GRAPHS OF PARENT FUNCTIONS

Linear Function

$f(x) = mx + b$

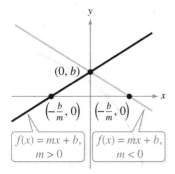

Domain: $(-\infty, \infty)$
Range $(m \neq 0)$: $(-\infty, \infty)$
x-intercept: $(-b/m, 0)$
y-intercept: $(0, b)$
Increasing when $m > 0$
Decreasing when $m < 0$

Absolute Value Function

$f(x) = |x| = \begin{cases} x, & x \geq 0 \\ -x, & x < 0 \end{cases}$

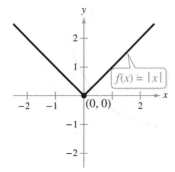

Domain: $(-\infty, \infty)$
Range: $[0, \infty)$
Intercept: $(0, 0)$
Decreasing on $(-\infty, 0)$
Increasing on $(0, \infty)$
Even function
y-axis symmetry

Square Root Function

$f(x) = \sqrt{x}$

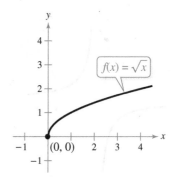

Domain: $[0, \infty)$
Range: $[0, \infty)$
Intercept: $(0, 0)$
Increasing on $(0, \infty)$

Greatest Integer Function

$f(x) = [\![x]\!]$

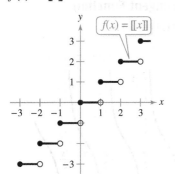

Domain: $(-\infty, \infty)$
Range: the set of integers
x-intercepts: in the interval $[0, 1)$
y-intercept: $(0, 0)$
Constant between each pair of
    consecutive integers
Jumps vertically one unit at
    each integer value

Quadratic (Squaring) Function

$f(x) = ax^2$

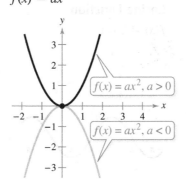

Domain: $(-\infty, \infty)$
Range $(a > 0)$: $[0, \infty)$
Range $(a < 0)$: $(-\infty, 0]$
Intercept: $(0, 0)$
Decreasing on $(-\infty, 0)$ for $a > 0$
Increasing on $(0, \infty)$ for $a > 0$
Increasing on $(-\infty, 0)$ for $a < 0$
Decreasing on $(0, \infty)$ for $a < 0$
Even function
y-axis symmetry
Relative minimum $(a > 0)$,
    relative maximum $(a < 0)$,
    or vertex: $(0, 0)$

Cubic Function

$f(x) = x^3$

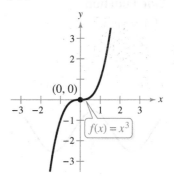

Domain: $(-\infty, \infty)$
Range: $(-\infty, \infty)$
Intercept: $(0, 0)$
Increasing on $(-\infty, \infty)$
Odd function
Origin symmetry

Rational (Reciprocal) Function

$f(x) = \dfrac{1}{x}$

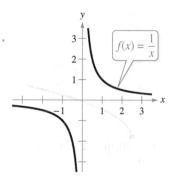

Domain: $(-\infty, 0) \cup (0, \infty)$
Range: $(-\infty, 0) \cup (0, \infty)$
No intercepts
Decreasing on $(-\infty, 0)$ and $(0, \infty)$
Odd function
Origin symmetry
Vertical asymptote: y-axis
Horizontal asymptote: x-axis

Exponential Function

$f(x) = a^x, \; a > 1$

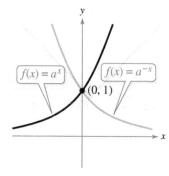

Domain: $(-\infty, \infty)$
Range: $(0, \infty)$
Intercept: $(0, 1)$
Increasing on $(-\infty, \infty)$
   for $f(x) = a^x$
Decreasing on $(-\infty, \infty)$
   for $f(x) = a^{-x}$
Horizontal asymptote: x-axis
Continuous

Logarithmic Function

$f(x) = \log_a x, \; a > 1$

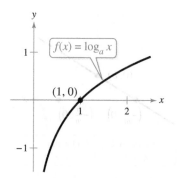

Domain: $(0, \infty)$
Range: $(-\infty, \infty)$
Intercept: $(1, 0)$
Increasing on $(0, \infty)$
Vertical asymptote: y-axis
Continuous
Reflection of graph of $f(x) = a^x$
   in the line $y = x$

Sine Function

$f(x) = \sin x$

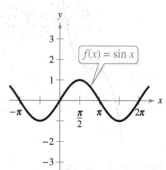

Domain: $(-\infty, \infty)$
Range: $[-1, 1]$
Period: 2π
x-intercepts: $(n\pi, 0)$
y-intercept: $(0, 0)$
Odd function
Origin symmetry

Cosine Function

$f(x) = \cos x$

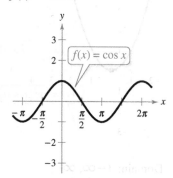

Domain: $(-\infty, \infty)$
Range: $[-1, 1]$
Period: 2π
x-intercepts: $\left(\dfrac{\pi}{2} + n\pi, 0\right)$
y-intercept: $(0, 1)$
Even function
y-axis symmetry

Tangent Function

$f(x) = \tan x$

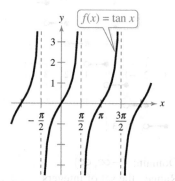

Domain: all $x \neq \dfrac{\pi}{2} + n\pi$
Range: $(-\infty, \infty)$
Period: π
x-intercepts: $(n\pi, 0)$
y-intercept: $(0, 0)$
Vertical asymptotes:
$$x = \dfrac{\pi}{2} + n\pi$$
Odd function
Origin symmetry

Cosecant Function

$f(x) = \csc x$

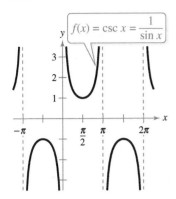

Domain: all $x \neq n\pi$
Range: $(-\infty, -1] \cup [1, \infty)$
Period: 2π
No intercepts
Vertical asymptotes: $x = n\pi$
Odd function
Origin symmetry

Secant Function

$f(x) = \sec x$

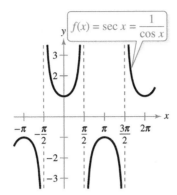

Domain: all $x \neq \dfrac{\pi}{2} + n\pi$
Range: $(-\infty, -1] \cup [1, \infty)$
Period: 2π
y-intercept: $(0, 1)$
Vertical asymptotes:

$$x = \frac{\pi}{2} + n\pi$$

Even function
y-axis symmetry

Cotangent Function

$f(x) = \cot x$

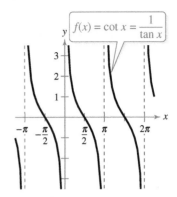

Domain: all $x \neq n\pi$
Range: $(-\infty, \infty)$
Period: π

x-intercepts: $\left(\dfrac{\pi}{2} + n\pi, 0\right)$

Vertical asymptotes: $x = n\pi$
Odd function
Origin symmetry

Inverse Sine Function

$f(x) = \arcsin x$

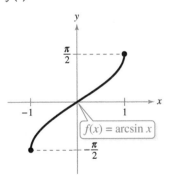

Domain: $[-1, 1]$

Range: $\left[-\dfrac{\pi}{2}, \dfrac{\pi}{2}\right]$

Intercept: $(0, 0)$
Odd function
Origin symmetry

Inverse Cosine Function

$f(x) = \arccos x$

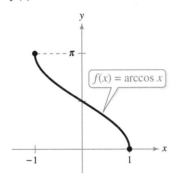

Domain: $[-1, 1]$
Range: $[0, \pi]$

y-intercept: $\left(0, \dfrac{\pi}{2}\right)$

Inverse Tangent Function

$f(x) = \arctan x$

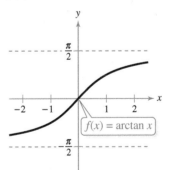

Domain: $(-\infty, \infty)$

Range: $\left(-\dfrac{\pi}{2}, \dfrac{\pi}{2}\right)$

Intercept: $(0, 0)$
Horizontal asymptotes:

$$y = \pm\frac{\pi}{2}$$

Odd function
Origin symmetry

Cosecant Function
$f(x) = \csc x$

Domain: all $x \neq n\pi$
Range: $(-\infty, -1] \cup [1, \infty)$
Period: 2π
No intercepts
Vertical asymptotes: $x = n\pi$
Odd function
Origin symmetry

Secant Function
$f(x) = \sec x$

Domain: all $x \neq \frac{\pi}{2} + n\pi$
Range: $(-\infty, -1] \cup [1, \infty)$
Period: 2π
Intercept: $(0, 1)$
Vertical asymptotes:
$x = \frac{\pi}{2} + n\pi$
Even function
y-axis symmetry

Cotangent Function
$f(x) = \cot x$

Domain: all $x \neq n\pi$
Range: $(-\infty, \infty)$
Period: π
x-intercepts: $\left(\frac{\pi}{2} + n\pi, 0\right)$
Vertical asymptotes: $x = n\pi$
Odd function
Origin symmetry

Inverse Sine Function
$f(x) = \arcsin x$

Domain: $[-1, 1]$
Range: $\left[-\frac{\pi}{2}, \frac{\pi}{2}\right]$
Intercept: $(0, 0)$
Odd function
Origin symmetry

Inverse Cosine Function
$f(x) = \arccos x$

Domain: $[-1, 1]$
Range: $[0, \pi]$
y-intercept: $\left(0, \frac{\pi}{2}\right)$

Inverse Tangent Function
$f(x) = \arctan x$

Domain: $(-\infty, \infty)$
Range: $\left(-\frac{\pi}{2}, \frac{\pi}{2}\right)$
Intercept: $(0, 0)$
Horizontal asymptotes:
$y = \pm\frac{\pi}{2}$
Odd function
Origin symmetry

Precalculus

Hudson Valley Community College

Custom Edition

Larson

 CENGAGE

Australia • Brazil • Mexico • Singapore • United Kingdom • United States

Precalculus, Hudson Valley Community College

Precalculus, 10th Edition
Ron Larson

© 2018 Cengage Learning. All rights reserved.

For product information and technology assistance, contact us at **Cengage Learning Customer & Sales Support, 1-800-354-9706.**

For permission to use material from this text or product, submit all requests online at **www.cengage.com/permissions.**
Further permissions questions can be emailed to **permissionrequest@cengage.com.**

This book contains select works from existing Cengage learning resources and was produced by Cengage learning Custom Solutions for collegiate use. As such, those adopting and/or contributing to this work are responsible for editorial content accuracy, continuity and completeness.

Compilation © 2017 Cengage Learning

ISBN: 9781337698894

Cengage Learning
20 Channel Street
Boston, MA 02210
USA

Cengage Learning is a leading provider of customized learning solutions with employees residing in nearly 40 different countries and sales in more than 125 countries around the world. Find your local representative at: **www.cengage.com.**

Cengage Learning products are represented in Canada by Nelson Education, Ltd.

For your course and learning solutions, visit **www.cengage.com.**

Purchase any of our products at your local college store or at our preferred online store **www.cengagebrain.com.**

Visit our custom book building website at **www.compose.cengage.com.**

Brief Contents

Preface

Welcome to *Precalculus,* Tenth Edition. We are excited to offer you a new edition with even more resources that will help you understand and master precalculus. This textbook includes features and resources that continue to make *Precalculus* a valuable learning tool for students and a trustworthy teaching tool for instructors.

Precalculus provides the clear instruction, precise mathematics, and thorough coverage that you expect for your course. Additionally, this new edition provides you with **free** access to three companion websites:

- **CalcView.com**—video solutions to selected exercises
- **CalcChat.com**—worked-out solutions to odd-numbered exercises and access to online tutors
- **LarsonPrecalculus.com**—companion website with resources to supplement your learning

These websites will help enhance and reinforce your understanding of the material presented in this text and prepare you for future mathematics courses. CalcView® and CalcChat® are also available as free mobile apps.

Features

NEW 📺 CalcView®

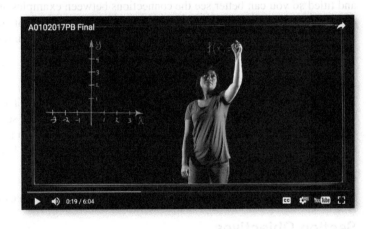

The website *CalcView.com* contains video solutions of selected exercises. Watch instructors progress step-by-step through solutions, providing guidance to help you solve the exercises. The CalcView mobile app is available for free at the Apple® App Store® or Google Play™ store. The app features an embedded QR Code® reader that can be used to scan the on-page codes 🔲 and go directly to the videos. You can also access the videos at *CalcView.com*.

CalcChat®
By Ron Larson

Free Easy Access Study Guide and Tutoring for Calculus Students

Easy Access Study Guide
24/7 Step-by-step solutions to odd-numbered exercises

Calculus & Linear Algebra	Precalculus & College Algebra	Applied Series
CALCULUS	PRECALCULUS	CALCULUS
Calculus 11e Calculus for AP 1e Elementary Linear Algebra 8e Calculus 10e Calculus ETF 6e Calculus 9e	Precalculus 10e College Algebra 10e Trigonometry 10e Precalculus with Limits A Graphing Approach 7e Precalculus Real Math Real People 7e College Algebra Real Math Real People 7e	Calculus an Applied Approach 10e Calculus An Applied Approach 9e Brief Calculus An Applied Approach 9e College Algebra with Applications for Business and the College Algebra and Calculus an Applied Approach 2e
VIEW	VIEW	VIEW

UPDATED CalcChat®

In each exercise set, be sure to notice the reference to *CalcChat.com*. This website provides free step-by-step solutions to all odd-numbered exercises in many of our textbooks. Additionally, you can chat with a tutor, at no charge, during the hours posted at the site. For over 14 years, hundreds of thousands of students have visited this site for help. The CalcChat mobile app is also available as a free download at the Apple® App Store® or Google Play™ store and features an embedded QR Code® reader.

App Store is a service mark of Apple Inc. Google Play is a trademark of Google Inc.
QR Code is a registered trademark of Denso Wave Incorporated.

REVISED LarsonPrecalculus.com

All companion website features have been
updated based on this revision, plus we have
added a new Collaborative Project feature. Access
to these features is free. You can view and listen to
worked-out solutions of Checkpoint problems in
English or Spanish, explore examples, download
data sets, watch lesson videos, and much more.

NEW Collaborative Project

You can find these extended group projects at
LarsonPrecalculus.com. Check your understanding
of the chapter concepts by solving in-depth, real-life
problems. These collaborative projects provide an
interesting and engaging way for you and other
students to work together and investigate ideas.

REVISED Exercise Sets

The exercise sets have been carefully and extensively examined to ensure they are rigorous
and relevant, and include topics our users have suggested. The exercises have been reorganized
and titled so you can better see the connections between examples and exercises. Multi-step,
real-life exercises reinforce problem-solving skills and mastery of concepts by giving you the
opportunity to apply the concepts in real-life situations. Error Analysis exercises have been
added throughout the text to help you identify common mistakes.

Table of Contents Changes

Based on market research and feedback from users, Section 6.5, The Complex Plane, has
been added. In addition, examples on finding the magnitude of a scalar multiple (Section 6.3),
multiplying in the complex plane (Section 6.6), using matrices to transform vectors
(Section 8.2), and further applications of 2×2 matrices (Section 8.5) have been added.

Chapter Opener

Each Chapter Opener highlights real-life applications used in the examples and exercises.

Section Objectives

A bulleted list of learning objectives provides you the opportunity
to preview what will be presented in the upcoming section.

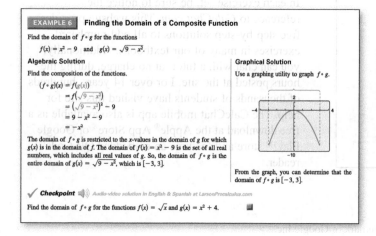

Side-By-Side Examples

Throughout the text, we present solutions to many
examples from multiple perspectives—algebraically,
graphically, and numerically. The side-by-side
format of this pedagogical feature helps you to see
that a problem can be solved in more than one way
and to see that different methods yield the same
result. The side-by-side format also addresses many
different learning styles.

Remarks

These hints and tips reinforce or expand upon
concepts, help you learn how to study mathematics,
caution you about common errors, address special
cases, or show alternative or additional steps to a
solution of an example.

Checkpoints

Accompanying every example, the Checkpoint problems encourage immediate practice and check your understanding of the concepts presented in the example. View and listen to worked-out solutions of the Checkpoint problems in English or Spanish at *LarsonPrecalculus.com*.

Technology

The technology feature gives suggestions for effectively using tools such as calculators, graphing utilities, and spreadsheet programs to help deepen your understanding of concepts, ease lengthy calculations, and provide alternate solution methods for verifying answers obtained by hand.

Historical Notes

These notes provide helpful information regarding famous mathematicians and their work.

Algebra of Calculus

Throughout the text, special emphasis is given to the algebraic techniques used in calculus. Algebra of Calculus examples and exercises are integrated throughout the text and are identified by the symbol ∫.

Summarize

The Summarize feature at the end of each section helps you organize the lesson's key concepts into a concise summary, providing you with a valuable study tool.

Vocabulary Exercises

The vocabulary exercises appear at the beginning of the exercise set for each section. These problems help you review previously learned vocabulary terms that you will use in solving the section exercises.

▷ TECHNOLOGY Use a graphing utility to check the result of Example 2. To do this, enter

$$Y1 = -(\sin(X))^3$$

and

$$Y2 = \sin(X)(\cos(X))^2$$
$$- \sin(X).$$

Select the *line* style for Y1 and the *path* style for Y2, then graph both equations in the same viewing window. The two graphs *appear* to coincide, so it is reasonable to assume that their expressions are equivalent. Note that the actual equivalence of the expressions can only be verified algebraically, as in Example 2. This graphical approach is only to check your work.

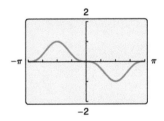

---

92. HOW DO YOU SEE IT? The graph represents the height h of a projectile after t seconds.

(a) Explain why h is a function of t.

(b) Approximate the height of the projectile after 0.5 second and after 1.25 seconds.

(c) Approximate the domain of h.

(d) Is t a function of h? Explain.

How Do You See It?

The How Do You See It? feature in each section presents a real-life exercise that you will solve by visual inspection using the concepts learned in the lesson. This exercise is excellent for classroom discussion or test preparation.

Project

The projects at the end of selected sections involve in-depth applied exercises in which you will work with large, real-life data sets, often creating or analyzing models. These projects are offered online at *LarsonPrecalculus.com*.

Chapter Summary

The Chapter Summary includes explanations and examples of the objectives taught in each chapter.

Instructor Resources

Annotated Instructor's Edition / ISBN-13: 978-1-337-27976-5
This is the complete student text plus point-of-use annotations for the instructor, including extra projects, classroom activities, teaching strategies, and additional examples. Answers to even-numbered text exercises, Vocabulary Checks, and Explorations are also provided.

Complete Solutions Manual (on instructor companion site)
This manual contains solutions to all exercises from the text, including Chapter Review Exercises and Chapter Tests, and Practice Tests with solutions.

Cengage Learning Testing Powered by Cognero (login.cengage.com)
CLT is a flexible online system that allows you to author, edit, and manage test bank content; create multiple test versions in an instant; and deliver tests from your LMS, your classroom, or wherever you want. This is available online via *www.cengage.com/login.*

Instructor Companion Site
Everything you need for your course in one place! This collection of book-specific lecture and class tools is available online via *www.cengage.com/login.* Access and download PowerPoint® presentations, images, the instructor's manual, and more.

Test Bank (on instructor companion site)
This contains text-specific multiple-choice and free response test forms.

Lesson Plans (on instructor companion site)
This manual provides suggestions for activities and lessons with notes on time allotment in order to ensure timeliness and efficiency during class.

MindTap for Mathematics
MindTap® is the digital learning solution that helps instructors engage and transform today's students into critical thinkers. Through paths of dynamic assignments and applications that you can personalize, real-time course analytics and an accessible reader, MindTap helps you turn cookie cutter into cutting edge, apathy into engagement, and memorizers into higher-level thinkers.

Student Study and Solutions Manual / ISBN-13: 978-1-337-28078-5
This guide offers step-by-step solutions for all odd-numbered text exercises,
Chapter Tests, and Cumulative Tests. It also contains Practice Tests.

Note-Taking Guide / ISBN-13: 978-1-337-28077-8
This is an innovative study aid, in the form of a notebook organizer, that helps
students develop a section-by-section summary of key concepts.

CengageBrain.com
To access additional course materials, please visit *www.cengagebrain.com*. At the
CengageBrain.com home page, search for the ISBN of your title (from the back cover
of your book) using the search box at the top of the page. This will take you to the
product page where these resources can be found.

MindTap for Mathematics
MindTap® provides you with the tools you need to better manage your limited
time—you can complete assignments whenever and wherever you are ready to learn
with course material specially customized for you by your instructor and streamlined
in one proven, easy-to-use interface. With an array of tools and apps—from note
taking to flashcards—you'll get a true understanding of course concepts, helping you
to achieve better grades and setting the groundwork for your future courses. This
access code entitles you to one term of usage.

Acknowledgments

I would like to thank the many people who have helped me prepare the text and the supplements package. Their encouragement, criticisms, and suggestions have been invaluable.

Thank you to all of the instructors who took the time to review the changes in this edition and to provide suggestions for improving it. Without your help, this book would not be possible.

Reviewers of the Tenth Edition

Gurdial Arora, *Xavier University of Louisiana*
Russell C. Chappell, *Twinsburg High School, Ohio*
Darlene Martin, *Lawson State Community College*
John Fellers, *North Allegheny School District*
Professor Steven Sikes, *Collin College*
Ann Slate, *Surry Community College*
John Elias, *Glenda Dawson High School*
Kathy Wood, *Lansing Catholic High School*
Darin Bauguess, *Surry Community College*
Brianna Kurtz, *Daytona State College*

Reviewers of the Previous Editions

Timothy Andrew Brown, *South Georgia College;* Blair E. Caboot, *Keystone College;* Shannon Cornell, *Amarillo College;* Gayla Dance, *Millsaps College;* Paul Finster, *El Paso Community College;* Paul A. Flasch, *Pima Community College West Campus;* Vadas Gintautas, *Chatham University;* Lorraine A. Hughes, *Mississippi State University;* Shu-Jen Huang, *University of Florida;* Renyetta Johnson, *East Mississippi Community College;* George Keihany, *Fort Valley State University;* Mulatu Lemma, *Savannah State University;* William Mays Jr., *Salem Community College;* Marcella Melby, *University of Minnesota;* Jonathan Prewett, *University of Wyoming;* Denise Reid, *Valdosta State University;* David L. Sonnier, *Lyon College;* David H. Tseng, *Miami Dade College—Kendall Campus;* Kimberly Walters, *Mississippi State University;* Richard Weil, *Brown College;* Solomon Willis, *Cleveland Community College;* Bradley R. Young, *Darton College*

My thanks to Robert Hostetler, The Behrend College, The Pennsylvania State University, and David Heyd, The Behrend College, The Pennsylvania State University, for their significant contributions to previous editions of this text.

I would also like to thank the staff at Larson Texts, Inc. who assisted with proofreading the manuscript, preparing and proofreading the art package, and checking and typesetting the supplements.

On a personal level, I am grateful to my spouse, Deanna Gilbert Larson, for her love, patience, and support. Also, a special thanks goes to R. Scott O'Neil. If you have suggestions for improving this text, please feel free to write to me. Over the past two decades, I have received many useful comments from both instructors and students, and I value these comments very highly.

Ron Larson, Ph.D.
Professor of Mathematics
Penn State University
www.RonLarson.com

PRECALCULUS

with

CalcChat® and CalcView®

10E

1 Functions and Their Graphs

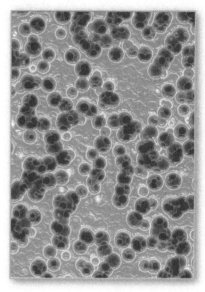

Snowstorm *(Exercise 47, page 66)*

Bacteria *(Example 8, page 80)*

Average Speed *(Example 7, page 54)*

Americans with Disabilities Act *(page 28)*

Alternative-Fuel Stations
(Example 10, page 42)

1.1 Rectangular Coordinates

The Cartesian plane can help you visualize relationships between two variables. For example, in Exercise 37 on page 9, given how far north and west one city is from another, plotting points to represent the cities can help you visualize these distances and determine the flying distance between the cities.

■ Plot points in the Cartesian plane.
■ Use the Distance Formula to find the distance between two points.
■ Use the Midpoint Formula to find the midpoint of a line segment.
■ Use a coordinate plane to model and solve real-life problems.

The Cartesian Plane

Just as you can represent real numbers by points on a real number line, you can represent ordered pairs of real numbers by points in a plane called the **rectangular coordinate system,** or the **Cartesian plane,** named after the French mathematician René Descartes (1596–1650).

Two real number lines intersecting at right angles form the Cartesian plane, as shown in Figure 1.1. The horizontal real number line is usually called the **x-axis,** and the vertical real number line is usually called the **y-axis.** The point of intersection of these two axes is the **origin,** and the two axes divide the plane into four **quadrants.**

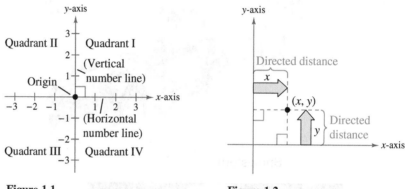

Figure 1.1   **Figure 1.2**

Each point in the plane corresponds to an **ordered pair** (x, y) of real numbers x and y, called **coordinates** of the point. The **x-coordinate** represents the directed distance from the y-axis to the point, and the **y-coordinate** represents the directed distance from the x-axis to the point, as shown in Figure 1.2.

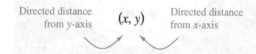

The notation (x, y) denotes both a point in the plane and an open interval on the real number line. The context will tell you which meaning is intended.

EXAMPLE 1   Plotting Points in the Cartesian Plane

Plot the points $(-1, 2)$, $(3, 4)$, $(0, 0)$, $(3, 0)$, and $(-2, -3)$.

Solution To plot the point $(-1, 2)$, imagine a vertical line through -1 on the x-axis and a horizontal line through 2 on the y-axis. The intersection of these two lines is the point $(-1, 2)$. Plot the other four points in a similar way, as shown in Figure 1.3.

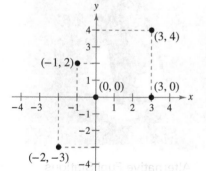

Figure 1.3

✓ **Checkpoint** ◀))) *Audio-video solution in English & Spanish at LarsonPrecalculus.com*

Plot the points $(-3, 2)$, $(4, -2)$, $(3, 1)$, $(0, -2)$, and $(-1, -2)$.

The beauty of a rectangular coordinate system is that it allows you to *see* relationships between two variables. It would be difficult to overestimate the importance of Descartes's introduction of coordinates in the plane. Today, his ideas are in common use in virtually every scientific and business-related field.

| EXAMPLE 2 | **Sketching a Scatter Plot** |

The table shows the numbers N (in millions) of subscribers to a cellular telecommunication service in the United States from 2005 through 2014, where t represents the year. Sketch a scatter plot of the data. *(Source: CTIA-The Wireless Association)*

Solution   To sketch a *scatter plot* of the data shown in the table, represent each pair of values by an ordered pair (t, N) and plot the resulting points. For example, let $(2005, 207.9)$ represent the first pair of values. Note that in the scatter plot below, the break in the t-axis indicates omission of the years before 2005, and the break in the N-axis indicates omission of the numbers less than 150 million.

DATA	Year, t	Subscribers, N
	2005	207.9
	2006	233.0
	2007	255.4
	2008	270.3
	2009	285.6
	2010	296.3
	2011	316.0
	2012	326.5
	2013	335.7
	2014	355.4

Spreadsheet at LarsonPrecalculus.com

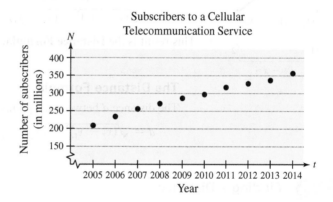

✓ *Checkpoint* ◀))) *Audio-video solution in English & Spanish at LarsonPrecalculus.com*

The table shows the numbers N (in thousands) of cellular telecommunication service employees in the United States from 2005 through 2014, where t represents the year. Sketch a scatter plot of the data. *(Source: CTIA-The Wireless Association)*

DATA	t	N
	2005	233.1
	2006	253.8
	2007	266.8
	2008	268.5
	2009	249.2
	2010	250.4
	2011	238.1
	2012	230.1
	2013	230.4
	2014	232.2

Spreadsheet at LarsonPrecalculus.com

▷ TECHNOLOGY The scatter plot in Example 2 is only one way to represent the data graphically. You could also represent the data using a bar graph or a line graph. Use a graphing utility to represent the data given in Example 2 graphically.

In Example 2, you could let $t = 1$ represent the year 2005. In that case, there would not be a break in the horizontal axis, and the labels 1 through 10 (instead of 2005 through 2014) would be on the tick marks.

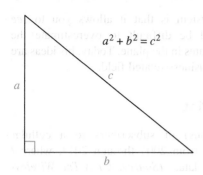

Figure 1.4

The Pythagorean Theorem and The Distance Formula

The Pythagorean Theorem is used extensively throughout this course.

Pythagorean Theorem

For a right triangle with hypotenuse length c and sides lengths a and b, you have $a^2 + b^2 = c^2$, as shown in Figure 1.4. (The converse is also true. That is, if $a^2 + b^2 = c^2$, then the triangle is a right triangle.)

Using the points (x_1, y_1) and (x_2, y_2), you can form a right triangle, as shown in Figure 1.5. The length of the hypotenuse of the right triangle is the distance d between the two points. The length of the vertical side of the triangle is $|y_2 - y_1|$ and the length of the horizontal side is $|x_2 - x_1|$. By the Pythagorean Theorem,

$$d^2 = |x_2 - x_1|^2 + |y_2 - y_1|^2$$
$$d = \sqrt{|x_2 - x_1|^2 + |y_2 - y_1|^2}$$
$$= \sqrt{(x_2 - x_1)^2 + (y_2 - y_1)^2}.$$

This result is the **Distance Formula.**

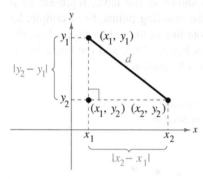

Figure 1.5

The Distance Formula

The distance d between the points (x_1, y_1) and (x_2, y_2) in the plane is

$$d = \sqrt{(x_2 - x_1)^2 + (y_2 - y_1)^2}.$$

EXAMPLE 3   **Finding a Distance**

Find the distance between the points $(-2, 1)$ and $(3, 4)$.

Algebraic Solution

Let $(x_1, y_1) = (-2, 1)$ and $(x_2, y_2) = (3, 4)$. Then apply the Distance Formula.

$$d = \sqrt{(x_2 - x_1)^2 + (y_2 - y_1)^2} \qquad \text{Distance Formula}$$
$$= \sqrt{[3 - (-2)]^2 + (4 - 1)^2} \qquad \text{Substitute for } x_1, y_1, x_2, \text{ and } y_2.$$
$$= \sqrt{(5)^2 + (3)^2} \qquad \text{Simplify.}$$
$$= \sqrt{34} \qquad \text{Simplify.}$$
$$\approx 5.83 \qquad \text{Use a calculator.}$$

So, the distance between the points is about 5.83 units.

Check

$$d^2 \overset{?}{=} 5^2 + 3^2 \qquad \text{Pythagorean Theorem}$$
$$\left(\sqrt{34}\right)^2 \overset{?}{=} 5^2 + 3^2 \qquad \text{Substitute for } d.$$
$$34 = 34 \qquad \text{Distance checks.} ✓$$

Graphical Solution

Use centimeter graph paper to plot the points $A(-2, 1)$ and $B(3, 4)$. Carefully sketch the line segment from A to B. Then use a centimeter ruler to measure the length of the segment.

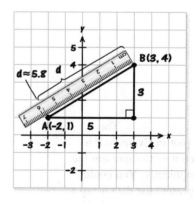

The line segment measures about 5.8 centimeters. So, the distance between the points is about 5.8 units.

✓ **Checkpoint**))) *Audio-video solution in English & Spanish at LarsonPrecalculus.com*

Find the distance between the points $(3, 1)$ and $(-3, 0)$.

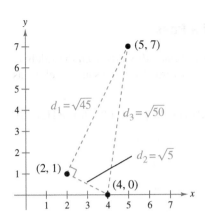

$d_1 = \sqrt{45}$

$d_3 = \sqrt{50}$

$d_2 = \sqrt{5}$

Figure 1.6

▷ **ALGEBRA HELP** To review the techniques for evaluating a radical, see Appendix A.2.

EXAMPLE 4 **Verifying a Right Triangle**

Show that the points

$$(2, 1), \quad (4, 0), \quad \text{and} \quad (5, 7)$$

are vertices of a right triangle.

Solution The three points are plotted in Figure 1.6. Using the Distance Formula, the lengths of the three sides are

$$d_1 = \sqrt{(5-2)^2 + (7-1)^2} = \sqrt{9+36} = \sqrt{45},$$

$$d_2 = \sqrt{(4-2)^2 + (0-1)^2} = \sqrt{4+1} = \sqrt{5}, \text{ and}$$

$$d_3 = \sqrt{(5-4)^2 + (7-0)^2} = \sqrt{1+49} = \sqrt{50}.$$

Because $(d_1)^2 + (d_2)^2 = 45 + 5 = 50 = (d_3)^2$, you can conclude by the converse of the Pythagorean Theorem that the triangle is a right triangle.

✓ **Checkpoint** ◀))) *Audio-video solution in English & Spanish at LarsonPrecalculus.com*

Show that the points $(2, -1)$, $(5, 5)$, and $(6, -3)$ are vertices of a right triangle.

The Midpoint Formula

To find the **midpoint** of the line segment that joins two points in a coordinate plane, find the average values of the respective coordinates of the two endpoints using the **Midpoint Formula.**

The Midpoint Formula

The midpoint of the line segment joining the points (x_1, y_1) and (x_2, y_2) is

$$\text{Midpoint} = \left(\frac{x_1 + x_2}{2}, \frac{y_1 + y_2}{2} \right).$$

For a proof of the Midpoint Formula, see Proofs in Mathematics on page 110.

EXAMPLE 5 **Finding the Midpoint of a Line Segment**

Find the midpoint of the line segment joining the points

$$(-5, -3) \quad \text{and} \quad (9, 3).$$

Solution Let $(x_1, y_1) = (-5, -3)$ and $(x_2, y_2) = (9, 3)$.

$$\text{Midpoint} = \left(\frac{x_1 + x_2}{2}, \frac{y_1 + y_2}{2} \right) \qquad \text{Midpoint Formula}$$

$$= \left(\frac{-5 + 9}{2}, \frac{-3 + 3}{2} \right) \qquad \text{Substitute for } x_1, y_1, x_2, \text{ and } y_2.$$

$$= (2, 0) \qquad \text{Simplify.}$$

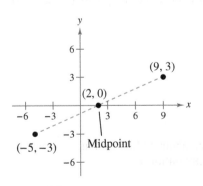

Figure 1.7

The midpoint of the line segment is $(2, 0)$, as shown in Figure 1.7.

✓ **Checkpoint** ◀))) *Audio-video solution in English & Spanish at LarsonPrecalculus.com*

Find the midpoint of the line segment joining the points

$$(-2, 8) \quad \text{and} \quad (4, -10).$$

Applications

EXAMPLE 6   **Finding the Length of a Pass**

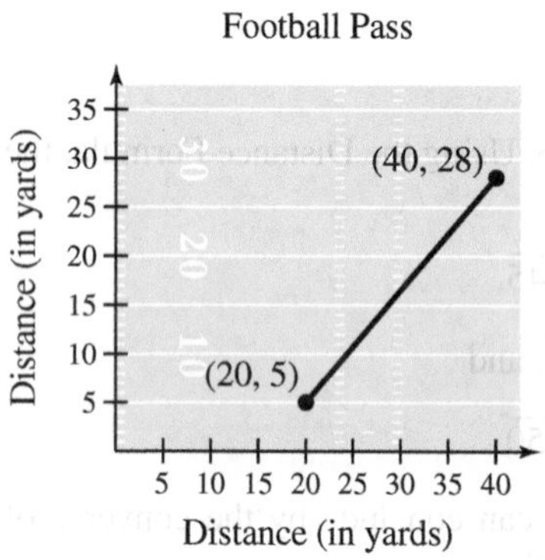

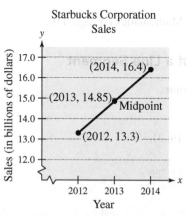

Football Pass

Figure 1.8

A football quarterback throws a pass from the 28-yard line, 40 yards from the sideline. A wide receiver catches the pass on the 5-yard line, 20 yards from the same sideline, as shown in Figure 1.8. How long is the pass?

Solution   The length of the pass is the distance between the points (40, 28) and (20, 5).

$$d = \sqrt{(x_2 - x_1)^2 + (y_2 - y_1)^2}$$ Distance Formula

$$= \sqrt{(40 - 20)^2 + (28 - 5)^2}$$ Substitute for x_1, y_1, x_2, and y_2.

$$= \sqrt{20^2 + 23^2}$$ Simplify.

$$= \sqrt{400 + 529}$$ Simplify.

$$= \sqrt{929}$$ Simplify.

$$\approx 30$$ Use a calculator.

So, the pass is about 30 yards long.

✓ *Checkpoint* 🔊)) *Audio-video solution in English & Spanish at LarsonPrecalculus.com*

A football quarterback throws a pass from the 10-yard line, 10 yards from the sideline. A wide receiver catches the pass on the 32-yard line, 25 yards from the same sideline. How long is the pass?

In Example 6, the scale along the goal line does not normally appear on a football field. However, when you use coordinate geometry to solve real-life problems, you are free to place the coordinate system in any way that helps you solve the problem.

EXAMPLE 7   **Estimating Annual Sales**

Starbucks Corporation had annual sales of approximately $13.3 billion in 2012 and $16.4 billion in 2014. Without knowing any additional information, what would you estimate the 2013 sales to have been?   *(Source: Starbucks Corporation)*

Starbucks Corporation Sales

Figure 1.9

Solution   Assuming that sales followed a linear pattern, you can estimate the 2013 sales by finding the midpoint of the line segment connecting the points (2012, 13.3) and (2014, 16.4).

$$\text{Midpoint} = \left(\frac{x_1 + x_2}{2}, \frac{y_1 + y_2}{2} \right)$$ Midpoint Formula

$$= \left(\frac{2012 + 2014}{2}, \frac{13.3 + 16.4}{2} \right)$$ Substitute for x_1, x_2, y_1, and y_2.

$$= (2013, 14.85)$$ Simplify.

So, you would estimate the 2013 sales to have been about $14.85 billion, as shown in Figure 1.9. (The actual 2013 sales were about $14.89 billion.)

✓ *Checkpoint* 🔊)) *Audio-video solution in English & Spanish at LarsonPrecalculus.com*

Yahoo! Inc. had annual revenues of approximately $5.0 billon in 2012 and $4.6 billion in 2014. Without knowing any additional information, what would you estimate the 2013 revenue to have been?   *(Source: Yahoo! Inc.)*

Much of computer graphics, including this computer-generated tessellation, consists of transformations of points in a coordinate plane. Example 8 illustrates one type of transformation called a translation. Other types include reflections, rotations, and stretches.

EXAMPLE 8   **Translating Points in the Plane**

See LarsonPrecalculus.com for an interactive version of this type of example.

The triangle in Figure 1.10 has vertices at the points $(-1, 2)$, $(1, -2)$, and $(2, 3)$. Shift the triangle three units to the right and two units up and find the coordinates of the vertices of the shifted triangle shown in Figure 1.11.

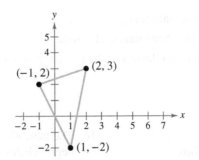

Figure 1.10

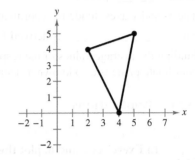

Figure 1.11

Solution   To shift the vertices three units to the right, add 3 to each of the x-coordinates. To shift the vertices two units up, add 2 to each of the y-coordinates.

Original Point	Translated Point
$(-1, 2)$	$(-1 + 3, 2 + 2) = (2, 4)$
$(1, -2)$	$(1 + 3, -2 + 2) = (4, 0)$
$(2, 3)$	$(2 + 3, 3 + 2) = (5, 5)$

✓ *Checkpoint*  *Audio-video solution in English & Spanish at LarsonPrecalculus.com*

Find the coordinates of the vertices of the parallelogram shown after translating it two units to the left and four units down.

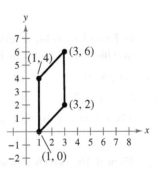

The figures in Example 8 were not really essential to the solution. Nevertheless, you should develop the habit of including sketches with your solutions because they serve as useful problem-solving tools.

Summarize (Section 1.1)

1. Describe the Cartesian plane *(page 2)*. For examples of plotting points in the Cartesian plane, see Examples 1 and 2.

2. State the Distance Formula *(page 4)*. For examples of using the Distance Formula to find the distance between two points, see Examples 3 and 4.

3. State the Midpoint Formula *(page 5)*. For an example of using the Midpoint Formula to find the midpoint of a line segment, see Example 5.

4. Describe examples of how to use a coordinate plane to model and solve real-life problems *(pages 6 and 7, Examples 6–8)*.

1.1 Exercises

See **CalcChat.com** for tutorial help and worked-out solutions to odd-numbered exercises.

Vocabulary: Fill in the blanks.

1. An ordered pair of real numbers can be represented in a plane called the rectangular coordinate system or the _____ plane.
2. The x- and y-axes divide the coordinate plane into four _____.
3. The _____ _____ is derived from the Pythagorean Theorem.
4. Finding the average values of the respective coordinates of the two endpoints of a line segment in a coordinate plane is also known as using the _____ _____.

Skills and Applications

 Plotting Points in the Cartesian Plane
In Exercises 5 and 6, plot the points.

5. $(2, 4)$, $(3, -1)$, $(-6, 2)$, $(-4, 0)$, $(-1, -8)$, $(1.5, -3.5)$
6. $(1, -5)$, $(-2, -7)$, $(3, 3)$, $(-2, 4)$, $(0, 5)$, $\left(\frac{2}{3}, \frac{5}{2}\right)$

Finding the Coordinates of a Point   In Exercises 7 and 8, find the coordinates of the point.

7. The point is three units to the left of the y-axis and four units above the x-axis.
8. The point is on the x-axis and 12 units to the left of the y-axis.

 Determining Quadrant(s) for a Point
In Exercises 9–14, determine the quadrant(s) in which (x, y) could be located.

9. $x > 0$ and $y < 0$
10. $x < 0$ and $y < 0$
11. $x = -4$ and $y > 0$
12. $x < 0$ and $y = 7$
13. $x + y = 0$, $x \neq 0$, $y \neq 0$
14. $xy > 0$

 Sketching a Scatter Plot   In Exercises 15 and 16, sketch a scatter plot of the data shown in the table.

15. The table shows the number y of Wal-Mart stores for each year x from 2008 through 2014. *(Source: Wal-Mart Stores, Inc.)*

DATA Year, x	Number of Stores, y
2008	7720
2009	8416
2010	8970
2011	10,130
2012	10,773
2013	10,942
2014	11,453

Spreadsheet at LarsonPrecalculus.com

16. The table shows the lowest temperature on record y (in degrees Fahrenheit) in Duluth, Minnesota, for each month x, where $x = 1$ represents January.   *(Source: NOAA)*

DATA Month, x	Temperature, y
1	−39
2	−39
3	−29
4	−5
5	17
6	27
7	35
8	32
9	22
10	8
11	−23
12	−34

Spreadsheet at LarsonPrecalculus.com

 Finding a Distance   In Exercises 17–22, find the distance between the points.

17. $(-2, 6)$, $(3, -6)$
18. $(8, 5)$, $(0, 20)$
19. $(1, 4)$, $(-5, -1)$
20. $(1, 3)$, $(3, -2)$
21. $\left(\frac{1}{2}, \frac{4}{3}\right)$, $(2, -1)$
22. $(9.5, -2.6)$, $(-3.9, 8.2)$

 Verifying a Right Triangle   In Exercises 23 and 24, (a) find the length of each side of the right triangle, and (b) show that these lengths satisfy the Pythagorean Theorem.

23.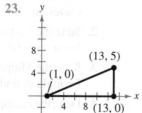

24.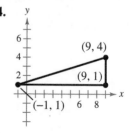

Verifying a Polygon In Exercises 25–28, show that the points form the vertices of the polygon.

25. Right triangle: $(4, 0), (2, 1), (-1, -5)$

26. Right triangle: $(-1, 3), (3, 5), (5, 1)$

27. Isosceles triangle: $(1, -3), (3, 2), (-2, 4)$

28. Isosceles triangle: $(2, 3), (4, 9), (-2, 7)$

Plotting, Distance, and Midpoint In Exercises 29–36, (a) plot the points, (b) find the distance between the points, and (c) find the midpoint of the line segment joining the points.

29. $(6, -3), (6, 5)$

30. $(1, 4), (8, 4)$

31. $(1, 1), (9, 7)$

32. $(1, 12), (6, 0)$

33. $(-1, 2), (5, 4)$

34. $(2, 10), (10, 2)$

35. $(-16.8, 12.3), (5.6, 4.9)$

36. $\left(\frac{1}{2}, 1\right), \left(-\frac{5}{2}, \frac{4}{3}\right)$

• • 37. Flying Distance • • • • • • • • • • • •

An airplane flies from Naples, Italy, in a straight line to Rome, Italy, which is 120 kilometers north and 150 kilometers west of Naples. How far does the plane fly?

38. Sports A soccer player passes the ball from a point that is 18 yards from the endline and 12 yards from the sideline. A teammate who is 42 yards from the same endline and 50 yards from the same sideline receives the pass. (See figure.) How long is the pass?

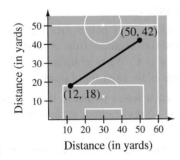

39. Sales The Coca-Cola Company had sales of $35,123 million in 2010 and $45,998 million in 2014. Use the Midpoint Formula to estimate the sales in 2012. Assume that the sales followed a linear pattern. *(Source: The Coca-Cola Company)*

40. Revenue per Share The revenue per share for Twitter, Inc. was $1.17 in 2013 and $3.25 in 2015. Use the Midpoint Formula to estimate the revenue per share in 2014. Assume that the revenue per share followed a linear pattern. *(Source: Twitter, Inc.)*

Translating Points in the Plane In Exercises 41–44, find the coordinates of the vertices of the polygon after the given translation to a new position in the plane.

41.

42.

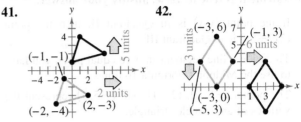

43. Original coordinates of vertices: $(-7, -2), (-2, 2), (-2, -4), (-7, -4)$

Shift: eight units up, four units to the right

44. Original coordinates of vertices: $(5, 8), (3, 6), (7, 6)$

Shift: 6 units down, 10 units to the left

45. Minimum Wage Use the graph below, which shows the minimum wages in the United States (in dollars) from 1950 through 2015. *(Source: U.S. Department of Labor)*

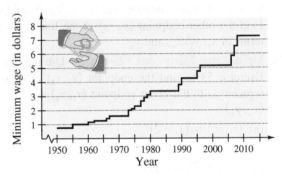

(a) Which decade shows the greatest increase in the minimum wage?

(b) Approximate the percent increases in the minimum wage from 1985 to 2000 and from 2000 to 2015.

(c) Use the percent increase from 2000 to 2015 to predict the minimum wage in 2030.

(d) Do you believe that your prediction in part (c) is reasonable? Explain.

46. Exam Scores The table shows the mathematics entrance test scores x and the final examination scores y in an algebra course for a sample of 10 students.

x	22	29	35	40	44	48	53	58	65	76
y	53	74	57	66	79	90	76	93	83	99

(a) Sketch a scatter plot of the data.

(b) Find the entrance test score of any student with a final exam score in the 80s.

(c) Does a higher entrance test score imply a higher final exam score? Explain.

Exploration

True or False? In Exercises 47–50, determine whether the statement is true or false. Justify your answer.

47. If the point (x, y) is in Quadrant II, then the point $(2x, -3y)$ is in Quadrant III.

48. To divide a line segment into 16 equal parts, you have to use the Midpoint Formula 16 times.

49. The points $(-8, 4)$, $(2, 11)$, and $(-5, 1)$ represent the vertices of an isosceles triangle.

50. If four points represent the vertices of a polygon, and the four side lengths are equal, then the polygon must be a square.

51. **Think About It** When plotting points on the rectangular coordinate system, when should you use different scales for the x- and y-axes? Explain.

52. **Think About It** What is the y-coordinate of any point on the x-axis? What is the x-coordinate of any point on the y-axis?

53. **Using the Midpoint Formula** A line segment has (x_1, y_1) as one endpoint and (x_m, y_m) as its midpoint. Find the other endpoint (x_2, y_2) of the line segment in terms of $x_1, y_1, x_m,$ and y_m.

54. **Using the Midpoint Formula** Use the result of Exercise 53 to find the endpoint (x_2, y_2) of each line segment with the given endpoint (x_1, y_1) and midpoint (x_m, y_m).
    (a) $(x_1, y_1) = (1, -2)$
    $(x_m, y_m) = (4, -1)$
    (b) $(x_1, y_1) = (-5, 11)$
    $(x_m, y_m) = (2, 4)$

55. **Using the Midpoint Formula** Use the Midpoint Formula three times to find the three points that divide the line segment joining (x_1, y_1) and (x_2, y_2) into four equal parts.

56. **Using the Midpoint Formula** Use the result of Exercise 55 to find the points that divide each line segment joining the given points into four equal parts.
    (a) $(x_1, y_1) = (1, -2)$
    $(x_2, y_2) = (4, -1)$
    (b) $(x_1, y_1) = (-2, -3)$
    $(x_2, y_2) = (0, 0)$

57. **Proof** Prove that the diagonals of the parallelogram in the figure intersect at their midpoints.

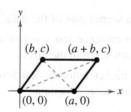

58. **HOW DO YOU SEE IT?** Use the plot of the point (x_0, y_0) in the figure. Match the transformation of the point with the correct plot. Explain. [The plots are labeled (i), (ii), (iii), and (iv).]

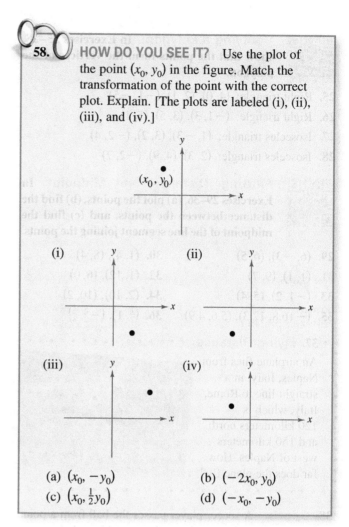

    (a) $(x_0, -y_0)$ (b) $(-2x_0, y_0)$
    (c) $\left(x_0, \frac{1}{2}y_0\right)$ (d) $(-x_0, -y_0)$

59. **Collinear Points** Three or more points are collinear when they all lie on the same line. Use the steps below to determine whether the set of points $\{A(2, 3), B(2, 6), C(6, 3)\}$ and the set of points $\{A(8, 3), B(5, 2), C(2, 1)\}$ are collinear.

    (a) For each set of points, use the Distance Formula to find the distances from A to B, from B to C, and from A to C. What relationship exists among these distances for each set of points?

    (b) Plot each set of points in the Cartesian plane. Do all the points of either set appear to lie on the same line?

    (c) Compare your conclusions from part (a) with the conclusions you made from the graphs in part (b). Make a general statement about how to use the Distance Formula to determine collinearity.

60. **Make a Conjecture**

    (a) Use the result of Exercise 58(a) to make a conjecture about the new location of a point when the sign of the y-coordinate is changed.

    (b) Use the result of Exercise 58(d) to make a conjecture about the new location of a point when the signs of both x- and y-coordinates are changed.

1.2 Graphs of Equations

The graph of an equation can help you visualize relationships between real-life quantities. For example, in Exercise 85 on page 21, you will use a graph to analyze life expectancy.

- ▣ Sketch graphs of equations.
- ▣ Find *x*- and *y*-intercepts of graphs of equations.
- ▣ Use symmetry to sketch graphs of equations.
- ▣ Write equations of circles.
- ▣ Use graphs of equations to solve real-life problems.

The Graph of an Equation

In Section 1.1, you used a coordinate system to graphically represent the relationship between two quantities as points in a coordinate plane.

Frequently, a relationship between two quantities is expressed as an **equation in two variables.** For example, $y = 7 - 3x$ is an equation in x and y. An ordered pair (a, b) is a **solution** or **solution point** of an equation in x and y when the substitutions $x = a$ and $y = b$ result in a true statement. For example, $(1, 4)$ is a solution of $y = 7 - 3x$ because $4 = 7 - 3(1)$ is a true statement.

In this section, you will review some basic procedures for sketching the graph of an equation in two variables. The **graph of an equation** is the set of all points that are solutions of the equation.

EXAMPLE 1   **Determining Solution Points**

Determine whether (a) $(2, 13)$ and (b) $(-1, -3)$ lie on the graph of $y = 10x - 7$.

Solution

a.

$$y = 10x - 7 \qquad \text{Write original equation.}$$

$$13 \overset{?}{=} 10(2) - 7 \qquad \text{Substitute 2 for } x \text{ and 13 for } y.$$

$$13 = 13 \qquad (2, 13) \text{ is a solution. } \checkmark$$

The point $(2, 13)$ *does* lie on the graph of $y = 10x - 7$ because it is a solution point of the equation.

b.

$$y = 10x - 7 \qquad \text{Write original equation.}$$

$$-3 \overset{?}{=} 10(-1) - 7 \qquad \text{Substitute } -1 \text{ for } x \text{ and } -3 \text{ for } y.$$

$$-3 \neq -17 \qquad (-1, -3) \text{ is not a solution.}$$

The point $(-1, -3)$ *does not* lie on the graph of $y = 10x - 7$ because it is *not* a solution point of the equation.

> ▷ **ALGEBRA HELP** When evaluating an expression or an equation, remember to follow the Basic Rules of Algebra. To review these rules, see Appendix A.1.

✓ *Checkpoint* *Audio-video solution in English & Spanish at LarsonPrecalculus.com*

Determine whether (a) $(3, -5)$ and (b) $(-2, 26)$ lie on the graph of $y = 14 - 6x$. ▪

The basic technique used for sketching the graph of an equation is the **point-plotting method.**

The Point-Plotting Method of Graphing

1. When possible, isolate one of the variables.
2. Construct a table of values showing several solution points.
3. Plot these points in a rectangular coordinate system.
4. Connect the points with a smooth curve or line.

It is important to use negative values, zero, and positive values for x (if possible) when constructing a table.

EXAMPLE 2   **Sketching the Graph of an Equation**

Sketch the graph of

$$3x + y = 7.$$

Solution

First, isolate the variable y.

$$y = -3x + 7 \qquad \text{Solve equation for } y.$$

Next, construct a table of values that consists of several solution points of the equation. For example, when $x = -3$,

$$y = -3(-3) + 7 = 16$$

which implies that $(-3, 16)$ is a solution point of the equation.

x	$y = -3x + 7$	(x, y)
-3	16	$(-3, 16)$
-2	13	$(-2, 13)$
-1	10	$(-1, 10)$
0	7	$(0, 7)$
1	4	$(1, 4)$
2	1	$(2, 1)$
3	-2	$(3, -2)$

From the table, it follows that

$$(-3, 16), \ (-2, 13), \ (-1, 10), \ (0, 7), \ (1, 4), \ (2, 1), \ \text{and} \ (3, -2)$$

are solution points of the equation. Plot these points and connect them with a line, as shown below.

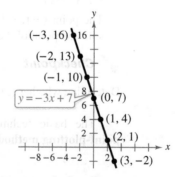

✓ **Checkpoint** ◀))) *Audio-video solution in English & Spanish at LarsonPrecalculus.com*

Sketch the graph of each equation.

a. $3x + y = 2$

b. $-2x + y = 1$

EXAMPLE 3   **Sketching the Graph of an Equation**

See LarsonPrecalculus.com for an interactive version of this type of example.

Sketch the graph of

$$y = x^2 - 2.$$

Solution

The equation is already solved for y, so begin by constructing a table of values.

x	-2	-1	0	1	2	3
$y = x^2 - 2$	2	-1	-2	-1	2	7
(x, y)	$(-2, 2)$	$(-1, -1)$	$(0, -2)$	$(1, -1)$	$(2, 2)$	$(3, 7)$

Next, plot the points given in the table, as shown in Figure 1.12. Finally, connect the points with a smooth curve, as shown in Figure 1.13.

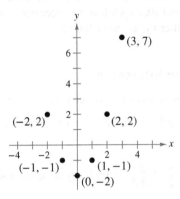

Figure 1.12   **Figure 1.13**

> **· · REMARK** One of your goals in this course is to learn to classify the basic shape of a graph from its equation. For instance, you will learn that the *linear equation* in Example 2 can be written in the form
>
> $$y = mx + b$$
>
> and its graph is a line. Similarly, the *quadratic equation* in Example 3 has the form
>
> $$y = ax^2 + bx + c$$
>
> and its graph is a parabola.

✓ **Checkpoint** 🔊))) *Audio-video solution in English & Spanish at LarsonPrecalculus.com*

Sketch the graph of each equation.

a. $y = x^2 + 3$ **b.** $y = 1 - x^2$

The point-plotting method demonstrated in Examples 2 and 3 is straightforward, but it has shortcomings. For instance, with too few solution points, it is possible to misrepresent the graph of an equation. To illustrate, when you only plot the four points

$$(-2, 2), (-1, -1), (1, -1), \quad \text{and} \quad (2, 2)$$

in Example 3, any one of the three graphs below is reasonable.

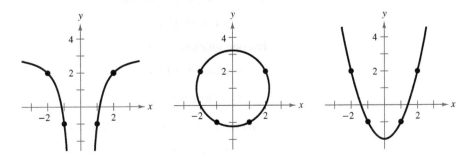

▷ TECHNOLOGY To graph
an equation involving x and y
on a graphing utility, use the
procedure below.

1. If necessary, rewrite the
   equation so that y is isolated
   on the left side.

2. Enter the equation in the
   graphing utility.

3. Determine a *viewing window*
   that shows all important
   features of the graph.

4. Graph the equation.

Intercepts of a Graph

Solution points of an equation that have zero as either the x-coordinate or the y-coordinate are called **intercepts.** They are the points at which the graph intersects or touches the x- or y-axis. It is possible for a graph to have no intercepts, one intercept, or several intercepts, as shown in the graphs below.

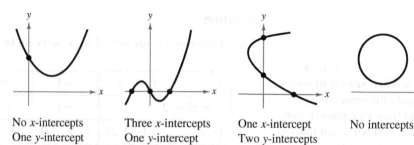

No x-intercepts
One y-intercept

Three x-intercepts
One y-intercept

One x-intercept
Two y-intercepts

No intercepts

Note that an x-intercept can be written as the ordered pair $(a, 0)$ and a y-intercept can be written as the ordered pair $(0, b)$. Sometimes it is convenient to denote the x-intercept as the x-coordinate a of the point $(a, 0)$ or the y-intercept as the y-coordinate b of the point $(0, b)$. Unless it is necessary to make a distinction, the term *intercept* will refer to either the point or the coordinate.

Finding Intercepts

1. To find x-intercepts, let y be zero and solve the equation for x.

2. To find y-intercepts, let x be zero and solve the equation for y.

EXAMPLE 4   **Finding x- and y-Intercepts**

Find the x- and y-intercepts of the graph of

$$y = x^3 - 4x.$$

Solution

To find the x-intercepts of the graph of $y = x^3 - 4x$, let $y = 0$. Then

$$0 = x^3 - 4x$$
$$= x(x^2 - 4)$$

has the solutions $x = 0$ and $x = \pm 2$.

x-intercepts: $(0, 0), (2, 0), (-2, 0)$ See figure.

To find the y-intercept of the graph of $y = x^3 - 4x$, let $x = 0$. Then

$$y = (0)^3 - 4(0)$$

has one solution, $y = 0$.

y-intercept: $(0, 0)$ See figure.

✓ **Checkpoint** ◀))) *Audio-video solution in English & Spanish at LarsonPrecalculus.com*

Find the x- and y-intercepts of the graph of

$$y = -x^2 - 5x.$$

Symmetry

Graphs of equations can have **symmetry** with respect to one of the coordinate axes or with respect to the origin. Symmetry with respect to the x-axis means that when you fold the Cartesian plane along the x-axis, the portion of the graph above the x-axis coincides with the portion below the x-axis. Symmetry with respect to the y-axis or the origin can be described in a similar manner. The graphs below show these three types of symmetry.

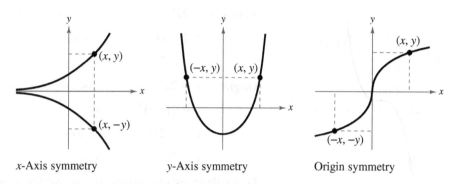

x-Axis symmetry y-Axis symmetry   Origin symmetry

Knowing the symmetry of a graph *before* attempting to sketch it is helpful, because then you need only half as many solution points to sketch the graph. Graphical and algebraic tests for these three basic types of symmetry are described below.

Graphical Tests for Symmetry

1. A graph is **symmetric with respect to the x-axis** if, whenever (x, y) is on the graph, $(x, -y)$ is also on the graph.

2. A graph is **symmetric with respect to the y-axis** if, whenever (x, y) is on the graph, $(-x, y)$ is also on the graph.

3. A graph is **symmetric with respect to the origin** if, whenever (x, y) is on the graph, $(-x, -y)$ is also on the graph.

For example, the graph of $y = x^2 - 2$ is symmetric with respect to the y-axis because (x, y) and $(-x, y)$ are on the graph of $y = x^2 - 2$. (See the table below and Figure 1.14.)

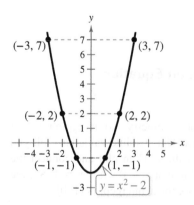

y-Axis symmetry

Figure 1.14

x	-3	-2	-1	1	2	3
y	7	2	-1	-1	2	7
(x, y)	$(-3, 7)$	$(-2, 2)$	$(-1, -1)$	$(1, -1)$	$(2, 2)$	$(3, 7)$

Algebraic Tests for Symmetry

1. The graph of an equation is symmetric with respect to the x-axis when replacing y with $-y$ yields an equivalent equation.

2. The graph of an equation is symmetric with respect to the y-axis when replacing x with $-x$ yields an equivalent equation.

3. The graph of an equation is symmetric with respect to the origin when replacing x with $-x$ and y with $-y$ yields an equivalent equation.

EXAMPLE 5 **Testing for Symmetry**

Test $y = 2x^3$ for symmetry with respect to both axes and the origin.

Solution

x-Axis:

	$y = 2x^3$	Write original equation.
	$-y = 2x^3$	Replace y with $-y$. Result is *not* an equivalent equation.

y-Axis:

	$y = 2x^3$	Write original equation.
	$y = 2(-x)^3$	Replace x with $-x$.
	$y = -2x^3$	Simplify. Result is *not* an equivalent equation.

Origin:

	$y = 2x^3$	Write original equation.
	$-y = 2(-x)^3$	Replace y with $-y$ and x with $-x$.
	$-y = -2x^3$	Simplify.
	$y = 2x^3$	Simplify. Result is an equivalent equation.

Of the three tests for symmetry, the test for origin symmetry is the only one satisfied. So, the graph of $y = 2x^3$ is symmetric with respect to the origin (see Figure 1.15).

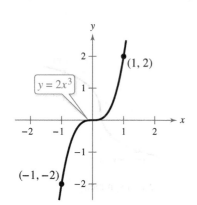

Figure 1.15

✓ *Checkpoint* ◀))) *Audio-video solution in English & Spanish at LarsonPrecalculus.com*

Test $y^2 = 6 - x$ for symmetry with respect to both axes and the origin.

EXAMPLE 6 **Using Symmetry as a Sketching Aid**

Use symmetry to sketch the graph of $x - y^2 = 1$.

Solution Of the three tests for symmetry, the test for x-axis symmetry is the only one satisfied, because $x - (-y)^2 = 1$ is equivalent to $x - y^2 = 1$. So, the graph is symmetric with respect to the x-axis. Find solution points above (or below) the x-axis and then use symmetry to obtain the graph, as shown in Figure 1.16.

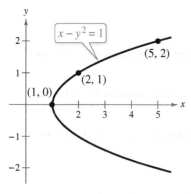

Figure 1.16

✓ *Checkpoint* ◀))) *Audio-video solution in English & Spanish at LarsonPrecalculus.com*

Use symmetry to sketch the graph of $y = x^2 - 4$.

EXAMPLE 7 **Sketching the Graph of an Equation**

Sketch the graph of $y = |x - 1|$.

Solution This equation fails all three tests for symmetry, so its graph is not symmetric with respect to either axis or to the origin. The absolute value bars tell you that y is always nonnegative. Construct a table of values. Then plot and connect the points, as shown in Figure 1.17. Notice from the table that $x = 0$ when $y = 1$. So, the y-intercept is $(0, 1)$. Similarly, $y = 0$ when $x = 1$. So, the x-intercept is $(1, 0)$.

x	-2	-1	0	1	2	3	4
$y = |x - 1|$	3	2	1	0	1	2	3
(x, y)	$(-2, 3)$	$(-1, 2)$	$(0, 1)$	$(1, 0)$	$(2, 1)$	$(3, 2)$	$(4, 3)$

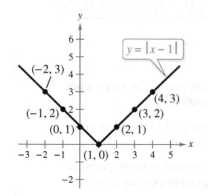

Figure 1.17

✓ *Checkpoint* ◀))) *Audio-video solution in English & Spanish at LarsonPrecalculus.com*

Sketch the graph of $y = |x - 2|$.

Circles

A **circle** is a set of points (x, y) in a plane that are the same distance r from a point called the center, (h, k), as shown at the right. By the Distance Formula,

$$\sqrt{(x - h)^2 + (y - k)^2} = r.$$

By squaring each side of this equation, you obtain the **standard form of the equation of a circle.** For example, for a circle with its center at $(h, k) = (1, 3)$ and radius $r = 4$,

$$\sqrt{(x - 1)^2 + (y - 3)^2} = 4 \qquad \text{Substitute for } h, k, \text{ and } r.$$
$$(x - 1)^2 + (y - 3)^2 = 16. \qquad \text{Square each side.}$$

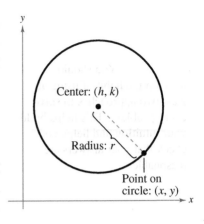

Standard Form of the Equation of a Circle

A point (x, y) lies on the circle of **radius** r and **center** (h, k) if and only if

$$(x - h)^2 + (y - k)^2 = r^2.$$

From this result, the standard form of the equation of a circle with radius *r and center at the origin,* $(h, k) = (0, 0)$, is

$$x^2 + y^2 = r^2. \qquad \text{Circle with radius } r \text{ and center at origin}$$

EXAMPLE 8   Writing the Equation of a Circle

The point $(3, 4)$ lies on a circle whose center is at $(-1, 2)$, as shown in Figure 1.18. Write the standard form of the equation of this circle.

Solution

The radius of the circle is the distance between $(-1, 2)$ and $(3, 4)$.

$$r = \sqrt{(x - h)^2 + (y - k)^2} \qquad \text{Distance Formula}$$
$$= \sqrt{[3 - (-1)]^2 + (4 - 2)^2} \qquad \text{Substitute for } x, y, h, \text{ and } k.$$
$$= \sqrt{4^2 + 2^2} \qquad \text{Simplify.}$$
$$= \sqrt{16 + 4} \qquad \text{Simplify.}$$
$$= \sqrt{20} \qquad \text{Radius}$$

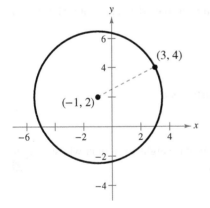

Figure 1.18

Using $(h, k) = (-1, 2)$ and $r = \sqrt{20}$, the equation of the circle is

$$(x - h)^2 + (y - k)^2 = r^2 \qquad \text{Equation of circle}$$
$$[x - (-1)]^2 + (y - 2)^2 = \left(\sqrt{20}\right)^2 \qquad \text{Substitute for } h, k, \text{ and } r.$$
$$(x + 1)^2 + (y - 2)^2 = 20. \qquad \text{Standard form}$$

✓ **Checkpoint** ◀))) Audio-video solution in English & Spanish at LarsonPrecalculus.com

The point $(1, -2)$ lies on a circle whose center is at $(-3, -5)$. Write the standard form of the equation of this circle.

To find h and k from the standard form of the equation of a circle, you may want to rewrite one or both of the quantities in parentheses. For example, $x + 1 = x - (-1)$.

Application

In this course, you will learn that there are many ways to approach a problem. Example 9 illustrates three common approaches.

A numerical approach: Construct and use a table.

A graphical approach: Draw and use a graph.

An algebraic approach: Use the rules of algebra.

EXAMPLE 9 **Maximum Weight**

The maximum weight y (in pounds) for a man in the United States Marine Corps can be approximated by the mathematical model

$$y = 0.040x^2 - 0.11x + 3.9, \quad 58 \le x \le 80$$

where x is the man's height (in inches). *(Source: U.S. Department of Defense)*

a. Construct a table of values that shows the maximum weights for men with heights of 62, 64, 66, 68, 70, 72, 74, and 76 inches.

b. Use the table of values to sketch a graph of the model. Then use the graph to estimate *graphically* the maximum weight for a man whose height is 71 inches.

c. Use the model to confirm *algebraically* the estimate you found in part (b).

Solution

a. Use a calculator to construct a table, as shown at the left.

b. Use the table of values to sketch the graph of the equation, as shown in Figure 1.19. From the graph, you can estimate that a height of 71 inches corresponds to a weight of about 198 pounds.

c. To confirm algebraically the estimate you found in part (b), substitute 71 for x in the model.

$$y = 0.040(71)^2 - 0.11(71) + 3.9$$
$$\approx 197.7$$

So, the graphical estimate of 198 pounds is fairly good.

DATA	Height, x	Weight, y
	62	150.8
	64	160.7
	66	170.9
	68	181.4
	70	192.2
	72	203.3
	74	214.8
	76	226.6

Spreadsheet at LarsonPrecalculus.com

✓ *Checkpoint* ◀))) Audio-video solution in English & Spanish at LarsonPrecalculus.com

Use Figure 1.19 to estimate *graphically* the maximum weight for a man whose height is 75 inches. Then confirm the estimate *algebraically*. ∎

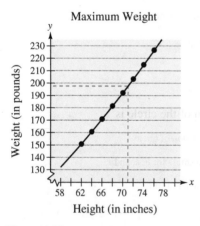

Maximum Weight

Figure 1.19

Summarize (Section 1.2)

1. Explain how to sketch the graph of an equation *(page 11)*. For examples of sketching graphs of equations, see Examples 2 and 3.

2. Explain how to find the x- and y-intercepts of a graph *(page 14)*. For an example of finding x- and y-intercepts, see Example 4.

3. Explain how to use symmetry to graph an equation *(page 15)*. For an example of using symmetry to graph an equation, see Example 6.

4. State the standard form of the equation of a circle *(page 17)*. For an example of writing the standard form of the equation of a circle, see Example 8.

5. Describe an example of how to use the graph of an equation to solve a real-life problem *(page 18, Example 9)*.

1.2 Exercises

See **CalcChat.com** for tutorial help and worked-out solutions to odd-numbered exercises.

Vocabulary: **Fill in the blanks.**

1. An ordered pair (a, b) is a _____ of an equation in x and y when the substitutions $x = a$ and $y = b$ result in a true statement.

2. The set of all solution points of an equation is the _____ of the equation.

3. The points at which a graph intersects or touches an axis are the _____ of the graph.

4. A graph is symmetric with respect to the _____ if, whenever (x, y) is on the graph, $(-x, y)$ is also on the graph.

5. The equation $(x - h)^2 + (y - k)^2 = r^2$ is the standard form of the equation of a _____ with center _____ and radius _____.

6. When you construct and use a table to solve a problem, you are using a _____ approach.

Skills and Applications

 Determining Solution Points In Exercises 7–14, determine whether each point lies on the graph of the equation.

Equation	Points	
7. $y = \sqrt{x + 4}$	(a) $(0, 2)$	(b) $(5, 3)$
8. $y = \sqrt{5 - x}$	(a) $(1, 2)$	(b) $(5, 0)$
9. $y = x^2 - 3x + 2$	(a) $(2, 0)$	(b) $(-2, 8)$
10. $y = 3 - 2x^2$	(a) $(-1, 1)$	(b) $(-2, 11)$
11. $y = 4 - |x - 2|$	(a) $(1, 5)$	(b) $(6, 0)$
12. $y = |x - 1| + 2$	(a) $(2, 3)$	(b) $(-1, 0)$
13. $x^2 + y^2 = 20$	(a) $(3, -2)$	(b) $(-4, 2)$
14. $2x^2 + 5y^2 = 8$	(a) $(6, 0)$	(b) $(0, 4)$

 Sketching the Graph of an Equation In Exercises 15–18, complete the table. Use the resulting solution points to sketch the graph of the equation.

15. $y = -2x + 5$

x	-1	0	1	2	$\frac{5}{2}$
y					
(x, y)					

16. $y + 1 = \frac{3}{4}x$

x	-2	0	1	$\frac{4}{3}$	2
y					
(x, y)					

17. $y + 3x = x^2$

x	-1	0	1	2	3
y					
(x, y)					

18. $y = 5 - x^2$

x	-2	-1	0	1	2
y					
(x, y)					

 Identifying x- and y-Intercepts In Exercises 19–22, identify the x- and y-intercepts of the graph. Verify your results algebraically.

19. $y = (x - 3)^2$

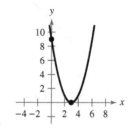

20. $y = 16 - 4x^2$

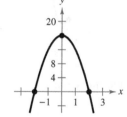

21. $y = |x + 2|$

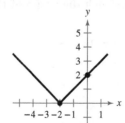

22. $y^2 = 4 - x$

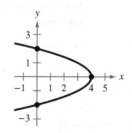

Finding x- and y-Intercepts In Exercises 23–32, find the x- and y-intercepts of the graph of the equation.

23. $y = 5x - 6$
24. $y = 8 - 3x$
25. $y = \sqrt{x + 4}$
26. $y = \sqrt{2x - 1}$
27. $y = |3x - 7|$
28. $y = -|x + 10|$
29. $y = 2x^3 - 4x^2$
30. $y = x^4 - 25$
31. $y^2 = 6 - x$
32. $y^2 = x + 1$

Testing for Symmetry In Exercises 33–40, use the algebraic tests to check for symmetry with respect to both axes and the origin.

33. $x^2 - y = 0$

34. $x - y^2 = 0$

35. $y = x^3$

36. $y = x^4 - x^2 + 3$

37. $y = \dfrac{x}{x^2 + 1}$

38. $y = \dfrac{1}{x^2 + 1}$

39. $xy^2 + 10 = 0$

40. $xy = 4$

Using Symmetry as a Sketching Aid In Exercises 41–44, assume that the graph has the given type of symmetry. Complete the graph of the equation. To print an enlarged copy of the graph, go to *MathGraphs.com*.

41.

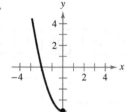

y-Axis symmetry

42.

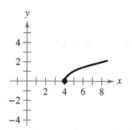

x-Axis symmetry

43.

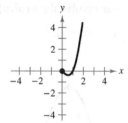

Origin symmetry

44.

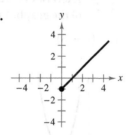

y-Axis symmetry

Sketching the Graph of an Equation In Exercises 45–56, find any intercepts and test for symmetry. Then sketch the graph of the equation.

45. $y = -3x + 1$

46. $y = 2x - 3$

47. $y = x^2 - 2x$

48. $y = -x^2 - 2x$

49. $y = x^3 + 3$

50. $y = x^3 - 1$

51. $y = \sqrt{x - 3}$

52. $y = \sqrt{1 - x}$

53. $y = |x - 6|$

54. $y = 1 - |x|$

55. $x = y^2 - 1$

56. $x = y^2 - 5$

$\mathrel{\rlap{/}{\triangle}}$ **Using Technology** In Exercises 57–66, use a graphing utility to graph the equation. Use a standard setting. Approximate any intercepts.

57. $y = 3 - \frac{1}{2}x$

58. $y = \frac{2}{3}x - 1$

59. $y = x^2 - 4x + 3$

60. $y = x^2 + x - 2$

61. $y = \dfrac{2x}{x - 1}$

62. $y = \dfrac{4}{x^2 + 1}$

63. $y = \sqrt[3]{x + 1}$

64. $y = x\sqrt{x + 6}$

65. $y = |x + 3|$

66. $y = 2 - |x|$

Writing the Equation of a Circle In Exercises 67–74, write the standard form of the equation of the circle with the given characteristics.

67. Center: $(0, 0)$; Radius: 3

68. Center: $(0, 0)$; Radius: 7

69. Center: $(-4, 5)$; Radius: 2

70. Center: $(1, -3)$; Radius: $\sqrt{11}$

71. Center: $(3, 8)$; Solution point: $(-9, 13)$

72. Center: $(-2, -6)$; Solution point: $(1, -10)$

73. Endpoints of a diameter: $(3, 2), (-9, -8)$

74. Endpoints of a diameter: $(11, -5), (3, 15)$

Sketching a Circle In Exercises 75–80, find the center and radius of the circle with the given equation. Then sketch the circle.

75. $x^2 + y^2 = 25$

76. $x^2 + y^2 = 16$

77. $(x - 1)^2 + (y + 3)^2 = 9$

78. $x^2 + (y - 1)^2 = 1$

79. $\left(x - \frac{1}{2}\right)^2 + \left(y - \frac{1}{2}\right)^2 = \frac{9}{4}$

80. $(x - 2)^2 + (y + 3)^2 = \frac{16}{9}$

81. **Depreciation** A hospital purchases a new magnetic resonance imaging (MRI) machine for $1.2 million. The depreciated value y (reduced value) after t years is given by $y = 1{,}200{,}000 - 80{,}000t$, $0 \le t \le 10$. Sketch the graph of the equation.

82. **Depreciation** You purchase an all-terrain vehicle (ATV) for $9500. The depreciated value y (reduced value) after t years is given by $y = 9500 - 1000t$, $0 \le t \le 6$. Sketch the graph of the equation.

$\mathrel{\rlap{/}{\triangle}}$ 83. **Geometry** A regulation NFL playing field of length x and width y has a perimeter of $346\frac{2}{3}$ or $\frac{1040}{3}$ yards.

(a) Draw a rectangle that gives a visual representation of the problem. Use the specified variables to label the sides of the rectangle.

(b) Show that the width of the rectangle is $y = \frac{520}{3} - x$ and its area is $A = x\left(\frac{520}{3} - x\right)$.

(c) Use a graphing utility to graph the area equation. Be sure to adjust your window settings.

(d) From the graph in part (c), estimate the dimensions of the rectangle that yield a maximum area.

(e) Use your school's library, the Internet, or some other reference source to find the actual dimensions and area of a regulation NFL playing field and compare your findings with the results of part (d).

The symbol $\mathrel{\rlap{/}{\triangle}}$ indicates an exercise or a part of an exercise in which you are instructed to use a graphing utility.

84. Architecture The arch support of a bridge is modeled by $y = -0.0012x^2 + 300$, where x and y are measured in feet and the x-axis represents the ground.

(a) Use a graphing utility to graph the equation.

(b) Find one x-intercept of the graph. Explain how to use the intercept and the symmetry of the graph to find the width of the arch support.

• • 85. Population Statistics • • • • • • • • • • • •

The table shows the life expectancies of a child (at birth) in the United States for selected years from 1940 through 2010. *(Source: U.S. National Center for Health Statistics)*

DATA	Year	Life Expectancy, y
	1940	62.9
	1950	68.2
	1960	69.7
	1970	70.8
	1980	73.7
	1990	75.4
	2000	76.8
	2010	78.7

Spreadsheet at LarsonPrecalculus.com

A model for the life expectancy during this period is

$$y = \frac{63.6 + 0.97t}{1 + 0.01t}, \quad 0 \le t \le 70$$

where y represents the life expectancy and t is the time in years, with $t = 0$ corresponding to 1940.

(a) Use a graphing utility to graph the data from the table and the model in the same viewing window. How well does the model fit the data? Explain.

(b) Determine the life expectancy in 1990 both graphically and algebraically.

(c) Use the graph to determine the year when life expectancy was approximately 70.1. Verify your answer algebraically.

(d) Find the y-intercept of the graph of the model. What does it represent in the context of the problem?

(e) Do you think this model can be used to predict the life expectancy of a child 50 years from now? Explain.

86. Electronics The resistance y (in ohms) of 1000 feet of solid copper wire at 68 degrees Fahrenheit is

$$y = \frac{10{,}370}{x^2}$$

where x is the diameter of the wire in mils (0.001 inch).

(a) Complete the table.

x	5	10	20	30	40	50
y						

x	60	70	80	90	100
y					

(b) Use the table of values in part (a) to sketch a graph of the model. Then use your graph to estimate the resistance when $x = 85.5$.

(c) Use the model to confirm algebraically the estimate you found in part (b).

(d) What can you conclude about the relationship between the diameter of the copper wire and the resistance?

Exploration

True or False? In Exercises 87–89, determine whether the statement is true or false. Justify your answer.

87. The graph of a linear equation cannot be symmetric with respect to the origin.

88. The graph of a linear equation can have either no x-intercepts or only one x-intercept.

89. A circle can have a total of zero, one, two, three, or four x- and y-intercepts.

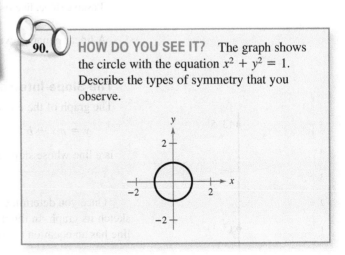

90. **HOW DO YOU SEE IT?** The graph shows the circle with the equation $x^2 + y^2 = 1$. Describe the types of symmetry that you observe.

91. Think About It Find a and b when the graph of $y = ax^2 + bx^3$ is symmetric with respect to (a) the y-axis and (b) the origin. (There are many correct answers.)

1.3 Linear Equations in Two Variables

Linear equations in two variables can help you model and solve real-life problems. For example, in Exercise 90 on page 33, you will use a surveyor's measurements to find a linear equation that models a mountain road.

- Use slope to graph linear equations in two variables.
- Find the slope of a line given two points on the line.
- Write linear equations in two variables.
- Use slope to identify parallel and perpendicular lines.
- Use slope and linear equations in two variables to model and solve real-life problems.

Using Slope

The simplest mathematical model for relating two variables is the **linear equation in two variables** $y = mx + b$. The equation is called *linear* because its graph is a line. (In mathematics, the term *line* means *straight line*.) By letting $x = 0$, you obtain

$$y = m(0) + b = b.$$

So, the line crosses the y-axis at $y = b$, as shown in the figures below. In other words, the y-intercept is $(0, b)$. The steepness, or *slope,* of the line is m.

$$y = mx + b$$

Slope ⎦   ⎦ y-Intercept

The **slope** of a nonvertical line is the number of units the line rises (or falls) vertically for each unit of horizontal change from left to right, as shown below.

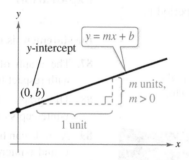

Positive slope, line rises

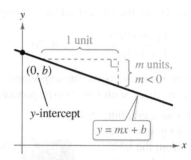

Negative slope, line falls

A linear equation written in **slope-intercept form** has the form $y = mx + b$.

> ### The Slope-Intercept Form of the Equation of a Line
> The graph of the equation
> $$y = mx + b$$
> is a line whose slope is m and whose y-intercept is $(0, b)$.

Once you determine the slope and the y-intercept of a line, it is relatively simple to sketch its graph. In the next example, note that none of the lines is vertical. A vertical line has an equation of the form

$$x = a. \qquad \text{Vertical line}$$

The equation of a vertical line cannot be written in the form $y = mx + b$ because the slope of a vertical line is undefined (see Figure 1.20).

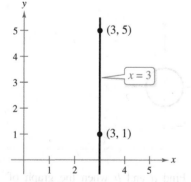

Slope is undefined.
Figure 1.20

EXAMPLE 1   **Graphing Linear Equations**

See LarsonPrecalculus.com for an interactive version of this type of example.

Sketch the graph of each linear equation.

a. $y = 2x + 1$

b. $y = 2$

c. $x + y = 2$

Solution

a. Because $b = 1$, the y-intercept is $(0, 1)$. Moreover, the slope is $m = 2$, so the line *rises* two units for each unit the line moves to the right (see figure).

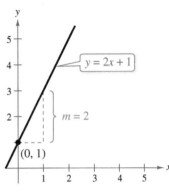

When m is positive, the line rises.

b. By writing this equation in the form $y = (0)x + 2$, you find that the y-intercept is $(0, 2)$ and the slope is $m = 0$. A slope of 0 implies that the line is horizontal—that is, it does not rise *or* fall (see figure).

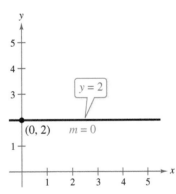

When m is 0, the line is horizontal.

c. By writing this equation in slope-intercept form

$$x + y = 2 \qquad \text{Write original equation.}$$
$$y = -x + 2 \qquad \text{Subtract } x \text{ from each side.}$$
$$y = (-1)x + 2 \qquad \text{Write in slope-intercept form.}$$

you find that the y-intercept is $(0, 2)$. Moreover, the slope is $m = -1$, so the line *falls* one unit for each unit the line moves to the right (see figure).

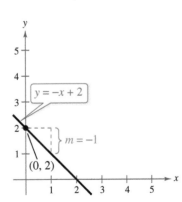

When m is negative, the line falls.

 Checkpoint ◄))) *Audio-video solution in English & Spanish at LarsonPrecalculus.com*

Sketch the graph of each linear equation.

a. $y = 3x + 2$ **b.** $y = -3$ **c.** $4x + y = 5$

Finding the Slope of a Line

Given an equation of a line, you can find its slope by writing the equation in slope-intercept form. When you are not given an equation, you can still find the slope by using two points on the line. For example, consider the line passing through the points (x_1, y_1) and (x_2, y_2) in the figure below.

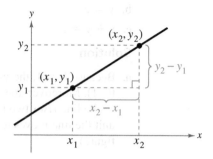

As you move from left to right along this line, a change of $(y_2 - y_1)$ units in the vertical direction corresponds to a change of $(x_2 - x_1)$ units in the horizontal direction.

$$y_2 - y_1 = \text{change in } y = \text{rise}$$

and

$$x_2 - x_1 = \text{change in } x = \text{run}$$

The ratio of $(y_2 - y_1)$ to $(x_2 - x_1)$ represents the slope of the line that passes through the points (x_1, y_1) and (x_2, y_2).

$$\text{Slope} = \frac{\text{change in } y}{\text{change in } x} = \frac{\text{rise}}{\text{run}} = \frac{y_2 - y_1}{x_2 - x_1}$$

The Slope of a Line Passing Through Two Points

The **slope** m of the nonvertical line through (x_1, y_1) and (x_2, y_2) is

$$m = \frac{y_2 - y_1}{x_2 - x_1}$$

where $x_1 \neq x_2$.

When using the formula for slope, the *order of subtraction* is important. Given two points on a line, you are free to label either one of them as (x_1, y_1) and the other as (x_2, y_2). However, once you do this, you must form the numerator and denominator using the same order of subtraction.

$$m = \frac{y_2 - y_1}{x_2 - x_1} \qquad m = \frac{y_1 - y_2}{x_1 - x_2} \qquad m = \frac{y_2 - y_1}{x_1 - x_2}$$

Correct   Correct   Incorrect

For example, the slope of the line passing through the points $(3, 4)$ and $(5, 7)$ can be calculated as

$$m = \frac{7 - 4}{5 - 3} = \frac{3}{2}$$

or as

$$m = \frac{4 - 7}{3 - 5} = \frac{-3}{-2} = \frac{3}{2}.$$

EXAMPLE 2   **Finding the Slope of a Line Through Two Points**

Find the slope of the line passing through each pair of points.

a. $(-2, 0)$ and $(3, 1)$ **b.** $(-1, 2)$ and $(2, 2)$

c. $(0, 4)$ and $(1, -1)$ **d.** $(3, 4)$ and $(3, 1)$

Solution

a. Letting $(x_1, y_1) = (-2, 0)$ and $(x_2, y_2) = (3, 1)$, you find that the slope is

$$m = \frac{y_2 - y_1}{x_2 - x_1} = \frac{1 - 0}{3 - (-2)} = \frac{1}{5}.$$ See Figure 1.21.

b. The slope of the line passing through $(-1, 2)$ and $(2, 2)$ is

$$m = \frac{2 - 2}{2 - (-1)} = \frac{0}{3} = 0.$$ See Figure 1.22.

c. The slope of the line passing through $(0, 4)$ and $(1, -1)$ is

$$m = \frac{-1 - 4}{1 - 0} = \frac{-5}{1} = -5.$$ See Figure 1.23.

d. The slope of the line passing through $(3, 4)$ and $(3, 1)$ is

$$m = \frac{1 - 4}{3 - 3} = \frac{-3}{0}.$$ See Figure 1.24.

Division by 0 is undefined, so the slope is undefined and the line is vertical.

· · **REMARK** In Figures
1.21 through 1.24, note the
relationships between slope
and the orientation of the line.

a. Positive slope: line rises
from left to right

b. Zero slope: line is horizontal

c. Negative slope: line falls
from left to right

d. Undefined slope: line is
vertical

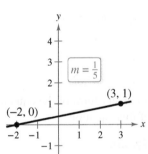

Figure 1.21

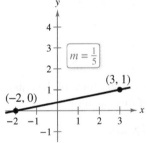

Figure 1.22

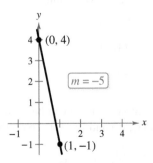

Figure 1.23

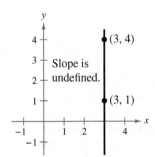

Figure 1.24

✓ *Checkpoint*))) *Audio-video solution in English & Spanish at LarsonPrecalculus.com*

Find the slope of the line passing through each pair of points.

a. $(-5, -6)$ and $(2, 8)$ **b.** $(4, 2)$ and $(2, 5)$

c. $(0, 0)$ and $(0, -6)$ **d.** $(0, -1)$ and $(3, -1)$

Writing Linear Equations in Two Variables

If (x_1, y_1) is a point on a line of slope m and (x, y) is *any other* point on the line, then

$$\frac{y - y_1}{x - x_1} = m.$$

This equation in the variables x and y can be rewritten in the **point-slope form** of the equation of a line

$$y - y_1 = m(x - x_1).$$

> ### Point-Slope Form of the Equation of a Line
> The equation of the line with slope m passing through the point (x_1, y_1) is
> $$y - y_1 = m(x - x_1).$$

The point-slope form is useful for *finding* the equation of a line. You should remember this form.

EXAMPLE 3   Using the Point-Slope Form

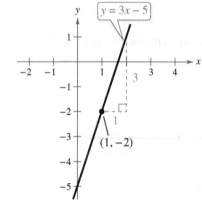

Figure 1.25

Find the slope-intercept form of the equation of the line that has a slope of 3 and passes through the point $(1, -2)$.

Solution   Use the point-slope form with $m = 3$ and $(x_1, y_1) = (1, -2)$.

$$y - y_1 = m(x - x_1) \qquad \text{Point-slope form}$$
$$y - (-2) = 3(x - 1) \qquad \text{Substitute for } m, x_1, \text{ and } y_1.$$
$$y + 2 = 3x - 3 \qquad \text{Simplify.}$$
$$y = 3x - 5 \qquad \text{Write in slope-intercept form.}$$

The slope-intercept form of the equation of the line is $y = 3x - 5$. Figure 1.25 shows the graph of this equation.

✓ **Checkpoint** ◀))) *Audio-video solution in English & Spanish at LarsonPrecalculus.com*

Find the slope-intercept form of the equation of the line that has the given slope and passes through the given point.

a. $m = 2$, $(3, -7)$

b. $m = -\frac{2}{3}$, $(1, 1)$

c. $m = 0$, $(1, 1)$

• **REMARK**   When you find an equation of the line that passes through two given points, you only need to substitute the coordinates of one of the points in the point-slope form. It does not matter which point you choose because both points will yield the same result.

The point-slope form can be used to find an equation of the line passing through two points (x_1, y_1) and (x_2, y_2). To do this, first find the slope of the line.

$$m = \frac{y_2 - y_1}{x_2 - x_1}, \quad x_1 \neq x_2$$

Then use the point-slope form to obtain the equation.

$$y - y_1 = \frac{y_2 - y_1}{x_2 - x_1}(x - x_1) \qquad \text{Two-point form}$$

This is sometimes called the **two-point form** of the equation of a line.

Parallel and Perpendicular Lines

Slope can tell you whether two nonvertical lines in a plane are parallel, perpendicular, or neither.

Parallel and Perpendicular Lines

1. Two distinct nonvertical lines are **parallel** if and only if their slopes are equal. That is,

$$m_1 = m_2.$$

2. Two nonvertical lines are **perpendicular** if and only if their slopes are negative reciprocals of each other. That is,

$$m_1 = \frac{-1}{m_2}.$$

EXAMPLE 4 **Finding Parallel and Perpendicular Lines**

Find the slope-intercept form of the equations of the lines that pass through the point $(2, -1)$ and are (a) parallel to and (b) perpendicular to the line $2x - 3y = 5$.

Solution Write the equation of the given line in slope-intercept form.

$2x - 3y = 5$	Write original equation.
$-3y = -2x + 5$	Subtract $2x$ from each side.
$y = \frac{2}{3}x - \frac{5}{3}$	Write in slope-intercept form.

Notice that the line has a slope of $m = \frac{2}{3}$.

a. Any line parallel to the given line must also have a slope of $\frac{2}{3}$. Use the point-slope form with $m = \frac{2}{3}$ and $(x_1, y_1) = (2, -1)$.

$y - (-1) = \frac{2}{3}(x - 2)$	Write in point-slope form.
$3(y + 1) = 2(x - 2)$	Multiply each side by 3.
$3y + 3 = 2x - 4$	Distributive Property
$y = \frac{2}{3}x - \frac{7}{3}$	Write in slope-intercept form.

Notice the similarity between the slope-intercept form of this equation and the slope-intercept form of the given equation.

b. Any line perpendicular to the given line must have a slope of $-\frac{3}{2}$ $\left(\text{because } -\frac{3}{2} \text{ is the negative reciprocal of } \frac{2}{3}\right)$. Use the point-slope form with $m = -\frac{3}{2}$ and $(x_1, y_1) = (2, -1)$.

$y - (-1) = -\frac{3}{2}(x - 2)$	Write in point-slope form.
$2(y + 1) = -3(x - 2)$	Multiply each side by 2.
$2y + 2 = -3x + 6$	Distributive Property
$y = -\frac{3}{2}x + 2$	Write in slope-intercept form.

The graphs of all three equations are shown in Figure 1.26.

✓ Checkpoint *Audio-video solution in English & Spanish at LarsonPrecalculus.com*

Find the slope-intercept form of the equations of the lines that pass through the point $(-4, 1)$ and are (a) parallel to and (b) perpendicular to the line $5x - 3y = 8$. ◼

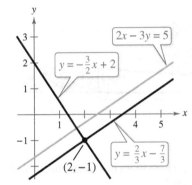

Figure 1.26

▷ **TECHNOLOGY** On a graphing utility, lines will not appear to have the correct slope unless you use a viewing window that has a square setting. For instance, graph the lines in Example 4 using the standard setting $-10 \leq x \leq 10$ and $-10 \leq y \leq 10$. Then reset the viewing window with the square setting $-9 \leq x \leq 9$ and $-6 \leq y \leq 6$. On which setting do the lines $y = \frac{2}{3}x - \frac{5}{3}$ and $y = -\frac{3}{2}x + 2$ appear to be perpendicular?

Applications

In real-life problems, the slope of a line can be interpreted as either a *ratio* or a *rate*. When the *x*-axis and *y*-axis have the same unit of measure, the slope has no units and is a **ratio**. When the *x*-axis and *y*-axis have different units of measure, the slope is a **rate** or **rate of change.**

EXAMPLE 5 Using Slope as a Ratio

The maximum recommended slope of a wheelchair ramp is $\frac{1}{12}$. A business installs a wheelchair ramp that rises 22 inches over a horizontal length of 24 feet. Is the ramp steeper than recommended? *(Source: ADA Standards for Accessible Design)*

Solution The horizontal length of the ramp is 24 feet or 12(24) = 288 inches (see figure). So, the slope of the ramp is

$$\text{Slope} = \frac{\text{vertical change}}{\text{horizontal change}} = \frac{22 \text{ in.}}{288 \text{ in.}} \approx 0.076.$$

Because $\frac{1}{12} \approx 0.083$, the slope of the ramp is not steeper than recommended.

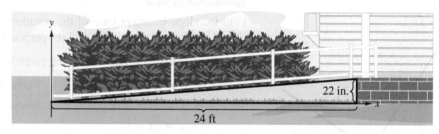

✓ *Checkpoint* 🔊))) *Audio-video solution in English & Spanish at LarsonPrecalculus.com*

The business in Example 5 installs a second ramp that rises 36 inches over a horizontal length of 32 feet. Is the ramp steeper than recommended?

EXAMPLE 6 Using Slope as a Rate of Change

A kitchen appliance manufacturing company determines that the total cost *C* (in dollars) of producing *x* units of a blender is given by

$$C = 25x + 3500. \qquad \text{Cost equation}$$

Interpret the *y*-intercept and slope of this line.

Solution The *y*-intercept (0, 3500) tells you that the cost of producing 0 units is \$3500. This is the *fixed cost* of production—it includes costs that must be paid regardless of the number of units produced. The slope of *m* = 25 tells you that the cost of producing each unit is \$25, as shown in Figure 1.27. Economists call the cost per unit the *marginal cost*. When the production increases by one unit, the "margin," or extra amount of cost, is \$25. So, the cost increases at a rate of \$25 per unit.

✓ *Checkpoint* 🔊))) *Audio-video solution in English & Spanish at LarsonPrecalculus.com*

An accounting firm determines that the value *V* (in dollars) of a copier *t* years after its purchase is given by

$$V = -300t + 1500.$$

Interpret the *y*-intercept and slope of this line.

The Americans with Disabilities Act (ADA) became law on July 26, 1990. It is the most comprehensive formulation of rights for persons with disabilities in U.S. (and world) history.

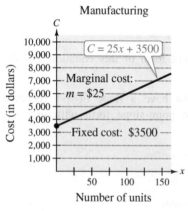

Manufacturing

C = 25*x* + 3500

Marginal cost: *m* = \$25

Fixed cost: \$3500

Cost (in dollars) vs. Number of units

Production cost
Figure 1.27

Businesses can deduct most of their expenses in the same year they occur. One exception is the cost of property that has a useful life of more than 1 year. Such costs must be *depreciated* (decreased in value) over the useful life of the property. Depreciating the *same amount* each year is called *linear* or *straight-line depreciation*. The *book value* is the difference between the original value and the total amount of depreciation accumulated to date.

EXAMPLE 7   Straight-Line Depreciation

A college purchased exercise equipment worth $12,000 for the new campus fitness center. The equipment has a useful life of 8 years. The salvage value at the end of 8 years is $2000. Write a linear equation that describes the book value of the equipment each year.

Solution   Let V represent the value of the equipment at the end of year t. Represent the initial value of the equipment by the data point $(0, 12,000)$ and the salvage value of the equipment by the data point $(8, 2000)$. The slope of the line is

$$m = \frac{2000 - 12,000}{8 - 0} = -\$1250$$

which represents the annual depreciation in *dollars per year*. Using the point-slope form, write an equation of the line.

$$V - 12,000 = -1250(t - 0) \qquad \text{Write in point-slope form.}$$
$$V = -1250t + 12,000 \qquad \text{Write in slope-intercept form.}$$

The table shows the book value at the end of each year, and Figure 1.28 shows the graph of the equation.

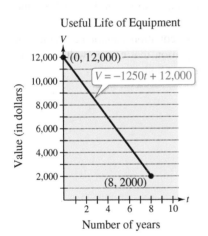

Useful Life of Equipment

Straight-line depreciation

Figure 1.28

Year, t	Value, V
0	12,000
1	10,750
2	9500
3	8250
4	7000
5	5750
6	4500
7	3250
8	2000

✓ **Checkpoint**  *Audio-video solution in English & Spanish at LarsonPrecalculus.com*

A manufacturing firm purchases a machine worth $24,750. The machine has a useful life of 6 years. After 6 years, the machine will have to be discarded and replaced, because it will have no salvage value. Write a linear equation that describes the book value of the machine each year.

In many real-life applications, the two data points that determine the line are often given in a disguised form. Note how the data points are described in Example 7.

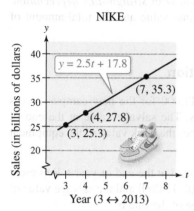

NIKE

Figure 1.29

EXAMPLE 8 **Predicting Sales**

The sales for NIKE were approximately $25.3 billion in 2013 and $27.8 billion in 2014. Using only this information, write a linear equation that gives the sales in terms of the year. Then predict the sales in 2017. *(Source: NIKE Inc.)*

Solution Let $t = 3$ represent 2013. Then the two given values are represented by the data points $(3, 25.3)$ and $(4, 27.8)$ The slope of the line through these points is

$$m = \frac{27.8 - 25.3}{4 - 3} = 2.5.$$

Use the point-slope form to write an equation that relates the sales y and the year t.

$$y - 25.3 = 2.5(t - 3) \qquad \text{Write in point-slope form.}$$
$$y = 2.5t + 17.8 \qquad \text{Write in slope-intercept form.}$$

According to this equation, the sales in 2017 will be

$$y = 2.5(7) + 17.8 = 17.5 + 17.8 = \$35.3 \text{ billion. (See Figure 1.29.)}$$

✓ *Checkpoint* 🔊))  Audio-video solution in English & Spanish at LarsonPrecalculus.com

The sales for Foot Locker were approximately $6.5 billion in 2013 and $7.2 billion in 2014. Repeat Example 8 using this information. *(Source: Foot Locker)*

The prediction method illustrated in Example 8 is called **linear extrapolation.** Note in Figure 1.30 that an extrapolated point does not lie between the given points. When the estimated point lies between two given points, as shown in Figure 1.31, the procedure is called **linear interpolation.**

The slope of a vertical line is undefined, so its equation cannot be written in slope-intercept form. However, every line has an equation that can be written in the **general form** $Ax + By + C = 0$, where A and B are not both zero.

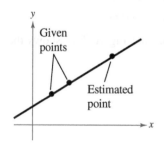

Linear extrapolation
Figure 1.30

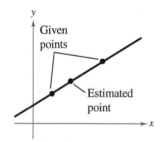

Linear interpolation
Figure 1.31

Summary of Equations of Lines

1. General form: $\qquad Ax + By + C = 0$
2. Vertical line: $\qquad x = a$
3. Horizontal line: $\qquad y = b$
4. Slope-intercept form: $y = mx + b$
5. Point-slope form: $\qquad y - y_1 = m(x - x_1)$
6. Two-point form: $\qquad y - y_1 = \dfrac{y_2 - y_1}{x_2 - x_1}(x - x_1)$

Summarize (Section 1.3)

1. Explain how to use slope to graph a linear equation in two variables *(page 22)* and how to find the slope of a line passing through two points *(page 24)*. For examples of using and finding slopes, see Examples 1 and 2.

2. State the point-slope form of the equation of a line *(page 26)*. For an example of using point-slope form, see Example 3.

3. Explain how to use slope to identify parallel and perpendicular lines *(page 27)*. For an example of finding parallel and perpendicular lines, see Example 4.

4. Describe examples of how to use slope and linear equations in two variables to model and solve real-life problems *(pages 28–30, Examples 5–8)*.

1.3 Exercises

See **CalcChat.com** for tutorial help and worked-out solutions to odd-numbered exercises.

Vocabulary: Fill in the blanks.

1. The simplest mathematical model for relating two variables is the _____ equation in two variables $y = mx + b$.

2. For a line, the ratio of the change in y to the change in x is the _____ of the line.

3. The _____-_____ form of the equation of a line with slope m passing through the point (x_1, y_1) is $y - y_1 = m(x - x_1)$.

4. Two distinct nonvertical lines are _____ if and only if their slopes are equal.

5. Two nonvertical lines are _____ if and only if their slopes are negative reciprocals of each other.

6. When the x-axis and y-axis have different units of measure, the slope can be interpreted as a _____.

7. _____ _____ is the prediction method used to estimate a point on a line when the point does not lie between the given points.

8. Every line has an equation that can be written in _____ form.

Skills and Applications

Identifying Lines In Exercises 9 and 10, identify the line that has each slope.

9. (a) $m = \frac{2}{3}$ 10. (a) $m = 0$

   (b) m is undefined.   (b) $m = -\frac{3}{4}$

   (c) $m = -2$ (c) $m = 1$

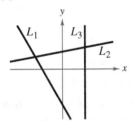   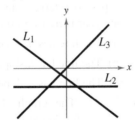

Sketching Lines In Exercises 11 and 12, sketch the lines through the point with the given slopes on the same set of coordinate axes.

Point	Slopes
11. $(2, 3)$	(a) 0 (b) 1
	(c) 2 (d) -3
12. $(-4, 1)$	(a) 3 (b) -3
	(c) $\frac{1}{2}$ (d) Undefined

Estimating the Slope of a Line In Exercises 13 and 14, estimate the slope of the line.

13. 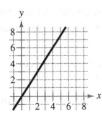   14.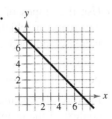

 Graphing a Linear Equation In Exercises 15–24, find the slope and y-intercept (if possible) of the line. Sketch the line.

15. $y = 5x + 3$ 16. $y = -x - 10$

17. $y = -\frac{3}{4}x - 1$ 18. $y = \frac{2}{3}x + 2$

19. $y - 5 = 0$ 20. $x + 4 = 0$

21. $5x - 2 = 0$ 22. $3y + 5 = 0$

23. $7x - 6y = 30$ 24. $2x + 3y = 9$

 Finding the Slope of a Line Through Two Points In Exercises 25–34, find the slope of the line passing through the pair of points.

25. $(0, 9), (6, 0)$ 26. $(10, 0), (0, -5)$

27. $(-3, -2), (1, 6)$ 28. $(2, -1), (-2, 1)$

29. $(5, -7), (8, -7)$ 30. $(-2, 1), (-4, -5)$

31. $(-6, -1), (-6, 4)$ 32. $(0, -10), (-4, 0)$

33. $(4.8, 3.1), (-5.2, 1.6)$

34. $\left(\frac{11}{2}, -\frac{4}{3}\right), \left(-\frac{3}{2}, -\frac{1}{3}\right)$

Using the Slope and a Point In Exercises 35–42, use the slope of the line and the point on the line to find three additional points through which the line passes. (There are many correct answers.)

35. $m = 0$, $(5, 7)$ 36. $m = 0$, $(3, -2)$

37. $m = 2$, $(-5, 4)$ 38. $m = -2$, $(0, -9)$

39. $m = -\frac{1}{3}$, $(4, 5)$ 40. $m = \frac{1}{4}$, $(3, -4)$

41. m is undefined, $(-4, 3)$

42. m is undefined, $(2, 14)$

Using the Point-Slope Form In Exercises 43–54, find the slope-intercept form of the equation of the line that has the given slope and passes through the given point. Sketch the line.

43. $m = 3$, $(0, -2)$ 44. $m = -1$, $(0, 10)$

45. $m = -2$, $(-3, 6)$ 46. $m = 4$, $(0, 0)$

47. $m = -\frac{1}{3}$, $(4, 0)$ 48. $m = \frac{1}{4}$, $(8, 2)$

49. $m = -\frac{1}{2}$, $(2, -3)$ 50. $m = \frac{3}{4}$, $(-2, -5)$

51. $m = 0$, $\left(4, \frac{5}{2}\right)$ 52. $m = 6$, $\left(2, \frac{3}{2}\right)$

53. $m = 5$, $(-5.1, 1.8)$ 54. $m = 0$, $(-2.5, 3.25)$

Finding an Equation of a Line In Exercises 55–64, find an equation of the line passing through the pair of points. Sketch the line.

55. $(5, -1), (-5, 5)$ 56. $(4, 3), (-4, -4)$

57. $(-7, 2), (-7, 5)$ 58. $(-6, -3), (2, -3)$

59. $\left(2, \frac{1}{2}\right), \left(\frac{1}{2}, \frac{5}{4}\right)$ 60. $(1, 1), \left(6, -\frac{2}{3}\right)$

61. $(1, 0.6), (-2, -0.6)$ 62. $(-8, 0.6), (2, -2.4)$

63. $(2, -1), \left(\frac{1}{3}, -1\right)$ 64. $\left(\frac{7}{3}, -8\right), \left(\frac{7}{3}, 1\right)$

Parallel and Perpendicular Lines In Exercises 65–68, determine whether the lines are parallel, perpendicular, or neither.

65. $L_1: y = -\frac{2}{3}x - 3$ 66. $L_1: y = \frac{1}{4}x - 1$

    $L_2: y = -\frac{2}{3}x + 4$ $L_2: y = 4x + 7$

67. $L_1: y = \frac{1}{2}x - 3$ 68. $L_1: y = -\frac{4}{5}x - 5$

    $L_2: y = -\frac{1}{2}x + 1$ $L_2: y = \frac{5}{4}x + 1$

Parallel and Perpendicular Lines In Exercises 69–72, determine whether the lines L_1 and L_2 passing through the pairs of points are parallel, perpendicular, or neither.

69. $L_1: (0, -1), (5, 9)$ 70. $L_1: (-2, -1), (1, 5)$

    $L_2: (0, 3), (4, 1)$ $L_2: (1, 3), (5, -5)$

71. $L_1: (-6, -3), (2, -3)$ 72. $L_1: (4, 8), (-4, 2)$

    $L_2: \left(3, -\frac{1}{2}\right), \left(6, -\frac{1}{2}\right)$ $L_2: (3, -5), \left(-1, \frac{1}{3}\right)$

 Finding Parallel and Perpendicular Lines In Exercises 73–80, find equations of the lines that pass through the given point and are (a) parallel to and (b) perpendicular to the given line.

73. $4x - 2y = 3$, $(2, 1)$ 74. $x + y = 7$, $(-3, 2)$

75. $3x + 4y = 7$, $\left(-\frac{2}{3}, \frac{7}{8}\right)$ 76. $5x + 3y = 0$, $\left(\frac{7}{8}, \frac{3}{4}\right)$

77. $y + 5 = 0$, $(-2, 4)$

78. $x - 4 = 0$, $(3, -2)$

79. $x - y = 4$, $(2.5, 6.8)$

80. $6x + 2y = 9$, $(-3.9, -1.4)$

Using Intercept Form In Exercises 81–86, use the *intercept form* to find the general form of the equation of the line with the given intercepts. The intercept form of the equation of a line with intercepts $(a, 0)$ and $(0, b)$ is

$$\frac{x}{a} + \frac{y}{b} = 1, \quad a \neq 0, \quad b \neq 0.$$

81. x-intercept: $(3, 0)$

    y-intercept: $(0, 5)$

82. x-intercept: $(-3, 0)$

    y-intercept: $(0, 4)$

83. x-intercept: $\left(-\frac{1}{6}, 0\right)$

    y-intercept: $\left(0, -\frac{2}{3}\right)$

84. x-intercept: $\left(\frac{2}{3}, 0\right)$

    y-intercept: $(0, -2)$

85. Point on line: $(1, 2)$

    x-intercept: $(c, 0)$, $c \neq 0$

    y-intercept: $(0, c)$, $c \neq 0$

86. Point on line: $(-3, 4)$

    x-intercept: $(d, 0)$, $d \neq 0$

    y-intercept: $(0, d)$, $d \neq 0$

87. **Sales** The slopes of lines representing annual sales y in terms of time x in years are given below. Use the slopes to interpret any change in annual sales for a one-year increase in time.

    (a) The line has a slope of $m = 135$.

    (b) The line has a slope of $m = 0$.

    (c) The line has a slope of $m = -40$.

88. **Sales** The graph shows the sales (in billions of dollars) for Apple Inc. in the years 2009 through 2015. *(Source: Apple Inc.)*

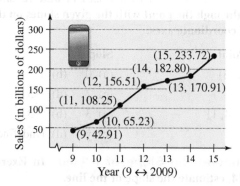

    (a) Use the slopes of the line segments to determine the years in which the sales showed the greatest increase and the least increase.

    (b) Find the slope of the line segment connecting the points for the years 2009 and 2015.

    (c) Interpret the meaning of the slope in part (b) in the context of the problem.

89. Road Grade You are driving on a road that has a 6% uphill grade. This means that the slope of the road is $\frac{6}{100}$. Approximate the amount of vertical change in your position when you drive 200 feet.

90. Road Grade

From the top of a mountain road, a surveyor takes several horizontal measurements x and several vertical measurements y, as shown in the table (x and y are measured in feet).

x	300	600	900	1200
y	-25	-50	-75	-100

x	1500	1800	2100
y	-125	-150	-175

(a) Sketch a scatter plot of the data.

(b) Use a straightedge to sketch the line that you think best fits the data.

(c) Find an equation for the line you sketched in part (b).

(d) Interpret the meaning of the slope of the line in part (c) in the context of the problem.

(e) The surveyor needs to put up a road sign that indicates the steepness of the road. For example, a surveyor would put up a sign that states "8% grade" on a road with a downhill grade that has a slope of $-\frac{8}{100}$. What should the sign state for the road in this problem?

Rate of Change **In Exercises 91 and 92, you are given the dollar value of a product in 2016 and the rate at which the value of the product is expected to change during the next 5 years. Use this information to write a linear equation that gives the dollar value V of the product in terms of the year t. (Let $t = 16$ represent 2016.)**

	2016 Value	Rate
91.	$3000	$150 decrease per year
92.	$200	$6.50 increase per year

93. Cost The cost C of producing n computer laptop bags is given by

$$C = 1.25n + 15{,}750, \quad n > 0.$$

Explain what the C-intercept and the slope represent.

94. Monthly Salary A pharmaceutical salesperson receives a monthly salary of $5000 plus a commission of 7% of sales. Write a linear equation for the salesperson's monthly wage W in terms of monthly sales S.

95. Depreciation A sandwich shop purchases a used pizza oven for $875. After 5 years, the oven will have to be discarded and replaced. Write a linear equation giving the value V of the equipment during the 5 years it will be in use.

96. Depreciation A school district purchases a high-volume printer, copier, and scanner for $24,000. After 10 years, the equipment will have to be replaced. Its value at that time is expected to be $2000. Write a linear equation giving the value V of the equipment during the 10 years it will be in use.

97. Temperature Conversion Write a linear equation that expresses the relationship between the temperature in degrees Celsius C and degrees Fahrenheit F. Use the fact that water freezes at 0°C (32°F) and boils at 100°C (212°F).

98. Neurology The average weight of a male child's brain is 970 grams at age 1 and 1270 grams at age 3. *(Source: American Neurological Association)*

(a) Assuming that the relationship between brain weight y and age t is linear, write a linear model for the data.

(b) What is the slope and what does it tell you about brain weight?

(c) Use your model to estimate the average brain weight at age 2.

(d) Use your school's library, the Internet, or some other reference source to find the actual average brain weight at age 2. How close was your estimate?

(e) Do you think your model could be used to determine the average brain weight of an adult? Explain.

99. Cost, Revenue, and Profit A roofing contractor purchases a shingle delivery truck with a shingle elevator for $42,000. The vehicle requires an average expenditure of $9.50 per hour for fuel and maintenance, and the operator is paid $11.50 per hour.

(a) Write a linear equation giving the total cost C of operating this equipment for t hours. (Include the purchase cost of the equipment.)

(b) Assuming that customers are charged $45 per hour of machine use, write an equation for the revenue R obtained from t hours of use.

(c) Use the formula for profit $P = R - C$ to write an equation for the profit obtained from t hours of use.

(d) Use the result of part (c) to find the break-even point—that is, the number of hours this equipment must be used to yield a profit of 0 dollars.

100. Geometry The length and width of a rectangular garden are 15 meters and 10 meters, respectively. A walkway of width x surrounds the garden.

(a) Draw a diagram that gives a visual representation of the problem.

(b) Write the equation for the perimeter y of the walkway in terms of x.

(c) Use a graphing utility to graph the equation for the perimeter.

(d) Determine the slope of the graph in part (c). For each additional one-meter increase in the width of the walkway, determine the increase in its perimeter.

Exploration

True or False? In Exercises 101 and 102, determine whether the statement is true or false. Justify your answer.

101. A line with a slope of $-\frac{5}{7}$ is steeper than a line with a slope of $-\frac{6}{7}$.

102. The line through $(-8, 2)$ and $(-1, 4)$ and the line through $(0, -4)$ and $(-7, 7)$ are parallel.

103. Right Triangle Explain how you can use slope to show that the points $A(-1, 5)$, $B(3, 7)$, and $C(5, 3)$ are the vertices of a right triangle.

104. Vertical Line Explain why the slope of a vertical line is undefined.

105. Error Analysis Describe the error.

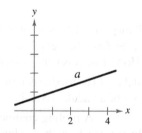   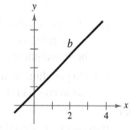

Line b has a greater slope than line a. ✗

106. Perpendicular Segments Find d_1 and d_2 in terms of m_1 and m_2, respectively (see figure). Then use the Pythagorean Theorem to find a relationship between m_1 and m_2.

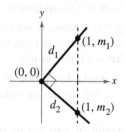

107. Think About It Is it possible for two lines with positive slopes to be perpendicular? Explain.

108. Slope and Steepness The slopes of two lines are -4 and $\frac{5}{2}$. Which is steeper? Explain.

109. Comparing Slopes Use a graphing utility to compare the slopes of the lines $y = mx$, where $m = 0.5, 1, 2,$ and 4. Which line rises most quickly? Now, let $m = -0.5, -1, -2,$ and -4. Which line falls most quickly? Use a square setting to obtain a true geometric perspective. What can you conclude about the slope and the "rate" at which the line rises or falls?

110. HOW DO YOU SEE IT? Match the description of the situation with its graph. Also determine the slope and y-intercept of each graph and interpret the slope and y-intercept in the context of the situation. [The graphs are labeled (i), (ii), (iii), and (iv).]

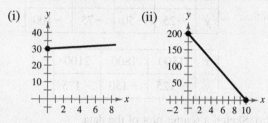

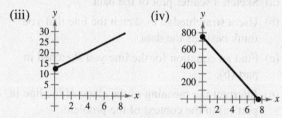

(a) A person is paying $20 per week to a friend to repay a $200 loan.

(b) An employee receives $12.50 per hour plus $2 for each unit produced per hour.

(c) A sales representative receives $30 per day for food plus $0.32 for each mile traveled.

(d) A computer that was purchased for $750 depreciates $100 per year.

Finding a Relationship for Equidistance In Exercises 111–114, find a relationship between x and y such that (x, y) is equidistant (the same distance) from the two points.

111. $(4, -1), (-2, 3)$ **112.** $(6, 5), (1, -8)$

113. $\left(3, \frac{5}{2}\right), (-7, 1)$ **114.** $\left(-\frac{1}{2}, -4\right), \left(\frac{7}{2}, \frac{5}{4}\right)$

Project: Bachelor's Degrees To work an extended application analyzing the numbers of bachelor's degrees earned by women in the United States from 2002 through 2013, visit this text's website at *LarsonPrecalculus.com*. (*Source: National Center for Education Statistics*)

1.4 Functions

Functions are used to model and solve real-life problems. For example, in Exercise 70 on page 47, you will use a function that models the force of water against the face of a dam.

- Determine whether relations between two variables are functions, and use function notation.
- Find the domains of functions.
- Use functions to model and solve real-life problems.
- Evaluate difference quotients.

Introduction to Functions and Function Notation

Many everyday phenomena involve two quantities that are related to each other by some rule of correspondence. The mathematical term for such a rule of correspondence is a **relation.** In mathematics, equations and formulas often represent relations. For example, the simple interest I earned on $1000 for 1 year is related to the annual interest rate r by the formula $I = 1000r$.

The formula $I = 1000r$ represents a special kind of relation that matches each item from one set with *exactly one* item from a different set. Such a relation is a **function.**

Definition of Function

A **function** f from a set A to a set B is a relation that assigns to each element x in the set A exactly one element y in the set B. The set A is the **domain** (or set of inputs) of the function f, and the set B contains the **range** (or set of outputs).

To help understand this definition, look at the function below, which relates the time of day to the temperature.

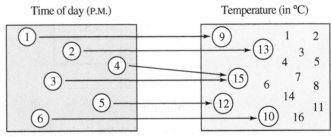

Set A is the domain.
Inputs: 1, 2, 3, 4, 5, 6

Set B contains the range.
Outputs: 9, 10, 12, 13, 15

The ordered pairs below can represent this function. The first coordinate (x-value) is the input and the second coordinate (y-value) is the output.

$$\{(1, 9), (2, 13), (3, 15), (4, 15), (5, 12), (6, 10)\}$$

Characteristics of a Function from Set *A* to Set *B*

1. Each element in A must be matched with an element in B.

2. Some elements in B may not be matched with any element in A.

3. Two or more elements in A may be matched with the same element in B.

4. An element in A (the domain) cannot be matched with two different elements in B.

Here are four common ways to represent functions.

Four Ways to Represent a Function

1. *Verbally* by a sentence that describes how the input variable is related to the output variable

2. *Numerically* by a table or a list of ordered pairs that matches input values with output values

3. *Graphically* by points in a coordinate plane in which the horizontal positions represent the input values and the vertical positions represent the output values

4. *Algebraically* by an equation in two variables

To determine whether a relation is a function, you must decide whether each input value is matched with exactly one output value. When any input value is matched with two or more output values, the relation is not a function.

EXAMPLE 1   **Testing for Functions**

Determine whether the relation represents y as a function of x.

a. The input value x is the number of representatives from a state, and the output value y is the number of senators.

b.

Input, x	Output, y
2	11
2	10
3	8
4	5
5	1

c.

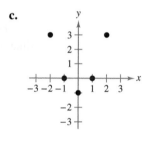

Solution

a. This verbal description *does* describe y as a function of x. Regardless of the value of x, the value of y is always 2. This is an example of a *constant function*.

b. This table *does not* describe y as a function of x. The input value 2 is matched with two different y-values.

c. The graph *does* describe y as a function of x. Each input value is matched with exactly one output value.

✓ **Checkpoint**  Audio-video solution in English & Spanish at LarsonPrecalculus.com

Determine whether the relation represents y as a function of x.

a. *Domain, x Range, y*

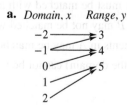

b.

Input, x	0	1	2	3	4
Output, y	-4	-2	0	2	4

Representing functions by sets of ordered pairs is common in *discrete mathematics*. In algebra, however, it is more common to represent functions by equations or formulas involving two variables. For example, the equation

$$y = x^2 \qquad\qquad \text{\textit{y} is a function of \textit{x}.}$$

represents the variable y as a function of the variable x. In this equation, x is the **independent variable** and y is the **dependent variable.** The domain of the function is the set of all values taken on by the independent variable x, and the range of the function is the set of all values taken on by the dependent variable y.

HISTORICAL NOTE

Many consider Leonhard Euler (1707–1783), a Swiss mathematician, to be the most prolific and productive mathematician in history. One of his greatest influences on mathematics was his use of symbols, or notation. Euler introduced the function notation $y = f(x)$.

EXAMPLE 2   Testing for Functions Represented Algebraically

See LarsonPrecalculus.com for an interactive version of this type of example.

Determine whether each equation represents y as a function of x.

a. $x^2 + y = 1$

b. $-x + y^2 = 1$

Solution   To determine whether y is a function of x, solve for y in terms of x.

a. Solving for y yields

$$x^2 + y = 1 \qquad\qquad \text{Write original equation.}$$
$$y = 1 - x^2. \qquad\qquad \text{Solve for \textit{y}.}$$

To each value of x there corresponds exactly one value of y. So, y is a function of x.

b. Solving for y yields

$$-x + y^2 = 1 \qquad\qquad \text{Write original equation.}$$
$$y^2 = 1 + x \qquad\qquad \text{Add \textit{x} to each side.}$$
$$y = \pm\sqrt{1 + x}. \qquad\qquad \text{Solve for \textit{y}.}$$

The $\pm$ indicates that to a given value of x there correspond two values of y. So, y is not a function of x.

✓ **Checkpoint**  Audio-video solution in English & Spanish at LarsonPrecalculus.com

Determine whether each equation represents y as a function of x.

a. $x^2 + y^2 = 8$ **b.** $y - 4x^2 = 36$

When using an equation to represent a function, it is convenient to name the function for easy reference. For example, the equation $y = 1 - x^2$ describes y as a function of x. By renaming this function "f," you can write the input, output, and equation using **function notation.**

Input	Output	Equation
x	$f(x)$	$f(x) = 1 - x^2$

The symbol $f(x)$ is read as *the value of f at x* or simply *f of x.* The symbol $f(x)$ corresponds to the y-value for a given x. So, $y = f(x)$. Keep in mind that f is the *name* of the function, whereas $f(x)$ is the *value* of the function at x. For example, the function $f(x) = 3 - 2x$ has *function values* denoted by $f(-1)$, $f(0)$, $f(2)$, and so on. To find these values, substitute the specified input values into the given equation.

$$\text{For } x = -1, \qquad f(-1) = 3 - 2(-1) = 3 + 2 = 5.$$
$$\text{For } x = 0, \qquad f(0) = 3 - 2(0) = 3 - 0 = 3.$$
$$\text{For } x = 2, \qquad f(2) = 3 - 2(2) = 3 - 4 = -1.$$

Although it is often convenient to use f as a function name and x as the independent variable, other letters may be used as well. For example,

$$f(x) = x^2 - 4x + 7, \quad f(t) = t^2 - 4t + 7, \quad \text{and} \quad g(s) = s^2 - 4s + 7$$

all define the same function. In fact, the role of the independent variable is that of a "placeholder." Consequently, the function can be described by

$$f(\blacksquare) = (\blacksquare)^2 - 4(\blacksquare) + 7.$$

EXAMPLE 3   Evaluating a Function

Let $g(x) = -x^2 + 4x + 1$. Find each function value.

a. $g(2)$ **b.** $g(t)$ **c.** $g(x + 2)$

Solution

a. Replace x with 2 in $g(x) = -x^2 + 4x + 1$.

$$g(2) = -(2)^2 + 4(2) + 1$$
$$= -4 + 8 + 1$$
$$= 5$$

b. Replace x with t.

$$g(t) = -(t)^2 + 4(t) + 1$$
$$= -t^2 + 4t + 1$$

c. Replace x with $x + 2$.

$$g(x + 2) = -(x + 2)^2 + 4(x + 2) + 1$$
$$= -(x^2 + 4x + 4) + 4x + 8 + 1$$
$$= -x^2 - 4x - 4 + 4x + 8 + 1$$
$$= -x^2 + 5$$

··**REMARK**   In Example 3(c), note that $g(x + 2)$ is not equal to $g(x) + g(2)$. In general, $g(u + v) \neq g(u) + g(v)$.

✓ **Checkpoint** 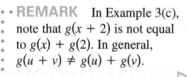 Audio-video solution in English & Spanish at LarsonPrecalculus.com

Let $f(x) = 10 - 3x^2$. Find each function value.

a. $f(2)$ **b.** $f(-4)$ **c.** $f(x - 1)$

A function defined by two or more equations over a specified domain is called a **piecewise-defined function.**

EXAMPLE 4   A Piecewise-Defined Function

Evaluate the function when $x = -1$, 0, and 1.

$$f(x) = \begin{cases} x^2 + 1, & x < 0 \\ x - 1, & x \geq 0 \end{cases}$$

Solution   Because $x = -1$ is less than 0, use $f(x) = x^2 + 1$ to obtain $f(-1) = (-1)^2 + 1 = 2$. For $x = 0$, use $f(x) = x - 1$ to obtain $f(0) = (0) - 1 = -1$. For $x = 1$, use $f(x) = x - 1$ to obtain $f(1) = (1) - 1 = 0$.

✓ **Checkpoint** 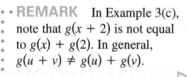 Audio-video solution in English & Spanish at LarsonPrecalculus.com

Evaluate the function given in Example 4 when $x = -2$, 2, and 3.

EXAMPLE 5   **Finding Values for Which $f(x) = 0$**

Find all real values of x for which $f(x) = 0$.

a. $f(x) = -2x + 10$ **b.** $f(x) = x^2 - 5x + 6$

Solution   For each function, set $f(x) = 0$ and solve for x.

a. $-2x + 10 = 0$ Set $f(x)$ equal to 0.

$\qquad -2x = -10$ Subtract 10 from each side.

$\qquad\quad x = 5$ Divide each side by -2.

So, $f(x) = 0$ when $x = 5$.

b. $x^2 - 5x + 6 = 0$ Set $f(x)$ equal to 0.

$(x - 2)(x - 3) = 0$ Factor.

$x - 2 = 0 \implies x = 2$ Set 1st factor equal to 0 and solve.

$x - 3 = 0 \implies x = 3$ Set 2nd factor equal to 0 and solve.

So, $f(x) = 0$ when $x = 2$ or $x = 3$.

✓ *Checkpoint* *Audio-video solution in English & Spanish at LarsonPrecalculus.com*

Find all real values of x for which $f(x) = 0$, where $f(x) = x^2 - 16$.

EXAMPLE 6   **Finding Values for Which $f(x) = g(x)$**

Find the values of x for which $f(x) = g(x)$.

a. $f(x) = x^2 + 1$ and $g(x) = 3x - x^2$

b. $f(x) = x^2 - 1$ and $g(x) = -x^2 + x + 2$

Solution

a. $\qquad x^2 + 1 = 3x - x^2$ Set $f(x)$ equal to $g(x)$.

$2x^2 - 3x + 1 = 0$ Write in general form.

$(2x - 1)(x - 1) = 0$ Factor.

$2x - 1 = 0 \implies x = \frac{1}{2}$ Set 1st factor equal to 0 and solve.

$x - 1 = 0 \implies x = 1$ Set 2nd factor equal to 0 and solve.

So, $f(x) = g(x)$ when $x = \dfrac{1}{2}$ or $x = 1$.

b. $\qquad x^2 - 1 = -x^2 + x + 2$ Set $f(x)$ equal to $g(x)$.

$2x^2 - x - 3 = 0$ Write in general form.

$(2x - 3)(x + 1) = 0$ Factor.

$2x - 3 = 0 \implies x = \frac{3}{2}$ Set 1st factor equal to 0 and solve.

$x + 1 = 0 \implies x = -1$ Set 2nd factor equal to 0 and solve.

So, $f(x) = g(x)$ when $x = \dfrac{3}{2}$ or $x = -1$.

✓ *Checkpoint* *Audio-video solution in English & Spanish at LarsonPrecalculus.com*

Find the values of x for which $f(x) = g(x)$, where $f(x) = x^2 + 6x - 24$ and $g(x) = 4x - x^2$.

The Domain of a Function

▷ TECHNOLOGY Use a
graphing utility to graph the
functions $y = \sqrt{4 - x^2}$ and
$y = \sqrt{x^2 - 4}$. What is the
domain of each function?
Do the domains of these two
functions overlap? If so, for
what values do the domains
overlap?

The domain of a function can be described explicitly or it can be *implied* by the expression used to define the function. The **implied domain** is the set of all real numbers for which the expression is defined. For example, the function

$$f(x) = \frac{1}{x^2 - 4} \qquad \text{Domain excludes } x\text{-values that result in division by zero.}$$

has an implied domain consisting of all real x other than $x = \pm 2$. These two values are excluded from the domain because division by zero is undefined. Another common type of implied domain is that used to avoid even roots of negative numbers. For example, the function

$$f(x) = \sqrt{x} \qquad \text{Domain excludes } x\text{-values that result in even roots of negative numbers.}$$

is defined only for $x \geq 0$. So, its implied domain is the interval $[0, \infty)$. In general, the domain of a function *excludes* values that cause division by zero *or* that result in the even root of a negative number.

EXAMPLE 7   Finding the Domains of Functions

Find the domain of each function.

a. f: $\{(-3, 0), (-1, 4), (0, 2), (2, 2), (4, -1)\}$ **b.** $g(x) = \dfrac{1}{x + 5}$

c. Volume of a sphere: $V = \frac{4}{3}\pi r^3$ **d.** $h(x) = \sqrt{4 - 3x}$

Solution

a. The domain of f consists of all first coordinates in the set of ordered pairs.

  Domain $= \{-3, -1, 0, 2, 4\}$

b. Excluding x-values that yield zero in the denominator, the domain of g is the set of all real numbers x except $x = -5$.

c. This function represents the volume of a sphere, so the values of the radius r must be positive. The domain is the set of all real numbers r such that $r > 0$.

d. This function is defined only for x-values for which

  $4 - 3x \geq 0$.

  By solving this inequality, you can conclude that $x \leq \frac{4}{3}$. So, the domain is the interval $\left(-\infty, \frac{4}{3}\right]$.

✓ *Checkpoint* 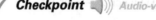 Audio-video solution in English & Spanish at LarsonPrecalculus.com

Find the domain of each function.

a. f: $\{(-2, 2), (-1, 1), (0, 3), (1, 1), (2, 2)\}$ **b.** $g(x) = \dfrac{1}{3 - x}$

c. Circumference of a circle: $C = 2\pi r$ **d.** $h(x) = \sqrt{x - 16}$

In Example 7(c), note that the domain of a function may be implied by the physical context. For example, from the equation

$$V = \frac{4}{3}\pi r^3$$

you have no reason to restrict r to positive values, but the physical context implies that a sphere cannot have a negative or zero radius.

Applications

EXAMPLE 8 **The Dimensions of a Container**

You work in the marketing department of a soft-drink company and are experimenting with a new can for iced tea that is slightly narrower and taller than a standard can. For your experimental can, the ratio of the height to the radius is 4.

a. Write the volume of the can as a function of the radius r.

b. Write the volume of the can as a function of the height h.

Solution

a. $V(r) = \pi r^2 h = \pi r^2 (4r) = 4\pi r^3$ Write V as a function of r.

b. $V(h) = \pi r^2 h = \pi \left(\dfrac{h}{4}\right)^2 h = \dfrac{\pi h^3}{16}$ Write V as a function of h.

✓ *Checkpoint*)) *Audio-video solution in English & Spanish at LarsonPrecalculus.com*

For the experimental can described in Example 8, write the *surface area* as a function of (a) the radius r and (b) the height h.

EXAMPLE 9 **The Path of a Baseball**

A batter hits a baseball at a point 3 feet above ground at a velocity of 100 feet per second and an angle of 45°. The path of the baseball is given by the function

$$f(x) = -0.0032x^2 + x + 3$$

where $f(x)$ is the height of the baseball (in feet) and x is the horizontal distance from home plate (in feet). Will the baseball clear a 10-foot fence located 300 feet from home plate?

Algebraic Solution

Find the height of the baseball when $x = 300$.

$$f(x) = -0.0032x^2 + x + 3$$ Write original function.

$$f(300) = -0.0032(300)^2 + 300 + 3$$ Substitute 300 for x.

$$= 15$$ Simplify.

When $x = 300$, the height of the baseball is 15 feet. So, the baseball will clear a 10-foot fence.

Graphical Solution

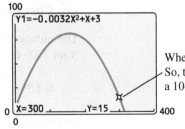

When $x = 300$, $y = 15$. So, the ball will clear a 10-foot fence.

✓ *Checkpoint*)) *Audio-video solution in English & Spanish at LarsonPrecalculus.com*

A second baseman throws a baseball toward the first baseman 60 feet away. The path of the baseball is given by the function

$$f(x) = -0.004x^2 + 0.3x + 6$$

where $f(x)$ is the height of the baseball (in feet) and x is the horizontal distance from the second baseman (in feet). The first baseman can reach 8 feet high. Can the first baseman catch the baseball without jumping?

Flexible-fuel vehicles are designed to operate on gasoline, E85, or a mixture of the two fuels. The concentration of ethanol in E85 fuel ranges from 51% to 83%, depending on where and when the E85 is produced.

EXAMPLE 10 **Alternative-Fuel Stations**

The number S of fuel stations that sold E85 (a gasoline-ethanol blend) in the United States increased in a linear pattern from 2008 through 2011, and then increased in a different linear pattern from 2012 through 2015, as shown in the bar graph. These two patterns can be approximated by the function

$$S(t) = \begin{cases} 260.8t - 439, & 8 \le t \le 11 \\ 151.2t + 714, & 12 \le t \le 15 \end{cases}$$

where t represents the year, with $t = 8$ corresponding to 2008. Use this function to approximate the number of stations that sold E85 each year from 2008 to 2015. *(Source: Alternative Fuels Data Center)*

Number of Stations Selling E85 in the U.S.

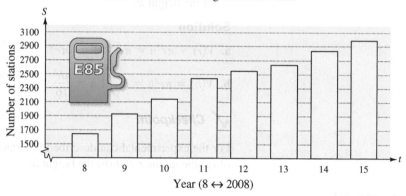

Solution From 2008 through 2011, use $S(t) = 260.8t - 439$.

1647	1908	2169	2430
2008	2009	2010	2011

From 2012 to 2015, use $S(t) = 151.2t + 714$.

2528	2680	2831	2982
2012	2013	2014	2015

✓ **Checkpoint** 🔊))) *Audio-video solution in English & Spanish at LarsonPrecalculus.com*

The number S of fuel stations that sold compressed natural gas in the United States from 2009 to 2015 can be approximated by the function

$$S(t) = \begin{cases} 69t + 151, & 9 \le t \le 11 \\ 160t - 803, & 12 \le t \le 15 \end{cases}$$

where t represents the year, with $t = 9$ corresponding to 2009. Use this function to approximate the number of stations that sold compressed natural gas each year from 2009 through 2015. *(Source: Alternative Fuels Data Center)* ■

Difference Quotients

One of the basic definitions in calculus uses the ratio

$$\frac{f(x + h) - f(x)}{h}, \quad h \ne 0.$$

This ratio is a **difference quotient,** as illustrated in Example 11.

EXAMPLE 11   **Evaluating a Difference Quotient**

• •REMARK You may find it easier to calculate the difference quotient in Example 11 by first finding $f(x + h)$, and then substituting the resulting expression into the difference quotient

$$\frac{f(x + h) - f(x)}{h}.$$

For $f(x) = x^2 - 4x + 7$, find $\dfrac{f(x + h) - f(x)}{h}$.

Solution

$$\frac{f(x + h) - f(x)}{h} = \frac{[(x + h)^2 - 4(x + h) + 7] - (x^2 - 4x + 7)}{h}$$

$$= \frac{x^2 + 2xh + h^2 - 4x - 4h + 7 - x^2 + 4x - 7}{h}$$

$$= \frac{2xh + h^2 - 4h}{h} = \frac{h(2x + h - 4)}{h} = 2x + h - 4, \quad h \neq 0$$

✓ *Checkpoint* ◀))) *Audio-video solution in English & Spanish at LarsonPrecalculus.com*

For $f(x) = x^2 + 2x - 3$, find $\dfrac{f(x + h) - f(x)}{h}$.

Summary of Function Terminology

Function: A **function** is a relationship between two variables such that to each value of the independent variable there corresponds exactly one value of the dependent variable.

Function notation: $y = f(x)$

   f is the *name* of the function.

   y is the **dependent variable.**

   x is the **independent variable.**

   $f(x)$ is the *value of the function at x.*

Domain: The **domain** of a function is the set of all values (inputs) of the independent variable for which the function is defined. If x is in the domain of f, then f is *defined* at x. If x is not in the domain of f, then f is *undefined* at x.

Range: The **range** of a function is the set of all values (outputs) taken on by the dependent variable (that is, the set of all function values).

Implied domain: If f is defined by an algebraic expression and the domain is not specified, then the **implied domain** consists of all real numbers for which the expression is defined.

Summarize   (Section 1.4)

1. State the definition of a function and describe function notation *(pages 35–39)*. For examples of determining functions and using function notation, see Examples 1–6.

2. State the definition of the implied domain of a function *(page 40)*. For an example of finding the domains of functions, see Example 7.

3. Describe examples of how functions can model real-life problems *(pages 41 and 42, Examples 8–10)*.

4. State the definition of a difference quotient *(page 42)*. For an example of evaluating a difference quotient, see Example 11.

1.4   Exercises

See CalcChat.com for tutorial help and worked-out solutions to odd-numbered exercises.

Vocabulary: Fill in the blanks.

1. A relation that assigns to each element x from a set of inputs, or _____, exactly one element y in a set of outputs, or _____, is a _____.

2. For an equation that represents y as a function of x, the set of all values taken on by the _____ variable x is the domain, and the set of all values taken on by the _____ variable y is the range.

3. If the domain of the function f is not given, then the set of values of the independent variable for which the expression is defined is the _____ _____.

4. One of the basic definitions in calculus uses the ratio $\dfrac{f(x + h) - f(x)}{h}$, $h \neq 0$. This ratio is a _____ _____.

Skills and Applications

Testing for Functions In Exercises 5–8, determine whether the relation represents y as a function of x.

5. *Domain, x* *Range, y*

6. *Domain, x* *Range, y*

7.

Input, x	10	7	4	7	10
Output, y	3	6	9	12	15

8.

Input, x	-2	0	2	4	6
Output, y	1	1	1	1	1

Testing for Functions In Exercises 9 and 10, which sets of ordered pairs represent functions from A to B? Explain.

9. $A = \{0, 1, 2, 3\}$ and $B = \{-2, -1, 0, 1, 2\}$
   (a) $\{(0, 1), (1, -2), (2, 0), (3, 2)\}$
   (b) $\{(0, -1), (2, 2), (1, -2), (3, 0), (1, 1)\}$
   (c) $\{(0, 0), (1, 0), (2, 0), (3, 0)\}$
   (d) $\{(0, 2), (3, 0), (1, 1)\}$

10. $A = \{a, b, c\}$ and $B = \{0, 1, 2, 3\}$
    (a) $\{(a, 1), (c, 2), (c, 3), (b, 3)\}$
    (b) $\{(a, 1), (b, 2), (c, 3)\}$
    (c) $\{(1, a), (0, a), (2, c), (3, b)\}$
    (d) $\{(c, 0), (b, 0), (a, 3)\}$

Testing for Functions Represented Algebraically In Exercises 11–18, determine whether the equation represents y as a function of x.

11. $x^2 + y^2 = 4$

12. $x^2 - y = 9$

13. $y = \sqrt{16 - x^2}$

14. $y = \sqrt{x + 5}$

15. $y = 4 - |x|$

16. $|y| = 4 - x$

17. $y = -75$

18. $x - 1 = 0$

Evaluating a Function In Exercises 19–30, find each function value, if possible.

19. $f(x) = 3x - 5$
    (a) $f(1)$ (b) $f(-3)$ (c) $f(x + 2)$

20. $V(r) = \frac{4}{3}\pi r^3$
    (a) $V(3)$ (b) $V\left(\frac{3}{2}\right)$ (c) $V(2r)$

21. $g(t) = 4t^2 - 3t + 5$
    (a) $g(2)$ (b) $g(t - 2)$ (c) $g(t) - g(2)$

22. $h(t) = -t^2 + t + 1$
    (a) $h(2)$ (b) $h(-1)$ (c) $h(x + 1)$

23. $f(y) = 3 - \sqrt{y}$
    (a) $f(4)$ (b) $f(0.25)$ (c) $f(4x^2)$

24. $f(x) = \sqrt{x + 8} + 2$
    (a) $f(-8)$ (b) $f(1)$ (c) $f(x - 8)$

25. $q(x) = 1/(x^2 - 9)$
    (a) $q(0)$ (b) $q(3)$ (c) $q(y + 3)$

26. $q(t) = (2t^2 + 3)/t^2$
    (a) $q(2)$ (b) $q(0)$ (c) $q(-x)$

27. $f(x) = |x|/x$
    (a) $f(2)$ (b) $f(-2)$ (c) $f(x - 1)$

28. $f(x) = |x| + 4$
    (a) $f(2)$ (b) $f(-2)$ (c) $f(x^2)$

29. $f(x) = \begin{cases} 2x + 1, & x < 0 \\ 2x + 2, & x \geq 0 \end{cases}$
    (a) $f(-1)$ (b) $f(0)$ (c) $f(2)$

30. $f(x) = \begin{cases} -3x - 3, & x < -1 \\ x^2 + 2x - 1, & x \geq -1 \end{cases}$
    (a) $f(-2)$ (b) $f(-1)$ (c) $f(1)$

Evaluating a Function In Exercises 31–34, complete the table.

31. $f(x) = -x^2 + 5$

x	-2	-1	0	1	2
$f(x)$					

32. $h(t) = \frac{1}{2}|t + 3|$

t	-5	-4	-3	-2	-1
$h(t)$					

33. $f(x) = \begin{cases} -\frac{1}{2}x + 4, & x \le 0 \\ (x - 2)^2, & x > 0 \end{cases}$

x	-2	-1	0	1	2
$f(x)$					

34. $f(x) = \begin{cases} 9 - x^2, & x < 3 \\ x - 3, & x \ge 3 \end{cases}$

x	1	2	3	4	5
$f(x)$					

 Finding Values for Which $f(x) = 0$ In Exercises 35–42, find all real values of x for which $f(x) = 0$.

35. $f(x) = 15 - 3x$ **36.** $f(x) = 4x + 6$

37. $f(x) = \dfrac{3x - 4}{5}$ **38.** $f(x) = \dfrac{12 - x^2}{8}$

39. $f(x) = x^2 - 81$ **40.** $f(x) = x^2 - 6x - 16$

41. $f(x) = x^3 - x$

42. $f(x) = x^3 - x^2 - 3x + 3$

 Finding Values for Which $f(x) = g(x)$ In Exercises 43–46, find the value(s) of x for which $f(x) = g(x)$.

43. $f(x) = x^2$, $g(x) = x + 2$

44. $f(x) = x^2 + 2x + 1$, $g(x) = 5x + 19$

45. $f(x) = x^4 - 2x^2$, $g(x) = 2x^2$

46. $f(x) = \sqrt{x} - 4$, $g(x) = 2 - x$

 Finding the Domain of a Function In Exercises 47–56, find the domain of the function.

47. $f(x) = 5x^2 + 2x - 1$

48. $g(x) = 1 - 2x^2$

49. $g(y) = \sqrt{y + 6}$

50. $f(t) = \sqrt[3]{t + 4}$

51. $g(x) = \dfrac{1}{x} - \dfrac{3}{x + 2}$ **52.** $h(x) = \dfrac{6}{x^2 - 4x}$

53. $f(s) = \dfrac{\sqrt{s - 1}}{s - 4}$ **54.** $f(x) = \dfrac{\sqrt{x + 6}}{6 + x}$

55. $f(x) = \dfrac{x - 4}{\sqrt{x}}$

56. $f(x) = \dfrac{x + 2}{\sqrt{x - 10}}$

57. Maximum Volume An open box of maximum volume is made from a square piece of material 24 centimeters on a side by cutting equal squares from the corners and turning up the sides (see figure).

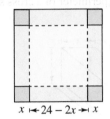 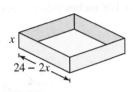

(a) The table shows the volumes V (in cubic centimeters) of the box for various heights x (in centimeters). Use the table to estimate the maximum volume.

Height, x	1	2	3	4	5	6
Volume, V	484	800	972	1024	980	864

(b) Plot the points (x, V) from the table in part (a). Does the relation defined by the ordered pairs represent V as a function of x?

(c) Given that V is a function of x, write the function and determine its domain.

58. Maximum Profit The cost per unit in the production of an MP3 player is $60. The manufacturer charges $90 per unit for orders of 100 or less. To encourage large orders, the manufacturer reduces the charge by $0.15 per MP3 player for each unit ordered in excess of 100 (for example, the charge is reduced to $87 per MP3 player for an order size of 120).

(a) The table shows the profits P (in dollars) for various numbers of units ordered, x. Use the table to estimate the maximum profit.

Units, x	130	140	150	160	170
Profit, P	3315	3360	3375	3360	3315

(b) Plot the points (x, P) from the table in part (a). Does the relation defined by the ordered pairs represent P as a function of x?

(c) Given that P is a function of x, write the function and determine its domain. (*Note:* $P = R - C$, where R is revenue and C is cost.)

59. Geometry Write the area A of a square as a function of its perimeter P.

60. Geometry Write the area A of a circle as a function of its circumference C.

61. Path of a Ball You throw a baseball to a child 25 feet away. The height y (in feet) of the baseball is given by

$$y = -\frac{1}{10}x^2 + 3x + 6$$

where x is the horizontal distance (in feet) from where you threw the ball. Can the child catch the baseball while holding a baseball glove at a height of 5 feet?

62. Postal Regulations A rectangular package has a combined length and girth (perimeter of a cross section) of 108 inches (see figure).

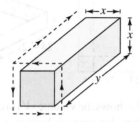

(a) Write the volume V of the package as a function of x. What is the domain of the function?

(b) Use a graphing utility to graph the function. Be sure to use an appropriate window setting.

(c) What dimensions will maximize the volume of the package? Explain.

63. Geometry A right triangle is formed in the first quadrant by the x- and y-axes and a line through the point $(2, 1)$ (see figure). Write the area A of the triangle as a function of x, and determine the domain of the function.

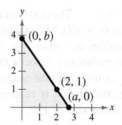

64. Geometry A rectangle is bounded by the x-axis and the semicircle $y = \sqrt{36 - x^2}$ (see figure). Write the area A of the rectangle as a function of x, and graphically determine the domain of the function.

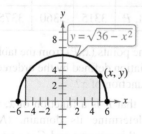

65. Pharmacology The percent p of prescriptions filled with generic drugs at CVS Pharmacies from 2008 through 2014 (see figure) can be approximated by the model

$$p(t) = \begin{cases} 2.77t + 45.2, & 8 \le t \le 11 \\ 1.95t + 55.9, & 12 \le t \le 14 \end{cases}$$

where t represents the year, with $t = 8$ corresponding to 2008. Use this model to find the percent of prescriptions filled with generic drugs in each year from 2008 through 2014. (*Source: CVS Health*)

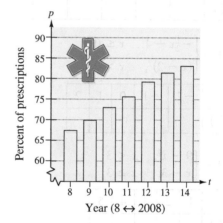

66. Median Sale Price The median sale price p (in thousands of dollars) of an existing one-family home in the United States from 2002 through 2014 (see figure) can be approximated by the model

$$p(t) = \begin{cases} -0.757t^2 + 20.80t + 127.2, & 2 \le t \le 6 \\ 3.879t^2 - 82.50t + 605.8, & 7 \le t \le 11 \\ -4.171t^2 + 124.34t - 714.2, & 12 \le t \le 14 \end{cases}$$

where t represents the year, with $t = 2$ corresponding to 2002. Use this model to find the median sale price of an existing one-family home in each year from 2002 through 2014. (*Source: National Association of Realtors*)

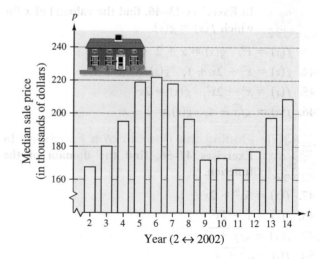

67. Cost, Revenue, and Profit A company produces a product for which the variable cost is $12.30 per unit and the fixed costs are $98,000. The product sells for $17.98. Let x be the number of units produced and sold.

(a) The total cost for a business is the sum of the variable cost and the fixed costs. Write the total cost C as a function of the number of units produced.

(b) Write the revenue R as a function of the number of units sold.

(c) Write the profit P as a function of the number of units sold. (*Note: $P = R - C$*)

68. Average Cost The inventor of a new game believes that the variable cost for producing the game is $0.95 per unit and the fixed costs are $6000. The inventor sells each game for $1.69. Let x be the number of games produced.

(a) The total cost for a business is the sum of the variable cost and the fixed costs. Write the total cost C as a function of the number of games produced.

(b) Write the average cost per unit $\overline{C} = \dfrac{C}{x}$ as a function of x.

69. Height of a Balloon A balloon carrying a transmitter ascends vertically from a point 3000 feet from the receiving station.

(a) Draw a diagram that gives a visual representation of the problem. Let h represent the height of the balloon and let d represent the distance between the balloon and the receiving station.

(b) Write the height of the balloon as a function of d. What is the domain of the function?

70. Physics

The function $F(y) = 149.76\sqrt{10}\,y^{5/2}$ estimates the force F (in tons) of water against the face of a dam, where y is the depth of the water (in feet).

(a) Complete the table. What can you conclude from the table?

y	5	10	20	30	40
$F(y)$					

(b) Use the table to approximate the depth at which the force against the dam is 1,000,000 tons.

(c) Find the depth at which the force against the dam is 1,000,000 tons algebraically.

71. Transportation For groups of 80 or more people, a charter bus company determines the rate per person according to the formula

$$\text{Rate} = 8 - 0.05(n - 80), \quad n \geq 80$$

where the rate is given in dollars and n is the number of people.

(a) Write the revenue R for the bus company as a function of n.

(b) Use the function in part (a) to complete the table. What can you conclude?

n	90	100	110	120	130	140	150
$R(n)$							

72. E-Filing The table shows the numbers of tax returns (in millions) made through e-file from 2007 through 2014. Let $f(t)$ represent the number of tax returns made through e-file in the year t. (*Source: eFile*)

Year	Number of Tax Returns Made Through E-File
2007	80.0
2008	89.9
2009	95.0
2010	98.7
2011	112.2
2012	112.1
2013	114.4
2014	125.8

DATA — Spreadsheet at LarsonPrecalculus.com

(a) Find $\dfrac{f(2014) - f(2007)}{2014 - 2007}$ and interpret the result in the context of the problem.

(b) Make a scatter plot of the data.

(c) Find a linear model for the data algebraically. Let N represent the number of tax returns made through e-file and let $t = 7$ correspond to 2007.

(d) Use the model found in part (c) to complete the table.

t	7	8	9	10	11	12	13	14
N								

(e) Compare your results from part (d) with the actual data.

(f) Use a graphing utility to find a linear model for the data. Let $x = 7$ correspond to 2007. How does the model you found in part (c) compare with the model given by the graphing utility?

Evaluating a Difference Quotient In Exercises 73–80, find the difference quotient and simplify your answer.

73. $f(x) = x^2 - 2x + 4,$ $\dfrac{f(2+h) - f(2)}{h},$ $h \neq 0$

74. $f(x) = 5x - x^2,$ $\dfrac{f(5+h) - f(5)}{h},$ $h \neq 0$

75. $f(x) = x^3 + 3x,$ $\dfrac{f(x+h) - f(x)}{h},$ $h \neq 0$

76. $f(x) = 4x^3 - 2x,$ $\dfrac{f(x+h) - f(x)}{h},$ $h \neq 0$

77. $g(x) = \dfrac{1}{x^2},$ $\dfrac{g(x) - g(3)}{x - 3},$ $x \neq 3$

78. $f(t) = \dfrac{1}{t - 2},$ $\dfrac{f(t) - f(1)}{t - 1},$ $t \neq 1$

79. $f(x) = \sqrt{5x},$ $\dfrac{f(x) - f(5)}{x - 5},$ $x \neq 5$

80. $f(x) = x^{2/3} + 1,$ $\dfrac{f(x) - f(8)}{x - 8},$ $x \neq 8$

Modeling Data In Exercises 81–84, determine which of the following functions

$$f(x) = cx, \quad g(x) = cx^2, \quad h(x) = c\sqrt{|x|}, \quad \text{and} \quad r(x) = \dfrac{c}{x}$$

can be used to model the data and determine the value of the constant c that will make the function fit the data in the table.

81.

x	-4	-1	0	1	4
y	-32	-2	0	-2	-32

82.

x	-4	-1	0	1	4
y	-1	$-\frac{1}{4}$	0	$\frac{1}{4}$	1

83.

x	-4	-1	0	1	4
y	-8	-32	Undefined	32	8

84.

x	-4	-1	0	1	4
y	6	3	0	3	6

Exploration

True or False? In Exercises 85–88, determine whether the statement is true or false. Justify your answer.

85. Every relation is a function.

86. Every function is a relation.

87. For the function

$$f(x) = x^4 - 1$$

the domain is $(-\infty, \infty)$ and the range is $(0, \infty)$.

88. The set of ordered pairs $\{(-8, -2), (-6, 0), (-4, 0), (-2, 2), (0, 4), (2, -2)\}$ represents a function.

89. Error Analysis Describe the error.

The functions

$$f(x) = \sqrt{x - 1} \quad \text{and} \quad g(x) = \dfrac{1}{\sqrt{x - 1}}$$

have the same domain, which is the set of all real numbers x such that $x \geq 1$.

90. Think About It Consider

$$f(x) = \sqrt{x - 2} \quad \text{and} \quad g(x) = \sqrt[3]{x - 2}.$$

Why are the domains of f and g different?

91. Think About It Given $f(x) = x^2$, is f the independent variable? Why or why not?

92. **HOW DO YOU SEE IT?** The graph represents the height h of a projectile after t seconds.

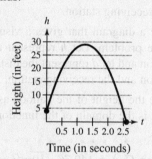

(a) Explain why h is a function of t.

(b) Approximate the height of the projectile after 0.5 second and after 1.25 seconds.

(c) Approximate the domain of h.

(d) Is t a function of h? Explain.

Think About It In Exercises 93 and 94, determine whether the statements use the word *function* in ways that are mathematically correct. Explain.

93. (a) The sales tax on a purchased item is a function of the selling price.

(b) Your score on the next algebra exam is a function of the number of hours you study the night before the exam.

94. (a) The amount in your savings account is a function of your salary.

(b) The speed at which a free-falling baseball strikes the ground is a function of the height from which it was dropped.

1.5 Analyzing Graphs of Functions

- ■ Use the Vertical Line Test for functions.
- ■ Find the zeros of functions.
- ■ Determine intervals on which functions are increasing or decreasing.
- ■ Determine relative minimum and relative maximum values of functions.
- ■ Determine the average rate of change of a function.
- ■ Identify even and odd functions.

The Graph of a Function

In Section 1.4, you studied functions from an algebraic point of view. In this section, you will study functions from a graphical perspective.

The **graph of a function** f is the collection of ordered pairs $(x, f(x))$ such that x is in the domain of f. As you study this section, remember that

x = the directed distance from the y-axis

$y = f(x)$ = the directed distance from the x-axis

as shown in the figure at the right.

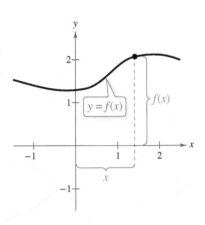

Graphs of functions can help you visualize relationships between variables in real life. For example, in Exercise 90 on page 59, you will use the graph of a function to visually represent the temperature in a city over a 24-hour period.

EXAMPLE 1   **Finding the Domain and Range of a Function**

Use the graph of the function f, shown in Figure 1.32, to find (a) the domain of f, (b) the function values $f(-1)$ and $f(2)$, and (c) the range of f.

Solution

a. The closed dot at $(-1, 1)$ indicates that $x = -1$ is in the domain of f, whereas the open dot at $(5, 2)$ indicates that $x = 5$ is not in the domain. So, the domain of f is all x in the interval $[-1, 5)$.

b. One point on the graph of f is $(-1, 1)$, so $f(-1) = 1$. Another point on the graph of f is $(2, -3)$, so $f(2) = -3$.

c. The graph does not extend below $f(2) = -3$ or above $f(0) = 3$, so the range of f is the interval $[-3, 3]$.

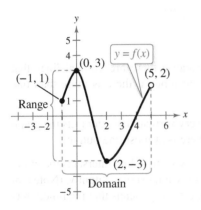

Figure 1.32

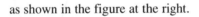

REMARK The use of dots (open or closed) at the extreme left and right points of a graph indicates that the graph does not extend beyond these points. If such dots are not on the graph, then assume that the graph extends beyond these points.

✓ **Checkpoint** ◀))) Audio-video solution in English & Spanish at LarsonPrecalculus.com

Use the graph of the function f to find (a) the domain of f, (b) the function values $f(0)$ and $f(3)$, and (c) the range of f.

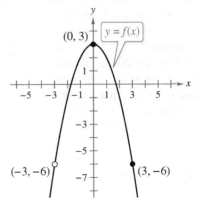

By the definition of a function, at most one y-value corresponds to a given x-value. So, no two points on the graph of a function have the same x-coordinate, or lie on the same vertical line. It follows, then, that a vertical line can intersect the graph of a function at most once. This observation provides a convenient visual test called the **Vertical Line Test** for functions.

Vertical Line Test for Functions

A set of points in a coordinate plane is the graph of y as a function of x if and only if no *vertical* line intersects the graph at more than one point.

EXAMPLE 2 **Vertical Line Test for Functions**

Use the Vertical Line Test to determine whether each graph represents y as a function of x.

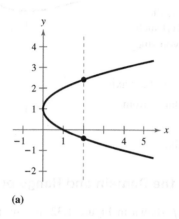
(a)

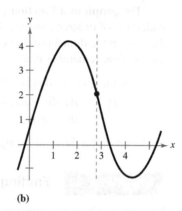
(b)

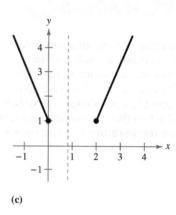
(c)

Solution

a. This *is not* a graph of y as a function of x, because there are vertical lines that intersect the graph twice. That is, for a particular input x, there is more than one output y.

b. This *is* a graph of y as a function of x, because every vertical line intersects the graph at most once. That is, for a particular input x, there is at most one output y.

c. This *is* a graph of y as a function of x, because every vertical line intersects the graph at most once. That is, for a particular input x, there is at most one output y. (Note that when a vertical line does not intersect the graph, it simply means that the function is undefined for that particular value of x.)

✓ **Checkpoint** ◀))) Audio-video solution in English & Spanish at LarsonPrecalculus.com

Use the Vertical Line Test to determine whether the graph represents y as a function of x.

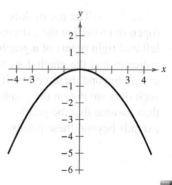

▷ **TECHNOLOGY** Most graphing utilities graph functions of x more easily than other types of equations. For example, the graph shown in (a) above represents the equation $x - (y - 1)^2 = 0$. To duplicate this graph using a graphing utility, you must first solve the equation for y to obtain $y = 1 \pm \sqrt{x}$, and then graph the two equations $y_1 = 1 + \sqrt{x}$ and $y_2 = 1 - \sqrt{x}$ in the same viewing window.

▷ **ALGEBRA HELP**   The
solution to Example 3 involves
solving equations. To review
the techniques for solving
equations, see Appendix A.5.

Zeros of a Function

If the graph of a function of x has an x-intercept at $(a, 0)$, then a is a **zero** of the function.

> **Zeros of a Function**
>
> The **zeros of a function** $y = f(x)$ are the x-values for which $f(x) = 0$.

EXAMPLE 3   **Finding the Zeros of Functions**

Find the zeros of each function algebraically.

a. $f(x) = 3x^2 + x - 10$

b. $g(x) = \sqrt{10 - x^2}$

c. $h(t) = \dfrac{2t - 3}{t + 5}$

Solution   To find the zeros of a function, set the function equal to zero and solve for the independent variable.

a.

$3x^2 + x - 10 = 0$	Set $f(x)$ equal to 0.
$(3x - 5)(x + 2) = 0$	Factor.
$3x - 5 = 0 \implies x = \frac{5}{3}$	Set 1st factor equal to 0 and solve.
$x + 2 = 0 \implies x = -2$	Set 2nd factor equal to 0 and solve.

The zeros of f are $x = \frac{5}{3}$ and $x = -2$. In Figure 1.33, note that the graph of f has $\left(\frac{5}{3}, 0\right)$ and $(-2, 0)$ as its x-intercepts.

b.

$\sqrt{10 - x^2} = 0$	Set $g(x)$ equal to 0.
$10 - x^2 = 0$	Square each side.
$10 = x^2$	Add x^2 to each side.
$\pm\sqrt{10} = x$	Extract square roots.

The zeros of g are $x = -\sqrt{10}$ and $x = \sqrt{10}$. In Figure 1.34, note that the graph of g has $\left(-\sqrt{10}, 0\right)$ and $\left(\sqrt{10}, 0\right)$ as its x-intercepts.

c.

$\dfrac{2t - 3}{t + 5} = 0$	Set $h(t)$ equal to 0.
$2t - 3 = 0$	Multiply each side by $t + 5$.
$2t = 3$	Add 3 to each side.
$t = \dfrac{3}{2}$	Divide each side by 2.

The zero of h is $t = \frac{3}{2}$. In Figure 1.35, note that the graph of h has $\left(\frac{3}{2}, 0\right)$ as its t-intercept.

✓ **Checkpoint** ◀))) *Audio-video solution in English & Spanish at LarsonPrecalculus.com*

Find the zeros of each function.

a. $f(x) = 2x^2 + 13x - 24$ **b.** $g(t) = \sqrt{t - 25}$ **c.** $h(x) = \dfrac{x^2 - 2}{x - 1}$

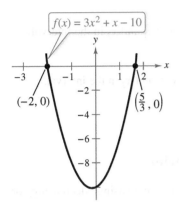

Zeros of f: $x = -2$, $x = \frac{5}{3}$
Figure 1.33

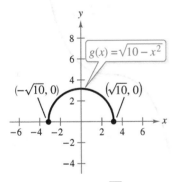

Zeros of g: $x = \pm\sqrt{10}$
Figure 1.34

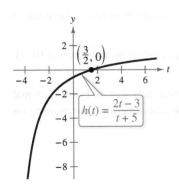

Zero of h: $t = \frac{3}{2}$
Figure 1.35

Increasing and Decreasing Functions

The more you know about the graph of a function, the more you know about the function itself. Consider the graph shown in Figure 1.36. As you move from *left to right*, this graph falls from $x = -2$ to $x = 0$, is constant from $x = 0$ to $x = 2$, and rises from $x = 2$ to $x = 4$.

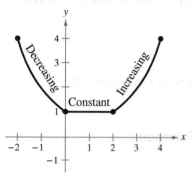

Figure 1.36

Increasing, Decreasing, and Constant Functions

A function f is **increasing** on an interval when, for any x_1 and x_2 in the interval,

$$x_1 < x_2 \quad \text{implies} \quad f(x_1) < f(x_2).$$

A function f is **decreasing** on an interval when, for any x_1 and x_2 in the interval,

$$x_1 < x_2 \quad \text{implies} \quad f(x_1) > f(x_2).$$

A function f is **constant** on an interval when, for any x_1 and x_2 in the interval,

$$f(x_1) = f(x_2).$$

EXAMPLE 4 **Describing Function Behavior**

Determine the open intervals on which each function is increasing, decreasing, or constant.

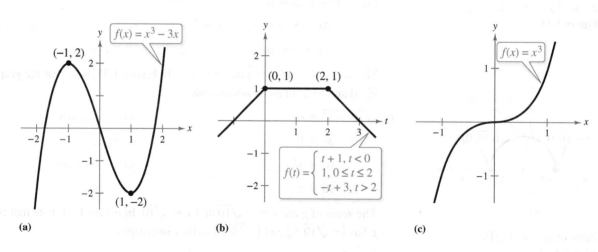

(a)   (b)   (c)

Solution

a. This function is increasing on the interval $(-\infty, -1)$, decreasing on the interval $(-1, 1)$, and increasing on the interval $(1, \infty)$.

b. This function is increasing on the interval $(-\infty, 0)$, constant on the interval $(0, 2)$, and decreasing on the interval $(2, \infty)$.

c. This function may appear to be constant on an interval near $x = 0$, but for all real values of x_1 and x_2, if $x_1 < x_2$, then $(x_1)^3 < (x_2)^3$. So, the function is increasing on the interval $(-\infty, \infty)$.

✓ **Checkpoint**  Audio-video solution in English & Spanish at LarsonPrecalculus.com

Graph the function

$$f(x) = x^3 + 3x^2 - 1.$$

Then determine the open intervals on which the function is increasing, decreasing, or constant.

Relative Minimum and Relative Maximum Values

The points at which a function changes its increasing, decreasing, or constant behavior are helpful in determining the **relative minimum** or **relative maximum** values of the function.

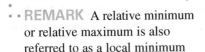

REMARK A relative minimum or relative maximum is also referred to as a local minimum or local maximum.

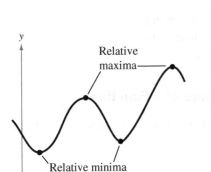

Figure 1.37

Definitions of Relative Minimum and Relative Maximum

A function value $f(a)$ is a **relative minimum** of f when there exists an interval (x_1, x_2) that contains a such that

$$x_1 < x < x_2 \quad \text{implies} \quad f(a) \le f(x).$$

A function value $f(a)$ is a **relative maximum** of f when there exists an interval (x_1, x_2) that contains a such that

$$x_1 < x < x_2 \quad \text{implies} \quad f(a) \ge f(x).$$

Figure 1.37 shows several different examples of relative minima and relative maxima. In Section 2.1, you will study a technique for finding the *exact point* at which a second-degree polynomial function has a relative minimum or relative maximum. For the time being, however, you can use a graphing utility to find reasonable approximations of these points.

EXAMPLE 5 **Approximating a Relative Minimum**

Use a graphing utility to approximate the relative minimum of the function

$$f(x) = 3x^2 - 4x - 2.$$

Solution The graph of f is shown in Figure 1.38. By using the *zoom* and *trace* features or the *minimum* feature of a graphing utility, you can approximate that the relative minimum of the function occurs at the point

$$(0.67, -3.33).$$

So, the relative minimum is approximately -3.33. Later, in Section 2.1, you will learn how to determine that the exact point at which the relative minimum occurs is $\left(\frac{2}{3}, -\frac{10}{3}\right)$ and the exact relative minimum is $-\frac{10}{3}$.

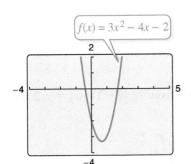

Figure 1.38

✓ **Checkpoint**  Audio-video solution in English & Spanish at LarsonPrecalculus.com

Use a graphing utility to approximate the relative maximum of the function

$$f(x) = -4x^2 - 7x + 3.$$

You can also use the *table* feature of a graphing utility to numerically approximate the relative minimum of the function in Example 5. Using a table that begins at 0.6 and increments the value of x by 0.01, you can approximate that the minimum of

$$f(x) = 3x^2 - 4x - 2$$

occurs at the point $(0.67, -3.33)$.

▷ **TECHNOLOGY** When you use a graphing utility to approximate the x- and y-values of the point where a relative minimum or relative maximum occurs, the *zoom* feature will often produce graphs that are nearly flat. To overcome this problem, manually change the vertical setting of the viewing window. The graph will stretch vertically when the values of Ymin and Ymax are closer together.

Average Rate of Change

In Section 1.3, you learned that the slope of a line can be interpreted as a *rate of change*. For a nonlinear graph, the **average rate of change** between any two points $(x_1, f(x_1))$ and $(x_2, f(x_2))$ is the slope of the line through the two points (see Figure 1.39). The line through the two points is called a **secant line,** and the slope of this line is denoted as m_{sec}.

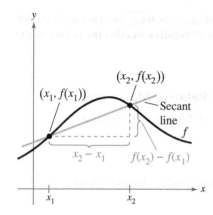

Figure 1.39

$$\text{Average rate of change of } f \text{ from } x_1 \text{ to } x_2 = \frac{f(x_2) - f(x_1)}{x_2 - x_1}$$

$$= \frac{\text{change in } y}{\text{change in } x}$$

$$= m_{sec}$$

EXAMPLE 6 **Average Rate of Change of a Function**

Find the average rates of change of $f(x) = x^3 - 3x$ (a) from $x_1 = -2$ to $x_2 = -1$ and (b) from $x_1 = 0$ to $x_2 = 1$ (see Figure 1.40).

Solution

a. The average rate of change of f from $x_1 = -2$ to $x_2 = -1$ is

$$\frac{f(x_2) - f(x_1)}{x_2 - x_1} = \frac{f(-1) - f(-2)}{-1 - (-2)} = \frac{2 - (-2)}{1} = 4. \qquad \text{Secant line has positive slope.}$$

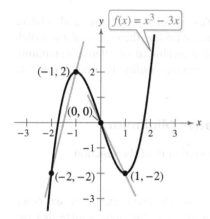

Figure 1.40

b. The average rate of change of f from $x_1 = 0$ to $x_2 = 1$ is

$$\frac{f(x_2) - f(x_1)}{x_2 - x_1} = \frac{f(1) - f(0)}{1 - 0} = \frac{-2 - 0}{1} = -2. \qquad \text{Secant line has negative slope.}$$

✓ **Checkpoint**  *Audio-video solution in English & Spanish at LarsonPrecalculus.com*

Find the average rates of change of $f(x) = x^2 + 2x$ (a) from $x_1 = -3$ to $x_2 = -2$ and (b) from $x_1 = -2$ to $x_2 = 0$.

EXAMPLE 7 **Finding Average Speed**

The distance s (in feet) a moving car is from a stoplight is given by the function

$$s(t) = 20t^{3/2}$$

where t is the time (in seconds). Find the average speed of the car (a) from $t_1 = 0$ to $t_2 = 4$ seconds and (b) from $t_1 = 4$ to $t_2 = 9$ seconds.

Solution

a. The average speed of the car from $t_1 = 0$ to $t_2 = 4$ seconds is

$$\frac{s(t_2) - s(t_1)}{t_2 - t_1} = \frac{s(4) - s(0)}{4 - 0} = \frac{160 - 0}{4} = 40 \text{ feet per second.}$$

b. The average speed of the car from $t_1 = 4$ to $t_2 = 9$ seconds is

$$\frac{s(t_2) - s(t_1)}{t_2 - t_1} = \frac{s(9) - s(4)}{9 - 4} = \frac{540 - 160}{5} = 76 \text{ feet per second.}$$

Average speed is an average rate of change.

✓ **Checkpoint** *Audio-video solution in English & Spanish at LarsonPrecalculus.com*

In Example 7, find the average speed of the car (a) from $t_1 = 0$ to $t_2 = 1$ second and (b) from $t_1 = 1$ second to $t_2 = 4$ seconds.

Even and Odd Functions

In Section 1.2, you studied different types of symmetry of a graph. In the terminology of functions, a function is said to be **even** when its graph is symmetric with respect to the *y*-axis and **odd** when its graph is symmetric with respect to the origin. The symmetry tests in Section 1.2 yield the tests for even and odd functions below.

Tests for Even and Odd Functions

A function $y = f(x)$ is **even** when, for each x in the domain of f, $f(-x) = f(x)$.

A function $y = f(x)$ is **odd** when, for each x in the domain of f, $f(-x) = -f(x)$.

EXAMPLE 8   **Even and Odd Functions**

See LarsonPrecalculus.com for an interactive version of this type of example.

a. The function $g(x) = x^3 - x$ is odd because $g(-x) = -g(x)$, as follows.

$$g(-x) = (-x)^3 - (-x)$$ Substitute $-x$ for x.

$$= -x^3 + x$$ Simplify.

$$= -(x^3 - x)$$ Distributive Property

$$= -g(x)$$ Test for odd function

b. The function $h(x) = x^2 + 1$ is even because $h(-x) = h(x)$, as follows.

$$h(-x) = (-x)^2 + 1 = x^2 + 1 = h(x)$$ Test for even function

Figure 1.41 shows the graphs and symmetry of these two functions.

✓ *Checkpoint* ◀))) Audio-video solution in English & Spanish at LarsonPrecalculus.com

Determine whether each function is even, odd, or neither. Then describe the symmetry.

a. $f(x) = 5 - 3x$ **b.** $g(x) = x^4 - x^2 - 1$ **c.** $h(x) = 2x^3 + 3x$

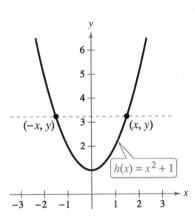

(a) Symmetric to origin: Odd Function

(b) Symmetric to *y*-axis: Even Function

Figure 1.41

Summarize   (Section 1.5)

1. State the Vertical Line Test for functions *(page 50)*. For an example of using the Vertical Line Test, see Example 2.

2. Explain how to find the zeros of a function *(page 51)*. For an example of finding the zeros of functions, see Example 3.

3. Explain how to determine intervals on which functions are increasing or decreasing *(page 52)*. For an example of describing function behavior, see Example 4.

4. Explain how to determine relative minimum and relative maximum values of functions *(page 53)*. For an example of approximating a relative minimum, see Example 5.

5. Explain how to determine the average rate of change of a function *(page 54)*. For examples of determining average rates of change, see Examples 6 and 7.

6. State the definitions of an even function and an odd function *(page 55)*. For an example of identifying even and odd functions, see Example 8.

1.5 Exercises

See **CalcChat.com** for tutorial help and worked-out solutions to odd-numbered exercises.

Vocabulary: Fill in the blanks.

1. The _____ _____ _____ is used to determine whether a graph represents y as a function of x.

2. The _____ of a function $y = f(x)$ are the values of x for which $f(x) = 0$.

3. A function f is _____ on an interval when, for any x_1 and x_2 in the interval, $x_1 < x_2$ implies $f(x_1) > f(x_2)$.

4. A function value $f(a)$ is a relative _____ of f when there exists an interval (x_1, x_2) containing a such that $x_1 < x < x_2$ implies $f(a) \geq f(x)$.

5. The _____ _____ _____ _____ between any two points $(x_1, f(x_1))$ and $(x_2, f(x_2))$ is the slope of the line through the two points, and this line is called the _____ line.

6. A function f is _____ when, for each x in the domain of f, $f(-x) = -f(x)$.

Skills and Applications

 Domain, Range, and Values of a Function In Exercises 7–10, use the graph of the function to find the domain and range of f and each function value.

7. (a) $f(-1)$ (b) $f(0)$ 8. (a) $f(-1)$ (b) $f(0)$
    (c) $f(1)$ (d) $f(2)$ (c) $f(1)$ (d) $f(3)$

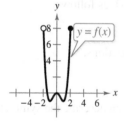 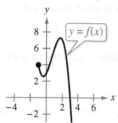

9. (a) $f(2)$ (b) $f(1)$ 10. (a) $f(-2)$ (b) $f(1)$
    (c) $f(3)$ (d) $f(-1)$ (c) $f(0)$ (d) $f(2)$

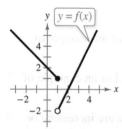 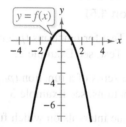

 Vertical Line Test for Functions In Exercises 11–14, use the Vertical Line Test to determine whether the graph represents y as a function of x. To print an enlarged copy of the graph, go to *MathGraphs.com*.

11. 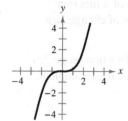   12.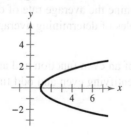

13. 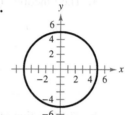   14.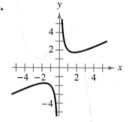

Finding the Zeros of a Function In Exercises 15–26, find the zeros of the function algebraically.

15. $f(x) = 3x + 18$

16. $f(x) = 15 - 2x$

17. $f(x) = 2x^2 - 7x - 30$

18. $f(x) = 3x^2 + 22x - 16$

19. $f(x) = \dfrac{x + 3}{2x^2 - 6}$

20. $f(x) = \dfrac{x^2 - 9x + 14}{4x}$

21. $f(x) = \frac{1}{3}x^3 - 2x$

22. $f(x) = -25x^4 + 9x^2$

23. $f(x) = x^3 - 4x^2 - 9x + 36$

24. $f(x) = 4x^3 - 24x^2 - x + 6$

25. $f(x) = \sqrt{2x - 1}$

26. $f(x) = \sqrt{3x + 2}$

Graphing and Finding Zeros In Exercises 27–32, (a) use a graphing utility to graph the function and find the zeros of the function and (b) verify your results from part (a) algebraically.

27. $f(x) = x^2 - 6x$

28. $f(x) = 2x^2 - 13x - 7$

29. $f(x) = \sqrt{2x + 11}$

30. $f(x) = \sqrt{3x - 14} - 8$

31. $f(x) = \dfrac{3x - 1}{x - 6}$

32. $f(x) = \dfrac{2x^2 - 9}{3 - x}$

Describing Function Behavior In Exercises 33–40, determine the open intervals on which the function is increasing, decreasing, or constant.

33. $f(x) = -\frac{1}{2}x^3$

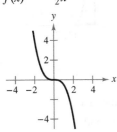

34. $f(x) = x^2 - 4x$

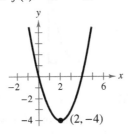

35. $f(x) = \sqrt{x^2 - 1}$

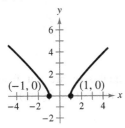

36. $f(x) = x^3 - 3x^2 + 2$

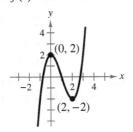

37. $f(x) = |x + 1| + |x - 1|$

38. $f(x) = \dfrac{x^2 + x + 1}{x + 1}$

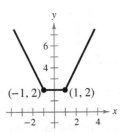

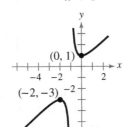

39. $f(x) = \begin{cases} 2x + 1, & x \le -1 \\ x^2 - 2, & x > -1 \end{cases}$

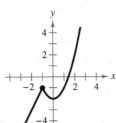

40. $f(x) = \begin{cases} x + 3, & x \le 0 \\ 3, & 0 < x \le 2 \\ 2x + 1, & x > 2 \end{cases}$

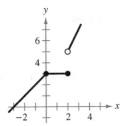

Describing Function Behavior In Exercises 41–48, use a graphing utility to graph the function and visually determine the open intervals on which the function is increasing, decreasing, or constant. Use a table of values to verify your results.

41. $f(x) = 3$
42. $g(x) = x$
43. $g(x) = \frac{1}{2}x^2 - 3$
44. $f(x) = 3x^4 - 6x^2$
45. $f(x) = \sqrt{1 - x}$
46. $f(x) = x\sqrt{x + 3}$
47. $f(x) = x^{3/2}$
48. $f(x) = x^{2/3}$

 Approximating Relative Minima or Maxima In Exercises 49–54, use a graphing utility to approximate (to two decimal places) any relative minima or maxima of the function.

49. $f(x) = x(x + 3)$
50. $f(x) = -x^2 + 3x - 2$
51. $h(x) = x^3 - 6x^2 + 15$
52. $f(x) = x^3 - 3x^2 - x + 1$
53. $h(x) = (x - 1)\sqrt{x}$
54. $g(x) = x\sqrt{4 - x}$

 Graphical Reasoning In Exercises 55–60, graph the function and determine the interval(s) for which $f(x) \ge 0$.

55. $f(x) = 4 - x$
56. $f(x) = 4x + 2$
57. $f(x) = 9 - x^2$
58. $f(x) = x^2 - 4x$
59. $f(x) = \sqrt{x - 1}$
60. $f(x) = |x + 5|$

 Average Rate of Change of a Function In Exercises 61–64, find the average rate of change of the function from x_1 to x_2.

Function	x-Values
61. $f(x) = -2x + 15$	$x_1 = 0, x_2 = 3$
62. $f(x) = x^2 - 2x + 8$	$x_1 = 1, x_2 = 5$
63. $f(x) = x^3 - 3x^2 - x$	$x_1 = -1, x_2 = 2$
64. $f(x) = -x^3 + 6x^2 + x$	$x_1 = 1, x_2 = 6$

65. Research and Development The amounts (in billions of dollars) the U.S. federal government spent on research and development for defense from 2010 through 2014 can be approximated by the model

$$y = 0.5079t^2 - 8.168t + 95.08$$

where t represents the year, with $t = 0$ corresponding to 2010. (*Source: American Association for the Advancement of Science*)

(a) Use a graphing utility to graph the model.

(b) Find the average rate of change of the model from 2010 to 2014. Interpret your answer in the context of the problem.

66. Finding Average Speed Use the information in Example 7 to find the average speed of the car from $t_1 = 0$ to $t_2 = 9$ seconds. Explain why the result is less than the value obtained in part (b) of Example 7.

Physics In Exercises 67–70, (a) use the position equation $s = -16t^2 + v_0t + s_0$ to write a function that represents the situation, (b) use a graphing utility to graph the function, (c) find the average rate of change of the function from t_1 to t_2, (d) describe the slope of the secant line through t_1 and t_2, (e) find the equation of the secant line through t_1 and t_2, and (f) graph the secant line in the same viewing window as your position function.

67. An object is thrown upward from a height of 6 feet at a velocity of 64 feet per second.

$t_1 = 0, t_2 = 3$

68. An object is thrown upward from a height of 6.5 feet at a velocity of 72 feet per second.

$t_1 = 0, t_2 = 4$

69. An object is thrown upward from ground level at a velocity of 120 feet per second.

$t_1 = 3, t_2 = 5$

70. An object is dropped from a height of 80 feet.

$t_1 = 1, t_2 = 2$

Even, Odd, or Neither? In Exercises 71–76, determine whether the function is even, odd, or neither. Then describe the symmetry.

71. $f(x) = x^6 - 2x^2 + 3$ **72.** $g(x) = x^3 - 5x$

73. $h(x) = x\sqrt{x+5}$ **74.** $f(x) = x\sqrt{1-x^2}$

75. $f(s) = 4s^{3/2}$ **76.** $g(s) = 4s^{2/3}$

Even, Odd, or Neither? In Exercises 77–82, sketch a graph of the function and determine whether it is even, odd, or neither. Verify your answer algebraically.

77. $f(x) = -9$ **78.** $f(x) = 5 - 3x$

79. $f(x) = -|x - 5|$ **80.** $h(x) = x^2 - 4$

81. $f(x) = \sqrt[3]{4x}$ **82.** $f(x) = \sqrt[3]{x - 4}$

Height of a Rectangle In Exercises 83 and 84, write the height h of the rectangle as a function of x.

83.

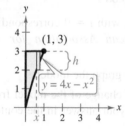

84.

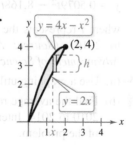

Length of a Rectangle In Exercises 85 and 86, write the length L of the rectangle as a function of y.

85.

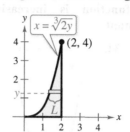

86.

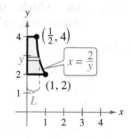

87. Error Analysis Describe the error.

The function $f(x) = 2x^3 - 5$ is odd because $f(-x) = -f(x)$, as follows.

$f(-x) = 2(-x)^3 - 5$

$= -2x^3 - 5$

$= -(2x^3 - 5)$

$= -f(x)$

88. Geometry Corners of equal size are cut from a square with sides of length 8 meters (see figure).

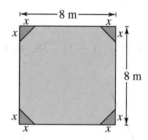

(a) Write the area A of the resulting figure as a function of x. Determine the domain of the function.

(b) Use a graphing utility to graph the area function over its domain. Use the graph to find the range of the function.

(c) Identify the figure that results when x is the maximum value in the domain of the function. What would be the length of each side of the figure?

89. Coordinate Axis Scale Each function described below models the specified data for the years 2006 through 2016, with $t = 6$ corresponding to 2006. Estimate a reasonable scale for the vertical axis (e.g., hundreds, thousands, millions, etc.) of the graph and justify your answer. (There are many correct answers.)

(a) $f(t)$ represents the average salary of college professors.

(b) $f(t)$ represents the U.S. population.

(c) $f(t)$ represents the percent of the civilian workforce that is unemployed.

(d) $f(t)$ represents the number of games a college football team wins.

90. Temperature

The table shows the temperatures y (in degrees Fahrenheit) in a city over a 24-hour period. Let x represent the time of day, where $x = 0$ corresponds to 6 A.M.

DATA	Time, x	Temperature, y
	0	34
	2	50
	4	60
	6	64
	8	63
	10	59
	12	53
	14	46
	16	40
	18	36
	20	34
	22	37
	24	45

Spreadsheet at LarsonPrecalculus.com

These data can be approximated by the model

$y = 0.026x^3 - 1.03x^2 + 10.2x + 34, \quad 0 \le x \le 24$.

(a) Use a graphing utility to create a scatter plot of the data. Then graph the model in the same viewing window.

(b) How well does the model fit the data?

(c) Use the graph to approximate the times when the temperature was increasing and decreasing.

(d) Use the graph to approximate the maximum and minimum temperatures during this 24-hour period.

(e) Could this model predict the temperatures in the city during the next 24-hour period? Why or why not?

Exploration

True or False? In Exercises 91–93, determine whether the statement is true or false. Justify your answer.

91. A function with a square root cannot have a domain that is the set of real numbers.

92. It is possible for an odd function to have the interval $[0, \infty)$ as its domain.

93. It is impossible for an even function to be increasing on its entire domain.

94. **HOW DO YOU SEE IT?** Use the graph of the function to answer parts (a)–(e).

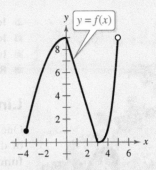

(a) Find the domain and range of f.

(b) Find the zero(s) of f.

(c) Determine the open intervals on which f is increasing, decreasing, or constant.

(d) Approximate any relative minimum or relative maximum values of f.

(e) Is f even, odd, or neither?

Think About It In Exercises 95 and 96, find the coordinates of a second point on the graph of a function f when the given point is on the graph and the function is (a) even and (b) odd.

95. $\left(-\frac{5}{3}, -7\right)$ **96.** $(2a, 2c)$

97. Writing Use a graphing utility to graph each function. Write a paragraph describing any similarities and differences you observe among the graphs.

(a) $y = x$ (b) $y = x^2$ (c) $y = x^3$

(d) $y = x^4$ (e) $y = x^5$ (f) $y = x^6$

98. Graphical Reasoning Graph each of the functions with a graphing utility. Determine whether each function is even, odd, or neither.

$f(x) = x^2 - x^4$ $g(x) = 2x^3 + 1$

$h(x) = x^5 - 2x^3 + x$ $j(x) = 2 - x^6 - x^8$

$k(x) = x^5 - 2x^4 + x - 2$ $p(x) = x^9 + 3x^5 - x^3 + x$

What do you notice about the equations of functions that are odd? What do you notice about the equations of functions that are even? Can you describe a way to identify a function as odd or even by inspecting the equation? Can you describe a way to identify a function as neither odd nor even by inspecting the equation?

99. Even, Odd, or Neither? Determine whether g is even, odd, or neither when f is an even function. Explain.

(a) $g(x) = -f(x)$ (b) $g(x) = f(-x)$

(c) $g(x) = f(x) - 2$ (d) $g(x) = f(x - 2)$

1.6 A Library of Parent Functions

■ **Identify and graph linear and squaring functions.**
■ **Identify and graph cubic, square root, and reciprocal functions.**
■ **Identify and graph step and other piecewise-defined functions.**
■ **Recognize graphs of parent functions.**

Linear and Squaring Functions

One of the goals of this text is to enable you to recognize the basic shapes of the graphs of different types of functions. For example, you know that the graph of the **linear function** $f(x) = ax + b$ is a line with slope $m = a$ and y-intercept at $(0, b)$. The graph of a linear function has the characteristics below.

- The domain of the function is the set of all real numbers.
- When $m \neq 0$, the range of the function is the set of all real numbers.
- The graph has an x-intercept at $(-b/m, 0)$ and a y-intercept at $(0, b)$.
- The graph is increasing when $m > 0$, decreasing when $m < 0$, and constant when $m = 0$.

Piecewise-defined functions model many real-life situations. For example, in Exercise 47 on page 66, you will write a piecewise-defined function to model the depth of snow during a snowstorm.

EXAMPLE 1 **Writing a Linear Function**

Write the linear function f for which $f(1) = 3$ and $f(4) = 0$.

Solution To find the equation of the line that passes through $(x_1, y_1) = (1, 3)$ and $(x_2, y_2) = (4, 0)$, first find the slope of the line.

$$m = \frac{y_2 - y_1}{x_2 - x_1} = \frac{0 - 3}{4 - 1} = \frac{-3}{3} = -1$$

Next, use the point-slope form of the equation of a line.

$y - y_1 = m(x - x_1)$	Point-slope form
$y - 3 = -1(x - 1)$	Substitute for x_1, y_1, and m.
$y = -x + 4$	Simplify.
$f(x) = -x + 4$	Function notation

The figure below shows the graph of this function.

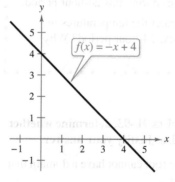

$f(x) = -x + 4$

✓ **Checkpoint** ◀))) *Audio-video solution in English & Spanish at LarsonPrecalculus.com*

Write the linear function f for which $f(-2) = 6$ and $f(4) = -9$.

There are two special types of linear functions, the **constant function** and the **identity function.** A constant function has the form

$$f(x) = c$$

and has a domain of all real numbers with a range consisting of a single real number c. The graph of a constant function is a horizontal line, as shown in Figure 1.42. The identity function has the form

$$f(x) = x.$$

Its domain and range are the set of all real numbers. The identity function has a slope of $m = 1$ and a y-intercept at $(0, 0)$. The graph of the identity function is a line for which each x-coordinate equals the corresponding y-coordinate. The graph is always increasing, as shown in Figure 1.43.

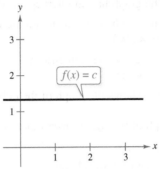

Figure 1.42

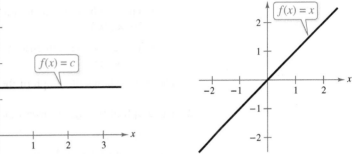

Figure 1.43

The graph of the **squaring function**

$$f(x) = x^2$$

is a U-shaped curve with the characteristics below.

- The domain of the function is the set of all real numbers.
- The range of the function is the set of all nonnegative real numbers.
- The function is even.
- The graph has an intercept at $(0, 0)$.
- The graph is decreasing on the interval $(-\infty, 0)$ and increasing on the interval $(0, \infty)$.
- The graph is symmetric with respect to the y-axis.
- The graph has a relative minimum at $(0, 0)$.

The figure below shows the graph of the squaring function.

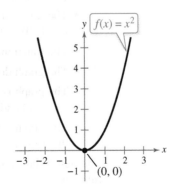

Cubic, Square Root, and Reciprocal Functions

Here are the basic characteristics of the graphs of the **cubic, square root,** and **reciprocal functions.**

1. The graph of the *cubic* function

$$f(x) = x^3$$

has the characteristics below.

- The domain of the function is the set of all real numbers.
- The range of the function is the set of all real numbers.
- The function is odd.
- The graph has an intercept at $(0, 0)$.
- The graph is increasing on the interval $(-\infty, \infty)$.
- The graph is symmetric with respect to the origin.

The figure shows the graph of the cubic function.

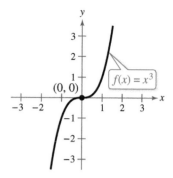

Cubic function

2. The graph of the *square root* function

$$f(x) = \sqrt{x}$$

has the characteristics below.

- The domain of the function is the set of all nonnegative real numbers.
- The range of the function is the set of all nonnegative real numbers.
- The graph has an intercept at $(0, 0)$.
- The graph is increasing on the interval $(0, \infty)$.

The figure shows the graph of the square root function.

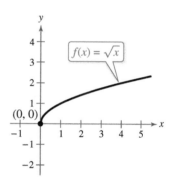

Square root function

3. The graph of the *reciprocal* function

$$f(x) = \frac{1}{x}$$

has the characteristics below.

- The domain of the function is $(-\infty, 0) \cup (0, \infty)$.
- The range of the function is $(-\infty, 0) \cup (0, \infty)$.
- The function is odd.
- The graph does not have any intercepts.
- The graph is decreasing on the intervals $(-\infty, 0)$ and $(0, \infty)$.
- The graph is symmetric with respect to the origin.

The figure shows the graph of the reciprocal function.

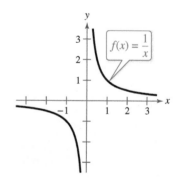

Reciprocal function

Step and Piecewise-Defined Functions

Functions whose graphs resemble sets of stairsteps are known as **step functions.** One common type of step function is the **greatest integer function,** denoted by $[\![x]\!]$ and defined as

$$f(x) = [\![x]\!] = \textit{the greatest integer less than or equal to x.}$$

Here are several examples of evaluating the greatest integer function.

$$[\![-1]\!] = (\text{greatest integer} \leq -1) = -1$$
$$\left[\!\!\left[-\tfrac{1}{2}\right]\!\!\right] = \left(\text{greatest integer} \leq -\tfrac{1}{2}\right) = -1$$
$$\left[\!\!\left[\tfrac{1}{10}\right]\!\!\right] = \left(\text{greatest integer} \leq \tfrac{1}{10}\right) = 0$$
$$[\![1.5]\!] = (\text{greatest integer} \leq 1.5) = 1$$
$$[\![1.9]\!] = (\text{greatest integer} \leq 1.9) = 1$$

The graph of the greatest integer function

$$f(x) = [\![x]\!]$$

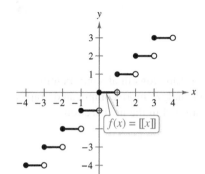

Figure 1.44

has the characteristics below, as shown in Figure 1.44.

- The domain of the function is the set of all real numbers.
- The range of the function is the set of all integers.
- The graph has a y-intercept at $(0, 0)$ and x-intercepts in the interval $[0, 1)$.
- The graph is constant between each pair of consecutive integer values of x.
- The graph jumps vertically one unit at each integer value of x.

▷ TECHNOLOGY Most graphing utilities display graphs in *connected* mode, which works well for graphs that do not have breaks. For graphs that do have breaks, such as the graph of the greatest integer function, it may be better to use *dot* mode. Graph the greatest integer function [often called Int(x)] in *connected* and *dot* modes, and compare the two results.

EXAMPLE 2 Evaluating a Step Function

Evaluate the function $f(x) = [\![x]\!] + 1$ when $x = -1, 2,$ and $\tfrac{3}{2}$.

Solution For $x = -1$, the greatest integer ≤ -1 is -1, so

$$f(-1) = [\![-1]\!] + 1 = -1 + 1 = 0.$$

For $x = 2$, the greatest integer ≤ 2 is 2, so

$$f(2) = [\![2]\!] + 1 = 2 + 1 = 3.$$

For $x = \tfrac{3}{2}$, the greatest integer $\leq \tfrac{3}{2}$ is 1, so

$$f\!\left(\tfrac{3}{2}\right) = \left[\!\!\left[\tfrac{3}{2}\right]\!\!\right] + 1 = 1 + 1 = 2.$$

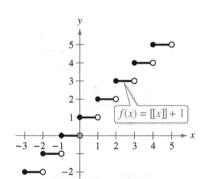

Figure 1.45

Verify your answers by examining the graph of $f(x) = [\![x]\!] + 1$ shown in Figure 1.45.

✓ **Checkpoint** ◀))) Audio-video solution in English & Spanish at LarsonPrecalculus.com

Evaluate the function $f(x) = [\![x + 2]\!]$ when $x = -\tfrac{3}{2}, 1,$ and $-\tfrac{5}{2}$. ■

Recall from Section 1.4 that a piecewise-defined function is defined by two or more equations over a specified domain. To graph a piecewise-defined function, graph each equation separately over the specified domain, as shown in Example 3.

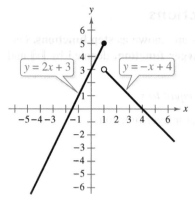

Figure 1.46

EXAMPLE 3   **Graphing a Piecewise-Defined Function**

See LarsonPrecalculus.com for an interactive version of this type of example.

Sketch the graph of $f(x) = \begin{cases} 2x + 3, & x \le 1 \\ -x + 4, & x > 1 \end{cases}$.

Solution   This piecewise-defined function consists of two linear functions. At $x = 1$ and to the left of $x = 1$, the graph is the line $y = 2x + 3$, and to the right of $x = 1$, the graph is the line $y = -x + 4$, as shown in Figure 1.46. Notice that the point $(1, 5)$ is a solid dot and the point $(1, 3)$ is an open dot. This is because $f(1) = 2(1) + 3 = 5$.

✓ **Checkpoint**))) Audio-video solution in English & Spanish at LarsonPrecalculus.com

Sketch the graph of $f(x) = \begin{cases} -\frac{1}{2}x - 6, & x \le -4 \\ x + 5, & x > -4 \end{cases}$.

Parent Functions

The graphs below represent the most commonly used functions in algebra. Familiarity with the characteristics of these graphs will help you analyze more complicated graphs obtained from these graphs by the transformations studied in the next section.

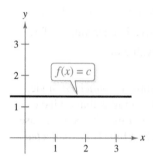

(a) Constant Function

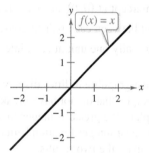

(b) Identity Function

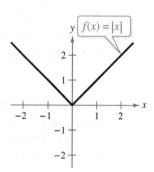

(c) Absolute Value Function

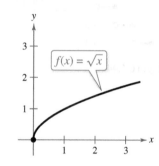

(d) Square Root Function

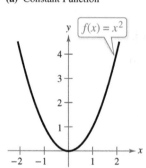

(e) Squaring Function

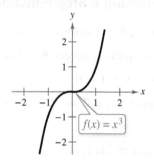

(f) Cubic Function

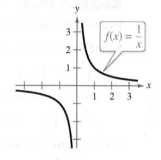

(g) Reciprocal Function

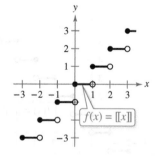

(h) Greatest Integer Function

Summarize   (Section 1.6)

1. Explain how to identify and graph linear and squaring functions *(pages 60 and 61)*. For an example involving a linear function, see Example 1.

2. Explain how to identify and graph cubic, square root, and reciprocal functions *(page 62)*.

3. Explain how to identify and graph step and other piecewise-defined functions *(page 63)*. For examples involving these functions, see Examples 2 and 3.

4. Identify and sketch the graphs of parent functions *(page 64)*.

1.6 Exercises

See **CalcChat.com** for tutorial help and worked-out solutions to odd-numbered exercises.

Vocabulary

In Exercises 1–9, write the most specific name of the function.

1. $f(x) = [\![x]\!]$

2. $f(x) = x$

3. $f(x) = 1/x$

4. $f(x) = x^2$

5. $f(x) = \sqrt{x}$

6. $f(x) = c$

7. $f(x) = |x|$

8. $f(x) = x^3$

9. $f(x) = ax + b$

10. Fill in the blank: The constant function and the identity function are two special types of _____ functions.

Skills and Applications

 Writing a Linear Function **In Exercises 11–14, (a) write the linear function f that has the given function values and (b) sketch the graph of the function.**

11. $f(1) = 4$, $f(0) = 6$ **12.** $f(-3) = -8$, $f(1) = 2$

13. $f\left(\frac{1}{2}\right) = -\frac{5}{3}$, $f(6) = 2$ **14.** $f\left(\frac{3}{5}\right) = \frac{1}{2}$, $f(4) = 9$

Graphing a Function **In Exercises 15–26, use a graphing utility to graph the function. Be sure to choose an appropriate viewing window.**

15. $f(x) = 2.5x - 4.25$ **16.** $f(x) = \frac{5}{6} - \frac{2}{3}x$

17. $g(x) = x^2 + 3$ **18.** $f(x) = -2x^2 - 1$

19. $f(x) = x^3 - 1$ **20.** $f(x) = (x - 1)^3 + 2$

21. $f(x) = \sqrt{x} + 4$ **22.** $h(x) = \sqrt{x + 2} + 3$

23. $f(x) = \dfrac{1}{x - 2}$ **24.** $k(x) = 3 + \dfrac{1}{x + 3}$

25. $g(x) = |x| - 5$ **26.** $f(x) = |x - 1|$

 Evaluating a Step Function **In Exercises 27–30, evaluate the function for the given values.**

27. $f(x) = [\![x]\!]$

(a) $f(2.1)$ (b) $f(2.9)$ (c) $f(-3.1)$ (d) $f\left(\frac{7}{2}\right)$

28. $h(x) = [\![x + 3]\!]$

(a) $h(-2)$ (b) $h\left(\frac{1}{2}\right)$ (c) $h(4.2)$ (d) $h(-21.6)$

29. $k(x) = [\![2x + 1]\!]$

(a) $k\left(\frac{1}{3}\right)$ (b) $k(-2.1)$ (c) $k(1.1)$ (d) $k\left(\frac{2}{3}\right)$

30. $g(x) = -7[\![x + 4]\!] + 6$

(a) $g\left(\frac{1}{8}\right)$ (b) $g(9)$ (c) $g(-4)$ (d) $g\left(\frac{3}{2}\right)$

Graphing a Step Function **In Exercises 31–34, sketch the graph of the function.**

31. $g(x) = -[\![x]\!]$ **32.** $g(x) = 4[\![x]\!]$

33. $g(x) = [\![x]\!] - 1$ **34.** $g(x) = [\![x - 3]\!]$

 Graphing a Piecewise-Defined Function **In Exercises 35–40, sketch the graph of the function.**

35. $g(x) = \begin{cases} x + 6, & x \le -4 \\ \frac{1}{2}x - 4, & x > -4 \end{cases}$

36. $f(x) = \begin{cases} 4 + x, & x \le 2 \\ x^2 + 2, & x > 2 \end{cases}$

37. $f(x) = \begin{cases} 1 - (x - 1)^2, & x \le 2 \\ \sqrt{x - 2}, & x > 2 \end{cases}$

38. $f(x) = \begin{cases} \sqrt{4 + x}, & x < 0 \\ \sqrt{4 - x}, & x \ge 0 \end{cases}$

39. $h(x) = \begin{cases} 4 - x^2, & x < -2 \\ 3 + x, & -2 \le x < 0 \\ x^2 + 1, & x \ge 0 \end{cases}$

40. $k(x) = \begin{cases} 2x + 1, & x \le -1 \\ 2x^2 - 1, & -1 < x \le 1 \\ 1 - x^2, & x > 1 \end{cases}$

Graphing a Function **In Exercises 41 and 42, (a) use a graphing utility to graph the function and (b) state the domain and range of the function.**

41. $s(x) = 2\left(\frac{1}{4}x - \left[\!\!\left[\frac{1}{4}x\right]\!\!\right]\right)$ **42.** $k(x) = 4\left(\frac{1}{2}x - \left[\!\!\left[\frac{1}{2}x\right]\!\!\right]\right)^2$

43. Wages A mechanic's pay is $14 per hour for regular time and time-and-a-half for overtime. The weekly wage function is

$$W(h) = \begin{cases} 14h, & 0 < h \le 40 \\ 21(h - 40) + 560, & h > 40 \end{cases}$$

where h is the number of hours worked in a week.

(a) Evaluate $W(30)$, $W(40)$, $W(45)$, and $W(50)$.

(b) The company decreases the regular work week to 36 hours. What is the new weekly wage function?

(c) The company increases the mechanic's pay to $16 per hour. What is the new weekly wage function? Use a regular work week of 40 hours.

44. Revenue The table shows the monthly revenue y (in thousands of dollars) of a landscaping business for each month of the year 2016, with $x = 1$ representing January.

Month, x	Revenue, y
1	5.2
2	5.6
3	6.6
4	8.3
5	11.5
6	15.8
7	12.8
8	10.1
9	8.6
10	6.9
11	4.5
12	2.7

DATA · Spreadsheet at LarsonPrecalculus.com

A mathematical model that represents these data is

$$f(x) = \begin{cases} -1.97x + 26.3 \\ 0.505x^2 - 1.47x + 6.3 \end{cases}.$$

(a) Use a graphing utility to graph the model. What is the domain of each part of the piecewise-defined function? How can you tell?

(b) Find $f(5)$ and $f(11)$ and interpret your results in the context of the problem.

(c) How do the values obtained from the model in part (b) compare with the actual data values?

45. Fluid Flow The intake pipe of a 100-gallon tank has a flow rate of 10 gallons per minute, and two drainpipes have flow rates of 5 gallons per minute each. The figure shows the volume V of fluid in the tank as a function of time t. Determine whether the input pipe and each drainpipe are open or closed in specific subintervals of the 1 hour of time shown in the graph. (There are many correct answers.)

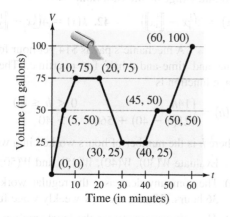

46. Delivery Charges The cost of mailing a package weighing up to, but not including, 1 pound is $2.72. Each additional pound or portion of a pound costs $0.50.

(a) Use the greatest integer function to create a model for the cost C of mailing a package weighing x pounds, where $x > 0$.

(b) Sketch the graph of the function.

· · 47. Snowstorm · · · · · · · · · · · ·

During a nine-hour snowstorm, it snows at a rate of 1 inch per hour for the first 2 hours, at a rate of 2 inches per hour for the next 6 hours, and at a rate of 0.5 inch per hour for the final hour. Write and graph a piecewise-defined function that gives the depth of the snow during the snowstorm. How many inches of snow accumulated from the storm?

48. **HOW DO YOU SEE IT?** For each graph of f shown below, answer parts (a)–(d).

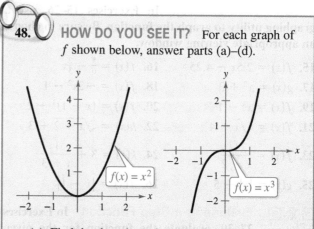

(a) Find the domain and range of f.

(b) Find the x- and y-intercepts of the graph of f.

(c) Determine the open intervals on which f is increasing, decreasing, or constant.

(d) Determine whether f is even, odd, or neither. Then describe the symmetry.

Exploration

True or False? **In Exercises 49 and 50, determine whether the statement is true or false. Justify your answer.**

49. A piecewise-defined function will always have at least one x-intercept or at least one y-intercept.

50. A linear equation will always have an x-intercept and a y-intercept.

1.7 Transformations of Functions

Transformations of functions model many real-life applications. For example, in Exercise 61 on page 74, you will use a transformation of a function to model the number of horsepower required to overcome wind drag on an automobile.

- Use vertical and horizontal shifts to sketch graphs of functions.
- Use reflections to sketch graphs of functions.
- Use nonrigid transformations to sketch graphs of functions.

Shifting Graphs

Many functions have graphs that are transformations of the parent graphs summarized in Section 1.6. For example, you obtain the graph of

$$h(x) = x^2 + 2$$

by shifting the graph of $f(x) = x^2$ *up* two units, as shown in Figure 1.47. In function notation, h and f are related as follows.

$$h(x) = x^2 + 2 = f(x) + 2 \qquad \text{Upward shift of two units}$$

Similarly, you obtain the graph of

$$g(x) = (x - 2)^2$$

by shifting the graph of $f(x) = x^2$ to the *right* two units, as shown in Figure 1.48. In this case, the functions g and f have the following relationship.

$$g(x) = (x - 2)^2 = f(x - 2) \qquad \text{Right shift of two units}$$

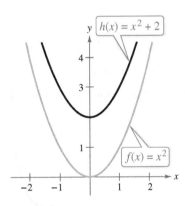

Figure 1.47

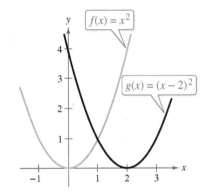

Figure 1.48

The list below summarizes this discussion about horizontal and vertical shifts.

· · REMARK In items 3 and 4, be sure you see that $h(x) = f(x - c)$ corresponds to a *right* shift and $h(x) = f(x + c)$ corresponds to a *left* shift for $c > 0$.

Vertical and Horizontal Shifts

Let c be a positive real number. **Vertical and horizontal shifts** in the graph of $y = f(x)$ are represented as follows.

1. Vertical shift c units *up:* $\qquad h(x) = f(x) + c$
2. Vertical shift c units *down:* $\qquad h(x) = f(x) - c$
3. Horizontal shift c units to the *right:* $h(x) = f(x - c)$
4. Horizontal shift c units to the *left:* $\quad h(x) = f(x + c)$

Some graphs are obtained from combinations of vertical and horizontal shifts, as demonstrated in Example 1(b). Vertical and horizontal shifts generate a *family of functions,* each with the same shape but at a different location in the plane.

EXAMPLE 1   **Shifting the Graph of a Function**

Use the graph of $f(x) = x^3$ to sketch the graph of each function.

a. $g(x) = x^3 - 1$

b. $h(x) = (x + 2)^3 + 1$

Solution

a. Relative to the graph of $f(x) = x^3$, the graph of

$$g(x) = x^3 - 1$$

is a downward shift of one unit, as shown below.

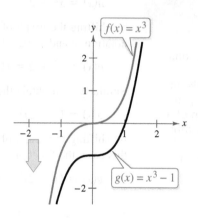

b. Relative to the graph of $f(x) = x^3$, the graph of

$$h(x) = (x + 2)^3 + 1$$

is a left shift of two units and an upward shift of one unit, as shown below.

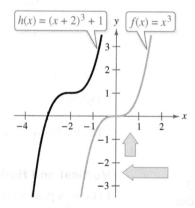

✓ **Checkpoint** ◀))) *Audio-video solution in English & Spanish at LarsonPrecalculus.com*

Use the graph of $f(x) = x^3$ to sketch the graph of each function.

a. $h(x) = x^3 + 5$

b. $g(x) = (x - 3)^3 + 2$

In Example 1(a), note that $g(x) = f(x) - 1$ and in Example 1(b), $h(x) = f(x + 2) + 1$. In Example 1(b), you obtain the same result whether the vertical shift precedes the horizontal shift or the horizontal shift precedes the vertical shift.

Reflecting Graphs

Another common type of transformation is a **reflection.** For example, if you consider the x-axis to be a mirror, then the graph of $h(x) = -x^2$ is the mirror image (or reflection) of the graph of $f(x) = x^2$, as shown in Figure 1.49.

Figure 1.49

Reflections in the Coordinate Axes

Reflections in the coordinate axes of the graph of $y = f(x)$ are represented as follows.

1. Reflection in the x-axis: $h(x) = -f(x)$

2. Reflection in the y-axis: $h(x) = f(-x)$

EXAMPLE 2   **Writing Equations from Graphs**

The graph of the function

$$f(x) = x^4$$

is shown in Figure 1.50. Each graph below is a transformation of the graph of f. Write an equation for the function represented by each graph.

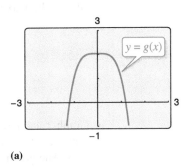

(a)

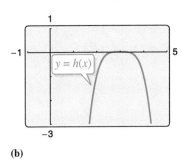

(b)

Figure 1.50

Solution

a. The graph of g is a reflection in the x-axis *followed by* an upward shift of two units of the graph of $f(x) = x^4$. So, an equation for g is

$$g(x) = -x^4 + 2.$$

b. The graph of h is a right shift of three units *followed by* a reflection in the x-axis of the graph of $f(x) = x^4$. So, an equation for h is

$$h(x) = -(x - 3)^4.$$

✓ **Checkpoint** 🔊))) *Audio-video solution in English & Spanish at LarsonPrecalculus.com*

The graph is a transformation of the graph of $f(x) = x^4$. Write an equation for the function represented by the graph.

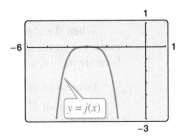

EXAMPLE 3 **Reflections and Shifts**

Compare the graph of each function with the graph of $f(x) = \sqrt{x}$.

a. $g(x) = -\sqrt{x}$ **b.** $h(x) = \sqrt{-x}$ **c.** $k(x) = -\sqrt{x+2}$

Algebraic Solution

a. The graph of g is a reflection of the graph of f in the x-axis because

$$g(x) = -\sqrt{x}$$
$$= -f(x).$$

b. The graph of h is a reflection of the graph of f in the y-axis because

$$h(x) = \sqrt{-x}$$
$$= f(-x).$$

c. The graph of k is a left shift of two units followed by a reflection in the x-axis because

$$k(x) = -\sqrt{x+2}$$
$$= -f(x+2).$$

Graphical Solution

a. Graph f and g on the same set of coordinate axes. The graph of g is a reflection of the graph of f in the x-axis.

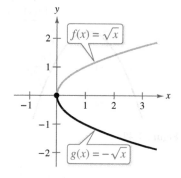

b. Graph f and h on the same set of coordinate axes. The graph of h is a reflection of the graph of f in the y-axis.

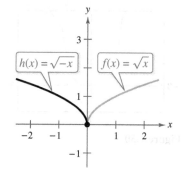

c. Graph f and k on the same set of coordinate axes. The graph of k is a left shift of two units followed by a reflection in the x-axis of the graph of f.

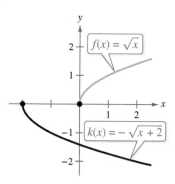

✓ *Checkpoint* ◀))) *Audio-video solution in English & Spanish at LarsonPrecalculus.com*

Compare the graph of each function with the graph of

$$f(x) = \sqrt{x-1}.$$

a. $g(x) = -\sqrt{x-1}$ **b.** $h(x) = \sqrt{-x-1}$

When sketching the graphs of functions involving square roots, remember that you must restrict the domain to exclude negative numbers inside the radical. For instance, here are the domains of the functions in Example 3.

Domain of $g(x) = -\sqrt{x}$: $x \geq 0$

Domain of $h(x) = \sqrt{-x}$: $x \leq 0$

Domain of $k(x) = -\sqrt{x+2}$: $x \geq -2$

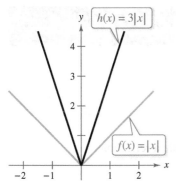

Figure 1.51

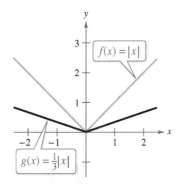

Figure 1.52

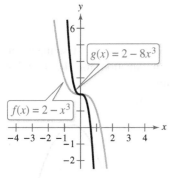

Figure 1.53

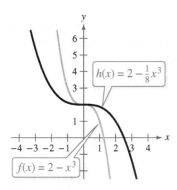

Figure 1.54

Nonrigid Transformations

Horizontal shifts, vertical shifts, and reflections are **rigid transformations** because the basic shape of the graph is unchanged. These transformations change only the *position* of the graph in the coordinate plane. **Nonrigid transformations** are those that cause a *distortion*—a change in the shape of the original graph. For example, a nonrigid transformation of the graph of $y = f(x)$ is represented by $g(x) = cf(x)$, where the transformation is a **vertical stretch** when $c > 1$ and a **vertical shrink** when $0 < c < 1$. Another nonrigid transformation of the graph of $y = f(x)$ is represented by $h(x) = f(cx)$, where the transformation is a **horizontal shrink** when $c > 1$ and a **horizontal stretch** when $0 < c < 1$.

EXAMPLE 4   **Nonrigid Transformations**

Compare the graph of each function with the graph of $f(x) = |x|$.

a. $h(x) = 3|x|$ **b.** $g(x) = \frac{1}{3}|x|$

Solution

a. Relative to the graph of $f(x) = |x|$, the graph of $h(x) = 3|x| = 3f(x)$ is a vertical stretch (each y-value is multiplied by 3). (See Figure 1.51.)

b. Similarly, the graph of $g(x) = \frac{1}{3}|x| = \frac{1}{3}f(x)$ is a vertical shrink $\left(\text{each } y\text{-value is multiplied by } \frac{1}{3}\right)$ of the graph of f. (See Figure 1.52.)

✓ *Checkpoint* 

Compare the graph of each function with the graph of $f(x) = x^2$.

a. $g(x) = 4x^2$ **b.** $h(x) = \frac{1}{4}x^2$

EXAMPLE 5   **Nonrigid Transformations**

See LarsonPrecalculus.com for an interactive version of this type of example.

Compare the graph of each function with the graph of $f(x) = 2 - x^3$.

a. $g(x) = f(2x)$ **b.** $h(x) = f\left(\frac{1}{2}x\right)$

Solution

a. Relative to the graph of $f(x) = 2 - x^3$, the graph of $g(x) = f(2x) = 2 - (2x)^3 = 2 - 8x^3$ is a horizontal shrink $(c > 1)$. (See Figure 1.53.)

b. Similarly, the graph of $h(x) = f\left(\frac{1}{2}x\right) = 2 - \left(\frac{1}{2}x\right)^3 = 2 - \frac{1}{8}x^3$ is a horizontal stretch $(0 < c < 1)$ of the graph of f. (See Figure 1.54.)

✓ *Checkpoint*  Audio-video solution in English & Spanish at LarsonPrecalculus.com

Compare the graph of each function with the graph of $f(x) = x^2 + 3$.

a. $g(x) = f(2x)$ **b.** $h(x) = f\left(\frac{1}{2}x\right)$

Summarize   (Section 1.7)

1. Explain how to shift the graph of a function vertically and horizontally *(page 67)*. For an example of shifting the graph of a function, see Example 1.

2. Explain how to reflect the graph of a function in the x-axis and in the y-axis *(page 69)*. For examples of reflecting graphs of functions, see Examples 2 and 3.

3. Describe nonrigid transformations of the graph of a function *(page 71)*. For examples of nonrigid transformations, see Examples 4 and 5.

1.7 Exercises

Vocabulary

In Exercises 1–3, fill in the blanks.

1. Horizontal shifts, vertical shifts, and reflections are _____ transformations.

2. A reflection in the x-axis of the graph of $y = f(x)$ is represented by $h(x) =$ _____, while a reflection in the y-axis of the graph of $y = f(x)$ is represented by $h(x) =$ _____.

3. A nonrigid transformation of the graph of $y = f(x)$ represented by $g(x) = cf(x)$ is a _____ _____ when $c > 1$ and a _____ _____ when $0 < c < 1$.

4. Match each function h with the transformation it represents, where $c > 0$.
    (a) $h(x) = f(x) + c$ (i) A horizontal shift of f, c units to the right
    (b) $h(x) = f(x) - c$ (ii) A vertical shift of f, c units down
    (c) $h(x) = f(x + c)$ (iii) A horizontal shift of f, c units to the left
    (d) $h(x) = f(x - c)$ (iv) A vertical shift of f, c units up

Skills and Applications

5. **Shifting the Graph of a Function** For each function, sketch the graphs of the function when $c = -2$, -1, 1, and 2 on the same set of coordinate axes.
    (a) $f(x) = |x| + c$ (b) $f(x) = |x - c|$

6. **Shifting the Graph of a Function** For each function, sketch the graphs of the function when $c = -3$, -2, 2, and 3 on the same set of coordinate axes.
    (a) $f(x) = \sqrt{x} + c$ (b) $f(x) = \sqrt{x - c}$

7. **Shifting the Graph of a Function** For each function, sketch the graphs of the function when $c = -4$, -1, 2, and 5 on the same set of coordinate axes.
    (a) $f(x) = [\![x]\!] + c$ (b) $f(x) = [\![x + c]\!]$

8. **Shifting the Graph of a Function** For each function, sketch the graphs of the function when $c = -3$, -2, 1, and 2 on the same set of coordinate axes.
    (a) $f(x) = \begin{cases} x^2 + c, & x < 0 \\ -x^2 + c, & x \ge 0 \end{cases}$

    (b) $f(x) = \begin{cases} (x + c)^2, & x < 0 \\ -(x + c)^2, & x \ge 0 \end{cases}$

Sketching Transformations **In Exercises 9 and 10, use the graph of f to sketch each graph. To print an enlarged copy of the graph, go to *MathGraphs.com*.**

9. (a) $y = f(-x)$
    (b) $y = f(x) + 4$
    (c) $y = 2f(x)$
    (d) $y = -f(x - 4)$
    (e) $y = f(x) - 3$
    (f) $y = -f(x) - 1$
    (g) $y = f(2x)$

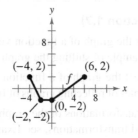

10. (a) $y = f(x - 5)$
    (b) $y = -f(x) + 3$
    (c) $y = \frac{1}{3}f(x)$
    (d) $y = -f(x + 1)$
    (e) $y = f(-x)$
    (f) $y = f(x) - 10$
    (g) $y = f\left(\frac{1}{3}x\right)$

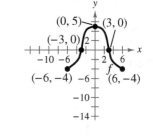

11. **Writing Equations from Graphs** Use the graph of $f(x) = x^2$ to write an equation for the function represented by each graph.
    (a)   (b)

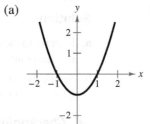

12. **Writing Equations from Graphs** Use the graph of $f(x) = x^3$ to write an equation for the function represented by each graph.
    (a)   (b)

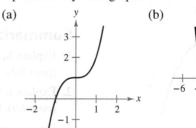

13. Writing Equations from Graphs Use the graph of $f(x) = |x|$ to write an equation for the function represented by each graph.

(a)   (b)

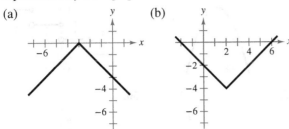

14. Writing Equations from Graphs Use the graph of $f(x) = \sqrt{x}$ to write an equation for the function represented by each graph.

(a)   (b)

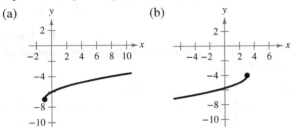

 Writing Equations from Graphs In Exercises 15–20, identify the parent function and the transformation represented by the graph. Write an equation for the function represented by the graph.

15.   16.

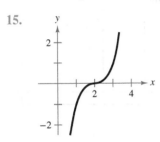

17.   18.

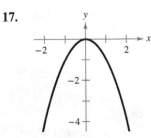

19.   20.

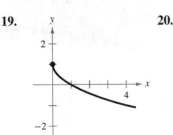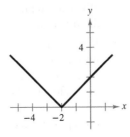

 Describing Transformations In Exercises 21–38, g is related to one of the parent functions described in Section 1.6. (a) Identify the parent function f. (b) Describe the sequence of transformations from f to g. (c) Sketch the graph of g. (d) Use function notation to write g in terms of f.

21. $g(x) = x^2 + 6$ 22. $g(x) = x^2 - 2$

23. $g(x) = -(x - 2)^3$ 24. $g(x) = -(x + 1)^3$

25. $g(x) = -3 - (x + 1)^2$

26. $g(x) = 4 - (x - 2)^2$

27. $g(x) = |x - 1| + 2$ 28. $g(x) = |x + 3| - 2$

29. $g(x) = 2\sqrt{x}$ 30. $g(x) = \frac{1}{2}\sqrt{x}$

31. $g(x) = 2[\![x]\!] - 1$ 32. $g(x) = -[\![x]\!] + 1$

33. $g(x) = |2x|$ 34. $g(x) = \left|\frac{1}{2}x\right|$

35. $g(x) = -2x^2 + 1$ 36. $g(x) = \frac{1}{2}x^2 - 2$

37. $g(x) = 3|x - 1| + 2$

38. $g(x) = -2|x + 1| - 3$

 Writing an Equation from a Description In Exercises 39–46, write an equation for the function whose graph is described.

39. The shape of $f(x) = x^2$, but shifted three units to the right and seven units down

40. The shape of $f(x) = x^2$, but shifted two units to the left, nine units up, and then reflected in the x-axis

41. The shape of $f(x) = x^3$, but shifted 13 units to the right

42. The shape of $f(x) = x^3$, but shifted six units to the left, six units down, and then reflected in the y-axis

43. The shape of $f(x) = |x|$, but shifted 12 units up and then reflected in the x-axis

44. The shape of $f(x) = |x|$, but shifted four units to the left and eight units down

45. The shape of $f(x) = \sqrt{x}$, but shifted six units to the left and then reflected in both the x-axis and the y-axis

46. The shape of $f(x) = \sqrt{x}$, but shifted nine units down and then reflected in both the x-axis and the y-axis

47. Writing Equations from Graphs Use the graph of $f(x) = x^2$ to write an equation for the function represented by each graph.

(a)   (b)

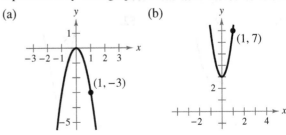

48. Writing Equations from Graphs   Use the graph of

$$f(x) = x^3$$

to write an equation for the function represented by each graph.

(a)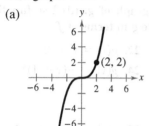

(b)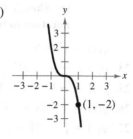

49. Writing Equations from Graphs   Use the graph of

$$f(x) = |x|$$

to write an equation for the function represented by each graph.

(a)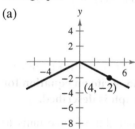

(b)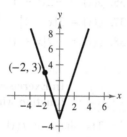

50. Writing Equations from Graphs   Use the graph of

$$f(x) = \sqrt{x}$$

to write an equation for the function represented by each graph.

(a)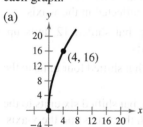

(b)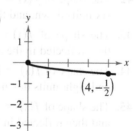

Writing Equations from Graphs   In Exercises 51–56, identify the parent function and the transformation represented by the graph. Write an equation for the function represented by the graph. Then use a graphing utility to verify your answer.

51.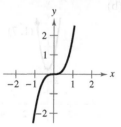

52.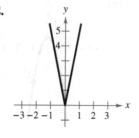

53.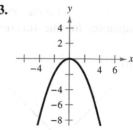

54.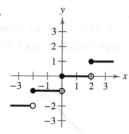

55.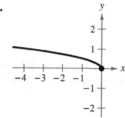

56.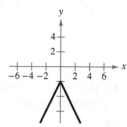

Writing Equations from Graphs   In Exercises 57–60, write an equation for the transformation of the parent function.

57.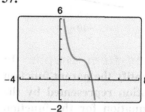

58.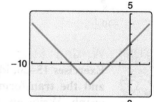

59.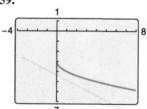

60.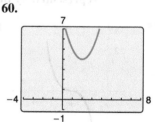

61. Automobile Aerodynamics

The horsepower H required to overcome wind drag on a particular automobile is given by

$$H(x) = 0.00004636x^3$$

where x is the speed of the car (in miles per hour).

(a) Use a graphing utility to graph the function.

(b) Rewrite the horsepower function so that x represents the speed in kilometers per hour. [Find $H(x/1.6)$.] Identify the type of transformation applied to the graph of the horsepower function.

62. Households The number N (in millions) of households in the United States from 2000 through 2014 can be approximated by

$$N(x) = -0.023(x - 33.12)^2 + 131, \quad 0 \le t \le 14$$

where t represents the year, with $t = 0$ corresponding to 2000. *(Source: U.S. Census Bureau)*

(a) Describe the transformation of the parent function $f(x) = x^2$. Then use a graphing utility to graph the function over the specified domain.

(b) Find the average rate of change of the function from 2000 to 2014. Interpret your answer in the context of the problem.

(c) Use the model to predict the number of households in the United States in 2022. Does your answer seem reasonable? Explain.

Exploration

True or False? In Exercises 63–66, determine whether the statement is true or false. Justify your answer.

63. The graph of $y = f(-x)$ is a reflection of the graph of $y = f(x)$ in the x-axis.

64. The graph of $y = -f(x)$ is a reflection of the graph of $y = f(x)$ in the y-axis.

65. The graphs of $f(x) = |x| + 6$ and $f(x) = |-x| + 6$ are identical.

66. If the graph of the parent function $f(x) = x^2$ is shifted six units to the right, three units up, and reflected in the x-axis, then the point $(-2, 19)$ will lie on the graph of the transformation.

67. Finding Points on a Graph The graph of $y = f(x)$ passes through the points $(0, 1)$, $(1, 2)$, and $(2, 3)$. Find the corresponding points on the graph of $y = f(x + 2) - 1$.

68. Think About It Two methods of graphing a function are plotting points and translating a parent function as shown in this section. Which method of graphing do you prefer to use for each function? Explain.

(a) $f(x) = 3x^2 - 4x + 1$ (b) $f(x) = 2(x - 1)^2 - 6$

69. Error Analysis Describe the error.

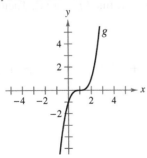

The graph of g is a right shift of one unit of the graph of $f(x) = x^3$. So, an equation for g is $g(x) = (x + 1)^3$.

70. **HOW DO YOU SEE IT?** Use the graph of $y = f(x)$ to find the open intervals on which the graph of each transformation is increasing and decreasing. If not possible, state the reason.

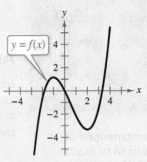

(a) $y = f(-x)$ (b) $y = -f(x)$ (c) $y = \frac{1}{2}f(x)$

(d) $y = -f(x - 1)$ (e) $y = f(x - 2) + 1$

71. Describing Profits Management originally predicted that the profits from the sales of a new product could be approximated by the graph of the function f shown. The actual profits are represented by the graph of the function g along with a verbal description. Use the concepts of transformations of graphs to write g in terms of f.

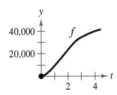

(a) The profits were only three-fourths as large as expected.

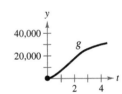

(b) The profits were consistently $10,000 greater than predicted.

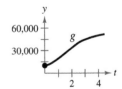

(c) There was a two-year delay in the introduction of the product. After sales began, profits grew as expected.

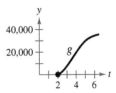

72. Reversing the Order of Transformations Reverse the order of transformations in Example 2(a). Do you obtain the same graph? Do the same for Example 2(b). Do you obtain the same graph? Explain.

1.8 Combinations of Functions: Composite Functions

- Add, subtract, multiply, and divide functions.
- Find the composition of one function with another function.
- Use combinations and compositions of functions to model and solve real-life problems.

Arithmetic Combinations of Functions

Arithmetic combinations of functions are used to model and solve real-life problems. For example, in Exercise 60 on page 82, you will use arithmetic combinations of functions to analyze numbers of pets in the United States.

Just as two real numbers can be combined by the operations of addition, subtraction, multiplication, and division to form other real numbers, two *functions* can be combined to create new functions. For example, the functions $f(x) = 2x - 3$ and $g(x) = x^2 - 1$ can be combined to form the sum, difference, product, and quotient of f and g.

$$f(x) + g(x) = (2x - 3) + (x^2 - 1) = x^2 + 2x - 4 \qquad \text{Sum}$$
$$f(x) - g(x) = (2x - 3) - (x^2 - 1) = -x^2 + 2x - 2 \qquad \text{Difference}$$
$$f(x)g(x) = (2x - 3)(x^2 - 1) = 2x^3 - 3x^2 - 2x + 3 \qquad \text{Product}$$
$$\frac{f(x)}{g(x)} = \frac{2x - 3}{x^2 - 1}, \quad x \neq \pm 1 \qquad \text{Quotient}$$

The domain of an **arithmetic combination** of functions f and g consists of all real numbers that are common to the domains of f and g. In the case of the quotient $f(x)/g(x)$, there is the further restriction that $g(x) \neq 0$.

Sum, Difference, Product, and Quotient of Functions

Let f and g be two functions with overlapping domains. Then, for all x common to both domains, the *sum, difference, product,* and *quotient* of f and g are defined as follows.

1. Sum: $(f + g)(x) = f(x) + g(x)$

2. Difference: $(f - g)(x) = f(x) - g(x)$

3. Product: $(fg)(x) = f(x) \cdot g(x)$

4. Quotient: $\left(\dfrac{f}{g}\right)(x) = \dfrac{f(x)}{g(x)}, \quad g(x) \neq 0$

EXAMPLE 1 **Finding the Sum of Two Functions**

Given $f(x) = 2x + 1$ and $g(x) = x^2 + 2x - 1$, find $(f + g)(x)$. Then evaluate the sum when $x = 3$.

Solution The sum of f and g is

$$(f + g)(x) = f(x) + g(x) = (2x + 1) + (x^2 + 2x - 1) = x^2 + 4x.$$

When $x = 3$, the value of this sum is

$$(f + g)(3) = 3^2 + 4(3) = 21.$$

✓ *Checkpoint* ◀))) *Audio-video solution in English & Spanish at LarsonPrecalculus.com*

Given $f(x) = x^2$ and $g(x) = 1 - x$, find $(f + g)(x)$. Then evaluate the sum when $x = 2$. ■

EXAMPLE 2 **Finding the Difference of Two Functions**

Given $f(x) = 2x + 1$ and $g(x) = x^2 + 2x - 1$, find $(f - g)(x)$. Then evaluate the difference when $x = 2$.

Solution The difference of f and g is

$$(f - g)(x) = f(x) - g(x) = (2x + 1) - (x^2 + 2x - 1) = -x^2 + 2.$$

When $x = 2$, the value of this difference is

$$(f - g)(2) = -(2)^2 + 2 = -2.$$

✓ *Checkpoint* ◀))) *Audio-video solution in English & Spanish at LarsonPrecalculus.com*

Given $f(x) = x^2$ and $g(x) = 1 - x$, find $(f - g)(x)$. Then evaluate the difference when $x = 3$.

EXAMPLE 3 **Finding the Product of Two Functions**

Given $f(x) = x^2$ and $g(x) = x - 3$, find $(fg)(x)$. Then evaluate the product when $x = 4$.

Solution The product of f and g is

$$(fg)(x) = f(x)g(x) = (x^2)(x - 3) = x^3 - 3x^2.$$

When $x = 4$, the value of this product is

$$(fg)(4) = 4^3 - 3(4)^2 = 16.$$

✓ *Checkpoint* ◀))) *Audio-video solution in English & Spanish at LarsonPrecalculus.com*

Given $f(x) = x^2$ and $g(x) = 1 - x$, find $(fg)(x)$. Then evaluate the product when $x = 3$.

In Examples 1–3, both f and g have domains that consist of all real numbers. So, the domains of $f + g$, $f - g$, and fg are also the set of all real numbers. Remember to consider any restrictions on the domains of f and g when forming the sum, difference, product, or quotient of f and g.

EXAMPLE 4 **Finding the Quotients of Two Functions**

Find $(f/g)(x)$ and $(g/f)(x)$ for the functions $f(x) = \sqrt{x}$ and $g(x) = \sqrt{4 - x^2}$. Then find the domains of f/g and g/f.

Solution The quotient of f and g is

$$\left(\frac{f}{g}\right)(x) = \frac{f(x)}{g(x)} = \frac{\sqrt{x}}{\sqrt{4 - x^2}}$$

and the quotient of g and f is

$$\left(\frac{g}{f}\right)(x) = \frac{g(x)}{f(x)} = \frac{\sqrt{4 - x^2}}{\sqrt{x}}.$$

• • **REMARK** Note that the domain of f/g includes $x = 0$, but not $x = 2$, because $x = 2$ yields a zero in the denominator, whereas the domain of g/f includes $x = 2$, but not $x = 0$, because $x = 0$ yields a zero in the denominator.

The domain of f is $[0, \infty)$ and the domain of g is $[-2, 2]$. The intersection of these domains is $[0, 2]$. So, the domains of f/g and g/f are as follows.

Domain of f/g: $[0, 2)$ Domain of g/f: $(0, 2]$

✓ *Checkpoint* ◀))) *Audio-video solution in English & Spanish at LarsonPrecalculus.com*

Find $(f/g)(x)$ and $(g/f)(x)$ for the functions $f(x) = \sqrt{x - 3}$ and $g(x) = \sqrt{16 - x^2}$. Then find the domains of f/g and g/f.

Composition of Functions

Another way of combining two functions is to form the **composition** of one with the other. For example, if $f(x) = x^2$ and $g(x) = x + 1$, then the composition of f with g is

$$f(g(x)) = f(x + 1)$$
$$= (x + 1)^2.$$

This composition is denoted as $f \circ g$ and reads as "f composed with g."

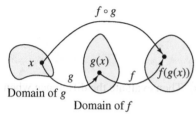

$f \circ g$

Domain of g

Domain of f

Figure 1.55

Definition of Composition of Two Functions

The **composition** of the function f with the function g is

$$(f \circ g)(x) = f(g(x)).$$

The domain of $f \circ g$ is the set of all x in the domain of g such that $g(x)$ is in the domain of f. (See Figure 1.55.)

EXAMPLE 5 **Compositions of Functions**

See LarsonPrecalculus.com for an interactive version of this type of example.

Given $f(x) = x + 2$ and $g(x) = 4 - x^2$, find the following.

a. $(f \circ g)(x)$ **b.** $(g \circ f)(x)$ **c.** $(g \circ f)(-2)$

Solution

a. The composition of f with g is as shown.

$$(f \circ g)(x) = f(g(x)) \qquad \text{Definition of } f \circ g$$
$$= f(4 - x^2) \qquad \text{Definition of } g(x)$$
$$= (4 - x^2) + 2 \qquad \text{Definition of } f(x)$$
$$= -x^2 + 6 \qquad \text{Simplify.}$$

b. The composition of g with f is as shown.

$$(g \circ f)(x) = g(f(x)) \qquad \text{Definition of } g \circ f$$
$$= g(x + 2) \qquad \text{Definition of } f(x)$$
$$= 4 - (x + 2)^2 \qquad \text{Definition of } g(x)$$
$$= 4 - (x^2 + 4x + 4) \qquad \text{Expand.}$$
$$= -x^2 - 4x \qquad \text{Simplify.}$$

Note that, in this case, $(f \circ g)(x) \neq (g \circ f)(x)$.

c. Evaluate the result of part (b) when $x = -2$.

$$(g \circ f)(-2) = -(-2)^2 - 4(-2) \qquad \text{Substitute.}$$
$$= -4 + 8 \qquad \text{Simplify.}$$
$$= 4 \qquad \text{Simplify.}$$

✓ Checkpoint))) *Audio-video solution in English & Spanish at LarsonPrecalculus.com*

Given $f(x) = 2x + 5$ and $g(x) = 4x^2 + 1$, find the following.

a. $(f \circ g)(x)$ **b.** $(g \circ f)(x)$ **c.** $(f \circ g)\left(-\frac{1}{2}\right)$

REMARK The tables of values below help illustrate the composition $(f \circ g)(x)$ in Example 5(a).

x	0	1	2	3
$g(x)$	4	3	0	-5

$g(x)$	4	3	0	-5
$f(g(x))$	6	5	2	-3

x	0	1	2	3
$f(g(x))$	6	5	2	-3

Note that the first two tables are combined (or "composed") to produce the values in the third table.

EXAMPLE 6 **Finding the Domain of a Composite Function**

Find the domain of $f \circ g$ for the functions

$$f(x) = x^2 - 9 \quad \text{and} \quad g(x) = \sqrt{9 - x^2}.$$

Algebraic Solution

Find the composition of the functions.

$$
\begin{aligned}
(f \circ g)(x) &= f(g(x)) \\
&= f\left(\sqrt{9 - x^2}\right) \\
&= \left(\sqrt{9 - x^2}\right)^2 - 9 \\
&= 9 - x^2 - 9 \\
&= -x^2
\end{aligned}
$$

The domain of $f \circ g$ is restricted to the x-values in the domain of g for which $g(x)$ is in the domain of f. The domain of $f(x) = x^2 - 9$ is the set of all real numbers, which includes all real values of g. So, the domain of $f \circ g$ is the entire domain of $g(x) = \sqrt{9 - x^2}$, which is $[-3, 3]$.

Graphical Solution

Use a graphing utility to graph $f \circ g$.

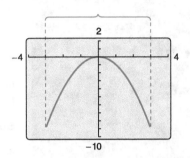

From the graph, you can determine that the domain of $f \circ g$ is $[-3, 3]$.

 ✓ *Checkpoint* 🔊))) *Audio-video solution in English & Spanish at LarsonPrecalculus.com*

Find the domain of $f \circ g$ for the functions $f(x) = \sqrt{x}$ and $g(x) = x^2 + 4$.

In Examples 5 and 6, you formed the composition of two given functions. In calculus, it is also important to be able to identify two functions that make up a given composite function. For example, the function $h(x) = (3x - 5)^3$ is the composition of $f(x) = x^3$ and $g(x) = 3x - 5$. That is,

$$h(x) = (3x - 5)^3 = [g(x)]^3 = f(g(x)).$$

Basically, to "decompose" a composite function, look for an "inner" function and an "outer" function. In the function h above, $g(x) = 3x - 5$ is the inner function and $f(x) = x^3$ is the outer function.

EXAMPLE 7 **Decomposing a Composite Function**

Write the function $h(x) = \dfrac{1}{(x - 2)^2}$ as a composition of two functions.

Solution   Consider $g(x) = x - 2$ as the inner function and $f(x) = \dfrac{1}{x^2} = x^{-2}$ as the outer function. Then write

$$
\begin{aligned}
h(x) &= \frac{1}{(x - 2)^2} \\
&= (x - 2)^{-2} \\
&= f(x - 2) \\
&= f(g(x)).
\end{aligned}
$$

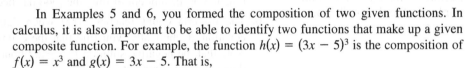 ✓ *Checkpoint* 🔊))) *Audio-video solution in English & Spanish at LarsonPrecalculus.com*

Write the function $h(x) = \dfrac{\sqrt[3]{8 - x}}{5}$ as a composition of two functions.

Application

| EXAMPLE 8 | Bacteria Count |

The number N of bacteria in a refrigerated food is given by

$$N(T) = 20T^2 - 80T + 500, \quad 2 \le T \le 14$$

where T is the temperature of the food in degrees Celsius. When the food is removed from refrigeration, the temperature of the food is given by

$$T(t) = 4t + 2, \quad 0 \le t \le 3$$

where t is the time in hours.

a. Find and interpret $(N \circ T)(t)$.

b. Find the time when the bacteria count reaches 2000.

Solution

a. $(N \circ T)(t) = N(T(t))$

$$= 20(4t + 2)^2 - 80(4t + 2) + 500$$

$$= 20(16t^2 + 16t + 4) - 320t - 160 + 500$$

$$= 320t^2 + 320t + 80 - 320t - 160 + 500$$

$$= 320t^2 + 420$$

The composite function $N \circ T$ represents the number of bacteria in the food as a function of the amount of time the food has been out of refrigeration.

b. The bacteria count reaches 2000 when $320t^2 + 420 = 2000$. By solving this equation algebraically, you find that the count reaches 2000 when $t \approx 2.2$ hours. Note that the negative solution $t \approx -2.2$ hours is rejected because it is not in the domain of the composite function.

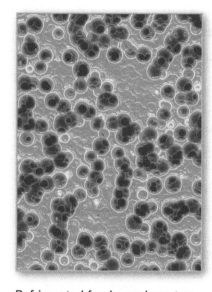

Refrigerated foods can have two types of bacteria: pathogenic bacteria, which can cause foodborne illness, and spoilage bacteria, which give foods an unpleasant look, smell, taste, or texture.

✓ **Checkpoint**   *Audio-video solution in English & Spanish at LarsonPrecalculus.com*

The number N of bacteria in a refrigerated food is given by

$$N(T) = 8T^2 - 14T + 200, \quad 2 \le T \le 12$$

where T is the temperature of the food in degrees Celsius. When the food is removed from refrigeration, the temperature of the food is given by

$$T(t) = 2t + 2, \quad 0 \le t \le 5$$

where t is the time in hours.

a. Find $(N \circ T)(t)$.

b. Find the time when the bacteria count reaches 1000.

Summarize   (Section 1.8)

1. Explain how to add, subtract, multiply, and divide functions *(page 76)*. For examples of finding arithmetic combinations of functions, see Examples 1–4.

2. Explain how to find the composition of one function with another function *(page 78)*. For examples that use compositions of functions, see Examples 5–7.

3. Describe a real-life example that uses a composition of functions *(page 80, Example 8)*.

1.8 Exercises

See **CalcChat.com** for tutorial help and worked-out solutions to odd-numbered exercises.

Vocabulary: Fill in the blanks.

1. Two functions f and g can be combined by the arithmetic operations of _____, _____, _____, and _____ to create new functions.

2. The _____ of the function f with the function g is $(f \circ g)(x) = f(g(x))$.

Skills and Applications

Graphing the Sum of Two Functions In Exercises 3 and 4, use the graphs of f and g to graph $h(x) = (f + g)(x)$. To print an enlarged copy of the graph, go to *MathGraphs.com*.

3.
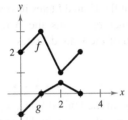

4.
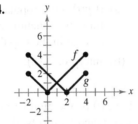

 Finding Arithmetic Combinations of Functions In Exercises 5–12, find (a) $(f + g)(x)$, (b) $(f - g)(x)$, (c) $(fg)(x)$, and (d) $(f/g)(x)$. What is the domain of f/g?

5. $f(x) = x + 2$, $g(x) = x - 2$

6. $f(x) = 2x - 5$, $g(x) = 2 - x$

7. $f(x) = x^2$, $g(x) = 4x - 5$

8. $f(x) = 3x + 1$, $g(x) = x^2 - 16$

9. $f(x) = x^2 + 6$, $g(x) = \sqrt{1 - x}$

10. $f(x) = \sqrt{x^2 - 4}$, $g(x) = \dfrac{x^2}{x^2 + 1}$

11. $f(x) = \dfrac{x}{x + 1}$, $g(x) = x^3$

12. $f(x) = \dfrac{2}{x}$, $g(x) = \dfrac{1}{x^2 - 1}$

 Evaluating an Arithmetic Combination of Functions In Exercises 13–24, evaluate the function for $f(x) = x + 3$ and $g(x) = x^2 - 2$.

13. $(f + g)(2)$

14. $(f + g)(-1)$

15. $(f - g)(0)$

16. $(f - g)(1)$

17. $(f - g)(3t)$

18. $(f + g)(t - 2)$

19. $(fg)(6)$

20. $(fg)(-6)$

21. $(f/g)(5)$

22. $(f/g)(0)$

23. $(f/g)(-1) - g(3)$

24. $(fg)(5) + f(4)$

 Graphical Reasoning In Exercises 25–28, use a graphing utility to graph f, g, and $f + g$ in the same viewing window. Which function contributes most to the magnitude of the sum when $0 \le x \le 2$? Which function contributes most to the magnitude of the sum when $x > 6$?

25. $f(x) = 3x$, $g(x) = -\dfrac{x^3}{10}$

26. $f(x) = \dfrac{x}{2}$, $g(x) = \sqrt{x}$

27. $f(x) = 3x + 2$, $g(x) = -\sqrt{x + 5}$

28. $f(x) = x^2 - \dfrac{1}{2}$, $g(x) = -3x^2 - 1$

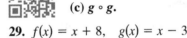 **Finding Compositions of Functions** In Exercises 29–34, find (a) $f \circ g$, (b) $g \circ f$, and (c) $g \circ g$.

29. $f(x) = x + 8$, $g(x) = x - 3$

30. $f(x) = -4x$, $g(x) = x + 7$

31. $f(x) = x^2$, $g(x) = x - 1$

32. $f(x) = 3x$, $g(x) = x^4$

33. $f(x) = \sqrt[3]{x - 1}$, $g(x) = x^3 + 1$

34. $f(x) = x^3$, $g(x) = \dfrac{1}{x}$

 Finding Domains of Functions and Composite Functions In Exercises 35–42, find (a) $f \circ g$ and (b) $g \circ f$. Find the domain of each function and of each composite function.

35. $f(x) = \sqrt{x + 4}$, $g(x) = x^2$

36. $f(x) = \sqrt[3]{x - 5}$, $g(x) = x^3 + 1$

37. $f(x) = x^3$, $g(x) = x^{2/3}$

38. $f(x) = x^5$, $g(x) = \sqrt[4]{x}$

39. $f(x) = |x|$, $g(x) = x + 6$

40. $f(x) = |x - 4|$, $g(x) = 3 - x$

41. $f(x) = \dfrac{1}{x}$, $g(x) = x + 3$

42. $f(x) = \dfrac{3}{x^2 - 1}$, $g(x) = x + 1$

Graphing Combinations of Functions In Exercises 43 and 44, on the same set of coordinate axes, (a) graph the functions f, g, and $f + g$ and (b) graph the functions f, g, and $f \circ g$.

43. $f(x) = \frac{1}{2}x$, $g(x) = x - 4$

44. $f(x) = x + 3$, $g(x) = x^2$

 Evaluating Combinations of Functions In Exercises 45–48, use the graphs of f and g to evaluate the functions.

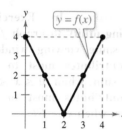 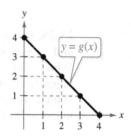

45. (a) $(f + g)(3)$ (b) $(f/g)(2)$

46. (a) $(f - g)(1)$ (b) $(fg)(4)$

47. (a) $(f \circ g)(2)$ (b) $(g \circ f)(2)$

48. (a) $(f \circ g)(1)$ (b) $(g \circ f)(3)$

 Decomposing a Composite Function In Exercises 49–56, find two functions f and g such that $(f \circ g)(x) = h(x)$. (There are many correct answers.)

49. $h(x) = (2x + 1)^2$ **50.** $h(x) = (1 - x)^3$

51. $h(x) = \sqrt[3]{x^2 - 4}$ **52.** $h(x) = \sqrt{9 - x}$

53. $h(x) = \dfrac{1}{x + 2}$ **54.** $h(x) = \dfrac{4}{(5x + 2)^2}$

55. $h(x) = \dfrac{-x^2 + 3}{4 - x^2}$

56. $h(x) = \dfrac{27x^3 + 6x}{10 - 27x^3}$

57. Stopping Distance The research and development department of an automobile manufacturer determines that when a driver is required to stop quickly to avoid an accident, the distance (in feet) the car travels during the driver's reaction time is given by $R(x) = \frac{3}{4}x$, where x is the speed of the car in miles per hour. The distance (in feet) the car travels while the driver is braking is given by $B(x) = \frac{1}{15}x^2$.

(a) Find the function that represents the total stopping distance T.

(b) Graph the functions R, B, and T on the same set of coordinate axes for $0 \le x \le 60$.

(c) Which function contributes most to the magnitude of the sum at higher speeds? Explain.

58. Business The annual cost C (in thousands of dollars) and revenue R (in thousands of dollars) for a company each year from 2010 through 2016 can be approximated by the models

$$C = 254 - 9t + 1.1t^2 \quad \text{and} \quad R = 341 + 3.2t$$

where t is the year, with $t = 10$ corresponding to 2010.

(a) Write a function P that represents the annual profit of the company.

(b) Use a graphing utility to graph C, R, and P in the same viewing window.

59. Vital Statistics Let $b(t)$ be the number of births in the United States in year t, and let $d(t)$ represent the number of deaths in the United States in year t, where $t = 10$ corresponds to 2010.

(a) If $p(t)$ is the population of the United States in year t, find the function $c(t)$ that represents the percent change in the population of the United States.

(b) Interpret $c(16)$.

60. Pets

Let $d(t)$ be the number of dogs in the United States in year t, and let $c(t)$ be the number of cats in the United States in year t, where $t = 10$ corresponds to 2010.

(a) Find the function $p(t)$ that represents the total number of dogs and cats in the United States.

(b) Interpret $p(16)$.

(c) Let $n(t)$ represent the population of the United States in year t, where $t = 10$ corresponds to 2010. Find and interpret

$$h(t) = p(t)/n(t).$$

61. Geometry A square concrete foundation is a base for a cylindrical tank (see figure).

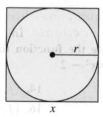

(a) Write the radius r of the tank as a function of the length x of the sides of the square.

(b) Write the area A of the circular base of the tank as a function of the radius r.

(c) Find and interpret $(A \circ r)(x)$.

62. Biology The number N of bacteria in a refrigerated food is given by

$$N(T) = 10T^2 - 20T + 600, \quad 2 \le T \le 20$$

where T is the temperature of the food in degrees Celsius. When the food is removed from refrigeration, the temperature of the food is given by

$$T(t) = 3t + 2, \quad 0 \le t \le 6$$

where t is the time in hours.

(a) Find and interpret $(N \circ T)(t)$.

(b) Find the bacteria count after 0.5 hour.

(c) Find the time when the bacteria count reaches 1500.

63. Salary You are a sales representative for a clothing manufacturer. You are paid an annual salary, plus a bonus of 3% of your sales over $500,000. Consider the two functions $f(x) = x - 500,000$ and $g(x) = 0.03x$. When x is greater than $500,000, which of the following represents your bonus? Explain.

(a) $f(g(x))$

(b) $g(f(x))$

64. Consumer Awareness The suggested retail price of a new hybrid car is p dollars. The dealership advertises a factory rebate of $2000 and a 10% discount.

(a) Write a function R in terms of p giving the cost of the hybrid car after receiving the rebate from the factory.

(b) Write a function S in terms of p giving the cost of the hybrid car after receiving the dealership discount.

(c) Find and interpret $(R \circ S)(p)$ and $(S \circ R)(p)$.

(d) Find $(R \circ S)(25,795)$ and $(S \circ R)(25,795)$. Which yields the lower cost for the hybrid car? Explain.

Exploration

True or False? In Exercises 65 and 66, determine whether the statement is true or false. Justify your answer.

65. If $f(x) = x + 1$ and $g(x) = 6x$, then

$$(f \circ g)(x) = (g \circ f)(x).$$

66. When you are given two functions f and g and a constant c, you can find $(f \circ g)(c)$ if and only if $g(c)$ is in the domain of f.

Siblings In Exercises 67 and 68, three siblings are three different ages. The oldest is twice the age of the middle sibling, and the middle sibling is six years older than one-half the age of the youngest.

67. (a) Write a composite function that gives the oldest sibling's age in terms of the youngest. Explain how you arrived at your answer.

(b) If the oldest sibling is 16 years old, find the ages of the other two siblings.

68. (a) Write a composite function that gives the youngest sibling's age in terms of the oldest. Explain how you arrived at your answer.

(b) If the youngest sibling is 2 years old, find the ages of the other two siblings.

69. Proof Prove that the product of two odd functions is an even function, and that the product of two even functions is an even function.

70. Conjecture Use examples to hypothesize whether the product of an odd function and an even function is even or odd. Then prove your hypothesis.

71. Writing Functions Write two unique functions f and g such that $(f \circ g)(x) = (g \circ f)(x)$ and f and g are (a) linear functions and (b) polynomial functions with degrees greater than one.

72. HOW DO YOU SEE IT? The graphs labeled $L_1, L_2, L_3,$ and L_4 represent four different pricing discounts, where p is the original price (in dollars) and S is the sale price (in dollars). Match each function with its graph. Describe the situations in parts (c) and (d).

(a) $f(p)$: A 50% discount is applied.

(b) $g(p)$: A $5 discount is applied.

(c) $(g \circ f)(p)$

(d) $(f \circ g)(p)$

73. Proof

(a) Given a function f, prove that g is even and h is odd, where $g(x) = \frac{1}{2}[f(x) + f(-x)]$ and

$$h(x) = \frac{1}{2}[f(x) - f(-x)].$$

(b) Use the result of part (a) to prove that any function can be written as a sum of even and odd functions. [*Hint:* Add the two equations in part (a).]

(c) Use the result of part (b) to write each function as a sum of even and odd functions.

$$f(x) = x^2 - 2x + 1, \quad k(x) = \frac{1}{x + 1}$$

1.9 Inverse Functions

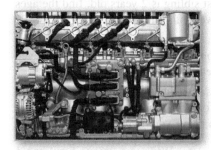

Inverse functions can help you model and solve real-life problems. For example, in Exercise 90 on page 92, you will write an inverse function and use it to determine the percent load interval for a diesel engine.

- Find inverse functions informally and verify that two functions are inverse functions of each other.
- Use graphs to verify that two functions are inverse functions of each other.
- Use the Horizontal Line Test to determine whether functions are one-to-one.
- Find inverse functions algebraically.

Inverse Functions

Recall from Section 1.4 that a set of ordered pairs can represent a function. For example, the function $f(x) = x + 4$ from the set $A = \{1, 2, 3, 4\}$ to the set $B = \{5, 6, 7, 8\}$ can be written as

$$f(x) = x + 4: \{(1, 5), (2, 6), (3, 7), (4, 8)\}.$$

In this case, by interchanging the first and second coordinates of each of the ordered pairs, you form the **inverse function** of f, which is denoted by f^{-1}. It is a function from the set B to the set A, and can be written as

$$f^{-1}(x) = x - 4: \{(5, 1), (6, 2), (7, 3), (8, 4)\}.$$

Note that the domain of f is equal to the range of f^{-1}, and vice versa, as shown in the figure below. Also note that the functions f and f^{-1} have the effect of "undoing" each other. In other words, when you form the composition of f with f^{-1} or the composition of f^{-1} with f, you obtain the identity function.

$$f(f^{-1}(x)) = f(x - 4) = (x - 4) + 4 = x$$
$$f^{-1}(f(x)) = f^{-1}(x + 4) = (x + 4) - 4 = x$$

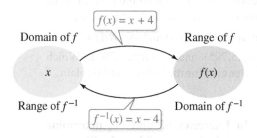

Domain of f $f(x) = x + 4$ Range of f

x $f(x)$

Range of f^{-1} $f^{-1}(x) = x - 4$ Domain of f^{-1}

EXAMPLE 1   **Finding an Inverse Function Informally**

Find the inverse function of $f(x) = 4x$. Then verify that both $f(f^{-1}(x))$ and $f^{-1}(f(x))$ are equal to the identity function.

Solution The function f *multiplies* each input by 4. To "undo" this function, you need to *divide* each input by 4. So, the inverse function of $f(x) = 4x$ is

$$f^{-1}(x) = \frac{x}{4}.$$

Verify that $f(f^{-1}(x)) = x$ and $f^{-1}(f(x)) = x$.

$$f(f^{-1}(x)) = f\left(\frac{x}{4}\right) = 4\left(\frac{x}{4}\right) = x \qquad f^{-1}(f(x)) = f^{-1}(4x) = \frac{4x}{4} = x$$

✓ **Checkpoint** ◀))) *Audio-video solution in English & Spanish at LarsonPrecalculus.com*

Find the inverse function of $f(x) = \frac{1}{5}x$. Then verify that both $f(f^{-1}(x))$ and $f^{-1}(f(x))$ are equal to the identity function. ∎

Definition of Inverse Function

Let f and g be two functions such that

$$f(g(x)) = x \quad \text{for every } x \text{ in the domain of } g$$

and

$$g(f(x)) = x \quad \text{for every } x \text{ in the domain of } f.$$

Under these conditions, the function g is the **inverse function** of the function f. The function g is denoted by f^{-1} (read "f-inverse"). So,

$$f(f^{-1}(x)) = x \quad \text{and} \quad f^{-1}(f(x)) = x.$$

The domain of f must be equal to the range of f^{-1}, and the range of f must be equal to the domain of f^{-1}.

Do not be confused by the use of -1 to denote the inverse function f^{-1}. In this text, whenever f^{-1} is written, it *always* refers to the inverse function of the function f and *not* to the reciprocal of $f(x)$.

If the function g is the inverse function of the function f, then it must also be true that the function f is the inverse function of the function g. So, it is correct to say that the functions f and g are *inverse functions of each other.*

EXAMPLE 2 **Verifying Inverse Functions**

Which of the functions is the inverse function of $f(x) = \dfrac{5}{x-2}$?

$$g(x) = \frac{x-2}{5} \qquad h(x) = \frac{5}{x} + 2$$

Solution By forming the composition of f with g, you have

$$f(g(x)) = f\left(\frac{x-2}{5}\right) = \frac{5}{\left(\dfrac{x-2}{5}\right) - 2} = \frac{25}{x-12} \neq x.$$

This composition is not equal to the identity function x, so g *is not* the inverse function of f. By forming the composition of f with h, you have

$$f(h(x)) = f\left(\frac{5}{x} + 2\right) = \frac{5}{\left(\dfrac{5}{x} + 2\right) - 2} = \frac{5}{\left(\dfrac{5}{x}\right)} = x.$$

So, it appears that h *is* the inverse function of f. Confirm this by showing that the composition of h with f is also equal to the identity function.

$$h(f(x)) = h\left(\frac{5}{x-2}\right) = \frac{5}{\left(\dfrac{5}{x-2}\right)} + 2 = x - 2 + 2 = x$$

Check to see that the domain of f is the same as the range of h and vice versa.

✓ *Checkpoint*))) *Audio-video solution in English & Spanish at LarsonPrecalculus.com*

Which of the functions is the inverse function of $f(x) = \dfrac{x-4}{7}$?

$$g(x) = 7x + 4 \qquad h(x) = \frac{7}{x-4}$$

The Graph of an Inverse Function

The graphs of a function f and its inverse function f^{-1} are related to each other in this way: If the point (a, b) lies on the graph of f, then the point (b, a) must lie on the graph of f^{-1}, and vice versa. This means that the graph of f^{-1} is a *reflection* of the graph of f in the line $y = x$, as shown in Figure 1.56.

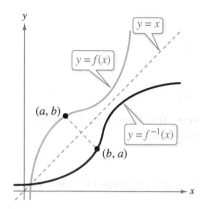

Figure 1.56

EXAMPLE 3   **Verifying Inverse Functions Graphically**

Verify graphically that the functions $f(x) = 2x - 3$ and $g(x) = \frac{1}{2}(x + 3)$ are inverse functions of each other.

Solution   Sketch the graphs of f and g on the same rectangular coordinate system, as shown in Figure 1.57. It appears that the graphs are reflections of each other in the line $y = x$. Further verify this reflective property by testing a few points on each graph. Note that for each point (a, b) on the graph of f, the point (b, a) is on the graph of g.

Graph of $f(x) = 2x - 3$	Graph of $g(x) = \frac{1}{2}(x + 3)$
$(-1, -5)$	$(-5, -1)$
$(0, -3)$	$(-3, 0)$
$(1, -1)$	$(-1, 1)$
$(2, 1)$	$(1, 2)$
$(3, 3)$	$(3, 3)$

The graphs of f and g are reflections of each other in the line $y = x$. So, f and g are inverse functions of each other.

✓ **Checkpoint** 🔊))) *Audio-video solution in English & Spanish at LarsonPrecalculus.com*

Verify graphically that the functions $f(x) = 4x - 1$ and $g(x) = \frac{1}{4}(x + 1)$ are inverse functions of each other.

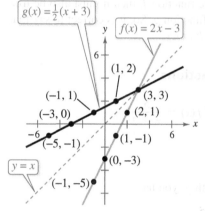

Figure 1.57

EXAMPLE 4   **Verifying Inverse Functions Graphically**

Verify graphically that the functions $f(x) = x^2$ $(x \geq 0)$ and $g(x) = \sqrt{x}$ are inverse functions of each other.

Solution   Sketch the graphs of f and g on the same rectangular coordinate system, as shown in Figure 1.58. It appears that the graphs are reflections of each other in the line $y = x$. Test a few points on each graph.

Graph of $f(x) = x^2$, $x \geq 0$	Graph of $g(x) = \sqrt{x}$
$(0, 0)$	$(0, 0)$
$(1, 1)$	$(1, 1)$
$(2, 4)$	$(4, 2)$
$(3, 9)$	$(9, 3)$

The graphs of f and g are reflections of each other in the line $y = x$. So, f and g are inverse functions of each other.

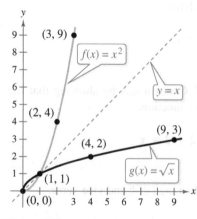

Figure 1.58

✓ **Checkpoint** 🔊))) *Audio-video solution in English & Spanish at LarsonPrecalculus.com*

Verify graphically that the functions $f(x) = x^2 + 1$ $(x \geq 0)$ and $g(x) = \sqrt{x - 1}$ are inverse functions of each other.

One-to-One Functions

The reflective property of the graphs of inverse functions gives you a graphical test for determining whether a function has an inverse function. This test is the **Horizontal Line Test** for inverse functions.

> **Horizontal Line Test for Inverse Functions**
>
> A function f has an inverse function if and only if no *horizontal* line intersects the graph of f at more than one point.

If no horizontal line intersects the graph of f at more than one point, then no y-value corresponds to more than one x-value. This is the essential characteristic of **one-to-one functions.**

> **One-to-One Functions**
>
> A function f is **one-to-one** when each value of the dependent variable corresponds to exactly one value of the independent variable. A function f has an inverse function if and only if f is one-to-one.

Consider the table of values for the function $f(x) = x^2$ on the left. The output $f(x) = 4$ corresponds to two inputs, $x = -2$ and $x = 2$, so f is not one-to-one. In the table on the right, x and y are interchanged. Here $x = 4$ corresponds to both $y = -2$ and $y = 2$, so this table does not represent a function. So, $f(x) = x^2$ is not one-to-one and does not have an inverse function.

x	$f(x) = x^2$
-2	4
-1	1
0	0
1	1
2	4
3	9

x	y
4	-2
1	-1
0	0
1	1
4	2
9	3

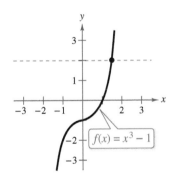

Figure 1.59

Figure 1.60

EXAMPLE 5   Applying the Horizontal Line Test

See LarsonPrecalculus.com for an interactive version of this type of example.

a. The graph of the function $f(x) = x^3 - 1$ is shown in Figure 1.59. No horizontal line intersects the graph of f at more than one point, so f *is* a one-to-one function and *does* have an inverse function.

b. The graph of the function $f(x) = x^2 - 1$ is shown in Figure 1.60. It is possible to find a horizontal line that intersects the graph of f at more than one point, so f *is not* a one-to-one function and *does not* have an inverse function.

✓ **Checkpoint** ◄))) *Audio-video solution in English & Spanish at LarsonPrecalculus.com*

Use the graph of f to determine whether the function has an inverse function.

a. $f(x) = \frac{1}{2}(3 - x)$ **b.** $f(x) = |x|$

Finding Inverse Functions Algebraically

•• **REMARK** Note what
happens when you try to
find the inverse function of a
function that is not one-to-one.

$$f(x) = x^2 + 1 \quad \text{Original function}$$

$$y = x^2 + 1 \quad \begin{array}{l}\text{Replace}\\ f(x) \text{ with } y.\end{array}$$

$$x = y^2 + 1 \quad \begin{array}{l}\text{Interchange}\\ x \text{ and } y.\end{array}$$

$$x - 1 = y^2 \quad \begin{array}{l}\text{Isolate}\\ y\text{-term.}\end{array}$$

$$y = \pm\sqrt{x - 1} \quad \begin{array}{l}\text{Solve}\\ \text{for } y.\end{array}$$

You obtain two y-values for
each x.

For relatively simple functions (such as the one in Example 1), you can find inverse functions by inspection. For more complicated functions, however, it is best to use the guidelines below. The key step in these guidelines is Step 3—interchanging the roles of x and y. This step corresponds to the fact that inverse functions have ordered pairs with the coordinates reversed.

Finding an Inverse Function

1. Use the Horizontal Line Test to decide whether f has an inverse function.

2. In the equation for $f(x)$, replace $f(x)$ with y.

3. Interchange the roles of x and y, and solve for y.

4. Replace y with $f^{-1}(x)$ in the new equation.

5. Verify that f and f^{-1} are inverse functions of each other by showing that the domain of f is equal to the range of f^{-1}, the range of f is equal to the domain of f^{-1}, and $f(f^{-1}(x)) = x$ and $f^{-1}(f(x)) = x$.

EXAMPLE 6 **Finding an Inverse Function Algebraically**

Find the inverse function of

$$f(x) = \frac{5 - x}{3x + 2}.$$

Solution The graph of f is shown in Figure 1.61. This graph passes the Horizontal Line Test. So, you know that f is one-to-one and has an inverse function.

$$f(x) = \frac{5 - x}{3x + 2} \qquad \text{Write original function.}$$

$$y = \frac{5 - x}{3x + 2} \qquad \text{Replace } f(x) \text{ with } y.$$

$$x = \frac{5 - y}{3y + 2} \qquad \text{Interchange } x \text{ and } y.$$

$$x(3y + 2) = 5 - y \qquad \text{Multiply each side by } 3y + 2.$$

$$3xy + 2x = 5 - y \qquad \text{Distributive Property}$$

$$3xy + y = 5 - 2x \qquad \text{Collect terms with } y.$$

$$y(3x + 1) = 5 - 2x \qquad \text{Factor.}$$

$$y = \frac{5 - 2x}{3x + 1} \qquad \text{Solve for } y.$$

$$f^{-1}(x) = \frac{5 - 2x}{3x + 1} \qquad \text{Replace } y \text{ with } f^{-1}(x).$$

Check that $f(f^{-1}(x)) = x$ and $f^{-1}(f(x)) = x$.

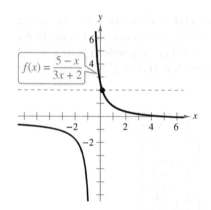

$$f(x) = \frac{5 - x}{3x + 2}$$

Figure 1.61

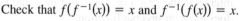 **Checkpoint**  Audio-video solution in English & Spanish at LarsonPrecalculus.com

Find the inverse function of

$$f(x) = \frac{5 - 3x}{x + 2}.$$

EXAMPLE 7   **Finding an Inverse Function Algebraically**

Find the inverse function of

$$f(x) = \sqrt{2x - 3}.$$

Solution The graph of f is shown in the figure below. This graph passes the Horizontal Line Test. So, you know that f is one-to-one and has an inverse function.

$f(x) = \sqrt{2x - 3}$	Write original function.
$y = \sqrt{2x - 3}$	Replace $f(x)$ with y.
$x = \sqrt{2y - 3}$	Interchange x and y.
$x^2 = 2y - 3$	Square each side.
$2y = x^2 + 3$	Isolate y-term.
$y = \dfrac{x^2 + 3}{2}$	Solve for y.
$f^{-1}(x) = \dfrac{x^2 + 3}{2}, \; x \geq 0$	Replace y with $f^{-1}(x)$.

The graph of f^{-1} in the figure is the reflection of the graph of f in the line $y = x$. Note that the range of f is the interval $[0, \infty)$, which implies that the domain of f^{-1} is the interval $[0, \infty)$. Moreover, the domain of f is the interval $\left[\frac{3}{2}, \infty\right)$, which implies that the range of f^{-1} is the interval $\left[\frac{3}{2}, \infty\right)$. Verify that $f(f^{-1}(x)) = x$ and $f^{-1}(f(x)) = x$.

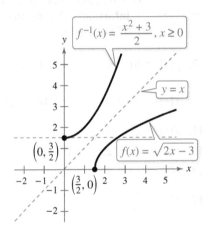

✓ *Checkpoint* ◀))) Audio-video solution in English & Spanish at LarsonPrecalculus.com

Find the inverse function of

$$f(x) = \sqrt[3]{10 + x}.$$

Summarize (Section 1.9)

1. State the definition of an inverse function *(page 85)*. For examples of finding inverse functions informally and verifying inverse functions, see Examples 1 and 2.

2. Explain how to use graphs to verify that two functions are inverse functions of each other *(page 86)*. For examples of verifying inverse functions graphically, see Examples 3 and 4.

3. Explain how to use the Horizontal Line Test to determine whether a function is one-to-one *(page 87)*. For an example of applying the Horizontal Line Test, see Example 5.

4. Explain how to find an inverse function algebraically *(page 88)*. For examples of finding inverse functions algebraically, see Examples 6 and 7.

1.9 Exercises

See **CalcChat.com** for tutorial help and worked-out solutions to odd-numbered exercises.

Vocabulary: Fill in the blanks.

1. If $f(g(x))$ and $g(f(x))$ both equal x, then the function g is the _____ function of the function f.
2. The inverse function of f is denoted by _____.
3. The domain of f is the _____ of f^{-1}, and the _____ of f^{-1} is the range of f.
4. The graphs of f and f^{-1} are reflections of each other in the line _____.
5. A function f is _____ when each value of the dependent variable corresponds to exactly one value of the independent variable.
6. A graphical test for the existence of an inverse function of f is the _____ Line Test.

Skills and Applications

 Finding an Inverse Function Informally
In Exercises 7–14, find the inverse function of f informally. Verify that $f(f^{-1}(x)) = x$ and $f^{-1}(f(x)) = x$.

7. $f(x) = 6x$
8. $f(x) = \dfrac{1}{3}x$
9. $f(x) = 3x + 1$
10. $f(x) = \dfrac{x-3}{2}$
11. $f(x) = x^2 - 4, \ x \geq 0$
12. $f(x) = x^2 + 2, \ x \geq 0$
13. $f(x) = x^3 + 1$
14. $f(x) = \dfrac{x^5}{4}$

 Verifying Inverse Functions In Exercises 15–18, verify that f and g are inverse functions algebraically.

15. $f(x) = \dfrac{x-9}{4}, \quad g(x) = 4x + 9$
16. $f(x) = -\dfrac{3}{2}x - 4, \quad g(x) = -\dfrac{2x+8}{3}$
17. $f(x) = \dfrac{x^3}{4}, \quad g(x) = \sqrt[3]{4x}$
18. $f(x) = x^3 + 5, \quad g(x) = \sqrt[3]{x-5}$

Sketching the Graph of an Inverse Function In Exercises 19 and 20, use the graph of the function to sketch the graph of its inverse function $y = f^{-1}(x)$.

19.
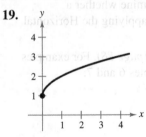

20.
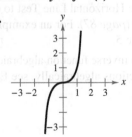

 Verifying Inverse Functions In Exercises 21–32, verify that f and g are inverse functions (a) algebraically and (b) graphically.

21. $f(x) = x - 5, \quad g(x) = x + 5$
22. $f(x) = 2x, \quad g(x) = \dfrac{x}{2}$
23. $f(x) = 7x + 1, \quad g(x) = \dfrac{x-1}{7}$
24. $f(x) = 3 - 4x, \quad g(x) = \dfrac{3-x}{4}$
25. $f(x) = x^3, \quad g(x) = \sqrt[3]{x}$
26. $f(x) = \dfrac{x^3}{3}, \quad g(x) = \sqrt[3]{3x}$
27. $f(x) = \sqrt{x+5}, \quad g(x) = x^2 - 5, \quad x \geq 0$
28. $f(x) = 1 - x^3, \quad g(x) = \sqrt[3]{1-x}$
29. $f(x) = \dfrac{1}{x}, \quad g(x) = \dfrac{1}{x}$
30. $f(x) = \dfrac{1}{1+x}, \quad x \geq 0, \quad g(x) = \dfrac{1-x}{x}, \quad 0 < x \leq 1$
31. $f(x) = \dfrac{x-1}{x+5}, \quad g(x) = -\dfrac{5x+1}{x-1}$
32. $f(x) = \dfrac{x+3}{x-2}, \quad g(x) = \dfrac{2x+3}{x-1}$

Using a Table to Determine an Inverse Function
In Exercises 33 and 34, does the function have an inverse function?

33.

x	-1	0	1	2	3	4
$f(x)$	-2	1	2	1	-2	-6

34.

x	-3	-2	-1	0	2	3
$f(x)$	10	6	4	1	-3	-10

Using a Table to Find an Inverse Function In Exercises 35 and 36, use the table of values for $y = f(x)$ to complete a table for $y = f^{-1}(x)$.

35.

x	-1	0	1	2	3	4
$f(x)$	3	5	7	9	11	13

36.

x	-3	-2	-1	0	1	2
$f(x)$	10	5	0	-5	-10	-15

Applying the Horizontal Line Test In Exercises 37–40, does the function have an inverse function?

37.
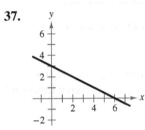

38.
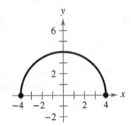

39.
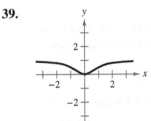

40.
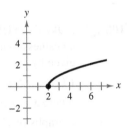

 Applying the Horizontal Line Test In Exercises 41–44, use a graphing utility to graph the function, and use the Horizontal Line Test to determine whether the function has an inverse function.

41. $g(x) = (x + 3)^2 + 2$ 42. $f(x) = \frac{1}{5}(x + 2)^3$

43. $f(x) = x\sqrt{9 - x^2}$ 44. $h(x) = |x| - |x - 4|$

 Finding and Analyzing Inverse Functions In Exercises 45–54, (a) find the inverse function of f, (b) graph both f and f^{-1} on the same set of coordinate axes, (c) describe the relationship between the graphs of f and f^{-1}, and (d) state the domains and ranges of f and f^{-1}.

45. $f(x) = x^5 - 2$ 46. $f(x) = x^3 + 8$

47. $f(x) = \sqrt{4 - x^2}, \quad 0 \le x \le 2$

48. $f(x) = x^2 - 2, \quad x \le 0$

49. $f(x) = \dfrac{4}{x}$ 50. $f(x) = -\dfrac{2}{x}$

51. $f(x) = \dfrac{x + 1}{x - 2}$ 52. $f(x) = \dfrac{x - 2}{3x + 5}$

53. $f(x) = \sqrt[3]{x - 1}$ 54. $f(x) = x^{3/5}$

 Finding an Inverse Function In Exercises 55–70, determine whether the function has an inverse function. If it does, find the inverse function.

55. $f(x) = x^4$ 56. $f(x) = \dfrac{1}{x^2}$

57. $g(x) = \dfrac{x + 1}{6}$ 58. $f(x) = 3x + 5$

59. $p(x) = -4$ 60. $f(x) = 0$

61. $f(x) = (x + 3)^2, \quad x \ge -3$

62. $q(x) = (x - 5)^2$

63. $f(x) = \begin{cases} x + 3, & x < 0 \\ 6 - x, & x \ge 0 \end{cases}$

64. $f(x) = \begin{cases} -x, & x \le 0 \\ x^2 - 3x, & x > 0 \end{cases}$

65. $h(x) = |x + 1| - 1$

66. $f(x) = |x - 2|, \quad x \le 2$

67. $f(x) = \sqrt{2x + 3}$

68. $f(x) = \sqrt{x - 2}$

69. $f(x) = \dfrac{6x + 4}{4x + 5}$

70. $f(x) = \dfrac{5x - 3}{2x + 5}$

Restricting the Domain In Exercises 71–78, restrict the domain of the function f so that the function is one-to-one and has an inverse function. Then find the inverse function f^{-1}. State the domains and ranges of f and f^{-1}. Explain your results. (There are many correct answers.)

71. $f(x) = |x + 2|$ 72. $f(x) = |x - 5|$

73. $f(x) = (x + 6)^2$ 74. $f(x) = (x - 4)^2$

75. $f(x) = -2x^2 + 5$

76. $f(x) = \frac{1}{2}x^2 - 1$

77. $f(x) = |x - 4| + 1$

78. $f(x) = -|x - 1| - 2$

Composition with Inverses In Exercises 79–84, use the functions $f(x) = \frac{1}{8}x - 3$ and $g(x) = x^3$ to find the value or function.

79. $(f^{-1} \circ g^{-1})(1)$ 80. $(g^{-1} \circ f^{-1})(-3)$

81. $(f^{-1} \circ f^{-1})(4)$ 82. $(g^{-1} \circ g^{-1})(-1)$

83. $(f \circ g)^{-1}$ 84. $g^{-1} \circ f^{-1}$

Composition with Inverses In Exercises 85–88, use the functions $f(x) = x + 4$ and $g(x) = 2x - 5$ to find the function.

85. $g^{-1} \circ f^{-1}$ 86. $f^{-1} \circ g^{-1}$

87. $(f \circ g)^{-1}$ 88. $(g \circ f)^{-1}$

89. Hourly Wage Your wage is $10.00 per hour plus $0.75 for each unit produced per hour. So, your hourly wage y in terms of the number of units produced x is $y = 10 + 0.75x$.

(a) Find the inverse function. What does each variable represent in the inverse function?

(b) Determine the number of units produced when your hourly wage is $24.25.

90. Diesel Mechanics

The function

$$y = 0.03x^2 + 245.50, \quad 0 < x < 100$$

approximates the exhaust temperature y in degrees Fahrenheit, where x is the percent load for a diesel engine.

(a) Find the inverse function. What does each variable represent in the inverse function?

(b) Use a graphing utility to graph the inverse function.

(c) The exhaust temperature of the engine must not exceed 500 degrees Fahrenheit. What is the percent load interval?

Exploration

True or False? **In Exercises 91 and 92, determine whether the statement is true or false. Justify your answer.**

91. If f is an even function, then f^{-1} exists.

92. If the inverse function of f exists and the graph of f has a y-intercept, then the y-intercept of f is an x-intercept of f^{-1}.

Creating a Table **In Exercises 93 and 94, use the graph of the function f to create a table of values for the given points. Then create a second table that can be used to find f^{-1}, and sketch the graph of f^{-1}, if possible.**

93.

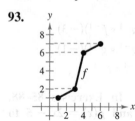

94.

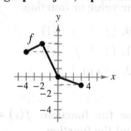

95. Proof Prove that if f and g are one-to-one functions, then $(f \circ g)^{-1}(x) = (g^{-1} \circ f^{-1})(x)$.

96. Proof Prove that if f is a one-to-one odd function, then f^{-1} is an odd function.

97. Think About It The function $f(x) = k(2 - x - x^3)$ has an inverse function, and $f^{-1}(3) = -2$. Find k.

98. Think About It Consider the functions $f(x) = x + 2$ and $f^{-1}(x) = x - 2$. Evaluate $f(f^{-1}(x))$ and $f^{-1}(f(x))$ for the given values of x. What can you conclude about the functions?

x	-10	0	7	45
$f(f^{-1}(x))$				
$f^{-1}(f(x))$				

99. Think About It Restrict the domain of

$$f(x) = x^2 + 1$$

to $x \geq 0$. Use a graphing utility to graph the function. Does the restricted function have an inverse function? Explain.

100. **HOW DO YOU SEE IT?** The cost C for a business to make personalized T-shirts is given by

$$C(x) = 7.50x + 1500$$

where x represents the number of T-shirts.

(a) The graphs of C and C^{-1} are shown below. Match each function with its graph.

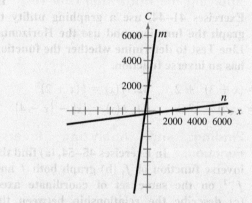

(b) Explain what $C(x)$ and $C^{-1}(x)$ represent in the context of the problem.

One-to-One Function Representation **In Exercises 101 and 102, determine whether the situation can be represented by a one-to-one function. If so, write a statement that best describes the inverse function.**

101. The number of miles n a marathon runner has completed in terms of the time t in hours

102. The depth of the tide d at a beach in terms of the time t over a 24-hour period

1.10 Mathematical Modeling and Variation

Mathematical models have a wide variety of real-life applications. For example, in Exercise 71 on page 103, you will use variation to model ocean temperatures at various depths.

- Use mathematical models to approximate sets of data points.
- Use the *regression* feature of a graphing utility to find equations of least squares regression lines.
- Write mathematical models for direct variation.
- Write mathematical models for direct variation as an *n*th power.
- Write mathematical models for inverse variation.
- Write mathematical models for combined variation.
- Write mathematical models for joint variation.

Introduction

In this section, you will study two techniques for fitting models to data: *least squares regression* and *direct and inverse variation*.

EXAMPLE 1   Using a Mathematical Model

The table shows the populations y (in millions) of the United States from 2008 through 2015.   *(Source: U.S. Census Bureau)*

Year	2008	2009	2010	2011	2012	2013	2014	2015
Population, y	304.1	306.8	309.3	311.7	314.1	316.5	318.9	321.2

Spreadsheet at LarsonPrecalculus.com

A linear model that approximates the data is

$$y = 2.43t + 284.9, \quad 8 \le t \le 15$$

where t represents the year, with $t = 8$ corresponding to 2008. Plot the actual data *and* the model on the same graph. How closely does the model represent the data?

Solution   Figure 1.62 shows the actual data and the model plotted on the same graph. From the graph, it appears that the model is a "good fit" for the actual data. To see how well the model fits, compare the actual values of y with the values of y found using the model. The values found using the model are labeled y^* in the table below.

t	8	9	10	11	12	13	14	15
y	304.1	306.8	309.3	311.7	314.1	316.5	318.9	321.2
y^*	304.3	306.8	309.2	311.6	314.1	316.5	318.9	321.4

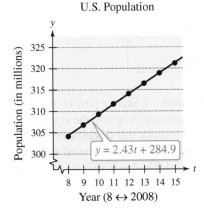

Figure 1.62

✓ *Checkpoint*))) Audio-video solution in English & Spanish at LarsonPrecalculus.com

The ordered pairs below give the median sales prices y (in thousands of dollars) of new homes sold in a neighborhood from 2009 through 2016. *(Spreadsheet at LarsonPrecalculus.com)*

DATA

(2009, 179.4)   (2011, 191.0)   (2013, 202.6)   (2015, 214.9)

(2010, 185.4)   (2012, 196.7)   (2014, 208.7)   (2016, 221.4)

A linear model that approximates the data is $y = 5.96t + 125.5$, $9 \le t \le 16$, where t represents the year, with $t = 9$ corresponding to 2009. Plot the actual data *and* the model on the same graph. How closely does the model represent the data?

Least Squares Regression and Graphing Utilities

So far in this text, you have worked with many different types of mathematical models that approximate real-life data. In some instances the model was given (as in Example 1), whereas in other instances you found the model using algebraic techniques or a graphing utility.

To find a model that approximates a set of data most accurately, statisticians use a measure called the **sum of the squared differences,** which is the sum of the squares of the differences between actual data values and model values. The "best-fitting" linear model, called the **least squares regression line,** is the one with the least sum of the squared differences.

Recall that you can approximate this line visually by plotting the data points and drawing the line that appears to best fit the data—or you can enter the data points into a graphing utility or software program and use the *linear regression* feature.

When you use the *regression* feature of a graphing utility or software program, an "*r*-value" may be output. This is the **correlation coefficient** of the data and gives a measure of how well the model fits the data. The closer the value of $|r|$ is to 1, the better the fit.

EXAMPLE 2   **Finding a Least Squares Regression Line**

See LarsonPrecalculus.com for an interactive version of this type of example.

The table shows the numbers E (in millions) of Medicare private health plan enrollees from 2008 through 2015. Construct a scatter plot that represents the data and find the equation of the least squares regression line for the data. *(Source: U.S. Centers for Medicare and Medicaid Services)*

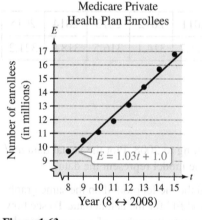

Medicare Private
Health Plan Enrollees

$E = 1.03t + 1.0$

Year (8 ↔ 2008)

Figure 1.63

DATA	Year	Enrollees, E
	2008	9.7
	2009	10.5
	2010	11.1
	2011	11.9
	2012	13.1
	2013	14.4
	2014	15.7
	2015	16.8

Spreadsheet at
LarsonPrecalculus.com

t	E	E^*
8	9.7	9.2
9	10.5	10.3
10	11.1	11.3
11	11.9	12.3
12	13.1	13.4
13	14.4	14.4
14	15.7	15.4
15	16.8	16.5

Solution Let $t = 8$ represent 2008. Figure 1.63 shows a scatter plot of the data. Using the *regression* feature of a graphing utility or software program, the equation of the least squares regression line is $E = 1.03t + 1.0$. To check this model, compare the actual E-values with the E-values found using the model, which are labeled E^* in the table at the left. The correlation coefficient for this model is $r \approx 0.992$, so the model is a good fit.

✓ *Checkpoint* ◀))) Audio-video solution in English & Spanish at LarsonPrecalculus.com

The ordered pairs below give the numbers E (in millions) of Medicare Advantage enrollees in health maintenance organization plans from 2008 through 2015. *(Spreadsheet at LarsonPrecalculus.com)* Construct a scatter plot that represents the data and find the equation of the least squares regression line for the data. *(Source: U.S. Centers for Medicare and Medicaid Services)*

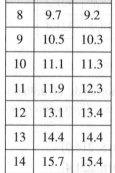

| (2008, 6.3) | (2010, 7.2) | (2012, 8.5) | (2014, 10.1) |
| (2009, 6.7) | (2011, 7.7) | (2013, 9.3) | (2015, 10.7) |

Direct Variation

There are two basic types of linear models. The more general model has a nonzero
y-intercept.

$$y = mx + b, \quad b \neq 0$$

The simpler model

$$y = kx$$

has a y-intercept of zero. In the simpler model, y **varies directly** as x, or is **directly
proportional** to x.

Direct Variation

The statements below are equivalent.

1. y **varies directly** as x.
2. y is **directly proportional** to x.
3. $y = kx$ for some nonzero constant k.

k is the **constant of variation** or the **constant of proportionality.**

EXAMPLE 3   **Direct Variation**

In Pennsylvania, the state income tax is directly proportional to *gross income*. You
work in Pennsylvania and your state income tax deduction is $46.05 for a gross
monthly income of $1500. Find a mathematical model that gives the Pennsylvania state
income tax in terms of gross income.

Solution

*Verbal
model*:   State income tax $=$ k $\cdot$ Gross income

Labels: State income tax $= y$ (dollars)
              Gross income $= x$ (dollars)
              Income tax rate $= k$ (percent in decimal form)

Equation: $y = kx$

To find the state income tax rate k, substitute the given information into the equation
$y = kx$ and solve.

$$y = kx \qquad \text{Write direct variation model.}$$

$$46.05 = k(1500) \qquad \text{Substitute 46.05 for } y \text{ and 1500 for } x.$$

$$0.0307 = k \qquad \text{Divide each side by 1500.}$$

So, the equation (or model) for state income tax in Pennsylvania is

$$y = 0.0307x.$$

In other words, Pennsylvania has a state income tax rate of 3.07% of gross income.
Figure 1.64 shows the graph of this equation.

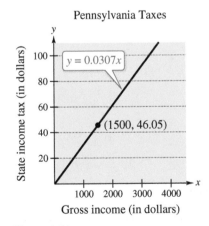

Pennsylvania Taxes

$y = 0.0307x$

(1500, 46.05)

State income tax (in dollars)

Gross income (in dollars)

Figure 1.64

✓ *Checkpoint*  *Audio-video solution in English & Spanish at LarsonPrecalculus.com*

The simple interest on an investment is directly proportional to the amount of the
investment. For example, an investment of $2500 earns $187.50 after 1 year. Find a
mathematical model that gives the interest I after 1 year in terms of the amount
invested P.

Direct Variation as an *n*th Power

Another type of direct variation relates one variable to a *power* of another variable. For example, in the formula for the area of a circle

$$A = \pi r^2$$

the area A is directly proportional to the square of the radius r. Note that for this formula, π is the constant of proportionality.

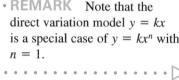

•• **REMARK** Note that the direct variation model $y = kx$ is a special case of $y = kx^n$ with $n = 1$.

> **Direct Variation as an *n*th Power**
>
> The statements below are equivalent.
>
> 1. y **varies directly as the *n*th power** of x.
> 2. y is **directly proportional to the *n*th power** of x.
> 3. $y = kx^n$ for some nonzero constant k.

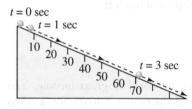

$t = 0$ sec
$t = 1$ sec
10 20 30 40 50 60 70
$t = 3$ sec

Figure 1.65

EXAMPLE 4 **Direct Variation as an *n*th Power**

The distance a ball rolls down an inclined plane is directly proportional to the square of the time it rolls. During the first second, the ball rolls 8 feet. (See Figure 1.65.)

a. Write an equation relating the distance traveled to the time.

b. How far does the ball roll during the first 3 seconds?

Solution

a. Letting d be the distance (in feet) the ball rolls and letting t be the time (in seconds), you have

$$d = kt^2.$$

Now, $d = 8$ when $t = 1$, so you have

$d = kt^2$	Write direct variation model.
$8 = k(\ )^2$	Substitute 8 for d and 1 for t.
$8 = k$	Simplify.

and, the equation relating distance to time is

$$d = 8t^2.$$

b. When $t = 3$, the distance traveled is

$d = 8(3)^2$	Substitute 3 for t.
$= 8(9)$	Simplify.
$= 72$ feet.	Simplify.

So, the ball rolls 72 feet during the first 3 seconds.

✓ *Checkpoint* ◗))) Audio-video solution in English & Spanish at LarsonPrecalculus.com

Neglecting air resistance, the distance s an object falls varies directly as the square of the duration t of the fall. An object falls a distance of 144 feet in 3 seconds. How far does it fall in 6 seconds?

In Examples 3 and 4, the direct variations are such that an *increase* in one variable corresponds to an *increase* in the other variable. You should not, however, assume that this always occurs with direct variation. For example, for the model $y = -3x$, an increase in x results in a *decrease* in y, and yet y is said to vary directly as x.

Inverse Variation

Inverse Variation

The statements below are equivalent.

1. y **varies inversely** as x.

2. y is **inversely proportional** to x.

3. $y = \dfrac{k}{x}$ for some nonzero constant k.

If x and y are related by an equation of the form $y = k/x^n$, then y varies inversely as the nth power of x (or y is inversely proportional to the nth power of x).

EXAMPLE 5 **Inverse Variation**

A company has found that the demand for one of its products varies inversely as the price of the product. When the price is \$6.25, the demand is 400 units. Approximate the demand when the price is \$5.75.

Solution

Let p be the price and let x be the demand. The demand varies inversely as the price, so you have

$$x = \frac{k}{p}.$$

Now, $x = 400$ when $p = 6.25$, so you have

$$x = \frac{k}{p} \qquad \text{Write inverse variation model.}$$

$$400 = \frac{k}{6.25} \qquad \text{Substitute 400 for } x \text{ and 6.25 for } p.$$

$$(400)(6.25) = k \qquad \text{Multiply each side by 6.25.}$$

$$2500 = k \qquad \text{Simplify.}$$

and the equation relating price and demand is

$$x = \frac{2500}{p}.$$

When $p = 5.75$, the demand is

$$x = \frac{2500}{p} \qquad \text{Write inverse variation model.}$$

$$= \frac{2500}{5.75} \qquad \text{Substitute 5.75 for } p.$$

$$\approx 435 \text{ units.} \qquad \text{Simplify.}$$

So, the demand for the product is about 435 units when the price is \$5.75.

Supply and demand are fundamental concepts in economics. The law of demand states that, all other factors remaining equal, the lower the price of the product, the higher the quantity demanded. The law of supply states that the higher the price of the product, the higher the quantity supplied. *Equilibrium* occurs when the demand and the supply are the same.

✓ *Checkpoint* ◀))) *Audio-video solution in English & Spanish at LarsonPrecalculus.com*

The company in Example 5 has found that the demand for another of its products also varies inversely as the price of the product. When the price is \$2.75, the demand is 600 units. Approximate the demand when the price is \$3.25.

Combined Variation

Some applications of variation involve problems with *both* direct and inverse variations in the same model. These types of models have **combined variation.**

EXAMPLE 6   Combined Variation

A gas law states that the volume of an enclosed gas varies inversely as the pressure (Figure 1.66) *and* directly as the temperature. The pressure of a gas is 0.75 kilogram per square centimeter when the temperature is 294 K and the volume is 8000 cubic centimeters.

a. Write an equation relating pressure, temperature, and volume.

b. Find the pressure when the temperature is 300 K and the volume is 7000 cubic centimeters.

Solution

a. Volume V varies directly as temperature T and inversely as pressure P, so you have

$$V = \frac{kT}{P}.$$

Now, $P = 0.75$ when $T = 294$ and $V = 8000$, so you have

$$V = \frac{kT}{P} \qquad \text{Write combined variation model.}$$

$$8000 = \frac{k(294)}{0.75} \qquad \text{Substitute 8000 for } V, 294 \text{ for } T, \text{ and } 0.75 \text{ for } P.$$

$$\frac{6000}{294} = k \qquad \text{Simplify.}$$

$$\frac{1000}{49} = k \qquad \text{Simplify.}$$

and the equation relating pressure, temperature, and volume is

$$V = \frac{1000}{49}\left(\frac{T}{P}\right).$$

b. Isolate P on one side of the equation by multiplying each side by P and dividing each side by V to obtain $P = \frac{1000}{49}\left(\frac{T}{V}\right)$. When $T = 300$ and $V = 7000$, the pressure is

$$P = \frac{1000}{49}\left(\frac{T}{V}\right) \qquad \text{Combined variation model solved for } P.$$

$$= \frac{1000}{49}\left(\frac{300}{7000}\right) \qquad \text{Substitute 300 for } T \text{ and 7000 for } V.$$

$$= \frac{300}{343} \qquad \text{Simplify.}$$

$$\approx 0.87 \text{ kilogram per square centimeter.} \qquad \text{Simplify.}$$

So, the pressure is about 0.87 kilogram per square centimeter when the temperature is 300 K and the volume is 7000 cubic centimeters.

✓ **Checkpoint** 🔊))) *Audio-video solution in English & Spanish at LarsonPrecalculus.com*

The resistance of a copper wire carrying an electrical current is directly proportional to its length and inversely proportional to its cross-sectional area. A copper wire with a diameter of 0.0126 inch has a resistance of 64.9 ohms per thousand feet. What length of 0.0201-inch-diameter copper wire will produce a resistance of 33.5 ohms?

P_1

P_2

V_1

V_2

If $P_2 > P_1$, then $V_2 < V_1$.

If the temperature is held constant and pressure increases, then the volume *decreases.*

Figure 1.66

Joint Variation

> ### Joint Variation
>
> The statements below are equivalent.
>
> **1.** z **varies jointly** as x and y.
>
> **2.** z is **jointly proportional** to x and y.
>
> **3.** $z = kxy$ for some nonzero constant k.

If x, y, and z are related by an equation of the form $z = kx^n y^m$, then z varies jointly as the nth power of x and the mth power of y.

EXAMPLE 7   Joint Variation

The *simple* interest for an investment is jointly proportional to the time and the principal. After one quarter (3 months), the interest on a principal of $5000 is $43.75. (a) Write an equation relating the interest, principal, and time. (b) Find the interest after three quarters.

Solution

a. Interest I (in dollars) is jointly proportional to principal P (in dollars) and time t (in years), so you have

$$I = kPt.$$

For $I = 43.75$, $P = 5000$, and $t = \frac{3}{12} = \frac{1}{4}$, you have $43.75 = k(5000)\left(\frac{1}{4}\right)$, which implies that $k = 4(43.75)/5000 = 0.035$. So, the equation relating interest, principal, and time is

$$I = 0.035Pt$$

which is the familiar equation for simple interest where the constant of proportionality, 0.035, represents an annual interest rate of 3.5%.

b. When $P = \$5000$ and $t = \frac{3}{4}$, the interest is $I = (0.035)(5000)\left(\frac{3}{4}\right) = \$131.25$.

✓ *Checkpoint*  *Audio-video solution in English & Spanish at LarsonPrecalculus.com*

The kinetic energy E of an object varies jointly with the object's mass m and the square of the object's velocity v. An object with a mass of 50 kilograms traveling at 16 meters per second has a kinetic energy of 6400 joules. What is the kinetic energy of an object with a mass of 70 kilograms traveling at 20 meters per second?

> ### Summarize   (Section 1.10)
>
> **1.** Explain how to use a mathematical model to approximate a set of data points *(page 93)*. For an example of using a mathematical model to approximate a set of data points, see Example 1.
>
> **2.** Explain how to use the *regression* feature of a graphing utility to find the equation of a least squares regression line *(page 94)*. For an example of finding the equation of a least squares regression line, see Example 2.
>
> **3.** Explain how to write mathematical models for direct variation, direct variation as an nth power, inverse variation, combined variation, and joint variation *(pages 95–99)*. For examples of these types of variation, see Examples 3–7.

1.10   Exercises   See CalcChat.com for tutorial help and worked-out solutions to odd-numbered exercises.

Vocabulary: **Fill in the blanks.**

1. Two techniques for fitting models to data are direct and inverse _____ and least squares _____.

2. Statisticians use a measure called the _____ of the _____ _____ to find a model that approximates a set of data most accurately.

3. The linear model with the least sum of the squared differences is called the _____ _____ _____ line.

4. An *r*-value, or _____ _____, of a set of data gives a measure of how well a model fits the data.

5. The direct variation model $y = kx^n$ can be described as "*y* varies directly as the *n*th power of *x*," or "*y* is _____ _____ to the *n*th power of *x*."

6. The mathematical model $y = \dfrac{2}{x}$ is an example of _____ variation.

7. Mathematical models that involve both direct and inverse variation have _____ variation.

8. The joint variation model $z = kxy$ can be described as "*z* varies jointly as *x* and *y*," or "*z* is _____ _____ to *x* and *y*."

Skills and Applications

Mathematical Models   In Exercises 9 and 10, (a) plot the actual data and the model of the same graph and (b) describe how closely the model represents the data. If the model does not closely represent the data, suggest another type of model that may be a better fit.

9. The ordered pairs below give the civilian noninstitutional U.S. populations *y* (in millions of people) 16 years of age and over not in the civilian labor force from 2006 through 2014. (*Spreadsheet at LarsonPrecalculus.com*)

DATA
(2006, 77.4)	(2011, 86.0)
(2007, 78.7)	(2012, 88.3)
(2008, 79.5)	(2013, 90.3)
(2009, 81.7)	(2014, 92.0)
(2010, 83.9)	

A model for the data is $y = 1.92t + 65.0$, $6 \le t \le 14$, where *t* represents the years, with *t* = 6 corresponding to 2006.   (*Source: U.S. Bureau of Labor Statistics*)

10. The ordered pairs below give the revenues *y* (in billions of dollars) for Activision Blizzard, Inc., from 2008 through 2014. (*Spreadsheet at LarsonPrecalculus.com*)

DATA
(2008, 3.03)	(2012, 4.86)
(2009, 4.28)	(2013, 4.58)
(2010, 4.45)	(2014, 4.41)
(2011, 4.76)	

A model for the data is $y = 0.184t + 2.32$, $8 \le t \le 14$, where *t* represents the year, with *t* = 8 corresponding to 2008.   (*Source: Activision Blizzard, Inc.*)

Sketching a Line   In Exercises 11–16, sketch the line that you think best approximates the data in the scatter plot. Then find an equation of the line. To print an enlarged copy of the graph, go to *MathGraphs.com*.

11.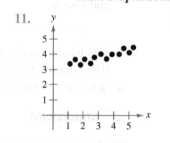

12.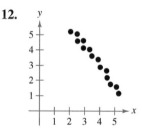

13.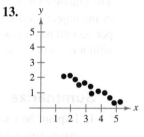

14.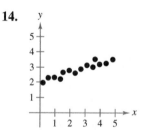

15.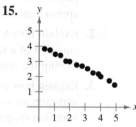

16.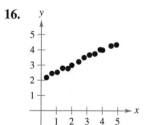

17. Sports The ordered pairs below give the winning times (in seconds) of the women's 100-meter freestyle in the Olympics from 1984 through 2012. *(Spreadsheet at LarsonPrecalculus.com) (Source: International Olympic Committee)*

DATA
(1984, 55.92)	(2000, 53.83)
(1988, 54.93)	(2004, 53.84)
(1992, 54.64)	(2008, 53.12)
(1996, 54.50)	(2012, 53.00)

(a) Sketch a scatter plot of the data. Let y represent the winning time (in seconds) and let $t = 84$ represent 1984.

(b) Sketch the line that you think best approximates the data and find an equation of the line.

(c) Use the *regression* feature of a graphing utility to find the equation of the least squares regression line that fits the data.

(d) Compare the linear model you found in part (b) with the linear model you found in part (c).

18. Broadway The ordered pairs below give the starting year and gross ticket sales S (in millions of dollars) for each Broadway season in New York City from 1997 through 2014. *(Spreadsheet at LarsonPrecalculus.com) (Source: The Broadway League)*

DATA
(1997, 558)	(2003, 771)	(2009, 1020)
(1998, 588)	(2004, 769)	(2010, 1081)
(1999, 603)	(2005, 862)	(2011, 1139)
(2000, 666)	(2006, 939)	(2012, 1139)
(2001, 643)	(2007, 938)	(2013, 1269)
(2002, 721)	(2008, 943)	(2014, 1365)

(a) Use a graphing utility to create a scatter plot of the data. Let $t = 7$ represent 1997.

(b) Use the *regression* feature of the graphing utility to find the equation of the least squares regression line that fits the data.

(c) Use the graphing utility to graph the scatter plot you created in part (a) and the model you found in part (b) in the same viewing window. How closely does the model represent the data?

(d) Use the model to predict the gross ticket sales during the season starting in 2021.

(e) Interpret the meaning of the slope of the linear model in the context of the problem.

 Direct Variation In Exercises 19–24, find a direct variation model that relates y and x.

19. $x = 2, y = 14$ **20.** $x = 5, y = 12$

21. $x = 5, y = 1$ **22.** $x = -24, y = 3$

23. $x = 4, y = 8\pi$ **24.** $x = \pi, y = -1$

 Direct Variation as an nth Power In Exercises 25–28, use the given values of k and n to complete the table for the direct variation model $y = kx^n$. Plot the points in a rectangular coordinate system.

x	2	4	6	8	10
$y = kx^n$					

25. $k = 1, n = 2$ **26.** $k = 2, n = 2$

27. $k = \frac{1}{2}, n = 3$ **28.** $k = \frac{1}{4}, n = 3$

Inverse Variation as an nth Power In Exercises 29–32, use the given values of k and n to complete the table for the inverse variation model $y = k/x^n$. Plot the points in a rectangular coordinate system.

x	2	4	6	8	10
$y = k/x^n$					

29. $k = 2, n = 1$ **30.** $k = 5, n = 1$

31. $k = 10, n = 2$ **32.** $k = 20, n = 2$

Think About It In Exercises 33 and 34, use the graph to determine whether y varies directly as some power of x or inversely as some power of x. Explain.

33.

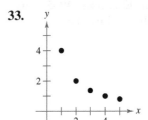

34.

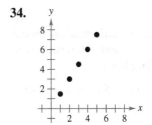

 Determining Variation In Exercises 35–38, determine whether the variation model represented by the ordered pairs (x, y) is of the form $y = kx$ or $y = k/x$, and find k. Then write a model that relates y and x.

35. $(5, 1), \left(10, \frac{1}{2}\right), \left(15, \frac{1}{3}\right), \left(20, \frac{1}{4}\right), \left(25, \frac{1}{5}\right)$

36. $(5, 2), (10, 4), (15, 6), (20, 8), (25, 10)$

37. $(5, -3.5), (10, -7), (15, -10.5), (20, -14), (25, -17.5)$

38. $(5, 24), (10, 12), (15, 8), (20, 6), \left(25, \frac{24}{5}\right)$

Finding a Mathematical Model In Exercises 39–48, find a mathematical model for the verbal statement.

39. A varies directly as the square of r.

40. V varies directly as the cube of l.

41. y varies inversely as the square of x.

42. h varies inversely as the square root of s.

43. F varies directly as g and inversely as r^2.

44. z varies jointly as the square of x and the cube of y.

45. *Newton's Law of Cooling:* The rate of change R of the temperature of an object is directly proportional to the difference between the temperature T of the object and the temperature T_e of the environment.

46. *Boyle's Law:* For a constant temperature, the pressure P of a gas is inversely proportional to the volume V of the gas.

47. *Direct Current:* The electric power P of a direct current circuit is jointly proportional to the voltage V and the electric current I.

48. *Newton's Law of Universal Gravitation:* The gravitational attraction F between two objects of masses m_1 and m_2 is jointly proportional to the masses and inversely proportional to the square of the distance r between the objects.

Describing a Formula In Exercises 49–52, use variation terminology to describe the formula.

49. $y = 2x^2$

50. $t = \dfrac{72}{r}$

51. $A = \frac{1}{2}bh$

52. $K = \frac{1}{2}mv^2$

 Finding a Mathematical Model In Exercises 53–60, find a mathematical model that represents the statement. (Determine the constant of proportionality.)

53. y is directly proportional to x. ($y = 54$ when $x = 3$.)

54. A varies directly as r^2. ($A = 9\pi$ when $r = 3$.)

55. y varies inversely as x. ($y = 3$ when $x = 25$.)

56. y is inversely proportional to x^3. ($y = 7$ when $x = 2$.)

57. z varies jointly as x and y. ($z = 64$ when $x = 4$ and $y = 8$.)

58. F is jointly proportional to r and the third power of s. ($F = 4158$ when $r = 11$ and $s = 3$.)

59. P varies directly as x and inversely as the square of y. ($P = \frac{28}{3}$ when $x = 42$ and $y = 9$.)

60. z varies directly as the square of x and inversely as y. ($z = 6$ when $x = 6$ and $y = 4$.)

61. Simple Interest The simple interest on an investment is directly proportional to the amount of the investment. An investment of \$3250 earns \$113.75 after 1 year. Find a mathematical model that gives the interest I after 1 year in terms of the amount invested P.

62. Simple Interest The simple interest on an investment is directly proportional to the amount of the investment. An investment of \$6500 earns \$211.25 after 1 year. Find a mathematical model that gives the interest I after 1 year in terms of the amount invested P.

63. Measurement Use the fact that 13 inches is approximately the same length as 33 centimeters to find a mathematical model that relates centimeters y to inches x. Then use the model to find the numbers of centimeters in 10 inches and 20 inches.

64. Measurement Use the fact that 14 gallons is approximately the same amount as 53 liters to find a mathematical model that relates liters y to gallons x. Then use the model to find the numbers of liters in 5 gallons and 25 gallons.

Hooke's Law In Exercises 65–68, use Hooke's Law, which states that the distance a spring stretches (or compresses) from its natural, or equilibrium, length varies directly as the applied force on the spring.

65. A force of 220 newtons stretches a spring 0.12 meter. What force stretches the spring 0.16 meter?

66. A force of 265 newtons stretches a spring 0.15 meter.

(a) What force stretches the spring 0.1 meter?

(b) How far does a force of 90 newtons stretch the spring?

67. The coiled spring of a toy supports the weight of a child. The weight of a 25-pound child compresses the spring a distance of 1.9 inches. The toy does not work properly when a weight compresses the spring more than 3 inches. What is the maximum weight for which the toy works properly?

68. An overhead garage door has two springs, one on each side of the door. A force of 15 pounds is required to stretch each spring 1 foot. Because of a pulley system, the springs stretch only one-half the distance the door travels. The door moves a total of 8 feet, and the springs are at their natural lengths when the door is open. Find the combined lifting force applied to the door by the springs when the door is closed.

69. Ecology The diameter of the largest particle that a stream can move is approximately directly proportional to the square of the velocity of the stream. When the velocity is $\frac{1}{4}$ mile per hour, the stream can move coarse sand particles about 0.02 inch in diameter. Approximate the velocity required to carry particles 0.12 inch in diameter.

70. Work The work W required to lift an object varies jointly with the object's mass m and the height h that the object is lifted. The work required to lift a 120-kilogram object 1.8 meters is 2116.8 joules. Find the amount of work required to lift a 100-kilogram object 1.5 meters.

• • **71. Ocean Temperatures** • • • • • • • • • • • • • •

The ordered pairs below give the average water temperatures C (in degrees Celsius) at several depths d (in meters) in the Indian Ocean. *(Spreadsheet at LarsonPrecalculus.com)* *(Source: NOAA)*

DATA
(1000, 4.85)   (2500, 1.888)
(1500, 3.525)   (3000, 1.583)
(2000, 2.468)   (3500, 1.422)

(a) Sketch a scatter plot of the data.

(b) Determine whether a direct variation model or an inverse variation model better fits the data.

(c) Find k for each pair of coordinates. Then find the mean value of k to find the constant of proportionality for the model you chose in part (b).

(d) Use your model to approximate the depth at which the water temperature is 3°C.

72. Light Intensity The ordered pairs below give the intensities y (in microwatts per square centimeter) of the light measured by a light probe located x centimeters from a light source. *(Spreadsheet at LarsonPrecalculus.com)*

DATA
(30, 0.1881)   (38, 0.1172)   (46, 0.0775)
(34, 0.1543)   (42, 0.0998)   (50, 0.0645)

A model that approximates the data is $y = 171.33/x^2$.

(a) Use a graphing utility to plot the data points and the model in the same viewing window.

(b) Use the model to approximate the light intensity 25 centimeters from the light source.

73. Music The fundamental frequency (in hertz) of a piano string is directly proportional to the square root of its tension and inversely proportional to its length and the square root of its mass density. A string has a frequency of 100 hertz. Find the frequency of a string with each property.

(a) Four times the tension

(b) Twice the length

(c) Four times the tension and twice the length

74. Beam Load The maximum load that a horizontal beam can safely support varies jointly as the width of the beam and the square of its depth and inversely as the length of the beam. Determine how each change affects the beam's maximum load.

(a) Doubling the width

(b) Doubling the depth

(c) Halving the length

(d) Halving the width and doubling the length

Exploration

True or False? **In Exercises 75 and 76, decide whether the statement is true or false. Justify your answer.**

75. If y is directly proportional to x and x is directly proportional to z, then y is directly proportional to z.

76. If y is inversely proportional to x and x is inversely proportional to z, then y is inversely proportional to z.

77. Error Analysis Describe the error.

In the equation for the surface area of a sphere, $S = 4\pi r^2$, the surface area S varies jointly with π and the square of the radius r.

78. **HOW DO YOU SEE IT?** Discuss how well a linear model approximates the data shown in each scatter plot.

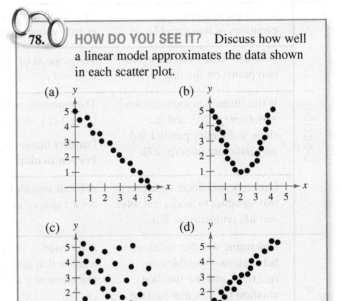

79. Think About It Let $y = 2x + 2$ and $t = x + 1$. What kind of variation do y and t have? Explain.

Project: Fraud and Identity Theft To work an extended application analyzing the numbers of fraud complaints and identity theft victims in the United States in 2014, visit this text's website at *LarsonPrecalculus.com*. *(Source: U.S. Federal Trade Commission)*

Chapter Summary

	What Did You Learn?	**Explanation/Examples**	**Review Exercises**
Section 1.1	Plot points in the Cartesian plane *(p. 2)*, use the Distance Formula *(p. 4)* and the Midpoint Formula *(p. 5)*, and use a coordinate plane to model and solve real-life problems *(p. 6)*.	For an ordered pair (x, y), the x-coordinate is the directed distance from the y-axis to the point, and the y-coordinate is the directed distance from the x-axis to the point. The coordinate plane can be used to estimate the annual sales of a company. (See Example 7.)	1–6
Section 1.2	Sketch graphs of equations *(p. 11)*, find x- and y-intercepts *(p. 14)*, and use symmetry to sketch graphs of equations *(p. 15)*.	To find x-intercepts, let y be zero and solve for x. To find y-intercepts, let x be zero and solve for y. Graphs can have symmetry with respect to one of the coordinate axes or with respect to the origin.	7–22
	Write equations of circles *(p. 17)*.	A point (x, y) lies on the circle of radius r and center (h, k) if and only if $(x - h)^2 + (y - k)^2 = r^2$.	23–27
	Use graphs of equations to solve real-life problems *(p. 18)*.	The graph of an equation can be used to estimate the maximum weight for a man in the U.S. Marine Corps. (See Example 9.)	28
Section 1.3	Use slope to graph linear equations in two variables *(p. 22)*.	The graph of the equation $y = mx + b$ is a line whose slope is m and whose y-intercept is $(0, b)$.	29–32
	Find the slope of a line given two points on the line *(p. 24)*.	The slope m of the nonvertical line through (x_1, y_1) and (x_2, y_2) is $m = (y_2 - y_1)/(x_2 - x_1)$, where $x_1 \neq x_2$.	33, 34
	Write linear equations in two variables *(p. 26)*, and use slope to identify parallel and perpendicular lines *(p. 27)*.	The equation of the line with slope m passing through the point (x_1, y_1) is $y - y_1 = m(x - x_1)$. **Parallel lines:** $m_1 = m_2$ **Perpendicular lines:** $m_1 = -1/m_2$	35–40
	Use slope and linear equations in two variables to model and solve real-life problems *(p. 28)*.	A linear equation in two variables can help you describe the book value of exercise equipment each year. (See Example 7.)	41, 42
Section 1.4	Determine whether relations between two variables are functions and use function notation *(p. 35)*, and find the domains of functions *(p. 40)*.	A function f from a set A (domain) to a set B (range) is a relation that assigns to each element x in the set A exactly one element y in the set B. **Equation:** $f(x) = 5 - x^2$ **$f(2)$:** $f(2) = 5 - 2^2 = 1$ **Domain of $f(x) = 5 - x^2$:** All real numbers	43–50
	Use functions to model and solve real-life problems *(p. 41)*.	A function can model the path of a baseball. (See Example 9.)	51, 52
	Evaluate difference quotients *(p. 42)*.	**Difference quotient:** $\dfrac{f(x + h) - f(x)}{h},\ h \neq 0$	53, 54
Section 1.5	Use the Vertical Line Test for functions *(p. 50)*.	A set of points in a coordinate plane is the graph of y as a function of x if and only if no *vertical* line intersects the graph at more than one point.	55, 56
	Find the zeros of functions *(p. 51)*.	**Zeros of $y = f(x)$:** x-values for which $f(x) = 0$	57, 58

	What Did You Learn?	**Explanation/Examples**	**Review Exercises**
Section 1.5	Determine intervals on which functions are increasing or decreasing *(p. 52)*, relative minimum and maximum values of functions *(p. 53)*, and the average rate of change of a function *(p. 54)*.	To determine whether a function is increasing, decreasing, or constant on an interval, determine whether the graph of the function rises, falls, or is constant from left to right. The points at which the behavior of a function changes can help determine relative minimum or relative maximum values. The average rate of change between any two points is the slope of the line (secant line) through the two points.	59–64
Section 1.5	Identify even and odd functions *(p. 55)*.	**Even:** For each x in the domain of f, $f(-x) = f(x)$. **Odd:** For each x in the domain of f, $f(-x) = -f(x)$.	65, 66
Section 1.6	Identify and graph different types of functions *(pp. 60, 62–64)*, and recognize graphs of parent functions *(p. 64)*.	**Linear:** $f(x) = ax + b$; **Squaring:** $f(x) = x^2$; **Cubic:** $f(x) = x^3$; **Square Root:** $f(x) = \sqrt{x}$; **Reciprocal:** $f(x) = 1/x$ Eight of the most commonly used functions in algebra are shown on page 64.	67–70
Section 1.7	Use vertical and horizontal shifts *(p. 67)*, reflections *(p. 69)*, and nonrigid transformations *(p. 71)* to sketch graphs of functions.	**Vertical shifts:** $h(x) = f(x) + c$ or $h(x) = f(x) - c$ **Horizontal shifts:** $h(x) = f(x - c)$ or $h(x) = f(x + c)$ **Reflection in *x*-axis:** $h(x) = -f(x)$ **Reflection in *y*-axis:** $h(x) = f(-x)$ **Nonrigid transformations:** $h(x) = cf(x)$ or $h(x) = f(cx)$	71–80
Section 1.8	Add, subtract, multiply, and divide functions *(p. 76)*, find compositions of functions *(p. 78)*, and use combinations and compositions of functions to model and solve real-life problems *(p. 80)*.	$(f + g)(x) = f(x) + g(x)$ $\quad$ $(f - g)(x) = f(x) - g(x)$ $(fg)(x) = f(x) \cdot g(x)$ $\quad\quad$ $(f/g)(x) = f(x)/g(x)$, $g(x) \neq 0$ The composition of the function f with the function g is $(f \circ g)(x) = f(g(x))$. A composite function can be used to represent the number of bacteria in food as a function of the amount of time the food has been out of refrigeration. (See Example 8.)	81–86
Section 1.9	Find inverse functions informally and verify that two functions are inverse functions of each other *(p. 84)*.	Let f and g be two functions such that $f(g(x)) = x$ for every x in the domain of g and $g(f(x)) = x$ for every x in the domain of f. Under these conditions, the function g is the inverse function of the function f.	87, 88
Section 1.9	Use graphs to verify inverse functions *(p. 86)*, use the Horizontal Line Test *(p. 87)*, and find inverse functions algebraically *(p. 88)*.	If the point (a, b) lies on the graph of f, then the point (b, a) must lie on the graph of f^{-1}, and vice versa. In short, the graph of f^{-1} is a reflection of the graph of f in the line $y = x$. To find an inverse function, replace $f(x)$ with y, interchange the roles of x and y, solve for y, and then replace y with $f^{-1}(x)$.	89–94
Section 1.10	Use mathematical models to approximate sets of data points *(p. 93)*, and use the *regression* feature of a graphing utility to find equations of least squares regression lines *(p. 94)*.	To see how well a model fits a set of data, compare the actual values of y with the model values. (See Example 1.) The sum of the squared differences is the sum of the squares of the differences between actual data values and model values. The least squares regression line is the linear model with the least sum of the squared differences.	95
Section 1.10	Write mathematical models for direct variation, direct variation as an *n*th power, inverse variation, combined variation, and joint variation *(pp. 95–99)*.	**Direct variation:** $y = kx$ for some nonzero constant k. **Direct variation as an *n*th power:** $y = kx^n$ for some nonzero constant k. **Inverse variation:** $y = k/x$ for some nonzero constant k. **Joint variation:** $z = kxy$ for some nonzero constant k.	96, 97

Review Exercises
See **CalcChat.com** for tutorial help and worked-out solutions to odd-numbered exercises.

1.1 **Plotting Points in the Cartesian Plane** **In Exercises 1 and 2, plot the points.**

1. $(5, 5), (-2, 0), (-3, 6), (-1, -7)$
2. $(0, 6), (8, 1), (5, -4), (-3, -3)$

Determining Quadrant(s) for a Point **In Exercises 3 and 4, determine the quadrant(s) in which (x, y) could be located.**

3. $x > 0$ and $y = -2$
4. $xy = 4$

5. **Plotting, Distance, and Midpoint** Plot the points $(-2, 6)$ and $(4, -3)$. Then find the distance between the points and the midpoint of the line segment joining the points.

6. **Sales** Barnes & Noble had annual sales of $6.8 billion in 2013 and $6.1 billion in 2015. Use the Midpoint Formula to estimate the sales in 2014. Assume that the annual sales follow a linear pattern. *(Source: Barnes & Noble, Inc.)*

1.2 **Sketching the Graph of an Equation** **In Exercises 7–10, construct a table of values that consists of several points of the equation. Use the resulting solution points to sketch the graph of the equation.**

7. $y = 3x - 5$
8. $y = -\frac{1}{2}x + 2$
9. $y = x^2 - 3x$
10. $y = 2x^2 - x - 9$

Finding x- and y-Intercepts **In Exercises 11–14, find the x- and y-intercepts of the graph of the equation.**

11. $y = 2x + 7$
12. $y = |x + 1| - 3$
13. $y = (x - 3)^2 - 4$
14. $y = x\sqrt{4 - x^2}$

Intercepts, Symmetry, and Graphing **In Exercises 15–22, find any intercepts and test for symmetry. Then sketch the graph of the equation.**

15. $y = -4x + 1$
16. $y = 5x - 6$
17. $y = 6 - x^2$
18. $y = x^2 - 12$
19. $y = x^3 + 5$
20. $y = -6 - x^3$
21. $y = \sqrt{x + 5}$
22. $y = |x| + 9$

Sketching a Circle **In Exercises 23–26, find the center and radius of the circle with the given equation. Then sketch the circle.**

23. $x^2 + y^2 = 9$
24. $x^2 + y^2 = 4$
25. $(x + 2)^2 + y^2 = 16$
26. $x^2 + (y - 8)^2 = 81$

27. **Writing the Equation of a Circle** Write the standard form of the equation of the circle for which the endpoints of a diameter are $(0, 0)$ and $(4, -6)$.

28. **Physics** The force F (in pounds) required to stretch a spring x inches from its natural length (see figure) is

$$F = \frac{5}{4}x, \quad 0 \le x \le 20.$$

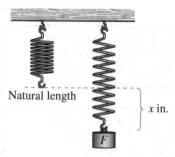

Natural length

x in.

F

(a) Use the model to complete the table.

x	0	4	8	12	16	20
Force, F						

(b) Sketch a graph of the model.

(c) Use the graph to estimate the force necessary to stretch the spring 10 inches.

1.3 **Graphing a Linear Equation** **In Exercises 29–32, find the slope and y-intercept (if possible) of the line. Sketch the line.**

29. $y = -\frac{1}{2}x + 1$
30. $2x - 3y = 6$
31. $y = 1$
32. $x = -6$

Finding the Slope of a Line Through Two Points **In Exercises 33 and 34, find the slope of the line passing through the pair of points.**

33. $(5, -2), (-1, 4)$
34. $(-1, 6), (3, -2)$

Using the Point-Slope Form **In Exercises 35 and 36, find the slope-intercept form of the equation of the line that has the given slope and passes through the given point. Sketch the line.**

35. $m = \frac{1}{3}, \quad (6, -5)$
36. $m = -\frac{3}{4}, \quad (-4, -2)$

Finding an Equation of a Line **In Exercises 37 and 38, find an equation of the line passing through the pair of points. Sketch the line.**

37. $(-6, 4), (4, 9)$
38. $(-9, -3), (-3, -5)$

Finding Parallel and Perpendicular Lines In Exercises 39 and 40, find equations of the lines that pass through the given point and are (a) parallel to and (b) perpendicular to the given line.

39. $5x - 4y = 8$, $(3, -2)$

40. $2x + 3y = 5$, $(-8, 3)$

41. Sales A discount outlet offers a 20% discount on all items. Write a linear equation giving the sale price S for an item with a list price L.

42. Hourly Wage A manuscript translator charges a starting fee of $50 plus $2.50 per page translated. Write a linear equation for the amount A earned for translating p pages.

1.4 Testing for Functions Represented Algebraically In Exercises 43–46, determine whether the equation represents y as a function of x.

43. $16x - y^4 = 0$

44. $2x - y - 3 = 0$

45. $y = \sqrt{1 - x}$

46. $|y| = x + 2$

Evaluating a Function In Exercises 47 and 48, find each function value.

47. $f(x) = x^2 + 1$

    (a) $f(2)$

    (b) $f(-4)$

    (c) $f(t^2)$

    (d) $f(t + 1)$

48. $h(x) = |x - 2|$

    (a) $h(-4)$

    (b) $h(-2)$

    (c) $h(0)$

    (d) $h(-x + 2)$

Finding the Domain of a Function In Exercises 49 and 50, find the domain of the function.

49. $f(x) = \sqrt{25 - x^2}$

50. $h(x) = \dfrac{x}{x^2 - x - 6}$

Physics In Exercises 51 and 52, the velocity of a ball projected upward from ground level is given by $v(t) = -32t + 48$, where t is the time in seconds and v is the velocity in feet per second.

51. Find the velocity when $t = 1$.

52. Find the time when the ball reaches its maximum height. [*Hint:* Find the time when $v(t) = 0$.]

Evaluating a Difference Quotient In Exercises 53 and 54, find the difference quotient and simplify your answer.

53. $f(x) = 2x^2 + 3x - 1$, $\dfrac{f(x + h) - f(x)}{h}$, $h \neq 0$

54. $f(x) = x^3 - 5x^2 + x$, $\dfrac{f(x + h) - f(x)}{h}$, $h \neq 0$

1.5 Vertical Line Test for Functions In Exercises 55 and 56, use the Vertical Line Test to determine whether the graph represents y as a function of x. To print an enlarged copy of the graph, go to *MathGraphs.com*.

55.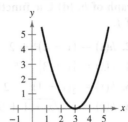

56.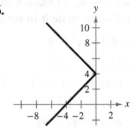

Finding the Zeros of a Function In Exercises 57 and 58, find the zeros of the function algebraically.

57. $f(x) = 3x^2 - 16x + 21$

58. $f(x) = 5x^2 + 4x - 1$

Describing Function Behavior In Exercises 59 and 60, use a graphing utility to graph the function and visually determine the open intervals on which the function is increasing, decreasing, or constant.

59. $f(x) = |x| + |x + 1|$ **60.** $f(x) = (x^2 - 4)^2$

Approximating Relative Minima or Maxima In Exercises 61 and 62, use a graphing utility to approximate (to two decimal places) any relative minima or maxima of the function.

61. $f(x) = -x^2 + 2x + 1$

62. $f(x) = x^3 - 4x^2 - 1$

Average Rate of Change of a Function In Exercises 63 and 64, find the average rate of change of the function from x_1 to x_2.

63. $f(x) = -x^2 + 8x - 4$, $x_1 = 0, x_2 = 4$

64. $f(x) = x^3 + 2x + 1$, $x_1 = 1, x_2 = 3$

Even, Odd, or Neither? In Exercises 65 and 66, determine whether the function is even, odd, or neither. Then describe the symmetry.

65. $f(x) = x^4 - 20x^2$ **66.** $f(x) = 2x\sqrt{x^2 + 3}$

1.6 Writing a Linear Function In Exercises 67 and 68, (a) write the linear function f that has the given function values and (b) sketch the graph of the function.

67. $f(2) = -6$, $f(-1) = 3$

68. $f(0) = -5$, $f(4) = -8$

Graphing a Function In Exercises 69 and 70, sketch the graph of the function.

69. $g(x) = [\![x]\!] - 2$

70. $f(x) = \begin{cases} 5x - 3, & x \geq -1 \\ -4x + 5, & x < -1 \end{cases}$

1.7 Describing Transformations In Exercises 71–80, h is related to one of the parent functions described in this chapter. (a) Identify the parent function f. (b) Describe the sequence of transformations from f to h. (c) Sketch the graph of h. (d) Use function notation to write h in terms of f.

71. $h(x) = x^2 - 9$

72. $h(x) = (x - 2)^3 + 2$

73. $h(x) = -\sqrt{x} + 4$

74. $h(x) = |x + 3| - 5$

75. $h(x) = -(x + 2)^2 + 3$

76. $h(x) = \frac{1}{2}(x - 1)^2 - 2$

77. $h(x) = -[\![x]\!] + 6$

78. $h(x) = -\sqrt{x + 1} + 9$

79. $h(x) = 5[\![x - 9]\!]$

80. $h(x) = -\frac{1}{3}x^3$

1.8 Finding Arithmetic Combinations of Functions In Exercises 81 and 82, find (a) $(f + g)(x)$, (b) $(f - g)(x)$, (c) $(fg)(x)$, and (d) $(f/g)(x)$. What is the domain of f/g?

81. $f(x) = x^2 + 3, \quad g(x) = 2x - 1$

82. $f(x) = x^2 - 4, \quad g(x) = \sqrt{3 - x}$

Finding Domains of Functions and Composite Functions In Exercises 83 and 84, find (a) $f \circ g$ and (b) $g \circ f$. Find the domain of each function and of each composite function.

83. $f(x) = \frac{1}{3}x - 3, \quad g(x) = 3x + 1$

84. $f(x) = x^3 - 4, \quad g(x) = \sqrt[3]{x + 7}$

Retail In Exercises 85 and 86, the price of a washing machine is x dollars. The function

$$f(x) = x - 100$$

gives the price of the washing machine after a $100 rebate. The function

$$g(x) = 0.95x$$

gives the price of the washing machine after a 5% discount.

85. Find and interpret $(f \circ g)(x)$.

86. Find and interpret $(g \circ f)(x)$.

1.9 Finding an Inverse Function Informally In Exercises 87 and 88, find the inverse function of f informally. Verify that $f(f^{-1}(x)) = x$ and $f^{-1}(f(x)) = x$.

87. $f(x) = 3x + 8$

88. $f(x) = \dfrac{x - 4}{5}$

Applying the Horizontal Line Test In Exercises 89 and 90, use a graphing utility to graph the function, and use the Horizontal Line Test to determine whether the function has an inverse function.

89. $f(x) = (x - 1)^2$

90. $h(t) = \dfrac{2}{t - 3}$

Finding and Analyzing Inverse Functions In Exercises 91 and 92, (a) find the inverse function of f, (b) graph both f and f^{-1} on the same set of coordinate axes, (c) describe the relationship between the graphs of f and f^{-1}, and (d) state the domains and ranges of f and f^{-1}.

91. $f(x) = \frac{1}{2}x - 3$

92. $f(x) = \sqrt{x + 1}$

Restricting the Domain In Exercises 93 and 94, restrict the domain of the function f to an interval on which the function is increasing, and find f^{-1} on that interval.

93. $f(x) = 2(x - 4)^2$

94. $f(x) = |x - 2|$

1.10

95. Agriculture The ordered pairs below give the amount B (in millions of pounds) of beef produced on private farms each year from 2007 through 2014. *(Spreadsheet at LarsonPrecalculus.com)* *(Source: United States Department of Agriculture)*

DATA
(2007, 102.7)	(2010, 84.2)	(2013, 70.4)
(2008, 95.9)	(2011, 75.0)	(2014, 67.9)
(2009, 90.2)	(2012, 76.3)	

(a) Use a graphing utility to create a scatter plot of the data. Let t represent the year, with $t = 7$ corresponding to 2007.

(b) Use the *regression* feature of the graphing utility to find the equation of the least squares regression line that fits the data. Then graph the model and the scatter plot you found in part (a) in the same viewing window. How closely does the model represent the data?

96. Travel Time The travel time between two cities is inversely proportional to the average speed. A train travels between the cities in 3 hours at an average speed of 65 miles per hour. How long does it take to travel between the cities at an average speed of 80 miles per hour?

97. Cost The cost of constructing a wooden box with a square base varies jointly as the height of the box and the square of the width of the box. Constructing a box of height 16 inches and of width 6 inches costs $28.80. How much does it cost to construct a box of height 14 inches and of width 8 inches?

Exploration

True or False? In Exercises 98 and 99, determine whether the statement is true or false. Justify your answer.

98. Relative to the graph of $f(x) = \sqrt{x}$, the graph of the function $h(x) = -\sqrt{x + 9} - 13$ is shifted 9 units to the left and 13 units down, then reflected in the x-axis.

99. If f and g are two inverse functions, then the domain of g is equal to the range of f.

Chapter Test

See **CalcChat.com** for tutorial help and worked-out solutions to odd-numbered exercises.

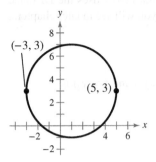

Figure for 6

Take this test as you would take a test in class. When you are finished, check your work against the answers given in the back of the book.

1. Plot the points $(-2, 5)$ and $(6, 0)$. Then find the distance between the points and the midpoint of the line segment joining the points.

2. A cylindrical can has a radius of 4 centimeters. Write the volume V of the can as a function of the height h.

In Exercises 3–5, find any intercepts and test for symmetry. Then sketch the graph of the equation.

3. $y = 3 - 5x$ 4. $y = 4 - |x|$ 5. $y = x^2 - 1$

6. Write the standard form of the equation of the circle shown at the left.

In Exercises 7 and 8, find an equation of the line passing through the pair of points. Sketch the line.

7. $(-2, 5), (1, -7)$ 8. $(-4, -7), \left(1, \frac{4}{3}\right)$

9. Find equations of the lines that pass through the point $(0, 4)$ and are (a) parallel to and (b) perpendicular to the line $5x + 2y = 3$.

10. Let $f(x) = \dfrac{\sqrt{x + 9}}{x^2 - 81}$. Find (a) $f(7)$, (b) $f(-5)$, and (c) $f(x - 9)$.

11. Find the domain of $f(x) = 10 - \sqrt{3 - x}$.

In Exercises 12–14, (a) find the zeros of the function, (b) use a graphing utility to graph the function, (c) approximate the open intervals on which the function is increasing, decreasing, or constant, and (d) determine whether the function is even, odd, or neither.

12. $f(x) = |x + 5|$ 13. $f(x) = 4x\sqrt{3 - x}$ 14. $f(x) = 2x^6 + 5x^4 - x^2$

15. Sketch the graph of $f(x) = \begin{cases} 3x + 7, & x \le -3 \\ 4x^2 - 1, & x > -3 \end{cases}$.

In Exercises 16–18, (a) identify the parent function f in the transformation, (b) describe the sequence of transformations from f to h, and (c) sketch the graph of h.

16. $h(x) = 4[\![x]\!]$ 17. $h(x) = \sqrt{x + 5} + 8$ 18. $h(x) = -2(x - 5)^3 + 3$

In Exercises 19 and 20, find (a) $(f + g)(x)$, (b) $(f - g)(x)$, (c) $(fg)(x)$, (d) $(f/g)(x)$, (e) $(f \circ g)(x)$, and (f) $(g \circ f)(x)$.

19. $f(x) = 3x^2 - 7, \quad g(x) = -x^2 - 4x + 5$ 20. $f(x) = 1/x, \quad g(x) = 2\sqrt{x}$

In Exercises 21–23, determine whether the function has an inverse function. If it does, find the inverse function.

21. $f(x) = x^3 + 8$ 22. $f(x) = |x^2 - 3| + 6$ 23. $f(x) = 3x\sqrt{x}$

In Exercises 24–26, find the mathematical model that represents the statement. (Determine the constant of proportionality.)

24. v varies directly as the square root of s. ($v = 24$ when $s = 16$.)

25. A varies jointly as x and y. ($A = 500$ when $x = 15$ and $y = 8$.)

26. b varies inversely as a. ($b = 32$ when $a = 1.5$.)

Proofs in Mathematics ■ ■ ■ ■ ■ ■ ■ ■ ■ ■ ■ ■ ■ ■

What does the word *proof* mean to you? In mathematics, the word *proof* means a valid argument. When you prove a statement or theorem, you must use facts, definitions, and accepted properties in a logical order. You can also use previously proved theorems in your proof. For example, the proof of the Midpoint Formula below uses the Distance Formula. There are several different proof methods, which you will see in later chapters.

The Midpoint Formula   *(p.5)*

The midpoint of the line segment joining the points (x_1, y_1) and (x_2, y_2) is

$$\text{Midpoint} = \left(\frac{x_1 + x_2}{2}, \frac{y_1 + y_2}{2}\right).$$

Proof

Using the figure, you must show that $d_1 = d_2$ and $d_1 + d_2 = d_3$.

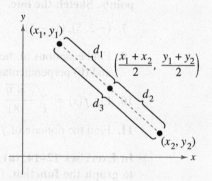

By the Distance Formula, you obtain

$$d_1 = \sqrt{\left(\frac{x_1 + x_2}{2} - x_1\right)^2 + \left(\frac{y_1 + y_2}{2} - y_1\right)^2}$$

$$= \sqrt{\left(\frac{x_2 - x_1}{2}\right)^2 + \left(\frac{y_2 - y_1}{2}\right)^2}$$

$$= \frac{1}{2}\sqrt{(x_2 - x_1)^2 + (y_2 - y_1)^2},$$

$$d_2 = \sqrt{\left(x_2 - \frac{x_1 + x_2}{2}\right)^2 + \left(y_2 - \frac{y_1 + y_2}{2}\right)^2}$$

$$= \sqrt{\left(\frac{x_2 - x_1}{2}\right)^2 + \left(\frac{y_2 - y_1}{2}\right)^2}$$

$$= \frac{1}{2}\sqrt{(x_2 - x_1)^2 + (y_2 - y_1)^2},$$

and

$$d_3 = \sqrt{(x_2 - x_1)^2 + (y_2 - y_1)^2}.$$

So, it follows that $d_1 = d_2$ and $d_1 + d_2 = d_3$.   ■

P.S. Problem Solving ▪ ▪ ▪ ▪ ▪ ▪ ▪ ▪ ▪ ▪ ▪ ▪ ▪ ▪ ▪

1. **Monthly Wages** As a salesperson, you receive a monthly salary of \$2000, plus a commission of 7% of sales. You receive an offer for a new job at \$2300 per month, plus a commission of 5% of sales.

   (a) Write a linear equation for your current monthly wage W_1 in terms of your monthly sales S.

   (b) Write a linear equation for the monthly wage W_2 of your new job offer in terms of the monthly sales S.

   (c) Use a graphing utility to graph both equations in the same viewing window. Find the point of intersection. What does the point of intersection represent?

   (d) You expect sales of \$20,000 per month. Should you change jobs? Explain.

2. **Cellphone Keypad** For the numbers 2 through 9 on a cellphone keypad (see figure), consider two relations: one mapping numbers onto letters, and the other mapping letters onto numbers. Are both relations functions? Explain.

1	2 ABC	3 DEF
4 GHI	5 JKL	6 MNO
7 PQRS	8 TUV	9 WXYZ
*	0	#

3. **Sums and Differences of Functions** What can be said about the sum and difference of each pair of functions?

   (a) Two even functions

   (b) Two odd functions

   (c) An odd function and an even function

4. **Inverse Functions** The functions

   $$f(x) = x \quad \text{and} \quad g(x) = -x$$

   are their own inverse functions. Graph each function and explain why this is true. Graph other linear functions that are their own inverse functions. Find a formula for a family of linear functions that are their own inverse functions.

5. **Proof** Prove that a function of the form

   $$y = a_{2n}x^{2n} + a_{2n-2}x^{2n-2} + \cdots + a_2 x^2 + a_0$$

   is an even function.

6. **Miniature Golf** A golfer is trying to make a hole-in-one on the miniature golf green shown. The golf ball is at the point $(2.5, 2)$ and the hole is at the point $(9.5, 2)$. The golfer wants to bank the ball off the side wall of the green at the point (x, y). Find the coordinates of the point (x, y). Then write an equation for the path of the ball.

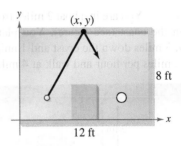

Figure for 6

7. **Titanic** At 2:00 P.M. on April 11, 1912, the *Titanic* left Cobh, Ireland, on her voyage to New York City. At 11:40 P.M. on April 14, the *Titanic* struck an iceberg and sank, having covered only about 2100 miles of the approximately 3400-mile trip.

   (a) What was the total duration of the voyage in hours?

   (b) What was the average speed in miles per hour?

   (c) Write a function relating the distance of the *Titanic* from New York City and the number of hours traveled. Find the domain and range of the function.

   (d) Graph the function in part (c).

8. **Average Rate of Change** Consider the function $f(x) = -x^2 + 4x - 3$. Find the average rate of change of the function from x_1 to x_2.

   (a) $x_1 = 1, x_2 = 2$

   (b) $x_1 = 1, x_2 = 1.5$

   (c) $x_1 = 1, x_2 = 1.25$

   (d) $x_1 = 1, x_2 = 1.125$

   (e) $x_1 = 1, x_2 = 1.0625$

   (f) Does the average rate of change seem to be approaching one value? If so, state the value.

   (g) Find the equations of the secant lines through the points $(x_1, f(x_1))$ and $(x_2, f(x_2))$ for parts (a)–(e).

   (h) Find the equation of the line through the point $(1, f(1))$ using your answer from part (f) as the slope of the line.

9. **Inverse of a Composition** Consider the functions $f(x) = 4x$ and $g(x) = x + 6$.

   (a) Find $(f \circ g)(x)$.

   (b) Find $(f \circ g)^{-1}(x)$.

   (c) Find $f^{-1}(x)$ and $g^{-1}(x)$.

   (d) Find $(g^{-1} \circ f^{-1})(x)$ and compare the result with that of part (b).

   (e) Repeat parts (a) through (d) for $f(x) = x^3 + 1$ and $g(x) = 2x$.

   (f) Write two one-to-one functions f and g, and repeat parts (a) through (d) for these functions.

   (g) Make a conjecture about $(f \circ g)^{-1}(x)$ and $(g^{-1} \circ f^{-1})(x)$.

10. Trip Time You are in a boat 2 miles from the nearest point on the coast (see figure). You plan to travel to point Q, 3 miles down the coast and 1 mile inland. You row at 2 miles per hour and walk at 4 miles per hour.

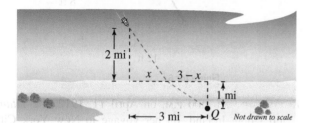

(a) Write the total time T (in hours) of the trip as a function of the distance x (in miles).

(b) Determine the domain of the function.

(c) Use a graphing utility to graph the function. Be sure to choose an appropriate viewing window.

(d) Find the value of x that minimizes T.

(e) Write a brief paragraph interpreting these values.

11. Heaviside Function The **Heaviside function**

$$H(x) = \begin{cases} 1, & x \geq 0 \\ 0, & x < 0 \end{cases}$$

is widely used in engineering applications. (See figure.) To print an enlarged copy of the graph, go to *MathGraphs.com*.

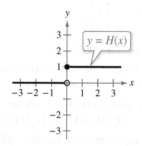

Sketch the graph of each function by hand.

(a) $H(x) - 2$

(b) $H(x - 2)$

(c) $-H(x)$

(d) $H(-x)$

(e) $\frac{1}{2}H(x)$

(f) $-H(x - 2) + 2$

12. Repeated Composition Let $f(x) = \dfrac{1}{1 - x}$.

(a) Find the domain and range of f.

(b) Find $f(f(x))$. What is the domain of this function?

(c) Find $f(f(f(x)))$. Is the graph a line? Why or why not?

13. Associative Property with Compositions Show that the Associative Property holds for compositions of functions—that is,

$$(f \circ (g \circ h))(x) = ((f \circ g) \circ h)(x).$$

14. Graphical Reasoning Use the graph of the function f to sketch the graph of each function. To print an enlarged copy of the graph, go to *MathGraphs.com*.

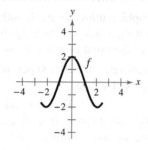

(a) $f(x + 1)$

(b) $f(x) + 1$

(c) $2f(x)$

(d) $f(-x)$

(e) $-f(x)$

(f) $|f(x)|$

(g) $f(|x|)$

15. Graphical Reasoning Use the graphs of f and f^{-1} to complete each table of function values.

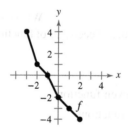
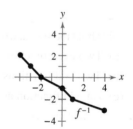

(a)

x	-4	-2	0	4
$(f(f^{-1}(x)))$				

(b)

x	-3	-2	0	1
$(f + f^{-1})(x)$				

(c)

x	-3	-2	0	1
$(f \cdot f^{-1})(x)$				

(d)

x	-4	-3	0	4
$|f^{-1}(x)|$				

2 Polynomial and Rational Functions

Candle Making Kits *(Example 12, page 161)*

Electrical Circuit
(Example 87, page 151)

Lyme Disease *(Exercise 82, page 144)*

Tree Growth
(Exercise 98, page 135)

Path of a Diver *(Exercise 67, page 121)*

2.1 Quadratic Functions and Models

- Analyze graphs of quadratic functions.
- Write quadratic functions in standard form and use the results to sketch their graphs.
- Find minimum and maximum values of quadratic functions in real-life applications.

The Graph of a Quadratic Function

In this and the next section, you will study graphs of polynomial functions. Section 1.6 introduced basic functions such as linear, constant, and squaring functions.

$f(x) = ax + b$ Linear function

$f(x) = c$ Constant function

$f(x) = x^2$ Squaring function

These are examples of **polynomial functions.**

Quadratic functions have many real-life applications. For example, in Exercise 67 on page 121, you will use a quadratic function that models the path of a diver.

Definition of a Polynomial Function

Let n be a nonnegative integer and let $a_n, a_{n-1}, \ldots, a_2, a_1, a_0$ be real numbers with $a_n \neq 0$. The function

$$f(x) = a_n x^n + a_{n-1} x^{n-1} + \cdots + a_2 x^2 + a_1 x + a_0$$

is a **polynomial function of x with degree n.**

Polynomial functions are classified by degree. For example, a constant function $f(x) = c$ with $c \neq 0$ has degree 0, and a linear function $f(x) = ax + b$ with $a \neq 0$ has degree 1. In this section, you will study **quadratic functions,** which are second-degree polynomial functions.

For example, each function listed below is a quadratic function.

$f(x) = x^2 + 6x + 2$

$g(x) = 2(x + 1)^2 - 3$

$h(x) = 9 + \frac{1}{4}x^2$

$k(x) = (x - 2)(x + 1)$

Note that the squaring function is a simple quadratic function.

Definition of a Quadratic Function

Let a, b, and c be real numbers with $a \neq 0$. The function

$$f(x) = ax^2 + bx + c$$ Quadratic function

is a **quadratic function.**

Time, t	Height, h
0	6
4	774
8	1030
12	774
16	6

Often, quadratic functions can model real-life data. For example, the table at the left shows the heights h (in feet) of a projectile fired from an initial height of 6 feet with an initial velocity of 256 feet per second at selected values of time t (in seconds). A quadratic model for the data in the table is

$$h(t) = -16t^2 + 256t + 6, \quad 0 \leq t \leq 16.$$

The graph of a quadratic function is a "U"-shaped curve called a **parabola.** Parabolas occur in many real-life applications—including those that involve reflective properties of satellite dishes and flashlight reflectors. You will study these properties in Section 10.2.

All parabolas are symmetric with respect to a line called the **axis of symmetry,** or simply the **axis** of the parabola. The point where the axis intersects the parabola is the **vertex** of the parabola. When the leading coefficient is positive, the graph of

$$f(x) = ax^2 + bx + c$$

is a parabola that opens upward. When the leading coefficient is negative, the graph is a parabola that opens downward. The next two figures show the axes and vertices of parabolas for cases where $a > 0$ and $a < 0$.

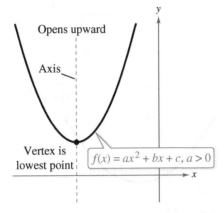

Leading coefficient is positive.

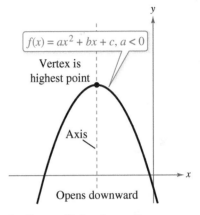

Leading coefficient is negative.

The simplest type of quadratic function is one in which $b = c = 0$. In this case, the function has the form $f(x) = ax^2$. Its graph is a parabola whose vertex is $(0, 0)$. When $a > 0$, the vertex is the point with the *minimum* y-value on the graph, and when $a < 0$, the vertex is the point with the *maximum* y-value on the graph, as shown in the figures below.

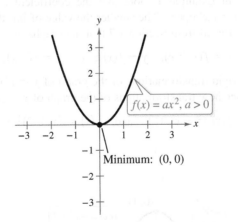

Leading coefficient is positive.

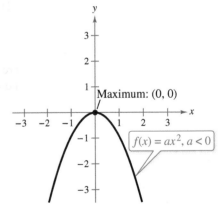

Leading coefficient is negative.

When sketching the graph of $f(x) = ax^2$, it is helpful to use the graph of $y = x^2$ as a reference, as suggested in Section 1.7. There you learned that when $a > 1$, the graph of $y = af(x)$ is a vertical stretch of the graph of $y = f(x)$. When $0 < a < 1$, the graph of $y = af(x)$ is a vertical shrink of the graph of $y = f(x)$. Example 1 demonstrates this again.

EXAMPLE 1   **Sketching Graphs of Quadratic Functions**

See LarsonPrecalculus.com for an interactive version of this type of example.

Sketch the graph of each quadratic function and compare it with the graph of $y = x^2$.

a. $f(x) = \frac{1}{3}x^2$ **b.** $g(x) = 2x^2$

Solution

▷ **ALGEBRA HELP** To
review techniques for
shifting, reflecting, stretching,
and shrinking graphs, see
Section 1.7.

a. Compared with $y = x^2$, each output of $f(x) = \frac{1}{3}x^2$ "shrinks" by a factor of $\frac{1}{3}$, producing the broader parabola shown in Figure 2.1.

b. Compared with $y = x^2$, each output of $g(x) = 2x^2$ "stretches" by a factor of 2, producing the narrower parabola shown in Figure 2.2.

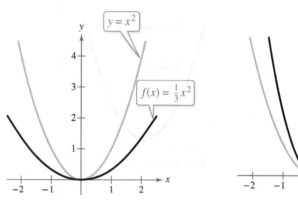

Figure 2.1   Figure 2.2

✓ *Checkpoint* 🔊))) *Audio-video solution in English & Spanish at LarsonPrecalculus.com*

Sketch the graph of each quadratic function and compare it with the graph of $y = x^2$.

a. $f(x) = \frac{1}{4}x^2$ **b.** $g(x) = -\frac{1}{6}x^2$ **c.** $h(x) = \frac{5}{2}x^2$ **d.** $k(x) = -4x^2$ ▪

In Example 1, note that the coefficient a determines how wide the parabola $f(x) = ax^2$ opens. The smaller the value of $|a|$, the wider the parabola opens. Recall from Section 1.7 that the graphs of

$$y = f(x \pm c), \quad y = f(x) \pm c, \quad y = f(-x), \quad \text{and} \quad y = -f(x)$$

are rigid transformations of the graph of $y = f(x)$. For example, in the figures below, notice how transformations of the graph of $y = x^2$ can produce the graphs of

$$f(x) = -x^2 + 1 \quad \text{and} \quad g(x) = (x + 2)^2 - 3.$$

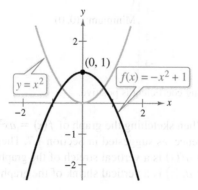

Reflection in *x*-axis followed by
an upward shift of one unit

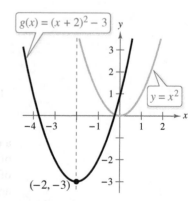

Left shift of two units followed by
a downward shift of three units

The Standard Form of a Quadratic Function

The **standard form** of a quadratic function is $f(x) = a(x - h)^2 + k$. This form is especially convenient for sketching a parabola because it identifies the vertex of the parabola as (h, k).

> **REMARK** The standard form of a quadratic function identifies four basic transformations of the graph of $y = x^2$.
>
> **a.** The factor a produces a vertical stretch or shrink.
>
> **b.** When $a < 0$, the factor a also produces a reflection in the x-axis.
>
> **c.** The factor $(x - h)^2$ represents a horizontal shift of h units.
>
> **d.** The term k represents a vertical shift of k units.

Standard Form of a Quadratic Function

The quadratic function

$$f(x) = a(x - h)^2 + k, \quad a \neq 0$$

is in **standard form**. The graph of f is a parabola whose axis is the vertical line $x = h$ and whose vertex is the point (h, k). When $a > 0$, the parabola opens upward, and when $a < 0$, the parabola opens downward.

To graph a parabola, it is helpful to begin by writing the quadratic function in standard form using the process of completing the square, as illustrated in Example 2. In this example, notice that when completing the square, you *add and subtract* the square of half the coefficient of x within the parentheses instead of adding the value to each side of the equation as is done in Appendix A.5.

EXAMPLE 2   **Using Standard Form to Graph a Parabola**

Sketch the graph of $f(x) = 2x^2 + 8x + 7$. Identify the vertex and the axis of the parabola.

> ▷ **ALGEBRA HELP** To review techniques for completing the square, see Appendix A.5.

Solution   Begin by writing the quadratic function in standard form. Notice that the first step in completing the square is to factor out any coefficient of x^2 that is not 1.

$$
\begin{aligned}
f(x) &= 2x^2 + 8x + 7 && \text{Write original function.}\\
&= 2(x^2 + 4x) + 7 && \text{Factor 2 out of } x\text{-terms.}\\
&= 2(x^2 + 4x + 4 - 4) + 7 && \text{Add and subtract 4 within parentheses.}\\
&\qquad\qquad \underset{(4/2)^2}{\big\uparrow} \\
&= 2(x^2 + 4x + 4) - 2(4) + 7 && \text{Distributive Property}\\
&= 2(x^2 + 4x + 4) - 8 + 7 && \text{Simplify.}\\
&= 2(x + 2)^2 - 1 && \text{Write in standard form.}
\end{aligned}
$$

The graph of f is a parabola that opens upward and has its vertex at $(-2, -1)$. This corresponds to a left shift of two units and a downward shift of one unit relative to the graph of $y = 2x^2$, as shown in the figure. The axis of the parabola is the vertical line through the vertex, $x = -2$, also shown in the figure.

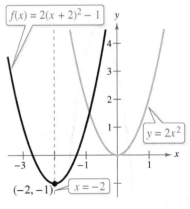

✓ *Checkpoint* ◁))) *Audio-video solution in English & Spanish at LarsonPrecalculus.com*

Sketch the graph of $f(x) = 3x^2 - 6x + 4$. Identify the vertex and the axis of the parabola.

▷ **ALGEBRA HELP** To review techniques for solving quadratic equations, see Appendix A.5.

To find the x-intercepts of the graph of $f(x) = ax^2 + bx + c$, you must solve the equation $ax^2 + bx + c = 0$. When $ax^2 + bx + c$ does not factor, use completing the square or the Quadratic Formula to find the x-intercepts. Remember, however, that a parabola may not have x-intercepts.

EXAMPLE 3 **Finding the Vertex and x-Intercepts of a Parabola**

Sketch the graph of $f(x) = -x^2 + 6x - 8$. Identify the vertex and x-intercepts.

Solution

$$f(x) = -x^2 + 6x - 8 \qquad \text{Write original function.}$$
$$= -(x^2 - 6x) - 8 \qquad \text{Factor } -1 \text{ out of } x\text{-terms.}$$
$$= -(x^2 - 6x + 9 - 9) - 8 \qquad \text{Add and subtract 9 within parentheses.}$$

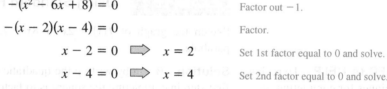

$$= -(x^2 - 6x + 9) - (-9) - 8 \qquad \text{Distributive Property}$$
$$= -(x - 3)^2 + 1 \qquad \text{Write in standard form.}$$

The graph of f is a parabola that opens downward with vertex $(3, 1)$. Next, find the x-intercepts of the graph.

$$-(x^2 - 6x + 8) = 0 \qquad \text{Factor out } -1.$$
$$-(x - 2)(x - 4) = 0 \qquad \text{Factor.}$$
$$x - 2 = 0 \ \Longrightarrow \ x = 2 \qquad \text{Set 1st factor equal to 0 and solve.}$$
$$x - 4 = 0 \ \Longrightarrow \ x = 4 \qquad \text{Set 2nd factor equal to 0 and solve.}$$

So, the x-intercepts are $(2, 0)$ and $(4, 0)$, as shown in Figure 2.3.

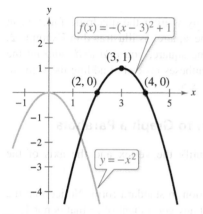

Figure 2.3

✓ **Checkpoint** *Audio-video solution in English & Spanish at LarsonPrecalculus.com*

Sketch the graph of $f(x) = x^2 - 4x + 3$. Identify the vertex and x-intercepts.

EXAMPLE 4 **Writing a Quadratic Function**

Write the standard form of the quadratic function whose graph is a parabola with vertex $(1, 2)$ and that passes through the point $(3, -6)$.

Solution The vertex is $(h, k) = (1, 2)$, so the equation has the form

$$f(x) = a(x - 1)^2 + 2. \qquad \text{Substitute for } h \text{ and } k \text{ in standard form.}$$

The parabola passes through the point $(3, -6)$, so it follows that $f(3) = -6$. So,

$$f(x) = a(x - 1)^2 + 2 \qquad \text{Write in standard form.}$$
$$-6 = a(3 - 1)^2 + 2 \qquad \text{Substitute 3 for } x \text{ and } -6 \text{ for } f(x).$$
$$-6 = 4a + 2 \qquad \text{Simplify.}$$
$$-8 = 4a \qquad \text{Subtract 2 from each side.}$$
$$-2 = a. \qquad \text{Divide each side by 4.}$$

The function in standard form is $f(x) = -2(x - 2)^2 + 2$. Figure 2.4 shows the graph of f.

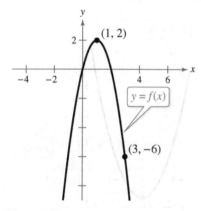

Figure 2.4

✓ **Checkpoint**  *Audio-video solution in English & Spanish at LarsonPrecalculus.com*

Write the standard form of the quadratic function whose graph is a parabola with vertex $(-4, 11)$ and that passes through the point $(-6, 15)$.

Finding Minimum and Maximum Values

Many applications involve finding the maximum or minimum value of a quadratic function. By completing the square within the quadratic function $f(x) = ax^2 + bx + c$, you can rewrite the function in standard form (see Exercise 79).

$$f(x) = a\left(x + \frac{b}{2a}\right)^2 + \left(c - \frac{b^2}{4a}\right) \qquad \text{Standard form}$$

So, the vertex of the graph of f is $\left(-\dfrac{b}{2a}, f\left(-\dfrac{b}{2a}\right)\right)$.

Minimum and Maximum Values of Quadratic Functions

Consider the function $f(x) = ax^2 + bx + c$ with vertex $\left(-\dfrac{b}{2a}, f\left(-\dfrac{b}{2a}\right)\right)$.

1. When $a > 0$, f has a *minimum* at $x = -\dfrac{b}{2a}$. The minimum value is $f\left(-\dfrac{b}{2a}\right)$.

2. When $a < 0$, f has a *maximum* at $x = -\dfrac{b}{2a}$. The maximum value is $f\left(-\dfrac{b}{2a}\right)$.

EXAMPLE 5 **Maximum Height of a Baseball**

The path of a baseball after being hit is modeled by $f(x) = -0.0032x^2 + x + 3$, where $f(x)$ is the height of the baseball (in feet) and x is the horizontal distance from home plate (in feet). What is the maximum height of the baseball?

Algebraic Solution

For this quadratic function, you have

$$f(x) = ax^2 + bx + c = -0.0032x^2 + x + 3$$

which shows that $a = -0.0032$ and $b = 1$. Because $a < 0$, the function has a maximum at $x = -b/(2a)$. So, the baseball reaches its maximum height when it is

$$x = -\frac{b}{2a} = -\frac{1}{2(-0.0032)} = 156.25 \text{ feet}$$

from home plate. At this distance, the maximum height is

$$f(156.25) = -0.0032(156.25)^2 + 156.25 + 3 = 81.125 \text{ feet}.$$

Graphical Solution

The maximum height is $y = 81.125$ feet at $x = 156.25$ feet.

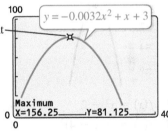

✓ **Checkpoint** 🔊))) *Audio-video solution in English & Spanish at LarsonPrecalculus.com*

Rework Example 5 when the path of the baseball is modeled by

$$f(x) = -0.007x^2 + x + 4.$$

Summarize (Section 2.1)

1. State the definition of a quadratic function and describe its graph *(pages 114–116)*. For an example of sketching graphs of quadratic functions, see Example 1.

2. State the standard form of a quadratic function *(page 117)*. For examples that use the standard form of a quadratic function, see Examples 2–4.

3. Explain how to find the minimum or maximum value of a quadratic function *(page 119)*. For a real-life application, see Example 5.

2.1 Exercises

See **CalcChat.com** for tutorial help and worked-out solutions to odd-numbered exercises.

Vocabulary: Fill in the blanks.

1. Linear, constant, and squaring functions are examples of _____ functions.

2. A polynomial function of x with degree n has the form $f(x) = a_n x^n + a_{n-1} x^{n-1} + \cdots + a_1 x + a_0$ ($a_n \neq 0$), where n is a _____ _____ and $a_n, a_{n-1}, \ldots, a_1, a_0$ are _____ numbers.

3. A _____ function is a second-degree polynomial function, and its graph is called a _____.

4. When the graph of a quadratic function opens downward, its leading coefficient is _____ and the vertex of the graph is a _____.

Skills and Applications

Matching In Exercises 5–8, match the quadratic function with its graph. [The graphs are labeled (a), (b), (c), and (d).]

(a)

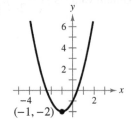

(b)

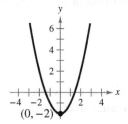

(c)

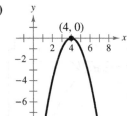

(d)

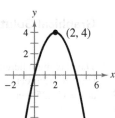

5. $f(x) = x^2 - 2$ 6. $f(x) = (x + 1)^2 - 2$
7. $f(x) = -(x - 4)^2$ 8. $f(x) = 4 - (x - 2)^2$

Sketching Graphs of Quadratic Functions In Exercises 9–12, sketch the graph of each quadratic function and compare it with the graph of $y = x^2$.

9. (a) $f(x) = \frac{1}{2}x^2$ (b) $g(x) = -\frac{1}{8}x^2$
   (c) $h(x) = \frac{3}{2}x^2$ (d) $k(x) = -3x^2$

10. (a) $f(x) = x^2 + 1$ (b) $g(x) = x^2 - 1$
    (c) $h(x) = x^2 + 3$ (d) $k(x) = x^2 - 3$

11. (a) $f(x) = (x - 1)^2$ (b) $g(x) = (3x)^2 + 1$
    (c) $h(x) = \left(\frac{1}{3}x\right)^2 - 3$ (d) $k(x) = (x + 3)^2$

12. (a) $f(x) = -\frac{1}{2}(x - 2)^2 + 1$
    (b) $g(x) = \left[\frac{1}{2}(x - 1)\right]^2 - 3$
    (c) $h(x) = -\frac{1}{2}(x + 2)^2 - 1$
    (d) $k(x) = [2(x + 1)]^2 + 4$

Using Standard Form to Graph a Parabola In Exercises 13–26, write the quadratic function in standard form and sketch its graph. Identify the vertex, axis of symmetry, and x-intercept(s).

13. $f(x) = x^2 - 6x$ 14. $g(x) = x^2 - 8x$
15. $h(x) = x^2 - 8x + 16$ 16. $g(x) = x^2 + 2x + 1$
17. $f(x) = x^2 - 6x + 2$ 18. $f(x) = x^2 + 16x + 61$
19. $f(x) = x^2 - 8x + 21$ 20. $f(x) = x^2 + 12x + 40$
21. $f(x) = x^2 - x + \frac{5}{4}$ 22. $f(x) = x^2 + 3x + \frac{1}{4}$
23. $f(x) = -x^2 + 2x + 5$ 24. $f(x) = -x^2 - 4x + 1$
25. $h(x) = 4x^2 - 4x + 21$ 26. $f(x) = 2x^2 - x + 1$

Using Technology In Exercises 27–34, use a graphing utility to graph the quadratic function. Identify the vertex, axis of symmetry, and x-intercept(s). Then check your results algebraically by writing the quadratic function in standard form.

27. $f(x) = -(x^2 + 2x - 3)$ 28. $f(x) = -(x^2 + x - 30)$
29. $g(x) = x^2 + 8x + 11$ 30. $f(x) = x^2 + 10x + 14$
31. $f(x) = -2x^2 + 12x - 18$
32. $f(x) = -4x^2 + 24x - 41$
33. $g(x) = \frac{1}{2}(x^2 + 4x - 2)$
34. $f(x) = \frac{3}{5}(x^2 + 6x - 5)$

Writing a Quadratic Function In Exercises 35 and 36, write the standard form of the quadratic function whose graph is the parabola shown.

35.

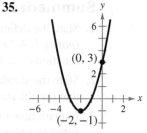

36.

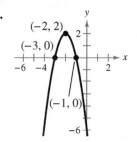

Writing a Quadratic Function In Exercises 37–46, write the standard form of the quadratic function whose graph is a parabola with the given vertex and that passes through the given point.

37. Vertex: $(-2, 5)$; point: $(0, 9)$
38. Vertex: $(-3, -10)$; point: $(0, 8)$
39. Vertex: $(1, -2)$; point: $(-1, 14)$
40. Vertex: $(2, 3)$; point: $(0, 2)$
41. Vertex: $(5, 12)$; point: $(7, 15)$
42. Vertex: $(-2, -2)$; point: $(-1, 0)$
43. Vertex: $\left(-\frac{1}{4}, \frac{3}{2}\right)$; point: $(-2, 0)$
44. Vertex: $\left(\frac{5}{2}, -\frac{3}{4}\right)$; point: $(-2, 4)$
45. Vertex: $\left(-\frac{5}{2}, 0\right)$; point: $\left(-\frac{7}{2}, -\frac{16}{3}\right)$
46. Vertex: $(6, 6)$; point: $\left(\frac{61}{10}, \frac{3}{2}\right)$

Graphical Reasoning In Exercises 47–50, determine the x-intercept(s) of the graph visually. Then find the x-intercept(s) algebraically to confirm your results.

47. $y = x^2 - 2x - 3$ 48. $y = x^2 - 4x - 5$

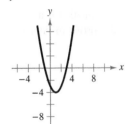
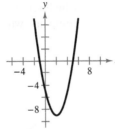

49. $y = 2x^2 + 5x - 3$ 50. $y = -2x^2 + 5x + 3$

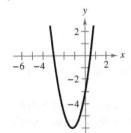
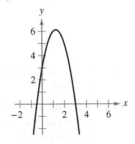

 Using Technology In Exercises 51–56, use a graphing utility to graph the quadratic function. Find the x-intercept(s) of the graph and compare them with the solutions of the corresponding quadratic equation when $f(x) = 0$.

51. $f(x) = x^2 - 4x$
52. $f(x) = -2x^2 + 10x$
53. $f(x) = x^2 - 9x + 18$
54. $f(x) = x^2 - 8x - 20$
55. $f(x) = 2x^2 - 7x - 30$
56. $f(x) = \frac{7}{10}(x^2 + 12x - 45)$

Finding Quadratic Functions In Exercises 57–62, find two quadratic functions, one that opens upward and one that opens downward, whose graphs have the given x-intercepts. (There are many correct answers.)

57. $(-3, 0), (3, 0)$ 58. $(-5, 0), (5, 0)$
59. $(-1, 0), (4, 0)$ 60. $(-2, 0), (3, 0)$
61. $(-3, 0), \left(-\frac{1}{2}, 0\right)$ 62. $\left(-\frac{3}{2}, 0\right), (-5, 0)$

Number Problems In Exercises 63–66, find two positive real numbers whose product is a maximum.

63. The sum is 110.
64. The sum is S.
65. The sum of the first and twice the second is 24.
66. The sum of the first and three times the second is 42.

67. **Path of a Diver**
The path of a diver is modeled by

$$f(x) = -\frac{4}{9}x^2 + \frac{24}{9}x + 12$$

where $f(x)$ is the height (in feet) and x is the horizontal distance (in feet) from the end of the diving board. What is the maximum height of the diver?

68. **Height of a Ball** The path of a punted football is modeled by

$$f(x) = -\frac{16}{2025}x^2 + \frac{9}{5}x + 1.5$$

where $f(x)$ is the height (in feet) and x is the horizontal distance (in feet) from the point at which the ball is punted.

(a) How high is the ball when it is punted?
(b) What is the maximum height of the punt?
(c) How long is the punt?

69. **Minimum Cost** A manufacturer of lighting fixtures has daily production costs of $C = 800 - 10x + 0.25x^2$, where C is the total cost (in dollars) and x is the number of units produced. What daily production number yields a minimum cost?

70. **Maximum Profit** The profit P (in hundreds of dollars) that a company makes depends on the amount x (in hundreds of dollars) the company spends on advertising according to the model $P = 230 + 20x - 0.5x^2$. What expenditure for advertising yields a maximum profit?

71. Maximum Revenue The total revenue R earned (in thousands of dollars) from manufacturing handheld video games is given by $R(p) = -25p^2 + 1200p$, where p is the price per unit (in dollars).

(a) Find the revenues when the prices per unit are $20, $25, and $30.

(b) Find the unit price that yields a maximum revenue. What is the maximum revenue? Explain.

72. Maximum Revenue The total revenue R earned per day (in dollars) from a pet-sitting service is given by $R(p) = -12p^2 + 150p$, where p is the price charged per pet (in dollars).

(a) Find the revenues when the prices per pet are $4, $6, and $8.

(b) Find the unit price that yields a maximum revenue. What is the maximum revenue? Explain.

73. Maximum Area A rancher has 200 feet of fencing to enclose two adjacent rectangular corrals (see figure).

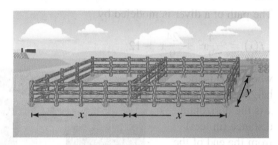

(a) Write the area A of the corrals as a function of x.

(b) What dimensions produce a maximum enclosed area?

74. Maximum Area A Norman window is constructed by adjoining a semicircle to the top of an ordinary rectangular window (see figure). The perimeter of the window is 16 feet.

(a) Write the area A of the window as a function of x.

(b) What dimensions produce a window of maximum area?

Exploration

True or False? In Exercises 75 and 76, determine whether the statement is true or false. Justify your answer.

75. The graph of $f(x) = -12x^2 - 1$ has no x-intercepts.

76. The graphs of $f(x) = -4x^2 - 10x + 7$ and $g(x) = 12x^2 + 30x + 1$ have the same axis of symmetry.

Think About It In Exercises 77 and 78, find the values of b such that the function has the given maximum or minimum value.

77. $f(x) = -x^2 + bx - 75$; Maximum value: 25

78. $f(x) = x^2 + bx - 25$; Minimum value: -50

79. Verifying the Vertex Write the quadratic function

$$f(x) = ax^2 + bx + c$$

in standard form to verify that the vertex occurs at

$$\left(-\frac{b}{2a}, f\left(-\frac{b}{2a}\right)\right).$$

80. **HOW DO YOU SEE IT?** The graph shows a quadratic function of the form

$$P(t) = at^2 + bt + c$$

which represents the yearly profit for a company, where $P(t)$ is the profit in year t.

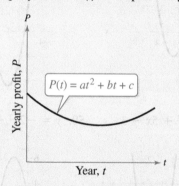

$P(t) = at^2 + bt + c$

(a) Is the value of a positive, negative, or zero? Explain.

(b) Write an expression in terms of a and b that represents the year t when the company made the least profit.

(c) The company made the same yearly profits in 2008 and 2016. Estimate the year in which the company made the least profit.

81. Proof Assume that the function

$$f(x) = ax^2 + bx + c, \quad a \neq 0$$

has two real zeros. Prove that the x-coordinate of the vertex of the graph is the average of the zeros of f. (*Hint:* Use the Quadratic Formula.)

Project: Height of a Basketball To work an extended application analyzing the height of a dropped basketball, visit this text's website at *LarsonPrecalculus.com*.

2.2 Polynomial Functions of Higher Degree

- ▣ Use transformations to sketch graphs of polynomial functions.
- ▣ Use the Leading Coefficient Test to determine the end behaviors of graphs of polynomial functions.
- ▣ Find real zeros of polynomial functions and use them as sketching aids.
- ▣ Use the Intermediate Value Theorem to help locate real zeros of polynomial functions.

Polynomial functions have many real-life applications. For example, in Exercise 98 on page 135, you will use a polynomial function to analyze the growth of a red oak tree.

Graphs of Polynomial Functions

In this section, you will study basic features of the graphs of polynomial functions. One feature is that the graph of a polynomial function is **continuous.** Essentially, this means that the graph of a polynomial function has no breaks, holes, or gaps, as shown in Figure 2.5(a). The graph shown in Figure 2.5(b) is an example of a piecewise-defined function that is not continuous.

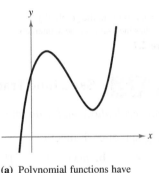

(a) Polynomial functions have continuous graphs.

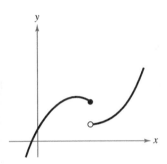

(b) Functions with graphs that are not continuous are not polynomial functions.

Figure 2.5

Another feature of the graph of a polynomial function is that it has only smooth, rounded turns, as shown in Figure 2.6(a). The graph of a polynomial function cannot have a sharp turn, such as the one shown in Figure 2.6(b).

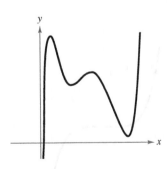

(a) Polynomial functions have graphs with smooth, rounded turns.

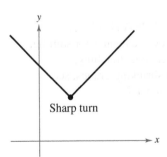

Sharp turn

(b) Functions with graphs that have sharp turns are not polynomial functions.

Figure 2.6

Sketching graphs of polynomial functions of degree greater than 2 is often more involved than sketching graphs of polynomial functions of degree 0, 1, or 2. However, using the features presented in this section, along with your knowledge of point plotting, intercepts, and symmetry, you should be able to make reasonably accurate sketches by hand.

▷

REMARK For functions of the form $f(x) = x^n$, if n is even, then the graph of the function is symmetric with respect to the y-axis, and if n is odd, then the graph of the function is symmetric with respect to the origin.

The polynomial functions that have the simplest graphs are monomial functions of the form $f(x) = x^n$, where n is an integer greater than zero. When n is *even*, the graph is similar to the graph of $f(x) = x^2$, and when n is *odd*, the graph is similar to the graph of $f(x) = x^3$, as shown in Figure 2.7. Moreover, the greater the value of n, the flatter the graph near the origin. Polynomial functions of the form $f(x) = x^n$ are often referred to as **power functions.**

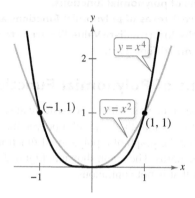

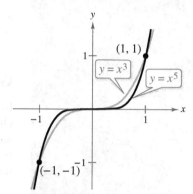

(a) When n is even, the graph of $y = x^n$ touches the x-axis at the x-intercept.

(b) When n is odd, the graph of $y = x^n$ crosses the x-axis at the x-intercept.

Figure 2.7

EXAMPLE 1 **Sketching Transformations of Monomial Functions**

See LarsonPrecalculus.com for an interactive version of this type of example.

Sketch the graph of each function.

a. $f(x) = -x^5$ **b.** $h(x) = (x + 1)^4$

Solution

a. The degree of $f(x) = -x^5$ is odd, so its graph is similar to the graph of $y = x^3$. In Figure 2.8, note that the negative coefficient has the effect of reflecting the graph in the x-axis.

b. The degree of $h(x) = (x + 1)^4$ is even, so its graph is similar to the graph of $y = x^2$. In Figure 2.9, note that the graph of h is a left shift by one unit of the graph of $y = x^4$.

▷ **ALGEBRA HELP** To review techniques for shifting, reflecting, stretching, and shrinking graphs, see Section 1.7.

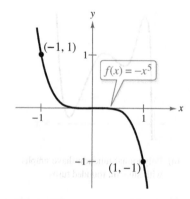

Figure 2.8

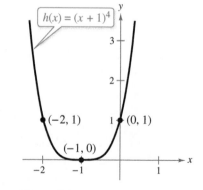

Figure 2.9

✓ *Checkpoint* 🔊))) Audio-video solution in English & Spanish at LarsonPrecalculus.com

Sketch the graph of each function.

a. $f(x) = (x + 5)^4$ **b.** $g(x) = x^4 - 7$

c. $h(x) = 7 - x^4$ **d.** $k(x) = \frac{1}{4}(x - 3)^4$

The Leading Coefficient Test

In Example 1, note that both graphs eventually rise or fall without bound as x moves to the left or to the right. A polynomial function's degree (even or odd) and its leading coefficient (positive or negative) determine whether the graph of the function eventually rises or falls, as described in the **Leading Coefficient Test.**

Leading Coefficient Test

As x moves without bound to the left or to the right, the graph of the polynomial function

$$f(x) = a_n x^n + \cdots + a_1 x + a_0, \quad a_n \neq 0$$

eventually rises or falls in the manner described below.

1. When n is *odd:*

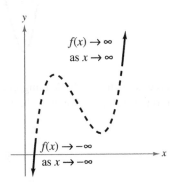

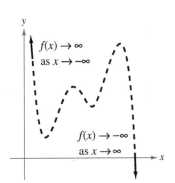

If the leading coefficient is positive ($a_n > 0$), then the graph falls to the left and rises to the right.

If the leading coefficient is negative ($a_n < 0$), then the graph rises to the left and falls to the right.

2. When n is *even:*

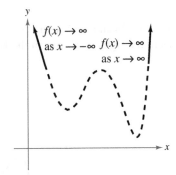

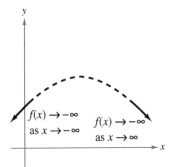

If the leading coefficient is positive ($a_n > 0$), then the graph rises to the left and to the right.

If the leading coefficient is negative ($a_n < 0$), then the graph falls to the left and to the right.

The dashed portions of the graphs indicate that the test determines *only* the right-hand and left-hand behavior of the graph.

> •••••••••••••••••••▷
> •• **REMARK** The notation
> "$f(x) \to -\infty$ as $x \to -\infty$"
> means that the graph falls to the
> left. The notation "$f(x) \to \infty$ as
> $x \to \infty$" means that the graph
> rises to the right. Identify and
> interpret similar notation for the
> other two possible types of end
> behavior given in the Leading
> Coefficient Test.

As you continue to study polynomial functions and their graphs, you will notice that the degree of a polynomial plays an important role in determining other characteristics of the polynomial function and its graph.

EXAMPLE 2 **Applying the Leading Coefficient Test**

Describe the left-hand and right-hand behavior of the graph of each function.

a. $f(x) = -x^3 + 4x$ **b.** $f(x) = x^4 - 5x^2 + 4$ **c.** $f(x) = x^5 - x$

Solution

a. The degree is odd and the leading coefficient is negative, so the graph rises to the left and falls to the right, as shown in the figure below.

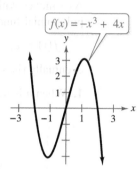

$$f(x) = -x^3 + 4x$$

b. The degree is even and the leading coefficient is positive, so the graph rises to the left and to the right, as shown in the figure below.

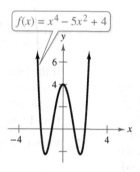

$$f(x) = x^4 - 5x^2 + 4$$

c. The degree is odd and the leading coefficient is positive, so the graph falls to the left and rises to the right, as shown in the figure below.

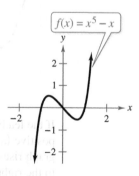

$$f(x) = x^5 - x$$

✓ *Checkpoint* ◀))) *Audio-video solution in English & Spanish at LarsonPrecalculus.com*

Describe the left-hand and right-hand behavior of the graph of each function.

a. $f(x) = \frac{1}{4}x^3 - 2x$ **b.** $f(x) = -3.6x^5 + 5x^3 - 1$

In Example 2, note that the Leading Coefficient Test tells you only whether the graph *eventually* rises or falls to the left or to the right. You must use other tests to determine other characteristics of the graph, such as intercepts and minimum and maximum points.

Real Zeros of Polynomial Functions

It is possible to show that for a polynomial function f of degree n, the two statements below are true.

· · · · · · · · · · · · · ▷

REMARK Remember that the *zeros* of a function of x are the x-values for which the function is zero.

1. The function f has, at most, n real zeros. (You will study this result in detail in the discussion of the Fundamental Theorem of Algebra in Section 2.5.)

2. The graph of f has, at most, $n - 1$ turning points. (Turning points, also called relative minima or relative maxima, are points at which the graph changes from increasing to decreasing or vice versa.)

Finding the zeros of a polynomial function is an important problem in algebra. There is a strong interplay between graphical and algebraic approaches to this problem.

Real Zeros of Polynomial Functions

When f is a polynomial function and a is a real number, the statements listed below are equivalent.

1. $x = a$ is a *zero* of the function f.
2. $x = a$ is a *solution* of the polynomial equation $f(x) = 0$.
3. $(x - a)$ is a *factor* of the polynomial $f(x)$.
4. $(a, 0)$ is an *x-intercept* of the graph of f.

EXAMPLE 3 **Finding Real Zeros of a Polynomial Function**

Find all real zeros of $f(x) = -2x^4 + 2x^2$. Then determine the maximum possible number of turning points of the graph of the function.

Solution To find the real zeros of the function, set $f(x)$ equal to zero and then solve for x.

$$-2x^4 + 2x^2 = 0 \qquad \text{Set } f(x) \text{ equal to 0.}$$
$$-2x^2(x^2 - 1) = 0 \qquad \text{Remove common monomial factor.}$$
$$-2x^2(x - 1)(x + 1) = 0 \qquad \text{Factor completely.}$$

So, the real zeros are $x = 0$, $x = 1$, and $x = -1$, and the corresponding x-intercepts occur at $(0, 0)$, $(1, 0)$, and $(-1, 0)$. The function is a fourth-degree polynomial, so the graph of f can have at most $4 - 1 = 3$ turning points. In this case, the graph of f has three turning points. Figure 2.10 shows the graph of f.

✓ *Checkpoint*  Audio-video solution in English & Spanish at LarsonPrecalculus.com

Find all real zeros of $f(x) = x^3 - 12x^2 + 36x$. Then determine the maximum possible number of turning points of the graph of the function. ◾

In Example 3, note that the factor $-2x^2$ yields the *repeated* zero $x = 0$. The exponent is even, so the graph touches the x-axis at $x = 0$.

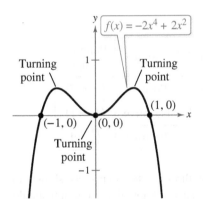

Turning point

Turning point

$f(x) = -2x^4 + 2x^2$

$(-1, 0)$ $(0, 0)$ $(1, 0)$

Turning point

Figure 2.10

▷ **ALGEBRA HELP** The solution to Example 3 uses polynomial factoring. To review the techniques for factoring polynomials, see Appendix A.3.

Repeated Zeros

A factor $(x - a)^k$, $k > 1$, yields a **repeated zero** $x = a$ of **multiplicity** k.

1. When k is odd, the graph *crosses* the x-axis at $x = a$.
2. When k is even, the graph *touches* the x-axis (but does not cross the x-axis) at $x = a$.

To graph polynomial functions, use the fact that a polynomial function can change signs only at its zeros. Between two consecutive zeros, a polynomial must be entirely positive or entirely negative. (This follows from the Intermediate Value Theorem, which you will study later in this section.) This means that when you put the real zeros of a polynomial function in order, they divide the real number line into intervals in which the function has no sign changes. These resulting intervals are **test intervals** in which you choose a representative x-value to determine whether the value of the polynomial function is positive (the graph lies above the x-axis) or negative (the graph lies below the x-axis).

▷ TECHNOLOGY Example 4 uses an *algebraic approach* to describe the graph of the function. A graphing utility can complement this approach. Remember to find a viewing window that shows all significant features of the graph. For instance, viewing window (a) illustrates all of the significant features of the function in Example 4, but viewing window (b) does not.

(a)

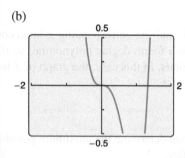

(b)

· · · · · · · · · · · · · ▷

· · REMARK   If you are unsure of the shape of a portion of the graph of a polynomial function, then plot some additional points. For instance, in Example 4, it is helpful to plot the additional point $\left(\frac{1}{2}, -\frac{5}{16}\right)$, as shown in Figure 2.12.

EXAMPLE 4   **Sketching the Graph of a Polynomial Function**

Sketch the graph of $f(x) = 3x^4 - 4x^3$.

Solution

1. *Apply the Leading Coefficient Test.*   The leading coefficient is positive and the degree is even, so you know that the graph eventually rises to the left and to the right (see Figure 2.11).

2. *Find the Real Zeros of the Function.*   Factoring $f(x) = 3x^4 - 4x^3$ as $f(x) = x^3(3x - 4)$ shows that the real zeros of f are $x = 0$ and $x = \frac{4}{3}$ (both of odd multiplicity). So, the x-intercepts occur at $(0, 0)$ and $\left(\frac{4}{3}, 0\right)$. Add these points to your graph, as shown in Figure 2.11.

3. *Plot a Few Additional Points.*   Use the zeros of the polynomial to find the test intervals. In each test interval, choose a representative x-value and evaluate the polynomial function, as shown in the table

Test Interval	Representative x-Value	Value of f	Sign	Point on Graph
$(-\infty, 0)$	-1	$f(-1) = 7$	Positive	$(-1, 7)$
$\left(0, \frac{4}{3}\right)$	1	$f(1) = -1$	Negative	$(1, -1)$
$\left(\frac{4}{3}, \infty\right)$	$\frac{3}{2}$	$f\left(\frac{3}{2}\right) = \frac{27}{16}$	Positive	$\left(\frac{3}{2}, \frac{27}{16}\right)$

4. *Draw the Graph.*   Draw a continuous curve through the points, as shown in Figure 2.12. Both zeros are of odd multiplicity, so you know that the graph should cross the x-axis at $x = 0$ and $x = \frac{4}{3}$.

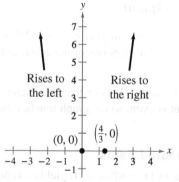

Figure 2.11

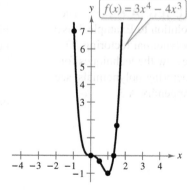

Figure 2.12

✓ *Checkpoint* ◄))) Audio-video solution in English & Spanish at LarsonPrecalculus.com

Sketch the graph of $f(x) = 2x^3 - 6x^2$.

A polynomial function is in **standard form** when its terms are in descending order of exponents from left to right. To avoid making a mistake when applying the Leading Coefficient Test, write the polynomial function in standard form first, if necessary.

EXAMPLE 5   Sketching the Graph of a Polynomial Function

Sketch the graph of $f(x) = -\frac{9}{2}x + 6x^2 - 2x^3$.

Solution

1. *Write in Standard Form and Apply the Leading Coefficient Test.* In standard form, the polynomial function is $f(x) = -2x^3 + 6x^2 - \frac{9}{2}x$. The leading coefficient is negative and the degree is odd, so you know that the graph eventually rises to the left and falls to the right (see Figure 2.13).

2. *Find the Real Zeros of the Function.* Factoring

$$f(x) = -2x^3 + 6x^2 - \frac{9}{2}x$$
$$= -\frac{1}{2}x(4x^2 - 12x + 9)$$
$$= -\frac{1}{2}x(2x - 3)^2$$

shows that the real zeros of f are $x = 0$ (odd multiplicity) and $x = \frac{3}{2}$ (even multiplicity). So, the x-intercepts occur at $(0, 0)$ and $\left(\frac{3}{2}, 0\right)$. Add these points to your graph, as shown in Figure 2.13.

3. *Plot a Few Additional Points.* Use the zeros of the polynomial to find the test intervals. In each test interval, choose a representative x-value and evaluate the polynomial function, as shown in the table.

> **REMARK** Observe in Example 5 that the sign of $f(x)$ is positive to the left of and negative to the right of the zero $x = 0$. Similarly, the sign of $f(x)$ is negative to the left and to the right of the zero $x = \frac{3}{2}$. This illustrates that (1) if the zero of a polynomial function is of *odd* multiplicity, then the graph crosses the x-axis at that zero, and (2) if the zero is of *even* multiplicity, then the graph touches the x-axis at that zero.

Test Interval	Representative x-Value	Value of f	Sign	Point on Graph
$(-\infty, 0)$	$-\frac{1}{2}$	$f\left(-\frac{1}{2}\right) = 4$	Positive	$\left(-\frac{1}{2}, 4\right)$
$\left(0, \frac{3}{2}\right)$	$\frac{1}{2}$	$f\left(\frac{1}{2}\right) = -1$	Negative	$\left(\frac{1}{2}, -1\right)$
$\left(\frac{3}{2}, \infty\right)$	2	$f(2) = -1$	Negative	$(2, -1)$

4. *Draw the Graph.* Draw a continuous curve through the points, as shown in Figure 2.14. From the multiplicities of the zeros, you know that the graph crosses the x-axis at $(0, 0)$ but does not cross the x-axis at $\left(\frac{3}{2}, 0\right)$.

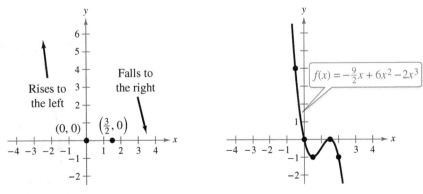

Figure 2.13   **Figure 2.14**

✓ *Checkpoint* 🔊))) Audio-video solution in English & Spanish at LarsonPrecalculus.com

Sketch the graph of $f(x) = -\frac{1}{4}x^4 + \frac{3}{2}x^3 - \frac{9}{4}x^2$.

The Intermediate Value Theorem

The **Intermediate Value Theorem** implies that if

$$(a, f(a)) \quad \text{and} \quad (b, f(b))$$

are two points on the graph of a polynomial function such that $f(a) \neq f(b)$, then for any number d between $f(a)$ and $f(b)$ there must be a number c between a and b such that $f(c) = d$. (See figure below.)

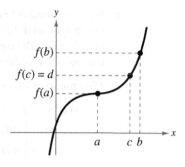

Intermediate Value Theorem

Let a and b be real numbers such that $a < b$. If f is a polynomial function such that $f(a) \neq f(b)$, then, in the interval $[a, b]$, f takes on every value between $f(a)$ and $f(b)$.

- ▷

REMARK Note that $f(a)$ and $f(b)$ must be of opposite signs in order to guarantee that a zero exists between them. If $f(a)$ and $f(b)$ are of the same sign, then it is inconclusive whether a zero exists between them.

One application of the Intermediate Value Theorem is in helping you locate real zeros of a polynomial function. If there exists a value $x = a$ at which a polynomial function is negative, and another value $x = b$ at which it is positive (or if it is positive when $x = a$ and negative when $x = b$), then the function has at least one real zero between these two values. For example, the function

$$f(x) = x^3 + x^2 + 1$$

is negative when $x = -2$ and positive when $x = -1$. So, it follows from the Intermediate Value Theorem that f must have a real zero somewhere between -2 and -1, as shown in the figure below.

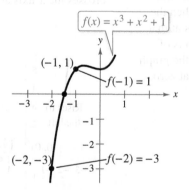

The function f must have a real zero somewhere between -2 and -1.

By continuing this line of reasoning, it is possible to approximate real zeros of a polynomial function to any desired accuracy. Example 6 further demonstrates this concept.

▷ TECHNOLOGY Using
the *table* feature of a
graphing utility can help
you approximate real zeros
of polynomial functions.
For instance, in Example 6,
construct a table that shows
function values for integer
values of *x*. Scrolling through
the table, notice that $f(-1)$ and
$f(0)$ differ in sign.

X	Y1
-2	-11
-1	-1
0	1
1	1
2	5
3	19
4	49

X=0

So, by the Intermediate
Value Theorem, the function
has a real zero between -1
and 0. Adjust your table to
show function values for
$-1 \le x \le 0$ using increments
of 0.1. Scrolling through this
table, notice that $f(-0.8)$ and
$f(-0.7)$ differ in sign.

X	Y1
-1	-1
-.9	-.539
-.8	-.152
-.7	.167
-.6	.424
-.5	.625
-.4	.776

X=-.7

So, the function has a real
zero between -0.8 and -0.7.
Repeating this process with
smaller increments, you should
obtain $x \approx -0.755$ as the real
zero of the function to three
decimal places, as stated in
Example 6. Use the *zero* or
root feature of the graphing
utility to confirm this result.

EXAMPLE 6 **Using the Intermediate Value Theorem**

Use the Intermediate Value Theorem to approximate the real zero of

$$f(x) = x^3 - x^2 + 1.$$

Solution Begin by computing a few function values.

x	-2	-1	0	1
$f(x)$	-11	-1	1	1

The value $f(-1)$ is negative and $f(0)$ is positive, so by the Intermediate Value Theorem, the function has a real zero between -1 and 0. To pinpoint this zero more closely, divide the interval $[-1, 0]$ into tenths and evaluate the function at each point. When you do this, you will find that

$$f(-0.8) = -0.152$$

and

$$f(-0.7) = 0.167.$$

So, f must have a real zero between -0.8 and -0.7, as shown in the figure. For a more accurate approximation, compute function values between $f(-0.8)$ and $f(-0.7)$ and apply the Intermediate Value Theorem again. Continue this process to verify that

$$x \approx -0.755$$

is an approximation (to the nearest thousandth) of the real zero of f.

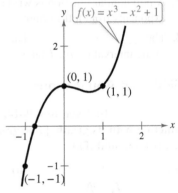

The function f has a real zero between -0.8 and -0.7.

✓ *Checkpoint* Audio-video solution in English & Spanish at LarsonPrecalculus.com

Use the Intermediate Value Theorem to approximate the real zero of

$$f(x) = x^3 - 3x^2 - 2.$$

Summarize (Section 2.2)

1. Explain how to use transformations to sketch graphs of polynomial functions *(page 124)*. For an example of sketching transformations of monomial functions, see Example 1.

2. Explain how to apply the Leading Coefficient Test *(page 125)*. For an example of applying the Leading Coefficient Test, see Example 2.

3. Explain how to find real zeros of polynomial functions and use them as sketching aids *(page 127)*. For examples involving finding real zeros of polynomial functions, see Examples 3–5.

4. Explain how to use the Intermediate Value Theorem to help locate real zeros of polynomial functions *(page 130)*. For an example of using the Intermediate Value Theorem, see Example 6.

2.2 Exercises

See CalcChat.com for tutorial help and worked-out solutions to odd-numbered exercises.

Vocabulary: Fill in the blanks.

1. The graph of a polynomial function is _____, which means that the graph has no breaks, holes, or gaps.

2. The _____ _____ _____ is used to determine the left-hand and right-hand behavior of the graph of a polynomial function.

3. A polynomial function of degree n has at most _____ real zeros and at most _____ turning points.

4. When $x = a$ is a zero of a polynomial function f, the three statements below are true.
    (a) $x = a$ is a _____ of the polynomial equation $f(x) = 0$.
    (b) _____ is a factor of the polynomial $f(x)$.
    (c) $(a, 0)$ is an _____ of the graph of f.

5. When a real zero $x = a$ of a polynomial function f is of even multiplicity, the graph of f _____ the x-axis at $x = a$, and when it is of odd multiplicity, the graph of f _____ the x-axis at $x = a$.

6. A factor $(x - a)^k$, $k > 1$, yields a _____ _____ $x = a$ of _____ k.

7. A polynomial function is written in _____ form when its terms are written in descending order of exponents from left to right.

8. The _____ _____ Theorem states that if f is a polynomial function such that $f(a) \neq f(b)$, then, in the interval $[a, b]$, f takes on every value between $f(a)$ and $f(b)$.

Skills and Applications

Matching In Exercises 9–14, match the polynomial function with its graph. [The graphs are labeled (a), (b), (c), (d), (e), and (f).]

9. $f(x) = -2x^2 - 5x$
10. $f(x) = 2x^3 - 3x + 1$
11. $f(x) = -\frac{1}{4}x^4 + 3x^2$
12. $f(x) = -\frac{1}{3}x^3 + x^2 - \frac{4}{3}$
13. $f(x) = x^4 + 2x^3$
14. $f(x) = \frac{1}{5}x^5 - 2x^3 + \frac{9}{5}x$

(a)

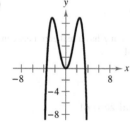

(b)

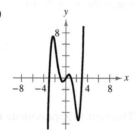

(c)

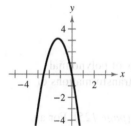

(d)

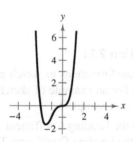

(e)

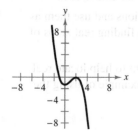

(f)

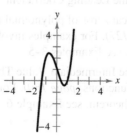

Sketching Transformations of Monomial Functions In Exercises 15–18, sketch the graph of $y = x^n$ and each transformation.

15. $y = x^3$
    (a) $f(x) = (x - 4)^3$
    (b) $f(x) = x^3 - 4$
    (c) $f(x) = -\frac{1}{4}x^3$
    (d) $f(x) = (x - 4)^3 - 4$

16. $y = x^5$
    (a) $f(x) = (x + 1)^5$
    (b) $f(x) = x^5 + 1$
    (c) $f(x) = 1 - \frac{1}{2}x^5$
    (d) $f(x) = -\frac{1}{2}(x + 1)^5$

17. $y = x^4$
    (a) $f(x) = (x + 3)^4$
    (b) $f(x) = x^4 - 3$
    (c) $f(x) = 4 - x^4$
    (d) $f(x) = \frac{1}{2}(x - 1)^4$
    (e) $f(x) = (2x)^4 + 1$
    (f) $f(x) = \left(\frac{1}{2}x\right)^4 - 2$

18. $y = x^6$
    (a) $f(x) = (x - 5)^6$
    (b) $f(x) = \frac{1}{8}x^6$
    (c) $f(x) = (x + 3)^6 - 4$
    (d) $f(x) = -\frac{1}{4}x^6 + 1$
    (e) $f(x) = \left(\frac{1}{4}x\right)^6 - 2$
    (f) $f(x) = (2x)^6 - 1$

Applying the Leading Coefficient Test In Exercises 19–28, describe the left-hand and right-hand behavior of the graph of the polynomial function.

19. $f(x) = 12x^3 + 4x$ **20.** $f(x) = 2x^2 - 3x + 1$
21. $g(x) = 5 - \frac{7}{2}x - 3x^2$ **22.** $h(x) = 1 - x^6$
23. $h(x) = 6x - 9x^3 + x^2$ **24.** $g(x) = 8 + \frac{1}{4}x^5 - x^4$
25. $f(x) = 9.8x^6 - 1.2x^3$
26. $h(x) = 1 - 0.5x^5 - 2.7x^3$
27. $f(s) = -\frac{7}{8}(s^3 + 5s^2 - 7s + 1)$
28. $h(t) = -\frac{4}{3}(t - 6t^3 + 2t^4 + 9)$

Using Technology In Exercises 29–32, use a graphing utility to graph the functions f and g in the same viewing window. Zoom out sufficiently far to show that the left-hand and right-hand behaviors of f and g appear identical.

29. $f(x) = 3x^3 - 9x + 1, \quad g(x) = 3x^3$
30. $f(x) = -\frac{1}{3}(x^3 - 3x + 2), \quad g(x) = -\frac{1}{3}x^3$
31. $f(x) = -(x^4 - 4x^3 + 16x), \quad g(x) = -x^4$
32. $f(x) = 3x^4 - 6x^2, \quad g(x) = 3x^4$

 Finding Real Zeros of a Polynomial Function In Exercises 33–48, (a) find all real zeros of the polynomial function, (b) determine whether the multiplicity of each zero is even or odd, (c) determine the maximum possible number of turning points of the graph of the function, and (d) use a graphing utility to graph the function and verify your answers.

33. $f(x) = x^2 - 36$ **34.** $f(x) = 81 - x^2$
35. $h(t) = t^2 - 6t + 9$ **36.** $f(x) = x^2 + 10x + 25$
37. $f(x) = \frac{1}{3}x^2 + \frac{1}{3}x - \frac{2}{3}$ **38.** $f(x) = \frac{1}{2}x^2 + \frac{5}{2}x - \frac{3}{2}$
39. $g(x) = 5x(x^2 - 2x - 1)$ **40.** $f(t) = t^2(3t^2 - 10t + 7)$
41. $f(x) = 3x^3 - 12x^2 + 3x$
42. $f(x) = x^4 - x^3 - 30x^2$
43. $g(t) = t^5 - 6t^3 + 9t$ **44.** $f(x) = x^5 + x^3 - 6x$
45. $f(x) = 3x^4 + 9x^2 + 6$ **46.** $f(t) = 2t^4 - 2t^2 - 40$
47. $g(x) = x^3 + 3x^2 - 4x - 12$
48. $f(x) = x^3 - 4x^2 - 25x + 100$

Using Technology In Exercises 49–52, (a) use a graphing utility to graph the function, (b) use the graph to approximate any x-intercepts of the graph, (c) find any real zeros of the function algebraically, and (d) compare the results of part (c) with those of part (b).

49. $y = 4x^3 - 20x^2 + 25x$
50. $y = 4x^3 + 4x^2 - 8x - 8$
51. $y = x^5 - 5x^3 + 4x$ **52.** $y = \frac{1}{5}x^5 - \frac{9}{5}x^3$

 Finding a Polynomial Function In Exercises 53–62, find a polynomial function that has the given zeros. (There are many correct answers.)

53. $0, 7$ **54.** $-2, 5$
55. $0, -2, -4$ **56.** $0, 1, 6$
57. $4, -3, 3, 0$ **58.** $-2, -1, 0, 1, 2$
59. $1 + \sqrt{2}, 1 - \sqrt{2}$ **60.** $4 + \sqrt{3}, 4 - \sqrt{3}$
61. $2, 2 + \sqrt{5}, 2 - \sqrt{5}$ **62.** $3, 2 + \sqrt{7}, 2 - \sqrt{7}$

 Finding a Polynomial Function In Exercises 63–70, find a polynomial of degree n that has the given zero(s). (There are many correct answers.)

	Zero(s)	Degree
63.	$x = -3$	$n = 2$
64.	$x = -\sqrt{2}, \sqrt{2}$	$n = 2$
65.	$x = -5, 0, 1$	$n = 3$
66.	$x = -2, 6$	$n = 3$
67.	$x = -5, 1, 2$	$n = 4$
68.	$x = -4, -1$	$n = 4$
69.	$x = 0, -\sqrt{3}, \sqrt{3}$	$n = 5$
70.	$x = -1, 4, 7, 8$	$n = 5$

 Sketching the Graph of a Polynomial Function In Exercises 71–84, sketch the graph of the function by (a) applying the Leading Coefficient Test, (b) finding the real zeros of the polynomial, (c) plotting sufficient solution points, and (d) drawing a continuous curve through the points.

71. $f(t) = \frac{1}{4}(t^2 - 2t + 15)$ **72.** $g(x) = -x^2 + 10x - 16$
73. $f(x) = x^3 - 25x$ **74.** $g(x) = -9x^2 + x^4$
75. $f(x) = -8 + \frac{1}{2}x^4$ **76.** $f(x) = 8 - x^3$
77. $f(x) = 3x^3 - 15x^2 + 18x$
78. $f(x) = -4x^3 + 4x^2 + 15x$
79. $f(x) = -5x^2 - x^3$ **80.** $f(x) = -48x^2 + 3x^4$
81. $f(x) = 9x^2(x + 2)^3$ **82.** $h(x) = \frac{1}{3}x^3(x - 4)^2$
83. $g(t) = -\frac{1}{4}(t - 2)^2(t + 2)^2$
84. $g(x) = \frac{1}{10}(x + 1)^2(x - 3)^3$

Using Technology In Exercises 85–88, use a graphing utility to graph the function. Use the *zero* or *root* feature to approximate the real zeros of the function. Then determine whether the multiplicity of each zero is even or odd.

85. $f(x) = x^3 - 16x$ **86.** $f(x) = \frac{1}{4}x^4 - 2x^2$
87. $g(x) = \frac{1}{5}(x + 1)^2(x - 3)(2x - 9)$
88. $h(x) = \frac{1}{5}(x + 2)^2(3x - 5)^2$

Using the Intermediate Value Theorem
In Exercises 89–92, (a) use the Intermediate Value Theorem and the *table* feature of a graphing utility to find intervals one unit in length in which the polynomial function is guaranteed to have a zero. (b) Adjust the table to approximate the zeros of the function to the nearest thousandth.

89. $f(x) = x^3 - 3x^2 + 3$

90. $f(x) = 0.11x^3 - 2.07x^2 + 9.81x - 6.88$

91. $g(x) = 3x^4 + 4x^3 - 3$ **92.** $h(x) = x^4 - 10x^2 + 3$

93. Maximum Volume You construct an open box from a square piece of material, 36 inches on a side, by cutting equal squares with sides of length x from the corners and turning up the sides (see figure).

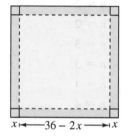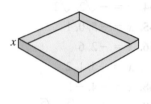

(a) Write a function V that represents the volume of the box.

(b) Determine the domain of the function V.

(c) Use a graphing utility to construct a table that shows the box heights x and the corresponding volumes $V(x)$. Use the table to estimate the dimensions that produce a maximum volume.

(d) Use the graphing utility to graph V and use the graph to estimate the value of x for which $V(x)$ is a maximum. Compare your result with that of part (c).

94. Maximum Volume You construct an open box with locking tabs from a square piece of material, 24 inches on a side, by cutting equal sections from the corners and folding along the dashed lines (see figure).

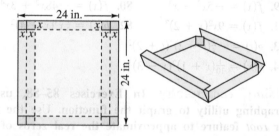

(a) Write a function V that represents the volume of the box.

(b) Determine the domain of the function V.

(c) Sketch a graph of the function and estimate the value of x for which $V(x)$ is a maximum.

95. Revenue The revenue R (in millions of dollars) for a software company from 2003 through 2016 can be modeled by

$$R = 6.212t^3 - 152.87t^2 + 990.2t - 414, \quad 3 \le t \le 16$$

where t represents the year, with $t = 3$ corresponding to 2003.

(a) Use a graphing utility to approximate any relative minima or maxima of the model over its domain.

(b) Use the graphing utility to approximate the intervals on which the revenue for the company is increasing and decreasing over its domain.

(c) Use the results of parts (a) and (b) to describe the company's revenue during this time period.

96. Revenue The revenue R (in millions of dollars) for a construction company from 2003 through 2010 can be modeled by

$$R = 0.1104t^4 - 5.152t^3 + 88.20t^2 - 654.8t + 1907, \quad 7 \le t \le 16$$

where t represents the year, with $t = 7$ corresponding to 2007.

(a) Use a graphing utility to approximate any relative minima or maxima of the model over its domain.

(b) Use the graphing utility to approximate the intervals on which the revenue for the company is increasing and decreasing over its domain.

(c) Use the results of parts (a) and (b) to describe the company's revenue during this time period.

97. Revenue The revenue R (in millions of dollars) for a beverage company is related to its advertising expense by the function

$$R = \frac{1}{100{,}000}(-x^3 + 600x^2), \quad 0 \le x \le 400$$

where x is the amount spent on advertising (in tens of thousands of dollars). Use the graph of this function to estimate the point on the graph at which the function is increasing most rapidly. This point is called the *point of diminishing returns* because any expense above this amount will yield less return per dollar invested in advertising.

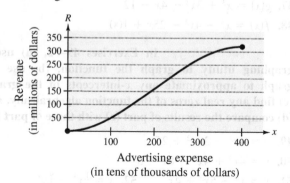

98. Arboriculture

The growth of a red oak tree is approximated by the function

$$G = -0.003t^3 + 0.137t^2 + 0.458t - 0.839,$$

$$2 \le t \le 34$$

where G is the height of the tree (in feet) and t is its age (in years).

(a) Use a graphing utility to graph the function.

(b) Estimate the age of the tree when it is growing most rapidly. This point is called the *point of diminishing returns* because the increase in size will be less with each additional year.

(c) Using calculus, the point of diminishing returns can be found by finding the vertex of the parabola

$$y = -0.009t^2 + 0.274t + 0.458.$$

Find the vertex of this parabola.

(d) Compare your results from parts (b) and (c).

Exploration

True or False? **In Exercises 99–102, determine whether the statement is true or false. Justify your answer.**

99. If the graph of a polynomial function falls to the right, then its leading coefficient is negative.

100. A fifth-degree polynomial function can have five turning points in its graph.

101. It is possible for a polynomial with an even degree to have a range of $(-\infty, \infty)$.

102. If f is a polynomial function of x such that $f(2) = -6$ and $f(6) = 6$, then f has at most one real zero between $x = 2$ and $x = 6$.

103. Modeling Polynomials Sketch the graph of a fourth-degree polynomial function that has a zero of multiplicity 2 and a negative leading coefficient. Sketch the graph of another polynomial function with the same characteristics except that the leading coefficient is positive.

104. Modeling Polynomials Sketch the graph of a fifth-degree polynomial function that has a zero of multiplicity 2 and a negative leading coefficient. Sketch the graph of another polynomial function with the same characteristics except that the leading coefficient is positive.

105. Graphical Reasoning Sketch the graph of the function $f(x) = x^4$. Explain how the graph of each function g differs (if it does) from the graph of f. Determine whether g is even, odd, or neither.

(a) $g(x) = f(x) + 2$ (b) $g(x) = f(x + 2)$

(c) $g(x) = f(-x)$ (d) $g(x) = -f(x)$

(e) $g(x) = f\left(\frac{1}{2}x\right)$ (f) $g(x) = \frac{1}{2}f(x)$

(g) $g(x) = f(x^{3/4})$ (h) $g(x) = (f \circ f)(x)$

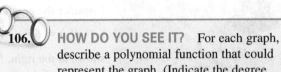

106. HOW DO YOU SEE IT? For each graph, describe a polynomial function that could represent the graph. (Indicate the degree of the function and the sign of its leading coefficient.)

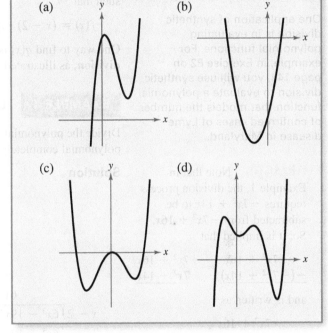

107. Think About It Use a graphing utility to graph the functions

$$y_1 = -\frac{1}{3}(x - 2)^5 + 1 \quad \text{and} \quad y_2 = \frac{3}{5}(x + 2)^5 - 3.$$

(a) Determine whether the graphs of y_1 and y_2 are increasing or decreasing. Explain.

(b) Will the graph of

$$g(x) = a(x - h)^5 + k$$

always be strictly increasing or strictly decreasing? If so, is this behavior determined by a, h, or k? Explain.

(c) Use a graphing utility to graph

$$f(x) = x^5 - 3x^2 + 2x + 1.$$

Use a graph and the result of part (b) to determine whether f can be written in the form $f(x) = a(x - h)^5 + k$. Explain.

2.3 Polynomial and Synthetic Division

- Use long division to divide polynomials by other polynomials.
- Use synthetic division to divide polynomials by binomials of the form $(x - k)$.
- Use the Remainder Theorem and the Factor Theorem.

One application of synthetic division is in evaluating polynomial functions. For example, in Exercise 82 on page 144, you will use synthetic division to evaluate a polynomial function that models the number of confirmed cases of Lyme disease in Maryland.

Long Division of Polynomials

Consider the graph of

$$f(x) = 6x^3 - 19x^2 + 16x - 4$$

shown at the right. Notice that one of the zeros of f is $x = 2$. This means that $(x - 2)$ is a factor of $f(x)$, and there exists a second-degree polynomial $q(x)$ such that

$$f(x) = (x - 2) \cdot q(x).$$

One way to find $q(x)$ is to use **long division,** as illustrated in Example 1.

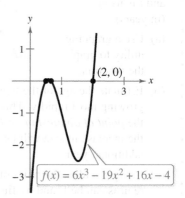

$f(x) = 6x^3 - 19x^2 + 16x - 4$

EXAMPLE 1   **Long Division of Polynomials**

Divide the polynomial $6x^3 - 19x^2 + 16x - 4$ by $x - 2$, and use the result to factor the polynomial completely.

Solution

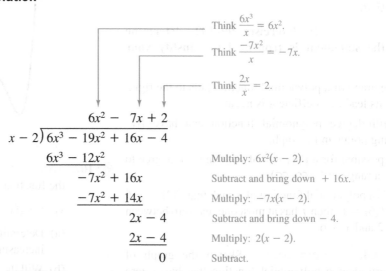

Think $\dfrac{6x^3}{x} = 6x^2$.

Think $\dfrac{-7x^2}{x} = -7x$.

Think $\dfrac{2x}{x} = 2$.

$$
\begin{array}{r}
6x^2 - 7x + 2 \\
x - 2 \overline{\smash{)}\ 6x^3 - 19x^2 + 16x - 4} \\
\underline{6x^3 - 12x^2} \\
-7x^2 + 16x \\
\underline{-7x^2 + 14x} \\
2x - 4 \\
\underline{2x - 4} \\
0
\end{array}
$$

Multiply: $6x^2(x - 2)$.
Subtract and bring down $+ 16x$.
Multiply: $-7x(x - 2)$.
Subtract and bring down $- 4$.
Multiply: $2(x - 2)$.
Subtract.

From this division, you have shown that

$$6x^3 - 19x^2 + 16x - 4 = (x - 2)(6x^2 - 7x + 2)$$

and by factoring the quadratic $6x^2 - 7x + 2$, you have

$$6x^3 - 19x^2 + 16x - 4 = (x - 2)(2x - 1)(3x - 2).$$

✓ **Checkpoint** ◀))) *Audio-video solution in English & Spanish at LarsonPrecalculus.com*

Divide the polynomial $9x^3 + 36x^2 - 49x - 196$ by $x + 4$, and use the result to factor the polynomial completely.

• • REMARK   Note that in Example 1, the division process requires $-7x^2 + 14x$ to be subtracted from $-7x^2 + 16x$. So, it is implied that

$$\dfrac{-7x^2 + 16x}{-(-7x^2 + 14x)} = \dfrac{-7x^2 + 16x}{7x^2 - 14x}$$

and is written as

$$
\begin{array}{r}
-7x^2 + 16x \\
\underline{-7x^2 + 14x} \\
2x.
\end{array}
$$

▷

• • REMARK   Note that the factorization found in Example 1 agrees with the graph of f above. The three x-intercepts occur at $(2, 0)$, $\left(\frac{1}{2}, 0\right)$, and $\left(\frac{2}{3}, 0\right)$.

▷

In Example 1, $x - 2$ is a factor of the polynomial

$$6x^3 - 19x^2 + 16x - 4$$

and the long division process produces a remainder of zero. Often, long division will produce a nonzero remainder. For example, when you divide $x^2 + 3x + 5$ by $x + 1$, you obtain a remainder of 3.

$$
\begin{array}{r}
x + 2 \quad \longleftarrow \text{Quotient} \\
x + 1 \,\overline{)\, x^2 + 3x + 5} \quad \longleftarrow \text{Dividend} \\
\underline{x^2 + \ x} \qquad\qquad \\
2x + 5 \qquad \\
\underline{2x + 2} \qquad \\
3 \quad \longleftarrow \text{Remainder}
\end{array}
$$

Divisor $\longrightarrow$

In fractional form, you can write this result as

$$
\underbrace{\frac{x^3 + 3x + 5}{\underbrace{x + 1}_{\text{Divisor}}}}_{\text{Dividend}} = \overbrace{x + 2}^{\text{Quotient}} + \frac{\overset{\text{Remainder}}{3}}{\underbrace{x + 1}_{\text{Divisor}}}.
$$

This implies that

$$x^2 + 3x + 5 = (x + 1)(x + 2) + 3 \qquad \text{Multiply each side by } (x + 1).$$

which illustrates a theorem called the **Division Algorithm.**

The Division Algorithm

If $f(x)$ and $d(x)$ are polynomials such that $d(x) \neq 0$, and the degree of $d(x)$ is less than or equal to the degree of $f(x)$, then there exist unique polynomials $q(x)$ and $r(x)$ such that

$$f(x) = d(x)q(x) + r(x)$$

$\underset{\text{Dividend}}{\uparrow} \quad \underset{\text{Divisor}}{\uparrow}\,\underset{\text{Quotient}}{\uparrow} \quad \underset{\text{Remainder}}{\uparrow}$

where $r(x) = 0$ or the degree of $r(x)$ is less than the degree of $d(x)$. If the remainder $r(x)$ is zero, then $d(x)$ *divides evenly* into $f(x)$.

Another way to write the Division Algorithm is

$$\frac{f(x)}{d(x)} = q(x) + \frac{r(x)}{d(x)}.$$

In the Division Algorithm, the rational expression $f(x)/d(x)$ is **improper** because the degree of $f(x)$ is greater than or equal to the degree of $d(x)$. On the other hand, the rational expression $r(x)/d(x)$ is **proper** because the degree of $r(x)$ is less than the degree of $d(x)$.

If necessary, follow these steps before you apply the Division Algorithm.

1. Write the terms of the dividend and divisor in descending powers of the variable.

2. Insert placeholders with zero coefficients for missing powers of the variable.

Note how Examples 2 and 3 apply these steps.

EXAMPLE 2 **Long Division of Polynomials**

Divide $x^3 - 1$ by $x - 1$. Check the result.

Solution There is no x^2-term or x-term in the dividend $x^3 - 1$, so you need to rewrite the dividend as $x^3 + 0x^2 + 0x - 1$ before you apply the Division Algorithm.

$$
\begin{array}{r}
x^2 + x + 1 \\
x - 1 \overline{)\, x^3 + 0x^2 + 0x - 1} \\
\underline{x^3 - x^2} \\
x^2 + 0x \\
\underline{x^2 - x} \\
x - 1 \\
\underline{x - 1} \\
0
\end{array}
$$

Multiply: $x^2(x - 1)$.

Subtract and bring down $0x$.

Multiply: $x(x - 1)$.

Subtract and bring down -1.

Multiply: $1(x - 1)$.

Subtract.

So, $x - 1$ divides evenly into $x^3 - 1$, and you can write

$$\frac{x^3 - 1}{x - 1} = x^2 + x + 1, \quad x \ne 1.$$

Check the result by multiplying.

$$(x - 1)(x^2 + x + 1) = x^3 + x^2 + x - x^2 - x - 1$$
$$= x^3 - 1$$

✓ **Checkpoint** ◀))) *Audio-video solution in English & Spanish at LarsonPrecalculus.com*

Divide $x^3 - 2x^2 - 9$ by $x - 3$. Check the result.

EXAMPLE 3 **Long Division of Polynomials**

See LarsonPrecalculus.com for an interactive version of this type of example.

Divide $-5x^2 - 2 + 3x + 2x^4 + 4x^3$ by $2x - 3 + x^2$. Check the result.

Solution Write the terms of the dividend and divisor in descending powers of x.

$$
\begin{array}{r}
2x^2 \qquad\; + 1 \\
x^2 + 2x - 3 \overline{)\, 2x^4 + 4x^3 - 5x^2 + 3x - 2} \\
\underline{2x^4 + 4x^3 - 6x^2} \\
x^2 + 3x - 2 \\
\underline{x^2 + 2x - 3} \\
x + 1
\end{array}
$$

Multiply: $2x^2(x^2 + 2x - 3)$.

Subtract and bring down $3x - 2$.

Multiply: $1(x^2 + 2x - 3)$.

Subtract.

Note that the first subtraction eliminated two terms from the dividend. When this happens, the quotient skips a term. You can write the result as

$$\frac{2x^4 + 4x^3 - 5x^2 + 3x - 2}{x^2 + 2x - 3} = 2x^2 + 1 + \frac{x + 1}{x^2 + 2x - 3}.$$

Check the result by multiplying.

$$(x^2 + 2x - 3)(2x^2 + 1) + x + 1 = 2x^4 + x^2 + 4x^3 + 2x - 6x^2 - 3 + x + 1$$
$$= 2x^4 + 4x^3 - 5x^2 + 3x - 2$$

✓ **Checkpoint** ◀))) *Audio-video solution in English & Spanish at LarsonPrecalculus.com*

Divide $-x^3 + 9x + 6x^4 - x^2 - 3$ by $1 + 3x$. Check the result.

Synthetic Division

For long division of polynomials by divisors of the form $x - k$, there is a shortcut called **synthetic division.** The pattern for synthetic division of a cubic polynomial is summarized below. (The pattern for higher-degree polynomials is similar.)

Synthetic Division (for a Cubic Polynomial)

To divide $ax^3 + bx^2 + cx + d$ by $x - k$, use this pattern.

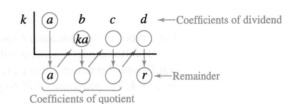

Vertical pattern: Add terms in columns.
Diagonal pattern: Multiply results by k.

This algorithm for synthetic division works only for divisors of the form $x - k$. Remember that $x + k = x - (-k)$.

EXAMPLE 4   **Using Synthetic Division**

Use synthetic division to divide

$$x^4 - 10x^2 - 2x + 4 \quad \text{by} \quad x + 3.$$

Solution   Begin by setting up an array. Include a zero for the missing x^3-term in the dividend.

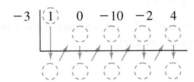

Then, use the synthetic division pattern by adding terms in columns and multiplying the results by -3.

$$
\begin{array}{r|rrrrr}
\text{Divisor: } x + 3 & \multicolumn{5}{l}{\text{Dividend: } x^4 - 10x^2 - 2x + 4} \\
\hline
-3 & 1 & 0 & -10 & -2 & 4 \\
 & & -3 & 9 & 3 & -3 \\
\hline
 & 1 & -3 & -1 & 1 & 1 \quad \longleftarrow \text{Remainder: } 1
\end{array}
$$

Quotient: $x^3 - 3x^2 - x + 1$

So, you have

$$\frac{x^4 - 10x^2 - 2x + 4}{x + 3} = x^3 - 3x^2 - x + 1 + \frac{1}{x + 3}.$$

✓ *Checkpoint* ◀))) *Audio-video solution in English & Spanish at LarsonPrecalculus.com*

Use synthetic division to divide $5x^3 + 8x^2 - x + 6$ by $x + 2$. ∎

The Remainder and Factor Theorems

The remainder obtained in the synthetic division process has an important interpretation, as described in the **Remainder Theorem.**

The Remainder Theorem

If a polynomial $f(x)$ is divided by $x - k$, then the remainder is

$r = f(k)$.

For a proof of the Remainder Theorem, see Proofs in Mathematics on page 193.

The Remainder Theorem tells you that synthetic division can be used to evaluate a polynomial function. That is, to evaluate a polynomial $f(x)$ when $x = k$, divide $f(x)$ by $x - k$. The remainder will be $f(k)$, as illustrated in Example 5.

EXAMPLE 5 **Using the Remainder Theorem**

Use the Remainder Theorem to evaluate

$$f(x) = 3x^3 + 8x^2 + 5x - 7$$

when $x = -2$. Check your answer.

Solution Using synthetic division gives the result below.

$$
\begin{array}{r|rrrr}
-2 & 3 & 8 & 5 & -7 \\
   &   & -6 & -4 & -2 \\
\hline
   & 3 & 2 & 1 & -9
\end{array}
$$

The remainder is $r = -9$, so

$$f(-2) = -9. \qquad {\scriptstyle r = f(k)}$$

This means that $(-2, -9)$ is a point on the graph of f. Check this by substituting $x = -2$ in the original function.

Check

$$
\begin{aligned}
f(-2) &= 3(-2)^3 + 8(-2)^2 + 5(-2) - 7 \\
&= 3(-8) + 8(4) - 10 - 7 \\
&= -24 + 32 - 10 - 7 \\
&= -9
\end{aligned}
$$

✓ **Checkpoint** ◄))) *Audio-video solution in English & Spanish at LarsonPrecalculus.com*

Use the Remainder Theorem to find each function value given

$$f(x) = 4x^3 + 10x^2 - 3x - 8.$$

Check your answer.

a. $f(-1)$ **b.** $f(4)$

c. $f\left(\frac{1}{2}\right)$ **d.** $f(-3)$

▷ **TECHNOLOGY** One way to evaluate a function with your graphing utility is to enter the function in the equation editor and use the *table* feature in *ask* mode. When you enter values in the X column of a table in *ask* mode, the corresponding function values are displayed in the function column.

Another important theorem is the **Factor Theorem,** stated below.

The Factor Theorem

A polynomial $f(x)$ has a factor $(x - k)$ if and only if $f(k) = 0$.

For a proof of the Factor Theorem, see Proofs in Mathematics on page 193.

Using the Factor Theorem, you can test whether a polynomial has $(x - k)$ as a factor by evaluating the polynomial at $x = k$. If the result is 0, then $(x - k)$ is a factor.

EXAMPLE 6 **Factoring a Polynomial: Repeated Division**

Show that $(x - 2)$ and $(x + 3)$ are factors of

$$f(x) = 2x^4 + 7x^3 - 4x^2 - 27x - 18.$$

Then find the remaining factors of $f(x)$.

Algebraic Solution

Using synthetic division with the factor $(x - 2)$ gives the result below.

$$
\begin{array}{r|rrrrr}
2 & 2 & 7 & -4 & -27 & -18 \\
  &   & 4 & 22 & 36 & 18 \\
\hline
  & 2 & 11 & 18 & 9 & 0
\end{array}
$$

0 remainder, so $f(2) = 0$ and $(x - 2)$ is a factor.

Take the result of this division and perform synthetic division again using the factor $(x + 3)$.

$$
\begin{array}{r|rrrr}
-3 & 2 & 11 & 18 & 9 \\
   &   & -6 & -15 & -9 \\
\hline
   & 2 & 5 & 3 & 0
\end{array}
$$

$\underbrace{}_{2x^2 + 5x + 3}$

0 remainder, so $f(-3) = 0$ and $(x + 3)$ is a factor.

The resulting quadratic expression factors as

$$2x^2 + 5x + 3 = (2x + 3)(x + 1)$$

so the complete factorization of $f(x)$ is

$$f(x) = (x - 2)(x + 3)(2x + 3)(x + 1).$$

Graphical Solution

The graph of $f(x) = 2x^4 + 7x^3 - 4x^2 - 27x - 18$ has four x-intercepts (see figure). These occur at $x = -3$, $x = -\frac{3}{2}$, $x = -1$, and $x = 2$. (Check this algebraically.) This implies that $(x + 3)$, $\left(x + \frac{3}{2}\right)$, $(x + 1)$, and $(x - 2)$ are factors of $f(x)$. [Note that $\left(x + \frac{3}{2}\right)$ and $(2x + 3)$ are equivalent factors because they both yield the same zero, $x = -\frac{3}{2}$.]

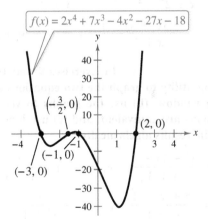

$f(x) = 2x^4 + 7x^3 - 4x^2 - 27x - 18$

$\left(-\frac{3}{2}, 0\right)$

$(2, 0)$

$(-1, 0)$

$(-3, 0)$

✓ **Checkpoint** 🔊)) *Audio-video solution in English & Spanish at LarsonPrecalculus.com*

Show that $(x + 3)$ is a factor of $f(x) = x^3 - 19x - 30$. Then find the remaining factors of $f(x)$.

Summarize   (Section 2.3)

1. Explain how to use long division to divide two polynomials (*pages 136 and 137*). For examples of long division of polynomials, see Examples 1–3.

2. Describe the algorithm for synthetic division (*page 139*). For an example of synthetic division, see Example 4.

3. State the Remainder Theorem and the Factor Theorem (*pages 140 and 141*). For an example of using the Remainder Theorem, see Example 5. For an example of using the Factor Theorem, see Example 6.

2.3 Exercises

See **CalcChat.com** for tutorial help and worked-out solutions to odd-numbered exercises.

Vocabulary

1. Two forms of the Division Algorithm are shown below. Identify and label each term or function.

$$f(x) = d(x)q(x) + r(x) \qquad \frac{f(x)}{d(x)} = q(x) + \frac{r(x)}{d(x)}$$

In Exercises 2–6, fill in the blanks.

2. In the Division Algorithm, the rational expression $r(x)/d(x)$ is _____ because the degree of $r(x)$ is less than the degree of $d(x)$.

3. In the Division Algorithm, the rational expression $f(x)/d(x)$ is _____ because the degree of $f(x)$ is greater than or equal to the degree of $d(x)$.

4. A shortcut for long division of polynomials is _____ _____, in which the divisor must be of the form $x - k$.

5. The _____ Theorem states that a polynomial $f(x)$ has a factor $(x - k)$ if and only if $f(k) = 0$.

6. The _____ Theorem states that if a polynomial $f(x)$ is divided by $x - k$, then the remainder is $r = f(k)$.

Skills and Applications

Using the Division Algorithm **In Exercises 7 and 8, use long division to verify that $y_1 = y_2$.**

7. $y_1 = \dfrac{x^2}{x + 2}$, $\quad y_2 = x - 2 + \dfrac{4}{x + 2}$

8. $y_1 = \dfrac{x^3 - 3x^2 + 4x - 1}{x + 3}$, $\quad y_2 = x^2 - 6x + 22 - \dfrac{67}{x + 3}$

Using Technology **In Exercises 9 and 10, (a) use a graphing utility to graph the two equations in the same viewing window, (b) use the graphs to verify that the expressions are equivalent, and (c) use long division to verify the results algebraically.**

9. $y_1 = \dfrac{x^2 + 2x - 1}{x + 3}$, $\quad y_2 = x - 1 + \dfrac{2}{x + 3}$

10. $y_1 = \dfrac{x^4 + x^2 - 1}{x^2 + 1}$, $\quad y_2 = x^2 - \dfrac{1}{x^2 + 1}$

Long Division of Polynomials **In Exercises 11–24, use long division to divide.**

11. $(2x^2 + 10x + 12) \div (x + 3)$

12. $(5x^2 - 17x - 12) \div (x - 4)$

13. $(4x^3 - 7x^2 - 11x + 5) \div (4x + 5)$

14. $(6x^3 - 16x^2 + 17x - 6) \div (3x - 2)$

15. $(x^4 + 5x^3 + 6x^2 - x - 2) \div (x + 2)$

16. $(x^3 + 4x^2 - 3x - 12) \div (x - 3)$

17. $(6x + 5) \div (x + 1)$ **18.** $(9x - 4) \div (3x + 2)$

19. $(x^3 - 9) \div (x^2 + 1)$ **20.** $(x^5 + 7) \div (x^4 - 1)$

21. $(3x + 2x^3 - 9 - 8x^2) \div (x^2 + 1)$

22. $(5x^3 - 16 - 20x + x^4) \div (x^2 - x - 3)$

23. $\dfrac{x^4}{(x - 1)^3}$ **24.** $\dfrac{2x^3 - 4x^2 - 15x + 5}{(x - 1)^2}$

Using Synthetic Division **In Exercises 25–44, use synthetic division to divide.**

25. $(2x^3 - 10x^2 + 14x - 24) \div (x - 4)$

26. $(5x^3 + 18x^2 + 7x - 6) \div (x + 3)$

27. $(6x^3 + 7x^2 - x + 26) \div (x - 3)$

28. $(2x^3 + 12x^2 + 14x - 3) \div (x + 4)$

29. $(4x^3 - 9x + 8x^2 - 18) \div (x + 2)$

30. $(9x^3 - 16x - 18x^2 + 32) \div (x - 2)$

31. $(-x^3 + 75x - 250) \div (x + 10)$

32. $(3x^3 - 16x^2 - 72) \div (x - 6)$

33. $(x^3 - 3x^2 + 5) \div (x - 4)$

34. $(5x^3 + 6x + 8) \div (x + 2)$

35. $\dfrac{10x^4 - 50x^3 - 800}{x - 6}$ **36.** $\dfrac{x^5 - 13x^4 - 120x + 80}{x + 3}$

37. $\dfrac{x^3 + 512}{x + 8}$ **38.** $\dfrac{x^3 - 729}{x - 9}$

39. $\dfrac{-3x^4}{x - 2}$ **40.** $\dfrac{-2x^5}{x + 2}$

41. $\dfrac{180x - x^4}{x - 6}$ **42.** $\dfrac{5 - 3x + 2x^2 - x^3}{x + 1}$

43. $\dfrac{4x^3 + 16x^2 - 23x - 15}{x + \frac{1}{2}}$ **44.** $\dfrac{3x^3 - 4x^2 + 5}{x - \frac{3}{2}}$

Using the Remainder Theorem In Exercises 45–50, write the function in the form $f(x) = (x - k)q(x) + r$ for the given value of k, and demonstrate that $f(k) = r$.

45. $f(x) = x^3 - x^2 - 10x + 7, \quad k = 3$

46. $f(x) = x^3 - 4x^2 - 10x + 8, \quad k = -2$

47. $f(x) = 15x^4 + 10x^3 - 6x^2 + 14, \quad k = -\frac{2}{3}$

48. $f(x) = 10x^3 - 22x^2 - 3x + 4, \quad k = \frac{1}{5}$

49. $f(x) = -4x^3 + 6x^2 + 12x + 4, \quad k = 1 - \sqrt{3}$

50. $f(x) = -3x^3 + 8x^2 + 10x - 8, \quad k = 2 + \sqrt{2}$

 Using the Remainder Theorem In Exercises 51–54, use the Remainder Theorem and synthetic division to find each function value. Verify your answers using another method.

51. $f(x) = 2x^3 - 7x + 3$

    (a) $f(1)$ (b) $f(-2)$ (c) $f(3)$ (d) $f(2)$

52. $g(x) = 2x^6 + 3x^4 - x^2 + 3$

    (a) $g(2)$ (b) $g(1)$ (c) $g(3)$ (d) $g(-1)$

53. $h(x) = x^3 - 5x^2 - 7x + 4$

    (a) $h(3)$ (b) $h\left(\frac{1}{2}\right)$ (c) $h(-2)$ (d) $h(-5)$

54. $f(x) = 4x^4 - 16x^3 + 7x^2 + 20$

    (a) $f(1)$ (b) $f(-2)$ (c) $f(5)$ (d) $f(-10)$

Using the Factor Theorem In Exercises 55–62, use synthetic division to show that x is a solution of the third-degree polynomial equation, and use the result to factor the polynomial completely. List all real solutions of the equation.

55. $x^3 + 6x^2 + 11x + 6 = 0, \quad x = -3$

56. $x^3 - 52x - 96 = 0, \quad x = -6$

57. $2x^3 - 15x^2 + 27x - 10 = 0, \quad x = \frac{1}{2}$

58. $48x^3 - 80x^2 + 41x - 6 = 0, \quad x = \frac{2}{3}$

59. $x^3 + 2x^2 - 3x - 6 = 0, \quad x = \sqrt{3}$

60. $x^3 + 2x^2 - 2x - 4 = 0, \quad x = \sqrt{2}$

61. $x^3 - 3x^2 + 2 = 0, \quad x = 1 + \sqrt{3}$

62. $x^3 - x^2 - 13x - 3 = 0, \quad x = 2 - \sqrt{5}$

 Factoring a Polynomial In Exercises 63–70, (a) verify the given factors of $f(x)$, (b) find the remaining factor(s) of $f(x)$, (c) use your results to write the complete factorization of $f(x)$, (d) list all real zeros of f, and (e) confirm your results by using a graphing utility to graph the function.

Function	Factors
63. $f(x) = 2x^3 + x^2 - 5x + 2$	$(x + 2), (x - 1)$
64. $f(x) = 3x^3 - x^2 - 8x - 4$	$(x + 1), (x - 2)$

Function	Factors
65. $f(x) = x^4 - 8x^3 + 9x^2$ $+ 38x - 40$	$(x - 5), (x + 2)$
66. $f(x) = 8x^4 - 14x^3 - 71x^2$ $- 10x + 24$	$(x + 2), (x - 4)$
67. $f(x) = 6x^3 + 41x^2 - 9x - 14$	$(2x + 1), (3x - 2)$
68. $f(x) = 10x^3 - 11x^2 - 72x + 45$	$(2x + 5), (5x - 3)$
69. $f(x) = 2x^3 - x^2 - 10x + 5$	$(2x - 1), \left(x + \sqrt{5}\right)$
70. $f(x) = x^3 + 3x^2 - 48x - 144$	$\left(x + 4\sqrt{3}\right), (x + 3)$

Approximating Zeros In Exercises 71–76, (a) use the *zero* or *root* feature of a graphing utility to approximate the zeros of the function accurate to three decimal places, (b) determine the exact value of one of the zeros, and (c) use synthetic division to verify your result from part (b), and then factor the polynomial completely.

71. $f(x) = x^3 - 2x^2 - 5x + 10$

72. $g(x) = x^3 + 3x^2 - 2x - 6$

73. $h(t) = t^3 - 2t^2 - 7t + 2$

74. $f(s) = s^3 - 12s^2 + 40s - 24$

75. $h(x) = x^5 - 7x^4 + 10x^3 + 14x^2 - 24x$

76. $g(x) = 6x^4 - 11x^3 - 51x^2 + 99x - 27$

Simplifying Rational Expressions In Exercises 77–80, simplify the rational expression by using long division or synthetic division.

77. $\dfrac{x^3 + x^2 - 64x - 64}{x + 8}$

78. $\dfrac{4x^3 - 8x^2 + x + 3}{2x - 3}$

79. $\dfrac{x^4 + 6x^3 + 11x^2 + 6x}{x^2 + 3x + 2}$

80. $\dfrac{x^4 + 9x^3 - 5x^2 - 36x + 4}{x^2 - 4}$

81. Profit A company that produces calculators estimates that the profit P (in dollars) from selling a specific model of calculator is given by

$$P = -152x^3 + 7545x^2 - 169,625, \quad 0 \le x \le 45$$

where x is the advertising expense (in tens of thousands of dollars). For this model of calculator, an advertising expense of \$400,000 ($x = 40$) results in a profit of \$2,174,375.

(a) Use a graphing utility to graph the profit function.

(b) Use the graph from part (a) to estimate another amount the company can spend on advertising that results in the same profit.

(c) Use synthetic division to confirm the result of part (b) algebraically.

82. Lyme Disease

The numbers N of confirmed cases of Lyme disease in Maryland from 2007 through 2014 are shown in the table, where t represents the year, with $t = 7$ corresponding to 2007. (*Source: Centers for Disease Control and Prevention*)

DATA	Year, t	Number, N
Spreadsheet at LarsonPrecalculus.com	7	2576
	8	1746
	9	1466
	10	1163
	11	938
	12	1113
	13	801
	14	957

(a) Use a graphing utility to create a scatter plot of the data.

(b) Use the *regression* feature of the graphing utility to find a *quartic* model for the data. (A quartic model has the form $at^4 + bt^3 + ct^2 + dt + e$, where a, b, c, d, and e are constant and t is variable.) Graph the model in the same viewing window as the scatter plot.

(c) Use the model to create a table of estimated values of N. Compare the model with the original data.

(d) Use synthetic division to confirm algebraically your estimated value for the year 2014.

Exploration

True or False? In Exercises 83–86, determine whether the statement is true or false. Justify your answer.

83. If $(7x + 4)$ is a factor of some polynomial function $f(x)$, then $\frac{4}{7}$ is a zero of f.

84. $(2x - 1)$ is a factor of the polynomial
$$6x^6 + x^5 - 92x^4 + 45x^3 + 184x^2 + 4x - 48.$$

85. The rational expression $\dfrac{x^3 + 2x^2 - 7x + 4}{x^2 - 4x - 12}$ is improper.

86. The equation
$$\frac{x^3 - 3x^2 + 4}{x + 1} = x^2 - 4x + 4$$
is true for all values of x.

Think About It In Exercises 87 and 88, perform the division. Assume that n is a positive integer.

87. $\dfrac{x^{3n} + 9x^{2n} + 27x^n + 27}{x^n + 3}$ **88.** $\dfrac{x^{3n} - 3x^{2n} + 5x^n - 6}{x^n - 2}$

89. Error Analysis Describe the error.

Use synthetic division to find the remainder when $x^2 + 3x - 5$ is divided by $x + 1$.

$$
\begin{array}{r|rrr}
1 & 1 & 3 & -5 \\
  &   & 1 & 4 \\
\hline
  & 1 & 4 & -1 \\
\end{array}
$$ ← Remainder: -1

90. **HOW DO YOU SEE IT?** The graph below shows a company's estimated profits for different advertising expenses. The company's actual profit was \$936,660 for an advertising expense of \$300,000.

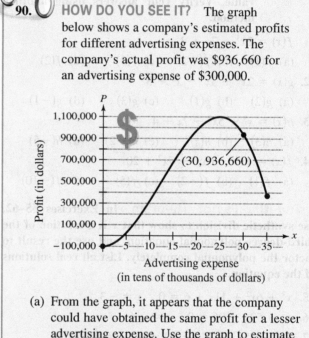

(a) From the graph, it appears that the company could have obtained the same profit for a lesser advertising expense. Use the graph to estimate this expense.

(b) The company's model is
$$P = -140.75x^3 + 5348.3x^2 - 76,560,$$
$$0 \le x \le 35$$
where P is the profit (in dollars) and x is the advertising expense (in tens of thousands of dollars). Explain how you could verify the lesser expense from part (a) algebraically.

Exploration In Exercises 91 and 92, find the constant c such that the denominator will divide evenly into the numerator.

91. $\dfrac{x^3 + 4x^2 - 3x + c}{x - 5}$ **92.** $\dfrac{x^5 - 2x^2 + x + c}{x + 2}$

93. Think About It Find the value of k such that $x - 4$ is a factor of $x^3 - kx^2 + 2kx - 8$.

2.4 Complex Numbers

Complex numbers are often used in electrical engineering. For example, in Exercise 87 on page 151, you will use complex numbers to find the impedance of an electrical circuit.

- Use the imaginary unit *i* to write complex numbers.
- Add, subtract, and multiply complex numbers.
- Use complex conjugates to write the quotient of two complex numbers in standard form.
- Find complex solutions of quadratic equations.

The Imaginary Unit *i*

You have learned that some quadratic equations have no real solutions. For example, the quadratic equation

$$x^2 + 1 = 0$$

has no real solution because there is no real number *x* that can be squared to produce -1. To overcome this deficiency, mathematicians created an expanded system of numbers using the **imaginary unit *i*,** defined as

$$i = \sqrt{-1} \qquad \text{Imaginary unit}$$

where $i^2 = -1$. By adding real numbers to real multiples of this imaginary unit, you obtain the set of **complex numbers.** Each complex number can be written in the **standard form *a* + *bi*.** For example, the standard form of the complex number $-5 + \sqrt{-9}$ is $-5 + 3i$ because

$$-5 + \sqrt{-9} = -5 + \sqrt{3^2(-1)} = -5 + 3\sqrt{-1} = -5 + 3i.$$

Definition of a Complex Number

Let *a* and *b* be real numbers. The number *a* + *bi* is a **complex number** written in **standard form.** The real number *a* is the **real part** and the number *bi* (where *b* is a real number) is the **imaginary part** of the complex number.

When $b = 0$, the number *a* + *bi* is a real number. When $b \neq 0$, the number *a* + *bi* is an **imaginary number.** A number of the form *bi*, where $b \neq 0$, is a **pure imaginary number.**

Every real number *a* can be written as a complex number using $b = 0$. That is, for every real number *a*, $a = a + 0i$. So, the set of real numbers is a subset of the set of complex numbers, as shown in the figure below.

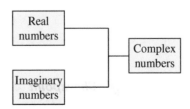

Equality of Complex Numbers

Two complex numbers *a* + *bi* and *c* + *di*, written in standard form, are equal to each other

$$a + bi = c + di \qquad \text{Equality of two complex numbers}$$

if and only if $a = c$ and $b = d$.

Operations with Complex Numbers

To add (or subtract) two complex numbers, add (or subtract) the real and imaginary parts of the numbers separately.

Addition and Subtraction of Complex Numbers

For two complex numbers $a + bi$ and $c + di$ written in standard form, the sum and difference are

$Sum:$ $(a + bi) + (c + di) = (a + c) + (b + d)i$

$Difference:$ $(a + bi) - (c + di) = (a - c) + (b - d)i.$

The **additive identity** in the complex number system is zero (the same as in the real number system). Furthermore, the **additive inverse** of the complex number $a + bi$ is

$-(a + bi) = -a - bi.$ Additive inverse

So, you have $(a + bi) + (-a - bi) = 0 + 0i = 0$.

| EXAMPLE 1 | **Adding and Subtracting Complex Numbers** |

a. $(4 + 7i) + (1 - 6i) = 4 + 7i + 1 - 6i$ Remove parentheses.

$= (4 + 1) + (7 - 6)i$ Group like terms.

$= 5 + i$ Write in standard form.

b. $(1 + 2i) + (3 - 2i) = 1 + 2i + 3 - 2i$ Remove parentheses.

$= (1 + 3) + (2 - 2)i$ Group like terms.

$= 4 + 0i$ Simplify.

$= 4$ Write in standard form.

c. $3i - (-2 + 3i) - (2 + 5i) = 3i + 2 - 3i - 2 - 5i$

$= (2 - 2) + (3 - 3 - 5)i$

$= 0 - 5i$

$= -5i$

d. $(3 + 2i) + (4 - i) - (7 + i) = 3 + 2i + 4 - i - 7 - i$

$= (3 + 4 - 7) + (2 - 1 - 1)i$

$= 0 + 0i$

$= 0$

> **REMARK** Note that the sum of two complex numbers can be a real number.

✓ *Checkpoint*))) *Audio-video solution in English & Spanish at LarsonPrecalculus.com*

Perform each operation and write the result in standard form.

a. $(7 + 3i) + (5 - 4i)$

b. $(3 + 4i) - (5 - 3i)$

c. $2i + (-3 - 4i) - (-3 - 3i)$

d. $(5 - 3i) + (3 + 5i) - (8 + 2i)$

Many of the properties of real numbers are valid for complex numbers as well. Here are some examples.

Associative Properties of Addition and Multiplication

Commutative Properties of Addition and Multiplication

Distributive Property of Multiplication Over Addition

Note the use of these properties when multiplying two complex numbers.

$$(a + bi)(c + di) = a(c + di) + bi(c + di) \qquad \text{Distributive Property}$$

$$= ac + (ad)i + (bc)i + (bd)i^2 \qquad \text{Distributive Property}$$

$$= ac + (ad)i + (bc)i + (bd)(-1) \qquad i^2 = -1$$

$$= ac - bd + (ad)i + (bc)i \qquad \text{Commutative Property}$$

$$= (ac - bd) + (ad + bc)i \qquad \text{Associative Property}$$

▷ **ALGEBRA HELP** To review the FOIL method, see Appendix A.3.

The procedure shown above is similar to multiplying two binomials and combining like terms, as in the FOIL method. So, you do not need to memorize this procedure.

EXAMPLE 2 **Multiplying Complex Numbers**

See LarsonPrecalculus.com for an interactive version of this type of example.

a. $4(-2 + 3i) = 4(-2) + 4(3i) \qquad$ Distributive Property

$$= -8 + 12i \qquad \text{Simplify.}$$

b. $(2 - i)(4 + 3i) = 8 + 6i - 4i - 3i^2 \qquad$ FOIL Method

$$= 8 + 6i - 4i - 3(-1) \qquad i^2 = -1$$

$$= (8 + 3) + (6 - 4)i \qquad \text{Group like terms.}$$

$$= 11 + 2i \qquad \text{Write in standard form.}$$

c. $(3 + 2i)(3 - 2i) = 9 - 6i + 6i - 4i^2 \qquad$ FOIL Method

$$= 9 - 6i + 6i - 4(-1) \qquad i^2 = -1$$

$$= 9 + 4 \qquad \text{Simplify.}$$

$$= 13 \qquad \text{Write in standard form.}$$

d. $(3 + 2i)^2 = (3 + 2i)(3 + 2i) \qquad$ Square of a binomial

$$= 9 + 6i + 6i + 4i^2 \qquad \text{FOIL Method}$$

$$= 9 + 6i + 6i + 4(-1) \qquad i^2 = -1$$

$$= 9 + 12i - 4 \qquad \text{Simplify.}$$

$$= 5 + 12i \qquad \text{Write in standard form.}$$

✓ **Checkpoint** ◀))) Audio-video solution in English & Spanish at LarsonPrecalculus.com

Perform each operation and write the result in standard form.

a. $-5(3 - 2i)$

b. $(2 - 4i)(3 + 3i)$

c. $(4 + 5i)(4 - 5i)$

d. $(4 + 2i)^2$

Complex Conjugates

Notice in Example 2(c) that the product of two complex numbers can be a real number. This occurs with pairs of complex numbers of the form $a + bi$ and $a - bi$, called **complex conjugates.**

$$(a + bi)(a - bi) = a^2 - abi + abi - b^2i^2$$
$$= a^2 - b^2(-1)$$
$$= a^2 + b^2$$

> **REMARK** Recall that the product of $a - b\sqrt{m}$ or $a + b\sqrt{m}$ and its conjugate is rational. Similarly, the product of a complex number and its conjugate is real.

EXAMPLE 3 **Multiplying Conjugates**

Multiply each complex number by its complex conjugate.

a. $1 + i$ **b.** $4 - 3i$

Solution

a. The complex conjugate of $1 + i$ is $1 - i$.

$$(1 + i)(1 - i) = 1^2 - i^2 = 1 - (-1) = 2$$

b. The complex conjugate of $4 - 3i$ is $4 + 3i$.

$$(4 - 3i)(4 + 3i) = 4^2 - (3i)^2 = 16 - 9i^2 = 16 - 9(-1) = 25$$

✓ **Checkpoint** Audio-video solution in English & Spanish at LarsonPrecalculus.com

Multiply each complex number by its complex conjugate.

a. $3 + 6i$ **b.** $2 - 5i$

To write the quotient of $a + bi$ and $c + di$ in standard form, where c and d are not both zero, multiply the numerator and denominator by the complex conjugate of the *denominator* to obtain

$$\frac{a + bi}{c + di} = \frac{a + bi}{c + di}\left(\frac{c - di}{c - di}\right) = \frac{(ac + bd) + (bc - ad)i}{c^2 + d^2} = \frac{ac + bd}{c^2 + d^2} + \left(\frac{bc - ad}{c^2 + d^2}\right)i.$$

> **REMARK** Note that when you multiply a quotient of complex numbers by
> $$\frac{c - di}{c - di}$$
> you are multiplying the quotient by a form of 1. So, you are not changing the original expression, you are only writing an equivalent expression.

EXAMPLE 4 **A Quotient of Complex Numbers in Standard Form**

$$\frac{2 + 3i}{4 - 2i} = \frac{2 + 3i}{4 - 2i}\left(\frac{4 + 2i}{4 + 2i}\right)$$ Multiply numerator and denominator by complex conjugate of denominator.

$$= \frac{8 + 4i + 12i + 6i^2}{16 - 4i^2}$$ Expand.

$$= \frac{8 - 6 + 16i}{16 + 4}$$ $i^2 = -1$

$$= \frac{2 + 16i}{20}$$ Simplify.

$$= \frac{1}{10} + \frac{4}{5}i$$ Write in standard form.

✓ **Checkpoint** Audio-video solution in English & Spanish at LarsonPrecalculus.com

Write $\dfrac{2 + i}{2 - i}$ in standard form.

Complex Solutions of Quadratic Equations

You can write a number such as $\sqrt{-3}$ in standard form by factoring out $i = \sqrt{-1}$.

$$\sqrt{-3} = \sqrt{3(-1)} = \sqrt{3}\sqrt{-1} = \sqrt{3}i$$

The number $\sqrt{3}i$ is the *principal square root* of -3.

> **REMARK** The definition of principal square root uses the rule
>
> $$\sqrt{ab} = \sqrt{a}\sqrt{b}$$
>
> for $a > 0$ and $b < 0$. This rule is not valid when *both* a and b are negative. For example,
>
> $$\sqrt{-5}\sqrt{-5} = \sqrt{5(-1)}\sqrt{5(-1)}$$
> $$= \sqrt{5}i\sqrt{5}i$$
> $$= \sqrt{25}i^2$$
> $$= 5i^2$$
> $$= -5$$
>
> whereas
>
> $$\sqrt{(-5)(-5)} = \sqrt{25} = 5.$$
>
> Be sure to convert complex numbers to standard form *before* performing any operations.

Principal Square Root of a Negative Number

When a is a positive real number, the **principal square root** of $-a$ is defined as

$$\sqrt{-a} = \sqrt{a}i.$$

EXAMPLE 5 **Writing Complex Numbers in Standard Form**

a. $\sqrt{-3}\sqrt{-12} = \sqrt{3}i\sqrt{12}i = \sqrt{36}i^2 = 6(-1) = -6$

b. $\sqrt{-48} - \sqrt{-27} = \sqrt{48}i - \sqrt{27}i = 4\sqrt{3}i - 3\sqrt{3}i = \sqrt{3}i$

c. $\left(-1 + \sqrt{-3}\right)^2 = \left(-1 + \sqrt{3}i\right)^2 = 1 - 2\sqrt{3}i + 3(-1)$
$$= -2 - 2\sqrt{3}i$$

✓ **Checkpoint** ◀)) Audio-video solution in English & Spanish at LarsonPrecalculus.com

Write $\sqrt{-14}\sqrt{-2}$ in standard form.

EXAMPLE 6 **Complex Solutions of a Quadratic Equation**

Solve $3x^2 - 2x + 5 = 0$.

Solution

$$x = \frac{-(-2) \pm \sqrt{(-2)^2 - 4(3)(5)}}{2(3)} \qquad \text{Quadratic Formula}$$

$$= \frac{2 \pm \sqrt{-56}}{6} \qquad \text{Simplify.}$$

$$= \frac{2 \pm 2\sqrt{14}i}{6} \qquad \text{Write } \sqrt{-56} \text{ in standard form.}$$

$$= \frac{1}{3} \pm \frac{\sqrt{14}}{3}i \qquad \text{Write solution in standard form.}$$

> ▷ **ALGEBRA HELP** To review the Quadratic Formula, see Appendix A.5.

✓ **Checkpoint** ◀)) Audio-video solution in English & Spanish at LarsonPrecalculus.com

Solve $8x^2 + 14x + 9 = 0$.

Summarize (Section 2.4)

1. Explain how to write complex numbers using the imaginary unit i *(page 145)*.

2. Explain how to add, subtract, and multiply complex numbers *(pages 146 and 147, Examples 1 and 2)*.

3. Explain how to use complex conjugates to write the quotient of two complex numbers in standard form *(page 148, Example 4)*.

4. Explain how to find complex solutions of a quadratic equation *(page 149, Example 6)*.

2.4 Exercises

See CalcChat.com for tutorial help and worked-out solutions to odd-numbered exercises.

Vocabulary: Fill in the blanks.

1. A _____ number has the form $a + bi$, where $a \neq 0$, $b = 0$.
2. An _____ number has the form $a + bi$, where $a \neq 0$, $b \neq 0$.
3. A _____ _____ number has the form $a + bi$, where $a = 0$, $b \neq 0$.
4. The imaginary unit i is defined as $i =$ _____, where $i^2 =$ _____.
5. When a is a positive real number, the _____ _____ root of $-a$ is defined as $\sqrt{-a} = \sqrt{a}\,i$.
6. The numbers $a + bi$ and $a - bi$ are called _____ _____, and their product is a real number $a^2 + b^2$.

Skills and Applications

Equality of Complex Numbers In Exercises 7–10, find real numbers a and b such that the equation is true.

7. $a + bi = 9 + 8i$
8. $a + bi = 10 - 5i$
9. $(a - 2) + (b + 1)i = 6 + 5i$
10. $(a + 2) + (b - 3)i = 4 + 7i$

 Writing a Complex Number in Standard Form In Exercises 11–22, write the complex number in standard form.

11. $2 + \sqrt{-25}$
12. $4 + \sqrt{-49}$
13. $1 - \sqrt{-12}$
14. $2 - \sqrt{-18}$
15. $\sqrt{-40}$
16. $\sqrt{-27}$
17. 23
18. 50
19. $-6i + i^2$
20. $-2i^2 + 4i$
21. $\sqrt{-0.04}$
22. $\sqrt{-0.0025}$

Adding or Subtracting Complex Numbers In Exercises 23–30, perform the operation and write the result in standard form.

23. $(5 + i) + (2 + 3i)$
24. $(13 - 2i) + (-5 + 6i)$
25. $(9 - i) - (8 - i)$
26. $(3 + 2i) - (6 + 13i)$
27. $\left(-2 + \sqrt{-8}\right) + \left(5 - \sqrt{-50}\right)$
28. $\left(8 + \sqrt{-18}\right) - \left(4 + 3\sqrt{2}\,i\right)$
29. $13i - (14 - 7i)$
30. $25 + (-10 + 11i) + 15i$

Multiplying Complex Numbers In Exercises 31–38, perform the operation and write the result in standard form.

31. $(1 + i)(3 - 2i)$
32. $(7 - 2i)(3 - 5i)$
33. $12i(1 - 9i)$
34. $-8i(9 + 4i)$
35. $\left(\sqrt{2} + 3i\right)\left(\sqrt{2} - 3i\right)$
36. $\left(4 + \sqrt{7}\,i\right)\left(4 - \sqrt{7}\,i\right)$
37. $(6 + 7i)^2$
38. $(5 - 4i)^2$

Multiplying Conjugates In Exercises 39–46, write the complex conjugate of the complex number. Then multiply the number by its complex conjugate.

39. $9 + 2i$
40. $8 - 10i$
41. $-1 - \sqrt{5}\,i$
42. $-3 + \sqrt{2}\,i$
43. $\sqrt{-20}$
44. $\sqrt{-15}$
45. $\sqrt{6}$
46. $1 + \sqrt{8}$

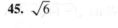 **A Quotient of Complex Numbers in Standard Form** In Exercises 47–54, write the quotient in standard form.

47. $\dfrac{2}{4 - 5i}$
48. $\dfrac{13}{1 - i}$
49. $\dfrac{5 + i}{5 - i}$
50. $\dfrac{6 - 7i}{1 - 2i}$
51. $\dfrac{9 - 4i}{i}$
52. $\dfrac{8 + 16i}{2i}$
53. $\dfrac{3i}{(4 - 5i)^2}$
54. $\dfrac{5i}{(2 + 3i)^2}$

 Performing Operations with Complex Numbers In Exercises 55–58, perform the operation and write the result in standard form.

55. $\dfrac{2}{1 + i} - \dfrac{3}{1 - i}$
56. $\dfrac{2i}{2 + i} + \dfrac{5}{2 - i}$
57. $\dfrac{i}{3 - 2i} + \dfrac{2i}{3 + 8i}$
58. $\dfrac{1 + i}{i} - \dfrac{3}{4 - i}$

 Writing a Complex Number in Standard Form In Exercises 59–66, write the complex number in standard form.

59. $\sqrt{-6}\,\sqrt{-2}$
60. $\sqrt{-5}\,\sqrt{-10}$
61. $\left(\sqrt{-15}\right)^2$
62. $\left(\sqrt{-75}\right)^2$
63. $\sqrt{-8} + \sqrt{-50}$
64. $\sqrt{-45} - \sqrt{-5}$
65. $\left(3 + \sqrt{-5}\right)\left(7 - \sqrt{-10}\right)$
66. $\left(2 - \sqrt{-6}\right)^2$

Complex Solutions of a Quadratic Equation In Exercises 67–76, use the Quadratic Formula to solve the quadratic equation.

67. $x^2 - 2x + 2 = 0$ 68. $x^2 + 6x + 10 = 0$

69. $4x^2 + 16x + 17 = 0$ 70. $9x^2 - 6x + 37 = 0$

71. $4x^2 + 16x + 21 = 0$ 72. $16t^2 - 4t + 3 = 0$

73. $\frac{3}{2}x^2 - 6x + 9 = 0$ 74. $\frac{7}{8}x^2 - \frac{3}{4}x + \frac{5}{16} = 0$

75. $1.4x^2 - 2x + 10 = 0$ 76. $4.5x^2 - 3x + 12 = 0$

Simplifying a Complex Number In Exercises 77–86, simplify the complex number and write it in standard form.

77. $-6i^3 + i^2$ 78. $4i^2 - 2i^3$

79. $-14i^5$ 80. $(-i)^3$

81. $\left(\sqrt{-72}\right)^3$ 82. $\left(\sqrt{-2}\right)^6$

83. $\dfrac{1}{i^3}$ 84. $\dfrac{1}{(2i)^3}$

85. $(3i)^4$ 86. $(-i)^6$

• • **87. Impedance of a Circuit** • • • • • • • • • •

The opposition to current in an electrical circuit is called its impedance. The impedance z in a parallel circuit with two pathways satisfies the equation

$$\frac{1}{z} = \frac{1}{z_1} + \frac{1}{z_2}$$

where z_1 is the impedance (in ohms) of pathway 1 and z_2 is the impedance (in ohms) of pathway 2.

(a) The impedance of each pathway in a parallel circuit is found by adding the impedances of all components in the pathway. Use the table to find z_1 and z_2

	Resistor	Inductor	Capacitor
Symbol	—⩗⩗⩗— $a\ \Omega$	—⟋⟋⟋⟋— $b\ \Omega$	—⊣⊢— $c\ \Omega$
Impedance	a	bi	$-ci$

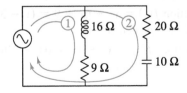

(b) Find the impedance z.

88. Cube of a Complex Number Cube each complex number.

(a) $-1 + \sqrt{3}i$ (b) $-1 - \sqrt{3}i$

Exploration

True or False? In Exercises 89–92, determine whether the statement is true or false. Justify your answer.

89. The sum of two complex numbers is always a real number.

90. There is no complex number that is equal to its complex conjugate.

91. $-i\sqrt{6}$ is a solution of $x^4 - x^2 + 14 = 56$.

92. $i^{44} + i^{150} - i^{74} - i^{109} + i^{61} = -1$

93. Pattern Recognition Find the missing values.

$i^1 = i$	$i^2 = -1$	$i^3 = -i$	$i^4 = 1$
$i^5 = $	$i^6 = $	$i^7 = $	$i^8 = $
$i^9 = $	$i^{10} = $	$i^{11} = $	$i^{12} = $

What pattern do you see? Write a brief description of how you would find i raised to any positive integer power.

94. **HOW DO YOU SEE IT?** The coordinate system shown below is called the complex plane. In the complex plane, the point (a, b) corresponds to the complex number $a + bi$.

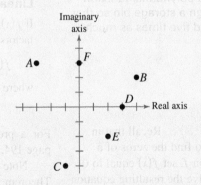

Match each complex number with its corresponding point.

(i) 3   (ii) $3i$ (iii) $4 + 2i$

(iv) $2 - 2i$ (v) $-3 + 3i$ (vi) $-1 - 4i$

95. Error Analysis Describe the error.

$$\sqrt{-6}\sqrt{-6} = \sqrt{(-6)(-6)} = \sqrt{36} = 6 \quad \times$$

96. Proof Prove that the complex conjugate of the product of two complex numbers $a_1 + b_1i$ and $a_2 + b_2i$ is the product of their complex conjugates.

97. Proof Prove that the complex conjugate of the sum of two complex numbers $a_1 + b_1i$ and $a_2 + b_2i$ is the sum of their complex conjugates.

2.5 Zeros of Polynomial Functions

Finding zeros of polynomial functions is an important part of solving many real-life problems. For example, in Exercise 105 on page 164, you will use the zeros of a polynomial function to redesign a storage bin so that it can hold five times as much food.

- Use the Fundamental Theorem of Algebra to determine numbers of zeros of polynomial functions.
- Find rational zeros of polynomial functions.
- Find complex zeros using conjugate pairs.
- Find zeros of polynomials by factoring.
- Use Descartes's Rule of Signs and the Upper and Lower Bound Rules to find zeros of polynomials.
- Find zeros of polynomials in real-life applications.

The Fundamental Theorem of Algebra

In the complex number system, every nth-degree polynomial function has *precisely n* zeros. This important result is derived from the **Fundamental Theorem of Algebra,** first proved by German mathematician Carl Friedrich Gauss (1777–1855).

> ### The Fundamental Theorem of Algebra
> If $f(x)$ is a polynomial of degree n, where $n > 0$, then f has at least one zero in the complex number system.

Using the Fundamental Theorem of Algebra and the equivalence of zeros and factors, you obtain the **Linear Factorization Theorem.**

> ### Linear Factorization Theorem
> If $f(x)$ is a polynomial of degree n, where $n > 0$, then $f(x)$ has precisely n linear factors
>
> $$f(x) = a_n(x - c_1)(x - c_2) \cdots (x - c_n)$$
>
> where $c_1, c_2, \ldots, c_n$ are complex numbers.

··REMARK Recall that in order to find the zeros of a function f, set $f(x)$ equal to 0 and solve the resulting equation for x. For instance, the function in Example 1(a) has a zero at $x = 2$ because

$$x - 2 = 0$$
$$x = 2.$$

For a proof of the Linear Factorization Theorem, see Proofs in Mathematics on page 194.

Note that the Fundamental Theorem of Algebra and the Linear Factorization Theorem tell you only that the zeros or factors of a polynomial exist, not how to find them. Such theorems are called **existence theorems.**

EXAMPLE 1 **Zeros of Polynomial Functions**

See LarsonPrecalculus.com for an interactive version of this type of example.

a. The first-degree polynomial function $f(x) = x - 2$ has exactly *one* zero: $x = 2$.

b. The second-degree polynomial function $f(x) = x^2 - 6x + 9 = (x - 3)(x - 3)$ has exactly *two* zeros: $x = 3$ and $x = 3$ (a *repeated zero*).

c. The third-degree polynomial function $f(x) = x^3 + 4x = x(x - 2i)(x + 2i)$ has exactly *three* zeros: $x = 0$, $x = 2i$, and $x = -2i$.

✓ **Checkpoint**))) Audio-video solution in English & Spanish at LarsonPrecalculus.com

Determine the number of zeros of the polynomial function $f(x) = x^4 - 1$.

The Rational Zero Test

The **Rational Zero Test** relates the possible rational zeros of a polynomial (having integer coefficients) to the leading coefficient and to the constant term of the polynomial.

The Rational Zero Test

If the polynomial

$$f(x) = a_n x^n + a_{n-1} x^{n-1} + \cdots + a_2 x^2 + a_1 x + a_0$$

has *integer* coefficients, then every rational zero of f has the form

$$\text{Rational zero} = \frac{p}{q}$$

where p and q have no common factors other than 1, and

   $p = $ a factor of the constant term a_0

   $q = $ a factor of the leading coefficient a_n.

Although they were not contemporaries, French mathematician Jean Le Rond d'Alembert (1717–1783) worked independently of Carl Friedrich Gauss in trying to prove the Fundamental Theorem of Algebra. His efforts were such that, in France, the Fundamental Theorem of Algebra is frequently known as d'Alembert's Theorem.

To use the Rational Zero Test, you should first list all rational numbers whose numerators are factors of the constant term and whose denominators are factors of the leading coefficient.

$$\text{Possible rational zeros: } \frac{\text{Factors of constant term}}{\text{Factors of leading coefficient}}$$

Having formed this list of *possible rational zeros,* use a trial-and-error method to determine which, if any, are actual zeros of the polynomial. Note that when the leading coefficient is 1, the possible rational zeros are simply the factors of the constant term.

EXAMPLE 2   **Rational Zero Test with Leading Coefficient of 1**

Find (if possible) the rational zeros of

$$f(x) = x^3 + x + 1.$$

Solution   The leading coefficient is 1, so the possible rational zeros are the factors of the constant term.

   Possible rational zeros: 1 and -1

Testing these possible zeros shows that neither works.

$$f(1) = (1)^3 + 1 + 1$$
$$= 3$$
$$f(-1) = (-1)^3 + (-1) + 1$$
$$= -1$$

So, the given polynomial has *no* rational zeros. Note from the graph of f in Figure 2.15 that f does have one real zero between -1 and 0. However, by the Rational Zero Test, you know that this real zero is *not* a rational number.

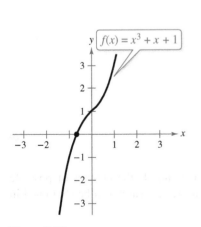

$f(x) = x^3 + x + 1$

Figure 2.15

✓ *Checkpoint* ◀))) Audio-video solution in English & Spanish at LarsonPrecalculus.com

Find (if possible) the rational zeros of

$$f(x) = x^3 + 2x^2 + 6x - 4.$$

EXAMPLE 3 **Rational Zero Test with Leading Coefficient of 1**

Find the rational zeros of

$$f(x) = x^4 - x^3 + x^2 - 3x - 6.$$

Solution The leading coefficient is 1, so the possible rational zeros are the factors of the constant term.

Possible rational zeros: $\pm 1, \pm 2, \pm 3, \pm 6$

By applying synthetic division successively, you find that $x = -1$ and $x = 2$ are the only two rational zeros.

$$
\begin{array}{r|rrrrr}
-1 & 1 & -1 & 1 & -3 & -6 \\
   &   & -1 & 2 & -3 & 6 \\
\hline
   & 1 & -2 & 3 & -6 & 0
\end{array}
$$
$\longrightarrow$ 0 remainder, so $x = -1$ is a zero.

$$
\begin{array}{r|rrrr}
2 & 1 & -2 & 3 & -6 \\
  &   & 2 & 0 & 6 \\
\hline
  & 1 & 0 & 3 & 0
\end{array}
$$
$\longrightarrow$ 0 remainder, so $x = 2$ is a zero.

So, $f(x)$ factors as

$$f(x) = (x + 1)(x - 2)(x^2 + 3).$$

The factor $(x^2 + 3)$ produces no real zeros, so $x = -1$ and $x = 2$ are the only *real* zeros of f. The figure below verifies this.

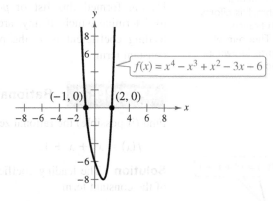

✓ **Checkpoint** ◀))) *Audio-video solution in English & Spanish at LarsonPrecalculus.com*

Find the rational zeros of

$$f(x) = x^3 - 15x^2 + 75x - 125.$$

When the leading coefficient of a polynomial is not 1, the number of possible rational zeros can increase dramatically. In such cases, the search can be shortened in several ways.

1. A graphing utility can help to speed up the calculations.

2. A graph can give good estimates of the locations of the zeros.

3. The Intermediate Value Theorem, along with a table of values, can give approximations of the zeros.

4. Synthetic division can be used to test the possible rational zeros.

After finding the first zero, the search becomes simpler by working with the lower-degree polynomial obtained in synthetic division, as shown in Example 3.

REMARK When there are few possible rational zeros, as in Example 2, it may be quicker to test the zeros by evaluating the function. When there are more possible rational zeros, as in Example 3, it may be quicker to use a different approach to test the zeros, such as using synthetic division or sketching a graph.

EXAMPLE 4 **Using the Rational Zero Test**

Find the rational zeros of $f(x) = 2x^3 + 3x^2 - 8x + 3$.

Solution The leading coefficient is 2 and the constant term is 3.

$$\textit{Possible rational zeros: } \frac{\text{Factors of 3}}{\text{Factors of 2}} = \frac{\pm 1, \pm 3}{\pm 1, \pm 2} = \pm 1, \pm 3, \pm \frac{1}{2}, \pm \frac{3}{2}$$

By synthetic division, $x = 1$ is a rational zero.

$$
\begin{array}{r|rrrr}
1 & 2 & 3 & -8 & 3 \\
  &   & 2 & 5 & -3 \\
\hline
  & 2 & 5 & -3 & 0 \\
\end{array}
$$

So, $f(x)$ factors as

$$f(x) = (x - 1)(2x^2 + 5x - 3)$$
$$= (x - 1)(2x - 1)(x + 3)$$

which shows that the rational zeros of f are $x = 1$, $x = \frac{1}{2}$, and $x = -3$.

✓ **Checkpoint**  Audio-video solution in English & Spanish at LarsonPrecalculus.com

Find the rational zeros of

$$f(x) = 2x^3 + x^2 - 13x + 6.$$

Recall from Section 2.2 that if $x = a$ is a zero of the polynomial function f, then $x = a$ is a solution of the polynomial equation $f(x) = 0$.

EXAMPLE 5 **Solving a Polynomial Equation**

Find all real solutions of $-10x^3 + 15x^2 + 16x - 12 = 0$.

Solution The leading coefficient is -10 and the constant term is -12.

$$\textit{Possible rational solutions: } \frac{\text{Factors of } -12}{\text{Factors of } -10} = \frac{\pm 1, \pm 2, \pm 3, \pm 4, \pm 6, \pm 12}{\pm 1, \pm 2, \pm 5, \pm 10}$$

With so many possibilities (32, in fact), it is worth your time to sketch a graph. In Figure 2.16, three reasonable solutions appear to be $x = -\frac{6}{5}$, $x = \frac{1}{2}$, and $x = 2$. Testing these by synthetic division shows that $x = 2$ is the only rational solution. So, you have

$$(x - 2)(-10x^2 - 5x + 6) = 0.$$

Using the Quadratic Formula to solve $-10x^2 - 5x + 6 = 0$, you find that the two additional solutions are irrational numbers.

$$x = \frac{5 + \sqrt{265}}{-20} \approx -1.0639$$

and

$$x = \frac{5 - \sqrt{265}}{-20} \approx 0.5639$$

✓ **Checkpoint**  Audio-video solution in English & Spanish at LarsonPrecalculus.com

Find all real solutions of

$$-2x^3 - 5x^2 + 15x + 18 = 0.$$

▷ **REMARK** Remember that when you find the rational zeros of a polynomial function with many possible rational zeros, as in Example 4, you must use trial and error. There is no quick algebraic method to determine which of the possibilities is an actual zero; however, sketching a graph may be helpful.

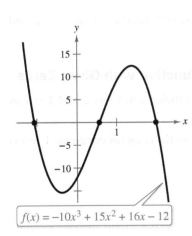

$$f(x) = -10x^3 + 15x^2 + 16x - 12$$

Figure 2.16

▷ **ALGEBRA HELP** To review the Quadratic Formula, see Appendix A.5.

Conjugate Pairs

In Example 1(c), note that the two complex zeros $2i$ and $-2i$ are complex conjugates. That is, they are of the forms $a + bi$ and $a - bi$.

Complex Zeros Occur in Conjugate Pairs

Let f be a polynomial function that has *real coefficients*. If $a + bi$, where $b \neq 0$, is a zero of the function, then the complex conjugate $a - bi$ is also a zero of the function.

Be sure you see that this result is true only when the polynomial function has *real coefficients*. For example, the result applies to the function $f(x) = x^2 + 1$, but not to the function $g(x) = x - i$.

EXAMPLE 6   **Finding a Polynomial Function with Given Zeros**

Find a fourth-degree polynomial function f with real coefficients that has -1, -1, and $3i$ as zeros.

Solution   You are given that $3i$ is a zero of f *and* the polynomial has real coefficients, so you know that the complex conjugate $-3i$ must also be a zero. Using the Linear Factorization Theorem, write $f(x)$ as

$$f(x) = a(x + 1)(x + 1)(x - 3i)(x + 3i).$$

For simplicity, let $a = 1$ to obtain

$$f(x) = (x^2 + 2x + 1)(x^2 + 9) = x^4 + 2x^3 + 10x^2 + 18x + 9.$$

✓ **Checkpoint** ◀))) *Audio-video solution in English & Spanish at LarsonPrecalculus.com*

Find a fourth-degree polynomial function f with real coefficients that has 2, -2, and $-7i$ as zeros.

EXAMPLE 7   **Finding a Polynomial Function with Given Zeros**

Find the cubic polynomial function f with real coefficients that has 2 and $1 - i$ as zeros, and $f(1) = 3$.

Solution   You are given that $1 - i$ is a zero of f, so the complex conjugate $1 + i$ is also a zero.

$$
\begin{aligned}
f(x) &= a(x - 2)[x - (1 - i)][x - (1 + i)] \\
&= a(x - 2)[(x - 1) + i][(x - 1) - i] \\
&= a(x - 2)[(x - 1)^2 + 1] \\
&= a(x - 2)(x^2 - 2x + 2) \\
&= a(x^3 - 4x^2 + 6x - 4)
\end{aligned}
$$

To find the value of a, use the fact that $f(1) = 3$ to obtain

$$a[(1)^3 - 4(1)^2 + 6(1) - 4] = 3.$$

So, $a = -3$ and

$$f(x) = -3(x^3 - 4x^2 + 6x - 4) = -3x^3 + 12x^2 - 18x + 12.$$

✓ **Checkpoint** ◀))) *Audio-video solution in English & Spanish at LarsonPrecalculus.com*

Find the *quartic* (fourth-degree) polynomial function f with real coefficients that has 1, -2, and $2i$ as zeros, and $f(-1) = 10$.

Factoring a Polynomial

The Linear Factorization Theorem states that you can write any nth-degree polynomial as the product of n linear factors.

$$f(x) = a_n(x - c_1)(x - c_2)(x - c_3) \cdots (x - c_n)$$

This result includes the possibility that some of the values of c_i are imaginary. The theorem below states that you can write $f(x)$ as the product of linear and quadratic factors with real coefficients. For a proof of this theorem, see Proofs in Mathematics on page 194.

Factors of a Polynomial

Every polynomial of degree $n > 0$ with real coefficients can be written as the product of linear and quadratic factors with real coefficients, where the quadratic factors have no real zeros.

A quadratic factor with no real zeros is *prime* or **irreducible over the reals.** Note that this is not the same as being *irreducible over the rationals.* For example, the quadratic $x^2 + 1 = (x - i)(x + i)$ is irreducible over the reals (and therefore over the rationals). On the other hand, the quadratic $x^2 - 2 = \left(x - \sqrt{2}\right)\left(x + \sqrt{2}\right)$ is irreducible over the rationals but *reducible* over the reals.

▷ TECHNOLOGY Another way to find the real zeros of the function in Example 8 is to use a graphing utility to graph the function (see figure).

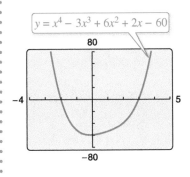

Then use the *zero* or *root* feature of the graphing utility to determine that $x = -2$ and $x = 3$ are the real zeros.

EXAMPLE 8 **Finding the Zeros of a Polynomial Function**

Find all the zeros of $f(x) = x^4 - 3x^3 + 6x^2 + 2x - 60$ given that $1 + 3i$ is a zero of f.

Solution Complex zeros occur in conjugate pairs, so you know that $1 - 3i$ is also a zero of f. This means that both $[x - (1 + 3i)]$ and $[x - (1 - 3i)]$ are factors of $f(x)$. Multiplying these two factors produces

$$[x - (1 + 3i)][x - (1 - 3i)] = [(x - 1) - 3i][(x - 1) + 3i]$$
$$= (x - 1)^2 - 9i^2$$
$$= x^2 - 2x + 10.$$

Using long division, divide $x^2 - 2x + 10$ into $f(x)$.

$$
\begin{array}{r}
x^2 - x - 6 \\
x^2 - 2x + 10 \overline{)\, x^4 - 3x^3 + 6x^2 + 2x - 60} \\
\underline{x^4 - 2x^3 + 10x^2} \\
-x^3 - 4x^2 + 2x \\
\underline{-x^3 + 2x^2 - 10x} \\
-6x^2 + 12x - 60 \\
\underline{-6x^2 + 12x - 60} \\
0
\end{array}
$$

So, you have

$$f(x) = (x^2 - 2x + 10)(x^2 - x - 6) = (x^2 - 2x + 10)(x - 3)(x + 2)$$

and can conclude that the zeros of f are $x = 1 + 3i$, $x = 1 - 3i$, $x = 3$, and $x = -2$.

▷ ALGEBRA HELP To review the techniques for polynomial long division, see Section 2.3.

✓ **Checkpoint** 🔊))) *Audio-video solution in English & Spanish at LarsonPrecalculus.com*

Find all the zeros of $f(x) = 3x^3 - 2x^2 + 48x - 32$ given that $4i$ is a zero of f. ∎

In Example 8, without knowing that $1 + 3i$ is a zero of f, it is still possible to find all the zeros of the function. You can first use synthetic division to find the real zeros -2 and 3. Then, factor the polynomial as

$$(x + 2)(x - 3)(x^2 - 2x + 10).$$

Finally, use the Quadratic Formula to solve $x^2 - 2x + 10 = 0$ to obtain the zeros $1 + 3i$ and $1 - 3i$.

In Example 9, you will find all the zeros, including the imaginary zeros, of a fifth-degree polynomial function.

EXAMPLE 9   Finding the Zeros of a Polynomial Function

Write

$$f(x) = x^5 + x^3 + 2x^2 - 12x + 8$$

as the product of linear factors and list all the zeros of the function.

Solution   The leading coefficient is 1, so the possible rational zeros are the factors of the constant term.

Possible rational zeros: ± 1, ± 2, ± 4, and ± 8

By synthetic division, $x = 1$ and $x = -2$ are zeros.

$$
\begin{array}{r|rrrrrr}
1 & 1 & 0 & 1 & 2 & -12 & 8 \\
  &   & 1 & 1 & 2 & 4 & -8 \\
\hline
  & 1 & 1 & 2 & 4 & -8 & 0
\end{array} \longrightarrow \text{1 is a zero.}
$$

$$
\begin{array}{r|rrrrr}
-2 & 1 & 1 & 2 & 4 & -8 \\
   &   & -2 & 2 & -8 & 8 \\
\hline
   & 1 & -1 & 4 & -4 & 0
\end{array} \longrightarrow \text{-2 is a zero.}
$$

So, you have

$$
\begin{aligned}
f(x) &= x^5 + x^3 + 2x^2 - 12x + 8 \\
&= (x - 1)(x + 2)(x^3 - x^2 + 4x - 4).
\end{aligned}
$$

Factoring by grouping,

$$x^3 - x^2 + 4x - 4 = (x - 1)(x^2 + 4)$$

and by factoring $x^2 + 4$ as

$$x^2 + 4 = (x - 2i)(x + 2i)$$

you obtain

$$f(x) = (x - 1)(x - 1)(x + 2)(x - 2i)(x + 2i)$$

which gives all five zeros of f.

$$x = 1, \quad x = 1, \quad x = -2, \quad x = 2i, \quad \text{and} \quad x = -2i$$

Figure 2.17 shows the graph of f. Notice that the *real* zeros are the only ones that appear as x-intercepts and that the real zero $x = 1$ is repeated.

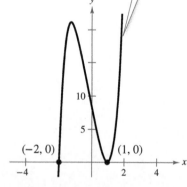

$f(x) = x^5 + x^3 + 2x^2 - 12x + 8$

$(-2, 0)$ $(1, 0)$

Figure 2.17

✓ **Checkpoint**))) *Audio-video solution in English & Spanish at LarsonPrecalculus.com*

Write

$$f(x) = x^4 + 8x^2 - 9$$

as the product of linear factors and list all the zeros of the function.

Other Tests for Zeros of Polynomials

You know that an nth-degree polynomial function can have *at most n* real zeros. Of course, many nth-degree polynomial functions do not have that many real zeros. For example, $f(x) = x^2 + 1$ has no real zeros, and $f(x) = x^3 + 1$ has only one real zero. The theorem below, called **Descartes's Rule of Signs,** uses variations in sign to analyze the number of real zeros of a polynomial. A **variation in sign** means that two consecutive nonzero coefficients have opposite signs.

Descartes's Rule of Signs

Let $f(x) = a_n x^n + a_{n-1}x^{n-1} + \cdots + a_2 x^2 + a_1 x + a_0$ be a polynomial with real coefficients and $a_0 \neq 0$.

1. The number of *positive real zeros* of f is either equal to the number of variations in sign of $f(x)$ or less than that number by an even integer.

2. The number of *negative real zeros* of f is either equal to the number of variations in sign of $f(-x)$ or less than that number by an even integer.

When using Descartes's Rule of Signs, count a zero of multiplicity k as k zeros. For example, the polynomial $x^3 - 3x + 2$ has two variations in sign, and so it has either two positive or no positive real zeros. This polynomial factors as

$$x^3 - 3x + 2 = (x - 1)(x - 1)(x + 2)$$

so the two positive real zeros are $x = 1$ of multiplicity 2.

EXAMPLE 10   **Using Descartes's Rule of Signs**

Determine the possible numbers of positive and negative real zeros of

$$f(x) = 3x^3 - 5x^2 + 6x - 4.$$

Solution   The original polynomial has *three* variations in sign.

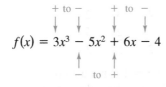

The polynomial

$$f(-x) = 3(-x)^3 - 5(-x)^2 + 6(-x) - 4$$
$$= -3x^3 - 5x^2 - 6x - 4$$

has no variations in sign. So, from Descartes's Rule of Signs, the polynomial

$$f(x) = 3x^3 - 5x^2 + 6x - 4$$

has either three positive real zeros or one positive real zero, and has no negative real zeros. Figure 2.18 shows that the function has only one real zero, $x = 1$.

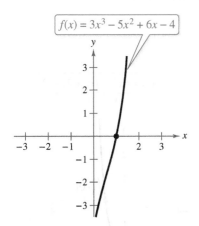

Figure 2.18

✓ *Checkpoint*

Determine the possible numbers of positive and negative real zeros of

$$f(x) = 2x^3 + 5x^2 + x + 8.$$

Another test for zeros of a polynomial function is related to the sign pattern in the last row of the synthetic division array. This test can give you an upper or lower bound for the real zeros of f. A real number c is an **upper bound** for the real zeros of f when no zeros are greater than c. Similarly, c is a **lower bound** when no real zeros of f are less than c.

Upper and Lower Bound Rules

Let $f(x)$ be a polynomial with real coefficients and a positive leading coefficient. Divide $f(x)$ by $x - c$ using synthetic division.

1. If $c > 0$ and each number in the last row is either positive or zero, then c is an **upper bound** for the real zeros of f.

2. If $c < 0$ and the numbers in the last row are alternately positive and negative (zero entries count as positive or negative), then c is a **lower bound** for the real zeros of f.

EXAMPLE 11   Finding Real Zeros of a Polynomial Function

Find all real zeros of

$$f(x) = 6x^3 - 4x^2 + 3x - 2.$$

Solution   List the possible rational zeros of f.

$$\frac{\text{Factors of } -2}{\text{Factors of } 6} = \frac{\pm 1, \pm 2}{\pm 1, \pm 2, \pm 3, \pm 6} = \pm 1, \pm \frac{1}{2}, \pm \frac{1}{3}, \pm \frac{1}{6}, \pm \frac{2}{3}, \pm 2$$

The original polynomial $f(x)$ has three variations in sign. The polynomial

$$f(-x) = 6(-x)^3 - 4(-x)^2 + 3(-x) - 2$$
$$= -6x^3 - 4x^2 - 3x - 2$$

has no variations in sign. So, by Descartes's Rule of Signs, there are three positive real zeros or one positive real zero, and no negative real zeros. Test $x = 1$.

$$\begin{array}{r|rrrr} 1 & 6 & -4 & 3 & -2 \\ & & 6 & 2 & 5 \\ \hline & 6 & 2 & 5 & 3 \end{array}$$

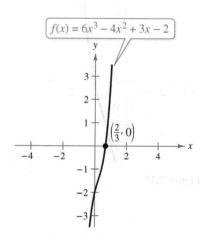

$f(x) = 6x^3 - 4x^2 + 3x - 2$

This shows that $x = 1$ is not a zero. However, the last row has all positive entries, telling you that $x = 1$ is an upper bound for the real zeros. So, restrict the search to zeros between 0 and 1. By trial and error, $x = \frac{2}{3}$ is a zero, and factoring,

$$f(x) = \left(x - \frac{2}{3}\right)(6x^2 + 3).$$

The factor $6x^2 + 3$ has no real zeros, so it follows that $x = \frac{2}{3}$ is the only real zero, as verified in the graph of f at the right.

✓ **Checkpoint** ◄))) *Audio-video solution in English & Spanish at LarsonPrecalculus.com*

Find all real zeros of $f(x) = 8x^3 - 4x^2 + 6x - 3$.

Application

EXAMPLE 12   **Using a Polynomial Model**

You design candle making kits. Each kit contains 25 cubic inches of candle wax and a mold for making a pyramid-shaped candle. You want the height of the candle to be 2 inches less than the length of each side of the candle's square base. What should the dimensions of your candle mold be?

Solution   The volume of a pyramid is $V = \frac{1}{3}Bh$, where B is the area of the base and h is the height. The area of the base is x^2 and the height is $(x - 2)$. So, the volume of the pyramid is $V = \frac{1}{3}x^2(x - 2)$. Substitute 25 for the volume and solve for x.

$25 = \frac{1}{3}x^2(x - 2)$ Substitute 25 for V.

$75 = x^3 - 2x^2$ Multiply each side by 3, and distribute x^2.

$0 = x^3 - 2x^2 - 75$ Write in general form.

The possible rational solutions are $x = \pm 1, \pm 3, \pm 5, \pm 15, \pm 25, \pm 75$. Note that in this case it makes sense to consider only positive x-values. Use synthetic division to test some of the possible solutions and determine that $x = 5$ is a solution.

$$
\begin{array}{r|rrrr}
5 & 1 & -2 & 0 & -75 \\
  &   & 5 & 15 & 75 \\
\hline
  & 1 & 3 & 15 & 0
\end{array}
$$

The other two solutions, which satisfy $x^2 + 3x + 15 = 0$, are imaginary, so discard them and conclude that the base of the candle mold should be 5 inches by 5 inches and the height should be $5 - 2 = 3$ inches.

✓ *Checkpoint*  *Audio-video solution in English & Spanish at LarsonPrecalculus.com*

Rework Example 12 when each kit contains 147 cubic inches of candle wax and you want the height of the pyramid-shaped candle to be 2 inches more than the length of each side of the candle's square base.

Before concluding this section, here is an additional hint that can help you find the zeros of a polynomial function. When the terms of $f(x)$ have a common monomial factor, you should factor it out before applying the tests in this section. For example, writing $f(x) = x^4 - 5x^3 + 3x^2 + x = x(x^3 - 5x^2 + 3x + 1)$ shows that $x = 0$ is a zero of f. Obtain the remaining zeros by analyzing the cubic factor.

Summarize   (Section 2.5)

1. State the Fundamental Theorem of Algebra and the Linear Factorization Theorem (*page 152, Example 1*).

2. Explain how to use the Rational Zero Test (*page 153, Examples 2–5*).

3. Explain how to use complex conjugates when analyzing a polynomial function (*page 156, Examples 6 and 7*).

4. Explain how to find the zeros of a polynomial function (*page 157, Examples 8 and 9*).

5. State Descartes's Rule of Signs and the Upper and Lower Bound Rules (*pages 159 and 160, Examples 10 and 11*).

6. Describe a real-life application of finding the zeros of a polynomial function (*page 161, Example 12*).

2.5 Exercises

See CalcChat.com for tutorial help and worked-out solutions to odd-numbered exercises.

Vocabulary: **Fill in the blanks.**

1. The _____ _____ of _____ states that if $f(x)$ is a polynomial of degree n ($n > 0$), then f has at least one zero in the complex number system.

2. The _____ _____ _____ states that if $f(x)$ is a polynomial of degree n ($n > 0$), then $f(x)$ has precisely n linear factors, $f(x) = a_n(x - c_1)(x - c_2) \cdots (x - c_n)$, where $c_1, c_2, \ldots, c_n$ are complex numbers.

3. The test that gives a list of the possible rational zeros of a polynomial function is the _____ _____ Test.

4. If $a + bi$, where $b \neq 0$, is a complex zero of a polynomial with real coefficients, then so is its _____ _____, $a - bi$.

5. Every polynomial of degree $n > 0$ with real coefficients can be written as the product of _____ and _____ factors with real coefficients, where the _____ factors have no real zeros.

6. A quadratic factor that cannot be factored further as a product of linear factors containing real numbers is _____ over the _____.

7. The theorem that can be used to determine the possible numbers of positive and negative real zeros of a function is called _____ _____ of _____.

8. A real number c is a _____ bound for the real zeros of f when no real zeros are less than c, and is a _____ bound when no real zeros are greater than c.

Skills and Applications

 Zeros of Polynomial Functions In Exercises 9–14, determine the number of zeros of the polynomial function.

9. $f(x) = x^3 + 2x^2 + 1$ 10. $f(x) = x^4 - 3x$

11. $g(x) = x^4 - x^5$ 12. $f(x) = x^3 - x^6$

13. $f(x) = (x + 5)^2$

14. $h(t) = (t - 1)^2 - (t + 1)^2$

Using the Rational Zero Test In Exercises 15–18, use the Rational Zero Test to list the possible rational zeros of f. Verify that the zeros of f shown in the graph are contained in the list.

15. $f(x) = x^3 + 2x^2 - x - 2$

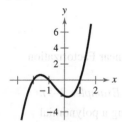

16. $f(x) = x^3 - 4x^2 - 4x + 16$

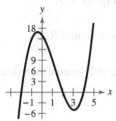

17. $f(x) = 2x^4 - 17x^3 + 35x^2 + 9x - 45$

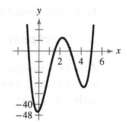

18. $f(x) = 4x^5 - 8x^4 - 5x^3 + 10x^2 + x - 2$

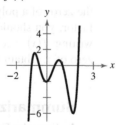

 Using the Rational Zero Test In Exercises 19–28, find (if possible) the rational zeros of the function.

19. $f(x) = x^3 - 7x - 6$ 20. $f(x) = x^3 - 13x + 12$

21. $g(t) = t^3 - 4t^2 + 4$ 22. $h(x) = x^3 - 19x + 30$

23. $h(t) = t^3 + 8t^2 + 13t + 6$

24. $g(x) = x^3 + 8x^2 + 12x + 18$

25. $C(x) = 2x^3 + 3x^2 - 1$

26. $f(x) = 3x^3 - 19x^2 + 33x - 9$

27. $g(x) = 9x^4 - 9x^3 - 58x^2 + 4x + 24$

28. $f(x) = 2x^4 - 15x^3 + 23x^2 + 15x - 25$

Solving a Polynomial Equation In Exercises 29–32, find all real solutions of the polynomial equation.

29. $-5x^3 + 11x^2 - 4x - 2 = 0$

30. $8x^3 + 10x^2 - 15x - 6 = 0$

31. $x^4 + 6x^3 + 3x^2 - 16x + 6 = 0$

32. $x^4 + 8x^3 + 14x^2 - 17x - 42 = 0$

Using the Rational Zero Test In Exercises 33–36, (a) list the possible rational zeros of f, (b) sketch the graph of f so that some of the possible zeros in part (a) can be disregarded, and then (c) determine all real zeros of f.

33. $f(x) = x^3 + x^2 - 4x - 4$

34. $f(x) = -3x^3 + 20x^2 - 36x + 16$

35. $f(x) = -4x^3 + 15x^2 - 8x - 3$

36. $f(x) = 4x^3 - 12x^2 - x + 15$

Using the Rational Zero Test In Exercises 37–40, (a) list the possible rational zeros of f, (b) use a graphing utility to graph f so that some of the possible zeros in part (a) can be disregarded, and then (c) determine all real zeros of f.

37. $f(x) = -2x^4 + 13x^3 - 21x^2 + 2x + 8$

38. $f(x) = 4x^4 - 17x^2 + 4$

39. $f(x) = 32x^3 - 52x^2 + 17x + 3$

40. $f(x) = 4x^3 + 7x^2 - 11x - 18$

 Finding a Polynomial Function with Given Zeros In Exercises 41–46, find a polynomial function with real coefficients that has the given zeros. (There are many correct answers.)

41. $1, 5i$

42. $4, -3i$

43. $2, 2, 1 + i$

44. $-1, 5, 3 - 2i$

45. $\frac{2}{3}, -1, 3 + \sqrt{2}i$

46. $-\frac{5}{2}, -5, 1 + \sqrt{3}i$

 Finding a Polynomial Function with Given Zeros In Exercises 47–50, find the polynomial function f with real coefficients that has the given degree, zeros, and solution point.

	Degree	Zeros	Solution Point
47.	4	$-2, 1, i$	$f(0) = -4$
48.	4	$-1, 2, \sqrt{2}i$	$f(1) = 12$
49.	3	$-3, 1 + \sqrt{3}i$	$f(-2) = 12$
50.	3	$-2, 1 - \sqrt{2}i$	$f(-1) = -12$

Factoring a Polynomial In Exercises 51–54, write the polynomial (a) as the product of factors that are irreducible over the *rationals*, (b) as the product of linear and quadratic factors that are irreducible over the *reals*, and (c) in completely factored form.

51. $f(x) = x^4 + 2x^2 - 8$

52. $f(x) = x^4 + 6x^2 - 27$

53. $f(x) = x^4 - 2x^3 - 3x^2 + 12x - 18$
    (*Hint:* One factor is $x^2 - 6$.)

54. $f(x) = x^4 - 3x^3 - x^2 - 12x - 20$
    (*Hint:* One factor is $x^2 + 4$.)

 Finding the Zeros of a Polynomial Function In Exercises 55–60, use the given zero to find all the zeros of the function.

Function	Zero
55. $f(x) = x^3 - x^2 + 4x - 4$	$2i$
56. $f(x) = 2x^3 + 3x^2 + 18x + 27$	$3i$
57. $g(x) = x^3 - 8x^2 + 25x - 26$	$3 + 2i$
58. $g(x) = x^3 + 9x^2 + 25x + 17$	$-4 + i$
59. $h(x) = x^4 - 6x^3 + 14x^2 - 18x + 9$	$1 - \sqrt{2}i$
60. $h(x) = x^4 + x^3 - 3x^2 - 13x + 14$	$-2 + \sqrt{3}i$

 Finding the Zeros of a Polynomial Function In Exercises 61–72, write the polynomial as the product of linear factors and list all the zeros of the function.

61. $f(x) = x^2 + 36$ **62.** $f(x) = x^2 + 49$

63. $h(x) = x^2 - 2x + 17$ **64.** $g(x) = x^2 + 10x + 17$

65. $f(x) = x^4 - 16$ **66.** $f(y) = y^4 - 256$

67. $f(z) = z^2 - 2z + 2$

68. $h(x) = x^3 - 3x^2 + 4x - 2$

69. $g(x) = x^3 - 3x^2 + x + 5$

70. $f(x) = x^3 - x^2 + x + 39$

71. $g(x) = x^4 - 4x^3 + 8x^2 - 16x + 16$

72. $h(x) = x^4 + 6x^3 + 10x^2 + 6x + 9$

Finding the Zeros of a Polynomial Function In Exercises 73–78, find all the zeros of the function. When there is an extended list of possible rational zeros, use a graphing utility to graph the function in order to disregard any of the possible rational zeros that are obviously not zeros of the function.

73. $f(x) = x^3 + 24x^2 + 214x + 740$

74. $f(s) = 2s^3 - 5s^2 + 12s - 5$

75. $f(x) = 16x^3 - 20x^2 - 4x + 15$

76. $f(x) = 9x^3 - 15x^2 + 11x - 5$

77. $f(x) = 2x^4 + 5x^3 + 4x^2 + 5x + 2$

78. $g(x) = x^5 - 8x^4 + 28x^3 - 56x^2 + 64x - 32$

Using Descartes's Rule of Signs In Exercises 79–86, use Descartes's Rule of Signs to determine the possible numbers of positive and negative real zeros of the function.

79. $g(x) = 2x^3 - 3x^2 - 3$ **80.** $h(x) = 4x^2 - 8x + 3$

81. $h(x) = 2x^3 + 3x^2 + 1$ **82.** $h(x) = 2x^4 - 3x - 2$

83. $g(x) = 6x^4 + 2x^3 - 3x^2 + 2$

84. $f(x) = 4x^3 - 3x^2 - 2x - 1$

85. $f(x) = 5x^3 + x^2 - x + 5$

86. $f(x) = 3x^3 - 2x^2 - x + 3$

Verifying Upper and Lower Bounds In Exercises 87–90, use synthetic division to verify the upper and lower bounds of the real zeros of f.

87. $f(x) = x^3 + 3x^2 - 2x + 1$

    (a) Upper: $x = 1$ (b) Lower: $x = -4$

88. $f(x) = x^3 - 4x^2 + 1$

    (a) Upper: $x = 4$ (b) Lower: $x = -1$

89. $f(x) = x^4 - 4x^3 + 16x - 16$

    (a) Upper: $x = 5$ (b) Lower: $x = -3$

90. $f(x) = 2x^4 - 8x + 3$

    (a) Upper: $x = 3$ (b) Lower: $x = -4$

Finding Real Zeros of a Polynomial Function In Exercises 91–94, find all real zeros of the function.

91. $f(x) = 16x^3 - 12x^2 - 4x + 3$

92. $f(z) = 12z^3 - 4z^2 - 27z + 9$

93. $f(y) = 4y^3 + 3y^2 + 8y + 6$

94. $g(x) = 3x^3 - 2x^2 + 15x - 10$

Finding the Rational Zeros of a Polynomial In Exercises 95–98, find the rational zeros of the polynomial function.

95. $P(x) = x^4 - \frac{25}{4}x^2 + 9 = \frac{1}{4}(4x^4 - 25x^2 + 36)$

96. $f(x) = x^3 - \frac{3}{2}x^2 - \frac{23}{2}x + 6$

    $= \frac{1}{2}(2x^3 - 3x^2 - 23x + 12)$

97. $f(x) = x^3 - \frac{1}{4}x^2 - x + \frac{1}{4} = \frac{1}{4}(4x^3 - x^2 - 4x + 1)$

98. $f(z) = z^3 + \frac{11}{6}z^2 - \frac{1}{2}z - \frac{1}{3} = \frac{1}{6}(6z^3 + 11z^2 - 3z - 2)$

Rational and Irrational Zeros In Exercises 99–102, match the cubic function with the numbers of rational and irrational zeros.

(a) **Rational zeros: 0; irrational zeros: 1**

(b) **Rational zeros: 3; irrational zeros: 0**

(c) **Rational zeros: 1; irrational zeros: 2**

(d) **Rational zeros: 1; irrational zeros: 0**

99. $f(x) = x^3 - 1$ **100.** $f(x) = x^3 - 2$

101. $f(x) = x^3 - x$ **102.** $f(x) = x^3 - 2x$

103. Geometry You want to make an open box from a rectangular piece of material, 15 centimeters by 9 centimeters, by cutting equal squares from the corners and turning up the sides.

    (a) Let x represent the side length of each of the squares removed. Draw a diagram showing the squares removed from the original piece of material and the resulting dimensions of the open box.

    (b) Use the diagram to write the volume V of the box as a function of x. Determine the domain of the function.

    (c) Sketch the graph of the function and approximate the dimensions of the box that yield a maximum volume.

    (d) Find values of x such that $V = 56$. Which of these values is a physical impossibility in the construction of the box? Explain.

104. Geometry A rectangular package to be sent by a delivery service (see figure) has a combined length and girth (perimeter of a cross section) of 120 inches.

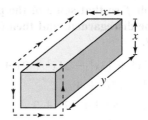

    (a) Use the diagram to write the volume V of the package as a function of x.

    (b) Use a graphing utility to graph the function and approximate the dimensions of the package that yield a maximum volume.

    (c) Find values of x such that $V = 13,500$. Which of these values is a physical impossibility in the construction of the package? Explain.

105. Geometry

A bulk food storage bin with dimensions 2 feet by 3 feet by 4 feet needs to be increased in size to hold five times as much food as the current bin.

    (a) Assume each dimension is increased by the same amount. Write a function that represents the volume V of the new bin.

    (b) Find the dimensions of the new bin.

106. Cost The ordering and transportation cost C (in thousands of dollars) for machine parts is given by

$$C(x) = 100\left(\frac{200}{x^2} + \frac{x}{x + 30}\right), \quad x \geq 1$$

where x is the order size (in hundreds). In calculus, it can be shown that the cost is a minimum when

$$3x^3 - 40x^2 - 2400x - 36,000 = 0.$$

Use a graphing utility to approximate the optimal order size to the nearest hundred units.

Exploration

True or False? **In Exercises 107 and 108, decide whether the statement is true or false. Justify your answer.**

107. It is possible for a third-degree polynomial function with integer coefficients to have no real zeros.

108. If $x = -i$ is a zero of the function

$$f(x) = x^3 + ix^2 + ix - 1$$

then $x = i$ must also be a zero of f.

Think About It **In Exercises 109–114, determine (if possible) the zeros of the function g when the function f has zeros at $x = r_1$, $x = r_2$, and $x = r_3$.**

109. $g(x) = -f(x)$
110. $g(x) = 3f(x)$
111. $g(x) = f(x - 5)$
112. $g(x) = f(2x)$
113. $g(x) = 3 + f(x)$
114. $g(x) = f(-x)$

115. Think About It A cubic polynomial function f has real zeros -2, $\frac{1}{2}$, and 3, and its leading coefficient is negative. Write an equation for f and sketch its graph. How many different polynomial functions are possible for f?

116. Think About It Sketch the graph of a fifth-degree polynomial function whose leading coefficient is positive and that has a zero at $x = 3$ of multiplicity 2.

Writing an Equation **In Exercises 117 and 118, the graph of a cubic polynomial function $y = f(x)$ is shown. One of the zeros is $1 + i$. Write an equation for f.**

117.

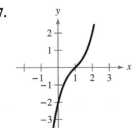

118.

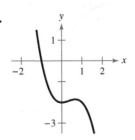

119. Error Analysis Describe the error.

The graph of a quartic (fourth-degree) polynomial $y = f(x)$ is shown. One of the zeros is i.

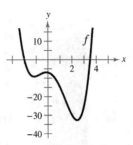

The function is $f(x) = (x + 2)(x - 3.5)(x - i)$. ✗

120. **HOW DO YOU SEE IT?** Use the information in the table to answer each question.

Interval	Value of $f(x)$
$(-\infty, -2)$	Positive
$(-2, 1)$	Negative
$(1, 4)$	Negative
$(4, \infty)$	Positive

(a) What are the three real zeros of the polynomial function f?

(b) What can be said about the behavior of the graph of f at $x = 1$?

(c) What is the least possible degree of f? Explain. Can the degree of f ever be odd? Explain.

(d) Is the leading coefficient of f positive or negative? Explain.

(e) Sketch a graph of a function that exhibits the behavior described in the table.

121. Think About It Let $y = f(x)$ be a quartic (fourth-degree) polynomial with leading coefficient $a = 1$ and

$$f(i) = f(2i) = 0.$$

Write an equation for f.

122. Think About It Let $y = f(x)$ be a cubic polynomial with leading coefficient $a = -1$ and

$$f(2) = f(i) = 0.$$

Write an equation for f.

123. Writing an Equation Write the equation for a quadratic function f (with integer coefficients) that has the given zeros. Assume that b is a positive integer.

(a) $\pm\sqrt{b}i$ (b) $a \pm bi$

2.6 Rational Functions

Rational functions have many real-life applications. For example, in Exercise 69 on page 176, you will use a rational function to determine the cost of supplying recycling bins to the population of a rural township.

- Find domains of rational functions.
- Find vertical and horizontal asymptotes of graphs of rational functions.
- Sketch graphs of rational functions.
- Sketch graphs of rational functions that have slant asymptotes.
- Use rational functions to model and solve real-life problems.

Introduction

A **rational function** is a quotient of polynomial functions. It can be written in the form

$$f(x) = \frac{N(x)}{D(x)}$$

where $N(x)$ and $D(x)$ are polynomials and $D(x)$ is not the zero polynomial.

The *domain* of a rational function of x includes all real numbers except x-values that make the denominator zero. Much of the discussion of rational functions will focus on the behavior of their graphs near x-values excluded from the domain.

EXAMPLE 1   Finding the Domain of a Rational Function

See LarsonPrecalculus.com for an interactive version of this type of example.

Find the domain of $f(x) = \dfrac{1}{x}$ and discuss the behavior of f near any excluded x-values.

Solution   The denominator is zero when $x = 0$, so the domain of f is all real numbers except $x = 0$. To determine the behavior of f near this excluded value, evaluate $f(x)$ to the left and right of $x = 0$, as shown in the tables below.

x	-1	-0.5	-0.1	-0.01	-0.001	$\rightarrow 0$
$f(x)$	-1	-2	-10	-100	-1000	$\rightarrow -\infty$

x	$0 \leftarrow$	0.001	0.01	0.1	0.5	1
$f(x)$	$\infty \leftarrow$	1000	100	10	2	1

Note that as x approaches 0 *from the left*, $f(x)$ decreases without bound. In contrast, as x approaches 0 *from the right*, $f(x)$ increases without bound. The graph of f is shown below.

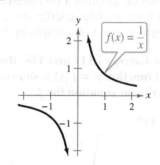

• •REMARK   Recall from Section 1.6 that the rational function

$$f(x) = \frac{1}{x}$$

is also referred to as the *reciprocal function*.

✓ Checkpoint ◉))) *Audio-video solution in English & Spanish at LarsonPrecalculus.com*

Find the domain of $f(x) = \dfrac{3x}{x - 1}$ and discuss the behavior of f near any excluded x-values.

Sunsetman/Shutterstock.com

Vertical and Horizontal Asymptotes

In Example 1, the behavior of f near $x = 0$ is as denoted below.

$$\underbrace{f(x) \to -\infty \text{ as } x \to 0^-}_{\substack{f(x) \text{ decreases without bound} \\ \text{as } x \text{ approaches 0 from the left.}}} \qquad \underbrace{f(x) \to \infty \text{ as } x \to 0^+}_{\substack{f(x) \text{ increases without bound} \\ \text{as } x \text{ approaches 0 from the right.}}}$$

The line $x = 0$ is a **vertical asymptote** of the graph of f, as shown in Figure 2.19. Notice that the graph of f also has a **horizontal asymptote**—the line $y = 0$. The behavior of f near $y = 0$ is as denoted below.

$$\underbrace{f(x) \to 0 \text{ as } x \to -\infty}_{\substack{f(x) \text{ approaches 0 as } x \\ \text{decreases without bound.}}} \qquad \underbrace{f(x) \to 0 \text{ as } x \to \infty}_{\substack{f(x) \text{ approaches 0 as } x \\ \text{increases without bound.}}}$$

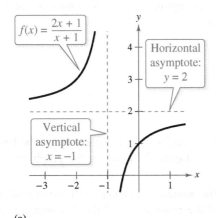

Figure 2.19

Definitions of Vertical and Horizontal Asymptotes

1. The line $x = a$ is a **vertical asymptote** of the graph of f when

$$f(x) \to \infty \quad \text{or} \quad f(x) \to -\infty$$

as $x \to a$, either from the right or from the left.

2. The line $y = b$ is a **horizontal asymptote** of the graph of f when

$$f(x) \to b$$

as $x \to \infty$ or $x \to -\infty$.

Eventually (as $x \to \infty$ or $x \to -\infty$), the distance between the horizontal asymptote and the points on the graph must approach zero. Figure 2.20 shows the vertical and horizontal asymptotes of the graphs of three rational functions.

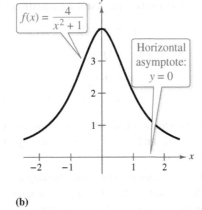

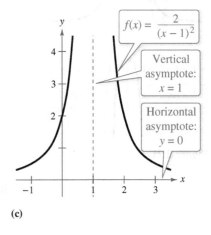

(a)   **(b)**   **(c)**

Figure 2.20

Verify numerically the horizontal asymptotes shown in Figure 2.20. For example, to show that the line $y = 2$ is the horizontal asymptote of the graph of

$$f(x) = \frac{2x + 1}{x + 1}$$

create a table that shows the value of $f(x)$ as x increases and decreases without bound.

The graphs of $f(x) = \dfrac{1}{x}$ in Figure 2.19 and $f(x) = \dfrac{2x + 1}{x + 1}$ in Figure 2.20(a) are **hyperbolas.** You will study hyperbolas in Chapter 10.

Vertical and Horizontal Asymptotes

Let f be the rational function

$$f(x) = \frac{N(x)}{D(x)} = \frac{a_n x^n + a_{n-1} x^{n-1} + \cdots + a_1 x + a_0}{b_m x^m + b_{m-1} x^{m-1} + \cdots + b_1 x + b_0}$$

where $N(x)$ and $D(x)$ have no common factors.

1. The graph of f has *vertical* asymptotes at the zeros of $D(x)$.

2. The graph of f has at most one *horizontal* asymptote determined by comparing the degrees of $N(x)$ and $D(x)$.

   a. When $n < m$, the graph of f has the line $y = 0$ (the x-axis) as a horizontal asymptote.

   b. When $n = m$, the graph of f has the line $y = \dfrac{a_n}{b_m}$ (ratio of the leading coefficients) as a horizontal asymptote.

   c. When $n > m$, the graph of f has no horizontal asymptote.

EXAMPLE 2   **Finding Vertical and Horizontal Asymptotes**

Find all vertical and horizontal asymptotes of the graph of each rational function.

a. $f(x) = \dfrac{2x^2}{x^2 - 1}$ **b.** $f(x) = \dfrac{x^2 + x - 2}{x^2 - x - 6}$

Solution

a. For this rational function, the degree of the numerator is *equal* to the degree of the denominator. The leading coefficient of the numerator is 2 and the leading coefficient of the denominator is 1, so the graph has the line $y = 2/1 = 2$ as a horizontal asymptote. To find any vertical asymptotes, set the denominator equal to zero and solve the resulting equation for x.

$$x^2 - 1 = 0 \qquad \text{Set denominator equal to zero.}$$
$$(x + 1)(x - 1) = 0 \qquad \text{Factor.}$$
$$x + 1 = 0 \implies x = -1 \qquad \text{Set 1st factor equal to 0.}$$
$$x - 1 = 0 \implies x = 1 \qquad \text{Set 2nd factor equal to 0.}$$

This equation has two real solutions, $x = -1$ and $x = 1$, so the graph has the lines $x = -1$ and $x = 1$ as vertical asymptotes. Figure 2.21 shows the graph of this function.

b. For this rational function, the degree of the numerator is equal to the degree of the denominator. The leading coefficients of the numerator and the denominator are both 1, so the graph has the line $y = 1/1 = 1$ as a horizontal asymptote. To find any vertical asymptotes, first factor the numerator and denominator as follows.

$$f(x) = \frac{x^2 + x - 2}{x^2 - x - 6} = \frac{(x - 1)(x + 2)}{(x + 2)(x - 3)} = \frac{x - 1}{x - 3}, \quad x \ne -2$$

Setting the denominator $x - 3$ (of the simplified function) equal to zero, you find that the graph has the line $x = 3$ as a vertical asymptote.

✓ **Checkpoint** ◀))) *Audio-video solution in English & Spanish at LarsonPrecalculus.com*

Find all vertical and horizontal asymptotes of the graph of $f(x) = \dfrac{3x^2 + 7x - 6}{x^2 + 4x + 3}$.

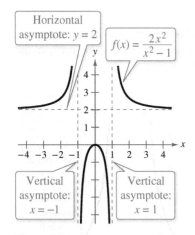

Horizontal asymptote: $y = 2$

$f(x) = \dfrac{2x^2}{x^2 - 1}$

Vertical asymptote: $x = -1$

Vertical asymptote: $x = 1$

Figure 2.21

▷

•• **REMARK**   There is a *hole* in the graph of f at $x = -2$. In Example 6, you will sketch the graph of a rational function that has a hole.

Sketching the Graph of a Rational Function

To sketch the graph of a rational function, use the following guidelines.

Guidelines for Graphing Rational Functions

Let $f(x) = \dfrac{N(x)}{D(x)}$, where $N(x)$ and $D(x)$ are polynomials and $D(x)$ is not the zero polynomial.

1. Simplify f, if possible. List any restrictions on the domain of f that are not implied by the simplified function.

2. Find and plot the y-intercept (if any) by evaluating $f(0)$.

3. Find the zeros of the numerator (if any). Then plot the corresponding x-intercepts.

4. Find the zeros of the denominator (if any). Then sketch the corresponding vertical asymptotes.

5. Find and sketch the horizontal asymptote (if any) by using the rule for finding the horizontal asymptote of a rational function on page 168.

6. Plot at least one point *between* and one point *beyond* each x-intercept and vertical asymptote.

7. Use smooth curves to complete the graph between and beyond the vertical asymptotes.

The concept of *test intervals* from Section 2.2 can be extended to graphing rational functions. Be aware, however, that although a polynomial function can change signs only at its zeros, a rational function can change signs both at its zeros and at its undefined values (the x-values for which its denominator is zero). So, to form the test intervals in which a rational function has no sign changes, arrange the x-values representing the zeros of both the numerator and the denominator of the rational function in increasing order.

You may also want to test for symmetry when graphing rational functions, especially for simple rational functions. Recall from Section 1.6 that the graph of the reciprocal function $f(x) = \dfrac{1}{x}$ is symmetric with respect to the origin.

▷ TECHNOLOGY Some graphing utilities have difficulty graphing rational functions with vertical asymptotes. In connected mode, the graphing utility may connect portions of the graph that are not supposed to be connected. For example, the graph on the left should consist of two unconnected portions—one to the left of $x = 2$ and the other to the right of $x = 2$. Changing the mode of the graphing utility to *dot* mode eliminates this problem. In *dot* mode, however, the graph is represented as a collection of dots (as shown in the graph on the right) rather than as a smooth curve.

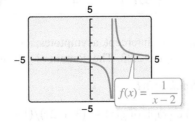
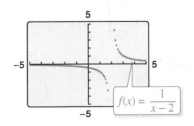

EXAMPLE 3 **Sketching the Graph of a Rational Function**

Sketch the graph of $g(x) = \dfrac{3}{x-2}$ and state its domain.

Solution

y-intercept:	$\left(0, -\frac{3}{2}\right)$, because $g(0) = -\frac{3}{2}$
x-intercept:	none, because there are no zeros of the numerator
Vertical asymptote:	$x = 2$, zero of denominator
Horizontal asymptote:	$y = 0$, because degree of $N(x) <$ degree of $D(x)$
Additional points:	

Test Interval	Representative x-Value	Value of g	Sign	Point on Graph
$(-\infty, 2)$	-4	$g(-4) = -\frac{1}{2}$	Negative	$\left(-4, -\frac{1}{2}\right)$
$(2, \infty)$	3	$g(3) = 3$	Positive	$(3, 3)$

By plotting the intercept, asymptotes, and a few additional points, you obtain the graph shown in Figure 2.22. The domain of g is all real numbers except $x = 2$.

✓ **Checkpoint** ◀))) *Audio-video solution in English & Spanish at LarsonPrecalculus.com*

Sketch the graph of $f(x) = \dfrac{1}{x+3}$ and state its domain.

EXAMPLE 4 **Sketching the Graph of a Rational Function**

Sketch the graph of $f(x) = (2x-1)/x$ and state its domain.

Solution

y-intercept:	none, because $x = 0$ is not in the domain
x-intercept:	$\left(\frac{1}{2}, 0\right)$, because $2x - 1 = 0$ when $x = \frac{1}{2}$
Vertical asymptote:	$x = 0$, zero of denominator
Horizontal asymptote:	$y = 2$, because degree of $N(x) =$ degree of $D(x)$
Additional points:	

Test Interval	Representative x-Value	Value of f	Sign	Point on Graph
$(-\infty, 0)$	-1	$f(-1) = 3$	Positive	$(-1, 3)$
$\left(0, \frac{1}{2}\right)$	$\frac{1}{4}$	$f\left(\frac{1}{4}\right) = -2$	Negative	$\left(\frac{1}{4}, -2\right)$
$\left(\frac{1}{2}, \infty\right)$	4	$f(4) = \frac{7}{4}$	Positive	$\left(4, \frac{7}{4}\right)$

By plotting the intercept, asymptotes, and a few additional points, you obtain the graph shown in Figure 2.23. The domain of f is all real numbers except $x = 0$.

✓ **Checkpoint** ◀))) *Audio-video solution in English & Spanish at LarsonPrecalculus.com*

Sketch the graph of $g(x) = (3 + 2x)/(1 + x)$ and state its domain.

• • **REMARK** You can use transformations to help you sketch graphs of rational functions. For instance, the graph of g in Example 3 is a vertical stretch and a right shift of the graph of $f(x) = 1/x$ because

$$g(x) = \frac{3}{x-2}$$

$$= 3\left(\frac{1}{x-2}\right)$$

$$= 3f(x-2).$$

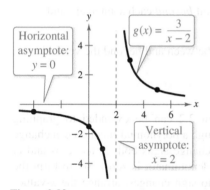

Figure 2.22

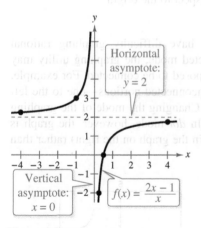

Figure 2.23

EXAMPLE 5 **Sketching the Graph of a Rational Function**

Sketch the graph of $f(x) = x/(x^2 - x - 2)$.

Solution Factoring the denominator, you have $f(x) = x/[(x + 1)(x - 2)]$.

Intercept: $(0, 0)$, because $f(0) = 0$

Vertical asymptotes: $x = -1$, $x = 2$, zeros of denominator

Horizontal asymptote: $y = 0$, because degree of $N(x) <$ degree of $D(x)$

Additional points:

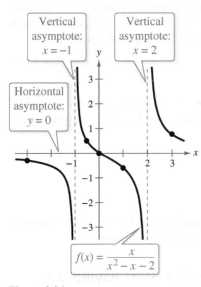

Vertical asymptote: $x = -1$

Vertical asymptote: $x = 2$

Horizontal asymptote: $y = 0$

$f(x) = \dfrac{x}{x^2 - x - 2}$

Figure 2.24

Test Interval	Representative x-Value	Value of f	Sign	Point on Graph
$(-\infty, -1)$	-3	$f(-3) = \frac{3}{10}$	Negative	$\left(-3, -\frac{3}{10}\right)$
$(-1, 0)$	$-\frac{1}{2}$	$f\left(-\frac{1}{2}\right) = \frac{2}{5}$	Positive	$\left(-\frac{1}{2}, \frac{2}{5}\right)$
$(0, 2)$	1	$f(1) = -\frac{1}{2}$	Negative	$\left(1, -\frac{1}{2}\right)$
$(2, \infty)$	3	$f(3) = \frac{3}{4}$	Positive	$\left(3, \frac{3}{4}\right)$

Figure 2.24 shows the graph of this function.

✓ *Checkpoint* 🔊))) *Audio-video solution in English & Spanish at LarsonPrecalculus.com*

Sketch the graph of $f(x) = 3x/(x^2 + x - 2)$.

•• **REMARK** If you are unsure of the shape of a portion of the graph of a rational function, then plot some additional points. Also note that when the numerator and the denominator of a rational function have a common factor, the graph of the function has a *hole* at the zero of the common factor. (See Example 6.)

EXAMPLE 6 **A Rational Function with Common Factors**

Sketch the graph of $f(x) = (x^2 - 9)/(x^2 - 2x - 3)$.

Solution By factoring the numerator and denominator, you have

$$f(x) = \frac{x^2 - 9}{x^2 - 2x - 3} = \frac{(x - 3)(x + 3)}{(x - 3)(x + 1)} = \frac{x + 3}{x + 1}, \quad x \neq 3.$$

y-intercept: $(0, 3)$, because $f(0) = 3$

x-intercept: $(-3, 0)$, because $x + 3 = 0$ when $x = -3$

Vertical asymptote: $x = -1$, zero of (simplified) denominator

Horizontal asymptote: $y = 1$, because degree of $N(x) =$ degree of $D(x)$

Additional points:

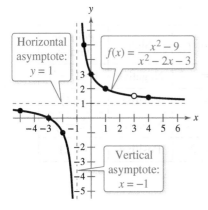

Horizontal asymptote: $y = 1$

$f(x) = \dfrac{x^2 - 9}{x^2 - 2x - 3}$

Vertical asymptote: $x = -1$

Hole at $x = 3$

Figure 2.25

Test Interval	Representative x-Value	Value of f	Sign	Point on Graph
$(-\infty, -3)$	-4	$f(-4) = \frac{1}{3}$	Positive	$\left(-4, \frac{1}{3}\right)$
$(-3, -1)$	-2	$f(-2) = -1$	Negative	$(-2, -1)$
$(-1, \infty)$	2	$f(2) = \frac{5}{3}$	Positive	$\left(2, \frac{5}{3}\right)$

Figure 2.25 shows the graph of this function. Notice that there is a hole in the graph at $x = 3$, because the numerator and denominator have a common factor of $x - 3$.

✓ *Checkpoint* 🔊))) *Audio-video solution in English & Spanish at LarsonPrecalculus.com*

Sketch the graph of $f(x) = (x^2 - 4)/(x^2 - x - 6)$.

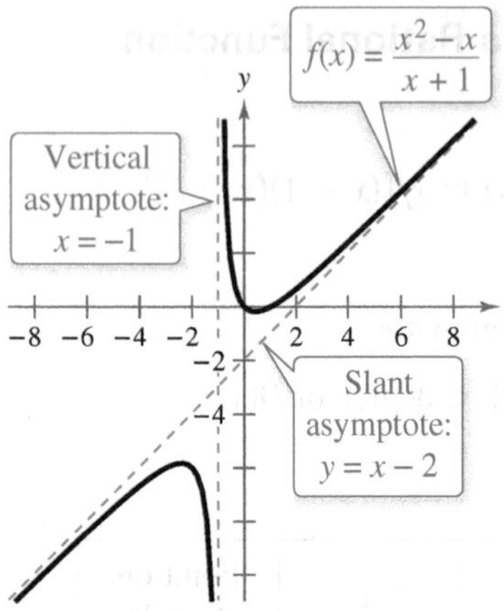

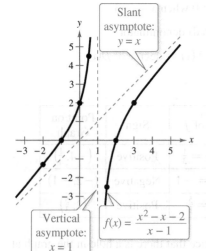

Figure 2.26

Slant Asymptotes

Consider a rational function whose denominator is of degree 1 or greater. If the degree of the numerator is exactly *one more* than the degree of the denominator, then the graph of the function has a **slant** (or **oblique**) **asymptote.** For example, the graph of

$$f(x) = \frac{x^2 - x}{x + 1}$$

has a slant asymptote, as shown in Figure 2.26. To find the equation of a slant asymptote, use long division. For example, by dividing $x + 1$ into $x^2 - x$, you obtain

$$f(x) = \frac{x^2 - x}{x + 1} = \underbrace{x - 2}_{\substack{\text{Slant asymptote} \\ (y = x - 2)}} + \frac{2}{x + 1}.$$

As x increases or decreases without bound, the remainder term $2/(x + 1)$ approaches 0, so the graph of f approaches the line $y = x - 2$, as shown in Figure 2.26.

EXAMPLE 7 A Rational Function with a Slant Asymptote

Sketch the graph of $f(x) = \dfrac{x^2 - x - 2}{x - 1}$.

Solution Factoring the numerator as $(x - 2)(x + 1)$ enables you to recognize the x-intercepts. Using long division

$$f(x) = \frac{x^2 - x - 2}{x - 1} = x - \frac{2}{x - 1}$$

enables you to recognize that the line $y = x$ is a slant asymptote of the graph.

y-intercept: (0, 2), because $f(0) = 2$

x-intercepts: (2, 0) and $(-1, 0)$, because $x - 2 = 0$ when $x = 2$ and $x + 1 = 0$ when $x = -1$

Vertical asymptote: $x = 1$, zero of denominator

Slant asymptote: $y = x$

Additional points:

Test Interval	Representative x-Value	Value of f	Sign	Point on Graph
$(-\infty, -1)$	-2	$f(-2) = -\frac{4}{3}$	Negative	$\left(-2, -\frac{4}{3}\right)$
$(-1, 1)$	$\frac{1}{2}$	$f\left(\frac{1}{2}\right) = \frac{9}{2}$	Positive	$\left(\frac{1}{2}, \frac{9}{2}\right)$
$(1, 2)$	$\frac{3}{2}$	$f\left(\frac{3}{2}\right) = -\frac{5}{2}$	Negative	$\left(\frac{3}{2}, -\frac{5}{2}\right)$
$(2, \infty)$	3	$f(3) = 2$	Positive	$(3, 2)$

Figure 2.27 shows the graph of the function.

✓ **Checkpoint** 🔊)) *Audio-video solution in English & Spanish at LarsonPrecalculus.com*

Sketch the graph of $f(x) = \dfrac{3x^2 + 1}{x}$.

Figure 2.27

Application

There are many examples of asymptotic behavior in real life. For instance, Example 8 shows how a vertical asymptote can help you to analyze the cost of removing pollutants from smokestack emissions.

EXAMPLE 8 **Cost-Benefit Model**

A utility company burns coal to generate electricity. The cost C (in dollars) of removing $p\%$ of the smokestack pollutants is given by

$$C = \frac{80,000p}{100 - p}, \quad 0 \le p \le 100.$$

You are a member of a state legislature considering a law that would require utility companies to remove 90% of the pollutants from their smokestack emissions. The current law requires 85% removal. How much additional cost would the utility company incur as a result of the new law?

Algebraic Solution

The current law requires 85% removal, so the current cost to the utility company is

$$C = \frac{80,000(85)}{100 - 85} \qquad \text{Evaluate } C \text{ when } p = 85.$$

$$\approx \$453,333.$$

The cost to remove 90% of the pollutants would be

$$C = \frac{80,000(90)}{100 - 90} \qquad \text{Evaluate } C \text{ when } p = 90.$$

$$= \$720,000.$$

So, the new law would require the utility company to spend an additional

$$720,000 - 453,333 = \$266,667. \qquad \begin{array}{l}\text{Subtract 85\% removal cost}\\\text{from 90\% removal cost.}\end{array}$$

Graphical Solution

Use a graphing utility to graph the function

$$y_1 = \frac{80,000x}{100 - x}$$

and use the *value* feature to approximate the values of y_1 when $x = 85$ and $x = 90$, as shown below. Note that the graph has a vertical asymptote at

$$x = 100.$$

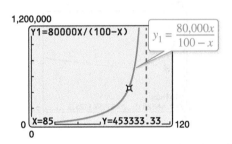

When $x = 85$, $y_1 \approx 453,333$.

When $x = 90$, $y_1 = 720,000$.

So, the new law would require the utility company to spend an additional

$$720,000 - 453,333 = \$266,667.$$

✓ *Checkpoint*))) *Audio-video solution in English & Spanish at LarsonPrecalculus.com*

The cost C (in millions of dollars) of removing $p\%$ of the industrial and municipal pollutants discharged into a river is given by

$$C = \frac{255p}{100 - p}, \quad 0 \le p < 100.$$

a. Find the costs of removing 20%, 45%, and 80% of the pollutants.

b. According to the model, is it possible to remove 100% of the pollutants? Explain.

EXAMPLE 9 **Finding a Minimum Area**

A rectangular page contains 48 square inches of print. The margins at the top and bottom of the page are each 1 inch deep. The margins on each side are $1\frac{1}{2}$ inches wide. What should the dimensions of the page be to use the least amount of paper?

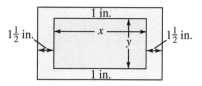

Figure 2.28

Graphical Solution

Let A be the area to be minimized. From Figure 2.28, you can write $A = (x + 3)(y + 2)$. The printed area inside the margins is given by $xy = 48$ or $y = 48/x$. To find the minimum area, rewrite the equation for A in terms of just one variable by substituting $48/x$ for y.

$$A = (x + 3)\left(\frac{48}{x} + 2\right) = \frac{(x + 3)(48 + 2x)}{x}, \quad x > 0$$

The graph of this rational function is shown below. Because x represents the width of the printed area, you need to consider only the portion of the graph for which x is positive. Use the *minimum* feature of a graphing utility to estimate that the minimum value of A occurs when $x \approx 8.5$ inches. The corresponding value of y is $48/8.5 \approx 5.6$ inches. So, the dimensions should be $x + 3 \approx 11.5$ inches by $y + 2 \approx 7.6$ inches.

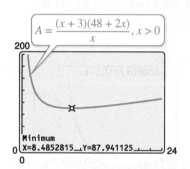

$$A = \frac{(x + 3)(48 + 2x)}{x}, x > 0$$

Numerical Solution

Let A be the area to be minimized. From Figure 2.28, you can write $A = (x + 3)(y + 2)$. The printed area inside the margins is given by $xy = 48$ or $y = 48/x$. To find the minimum area, rewrite the equation for A in terms of just one variable by substituting $48/x$ for y.

$$A = (x + 3)\left(\frac{48}{x} + 2\right) = \frac{(x + 3)(48 + 2x)}{x}, \quad x > 0$$

Use the *table* feature of a graphing utility to create a table of values for the function $y_1 = [(x + 3)(48 + 2x)]/x$ beginning at $x = 1$ and increasing by 1. The minimum value of y_1 occurs when x is somewhere between 8 and 9, as shown in Figure 2.29. To approximate the minimum value of y_1 to one decimal place, change the table to begin at $x = 8$ and increase by 0.1. The minimum value of y_1 occurs when $x \approx 8.5$, as shown in Figure 2.30. The corresponding value of y is $48/8.5 \approx 5.6$ inches. So, the dimensions should be $x + 3 \approx 11.5$ inches by $y + 2 \approx 7.6$ inches.

X	Y₁	
6	90	
7	88.571	
8	88	
9	88	
10	88.4	
11	89.091	
12	90	
X=8		

X	Y₁	
8.2	87.961	
8.3	87.949	
8.4	87.943	
8.5	87.941	
8.6	87.944	
8.7	87.952	
8.8	87.964	
X=8.5		

Figure 2.29   **Figure 2.30**

✓ **Checkpoint** ◄))) *Audio-video solution in English & Spanish at LarsonPrecalculus.com*

Rework Example 9 when the margins on each side are 2 inches wide and the page contains 40 square inches of print.

Summarize (Section 2.6)

1. State the definition of a rational function and describe the domain *(page 166)*. For an example of finding the domain of a rational function, see Example 1.

2. Explain how to find the vertical and horizontal asymptotes of the graph of a rational function *(page 168)*. For an example of finding vertical and horizontal asymptotes of graphs of rational functions, see Example 2.

3. Explain how to sketch the graph of a rational function *(page 169)*. For examples of sketching the graphs of rational functions, see Examples 3–6.

4. Explain how to determine whether the graph of a rational function has a slant asymptote *(page 172)*. For an example of sketching the graph of a rational function that has a slant asymptote, see Example 7.

5. Describe examples of how to use rational functions to model and solve real-life problems *(pages 173 and 174, Examples 8 and 9)*.

2.6 Exercises

Vocabulary: Fill in the blanks.

1. Functions of the form $f(x) = N(x)/D(x)$, where $N(x)$ and $D(x)$ are polynomials and $D(x)$ is not the zero polynomial, are called _____ _____.

2. When $f(x) \to \pm\infty$ as $x \to a$ from the left or the right, $x = a$ is a _____ _____ of the graph of f.

3. When $f(x) \to b$ as $x \to \pm\infty$, $y = b$ is a _____ _____ of the graph of f.

4. For the rational function $f(x) = N(x)/D(x)$, if the degree of $N(x)$ is exactly one more than the degree of $D(x)$, then the graph of f has a _____ (or oblique) _____.

Skills and Applications

Finding the Domain of a Rational Function In Exercises 5–8, find the domain of the function and discuss the behavior of f near any excluded x-values.

5. $f(x) = \dfrac{1}{x - 1}$

6. $f(x) = \dfrac{5x}{x + 2}$

7. $f(x) = \dfrac{3x^2}{x^2 - 1}$

8. $f(x) = \dfrac{2x}{x^2 - 4}$

Finding Vertical and Horizontal Asymptotes In Exercises 9–16, find all vertical and horizontal asymptotes of the graph of the function.

9. $f(x) = \dfrac{4}{x^2}$

10. $f(x) = \dfrac{1}{(x - 2)^3}$

11. $f(x) = \dfrac{5 + x}{5 - x}$

12. $f(x) = \dfrac{3 - 7x}{3 + 2x}$

13. $f(x) = \dfrac{x^3}{x^2 - x}$

14. $f(x) = \dfrac{4x^2}{x + 2}$

15. $f(x) = \dfrac{x^2 - 3x - 4}{2x^2 + x - 1}$

16. $f(x) = \dfrac{-4x^2 + 1}{x^2 + x + 3}$

Sketching the Graph of a Rational Function In Exercises 17–38, (a) state the domain of the function, (b) identify all intercepts, (c) find any vertical or horizontal asymptotes, and (d) plot additional solution points as needed to sketch the graph of the rational function.

17. $f(x) = \dfrac{1}{x + 1}$

18. $f(x) = \dfrac{1}{x - 3}$

19. $h(x) = \dfrac{-1}{x + 4}$

20. $g(x) = \dfrac{1}{6 - x}$

21. $C(x) = \dfrac{2x + 3}{x + 2}$

22. $P(x) = \dfrac{1 - 3x}{1 - x}$

23. $f(x) = \dfrac{x^2}{x^2 + 9}$

24. $f(t) = \dfrac{1 - 2t}{t}$

25. $g(s) = \dfrac{4s}{s^2 + 4}$

26. $f(x) = -\dfrac{x}{(x - 2)^2}$

27. $h(x) = \dfrac{2x}{x^2 - 3x - 4}$

28. $g(x) = \dfrac{3x}{x^2 + 2x - 3}$

29. $f(x) = \dfrac{x - 4}{x^2 - 16}$

30. $f(x) = \dfrac{x + 1}{x^2 - 1}$

31. $f(t) = \dfrac{t^2 - 1}{t - 1}$

32. $f(x) = \dfrac{x^2 - 36}{x + 6}$

33. $f(x) = \dfrac{x^2 - 25}{x^2 - 4x - 5}$

34. $f(x) = \dfrac{x^2 - 4}{x^2 - 3x + 2}$

35. $f(x) = \dfrac{x^2 + 3x}{x^2 + x - 6}$

36. $f(x) = \dfrac{5(x + 4)}{x^2 + x - 12}$

37. $f(x) = \dfrac{2x^2 - 5x - 3}{x^3 - 2x^2 - x + 2}$

38. $f(x) = \dfrac{x^2 - x - 2}{x^3 - 2x^2 - 5x + 6}$

Matching In Exercises 39–42, match the rational function with its graph. [The graphs are labeled (a)–(d).]

(a)
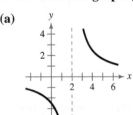

(b)
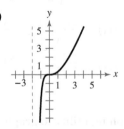

(c)
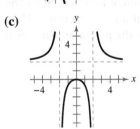

(d)
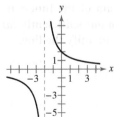

39. $f(x) = \dfrac{4}{x + 2}$

40. $f(x) = \dfrac{5}{x - 2}$

41. $f(x) = \dfrac{2x}{x^2 - 4}$

42. $f(x) = \dfrac{3x^3}{(x + 2)^2}$

Comparing Graphs of Functions **In Exercises 43–46**, (a) state the domains of f and g, (b) use a graphing utility to graph f and g in the same viewing window, and (c) explain why the graphing utility may not show the difference in the domains of f and g.

43. $f(x) = \dfrac{x^2 - 1}{x + 1}$, $\quad g(x) = x - 1$

44. $f(x) = \dfrac{x^2(x - 2)}{x^2 - 2x}$, $\quad g(x) = x$

45. $f(x) = \dfrac{x - 2}{x^2 - 2x}$, $\quad g(x) = \dfrac{1}{x}$

46. $f(x) = \dfrac{2x - 6}{x^2 - 7x + 12}$, $\quad g(x) = \dfrac{2}{x - 4}$

A Rational Function with a Slant Asymptote **In Exercises 47–60**, (a) state the domain of the function, (b) identify all intercepts, (c) find any vertical or slant asymptotes, and (d) plot additional solution points as needed to sketch the graph of the rational function.

47. $h(x) = \dfrac{x^2 - 4}{x}$

48. $g(x) = \dfrac{x^2 + 5}{x}$

49. $f(x) = \dfrac{2x^2 + 1}{x}$

50. $f(x) = \dfrac{-x^2 - 2}{x}$

51. $g(x) = \dfrac{x^2 + 1}{x}$

52. $h(x) = \dfrac{x^2}{x - 1}$

53. $f(t) = -\dfrac{t^2 + 1}{t + 5}$

54. $f(x) = \dfrac{x^2 + 1}{x + 1}$

55. $f(x) = \dfrac{x^3}{x^2 - 4}$

56. $g(x) = \dfrac{x^3}{2x^2 - 8}$

57. $f(x) = \dfrac{x^2 - x + 1}{x - 1}$

58. $f(x) = \dfrac{2x^2 - 5x + 5}{x - 2}$

59. $f(x) = \dfrac{2x^3 - x^2 - 2x + 1}{x^2 + 3x + 2}$

60. $f(x) = \dfrac{2x^3 + x^2 - 8x - 4}{x^2 - 3x + 2}$

 Using Technology **In Exercises 61–64**, use a graphing utility to graph the rational function. State the domain of the function and find any asymptotes. Then zoom out sufficiently far so that the graph appears as a line. Identify the line.

61. $f(x) = \dfrac{x^2 + 2x - 8}{x + 2}$

62. $f(x) = \dfrac{2x^2 + x}{x + 1}$

63. $g(x) = \dfrac{1 + 3x^2 - x^3}{x^2}$

64. $h(x) = \dfrac{12 - 2x - x^2}{2(4 + x)}$

Graphical Reasoning **In Exercises 65–68**, (a) use the graph to determine any x-intercepts of the graph of the rational function and (b) set $y = 0$ and solve the resulting equation to confirm your result in part (a).

65. $y = \dfrac{x + 1}{x - 3}$

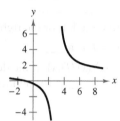

66. $y = \dfrac{2x}{x - 3}$

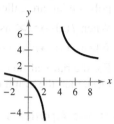

67. $y = \dfrac{1}{x} - x$

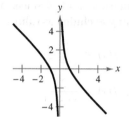

68. $y = x - 3 + \dfrac{2}{x}$

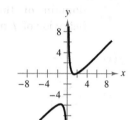

69. Recycling

The cost C (in dollars) of supplying recycling bins to $p\%$ of the population of a rural township is given by

$$C = \dfrac{25{,}000p}{100 - p}, \quad 0 \le p < 100.$$

(a) Use a graphing utility to graph the cost function.

(b) Find the costs of supplying bins to 15%, 50%, and 90% of the population.

(c) According to the model, is it possible to supply bins to 100% of the population? Explain.

70. Population Growth The game commission introduces 100 deer into newly acquired state game lands. The population N of the herd is modeled by

$$N = \dfrac{20(5 + 3t)}{1 + 0.04t}, \quad t \ge 0$$

where t is the time in years.

 (a) Use a graphing utility to graph this model.

(b) Find the populations when $t = 5$, $t = 10$, and $t = 25$.

(c) What is the limiting size of the herd as time increases?

71. Page Design   A rectangular page contains 64 square inches of print. The margins at the top and bottom of the page are each 1 inch deep. The margins on each side are $1\frac{1}{2}$ inches wide. What should the dimensions of the page be to use the least amount of paper?

72. Page Design   A page that is x inches wide and y inches high contains 30 square inches of print. The top and bottom margins are each 1 inch deep, and the margins on each side are 2 inches wide (see figure).

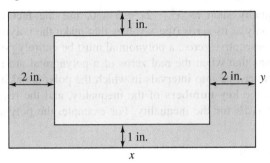

(a) Write a functions for the total area A of the page in terms of x.

(b) Determine the domain of the function based on the physical constraints of the problem.

(c) Use a graphing utility to graph the area function and approximate the dimensions of the page that use the least amount of paper.

73. Average Speed   A driver's average speed is 50 miles per hour on a round trip between two cities 100 miles apart. The average speeds for going and returning were x and y miles per hour, respectively.

(a) Show that $y = (25x)/(x - 25)$.

(b) Determine the vertical and horizontal asymptotes of the graph of the function.

(c) Use a graphing utility to graph the function.

(d) Complete the table.

x	30	35	40	45	50	55	60
y							

(e) Are the results in the table what you expected? Explain.

(f) Is it possible to average 20 miles per hour in one direction and still average 50 miles per hour on the round trip? Explain.

74. Medicine   The concentration C of a chemical in the bloodstream t hours after injection into muscle tissue is given by

$$C = \frac{3t^2 + t}{t^3 + 50}, \quad t > 0.$$

Use a graphing utility to graph the function. Determine the horizontal asymptote of the graph of the function and interpret its meaning in the context of the problem.

Exploration

True or False?   **In Exercises 75–77, determine whether the statement is true or false. Justify your answer.**

75. The graph of a polynomial function can have infinitely many vertical asymptotes.

76. The graph of a rational function can never cross one of its asymptotes.

77. The graph of a rational function can have a vertical asymptote, a horizontal asymptote, and a slant asymptote.

78. 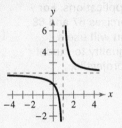 **HOW DO YOU SEE IT?**   The graph of a rational function

$$f(x) = \frac{N(x)}{D(x)}$$

is shown below. Determine which of the statements about the function is false. Justify your answer.

(a) $D(1) = 0$.

(b) The degree of $N(x)$ and $D(x)$ are equal.

(c) The ratio of the leading coefficients of $N(x)$ and $D(x)$ is 1.

79. Writing   Is every rational function a polynomial function? Is every polynomial function a rational function? Explain.

Writing a Rational Function   **In Exercises 80–82, write a rational function f whose graph has the specified characteristics. (There are many correct answers.)**

80. Vertical asymptote: None

Horizontal asymptote: $y = 2$

81. Vertical asymptotes: $x = -2$, $x = 1$

Horizontal asymptote: None

82. Vertical asymptote: $x = 2$

Slant asymptote: $y = x + 1$

Zero of the function: $x = -2$

Project: Department of Defense   To work an extended application analyzing the total numbers of military personnel on active duty from 1984 through 2014, visit this text's website at *LarsonPrecalculus.com*. (*Source: U.S. Department of Defense*)

2.7 Nonlinear Inequalities

Nonlinear inequalities have many real-life applications. For example, in Exercises 67 and 68 on page 186, you will use a polynomial inequality to model the height of a projectile.

■ Solve polynomial inequalities.
■ Solve rational inequalities.
■ Use nonlinear inequalities to model and solve real-life problems.

Polynomial Inequalities

To solve a polynomial inequality such as $x^2 - 2x - 3 < 0$, use the fact that a polynomial can change signs only at its *zeros* (the x-values that make the polynomial equal to zero). Between two consecutive zeros, a polynomial must be entirely positive or entirely negative. This means that when the real zeros of a polynomial are put in order, they divide the real number line into intervals in which the polynomial has no sign changes. These zeros are the **key numbers** of the inequality, and the resulting open intervals are the *test intervals* for the inequality. For example, the polynomial $x^2 - 2x - 3$ factors as

$$x^2 - 2x - 3 = (x + 1)(x - 3)$$

so it has two zeros,

$$x = -1 \quad \text{and} \quad x = 3.$$

These zeros divide the real number line into three test intervals:

$$(-\infty, -1), \quad (-1, 3), \quad \text{and} \quad (3, \infty). \quad \text{(See figure below.)}$$

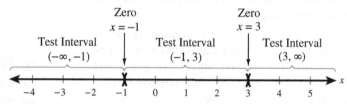

Three test intervals for $x^2 - 2x - 3$

To solve the inequality $x^2 - 2x - 3 < 0$, you need to test only one value from each of these test intervals. When a value from a test interval satisfies the original inequality, you can conclude that the interval is a solution of the inequality.

Use the same basic approach, generalized below, to find the solution set of any polynomial inequality.

· ·REMARK   The solution set of

$$x^2 - 2x - 3 < 0$$

discussed above, is the open interval $(-1, 3)$. Use Step 3 to verify this. By choosing the representative x-values $x = -2$, $x = 0$, and $x = 4$, you will find that the value of the polynomial is negative only in $(-1, 3)$.

Test Intervals for a Polynomial Inequality

To determine the intervals on which the values of a polynomial are entirely negative or entirely positive, use the steps below.

1. Find all real zeros of the polynomial, and arrange the zeros in increasing order. These zeros are the key numbers of the inequality.

2. Use the key numbers of the inequality to determine the test intervals.

3. Choose one representative x-value in each test interval and evaluate the polynomial at that value. When the value of the polynomial is negative, the polynomial has negative values for every x-value in the interval. When the value of the polynomial is positive, the polynomial has positive values for every x-value in the interval.

EXAMPLE 1   **Solving a Polynomial Inequality**

Solve $x^2 - x - 6 < 0$. Then graph the solution set.

▷ **ALGEBRA HELP** To review the techniques for factoring polynomials, see Appendix A.5.

Solution   Factoring the polynomial

$$x^2 - x - 6 = (x + 2)(x - 3)$$

shows that the key numbers are $x = -2$ and $x = 3$. So, the inequality's test intervals are

$$(-\infty, -2), \quad (-2, 3), \quad \text{and} \quad (3, \infty) \qquad \text{Test intervals}$$

In each test interval, choose a representative x-value and evaluate the polynomial.

Test Interval	x-Value	Polynomial Value	Conclusion
$(-\infty, -2)$	$x = -3$	$(-3)^2 - (-3) - 6 = 6$	Positive
$(-2, 3)$	$x = 0$	$(0)^2 - (0) - 6 = -6$	Negative
$(3, \infty)$	$x = 4$	$(4)^2 - (4) - 6 = 6$	Positive

The inequality is satisfied for all x-values in $(-2, 3)$. This implies that the solution set of the inequality

$$x^2 - x - 6 < 0$$

is the interval $(-2, 3)$, as shown on the number line below. Note that the original inequality contains a "less than" symbol. This means that the solution set does not contain the endpoints of the test interval $(-2, 3)$.

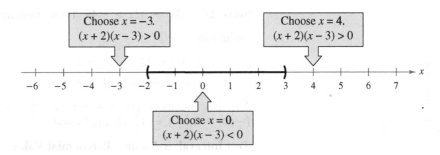

✓ *Checkpoint* ◀))) *Audio-video solution in English & Spanish at LarsonPrecalculus.com*

Solve $x^2 - x - 20 < 0$. Then graph the solution set.

As with linear inequalities, you can check the reasonableness of a solution by substituting x-values into the original inequality. For instance, to check the solution found in Example 1, substitute several x-values from the interval $(-2, 3)$ into the inequality

$$x^2 - x - 6 < 0.$$

Regardless of which x-values you choose, the inequality should be satisfied.

You can also use a graph to check the result of Example 1. Sketch the graph of

$$y = x^2 - x - 6$$

as shown in Figure 2.31. Notice that the graph is below the x-axis on the interval $(-2, 3)$.

In Example 1, the polynomial inequality is in general form (with the polynomial on one side and zero on the other). Whenever this is not the case, you should begin by writing the inequality in general form.

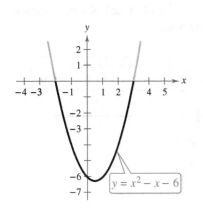

Figure 2.31

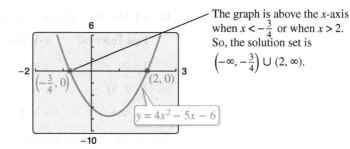

> **EXAMPLE 2** **Solving a Polynomial Inequality**

See LarsonPrecalculus.com for an interactive version of this type of example.

Solve $4x^2 - 5x > 6$.

Algebraic Solution

$$4x^2 - 5x - 6 > 0 \qquad \text{Write in general form.}$$

$$(x - 2)(4x + 3) > 0 \qquad \text{Factor.}$$

Key numbers: $x = -\frac{3}{4}$, $x = 2$

Test intervals: $\left(-\infty, -\frac{3}{4}\right), \left(-\frac{3}{4}, 2\right), (2, \infty)$

Test: Is $(x - 2)(4x + 3) > 0$?

Testing these intervals shows that the polynomial $4x^2 - 5x - 6$ is positive on the open intervals $\left(-\infty, -\frac{3}{4}\right)$ and $(2, \infty)$. So, the solution set of the inequality is $\left(-\infty, -\frac{3}{4}\right) \cup (2, \infty)$.

Graphical Solution

First write the polynomial inequality $4x^2 - 5x > 6$ as $4x^2 - 5x - 6 > 0$. Then use a graphing utility to graph $y = 4x^2 - 5x - 6$.

The graph is above the x-axis when $x < -\frac{3}{4}$ or when $x > 2$. So, the solution set is $\left(-\infty, -\frac{3}{4}\right) \cup (2, \infty)$.

✓ **Checkpoint** ◀ッ)) *Audio-video solution in English & Spanish at LarsonPrecalculus.com*

Solve $2x^2 + 3x < 5$ (a) algebraically and (b) graphically.

> **EXAMPLE 3** **Solving a Polynomial Inequality**

Solve $2x^3 - 3x^2 - 32x > -48$. Then graph the solution set.

Solution

$$2x^3 - 3x^2 - 32x + 48 > 0 \qquad \text{Write in general form.}$$

$$(x - 4)(x + 4)(2x - 3) > 0 \qquad \text{Factor by grouping.}$$

The key numbers are $x = -4$, $x = \frac{3}{2}$, and $x = 4$, and the test intervals are $(-\infty, -4), \left(-4, \frac{3}{2}\right), \left(\frac{3}{2}, 4\right)$, and $(4, \infty)$.

Test Interval	x-Value	Polynomial Value	Conclusion
$(-\infty, -4)$	$x = -5$	$2(-5)^3 - 3(-5)^2 - 32(-5) + 48 = -117$	Negative
$\left(-4, \frac{3}{2}\right)$	$x = 0$	$2(0)^3 - 3(0)^2 - 32(0) + 48 = 48$	Positive
$\left(\frac{3}{2}, 4\right)$	$x = 2$	$2(2)^3 - 3(2)^2 - 32(2) + 48 = -12$	Negative
$(4, \infty)$	$x = 5$	$2(5)^3 - 3(5)^2 - 32(5) + 48 = 63$	Positive

The inequality is satisfied on the open intervals $\left(-4, \frac{3}{2}\right)$ and $(4, \infty)$. So, the solution set is $\left(-4, \frac{3}{2}\right) \cup (4, \infty)$, as shown on the number line below.

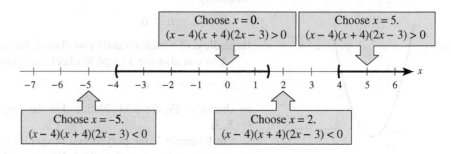

✓ **Checkpoint** ◀ッ)) *Audio-video solution in English & Spanish at LarsonPrecalculus.com*

Solve $3x^3 - x^2 - 12x > -4$. Then graph the solution set.

You may find it easier to determine the sign of a polynomial from its *factored* form. For instance, in Example 2, when you substitute the test value $x = 1$ into the factored form

$$(x - 2)(4x + 3)$$

the sign pattern of the factors is

$$(-)(+)$$

which yields a negative result. Use factored forms to determine the signs of the polynomials in other examples in this section.

When solving a polynomial inequality, be sure to account for the inequality symbol. For instance, in Example 2, note that the original inequality symbol is "greater than" and the solution consists of two open intervals. If the original inequality had been

$$4x^2 - 5x \geq 6$$

then the solution set would have been

$$\left(-\infty, -\tfrac{3}{4}\right] \cup [2, \infty).$$

Each of the polynomial inequalities in Examples 1, 2, and 3 has a solution set that consists of a single interval or the union of two intervals. When solving the exercises for this section, watch for unusual solution sets, as illustrated in Example 4.

EXAMPLE 4   Unusual Solution Sets

a. The solution set of

$$x^2 + 2x + 4 > 0$$

consists of the entire set of real numbers, $(-\infty, \infty)$. In other words, the value of the quadratic polynomial $x^2 + 2x + 4$ is positive for every real value of x.

b. The solution set of

$$x^2 + 2x + 1 \leq 0$$

consists of the single real number $\{-1\}$, because the inequality has only one key number, $x = -1$, and it is the only value that satisfies the inequality.

c. The solution set of

$$x^2 + 3x + 5 < 0$$

is empty. In other words, $x^2 + 3x + 5$ is not less than zero for any value of x.

d. The solution set of

$$x^2 - 4x + 4 > 0$$

consists of all real numbers except $x = 2$. This solution set can be written in interval notation as

$$(-\infty, 2) \cup (2, \infty).$$

✓ **Checkpoint** ◀))) Audio-video solution in English & Spanish at LarsonPrecalculus.com

What is unusual about the solution set of each inequality?

a. $x^2 + 6x + 9 < 0$

b. $x^2 + 4x + 4 \leq 0$

c. $x^2 - 6x + 9 > 0$

d. $x^2 - 2x + 1 \geq 0$

Rational Inequalities

The concepts of key numbers and test intervals can be extended to rational inequalities. Use the fact that the value of a rational expression can change sign at its *zeros* (the *x*-values for which its numerator is zero) and at its *undefined values* (the *x*-values for which its denominator is zero). These two types of numbers make up the *key numbers* of a rational inequality. When solving a rational inequality, begin by writing the inequality in general form, that is, with zero on the right side of the inequality.

EXAMPLE 5   Solving a Rational Inequality

•·**REMARK**   By writing 3 as $\frac{3}{1}$, you should be able to see that the least common denominator is $(x - 5)(1) = x - 5$. So, rewriting the general form as

$$\frac{2x - 7}{x - 5} - \frac{3(x - 5)}{x - 5} \le 0$$

and subtracting gives the result shown.

Solve $\dfrac{2x - 7}{x - 5} \le 3$. Then graph the solution set.

Solution

$$\frac{2x - 7}{x - 5} \le 3 \qquad \text{Write original inequality.}$$

$$\frac{2x - 7}{x - 5} - 3 \le 0 \qquad \text{Write in general form.}$$

$$\frac{2x - 7 - 3x + 15}{x - 5} \le 0 \qquad \text{Find the LCD and subtract fractions.}$$

$$\frac{-x + 8}{x - 5} \le 0 \qquad \text{Simplify.}$$

Key numbers: $x = 5$, $x = 8$ Zeros and undefined values of rational expression

Test intervals: $(-\infty, 5), (5, 8), (8, \infty)$

Test: Is $\dfrac{-x + 8}{x - 5} \le 0$?

Testing these intervals, as shown in the figure below, the inequality is satisfied on the open intervals $(-\infty, 5)$ and $(8, \infty)$. Moreover,

$$\frac{-x + 8}{x - 5} = 0$$

when $x = 8$, so the solution set is $(-\infty, 5) \cup [8, \infty)$. (Be sure to use a bracket to signify that $x = 8$ is included in the solution set.)

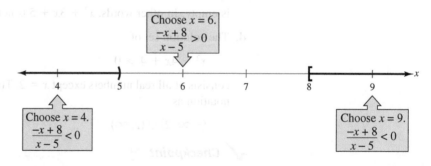

✓ **Checkpoint**))) *Audio-video solution in English & Spanish at LarsonPrecalculus.com*

Solve each inequality. Then graph the solution set.

a. $\dfrac{x - 2}{x - 3} \ge -3$

b. $\dfrac{4x - 1}{x - 6} > 3$

Applications

One common application of inequalities comes from business and involves profit, revenue, and cost. The formula that relates these three quantities is

$$\boxed{\text{Profit}} = \boxed{\text{Revenue}} - \boxed{\text{Cost}}$$

$$P = R - C.$$

EXAMPLE 6   Profit from a Product

The marketing department of a calculator manufacturer determines that the demand for a new model of calculator is

$$p = 100 - 0.00001x, \quad 0 \le x \le 10{,}000{,}000 \qquad \text{Demand equation}$$

where p is the price per calculator (in dollars) and x represents the number of calculators sold. (According to this model, no one would be willing to pay \$100 for the calculator. At the other extreme, the company could not *give* away more than 10 million calculators.) The revenue for selling x calculators is

$$R = xp = x(100 - 0.00001x). \qquad \text{Revenue equation}$$

The total cost of producing x calculators is \$10 per calculator plus a one-time development cost of \$2,500,000. So, the total cost is

$$C = 10x + 2{,}500{,}000. \qquad \text{Cost equation}$$

What prices can the company charge per calculator to obtain a profit of at least \$190,000,000?

Solution

Verbal model: $\boxed{\text{Profit}} = \boxed{\text{Revenue}} - \boxed{\text{Cost}}$

Equation: $P = R - C$

$$P = 100x - 0.00001x^2 - (10x + 2{,}500{,}000)$$

$$P = -0.00001x^2 + 90x - 2{,}500{,}000$$

To answer the question, solve the inequality

$$P \ge 190{,}000{,}000$$

$$-0.00001x^2 + 90x - 2{,}500{,}000 \ge 190{,}000{,}000.$$

Write the inequality in general form, find the key numbers and the test intervals, and then test a value in each test interval to find that the solution is

$$3{,}500{,}000 \le x \le 5{,}500{,}000$$

as shown in Figure 2.32. Substituting the x-values in the original demand equation shows that prices of

$$\$45.00 \le p \le \$65.00$$

yield a profit of at least \$190,000,000.

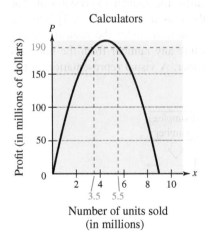

Calculators

Profit (in millions of dollars)

Number of units sold (in millions)

Figure 2.32

✓ *Checkpoint* ◀))) *Audio-video solution in English & Spanish at LarsonPrecalculus.com*

The revenue and cost equations for a product are

$$R = x(60 - 0.0001x) \quad \text{and} \quad C = 12x + 1{,}800{,}000$$

where R and C are measured in dollars and x represents the number of units sold. How many units must be sold to obtain a profit of at least \$3,600,000?

Another common application of inequalities is finding the domain of an expression that involves a square root, as shown in Example 7.

EXAMPLE 7   **Finding the Domain of an Expression**

Find the domain of $\sqrt{64 - 4x^2}$.

Algebraic Solution

Recall that the domain of an expression is the set of all x-values for which the expression is defined. The expression $\sqrt{64 - 4x^2}$ is defined only when $64 - 4x^2$ is nonnegative, so the inequality $64 - 4x^2 \geq 0$ gives the domain.

$$64 - 4x^2 \geq 0 \qquad \text{Write in general form.}$$
$$16 - x^2 \geq 0 \qquad \text{Divide each side by 4.}$$
$$(4 - x)(4 + x) \geq 0 \qquad \text{Write in factored form.}$$

The inequality has two key numbers: $x = -4$ and $x = 4$. Use these two numbers to test the inequality.

Key numbers: $x = -4, x = 4$

Test intervals: $(-\infty, -4), (-4, 4), (4, \infty)$

Test: Is $(4 - x)(4 + x) \geq 0$?

A test shows that the inequality is satisfied in the *closed interval* $[-4, 4]$. So, the domain of the expression $\sqrt{64 - 4x^2}$ is the closed interval $[-4, 4]$.

Graphical Solution

Begin by sketching the graph of the equation $y = \sqrt{64 - 4x^2}$, as shown below. The graph shows that the x-values extend from -4 to 4 (including -4 and 4). So, the domain of the expression $\sqrt{64 - 4x^2}$ is the closed interval $[-4, 4]$.

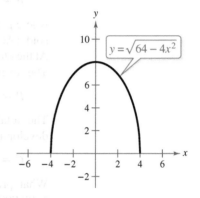

$y = \sqrt{64 - 4x^2}$

 **Checkpoint** *Audio-video solution in English & Spanish at LarsonPrecalculus.com*

Find the domain of $\sqrt{x^2 - 7x + 10}$.

You can check the reasonableness of the solution to Example 7 by choosing a representative x-value in the interval and evaluating the radical expression at that value. When you substitute any number from the closed interval $[-4, 4]$ into the expression $\sqrt{64 - 4x^2}$, you obtain a nonnegative number under the radical symbol that simplifies to a real number. When you substitute any number from the intervals $(-\infty, -4)$ and $(4, \infty)$, you obtain a complex number. A visual representation of the intervals is shown below.

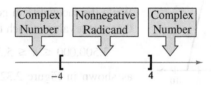

Summarize   (Section 2.7)

1. Explain how to solve a polynomial inequality *(page 178)*. For examples of solving polynomial inequalities, see Examples 1–4.

2. Explain how to solve a rational inequality *(page 182)*. For an example of solving a rational inequality, see Example 5.

3. Describe applications of polynomial inequalities *(pages 183 and 184, Examples 6 and 7)*.

2.7 Exercises

See CalcChat.com for tutorial help and worked-out solutions to odd-numbered exercises.

Vocabulary: Fill in the blanks.

1. Between two consecutive zeros, a polynomial must be entirely _____ or entirely _____.
2. To solve a polynomial inequality, find the _____ numbers of the inequality, and use these numbers to create _____ _____ for the inequality.
3. A rational expression can change sign at its _____ and its _____ _____.
4. The formula that relates cost, revenue, and profit is _____.

Skills and Applications

 Checking Solutions **In Exercises 5–8, determine whether each value of x is a solution of the inequality.**

Inequality	Values
5. $x^2 - 3 < 0$	(a) $x = 3$ (b) $x = 0$ (c) $x = \frac{3}{2}$ (d) $x = -5$
6. $x^2 - 2x - 8 \geq 0$	(a) $x = -2$ (b) $x = 0$ (c) $x = -4$ (d) $x = 1$
7. $\dfrac{x + 2}{x - 4} \geq 3$	(a) $x = 5$ (b) $x = 4$ (c) $x = -\frac{9}{2}$ (d) $x = \frac{9}{2}$
8. $\dfrac{3x^2}{x^2 + 4} < 1$	(a) $x = -2$ (b) $x = -1$ (c) $x = 0$ (d) $x = 3$

Finding Key Numbers **In Exercises 9–12, find the key numbers of the inequality.**

9. $x^2 - 3x - 18 > 0$
10. $9x^3 - 25x^2 \leq 0$
11. $\dfrac{1}{x - 5} + 1 \geq 0$
12. $\dfrac{x}{x + 2} - \dfrac{2}{x - 1} < 0$

 Solving a Polynomial Inequality **In Exercises 13–36, solve the inequality. Then graph the solution set.**

13. $2x^2 + 4x < 0$
14. $3x^2 - 9x \geq 0$
15. $x^2 < 9$
16. $x^2 \leq 25$
17. $(x + 2)^2 \leq 25$
18. $(x - 3)^2 \geq 1$
19. $x^2 + 6x + 1 \geq -7$
20. $x^2 - 8x + 2 < 11$
21. $x^2 + x < 6$
22. $x^2 + 2x > 3$
23. $x^2 < 3 - 2x$
24. $x^2 > 2x + 8$
25. $3x^2 - 11x > 20$
26. $-2x^2 + 6x \leq -15$
27. $x^3 - 3x^2 - x + 3 > 0$
28. $x^3 + 2x^2 - 4x \leq 8$
29. $-x^3 + 7x^2 + 9x > 63$
30. $2x^3 + 13x^2 - 8x \geq 52$
31. $4x^3 - 6x^2 < 0$
32. $4x^3 - 12x^2 > 0$
33. $x^3 - 4x \geq 0$
34. $2x^3 - x^4 \leq 0$
35. $(x - 1)^2(x + 2)^3 \geq 0$
36. $x^4(x - 3) \leq 0$

 Unusual Solution Sets **In Exercises 37–40, explain what is unusual about the solution set of the inequality.**

37. $4x^2 - 4x + 1 \leq 0$
38. $x^2 + 3x + 8 > 0$
39. $x^2 - 6x + 12 \leq 0$
40. $x^2 - 8x + 16 > 0$

Solving a Rational Inequality **In Exercises 41–52, solve the inequality. Then graph the solution set.**

41. $\dfrac{4x - 1}{x} > 0$
42. $\dfrac{x^2 - 1}{x} < 0$
43. $\dfrac{3x + 5}{x - 1} < 2$
44. $\dfrac{x + 12}{x + 2} \geq 3$
45. $\dfrac{2}{x + 5} > \dfrac{1}{x - 3}$
46. $\dfrac{5}{x - 6} > \dfrac{3}{x + 2}$
47. $\dfrac{1}{x - 3} \leq \dfrac{9}{4x + 3}$
48. $\dfrac{1}{x} \geq \dfrac{1}{x + 3}$
49. $\dfrac{x^2 + 2x}{x^2 - 9} \leq 0$
50. $\dfrac{x^2 + x - 6}{x} \geq 0$
51. $\dfrac{3}{x - 1} + \dfrac{2x}{x + 1} > -1$
52. $\dfrac{3x}{x - 1} \leq \dfrac{x}{x + 4} + 3$

Using Technology **In Exercises 53–60, use a graphing utility to graph the equation. Use the graph to approximate the values of x that satisfy each inequality.**

Equation	Inequalities
53. $y = -x^2 + 2x + 3$	(a) $y \leq 0$ (b) $y \geq 3$
54. $y = \frac{1}{2}x^2 - 2x + 1$	(a) $y \leq 0$ (b) $y \geq 7$
55. $y = \frac{1}{8}x^3 - \frac{1}{2}x$	(a) $y \geq 0$ (b) $y \leq 6$
56. $y = x^3 - x^2 - 16x + 16$	(a) $y \leq 0$ (b) $y \geq 36$
57. $y = \dfrac{3x}{x - 2}$	(a) $y \leq 0$ (b) $y \geq 6$
58. $y = \dfrac{2(x - 2)}{x + 1}$	(a) $y \leq 0$ (b) $y \geq 8$
59. $y = \dfrac{2x^2}{x^2 + 4}$	(a) $y \geq 1$ (b) $y \leq 2$
60. $y = \dfrac{5x}{x^2 + 4}$	(a) $y \geq 1$ (b) $y \leq 0$

Solving an Inequality   **In Exercises 61–66, solve the inequality. (Round your answers to two decimal places.)**

61. $0.3x^2 + 6.26 < 10.8$ **62.** $-1.3x^2 + 3.78 > 2.12$

63. $-0.5x^2 + 12.5x + 1.6 > 0$

64. $1.2x^2 + 4.8x + 3.1 < 5.3$

65. $\dfrac{1}{2.3x - 5.2} > 3.4$ **66.** $\dfrac{2}{3.1x - 3.7} > 5.8$

· · Height of a Projectile · · · · · · · · · · · · · · ·

In Exercises 67 and 68, use the position equation

$s = -16t^2 + v_0 t + s_0$

where s represents the height of an object (in feet), v_0 represents the initial velocity of the object (in feet per second), s_0 represents the initial height of the object (in feet), and t represents the time (in seconds).

67. A projectile is fired straight upward from ground level $(s_0 = 0)$ with an initial velocity of 160 feet per second.

(a) At what instant will it be back at ground level?

(b) When will the height exceed 384 feet?

68. A projectile is fired straight upward from ground level $(s_0 = 0)$ with an initial velocity of 128 feet per second.

(a) At what instant will it be back at ground level?

(b) When will the height be less than 128 feet?

69. Cost, Revenue, and Profit   The revenue and cost equations for a product are $R = x(75 - 0.0005x)$ and $C = 30x + 250,000$, where R and C are measured in dollars and x represents the number of units sold. How many units must be sold to obtain a profit of at least $750,000? What is the price per unit?

70. Cost, Revenue, and Profit   The revenue and cost equations for a product are $R = x(50 - 0.0002x)$ and $C = 12x + 150,000$, where R and C are measured in dollars and x represents the number of units sold. How many units must be sold to obtain a profit of at least $1,650,000? What is the price per unit?

Finding the Domain of an Expression   **In Exercises 71–76, find the domain of the expression. Use a graphing utility to verify your result.**

71. $\sqrt{4 - x^2}$ **72.** $\sqrt{x^2 - 9}$

73. $\sqrt{x^2 - 9x + 20}$ **74.** $\sqrt{49 - x^2}$

75. $\sqrt{\dfrac{x}{x^2 - 2x - 35}}$ **76.** $\sqrt{\dfrac{x}{x^2 - 9}}$

77. School Enrollment   The table shows the numbers N (in millions) of students enrolled in elementary and secondary schools in the United States from 2005 through 2014. *(Source: National Center for Education Statistics)*

DATA	Year	Number, N
	2005	49.11
	2006	49.32
	2007	49.29
	2008	49.27
	2009	49.36
	2010	49.48
	2011	49.52
	2012	49.77
	2013	49.94
	2014	49.99

Spreadsheet at LarsonPrecalculus.com

(a) Use a graphing utility to create a scatter plot of the data. Let t represent the year, with $t = 5$ corresponding to 2005.

(b) Use the *regression* feature of the graphing utility to find a *quartic* model for the data. (A quartic model has the form $at^4 + bt^3 + ct^2 + dt + e$, where a, b, c, d, and e are constant and t is variable.)

(c) Graph the model and the scatter plot in the same viewing window. How well does the model fit the data?

(d) According to the model, after 2014, when did the number of students enrolled in elementary and secondary schools fall below 48 million?

(e) Is the model valid for long-term predictions of student enrollment? Explain.

78. Safe Load   The maximum safe load uniformly distributed over a one-foot section of a two-inch-wide wooden beam can be approximated by the model

$\text{Load} = 168.5d^2 - 472.1$

where d is the depth of the beam.

(a) Evaluate the model for $d = 4$, $d = 6$, $d = 8$, $d = 10$, and $d = 12$. Use the results to create a bar graph.

(b) Determine the minimum depth of the beam that will safely support a load of 2000 pounds.

79. Geometry   A rectangular playing field with a perimeter of 100 meters is to have an area of at least 500 square meters. Within what bounds must the length of the rectangle lie?

80. Geometry A rectangular parking lot with a perimeter of 440 feet is to have an area of at least 8000 square feet. Within what bounds must the length of the rectangle lie?

81. Resistors When two resistors of resistances R_1 and R_2 are connected in parallel (see figure), the total resistance R satisfies the equation

$$\frac{1}{R} = \frac{1}{R_1} + \frac{1}{R_2}.$$

Find R_1 for a parallel circuit in which $R_2 = 2$ ohms and R must be at least 1 ohm.

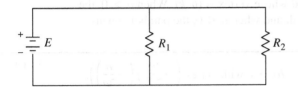

82. Teachers' Salaries The table shows the mean salaries S (in thousands of dollars) of public school classroom teachers in the United States from 2002 through 2013.

DATA	Year	Salary, S
	2002	44.7
	2003	45.7
	2004	46.5
	2005	47.5
	2006	49.1
	2007	51.1
	2008	52.8
	2009	54.3
	2010	55.2
	2011	56.1
	2012	55.4
	2013	56.4

Spreadsheet at LarsonPrecalculus.com

A model that approximates these data is

$$S = \frac{40.32 + 3.53t}{1 + 0.039t}, \quad 2 \le t \le 13$$

where t represents the year, with $t = 2$ corresponding to 2002. *(Source: National Center for Education Statistics)*

(a) Use a graphing utility to create a scatter plot of the data. Then graph the model in the same viewing window.

(b) How well does the model fit the data? Explain.

(c) Use the model to predict when the salary for classroom teachers will exceed $65,000.

(d) Is the model valid for long-term predictions of classroom teacher salaries? Explain.

Exploration

True or False? In Exercises 83 and 84, determine whether the statement is true or false. Justify your answer.

83. The zeros of the polynomial $x^3 - 2x^2 - 11x + 12$ divide the real number line into three test intervals.

84. The solution set of the inequality $\frac{3}{2}x^2 + 3x + 6 \ge 0$ is the entire set of real numbers.

85. Graphical Reasoning Use a graphing utility to verify the results in Example 4. For instance, the graph of $y = x^2 + 2x + 4$ is shown below. Notice that the y-values are greater than 0 for all values of x, as stated in Example 4(a). Use the graphing utility to graph $y = x^2 + 2x + 1$, $y = x^2 + 3x + 5$, and $y = x^2 - 4x + 4$. Explain how you can use the graphs to verify the results of parts (b), (c), and (d) of Example 4.

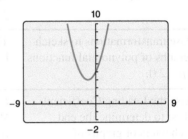

86. HOW DO YOU SEE IT? Consider the polynomial

$$(x - a)(x - b)$$

and the real number line shown below.

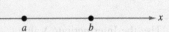

(a) Identify the points on the line at which the polynomial is zero.

(b) For each of the three subintervals of the real number line, write the sign of each factor and the sign of the product.

(c) At what x-values does the polynomial change signs?

Conjecture In Exercises 87–90, (a) find the interval(s) for b such that the equation has at least one real solution and (b) write a conjecture about the interval(s) based on the values of the coefficients.

87. $x^2 + bx + 9 = 0$

88. $x^2 + bx - 9 = 0$

89. $3x^2 + bx + 10 = 0$

90. $2x^2 + bx + 5 = 0$

Chapter Summary

	What Did You Learn?	Explanation/Examples	Review Exercises
Section 2.1	Analyze graphs of quadratic functions (p. 114).	Let a, b, and c be real numbers with $a \neq 0$. The function $f(x) = ax^2 + bx + c$ is a quadratic function. Its graph is a "U"-shaped curve called a parabola.	1, 2
	Write quadratic functions in standard form and use the results to sketch their graphs (p. 117).	The quadratic function $f(x) = a(x - h)^2 + k$, $a \neq 0$, is in standard form. The graph of f is a parabola whose axis is the vertical line $x = h$ and whose vertex is (h, k). When $a > 0$, the parabola opens upward, and when $a < 0$, the parabola opens downward.	3–8
	Find minimum and maximum values of quadratic functions in real-life applications (p. 119).	Consider $f(x) = ax^2 + bx + c$ with vertex $\left(-\dfrac{b}{2a}, f\left(-\dfrac{b}{2a}\right)\right)$. When $a > 0$, f has a *minimum* at $x = -b/(2a)$. When $a < 0$, f has a *maximum* at $x = -b/(2a)$.	9, 10
Section 2.2	Use transformations to sketch graphs of polynomial functions (p. 123).	The graph of a polynomial function is continuous (no breaks, holes, or gaps) and has only smooth, rounded turns.	11, 12
	Use the Leading Coefficient Test to determine the end behaviors of graphs of polynomial functions (p. 125).	Consider the graph of $f(x) = a_n x^n + \cdots + a_1 x + a_0$, $a_n \neq 0$. **When n is odd:** If $a_n > 0$, then the graph falls to the left and rises to the right. If $a_n < 0$, then the graph rises to the left and falls to the right. **When n is even:** If $a_n > 0$, then the graph rises to the left and to the right. If $a_n < 0$, then the graph falls to the left and to the right.	13–16
	Find real zeros of polynomial functions and use them as sketching aids (p. 127).	When f is a polynomial function and a is a real number, the following are equivalent: (1) $x = a$ is a *zero* of f, (2) $x = a$ is a *solution* of the equation $f(x) = 0$, (3) $(x - a)$ is a *factor* of the polynomial $f(x)$, and (4) $(a, 0)$ is an *x-intercept* of the graph of f.	17–20
	Use the Intermediate Value Theorem to help locate real zeros of polynomial functions (p. 130).	Let a and b be real numbers such that $a < b$. If f is a polynomial function such that $f(a) \neq f(b)$, then, in the interval $[a, b]$, f takes on every value between $f(a)$ and $f(b)$.	21, 22
Section 2.3	Use long division to divide polynomials by other polynomials (p. 136).	Dividend Quotient Remainder $$\dfrac{x^2 + 3x + 5}{x + 1} = x + 2 + \dfrac{3}{x + 1}$$ Divisor — $x + 1$ Divisor	23, 24
	Use synthetic division to divide polynomials by binomials of the form $(x - k)$ (p. 139).	Divisor: $x + 3$ Dividend: $x^4 - 10x^2 - 2x + 4$ $$-3 \begin{array}{|rrrrr} 1 & 0 & -10 & -2 & 4 \\ & -3 & 9 & 3 & -3 \\ \hline 1 & -3 & -1 & 1 & (1) \end{array}$$ Remainder: 1 Quotient: $x^3 - 3x^2 - x + 1$	25, 26
	Use the Remainder Theorem and the Factor Theorem (p. 140).	**The Remainder Theorem:** If a polynomial $f(x)$ is divided by $x - k$, then the remainder is $r = f(k)$. **The Factor Theorem:** A polynomial $f(x)$ has a factor $(x - k)$ if and only if $f(k) = 0$.	27, 28

	What Did You Learn?	**Explanation/Examples**	**Review Exercises**
Section 2.4	Use the imaginary unit i to write complex number *(p. 145)*.	When a and b are real numbers, $a + bi$ is a complex number. Two complex numbers $a + bi$ and $c + di$, written in standard form, are equal to each other if and only if $a = c$ and $b = d$.	29, 30
	Add, subtract, and multiply complex number *(p. 146)*.	**Sum:** $(a + bi) + (c + di) = (a + c) + (b + d)i$ **Difference:** $(a + bi) - (c + di) = (a - c) + (b - d)i$	31–34
	Use complex conjugates to write the quotient of two complex numbers in standard form *(p. 148)*.	To write $(a + bi)/(c + di)$ in standard form, where c and d are not both zero, multiply the numerator and denominator by the complex conjugate of the denominator, $c - di$.	35–38
	Find complex solutions of quadratic equations *(p. 149)*.	When a is a positive real number, the principal square root of $-a$ is defined as $\sqrt{-a} = \sqrt{a}i$.	39, 40
Section 2.5	Use the Fundamental Theorem of Algebra to determine the numbers of zeros of polynomial functions *(p. 152)*.	**The Fundamental Theorem of Algebra** If $f(x)$ is a polynomial of degree n, where $n > 0$, then f has at least one zero in the complex number system.	41, 42
	Find rational zeros of polynomial functions *(p. 153)*, and find complex zeros using conjugate pairs *(p. 156)*.	The Rational Zero Test relates the possible rational zeros of a polynomial to the leading coefficient and constant term. **Complex Zeros:** Let f be a polynomial function that has real coefficients. If $a + bi$, where $b \neq 0$, is a zero of the function, then the complex conjugate $a - bi$ is also a zero of the function.	43, 44
	Find zeros of polynomial by factoring *(p. 157)*, use Descartes's Rule of Signs and the Upper and Lower Bound Rules *(p. 159)*, and find zeros of polynomials in real-life applications *(p. 161)*.	Every polynomial of degree $n > 0$ with real coefficients can be written as the product of linear and quadratic factors with real coefficients, where the quadratic factors have no real zeros.	45–48
Section 2.6	Find domains *(p. 166)*, and vertical and horizontal asymptotes *(p. 167)*, of graphs of rational functions.	The domain of a rational function of x includes all real numbers except x-values that make the denominator zero. The line $x = a$ is a vertical asymptote of the graph of f when $f(x) \to \infty$ or $f(x) \to -\infty$ as $x \to a$, either from the right or from the left. The line $y = b$ is a horizontal asymptote of the graph of f when $f(x) \to b$ as $x \to \infty$ or $x \to -\infty$.	49, 50
	Sketch the graphs of rational functions *(p. 169)*, including functions with slant asymptotes *(p. 172)*.	Consider a rational function whose denominator is of degree 1 or greater. If the degree of the numerator is exactly *one more* than the degree of the denominator, then the graph of the function has a slant asymptote.	51–58
	Use rational functions to model and solve real-life problems *(p. 173)*.	A rational function can help you model the cost of removing a given percent of the smokestack pollutants at a utility company that burns coals. (See Example 8.)	59, 60
Section 2.7	Solve polynomial *(p. 178)*, and rational *(p. 182)* inequalities.	Use the concepts of key numbers and text intervals to solve both polynomial and rational inequalities.	61–64
	Use nonlinear inequalities to model and solve real-life problems *(p. 183)*.	A common application of nonlinear inequalities involves profit P, revenue R, and cost C. (See Example 6.)	65

Review Exercises
See **CalcChat.com** for tutorial help and worked-out solutions to odd-numbered exercises.

2.1 Sketching Graphs of Quadratic Functions
In Exercises 1 and 2, sketch the graph of each quadratic function and compare it with the graph of $y = x^2$.

1. (a) $g(x) = -2x^2$
   (b) $h(x) = x^2 + 2$

2. (a) $h(x) = (x - 3)^2$
   (b) $k(x) = \frac{1}{2}x - 1$

Using Standard Form to Graph a Parabola In Exercises 3–8, write the quadratic function in standard form and sketch its graph. Identify the vertex, axis of symmetry, and x-intercept(s).

3. $g(x) = x^2 - 2x$

4. $f(x) = x^2 + 8x + 10$

5. $h(x) = 3 + 4x - x^2$

6. $f(t) = -2t^2 + 4t + 1$

7. $h(x) = 4x^2 + 4x + 13$

8. $f(x) = \frac{1}{3}(x^2 + 5x - 4)$

9. Geometry The perimeter of a rectangle is 1000 meters.

(a) Write the width y as a function of the length x. Use the result to write the area A as a function of x.

(b) Of all possible rectangles with perimeters of 1000 meters, find the dimensions of the one with the maximum area.

10. Minimum Cost A soft-drink manufacturer has a daily production cost of $C = 70{,}000 - 120x + 0.055x^2$, where C is the total cost (in dollars) and x is the number of units produced. How many units should they produce each day to yield a minimum cost?

2.2 Sketching a Transformation of a Monomial Function In Exercises 11 and 12, sketch the graphs of $y = x^n$ and the transformation.

11. $y = x^4$, $f(x) = 6 - x^4$

12. $y = x^5$, $f(x) = \frac{1}{2}x^5 + 3$

Applying the Leading Coefficient Test In Exercises 13–16, describe the left-hand and right-hand behavior of the graph of the polynomial function.

13. $f(x) = -2x^2 - 5x + 12$
14. $f(x) = 4x - \frac{1}{2}x^3$
15. $g(x) = -3x^3 - 8x^4 + x^5$
16. $h(x) = 5 + 9x^6 - 6x^5$

Sketching the Graph of a Polynomial Function In Exercises 17–20, sketch the graph of the function by (a) applying the Leading Coefficient Test, (b) finding the real zeros of the polynomial, (c) plotting sufficient solution points, and (d) drawing a continuous curve through the points.

17. $g(x) = 2x^3 + 4x^2$

18. $h(x) = 3x^2 - x^4$

19. $f(x) = -x^3 + x^2 - 2$

20. $f(x) = x(x^3 + x^2 - 5x + 3)$

Using the Intermediate Value Theorem In Exercises 21 and 22, (a) use the Intermediate Value Theorem and the *table* feature of a graphing utility to find intervals one unit in length in which the polynomial function is guaranteed to have a zero. (b) Adjust the table to approximate the zeros of the function to the nearest thousandth. Use the *zero* or *root* feature of the graphing utility to verify your results.

21. $f(x) = 3x^3 - x^2 + 3$ **22.** $f(x) = x^4 - 5x - 1$

2.3 Long Division of Polynomials In Exercises 23 and 24, use long division to divide.

23. $\dfrac{30x^2 - 3x + 8}{5x - 3}$

24. $\dfrac{5x^3 - 21x^2 - 25x - 4}{x^2 - 5x - 1}$

Using Synthetic Division In Exercises 25 and 26, use synthetic division to divide.

25. $\dfrac{2x^3 - 25x^2 + 66x + 48}{x - 8}$

26. $\dfrac{x^4 - 2x^2 + 9x}{x + 3}$

Factoring a Polynomial In Exercises 27 and 28, (a) verify the given factor(s) of $f(x)$, (b) find the remaining factors of $f(x)$, (c) use your results to write the complete factorization of $f(x)$, (d) list all real zeros of f, and (e) confirm your results by using a graphing utility to graph the function.

Function	Factor(s)
27. $f(x) = 2x^3 + 11x^2 - 21x - 90$	$(x + 6)$
28. $f(x) = x^4 - 4x^3 - 7x^2 + 22x + 24$	$(x + 2), (x - 3)$

2.4 Writing a Complex Number in Standard Form In Exercises 29 and 30, write the complex number in standard form.

29. $4 + \sqrt{-9}$ **30.** $-5i + i^2$

Performing Operations with Complex Numbers In Exercises 31–34, perform the operation and write the result in standard form.

31. $(6 - 4i) + (-9 + i)$ **32.** $(7 - 2i) - (3 - 8i)$
33. $-3i(-2 + 5i)$ **34.** $(4 + i)(3 - 10i)$

Quotient of Complex Numbers in Standard Form In Exercises 35 and 36, write the quotient in standard form.

35. $\dfrac{4}{1 - 2i}$ **36.** $\dfrac{3 + 2i}{5 + i}$

Performing Operations with Complex Numbers
In Exercises 37 and 38, perform the operation and write the result in standard form.

37. $\dfrac{4}{2-3i} + \dfrac{2}{1+i}$ **38.** $\dfrac{1}{2+i} - \dfrac{5}{1+4i}$

Complex Solutions of a Quadratic Equation   In
Exercises 39 and 40, use the Quadratic Formula to solve the quadratic equation.

39. $x^2 - 2x + 10 = 0$ **40.** $6x^2 + 3x + 27 = 0$

2.5 **Zeros of Polynomial Functions**   **In Exercises 41 and 42, determine the number of zeros of the polynomial function.**

41. $g(x) = x^2 - 2x - 8$ **42.** $h(t) = t^2 - t^5$

Using the Rational Zero Test   In Exercises 43 and
44, find the rational zeros of the function.

43. $f(x) = 4x^3 - 27x^2 + 11x + 42$
44. $f(x) = x^4 + x^3 - 11x^2 + x - 12$

Finding the Zeros of a Polynomial Function   In
Exercises 45 and 46, write the polynomial as the product of linear factors and list all the zeros of the function.

45. $g(x) = x^3 - 7x^2 + 36$
46. $f(x) = x^4 + 8x^3 + 8x^2 - 72x - 153$

47. Using Descartes's Rule of Signs Use Descartes's Rule of Signs to determine the possible numbers of positive and negative real zeros of $h(x) = -2x^5 + 4x^3 - 2x^2 + 5$.

48. Verifying Upper and Lower Bounds Use synthetic division to verify the upper and lower bounds of the real zeros of $f(x) = 4x^3 - 3x^2 + 4x - 3$.
  (a) Upper: $x = 1$ (b) Lower: $x = -\frac{1}{4}$

2.6 **Finding Domain and Asymptotes**   **In Exercises 49 and 50, find the domain and the vertical and horizontal asymptotes of the graph of the rational function.**

49. $f(x) = \dfrac{3x}{x+10}$ **50.** $f(x) = \dfrac{8}{x^2 - 10x + 24}$

Sketching the Graph of a Rational Function   In
Exercises 51–58, (a) state the domain of the function, (b) identify all intercepts, (c) find any asymptotes, and (d) plot additional solution points as needed to sketch the graph of the rational function.

51. $f(x) = \dfrac{4}{x}$ **52.** $h(x) = \dfrac{x-4}{x-7}$

53. $f(x) = \dfrac{x}{x^2 - 16}$ **54.** $f(x) = \dfrac{-8x}{x^2 + 4}$

55. $f(x) = \dfrac{6x^2 - 11x + 3}{3x^2 - x}$ **56.** $f(x) = \dfrac{6x^2 - 7x + 2}{4x^2 - 1}$

57. $f(x) = \dfrac{2x^3}{x^2 + 1}$ **58.** $f(x) = \dfrac{2x^2 + 2}{x + 1}$

59. Seizure of Illegal Drugs The cost C (in millions of dollars) for the federal government to seize $p\%$ of an illegal drug as it enters the country is given by

$$C = \frac{528p}{100 - p}, \quad 0 \le p \le 100.$$

  (a) Use a graphing utility to graph the cost function.
  (b) Find the costs of seizing 25%, 50%, and 75% of the drug.
  (c) According to the model, it is possible to seize 100% of the drug? Explain.

60. Page Design A page that is x inches wide and y inches high contains 30 square inches of print. The top and bottom margins are each 2 inches deep, and the margins on each side are 2 inches wide.
  (a) Write a function for the total area A of the page in terms of x.
  (b) Determine the domain of the function based of the physical constraints of the problem.
  (c) Use a graphing utility to graph the area function and approximate the dimensions of the page that use the least amount of paper.

2.7 **Solving an Inequality**   **In Exercises 61–64, solve the inequality. Then graph the solution set.**

61. $12x^2 + 5x < 2$ **62.** $x^3 - 16x \ge 0$

63. $\dfrac{2}{x+1} \ge \dfrac{3}{x-1}$ **64.** $\dfrac{x^2 - 9x + 20}{x} < 0$

65. Biology A biologist introduces 200 ladybugs into a crop field. The population P of the ladybugs can be approximated by the model

$$P = \frac{1000(1 + 3t)}{5 + t}$$

where t is the time in days. Find the time required for the population to increase to at least 2000 ladybugs.

Exploration

True or False?   **In Exercises 66 and 67, determine whether the statement is true or false. Justify your answer.**

66. A fourth-degree polynomial with real coefficients can have -5, $-8i$, $4i$, and 5 as its zeros.

67. The domain of a rational function can never be the set of all real numbers.

68. Writing Describe what is meant by an asymptote of a graph.

Chapter Test

See CalcChat.com for tutorial help and worked-out solutions to odd-numbered exercises.

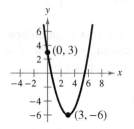

Figure for 2

Take this test as you would take a test in class. When you are finished, check your work against the answers given in the back of the book.

1. Sketch the graph of each quadratic function and compare it with the graph of $y = x^2$.

    (a) $g(x) = -x^2 + 4$ (b) $g(x) = \left(x - \frac{3}{2}\right)^2$

2. Write the standard form of the equation of the parabola shown at the left.

3. The path of a ball is modeled by the function $f(x) = -\frac{1}{20}x^2 + 3x + 5$, where $f(x)$ is the height (in feet) of the ball and x is the horizontal distance (in feet) from where the ball was thrown.

    (a) What is the maximum height of the ball?

    (b) Which number determines the height at which the ball was thrown? Does changing this value change the coordinates of the maximum height of the ball? Explain.

4. Describe the left-hand and right-hand behavior of the graph of the function $h(t) = -\frac{3}{4}t^5 + 2t^2$. Then sketch its graph.

5. Divide using long division.

$$\frac{3x^3 + 4x - 1}{x^2 + 1}$$

6. Divide using synthetic division.

$$\frac{2x^4 - 3x^2 + 4x - 1}{x + 2}$$

7. Use synthetic division to show that $x = \frac{5}{2}$ is a zero of the function

$$f(x) = 2x^3 - 5x^2 - 6x + 15.$$

Use the result to factor the polynomial function completely and list all the zeros of the function.

8. Perform each operation and write the result in standard form.

    (a) $\sqrt{-16} - 2(7 + 2i)$

    (b) $(5 - i)(3 + 4i)$

9. Write the quotient in standard form: $\dfrac{8}{1 + 2i}$.

In Exercises 10 and 11, find a polynomial function with real coefficients that has the given zeros. (There are many correct answers.)

10. $0, 2, 3i$ 11. $1, 1, 2 + \sqrt{3}i$

In Exercises 12 and 13, find all the zeros of the function.

12. $f(x) = 3x^3 + 14x^2 - 7x - 10$ 13. $f(x) = x^4 - 9x^2 - 22x - 24$

In Exercises 14–16, identify any intercepts and asymptotes of the graph of the function. Then sketch the graph of the function.

14. $h(x) = \dfrac{3}{x^2} - 1$ 15. $f(x) = \dfrac{2x^2 - 5x - 12}{x^2 - 16}$ 16. $g(x) = \dfrac{x^2 + 2}{x - 1}$

In Exercises 17 and 18, solve the inequality. The graph the solution set.

17. $2x^2 + 5x > 12$ 18. $\dfrac{2}{x} \le \dfrac{1}{x + 6}$

Proofs in Mathematics ▪ ▫ ▪ ▫ ▪ ▫ ▪ ▫ ▪ ▫ ▪ ▫ ▪ ▫ ▪ ▫ ▫ ▪

These two pages contain proofs of four important theorems about polynomial functions. The first two theorems are from Section 2.3, and the second two theorems are from Section 2.5.

The Remainder Theorem   *(p. 140)*

If a polynomial $f(x)$ is divided by $x - k$, then the remainder is

$$r = f(k).$$

Proof

Using the Division Algorithm with the divisor $(x - k)$, you have

$$f(x) = (x - k)q(x) + r(x).$$

Either $r(x) = 0$ or the degree of $r(x)$ is less than the degree of $x - k$, so you know that $r(x)$ must be a constant. That is, $r(x) = r$. Now, by evaluating $f(x)$ at $x = k$, you have

$$f(k) = (k - k)q(k) + r$$
$$= (0)q(k) + r$$
$$= r.$$

To be successful in algebra, it is important that you understand the connection among *factors* of a polynomial, *zeros* of a polynomial function, and *solutions* or *roots* of a polynomial equation. The Factor Theorem is the basis for this connection.

The Factor Theorem   *(p. 141)*

A polynomial $f(x)$ has a factor $(x - k)$ if and only if $f(k) = 0$.

Proof

Using the Division Algorithm with the factor $(x - k)$, you have

$$f(x) = (x - k)q(x) + r(x).$$

By the Remainder Theorem, $r(x) = r = f(k)$, and you have

$$f(x) = (x - k)q(x) + f(k)$$

where $q(x)$ is a polynomial of lesser degree than $f(x)$. If $f(k) = 0$, then

$$f(x) = (x - k)q(x)$$

and you see that $(x - k)$ is a factor of $f(x)$. Conversely, if $(x - k)$ is a factor of $f(x)$, then division of $f(x)$ by $(x - k)$ yields a remainder of 0. So, by the Remainder Theorem, you have $f(k) = 0$.

THE FUNDAMENTAL THEOREM OF ALGEBRA

The Fundamental Theorem of Algebra, which is closely related to the Linear Factorization Theorem, has a long and interesting history. In the early work with polynomial equations, the Fundamental Theorem of Algebra was thought to have been false, because imaginary solutions were not considered. In fact, in the very early work by mathematicians such as Abu al-Khwarizmi (c. 800 A.D.), negative solutions were also not considered.

Once imaginary numbers were considered, several mathematicians attempted to give a general proof of the Fundamental Theorem of Algebra. These included Jean Le Rond d'Alembert (1746), Leonhard Euler (1749), Joseph-Louis Lagrange (1772), and Pierre Simon Laplace (1795). The mathematician usually credited with the first complete and correct proof of the Fundamental Theorem of Algebra is Carl Friedrich Gauss, who published the proof in 1816.

Linear Factorization Theorem   *(p. 152)*

If $f(x)$ is a polynomial of degree n, where $n > 0$, then $f(x)$ has precisely n linear factors

$$f(x) = a_n(x - c_1)(x - c_2) \cdots (x - c_n)$$

where $c_1, c_2, \ldots, c_n$ are complex numbers.

Proof

Using the Fundamental Theorem of Algebra, you know that f must have at least one zero, c_1. Consequently, $(x - c_1)$ is a factor of $f(x)$, and you have

$$f(x) = (x - c_1)f_1(x).$$

If the degree of $f_1(x)$ is greater than zero, then you again apply the Fundamental Theorem of Algebra to conclude that f_1 must have a zero c_2, which implies that

$$f(x) = (x - c_1)(x - c_2)f_2(x).$$

It is clear that the degree of $f_1(x)$ is $n - 1$, that the degree of $f_2(x)$ is $n - 2$, and that you can repeatedly apply the Fundamental Theorem of Algebra n times until you obtain

$$f(x) = a_n(x - c_1)(x - c_2) \cdots (x - c_n)$$

where a_n is the leading coefficient of the polynomial $f(x)$.   ■

Factors of a Polynomial   *(p. 157)*

Every polynomial of degree $n > 0$ with real coefficients can be written as the product of linear and quadratic factors with real coefficients, where the quadratic factors have no real zeros.

Proof

To begin, use the Linear Factorization Theorem to conclude that $f(x)$ can be *completely* factored in the form

$$f(x) = d(x - c_1)(x - c_2)(x - c_3) \cdots (x - c_n).$$

If each c_i is real, then there is nothing more to prove. If any c_i is imaginary $(c_i = a + bi, b \neq 0)$, then you know that the conjugate $c_j = a - bi$ is also a zero, because the coefficients of $f(x)$ are real. By multiplying the corresponding factors, you obtain

$$
\begin{aligned}
(x - c_i)(x - c_j) &= [x - (a + bi)][x - (a - bi)] \\
&= [(x - a) - bi][(x - a) + bi] \\
&= (x - a)^2 + b^2 \\
&= x^2 - 2ax + (a^2 + b^2)
\end{aligned}
$$

where each coefficient is real.   ■

P.S. Problem Solving

1. Verifying the Remainder Theorem Show that if $f(x) = ax^3 + bx^2 + cx + d$, then $f(k) = r$, where $r = ak^3 + bk^2 + ck + d$, using long division. In other words, verify the Remainder Theorem for a third-degree polynomial function.

2. Babylonian Mathematics In 2000 B.C., the Babylonians solved polynomial equations by referring to tables of values. One such table gave the values of $y^3 + y^2$. To be able to use this table, the Babylonians sometimes used the method below to manipulate the equation.

$$ax^3 + bx^2 = c \qquad \text{Original equation}$$

$$\frac{a^3 x^3}{b^3} + \frac{a^2 x^2}{b^2} = \frac{a^2 c}{b^3} \qquad \text{Multiply each side by } \frac{a^2}{b^3}.$$

$$\left(\frac{ax}{b}\right)^3 + \left(\frac{ax}{b}\right)^2 = \frac{a^2 c}{b^3} \qquad \text{Rewrite.}$$

Then they would find $(a^2 c)/b^3$ in the $y^3 + y^2$ column of the table. They knew that the corresponding y-value was equal to $(ax)/b$, so they could conclude that $x = (by)/a$.

(a) Calculate $y^3 + y^2$ for $y = 1, 2, 3, \ldots, 10$. Record the values in a table.

(b) Use the table from part (a) and the method above to solve each equation.

    (i) $x^3 + x^2 = 252$

    (ii) $x^3 + 2x^2 = 288$

    (iii) $3x^3 + x^2 = 90$

    (iv) $2x^3 + 5x^2 = 2500$

    (v) $7x^3 + 6x^2 = 1728$

    (vi) $10x^3 + 3x^2 = 297$

(c) Using the methods from this chapter, verify your solution of each equation.

3. Finding Dimensions At a glassware factory, molten cobalt glass is poured into molds to make paperweights. Each mold is a rectangular prism whose height is 3 inches greater than the length of each side of the square base. A machine pours 20 cubic inches of liquid glass into each mold. What are the dimensions of the mold?

4. True or False? Determine whether the statement is true or false. If false, provide one or more reasons why the statement is false and correct the statement. Let $f(x) = ax^3 + bx^2 + cx + d$, $a \neq 0$, and let $f(2) = -1$. Then

$$\frac{f(x)}{x + 1} = q(x) + \frac{2}{x + 1}$$

where $q(x)$ is a second-degree polynomial.

5. Finding the Equation of a Parabola The parabola shown in the figure has an equation of the form $y = ax^2 + bx + c$. Find the equation of this parabola using each method.

(a) Find the equation analytically.

(b) Use the regression feature of a graphing utility to find the equation.

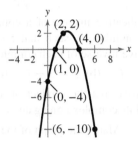

6. Finding the Slope of a Tangent Line One of the fundamental themes of calculus is to find the slope of the tangent line to a curve at a point. To see how this can be done, consider the point $(2, 4)$ on the graph of the quadratic function $f(x) = x^2$, as shown in the figure.

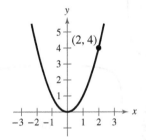

(a) Find the slope m_1 of the line joining $(2, 4)$ and $(3, 9)$. Is the slope of the tangent line at $(2, 4)$ greater than or less than the slope of the line through $(2, 4)$ and $(3, 9)$?

(b) Find the slope m_2 of the line joining $(2, 4)$ and $(1, 1)$. Is the slope of the tangent line at $(2, 4)$ greater than or less than the slope of the line through $(2, 4)$ and $(1, 1)$?

(c) Find the slope m_3 of the line joining $(2, 4)$ and $(2.1, 4.41)$. Is the slope of the tangent line at $(2, 4)$ greater than or less than the slope of the line through $(2, 4)$ and $(2.1, 4.41)$?

(d) Find the slope m_h of the line joining $(2, 4)$ and $(2 + h, f(2 + h))$ in terms of the nonzero number h.

(e) Evaluate the slope formula from part (d) for $h = -1$, 1, and 0.1. Compare these values with those in parts (a)–(c).

(f) What can you conclude the slope m_{tan} of the tangent line at $(2, 4)$ to be? Explain.

7. Writing Cubic Functions For each part, write a cubic function of the form $f(x) = (x - k)q(x) + r$ whose graph has the specified characteristics. (There are many correct answers.)

(a) Passes through the point $(2, 5)$ and rises to the right

(b) Passes through the point $(-3, 1)$ and falls to the right

8. Multiplicative Inverse of a Complex Number The multiplicative inverse of a complex number z is a complex number z_m such that $z \cdot z_m = 1$. Find the multiplicative inverse of each complex number.

(a) $z = 1 + i$ (b) $z = 3 - i$ (c) $z = -2 + 8i$

9. Proof Prove that the product of a complex number $a + bi$ and its complex conjugate is a real number.

10. Matching Match the graph of the rational function

$$f(x) = \frac{ax + b}{cx + d}$$

with the given conditions.

(a)
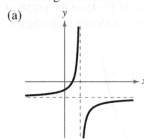

(b)
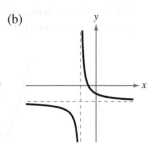

(c)
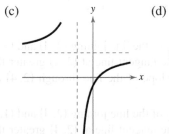

(d)
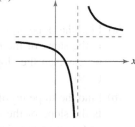

(i) $a > 0$ (ii) $a > 0$ (iii) $a < 0$ (iv) $a > 0$
   $b < 0$ $b > 0$ $b > 0$ $b < 0$
   $c > 0$ $c < 0$ $c > 0$ $c > 0$
   $d < 0$ $d < 0$ $d < 0$ $d > 0$

11. Effects of Values on a Graph Consider the function

$$f(x) = \frac{ax}{(x - b)^2}.$$

(a) Determine the effect on the graph of f when $b \neq 0$ and a is varied. Consider cases in which a is positive and a is negative.

(b) Determine the effect on the graph of f when $a \neq 0$ and b is varied.

12. Distinct Vision The endpoints of the interval over which distinct vision is possible are called the *near point* and *far point* of the eye (see figure). With increasing age, these points normally change. The table shows the approximate near points y (in inches) for various ages x (in years).

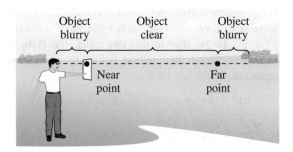

Age, x	Near Point, y
16	3.0
32	4.7
44	9.8
50	19.7
60	39.4

(a) Use the *regression* feature of a graphing utility to find a quadratic model for the data. Use the graphing utility to plot the data and graph the model in the same viewing window.

(b) Find a rational model for the data. Take the reciprocals of the near points to generate the points $(x, 1/y)$. Use the *regression* feature of the graphing utility to find a linear model for the data. The resulting line has the form

$$\frac{1}{y} = ax + b.$$

Solve for y. Use the graphing utility to plot the data and graph the model in the same viewing window.

(c) Use the *table* feature of the graphing utility to construct a table showing the predicted near point based on each model for each of the ages in the original table. How well do the models fit the original data?

(d) Use both models to estimate the near point for a person who is 25 years old. Which model is a better fit?

(e) Do you think either model can be used to predict the near point for a person who is 70 years old? Explain.

13. Zeros of a Cubic Function Can a cubic function with real coefficients have two real zeros and one complex zero? Explain.

3 Exponential and Logarithmic Functions

Beaver Population *(Exercise 83, page 234)*

Earthquakes
(Example 6, page 242)

Sound Intensity *(Exercises 79–82, page 224)*

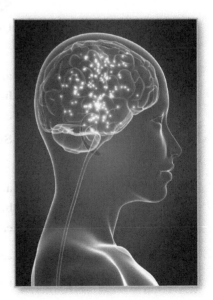

Human Memory Model
(Exercise 83, page 218)

Nuclear Reactor Accident *(Example 9, page 205)*

3.1 Exponential Functions and Their Graphs

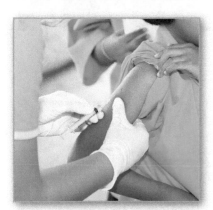

- Recognize and evaluate exponential functions with base *a*.
- Graph exponential functions and use the One-to-One Property.
- Recognize, evaluate, and graph exponential functions with base *e*.
- Use exponential functions to model and solve real-life problems.

Exponential Functions

So far, this text has dealt mainly with **algebraic functions,** which include polynomial functions and rational functions. In this chapter, you will study two types of nonalgebraic functions—*exponential functions* and *logarithmic functions*. These functions are examples of **transcendental functions.** This section will focus on exponential functions.

Exponential functions can help you model and solve real-life problems. For example, in Exercise 66 on page 208, you will use an exponential function to model the concentration of a drug in the bloodstream.

> ### Definition of Exponential Function
> The **exponential function** *f* with base *a* is denoted by
>
> $$f(x) = a^x$$
>
> where $a > 0$, $a \neq 1$, and *x* is any real number.

The base *a* of an exponential function cannot be 1 because $a = 1$ yields $f(x) = 1^x = 1$. This is a constant function, not an exponential function.

You have evaluated a^x for integer and rational values of *x*. For example, you know that $4^3 = 64$ and $4^{1/2} = 2$. However, to evaluate 4^x for any real number *x*, you need to interpret forms with *irrational* exponents. For the purposes of this text, it is sufficient to think of $a^{\sqrt{2}}$ (where $\sqrt{2} \approx 1.41421356$) as the number that has the successively closer approximations

$$a^{1.4}, a^{1.41}, a^{1.414}, a^{1.4142}, a^{1.41421}, \ldots.$$

EXAMPLE 1   Evaluating Exponential Functions

Use a calculator to evaluate each function at the given value of *x*.

Function	Value
a. $f(x) = 2^x$	$x = -3.1$
b. $f(x) = 2^{-x}$	$x = \pi$
c. $f(x) = 0.6^x$	$x = \frac{3}{2}$

Solution

Function Value	Calculator Keystrokes	Display
a. $f(-3.1) = 2^{-3.1}$	2 [∧] [(−)] 3.1 [ENTER]	0.1166291
b. $f(\pi) = 2^{-\pi}$	2 [∧] [(−)] π [ENTER]	0.1133147
c. $f\left(\frac{3}{2}\right) = (0.6)^{3/2}$	.6 [∧] [(] 3 [÷] 2 [)] [ENTER]	0.4647580

✓ *Checkpoint* ◄))) Audio-video solution in English & Spanish at LarsonPrecalculus.com

Use a calculator to evaluate $f(x) = 8^{-x}$ at $x = \sqrt{2}$.

When evaluating exponential functions with a calculator, it may be necessary to enclose fractional exponents in parentheses. Some calculators do not correctly interpret an exponent that consists of an expression unless parentheses are used.

Graphs of Exponential Functions

The graphs of all exponential functions have similar characteristics, as shown in Examples 2, 3, and 5.

EXAMPLE 2 **Graphs of $y = a^x$**

In the same coordinate plane, sketch the graph of each function.

a. $f(x) = 2^x$ **b.** $g(x) = 4^x$

Solution Begin by constructing a table of values.

x	-3	-2	-1	0	1	2
2^x	$\frac{1}{8}$	$\frac{1}{4}$	$\frac{1}{2}$	1	2	4
4^x	$\frac{1}{64}$	$\frac{1}{16}$	$\frac{1}{4}$	1	4	16

> ▷ **ALGEBRA HELP** To review the techniques for sketching the graph of an equation, see Section 1.2.

To sketch the graph of each function, plot the points from the table and connect them with a smooth curve, as shown in Figure 3.1. Note that both graphs are increasing. Moreover, the graph of $g(x) = 4^x$ is increasing more rapidly than the graph of $f(x) = 2^x$.

 **Checkpoint** Audio-video solution in English & Spanish at LarsonPrecalculus.com

In the same coordinate plane, sketch the graph of each function.

a. $f(x) = 3^x$ **b.** $g(x) = 9^x$

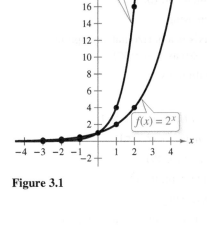

Figure 3.1

The table in Example 2 was evaluated by hand for integer values of x. You can also evaluate $f(x)$ and $g(x)$ for noninteger values of x by using a calculator.

EXAMPLE 3 **Graphs of $y = a^{-x}$**

In the same coordinate plane, sketch the graph of each function.

a. $F(x) = 2^{-x}$ **b.** $G(x) = 4^{-x}$

Solution Begin by constructing a table of values.

x	-2	-1	0	1	2	3
2^{-x}	4	2	1	$\frac{1}{2}$	$\frac{1}{4}$	$\frac{1}{8}$
4^{-x}	16	4	1	$\frac{1}{4}$	$\frac{1}{16}$	$\frac{1}{64}$

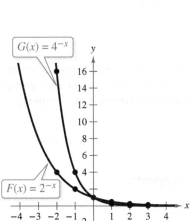

Figure 3.2

To sketch the graph of each function, plot the points from the table and connect them with a smooth curve, as shown in Figure 3.2. Note that both graphs are decreasing. Moreover, the graph of $G(x) = 4^{-x}$ is decreasing more rapidly than the graph of $F(x) = 2^{-x}$.

 Checkpoint Audio-video solution in English & Spanish at LarsonPrecalculus.com

In the same coordinate plane, sketch the graph of each function.

a. $f(x) = 3^{-x}$ **b.** $g(x) = 9^{-x}$

Note that it is possible to use one of the properties of exponents to rewrite the functions in Example 3 with positive exponents.

$$F(x) = 2^{-x} = \frac{1}{2^x} = \left(\frac{1}{2}\right)^x \quad \text{and} \quad G(x) = 4^{-x} = \frac{1}{4^x} = \left(\frac{1}{4}\right)^x$$

Comparing the functions in Examples 2 and 3, observe that

$$F(x) = 2^{-x} = f(-x) \quad \text{and} \quad G(x) = 4^{-x} = g(-x).$$

Consequently, the graph of F is a reflection (in the y-axis) of the graph of f. The graphs of G and g have the same relationship. The graphs in Figures 3.1 and 3.2 are typical of the exponential functions $y = a^x$ and $y = a^{-x}$. They have one y-intercept and one horizontal asymptote (the x-axis), and they are continuous. Here is a summary of the basic characteristics of the graphs of these exponential functions.

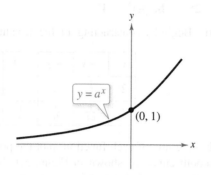

Graph of $y = a^x$, $a > 1$
- Domain: $(-\infty, \infty)$
- Range: $(0, \infty)$
- y-intercept: $(0, 1)$
- Increasing
- x-axis is a horizontal asymptote ($a^x \to 0$ as $x \to -\infty$).
- Continuous

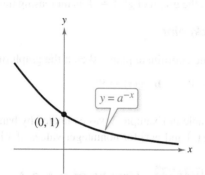

Graph of $y = a^{-x}$, $a > 1$
- Domain: $(-\infty, \infty)$
- Range: $(0, \infty)$
- y-intercept: $(0, 1)$
- Decreasing
- x-axis is a horizontal asymptote ($a^{-x} \to 0$ as $x \to \infty$).
- Continuous

Notice that the graph of an exponential function is always increasing or always decreasing, so the graph passes the Horizontal Line Test. Therefore, an exponential function is a one-to-one function. You can use the following **One-to-One Property** to solve simple exponential equations.

For $a > 0$ and $a \neq 1$, $a^x = a^y$ if and only if $x = y$.   One-to-One Property

EXAMPLE 4   **Using the One-to-One Property**

a. $9 = 3^{x+1}$ Original equation

$\quad 3^2 = 3^{x+1}$ $9 = 3^2$

$\quad 2 = x + 1$ One-to-One Property

$\quad 1 = x$ Solve for x.

b. $\left(\frac{1}{2}\right)^x = 8$ Original equation

$\quad 2^{-x} = 2^3$ $\left(\frac{1}{2}\right)^x = 2^{-x}, 8 = 2^3$

$\quad x = -3$ One-to-One Property

✓ **Checkpoint** ◀))) *Audio-video solution in English & Spanish at LarsonPrecalculus.com*

Use the One-to-One Property to solve the equation for x.

a. $8 = 2^{2x-1}$ **b.** $\left(\frac{1}{3}\right)^{-x} = 27$

In Example 5, notice how the graph of $y = a^x$ can be used to sketch the graphs of functions of the form $f(x) = b \pm a^{x+c}$.

▷ **ALGEBRA HELP** To review the techniques for transforming the graph of a function, see Section 1.7.

EXAMPLE 5

Transformations of Graphs of Exponential Functions

See LarsonPrecalculus.com for an interactive version of this type of example.

Describe the transformation of the graph of $f(x) = 3^x$ that yields each graph.

a.
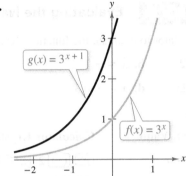

b.
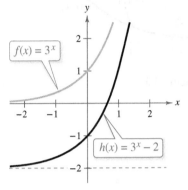

c.
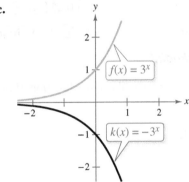

d.
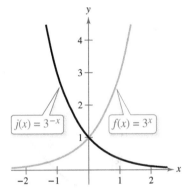

Solution

a. Because $g(x) = 3^{x+1} = f(x + 1)$, the graph of g is obtained by shifting the graph of f one unit to the *left*.

b. Because $h(x) = 3^x - 2 = f(x) - 2$, the graph of h is obtained by shifting the graph of f *down* two units.

c. Because $k(x) = -3^x = -f(x)$, the graph of k is obtained by *reflecting* the graph of f in the x-axis.

d. Because $j(x) = 3^{-x} = f(-x)$, the graph of j is obtained by *reflecting* the graph of f in the y-axis.

✓ *Checkpoint*  Audio-video solution in English & Spanish at LarsonPrecalculus.com

Describe the transformation of the graph of $f(x) = 4^x$ that yields the graph of each function.

a. $g(x) = 4^{x-2}$ b. $h(x) = 4^x + 3$ c. $k(x) = 4^{-x} - 3$

Note how each transformation in Example 5 affects the y-intercept and the horizontal asymptote.

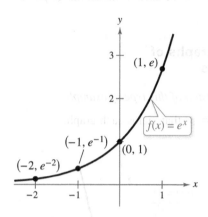

Figure 3.3

The Natural Base e

In many applications, the most convenient choice for a base is the irrational number

$$e \approx 2.718281828 . . . . .$$

This number is called the **natural base.** The function $f(x) = e^x$ is called the **natural exponential function.** Figure 3.3 shows its graph. Be sure you see that for the exponential function $f(x) = e^x$, e is the constant 2.718281828 . . . , whereas x is the variable.

EXAMPLE 6 **Evaluating the Natural Exponential Function**

Use a calculator to evaluate the function $f(x) = e^x$ at each value of x.

a. $x = -2$ **b.** $x = -1$

c. $x = 0.25$ **d.** $x = -0.3$

Solution

	Function Value	Calculator Keystrokes	Display
a.	$f(-2) = e^{-2}$	$\boxed{e^x}$ $\boxed{(-)}$ 2 $\boxed{\text{ENTER}}$	0.1353353
b.	$f(-1) = e^{-1}$	$\boxed{e^x}$ $\boxed{(-)}$ 1 $\boxed{\text{ENTER}}$	0.3678794
c.	$f(0.25) = e^{0.25}$	$\boxed{e^x}$ 0.25 $\boxed{\text{ENTER}}$	1.2840254
d.	$f(-0.3) = e^{-0.3}$	$\boxed{e^x}$ $\boxed{(-)}$ 0.3 $\boxed{\text{ENTER}}$	0.7408182

✓ *Checkpoint*  Audio-video solution in English & Spanish at LarsonPrecalculus.com

Use a calculator to evaluate the function $f(x) = e^x$ at each value of x.

a. $x = 0.3$

b. $x = -1.2$

c. $x = 6.2$

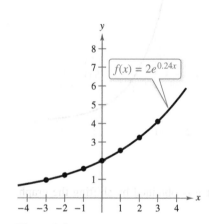

Figure 3.4

EXAMPLE 7 **Graphing Natural Exponential Functions**

Sketch the graph of each natural exponential function.

a. $f(x) = 2e^{0.24x}$

b. $g(x) = \frac{1}{2}e^{-0.58x}$

Solution Begin by using a graphing utility to construct a table of values.

x	-3	-2	-1	0	1	2	3
$f(x)$	0.974	1.238	1.573	2.000	2.542	3.232	4.109
$g(x)$	2.849	1.595	0.893	0.500	0.280	0.157	0.088

To graph each function, plot the points from the table and connect them with a smooth curve, as shown in Figures 3.4 and 3.5. Note that the graph in Figure 3.4 is increasing, whereas the graph in Figure 3.5 is decreasing.

✓ *Checkpoint*  Audio-video solution in English & Spanish at LarsonPrecalculus.com

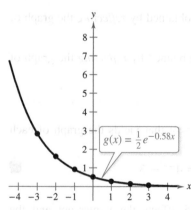

Figure 3.5

Sketch the graph of $f(x) = 5e^{0.17x}$.

Applications

One of the most familiar examples of exponential growth is an investment earning *continuously compounded interest*. The formula for *interest compounded n times per year* is

$$A = P\left(1 + \frac{r}{n}\right)^{nt}.$$

In this formula, A is the balance in the account, P is the principal (or original deposit), r is the annual interest rate (in decimal form), n is the number of compoundings per year, and t is the time in years. Exponential functions can be used to *develop* this formula and show how it leads to continuous compounding.

Consider a principal P invested at an annual interest rate r, compounded once per year. When the interest is added to the principal at the end of the first year, the new balance P_1 is

$$P_1 = P + Pr$$
$$= P(1 + r).$$

This pattern of multiplying the balance by $1 + r$ repeats each successive year, as shown here.

Year	Balance After Each Compounding
0	$P = P$
1	$P_1 = P(1 + r)$
2	$P_2 = P_1(1 + r) = P(1 + r)(1 + r) = P(1 + r)^2$
3	$P_3 = P_2(1 + r) = P(1 + r)^2(1 + r) = P(1 + r)^3$
$\vdots$	$\vdots$
t	$P_t = P(1 + r)^t$

To accommodate more frequent (quarterly, monthly, or daily) compounding of interest, let n be the number of compoundings per year and let t be the number of years. Then the rate per compounding is r/n, and the account balance after t years is

$$A = P\left(1 + \frac{r}{n}\right)^{nt}. \qquad \text{Amount (balance) with } n \text{ compoundings per year}$$

When the number of compoundings n increases without bound, the process approaches what is called **continuous compounding.** In the formula for n compoundings per year, let $m = n/r$. This yields a new expression.

$$A = P\left(1 + \frac{r}{n}\right)^{nt} \qquad \text{Amount with } n \text{ compoundings per year}$$

$$= P\left(1 + \frac{r}{mr}\right)^{mrt} \qquad \text{Substitute } mr \text{ for } n.$$

$$= P\left(1 + \frac{1}{m}\right)^{mrt} \qquad \text{Simplify.}$$

$$= P\left[\left(1 + \frac{1}{m}\right)^m\right]^{rt} \qquad \text{Property of exponents}$$

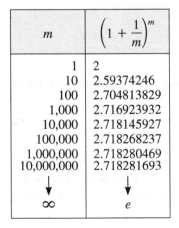

m	$\left(1 + \dfrac{1}{m}\right)^m$
1	2
10	2.59374246
100	2.704813829
1,000	2.716923932
10,000	2.718145927
100,000	2.718268237
1,000,000	2.718280469
10,000,000	2.718281693
$\downarrow$	$\downarrow$
∞	e

As m increases without bound (that is, as $m \to \infty$), the table at the left shows that $[1 + (1/m)]^m \to e$. This allows you to conclude that the formula for continuous compounding is

$$A = Pe^{rt}. \qquad \text{Substitute } e \text{ for } [1 + (1/m)]^m.$$

•• REMARK Be sure you see
that, when using the formulas
for compound interest, you must
write the annual interest rate in
decimal form. For example, you
must write 6% as 0.06.

⋯⋯⋯⋯⋯⋯⋯▷

Formulas for Compound Interest

After t years, the balance A in an account with principal P and annual interest rate r (in decimal form) is given by one of these two formulas.

1. For n compoundings per year: $A = P\left(1 + \dfrac{r}{n}\right)^{nt}$

2. For continuous compounding: $A = Pe^{rt}$

EXAMPLE 8   **Compound Interest**

You invest \$12,000 at an annual rate of 3%. Find the balance after 5 years for each type of compounding.

a. Quarterly

b. Monthly

c. Continuous

Solution

a. For quarterly compounding, use $n = 4$ to find the balance after 5 years.

$$A = P\left(1 + \frac{r}{n}\right)^{nt} \qquad \text{Formula for compound interest}$$

$$= 12{,}000\left(1 + \frac{0.03}{4}\right)^{4(5)} \qquad \text{Substitute for } P, r, n, \text{ and } t.$$

$$\approx 13{,}934.21 \qquad \text{Use a calculator.}$$

b. For monthly compounding, use $n = 12$ to find the balance after 5 years.

$$A = P\left(1 + \frac{r}{n}\right)^{nt} \qquad \text{Formula for compound interest}$$

$$= 12{,}000\left(1 + \frac{0.03}{12}\right)^{12(5)} \qquad \text{Substitute for } P, r, n, \text{ and } t.$$

$$\approx \$13{,}939.40 \qquad \text{Use a calculator.}$$

c. Use the formula for continuous compounding to find the balance after 5 years.

$$A = Pe^{rt} \qquad \text{Formula for continuous compounding}$$

$$= 12{,}000e^{0.03(5)} \qquad \text{Substitute for } P, r, \text{ and } t.$$

$$\approx \$13{,}942.01 \qquad \text{Use a calculator.}$$

✓ *Checkpoint*  Audio-video solution in English & Spanish at LarsonPrecalculus.com

You invest \$6000 at an annual rate of 4%. Find the balance after 7 years for each type of compounding.

a. Quarterly   **b.** Monthly   **c.** Continuous

In Example 8, note that continuous compounding yields more than quarterly and monthly compounding. This is typical of the two types of compounding. That is, for a given principal, interest rate, and time, continuous compounding will always yield a larger balance than compounding n times per year.

EXAMPLE 9 **Radioactive Decay**

The International Atomic Energy Authority ranks nuclear incidents and accidents by severity using a scale from 1 to 7 called the International Nuclear and Radiological Event Scale (INES). A level 7 ranking is the most severe. To date, the Chernobyl accident and an accident at Japan's Fukushima Daiichi power plant in 2011 are the only two disasters in history to be given an INES level 7 ranking.

In 1986, a nuclear reactor accident occurred in Chernobyl in what was then the Soviet Union. The explosion spread highly toxic radioactive chemicals, such as plutonium $\left(^{239}Pu\right)$, over hundreds of square miles, and the government evacuated the city and the surrounding area. To see why the city is now uninhabited, consider the model

$$P = 10\left(\frac{1}{2}\right)^{t/24,100}$$

which represents the amount of plutonium P that remains (from an initial amount of 10 pounds) after t years. Sketch the graph of this function over the interval from $t = 0$ to $t = 100,000$, where $t = 0$ represents 1986. How much of the 10 pounds will remain in the year 2020? How much of the 10 pounds will remain after 100,000 years?

Solution The graph of this function is shown in the figure at the right. Note from this graph that plutonium has a *half-life* of about 24,100 years. That is, after 24,100 years, *half* of the original amount will remain. After another 24,100 years, one-quarter of the original amount will remain, and so on. In the year 2020 $(t = 34)$, there will still be

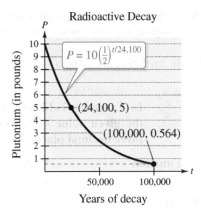

Radioactive Decay

$$P = 10\left(\frac{1}{2}\right)^{34/24,100}$$

$$\approx 10\left(\frac{1}{2}\right)^{0.0014108}$$

$$\approx 9.990 \text{ pounds}$$

of plutonium remaining. After 100,000 years, there will still be

$$P = 10\left(\frac{1}{2}\right)^{100,000/24,100}$$

$$\approx 0.564 \text{ pound}$$

of plutonium remaining.

 **Checkpoint**))) *Audio-video solution in English & Spanish at LarsonPrecalculus.com*

In Example 9, how much of the 10 pounds will remain in the year 2089? How much of the 10 pounds will remain after 125,000 years?

Summarize (Section 3.1)

1. State the definition of the exponential function f with base a *(page 198)*. For an example of evaluating exponential functions, see Example 1.

2. Describe the basic characteristics of the graphs of the exponential functions $y = a^x$ and $y = a^{-x}, a > 1$ *(page 200)*. For examples of graphing exponential functions, see Examples 2, 3, and 5.

3. State the definitions of the natural base and the natural exponential function *(page 202)*. For examples of evaluating and graphing natural exponential functions, see Examples 6 and 7.

4. Describe real-life applications involving exponential functions *(pages 204 and 205, Examples 8 and 9)*.

3.1 Exercises

See **CalcChat.com** for tutorial help and worked-out solutions to odd-numbered exercises.

Vocabulary: Fill in the blanks.

1. Polynomial and rational functions are examples of _____ functions.

2. Exponential and logarithmic functions are examples of nonalgebraic functions, also called _____ functions.

3. The _____ Property can be used to solve simple exponential equations.

4. The exponential function $f(x) = e^x$ is called the _____ _____ function, and the base e is called the _____ base.

5. To find the amount A in an account after t years with principal P and an annual interest rate r (in decimal form) compounded n times per year, use the formula _____.

6. To find the amount A in an account after t years with principal P and an annual interest rate r (in decimal form) compounded continuously, use the formula _____.

Skills and Applications

 Evaluating an Exponential Function In Exercises 7–12, evaluate the function at the given value of x. Round your result to three decimal places.

Function	Value
7. $f(x) = 0.9^x$	$x = 1.4$
8. $f(x) = 4.7^x$	$x = -\pi$
9. $f(x) = 3^x$	$x = \frac{2}{5}$
10. $f(x) = \left(\frac{2}{3}\right)^{5x}$	$x = \frac{3}{10}$
11. $f(x) = 5000(2^x)$	$x = -1.5$
12. $f(x) = 200(1.2)^{12x}$	$x = 24$

Matching an Exponential Function with Its Graph In Exercises 13–16, match the exponential function with its graph. [The graphs are labeled (a), (b), (c), and (d).]

(a)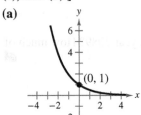

(b)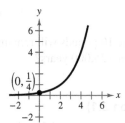

(c)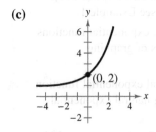

(d)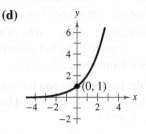

13. $f(x) = 2^x$

14. $f(x) = 2^x + 1$

15. $f(x) = 2^{-x}$

16. $f(x) = 2^{x-2}$

 Graphing an Exponential Function In Exercises 17–24, use a graphing utility to construct a table of values for the function. Then sketch the graph of the function.

17. $f(x) = 7^x$

18. $f(x) = 7^{-x}$

19. $f(x) = \left(\frac{1}{4}\right)^{-x}$

20. $f(x) = \left(\frac{1}{4}\right)^x$

21. $f(x) = 4^{x-1}$

22. $f(x) = 4^{x+1}$

23. $f(x) = 2^{x+1} + 3$

24. $f(x) = 3^{x-2} + 1$

 Using the One-to-One Property In Exercises 25–28, use the One-to-One Property to solve the equation for x.

25. $3^{x+1} = 27$

26. $2^{x-2} = 64$

27. $\left(\frac{1}{2}\right)^x = 32$

28. $5^{x-2} = \frac{1}{125}$

 Transformations of the Graph of an Exponential Function In Exercises 29–32, describe the transformation(s) of the graph of f that yield(s) the graph of g.

29. $f(x) = 3^x$, $g(x) = 3^x + 1$

30. $f(x) = \left(\frac{7}{2}\right)^x$, $g(x) = -\left(\frac{7}{2}\right)^{-x}$

31. $f(x) = 10^x$, $g(x) = 10^{-x+3}$

32. $f(x) = 0.3^x$, $g(x) = -0.3^x + 5$

 Evaluating a Natural Exponential Function In Exercises 33–36, evaluate the function at the given value of x. Round your result to three decimal places.

Function	Value
33. $f(x) = e^x$	$x = 1.9$
34. $f(x) = 1.5e^{x/2}$	$x = 240$
35. $f(x) = 5000e^{0.06x}$	$x = 6$
36. $f(x) = 250e^{0.05x}$	$x = 20$

Graphing a Natural Exponential Function In Exercises 37–40, use a graphing utility to construct a table of values for the function. Then sketch the graph of the function.

37. $f(x) = 3e^{x+4}$ **38.** $f(x) = 2e^{-1.5x}$

39. $f(x) = 2e^{x-2} + 4$ **40.** $f(x) = 2 + e^{x-5}$

Graphing a Natural Exponential Function In Exercises 41–44, use a graphing utility to graph the exponential function.

41. $s(t) = 2e^{0.5t}$ **42.** $s(t) = 3e^{-0.2t}$

43. $g(x) = 1 + e^{-x}$ **44.** $h(x) = e^{x-2}$

Using the One-to-One Property In Exercises 45–48, use the One-to-One Property to solve the equation for x.

45. $e^{3x+2} = e^3$ **46.** $e^{2x-1} = e^4$

47. $e^{x^2-3} = e^{2x}$ **48.** $e^{x^2+6} = e^{5x}$

Compound Interest In Exercises 49–52, complete the table by finding the balance A when P dollars is invested at rate r for t years and compounded n times per year.

n	1	2	4	12	365	Continuous
A						

49. $P = \$1500, r = 2\%, t = 10$ years

50. $P = \$2500, r = 3.5\%, t = 10$ years

51. $P = \$2500, r = 4\%, t = 20$ years

52. $P = \$1000, r = 6\%, t = 40$ years

Compound Interest In Exercises 53–56, complete the table by finding the balance A when \$12,000 is invested at rate r for t years, compounded continuously.

t	10	20	30	40	50
A					

53. $r = 4\%$ **54.** $r = 6\%$

55. $r = 6.5\%$ **56.** $r = 3.5\%$

57. Trust Fund On the day of a child's birth, a parent deposits \$30,000 in a trust fund that pays 5% interest, compounded continuously. Determine the balance in this account on the child's 25th birthday.

58. Trust Fund A philanthropist deposits \$5000 in a trust fund that pays 7.5% interest, compounded continuously. The balance will be given to the college from which the philanthropist graduated after the money has earned interest for 50 years. How much will the college receive?

59. Inflation Assuming that the annual rate of inflation averages 4% over the next 10 years, the approximate costs C of goods or services during any year in that decade can be modeled by $C(t) = P(1.04)^t$, where t is the time in years and P is the present cost. The price of an oil change for your car is presently \$29.88. Estimate the price 10 years from now.

60. Computer Virus The number V of computers infected by a virus increases according to the model $V(t) = 100e^{4.6052t}$, where t is the time in hours. Find the number of computers infected after (a) 1 hour, (b) 1.5 hours, and (c) 2 hours.

61. Population Growth The projected population of the United States for the years 2025 through 2055 can be modeled by $P = 307.58e^{0.0052t}$, where P is the population (in millions) and t is the time (in years), with $t = 25$ corresponding to 2025. *(Source: U.S. Census Bureau)*

(a) Use a graphing utility to graph the function for the years 2025 through 2055.

(b) Use the *table* feature of the graphing utility to create a table of values for the same time period as in part (a).

(c) According to the model, during what year will the population of the United States exceed 430 million?

62. Population The population P (in millions) of Italy from 2003 through 2015 can be approximated by the model $P = 57.59e^{0.0051t}$, where t represents the year, with $t = 3$ corresponding to 2003. *(Source: U.S. Census Bureau)*

(a) According to the model, is the population of Italy increasing or decreasing? Explain.

(b) Find the populations of Italy in 2003 and 2015.

(c) Use the model to predict the populations of Italy in 2020 and 2025.

63. Radioactive Decay Let Q represent a mass (in grams) of radioactive plutonium (^{239}Pu), whose half-life is 24,100 years. The quantity of plutonium present after t years is $Q = 16\left(\frac{1}{2}\right)^{t/24,100}$.

(a) Determine the initial quantity (when $t = 0$).

(b) Determine the quantity present after 75,000 years.

(c) Use a graphing utility to graph the function over the interval $t = 0$ to $t = 150,000$.

64. Radioactive Decay Let Q represent a mass (in grams) of carbon (^{14}C), whose half-life is 5715 years. The quantity of carbon 14 present after t years is $Q = 10\left(\frac{1}{2}\right)^{t/5715}$.

(a) Determine the initial quantity (when $t = 0$).

(b) Determine the quantity present after 2000 years.

(c) Sketch the graph of the function over the interval $t = 0$ to $t = 10,000$.

65. Depreciation The value of a wheelchair conversion van that originally cost $49,810 depreciates so that each year it is worth $\frac{7}{8}$ of its value for the previous year.

(a) Find a model for $V(t)$, the value of the van after t years.

(b) Determine the value of the van 4 years after it was purchased.

66. Chemistry

Immediately following an injection, the concentration of a drug in the bloodstream is 300 milligrams per milliliter. After t hours, the concentration is 75% of the level of the previous hour.

(a) Find a model for $C(t)$, the concentration of the drug after t hours.

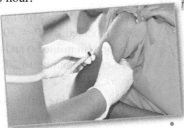

(b) Determine the concentration of the drug after 8 hours.

Exploration

True or False? In Exercises 67 and 68, determine whether the statement is true or false. Justify your answer.

67. The line $y = -2$ is an asymptote for the graph of $f(x) = 10^x - 2$.

68. $e = \dfrac{271,801}{99,990}$

Think About It In Exercises 69–72, use properties of exponents to determine which functions (if any) are the same.

69. $f(x) = 3^{x-2}$
$g(x) = 3^x - 9$
$h(x) = \frac{1}{9}(3^x)$

70. $f(x) = 4^x + 12$
$g(x) = 2^{2x+6}$
$h(x) = 64(4^x)$

71. $f(x) = 16(4^{-x})$
$g(x) = \left(\frac{1}{4}\right)^{x-2}$
$h(x) = 16(2^{-2x})$

72. $f(x) = e^{-x} + 3$
$g(x) = e^{3-x}$
$h(x) = -e^{x-3}$

73. Solving Inequalities Graph the functions $y = 3^x$ and $y = 4^x$ and use the graphs to solve each inequality.

(a) $4^x < 3^x$

(b) $4^x > 3^x$

74. Using Technology Use a graphing utility to graph each function. Use the graph to find where the function is increasing and decreasing, and approximate any relative maximum or minimum values.

(a) $f(x) = x^2 e^{-x}$

(b) $g(x) = x2^{3-x}$

75. Graphical Reasoning Use a graphing utility to graph $y_1 = [1 + (1/x)]^x$ and $y_2 = e$ in the same viewing window. Using the *trace* feature, explain what happens to the graph of y_1 as x increases.

76. Graphical Reasoning Use a graphing utility to graph

$$f(x) = \left(1 + \frac{0.5}{x}\right)^x \quad \text{and} \quad g(x) = e^{0.5}$$

in the same viewing window. What is the relationship between f and g as x increases and decreases without bound?

77. Comparing Graphs Use a graphing utility to graph each pair of functions in the same viewing window. Describe any similarities and differences in the graphs.

(a) $y_1 = 2^x, y_2 = x^2$

(b) $y_1 = 3^x, y_2 = x^3$

78. HOW DO YOU SEE IT? The figure shows the graphs of $y = 2^x$, $y = e^x$, $y = 10^x$, $y = 2^{-x}$, $y = e^{-x}$, and $y = 10^{-x}$. Match each function with its graph. [The graphs are labeled (a) through (f).] Explain your reasoning.

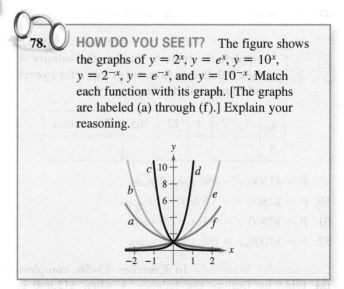

79. Think About It Which functions are exponential?

(a) $f(x) = 3x$

(b) $g(x) = 3x^2$

(c) $h(x) = 3^x$

(d) $k(x) = 2^{-x}$

80. Compound Interest Use the formula

$$A = P\left(1 + \frac{r}{n}\right)^{nt}$$

to calculate the balance A of an investment when $P = \$3000$, $r = 6\%$, and $t = 10$ years, and compounding is done (a) by the day, (b) by the hour, (c) by the minute, and (d) by the second. Does increasing the number of compoundings per year result in unlimited growth of the balance? Explain.

Project: Population per Square Mile To work an extended application analyzing the population per square mile of the United States, visit this text's website at *LarsonPrecalculus.com*. (*Source: U.S. Census Bureau*)

3.2 Logarithmic Functions and Their Graphs

- Recognize and evaluate logarithmic functions with base *a*.
- Graph logarithmic functions.
- Recognize, evaluate, and graph natural logarithmic functions.
- Use logarithmic functions to model and solve real-life problems.

Logarithmic Functions

In Section 3.1, you learned that the exponential function $f(x) = a^x$ is one-to-one. It follows that $f(x) = a^x$ must have an inverse function. This inverse function is the **logarithmic function with base *a*.**

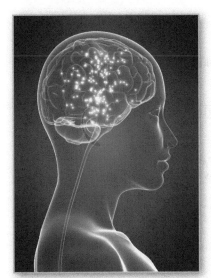

Logarithmic functions can often model scientific observations. For example, in Exercise 83 on page 218, you will use a logarithmic function that models human memory.

Definition of Logarithmic Function with Base *a*

For $x > 0$, $a > 0$, and $a \neq 1$,

$$y = \log_a x \text{ if and only if } x = a^y.$$

The function

$$f(x) = \log_a x \qquad \text{Read as "log base } a \text{ of } x."$$

is the **logarithmic function with base *a*.**

The equations $y = \log_a x$ and $x = a^y$ are equivalent. For example, $2 = \log_3 9$ is equivalent to $9 = 3^2$, and $5^3 = 125$ is equivalent to $\log_5 125 = 3$.

When evaluating logarithms, remember that *a logarithm is an exponent*. This means that $\log_a x$ is the exponent to which *a* must be raised to obtain *x*. For example, $\log_2 8 = 3$ because 2 raised to the third power is 8.

EXAMPLE 1 **Evaluating Logarithms**

Evaluate each logarithm at the given value of *x*.

a. $f(x) = \log_2 x, \quad x = 32$ **b.** $f(x) = \log_3 x, \quad x = 1$

c. $f(x) = \log_4 x, \quad x = 2$ **d.** $f(x) = \log_{10} x, \quad x = \frac{1}{100}$

Solution

a. $f(32) = \log_2 32 = 5$ because $2^5 = 32$.

b. $f(1) = \log_3 1 = 0$ because $3^0 = 1$.

c. $f(2) = \log_4 2 = \frac{1}{2}$ because $4^{1/2} = \sqrt{4} = 2$.

d. $f\left(\frac{1}{100}\right) = \log_{10} \frac{1}{100} = -2$ because $10^{-2} = \frac{1}{10^2} = \frac{1}{100}$.

✓ **Checkpoint**))) *Audio-video solution in English & Spanish at LarsonPrecalculus.com*

Evaluate each logarithm at the given value of *x*.

a. $f(x) = \log_6 x, x = 1$ **b.** $f(x) = \log_5 x, x = \frac{1}{125}$ **c.** $f(x) = \log_7 x, x = 343$

The logarithmic function with base 10 is called the **common logarithmic function.** It is denoted by $\log_{10}$ or simply log. On most calculators, it is denoted by $\boxed{\text{LOG}}$. Example 2 shows how to use a calculator to evaluate common logarithmic functions. You will learn how to use a calculator to calculate logarithms with any base in Section 3.3.

> **EXAMPLE 2** **Evaluating Common Logarithms on a Calculator**

Use a calculator to evaluate the function $f(x) = \log x$ at each value of x.

a. $x = 10$ **b.** $x = \frac{1}{3}$ **c.** $x = -2$

Solution

Function Value	Calculator Keystrokes	Display
a. $f(10) = \log 10$	LOG 10 ENTER	1
b. $f\left(\frac{1}{3}\right) = \log \frac{1}{3}$	LOG (1 ÷ 3) ENTER	-0.4771213
c. $f(-2) = \log(-2)$	LOG (−) 2 ENTER	ERROR

Note that the calculator displays an error message (or a complex number) when you try to evaluate $\log(-2)$. This occurs because there is no real number power to which 10 can be raised to obtain -2.

✓ **Checkpoint** ◀))) *Audio-video solution in English & Spanish at LarsonPrecalculus.com*

Use a calculator to evaluate the function $f(x) = \log x$ at each value of x.

a. $x = 275$ **b.** $x = -\frac{1}{2}$ **c.** $x = \frac{1}{2}$

The definition of the logarithmic function with base a leads to several properties.

Properties of Logarithms

1. $\log_a 1 = 0$ because $a^0 = 1$.

2. $\log_a a = 1$ because $a^1 = a$.

3. $\log_a a^x = x$ and $a^{\log_a x} = x$ Inverse Properties

4. If $\log_a x = \log_a y$, then $x = y$.   One-to-One Property

---

> **EXAMPLE 3** **Using Properties of Logarithms**

a. Simplify $\log_4 1$.   **b.** Simplify $\log_{\sqrt{7}} \sqrt{7}$.   **c.** Simplify $6^{\log_6 20}$.

Solution

a. $\log_4 1 = 0$ Property 1

b. $\log_{\sqrt{7}} \sqrt{7} = 1$ Property 2

c. $6^{\log_6 20} = 20$ Property 3 (Inverse Property)

✓ **Checkpoint** ◀))) *Audio-video solution in English & Spanish at LarsonPrecalculus.com*

a. Simplify $\log_9 9$.   **b.** Simplify $20^{\log_{20} 3}$.   **c.** Simplify $\log_{\sqrt{3}} 1$.

> **EXAMPLE 4** **Using the One-to-One Property**

a. $\log_3 x = \log_3 12$ Original equation

     $x = 12$ One-to-One Property

b. $\log(2x + 1) = \log 3x$ ⟹ $2x + 1 = 3x$ ⟹ $1 = x$

c. $\log_4(x^2 - 6) = \log_4 10$ ⟹ $x^2 - 6 = 10$ ⟹ $x^2 = 16$ ⟹ $x = \pm 4$

✓ **Checkpoint** ◀))) *Audio-video solution in English & Spanish at LarsonPrecalculus.com*

Solve $\log_5(x^2 + 3) = \log_5 12$ for x.

Graphs of Logarithmic Functions

To sketch the graph of $y = \log_a x$, use the fact that the graphs of inverse functions are reflections of each other in the line $y = x$.

EXAMPLE 5 **Graphing Exponential and Logarithmic Functions**

In the same coordinate plane, sketch the graph of each function.

a. $f(x) = 2^x$ **b.** $g(x) = \log_2 x$

Solution

a. For $f(x) = 2^x$, construct a table of values. By plotting these points and connecting them with a smooth curve, you obtain the graph shown in Figure 3.6.

x	-2	-1	0	1	2	3
$f(x) = 2^x$	$\frac{1}{4}$	$\frac{1}{2}$	1	2	4	8

b. Because $g(x) = \log_2 x$ is the inverse function of $f(x) = 2^x$, the graph of g is obtained by plotting the points $(f(x), x)$ and connecting them with a smooth curve. The graph of g is a reflection of the graph of f in the line $y = x$, as shown in Figure 3.6.

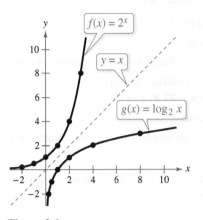

Figure 3.6

✓ *Checkpoint*  *Audio-video solution in English & Spanish at LarsonPrecalculus.com*

In the same coordinate plane, sketch the graphs of (a) $f(x) = 8^x$ and (b) $g(x) = \log_8 x$.

EXAMPLE 6 **Sketching the Graph of a Logarithmic Function**

Sketch the graph of $f(x) = \log x$. Identify the vertical asymptote.

Solution Begin by constructing a table of values. Note that some of the values can be obtained without a calculator by using the properties of logarithms. Others require a calculator.

	Without calculator				With calculator		
x	$\frac{1}{100}$	$\frac{1}{10}$	1	10	2	5	8
$f(x) = \log x$	-2	-1	0	1	0.301	0.699	0.903

Next, plot the points and connect them with a smooth curve, as shown in the figure below. The vertical asymptote is $x = 0$ (y-axis).

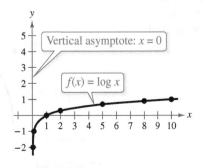

✓ *Checkpoint* *Audio-video solution in English & Spanish at LarsonPrecalculus.com*

Sketch the graph of $f(x) = \log_3 x$ by constructing a table of values without using a calculator. Identify the vertical asymptote.

The graph in Example 6 is typical for functions of the form $f(x) = \log_a x$, $a > 1$. They have one x-intercept and one vertical asymptote. Notice how slowly the graph rises for $x > 1$. Here are the basic characteristics of logarithmic graphs.

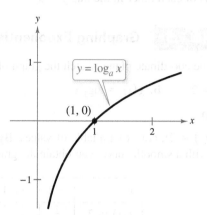

Graph of $y = \log_a x$, $a > 1$
- Domain: $(0, \infty)$
- Range: $(-\infty, \infty)$
- x-intercept: $(1, 0)$
- Increasing
- One-to-one, therefore has an inverse function
- y-axis is a vertical asymptote ($\log_a x \to -\infty$ as $x \to 0^+$).
- Continuous
- Reflection of graph of $y = a^x$ in the line $y = x$

Some basic characteristics of the graph of $f(x) = a^x$ are listed below to illustrate the inverse relation between $f(x) = a^x$ and $g(x) = \log_a x$.

- Domain: $(-\infty, \infty)$
- y-intercept: $(0, 1)$
- Range: $(0, \infty)$
- x-axis is a horizontal asymptote ($a^x \to 0$ as $x \to -\infty$).

The next example uses the graph of $y = \log_a x$ to sketch the graphs of functions of the form $f(x) = b \pm \log_a(x + c)$.

EXAMPLE 7   **Shifting Graphs of Logarithmic Functions**

See LarsonPrecalculus.com for an interactive version of this type of example.

Use the graph of $f(x) = \log x$ to sketch the graph of each function.

a. $g(x) = \log(x - 1)$ **b.** $h(x) = 2 + \log x$

Solution

a. Because $g(x) = \log(x - 1) = f(x - 1)$, the graph of g can be obtained by shifting the graph of f one unit to the right, as shown in Figure 3.7.

b. Because $h(x) = 2 + \log x = 2 + f(x)$, the graph of h can be obtained by shifting the graph of f two units up, as shown in Figure 3.8.

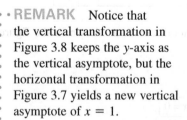

REMARK Notice that the vertical transformation in Figure 3.8 keeps the y-axis as the vertical asymptote, but the horizontal transformation in Figure 3.7 yields a new vertical asymptote of $x = 1$.

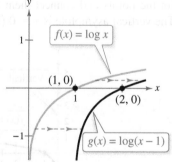

Figure 3.7

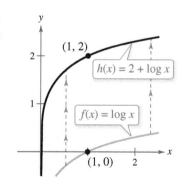

Figure 3.8

▷ **ALGEBRA HELP** To review the techniques for shifting, reflecting, and stretching graphs, see Section 1.7.

✓ *Checkpoint*  *Audio-video solution in English & Spanish at LarsonPrecalculus.com*

Use the graph of $f(x) = \log_3 x$ to sketch the graph of each function.

a. $g(x) = -1 + \log_3 x$ **b.** $h(x) = \log_3(x + 3)$

The Natural Logarithmic Function

By looking back at the graph of the natural exponential function introduced on page 202 in Section 3.1, you will see that $f(x) = e^x$ is one-to-one and so has an inverse function. This inverse function is called the **natural logarithmic function** and is denoted by the special symbol $\ln x$, read as "the natural log of x" or "el en of x."

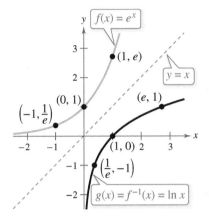

Reflection of graph of $f(x) = e^x$ in the line $y = x$
Figure 3.9

> **The Natural Logarithmic Function**
>
> The function
>
> $$f(x) = \log_e x = \ln x, \quad x > 0$$
>
> is called the **natural logarithmic function.**

The equations $y = \ln x$ and $x = e^y$ are equivalent. Note that the natural logarithm $\ln x$ is written without a base. The base is understood to be e.

Because the functions $f(x) = e^x$ and $g(x) = \ln x$ are inverse functions of each other, their graphs are reflections of each other in the line $y = x$, as shown in Figure 3.9.

▷ **TECHNOLOGY** On most calculators, the natural logarithm is denoted by ⬚LN⬚ as illustrated in Example 8.

EXAMPLE 8 **Evaluating the Natural Logarithmic Function**

Use a calculator to evaluate the function $f(x) = \ln x$ at each value of x.

a. $x = 2$

b. $x = 0.3$

c. $x = -1$

d. $x = 1 + \sqrt{2}$

Solution

Function Value	Calculator Keystrokes	Display
a. $f(2) = \ln 2$	⬚LN⬚ 2 ⬚ENTER⬚	0.6931472
b. $f(0.3) = \ln 0.3$	⬚LN⬚ .3 ⬚ENTER⬚	−1.2039728
c. $f(-1) = \ln(-1)$	⬚LN⬚ ⬚(−)⬚ 1 ⬚ENTER⬚	ERROR
d. $f(1 + \sqrt{2}) = \ln(1 + \sqrt{2})$	⬚LN⬚ ⬚(⬚ 1 ⬚+⬚ ⬚√⬚ 2 ⬚)⬚ ⬚ENTER⬚	0.8813736

✓ **Checkpoint** ◀))) *Audio-video solution in English & Spanish at LarsonPrecalculus.com*

Use a calculator to evaluate the function $f(x) = \ln x$ at each value of x.

a. $x = 0.01$ **b.** $x = 4$

c. $x = \sqrt{3} + 2$ **d.** $x = \sqrt{3} - 2$

• • • • • • • • • • • • • • • • • • ▷

•• **REMARK** In Example 8(c), be sure you see that $\ln(-1)$ gives an error message on most calculators. This occurs because the domain of $\ln x$ is the set of *positive real numbers* (see Figure 3.9). So, $\ln(-1)$ is undefined.

The properties of logarithms on page 210 are also valid for natural logarithms.

> **Properties of Natural Logarithms**
>
> **1.** $\ln 1 = 0$ because $e^0 = 1$.
>
> **2.** $\ln e = 1$ because $e^1 = e$.
>
> **3.** $\ln e^x = x$ and $e^{\ln x} = x$ Inverse Properties
>
> **4.** If $\ln x = \ln y$, then $x = y$.   One-to-One Property

EXAMPLE 9 **Using Properties of Natural Logarithms**

Use the properties of natural logarithms to simplify each expression.

a. $\ln \dfrac{1}{e}$ **b.** $e^{\ln 5}$ **c.** $\dfrac{\ln 1}{3}$ **d.** $2 \ln e$

Solution

a. $\ln \dfrac{1}{e} = \ln e^{-1} = -1$ Property 3 (Inverse Property)

b. $e^{\ln 5} = 5$ Property 3 (Inverse Property)

c. $\dfrac{\ln 1}{3} = \dfrac{0}{3} = 0$ Property 1

d. $2 \ln e = 2(1) = 2$ Property 2

✓ **Checkpoint**))) Audio-video solution in English & Spanish at LarsonPrecalculus.com

Use the properties of natural logarithms to simplify each expression.

a. $\ln e^{1/3}$ **b.** $5 \ln 1$ **c.** $\frac{3}{4} \ln e$ **d.** $e^{\ln 7}$

EXAMPLE 10 **Finding the Domains of Logarithmic Functions**

Find the domain of each function.

a. $f(x) = \ln(x - 2)$ **b.** $g(x) = \ln(2 - x)$ **c.** $h(x) = \ln x^2$

Solution

a. Because $\ln(x - 2)$ is defined only when

$$x - 2 > 0$$

it follows that the domain of f is $(2, \infty)$, as shown in Figure 3.10.

b. Because $\ln(2 - x)$ is defined only when

$$2 - x > 0$$

it follows that the domain of g is $(-\infty, 2)$, as shown in Figure 3.11.

c. Because $\ln x^2$ is defined only when

$$x^2 > 0$$

it follows that the domain of h is all real numbers except $x = 0$, as shown in Figure 3.12.

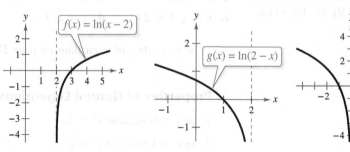

Figure 3.10   Figure 3.11   Figure 3.12

✓ **Checkpoint**))) Audio-video solution in English & Spanish at LarsonPrecalculus.com

Find the domain of $f(x) = \ln(x + 3)$.

Application

EXAMPLE 11   **Human Memory Model**

Students participating in a psychology experiment attended several lectures on a subject and took an exam. Every month for a year after the exam, the students took a retest to see how much of the material they remembered. The average scores for the group are given by the *human memory model* $f(t) = 75 - 6 \ln(t + 1)$, $0 \le t \le 12$, where t is the time in months.

a. What was the average score on the original exam ($t = 0$)?

b. What was the average score at the end of $t = 2$ months?

c. What was the average score at the end of $t = 6$ months?

Algebraic Solution

a. The original average score was

$$f(0) = 75 - 6 \ln(0 + 1)$$ Substitute 0 for t.

$$= 75 - 6 \ln 1$$ Simplify.

$$= 75 - 6(0)$$ Property of natural logarithms

$$= 75.$$ Solution

b. After 2 months, the average score was

$$f(2) = 75 - 6 \ln(2 + 1)$$ Substitute 2 for t.

$$= 75 - 6 \ln 3$$ Simplify.

$$\approx 75 - 6(1.0986)$$ Use a calculator.

$$\approx 68.41.$$ Solution

c. After 6 months, the average score was

$$f(6) = 75 - 6 \ln(6 + 1)$$ Substitute 6 for t.

$$= 75 - 6 \ln 7$$ Simplify.

$$\approx 75 - 6(1.9459)$$ Use a calculator.

$$\approx 63.32.$$ Solution

Graphical Solution

a.

When $t = 0$, $y = 75$. So, the original average score was 75.

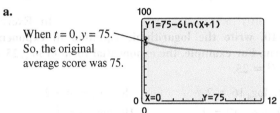

b.

When $t = 2$, $y \approx 68.41$. So, the average score after 2 months was about 68.41.

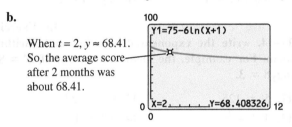

c.

When $t = 6$, $y \approx 63.32$. So, the average score after 6 months was about 63.32.

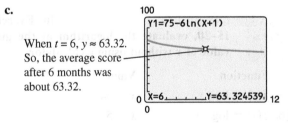

✓ *Checkpoint* ◀))) *Audio-video solution in English & Spanish at LarsonPrecalculus.com*

In Example 11, find the average score at the end of (a) $t = 1$ month, (b) $t = 9$ months, and (c) $t = 12$ months.

Summarize   (Section 3.2)

1. State the definition of the logarithmic function with base a *(page 209)* and make a list of the properties of logarithms *(page 210)*. For examples of evaluating logarithmic functions and using the properties of logarithms, see Examples 1–4.

2. Explain how to graph a logarithmic function *(pages 211 and 212)*. For examples of graphing logarithmic functions, see Examples 5–7.

3. State the definition of the natural logarithmic function and make a list of the properties of natural logarithms *(page 213)*. For examples of evaluating natural logarithmic functions and using the properties of natural logarithms, see Examples 8 and 9.

4. Describe a real-life application that uses a logarithmic function to model and solve a problem *(page 215, Example 11)*.

3.2 Exercises

Vocabulary: Fill in the blanks.

1. The inverse function of the exponential function $f(x) = a^x$ is the _____ function with base a.
2. The common logarithmic function has base _____.
3. The logarithmic function $f(x) = \ln x$ is the _____ logarithmic function and has base _____.
4. The Inverse Properties of logarithms state that $\log_a a^x = x$ and _____.
5. The One-to-One Property of natural logarithms states that if $\ln x = \ln y$, then _____.
6. The domain of the natural logarithmic function is the set of _____ _____ _____.

Skills and Applications

Writing an Exponential Equation In Exercises 7–10, write the logarithmic equation in exponential form. For example, the exponential form of $\log_5 25 = 2$ is $5^2 = 25$.

7. $\log_4 16 = 2$
8. $\log_9 \frac{1}{81} = -2$
9. $\log_{12} 12 = 1$
10. $\log_{32} 4 = \frac{2}{5}$

Writing a Logarithmic Equation In Exercises 11–14, write the exponential equation in logarithmic form. For example, the logarithmic form of $2^3 = 8$ is $\log_2 8 = 3$.

11. $5^3 = 125$
12. $9^{3/2} = 27$
13. $4^{-3} = \frac{1}{64}$
14. $24^0 = 1$

Evaluating a Logarithm In Exercises 15–20, evaluate the logarithm at the given value of x without using a calculator.

Function	Value
15. $f(x) = \log_2 x$	$x = 64$
16. $f(x) = \log_{25} x$	$x = 5$
17. $f(x) = \log_8 x$	$x = 1$
18. $f(x) = \log x$	$x = 10$
19. $g(x) = \log_a x$	$x = a^{-2}$
20. $g(x) = \log_b x$	$x = \sqrt{b}$

Evaluating a Common Logarithm on a Calculator In Exercises 21–24, use a calculator to evaluate $f(x) = \log x$ at the given value of x. Round your result to three decimal places.

21. $x = \frac{7}{8}$
22. $x = \frac{1}{500}$
23. $x = 12.5$
24. $x = 96.75$

Using Properties of Logarithms In Exercises 25–28, use the properties of logarithms to simplify the expression.

25. $\log_8 8$
26. $\log_\pi \pi^2$
27. $\log_{7.5} 1$
28. $5^{\log_5 3}$

Using the One-to-One Property In Exercises 29–32, use the One-to-One Property to solve the equation for x.

29. $\log_5(x + 1) = \log_5 6$
30. $\log_2(x - 3) = \log_2 9$
31. $\log 11 = \log(x^2 + 7)$
32. $\log(x^2 + 6x) = \log 27$

Graphing Exponential and Logarithmic Functions In Exercises 33–36, sketch the graphs of f and g in the same coordinate plane.

33. $f(x) = 7^x$, $g(x) = \log_7 x$
34. $f(x) = 5^x$, $g(x) = \log_5 x$
35. $f(x) = 6^x$, $g(x) = \log_6 x$
36. $f(x) = 10^x$, $g(x) = \log x$

Matching a Logarithmic Function with Its Graph In Exercises 37–40, use the graph of $g(x) = \log_3 x$ to match the given function with its graph. Then describe the relationship between the graphs of f and g. [The graphs are labeled (a), (b), (c), and (d).]

(a)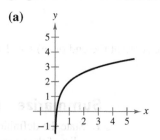

(b)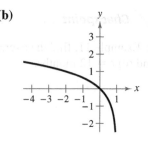

(c)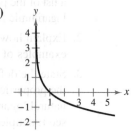

(d)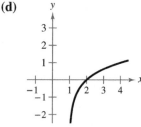

37. $f(x) = \log_3 x + 2$
38. $f(x) = \log_3(x - 1)$
39. $f(x) = \log_3(1 - x)$
40. $f(x) = -\log_3 x$

Sketching the Graph of a Logarithmic Function In Exercises 41–48, find the domain, x-intercept, and vertical asymptote of the logarithmic function and sketch its graph.

41. $f(x) = \log_4 x$ **42.** $g(x) = \log_6 x$

43. $y = \log_3 x + 1$

44. $h(x) = \log_4(x - 3)$

45. $f(x) = -\log_6(x + 2)$

46. $y = \log_5(x - 1) + 4$

47. $y = \log \dfrac{x}{7}$

48. $y = \log(-2x)$

Writing a Natural Exponential Equation In Exercises 49–52, write the logarithmic equation in exponential form.

49. $\ln \frac{1}{2} = -0.693 \ldots$ **50.** $\ln 7 = 1.945 \ldots$

51. $\ln 250 = 5.521 \ldots$ **52.** $\ln 1 = 0$

Writing a Natural Logarithmic Equation In Exercises 53–56, write the exponential equation in logarithmic form.

53. $e^2 = 7.3890 \ldots$ **54.** $e^{-3/4} = 0.4723 \ldots$

55. $e^{-4x} = \frac{1}{2}$ **56.** $e^{2x} = 3$

 Evaluating a Logarithmic Function In Exercises 57–60, use a calculator to evaluate the function at the given value of x. Round your result to three decimal places.

Function	Value
57. $f(x) = \ln x$	$x = 18.42$
58. $f(x) = 3 \ln x$	$x = 0.74$
59. $g(x) = 8 \ln x$	$x = \sqrt{5}$
60. $g(x) = -\ln x$	$x = \frac{1}{2}$

 Using Properties of Natural Logarithms In Exercises 61–66, use the properties of natural logarithms to simplify the expression.

61. $e^{\ln 4}$ **62.** $\ln \dfrac{1}{e^2}$

63. $2.5 \ln 1$ **64.** $\dfrac{\ln e}{\pi}$

65. $\ln e^{\ln e}$ **66.** $e^{\ln(1/e)}$

Graphing a Natural Logarithmic Function In Exercises 67–70, find the domain, x-intercept, and vertical asymptote of the logarithmic function and sketch its graph.

67. $f(x) = \ln(x - 4)$ **68.** $h(x) = \ln(x + 5)$

69. $g(x) = \ln(-x)$ **70.** $f(x) = \ln(3 - x)$

Graphing a Natural Logarithmic Function In Exercises 71–74, use a graphing utility to graph the function. Be sure to use an appropriate viewing window.

71. $f(x) = \ln(x - 1)$ **72.** $f(x) = \ln(x + 2)$

73. $f(x) = -\ln x + 8$ **74.** $f(x) = 3 \ln x - 1$

Using the One-to-One Property In Exercises 75–78, use the One-to-One Property to solve the equation for x.

75. $\ln(x + 4) = \ln 12$ **76.** $\ln(x - 7) = \ln 7$

77. $\ln(x^2 - x) = \ln 6$ **78.** $\ln(x^2 - 2) = \ln 23$

79. Monthly Payment   The model

$$t = 16.625 \ln \frac{x}{x - 750}, \quad x > 750$$

approximates the length of a home mortgage of $150,000 at 6% in terms of the monthly payment. In the model, t is the length of the mortgage in years and x is the monthly payment in dollars.

(a) Approximate the lengths of a $150,000 mortgage at 6% when the monthly payment is $897.72 and when the monthly payment is $1659.24.

(b) Approximate the total amounts paid over the term of the mortgage with a monthly payment of $897.72 and with a monthly payment of $1659.24. What amount of the total is interest costs in each case?

(c) What is the vertical asymptote for the model? Interpret its meaning in the context of the problem.

80. Telephone Service   The percent P of households in the United States with wireless-only telephone service from 2005 through 2014 can be approximated by the model

$$P = -3.42 + 1.297t \ln t, \quad 5 \le t \le 14$$

where t represents the year, with $t = 5$ corresponding to 2005. *(Source: National Center for Health Statistics)*

(a) Approximate the percents of households with wireless-only telephone service in 2008 and 2012.

(b) Use a graphing utility to graph the function.

(c) Can the model be used to predict the percent of households with wireless-only telephone service in 2020? in 2030? Explain.

81. Population   The time t (in years) for the world population to double when it is increasing at a continuous rate r (in decimal form) is given by $t = (\ln 2)/r$.

(a) Complete the table and interpret your results.

r	0.005	0.010	0.015	0.020	0.025	0.030
t						

(b) Use a graphing utility to graph the function.

82. Compound Interest A principal P, invested at $5\frac{1}{2}\%$ and compounded continuously, increases to an amount K times the original principal after t years, where $t = (\ln K)/0.055$.

(a) Complete the table and interpret your results.

K	1	2	4	6	8	10	12
t							

(b) Sketch a graph of the function.

83. Human Memory Model

Students in a mathematics class took an exam and then took a retest monthly with an equivalent exam. The average scores for the class are given by the human memory model

$$f(t) = 80 - 17 \log(t + 1), \quad 0 \le t \le 12$$

where t is the time in months.

(a) Use a graphing utility to graph the model over the specified domain.

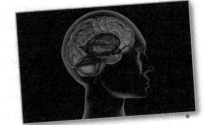

(b) What was the average score on the original exam ($t = 0$)?

(c) What was the average score after 4 months?

(d) What was the average score after 10 months?

84. Sound Intensity The relationship between the number of decibels β and the intensity of a sound I (in watts per square meter) is

$$\beta = 10 \log \frac{I}{10^{-12}}.$$

(a) Determine the number of decibels of a sound with an intensity of 1 watt per square meter.

(b) Determine the number of decibels of a sound with an intensity of 10^{-2} watt per square meter.

(c) The intensity of the sound in part (a) is 100 times as great as that in part (b). Is the number of decibels 100 times as great? Explain.

Exploration

True or False? **In Exercises 85 and 86, determine whether the statement is true or false. Justify your answer.**

85. The graph of $f(x) = \log_6 x$ is a reflection of the graph of $g(x) = 6^x$ in the x-axis.

86. The graph of $f(x) = \ln(-x)$ is a reflection of the graph of $h(x) = e^{-x}$ in the line $y = -x$.

87. Graphical Reasoning Use a graphing utility to graph f and g in the same viewing window and determine which is increasing at the greater rate as x approaches $+\infty$. What can you conclude about the rate of growth of the natural logarithmic function?

(a) $f(x) = \ln x,\quad g(x) = \sqrt{x}$

(b) $f(x) = \ln x,\quad g(x) = \sqrt[4]{x}$

88. HOW DO YOU SEE IT? The figure shows the graphs of $f(x) = 3^x$ and $g(x) = \log_3 x$. [The graphs are labeled m and n.]

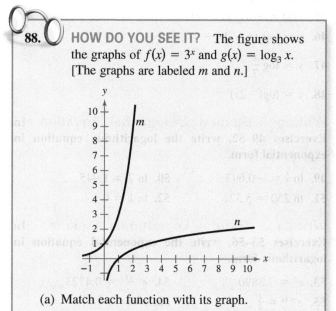

(a) Match each function with its graph.

(b) Given that $f(a) = b$, what is $g(b)$? Explain.

Error Analysis **In Exercises 89 and 90, describe the error.**

89.

x	1	2	8
y	0	1	3

From the table, you can conclude that y is an exponential function of x.

90.

x	1	2	5
y	2	4	32

From the table, you can conclude that y is a logarithmic function of x.

91. Numerical Analysis

(a) Complete the table for the function $f(x) = (\ln x)/x$.

x	1	5	10	10^2	10^4	10^6
$f(x)$						

(b) Use the table in part (a) to determine what value $f(x)$ approaches as x increases without bound.

(c) Use a graphing utility to confirm the result of part (b).

92. Writing Explain why $\log_a x$ is defined only for $0 < a < 1$ and $a > 1$.

3.3 Properties of Logarithms

Logarithmic functions have many real-life applications. For example, in Exercises 79–82 on page 224, you will use a logarithmic function that models the relationship between the number of decibels and the intensity of a sound.

- ◼ Use the change-of-base formula to rewrite and evaluate logarithmic expressions.
- ◼ Use properties of logarithms to evaluate or rewrite logarithmic expressions.
- ◼ Use properties of logarithms to expand or condense logarithmic expressions.
- ◼ Use logarithmic functions to model and solve real-life problems.

Change of Base

Most calculators have only two types of log keys, $\boxed{\text{LOG}}$ for common logarithms (base 10) and $\boxed{\text{LN}}$ for natural logarithms (base e). Although common logarithms and natural logarithms are the most frequently used, you may occasionally need to evaluate logarithms with other bases. To do this, use the **change-of-base formula.**

Change-of-Base Formula

Let a, b, and x be positive real numbers such that $a \neq 1$ and $b \neq 1$. Then $\log_a x$ can be converted to a different base as follows.

Base b	Base 10	Base e
$\log_a x = \dfrac{\log_b x}{\log_b a}$	$\log_a x = \dfrac{\log x}{\log a}$	$\log_a x = \dfrac{\ln x}{\ln a}$

One way to look at the change-of-base formula is that logarithms with base a are *constant multiples* of logarithms with base b. The constant multiplier is

$$\frac{1}{\log_b a}.$$

EXAMPLE 1   Changing Bases Using Common Logarithms

$$\log_4 25 = \frac{\log 25}{\log 4} \qquad \log_a x = \frac{\log x}{\log a}$$

$$\approx \frac{1.39794}{0.60206} \qquad \text{Use a calculator.}$$

$$\approx 2.3219 \qquad \text{Simplify.}$$

✓ **Checkpoint** ◀))) *Audio-video solution in English & Spanish at LarsonPrecalculus.com*

Evaluate $\log_2 12$ using the change-of-base formula and common logarithms.

EXAMPLE 2   Changing Bases Using Natural Logarithms

$$\log_4 25 = \frac{\ln 25}{\ln 4} \qquad \log_a x = \frac{\ln x}{\ln a}$$

$$\approx \frac{3.21888}{1.38629} \qquad \text{Use a calculator.}$$

$$\approx 2.3219 \qquad \text{Simplify.}$$

✓ **Checkpoint** ◀))) *Audio-video solution in English & Spanish at LarsonPrecalculus.com*

Evaluate $\log_2 12$ using the change-of-base formula and natural logarithms. ◼

Properties of Logarithms

You know from the preceding section that the logarithmic function with base a is the *inverse function* of the exponential function with base a. So, it makes sense that the properties of exponents have corresponding properties involving logarithms. For example, the exponential property $a^m a^n = a^{m+n}$ has the corresponding logarithmic property $\log_a(uv) = \log_a u + \log_a v$.

> •• REMARK There is no property that can be used to rewrite $\log_a(u \pm v)$. Specifically, $\log_a(u + v)$ is *not* equal to $\log_a u + \log_a v$.

Properties of Logarithms

Let a be a positive number such that $a \neq 1$, let n be a real number, and let u and v be positive real numbers.

	Logarithm with Base a	Natural Logarithm
1. Product Property:	$\log_a(uv) = \log_a u + \log_a v$	$\ln(uv) = \ln u + \ln v$
2. Quotient Property:	$\log_a \dfrac{u}{v} = \log_a u - \log_a v$	$\ln \dfrac{u}{v} = \ln u - \ln v$
3. Power Property:	$\log_a u^n = n \log_a u$	$\ln u^n = n \ln u$

For proofs of the properties listed above, see Proofs in Mathematics on page 256.

EXAMPLE 3   Using Properties of Logarithms

Write each logarithm in terms of $\ln 2$ and $\ln 3$.

a. $\ln 6$ **b.** $\ln \dfrac{2}{27}$

Solution

a. $\ln 6 = \ln(2 \cdot 3)$ Rewrite 6 as $2 \cdot 3$.

$\quad\quad = \ln 2 + \ln 3$ Product Property

b. $\ln \dfrac{2}{27} = \ln 2 - \ln 27$ Quotient Property

$\quad\quad\quad = \ln 2 - \ln 3^3$ Rewrite 27 as 3^3.

$\quad\quad\quad = \ln 2 - 3 \ln 3$ Power Property

✓ **Checkpoint**  Audio-video solution in English & Spanish at LarsonPrecalculus.com

Write each logarithm in terms of $\log 3$ and $\log 5$.

a. $\log 75$ **b.** $\log \dfrac{9}{125}$

EXAMPLE 4   Using Properties of Logarithms

Find the exact value of $\log_5 \sqrt[3]{5}$ without using a calculator.

Solution

$\log_5 \sqrt[3]{5} = \log_5 5^{1/3} = \tfrac{1}{3} \log_5 5 = \tfrac{1}{3}(1) = \tfrac{1}{3}$

✓ **Checkpoint** 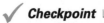 Audio-video solution in English & Spanish at LarsonPrecalculus.com

Find the exact value of $\ln e^6 - \ln e^2$ without using a calculator.

HISTORICAL NOTE

John Napier, a Scottish mathematician, developed logarithms as a way to simplify tedious calculations. Napier worked about 20 years on the development of logarithms before publishing his work is 1614. Napier only partially succeeded in his quest to simplify tedious calculations. Nonetheless, the development of logarithms was a step forward and received immediate recognition.

Rewriting Logarithmic Expressions

The properties of logarithms are useful for rewriting logarithmic expressions in forms that simplify the operations of algebra. This is true because these properties convert complicated products, quotients, and exponential forms into simpler sums, differences, and products, respectively.

EXAMPLE 5 **Expanding Logarithmic Expressions**

Expand each logarithmic expression.

a. $\log_4 5x^3y$ **b.** $\ln \dfrac{\sqrt{3x-5}}{7}$

Solution

a. $\log_4 5x^3y = \log_4 5 + \log_4 x^3 + \log_4 y$ Product Property

$= \log_4 5 + 3\log_4 x + \log_4 y$ Power Property

> ▷ **ALGEBRA HELP** To review rewriting radicals and rational exponents, see Appendix A.2.

b. $\ln \dfrac{\sqrt{3x-5}}{7} = \ln \dfrac{(3x-5)^{1/2}}{7}$ Rewrite using rational exponent.

$= \ln(3x-5)^{1/2} - \ln 7$ Quotient Property

$= \dfrac{1}{2}\ln(3x-5) - \ln 7$ Power Property

✓ *Checkpoint* ◀))) Audio-video solution in English & Spanish at LarsonPrecalculus.com

Expand the expression $\log_3 \dfrac{4x^2}{\sqrt{y}}$.

Example 5 uses the properties of logarithms to *expand* logarithmic expressions. Example 6 reverses this procedure and uses the properties of logarithms to *condense* logarithmic expressions.

EXAMPLE 6 **Condensing Logarithmic Expressions**

See LarsonPrecalculus.com for an interactive version of this type of example.

Condense each logarithmic expression.

a. $\dfrac{1}{2}\log x + 3\log(x+1)$ **b.** $2\ln(x+2) - \ln x$ **c.** $\dfrac{1}{3}[\log_2 x + \log_2(x+1)]$

Solution

a. $\dfrac{1}{2}\log x + 3\log(x+1) = \log x^{1/2} + \log(x+1)^3$ Power Property

$= \log\left[\sqrt{x}\,(x+1)^3\right]$ Product Property

b. $2\ln(x+2) - \ln x = \ln(x+2)^2 - \ln x$ Power Property

$= \ln \dfrac{(x+2)^2}{x}$ Quotient Property

c. $\dfrac{1}{3}[\log_2 x + \log_2(x+1)] = \dfrac{1}{3}\log_2[x(x+1)]$ Product Property

$= \log_2[x(x+1)]^{1/3}$ Power Property

$= \log_2 \sqrt[3]{x(x+1)}$ Rewrite with a radical.

✓ *Checkpoint* ◀))) Audio-video solution in English & Spanish at LarsonPrecalculus.com

Condense the expression $2[\log(x+3) - 2\log(x-2)]$.

Application

One way to determine a possible relationship between the x- and y-values of a set of nonlinear data is to take the natural logarithm of each x-value and each y-value. If the plotted points $(\ln x, \ln y)$ lie on a line, then x and y are related by the equation $\ln y = m \ln x$, where m is the slope of the line.

EXAMPLE 7 **Finding a Mathematical Model**

The table shows the mean distance x from the sun and the period y (the time it takes a planet to orbit the sun, in years) for each of the six planets that are closest to the sun. In the table, the mean distance is given in astronomical units (where one astronomical unit is defined as Earth's mean distance from the sun). The points from the table are plotted in Figure 3.13. Find an equation that relates y and x.

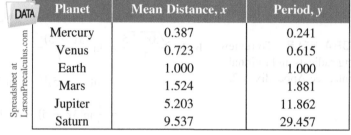

	Planet	Mean Distance, x	Period, y
DATA	Mercury	0.387	0.241
	Venus	0.723	0.615
	Earth	1.000	1.000
	Mars	1.524	1.881
	Jupiter	5.203	11.862
	Saturn	9.537	29.457

Spreadsheet at LarsonPrecalculus.com

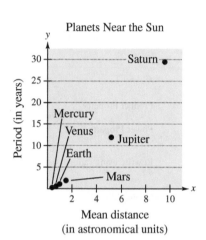

Planets Near the Sun

Period (in years)

Mean distance (in astronomical units)

Figure 3.13

Planet	ln x	ln y
Mercury	−0.949	−1.423
Venus	−0.324	−0.486
Earth	0.000	0.000
Mars	0.421	0.632
Jupiter	1.649	2.473
Saturn	2.255	3.383

Solution From Figure 3.13, it is not clear how to find an equation that relates y and x. To solve this problem, make a table of values giving the natural logarithms of all x- and y-values of the data (see the table at the left). Plot each point $(\ln x, \ln y)$. These points appear to lie on a line (see Figure 3.14). Choose two points to determine the slope of the line. Using the points $(0.421, 0.632)$ and $(0, 0)$, the slope of the line is

$$m = \frac{0.632 - 0}{0.421 - 0} \approx 1.5 = \frac{3}{2}.$$

By the point-slope form, the equation of the line is $Y = \frac{3}{2}X$, where $Y = \ln y$ and $X = \ln x$. So, an equation that relates y and x is $\ln y = \frac{3}{2} \ln x$.

✓ Checkpoint  Audio-video solution in English & Spanish at LarsonPrecalculus.com

Find a logarithmic equation that relates y and x for the following ordered pairs.

$(0.37, 0.51), (1.00, 1.00), (2.72, 1.95), (7.39, 3.79), (20.09, 7.39)$

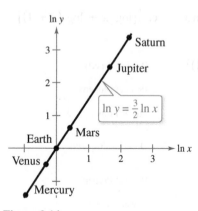

Figure 3.14

$\ln y = \frac{3}{2} \ln x$

Summarize (Section 3.3)

1. State the change-of-base formula *(page 219)*. For examples of using the change-of-base formula to rewrite and evaluate logarithmic expressions, see Examples 1 and 2.

2. Make a list of the properties of logarithms *(page 220)*. For examples of using the properties of logarithms to evaluate or rewrite logarithmic expressions, see Examples 3 and 4.

3. Explain how to use the properties of logarithms to expand or condense logarithmic expressions *(page 221)*. For examples of expanding and condensing logarithmic expressions, see Examples 5 and 6.

4. Describe an example of how to use a logarithmic function to model and solve a real-life problem *(page 222, Example 7)*.

3.3 Exercises

See **CalcChat.com** for tutorial help and worked-out solutions to odd-numbered exercises.

Vocabulary

In Exercises 1–3, fill in the blanks.

1. To evaluate a logarithm to any base, use the _____ formula.

2. The change-of-base formula for base e is $\log_a x = $ _____.

3. When you consider $\log_a x$ to be a constant multiple of $\log_b x$, the constant multiplier is _____.

4. Name the property of logarithms illustrated by each statement.

   (a) $\ln(uv) = \ln u + \ln v$ (b) $\log_a u^n = n \log_a u$ (c) $\ln \dfrac{u}{v} = \ln u - \ln v$

Skills and Applications

Changing Bases In Exercises 5–8, rewrite the logarithm as a ratio of (a) common logarithms and (b) natural logarithms.

5. $\log_5 16$

6. $\log_{1/5} 4$

7. $\log_x \dfrac{3}{10}$

8. $\log_{2.6} x$

Using the Change-of-Base Formula In Exercises 9–12, evaluate the logarithm using the change-of-base formula. Round your result to three decimal places.

9. $\log_3 17$

10. $\log_{0.4} 12$

11. $\log_\pi 0.5$

12. $\log_{2/3} 0.125$

Using Properties of Logarithms In Exercises 13–18, use the properties of logarithms to write the logarithm in terms of $\log_3 5$ and $\log_3 7$.

13. $\log_3 35$

14. $\log_3 \dfrac{5}{7}$

15. $\log_3 \dfrac{7}{25}$

16. $\log_3 175$

17. $\log_3 \dfrac{21}{5}$

18. $\log_3 \dfrac{45}{49}$

Using Properties of Logarithms In Exercises 19–32, find the exact value of the logarithmic expression without using a calculator. (If this is not possible, state the reason.)

19. $\log_3 9$

20. $\log_5 \dfrac{1}{125}$

21. $\log_6 \sqrt[3]{\dfrac{1}{6}}$

22. $\log_2 \sqrt[4]{8}$

23. $\log_2(-2)$

24. $\log_3(-27)$

25. $\ln \sqrt[4]{e^3}$

26. $\ln(1/\sqrt{e})$

27. $\ln e^2 + \ln e^5$

28. $2 \ln e^6 - \ln e^5$

29. $\log_5 75 - \log_5 3$

30. $\log_4 2 + \log_4 32$

31. $\log_4 8$

32. $\log_8 16$

Using Properties of Logarithms In Exercises 33–40, approximate the logarithm using the properties of logarithms, given $\log_b 2 \approx 0.3562$, $\log_b 3 \approx 0.5646$, and $\log_b 5 \approx 0.8271$.

33. $\log_b 10$

34. $\log_b \dfrac{2}{3}$

35. $\log_b 0.04$

36. $\log_b \sqrt{2}$

37. $\log_b 45$

38. $\log_b(3b^2)$

39. $\log_b(2b)^{-2}$

40. $\log_b \sqrt[3]{3b}$

Expanding a Logarithmic Expression In Exercises 41–60, use the properties of logarithms to expand the expression as a sum, difference, and/or constant multiple of logarithms. (Assume all variables are positive.)

41. $\ln 7x$

42. $\log_3 13z$

43. $\log_8 x^4$

44. $\ln(xy)^3$

45. $\log_5 \dfrac{5}{x}$

46. $\log_6 \dfrac{w^2}{v}$

47. $\ln \sqrt{z}$

48. $\ln \sqrt[3]{t}$

49. $\ln xyz^2$

50. $\log_4 11b^2c$

51. $\ln z(z-1)^2, \quad z > 1$

52. $\ln \dfrac{x^2 - 1}{x^3}, \quad x > 1$

53. $\log_2 \dfrac{\sqrt{a^2 - 4}}{7}, \quad a > 2$

54. $\ln \dfrac{3}{\sqrt{x^2 + 1}}$

55. $\log_5 \dfrac{x^2}{y^2 z^3}$

56. $\log_{10} \dfrac{xy^4}{z^5}$

57. $\ln \sqrt[3]{\dfrac{yz}{x^2}}$

58. $\log_2 x^4 \sqrt{\dfrac{y}{z^3}}$

59. $\ln \sqrt[4]{x^3(x^2 + 3)}$

60. $\ln \sqrt{x^2(x + 2)}$

Condensing a Logarithmic Expression
In Exercises 61–76, condense the expression to the logarithm of a single quantity.

61. $\ln 3 + \ln x$

62. $\log_5 8 - \log_5 t$

63. $\frac{2}{3} \log_7(z - 2)$

64. $-4 \ln 3x$

65. $\log_3 5x - 4 \log_3 x$

66. $2 \log_2 x + 4 \log_2 y$

67. $\log x + 2 \log(x + 1)$

68. $2 \ln 8 - 5 \ln(z - 4)$

69. $\log x - 2 \log y + 3 \log z$

70. $3 \log_3 x + \frac{1}{4} \log_3 y - 4 \log_3 z$

71. $\ln x - [\ln(x + 1) + \ln(x - 1)]$

72. $4[\ln z + \ln(z + 5)] - 2 \ln(z - 5)$

73. $\frac{1}{2}[2 \ln(x + 3) + \ln x - \ln(x^2 - 1)]$

74. $2[3 \ln x - \ln(x + 1) - \ln(x - 1)]$

75. $\frac{1}{3}[\log_8 y + 2 \log_8(y + 4)] - \log_8(y - 1)$

76. $\frac{1}{2}[\log_4(x + 1) + 2 \log_4(x - 1)] + 6 \log_4 x$

Comparing Logarithmic Quantities **In Exercises 77 and 78, determine which (if any) of the logarithmic expressions are equal. Justify your answer.**

77. $\dfrac{\log_2 32}{\log_2 4}$, $\log_2 \dfrac{32}{4}$, $\log_2 32 - \log_2 4$

78. $\log_7 \sqrt{70}$, $\log_7 35$, $\frac{1}{2} + \log_7 \sqrt{10}$

• • **Sound Intensity** • • • • • • • • • • • • • • • • • •

In Exercises 79–82, use the following information.
The relationship between the number of decibels β and the intensity of a sound I (in watts per square meter) is

$$\beta = 10 \log \frac{I}{10^{-12}}.$$

79. Use the properties of logarithms to write the formula in a simpler form. Then determine the number of decibels of a sound with an intensity of 10^{-6} watt per square meter.

80. Find the difference in loudness between an average office with an intensity of 1.26×10^{-7} watt per square meter and a broadcast studio with an intensity of 3.16×10^{-10} watt per square meter.

81. Find the difference in loudness between a vacuum cleaner with an intensity of 10^{-4} watt per square meter and rustling leaves with an intensity of 10^{-11} watt per square meter.

82. You and your roommate are playing your stereos at the same time and at the same intensity. How much louder is the music when both stereos are playing compared with just one stereo playing?

Curve Fitting **In Exercises 83–86, find a logarithmic equation that relates y and x.**

83.

x	1	2	3	4	5	6
y	1	1.189	1.316	1.414	1.495	1.565

84.

x	1	2	3	4	5	6
y	1	0.630	0.481	0.397	0.342	0.303

85.

x	1	2	3	4	5	6
y	2.5	2.102	1.9	1.768	1.672	1.597

86.

x	1	2	3	4	5	6
y	0.5	2.828	7.794	16	27.951	44.091

87. **Stride Frequency of Animals** Four-legged animals run with two different types of motion: trotting and galloping. An animal that is trotting has at least one foot on the ground at all times, whereas an animal that is galloping has all four feet off the ground at some point in its stride. The number of strides per minute at which an animal breaks from a trot to a gallop depends on the weight of the animal. Use the table to find a logarithmic equation that relates an animal's weight x (in pounds) and its lowest stride frequency while galloping y (in strides per minute).

DATA	Weight, x	Stride Frequency, y
	25	191.5
	35	182.7
	50	173.8
	75	164.2
	500	125.9
	1000	114.2

Spreadsheet at LarsonPrecalculus.com

88. **Nail Length** The approximate lengths and diameters (in inches) of bright common wire nails are shown in the table. Find a logarithmic equation that relates the diameter y of a bright common wire nail to its length x.

Length, x	Diameter, y
2	0.113
3	0.148
4	0.192
5	0.225
6	0.262

89. Comparing Models A cup of water at an initial temperature of 78°C is placed in a room at a constant temperature of 21°C. The temperature of the water is measured every 5 minutes during a half-hour period. The results are recorded as ordered pairs of the form (t, T), where t is the time (in minutes) and T is the temperature (in degrees Celsius).

$(0, 78.0°)$, $(5, 66.0°)$, $(10, 57.5°)$, $(15, 51.2°)$, $(20, 46.3°)$, $(25, 42.4°)$, $(30, 39.6°)$

(a) Subtract the room temperature from each of the temperatures in the ordered pairs. Use a graphing utility to plot the data points (t, T) and $(t, T - 21)$.

(b) An exponential model for the data $(t, T - 21)$ is $T - 21 = 54.4(0.964)^t$. Solve for T and graph the model. Compare the result with the plot of the original data.

(c) Use the graphing utility to plot the points $(t, \ln(T - 21))$ and observe that the points appear to be linear. Use the *regression* feature of the graphing utility to fit a line to these data. This resulting line has the form $\ln(T - 21) = at + b$, which is equivalent to $e^{\ln(T-21)} = e^{at+b}$. Solve for T, and verify that the result is equivalent to the model in part (b).

(d) Fit a rational model to the data. Take the reciprocals of the y-coordinates of the revised data points to generate the points

$$\left(t, \frac{1}{T - 21}\right).$$

Use the graphing utility to graph these points and observe that they appear to be linear. Use the *regression* feature of the graphing utility to fit a line to these data. The resulting line has the form

$$\frac{1}{T - 21} = at + b.$$

Solve for T, and use the graphing utility to graph the rational function and the original data points.

90. Writing Write a short paragraph explaining why the transformations of the data in Exercise 89 were necessary to obtain the models. Why did taking the logarithms of the temperatures lead to a linear scatter plot? Why did taking the reciprocals of the temperatures lead to a linear scatter plot?

Exploration

True or False? In Exercises 91–96, determine whether the statement is true or false given that $f(x) = \ln x$. Justify your answer.

91. $f(0) = 0$

92. $f(ax) = f(a) + f(x)$, $a > 0$, $x > 0$

93. $f(x - 2) = f(x) - f(2)$, $x > 2$

94. $\sqrt{f(x)} = \frac{1}{2}f(x)$

95. If $f(u) = 2f(v)$, then $v = u^2$.

96. If $f(x) < 0$, then $0 < x < 1$.

Using the Change-of-Base Formula In Exercises 97–100, use the change-of-base formula to rewrite the logarithm as a ratio of logarithms. Then use a graphing utility to graph the ratio.

97. $f(x) = \log_2 x$

98. $f(x) = \log_{1/2} x$

99. $f(x) = \log_{1/4} x$

100. $f(x) = \log_{11.8} x$

Error Analysis In Exercises 101 and 102, describe the error.

101. $(\ln e)^2 = 2(\ln e) = 2(1) = 2$ ✗

102. $\log_2 8 = \log_2(4 + 4)$
$= \log_2 4 + \log_2 4$
$= \log_2 2^2 + \log_2 2^2$
$= 2 + 2$
$= 4$

103. Graphical Reasoning Use a graphing utility to graph the functions $y_1 = \ln x - \ln(x - 3)$ and $y_2 = \ln \dfrac{x}{x - 3}$ in the same viewing window. Does the graphing utility show the functions with the same domain? If not, explain why some numbers are in the domain of one function but not the other.

104. HOW DO YOU SEE IT? The figure shows the graphs of $y = \ln x$, $y = \ln x^2$, $y = \ln 2x$, and $y = \ln 2$. Match each function with its graph. (The graphs are labeled A through D.) Explain.

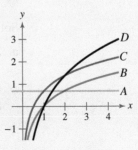

105. Think About It For which integers between 1 and 20 can you approximate natural logarithms, given the values $\ln 2 \approx 0.6931$, $\ln 3 \approx 1.0986$, and $\ln 5 \approx 1.6094$? Approximate these logarithms. (Do not use a calculator.)

3.4 Exponential and Logarithmic Equations

■ Solve simple exponential and logarithmic equations.
■ Solve more complicated exponential equations.
■ Solve more complicated logarithmic equations.
■ Use exponential and logarithmic equations to model and solve real-life problems.

Exponential and logarithmic equations have many life science applications. For example, Exercise 83 on page 234 uses an exponential function to model the beaver population in a given area.

Introduction

So far in this chapter, you have studied the definitions, graphs, and properties of exponential and logarithmic functions. In this section, you will study procedures for *solving equations* involving exponential and logarithmic expressions.

There are two basic strategies for solving exponential or logarithmic equations. The first is based on the One-to-One Properties and was used to solve simple exponential and logarithmic equations in Sections 3.1 and 3.2. The second is based on the Inverse Properties. For $a > 0$ and $a \neq 1$, the properties below are true for all x and y for which $\log_a x$ and $\log_a y$ are defined.

One-to-One Properties

$a^x = a^y$ if and only if $x = y$.

$\log_a x = \log_a y$ if and only if $x = y$.

Inverse Properties

$a^{\log_a x} = x$

$\log_a a^x = x$

EXAMPLE 1 **Solving Simple Equations**

Original Equation	Rewritten Equation	Solution	Property
a. $2^x = 32$	$2^x = 2^5$	$x = 5$	One-to-One
b. $\ln x - \ln 3 = 0$	$\ln x = \ln 3$	$x = 3$	One-to-One
c. $\left(\frac{1}{3}\right)^x = 9$	$3^{-x} = 3^2$	$x = -2$	One-to-One
d. $e^x = 7$	$\ln e^x = \ln 7$	$x = \ln 7$	Inverse
e. $\ln x = -3$	$e^{\ln x} = e^{-3}$	$x = e^{-3}$	Inverse
f. $\log x = -1$	$10^{\log x} = 10^{-1}$	$x = 10^{-1} = \frac{1}{10}$	Inverse
g. $\log_3 x = 4$	$3^{\log_3 x} = 3^4$	$x = 81$	Inverse

✓ **Checkpoint**  Audio-video solution in English & Spanish at LarsonPrecalculus.com

Solve each equation for x.

a. $2^x = 512$ **b.** $\log_6 x = 3$ **c.** $5 - e^x = 0$ **d.** $9^x = \frac{1}{3}$ ■

Strategies for Solving Exponential and Logarithmic Equations

1. Rewrite the original equation in a form that allows the use of the One-to-One Properties of exponential or logarithmic functions.

2. Rewrite an *exponential* equation in logarithmic form and apply the Inverse Property of logarithmic functions.

3. Rewrite a *logarithmic* equation in exponential form and apply the Inverse Property of exponential functions.

Solving Exponential Equations

EXAMPLE 2 **Solving Exponential Equations**

Solve each equation and approximate the result to three decimal places, if necessary.

a. $e^{-x^2} = e^{-3x-4}$ **b.** $3(2^x) = 42$

Solution

a.
$$e^{-x^2} = e^{-3x-4}$$ Write original equation.

$$-x^2 = -3x - 4$$ One-to-One Property

$$x^2 - 3x - 4 = 0$$ Write in general form.

$$(x + 1)(x - 4) = 0$$ Factor.

$$x + 1 = 0 \implies x = -1$$ Set 1st factor equal to 0.

$$x - 4 = 0 \implies x = 4$$ Set 2nd factor equal to 0.

The solutions are $x = -1$ and $x = 4$. Check these in the original equation.

b.
$$3(2^x) = 42$$ Write original equation.

$$2^x = 14$$ Divide each side by 3.

$$\log_2 2^x = \log_2 14$$ Take log (base 2) of each side.

$$x = \log_2 14$$ Inverse Property

$$x = \frac{\ln 14}{\ln 2} \approx 3.807$$ Change-of-base formula

The solution is $x = \log_2 14 \approx 3.807$. Check this in the original equation.

> **· · REMARK**
> Another way to solve
> Example 2(b) is by taking the
> natural log of each side and
> then applying the Power
> Property.
> $$3(2^x) = 42$$
> $$2^x = 14$$
> $$\ln 2^x = \ln 14$$
> $$x \ln 2 = \ln 14$$
> $$x = \frac{\ln 14}{\ln 2} \approx 3.807$$
> Notice that you obtain the same
> result as in Example 2(b).

✓ **Checkpoint**  *Audio-video solution in English & Spanish at LarsonPrecalculus.com*

Solve each equation and approximate the result to three decimal places, if necessary.

a. $e^{2x} = e^{x^2-8}$ **b.** $2(5^x) = 32$

In Example 2(b), the exact solution is $x = \log_2 14$, and the approximate solution is $x \approx 3.807$. An exact answer is preferred when the solution is an intermediate step in a larger problem. For a final answer, an approximate solution is more practical.

EXAMPLE 3 **Solving an Exponential Equation**

Solve $e^x + 5 = 60$ and approximate the result to three decimal places.

Solution

$$e^x + 5 = 60$$ Write original equation.

$$e^x = 55$$ Subtract 5 from each side.

$$\ln e^x = \ln 55$$ Take natural log of each side.

$$x = \ln 55 \approx 4.007$$ Inverse Property

> **· · REMARK** Remember that
> the natural logarithmic function
> has a base of e.

The solution is $x = \ln 55 \approx 4.007$. Check this in the original equation.

✓ **Checkpoint**  *Audio-video solution in English & Spanish at LarsonPrecalculus.com*

Solve $e^x - 7 = 23$ and approximate the result to three decimal places.

EXAMPLE 4 **Solving an Exponential Equation**

Solve $2(3^{2t-5}) - 4 = 11$ and approximate the result to three decimal places.

Solution

$2(3^{2t-5}) - 4 = 11$	Write original equation.
$2(3^{2t-5}) = 15$	Add 4 to each side.
$3^{2t-5} = \dfrac{15}{2}$	Divide each side by 2.
$\log_3 3^{2t-5} = \log_3 \dfrac{15}{2}$	Take log (base 3) of each side.
$2t - 5 = \log_3 \dfrac{15}{2}$	Inverse Property
$2t = 5 + \log_3 7.5$	Add 5 to each side.
$t = \dfrac{5}{2} + \dfrac{1}{2}\log_3 7.5$	Divide each side by 2.
$t \approx 3.417$	Use a calculator.

REMARK Remember that to evaluate a logarithm such as $\log_3 7.5$, you need to use the change-of-base formula.

$$\log_3 7.5 = \frac{\ln 7.5}{\ln 3} \approx 1.834$$

The solution is $t = \frac{5}{2} + \frac{1}{2}\log_3 7.5 \approx 3.417$. Check this in the original equation.

✓ **Checkpoint**  Audio-video solution in English & Spanish at LarsonPrecalculus.com

Solve $6(2^{t+5}) + 4 = 11$ and approximate the result to three decimal places.

When an equation involves two or more exponential expressions, you can still use a procedure similar to that demonstrated in Examples 2, 3, and 4. However, it may include additional algebraic techniques.

EXAMPLE 5 **Solving an Exponential Equation of Quadratic Type**

Solve $e^{2x} - 3e^x + 2 = 0$.

Algebraic Solution

$e^{2x} - 3e^x + 2 = 0$	Write original equation.
$(e^x)^2 - 3e^x + 2 = 0$	Write in quadratic form.
$(e^x - 2)(e^x - 1) = 0$	Factor.
$e^x - 2 = 0$	Set 1st factor equal to 0.
$x = \ln 2$	Solve for x.
$e^x - 1 = 0$	Set 2nd factor equal to 0.
$x = 0$	Solve for x.

The solutions are $x = \ln 2 \approx 0.693$ and $x = 0$. Check these in the original equation.

Graphical Solution

Use a graphing utility to graph $y = e^{2x} - 3e^x + 2$ and then find the zeros.

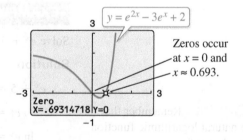

$y = e^{2x} - 3e^x + 2$

Zeros occur at $x = 0$ and $x \approx 0.693$.

Zero
X=.69314718 Y=0

So, the solutions are $x = 0$ and $x \approx 0.693$.

✓ **Checkpoint**  Audio-video solution in English & Spanish at LarsonPrecalculus.com

Solve $e^{2x} - 7e^x + 12 = 0$.

Solving Logarithmic Equations

To solve a logarithmic equation, write it in exponential form. This procedure is called *exponentiating* each side of an equation.

$\ln x = 3$ Logarithmic form

$e^{\ln x} = e^3$ Exponentiate each side.

$x = e^3$ Exponential form

EXAMPLE 6   Solving Logarithmic Equations

a. $\ln x = 2$ Original equation

$e^{\ln x} = e^2$ Exponentiate each side.

$x = e^2$ Inverse Property

b. $\log_3(5x - 1) = \log_3(x + 7)$ Original equation

$5x - 1 = x + 7$ One-to-One Property

$x = 2$ Solve for x.

c. $\log_6(3x + 14) - \log_6 5 = \log_6 2x$ Original equation

$\log_6\left(\dfrac{3x + 14}{5}\right) = \log_6 2x$ Quotient Property of Logarithms

$\dfrac{3x + 14}{5} = 2x$ One-to-One Property

$3x + 14 = 10x$ Multiply each side by 5.

$x = 2$ Solve for x.

✓ **Checkpoint** ◀))) *Audio-video solution in English & Spanish at LarsonPrecalculus.com*

Solve each equation.

a. $\ln x = \frac{2}{3}$ **b.** $\log_2(2x - 3) = \log_2(x + 4)$ **c.** $\log 4x - \log(12 + x) = \log 2$

EXAMPLE 7   Solving a Logarithmic Equation

Solve $5 + 2 \ln x = 4$ and approximate the result to three decimal places.

Algebraic Solution

$5 + 2 \ln x = 4$ Write original equation.

$2 \ln x = -1$ Subtract 5 from each side.

$\ln x = -\dfrac{1}{2}$ Divide each side by 2.

$e^{\ln x} = e^{-1/2}$ Exponentiate each side.

$x = e^{-1/2}$ Inverse Property

$x \approx 0.607$ Use a calculator.

Graphical Solution

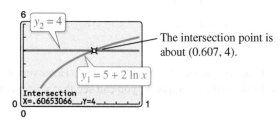

The intersection point is about $(0.607, 4)$.

So, the solution is $x \approx 0.607$.

✓ **Checkpoint** ◀))) *Audio-video solution in English & Spanish at LarsonPrecalculus.com*

Solve $7 + 3 \ln x = 5$ and approximate the result to three decimal places.

EXAMPLE 8   **Solving a Logarithmic Equation**

Solve $2 \log_5 3x = 4$.

Solution

$$2 \log_5 3x = 4 \qquad \text{Write original equation.}$$

$$\log_5 3x = 2 \qquad \text{Divide each side by 2.}$$

$$5^{\log_5 3x} = 5^2 \qquad \text{Exponentiate each side (base 5).}$$

$$3x = 25 \qquad \text{Inverse Property}$$

$$x = \frac{25}{3} \qquad \text{Divide each side by 3.}$$

The solution is $x = \frac{25}{3}$. Check this in the original equation.

✓ **Checkpoint** ◀))) *Audio-video solution in English & Spanish at LarsonPrecalculus.com*

Solve $3 \log_4 6x = 9$.

The domain of a logarithmic function generally does not include all real numbers, so you should be sure to check for extraneous solutions of logarithmic equations.

EXAMPLE 9   **Checking for Extraneous Solutions**

Solve

$$\log 5x + \log(x - 1) = 2.$$

Algebraic Solution

$$\log 5x + \log(x - 1) = 2 \qquad \text{Write original equation.}$$

$$\log[5x(x - 1)] = 2 \qquad \text{Product Property of Logarithms}$$

$$10^{\log(5x^2 - 5x)} = 10^2 \qquad \text{Exponentiate each side (base 10).}$$

$$5x^2 - 5x = 100 \qquad \text{Inverse Property}$$

$$x^2 - x - 20 = 0 \qquad \text{Write in general form.}$$

$$(x - 5)(x + 4) = 0 \qquad \text{Factor.}$$

$$x - 5 = 0 \qquad \text{Set 1st factor equal to 0.}$$

$$x = 5 \qquad \text{Solve for } x.$$

$$x + 4 = 0 \qquad \text{Set 2nd factor equal to 0.}$$

$$x = -4 \qquad \text{Solve for } x.$$

The solutions appear to be $x = 5$ and $x = -4$. However, when you check these in the original equation, you can see that $x = 5$ is the only solution.

Graphical Solution

First, rewrite the original equation as

$$\log 5x + \log(x - 1) - 2 = 0.$$

Then use a graphing utility to graph the equation

$$y = \log 5x + \log(x - 1) - 2$$

and find the zero(s).

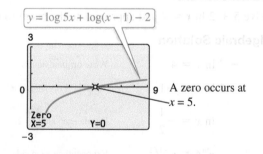

A zero occurs at $x = 5$.

So, the solution is $x = 5$.

✓ **Checkpoint** ◀))) *Audio-video solution in English & Spanish at LarsonPrecalculus.com*

Solve $\log x + \log(x - 9) = 1$.

In Example 9, the domain of $\log 5x$ is $x > 0$ and the domain of $\log(x - 1)$ is $x > 1$, so the domain of the original equation is $x > 1$. This means that the solution $x = -4$ is extraneous. The graphical solution verifies this conclusion.

Applications

EXAMPLE 10 **Doubling an Investment**

See LarsonPrecalculus.com for an interactive version of this type of example.

You invest $500 at an annual interest rate of 6.75%, compounded continuously. How long will it take your money to double?

Solution Using the formula for continuous compounding, the balance is

$$A = Pe^{rt}$$

$$A = 500e^{0.0675t}.$$

To find the time required for the balance to double, let $A = 1000$ and solve the resulting equation for t.

$500e^{0.0675t} = 1000$	Let $A = 1000$.
$e^{0.0675t} = 2$	Divide each side by 500.
$\ln e^{0.0675t} = \ln 2$	Take natural log of each side.
$0.0675t = \ln 2$	Inverse Property
$t = \dfrac{\ln 2}{0.0675}$	Divide each side by 0.0675.
$t \approx 10.27$	Use a calculator.

The balance in the account will double after approximately 10.27 years. This result is demonstrated graphically below.

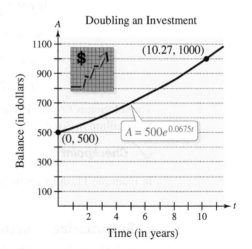

Doubling an Investment

Checkpoint Audio-video solution in English & Spanish at LarsonPrecalculus.com

You invest $500 at an annual interest rate of 5.25%, compounded continuously. How long will it take your money to double? Compare your result with that of Example 10.

In Example 10, an approximate answer of 10.27 years is given. Within the context of the problem, the exact solution

$$t = \frac{\ln 2}{0.0675}$$

does not make sense as an answer.

EXAMPLE 11 **Retail Sales**

The retail sales y (in billions of dollars) of e-commerce companies in the United States from 2009 through 2014 can be modeled by

$$y = -614 + 342.2 \ln t, \quad 9 \le t \le 14$$

where t represents the year, with $t = 9$ corresponding to 2009 (see figure). During which year did the sales reach $240 billion? (*Source: U.S. Census Bureau*)

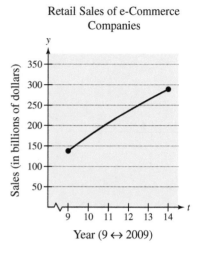

Retail Sales of e-Commerce Companies

Year (9 ↔ 2009)

Solution

$-614 + 342.2 \ln t = y$	Write original equation.
$-614 + 342.2 \ln t = 240$	Substitute 240 for y.
$342.2 \ln t = 854$	Add 614 to each side.
$\ln t = \dfrac{854}{342.2}$	Divide each side by 342.2.
$e^{\ln t} = e^{854/342.2}$	Exponentiate each side.
$t = e^{854/342.2}$	Inverse Property
$t \approx 12$	Use a calculator.

The solution is $t \approx 12$. Because $t = 9$ represents 2009, it follows that the sales reached $240 billion in 2012.

✓ *Checkpoint*))) Audio-video solution in English & Spanish at LarsonPrecalculus.com

In Example 11, during which year did the sales reach $180 billion?

Summarize (Section 3.4)

1. State the One-to-One Properties and the Inverse Properties that are used to solve simple exponential and logarithmic equations (*page 226*). For an example of solving simple exponential and logarithmic equations, see Example 1.

2. Describe strategies for solving exponential equations (*pages 227 and 228*). For examples of solving exponential equations, see Examples 2–5.

3. Describe strategies for solving logarithmic equations (*pages 229 and 230*). For examples of solving logarithmic equations, see Examples 6–9.

4. Describe examples of how to use exponential and logarithmic equations to model and solve real-life problems (*pages 231 and 232, Examples 10 and 11*).

3.4 Exercises

See **CalcChat.com** for tutorial help and worked-out solutions to odd-numbered exercises.

Vocabulary: Fill in the blanks.

1. To solve exponential and logarithmic equations, you can use the One-to-One and Inverse Properties below.

  (a) $a^x = a^y$ if and only if _____.

  (b) $\log_a x = \log_a y$ if and only if _____.

  (c) $a^{\log_a x} =$ _____

  (d) $\log_a a^x =$ _____

2. An _____ solution does not satisfy the original equation.

Skills and Applications

Determining Solutions In Exercises 3–6, determine whether each x-value is a solution (or an approximate solution) of the equation.

3. $4^{2x-7} = 64$

  (a) $x = 5$

  (b) $x = 2$

  (c) $x = \frac{1}{2}(\log_4 64 + 7)$

4. $4e^{x-1} = 60$

  (a) $x = 1 + \ln 15$

  (b) $x \approx 1.708$

  (c) $x = \ln 16$

5. $\log_2(x + 3) = 10$

  (a) $x = 1021$

  (b) $x = 17$

  (c) $x = 10^2 - 3$

6. $\ln(2x + 3) = 5.8$

  (a) $x = \frac{1}{2}(-3 + \ln 5.8)$

  (b) $x = \frac{1}{2}(-3 + e^{5.8})$

  (c) $x \approx 163.650$

 Solving a Simple Equation In Exercises 7–16, solve for x.

7. $4^x = 16$

8. $\left(\frac{1}{2}\right)^x = 32$

9. $\ln x - \ln 2 = 0$

10. $\log x - \log 10 = 0$

11. $e^x = 2$

12. $e^x = \frac{1}{3}$

13. $\ln x = -1$

14. $\log x = -2$

15. $\log_4 x = 3$

16. $\log_5 x = \frac{1}{2}$

Approximating a Point of Intersection In Exercises 17 and 18, approximate the point of intersection of the graphs of f and g. Then solve the equation $f(x) = g(x)$ algebraically to verify your approximation.

17. $f(x) = 2^x, g(x) = 8$

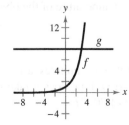

18. $f(x) = \log_3 x, g(x) = 2$

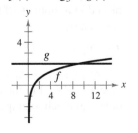

 Solving an Exponential Equation In Exercises 19–46, solve the exponential equation algebraically. Approximate the result to three decimal places, if necessary.

19. $e^x = e^{x^2-2}$

20. $e^{x^2-3} = e^{x-2}$

21. $4(3^x) = 20$

22. $4e^x = 91$

23. $e^x - 8 = 31$

24. $5^x + 8 = 26$

25. $3^{2x} = 80$

26. $4^{-3t} = 0.10$

27. $3^{2-x} = 400$

28. $7^{-3-x} = 242$

29. $8(10^{3x}) = 12$

30. $8(3^{6-x}) = 40$

31. $e^{3x} = 12$

32. $500e^{-2x} = 125$

33. $7 - 2e^x = 5$

34. $-14 + 3e^x = 11$

35. $6(2^{3x-1}) - 7 = 9$

36. $8(4^{6-2x}) + 13 = 41$

37. $3^x = 2^{x-1}$

38. $e^{x+1} = 2^{x+2}$

39. $4^x = 5^{x^2}$

40. $3^{x^2} = 7^{6-x}$

41. $e^{2x} - 4e^x - 5 = 0$

42. $e^{2x} - 5e^x + 6 = 0$

43. $\dfrac{1}{1 - e^x} = 5$

44. $\dfrac{100}{1 + e^{2x}} = 1$

45. $\left(1 + \dfrac{0.065}{365}\right)^{365t} = 4$

46. $\left(1 + \dfrac{0.10}{12}\right)^{12t} = 2$

 Solving a Logarithmic Equation In Exercises 47–62, solve the logarithmic equation algebraically. Approximate the result to three decimal places, if necessary.

47. $\ln x = -3$

48. $\ln x - 7 = 0$

49. $2.1 = \ln 6x$

50. $\log 3z = 2$

51. $3 - 4 \ln x = 11$

52. $3 + 8 \ln x = 7$

53. $6 \log_3 0.5x = 11$

54. $4 \log(x - 6) = 11$

55. $\ln x - \ln(x + 1) = 2$

56. $\ln x + \ln(x + 1) = 1$

57. $\ln(x + 5) = \ln(x - 1) - \ln(x + 1)$

58. $\ln(x + 1) - \ln(x - 2) = \ln x$

59. $\log(3x + 4) = \log(x - 10)$

60. $\log_2 x + \log_2(x + 2) = \log_2(x + 6)$

61. $\log_4 x - \log_4(x - 1) = \frac{1}{2}$

62. $\log 8x - \log(1 + \sqrt{x}) = 2$

Using Technology **In Exercises 63–70, use a graphing utility to graphically solve the equation. Approximate the result to three decimal places. Verify your result algebraically.**

63. $5^x = 212$

64. $6e^{1-x} = 25$

65. $8e^{-2x/3} = 11$

66. $e^{0.09t} = 3$

67. $3 - \ln x = 0$

68. $10 - 4\ln(x - 2) = 0$

69. $2\ln(x + 3) = 3$

70. $\ln(x + 1) = 2 - \ln x$

Compound Interest **In Exercises 71 and 72, you invest \$2500 in an account at interest rate r, compounded continuously. Find the time required for the amount to (a) double and (b) triple.**

71. $r = 0.025$

72. $r = 0.0375$

Algebra of Calculus **In Exercises 73–80, solve the equation algebraically. Round your result to three decimal places, if necessary. Verify your answer using a graphing utility.**

73. $2x^2 e^{2x} + 2xe^{2x} = 0$

74. $-x^2 e^{-x} + 2xe^{-x} = 0$

75. $-xe^{-x} + e^{-x} = 0$

76. $e^{-2x} - 2xe^{-2x} = 0$

77. $\dfrac{1 + \ln x}{2} = 0$

78. $\dfrac{1 - \ln x}{x^2} = 0$

79. $2x\ln x + x = 0$

80. $2x\ln\left(\dfrac{1}{x}\right) - x = 0$

81. Average Heights The percent m of American males between the ages of 20 and 29 who are under x inches tall is modeled by

$$m(x) = \frac{100}{1 + e^{-0.5536(x - 69.51)}}, \quad 64 \le x \le 78$$

and the percent f of American females between the ages of 20 and 29 who are under x inches tall is modeled by

$$f(x) = \frac{100}{1 + e^{-0.5834(x - 64.49)}}, \quad 60 \le x \le 78.$$

(Source: U.S. National Center for Health Statistics)

(a) Use the graph to determine any horizontal asymptotes of the graphs of the functions. Interpret the meaning in the context of the problem.

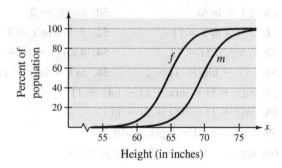

(b) What is the average height of each sex?

82. Demand The demand equation for a smartphone is

$$p = 5000\left(1 - \frac{4}{4 + e^{-0.002x}}\right).$$

Find the demand x for each price.

(a) $p = \$169$

(b) $p = \$299$

83. Ecology

The number N of beavers in a given area after x years can be approximated by

$$N = 5.5 \cdot 10^{0.23x}, \quad 0 \le x \le 10.$$

Use the model to approximate how many years it will take for the beaver population to reach 78.

84. Ecology The number N of trees of a given species per acre is approximated by the model

$$N = 3500(10^{-0.12x}), \quad 3 \le x \le 30$$

where x is the average diameter of the trees (in inches) 4.5 feet above the ground. Use the model to approximate the average diameter of the trees in a test plot when $N = 22$.

85. Population The population P (in thousands) of Alaska in the years 2005 through 2015 can be modeled by

$$P = 75\ln t + 540, \quad 5 \le t \le 15$$

where t represents the year, with $t = 5$ corresponding to 2005. During which year did the population of Alaska exceed 720 thousand? *(Source: U.S. Census Bureau)*

86. Population The population P (in thousands) of Montana in the years 2005 through 2015 can be modeled by

$$P = 81\ln t + 807, \quad 5 \le t \le 15$$

where t represents the year, with $t = 5$ corresponding to 2005. During which year did the population of Montana exceed 965 thousand? *(Source: U.S. Census Bureau)*

87. Temperature An object at a temperature of 80°C is placed in a room at 20°C. The temperature of the object is given by

$$T = 20 + 60e^{-0.06m}$$

where m represents the number of minutes after the object is placed in the room. How long does it take the object to reach a temperature of 70°C?

88. Temperature An object at a temperature of 160°C was removed from a furnace and placed in a room at 20°C. The temperature T of the object was measured each hour h and recorded in the table. A model for the data is

$$T = 20 + 140e^{-0.68h}.$$

DATA	Hour, h	Temperature, T
	0	160°
	1	90°
	2	56°
	3	38°
	4	29°
	5	24°

Spreadsheet at LarsonPrecalculus.com

(a) The figure below shows the graph of the model. Use the graph to identify the horizontal asymptote of the model and interpret the asymptote in the context of the problem.

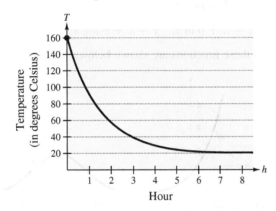

(b) Use the model to approximate the time it took for the object to reach a temperature of 100°C.

Exploration

True or False? In Exercises 89–92, rewrite each verbal statement as an equation. Then decide whether the statement is true or false. Justify your answer.

89. The logarithm of the product of two numbers is equal to the sum of the logarithms of the numbers.

90. The logarithm of the sum of two numbers is equal to the product of the logarithms of the numbers.

91. The logarithm of the difference of two numbers is equal to the difference of the logarithms of the numbers.

92. The logarithm of the quotient of two numbers is equal to the difference of the logarithms of the numbers.

93. Think About It Is it possible for a logarithmic equation to have more than one extraneous solution? Explain.

94. **HOW DO YOU SEE IT?** Solving $\log_3 x + \log_3(x - 8) = 2$ algebraically, the solutions appear to be $x = 9$ and $x = -1$. Use the graph of

$$y = \log_3 x + \log_3(x - 8) - 2$$

to determine whether each value is an actual solution of the equation. Explain.

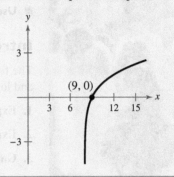

95. Finance You are investing P dollars at an annual interest rate of r, compounded continuously, for t years. Which change below results in the highest value of the investment? Explain.

(a) Double the amount you invest.

(b) Double your interest rate.

(c) Double the number of years.

96. Think About It Are the times required for the investments in Exercises 71 and 72 to quadruple twice as long as the times for them to double? Give a reason for your answer and verify your answer algebraically.

97. Effective Yield The *effective yield* of an investment plan is the percent increase in the balance after 1 year. Find the effective yield for each investment plan. Which investment plan has the greatest effective yield? Which investment plan will have the highest balance after 5 years?

(a) 7% annual interest rate, compounded annually

(b) 7% annual interest rate, compounded continuously

(c) 7% annual interest rate, compounded quarterly

(d) 7.25% annual interest rate, compounded quarterly

98. Graphical Reasoning Let $f(x) = \log_a x$ and $g(x) = a^x$, where $a > 1$.

(a) Let $a = 1.2$ and use a graphing utility to graph the two functions in the same viewing window. What do you observe? Approximate any points of intersection of the two graphs.

(b) Determine the value(s) of a for which the two graphs have one point of intersection.

(c) Determine the value(s) of a for which the two graphs have two points of intersection.

3.5 Exponential and Logarithmic Models

Exponential growth and decay models can often represent populations. For example, in Exercise 30 on page 244, you will use exponential growth and decay models to compare the populations of several countries.

- Recognize the five most common types of models involving exponential and logarithmic functions.
- Use exponential growth and decay functions to model and solve real-life problems.
- Use Gaussian functions to model and solve real-life problems.
- Use logistic growth functions to model and solve real-life problems.
- Use logarithmic functions to model and solve real-life problems.

Introduction

The five most common types of mathematical models involving exponential functions and logarithmic functions are listed below.

1. **Exponential growth model:** $y = ae^{bx}, \quad b > 0$

2. **Exponential decay model:** $y = ae^{-bx}, \quad b > 0$

3. **Gaussian model:** $y = ae^{-(x-b)^2/c}$

4. **Logistic growth model:** $y = \dfrac{a}{1 + be^{-rx}}$

5. **Logarithmic models:** $y = a + b \ln x, \quad y = a + b \log x$

The basic shapes of the graphs of these functions are shown below.

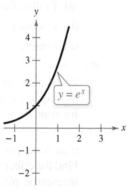

Exponential growth model

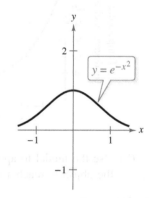

Gaussian model

Exponential decay model

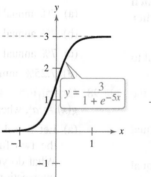

Logistic growth model

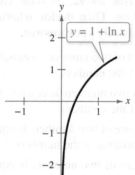

Natural logarithmic model

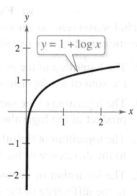

Common logarithmic model

You often gain insight into a situation modeled by an exponential or logarithmic function by identifying and interpreting the asymptotes of the graph of the function. Identify the asymptote(s) of the graph of each function shown above.

Exponential Growth and Decay

EXAMPLE 1 **Online Advertising**

The amounts S (in billions of dollars) spent in the United States on mobile online advertising in the years 2010 through 2014 are shown in the table. A scatter plot of the data is shown at the right. *(Source: IAB/Price Waterhouse Coopers)*

Year	2010	2011	2012	2013	2014
Advertising Spending	0.6	1.6	3.4	7.1	12.5

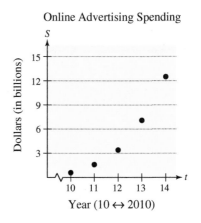

Online Advertising Spending

Year (10 ↔ 2010)

An exponential growth model that approximates the data is

$$S = 0.00036e^{0.7563t}, \quad 10 \le t \le 14$$

where t represents the year, with $t = 10$ corresponding to 2010. Compare the values found using the model with the amounts shown in the table. According to this model, in what year will the amount spent on mobile online advertising be approximately $65 billion?

Algebraic Solution

The table compares the actual amounts with the values found using the model.

Year	2010	2011	2012	2013	2014
Advertising Spending	0.6	1.6	3.4	7.1	12.5
Model	0.7	1.5	3.1	6.7	14.3

To find when the amount spent on mobile online advertising is about $65 billion, let $S = 65$ in the model and solve for t.

$0.00036e^{0.7563t} = S$	Write original model.
$0.00036e^{0.7563t} = 65$	Substitute 65 for S.
$e^{0.7563t} \approx 180{,}556$	Divide each side by 0.00036.
$\ln e^{0.7563t} \approx \ln 180{,}556$	Take natural log of each side.
$0.7563t \approx 12.1038$	Inverse Property
$t \approx 16$	Divide each side by 0.7563.

According to the model, the amount spent on mobile online advertising will be about $65 billion in 2016.

Graphical Solution

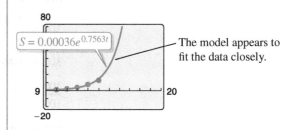

The model appears to fit the data closely.

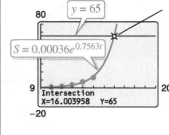

The intersection point of the model and the line $y = 65$ is about (16, 65). So, according to the model, the amount spent on mobile online advertising will be about $65 billion in 2016.

✓ **Checkpoint** ◀))) *Audio-video solution in English & Spanish at LarsonPrecalculus.com*

In Example 1, in what year will the amount spent on mobile online advertising be about $300 billion?

▷ TECHNOLOGY Some graphing utilities have an *exponential regression* feature that can help you find exponential models to represent data. If you have such a graphing utility, use it to find an exponential model for the data given in Example 1. How does your model compare with the model given in Example 1?

In Example 1, the exponential growth model is given. Sometimes you must find such a model. One technique for doing this is shown in Example 2.

EXAMPLE 2   Modeling Population Growth

In a research experiment, a population of fruit flies is increasing according to the law of exponential growth. After 2 days there are 100 flies, and after 4 days there are 300 flies. How many flies will there be after 5 days?

Solution   Let y be the number of flies at time t (in days). From the given information, you know that $y = 100$ when $t = 2$ and $y = 300$ when $t = 4$. Substituting this information into the model $y = ae^{bt}$ produces

$$100 = ae^{2b} \quad \text{and} \quad 300 = ae^{4b}.$$

To solve for b, solve for a in the first equation.

$100 = ae^{2b}$	Write first equation.
$\dfrac{100}{e^{2b}} = a$	Solve for a.

Then substitute the result into the second equation.

$300 = ae^{4b}$	Write second equation.
$300 = \left(\dfrac{100}{e^{2b}}\right)e^{4b}$	Substitute $\dfrac{100}{e^{2b}}$ for a.
$300 = 100e^{2b}$	Simplify.
$\dfrac{300}{100} = e^{2b}$	Divide each side by 100.
$\ln 3 = 2b$	Take natural log of each side.
$\dfrac{1}{2}\ln 3 = b$	Solve for b.

Now substitute $\frac{1}{2}\ln 3$ for b in the expression you found for a.

$a = \dfrac{100}{e^{2[(1/2)\ln 3]}}$	Substitute $\frac{1}{2}\ln 3$ for b.
$= \dfrac{100}{e^{\ln 3}}$	Simplify.
$= \dfrac{100}{3}$	Inverse Property
≈ 33.33	Divide.

So, with $a \approx 33.33$ and $b = \frac{1}{2}\ln 3 \approx 0.5493$, the exponential growth model is

$$y = 33.33e^{0.5493t}$$

as shown in Figure 3.15. After 5 days, the population will be

$$y = 33.33e^{0.5493(5)}$$
$$\approx 520 \text{ flies.}$$

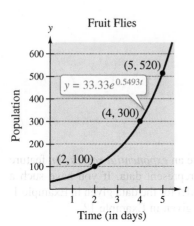

Fruit Flies

$y = 33.33e^{0.5493t}$

(5, 520)

(4, 300)

(2, 100)

Population

Time (in days)

Figure 3.15

✓ *Checkpoint*))) *Audio-video solution in English & Spanish at LarsonPrecalculus.com*

The number of bacteria in a culture is increasing according to the law of exponential growth. After 1 hour there are 100 bacteria, and after 2 hours there are 200 bacteria. How many bacteria will there be after 3 hours?

In living organic material, the ratio of the number of radioactive carbon isotopes (carbon-14) to the number of nonradioactive carbon isotopes (carbon-12) is about 1 to 10^{12}. When organic material dies, its carbon-12 content remains fixed, whereas its radioactive carbon-14 begins to decay with a half-life of about 5700 years. To estimate the age (the number of years since death) of organic material, scientists use the formula

$$R = \frac{1}{10^{12}}e^{-t/8223} \qquad \text{Carbon dating model}$$

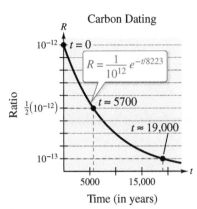

where R represents the ratio of carbon-14 to carbon-12 of organic material t years after death. The graph of R is shown at the right. Note that R decreases as t increases.

EXAMPLE 3   Carbon Dating

Estimate the age of a newly discovered fossil for which the ratio of carbon-14 to carbon-12 is $R = \dfrac{1}{10^{13}}$.

Algebraic Solution

In the carbon dating model, substitute the given value of R to obtain the following.

$$\frac{1}{10^{12}}e^{-t/8223} = R \qquad \text{Write original model.}$$

$$\frac{e^{-t/8223}}{10^{12}} = \frac{1}{10^{13}} \qquad \text{Substitute } \tfrac{1}{10^{13}} \text{ for } R.$$

$$e^{-t/8223} = \frac{1}{10} \qquad \text{Multiply each side by } 10^{12}.$$

$$\ln e^{-t/8223} = \ln \frac{1}{10} \qquad \text{Take natural log of each side.}$$

$$-\frac{t}{8223} \approx -2.3026 \qquad \text{Inverse Property}$$

$$t \approx 18{,}934 \qquad \text{Multiply each side by } -8223.$$

So, to the nearest thousand years, the age of the fossil is about 19,000 years.

Graphical Solution

Use a graphing utility to graph

$$y_1 = \frac{1}{10^{12}}e^{-x/8223} \quad \text{and} \quad y_2 = \frac{1}{10^{13}}$$

in the same viewing window.

Use the *intersect* feature to estimate that $x \approx 18{,}934$ when $y = 1/10^{13}$.

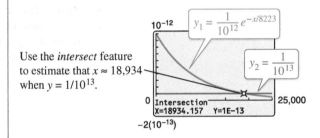

So, to the nearest thousand years, the age of the fossil is about 19,000 years.

✓ **Checkpoint** 🔊))) *Audio-video solution in English & Spanish at LarsonPrecalculus.com*

Estimate the age of a newly discovered fossil for which the ratio of carbon-14 to carbon-12 is $R = 1/10^{14}$.

The value of b in the exponential decay model $y = ae^{-bt}$ determines the *decay* of radioactive isotopes. For example, to find how much of an initial 10 grams of ^{226}Ra isotope with a half-life of 1599 years is left after 500 years, substitute this information into the model $y = ae^{-bt}$.

$$\frac{1}{2}(10) = 10e^{-b(1599)} \quad \Longrightarrow \quad \ln\frac{1}{2} = -1599b \quad \Longrightarrow \quad b = -\frac{\ln\frac{1}{2}}{1599}$$

Using the value of b found above and $a = 10$, the amount left is

$$y = 10e^{-[-\ln(1/2)/1599](500)} \approx 8.05 \text{ grams.}$$

Gaussian Models

As mentioned at the beginning of this section, Gaussian models are of the form

$$y = ae^{-(x-b)^2/c}.$$

This type of model is commonly used in probability and statistics to represent populations that are **normally distributed.** For *standard* normal distributions, the model takes the form

$$y = \frac{1}{\sqrt{2\pi}}e^{-x^2/2}.$$

The graph of a Gaussian model is called a **bell-shaped curve.** Use a graphing utility to graph the standard normal distribution curve. Can you see why it is called a bell-shaped curve?

The **average value** of a population can be found from the bell-shaped curve by observing where the maximum *y*-value of the function occurs. The *x*-value corresponding to the maximum *y*-value of the function represents the average value of the independent variable—in this case, *x*.

EXAMPLE 4   SAT Scores

See LarsonPrecalculus.com for an interactive version of this type of example.

In 2015, the SAT mathematics scores for college-bound seniors in the United States roughly followed the normal distribution

$$y = 0.0033e^{-(x-511)^2/28,800}, \quad 200 \le x \le 800$$

where *x* is the SAT score for mathematics. Use a graphing utility to graph this function and estimate the average SAT mathematics score.   *(Source: The College Board)*

Solution   The graph of the function is shown below. On this bell-shaped curve, the maximum value of the curve corresponds to the average score. Using the *maximum* feature of the graphing utility, you find that the average mathematics score for college-bound seniors in 2015 was about 511.

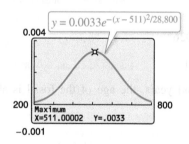

✓ **Checkpoint** ◀))) *Audio-video solution in English & Spanish at LarsonPrecalculus.com*

In 2015, the SAT critical reading scores for college-bound seniors in the United States roughly followed the normal distribution

$$y = 0.0034e^{-(x-495)^2/26,912}, \quad 200 \le x \le 800$$

where *x* is the SAT score for critical reading. Use a graphing utility to graph this function and estimate the average SAT critical reading score. *(Source: The College Board)*

In Example 4, note that 50% of the seniors who took the test earned scores greater than 511 (see Figure 3.16).

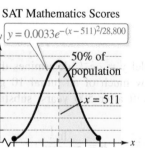

SAT Mathematics Scores

Figure 3.16

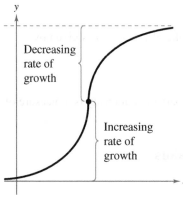

Figure 3.17

Logistic Growth Models

Some populations initially have rapid growth, followed by a declining rate of growth, as illustrated by the graph in Figure 3.17. One model for describing this type of growth pattern is the **logistic curve** given by the function

$$y = \frac{a}{1 + be^{-rx}}$$

where y is the population size and x is the time. An example is a bacteria culture that is initially allowed to grow under ideal conditions and then under less favorable conditions that inhibit growth. A logistic growth curve is also called a **sigmoidal curve.**

EXAMPLE 5   **Spread of a Virus**

On a college campus of 5000 students, one student returns from vacation with a contagious and long-lasting flu virus. The spread of the virus is modeled by

$$y = \frac{5000}{1 + 4999e^{-0.8t}}, \quad t \geq 0$$

where y is the total number of students infected after t days. The college will cancel classes when 40% or more of the students are infected.

a. How many students are infected after 5 days?

b. After how many days will the college cancel classes?

Algebraic Solution

a. After 5 days, the number of students infected is

$$y = \frac{5000}{1 + 4999e^{-0.8(5)}} = \frac{5000}{1 + 4999e^{-4}} \approx 54.$$

b. The college will cancel classes when the number of infected students is $(0.40)(5000) = 2000$.

$$2000 = \frac{5000}{1 + 4999e^{-0.8t}}$$

$$1 + 4999e^{-0.8t} = 2.5$$

$$e^{-0.8t} = \frac{1.5}{4999}$$

$$-0.8t = \ln \frac{1.5}{4999}$$

$$t = -\frac{1}{0.8} \ln \frac{1.5}{4999}$$

$$t \approx 10.14$$

So, after about 10 days, at least 40% of the students will be infected, and the college will cancel classes.

Graphical Solution

a.

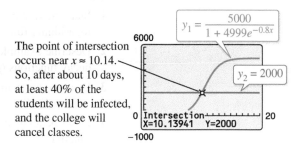

Use the *value* feature to estimate that $y \approx 54$ when $x = 5$. So, after 5 days, about 54 students are infected.

b. The college will cancel classes when the number of infected students is $(0.40)(5000) = 2000$. Use a graphing utility to graph

$$y_1 = \frac{5000}{1 + 4999e^{-0.8x}} \quad \text{and} \quad y_2 = 2000$$

in the same viewing window. Use the *intersect* feature of the graphing utility to find the point of intersection of the graphs.

The point of intersection occurs near $x \approx 10.14$. So, after about 10 days, at least 40% of the students will be infected, and the college will cancel classes.

 ***Checkpoint*** 🔊))) *Audio-video solution in English & Spanish at LarsonPrecalculus.com*

In Example 5, after how many days are 250 students infected?

Logarithmic Models

On the Richter scale, the magnitude R of an earthquake of intensity I is given by

$$R = \log \frac{I}{I_0}$$

where $I_0 = 1$ is the minimum intensity used for comparison. (Intensity is a measure of the wave energy of an earthquake.)

 EXAMPLE 6 **Magnitudes of Earthquakes**

Find the intensity of each earthquake.

a. Piedmont, California, in 2015: $R = 4.0$ **b.** Nepal in 2015: $R = 7.8$

Solution

a. Because $I_0 = 1$ and $R = 4.0$, you have

$$4.0 = \log \frac{I}{1}$$ Substitute 1 for I_0 and 4.0 for R.

$$10^{4.0} = 10^{\log I}$$ Exponentiate each side.

$$10^{4.0} = I$$ Inverse Property

$$10,000 = I.$$ Simplify.

b. For $R = 7.8$, you have

$$7.8 = \log \frac{I}{1}$$ Substitute 1 for I_0 and 7.8 for R.

$$10^{7.8} = 10^{\log I}$$ Exponentiate each side.

$$10^{7.8} = I$$ Inverse Property

$$63,000,000 \approx I.$$ Use a calculator.

On April 25, 2015, an earthquake of magnitude 7.8 struck in Nepal. The city of Kathmandu took extensive damage, including the collapse of the 203-foot Dharahara Tower, built by Nepal's first prime minister in 1832.

Note that an increase of 3.8 units on the Richter scale (from 4.0 to 7.8) represents an increase in intensity by a factor of $10^{7.8}/10^4 \approx 63,000,000/10,000 = 6300$. In other words, the intensity of the earthquake in Nepal was about 6300 times as great as that of the earthquake in Piedmont, California.

✓ **Checkpoint**   Audio-video solution in English & Spanish at LarsonPrecalculus.com

Find the intensities of earthquakes whose magnitudes are (a) $R = 6.0$ and (b) $R = 7.9$.

Summarize (Section 3.5)

1. State the five most common types of models involving exponential and logarithmic functions (*page 236*).
2. Describe examples of real-life applications that use exponential growth and decay functions (*pages 237–239, Examples 1–3*).
3. Describe an example of a real-life application that uses a Gaussian function (*page 240, Example 4*).
4. Describe an example of a real-life application that uses a logistic growth function (*page 241, Example 5*).
5. Describe an example of a real-life application that uses a logarithmic function (*page 242, Example 6*).

3.5 Exercises

See **CalcChat.com** for tutorial help and worked-out solutions to odd-numbered exercises.

Vocabulary: Fill in the blanks.

1. An exponential growth model has the form _____, and an exponential decay model has the form _____.
2. A logarithmic model has the form _____ or _____.
3. In probability and statistics, Gaussian models commonly represent populations that are _____ _____.
4. A logistic growth model has the form _____.

Skills and Applications

Solving for a Variable In Exercises 5 and 6, (a) solve for *P* and (b) solve for *t*.

5. $A = Pe^{rt}$

6. $A = P\left(1 + \dfrac{r}{n}\right)^{nt}$

 Compound Interest In Exercises 7–12, find the missing values assuming continuously compounded interest.

	Initial Investment	Annual % Rate	Time to Double	Amount After 10 Years
7.	$1000	3.5%		
8.	$750	$10\frac{1}{2}$%		
9.	$750		$7\frac{3}{4}$ yr	
10.	$500			$1505.00
11.		4.5%		$10,000.00
12.			12 yr	$2000.00

Compound Interest In Exercises 13 and 14, determine the principal *P* that must be invested at rate *r*, compounded monthly, so that $500,000 will be available for retirement in *t* years.

13. $r = 5\%, t = 10$

14. $r = 3\frac{1}{2}\%, t = 15$

Compound Interest In Exercises 15 and 16, determine the time necessary for *P* dollars to double when it is invested at interest rate *r* compounded (a) annually, (b) monthly, (c) daily, and (d) continuously.

15. $r = 10\%$

16. $r = 6.5\%$

17. **Compound Interest** Complete the table for the time *t* (in years) necessary for *P* dollars to triple when it is invested at an interest rate *r* compounded (a) continuously and (b) annually.

r	2%	4%	6%	8%	10%	12%
t						

18. **Modeling Data** Draw scatter plots of the data in Exercise 17. Use the *regression* feature of a graphing utility to find models for the data.

19. **Comparing Models** If $1 is invested over a 10-year period, then the balance *A* after *t* years is given by either $A = 1 + 0.075[\![t]\!]$ or $A = e^{0.07t}$ depending on whether the interest is simple interest at $7\frac{1}{2}\%$ or continuous compound interest at 7%. Graph each function on the same set of axes. Which grows at a greater rate? (Remember that $[\![t]\!]$ is the greatest integer function discussed in Section 1.6.)

20. **Comparing Models** If $1 is invested over a 10-year period, then the balance *A* after *t* years is given by either $A = 1 + 0.06[\![t]\!]$ or $A = [1 + (0.055/365)]^{[\![365t]\!]}$ depending on whether the interest is simple interest at 6% or compound interest at $5\frac{1}{2}\%$ compounded daily. Use a graphing utility to graph each function in the same viewing window. Which grows at a greater rate?

 Radioactive Decay In Exercises 21–24, find the missing value for the radioactive isotope.

Isotope	Half-life (years)	Initial Quantity	Amount After 1000 Years
21. ^{226}Ra	1599	10 g	
22. ^{14}C	5715	6.5 g	
23. ^{14}C	5715		2 g
24. ^{239}Pu	24,100		0.4 g

Finding an Exponential Model In Exercises 25–28, find the exponential model that fits the points shown in the graph or table.

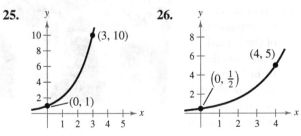

25.

26.

27.

x	0	4
y	5	1

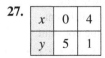

28.

x	0	3
y	1	$\frac{1}{4}$

29. Population The populations P (in thousands) of Horry County, South Carolina, from 1971 through 2014 can be modeled by

$$P = 76.6e^{0.0313t}$$

where t represents the year, with $t = 1$ corresponding to 1971. *(Source: U.S. Census Bureau)*

(a) Use the model to complete the table.

Year	Population
1980	
1990	
2000	
2010	

(b) According to the model, when will the population of Horry County reach 360,000?

(c) Do you think the model is valid for long-term predictions of the population? Explain.

• • 30. Population • • • • • • • • • • • • • • • • • • •

The table shows the mid-year populations (in millions) of five countries in 2015 and the projected populations (in millions) for the year 2025. *(Source: U.S. Census Bureau)*

Country	2015	2025
Bulgaria	7.2	6.7
Canada	35.1	37.6
China	1367.5	1407.0
United Kingdom	64.1	67.2
United States	321.4	347.3

(a) Find the exponential growth or decay model $y = ae^{bt}$ or $y = ae^{-bt}$ for the population of each country by letting $t = 15$ correspond to 2015. Use the model to predict the population of each country in 2035.

(b) You can see that the populations of the United States and the United Kingdom are growing at different rates. What constant in the equation $y = ae^{bt}$ gives the growth rate? Discuss the relationship between the different growth rates and the magnitude of the constant.

31. Website Growth The number y of hits a new website receives each month can be modeled by $y = 4080e^{kt}$, where t represents the number of months the website has been operating. In the website's third month, there were 10,000 hits. Find the value of k, and use this value to predict the number of hits the website will receive after 24 months.

32. Population The population P (in thousands) of Tallahassee, Florida, from 2000 through 2014 can be modeled by $P = 150.9e^{kt}$, where t represents the year, with $t = 0$ corresponding to 2000. In 2005, the population of Tallahassee was about 163,075. *(Source: U.S. Census Bureau)*

(a) Find the value of k. Is the population increasing or decreasing? Explain.

(b) Use the model to predict the populations of Tallahassee in 2020 and 2025. Are the results reasonable? Explain.

(c) According to the model, during what year will the population reach 200,000?

33. Bacteria Growth The number of bacteria in a culture is increasing according to the law of exponential growth. After 3 hours there are 100 bacteria, and after 5 hours there are 400 bacteria. How many bacteria will there be after 6 hours?

34. Bacteria Growth The number of bacteria in a culture is increasing according to the law of exponential growth. The initial population is 250 bacteria, and the population after 10 hours is double the population after 1 hour. How many bacteria will there be after 6 hours?

35. Depreciation A laptop computer that costs $575 new has a book value of $275 after 2 years.

(a) Find the linear model $V = mt + b$.

(b) Find the exponential model $V = ae^{kt}$.

(c) Use a graphing utility to graph the two models in the same viewing window. Which model depreciates faster in the first 2 years?

(d) Find the book values of the computer after 1 year and after 3 years using each model.

(e) Explain the advantages and disadvantages of using each model to a buyer and a seller.

36. Learning Curve The management at a plastics factory has found that the maximum number of units a worker can produce in a day is 30. The learning curve for the number N of units produced per day after a new employee has worked t days is modeled by $N = 30(1 - e^{kt})$. After 20 days on the job, a new employee produces 19 units.

(a) Find the learning curve for this employee. (*Hint:* First, find the value of k.)

(b) How many days does the model predict will pass before this employee is producing 25 units per day?

37. Carbon Dating The ratio of carbon-14 to carbon-12 in a piece of wood discovered in a cave is $R = 1/8^{14}$. Estimate the age of the piece of wood.

38. Carbon Dating The ratio of carbon-14 to carbon-12 in a piece of paper buried in a tomb is $R = 1/13^{11}$. Estimate the age of the piece of paper.

39. IQ Scores The IQ scores for a sample of students at a small college roughly follow the normal distribution

$$y = 0.0266e^{-(x-100)^2/450}, \quad 70 \le x \le 115$$

where x is the IQ score.

(a) Use a graphing utility to graph the function.

(b) From the graph in part (a), estimate the average IQ score of a student.

40. Education The amount of time (in hours per week) a student utilizes a math-tutoring center roughly follows the normal distribution

$$y = 0.7979e^{-(x-5.4)^2/0.5}, \quad 4 \le x \le 7$$

where x is the number of hours.

(a) Use a graphing utility to graph the function.

(b) From the graph in part (a), estimate the average number of hours per week a student uses the tutoring center.

41. Cell Sites A cell site is a site where electronic communications equipment is placed in a cellular network for the use of mobile phones. The numbers y of cell sites from 1985 through 2014 can be modeled by

$$y = \frac{320{,}110}{1 + 374e^{-0.252t}}$$

where t represents the year, with $t = 5$ corresponding to 1985. *(Source: CTIA-The Wireless Association)*

(a) Use the model to find the numbers of cell sites in the years 1998, 2003, and 2006.

(b) Use a graphing utility to graph the function.

(c) Use the graph to determine the year in which the number of cell sites reached 270,000.

(d) Confirm your answer to part (c) algebraically.

42. Population The population P (in thousands) of a city from 2000 through 2016 can be modeled by

$$P = \frac{2632}{1 + 0.083e^{0.050t}}$$

where t represents the year, with $t = 0$ corresponding to 2000.

(a) Use the model to find the populations of the city in the years 2000, 2005, 2010, and 2015.

(b) Use a graphing utility to graph the function.

(c) Use the graph to determine the year in which the population reached 2.2 million.

(d) Confirm your answer to part (c) algebraically.

43. Population Growth A conservation organization released 100 animals of an endangered species into a game preserve. The preserve has a carrying capacity of 1000 animals. The growth of the pack is modeled by the logistic curve

$$p(t) = \frac{1000}{1 + 9e^{-0.1656t}}$$

where t is measured in months (see figure).

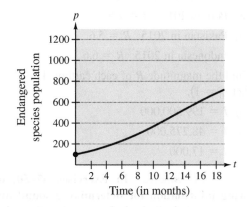

(a) Estimate the population after 5 months.

(b) After how many months is the population 500?

(c) Use a graphing utility to graph the function. Use the graph to determine the horizontal asymptotes, and interpret the meaning of the asymptotes in the context of the problem.

44. Sales After discontinuing all advertising for a tool kit in 2010, the manufacturer noted that sales began to drop according to the model

$$S = \frac{500{,}000}{1 + 0.1e^{kt}}$$

where S represents the number of units sold and t represents the year, with $t = 0$ corresponding to 2010 (see figure). In 2014, 300,000 units were sold.

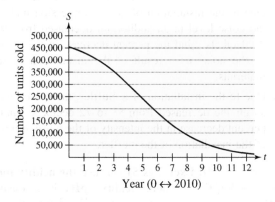

(a) Use the graph to estimate sales in 2020.

(b) Complete the model by solving for k.

(c) Use the model to estimate sales in 2020. Compare your results with that of part (a).

Geology In Exercises 45 and 46, use the Richter scale

$$R = \log \frac{I}{I_0}$$

for measuring the magnitude R of an earthquake.

45. Find the intensity I of an earthquake measuring R on the Richter scale (let $I_0 = 1$).

(a) Peru in 2015: $R = 7.6$

(b) Pakistan in 2015: $R = 5.6$

(c) Indonesia in 2015: $R = 6.6$

46. Find the magnitude R of each earthquake of intensity I (let $I_0 = 1$).

(a) $I = 199{,}500{,}000$

(b) $I = 48{,}275{,}000$

(c) $I = 17{,}000$

Intensity of Sound In Exercises 47–50, use the following information for determining sound intensity. The number of decibels β of a sound with an intensity of I watts per square meter is given by $\beta = 10 \log(I/I_0)$, where I_0 is an intensity of 10^{-12} watt per square meter, corresponding roughly to the faintest sound that can be heard by the human ear. In Exercises 47 and 48, find the number of decibels β of the sound.

47. (a) $I = 10^{-10}$ watt per m² (quiet room)

(b) $I = 10^{-5}$ watt per m² (busy street corner)

(c) $I = 10^{-8}$ watt per m² (quiet radio)

(d) $I = 10^{-3}$ watt per m² (loud car horn)

48. (a) $I = 10^{-11}$ watt per m² (rustle of leaves)

(b) $I = 10^{2}$ watt per m² (jet at 30 meters)

(c) $I = 10^{-4}$ watt per m² (door slamming)

(d) $I = 10^{-6}$ watt per m² (normal conversation)

49. Due to the installation of noise suppression materials, the noise level in an auditorium decreased from 93 to 80 decibels. Find the percent decrease in the intensity of the noise as a result of the installation of these materials.

50. Due to the installation of a muffler, the noise level of an engine decreased from 88 to 72 decibels. Find the percent decrease in the intensity of the noise as a result of the installation of the muffler.

pH Levels In Exercises 51–56, use the acidity model pH $= -\log[\text{H}^+]$, where acidity (pH) is a measure of the hydrogen ion concentration $[\text{H}^+]$ (measured in moles of hydrogen per liter) of a solution.

51. Find the pH when $[\text{H}^+] = 2.3 \times 10^{-5}$.

52. Find the pH when $[\text{H}^+] = 1.13 \times 10^{-5}$.

53. Compute $[\text{H}^+]$ for a solution in which pH $= 5.8$.

54. Compute $[\text{H}^+]$ for a solution in which pH $= 3.2$.

55. Apple juice has a pH of 2.9 and drinking water has a pH of 8.0. The hydrogen ion concentration of the apple juice is how many times the concentration of drinking water?

56. The pH of a solution decreases by one unit. By what factor does the hydrogen ion concentration increase?

57. **Forensics** At 8:30 A.M., a coroner went to the home of a person who had died during the night. In order to estimate the time of death, the coroner took the person's temperature twice. At 9:00 A.M. the temperature was 85.7°F, and at 11:00 A.M. the temperature was 82.8°F. From these two temperatures, the coroner was able to determine that the time elapsed since death and the body temperature were related by the formula

$$t = -10 \ln \frac{T - 70}{98.6 - 70}$$

where t is the time in hours elapsed since the person died and T is the temperature (in degrees Fahrenheit) of the person's body. (This formula comes from a general cooling principle called *Newton's Law of Cooling*. It uses the assumptions that the person had a normal body temperature of 98.6°F at death and that the room temperature was a constant 70°F.) Use the formula to estimate the time of death of the person.

58. **Home Mortgage** A \$120,000 home mortgage for 30 years at $7\frac{1}{2}$% has a monthly payment of \$839.06. Part of the monthly payment covers the interest charge on the unpaid balance, and the remainder of the payment reduces the principal. The amount paid toward the interest is

$$u = M - \left(M - \frac{Pr}{12}\right)\left(1 + \frac{r}{12}\right)^{12t}$$

and the amount paid toward the reduction of the principal is

$$v = \left(M - \frac{Pr}{12}\right)\left(1 + \frac{r}{12}\right)^{12t}.$$

In these formulas, P is the amount of the mortgage, r is the interest rate (in decimal form), M is the monthly payment, and t is the time in years.

(a) Use a graphing utility to graph each function in the same viewing window. (The viewing window should show all 30 years of mortgage payments.)

(b) In the early years of the mortgage, is the greater part of the monthly payment paid toward the interest or the principal? Approximate the time when the monthly payment is evenly divided between interest and principal reduction.

(c) Repeat parts (a) and (b) for a repayment period of 20 years ($M = \$966.71$). What can you conclude?

59. Home Mortgage The total interest u paid on a home mortgage of P dollars at interest rate r (in decimal form) for t years is

$$u = P\left[\frac{rt}{1 - \left(\dfrac{1}{1 + r/12}\right)^{12t}} - 1\right].$$

Consider a $120,000 home mortgage at $7\frac{1}{2}\%$.

(a) Use a graphing utility to graph the total interest function.

(b) Approximate the length of the mortgage for which the total interest paid is the same as the size of the mortgage. Is it possible that some people are paying twice as much in interest charges as the size of the mortgage?

60. Car Speed The table shows the time t (in seconds) required for a car to attain a speed of s miles per hour from a standing start.

Speed, s	Time, t
30	3.4
40	5.0
50	7.0
60	9.3
70	12.0
80	15.8
90	20.0

DATA — Spreadsheet at LarsonPrecalculus.com

Two models for these data are given below.

$t_1 = 40.757 + 0.556s - 15.817 \ln s$

$t_2 = 1.2259 + 0.0023s^2$

(a) Use the *regression* feature of a graphing utility to find a linear model t_3 and an exponential model t_4 for the data.

(b) Use the graphing utility to graph the data and each model in the same viewing window.

(c) Create a table comparing the data with estimates obtained from each model.

(d) Use the results of part (c) to find the sum of the absolute values of the differences between the data and the estimated values found using each model. Based on the four sums, which model do you think best fits the data? Explain.

Exploration

True or False? **In Exercises 61–64, determine whether the statement is true or false. Justify your answer.**

61. The domain of a logistic growth function cannot be the set of real numbers.

62. A logistic growth function will always have an x-intercept.

63. The graph of $f(x) = \dfrac{4}{1 + 6e^{-2x}} + 5$ is the graph of $g(x) = \dfrac{4}{1 + 6e^{-2x}}$ shifted to the right five units.

64. The graph of a Gaussian model will never have an x-intercept.

65. Writing Use your school's library, the Internet, or some other reference source to write a paper describing John Napier's work with logarithms.

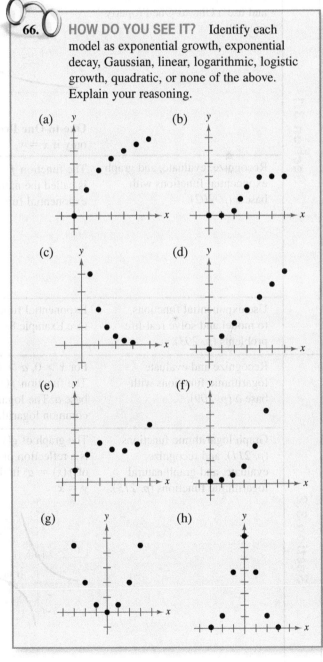

66. **HOW DO YOU SEE IT?** Identify each model as exponential growth, exponential decay, Gaussian, linear, logarithmic, logistic growth, quadratic, or none of the above. Explain your reasoning.

(a) (b) (c) (d) (e) (f) (g) (h)

Project: Sales per Share To work an extended application analyzing the sales per share for Kohl's Corporation from 1999 through 2014, visit this text's website at *LarsonPrecalculus.com*. (*Source: Kohl's Corporation*)

Chapter Summary

	What Did You Learn?	**Explanation/Examples**	**Review Exercises**
Section 3.1	Recognize and evaluate exponential functions with base a (p. 198).	The exponential function f with base a is denoted by $f(x) = a^x$, where $a > 0$, $a \neq 1$, and x is any real number.	1–6
	Graph exponential functions and use a One-to-One Property (p. 199).	**One-to-One Property:** For $a > 0$ and $a \neq 1$, $a^x = a^y$ if and only if $x = y$.	7–20
	Recognize, evaluate, and graph exponential functions with base e (p. 202).	The function $f(x) = e^x$ is called the natural exponential function.	21–28
	Use exponential functions to model and solve real-life problems (p. 203).	Exponential functions are used in compound interest formulas (see Example 8) and in radioactive decay models (see Example 9).	29–32
Section 3.2	Recognize and evaluate logarithmic functions with base a (p. 209).	For $x > 0$, $a > 0$, and $a \neq 1$, $y = \log_a x$ if and only if $x = a^y$. The function $f(x) = \log_a x$ is the logarithmic function with base a. The logarithmic function with base 10 is called the common logarithmic function. It is denoted by $\log_{10}$ or log.	33–44
	Graph logarithmic functions (p. 211), and recognize, evaluate, and graph natural logarithmic functions (p. 213).	The graph of $g(x) = \log_a x$ is a reflection of the graph of $f(x) = a^x$ in the line $y = x$. The function $g(x) = \ln x$, $x > 0$, is called the natural logarithmic function. Its graph is a reflection of the graph of $f(x) = e^x$ in the line $y = x$.	45–56
	Use logarithmic functions to model and solve real-life problems (p. 215).	A logarithmic function can model human memory. (See Example 11.)	57, 58

	What Did You Learn?	**Explanation/Examples**	**Review Exercises**
Section 3.3	Use the change-of-base formula to rewrite and evaluate logarithmic expressions (p. 219).	Let a, b, and x be positive real numbers such that $a \neq 1$ and $b \neq 1$. Then $\log_a x$ can be converted to a different base as follows. **Base b** $\qquad$ **Base 10** $\qquad$ **Base e** $\log_a x = \dfrac{\log_b x}{\log_b a}$ $\quad$ $\log_a x = \dfrac{\log x}{\log a}$ $\quad$ $\log_a x = \dfrac{\ln x}{\ln a}$	59–62
	Use properties of logarithms to evaluate, rewrite, expand, or condense logarithmic expressions (pp. 220–221).	Let a be a positive number such that $a \neq 1$, let n be a real number, and let u and v be positive real numbers. **1. Product Property:** $\log_a(uv) = \log_a u + \log_a v$ $\qquad\qquad\qquad\qquad \ln(uv) = \ln u + \ln v$ **2. Quotient Property:** $\log_a(u/v) = \log_a u - \log_a v$ $\qquad\qquad\qquad\qquad \ln(u/v) = \ln u - \ln v$ **3. Power Property:** $\log_a u^n = n \log_a u, \ln u^n = n \ln u$	63–78
	Use logarithmic functions to model and solve real-life problems (p. 222).	Logarithmic functions can help you find an equation that relates the periods of several planets and their distances from the sun. (See Example 7.)	79, 80
Section 3.4	Solve simple exponential and logarithmic equations (p. 226).	One-to-One Properties and Inverse Properties of exponential or logarithmic functions are used to solve exponential or logarithmic equations.	81–86
	Solve more complicated exponential equations (p. 227) and logarithmic equations (p. 229).	To solve more complicated equations, rewrite the equations to allow the use of the One-to-One Properties or Inverse Properties of exponential or logarithmic functions. (See Examples 2–9.)	87–102
	Use exponential and logarithmic equations to model and solve real-life problems (p. 231).	Exponential and logarithmic equations can help you determine how long it will take to double an investment (see Example 10) and find the year in which an industry had a given amount of sales (see Example 11).	103, 104
Section 3.5	Recognize the five most common types of models involving exponential and logarithmic functions (p. 236).	**1. Exponential growth model:** $y = ae^{bx}, \quad b > 0$ **2. Exponential decay model:** $y = ae^{-bx}, \quad b > 0$ **3. Gaussian model:** $y = ae^{-(x-b)^2/c}$ **4. Logistic growth model:** $y = \dfrac{a}{1 + be^{-rx}}$ **5. Logarithmic models:** $y = a + b \ln x, y = a + b \log x$	105–110
	Use exponential growth and decay functions to model and solve real-life problems (p. 237).	An exponential growth function can help you model a population of fruit flies (see Example 2), and an exponential decay function can help you estimate the age of a fossil (see Example 3).	111, 112
	Use Gaussian functions (p. 240), logistic growth functions (p. 241), and logarithmic functions (p. 242) to model and solve real-life problems.	A Gaussian function can help you model SAT mathematics scores for college-bound seniors. (See Example 4.) A logistic growth function can help you model the spread of a flu virus. (See Example 5.) A logarithmic function can help you find the intensity of an earthquake given its magnitude. (See Example 6.)	113–115

Review Exercises
See CalcChat.com for tutorial help and worked-out solutions to odd-numbered exercises.

3.1 Evaluating an Exponential Function In Exercises 1–6, evaluate the function at the given value of x. Round your result to three decimal places.

1. $f(x) = 0.3^x$, $x = 1.5$ **2.** $f(x) = 30^x$, $x = \sqrt{3}$

3. $f(x) = 2^x$, $x = \frac{2}{3}$ **4.** $f(x) = \left(\frac{1}{2}\right)^{2x}$, $x = \pi$

5. $f(x) = 7(0.2^x)$, $x = -\sqrt{11}$

6. $f(x) = -14(5^x)$, $x = -0.8$

Graphing an Exponential Function In Exercises 7–12, use a graphing utility to construct a table of values for the function. Then sketch the graph of the function.

7. $f(x) = 4^{-x} + 4$ **8.** $f(x) = 2.65^{x-1}$

9. $f(x) = 5^{x-2} + 4$ **10.** $f(x) = 2^{x-6} - 5$

11. $f(x) = \left(\frac{1}{2}\right)^{-x} + 3$ **12.** $f(x) = \left(\frac{1}{8}\right)^{x+2} - 5$

Using a One-to-One Property In Exercises 13–16, use a One-to-One Property to solve the equation for x.

13. $\left(\frac{1}{3}\right)^{x-3} = 9$ **14.** $3^{x+3} = \frac{1}{81}$

15. $e^{3x-5} = e^7$ **16.** $e^{8-2x} = e^{-3}$

Transforming the Graph of an Exponential Function In Exercises 17–20, describe the transformation of the graph of f that yields the graph of g.

17. $f(x) = 5^x$, $g(x) = 5^x + 1$

18. $f(x) = 6^x$, $g(x) = 6^{x+1}$

19. $f(x) = 3^x$, $g(x) = 1 - 3^x$

20. $f(x) = \left(\frac{1}{2}\right)^x$, $g(x) = -\left(\frac{1}{2}\right)^{x+2}$

Evaluating the Natural Exponential Function In Exercises 21–24, evaluate $f(x) = e^x$ at the given value of x. Round your result to three decimal places.

21. $x = 3.4$ **22.** $x = -2.5$

23. $x = \frac{3}{5}$ **24.** $x = \frac{2}{7}$

Graphing a Natural Exponential Function In Exercises 25–28, use a graphing utility to construct a table of values for the function. Then sketch the graph of the function.

25. $h(x) = e^{-x/2}$ **26.** $h(x) = 2 - e^{-x/2}$

27. $f(x) = e^{x+2}$ **28.** $s(t) = 4e^{t-1}$

29. Waiting Times The average time between new posts on a message board is 3 minutes. The probability F of waiting less than t minutes until the next post is approximated by the model $F(t) = 1 - e^{-t/3}$. A message has just been posted. Find the probability that the next post will be within (a) 1 minute, (b) 2 minutes, and (c) 5 minutes.

30. Depreciation After t years, the value V of a car that originally cost $23,970 is given by $V(t) = 23,970\left(\frac{3}{4}\right)^t$.

(a) Use a graphing utility to graph the function.

(b) Find the value of the car 2 years after it was purchased.

(c) According to the model, when does the car depreciate most rapidly? Is this realistic? Explain.

(d) According to the model, when will the car have no value?

Compound Interest In Exercises 31 and 32, complete the table by finding the balance A when P dollars is invested at rate r for t years and compounded n times per year.

n	1	2	4	12	365	Continuous
A						

31. $P = \$5000$, $r = 3\%$, $t = 10$ years

32. $P = \$4500$, $r = 2.5\%$, $t = 30$ years

3.2 Writing a Logarithmic Equation In Exercises 33–36, write the exponential equation in logarithmic form. For example, the logarithmic form of $2^3 = 8$ is $\log_2 8 = 3$.

33. $3^3 = 27$ **34.** $25^{3/2} = 125$

35. $e^{0.8} = 2.2255\ldots$ **36.** $e^0 = 1$

Evaluating a Logarithm In Exercises 37–40, evaluate the logarithm at the given value of x without using a calculator.

37. $f(x) = \log x$, $x = 1000$ **38.** $g(x) = \log_9 x$, $x = 3$

39. $g(x) = \log_2 x$, $x = \frac{1}{4}$ **40.** $f(x) = \log_3 x$, $x = \frac{1}{81}$

Using a One-to-One Property In Exercises 41–44, use a One-to-One Property to solve the equation for x.

41. $\log_4(x + 7) = \log_4 14$ **42.** $\log_8(3x - 10) = \log_8 5$

43. $\ln(x + 9) = \ln 4$ **44.** $\log(3x - 2) = \log 7$

Sketching the Graph of a Logarithmic Function In Exercises 45–48, find the domain, x-intercept, and vertical asymptote of the logarithmic function and sketch its graph.

45. $g(x) = \log_7 x$

46. $f(x) = \log \dfrac{x}{3}$

47. $f(x) = 4 - \log(x + 5)$

48. $f(x) = \log(x - 3) + 1$

Evaluating a Logarithmic Function In Exercises 49–52, use a calculator to evaluate the function at the given value of x. Round your result to three decimal places, if necessary.

49. $f(x) = \ln x$, $x = 22.6$ **50.** $f(x) = \ln x$, $x = e^{-12}$

51. $f(x) = \frac{1}{2} \ln x$, $x = \sqrt{e}$

52. $f(x) = 5 \ln x$, $x = 0.98$

Graphing a Natural Logarithmic Function In Exercises 53–56, find the domain, x-intercept, and vertical asymptote of the logarithmic function and sketch its graph.

53. $f(x) = \ln x + 6$ **54.** $f(x) = \ln x - 5$

55. $h(x) = \ln(x - 6)$ **56.** $f(x) = \ln(x + 4)$

57. Astronomy The formula $M = m - 5 \log(d/10)$ gives the distance d (in parsecs) from Earth to a star with apparent magnitude m and absolute magnitude M. The star Rasalhague has an apparent magnitude of 2.08 and an absolute magnitude of 1.3. Find the distance from Earth to Rasalhague.

58. Snow Removal The number of miles s of roads cleared of snow is approximated by the model

$$s = 25 - \frac{13 \ln(h/12)}{\ln 3}, \quad 2 \le h \le 15$$

where h is the depth (in inches) of the snow. Use this model to find s when $h = 10$ inches.

3.3 Using the Change-of-Base Formula In Exercises 59–62, evaluate the logarithm using the change-of-base formula (a) with common logarithms and (b) with natural logarithms. Round your results to three decimal places.

59. $\log_2 6$ **60.** $\log_{12} 200$

61. $\log_{1/2} 5$ **62.** $\log_4 0.75$

Using Properties of Logarithms In Exercises 63–66, use the properties of logarithms to write the logarithm in terms of $\log_2 3$ and $\log_2 5$.

63. $\log_2 \frac{5}{3}$ **64.** $\log_2 45$

65. $\log_2 \frac{9}{5}$ **66.** $\log_2 \frac{20}{9}$

Expanding a Logarithmic Expression In Exercises 67–72, use the properties of logarithms to expand the expression as a sum, difference, and/or constant multiple of logarithms. (Assume all variables are positive.)

67. $\log 7x^2$ **68.** $\log 11x^3$

69. $\log_3 \frac{9}{\sqrt{x}}$ **70.** $\log_7 \frac{\sqrt[3]{x}}{19}$

71. $\ln x^2 y^2 z$ **72.** $\ln\left(\frac{y-1}{3}\right)^2$, $y > 1$

Condensing a Logarithmic Expression In Exercises 73–78, condense the expression to the logarithm of a single quantity.

73. $\ln 7 + \ln x$

74. $\log_2 y - \log_2 3$

75. $\log x - \frac{1}{2} \log y$

76. $3 \ln x + 2 \ln(x + 1)$

77. $\frac{1}{2} \log_3 x - 2 \log_3 (y + 8)$

78. $5 \ln(x - 2) - \ln(x + 2) - 3 \ln x$

79. Climb Rate The time t (in minutes) for a small plane to climb to an altitude of h feet is modeled by

$$t = 50 \log[18{,}000/(18{,}000 - h)]$$

where 18,000 feet is the plane's absolute ceiling.

(a) Determine the domain of the function in the context of the problem.

(b) Use a graphing utility to graph the function and identify any asymptotes.

(c) As the plane approaches its absolute ceiling, what can be said about the time required to increase its altitude?

(d) Find the time it takes for the plane to climb to an altitude of 4000 feet.

80. Human Memory Model Students in a learning theory study took an exam and then retested monthly for 6 months with an equivalent exam. The data obtained in the study are given by the ordered pairs (t, s), where t is the time (in months) after the initial exam and s is the average score for the class. Use the data to find a logarithmic equation that relates t and s.

(1, 84.2), (2, 78.4), (3, 72.1),
(4, 68.5), (5, 67.1), (6, 65.3)

3.4 Solving a Simple Equation In Exercises 81–86, solve for x.

81. $5^x = 125$

82. $6^x = \frac{1}{216}$

83. $e^x = 3$

84. $\log x - \log 5 = 0$

85. $\ln x = 4$

86. $\ln x = -1.6$

Solving an Exponential Equation In Exercises 87–90, solve the exponential equation algebraically. Approximate the result to three decimal places.

87. $e^{4x} = e^{x^2 + 3}$

88. $e^{3x} = 25$

89. $2^x - 3 = 29$

90. $e^{2x} - 6e^x + 8 = 0$

Solving a Logarithmic Equation In Exercises 91–98, solve the logarithmic equation algebraically. Approximate the result to three decimal places.

91. $\ln 3x = 8.2$ **92.** $4 \ln 3x = 15$

93. $\ln x + \ln(x - 3) = 1$

94. $\ln(x + 2) - \ln x = 2$

95. $\log_8(x - 1) = \log_8(x - 2) - \log_8(x + 2)$

96. $\log_6(x + 2) - \log_6 x = \log_6(x + 5)$

97. $\log(1 - x) = -1$

98. $\log(-x - 4) = 2$

Using Technology In Exercises 99–102, use a graphing utility to graphically solve the equation. Approximate the result to three decimal places. Verify your result algebraically.

99. $25e^{-0.3x} = 12$

100. $2 = 5 - e^{x+7}$

101. $2 \ln(x + 3) - 3 = 0$

102. $2 \ln x - \ln(3x - 1) = 0$

103. Compound Interest You deposit $8500 in an account that pays 1.5% interest, compounded continuously. How long will it take for the money to triple?

104. Meteorology The speed of the wind S (in miles per hour) near the center of a tornado and the distance d (in miles) the tornado travels are related by the model $S = 93 \log d + 65$. On March 18, 1925, a large tornado struck portions of Missouri, Illinois, and Indiana with a wind speed at the center of about 283 miles per hour. Approximate the distance traveled by this tornado.

3.5 **Matching a Function with Its Graph** In Exercises 105–110, match the function with its graph. [The graphs are labeled (a), (b), (c), (d), (e), and (f).]

(a)

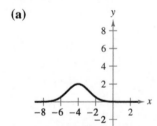

(b)

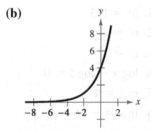

(c)   **(d)**

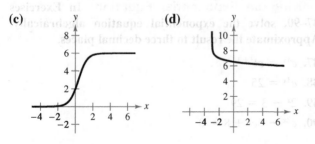

(e)

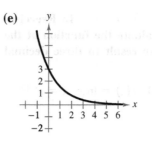

(f)

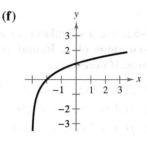

105. $y = 3e^{-2x/3}$ **106.** $y = 4e^{2x/3}$

107. $y = \ln(x + 3)$ **108.** $y = 7 - \log(x + 3)$

109. $y = 2e^{-(x+4)^2/3}$ **110.** $y = \dfrac{6}{1 + 2e^{-2x}}$

111. Finding an Exponential Model Find the exponential model $y = ae^{bx}$ that fits the points $(0, 2)$ and $(4, 3)$.

112. Wildlife Population A species of bat is in danger of becoming extinct. Five years ago, the total population of the species was 2000. Two years ago, the total population of the species was 1400. What was the total population of the species one year ago?

113. Test Scores The test scores for a biology test follow the normal distribution

$$y = 0.0499e^{-(x-71)^2/128}, \quad 40 \le x \le 100$$

where x is the test score. Use a graphing utility to graph the equation and estimate the average test score.

114. Typing Speed In a typing class, the average number N of words per minute typed after t weeks of lessons is

$$N = 157/(1 + 5.4e^{-0.12t}).$$

Find the time necessary to type (a) 50 words per minute and (b) 75 words per minute.

115. Sound Intensity The relationship between the number of decibels β and the intensity of a sound I (in watts per square meter) is

$$\beta = 10 \log(I/10^{-12}).$$

Find the intensity I for each decibel level β.

(a) $\beta = 60$ (b) $\beta = 135$ (c) $\beta = 1$

Exploration

116. Graph of an Exponential Function Consider the graph of $y = e^{kt}$. Describe the characteristics of the graph when k is positive and when k is negative.

True or False? In Exercises 117 and 118, determine whether the equation is true or false. Justify your answer.

117. $\log_b b^{2x} = 2x$

118. $\ln(x + y) = \ln x + \ln y$

Chapter Test

See **CalcChat.com** for tutorial help and worked-out solutions to odd-numbered exercises.

Take this test as you would take a test in class. When you are finished, check your work against the answers given in the back of the book.

In Exercises 1–4, evaluate the expression. Round your result to three decimal places.

1. $0.7^{2.5}$ **2.** $3^{-\pi}$ **3.** $e^{-7/10}$ **4.** $e^{3.1}$

In Exercises 5–7, use a graphing utility to construct a table of values for the function. Then sketch the graph of the function.

5. $f(x) = 10^{-x}$ **6.** $f(x) = -6^{x-2}$ **7.** $f(x) = 1 - e^{2x}$

8. Evaluate (a) $\log_7 7^{-0.89}$ and (b) $4.6 \ln e^2$.

In Exercises 9–11, find the domain, x-intercept, and vertical asymptote of the logarithmic function and sketch its graph.

9. $f(x) = 4 + \log x$ **10.** $f(x) = \ln(x - 4)$ **11.** $f(x) = 1 + \ln(x + 6)$

In Exercises 12–14, evaluate the logarithm using the change-of-base formula. Round your result to three decimal places.

12. $\log_5 35$ **13.** $\log_{16} 0.63$ **14.** $\log_{3/4} 24$

In Exercises 15–17, use the properties of logarithms to expand the expression as a sum, difference, and/or constant multiple of logarithms. (Assume all variables are positive.)

15. $\log_2 3a^4$ **16.** $\ln \dfrac{\sqrt{x}}{7}$ **17.** $\log \dfrac{10x^2}{y^3}$

In Exercises 18–20, condense the expression to the logarithm of a single quantity.

18. $\log_3 13 + \log_3 y$ **19.** $4 \ln x - 4 \ln y$

20. $3 \ln x - \ln(x + 3) + 2 \ln y$

In Exercises 21–26, solve the equation algebraically. Approximate the result to three decimal places, if necessary.

21. $5^x = \dfrac{1}{25}$ **22.** $3e^{-5x} = 132$

23. $\dfrac{1025}{8 + e^{4x}} = 5$ **24.** $\ln x = \dfrac{1}{2}$

25. $18 + 4 \ln x = 7$ **26.** $\log x + \log(x - 15) = 2$

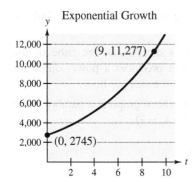

Figure for 27

27. Find the exponential growth model that fits the points shown in the graph.

28. The half-life of radioactive actinium (^{227}Ac) is 21.77 years. What percent of a present amount of radioactive actinium will remain after 19 years?

29. A model that can predict a child's height H (in centimeters) based on the child's age is $H = 70.228 + 5.104x + 9.222 \ln x$, $\frac{1}{4} \le x \le 6$, where x is the child's age in years. *(Source: Snapshots of Applications in Mathematics)*

    (a) Construct a table of values for the model. Then sketch the graph of the model.

    (b) Use the graph from part (a) to predict the height of a four-year-old child. Then confirm your prediction algebraically.

Cumulative Test for Chapters 1–3

See CalcChat.com for tutorial help and worked-out solutions to odd-numbered exercises.

Take this test as you would take a test in class. When you are finished, check your work against the answers given in the back of the book.

1. Plot the points $(-2, 5)$ and $(3, -1)$. Find the midpoint of the line segment joining the points and the distance between the points.

In Exercises 2–4, sketch the graph of the equation.

2. $x - 3y + 12 = 0$ 3. $y = x^2 - 9$ 4. $y = \sqrt{4 - x}$

5. Find the slope-intercept form of the equation of the line passing through $\left(-\frac{1}{2}, 1\right)$ and $(3, 8)$.

6. Explain why the graph at the left does not represent y as a function of x.

7. Let $f(x) = \dfrac{x}{x - 2}$. Find each function value, if possible.

    (a) $f(6)$ (b) $f(2)$ (c) $f(s + 2)$

8. Compare the graph of each function with the graph of $y = \sqrt[3]{x}$. (*Note:* It is not necessary to sketch the graphs.)

    (a) $r(x) = \frac{1}{2}\sqrt[3]{x}$ (b) $h(x) = \sqrt[3]{x} + 2$ (c) $g(x) = \sqrt[3]{x + 2}$

In Exercises 9 and 10, find (a) $(f + g)(x)$, (b) $(f - g)(x)$, (c) $(fg)(x)$, and (d) $(f/g)(x)$. What is the domain of f/g?

9. $f(x) = x - 4$, $g(x) = 3x + 1$
10. $f(x) = \sqrt{x - 1}$, $g(x) = x^2 + 1$

In Exercises 11 and 12, find (a) $f \circ g$ and (b) $g \circ f$. Find the domain of each composite function.

11. $f(x) = 2x^2$, $g(x) = \sqrt{x + 6}$
12. $f(x) = x - 2$, $g(x) = |x|$

13. Determine whether $h(x) = 3x - 4$ has an inverse function. If is does, find the inverse function.

14. The power P produced by a wind turbine varies directly as the cube of the wind speed S. A wind speed of 27 miles per hour produces a power output of 750 kilowatts. Find the output for a wind speed of 40 miles per hour.

15. Write the standard form of the quadratic function whose graph is a parabola with vertex $(-8, 5)$ and that passes through the point $(-4, -7)$.

In Exercises 16–18, sketch the graph of the function.

16. $h(x) = -x^2 + 10x - 21$

17. $f(t) = -\frac{1}{2}(t - 1)^2(t + 2)^2$

18. $g(s) = s^3 - 3s^2$

In Exercises 19–21, find all the zeros of the function.

19. $f(x) = x^3 + 2x^2 + 4x + 8$
20. $f(x) = x^4 + 4x^3 - 21x^2$
21. $f(x) = 2x^4 - 11x^3 + 30x^2 - 62x - 40$

Figure for 6

22. Use long division to divide: $\dfrac{6x^3 - 4x^2}{2x^2 + 1}$.

23. Use synthetic division to divide $3x^4 + 2x^2 - 5x + 3$ by $x - 2$.

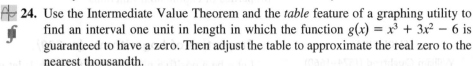 **24.** Use the Intermediate Value Theorem and the *table* feature of a graphing utility to find an interval one unit in length in which the function $g(x) = x^3 + 3x^2 - 6$ is guaranteed to have a zero. Then adjust the table to approximate the real zero to the nearest thousandth.

In Exercises 25–27, sketch the graph of the rational function. Identify all intercepts and find any asymptotes.

25. $f(x) = \dfrac{2x}{x^2 + 2x - 3}$

26. $f(x) = \dfrac{x^2 - 4}{x^2 + x - 2}$

27. $f(x) = \dfrac{x^3 - 2x^2 - 9x + 18}{x^2 + 4x + 3}$

In Exercises 28 and 29, solve the inequality. Then graph the solution set.

28. $2x^3 - 18x \le 0$

29. $\dfrac{1}{x + 1} \ge \dfrac{1}{x + 5}$

In Exercises 30 and 31, describe the transformations of the graph of f that yield the graph of g.

30. $f(x) = \left(\frac{2}{5}\right)^x, \quad g(x) = -\left(\frac{2}{5}\right)^{-x+3}$

31. $f(x) = 2.2^x, \quad g(x) = -2.2^x + 4$

In Exercises 32–35, use a calculator to evaluate the expression. Round your result to three decimal places.

32. $\log 98$

33. $\log \frac{6}{7}$

34. $\ln \sqrt{31}$

35. $\ln\left(\sqrt{30} - 4\right)$

36. Use the properties of logarithms to expand $\ln\left(\dfrac{x^2 - 25}{x^4}\right)$, where $x > 5$.

37. Condense $2 \ln x - \frac{1}{2} \ln(x + 5)$ to the logarithm of a single quantity.

In Exercises 38–40, solve the equation algebraically. Approximate the result to three decimal places.

38. $6e^{2x} = 72$ **39.** $e^{2x} - 13e^x + 42 = 0$ **40.** $\ln\sqrt{x + 2} = 3$

41. On the day a grandchild is born, a grandparent deposits \$2500 in a fund earning 7.5% interest, compounded continuously. Determine the balance in the account on the grandchild's 25th birthday.

42. The number N of bacteria in a culture is given by the model $N = 175e^{kt}$, where t is the time in hours. If $N = 420$ when $t = 8$, then estimate the time required for the population to double in size.

43. The population P (in millions) of Texas from 2001 through 2014 can be approximated by the model $P = 20.913e^{0.0184t}$, where t represents the year, with $t = 1$ corresponding to 2001. According to this model, when will the population reach 32 million? *(Source: U.S. Census Bureau)*

Proofs in Mathematics ■ ■ ■ ■ ■ ■ ■ ■ ■ ■ ■ ■ ■ ■ ■ ■

Each of the three properties of logarithms listed below can be proved by using properties of exponential functions.

Properties of Logarithms *(p. 220)*

Let a be a positive number such that $a \neq 1$, let n be a real number, and let u and v be positive real numbers.

	Logarithm with Base a	Natural Logarithm
1. Product Property:	$\log_a(uv) = \log_a u + \log_a v$	$\ln(uv) = \ln u + \ln v$
2. Quotient Property:	$\log_a \dfrac{u}{v} = \log_a u - \log_a v$	$\ln \dfrac{u}{v} = \ln u - \ln v$
3. Power Property:	$\log_a u^n = n \log_a u$	$\ln u^n = n \ln u$

Proof

Let

$$x = \log_a u \quad \text{and} \quad y = \log_a v.$$

The corresponding exponential forms of these two equations are

$$a^x = u \quad \text{and} \quad a^y = v.$$

To prove the Product Property, multiply u and v to obtain

$$uv = a^x a^y$$
$$= a^{x+y}.$$

The corresponding logarithmic form of $uv = a^{x+y}$ is $\log_a(uv) = x + y$. So,

$$\log_a(uv) = \log_a u + \log_a v.$$

To prove the Quotient Property, divide u by v to obtain

$$\frac{u}{v} = \frac{a^x}{a^y}$$
$$= a^{x-y}.$$

The corresponding logarithmic form of $\dfrac{u}{v} = a^{x-y}$ is $\log_a \dfrac{u}{v} = x - y$. So,

$$\log_a \frac{u}{v} = \log_a u - \log_a v.$$

To prove the Power Property, substitute a^x for u in the expression $\log_a u^n$.

$$\log_a u^n = \log_a(a^x)^n \qquad \text{Substitute } a^x \text{ for } u.$$
$$= \log_a a^{nx} \qquad \text{Property of Exponents}$$
$$= nx \qquad \text{Inverse Property}$$
$$= n \log_a u \qquad \text{Substitute } \log_a u \text{ for } x.$$

So, $\log_a u^n = n \log_a u$.

P.S. Problem Solving ▪ ▪ ▪ ▪ ▪ ▪ ▪ ▪ ▪ ▪ ▪ ▪ ▪ ▪ ▪

1. **Graphical Reasoning** Graph the exponential function $y = a^x$ for $a = 0.5$, 1.2, and 2.0. Which of these curves intersects the line $y = x$? Determine all positive numbers a for which the curve $y = a^x$ intersects the line $y = x$.

2. **Graphical Reasoning** Use a graphing utility to graph each of the functions $y_1 = e^x$, $y_2 = x^2$, $y_3 = x^3$, $y_4 = \sqrt{x}$, and $y_5 = |x|$. Which function increases at the greatest rate as x approaches ∞?

3. **Conjecture** Use the result of Exercise 2 to make a conjecture about the rate of growth of $y_1 = e^x$ and $y = x^n$, where n is a natural number and x approaches ∞.

4. **Implication of "Growing Exponentially"** Use the results of Exercises 2 and 3 to describe what is implied when it is stated that a quantity is growing exponentially.

5. **Exponential Function** Given the exponential function

   $$f(x) = a^x$$

   show that

   (a) $f(u + v) = f(u) \cdot f(v)$ and   (b) $f(2x) = [f(x)]^2$.

6. **Hyperbolic Functions** Given that

   $$f(x) = \frac{e^x + e^{-x}}{2} \quad \text{and} \quad g(x) = \frac{e^x - e^{-x}}{2}$$

   show that

   $$[f(x)]^2 - [g(x)]^2 = 1.$$

7. **Graphical Reasoning** Use a graphing utility to compare the graph of the function $y = e^x$ with the graph of each function. [$n!$ (read "n factorial") is defined as $n! = 1 \cdot 2 \cdot 3 \cdots (n - 1) \cdot n$.]

   (a) $y_1 = 1 + \dfrac{x}{1!}$

   (b) $y_2 = 1 + \dfrac{x}{1!} + \dfrac{x^2}{2!}$

   (c) $y_3 = 1 + \dfrac{x}{1!} + \dfrac{x^2}{2!} + \dfrac{x^3}{3!}$

8. **Identifying a Pattern** Identify the pattern of successive polynomials given in Exercise 7. Extend the pattern one more term and compare the graph of the resulting polynomial function with the graph of $y = e^x$. What do you think this pattern implies?

9. **Finding an Inverse Function** Graph the function

   $$f(x) = e^x - e^{-x}.$$

   From the graph, the function appears to be one-to-one. Assume that f has an inverse function and find $f^{-1}(x)$.

10. **Finding a Pattern for an Inverse Function** Find a pattern for $f^{-1}(x)$ when

    $$f(x) = \frac{a^x + 1}{a^x - 1}$$

    where $a > 0$, $a \ne 1$.

11. **Determining the Equation of a Graph** Determine whether the graph represents equation (a), (b), or (c). Explain your reasoning.

    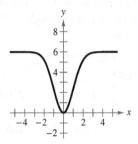

    (a) $y = 6e^{-x^2/2}$

    (b) $y = \dfrac{6}{1 + e^{-x/2}}$

    (c) $y = 6(1 - e^{-x^2/2})$

12. **Simple and Compound Interest** You have two options for investing \$500. The first earns 7% interest compounded annually, and the second earns 7% simple interest. The figure shows the growth of each investment over a 30-year period.

    (a) Determine which graph represents each type of investment. Explain your reasoning.

    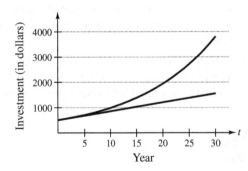

    (b) Verify your answer in part (a) by finding the equations that model the investment growth and by graphing the models.

    (c) Which option would you choose? Explain.

13. **Radioactive Decay** Two different samples of radioactive isotopes are decaying. The isotopes have initial amounts of c_1 and c_2 and half-lives of k_1 and k_2, respectively. Find an expression for the time t required for the samples to decay to equal amounts.

14. Bacteria Decay A lab culture initially contains 500 bacteria. Two hours later, the number of bacteria decreases to 200. Find the exponential decay model of the form

$$B = B_0 a^{kt}$$

that approximates the number of bacteria B in the culture after t hours.

15. Colonial Population The table shows the colonial population estimates of the American colonies for each decade from 1700 through 1780. *(Source: U.S. Census Bureau)*

DATA	Year	Population
	1700	250,900
	1710	331,700
	1720	466,200
	1730	629,400
	1740	905,600
	1750	1,170,800
	1760	1,593,600
	1770	2,148,100
	1780	2,780,400

Spreadsheet at LarsonPrecalculus.com

Let y represent the population in the year t, with $t = 0$ corresponding to 1700.

(a) Use the *regression* feature of a graphing utility to find an exponential model for the data.

(b) Use the *regression* feature of the graphing utility to find a quadratic model for the data.

(c) Use the graphing utility to plot the data and the models from parts (a) and (b) in the same viewing window.

(d) Which model is a better fit for the data? Would you use this model to predict the population of the United States in 2020? Explain your reasoning.

16. Ratio of Logarithms Show that

$$\frac{\log_a x}{\log_{a/b} x} = 1 + \log_a \frac{1}{b}.$$

17. Solving a Logarithmic Equation Solve

$(\ln x)^2 = \ln x^2$.

18. Graphical Reasoning Use a graphing utility to compare the graph of each function with the graph of $y = \ln x$.

(a) $y_1 = x - 1$

(b) $y_2 = (x - 1) - \frac{1}{2}(x - 1)^2$

(c) $y_3 = (x - 1) - \frac{1}{2}(x - 1)^2 + \frac{1}{3}(x - 1)^3$

19. Identifying a Pattern Identify the pattern of successive polynomials given in Exercise 18. Extend the pattern one more term and compare the graph of the resulting polynomial function with the graph of $y = \ln x$. What do you think the pattern implies?

20. Finding Slope and y-Intercept Take the natural log of each side of each equation below.

$$y = ab^x, \quad y = ax^b$$

(a) What are the slope and y-intercept of the line relating x and $\ln y$ for $y = ab^x$?

(b) What are the slope and y-intercept of the line relating $\ln x$ and $\ln y$ for $y = ax^b$?

Ventilation Rate In Exercises 21 and 22, use the model

$$y = 80.4 - 11 \ln x, \quad 100 \le x \le 1500$$

which approximates the minimum required ventilation rate in terms of the air space per child in a public school classroom. In the model, x is the air space (in cubic feet) per child and y is the ventilation rate (in cubic feet per minute) per child.

21. Use a graphing utility to graph the model and approximate the required ventilation rate when there are 300 cubic feet of air space per child.

22. In a classroom designed for 30 students, the air conditioning system can move 450 cubic feet of air per minute.

(a) Determine the ventilation rate per child in a full classroom.

(b) Estimate the air space required per child.

(c) Determine the minimum number of square feet of floor space required for the room when the ceiling height is 30 feet.

Using Technology In Exercises 23–26, (a) use a graphing utility to create a scatter plot of the data, (b) decide whether the data could best be modeled by a linear model, an exponential model, or a logarithmic model, (c) explain why you chose the model you did in part (b), (d) use the *regression* feature of the graphing utility to find the model you chose in part (b) for the data and graph the model with the scatter plot, and (e) determine how well the model you chose fits the data.

23. $(1, 2.0), (1.5, 3.5), (2, 4.0), (4, 5.8), (6, 7.0), (8, 7.8)$

24. $(1, 4.4), (1.5, 4.7), (2, 5.5), (4, 9.9), (6, 18.1), (8, 33.0)$

25. $(1, 7.5), (1.5, 7.0), (2, 6.8), (4, 5.0), (6, 3.5), (8, 2.0)$

26. $(1, 5.0), (1.5, 6.0), (2, 6.4), (4, 7.8), (6, 8.6), (8, 9.0)$

4 Trigonometry

Television Coverage *(Exercise 85, page 317)*

Waterslide Design
(Exercise 30, page 335)

Respiratory Cycle *(Exercise 80, page 306)*

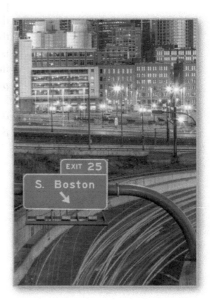

Temperature of a City
(Exercise 99, page 296)

Skateboard Ramp *(Example 10, page 283)*

4.1 Radian and Degree Measure

Angles and their measure have a wide variety of real-life applications. For example, in Exercise 68 on page 269, you will use angles and their measure to model the distance a cyclist travels.

- Describe angles.
- Use radian measure.
- Use degree measure.
- Use angles and their measure to model and solve real-life problems.

Angles

As derived from the Greek language, the word **trigonometry** means "measurement of triangles." Originally, trigonometry dealt with relationships among the sides and angles of triangles and was instrumental in the development of astronomy, navigation, and surveying. With the development of calculus and the physical sciences in the 17th century, a different perspective arose—one that viewed the classic trigonometric relationships as *functions* with the set of real numbers as their domains. Consequently, the applications of trigonometry expanded to include a vast number of physical phenomena, such as sound waves, planetary orbits, vibrating strings, pendulums, and orbits of atomic particles. This text incorporates *both* perspectives, starting with angles and their measure.

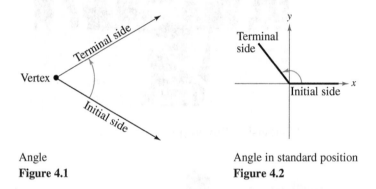

Angle
Figure 4.1

Angle in standard position
Figure 4.2

Rotating a ray (half-line) about its endpoint determines an **angle.** The starting position of the ray is the **initial side** of the angle, and the position after rotation is the **terminal side,** as shown in Figure 4.1. The endpoint of the ray is the **vertex** of the angle. This perception of an angle fits a coordinate system in which the origin is the vertex and the initial side coincides with the positive *x*-axis. Such an angle is in **standard position,** as shown in Figure 4.2. Counterclockwise rotation generates **positive angles** and clockwise rotation generates **negative angles,** as shown in Figure 4.3. Labels for angles can be Greek letters such as α (alpha), β (beta), and θ (theta) or uppercase letters such as *A*, *B*, and *C*. In Figure 4.4, note that angles α and β have the same initial and terminal sides. Such angles are **coterminal.**

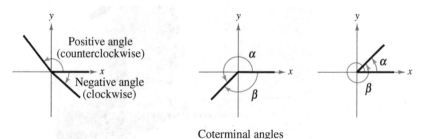

Figure 4.3

Coterminal angles
Figure 4.4

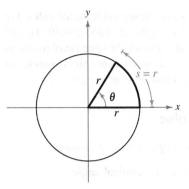

Arc length = radius when θ = 1 radian.
Figure 4.5

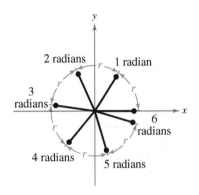

Figure 4.6

•• REMARK The phrase "θ lies in a quadrant" is an abbreviation for the phrase "the terminal side of θ lies in a quadrant." The terminal sides of the "quadrantal angles" 0, $\pi/2$, π, and $3\pi/2$ do not lie within quadrants. ▷

Radian Measure

The amount of rotation from the initial side to the terminal side determines the **measure of an angle.** One way to measure angles is in *radians*. This type of measure is especially useful in calculus. To define a radian, use a **central angle** of a circle, which is an angle whose vertex is the center of the circle, as shown in Figure 4.5.

Definition of a Radian

One **radian** (rad) is the measure of a central angle θ that intercepts an arc s equal in length to the radius r of the circle. (See Figure 4.5.) Algebraically, this means that

$$\theta = \frac{s}{r}$$

where θ is measured in radians. (Note that $\theta = 1$ when $s = r$.)

The circumference of a circle is $2\pi r$ units, so it follows that a central angle of one full revolution (counterclockwise) corresponds to an arc length of $s = 2\pi r$. Moreover, $2\pi \approx 6.28$, so there are just over six radius lengths in a full circle, as shown in Figure 4.6. The units of measure for s and r are the same, so the ratio s/r has no units—it is a real number.

The measure of an angle of one full revolution is $s/r = 2\pi r/r = 2\pi$ radians, so you can obtain the following.

$$\frac{1}{2} \text{ revolution} = \frac{2\pi}{2} = \pi \text{ radians} \qquad \frac{1}{4} \text{ revolution} = \frac{2\pi}{4} = \frac{\pi}{2} \text{ radians}$$

$$\frac{1}{6} \text{ revolution} = \frac{2\pi}{6} = \frac{\pi}{3} \text{ radians}$$

These and other common angles are shown below.

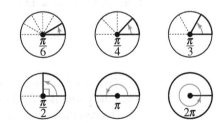

Recall that the four quadrants in a coordinate system are numbered I, II, III, and IV. The figure below shows which angles between 0 and 2π lie in each of the four quadrants. Note that angles between 0 and $\pi/2$ are **acute** angles and angles between $\pi/2$ and π are **obtuse** angles.

$$\theta = \frac{\pi}{2}$$

Quadrant II	Quadrant I
$\frac{\pi}{2} < \theta < \pi$	$0 < \theta < \frac{\pi}{2}$

$\theta = \pi$ ————————→ $\theta = 0$

Quadrant III	Quadrant IV
$\pi < \theta < \frac{3\pi}{2}$	$\frac{3\pi}{2} < \theta < 2\pi$

$$\theta = \frac{3\pi}{2}$$

Two angles are coterminal when they have the same initial and terminal sides. For example, the angles 0 and 2π are coterminal, as are the angles $\pi/6$ and $13\pi/6$. To find an angle that is coterminal to a given angle θ, add or subtract 2π (one revolution), as demonstrated in Example 1. A given angle θ has infinitely many coterminal angles. For example, $\theta = \pi/6$ is coterminal with $(\pi/6) + 2n\pi$, where n is an integer.

EXAMPLE 1   Finding Coterminal Angles

See LarsonPrecalculus.com for an interactive version of this type of example.

▷ **ALGEBRA HELP** To review operations involving fractions, see Appendix A.1.

a. For the positive angle $13\pi/6$, subtract 2π to obtain a coterminal angle.

$$\frac{13\pi}{6} - 2\pi = \frac{\pi}{6} \qquad \text{See Figure 4.7.}$$

b. For the negative angle $-2\pi/3$, add 2π to obtain a coterminal angle.

$$-\frac{2\pi}{3} + 2\pi = \frac{4\pi}{3} \qquad \text{See Figure 4.8.}$$

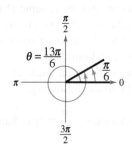

Figure 4.7   **Figure 4.8**

✓ **Checkpoint** ◀))) *Audio-video solution in English & Spanish at LarsonPrecalculus.com*

Determine two coterminal angles (one positive and one negative) for each angle.

a. $\theta = \dfrac{9\pi}{4}$ **b.** $\theta = -\dfrac{\pi}{3}$

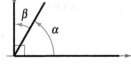

Complementary angles

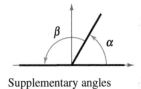

Supplementary angles
Figure 4.9

Two positive angles α and β are **complementary** (complements of each other) when their sum is $\pi/2$. Two positive angles are **supplementary** (supplements of each other) when their sum is π. (See Figure 4.9.)

EXAMPLE 2   Complementary and Supplementary Angles

a. The complement of $\dfrac{2\pi}{5}$ is $\dfrac{\pi}{2} - \dfrac{2\pi}{5} = \dfrac{5\pi}{10} - \dfrac{4\pi}{10} = \dfrac{\pi}{10}$.

The supplement of $\dfrac{2\pi}{5}$ is $\pi - \dfrac{2\pi}{5} = \dfrac{5\pi}{5} - \dfrac{2\pi}{5} = \dfrac{3\pi}{5}$.

b. There is no complement of $4\pi/5$ because $4\pi/5$ is greater than $\pi/2$. (Remember that complements are *positive* angles.) The supplement of $4\pi/5$ is

$$\pi - \frac{4\pi}{5} = \frac{5\pi}{5} - \frac{4\pi}{5} = \frac{\pi}{5}.$$

✓ **Checkpoint** ◀))) *Audio-video solution in English & Spanish at LarsonPrecalculus.com*

Find (if possible) the complement and supplement of (a) $\pi/6$ and (b) $5\pi/6$.

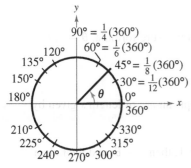

Figure 4.10

 $\frac{\pi}{6}$ 30° $\frac{\pi}{4}$ 45°

 $\frac{\pi}{3}$ 60° $\frac{\pi}{2}$ 90°

 π 180° 2π 360°

Figure 4.11

▷ **TECHNOLOGY** With calculators, it is convenient to use *decimal* degrees to denote fractional parts of degrees. Historically, however, fractional parts of degrees were expressed in *minutes* and *seconds*, using the prime (′) and double prime (″) notations, respectively. That is,

$$1' = \text{one minute} = \tfrac{1}{60}(1°)$$

$$1'' = \text{one second} = \tfrac{1}{3600}(1°).$$

For example, you would write an angle θ of 64 degrees, 32 minutes, and 47 seconds as $\theta = 64° \, 32' \, 47''$.

Many calculators have special keys for converting an angle in degrees, minutes, and seconds (D° M′ S″) to decimal degree form and vice versa.

Degree Measu

Another way to measur degree (1°) is equivale To measure angles, it is shown in Figure 4.10. half revolution corresp

One complete rev related by the equation

$$360° = 2\pi \text{ rad}$$

From these equations,

$$1° = \frac{\pi}{180} \text{ rad}$$

which lead to the con

Chapter 4 Trigonometry

264

Conversions Be

1. To convert degrees to radians, multiply degrees by $\dfrac{\pi \text{ rad}}{180°}$.

2. To convert radians to degrees, multiply radians by $\dfrac{180°}{\pi \text{ rad}}$.

To apply these two conversion rules, use the basic relationship $\pi \text{ rad} = 180°$. (See Figure 4.11.)

When no units of angle measure are specified, *radian measure is implied.* For example, $\theta = 2$ implies that $\theta = 2$ radians.

EXAMPLE 3   **Converting from Degrees to Radians**

a. $135° = (135 \text{ deg})\left(\dfrac{\pi \text{ rad}}{180 \text{ deg}}\right) = \dfrac{3\pi}{4}$ radians   Multiply by $\frac{\pi \text{ rad}}{180°}$.

b. $540° = (540 \text{ deg})\left(\dfrac{\pi \text{ rad}}{180 \text{ deg}}\right) = 3\pi$ radians   Multiply by $\frac{\pi \text{ rad}}{180°}$.

✓ **Checkpoint** Audio-video solution in English & Spanish at LarsonPrecalculus.com

Convert each degree measure to radian measure as a multiple of π. Do not use a calculator.

a. 60°   b. 320°

EXAMPLE 4   **Converting from Radians to Degrees**

a. $-\dfrac{\pi}{2} \text{ rad} = \left(-\dfrac{\pi}{2} \text{ rad}\right)\left(\dfrac{180 \text{ deg}}{\pi \text{ rad}}\right) = -90°$ Multiply by $\frac{180°}{\pi \text{ rad}}$.

b. $2 \text{ rad} = (2 \text{ rad})\left(\dfrac{180 \text{ deg}}{\pi \text{ rad}}\right) = \dfrac{360°}{\pi} \approx 114.59°$ Multiply by $\frac{180°}{\pi \text{ rad}}$.

✓ **Checkpoint** Audio-video solution in English & Spanish at LarsonPrecalculus.com

Convert each radian measure to degree measure. Do not use a calculator.

a. $\pi/6$ b. $5\pi/3$

measure arc length along a circle, use the radian measure formula, $\theta = s/r$.

Arc Length

For a circle of radius r, a central angle θ intercepts an arc of length s given by

$$s = r\theta \qquad \text{Length of circular arc}$$

where θ is measured in radians. Note that if $r = 1$, then $s = \theta$, and the radian measure of θ equals the arc length.

EXAMPLE 5 Finding Arc Length

A circle has a radius of 4 inches. Find the length of the arc intercepted by a central angle of $240°$, as shown in Figure 4.12.

Solution To use the formula $s = r\theta$, first convert $240°$ to radian measure.

$$240° = (240 \text{ deg})\left(\frac{\pi \text{ rad}}{180 \text{ deg}}\right)$$

$$= \frac{4\pi}{3} \text{ radians}$$

Then, using a radius of $r = 4$ inches, find the arc length.

$$s = r\theta \qquad \text{Length of circular arc}$$

$$= 4\left(\frac{4\pi}{3}\right) \qquad \text{Substitute for } r \text{ and } \theta.$$

$$\approx 16.76 \text{ inches} \qquad \text{Use a calculator.}$$

Note that the units for r determine the units for $r\theta$ because θ is in radian measure, which has no units.

✓ **Checkpoint** 🔊))) *Audio-video solution in English & Spanish at LarsonPrecalculus.com*

A circle has a radius of 27 inches. Find the length of the arc intercepted by a central angle of $160°$.

θ = 240°

r = 4

Figure 4.12

The formula for the length of a circular arc can be used to analyze the motion of a particle moving at a *constant speed* along a circular path.

•• **REMARK**

Linear speed measures how fast the particle moves, and angular speed measures how fast the angle changes. To establish a relationship between linear speed v and angular speed ω, divide each side of the formula for arc length by t, as shown.

$$s = r\theta$$

$$\frac{s}{t} = \frac{r\theta}{t}$$

$$v = r\omega$$

Linear and Angular Speeds

Consider a particle moving at a constant speed along a circular arc of radius r. If s is the length of the arc traveled in time t, then the **linear speed** v of the particle is

$$\text{Linear speed } v = \frac{\text{arc length}}{\text{time}} = \frac{s}{t}.$$

Moreover, if θ is the angle (in radian measure) corresponding to the arc length s, then the **angular speed** ω (the lowercase Greek letter omega) of the particle is

$$\text{Angular speed } \omega = \frac{\text{central angle}}{\text{time}} = \frac{\theta}{t}.$$

EXAMPLE 6 **Finding Linear Speed**

The second hand of a clock is 10.2 centimeters long, as shown at the right. Find the linear speed of the tip of the second hand as it passes around the clock face.

Solution In one revolution, the arc length traveled is

$$s = 2\pi r$$

$$= 2\pi(10.2) \qquad \text{Substitute for } r.$$

$$= 20.4\pi \text{ centimeters.}$$

The time required for the second hand to travel this distance is

$$t = 1 \text{ minute} = 60 \text{ seconds.}$$

So, the linear speed of the tip of the second hand is

$$v = \frac{s}{t}$$

$$= \frac{20.4\pi \text{ centimeters}}{60 \text{ seconds}}$$

$$\approx 1.07 \text{ centimeters per second.}$$

✓ **Checkpoint** ◀))) *Audio-video solution in English & Spanish at LarsonPrecalculus.com*

The second hand of a clock is 8 centimeters long. Find the linear speed of the tip of the second hand as it passes around the clock face.

Figure 4.13

EXAMPLE 7 **Finding Angular and Linear Speeds**

The blades of a wind turbine are 116 feet long (see Figure 4.13). The propeller rotates at 15 revolutions per minute.

a. Find the angular speed of the propeller in radians per minute.

b. Find the linear speed of the tips of the blades.

Solution

a. Each revolution corresponds to 2π radians, so the propeller turns $15(2\pi) = 30\pi$ radians per minute. In other words, the angular speed is

$$\omega = \frac{\theta}{t} = \frac{30\pi \text{ radians}}{1 \text{ minute}} = 30\pi \text{ radians per minute.}$$

b. The linear speed is

$$v = \frac{s}{t} = \frac{r\theta}{t} = \frac{116(30\pi) \text{ feet}}{1 \text{ minute}} \approx 10,933 \text{ feet per minute.}$$

✓ **Checkpoint** ◀))) *Audio-video solution in English & Spanish at LarsonPrecalculus.com*

The circular blade on a saw has a radius of 4 inches and it rotates at 2400 revolutions per minute.

a. Find the angular speed of the blade in radians per minute.

b. Find the linear speed of the edge of the blade.

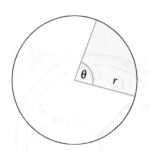

Figure 4.14

A **sector** of a circle is the region bounded by two radii of the circle and their intercepted arc (see Figure 4.14).

Area of a Sector of a Circle

For a circle of radius r, the area A of a sector of the circle with central angle θ is

$$A = \frac{1}{2}r^2\theta$$

where θ is measured in radians.

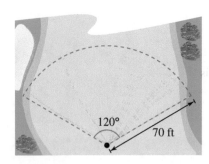

Figure 4.15

EXAMPLE 8   Area of a Sector of a Circle

A sprinkler on a golf course fairway sprays water over a distance of 70 feet and rotates through an angle of 120° (see Figure 4.15). Find the area of the fairway watered by the sprinkler.

Solution

First convert 120° to radian measure.

$$\theta = 120°$$

$$= (120 \text{ deg})\left(\frac{\pi \text{ rad}}{180 \text{ deg}}\right) \qquad \text{Multiply by } \frac{\pi \text{ rad}}{180°}.$$

$$= \frac{2\pi}{3} \text{ radians}$$

Then, using $\theta = 2\pi/3$ and $r = 70$, the area is

$$A = \frac{1}{2}r^2\theta \qquad \text{Formula for the area of a sector of a circle}$$

$$= \frac{1}{2}(70)^2\left(\frac{2\pi}{3}\right) \qquad \text{Substitute for } r \text{ and } \theta.$$

$$= \frac{4900\pi}{3} \qquad \text{Multiply.}$$

$$\approx 5131 \text{ square feet.} \qquad \text{Use a calculator.}$$

✓ *Checkpoint* ◄))) *Audio-video solution in English & Spanish at LarsonPrecalculus.com*

A sprinkler sprays water over a distance of 40 feet and rotates through an angle of 80°. Find the area watered by the sprinkler. ▪

Summarize   (Section 4.1)

1. Describe an angle *(page 260)*.
2. Explain how to use radian measure *(page 261)*. For examples involving radian measure, see Examples 1 and 2.
3. Explain how to use degree measure *(page 263)*. For examples involving degree measure, see Examples 3 and 4.
4. Describe real-life applications involving angles and their measure *(pages 264–266, Examples 5–8)*.

4.1 Exercises

Vocabulary: Fill in the blanks.

1. Two angles that have the same initial and terminal sides are _____.

2. One _____ is the measure of a central angle that intercepts an arc equal in length to the radius of the circle.

3. Two positive angles that have a sum of $\pi/2$ are _____ angles, and two positive angles that have a sum of π are _____ angles.

4. The angle measure that is equivalent to a rotation of $\frac{1}{360}$ of a complete revolution about an angle's vertex is one _____.

5. The _____ speed of a particle is the ratio of the arc length traveled to the elapsed time, and the _____ speed of a particle is the ratio of the change in the central angle to the elapsed time.

6. The area A of a sector of a circle with radius r and central angle θ, where θ is measured in radians, is given by the formula _____.

Skills and Applications

Estimating an Angle In Exercises 7–10, estimate the angle to the nearest one-half radian.

7. 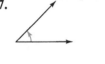   8.

9.   10.

Determining Quadrants In Exercises 11 and 12, determine the quadrant in which each angle lies.

11. (a) $\dfrac{\pi}{4}$ (b) $-\dfrac{5\pi}{4}$ 12. (a) $-\dfrac{\pi}{6}$ (b) $\dfrac{11\pi}{9}$

Sketching Angles In Exercises 13 and 14, sketch each angle in standard position.

13. (a) $\dfrac{\pi}{3}$ (b) $-\dfrac{2\pi}{3}$ 14. (a) $\dfrac{5\pi}{2}$ (b) 4

Finding Coterminal Angles In Exercises 15 and 16, determine two coterminal angles (one positive and one negative) for each angle. Give your answers in radians.

15. (a) $\dfrac{\pi}{6}$ (b) $-\dfrac{5\pi}{6}$ 16. (a) $\dfrac{2\pi}{3}$ (b) $-\dfrac{9\pi}{4}$

Complementary and Supplementary Angles In Exercises 17–20, find (if possible) the complement and supplement of each angle.

17. (a) $\dfrac{\pi}{12}$ (b) $\dfrac{11\pi}{12}$ 18. (a) $\dfrac{\pi}{3}$ (b) $\dfrac{\pi}{4}$

19. (a) 1   (b) 2   20. (a) 3   (b) 1.5

Estimating an Angle In Exercises 21–24, estimate the number of degrees in the angle.

21.    22.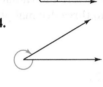

23.    24.

Determining Quadrants In Exercises 25 and 26, determine the quadrant in which each angle lies.

25. (a) $130°$ (b) $-8.3°$
26. (a) $-132° \, 50'$ (b) $3.4°$

Sketching Angles In Exercises 27 and 28, sketch each angle in standard position.

27. (a) $270°$ (b) $-120°$ 28. (a) $135°$ (b) $-750°$

Finding Coterminal Angles In Exercises 29 and 30, determine two coterminal angles (one positive and one negative) for each angle. Give your answers in degrees.

29. (a) $120°$ (b) $-210°$ 30. (a) $45°$ (b) $-420°$

Complementary and Supplementary Angles In Exercises 31–34, find (if possible) the complement and supplement of each angle.

31. (a) $18°$ (b) $85°$ 32. (a) $46°$ (b) $93°$
33. (a) $24°$ (b) $126°$ 34. (a) $130°$ (b) $170°$

Converting from Degrees to Radians In Exercises 35 and 36, convert each degree measure to radian measure as a multiple of π. Do not use a calculator.

35. (a) $120°$ (b) $-20°$
36. (a) $-60°$ (b) $144°$

 Converting from Radians to Degrees In Exercises 37 and 38, convert each radian measure to degree measure. Do not use a calculator.

37. (a) $\dfrac{3\pi}{2}$ (b) $-\dfrac{7\pi}{6}$

38. (a) $-\dfrac{7\pi}{12}$ (b) $\dfrac{5\pi}{4}$

Converting from Degrees to Radians In Exercises 39–42, convert the degree measure to radian measure. Round to three decimal places.

39. $45°$ 40. $-48.27°$
41. $-0.54°$ 42. $345°$

Converting from Radians to Degrees In Exercises 43–46, convert the radian measure to degree measure. Round to three decimal places, if necessary.

43. $\dfrac{5\pi}{11}$ 44. $\dfrac{15\pi}{8}$

45. -4.2π 46. -0.57

Converting to Decimal Degree Form In Exercises 47 and 48, convert each angle measure to decimal degree form.

47. (a) $54° \, 45'$ (b) $-128° \, 30'$
48. (a) $135° \, 10' \, 36''$ (b) $-408° \, 16' \, 20''$

Converting to D° M′ S″ Form In Exercises 49 and 50, convert each angle measure to D° M′ S″ form.

49. (a) $240.6°$ (b) $-145.8°$
50. (a) $345.12°$ (b) $-3.58°$

 Finding Arc Length In Exercises 51 and 52, find the length of the arc on a circle of radius r intercepted by a central angle θ.

51. $r = 15$ inches, $\theta = 120°$
52. $r = 3$ meters, $\theta = 150°$

Finding the Central Angle In Exercises 53 and 54, find the radian measure of the central angle of a circle of radius r that intercepts an arc of length s.

53. $r = 80$ kilometers, $s = 150$ kilometers
54. $r = 14$ feet, $s = 8$ feet

Finding the Central Angle In Exercises 55 and 56, find the radian measure of the central angle.

55.   56.

 Area of a Sector of a Circle In Exercises 57 and 58, find the area of the sector of a circle of radius r and central angle θ.

57. $r = 6$ inches, $\theta = \dfrac{\pi}{3}$ 58. $r = 2.5$ feet, $\theta = 225°$

Error Analysis In Exercises 59 and 60, describe the error.

59. $20° = (20 \; \text{deg})\left(\dfrac{180 \; \text{rad}}{\pi \; \text{deg}}\right) = \dfrac{3600}{\pi} \; \text{rad}$

60. A circle has a radius of 6 millimeters. The length of the arc intercepted by a central angle of $72°$ is

$$s = r\theta$$
$$= 6(72)$$
$$= 432 \text{ millimeters.}$$

Earth-Space Science In Exercises 61 and 62, find the distance between the cities. Assume that Earth is a sphere of radius 4000 miles and that the cities are on the same longitude (one city is due north of the other).

City	Latitude
61. Dallas, Texas	$32° \, 47' \, 9''$ N
Omaha, Nebraska	$41° \, 15' \, 50''$ N
62. San Francisco, California	$37° \, 47' \, 36''$ N
Seattle, Washington	$47° \, 37' \, 18''$ N

63. **Instrumentation** The pointer on a voltmeter is 6 centimeters in length (see figure). Find the number of degrees through which the pointer rotates when it moves 2.5 centimeters on the scale.

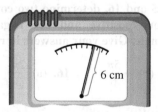

64. **Linear and Angular Speed** A $7\frac{1}{4}$-inch circular power saw blade rotates at 5200 revolutions per minute.

(a) Find the angular speed of the saw blade in radians per minute.

(b) Find the linear speed (in feet per minute) of the saw teeth as they contact the wood being cut.

65. Linear and Angular Speed A carousel with a 50-foot diameter makes 4 revolutions per minute.

(a) Find the angular speed of the carousel in radians per minute.

(b) Find the linear speed (in feet per minute) of the platform rim of the carousel.

66. Linear and Angular Speed A Blu-ray disc is approximately 12 centimeters in diameter. The drive motor of a Blu-ray player is able to rotate up to 10,000 revolutions per minute.

(a) Find the maximum angular speed (in radians per second) of a Blu-ray disc as it rotates.

(b) Find the maximum linear speed (in meters per second) of a point on the outermost track as the disc rotates.

67. Linear and Angular Speed A computerized spin balance machine rotates a 25-inch-diameter tire at 480 revolutions per minute.

(a) Find the road speed (in miles per hour) at which the tire is being balanced.

(b) At what rate should the spin balance machine be set so that the tire is being tested for 55 miles per hour?

• • 68. Speed of a Bicycle • • • • • • • • • • • • • •

The radii of the pedal sprocket, the wheel sprocket, and the wheel of the bicycle in the figure are 4 inches, 2 inches, and 14 inches, respectively. A cyclist pedals at a rate of 1 revolution per second.

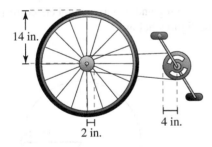

(a) Find the speed of the bicycle in feet per second and miles per hour.

(b) Use your result from part (a) to write a function for the distance d (in miles) a cyclist travels in terms of the number n of revolutions of the pedal sprocket.

(c) Write a function for the distance d (in miles) a cyclist travels in terms of the time t (in seconds). Compare this function with the function from part (b).

69. Area A sprinkler on a golf green is set to spray water over a distance of 15 meters and to rotate through an angle of 150°. Draw a diagram that shows the region that can be irrigated with the sprinkler. Find the area of the region.

70. Area A car's rear windshield wiper rotates 125°. The total length of the wiper mechanism is 25 inches and the length of the wiper blade is 14 inches. Find the area wiped by the wiper blade.

Exploration

True or False? **In Exercises 71–74, determine whether the statement is true or false. Justify your answer.**

71. An angle measure containing π must be in radian measure.

72. A measurement of 4 radians corresponds to two complete revolutions from the initial side to the terminal side of an angle.

73. The difference between the measures of two coterminal angles is always a multiple of 360° when expressed in degrees and is always a multiple of 2π radians when expressed in radians.

74. An angle that measures $-1260°$ lies in Quadrant III.

75. Writing When the radius of a circle increases and the magnitude of a central angle is held constant, how does the length of the intercepted arc change? Explain.

76. **HOW DO YOU SEE IT?** Determine which angles in the figure are coterminal angles with angle A. Explain.

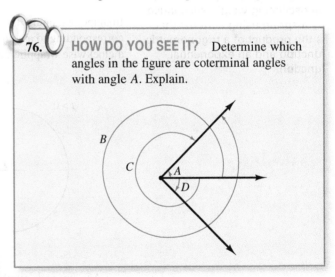

77. Think About It A fan motor turns at a given angular speed. How does the speed of the tips of the blades change when a fan of greater diameter is installed on the motor? Explain.

78. Think About It Is a degree or a radian the larger unit of measure? Explain.

79. Proof Prove that the area of a circular sector of radius r with central angle θ is $A = \frac{1}{2}\theta r^2$, where θ is measured in radians.

4.2 Trigonometric Functions: The Unit Circle

- Identify a unit circle and describe its relationship to real numbers.
- Evaluate trigonometric functions using the unit circle.
- Use domain and period to evaluate sine and cosine functions, and use a calculator to evaluate trigonometric functions.

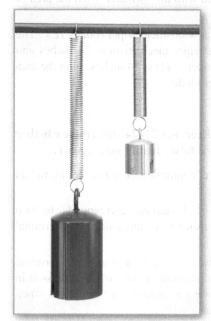

Trigonometric functions can help you analyze the movement of an oscillating weight. For example, in Exercise 50 on page 276, you will analyze the displacement of an oscillating weight suspended by a spring using a model that is the product of a trigonometric function and an exponential function.

The Unit Circle

The two historical perspectives of trigonometry incorporate different methods for introducing the trigonometric functions. One such perspective is based on the unit circle. Consider the **unit circle** given by

$$x^2 + y^2 = 1 \qquad \text{Unit circle}$$

as shown in the figure below.

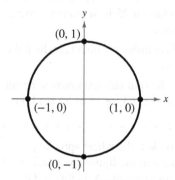

Imagine wrapping the real number line around this circle, with positive numbers corresponding to a counterclockwise wrapping and negative numbers corresponding to a clockwise wrapping, as shown in the figures below.

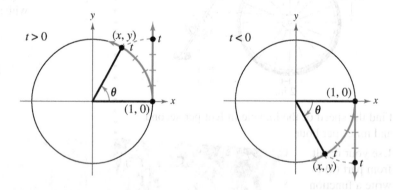

As the real number line wraps around the unit circle, each real number t corresponds to a point (x, y) on the circle. For example, the real number 0 corresponds to the point $(1, 0)$. Moreover, the unit circle has a circumference of 2π, so the real number 2π also corresponds to the point $(1, 0)$.

Each real number t also corresponds to a central angle θ (in standard position) whose radian measure is t. With this interpretation of t, the arc length formula

$$s = r\theta \quad (\text{with } r = 1)$$

indicates that the real number t is the (directional) length of the arc intercepted by the angle θ, given in radians.

The Trigonometric Functions

From the preceding discussion, the coordinates x and y are two functions of the real variable t. These coordinates are used to define the six trigonometric functions of a real number t.

| sine | cosecant | cosine | secant | tangent | cotangent |

Abbreviations for these six functions are sin, csc, cos, sec, tan, and cot, respectively.

> **Definitions of Trigonometric Functions**
>
> Let t be a real number and let (x, y) be the point on the unit circle corresponding to t.
>
> $$\sin t = y \qquad\qquad \cos t = x \qquad\qquad \tan t = \frac{y}{x}, \quad x \neq 0$$
>
> $$\csc t = \frac{1}{y}, \quad y \neq 0 \qquad \sec t = \frac{1}{x}, \quad x \neq 0 \qquad \cot t = \frac{x}{y}, \quad y \neq 0$$

• • REMARK Note that the functions in the second row are the *reciprocals* of the corresponding functions in the first row. ▷

In the definitions of the trigonometric functions, note that the tangent and secant are not defined when $x = 0$. For example, $t = \pi/2$ corresponds to $(x, y) = (0, 1)$, so $\tan(\pi/2)$ and $\sec(\pi/2)$ are *undefined*. Similarly, the cotangent and cosecant are not defined when $y = 0$. For example, $t = 0$ corresponds to $(x, y) = (1, 0)$, so $\cot 0$ and $\csc 0$ are *undefined*.

In Figure 4.16, the unit circle is divided into eight equal arcs, corresponding to t-values of

$$0, \frac{\pi}{4}, \frac{\pi}{2}, \frac{3\pi}{4}, \pi, \frac{5\pi}{4}, \frac{3\pi}{2}, \frac{7\pi}{4}, \text{ and } 2\pi.$$

Similarly, in Figure 4.17, the unit circle is divided into 12 equal arcs, corresponding to t-values of

$$0, \frac{\pi}{6}, \frac{\pi}{3}, \frac{\pi}{2}, \frac{2\pi}{3}, \frac{5\pi}{6}, \pi, \frac{7\pi}{6}, \frac{4\pi}{3}, \frac{3\pi}{2}, \frac{5\pi}{3}, \frac{11\pi}{6}, \text{ and } 2\pi.$$

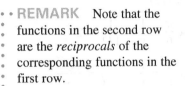

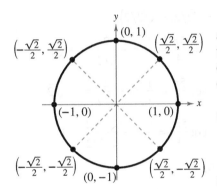

Figure 4.16

To verify the points on the unit circle in Figure 4.16, note that

$$\left(\frac{\sqrt{2}}{2}, \frac{\sqrt{2}}{2}\right)$$

lies on the line $y = x$. So, substituting x for y in the equation of the unit circle produces the following

$$x^2 + x^2 = 1 \implies 2x^2 = 1 \implies x^2 = \frac{1}{2} \implies x = \pm\frac{\sqrt{2}}{2}$$

Because the point is in the first quadrant and $y = x$, you have

$$x = \frac{\sqrt{2}}{2} \quad \text{and} \quad y = \frac{\sqrt{2}}{2}.$$

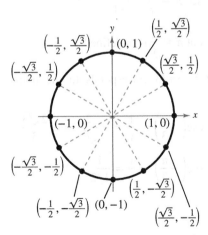

Figure 4.17

Similar reasoning can be used to verify the rest of the points in Figure 4.16 and the points in Figure 4.17.

Using the (x, y) coordinates in Figures 4.16 and 4.17, you can evaluate the trigonometric functions for these common t-values. Examples 1 and 2 demonstrate this procedure. You should study and learn these exact function values for common t-values because they will help you perform calculations in later sections.

EXAMPLE 1   **Evaluating Trigonometric Functions**

See LarsonPrecalculus.com for an interactive version of this type of example.

▷ **ALGEBRA HELP**   To review dividing fractions and rationalizing denominators, see Appendix A.1 and Appendix A.2, respectively.

Evaluate the six trigonometric functions at each real number.

a. $t = \dfrac{\pi}{6}$ **b.** $t = \dfrac{5\pi}{4}$ **c.** $t = \pi$ **d.** $t = -\dfrac{\pi}{3}$

Solution   For each t-value, begin by finding the corresponding point (x, y) on the unit circle. Then use the definitions of trigonometric functions listed on page 271.

a. $t = \pi/6$ corresponds to the point $(x, y) = \left(\sqrt{3}/2, \, 1/2\right)$.

$$\sin \frac{\pi}{6} = y = \frac{1}{2} \qquad\qquad \csc \frac{\pi}{6} = \frac{1}{y} = \frac{1}{1/2} = 2$$

$$\cos \frac{\pi}{6} = x = \frac{\sqrt{3}}{2} \qquad\qquad \sec \frac{\pi}{6} = \frac{1}{x} = \frac{2}{\sqrt{3}} = \frac{2\sqrt{3}}{3}$$

$$\tan \frac{\pi}{6} = \frac{y}{x} = \frac{1/2}{\sqrt{3}/2} = \frac{1}{\sqrt{3}} = \frac{\sqrt{3}}{3} \qquad\qquad \cot \frac{\pi}{6} = \frac{x}{y} = \frac{\sqrt{3}/2}{1/2} = \sqrt{3}$$

b. $t = 5\pi/4$ corresponds to the point $(x, y) = \left(-\sqrt{2}/2, \, -\sqrt{2}/2\right)$.

$$\sin \frac{5\pi}{4} = y = -\frac{\sqrt{2}}{2} \qquad\qquad \csc \frac{5\pi}{4} = \frac{1}{y} = -\frac{2}{\sqrt{2}} = -\sqrt{2}$$

$$\cos \frac{5\pi}{4} = x = -\frac{\sqrt{2}}{2} \qquad\qquad \sec \frac{5\pi}{4} = \frac{1}{x} = -\frac{2}{\sqrt{2}} = -\sqrt{2}$$

$$\tan \frac{5\pi}{4} = \frac{y}{x} = \frac{-\sqrt{2}/2}{-\sqrt{2}/2} = 1 \qquad\qquad \cot \frac{5\pi}{4} = \frac{x}{y} = \frac{-\sqrt{2}/2}{-\sqrt{2}/2} = 1$$

c. $t = \pi$ corresponds to the point $(x, y) = (-1, 0)$.

$$\sin \pi = y = 0 \qquad\qquad \csc \pi = \frac{1}{y} \text{ is undefined.}$$

$$\cos \pi = x = -1 \qquad\qquad \sec \pi = \frac{1}{x} = \frac{1}{-1} = -1$$

$$\tan \pi = \frac{y}{x} = \frac{0}{-1} = 0 \qquad\qquad \cot \pi = \frac{x}{y} \text{ is undefined.}$$

d. Moving *clockwise* around the unit circle, $t = -\pi/3$ corresponds to the point $(x, y) = \left(1/2, \, -\sqrt{3}/2\right)$.

$$\sin\left(-\frac{\pi}{3}\right) = y = -\frac{\sqrt{3}}{2} \qquad\qquad \csc\left(-\frac{\pi}{3}\right) = \frac{1}{y} = -\frac{2}{\sqrt{3}} = -\frac{2\sqrt{3}}{3}$$

$$\cos\left(-\frac{\pi}{3}\right) = x = \frac{1}{2} \qquad\qquad \sec\left(-\frac{\pi}{3}\right) = \frac{1}{x} = \frac{1}{1/2} = 2$$

$$\tan\left(-\frac{\pi}{3}\right) = \frac{y}{x} = \frac{-\sqrt{3}/2}{1/2} = -\sqrt{3}$$

$$\cot\left(-\frac{\pi}{3}\right) = \frac{x}{y} = \frac{1/2}{-\sqrt{3}/2} = -\frac{1}{\sqrt{3}} = -\frac{\sqrt{3}}{3}$$

✓ *Checkpoint*))) Audio-video solution in English & Spanish at LarsonPrecalculus.com

Evaluate the six trigonometric functions at each real number.

a. $t = \pi/2$ **b.** $t = 0$ **c.** $t = -5\pi/6$ **d.** $t = -3\pi/4$

Domain and Period of Sine and Cosine

The *domain* of the sine and cosine functions is the set of all real numbers. To determine the *range* of these two functions, consider the unit circle shown in Figure 4.18. You know that $\sin t = y$ and $\cos t = x$. Moreover, (x, y) is on the unit circle, so you also know that $-1 \le y \le 1$ and $-1 \le x \le 1$. This means that the values of sine and cosine also range between -1 and 1.

$$\begin{array}{ccc} -1 \le \ \ y \ \ \le 1 & & -1 \le \ \ x \ \ \le 1 \\ & \text{and} & \\ -1 \le \sin t \le 1 & & -1 \le \cos t \le 1 \end{array}$$

Adding 2π to each value of t in the interval $[0, 2\pi]$ results in a revolution around the unit circle, as shown in the figure below.

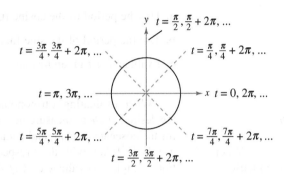

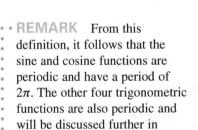

Figure 4.18

The values of $\sin(t + 2\pi)$ and $\cos(t + 2\pi)$ correspond to those of $\sin t$ and $\cos t$. Repeated revolutions (positive or negative) on the unit circle yield similar results. This leads to the general result

$$\sin(t + 2\pi n) = \sin t \quad \text{and} \quad \cos(t + 2\pi n) = \cos t$$

for any integer n and real number t. Functions that behave in such a repetitive (or cyclic) manner are **periodic.**

• • REMARK From this definition, it follows that the sine and cosine functions are periodic and have a period of 2π. The other four trigonometric functions are also periodic and will be discussed further in Section 4.6.

Definition of Periodic Function

A function f is **periodic** when there exists a positive real number c such that

$$f(t + c) = f(t)$$

for all t in the domain of f. The smallest number c for which f is periodic is the **period** of f.

Recall from Section 1.5 that a function f is *even* when $f(-t) = f(t)$ and is *odd* when $f(-t) = -f(t)$.

Even and Odd Trigonometric Functions

The cosine and secant functions are *even.*

$$\cos(-t) = \cos t \qquad\qquad \sec(-t) = \sec t$$

The sine, cosecant, tangent, and cotangent functions are *odd.*

$$\sin(-t) = -\sin t \qquad\qquad \csc(-t) = -\csc t$$

$$\tan(-t) = -\tan t \qquad\qquad \cot(-t) = -\cot t$$

EXAMPLE 2 **Evaluating Sine and Cosine**

a. Because $\dfrac{13\pi}{6} = 2\pi + \dfrac{\pi}{6}$, you have $\sin \dfrac{13\pi}{6} = \sin\left(2\pi + \dfrac{\pi}{6}\right) = \sin \dfrac{\pi}{6} = \dfrac{1}{2}$.

b. Because $-\dfrac{7\pi}{2} = -4\pi + \dfrac{\pi}{2}$, you have

$$\cos\left(-\dfrac{7\pi}{2}\right) = \cos\left(-4\pi + \dfrac{\pi}{2}\right) = \cos \dfrac{\pi}{2} = 0.$$

c. For $\sin t = \dfrac{4}{5}$, $\sin(-t) = -\dfrac{4}{5}$ because the sine function is odd.

✓ **Checkpoint** 🔊)) *Audio-video solution in English & Spanish at LarsonPrecalculus.com*

a. Use the period of the cosine function to evaluate $\cos(9\pi/2)$.

b. Use the period of the sine function to evaluate $\sin(-7\pi/3)$.

c. Evaluate $\cos t$ given that $\cos(-t) = 0.3$. ◼

When evaluating a trigonometric function with a calculator, set the calculator to the desired *mode* of measurement (*degree* or *radian*). Most calculators do not have keys for the cosecant, secant, and cotangent functions. To evaluate these functions, you can use the ⌊x⁻¹⌋ key with their respective reciprocal functions: sine, cosine, and tangent. For example, to evaluate $\csc(\pi/8)$, use the fact that

$$\csc \dfrac{\pi}{8} = \dfrac{1}{\sin(\pi/8)}$$

and enter the keystroke sequence below in *radian* mode.

⌊()⌋ ⌊SIN⌋ ⌊()⌋ ⌊π⌋ ⌊÷⌋ 8 ⌊()⌋ ⌊()⌋ ⌊x⁻¹⌋ ⌊ENTER⌋   Display 2.6131259

▷ TECHNOLOGY When evaluating trigonometric functions with a calculator, remember to enclose all fractional angle measures in parentheses. For example, to evaluate $\sin t$ for $t = \pi/6$, enter

⌊SIN⌋ ⌊()⌋ ⌊π⌋ ⌊÷⌋ 6 ⌊()⌋ ⌊ENTER⌋.

These keystrokes yield the correct value of 0.5. Note that some calculators automatically place a left parenthesis after trigonometric functions.

EXAMPLE 3 **Using a Calculator**

Function	Mode	Calculator Keystrokes	Display
a. $\sin \dfrac{2\pi}{3}$	Radian	⌊SIN⌋ ⌊()⌋ 2 ⌊π⌋ ⌊÷⌋ 3 ⌊()⌋ ⌊ENTER⌋	0.8660254
b. $\cot 1.5$	Radian	⌊()⌋ ⌊TAN⌋ ⌊()⌋ 1.5 ⌊()⌋ ⌊()⌋ ⌊x⁻¹⌋ ⌊ENTER⌋	0.0709148

✓ **Checkpoint** 🔊)) *Audio-video solution in English & Spanish at LarsonPrecalculus.com*

Use a calculator to evaluate (a) $\sin(5\pi/7)$ and (b) $\csc 2.0$. ◼

Summarize (Section 4.2)

1. Explain how to identify a unit circle and describe its relationship to real numbers *(page 270)*.

2. State the unit circle definitions of trigonometric functions *(page 271)*. For an example of evaluating trigonometric functions using the unit circle, see Example 1.

3. Explain how to use domain and period to evaluate sine and cosine functions *(page 273)*, and describe how to use a calculator to evaluate trigonometric functions *(page 274)*. For an example of using domain and period to evaluate sine and cosine functions, see Example 2. For an example of using a calculator to evaluate trigonometric functions, see Example 3.

4.2 Exercises

Vocabulary: Fill in the blanks.

1. Each real number t corresponds to a point (x, y) on the _____ _____.
2. A function f is _____ when there exists a positive real number c such that $f(t + c) = f(t)$ for all t in the domain of f.
3. The smallest number c for which a function f is periodic is the _____ of f.
4. A function f is _____ when $f(-t) = -f(t)$ and _____ when $f(-t) = f(t)$.

Skills and Applications

Evaluating Trigonometric Functions **In Exercises 5–8, find the exact values of the six trigonometric functions of the real number t.**

5.
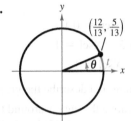

6.
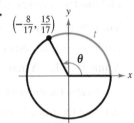

7.
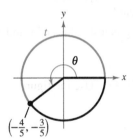

8.
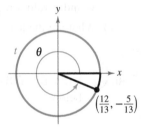

Finding a Point on the Unit Circle **In Exercises 9–12, find the point (x, y) on the unit circle that corresponds to the real number t.**

9. $t = \pi/2$
10. $t = \pi/4$
11. $t = 5\pi/6$
12. $t = 4\pi/3$

Evaluating Sine, Cosine, and Tangent In Exercises 13–22, evaluate (if possible) the sine, cosine, and tangent at the real number.

13. $t = \dfrac{\pi}{4}$
14. $t = \dfrac{\pi}{3}$

15. $t = -\dfrac{\pi}{6}$
16. $t = -\dfrac{\pi}{4}$

17. $t = -\dfrac{7\pi}{4}$
18. $t = -\dfrac{4\pi}{3}$

19. $t = \dfrac{11\pi}{6}$
20. $t = \dfrac{5\pi}{3}$

21. $t = -\dfrac{3\pi}{2}$
22. $t = -2\pi$

 Evaluating Trigonometric Functions In Exercises 23–30, evaluate (if possible) the six trigonometric functions at the real number.

23. $t = 2\pi/3$
24. $t = 5\pi/6$
25. $t = 4\pi/3$
26. $t = 7\pi/4$
27. $t = -5\pi/3$
28. $t = -3\pi/2$
29. $t = -\pi/2$
30. $t = -\pi$

 Using Period to Evaluate Sine and Cosine In Exercises 31–36, evaluate the trigonometric function using its period as an aid.

31. $\sin 4\pi$
32. $\cos 3\pi$
33. $\cos(7\pi/3)$
34. $\sin(9\pi/4)$
35. $\sin(19\pi/6)$
36. $\sin(-8\pi/3)$

 Using the Value of a Function In Exercises 37–42, use the given value to evaluate each function.

37. $\sin t = \dfrac{1}{2}$
   (a) $\sin(-t)$
   (b) $\csc(-t)$
38. $\sin(-t) = \dfrac{3}{8}$
   (a) $\sin t$
   (b) $\csc t$
39. $\cos(-t) = -\dfrac{1}{5}$
   (a) $\cos t$
   (b) $\sec(-t)$
40. $\cos t = -\dfrac{3}{4}$
   (a) $\cos(-t)$
   (b) $\sec(-t)$
41. $\sin t = \dfrac{4}{5}$
   (a) $\sin(\pi - t)$
   (b) $\sin(t + \pi)$
42. $\cos t = \dfrac{4}{5}$
   (a) $\cos(\pi - t)$
   (b) $\cos(t + \pi)$

Using a Calculator In Exercises 43–48, use a calculator to evaluate the trigonometric function. Round your answer to four decimal places. (Be sure the calculator is in the correct mode.)

43. $\sin 0.6$
44. $\cos(-2.8)$
45. $\tan(\pi/8)$
46. $\tan(5\pi/7)$
47. $\sec 3.1$
48. $\cot(-1.1)$

49. Harmonic Motion The displacement from equilibrium of an oscillating weight suspended by a spring is given by

$$y(t) = \frac{1}{2} \cos 6t$$

where y is the displacement in feet and t is the time in seconds. Find the displacement when (a) $t = 0$, (b) $t = \frac{1}{4}$, and (c) $t = \frac{1}{2}$.

50. Harmonic Motion

The displacement from equilibrium of an oscillating weight suspended by a spring and subject to the damping effect of friction is given by

$$y(t) = \frac{1}{2} e^{-t} \cos 6t$$

where y is the displacement in feet and t is the time in seconds.

(a) Complete the table

t	0	$\frac{1}{4}$	$\frac{1}{2}$	$\frac{3}{4}$	1
y					

(b) Use the *table* feature of a graphing utility to approximate the time when the weight reaches equilibrium.

(c) What appears to happen to the displacement as t increases?

Exploration

True of False? In Exercises 51–54, determine whether the statement is true or false. Justify your answer.

51. Because $\sin(-t) = -\sin t$, the sine of a negative angle is a negative number.

52. The real number 0 corresponds to the point $(0, 1)$ on the unit circle.

53. $\tan a = \tan(a - 6\pi)$

54. $\cos\left(-\frac{7\pi}{2}\right) = \cos\left(\pi + \frac{\pi}{2}\right)$

55. Conjecture Let (x_1, y_1) and (x_2, y_2) be points on the unit circle corresponding to $t = t_1$ and $t = \pi - t_1$, respectively.

(a) Identify the symmetry of the points (x_1, y_1) and (x_2, y_2).

(b) Make a conjecture about any relationship between $\sin t_1$ and $\sin(\pi - t_1)$.

(c) Make a conjecture about any relationship between $\cos t_1$ and $\cos(\pi - t_1)$.

56. Using the Unit Circle Use the unit circle to verify that the cosine and secant functions are even and that the sine, cosecant, tangent, and cotangent functions are odd.

57. Error Analysis Describe the error.

Your classmate uses a calculator to evaluate $\tan(\pi/2)$ and gets a result of 0.0274224385.

58. Verifying Expressions Are Not Equal Verify that

$$\sin(t_1 + t_2) \neq \sin t_1 + \sin t_2$$

by approximating $\sin 0.25$, $\sin 0.75$, and $\sin 1$.

59. Using Technology With a graphing utility in *radian* and *parametric* modes, enter the equations

$$X_{1T} = \cos T \quad \text{and} \quad Y_{1T} = \sin T$$

and use the settings below.

Tmin = 0, Tmax = 6.3, Tstep = 0.1

Xmin = −1.5, Xmax = 1.5, Xscl = 1

Ymin = −1, Ymax = 1, Yscl = 1

(a) Graph the entered equations and describe the graph.

(b) Use the *trace* feature to move the cursor around the graph. What do the t-values represent? What do the x- and y-values represent?

(c) What are the least and greatest values of x and y?

60. HOW DO YOU SEE IT? Use the figure below.

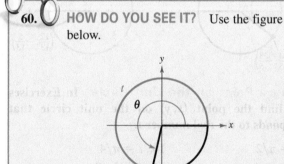

(a) Are all of the trigonometric functions of t defined? Explain.

(b) For those trigonometric functions that are defined, determine whether the sign of the trigonometric function is positive or negative. Explain.

61. Think About It Because $f(t) = \sin t$ is an odd function and $g(t) = \cos t$ is an even function, what can be said about the function $h(t) = f(t)g(t)$?

62. Think About It Because $f(t) = \sin t$ and $g(t) = \tan t$ are odd functions, what can be said about the function $h(t) = f(t)g(t)$?

4.3 Right Triangle Trigonometry

Right triangle trigonometry has many real-life applications. For example, in Exercise 72 on page 287, you will use right triangle trigonometry to analyze the height of a helium-filled balloon.

- Evaluate trigonometric functions of acute angles.
- Use fundamental trigonometric identities.
- Use trigonometric functions to model and solve real-life problems.

The Six Trigonometric Functions

This section introduces the trigonometric functions from a *right triangle* perspective. Consider the right triangle shown below, in which one acute angle is labeled θ. Relative to the angle θ, the three sides of the triangle are the **hypotenuse,** the **opposite side** (the side opposite the angle θ), and the **adjacent side** (the side adjacent to the angle θ).

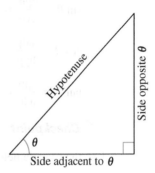

Using the lengths of these three sides, you can form six ratios that define the six trigonometric functions of the acute angle θ.

> **sine**   **cosecant**   **cosine**   **secant**   **tangent**   **cotangent**

In the definitions below,

$$0° < \theta < 90°$$

(θ lies in the first quadrant). For such angles, the value of each trigonometric function is *positive*.

Right Triangle Definitions of Trigonometric Functions

Let θ be an *acute* angle of a right triangle. The six trigonometric functions of the angle θ are defined below. (Note that the functions in the second row are the *reciprocals* of the corresponding functions in the first row.)

$$\sin\theta = \frac{\text{opp}}{\text{hyp}} \qquad \cos\theta = \frac{\text{adj}}{\text{hyp}} \qquad \tan\theta = \frac{\text{opp}}{\text{adj}}$$

$$\csc\theta = \frac{\text{hyp}}{\text{opp}} \qquad \sec\theta = \frac{\text{hyp}}{\text{adj}} \qquad \cot\theta = \frac{\text{adj}}{\text{opp}}$$

The abbreviations

   opp,   *adj,* and *hyp*

represent the lengths of the three sides of a right triangle.

   opp = the length of the side *opposite* θ

   adj = the length of the side *adjacent to* θ

   hyp = the length of the *hypotenuse*

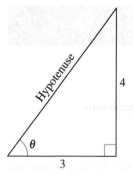

Figure 4.19

EXAMPLE 1   **Evaluating Trigonometric Functions**

See LarsonPrecalculus.com for an interactive version of this type of example.

Use the triangle in Figure 4.19 to find the values of the six trigonometric functions of θ.

Solution   By the Pythagorean Theorem, $(\text{hyp})^2 = (\text{opp})^2 + (\text{adj})^2$, it follows that

$$\text{hyp} = \sqrt{4^2 + 3^2}$$
$$= \sqrt{25}$$
$$= 5.$$

So, the six trigonometric functions of θ are

$$\sin \theta = \frac{\text{opp}}{\text{hyp}} = \frac{4}{5} \qquad \csc \theta = \frac{\text{hyp}}{\text{opp}} = \frac{5}{4}$$

$$\cos \theta = \frac{\text{adj}}{\text{hyp}} = \frac{3}{5} \qquad \sec \theta = \frac{\text{hyp}}{\text{adj}} = \frac{5}{3}$$

$$\tan \theta = \frac{\text{opp}}{\text{adj}} = \frac{4}{3} \qquad \cot \theta = \frac{\text{adj}}{\text{opp}} = \frac{3}{4}.$$

✓ *Checkpoint* *Audio-video solution in English & Spanish at LarsonPrecalculus.com*

Use the triangle below to find the values of the six trigonometric functions of θ.

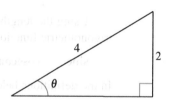

In Example 1, you were given the lengths of two sides of the right triangle, but not the angle θ. Often, you will be asked to find the trigonometric functions of a *given* acute angle θ. To do this, construct a right triangle having θ as one of its angles.

EXAMPLE 2   **Evaluating Trigonometric Functions of 45°**

Find the values of sin 45°, cos 45°, and tan 45°.

Solution   Construct a right triangle having 45° as one of its acute angles, as shown in Figure 4.20. Choose 1 as the length of the adjacent side. From geometry, you know that the other acute angle is also 45°. So, the triangle is isosceles and the length of the opposite side is also 1. By the Pythagorean Theorem, the length of the hypotenuse is $\sqrt{2}$.

$$\sin 45° = \frac{\text{opp}}{\text{hyp}} = \frac{1}{\sqrt{2}} = \frac{\sqrt{2}}{2}$$

$$\cos 45° = \frac{\text{adj}}{\text{hyp}} = \frac{1}{\sqrt{2}} = \frac{\sqrt{2}}{2}$$

$$\tan 45° = \frac{\text{opp}}{\text{adj}} = \frac{1}{1} = 1$$

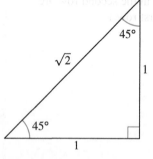

Figure 4.20

✓ *Checkpoint* *Audio-video solution in English & Spanish at LarsonPrecalculus.com*

Find the values of cot 45°, sec 45°, and csc 45°.

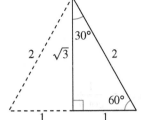

Figure 4.21

• • REMARK   The angles 30°, 45°, and 60° ($\pi/6$, $\pi/4$, and $\pi/3$ radians, respectively) occur frequently in trigonometry, so you should learn to construct the triangles shown in Figures 4.20 and 4.21.
• • • • • • • • • • • • • • • • • ▷

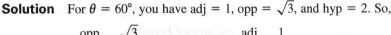 **EXAMPLE 3**   **Evaluating Trigonometric Functions of 30° and 60°**

Use the equilateral triangle shown in Figure 4.21 to find the values of sin 60°, cos 60°, sin 30°, and cos 30°.

Solution   For $\theta = 60°$, you have adj $= 1$, opp $= \sqrt{3}$, and hyp $= 2$. So,

$$\sin 60° = \frac{\text{opp}}{\text{hyp}} = \frac{\sqrt{3}}{2} \quad \text{and} \quad \cos 60° = \frac{\text{adj}}{\text{hyp}} = \frac{1}{2}.$$

For $\theta = 30°$, adj $= \sqrt{3}$, opp $= 1$, and hyp $= 2$. So,

$$\sin 30° = \frac{\text{opp}}{\text{hyp}} = \frac{1}{2} \quad \text{and} \quad \cos 30° = \frac{\text{adj}}{\text{hyp}} = \frac{\sqrt{3}}{2}.$$

✓ *Checkpoint* *Audio-video solution in English & Spanish at LarsonPrecalculus.com*

Use the equilateral triangle shown in Figure 4.21 to find the values of tan 60° and tan 30°.

Sines, Cosines, and Tangents of Special Angles

$$\sin 30° = \sin \frac{\pi}{6} = \frac{1}{2} \qquad \cos 30° = \cos \frac{\pi}{6} = \frac{\sqrt{3}}{2} \qquad \tan 30° = \tan \frac{\pi}{6} = \frac{\sqrt{3}}{3}$$

$$\sin 45° = \sin \frac{\pi}{4} = \frac{\sqrt{2}}{2} \qquad \cos 45° = \cos \frac{\pi}{4} = \frac{\sqrt{2}}{2} \qquad \tan 45° = \tan \frac{\pi}{4} = 1$$

$$\sin 60° = \sin \frac{\pi}{3} = \frac{\sqrt{3}}{2} \qquad \cos 60° = \cos \frac{\pi}{3} = \frac{1}{2} \qquad \tan 60° = \tan \frac{\pi}{3} = \sqrt{3}$$

Note that $\sin 30° = \frac{1}{2} = \cos 60°$. This occurs because 30° and 60° are complementary angles. In general, it can be shown from the right triangle definitions that *cofunctions of complementary angles are equal.* That is, if θ is an acute angle, then the relationships below are true.

$$\sin(90° - \theta) = \cos \theta \qquad \cos(90° - \theta) = \sin \theta \qquad \tan(90° - \theta) = \cot \theta$$

$$\cot(90° - \theta) = \tan \theta \qquad \sec(90° - \theta) = \csc \theta \qquad \csc(90° - \theta) = \sec \theta$$

To use a calculator to evaluate trigonometric functions of angles measured in degrees, remember to set the calculator to *degree* mode.

EXAMPLE 4   **Using a Calculator**

Use a calculator to evaluate sec 5° 40′ 12″.

Solution   Begin by converting to decimal degree form. $\left[\text{Recall that } 1' = \frac{1}{60}(1°) \text{ and } 1'' = \frac{1}{3600}(1°).\right]$

$$5° \, 40' \, 12'' = 5° + \left(\frac{40}{60}\right)° + \left(\frac{12}{3600}\right)° = 5.67°$$

Then, use a calculator to evaluate sec 5.67°.

Function	Calculator Keystrokes	Display
sec 5° 40′ 12″ = sec 5.67°	⬚ `COS` ⬚ 5.67 ⬚ ⬚ `x⁻¹` `ENTER`	1.0049166

✓ *Checkpoint* *Audio-video solution in English & Spanish at LarsonPrecalculus.com*

Use a calculator to evaluate csc 34° 30′ 36″.

Trigonometric Identities

Trigonometric identities are relationships between trigonometric functions.

Fundamental Trigonometric Identities

Reciprocal Identities

$$\sin\theta = \frac{1}{\csc\theta} \qquad \cos\theta = \frac{1}{\sec\theta} \qquad \tan\theta = \frac{1}{\cot\theta}$$

$$\csc\theta = \frac{1}{\sin\theta} \qquad \sec\theta = \frac{1}{\cos\theta} \qquad \cot\theta = \frac{1}{\tan\theta}$$

Quotient Identities

$$\tan\theta = \frac{\sin\theta}{\cos\theta} \qquad \cot\theta = \frac{\cos\theta}{\sin\theta}$$

Pythagorean Identities

$$\sin^2\theta + \cos^2\theta = 1$$

$$1 + \tan^2\theta = \sec^2\theta$$

$$1 + \cot^2\theta = \csc^2\theta$$

▷ • **REMARK** Do not confuse, for example, $\sin^2\theta$ with $\sin\theta^2$. With $\sin^2\theta$, you are squaring $\sin\theta$. With $\sin\theta^2$, you are squaring θ and then finding the sine.

Note that $\sin^2\theta$ represents $(\sin\theta)^2$, $\cos^2\theta$ represents $(\cos\theta)^2$, and so on.

EXAMPLE 5 Applying Trigonometric Identities

Let θ be an acute angle such that $\sin\theta = 0.6$. Find the value of (a) $\cos\theta$ and (b) $\tan\theta$ using trigonometric identities.

Solution

a. To find the value of $\cos\theta$, use the Pythagorean identity

$$\sin^2\theta + \cos^2\theta = 1.$$

So, you have

$(0.6)^2 + \cos^2\theta = 1$	Substitute 0.6 for $\sin\theta$.
$\cos^2\theta = 1 - (0.6)^2$	Subtract $(0.6)^2$ from each side.
$\cos^2\theta = 0.64$	Simplify.
$\cos\theta = \sqrt{0.64}$	Extract positive square root.
$\cos\theta = 0.8.$	Simplify.

b. Now, knowing the sine and cosine of θ, you can find the tangent of θ.

$$\tan\theta = \frac{\sin\theta}{\cos\theta} = \frac{0.6}{0.8} = 0.75$$

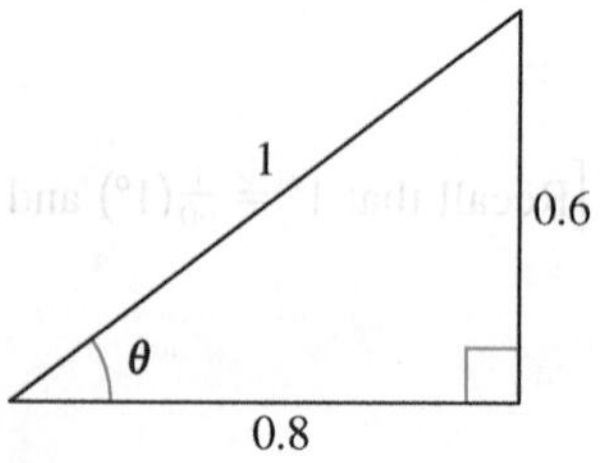

Use the definitions of $\cos\theta$ and $\tan\theta$ and the triangle shown in Figure 4.22 to check these results.

Figure 4.22

✓ *Checkpoint*  Audio-video solution in English & Spanish at LarsonPrecalculus.com

Let θ be an acute angle such that $\cos\theta = 0.96$. Find the value of (a) $\sin\theta$ and (b) $\tan\theta$ using trigonometric identities.

EXAMPLE 6 **Applying Trigonometric Identities**

Let θ be an acute angle such that $\tan \theta = \frac{1}{3}$. Find the value of (a) $\cot \theta$ and (b) $\sec \theta$ using trigonometric identities.

Solution

a. $\cot \theta = \dfrac{1}{\tan \theta}$ Reciprocal identity

       $= \dfrac{1}{1/3}$ Substitute $\frac{1}{3}$ for $\tan \theta$.

       $= 3$ Simplify.

b. $\sec^2 \theta = 1 + \tan^2 \theta$ Pythagorean identity

     $\sec^2 \theta = 1 + \left(\dfrac{1}{3}\right)^2$ Substitute $\frac{1}{3}$ for $\tan \theta$.

     $\sec^2 \theta = \dfrac{10}{9}$ Simplify.

     $\sec \theta = \dfrac{\sqrt{10}}{3}$ Extract positive square root and simplify.

Use the definitions of $\cot \theta$ and $\sec \theta$ and the triangle below to check these results.

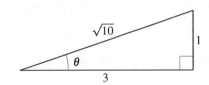

✓ **Checkpoint** 🔊))) *Audio-video solution in English & Spanish at LarsonPrecalculus.com*

Let θ be an acute angle such that $\tan \theta = 2$. Find the value of (a) $\cot \theta$ and (b) $\sec \theta$ using trigonometric identities.

EXAMPLE 7 **Using Trigonometric Identities**

Use trigonometric identities to transform the left side of the equation into the right side $(0 < \theta < \pi/2)$.

a. $\sin \theta \csc \theta = 1$ **b.** $(\csc \theta + \cot \theta)(\csc \theta - \cot \theta) = 1$

Solution

a. $\sin \theta \csc \theta = \left(\dfrac{1}{\cancel{\csc \theta}}\right) \cancel{\csc \theta} = 1$ Use a reciprocal identity and simplify.

b. $(\csc \theta + \cot \theta)(\csc \theta - \cot \theta)$

     $= \csc^2 \theta - \csc \theta \cot \theta + \csc \theta \cot \theta - \cot^2 \theta$ FOIL Method

     $= \csc^2 \theta - \cot^2 \theta$ Simplify.

     $= 1$ Pythagorean identity

✓ **Checkpoint** 🔊))) *Audio-video solution in English & Spanish at LarsonPrecalculus.com*

Use trigonometric identities to transform the left side of the equation into the right side $(0 < \theta < \pi/2)$.

a. $\tan \theta \csc \theta = \sec \theta$ **b.** $(\csc \theta + 1)(\csc \theta - 1) = \cot^2 \theta$

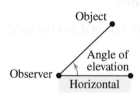

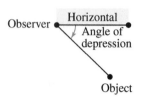

Figure 4.23

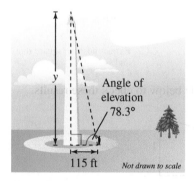

Figure 4.24

Applications Involving Right Triangles

Many applications of trigonometry involve **solving right triangles.** In this type of application, you are usually given one side of a right triangle and one of the acute angles and are asked to find one of the other sides, *or* you are given two sides and are asked to find one of the acute angles.

In Example 8, you are given the **angle of elevation,** which represents the angle from the horizontal upward to an object. In other applications you may be given the **angle of depression,** which represents the angle from the horizontal downward to an object. (See Figure 4.23.)

EXAMPLE 8 Solving a Right Triangle

A surveyor stands 115 feet from the base of the Washington Monument, as shown in Figure 4.24. The surveyor measures the angle of elevation to the top of the monument to be 78.3°. How tall is the Washington Monument?

Solution From Figure 4.24,

$$\tan 78.3° = \frac{\text{opp}}{\text{adj}} = \frac{y}{115}$$

where y is the height of the monument. So, the height of the Washington Monument is

$$y = 115 \tan 78.3°$$
$$\approx 115(4.8288)$$
$$\approx 555 \text{ feet.}$$

✓ **Checkpoint** ◀))) *Audio-video solution in English & Spanish at LarsonPrecalculus.com*

The angle of elevation to the top of a flagpole at a distance of 19 feet from its base is 64.6°. How tall is the flagpole?

EXAMPLE 9 Solving a Right Triangle

A lighthouse is 200 yards from a bike path along the edge of a lake. A walkway to the lighthouse is 400 yards long. (See Figure 4.25.) Find the acute angle θ between the bike path and the walkway.

Solution From Figure 4.25, the sine of the angle θ is

$$\sin \theta = \frac{\text{opp}}{\text{hyp}} = \frac{200}{400} = \frac{1}{2}.$$

You should recognize that $\theta = 30°$.

✓ **Checkpoint** ◀))) *Audio-video solution in English & Spanish at LarsonPrecalculus.com*

Find the acute angle θ between the two paths shown below.

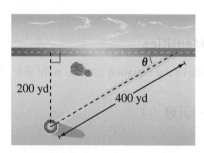

Figure 4.25

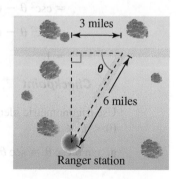

In Example 9, you were able to recognize that the special angle $\theta = 30°$ satisfies the equation $\sin \theta = \frac{1}{2}$. However, when θ is not a special angle, you can *estimate* its value. For example, to estimate the acute angle θ in the equation $\sin \theta = 0.6$, you could reason that $\sin 30° = \frac{1}{2} = 0.5000$ and $\sin 45° = 1/\sqrt{2} \approx 0.7071$, so θ lies somewhere between 30° and 45°. In a later section, you will study a method of determining a more precise value of θ.

EXAMPLE 10 **Solving a Right Triangle**

Find the length c and the height b of the skateboard ramp below.

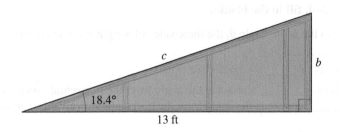

Solution From the figure,

$$\cos 18.4° = \frac{\text{adj}}{\text{hyp}} = \frac{13}{c}.$$

So, the length of the skateboard ramp is

$$c = \frac{13}{\cos 18.4°} \approx \frac{13}{0.9489} \approx 13.7 \text{ feet.}$$

Also from the figure,

$$\tan 18.4° = \frac{\text{opp}}{\text{adj}} = \frac{b}{13}.$$

So, the height is

$$b = 13 \tan 18.4° \approx 13(0.3327) \approx 4.3 \text{ feet.}$$

✓ *Checkpoint* ◀))) *Audio-video solution in English & Spanish at LarsonPrecalculus.com*

Find the length c and the horizontal length a of the loading ramp below.

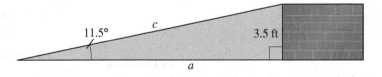

Skateboarders can go to a skatepark, which is a recreational environment built with many different types of ramps and rails.

Summarize (Section 4.3)

1. State the right triangle definitions of the six trigonometric functions *(page 277)*. For examples of evaluating trigonometric functions of acute angles, see Examples 1–4.

2. List the reciprocal, quotient, and Pythagorean identities *(page 280)*. For examples of using these identities, see Examples 5–7.

3. Describe real-life applications of trigonometric functions *(pages 282 and 283, Examples 8–10)*.

4.3 Exercises

See **CalcChat.com** for tutorial help and worked-out solutions to odd-numbered exercises.

Vocabulary

1. Match each trigonometric function with its right triangle definition.

   (a) sine   (b) cosine   (c) tangent   (d) cosecant   (e) secant   (f) cotangent

   (i) $\dfrac{\text{hypotenuse}}{\text{adjacent}}$ (ii) $\dfrac{\text{adjacent}}{\text{opposite}}$ (iii) $\dfrac{\text{hypotenuse}}{\text{opposite}}$ (iv) $\dfrac{\text{adjacent}}{\text{hypotenuse}}$ (v) $\dfrac{\text{opposite}}{\text{hypotenuse}}$ (vi) $\dfrac{\text{opposite}}{\text{adjacent}}$

In Exercises 2–4, fill in the blanks.

2. Relative to the acute angle θ, the three sides of a right triangle are the _____ side, the _____ side, and the _____.

3. Cofunctions of _____ angles are equal.

4. An angle of _____ represents the angle from the horizontal upward to an object, whereas an angle of _____ represents the angle from the horizontal downward to an object.

Skills and Applications

 Evaluating Trigonometric Functions In Exercises 5–10, find the exact values of the six trigonometric functions of the angle θ.

5.

6.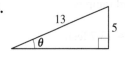

7.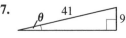

8.

9.

10.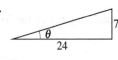

Evaluating Trigonometric Functions In Exercises 11–14, find the exact values of the six trigonometric functions of the angle θ for each of the two triangles. Explain why the function values are the same.

11.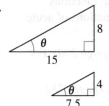

12.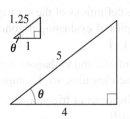

13.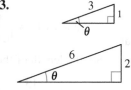

14.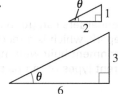

Evaluating Trigonometric Functions In Exercises 15–22, sketch a right triangle corresponding to the trigonometric function of the acute angle θ. Then find the exact values of the other five trigonometric functions of θ.

15. $\cos \theta = \frac{15}{17}$ **16.** $\sin \theta = \frac{3}{5}$

17. $\sec \theta = \frac{6}{5}$ **18.** $\tan \theta = \frac{4}{5}$

19. $\sin \theta = \frac{1}{5}$ **20.** $\sec \theta = \frac{17}{7}$

21. $\cot \theta = 3$ **22.** $\csc \theta = 9$

 Evaluating Trigonometric Functions of 30°, 45°, and 60° In Exercises 23–28, construct an appropriate triangle to find the missing values. ($0° \le \theta \le 90°, 0 \le \theta \le \pi/2$)

Function	θ (deg)	θ (rad)	Function Value
23. tan	30°		
24. cos	45°		
25. sin		$\dfrac{\pi}{4}$	
26. tan		$\dfrac{\pi}{3}$	
27. sec		$\dfrac{\pi}{4}$	
28. csc		$\dfrac{\pi}{6}$	

Using a Calculator In Exercises 29–36, use a calculator to evaluate each function. Round your answers to four decimal places. (Be sure the calculator is in the correct mode.)

29. (a) sin 20°   (b) cos 70°
30. (a) tan 23.5°   (b) cot 66.5°
31. (a) sin 14.21°   (b) csc 14.21°
32. (a) cot 79.56°   (b) sec 79.56°
33. (a) cos 4° 50′ 15″   (b) sec 4° 50′ 15″
34. (a) sec 42° 12′   (b) csc 48° 7′
35. (a) cot 17° 15′   (b) tan 17° 15′
36. (a) sec 56° 8′ 10″   (b) cos 56° 8′ 10″

 Applying Trigonometric Identities In Exercises 37–42, use the given function value(s) and the trigonometric identities to find the exact value of each indicated trigonometric function.

37. $\sin 60° = \dfrac{\sqrt{3}}{2}$, $\cos 60° = \dfrac{1}{2}$

   (a) sin 30°   (b) cos 30°
   (c) tan 60°   (d) cot 60°

38. $\sin 30° = \dfrac{1}{2}$, $\tan 30° = \dfrac{\sqrt{3}}{3}$

   (a) csc 30°   (b) cot 60°
   (c) cos 30°   (d) cot 30°

39. $\cos \theta = \frac{1}{3}$

   (a) $\sin \theta$ (b) $\tan \theta$
   (c) $\sec \theta$ (d) $\csc(90° - \theta)$

40. $\sec \theta = 5$

   (a) $\cos \theta$ (b) $\cot \theta$
   (c) $\cot(90° - \theta)$ (d) $\sin \theta$

41. $\cot \alpha = 3$

   (a) $\tan \alpha$ (b) $\csc \alpha$
   (c) $\cot(90° - \alpha)$ (d) $\sin \alpha$

42. $\cos \beta = \dfrac{\sqrt{7}}{4}$

   (a) $\sec \beta$ (b) $\sin \beta$
   (c) $\cot \beta$ (d) $\sin(90° - \beta)$

 Using Trigonometric Identities In Exercises 43–52, use trigonometric identities to transform the left side of the equation into the right side $(0 < \theta < \pi/2)$.

43. $\tan \theta \cot \theta = 1$
44. $\cos \theta \sec \theta = 1$
45. $\tan \alpha \cos \alpha = \sin \alpha$
46. $\cot \alpha \sin \alpha = \cos \alpha$
47. $(1 + \sin \theta)(1 - \sin \theta) = \cos^2 \theta$
48. $(1 + \cos \theta)(1 - \cos \theta) = \sin^2 \theta$
49. $(\sec \theta + \tan \theta)(\sec \theta - \tan \theta) = 1$
50. $\sin^2 \theta - \cos^2 \theta = 2 \sin^2 \theta - 1$
51. $\dfrac{\sin \theta}{\cos \theta} + \dfrac{\cos \theta}{\sin \theta} = \csc \theta \sec \theta$
52. $\dfrac{\tan \beta + \cot \beta}{\tan \beta} = \csc^2 \beta$

Finding Special Angles of a Triangle In Exercises 53–58, find each value of θ in degrees $(0° < \theta < 90°)$ and radians $(0 < \theta < \pi/2)$ without using a calculator.

53. (a) $\sin \theta = \frac{1}{2}$ (b) $\csc \theta = 2$
54. (a) $\cos \theta = \dfrac{\sqrt{2}}{2}$ (b) $\tan \theta = 1$
55. (a) $\sec \theta = 2$ (b) $\cot \theta = 1$
56. (a) $\tan \theta = \sqrt{3}$ (b) $\csc \theta = \sqrt{2}$
57. (a) $\csc \theta = \dfrac{2\sqrt{3}}{3}$ (b) $\sin \theta = \dfrac{\sqrt{2}}{2}$
58. (a) $\cot \theta = \dfrac{\sqrt{3}}{3}$ (b) $\sec \theta = \sqrt{2}$

Finding Side Lengths of a Triangle In Exercises 59–62, find the exact values of the indicated variables.

59. Find x and y.

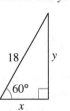

60. Find x and r.

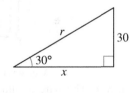

61. Find x and r.

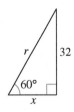

62. Find x and r.

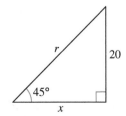

63. **Empire State Building** You are standing 45 meters from the base of the Empire State Building. You estimate that the angle of elevation to the top of the 86th floor (the observatory) is 82°. The total height of the building is another 123 meters above the 86th floor. What is the approximate height of the building? One of your friends is on the 86th floor. What is the distance between you and your friend?

64. Height of a Tower A six-foot person walks from the base of a broadcasting tower directly toward the tip of the shadow cast by the tower. When the person is 132 feet from the tower and 3 feet from the tip of the shadow, the person's shadow starts to appear beyond the tower's shadow.

(a) Draw a right triangle that gives a visual representation of the problem. Label the known quantities of the triangle and use a variable to represent the height of the tower.

(b) Use a trigonometric function to write an equation involving the unknown quantity.

(c) What is the height of the tower?

65. Angle of Elevation You are skiing down a mountain with a vertical height of 1250 feet. The distance from the top of the mountain to the base is 2500 feet. What is the angle of elevation from the base to the top of the mountain?

66. Biology A biologist wants to know the width w of a river to properly set instruments for an experiment. From point A, the biologist walks downstream 100 feet and sights to point C (see figure). From this sighting, it is determined that $\theta = 54°$. How wide is the river?

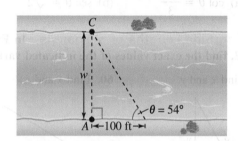

67. Guy Wire A guy wire runs from the ground to a cell tower. The wire is attached to the cell tower 150 feet above the ground. The angle formed between the wire and the ground is 43° (see figure).

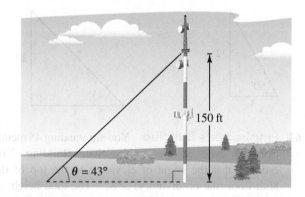

(a) How long is the guy wire?

(b) How far from the base of the tower is the guy wire anchored to the ground?

68. Height of a Mountain In traveling across flat land, you see a mountain directly in front of you. Its angle of elevation (to the peak) is 3.5°. After you drive 13 miles closer to the mountain, the angle of elevation is 9° (see figure). Approximate the height of the mountain.

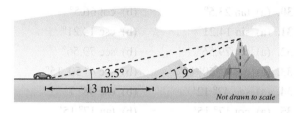

Not drawn to scale

69. Machine Shop Calculations A steel plate has the form of one-fourth of a circle with a radius of 60 centimeters. Two two-centimeter holes are drilled in the plate, positioned as shown in the figure. Find the coordinates of the center of each hole.

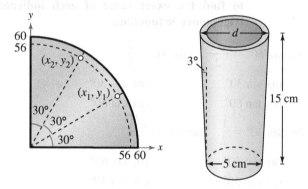

Figure for 69   Figure for 70

70. Machine Shop Calculations A tapered shaft has a diameter of 5 centimeters at the small end and is 15 centimeters long (see figure). The taper is 3°. Find the diameter d of the large end of the shaft.

71. Geometry Use a compass to sketch a quarter of a circle of radius 10 centimeters. Using a protractor, construct an angle of 20° in standard position (see figure). Drop a perpendicular line from the point of intersection of the terminal side of the angle and the arc of the circle. By actual measurement, calculate the coordinates (x, y) of the point of intersection and use these measurements to approximate the six trigonometric functions of a 20° angle.

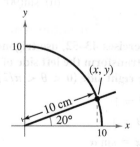

• • **72. Helium-Filled Balloon** • • • • • • • • • • • • •

A 20-meter line is used to tether a helium-filled balloon. The line makes an angle of approximately 85° with the ground because of a breeze.

(a) Draw a right triangle that gives a visual representation of the problem. Label the known quantities of the triangle and use a variable to represent the height of the balloon.

(b) Use a trigonometric function to write and solve an equation for the height of the balloon.

(c) The breeze becomes stronger and the angle the line makes with the ground decreases. How does this affect the triangle you drew in part (a)?

(d) Complete the table, which shows the heights (in meters) of the balloon for decreasing angle measures θ.

Angle, θ	80°	70°	60°	50°
Height				

Angle, θ	40°	30°	20°	10°
Height				

(e) As θ approaches 0°, how does this affect the height of the balloon? Draw a right triangle to explain your reasoning.

73. **Johnstown Inclined Plane** The Johnstown Inclined Plane in Pennsylvania is one of the longest and steepest hoists in the world. The railway cars travel a distance of 896.5 feet at an angle of approximately 35.4°, rising to a height of 1693.5 feet above sea level.

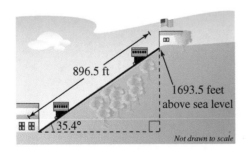

896.5 ft

1693.5 feet above sea level

35.4°

Not drawn to scale

(a) Find the vertical rise of the inclined plane.

(b) Find the elevation of the lower end of the inclined plane.

(c) The cars move up the mountain at a rate of 300 feet per minute. Find the rate at which they rise vertically.

74. **Error Analysis** Describe the error.

$$\cos 60° = \frac{\text{opp}}{\text{hyp}} = \frac{1}{2} \quad \times$$

Exploration

True or False? **In Exercises 75–80, determine whether the statement is true or false. Justify your answer.**

75. $\sin 60° \csc 60° = 1$ 76. $\sec 30° = \csc 30°$

77. $\sin 45° + \cos 45° = 1$ 78. $\cos 60° - \sin 30° = 0$

79. $\dfrac{\sin 60°}{\sin 30°} = \sin 2°$ 80. $\tan[(5°)^2] = \tan^2 5°$

81. **Think About It** You are given the value of $\tan \theta$. Is it possible to find the value of $\sec \theta$ without finding the measure of θ? Explain.

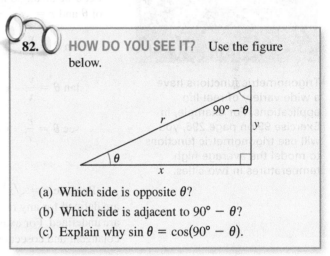

82. **HOW DO YOU SEE IT?** Use the figure below.

(a) Which side is opposite θ?

(b) Which side is adjacent to $90° - \theta$?

(c) Explain why $\sin \theta = \cos(90° - \theta)$.

83. **Think About It** Complete the table.

θ	0.1	0.2	0.3	0.4	0.5
$\sin \theta$					

(a) Is θ or $\sin \theta$ greater for θ in the interval $(0, 0.5]$?

(b) As θ approaches 0, how do θ and $\sin \theta$ compare? Explain.

84. **Think About It** Complete the table.

θ	0°	18°	36°	54°	72°	90°
$\sin \theta$						
$\cos \theta$						

(a) Discuss the behavior of the sine function for $0° \leq \theta \leq 90°$.

(b) Discuss the behavior of the cosine function for $0° \leq \theta \leq 90°$.

(c) Use the definitions of the sine and cosine functions to explain the results of parts (a) and (b).

4.4 Trigonometric Functions of Any Angle

Trigonometric functions have a wide variety of real-life applications. For example, in Exercise 99 on page 296, you will use trigonometric functions to model the average high temperatures in two cities.

- Evaluate trigonometric functions of any angle.
- Find reference angles.
- Evaluate trigonometric functions of real numbers.

Introduction

In Section 4.3, the definitions of trigonometric functions were restricted to acute angles. In this section, the definitions are extended to cover *any* angle. When θ is an *acute* angle, the definitions here coincide with those in the preceding section.

Definitions of Trigonometric Functions of Any Angle

Let θ be an angle in standard position with (x, y) a point on the terminal side of θ and $r = \sqrt{x^2 + y^2} \neq 0$.

$$\sin \theta = \frac{y}{r} \qquad \cos \theta = \frac{x}{r}$$

$$\tan \theta = \frac{y}{x}, \quad x \neq 0 \qquad \cot \theta = \frac{x}{y}, \quad y \neq 0$$

$$\sec \theta = \frac{r}{x}, \quad x \neq 0 \qquad \csc \theta = \frac{r}{y}, \quad y \neq 0$$

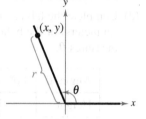

Because $r = \sqrt{x^2 + y^2}$ *cannot* be zero, it follows that the sine and cosine functions are defined for any real value of θ. However, when $x = 0$, the tangent and secant of θ are undefined. For example, the tangent of $90°$ is undefined. Similarly, when $y = 0$, the cotangent and cosecant of θ are undefined.

EXAMPLE 1 Evaluating Trigonometric Functions

Let $(-3, 4)$ be a point on the terminal side of θ. Find the sine, cosine, and tangent of θ.

Solution Referring to Figure 4.26, $x = -3$, $y = 4$, and

$$r = \sqrt{x^2 + y^2}$$
$$= \sqrt{(-3)^2 + 4^2}$$
$$= 5.$$

So, you have

$$\sin \theta = \frac{y}{r} = \frac{4}{5}$$

$$\cos \theta = \frac{x}{r} = -\frac{3}{5}$$

and

$$\tan \theta = \frac{y}{x} = -\frac{4}{3}.$$

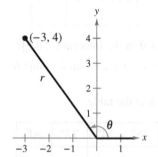

Figure 4.26

▷ **ALGEBRA HELP** The formula $r = \sqrt{x^2 + y^2}$ is an application of the Distance Formula. To review the Distance Formula, see Section 1.1.

✓ **Checkpoint** 🔊))) *Audio-video solution in English & Spanish at LarsonPrecalculus.com*

Let $(-2, 3)$ be a point on the terminal side of θ. Find the sine, cosine, and tangent of θ.

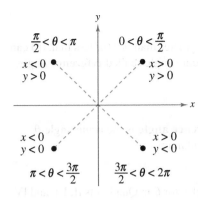

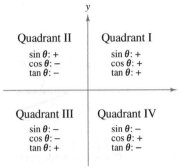

Figure 4.27

The *signs* of the trigonometric functions in the four quadrants can be determined from the definitions of the functions. For example, $\cos \theta = x/r$, so $\cos \theta$ is positive wherever $x > 0$, which is in Quadrants I and IV. (Remember, r is always positive.) Figure 4.27 shows this and other results. Use similar reasoning to verify the other results.

EXAMPLE 2   **Evaluating Trigonometric Functions**

Given $\tan \theta = -\frac{5}{4}$ and $\cos \theta > 0$, find $\sin \theta$ and $\sec \theta$.

Solution   Note that θ lies in Quadrant IV because that is the only quadrant in which the tangent is negative and the cosine is positive. Moreover, using

$$\tan \theta = \frac{y}{x} = -\frac{5}{4}$$

and the fact that y is negative in Quadrant IV, let $y = -5$ and $x = 4$. So, $r = \sqrt{16 + 25} = \sqrt{41}$ and you have the results below.

$$\sin \theta = \frac{y}{r}$$

$$= \frac{-5}{\sqrt{41}} \qquad \text{Exact value}$$

$$\approx -0.7809 \qquad \text{Approximate value}$$

$$\sec \theta = \frac{r}{x}$$

$$= \frac{\sqrt{41}}{4} \qquad \text{Exact value}$$

$$\approx 1.6008 \qquad \text{Approximate value}$$

✓ *Checkpoint*))) *Audio-video solution in English & Spanish at LarsonPrecalculus.com*

Given $\sin \theta = \frac{4}{5}$ and $\tan \theta < 0$, find $\cos \theta$ and $\tan \theta$.

EXAMPLE 3   **Trigonometric Functions of Quadrantal Angles**

Evaluate the cosine and tangent functions at the quadrantal angles 0, $\frac{\pi}{2}$, π, and $\frac{3\pi}{2}$.

Solution   To begin, choose a point on the terminal side of each angle, as shown in Figure 4.28. For each of the four points, $r = 1$ and you have the results below.

$$\cos 0 = \frac{x}{r} = \frac{1}{1} = 1 \qquad \tan 0 = \frac{y}{x} = \frac{0}{1} = 0 \qquad\qquad (x, y) = (1, 0)$$

$$\cos \frac{\pi}{2} = \frac{x}{r} = \frac{0}{1} = 0 \qquad \tan \frac{\pi}{2} = \frac{y}{x} = \frac{1}{0} \implies \text{undefined} \qquad (x, y) = (0, 1)$$

$$\cos \pi = \frac{x}{r} = \frac{-1}{1} = -1 \qquad \tan \pi = \frac{y}{x} = \frac{0}{-1} = 0 \qquad\qquad (x, y) = (-1, 0)$$

$$\cos \frac{3\pi}{2} = \frac{x}{r} = \frac{0}{1} = 0 \qquad \tan \frac{3\pi}{2} = \frac{y}{x} = \frac{-1}{0} \implies \text{undefined} \qquad (x, y) = (0, -1)$$

✓ *Checkpoint*))) *Audio-video solution in English & Spanish at LarsonPrecalculus.com*

Evaluate the sine and cotangent functions at the quadrantal angle $\frac{3\pi}{2}$.

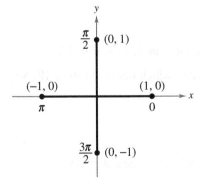

Figure 4.28

Reference Angles

The values of the trigonometric functions of angles greater than 90° (or less than 0°) can be determined from their values at corresponding acute angles called **reference angles.**

Definition of a Reference Angle

Let θ be an angle in standard position. Its **reference angle** is the acute angle θ' formed by the terminal side of θ and the horizontal axis.

The three figures below show the reference angles for θ in Quadrants II, III, and IV.

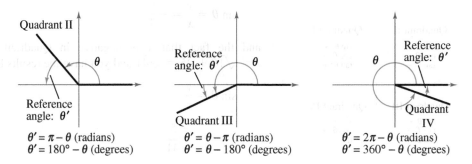

$\theta' = \pi - \theta$ (radians)
$\theta' = 180° - \theta$ (degrees)

$\theta' = \theta - \pi$ (radians)
$\theta' = \theta - 180°$ (degrees)

$\theta' = 2\pi - \theta$ (radians)
$\theta' = 360° - \theta$ (degrees)

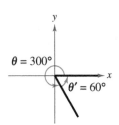

Figure 4.29

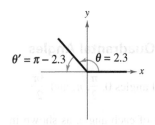

Figure 4.30

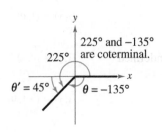

Figure 4.31

EXAMPLE 4   Finding Reference Angles

Find the reference angle θ'.

a. $\theta = 300°$ **b.** $\theta = 2.3$ **c.** $\theta = -135°$

Solution

a. Because 300° lies in Quadrant IV, the angle it makes with the x-axis is

$$\theta' = 360° - 300°$$
$$= 60°. \qquad \text{Degrees}$$

Figure 4.29 shows the angle $\theta = 300°$ and its reference angle $\theta' = 60°$.

b. Because 2.3 lies between $\pi/2 \approx 1.5708$ and $\pi \approx 3.1416$, it follows that it is in Quadrant II and its reference angle is

$$\theta' = \pi - 2.3$$
$$\approx 0.8416. \qquad \text{Radians}$$

Figure 4.30 shows the angle $\theta = 2.3$ and its reference angle $\theta' = \pi - 2.3$.

c. First, determine that $-135°$ is coterminal with 225°, which lies in Quadrant III. So, the reference angle is

$$\theta' = 225° - 180°$$
$$= 45°. \qquad \text{Degrees}$$

Figure 4.31 shows the angle $\theta = -135°$ and its reference angle $\theta' = 45°$.

✓ **Checkpoint**))) *Audio-video solution in English & Spanish at LarsonPrecalculus.com*

Find the reference angle θ'.

a. $\theta = 213°$ **b.** $\theta = \dfrac{14\pi}{9}$ **c.** $\theta = \dfrac{4\pi}{5}$

Trigonometric Functions of Real Numbers

To see how to use a reference angle to evaluate a trigonometric function, consider the point (x, y) on the terminal side of the angle θ, as shown at the right. You know that

$$\sin \theta = \frac{y}{r}$$

and

$$\tan \theta = \frac{y}{x}.$$

For the right triangle with acute angle θ' and sides of lengths $|x|$ and $|y|$, you have

$$\sin \theta' = \frac{\text{opp}}{\text{hyp}} = \frac{|y|}{r}$$

and

$$\tan \theta' = \frac{\text{opp}}{\text{adj}} = \frac{|y|}{|x|}.$$

$\text{opp} = |y|, \ \text{adj} = |x|$

So, it follows that $\sin \theta$ and $\sin \theta'$ are equal, *except possibly in sign*. The same is true for $\tan \theta$ and $\tan \theta'$ and for the other four trigonometric functions. In all cases, the quadrant in which θ lies determines the sign of the function value.

Evaluating Trigonometric Functions of Any Angle

To find the value of a trigonometric function of any angle θ:

1. Determine the function value of the associated reference angle θ'.

2. Depending on the quadrant in which θ lies, affix the appropriate sign to the function value.

•• REMARK   Learning the table of values at the right is worth the effort because doing so will increase both your efficiency and your confidence when working in trigonometry. Below is a pattern for the sine function that may help you remember the values.

θ	0°	30°	45°	60°	90°
$\sin \theta$	$\frac{\sqrt{0}}{2}$	$\frac{\sqrt{1}}{2}$	$\frac{\sqrt{2}}{2}$	$\frac{\sqrt{3}}{2}$	$\frac{\sqrt{4}}{2}$

Reverse the order to get cosine values of the same angles.

Using reference angles and the special angles discussed in the preceding section enables you to greatly extend the scope of *exact* trigonometric function values. For example, knowing the function values of 30° means that you know the function values of all angles for which 30° is a reference angle. For convenience, the table below shows the exact values of the sine, cosine, and tangent functions of special angles and quadrantal angles.

Trigonometric Values of Common Angles

θ (degrees)	0°	30°	45°	60°	90°	180°	270°
θ (radians)	0	$\frac{\pi}{6}$	$\frac{\pi}{4}$	$\frac{\pi}{3}$	$\frac{\pi}{2}$	π	$\frac{3\pi}{2}$
$\sin \theta$	0	$\frac{1}{2}$	$\frac{\sqrt{2}}{2}$	$\frac{\sqrt{3}}{2}$	1	0	-1
$\cos \theta$	1	$\frac{\sqrt{3}}{2}$	$\frac{\sqrt{2}}{2}$	$\frac{1}{2}$	0	-1	0
$\tan \theta$	0	$\frac{\sqrt{3}}{3}$	1	$\sqrt{3}$	Undef.	0	Undef.

EXAMPLE 5   **Using Reference Angles**

See LarsonPrecalculus.com for an interactive version of this type of example.

Evaluate each trigonometric function.

a. $\cos \dfrac{4\pi}{3}$ **b.** $\tan(-210°)$ **c.** $\csc \dfrac{11\pi}{4}$

Solution

a. Because $\theta = 4\pi/3$ lies in Quadrant III, the reference angle is

$$\theta' = \frac{4\pi}{3} - \pi = \frac{\pi}{3}$$

as shown at the right. The cosine is negative in Quadrant III, so

$$\cos \frac{4\pi}{3} = (-)\cos \frac{\pi}{3}$$

$$= -\frac{1}{2}.$$

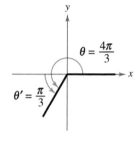

b. Because $-210° + 360° = 150°$, it follows that $-210°$ is coterminal with the second-quadrant angle $150°$. So, the reference angle is

$$\theta' = 180° - 150°$$

$$= 30°$$

as shown at the right. The tangent is negative in Quadrant II, so

$$\tan(-210°) = (-)\tan 30°$$

$$= -\frac{\sqrt{3}}{3}.$$

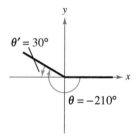

c. Because $(11\pi/4) - 2\pi = 3\pi/4$, it follows that $11\pi/4$ is coterminal with the second-quadrant angle $3\pi/4$. So, the reference angle is

$$\theta' = \pi - \frac{3\pi}{4} = \frac{\pi}{4}$$

as shown at the right. The cosecant is positive in Quadrant II, so

$$\csc \frac{11\pi}{4} = (+)\csc \frac{\pi}{4}$$

$$= \frac{1}{\sin(\pi/4)}$$

$$= \sqrt{2}.$$

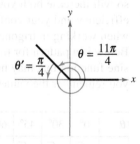

✓ *Checkpoint* *Audio-video solution in English & Spanish at LarsonPrecalculus.com*

Evaluate each trigonometric function.

a. $\sin \dfrac{7\pi}{4}$ **b.** $\cos(-120°)$ **c.** $\tan \dfrac{11\pi}{6}$

> EXAMPLE 6   **Using Trigonometric Identities**

Let θ be an angle in Quadrant II such that $\sin \theta = \frac{1}{3}$. Find (a) $\cos \theta$ and (b) $\tan \theta$ by using trigonometric identities.

Solution

REMARK The fundamental trigonometric identities listed in the preceding section (for an acute angle θ) are also valid when θ is any angle in the domain of the function.

a. Using the Pythagorean identity $\sin^2 \theta + \cos^2 \theta = 1$, you obtain

$$\left(\frac{1}{3}\right)^2 + \cos^2 \theta = 1 \quad \Longrightarrow \quad \cos^2 \theta = 1 - \frac{1}{9} = \frac{8}{9}.$$

You know that $\cos \theta < 0$ in Quadrant II, so use the negative root to obtain

$$\cos \theta = -\frac{\sqrt{8}}{\sqrt{9}} = -\frac{2\sqrt{2}}{3}.$$

b. Using the trigonometric identity $\tan \theta = \dfrac{\sin \theta}{\cos \theta}$, you obtain

$$\tan \theta = \frac{1/3}{-2\sqrt{2}/3} = -\frac{1}{2\sqrt{2}} = -\frac{\sqrt{2}}{4}.$$

✓ *Checkpoint* ◀))) *Audio-video solution in English & Spanish at LarsonPrecalculus.com*

Let θ be an angle in Quadrant III such that $\sin \theta = -\frac{4}{5}$. Find (a) $\cos \theta$ and (b) $\tan \theta$ by using trigonometric identities.

> EXAMPLE 7   **Using a Calculator**

Use a calculator to evaluate each trigonometric function.

a. $\cot 410°$ **b.** $\sin(-7)$ **c.** $\sec \dfrac{\pi}{9}$

Solution

Function	Mode	Calculator Keystrokes	Display
a. $\cot 410°$	Degree	(TAN (410)) x⁻¹ ENTER	0.8390996
b. $\sin(-7)$	Radian	SIN (((−) 7) ENTER	−0.6569866
c. $\sec(\pi/9)$	Radian	(COS (π ÷ 9)) x⁻¹ ENTER	1.0641778

✓ *Checkpoint* ◀)))  *Audio-video solution in English & Spanish at LarsonPrecalculus.com*

Use a calculator to evaluate each trigonometric function.

a. $\tan 119°$ **b.** $\csc 5$ **c.** $\cos \dfrac{\pi}{5}$

Summarize   (Section 4.4)

1. State the definitions of the trigonometric functions of any angle *(page 288)*. For examples of evaluating trigonometric functions, see Examples 1–3.

2. Explain how to use a reference angle *(page 290)*. For an example of finding reference angles, see Example 4.

3. Explain how to evaluate a trigonometric function of a real number *(page 291)*. For examples of evaluating trigonometric functions of real numbers, see Examples 5–7.

4.4 Exercises

Vocabulary: Fill in the blanks.

In Exercises 1–6, let θ be an angle in standard position with (x, y) a point on the terminal side of θ and $r = \sqrt{x^2 + y^2} \neq 0$.

1. $\sin \theta =$ _____

2. $\dfrac{r}{y} =$ _____

3. $\tan \theta =$ _____

4. $\sec \theta =$ _____

5. $\dfrac{x}{r} =$ _____

6. $\dfrac{x}{y} =$ _____

7. Because $r = \sqrt{x^2 + y^2}$ cannot be _____, the sine and cosine functions are _____ for any real value of θ.

8. The acute angle formed by the terminal side of an angle θ in standard position and the horizontal axis is the _____ angle of θ and is denoted by θ'.

Skills and Applications

Evaluating Trigonometric Functions
In Exercises 9–12, find the exact values of the six trigonometric functions of each angle θ.

9. (a) 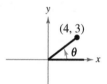 (b)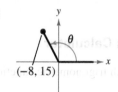

10. (a) 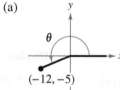 (b)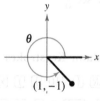

11. (a) 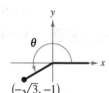 (b)

12. (a) 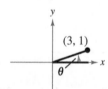 (b)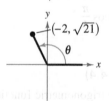

Evaluating Trigonometric Functions In Exercises 13–18, the point is on the terminal side of an angle in standard position. Find the exact values of the six trigonometric functions of the angle.

13. $(5, 12)$

14. $(8, 15)$

15. $(-5, -2)$

16. $(-4, 10)$

17. $(-5.4, 7.2)$

18. $\left(3\frac{1}{2}, -2\sqrt{15}\right)$

Determining a Quadrant In Exercises 19–22, determine the quadrant in which θ lies.

19. $\sin \theta > 0, \quad \cos \theta > 0$

20. $\sin \theta < 0, \quad \cos \theta < 0$

21. $\csc \theta > 0, \quad \tan \theta < 0$

22. $\sec \theta > 0, \quad \cot \theta < 0$

Evaluating Trigonometric Functions
In Exercises 23–32, find the exact values of the remaining trigonometric functions of θ satisfying the given conditions.

23. $\tan \theta = \frac{15}{8}, \quad \sin \theta > 0$

24. $\cos \theta = \frac{8}{17}, \quad \tan \theta < 0$

25. $\sin \theta = 0.6, \quad \theta$ lies in Quadrant II.

26. $\cos \theta = -0.8, \quad \theta$ lies in Quadrant III.

27. $\cot \theta = -3, \quad \cos \theta > 0$

28. $\csc \theta = 4, \quad \cot \theta < 0$

29. $\cos \theta = 0, \quad \csc \theta = 1$

30. $\sin \theta = 0, \quad \sec \theta = -1$

31. $\cot \theta$ is undefined, $\dfrac{\pi}{2} \leq \theta \leq \dfrac{3\pi}{2}$

32. $\tan \theta$ is undefined, $\pi \leq \theta \leq 2\pi$

An Angle Formed by a Line Through the Origin In Exercises 33–36, the terminal side of θ lies on the given line in the specified quadrant. Find the exact values of the six trigonometric functions of θ by finding a point on the line.

Line	Quadrant
33. $y = -x$	II
34. $y = \frac{1}{3}x$	III
35. $2x - y = 0$	I
36. $4x + 3y = 0$	IV

Trigonometric Function of a Quadrantal Angle In Exercises 37–46, evaluate the trigonometric function of the quadrantal angle, if possible.

37. $\sin 0$

38. $\csc \dfrac{3\pi}{2}$

39. $\sec \dfrac{3\pi}{2}$

40. $\sec \pi$

41. $\sin \dfrac{\pi}{2}$

42. $\cot 0$

43. $\csc \pi$

44. $\cot \dfrac{\pi}{2}$

45. $\cos \dfrac{9\pi}{2}$

46. $\tan\left(-\dfrac{\pi}{2}\right)$

 Finding a Reference Angle In Exercises 47–54, find the reference angle θ'. Sketch θ in standard position and label θ'.

47. $\theta = 160°$

48. $\theta = 309°$

49. $\theta = -125°$

50. $\theta = -215°$

51. $\theta = \dfrac{2\pi}{3}$

52. $\theta = \dfrac{7\pi}{6}$

53. $\theta = 4.8$

54. $\theta = 12.9$

 Using a Reference Angle In Exercises 55–68, evaluate the sine, cosine, and tangent of the angle without using a calculator.

55. $225°$

56. $300°$

57. $750°$

58. $675°$

59. $-120°$

60. $-570°$

61. $\dfrac{2\pi}{3}$

62. $\dfrac{3\pi}{4}$

63. $-\dfrac{\pi}{6}$

64. $-\dfrac{2\pi}{3}$

65. $\dfrac{11\pi}{4}$

66. $\dfrac{13\pi}{6}$

67. $-\dfrac{17\pi}{6}$

68. $-\dfrac{23\pi}{4}$

 Using a Trigonometric Identity In Exercises 69–74, use the function value to find the indicated trigonometric value in the specified quadrant.

	Function Value	Quadrant	Trigonometric Value
69.	$\sin \theta = -\dfrac{3}{5}$	IV	$\cos \theta$
70.	$\cot \theta = -3$	II	$\csc \theta$
71.	$\tan \theta = \dfrac{3}{2}$	III	$\sec \theta$
72.	$\csc \theta = -2$	IV	$\cot \theta$
73.	$\cos \theta = \dfrac{5}{8}$	I	$\csc \theta$
74.	$\sec \theta = -\dfrac{9}{4}$	III	$\cot \theta$

 Using a Calculator In Exercises 75–90, use a calculator to evaluate the trigonometric function. Round your answer to four decimal places. (Be sure the calculator is in the correct mode.)

75. $\sin 10°$

76. $\tan 304°$

77. $\cos(-110°)$

78. $\sin(-330°)$

79. $\cot 178°$

80. $\sec 72°$

81. $\csc 405°$

82. $\cot(-560°)$

83. $\tan \dfrac{\pi}{9}$

84. $\cos \dfrac{2\pi}{7}$

85. $\sec \dfrac{11\pi}{8}$

86. $\csc \dfrac{15\pi}{4}$

87. $\sin(-0.65)$

88. $\cos 1.35$

89. $\csc(-10)$

90. $\sec(-4.6)$

 Solving for θ In Exercises 91–96, find two solutions of each equation. Give your answers in degrees ($0° \le \theta < 360°$) and in radians ($0 \le \theta < 2\pi$). Do not use a calculator.

91. (a) $\sin \theta = \dfrac{1}{2}$

(b) $\sin \theta = -\dfrac{1}{2}$

92. (a) $\cos \theta = \dfrac{\sqrt{2}}{2}$

(b) $\cos \theta = -\dfrac{\sqrt{2}}{2}$

93. (a) $\cos \theta = \dfrac{1}{2}$

(b) $\sec \theta = 2$

94. (a) $\sin \theta = \dfrac{\sqrt{3}}{2}$

(b) $\csc \theta = \dfrac{2\sqrt{3}}{3}$

95. (a) $\tan \theta = 1$

(b) $\cot \theta = -\sqrt{3}$

96. (a) $\cot \theta = 0$

(b) $\sec \theta = -\sqrt{2}$

97. Distance An airplane, flying at an altitude of 6 miles, is on a flight path that passes directly over an observer (see figure). Let θ be the angle of elevation from the observer to the plane. Find the distance d from the observer to the plane when (a) $\theta = 30°$, (b) $\theta = 90°$, and (c) $\theta = 120°$.

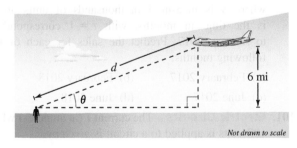

Not drawn to scale

98. Harmonic Motion The displacement from equilibrium of an oscillating weight suspended by a spring is given by $y(t) = 2 \cos 6t$, where y is the displacement in centimeters and t is the time in seconds. Find the displacement when (a) $t = 0$, (b) $t = \dfrac{1}{4}$, and (c) $t = \dfrac{1}{2}$.

• • 99. Temperature • • • • • • • • • • • • • • •

The table shows the average high temperatures (in degrees Fahrenheit) in Boston, Massachusetts (*B*), and Fairbanks, Alaska (*F*), for selected months in 2015. *(Source: U.S. Climate Data)*

Month	Boston, B	Fairbanks, F
January	33	1
March	41	31
June	72	71
August	83	62
November	56	17

Spreadsheet at LarsonPrecalculus.com

(a) Use the *regression* feature of a graphing utility to find a model of the form

$$y = a \sin(bt + c) + d$$

for each city. Let *t* represent the month, with *t* = 1 corresponding to January.

(b) Use the models from part (a) to estimate the monthly average high temperatures for the two cities in February, April, May, July, September, October, and December.

(c) Use a graphing utility to graph both models in the same viewing window. Compare the temperatures for the two cities.

100. Sales A company that produces snowboards forecasts monthly sales over the next 2 years to be

$$S = 23.1 + 0.442t + 4.3 \cos \frac{\pi t}{6}$$

where *S* is measured in thousands of units and *t* is the time in months, with *t* = 1 corresponding to January 2017. Predict the sales for each of the following months.

(a) February 2017   (b) February 2018

(c) June 2017   (d) June 2018

101. Electric Circuits The current *I* (in amperes) when 100 volts is applied to a circuit is given by

$$I = 5e^{-2t} \sin t$$

where *t* is the time (in seconds) after the voltage is applied. Approximate the current at *t* = 0.7 second after the voltage is applied.

102. **HOW DO YOU SEE IT?** Consider an angle in standard position with *r* = 12 centimeters, as shown in the figure. Describe the changes in the values of *x*, *y*, sin *θ*, cos *θ*, and tan *θ* as *θ* increases continuously from 0° to 90°.

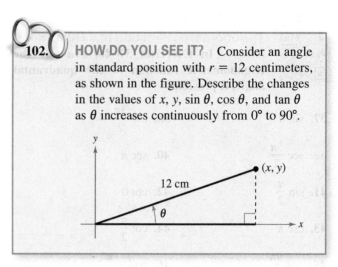

Exploration

True or False? In Exercises 103 and 104, determine whether the statement is true or false. Justify your answer.

103. In each of the four quadrants, the signs of the secant function and the sine function are the same.

104. The reference angle for an angle *θ* (in degrees) is the angle $\theta' = 360°n - \theta$, where *n* is an integer and $0° \le \theta' \le 360°$.

105. Writing Write a short essay explaining to a classmate how to evaluate the six trigonometric functions of any angle *θ* in standard position. Include an explanation of reference angles and how to use them, the signs of the functions in each of the four quadrants, and the trigonometric values of common angles. Include figures or diagrams in your essay.

106. Think About It The figure shows point *P*(*x*, *y*) on a unit circle and right triangle *OAP*.

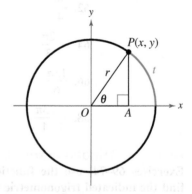

(a) Find sin *t* and cos *t* using the unit circle definitions of sine and cosine (from Section 4.2).

(b) What is the value of *r*? Explain.

(c) Use the definitions of sine and cosine given in this section to find sin *θ* and cos *θ*. Write your answers in terms of *x* and *y*.

(d) Based on your answers to parts (a) and (c), what can you conclude?

4.5 Graphs of Sine and Cosine Functions

■ Sketch the graphs of basic sine and cosine functions.
■ Use amplitude and period to help sketch the graphs of sine and cosine functions.
■ Sketch translations of the graphs of sine and cosine functions.
■ Use sine and cosine functions to model real-life data.

Graphs of sine and cosine functions have many scientific applications. For example, in Exercise 80 on page 306, you will use the graph of a sine function to analyze airflow during a respiratory cycle.

Basic Sine and Cosine Curves

In this section, you will study techniques for sketching the graphs of the sine and cosine functions. The graph of the sine function, shown in Figure 4.32, is a **sine curve.** In the figure, the black portion of the graph represents one period of the function and is **one cycle** of the sine curve. The gray portion of the graph indicates that the basic sine curve repeats indefinitely to the left and right. Figure 4.33 shows the graph of the cosine function.

Recall from Section 4.4 that the domain of the sine and cosine functions is the set of all real numbers. Moreover, the range of each function is the interval $[-1, 1]$, and each function has a period of 2π. This information is consistent with the basic graphs shown in Figures 4.32 and 4.33.

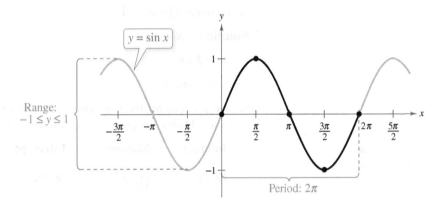

Figure 4.32

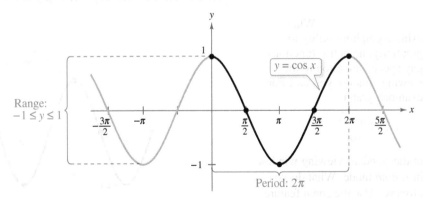

Figure 4.33

Note in Figures 4.32 and 4.33 that the sine curve is symmetric with respect to the *origin*, whereas the cosine curve is symmetric with respect to the *y-axis*. These properties of symmetry follow from the fact that the sine function is odd and the cosine function is even.

To sketch the graphs of the basic sine and cosine functions, it helps to note five **key points** in one period of each graph: the *intercepts, maximum points,* and *minimum points* (see graphs below).

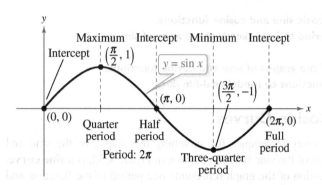

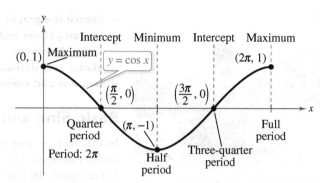

EXAMPLE 1 **Using Key Points to Sketch a Sine Curve**

See LarsonPrecalculus.com for an interactive version of this type of example.

Sketch the graph of

$$y = 2 \sin x$$

on the interval $[-\pi, 4\pi]$.

Solution Note that

$$y = 2 \sin x$$
$$= 2(\sin x).$$

So, the y-values for the key points have twice the magnitude of those on the graph of $y = \sin x$. Divide the period 2π into four equal parts to obtain the key points

Intercept	Maximum	Intercept	Minimum	Intercept
$(0, 0)$,	$\left(\dfrac{\pi}{2}, 2\right)$,	$(\pi, 0)$,	$\left(\dfrac{3\pi}{2}, -2\right)$, and	$(2\pi, 0)$.

By connecting these key points with a smooth curve and extending the curve in both directions over the interval $[-\pi, 4\pi]$, you obtain the graph below.

▷ **TECHNOLOGY** When using a graphing utility to graph trigonometric functions, pay special attention to the viewing window you use. For example, graph

$$y = \frac{\sin 10x}{10}$$

in the standard viewing window in *radian* mode. What do you observe? Use the *zoom* feature to find a viewing window that displays a good view of the graph.

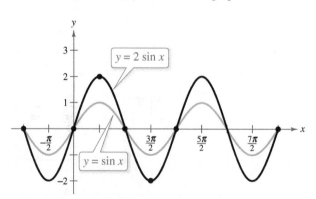

✓ **Checkpoint** ◀))) *Audio-video solution in English & Spanish at LarsonPrecalculus.com*

Sketch the graph of

$$y = 2 \cos x$$

on the interval $\left[-\dfrac{\pi}{2}, \dfrac{9\pi}{2}\right]$.

Amplitude and Period

In the rest of this section, you will study the effect of each of the constants a, b, c, and d on the graphs of equations of the forms

$$y = d + a \sin(bx - c)$$

and

$$y = d + a \cos(bx - c).$$

A quick review of the transformations you studied in Section 1.7 will help in this investigation.

The constant factor a in $y = a \sin x$ and $y = a \cos x$ acts as a *scaling factor*—a *vertical stretch* or *vertical shrink* of the basic curve. When $|a| > 1$, the basic curve is stretched, and when $0 < |a| < 1$, the basic curve is shrunk. The result is that the graphs of $y = a \sin x$ and $y = a \cos x$ range between $-a$ and a instead of between -1 and 1. The absolute value of a is the **amplitude** of the function. The range of the function for $a > 0$ is $-a \leq y \leq a$.

> ### Definition of the Amplitude of Sine and Cosine Curves
>
> The **amplitude** of $y = a \sin x$ and $y = a \cos x$ represents half the distance between the maximum and minimum values of the function and is given by
>
> Amplitude $= |a|$.

EXAMPLE 2 **Scaling: Vertical Shrinking and Stretching**

In the same coordinate plane, sketch the graph of each function.

a. $y = \dfrac{1}{2} \cos x$

b. $y = 3 \cos x$

Solution

a. The amplitude of $y = \frac{1}{2} \cos x$ is $\frac{1}{2}$, so the maximum value is $\frac{1}{2}$ and the minimum value is $-\frac{1}{2}$. Divide one cycle, $0 \leq x \leq 2\pi$, into four equal parts to obtain the key points

Maximum	Intercept	Minimum	Intercept	Maximum
$\left(0, \dfrac{1}{2}\right),$	$\left(\dfrac{\pi}{2}, 0\right),$	$\left(\pi, -\dfrac{1}{2}\right),$	$\left(\dfrac{3\pi}{2}, 0\right),$	and $\left(2\pi, \dfrac{1}{2}\right).$

b. A similar analysis shows that the amplitude of $y = 3 \cos x$ is 3, and the key points are

Maximum	Intercept	Minimum	Intercept	Maximum
$(0, 3),$	$\left(\dfrac{\pi}{2}, 0\right),$	$(\pi, -3),$	$\left(\dfrac{3\pi}{2}, 0\right),$	and $(2\pi, 3).$

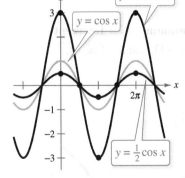

Figure 4.34

Figure 4.34 shows the graphs of these two functions. Notice that the graph of $y = \frac{1}{2} \cos x$ is a vertical *shrink* of the graph of $y = \cos x$ and the graph of $y = 3 \cos x$ is a vertical *stretch* of the graph of $y = \cos x$.

✓ **Checkpoint** ◀))) *Audio-video solution in English & Spanish at LarsonPrecalculus.com*

In the same coordinate plane, sketch the graph of each function.

a. $y = \dfrac{1}{3} \sin x$

b. $y = 3 \sin x$

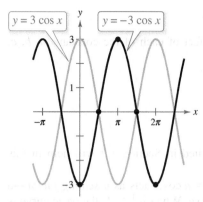

Figure 4.35

You know from Section 1.7 that the graph of $y = -f(x)$ is a **reflection** in the x-axis of the graph of $y = f(x)$. For example, the graph of $y = -3 \cos x$ is a reflection of the graph of $y = 3 \cos x$, as shown in Figure 4.35.

Next, consider the effect of the positive real number b on the graphs of $y = a \sin bx$ and $y = a \cos bx$. For example, compare the graphs of $y = a \sin x$ and $y = a \sin bx$. The graph of $y = a \sin x$ completes one cycle from $x = 0$ to $x = 2\pi$, so it follows that the graph of $y = a \sin bx$ completes one cycle from $x = 0$ to $x = 2\pi/b$.

Period of Sine and Cosine Functions

Let b be a positive real number. The **period** of $y = a \sin bx$ and $y = a \cos bx$ is given by

$$\text{Period} = \frac{2\pi}{b}.$$

Note that when $0 < b < 1$, the period of $y = a \sin bx$ is greater than 2π and represents a *horizontal stretch* of the basic curve. Similarly, when $b > 1$, the period of $y = a \sin bx$ is less than 2π and represents a *horizontal shrink* of the basic curve. These two statements are also true for $y = a \cos bx$. When b is negative, rewrite the function using the identity $\sin(-x) = -\sin x$ or $\cos(-x) = \cos x$.

EXAMPLE 3   Scaling: Horizontal Stretching

Sketch the graph of

$$y = \sin \frac{x}{2}.$$

Solution   The amplitude is 1. Moreover, $b = \frac{1}{2}$, so the period is

$$\frac{2\pi}{b} = \frac{2\pi}{\frac{1}{2}} = 4\pi. \qquad \text{Substitute for } b.$$

Now, divide the period-interval $[0, 4\pi]$ into four equal parts using the values π, 2π, and 3π to obtain the key points

Intercept	Maximum	Intercept	Minimum	Intercept
$(0, 0)$,	$(\pi, 1)$,	$(2\pi, 0)$,	$(3\pi, -1)$, and	$(4\pi, 0)$.

The graph is shown below.

- - - - - - - - - - - - ▷

··REMARK In general, to divide a period-interval into four equal parts, successively add "period/4," starting with the left endpoint of the interval. For example, for the period-interval $[-\pi/6, \pi/2]$ of length $2\pi/3$, you would successively add

$$\frac{2\pi/3}{4} = \frac{\pi}{6}$$

to obtain $-\pi/6$, 0, $\pi/6$, $\pi/3$, and $\pi/2$ as the x-values for the key points on the graph.

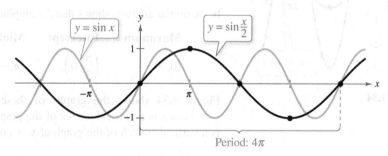

Period: 4π

✓ **Checkpoint** ◀))) Audio-video solution in English & Spanish at LarsonPrecalculus.com

Sketch the graph of

$$y = \cos \frac{x}{3}.$$

Translations of Sine and Cosine Curves

The constant c in the equations

$$y = a \sin(bx - c) \quad \text{and} \quad y = a \cos(bx - c)$$

results in *horizontal translations* (shifts) of the basic curves. For example, compare the graphs of $y = a \sin bx$ and $y = a \sin(bx - c)$. The graph of $y = a \sin(bx - c)$ completes one cycle from $bx - c = 0$ to $bx - c = 2\pi$. Solve for x to find that the interval for one cycle is

Left endpoint   Right endpoint

$$\overbrace{} \qquad \overbrace{}$$

$$\frac{c}{b} \le x \le \frac{c}{b} + \frac{2\pi}{b}.$$

$$\underbrace{}_{\text{Period}}$$

This implies that the period of $y = a \sin(bx - c)$ is $2\pi/b$, and the graph of $y = a \sin bx$ is shifted by an amount c/b. The number c/b is the **phase shift.**

Graphs of Sine and Cosine Functions

The graphs of $y = a \sin(bx - c)$ and $y = a \cos(bx - c)$ have the characteristics below. (Assume $b > 0$.)

$$\text{Amplitude} = |a| \qquad \text{Period} = \frac{2\pi}{b}$$

The left and right endpoints of a one-cycle interval can be determined by solving the equations $bx - c = 0$ and $bx - c = 2\pi$.

EXAMPLE 4   **Horizontal Translation**

Analyze the graph of $y = \dfrac{1}{2} \sin\left(x - \dfrac{\pi}{3}\right)$.

Algebraic Solution

The amplitude is $\frac{1}{2}$ and the period is $2\pi/1 = 2\pi$. Solving the equations

$$x - \frac{\pi}{3} = 0 \quad \implies \quad x = \frac{\pi}{3}$$

and

$$x - \frac{\pi}{3} = 2\pi \quad \implies \quad x = \frac{7\pi}{3}$$

shows that the interval $[\pi/3, 7\pi/3]$ corresponds to one cycle of the graph. Dividing this interval into four equal parts produces the key points

Intercept	Maximum	Intercept	Minimum	Intercept
$\left(\dfrac{\pi}{3}, 0\right),$	$\left(\dfrac{5\pi}{6}, \dfrac{1}{2}\right),$	$\left(\dfrac{4\pi}{3}, 0\right),$	$\left(\dfrac{11\pi}{6}, -\dfrac{1}{2}\right),$ and	$\left(\dfrac{7\pi}{3}, 0\right).$

Graphical Solution

Use a graphing utility set in *radian* mode to graph $y = (1/2) \sin[x - (\pi/3)]$, as shown in the figure below. Use the *minimum*, *maximum*, and *zero* or *root* features of the graphing utility to approximate the key points $(1.05, 0)$, $(2.62, 0.5)$, $(4.19, 0)$, $(5.76, -0.5)$, and $(7.33, 0)$.

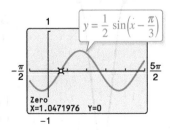

✓ **Checkpoint**))) *Audio-video solution in English & Spanish at LarsonPrecalculus.com*

Analyze the graph of $y = 2 \cos\left(x - \dfrac{\pi}{2}\right)$.

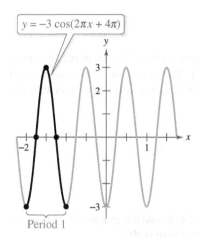

$y = -3 \cos(2\pi x + 4\pi)$

Figure 4.36

EXAMPLE 5   Horizontal Translation

Sketch the graph of

$$y = -3 \cos(2\pi x + 4\pi).$$

Solution   The amplitude is 3 and the period is $2\pi/2\pi = 1$. Solving the equations

$$2\pi x + 4\pi = 0$$
$$2\pi x = -4\pi$$
$$x = -2$$

and

$$2\pi x + 4\pi = 2\pi$$
$$2\pi x = -2\pi$$
$$x = -1$$

shows that the interval $[-2, -1]$ corresponds to one cycle of the graph. Dividing this interval into four equal parts produces the key points

Minimum	Intercept	Maximum	Intercept	Minimum
$(-2, -3),$	$\left(-\dfrac{7}{4}, 0\right),$	$\left(-\dfrac{3}{2}, 3\right),$	$\left(-\dfrac{5}{4}, 0\right),$ and	$(-1, -3).$

Figure 4.36 shows the graph.

✓ *Checkpoint*  Audio-video solution in English & Spanish at LarsonPrecalculus.com

Sketch the graph of

$$y = -\frac{1}{2} \sin(\pi x + \pi).$$

The constant d in the equations

$$y = d + a \sin(bx - c) \quad \text{and} \quad y = d + a \cos(bx - c)$$

results in *vertical translations* of the basic curves. The shift is d units up for $d > 0$ and d units down for $d < 0$. In other words, the graph oscillates about the horizontal line $y = d$ instead of about the x-axis.

EXAMPLE 6   Vertical Translation

Sketch the graph of

$$y = 2 + 3 \cos 2x.$$

Solution   The amplitude is 3 and the period is $2\pi/2 = \pi$. The key points over the interval $[0, \pi]$ are

$$(0, 5), \quad \left(\frac{\pi}{4}, 2\right), \quad \left(\frac{\pi}{2}, -1\right), \quad \left(\frac{3\pi}{2}, 2\right), \quad \text{and} \quad (\pi, 5).$$

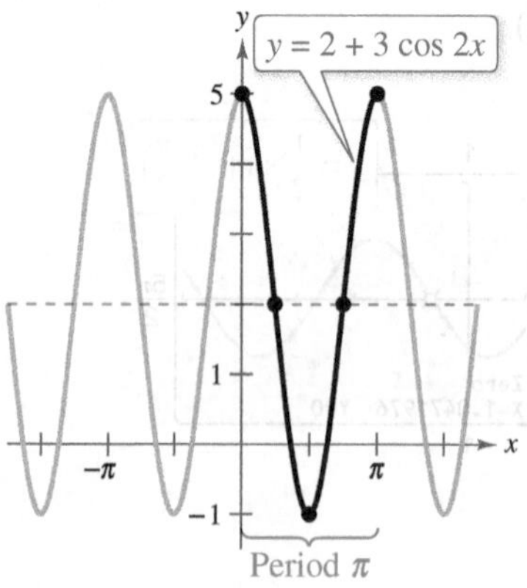

Figure 4.37 shows the graph. Compared with the graph of $f(x) = 3 \cos 2x$, the graph of $y = 2 + 3 \cos 2x$ is shifted up two units.

✓ *Checkpoint*  Audio-video solution in English & Spanish at LarsonPrecalculus.com

Sketch the graph of

$$y = 2 \cos x - 5.$$

$y = 2 + 3 \cos 2x$

Figure 4.37

Mathematical Modeling

DATA	Time, t	Depth, y
	0	3.4
	2	8.7
	4	11.3
	6	9.1
	8	3.8
	10	0.1
	12	1.2

Spreadsheet at LarsonPrecalculus.com

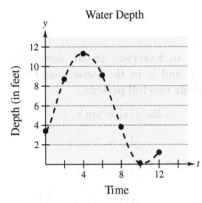

Figure 4.38

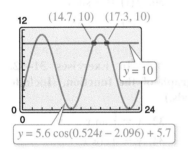

$y = 5.6 \cos(0.524t - 2.096) + 5.7$

Figure 4.39

EXAMPLE 7 **Finding a Trigonometric Model**

The table shows the depths (in feet) of the water at the end of a dock every two hours from midnight to noon, where $t = 0$ corresponds to midnight. (a) Use a trigonometric function to model the data. (b) Find the depths at 9 A.M. and 3 P.M. (c) A boat needs at least 10 feet of water to moor at the dock. During what times in the afternoon can it safely dock?

Solution

a. Begin by graphing the data, as shown in Figure 4.38. Use either a sine or cosine model. For example, a cosine model has the form $y = a \cos(bt - c) + d$. The difference between the maximum value and the minimum value is twice the amplitude of the function. So, the amplitude is

$$a = \tfrac{1}{2}[(\text{maximum depth}) - (\text{minimum depth})] = \tfrac{1}{2}(11.3 - 0.1) = 5.6.$$

The cosine function completes one half of a cycle between the times at which the maximum and minimum depths occur. So, the period p is

$$p = 2[(\text{time of min. depth}) - (\text{time of max. depth})] = 2(10 - 4) = 12$$

which implies that $b = 2\pi/p \approx 0.524$. The maximum depth occurs 4 hours after midnight, so consider the left endpoint to be $c/b = 4$, which means that $c \approx 4(0.524) = 2.096$. Moreover, the average depth is $\tfrac{1}{2}(11.3 + 0.1) = 5.7$, so it follows that $d = 5.7$. Substituting the values of a, b, c, and d into the cosine model yields $y = 5.6 \cos(0.524t - 2.096) + 5.7$.

b. The depths at 9 A.M. and 3 P.M. are

$$y = 5.6 \cos(0.524 \cdot 9 - 2.096) + 5.7 \approx 0.84 \text{ foot} \qquad \text{9 A.M.}$$

and

$$y = 5.6 \cos(0.524 \cdot 15 - 2.096) + 5.7 \approx 10.56 \text{ feet.} \qquad \text{3 P.M.}$$

c. Using a graphing utility, graph the model with the line $y = 10$. Using the *intersect* feature, determine that the depth is at least 10 feet between 2:42 P.M. ($t \approx 14.7$) and 5:18 P.M. ($t \approx 17.3$), as shown in Figure 4.39.

✓ **Checkpoint**  Audio-video solution in English & Spanish at LarsonPrecalculus.com

Find a sine model for the data in Example 7.

Summarize (Section 4.5)

1. Explain how to sketch the graphs of basic sine and cosine functions *(page 297)*. For an example of sketching the graph of a sine function, see Example 1.

2. Explain how to use amplitude and period to help sketch the graphs of sine and cosine functions *(pages 299 and 300)*. For examples of using amplitude and period to sketch graphs of sine and cosine functions, see Examples 2 and 3.

3. Explain how to sketch translations of the graphs of sine and cosine functions *(page 301)*. For examples of translating the graphs of sine and cosine functions, see Examples 4–6.

4. Give an example of using a sine or cosine function to model real-life data *(page 303, Example 7)*.

4.5 Exercises

Vocabulary: Fill in the blanks.

1. One period of a sine or cosine function is one _____ of the sine or cosine curve.

2. The _____ of a sine or cosine curve represents half the distance between the maximum and minimum values of the function.

3. For the function $y = a \sin(bx - c)$, $\dfrac{c}{b}$ represents the _____ _____ of one cycle of the graph of the function.

4. For the function $y = d + a \cos(bx - c)$, d represents a _____ _____ of the basic curve.

Skills and Applications

 Finding the Period and Amplitude In Exercises 5–12, find the period and amplitude.

5. $y = 2 \sin 5x$

6. $y = 3 \cos 2x$

7. $y = \dfrac{3}{4} \cos \dfrac{\pi x}{2}$

8. $y = -5 \sin \dfrac{\pi x}{3}$

9. $y = -\dfrac{1}{2} \sin \dfrac{5x}{4}$

10. $y = \dfrac{1}{4} \sin \dfrac{x}{6}$

11. $y = -\dfrac{5}{3} \cos \dfrac{\pi x}{12}$

12. $y = -\dfrac{2}{5} \cos 10\pi x$

Describing the Relationship Between Graphs In Exercises 13–24, describe the relationship between the graphs of f and g. Consider amplitude, period, and shifts.

13. $f(x) = \cos x$
    $g(x) = \cos 5x$

14. $f(x) = \sin x$
    $g(x) = 2 \sin x$

15. $f(x) = \cos 2x$
    $g(x) = -\cos 2x$

16. $f(x) = \sin 3x$
    $g(x) = \sin(-3x)$

17. $f(x) = \sin x$
    $g(x) = \sin(x - \pi)$

18. $f(x) = \cos x$
    $g(x) = \cos(x + \pi)$

19. $f(x) = \sin 2x$
    $g(x) = 3 + \sin 2x$

20. $f(x) = \cos 4x$
    $g(x) = -2 + \cos 4x$

21.
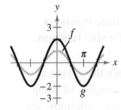

22.
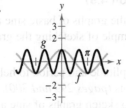

23.
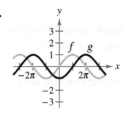

24.
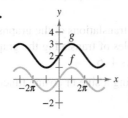

 Sketching Graphs of Sine or Cosine Functions In Exercises 25–30, sketch the graphs of f and g in the same coordinate plane. (Include two full periods.)

25. $f(x) = \sin x$
    $g(x) = \sin \dfrac{x}{3}$

26. $f(x) = \sin x$
    $g(x) = 4 \sin x$

27. $f(x) = \cos x$
    $g(x) = 2 + \cos x$

28. $f(x) = \cos x$
    $g(x) = \cos\left(x + \dfrac{\pi}{2}\right)$

29. $f(x) = -\cos x$
    $g(x) = -\cos(x - \pi)$

30. $f(x) = -\sin x$
    $g(x) = -3 \sin x$

 Sketching the Graph of a Sine or Cosine Function In Exercises 31–52, sketch the graph of the function. (Include two full periods.)

31. $y = 5 \sin x$

32. $y = \frac{1}{4} \sin x$

33. $y = \frac{1}{3} \cos x$

34. $y = 4 \cos x$

35. $y = \cos \dfrac{x}{2}$

36. $y = \sin 4x$

37. $y = \cos 2\pi x$

38. $y = \sin \dfrac{\pi x}{4}$

39. $y = -\sin \dfrac{2\pi x}{3}$

40. $y = 10 \cos \dfrac{\pi x}{6}$

41. $y = \cos\left(x - \dfrac{\pi}{2}\right)$

42. $y = \sin(x - 2\pi)$

43. $y = 3 \sin(x + \pi)$

44. $y = -4 \cos\left(x + \dfrac{\pi}{4}\right)$

45. $y = 2 - \sin \dfrac{2\pi x}{3}$

46. $y = -3 + 5 \cos \dfrac{\pi t}{12}$

47. $y = 2 + 5 \cos 6\pi x$

48. $y = 2 \sin 3x + 5$

49. $y = 3 \sin(x + \pi) - 3$

50. $y = -3 \sin(6x + \pi)$

51. $y = \dfrac{2}{3} \cos\left(\dfrac{x}{2} - \dfrac{\pi}{4}\right)$

52. $y = 4 \cos\left(\pi x + \dfrac{\pi}{2}\right) - 1$

Describing a Transformation In Exercises 53–58, g is related to a parent function $f(x) = \sin(x)$ or $f(x) = \cos(x)$. (a) Describe the sequence of transformations from f to g. (b) Sketch the graph of g. (c) Use function notation to write g in terms of f.

53. $g(x) = \sin(4x - \pi)$

54. $g(x) = \sin(2x + \pi)$

55. $g(x) = \cos\left(x - \dfrac{\pi}{2}\right) + 2$

56. $g(x) = 1 + \cos(x + \pi)$

57. $g(x) = 2\sin(4x - \pi) - 3$

58. $g(x) = 4 - \sin\left(2x + \dfrac{\pi}{2}\right)$

Graphing a Sine or Cosine Function In Exercises 59–64, use a graphing utility to graph the function. (Include two full periods.) Be sure to choose an appropriate viewing window.

59. $y = -2\sin(4x + \pi)$

60. $y = -4\sin\left(\dfrac{2}{3}x - \dfrac{\pi}{3}\right)$

61. $y = \cos\left(2\pi x - \dfrac{\pi}{2}\right) + 1$

62. $y = 3\cos\left(\dfrac{\pi x}{2} + \dfrac{\pi}{2}\right) - 2$

63. $y = -0.1\sin\left(\dfrac{\pi x}{10} + \pi\right)$

64. $y = \dfrac{1}{100}\cos 120\pi t$

Graphical Reasoning In Exercises 65–68, find a and d for the function $f(x) = a\cos x + d$ such that the graph of f matches the figure.

65.

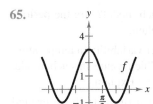

66.

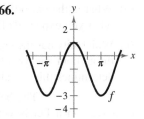

67.

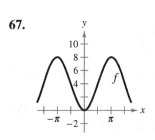

68.

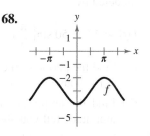

Graphical Reasoning In Exercises 69–72, find a, b, and c for the function $f(x) = a\sin(bx - c)$ such that the graph of f matches the figure.

69.

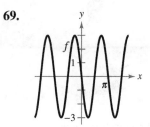

70.

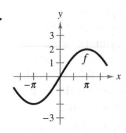

71.

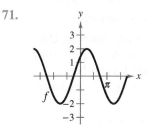

72.

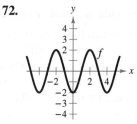

Using Technology In Exercises 73 and 74, use a graphing utility to graph y_1 and y_2 in the interval $[-2\pi, 2\pi]$. Use the graphs to find real numbers x such that $y_1 = y_2$.

73. $y_1 = \sin x$, $y_2 = -\dfrac{1}{2}$ **74.** $y_1 = \cos x$, $y_2 = -1$

Writing an Equation In Exercises 75–78, write an equation for a function with the given characteristics.

75. A sine curve with a period of π, an amplitude of 2, a right phase shift of $\pi/2$, and a vertical translation up 1 unit

76. A sine curve with a period of 4π, an amplitude of 3, a left phase shift of $\pi/4$, and a vertical translation down 1 unit

77. A cosine curve with a period of π, an amplitude of 1, a left phase shift of π, and a vertical translation down $\dfrac{3}{2}$ units

78. A cosine curve with a period of 4π, an amplitude of 3, a right phase shift of $\pi/2$, and a vertical translation up 2 units

79. Respiratory Cycle For a person exercising, the velocity v (in liters per second) of airflow during a respiratory cycle (the time from the beginning of one breath to the beginning of the next) is modeled by

$$v = 1.75\sin(\pi t/2)$$

where t is the time (in seconds). (Inhalation occurs when $v > 0$, and exhalation occurs when $v < 0$.)

(a) Find the time for one full respiratory cycle.

(b) Find the number of cycles per minute.

(c) Sketch the graph of the velocity function.

80. Respiratory Cycle

For a person at rest, the velocity v (in liters per second) of airflow during a respiratory cycle (the time from the beginning of one breath to the beginning of the next) is modeled by $v = 0.85 \sin(\pi t/3)$, where t is the time (in seconds).

(a) Find the time for one full respiratory cycle.

(b) Find the number of cycles per minute.

(c) Sketch the graph of the velocity function. Use the graph to confirm your answer in part (a) by finding two times when new breaths begin. (Inhalation occurs when $v > 0$, and exhalation occurs when $v < 0$.)

81. Biology   The function $P = 100 - 20 \cos(5\pi t/3)$ approximates the blood pressure P (in millimeters of mercury) at time t (in seconds) for a person at rest.

(a) Find the period of the function.

(b) Find the number of heartbeats per minute.

82. Piano Tuning   When tuning a piano, a technician strikes a tuning fork for the A above middle C and sets up a wave motion that can be approximated by $y = 0.001 \sin 880\pi t$, where t is the time (in seconds).

(a) What is the period of the function?

(b) The frequency f is given by $f = 1/p$. What is the frequency of the note?

83. Astronomy   The table shows the percent y (in decimal form) of the moon's face illuminated on day x in the year 2018, where $x = 1$ corresponds to January 1. *(Source: U.S. Naval Observatory)*

DATA	x	y
	1	1.0
	8	0.5
	16	0.0
	24	0.5
	31	1.0
	38	0.5

Spreadsheet at LarsonPrecalculus.com

(a) Create a scatter plot of the data.

(b) Find a trigonometric model for the data.

(c) Add the graph of your model in part (b) to the scatter plot. How well does the model fit the data?

(d) What is the period of the model?

(e) Estimate the percent of the moon's face illuminated on March 12, 2018.

84. Meteorology   The table shows the maximum daily high temperatures (in degrees Fahrenheit) in Las Vegas L and International Falls I for month t, where $t = 1$ corresponds to January. *(Source: National Climatic Data Center)*

DATA Month, t	Las Vegas, L	International Falls, I
1	57.1	13.8
2	63.0	22.4
3	69.5	34.9
4	78.1	51.5
5	87.8	66.6
6	98.9	74.2
7	104.1	78.6
8	101.8	76.3
9	93.8	64.7
10	80.8	51.7
11	66.0	32.5
12	57.3	18.1

Spreadsheet at LarsonPrecalculus.com

(a) A model for the temperatures in Las Vegas is

$$L(t) = 80.60 + 23.50 \cos\left(\frac{\pi t}{6} - 3.67\right).$$

Find a trigonometric model for the temperatures in International Falls.

(b) Use a graphing utility to graph the data points and the model for the temperatures in Las Vegas. How well does the model fit the data?

(c) Use the graphing utility to graph the data points and the model for the temperatures in International Falls. How well does the model fit the data?

(d) Use the models to estimate the average maximum temperature in each city. Which value in each model did you use? Explain.

(e) What is the period of each model? Are the periods what you expected? Explain.

(f) Which city has the greater variability in temperature throughout the year? Which value in each model determines this variability? Explain.

85. Ferris Wheel   The height h (in feet) above ground of a seat on a Ferris wheel at time t (in seconds) is modeled by

$$h(t) = 53 + 50 \sin\left(\frac{\pi}{10}t - \frac{\pi}{2}\right).$$

(a) Find the period of the model. What does the period tell you about the ride?

(b) Find the amplitude of the model. What does the amplitude tell you about the ride?

(c) Use a graphing utility to graph one cycle of the model.

86. Fuel Consumption The daily consumption C (in gallons) of diesel fuel on a farm is modeled by

$$C = 30.3 + 21.6 \sin\left(\frac{2\pi t}{365} + 10.9\right)$$

where t is the time (in days), with $t = 1$ corresponding to January 1.

(a) What is the period of the model? Is it what you expected? Explain.

(b) What is the average daily fuel consumption? Which value in the model did you use? Explain.

(c) Use a graphing utility to graph the model. Use the graph to approximate the time of the year when consumption exceeds 40 gallons per day.

Exploration

True or False? **In Exercises 87–89, determine whether the statement is true or false. Justify your answer.**

87. The graph of $g(x) = \sin(x + 2\pi)$ is a translation of the graph of $f(x) = \sin x$ exactly one period to the right, and the two graphs look identical.

88. The function $y = \frac{1}{2}\cos 2x$ has an amplitude that is twice that of the function $y = \cos x$.

89. The graph of $y = -\cos x$ is a reflection of the graph of $y = \sin[x + (\pi/2)]$ in the x-axis.

90. HOW DO YOU SEE IT? The figure below shows the graph of $y = \sin(x - c)$ for

$$c = -\frac{\pi}{4}, \quad 0, \quad \text{and} \quad \frac{\pi}{4}.$$

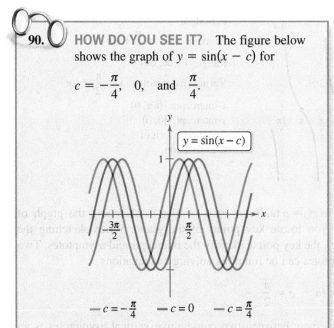

(a) How does the value of c affect the graph?

(b) Which graph is equivalent to that of

$$y = -\cos\left(x + \frac{\pi}{4}\right)?$$

Conjecture **In Exercises 91 and 92, graph f and g in the same coordinate plane. (Include two full periods.) Make a conjecture about the functions.**

91. $f(x) = \sin x, \quad g(x) = \cos\left(x - \frac{\pi}{2}\right)$

92. $f(x) = \sin x, \quad g(x) = -\cos\left(x + \frac{\pi}{2}\right)$

93. Writing Sketch the graph of $y = \cos bx$ for $b = \frac{1}{2}, 2$, and 3. How does the value of b affect the graph? How many complete cycles of the graph occur between 0 and 2π for each value of b?

94. Polynomial Approximations Using calculus, it can be shown that the sine and cosine functions can be approximated by the polynomials

$$\sin x \approx x - \frac{x^3}{3!} + \frac{x^5}{5!}$$

and

$$\cos x \approx 1 - \frac{x^2}{2!} + \frac{x^4}{4!}$$

where x is in radians.

(a) Use a graphing utility to graph the sine function and its polynomial approximation in the same viewing window. How do the graphs compare?

(b) Use the graphing utility to graph the cosine function and its polynomial approximation in the same viewing window. How do the graphs compare?

(c) Study the patterns in the polynomial approximations of the sine and cosine functions and predict the next term in each. Then repeat parts (a) and (b). How does the accuracy of the approximations change when an additional term is added?

95. Polynomial Approximations Use the polynomial approximations of the sine and cosine functions in Exercise 94 to approximate each function value. Compare the results with those given by a calculator. Is the error in the approximation the same in each case? Explain.

(a) $\sin\frac{1}{2}$ (b) $\sin 1$

(c) $\sin\frac{\pi}{6}$ (d) $\cos(-0.5)$

(e) $\cos 1$ (f) $\cos\frac{\pi}{4}$

Project: Meteorology To work an extended application analyzing the mean monthly temperature and mean monthly precipitation for Honolulu, Hawaii, visit this text's website at *LarsonPrecalculus.com*. (*Source: National Climatic Data Center*)

4.6 Graphs of Other Trigonometric Functions

Graphs of trigonometric functions have many real-life applications, such as in modeling the distance from a television camera to a unit in a parade, as in Exercise 85 on page 317.

■ Sketch the graphs of tangent functions.
■ Sketch the graphs of cotangent functions.
■ Sketch the graphs of secant and cosecant functions.
■ Sketch the graphs of damped trigonometric functions.

Graph of the Tangent Function

Recall that the tangent function is odd. That is, $\tan(-x) = -\tan x$. Consequently, the graph of $y = \tan x$ is symmetric with respect to the origin. You also know from the identity $\tan x = (\sin x)/(\cos x)$ that the tangent function is undefined for values at which $\cos x = 0$. Two such values are $x = \pm\pi/2 \approx \pm1.5708$. As shown in the table below, $\tan x$ increases without bound as x approaches $\pi/2$ from the left and decreases without bound as x approaches $-\pi/2$ from the right.

x	$-\dfrac{\pi}{2}$	-1.57	-1.5	$-\dfrac{\pi}{4}$	0	$\dfrac{\pi}{4}$	1.5	1.57	$\dfrac{\pi}{2}$
$\tan x$	Undef.	-1255.8	-14.1	-1	0	1	14.1	1255.8	Undef.

So, the graph of $y = \tan x$ (shown below) has *vertical asymptotes* at $x = \pi/2$ and $x = -\pi/2$. Moreover, the period of the tangent function is π, so vertical asymptotes also occur at $x = (\pi/2) + n\pi$, where n is an integer. The domain of the tangent function is the set of all real numbers other than $x = (\pi/2) + n\pi$, and the range is the set of all real numbers.

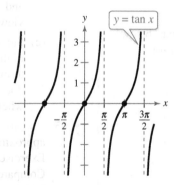

Period: π
Domain: all $x \neq \dfrac{\pi}{2} + n\pi$
Range: $(-\infty, \infty)$
Vertical asymptotes: $x = \dfrac{\pi}{2} + n\pi$
x-intercepts: $(n\pi, 0)$
y-intercept: $(0, 0)$
Symmetry: origin
Odd function

▷ **ALGEBRA HELP**
• To review odd and even functions, see Section 1.5.
• To review symmetry of a graph, see Section 1.2.
• To review fundamental trigonometric identities, see Section 4.3.
• To review asymptotes, see Section 2.6.
• To review domain and range of a function, see Section 1.4.
• To review intercepts of a graph, see Section 1.2.

Sketching the graph of $y = a \tan(bx - c)$ is similar to sketching the graph of $y = a \sin(bx - c)$ in that you locate key points of the graph. When sketching the graph of $y = a \tan(bx - c)$, the key points identify the intercepts and asymptotes. Two consecutive vertical asymptotes can be found by solving the equations

$$bx - c = -\frac{\pi}{2} \quad \text{and} \quad bx - c = \frac{\pi}{2}.$$

On the x-axis, the point halfway between two consecutive vertical asymptotes is an x-intercept of the graph. The period of the function $y = a \tan(bx - c)$ is the distance between two consecutive vertical asymptotes. The amplitude of a tangent function is not defined. After plotting two consecutive asymptotes and the x-intercept between them, plot additional points between the asymptotes and sketch one cycle. Finally, sketch one or two additional cycles to the left and right.

EXAMPLE 1 **Sketching the Graph of a Tangent Function**

Sketch the graph of $y = \tan \dfrac{x}{2}$.

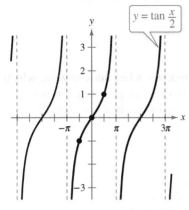

Figure 4.40

Solution

Solving the equations

$$\frac{x}{2} = -\frac{\pi}{2} \quad \text{and} \quad \frac{x}{2} = \frac{\pi}{2}$$

shows that two consecutive vertical asymptotes occur at $x = -\pi$ and $x = \pi$. Between these two asymptotes, find a few points, including the *x*-intercept, as shown in the table. Figure 4.40 shows three cycles of the graph.

x	$-\pi$	$-\dfrac{\pi}{2}$	0	$\dfrac{\pi}{2}$	π
$\tan \dfrac{x}{2}$	Undef.	-1	0	1	Undef.

✓ **Checkpoint** 🔊⟩⟩⟩ *Audio-video solution in English & Spanish at LarsonPrecalculus.com*

Sketch the graph of $y = \tan \dfrac{x}{4}$.

EXAMPLE 2 **Sketching the Graph of a Tangent Function**

Sketch the graph of $y = -3 \tan 2x$.

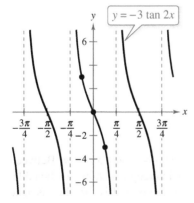

Figure 4.41

Solution

Solving the equations

$$2x = -\frac{\pi}{2} \quad \text{and} \quad 2x = \frac{\pi}{2}$$

shows that two consecutive vertical asymptotes occur at $x = -\pi/4$ and $x = \pi/4$. Between these two asymptotes, find a few points, including the *x*-intercept, as shown in the table. Figure 4.41 shows three cycles of the graph.

x	$-\dfrac{\pi}{4}$	$-\dfrac{\pi}{8}$	0	$\dfrac{\pi}{8}$	$\dfrac{\pi}{4}$
$-3 \tan 2x$	Undef.	3	0	-3	Undef.

✓ **Checkpoint** 🔊⟩⟩⟩ *Audio-video solution in English & Spanish at LarsonPrecalculus.com*

Sketch the graph of $y = \tan 2x$.

Compare the graphs in Examples 1 and 2. The graph of $y = a \tan(bx - c)$ increases between consecutive vertical asymptotes when $a > 0$ and decreases between consecutive vertical asymptotes when $a < 0$. In other words, the graph for $a < 0$ is a reflection in the *x*-axis of the graph for $a > 0$. Also, the period is greater when $0 < b < 1$ than when $b > 1$. In other words, compared with the case where $b = 1$, the period represents a horizontal stretch when $0 < b < 1$ and a horizontal shrink when $b > 1$.

Graph of the Cotangent Function

The graph of the cotangent function is similar to the graph of the tangent function. It also has a period of π. However, the identity

$$y = \cot x = \frac{\cos x}{\sin x}$$

shows that the cotangent function has vertical asymptotes when $\sin x$ is zero, which occurs at $x = n\pi$, where n is an integer. The graph of the cotangent function is shown below. Note that two consecutive vertical asymptotes of the graph of $y = a\cot(bx - c)$ can be found by solving the equations

$$bx - c = 0 \quad \text{and} \quad bx - c = \pi.$$

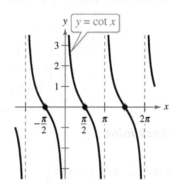

Period: π
Domain: all $x \neq n\pi$
Range: $(-\infty, \infty)$
Vertical asymptotes: $x = n\pi$
x-intercepts: $\left(\frac{\pi}{2} + n\pi, 0 \right)$
Symmetry: origin
Odd function

EXAMPLE 3   **Sketching the Graph of a Cotangent Function**

Sketch the graph of

$$y = 2\cot\frac{x}{3}.$$

Solution

Solving the equations

$$\frac{x}{3} = 0 \quad \text{and} \quad \frac{x}{3} = \pi$$

shows that two consecutive vertical asymptotes occur at $x = 0$ and $x = 3\pi$. Between these two asymptotes, find a few points, including the x-intercept, as shown in the table. Figure 4.42 shows three cycles of the graph. Note that the period is 3π, the distance between consecutive asymptotes.

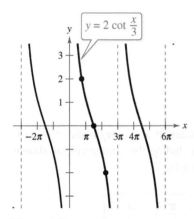

Figure 4.42

x	0	$\dfrac{3\pi}{4}$	$\dfrac{3\pi}{2}$	$\dfrac{9\pi}{4}$	3π
$2\cot\dfrac{x}{3}$	Undef.	2	0	-2	Undef.

✔ **Checkpoint** 🔊 *Audio-video solution in English & Spanish at LarsonPrecalculus.com*

Sketch the graph of

$$y = \cot\frac{x}{4}.$$

Graphs of the Reciprocal Functions

You can obtain the graphs of the cosecant and secant functions from the graphs of the sine and cosine functions, respectively, using the reciprocal identities

$$\csc x = \frac{1}{\sin x} \quad \text{and} \quad \sec x = \frac{1}{\cos x}.$$

For example, at a given value of x, the y-coordinate of $\sec x$ is the reciprocal of the y-coordinate of $\cos x$. Of course, when $\cos x = 0$, the reciprocal does not exist. Near such values of x, the behavior of the secant function is similar to that of the tangent function. In other words, the graphs of

$$\tan x = \frac{\sin x}{\cos x} \quad \text{and} \quad \sec x = \frac{1}{\cos x}$$

have vertical asymptotes where $\cos x = 0$, that is, at $x = (\pi/2) + n\pi$, where n is an integer. Similarly,

$$\cot x = \frac{\cos x}{\sin x} \quad \text{and} \quad \csc x = \frac{1}{\sin x}$$

have vertical asymptotes where $\sin x = 0$, that is, at $x = n\pi$, where n is an integer.

To sketch the graph of a secant or cosecant function, first make a sketch of its reciprocal function. For example, to sketch the graph of $y = \csc x$, first sketch the graph of $y = \sin x$. Then find reciprocals of the y-coordinates to obtain points on the graph of $y = \csc x$. You can use this procedure to obtain the graphs below.

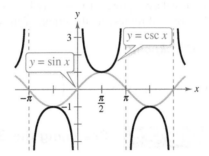

Period: 2π
Domain: all $x \neq n\pi$
Range: $(-\infty, -1] \cup [1, \infty)$
Vertical asymptotes: $x = n\pi$
No intercepts
Symmetry: origin
Odd function

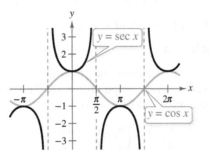

Period: 2π
Domain: all $x \neq \dfrac{\pi}{2} + n\pi$
Range: $(-\infty, -1] \cup [1, \infty)$
Vertical asymptotes: $x = \dfrac{\pi}{2} + n\pi$
y-intercept: $(0, 1)$
Symmetry: y-axis
Even function

In comparing the graphs of the cosecant and secant functions with those of the sine and cosine functions, respectively, note that the "hills" and "valleys" are interchanged. For example, a hill (or maximum point) on the sine curve corresponds to a valley (a relative minimum) on the cosecant curve, and a valley (or minimum point) on the sine curve corresponds to a hill (a relative maximum) on the cosecant curve, as shown in Figure 4.43. Additionally, x-intercepts of the sine and cosine functions become vertical asymptotes of the cosecant and secant functions, respectively (see Figure 4.43).

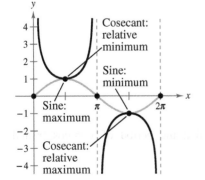

Figure 4.43

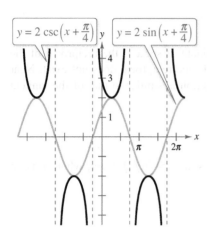

Figure 4.44

EXAMPLE 4 **Sketching the Graph of a Cosecant Function**

Sketch the graph of $y = 2 \csc\left(x + \dfrac{\pi}{4}\right)$.

Solution

Begin by sketching the graph of

$$y = 2 \sin\left(x + \frac{\pi}{4}\right).$$

For this function, the amplitude is 2 and the period is 2π. Solving the equations

$$x + \frac{\pi}{4} = 0 \quad \text{and} \quad x + \frac{\pi}{4} = 2\pi$$

shows that one cycle of the sine function corresponds to the interval from $x = -\pi/4$ to $x = 7\pi/4$. The gray curve in Figure 4.44 represents the graph of the sine function. At the midpoint and endpoints of this interval, the sine function is zero. So, the corresponding cosecant function

$$y = 2 \csc\left(x + \frac{\pi}{4}\right)$$

$$= 2\left(\frac{1}{\sin[x + (\pi/4)]}\right)$$

has vertical asymptotes at $x = -\pi/4$, $x = 3\pi/4$, $x = 7\pi/4$, and so on. The black curve in Figure 4.44 represents the graph of the cosecant function.

✓ *Checkpoint* 🔊))) *Audio-video solution in English & Spanish at LarsonPrecalculus.com*

Sketch the graph of $y = 2 \csc\left(x + \dfrac{\pi}{2}\right)$.

EXAMPLE 5 **Sketching the Graph of a Secant Function**

See LarsonPrecalculus.com for an interactive version of this type of example.

Sketch the graph of $y = \sec 2x$.

Solution

Begin by sketching the graph of $y = \cos 2x$, shown as the gray curve in Figure 4.45. Then, form the graph of $y = \sec 2x$, shown as the black curve in the figure. Note that the x-intercepts of $y = \cos 2x$

$$\left(-\frac{\pi}{4}, 0\right), \quad \left(\frac{\pi}{4}, 0\right), \quad \left(\frac{3\pi}{4}, 0\right), \dots$$

correspond to the vertical asymptotes

$$x = -\frac{\pi}{4}, \quad x = \frac{\pi}{4}, \quad x = \frac{3\pi}{4}, \dots$$

of the graph of $y = \sec 2x$. Moreover, notice that the period of $y = \cos 2x$ and $y = \sec 2x$ is π.

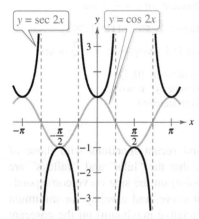

Figure 4.45

✓ *Checkpoint* 🔊))) *Audio-video solution in English & Spanish at LarsonPrecalculus.com*

Sketch the graph of $y = \sec \dfrac{x}{2}$.

Damped Trigonometric Graphs

You can graph a *product* of two functions using properties of the individual functions. For example, consider the function

$$f(x) = x \sin x$$

as the product of the functions $y = x$ and $y = \sin x$. Using properties of absolute value and the fact that $|\sin x| \le 1$, you have

$$0 \le |x||\sin x| \le |x|.$$

Consequently,

$$-|x| \le x \sin x \le |x|$$

which means that the graph of $f(x) = x \sin x$ lies between the lines $y = -x$ and $y = x$. Furthermore,

$$f(x) = x \sin x = \pm x \quad \text{at} \quad x = \frac{\pi}{2} + n\pi$$

and

$$f(x) = x \sin x = 0 \quad \text{at} \quad x = n\pi$$

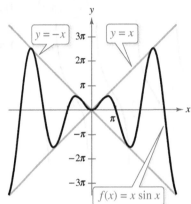

· · · · · · · · · · · · · · · · ▷

· · **REMARK** Do you see why the graph of $f(x) = x \sin x$ touches the lines $y = \pm x$ at $x = (\pi/2) + n\pi$ and why the graph has x-intercepts at $x = n\pi$? Recall that the sine function is equal to ± 1 at odd multiples of $\pi/2$ and is equal to 0 at multiples of π.

where n is an integer, so the graph of f touches the line $y = x$ or the line $y = -x$ at $x = (\pi/2) + n\pi$ and has x-intercepts at $x = n\pi$. A sketch of f is shown at the right. In the function $f(x) = x \sin x$, the factor x is called the **damping factor.**

EXAMPLE 6   **Damped Sine Curve**

Sketch the graph of $f(x) = e^{-x} \sin 3x$.

Solution

Consider f as the product of the two functions $y = e^{-x}$ and $y = \sin 3x$, each of which has the set of real numbers as its domain. For any real number x, you know that $e^{-x} > 0$ and $|\sin 3x| \le 1$. So,

$$e^{-x}|\sin 3x| \le e^{-x}$$

which means that

$$-e^{-x} \le e^{-x} \sin 3x \le e^{-x}.$$

Furthermore,

$$f(x) = e^{-x} \sin 3x = \pm e^{-x} \quad \text{at} \quad x = \frac{\pi}{6} + \frac{n\pi}{3}$$

and

$$f(x) = e^{-x} \sin 3x = 0 \quad \text{at} \quad x = \frac{n\pi}{3}$$

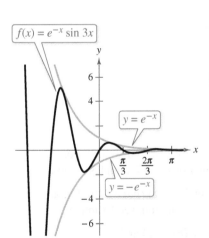

Figure 4.46

so the graph of f touches the curve $y = e^{-x}$ or the curve $y = -e^{-x}$ at $x = (\pi/6) + (n\pi/3)$ and has intercepts at $x = n\pi/3$. Figure 4.46 shows a sketch of f.

✓ **Checkpoint**))) *Audio-video solution in English & Spanish at LarsonPrecalculus.com*

Sketch the graph of $f(x) = e^x \sin 4x$.

Below is a summary of the characteristics of the six basic trigonometric functions.

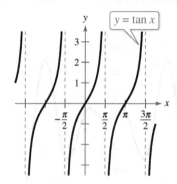

Domain: $(-\infty, \infty)$
Range: $[-1, 1]$
Period: 2π

$y = \sin x$

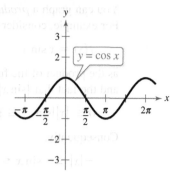

Domain: $(-\infty, \infty)$
Range: $[-1, 1]$
Period: 2π

$y = \cos x$

$y = \tan x$

Domain: all $x \neq \dfrac{\pi}{2} + n\pi$
Range: $(-\infty, \infty)$
Period: π

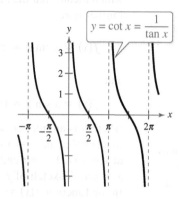

$y = \cot x = \dfrac{1}{\tan x}$

Domain: all $x \neq n\pi$
Range: $(-\infty, \infty)$
Period: π

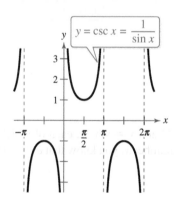

$y = \csc x = \dfrac{1}{\sin x}$

Domain: all $x \neq n\pi$
Range: $(-\infty, -1] \cup [1, \infty)$
Period: 2π

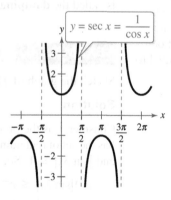

$y = \sec x = \dfrac{1}{\cos x}$

Domain: all $x \neq \dfrac{\pi}{2} + n\pi$
Range: $(-\infty, -1] \cup [1, \infty)$
Period: 2π

Summarize   (Section 4.6)

1. Explain how to sketch the graph of $y = a \tan(bx - c)$ *(page 308)*. For examples of sketching graphs of tangent functions, see Examples 1 and 2.

2. Explain how to sketch the graph of $y = a \cot(bx - c)$ *(page 310)*. For an example of sketching the graph of a cotangent function, see Example 3.

3. Explain how to sketch the graphs of $y = a \csc(bx - c)$ and $y = a \sec(bx - c)$ *(page 311)*. For examples of sketching graphs of cosecant and secant functions, see Examples 4 and 5.

4. Explain how to sketch the graph of a damped trigonometric function *(page 313)*. For an example of sketching the graph of a damped trigonometric function, see Example 6.

4.6 Exercises

Vocabulary: Fill in the blanks.

1. The tangent, cotangent, and cosecant functions are _____, so the graphs of these functions have symmetry with respect to the _____.
2. The graphs of the tangent, cotangent, secant, and cosecant functions have _____ asymptotes.
3. To sketch the graph of a secant or cosecant function, first make a sketch of its _____ function.
4. For the function $f(x) = g(x) \cdot \sin x$, $g(x)$ is called the _____ factor.
5. The period of $y = \tan x$ is _____.
6. The domain of $y = \cot x$ is all real numbers such that _____.
7. The range of $y = \sec x$ is _____.
8. The period of $y = \csc x$ is _____.

Skills and Applications

Matching In Exercises 9–14, match the function with its graph. State the period of the function. [The graphs are labeled (a), (b), (c), (d), (e), and (f).]

(a)
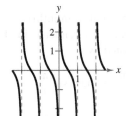

(b)
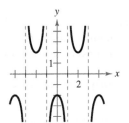

(c)
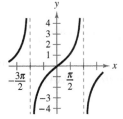

(d)
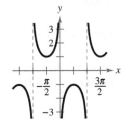

(e)
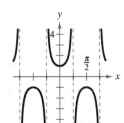

(f)
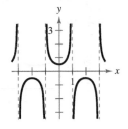

9. $y = \sec 2x$

10. $y = \tan \dfrac{x}{2}$

11. $y = \dfrac{1}{2} \cot \pi x$

12. $y = -\csc x$

13. $y = \dfrac{1}{2} \sec \dfrac{\pi x}{2}$

14. $y = -2 \sec \dfrac{\pi x}{2}$

Sketching the Graph of a Trigonometric Function In Exercises 15–38, sketch the graph of the function. (Include two full periods.)

15. $y = \dfrac{1}{3} \tan x$

16. $y = -\dfrac{1}{2} \tan x$

17. $y = -\dfrac{1}{2} \sec x$

18. $y = \dfrac{1}{4} \sec x$

19. $y = -2 \tan 3x$

20. $y = -3 \tan \pi x$

21. $y = \csc \pi x$

22. $y = 3 \csc 4x$

23. $y = \dfrac{1}{2} \sec \pi x$

24. $y = 2 \sec 3x$

25. $y = \csc \dfrac{x}{2}$

26. $y = \csc \dfrac{x}{3}$

27. $y = 3 \cot 2x$

28. $y = 3 \cot \dfrac{\pi x}{2}$

29. $y = \tan \dfrac{\pi x}{4}$

30. $y = \tan 4x$

31. $y = 2 \csc(x - \pi)$

32. $y = \csc(2x - \pi)$

33. $y = 2 \sec(x + \pi)$

34. $y = \tan(x + \pi)$

35. $y = -\sec \pi x + 1$

36. $y = -2 \sec 4x + 2$

37. $y = \dfrac{1}{4} \csc\left(x + \dfrac{\pi}{4}\right)$

38. $y = 2 \cot\left(x + \dfrac{\pi}{2}\right)$

Graphing a Trigonometric Function In Exercises 39–48, use a graphing utility to graph the function. (Include two full periods.)

39. $y = \tan \dfrac{x}{3}$

40. $y = -\tan 2x$

41. $y = -2 \sec 4x$

42. $y = \sec \pi x$

43. $y = \tan\left(x - \dfrac{\pi}{4}\right)$

44. $y = \dfrac{1}{4} \cot\left(x - \dfrac{\pi}{2}\right)$

45. $y = -\csc(4x - \pi)$

46. $y = 2 \sec(2x - \pi)$

47. $y = 0.1 \tan\left(\dfrac{\pi x}{4} + \dfrac{\pi}{4}\right)$

48. $y = \dfrac{1}{3} \sec\left(\dfrac{\pi x}{2} + \dfrac{\pi}{2}\right)$

Solving a Trigonometric Equation In Exercises 49–56, find the solutions of the equation in the interval $[-2\pi, 2\pi]$. Use a graphing utility to verify your results.

49. $\tan x = 1$

50. $\tan x = \sqrt{3}$

51. $\cot x = -\sqrt{3}$

52. $\cot x = 1$

53. $\sec x = -2$

54. $\sec x = 2$

55. $\csc x = \sqrt{2}$

56. $\csc x = -2$

 Even and Odd Trigonometric Functions In Exercises 57–64, use the graph of the function to determine whether the function is even, odd, or neither. Verify your answer algebraically.

57. $f(x) = \sec x$

58. $f(x) = \tan x$

59. $g(x) = \cot x$

60. $g(x) = \csc x$

61. $f(x) = x + \tan x$

62. $f(x) = x^2 - \sec x$

63. $g(x) = x \csc x$

64. $g(x) = x^2 \cot x$

Identifying Damped Trigonometric Functions In Exercises 65–68, match the function with its graph. Describe the behavior of the function as x approaches zero. [The graphs are labeled (a), (b), (c), and (d).]

(a)

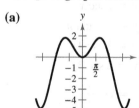

(b)

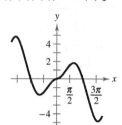

(c)

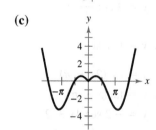

(d)

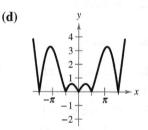

65. $f(x) = |x \cos x|$

66. $f(x) = x \sin x$

67. $g(x) = |x| \sin x$

68. $g(x) = |x| \cos x$

Conjecture In Exercises 69–72, graph the functions f and g. Use the graphs to make a conjecture about the relationship between the functions.

69. $f(x) = \sin x + \cos\left(x + \dfrac{\pi}{2}\right)$, $g(x) = 0$

70. $f(x) = \sin x - \cos\left(x + \dfrac{\pi}{2}\right)$, $g(x) = 2 \sin x$

71. $f(x) = \sin^2 x$, $g(x) = \frac{1}{2}(1 - \cos 2x)$

72. $f(x) = \cos^2 \dfrac{\pi x}{2}$, $g(x) = \frac{1}{2}(1 + \cos \pi x)$

Analyzing a Damped Trigonometric Graph In Exercises 73–76, use a graphing utility to graph the function and the damping factor of the function in the same viewing window. Describe the behavior of the function as x increases without bound.

73. $g(x) = e^{-x^2/2} \sin x$

74. $f(x) = e^{-x} \cos x$

75. $f(x) = 2^{-x/4} \cos \pi x$

76. $h(x) = 2^{-x^2/4} \sin x$

Analyzing a Trigonometric Graph In Exercises 77–82, use a graphing utility to graph the function. Describe the behavior of the function as x approaches zero.

77. $y = \dfrac{6}{x} + \cos x$, $x > 0$

78. $y = \dfrac{4}{x} + \sin 2x$, $x > 0$

79. $g(x) = \dfrac{\sin x}{x}$

80. $f(x) = \dfrac{1 - \cos x}{x}$

81. $f(x) = \sin \dfrac{1}{x}$

82. $h(x) = x \sin \dfrac{1}{x}$

83. Meteorology The normal monthly high temperatures H (in degrees Fahrenheit) in Erie, Pennsylvania, are approximated by

$$H(t) = 57.54 - 18.53 \cos \frac{\pi t}{6} - 14.03 \sin \frac{\pi t}{6}$$

and the normal monthly low temperatures L are approximated by

$$L(t) = 42.03 - 15.99 \cos \frac{\pi t}{6} - 14.32 \sin \frac{\pi t}{6}$$

where t is the time (in months), with $t = 1$ corresponding to January (see figure). *(Source: NOAA)*

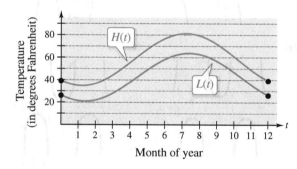

Month of year

(a) What is the period of each function?

(b) During what part of the year is the difference between the normal high and normal low temperatures greatest? When is it least?

(c) The sun is northernmost in the sky around June 21, but the graph shows the warmest temperatures at a later date. Approximate the lag time of the temperatures relative to the position of the sun.

84. Sales The projected monthly sales S (in thousands of units) of lawn mowers are modeled by

$$S = 74 + 3t - 40 \cos \frac{\pi t}{6}$$

where t is the time (in months), with $t = 1$ corresponding to January.

(a) Graph the sales function over 1 year.

(b) What are the projected sales for June?

85. Television Coverage

A television camera is on a reviewing platform 27 meters from the street on which a parade passes from left to right (see figure). Write the distance d from the camera to a unit in the parade as a function of the angle x, and graph the function over the interval $-\pi/2 < x < \pi/2$. (Consider x as negative when a unit in the parade approaches from the left.)

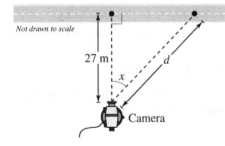

Not drawn to scale

27 m

x

d

Camera

86. Distance A plane flying at an altitude of 7 miles above a radar antenna passes directly over the radar antenna (see figure). Let d be the ground distance from the antenna to the point directly under the plane and let x be the angle of elevation to the plane from the antenna. (d is positive as the plane approaches the antenna.) Write d as a function of x and graph the function over the interval $0 < x < \pi$.

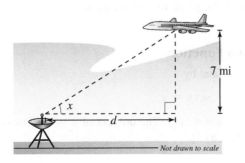

7 mi

x

d

Not drawn to scale

True or False? **In Exercises 87 and 88, determine whether the statement is true or false. Justify your answer.**

87. You can obtain the graph of $y = \csc x$ on a calculator by graphing the reciprocal of $y = \sin x$.

88. You can obtain the graph of $y = \sec x$ on a calculator by graphing a translation of the reciprocal of $y = \sin x$.

89. Think About It Consider the function $f(x) = x - \cos x$.

(a) Use a graphing utility to graph the function and verify that there exists a zero between 0 and 1. Use the graph to approximate the zero.

(b) Starting with $x_0 = 1$, generate a sequence x_1, x_2, x_3, . . . , where $x_n = \cos(x_{n-1})$. For example, $x_0 = 1, x_1 = \cos(x_0), x_2 = \cos(x_1), x_3 = \cos(x_2), \ldots$. What value does the sequence approach?

90. **HOW DO YOU SEE IT?** Determine which function each graph represents. Do not use a calculator. Explain.

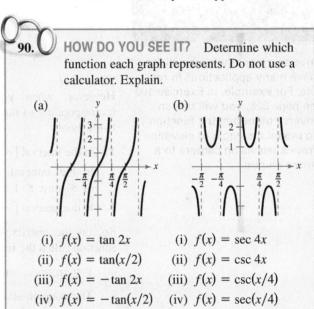

(i) $f(x) = \tan 2x$ (i) $f(x) = \sec 4x$

(ii) $f(x) = \tan(x/2)$ (ii) $f(x) = \csc 4x$

(iii) $f(x) = -\tan 2x$ (iii) $f(x) = \csc(x/4)$

(iv) $f(x) = -\tan(x/2)$ (iv) $f(x) = \sec(x/4)$

Graphical Reasoning **In Exercises 91 and 92, use a graphing utility to graph the function. Use the graph to determine the behavior of the function as $x \to c$. (*Note:* The notation $x \to c^+$ indicates that x approaches c from the right and $x \to c^-$ indicates that x approaches c from the left.)**

(a) $x \to 0^+$ (b) $x \to 0^-$ (c) $x \to \pi^+$ (d) $x \to \pi^-$

91. $f(x) = \cot x$ **92.** $f(x) = \csc x$

Graphical Reasoning **In Exercises 93 and 94, use a graphing utility to graph the function. Use the graph to determine the behavior of the function as $x \to c$.**

(a) $x \to (\pi/2)^+$ (b) $x \to (\pi/2)^-$

(c) $x \to (-\pi/2)^+$ (d) $x \to (-\pi/2)^-$

93. $f(x) = \tan x$ **94.** $f(x) = \sec x$

4.7 Inverse Trigonometric Functions

■ Evaluate and graph the inverse sine function.
■ Evaluate and graph other inverse trigonometric functions.
■ Evaluate compositions with inverse trigonometric functions.

Inverse Sine Function

Recall from Section 1.9 that for a function to have an inverse function, it must be one-to-one—that is, it must pass the Horizontal Line Test. Notice in Figure 4.47 that $y = \sin x$ does not pass the test because different values of x yield the same y-value.

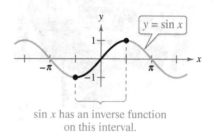

$y = \sin x$

$\sin x$ has an inverse function
on this interval.

Figure 4.47

Inverse trigonometric functions have many applications in real life. For example, in Exercise 100 on page 326, you will use an inverse trigonometric function to model the angle of elevation from a television camera to a space shuttle.

However, when you restrict the domain to the interval $-\pi/2 \le x \le \pi/2$ (corresponding to the black portion of the graph in Figure 4.47), the properties listed below hold.

1. On the interval $[-\pi/2, \pi/2]$, the function $y = \sin x$ is increasing.
2. On the interval $[-\pi/2, \pi/2]$, $y = \sin x$ takes on its full range of values, $-1 \le \sin x \le 1$.
3. On the interval $[-\pi/2, \pi/2]$, $y = \sin x$ is one-to-one.

So, on the restricted domain $-\pi/2 \le x \le \pi/2$, $y = \sin x$ has a unique inverse function called the **inverse sine function.** It is denoted by

$$y = \arcsin x \quad \text{or} \quad y = \sin^{-1} x.$$

The notation $\sin^{-1} x$ is consistent with the inverse function notation $f^{-1}(x)$. The arcsin x notation (read as "the arcsine of x") comes from the association of a central angle with its intercepted *arc length* on a unit circle. So, arcsin x means the angle (or arc) whose sine is x. Both notations, arcsin x and $\sin^{-1} x$ are commonly used in mathematics. You must remember that $\sin^{-1} x$ denotes the *inverse* sine function, *not* $1/\sin x$. The values of arcsin x lie in the interval

$$-\frac{\pi}{2} \le \arcsin x \le \frac{\pi}{2}.$$

Figure 4.48 on the next page shows the graph of $y = \arcsin x$.

REMARK When evaluating the inverse sine function, it helps to remember the phrase "the arcsine of x is the angle (or number) whose sine is x."

Definition of Inverse Sine Function

The **inverse sine function** is defined by

$$y = \arcsin x \quad \text{if and only if} \quad \sin y = x$$

where $-1 \le x \le 1$ and $-\pi/2 \le y \le \pi/2$. The domain of $y = \arcsin x$ is $[-1, 1]$, and the range is $[-\pi/2, \pi/2]$.

EXAMPLE 1 **Evaluating the Inverse Sine Function**

If possible, find the exact value of each expression.

a. $\arcsin\left(-\dfrac{1}{2}\right)$ **b.** $\sin^{-1}\dfrac{\sqrt{3}}{2}$ **c.** $\sin^{-1}2$

Solution

a. You know that $\sin\left(-\dfrac{\pi}{6}\right) = -\dfrac{1}{2}$ and $-\dfrac{\pi}{6}$ lies in $\left[-\dfrac{\pi}{2}, \dfrac{\pi}{2}\right]$, so

$$\arcsin\left(-\dfrac{1}{2}\right) = -\dfrac{\pi}{6}. \qquad \text{Angle whose sine is } -\tfrac{1}{2}$$

b. You know that $\sin\dfrac{\pi}{3} = \dfrac{\sqrt{3}}{2}$ and $\dfrac{\pi}{3}$ lies in $\left[-\dfrac{\pi}{2}, \dfrac{\pi}{2}\right]$, so

$$\sin^{-1}\dfrac{\sqrt{3}}{2} = \dfrac{\pi}{3}. \qquad \text{Angle whose sine is } \sqrt{3}/2$$

c. It is not possible to evaluate $y = \sin^{-1}x$ when $x = 2$ because there is no angle whose sine is 2. Remember that the domain of the inverse sine function is $[-1, 1]$.

✓ **Checkpoint** Audio-video solution in English & Spanish at LarsonPrecalculus.com

If possible, find the exact value of each expression.

a. $\arcsin 1$ **b.** $\sin^{-1}(-2)$

EXAMPLE 2 **Graphing the Arcsine Function**

See LarsonPrecalculus.com for an interactive version of this type of example.

Sketch the graph of $y = \arcsin x$.

Solution

By definition, the equations $y = \arcsin x$ and $\sin y = x$ are equivalent for $-\pi/2 \le y \le \pi/2$. So, their graphs are the same. From the interval $[-\pi/2, \pi/2]$, assign values to y in the equation $\sin y = x$ to make a table of values.

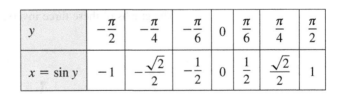

y	$-\dfrac{\pi}{2}$	$-\dfrac{\pi}{4}$	$-\dfrac{\pi}{6}$	0	$\dfrac{\pi}{6}$	$\dfrac{\pi}{4}$	$\dfrac{\pi}{2}$
$x = \sin y$	-1	$-\dfrac{\sqrt{2}}{2}$	$-\dfrac{1}{2}$	0	$\dfrac{1}{2}$	$\dfrac{\sqrt{2}}{2}$	1

Then plot the points and connect them with a smooth curve. Figure 4.48 shows the graph of $y = \arcsin x$. Note that it is the reflection (in the line $y = x$) of the black portion of the graph in Figure 4.47. Be sure you see that Figure 4.48 shows the *entire* graph of the inverse sine function. Remember that the domain of $y = \arcsin x$ is the closed interval $[-1, 1]$ and the range is the closed interval $[-\pi/2, \pi/2]$.

✓ **Checkpoint** Audio-video solution in English & Spanish at LarsonPrecalculus.com

Use a graphing utility to graph $f(x) = \sin x$, $g(x) = \arcsin x$, and $y = x$ in the same viewing window to verify geometrically that g is the inverse function of f. (Be sure to restrict the domain of f properly.) ∎

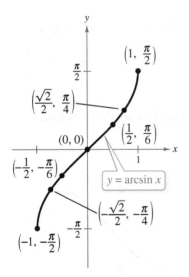

Figure 4.48

Other Inverse Trigonometric Functions

The cosine function is decreasing and one-to-one on the interval $0 \le x \le \pi$, as shown in the graph below.

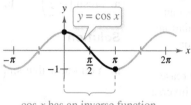

cos x has an inverse function
on this interval.

Consequently, on this interval the cosine function has an inverse function—the **inverse cosine function**—denoted by

$$y = \arccos x \quad \text{or} \quad y = \cos^{-1} x.$$

Similarly, to define an **inverse tangent function,** restrict the domain of $y = \tan x$ to the interval $(-\pi/2, \pi/2)$. The inverse tangent function is denoted by

$$y = \arctan x \quad \text{or} \quad y = \tan^{-1} x.$$

The list below summarizes the definitions of the three most common inverse trigonometric functions. Definitions of the remaining three are explored in Exercises 111–113.

Definitions of the Inverse Trigonometric Functions

Function	Domain	Range
$y = \arcsin x$ if and only if $\sin y = x$	$-1 \le x \le 1$	$-\dfrac{\pi}{2} \le y \le \dfrac{\pi}{2}$
$y = \arccos x$ if and only if $\cos y = x$	$-1 \le x \le 1$	$0 \le y \le \pi$
$y = \arctan x$ if and only if $\tan y = x$	$-\infty < x < \infty$	$-\dfrac{\pi}{2} < y < \dfrac{\pi}{2}$

The graphs of these three inverse trigonometric functions are shown below.

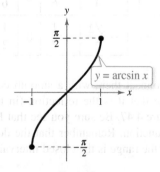

Domain: $[-1, 1]$
Range: $\left[-\dfrac{\pi}{2}, \dfrac{\pi}{2} \right]$
Intercept: $(0, 0)$
Symmetry: origin
Odd function

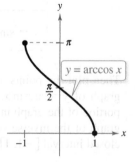

Domain: $[-1, 1]$
Range: $[0, \pi]$
y-intercept: $\left(0, \dfrac{\pi}{2} \right)$

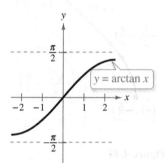

Domain: $(-\infty, \infty)$
Range: $\left(-\dfrac{\pi}{2}, \dfrac{\pi}{2} \right)$
Horizontal asymptotes: $y = \pm\dfrac{\pi}{2}$
Intercept: $(0, 0)$
Symmetry: origin
Odd function

EXAMPLE 3   **Evaluating Inverse Trigonometric Functions**

Find the exact value of each expression.

a. $\arccos \dfrac{\sqrt{2}}{2}$

b. $\arctan 0$

c. $\tan^{-1}(-1)$

Solution

a. You know that $\cos(\pi/4) = \sqrt{2}/2$ and $\pi/4$ lies in $[0, \pi]$, so

$$\arccos \frac{\sqrt{2}}{2} = \frac{\pi}{4}. \qquad \text{Angle whose cosine is } \sqrt{2}/2$$

b. You know that $\tan 0 = 0$ and 0 lies in $(-\pi/2, \pi/2)$, so

$$\arctan 0 = 0. \qquad \text{Angle whose tangent is } 0$$

c. You know that $\tan(-\pi/4) = -1$ and $-\pi/4$ lies in $(-\pi/2, \pi/2)$, so

$$\tan^{-1}(-1) = -\frac{\pi}{4}. \qquad \text{Angle whose tangent is } -1$$

✓ *Checkpoint*   *Audio-video solution in English & Spanish at LarsonPrecalculus.com*

Find the exact value of $\cos^{-1}(-1)$.

EXAMPLE 4   **Calculators and Inverse Trigonometric Functions**

Use a calculator to approximate the value of each expression, if possible.

a. $\arctan(-8.45)$

b. $\sin^{-1} 0.2447$

c. $\arccos 2$

Solution

Function	Mode	Calculator Keystrokes
a. $\arctan(-8.45)$	Radian	TAN⁻¹ ((−) 8.45) ENTER

From the display, it follows that $\arctan(-8.45) \approx -1.4530010$.

b. $\sin^{-1} 0.2447$	Radian	SIN⁻¹ (0.2447) ENTER

From the display, it follows that $\sin^{-1} 0.2447 \approx 0.2472103$.

c. $\arccos 2$	Radian	COS⁻¹ (2) ENTER

The calculator should display an *error message* because the domain of the inverse cosine function is $[-1, 1]$.

✓ *Checkpoint*  *Audio-video solution in English & Spanish at LarsonPrecalculus.com*

Use a calculator to approximate the value of each expression, if possible.

a. $\arctan 4.84$

b. $\arcsin(-1.1)$

c. $\arccos(-0.349)$

In Example 4, had you set the calculator to *degree* mode, the displays would have been in degrees rather than in radians. This convention is peculiar to calculators. By definition, the values of inverse trigonometric functions are *always in radians*.

Compositions with Inverse Trigonometric Functions

▷ **ALGEBRA HELP**   To review
compositions of functions, see
Section 1.8.

Recall from Section 1.9 that for all x in the domains of f and f^{-1}, inverse functions have the properties

$$f(f^{-1}(x)) = x \quad \text{and} \quad f^{-1}(f(x)) = x.$$

Inverse Properties of Trigonometric Functions

If $-1 \le x \le 1$ and $-\pi/2 \le y \le \pi/2$, then

$$\sin(\arcsin x) = x \quad \text{and} \quad \arcsin(\sin y) = y.$$

If $-1 \le x \le 1$ and $0 \le y \le \pi$, then

$$\cos(\arccos x) = x \quad \text{and} \quad \arccos(\cos y) = y.$$

If x is a real number and $-\pi/2 < y < \pi/2$, then

$$\tan(\arctan x) = x \quad \text{and} \quad \arctan(\tan y) = y.$$

Keep in mind that these inverse properties do not apply for arbitrary values of x and y. For example,

$$\arcsin\left(\sin \frac{3\pi}{2}\right) = \arcsin(-1) = -\frac{\pi}{2} \ne \frac{3\pi}{2}.$$

In other words, the property $\arcsin(\sin y) = y$ is not valid for values of y outside the interval $[-\pi/2, \pi/2]$.

EXAMPLE 5   **Using Inverse Properties**

If possible, find the exact value of each expression.

a. $\tan[\arctan(-5)]$ **b.** $\arcsin\left(\sin \dfrac{5\pi}{3}\right)$ **c.** $\cos(\cos^{-1} \pi)$

Solution

a. You know that -5 lies in the domain of the arctangent function, so the inverse property applies, and you have

$$\tan[\arctan(-5)] = -5.$$

b. In this case, $5\pi/3$ does not lie in the range of the arcsine function, $-\pi/2 \le y \le \pi/2$. However, $5\pi/3$ is coterminal with

$$\frac{5\pi}{3} - 2\pi = -\frac{\pi}{3}$$

which does lie in the range of the arcsine function, and you have

$$\arcsin\left(\sin \frac{5\pi}{3}\right) = \arcsin\left[\sin\left(-\frac{\pi}{3}\right)\right] = -\frac{\pi}{3}.$$

c. The expression $\cos(\cos^{-1} \pi)$ is not defined because $\cos^{-1} \pi$ is not defined. Remember that the domain of the inverse cosine function is $[-1, 1]$.

✓ **Checkpoint** 🔊))) Audio-video solution in English & Spanish at LarsonPrecalculus.com

If possible, find the exact value of each expression.

a. $\tan[\tan^{-1}(-14)]$ **b.** $\sin^{-1}\left(\sin \dfrac{7\pi}{4}\right)$ **c.** $\cos(\arccos 0.54)$

EXAMPLE 6 **Evaluating Compositions of Functions**

Find the exact value of each expression.

a. $\tan\left(\arccos \frac{2}{3}\right)$ **b.** $\cos\left[\arcsin\left(-\frac{3}{5}\right)\right]$

Solution

a. If you let $u = \arccos \frac{2}{3}$, then $\cos u = \frac{2}{3}$. The range of the inverse cosine function is $[0, \pi]$ and $\cos u$ is positive, so u is a *first*-quadrant angle. Sketch and label a right triangle with acute angle u, as shown in Figure 4.49. Consequently,

$$\tan\left(\arccos \frac{2}{3}\right) = \tan u = \frac{\text{opp}}{\text{adj}} = \frac{\sqrt{5}}{2}.$$

b. If you let $u = \arcsin\left(-\frac{3}{5}\right)$, then $\sin u = -\frac{3}{5}$. The range of the inverse sine function is $[-\pi/2, \pi/2]$ and $\sin u$ is negative, so u is a *fourth*-quadrant angle. Sketch and label a right triangle with acute angle u, as shown in Figure 4.50. Consequently,

$$\cos\left[\arcsin\left(-\frac{3}{5}\right)\right] = \cos u = \frac{\text{adj}}{\text{hyp}} = \frac{4}{5}.$$

✓ *Checkpoint* Audio-video solution in English & Spanish at *LarsonPrecalculus.com*

Find the exact value of $\cos\left[\arctan\left(-\frac{3}{4}\right)\right]$.

 EXAMPLE 7 **Some Problems from Calculus**

Write an algebraic expression that is equivalent to each expression.

a. $\sin(\arccos 3x)$, $0 \le x \le \frac{1}{3}$ **b.** $\cot(\arccos 3x)$, $0 \le x < \frac{1}{3}$

Solution

If you let $u = \arccos 3x$, then $\cos u = 3x$, where $-1 \le 3x \le 1$. Write

$$\cos u = \frac{\text{adj}}{\text{hyp}} = \frac{3x}{1}$$

and sketch a right triangle with acute angle u, as shown in Figure 4.51. From this triangle, convert each expression to algebraic form.

a. $\sin(\arccos 3x) = \sin u = \frac{\text{opp}}{\text{hyp}} = \sqrt{1 - 9x^2}$, $0 \le x \le \frac{1}{3}$

b. $\cot(\arccos 3x) = \cot u = \frac{\text{adj}}{\text{opp}} = \frac{3x}{\sqrt{1 - 9x^2}}$, $0 \le x < \frac{1}{3}$

✓ *Checkpoint* Audio-video solution in English & Spanish at *LarsonPrecalculus.com*

Write an algebraic expression that is equivalent to $\sec(\arctan x)$.

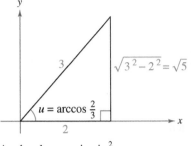

Angle whose cosine is $\frac{2}{3}$
Figure 4.49

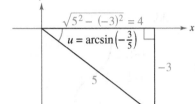

Angle whose sine is $-\frac{3}{5}$
Figure 4.50

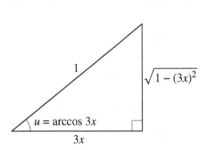

Angle whose cosine is $3x$
Figure 4.51

Summarize (Section 4.7)

1. State the definition of the inverse sine function *(page 318)*. For examples of evaluating and graphing the inverse sine function, see Examples 1 and 2.

2. State the definitions of the inverse cosine and inverse tangent functions *(page 320)*. For examples of evaluating inverse trigonometric functions, see Examples 3 and 4.

3. State the inverse properties of trigonometric functions *(page 322)*. For examples of finding compositions with inverse trigonometric functions, see Examples 5–7.

4.7 Exercises

See CalcChat.com for tutorial help and worked-out solutions to odd-numbered exercises.

Vocabulary: Fill in the blanks.

Function	Alternative Notation	Domain	Range
1. $y = \arcsin x$	_____	_____	$-\dfrac{\pi}{2} \le y \le \dfrac{\pi}{2}$
2. _____	$y = \cos^{-1} x$	$-1 \le x \le 1$	_____
3. $y = \arctan x$	_____	_____	_____

4. A trigonometric function has an _____ function only when its domain is restricted.

Skills and Applications

 Evaluating an Inverse Trigonometric Function In Exercises 5–18, find the exact value of the expression, if possible.

5. $\arcsin \dfrac{1}{2}$

6. $\arcsin 0$

7. $\arccos \dfrac{1}{2}$

8. $\arccos 0$

9. $\arctan \dfrac{\sqrt{3}}{3}$

10. $\arctan 1$

11. $\arcsin 3$

12. $\arctan \sqrt{3}$

13. $\tan^{-1}\left(-\sqrt{3}\right)$

14. $\cos^{-1}(-2)$

15. $\arccos\left(-\dfrac{1}{2}\right)$

16. $\arcsin \dfrac{\sqrt{2}}{2}$

17. $\sin^{-1}\left(-\dfrac{\sqrt{3}}{2}\right)$

18. $\tan^{-1}\left(-\dfrac{\sqrt{3}}{3}\right)$

 Graphing an Inverse Trigonometric Function In Exercises 19 and 20, use a graphing utility to graph f, g, and $y = x$ in the same viewing window to verify geometrically that g is the inverse function of f. (Be sure to restrict the domain of f properly.)

19. $f(x) = \cos x, \quad g(x) = \arccos x$

20. $f(x) = \tan x, \quad g(x) = \arctan x$

 **Calculators and Inverse Trigonometric Functions** In Exercises 21–36, use a calculator to approximate the value of the expression, if possible. Round your result to two decimal places.

21. $\arccos 0.37$

22. $\arcsin 0.65$

23. $\arcsin(-0.75)$

24. $\arccos(-0.7)$

25. $\arctan(-3)$

26. $\arctan 25$

27. $\sin^{-1} 1.36$

28. $\cos^{-1} 0.26$

29. $\arccos(-0.41)$

30. $\arcsin(-0.125)$

31. $\arctan 0.92$

32. $\arctan 2.8$

33. $\arcsin \dfrac{7}{8}$

34. $\arccos\left(-\dfrac{4}{3}\right)$

35. $\tan^{-1}\left(-\dfrac{95}{7}\right)$

36. $\tan^{-1}\left(-\sqrt{372}\right)$

Finding Missing Coordinates In Exercises 37 and 38, determine the missing coordinates of the points on the graph of the function.

37.
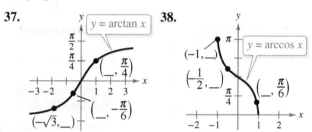

38.
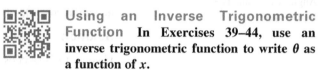

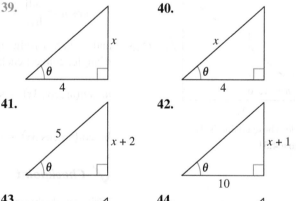 **Using an Inverse Trigonometric Function** In Exercises 39–44, use an inverse trigonometric function to write θ as a function of x.

39.

40.

41.

42.

43.

44.

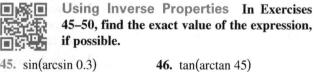 **Using Inverse Properties** In Exercises 45–50, find the exact value of the expression, if possible.

45. $\sin(\arcsin 0.3)$

46. $\tan(\arctan 45)$

47. $\cos\left[\arccos\left(-\sqrt{3}\right)\right]$

48. $\sin[\arcsin(-0.2)]$

49. $\arcsin[\sin(9\pi/4)]$

50. $\arccos[\cos(-3\pi/2)]$

Evaluating a Composition of Functions
In Exercises 51–62, find the exact value of the expression, if possible.

51. $\sin\left(\arctan\frac{3}{4}\right)$

52. $\cos\left(\arcsin\frac{4}{5}\right)$

53. $\cos(\tan^{-1}2)$

54. $\sin\left(\cos^{-1}\sqrt{5}\right)$

55. $\sec\left(\arcsin\frac{5}{13}\right)$

56. $\csc\left[\arctan\left(-\frac{5}{12}\right)\right]$

57. $\cot\left[\arctan\left(-\frac{3}{5}\right)\right]$

58. $\sec\left[\arccos\left(-\frac{3}{4}\right)\right]$

59. $\tan\left[\arccos\left(-\frac{2}{3}\right)\right]$

60. $\cot\left(\arctan\frac{5}{8}\right)$

61. $\csc\left(\cos^{-1}\frac{\sqrt{3}}{2}\right)$

62. $\tan\left[\sin^{-1}\left(-\frac{\sqrt{2}}{2}\right)\right]$

Writing an Expression   In Exercises 63–72, write an algebraic expression that is equivalent to the given expression.

63. $\cos(\arcsin 2x)$

64. $\sin(\arctan x)$

65. $\cot(\arctan x)$

66. $\sec(\arctan 3x)$

67. $\sin(\arccos x)$

68. $\csc[\arccos(x-1)]$

69. $\tan\left(\arccos\frac{x}{3}\right)$

70. $\cot\left(\arctan\frac{1}{x}\right)$

71. $\csc\left(\arctan\frac{x}{a}\right)$

72. $\cos\left(\arcsin\frac{x-h}{r}\right)$

Using Technology   In Exercises 73 and 74, use a graphing utility to graph f and g in the same viewing window to verify that the two functions are equal. Explain why they are equal. Identify any asymptotes of the graphs.

73. $f(x) = \sin(\arctan 2x), \quad g(x) = \dfrac{2x}{\sqrt{1+4x^2}}$

74. $f(x) = \tan\left(\arccos\dfrac{x}{2}\right), \quad g(x) = \dfrac{\sqrt{4-x^2}}{x}$

Completing an Equation   In Exercises 75–78, complete the equation.

75. $\arctan\dfrac{9}{x} = \arcsin\left(\boxed{}\right), \quad x > 0$

76. $\arcsin\dfrac{\sqrt{36-x^2}}{6} = \arccos\left(\boxed{}\right), \quad 0 \le x \le 6$

77. $\arccos\dfrac{3}{\sqrt{x^2-2x+10}} = \arcsin\left(\boxed{}\right)$

78. $\arccos\dfrac{x-2}{2} = \arctan\left(\boxed{}\right), \quad 2 < x < 4$

Sketching the Graph of a Function   In Exercises 79–84, sketch the graph of the function and compare the graph to the graph of the parent inverse trigonometric function.

79. $y = 2\arcsin x$

80. $f(x) = \arctan 2x$

81. $f(x) = \dfrac{\pi}{2} + \arctan x$

82. $g(t) = \arccos(t+2)$

83. $h(v) = \arccos\dfrac{v}{2}$

84. $f(x) = \arcsin\dfrac{x}{4}$

Graphing an Inverse Trigonometric Function   In Exercises 85–90, use a graphing utility to graph the function.

85. $f(x) = 2\arccos 2x$

86. $f(x) = \pi\arcsin 4x$

87. $f(x) = \arctan(2x-3)$

88. $f(x) = -3 + \arctan \pi x$

89. $f(x) = \pi - \sin^{-1}\dfrac{2}{3}$

90. $f(x) = \dfrac{\pi}{2} + \cos^{-1}\dfrac{1}{\pi}$

Using a Trigonometric Identity   In Exercises 91 and 92, write the function in terms of the sine function by using the identity

$$A\cos\omega t + B\sin\omega t = \sqrt{A^2+B^2}\,\sin\left(\omega t + \arctan\frac{A}{B}\right).$$

Use a graphing utility to graph both forms of the function. What does the graph imply?

91. $f(t) = 3\cos 2t + 3\sin 2t$

92. $f(t) = 4\cos \pi t + 3\sin \pi t$

Behavior of an Inverse Trigonometric Function In Exercises 93–98, fill in the blank. If not possible, state the reason. (*Note:* The notation $x \to c^+$ indicates that x approaches c from the right and $x \to c^-$ indicates that x approaches c from the left.)

93. As $x \to 1^-$, the value of $\arcsin x \to$ ▢ .

94. As $x \to 1^-$, the value of $\arccos x \to$ ▢ .

95. As $x \to \infty$, the value of $\arctan x \to$ ▢ .

96. As $x \to -1^+$, the value of $\arcsin x \to$ ▢ .

97. As $x \to -1^+$, the value of $\arccos x \to$ ▢ .

98. As $x \to -\infty$, the value of $\arctan x \to$ ▢ .

99. **Docking a Boat**   A boat is pulled in by means of a winch located on a dock 5 feet above the deck of the boat (see figure). Let θ be the angle of elevation from the boat to the winch and let s be the length of the rope from the winch to the boat.

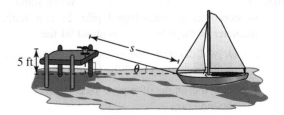

(a) Write θ as a function of s.

(b) Find θ when $s = 40$ feet and $s = 20$ feet.

100. Videography

A television camera at ground level films the lift-off of a space shuttle at a point 750 meters from the launch pad (see figure). Let θ be the angle of elevation to the shuttle and let s be the height of the shuttle.

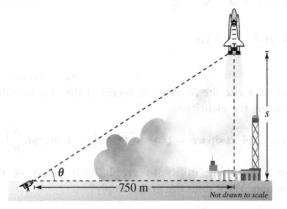

Not drawn to scale

(a) Write θ as a function of s.

(b) Find θ when $s = 300$ meters and $s = 1200$ meters.

101. Granular Angle of Repose Different types of granular substances naturally settle at different angles when stored in cone-shaped piles. This angle θ is called the *angle of repose* (see figure). When rock salt is stored in a cone-shaped pile 5.5 meters high, the diameter of the pile's base is about 17 meters.

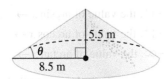

5.5 m

8.5 m

(a) Find the angle of repose for rock salt.

(b) How tall is a pile of rock salt that has a base diameter of 20 meters?

102. Granular Angle of Repose When shelled corn is stored in a cone-shaped pile 20 feet high, the diameter of the pile's base is about 94 feet.

(a) Draw a diagram that gives a visual representation of the problem. Label the known quantities.

(b) Find the angle of repose (see Exercise 101) for shelled corn.

(c) How tall is a pile of shelled corn that has a base diameter of 60 feet?

103. Photography A photographer takes a picture of a three-foot-tall painting hanging in an art gallery. The camera lens is 1 foot below the lower edge of the painting (see figure). The angle β subtended by the camera lens x feet from the painting is given by

$$\beta = \arctan \frac{3x}{x^2 + 4}, \quad x > 0.$$

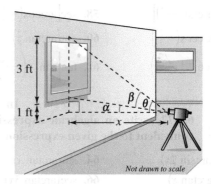

3 ft

1 ft

Not drawn to scale

(a) Use a graphing utility to graph β as a function of x.

(b) Use the graph to approximate the distance from the picture when β is maximum.

(c) Identify the asymptote of the graph and interpret its meaning in the context of the problem.

104. Angle of Elevation An airplane flies at an altitude of 6 miles toward a point directly over an observer. Consider θ and x as shown in the figure.

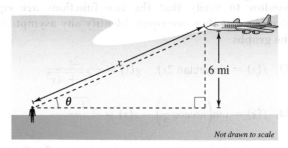

6 mi

Not drawn to scale

(a) Write θ as a function of x.

(b) Find θ when $x = 12$ miles and $x = 7$ miles.

105. Police Patrol A police car with its spotlight on is parked 20 meters from a warehouse. Consider θ and x as shown in the figure.

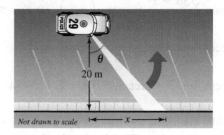

20 m

Not drawn to scale

(a) Write θ as a function of x.

(b) Find θ when $x = 5$ meters and $x = 12$ meters.

Exploration

True or False? **In Exercises 106–109, determine whether the statement is true or false. Justify your answer.**

106. $\sin \dfrac{5\pi}{6} = \dfrac{1}{2} \implies \arcsin \dfrac{1}{2} = \dfrac{5\pi}{6}$

107. $\tan\left(-\dfrac{\pi}{4}\right) = -1 \implies \arctan(-1) = -\dfrac{\pi}{4}$

108. $\arctan x = \dfrac{\arcsin x}{\arccos x}$ **109.** $\sin^{-1} x = \dfrac{1}{\sin x}$

110. HOW DO YOU SEE IT? Use the figure below to determine the value(s) of x for which each statement is true.

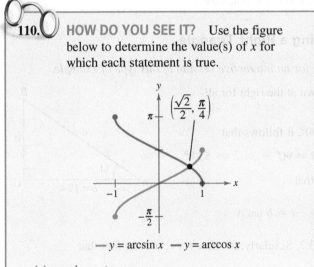

$$\left(\dfrac{\sqrt{2}}{2}, \dfrac{\pi}{4}\right)$$

— $y = \arcsin x$ — $y = \arccos x$

(a) $\arcsin x < \arccos x$

(b) $\arcsin x = \arccos x$

(c) $\arcsin x > \arccos x$

111. Inverse Cotangent Function Define the inverse cotangent function by restricting the domain of the cotangent function to the interval $(0, \pi)$, and sketch the graph of the inverse trigonometric function.

112. Inverse Secant Function Define the inverse secant function by restricting the domain of the secant function to the intervals $[0, \pi/2)$ and $(\pi/2, \pi]$, and sketch the graph of the inverse trigonometric function.

113. Inverse Cosecant Function Define the inverse cosecant function by restricting the domain of the cosecant function to the intervals $[-\pi/2, 0)$ and $(0, \pi/2]$, and sketch the graph of the inverse trigonometric function.

114. Writing Use the results of Exercises 111–113 to explain how to graph (a) the inverse cotangent function, (b) the inverse secant function, and (c) the inverse cosecant function on a graphing utility.

Evaluating an Inverse Trigonometric Function **In Exercises 115–120, use the results of Exercises 111–113 to find the exact value of the expression.**

115. $\operatorname{arcsec} \sqrt{2}$

116. $\operatorname{arcsec} 1$

117. $\operatorname{arccot}(-1)$

118. $\operatorname{arccot}\left(-\sqrt{3}\right)$

119. $\operatorname{arccsc}(-1)$

120. $\operatorname{arccsc} \dfrac{2\sqrt{3}}{3}$

Calculators and Inverse Trigonometric Functions **In Exercises 121–126, use the results of Exercises 111–113 and a calculator to approximate the value of the expression. Round your result to two decimal places.**

121. $\operatorname{arcsec} 2.54$

122. $\operatorname{arcsec}(-1.52)$

123. $\operatorname{arccsc}\left(-\dfrac{25}{3}\right)$

124. $\operatorname{arccsc}(-12)$

125. $\operatorname{arccot} 5.25$

126. $\operatorname{arccot}\left(-\dfrac{16}{7}\right)$

127. Area In calculus, it is shown that the area of the region bounded by the graphs of $y = 0$, $y = 1/(x^2 + 1)$, $x = a$, and $x = b$ (see figure) is given by

Area $= \arctan b - \arctan a.$

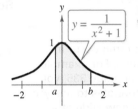

Find the area for each value of a and b.

(a) $a = 0, b = 1$ (b) $a = -1, b = 1$

(c) $a = 0, b = 3$ (d) $a = -1, b = 3$

128. Think About It Use a graphing utility to graph the functions $f(x) = \sqrt{x}$ and $g(x) = 6 \arctan x$. For $x > 0$, it appears that $g > f$. Explain how you know that there exists a positive real number a such that $g < f$ for $x > a$. Approximate the number a.

129. Think About It Consider the functions

$$f(x) = \sin x \quad \text{and} \quad f^{-1}(x) = \arcsin x.$$

(a) Use a graphing utility to graph the composite functions $f \circ f^{-1}$ and $f^{-1} \circ f$.

(b) Explain why the graphs in part (a) are not the graph of the line $y = x$. Why do the graphs of $f \circ f^{-1}$ and $f^{-1} \circ f$ differ?

130. Proof Prove each identity.

(a) $\arcsin(-x) = -\arcsin x$

(b) $\arctan(-x) = -\arctan x$

(c) $\arctan x + \arctan \dfrac{1}{x} = \dfrac{\pi}{2}, \quad x > 0$

(d) $\arcsin x + \arccos x = \dfrac{\pi}{2}$

(e) $\arcsin x = \arctan \dfrac{x}{\sqrt{1 - x^2}}$

4.8   Applications and Models

- ■ Solve real-life problems involving right triangles.
- ■ Solve real-life problems involving directional bearings.
- ■ Solve real-life problems involving harmonic motion.

Applications Involving Right Triangles

In this section, the three angles of a right triangle are denoted by A, B, and C (where C is the right angle), and the lengths of the sides opposite these angles are denoted by a, b, and c, respectively (where c is the hypotenuse).

Right triangles often occur in real-life situations. For example, in Exercise 30 on page 335, you will use right triangles to analyze the design of a new slide at a water park.

EXAMPLE 1   Solving a Right Triangle

See LarsonPrecalculus.com for an interactive version of this type of example.

Solve the right triangle shown at the right for all unknown sides and angles.

Solution   Because $C = 90°$, it follows that

$$A + B = 90° \quad \text{and} \quad B = 90° - 34.2° = 55.8°.$$

To solve for a, use the fact that

$$\tan A = \frac{\text{opp}}{\text{adj}} = \frac{a}{b} \implies a = b \tan A.$$

So, $a = 19.4 \tan 34.2° \approx 13.2$. Similarly, to solve for c, use the fact that

$$\cos A = \frac{\text{adj}}{\text{hyp}} = \frac{b}{c} \implies c = \frac{b}{\cos A}.$$

So, $c = \dfrac{19.4}{\cos 34.2°} \approx 23.5$.

✓ **Checkpoint**))) *Audio-video solution in English & Spanish at LarsonPrecalculus.com*

Solve the right triangle shown at the right for all unknown sides and angles.

EXAMPLE 2   Finding a Side of a Right Triangle

The height of a mountain is 5000 feet. The distance between its peak and that of an adjacent mountain is 25,000 feet. The angle of elevation between the two peaks is 27°. (See Figure 4.52.) What is the height of the adjacent mountain?

Solution   From the figure, $\sin A = a/c$, so

$$a = c \sin A = 25{,}000 \sin 27° \approx 11{,}350.$$

The height of the adjacent mountain is about $11{,}350 + 5000 = 16{,}350$ feet.

✓ **Checkpoint**))) *Audio-video solution in English & Spanish at LarsonPrecalculus.com*

A ladder that is 16 feet long leans against the side of a house. The angle of elevation of the ladder is 80°. Find the height from the top of the ladder to the ground. ■

Figure 4.52

EXAMPLE 3 **Finding a Side of a Right Triangle**

At a point 200 feet from the base of a building, the angle of elevation to the *bottom* of a smokestack is 35°, whereas the angle of elevation to the *top* is 53°, as shown in Figure 4.53. Find the height s of the smokestack alone.

Solution

This problem involves two right triangles. For the smaller right triangle, use the fact that

$$\tan 35° = \frac{a}{200}$$

to find that the height of the building is

$$a = 200 \tan 35°.$$

For the larger right triangle, use the equation

$$\tan 53° = \frac{a + s}{200}$$

to find that

$$a + s = 200 \tan 53°.$$

So, the height of the smokestack is

$$s = 200 \tan 53° - a$$

$$= 200 \tan 53° - 200 \tan 35°$$

$$\approx 125.4 \text{ feet}.$$

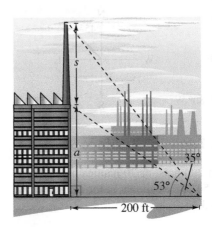

Figure 4.53

✓ *Checkpoint*))) *Audio-video solution in English & Spanish at LarsonPrecalculus.com*

At a point 65 feet from the base of a church, the angles of elevation to the bottom of the steeple and the top of the steeple are 35° and 43°, respectively. Find the height of the steeple.

EXAMPLE 4 **Finding an Angle of Depression**

A swimming pool is 20 meters long and 12 meters wide. The bottom of the pool is slanted so that the water depth is 1.3 meters at the shallow end and 4 meters at the deep end, as shown in Figure 4.54. Find the angle of depression (in degrees) of the bottom of the pool.

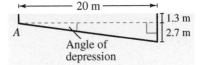

Figure 4.54

Solution   Using the tangent function,

$$\tan A = \frac{\text{opp}}{\text{adj}}$$

$$= \frac{2.7}{20}$$

$$= 0.135.$$

So, the angle of depression is

$$A = \arctan 0.135 \approx 0.13419 \text{ radian} \approx 7.69°.$$

✓ *Checkpoint*))) *Audio-video solution in English & Spanish at LarsonPrecalculus.com*

From the time a small airplane is 100 feet high and 1600 ground feet from its landing runway, the plane descends in a straight line to the runway. Determine the angle of descent (in degrees) of the plane.

. . . .

▷ •

••REMARK In *air navigation*, bearings are measured in degrees *clockwise* from north. The figures below illustrate examples of air navigation bearings

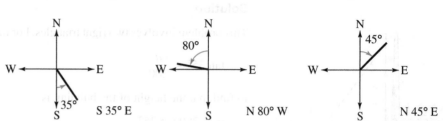

Trigonometry and Bearings

In surveying and navigation, directions can be given in terms of **bearings.** A bearing measures the acute angle that a path or line of sight makes with a fixed north-south line. For example, in the figures below, the bearing S 35° E means 35 degrees east of south, N 80° W means 80 degrees west of north, and N 45° E means 45 degrees east of north.

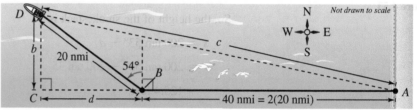

| EXAMPLE 5 | **Finding Directions in Terms of Bearings** |

A ship leaves port at noon and heads due west at 20 knots, or 20 nautical miles (nmi) per hour. At 2 P.M. the ship changes course to N 54° W, as shown in the figure below. Find the ship's bearing and distance from port at 3 P.M.

Solution

For triangle BCD, you have

$$B = 90° - 54° = 36°.$$

The two sides of this triangle are

$$b = 20 \sin 36° \quad \text{and} \quad d = 20 \cos 36°.$$

For triangle ACD, find angle A.

$$\tan A = \frac{b}{d + 40} = \frac{20 \sin 36°}{20 \cos 36° + 40} \approx 0.209$$

$$A \approx \arctan 0.209 \approx 0.20603 \text{ radian} \approx 11.80°$$

The angle with the north-south line is $90° - 11.80° = 78.20°$. So, the bearing of the ship is N 78.20° W. Finally, from triangle ACD, you have

$$\sin A = \frac{b}{c}$$

which yields

$$c = \frac{b}{\sin A} = \frac{20 \sin 36°}{\sin 11.80°} \approx 57.5 \text{ nautical miles.} \qquad \text{Distance from port}$$

✓ **Checkpoint**))) Audio-video solution in English & Spanish at LarsonPrecalculus.com

A sailboat leaves a pier heading due west at 8 knots. After 15 minutes, the sailboat changes course to N 16° W at 10 knots. Find the sailboat's bearing and distance from the pier after 12 minutes on this course.

Harmonic Motion

The periodic nature of the trigonometric functions is useful for describing the motion of a point on an object that vibrates, oscillates, rotates, or is moved by wave motion.

For example, consider a ball that is bobbing up and down on the end of a spring. Assume that the maximum distance the ball moves vertically upward or downward from its equilibrium (at rest) position is 10 centimeters (see figure). Assume further that the time it takes for the ball to move from its maximum displacement above zero to its maximum displacement below zero and back again is $t = 4$ seconds. With the ideal conditions of perfect elasticity and no friction or air resistance, the ball would continue to move up and down in a uniform and regular manner.

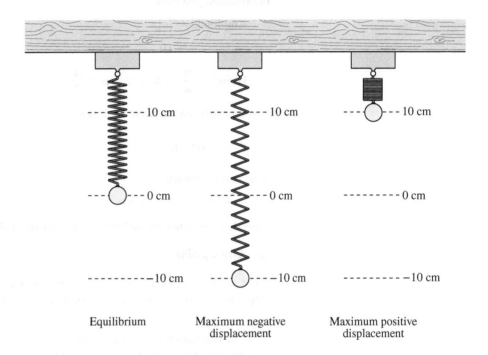

| Equilibrium | Maximum negative displacement | Maximum positive displacement |

The period (time for one complete cycle) of the motion is

Period = 4 seconds

the amplitude (maximum displacement from equilibrium) is

Amplitude = 10 centimeters

and the **frequency** (number of cycles per second) is

Frequency = $\dfrac{1}{4}$ cycle per second.

Motion of this nature can be described by a sine or cosine function and is called **simple harmonic motion.**

Definition of Simple Harmonic Motion

A point that moves on a coordinate line is in **simple harmonic motion** when its distance d from the origin at time t is given by either

$$d = a \sin \omega t \quad \text{or} \quad d = a \cos \omega t$$

where a and ω are real numbers such that $\omega > 0$. The motion has amplitude $|a|$, period $\dfrac{2\pi}{\omega}$, and frequency $\dfrac{\omega}{2\pi}$.

EXAMPLE 6 **Simple Harmonic Motion**

Write an equation for the simple harmonic motion of the ball described on the preceding page.

Solution

The spring is at equilibrium ($d = 0$) when $t = 0$, so use the equation

$$d = a \sin \omega t.$$

Moreover, the maximum displacement from zero is 10 and the period is 4. Using this information, you have

$$\text{Amplitude} = |a|$$

$$= 10$$

$$\text{Period} = \frac{2\pi}{\omega} = 4 \quad \Longrightarrow \quad \omega = \frac{\pi}{2}.$$

Consequently, an equation of motion is

$$d = 10 \sin \frac{\pi}{2} t.$$

Note that the choice of

$$a = 10 \quad \text{or} \quad a = -10$$

depends on whether the ball initially moves up or down.

✓ **Checkpoint** 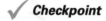 Audio-video solution in English & Spanish at LarsonPrecalculus.com

Write an equation for simple harmonic motion for which $d = 0$ when $t = 0$, the amplitude is 6 centimeters, and the period is 3 seconds. ■

One illustration of the relationship between sine waves and harmonic motion is in the wave motion that results when you drop a stone into a calm pool of water. The waves move outward in roughly the shape of sine (or cosine) waves, as shown at the right. Now suppose you are fishing in the same pool of water and your fishing bobber does not move horizontally. As the waves move outward from the dropped stone, the fishing bobber moves up and down in simple harmonic motion, as shown below.

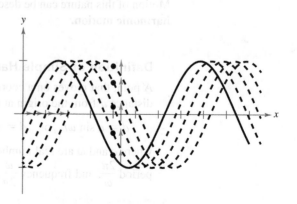

EXAMPLE 7 **Simple Harmonic Motion**

Consider the equation for simple harmonic motion $d = 6 \cos \dfrac{3\pi}{4}t$. Find (a) the maximum displacement, (b) the frequency, (c) the value of d when $t = 4$, and (d) the least positive value of t for which $d = 0$.

Algebraic Solution

The equation has the form $d = a \cos \omega t$, with $a = 6$ and $\omega = 3\pi/4$.

a. The maximum displacement (from the point of equilibrium) is the amplitude. So, the maximum displacement is 6.

b. Frequency $= \dfrac{\omega}{2\pi}$

$\qquad\quad = \dfrac{3\pi/4}{2\pi}$

$\qquad\quad = \dfrac{3}{8}$ cycle per unit of time

c. $d = 6 \cos\left[\dfrac{3\pi}{4}(4)\right] = 6 \cos 3\pi = 6(-1) = -6$

d. To find the least positive value of t for which $d = 0$, solve

$\qquad 6 \cos \dfrac{3\pi}{4}t = 0.$

First divide each side by 6 to obtain

$\qquad \cos \dfrac{3\pi}{4}t = 0.$

This equation is satisfied when

$\qquad \dfrac{3\pi}{4}t = \dfrac{\pi}{2}, \dfrac{3\pi}{2}, \dfrac{5\pi}{2}, \ldots$

Multiply these values by $4/(3\pi)$ to obtain

$\qquad t = \dfrac{2}{3}, 2, \dfrac{10}{3}, \ldots$

So, the least positive value of t is $t = \frac{2}{3}$.

Graphical Solution

Use a graphing utility set in *radian* mode.

a.

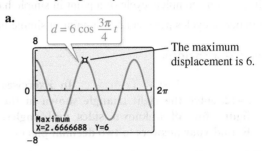

The maximum displacement is 6.

b.

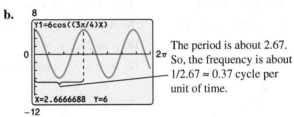

The period is about 2.67. So, the frequency is about $1/2.67 \approx 0.37$ cycle per unit of time.

c.

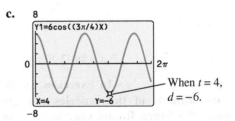

When $t = 4$, $d = -6$.

d. The least positive value of t for which $d = 0$ is $t \approx 0.67$.

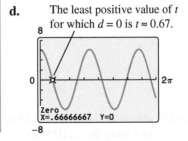

 Checkpoint)))) Audio-video solution in English & Spanish at LarsonPrecalculus.com

Rework Example 7 for the equation $d = 4 \cos 6\pi t$.

Summarize **(Section 4.8)**

1. Describe real-life applications of right triangles (*pages 328 and 329, Examples 1–4*).

2. Describe a real-life application of a directional bearing (*page 330, Example 5*).

3. Describe real-life applications of simple harmonic motion (*pages 332 and 333, Examples 6 and 7*).

4.8 Exercises

See **CalcChat.com** for tutorial help and worked-out solutions to odd-numbered exercises.

Vocabulary: Fill in the blanks.

1. A _____ measures the acute angle that a path or line of sight makes with a fixed north-south line.

2. A point that moves on a coordinate line is in simple _____ _____ when its distance d from the origin at time t is given by either $d = a \sin \omega t$ or $d = a \cos \omega t$.

3. The time for one complete cycle of a point in simple harmonic motion is its _____.

4. The number of cycles per second of a point in simple harmonic motion is its _____.

Skills and Applications

Solving a Right Triangle **In Exercises 5–12, solve the right triangle shown in the figure for all unknown sides and angles. Round your answers to two decimal places.**

5. $A = 60°$, $c = 12$

6. $B = 25°$, $b = 4$

7. $B = 72.8°$, $a = 4.4$

8. $A = 8.4°$, $a = 40.5$

9. $a = 3$, $b = 4$

10. $a = 25$, $c = 35$

11. $b = 15.70$, $c = 55.16$

12. $b = 1.32$, $c = 9.45$

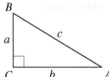

Figure for 5–12

Figure for 13–16

Finding an Altitude **In Exercises 13–16, find the altitude of the isosceles triangle shown in the figure. Round your answers to two decimal places.**

13. $\theta = 45°$, $b = 6$

14. $\theta = 22°$, $b = 14$

15. $\theta = 32°$, $b = 8$

16. $\theta = 27°$, $b = 11$

17. **Length** The sun is 25° above the horizon. Find the length of a shadow cast by a building that is 100 feet tall (see figure).

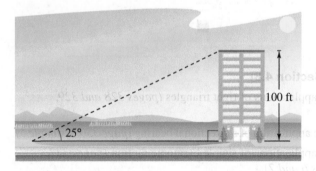

18. **Length** The sun is 20° above the horizon. Find the length of a shadow cast by a park statue that is 12 feet tall.

19. **Height** A ladder that is 20 feet long leans against the side of a house. The angle of elevation of the ladder is 80°. Find the height from the top of the ladder to the ground.

20. **Height** The length of a shadow of a tree is 125 feet when the angle of elevation of the sun is 33°. Approximate the height of the tree.

21. **Height** At a point 50 feet from the base of a church, the angles of elevation to the bottom of the steeple and the top of the steeple are 35° and 48°, respectively. Find the height of the steeple.

22. **Distance** An observer in a lighthouse 350 feet above sea level observes two ships directly offshore. The angles of depression to the ships are 4° and 6.5° (see figure). How far apart are the ships?

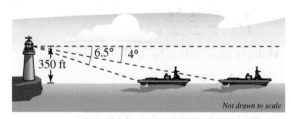

Not drawn to scale

23. **Distance** A passenger in an airplane at an altitude of 10 kilometers sees two towns directly to the east of the plane. The angles of depression to the towns are 28° and 55° (see figure). How far apart are the towns?

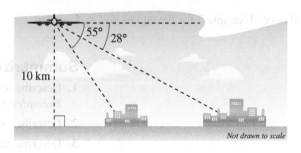

Not drawn to scale

24. Angle of Elevation The height of an outdoor basketball backboard is $12\frac{1}{2}$ feet, and the backboard casts a shadow 17 feet long.

(a) Draw a right triangle that gives a visual representation of the problem. Label the known and unknown quantities.

(b) Use a trigonometric function to write an equation involving the unknown angle of elevation.

(c) Find the angle of elevation.

25. Angle of Elevation An engineer designs a 75-foot cellular telephone tower. Find the angle of elevation to the top of the tower at a point on level ground 50 feet from its base.

26. Angle of Depression A cellular telephone tower that is 120 feet tall is placed on top of a mountain that is 1200 feet above sea level. What is the angle of depression from the top of the tower to a cell phone user who is 5 horizontal miles away and 400 feet above sea level?

27. Angle of Depression A Global Positioning System satellite orbits 12,500 miles above Earth's surface (see figure). Find the angle of depression from the satellite to the horizon. Assume the radius of Earth is 4000 miles.

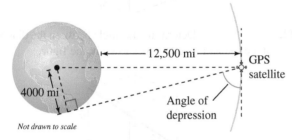

Not drawn to scale

28. Height You are holding one of the tethers attached to the top of a giant character balloon that is floating approximately 20 feet above ground level. You are standing approximately 100 feet ahead of the balloon (see figure).

Not drawn to scale

(a) Find an equation for the length l of the tether you are holding in terms of h, the height of the balloon from top to bottom.

(b) Find an equation for the angle of elevation θ from you to the top of the balloon.

(c) The angle of elevation to the top of the balloon is 35°. Find the height h of the balloon.

29. Altitude You observe a plane approaching overhead and assume that its speed is 550 miles per hour. The angle of elevation of the plane is 16° at one time and 57° one minute later. Approximate the altitude of the plane.

30. Waterslide Design

The designers of a water park have sketched a preliminary drawing of a new slide (see figure).

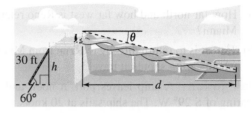

(a) Find the height h of the slide.

(b) Find the angle of depression θ from the top of the slide to the end of the slide at the ground in terms of the horizontal distance d a rider travels.

(c) Safety restrictions require the angle of depression to be no less than 25° and no more than 30°. Find an interval for how far a rider travels horizontally.

31. Speed Enforcement A police department has set up a speed enforcement zone on a straight length of highway. A patrol car is parked parallel to the zone, 200 feet from one end and 150 feet from the other end (see figure).

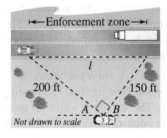

Not drawn to scale

(a) Find the length l of the zone and the measures of angles A and B (in degrees).

(b) Find the minimum amount of time (in seconds) it takes for a vehicle to pass through the zone without exceeding the posted speed limit of 35 miles per hour.

32. Airplane Ascent During takeoff, an airplane's angle of ascent is 18° and its speed is 260 feet per second.

(a) Find the plane's altitude after 1 minute.

(b) How long will it take for the plane to climb to an altitude of 10,000 feet?

33. Air Navigation An airplane flying at 550 miles per hour has a bearing of 52°. After flying for 1.5 hours, how far north and how far east will the plane have traveled from its point of departure?

34. Air Navigation A jet leaves Reno, Nevada, and heads toward Miami, Florida, at a bearing of 100°. The distance between the two cities is approximately 2472 miles.

(a) How far north and how far west is Reno relative to Miami?

(b) The jet is to return directly to Reno from Miami. At what bearing should it travel?

35. Navigation A ship leaves port at noon and has a bearing of S 29° W. The ship sails at 20 knots.

(a) How many nautical miles south and how many nautical miles west will the ship have traveled by 6:00 P.M.?

(b) At 6:00 P.M., the ship changes course to due west. Find the ship's bearing and distance from port at 7:00 P.M.

36. Navigation A privately owned yacht leaves a dock in Myrtle Beach, South Carolina, and heads toward Freeport in the Bahamas at a bearing of S 1.4° E. The yacht averages a speed of 20 knots over the 428-nautical-mile trip.

(a) How long will it take the yacht to make the trip?

(b) How far east and south is the yacht after 12 hours?

(c) A plane leaves Myrtle Beach to fly to Freeport. At what bearing should it travel?

37. Navigation A ship is 45 miles east and 30 miles south of port. The captain wants to sail directly to port. What bearing should the captain take?

38. Air Navigation An airplane is 160 miles north and 85 miles east of an airport. The pilot wants to fly directly to the airport. What bearing should the pilot take?

39. Surveying A surveyor wants to find the distance across a pond (see figure). The bearing from A to B is N 32° W. The surveyor walks 50 meters from A to C, and at the point C the bearing to B is N 68° W.

(a) Find the bearing from A to C.

(b) Find the distance from A to B.

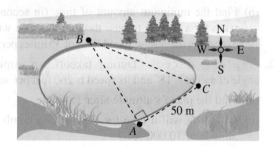

40. Location of a Fire Fire tower A is 30 kilometers due west of fire tower B. A fire is spotted from the towers, and the bearings from A and B are N 76° E and N 56° W, respectively (see figure). Find the distance d of the fire from the line segment AB.

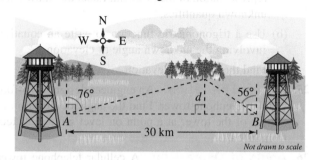

Not drawn to scale

41. Geometry Determine the angle between the diagonal of a cube and the diagonal of its base, as shown in the figure.

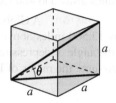

42. Geometry Determine the angle between the diagonal of a cube and its edge, as shown in the figure.

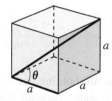

43. Geometry Find the length of the sides of a regular pentagon inscribed in a circle of radius 25 inches.

44. Geometry Find the length of the sides of a regular hexagon inscribed in a circle of radius 25 inches.

Simple Harmonic Motion In Exercises 45–48, find a model for simple harmonic motion satisfying the specified conditions.

Displacement ($t = 0$)	Amplitude	Period
45. 0	4 centimeters	2 seconds
46. 0	3 meters	6 seconds
47. 3 inches	3 inches	1.5 seconds
48. 2 feet	2 feet	10 seconds

49. Tuning Fork A point on the end of a tuning fork moves in simple harmonic motion described by $d = a \sin \omega t$. Find ω given that the tuning fork for middle C has a frequency of 262 vibrations per second.

50. Wave Motion A buoy oscillates in simple harmonic motion as waves go past. The buoy moves a total of 3.5 feet from its low point to its high point (see figure), and it returns to its high point every 10 seconds. Write an equation that describes the motion of the buoy where the high point corresponds to the time $t = 0$.

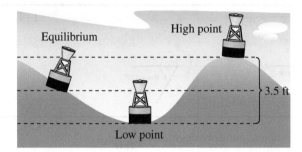

Simple Harmonic Motion **In Exercises 51–54, for the simple harmonic motion described by the trigonometric function, find (a) the maximum displacement, (b) the frequency, (c) the value of d when $t = 5$, and (d) the least positive value of t for which $d = 0$. Use a graphing utility to verify your results.**

51. $d = 9 \cos \dfrac{6\pi}{5} t$ **52.** $d = \dfrac{1}{2} \cos 20\pi t$

53. $d = \dfrac{1}{4} \sin 6\pi t$ **54.** $d = \dfrac{1}{64} \sin 792\pi t$

55. Oscillation of a Spring A ball that is bobbing up and down on the end of a spring has a maximum displacement of 3 inches. Its motion (in ideal conditions) is modeled by $y = \frac{1}{4} \cos 16t$, $t > 0$, where y is measured in feet and t is the time in seconds.

(a) Graph the function.

(b) What is the period of the oscillations?

(c) Determine the first time the weight passes the point of equilibrium $(y = 0)$.

56. Hours of Daylight The numbers of hours H of daylight in Denver, Colorado, on the 15th of each month starting with January are: 9.68, 10.72, 11.92, 13.25, 14.35, 14.97, 14.72, 13.73, 12.47, 11.18, 10.00, and 9.37. A model for the data is

$$H(t) = 12.13 + 2.77 \sin\left(\frac{\pi t}{6} - 1.60\right)$$

where t represents the month, with $t = 1$ corresponding to January. *(Source: United States Navy)*

(a) Use a graphing utility to graph the data and the model in the same viewing window.

(b) What is the period of the model? Is it what you expected? Explain.

(c) What is the amplitude of the model? What does it represent in the context of the problem?

57. Sales The table shows the average sales S (in millions of dollars) of an outerwear manufacturer for each month t, where $t = 1$ corresponds to January.

Time, t	1	2	3	4
Sales, S	13.46	11.15	8.00	4.85

Time, t	5	6	7	8
Sales, S	2.54	1.70	2.54	4.85

Time, t	9	10	11	12
Sales, S	8.00	11.15	13.46	14.30

(a) Create a scatter plot of the data.

(b) Find a trigonometric model that fits the data. Graph the model with your scatter plot. How well does the model fit the data?

(c) What is the period of the model? Do you think it is reasonable given the context? Explain.

(d) Interpret the meaning of the model's amplitude in the context of the problem.

Exploration

58. **HOW DO YOU SEE IT?** The graph below shows the displacement of an object in simple harmonic motion.

(a) What is the amplitude?

(b) What is the period?

(c) Is the equation of the simple harmonic motion of the form $d = a \sin \omega t$ or $d = a \cos \omega t$?

True or False? **In Exercises 59 and 60, determine whether the statement is true or false. Justify your answer.**

59. The Leaning Tower of Pisa is not vertical, but when you know the angle of elevation θ to the top of the tower as you stand d feet away from it, its height h can be found using the formula $h = d \tan \theta$.

60. The bearing N 24° E means 24 degrees north of east.

Chapter Summary

What Did You Learn?	Explanation/Examples	Review Exercises
Section 4.1 Describe angles *(p. 260)*.		1–4
Use radian measure *(p. 261)* and degree measure *(p. 263)*.	To convert degrees to radians, multiply degrees by $\frac{\pi \text{ rad}}{180°}$. To convert radians to degrees, multiply radians by $\frac{180°}{\pi \text{ rad}}$.	5–14
Use angles and their measure to model and solve real-life problems *(p. 264)*.	Angles and their measure can be used to find arc length and the area of a sector of a circle. (See Examples 5 and 8.)	15–18
Section 4.2 Identify a unit circle and describe its relationship to real numbers *(p. 270)*.		19–22
Evaluate trigonometric functions using the unit circle *(p. 271)*.	For the point (x, y) on the unit circle corresponding to a real number t: $\sin t = y$; $\cos t = x$; $\tan t = \frac{y}{x}$, $x \neq 0$; $\csc t = \frac{1}{y}$, $y \neq 0$; $\sec t = \frac{1}{x}$, $x \neq 0$; and $\cot t = \frac{x}{y}$, $y \neq 0$.	23, 24
Use domain and period to evaluate sine and cosine functions *(p. 273)*, and use a calculator to evaluate trigonometric functions *(p. 274)*.	Because $\frac{13\pi}{6} = 2\pi + \frac{\pi}{6}$, $\sin \frac{13\pi}{6} = \sin \frac{\pi}{6} = \frac{1}{2}$. $\sin \frac{3\pi}{8} \approx 0.9239$, $\cot(-1.2) \approx -0.3888$	25–32
Section 4.3 Evaluate trigonometric functions of acute angles *(p. 277)*.	$\sin \theta = \frac{\text{opp}}{\text{hyp}}$, $\quad \cos \theta = \frac{\text{adj}}{\text{hyp}}$, $\quad \tan \theta = \frac{\text{opp}}{\text{adj}}$ $\csc \theta = \frac{\text{hyp}}{\text{opp}}$, $\quad \sec \theta = \frac{\text{hyp}}{\text{adj}}$, $\quad \cot \theta = \frac{\text{adj}}{\text{opp}}$ $\csc 29° 15' = 1/\sin 29.25° \approx 2.0466$	33–38
Use fundamental trigonometric identities *(p. 280)*.	$\sin \theta = \frac{1}{\csc \theta}$, $\quad \tan \theta = \frac{\sin \theta}{\cos \theta}$, $\quad \sin^2 \theta + \cos^2 \theta = 1$	39, 40
Use trigonometric functions to model and solve real-life problems *(p. 282)*.	Trigonometric functions can be used to find the height of a monument, the angle between two paths, and the length and height of a ramp. (See Examples 8–10.)	41, 42

	What Did You Learn?	**Explanation/Examples**	**Review Exercises**
Section 4.4	Evaluate trigonometric functions of any angle *(p. 288)*.	Let $(3, 4)$ be a point on the terminal side of θ. Then $\sin \theta = \frac{4}{5}$, $\cos \theta = \frac{3}{5}$, and $\tan \theta = \frac{4}{3}$.	43–50
	Find reference angles *(p. 290)*.	Let θ be an angle in standard position. Its reference angle is the acute angle θ' formed by the terminal side of θ and the horizontal axis.	51–54
	Evaluate trigonometric functions of real numbers *(p. 291)*.	$\cos \frac{7\pi}{3} = \frac{1}{2}$ because $\theta' = \frac{7\pi}{3} - 2\pi = \frac{\pi}{3}$ and $\cos \frac{\pi}{3} = \frac{1}{2}$.	55–62
Section 4.5	Sketch the graphs of sine and cosine functions using amplitude and period *(p. 299)*.		63, 64
	Sketch translations of the graphs of sine and cosine functions *(p. 301)*.	For $y = d + a \sin(bx - c)$ and $y = d + a \cos(bx - c)$, the constant c results in horizontal translations and the constant d results in vertical translations. (See Examples 4–6.)	65–68
	Use sine and cosine functions to model real-life data *(p. 303)*.	A cosine function can be used to model the depth of the water at the end of a dock. (See Example 7.)	69, 70
Section 4.6	Sketch the graphs of tangent *(p. 308)*, cotangent *(p. 310)*, secant *(p. 311)*, and cosecant functions *(p. 311)*.		71–74
	Sketch the graphs of damped trigonometric functions *(p. 313)*.	In $f(x) = x \cos 2x$, the factor x is called the damping factor.	75, 76
Section 4.7	Evaluate and graph inverse trigonometric functions *(p. 318)*.	$\arcsin\left(\frac{1}{2}\right) = \frac{\pi}{6}$, $\cos^{-1}\left(-\frac{\sqrt{2}}{2}\right) = \frac{3\pi}{4}$, $\tan^{-1}\sqrt{3} = \frac{\pi}{3}$	77–86
	Evaluate compositions with inverse trigonometric functions *(p. 322)*.	$\sin(\sin^{-1} 0.4) = 0.4$, $\cos\left(\arctan \frac{5}{12}\right) = \frac{12}{13}$	87–92
Section 4.8	Solve real-life problems involving right triangles *(p. 328)*.	A trigonometric function can be used to find the height of a smokestack on top of a building. (See Example 3.)	93, 94
	Solve real-life problems involving directional bearings *(p. 330)*.	Trigonometric functions can be used to find a ship's bearing and distance from a port at a given time. (See Example 5.)	95
	Solve real-life problems involving harmonic motion *(p. 331)*.	Trigonometric functions can be used to describe the motion of a point on an object that vibrates, oscillates, rotates, or is moved by wave motion. (See Examples 6 and 7.)	96

4.1 **Using Radian or Degree Measure** In Exercises 1–4, (a) sketch the angle in standard position, (b) determine the quadrant in which the angle lies, and (c) determine two coterminal angles (one positive and one negative).

1. $\dfrac{15\pi}{4}$ **2.** $-\dfrac{4\pi}{3}$

3. $-110°$ **4.** $280°$

Converting from Degrees to Radians In Exercises 5–8, convert the degree measure to radian measure. Round to three decimal places.

5. $450°$ **6.** $190°$

7. $-16°$ **8.** $-112°$

Converting from Radians to Degrees In Exercises 9–12, convert the radian measure to degree measure. Round to three decimal places, if necessary.

9. $\dfrac{3\pi}{10}$ **10.** $-\dfrac{11\pi}{6}$

11. -3.5 **12.** 5.7

Converting to D° M′ S″ Form In Exercises 13 and 14, convert the angle measure to D° M′ S″ form.

13. $198.4°$ **14.** $-5.96°$

15. Arc Length Find the length of the arc on a circle of radius 20 inches intercepted by a central angle of $138°$.

16. Phonograph Phonograph records are vinyl discs that rotate on a turntable. A typical record album is 12 inches in diameter and plays at $33\frac{1}{3}$ revolutions per minute.

(a) Find the angular speed of a record album.

(b) Find the linear speed (in inches per minute) of the outer edge of a record album.

Area of a Sector of a Circle In Exercises 17 and 18, find the area of the sector of a circle of radius r and central angle θ.

Radius r	Central Angle θ
17. 20 inches	$150°$
18. 7.5 millimeters	$2\pi/3$ radians

4.2 **Finding a Point on the Unit Circle** In Exercises 19–22, find the point (x, y) on the unit circle that corresponds to the real number t.

19. $t = 2\pi/3$ **20.** $t = 7\pi/4$

21. $t = 7\pi/6$ **22.** $t = -4\pi/3$

Evaluating Trigonometric Functions In Exercises 23 and 24, evaluate (if possible) the six trigonometric functions at the real number.

23. $t = \dfrac{3\pi}{4}$ **24.** $t = -\dfrac{2\pi}{3}$

Using Period to Evaluate Sine and Cosine In Exercises 25–28, evaluate the trigonometric function using its period as an aid.

25. $\sin \dfrac{11\pi}{4}$ **26.** $\cos 4\pi$

27. $\cos\left(-\dfrac{17\pi}{6}\right)$ **28.** $\sin\left(-\dfrac{13\pi}{3}\right)$

Using a Calculator In Exercises 29–32, use a calculator to evaluate the trigonometric function. Round your answer to four decimal places. (Be sure the calculator is in the correct mode.)

29. $\sec \dfrac{12\pi}{5}$ **30.** $\sin\left(-\dfrac{\pi}{9}\right)$

31. $\tan 33$ **32.** $\csc 10.5$

4.3 **Evaluating Trigonometric Functions** In Exercises 33 and 34, find the exact values of the six trigonometric functions of the angle θ.

33. 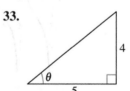   **34.**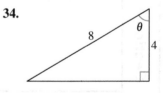

Using a Calculator In Exercises 35–38, use a calculator to evaluate the trigonometric function. Round your answer to four decimal places. (Be sure the calculator is in the correct mode.)

35. $\tan 33°$ **36.** $\sec 79.3°$

37. $\cot 15° \, 14'$ **38.** $\cos 78° \, 11' \, 58''$

Applying Trigonometric Identities In Exercises 39 and 40, use the given function value and the trigonometric identities to find the exact value of each indicated trigonometric function.

39. $\sin \theta = \frac{1}{3}$

(a) $\csc \theta$ (b) $\cos \theta$

(c) $\sec \theta$ (d) $\tan \theta$

40. $\csc \theta = 5$

(a) $\sin \theta$ (b) $\cot \theta$

(c) $\tan \theta$ (d) $\sec(90° - \theta)$

41. **Railroad Grade** A train travels 3.5 kilometers on a straight track with a grade of 1.2° (see figure). What is the vertical rise of the train in that distance?

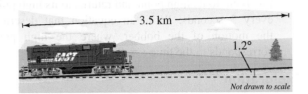

3.5 km

1.2°

Not drawn to scale

42. **Guy Wire** A guy wire runs from the ground to the top of a 25-foot telephone pole. The angle formed between the wire and the ground is 52°. How far from the base of the pole is the guy wire anchored to the ground? Assume the pole is perpendicular to the ground.

4.4 **Evaluating Trigonometric Functions** In Exercises 43–46, the point is on the terminal side of an angle in standard position. Find the exact values of the six trigonometric functions of the angle.

43. $(12, 16)$ **44.** $(3, -4)$

45. $(0.3, 0.4)$ **46.** $\left(-\frac{10}{3}, -\frac{2}{3}\right)$

Evaluating Trigonometric Functions In Exercises 47–50, find the exact values of the remaining five trigonometric functions of θ satisfying the given conditions.

47. $\sec \theta = \frac{6}{5}$, $\tan \theta < 0$

48. $\csc \theta = \frac{3}{2}$, $\cos \theta < 0$

49. $\cos \theta = -\frac{2}{5}$, $\sin \theta > 0$

50. $\sin \theta = -\frac{1}{2}$, $\cos \theta > 0$

Finding a Reference Angle In Exercises 51–54, find the reference angle θ'. Sketch θ in standard position and label θ'.

51. $\theta = 264°$ **52.** $\theta = 635°$

53. $\theta = -6\pi/5$ **54.** $\theta = 17\pi/3$

Using a Reference Angle In Exercises 55–58, evaluate the sine, cosine, and tangent of the angle without using a calculator.

55. $-150°$ **56.** $495°$

57. $\pi/3$ **58.** $-5\pi/4$

Using a Calculator In Exercises 59–62, use a calculator to evaluate the trigonometric function. Round your answer to four decimal places. (Be sure the calculator is in the correct mode.)

59. $\sin 106°$

60. $\tan 37°$

61. $\tan(-17\pi/15)$

62. $\cos(-25\pi/7)$

4.5 **Sketching the Graph of a Sine or Cosine Function** In Exercises 63–68, sketch the graph of the function. (Include two full periods.)

63. $y = \sin 6x$

64. $f(x) = -\cos 3x$

65. $y = 5 + \sin \pi x$

66. $y = -4 - \cos \pi x$

67. $g(t) = \frac{5}{2} \sin(t - \pi)$

68. $g(t) = 3 \cos(t + \pi)$

69. **Sound Waves** Sound waves can be modeled using sine functions of the form $y = a \sin bx$, where x is measured in seconds.

   (a) Write an equation of a sound wave whose amplitude is 2 and whose period is $\frac{1}{264}$ second.

   (b) What is the frequency of the sound wave described in part (a)?

70. **Meteorology** The times S of sunset (Greenwich Mean Time) at 40° north latitude on the 15th of each month starting with January are: 16:59, 17:35, 18:06, 18:38, 19:08, 19:30, 19:28, 18:57, 18:10, 17:21, 16:44, and 16:36. A model (in which minutes have been converted to the decimal parts of an hour) for the data is

$$S(t) = 18.10 - 1.41 \sin\left(\frac{\pi t}{6} + 1.55\right)$$

where t represents the month, with $t = 1$ corresponding to January. *(Source: NOAA)*

   (a) Use a graphing utility to graph the data and the model in the same viewing window.

   (b) What is the period of the model? Is it what you expected? Explain.

   (c) What is the amplitude of the model? What does it represent in the context of the problem?

4.6 **Sketching the Graph of a Trigonometric Function** In Exercises 71–74, sketch the graph of the function. (Include two full periods.)

71. $f(t) = \tan\left(t + \frac{\pi}{2}\right)$ **72.** $f(x) = \frac{1}{2} \cot x$

73. $f(x) = \frac{1}{2} \csc \frac{x}{2}$ **74.** $h(t) = \sec\left(t - \frac{\pi}{4}\right)$

Analyzing a Damped Trigonometric Graph In Exercises 75 and 76, use a graphing utility to graph the function and the damping factor of the function in the same viewing window. Describe the behavior of the function as x increases without bound.

75. $f(x) = x \cos x$

76. $g(x) = e^x \cos x$

4.7 Evaluating an Inverse Trigonometric Function In Exercises 77–80, find the exact value of the expression.

77. $\arcsin(-1)$

78. $\cos^{-1} 1$

79. $\operatorname{arccot} \sqrt{3}$

80. $\operatorname{arcsec}\left(-\sqrt{2}\right)$

Calculators and Inverse Trigonometric Functions In Exercises 81–84, use a calculator to approximate the value of the expression, if possible. Round your result to two decimal places.

81. $\tan^{-1}(-1.3)$

82. $\arccos 0.372$

83. $\operatorname{arccot} 15.5$

84. $\operatorname{arccsc}(-4.03)$

Graphing an Inverse Trigonometric Function In Exercises 85 and 86, use a graphing utility to graph the function.

85. $f(x) = \arctan(x/2)$

86. $f(x) = -\arcsin 2x$

Evaluating a Composition of Functions In Exercises 87–90, find the exact value of the expression.

87. $\cos\left(\arctan \frac{3}{4}\right)$

88. $\tan\left(\arccos \frac{3}{5}\right)$

89. $\sec\left(\arctan \frac{12}{5}\right)$

90. $\cot\left[\arcsin\left(-\frac{12}{13}\right)\right]$

Writing an Expression In Exercises 91 and 92, write an algebraic expression that is equivalent to the given expression.

91. $\tan[\arccos(x/2)]$

92. $\sec[\arcsin(x-1)]$

4.8

93. **Angle of Elevation** The height of a radio transmission tower is 70 meters, and it casts a shadow of length 30 meters. Draw a right triangle that gives a visual representation of the problem. Label the known and unknown quantities. Then find the angle of elevation.

94. **Height** A football lands at the edge of the roof of your school building. When you are 25 feet from the base of the building, the angle of elevation to the football is 21°. How high off the ground is the football?

95. **Air Navigation** From city A to city B, a plane flies 650 miles at a bearing of 48°. From city B to city C, the plane flies 810 miles at a bearing of 115°. Find the distance from city A to city C and the bearing from city A to city C.

96. **Wave Motion** A fishing bobber oscillates in simple harmonic motion because of the waves in a lake. The bobber moves a total of 1.5 inches from its low point to its high point and returns to its high point every 3 seconds. Write an equation that describes the motion of the bobber, where the high point corresponds to the time $t = 0$.

Exploration

True or False? In Exercises 97 and 98, determine whether the statement is true or false. Justify your answer.

97. $y = \sin \theta$ is not a function because $\sin 30° = \sin 150°$.

98. Because $\tan(3\pi/4) = -1$, $\arctan(-1) = 3\pi/4$.

99. **Writing** Describe the behavior of $f(\theta) = \sec \theta$ at the zeros of $g(\theta) = \cos \theta$. Explain.

100. **Conjecture**

(a) Use a graphing utility to complete the table.

θ	0.1	0.4	0.7	1.0	1.3
$\tan\left(\theta - \dfrac{\pi}{2}\right)$					
$-\cot \theta$					

(b) Make a conjecture about the relationship between $\tan[\theta - (\pi/2)]$ and $-\cot \theta$.

101. **Writing** When graphing the sine and cosine functions, determining the amplitude is part of the analysis. Explain why this is not true for the other four trigonometric functions.

102. **Oscillation of a Spring** A weight is suspended from a ceiling by a steel spring. The weight is lifted (positive direction) from the equilibrium position and released. The resulting motion of the weight is modeled by $y = Ae^{-kt} \cos bt = \frac{1}{5}e^{-t/10} \cos 6t$, where y is the distance (in feet) from equilibrium and t is the time (in seconds). The figure shows the graph of the function. For each of the following, describe the change in the graph without graphing the resulting function.

(a) A is changed from $\frac{1}{5}$ to $\frac{1}{3}$.

(b) k is changed from $\frac{1}{10}$ to $\frac{1}{3}$.

(c) b is changed from 6 to 9.

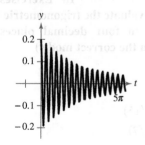

Take this test as you would take a test in class. When you are finished, check your work against the answers given in the back of the book.

1. Consider an angle that measures $\dfrac{5\pi}{4}$ radians.

   (a) Sketch the angle in standard position.

   (b) Determine two coterminal angles (one positive and one negative).

   (c) Convert the radian measure to degree measure.

2. A truck is moving at a rate of 105 kilometers per hour, and the diameter of each of its wheels is 1 meter. Find the angular speed of the wheels in radians per minute.

3. A water sprinkler sprays water on a lawn over a distance of 25 feet and rotates through an angle of 130°. Find the area of the lawn watered by the sprinkler.

4. Given that θ is an acute angle and $\tan \theta = \frac{3}{2}$, find the exact values of the other five trigonometric functions of θ.

5. Find the exact values of the six trigonometric functions of the angle θ shown in the figure.

6. Find the reference angle θ' of the angle $\theta = 205°$. Sketch θ in standard position and label θ'.

7. Determine the quadrant in which θ lies when $\sec \theta < 0$ and $\tan \theta > 0$.

8. Find two exact values of θ in degrees $(0° \le \theta < 360°)$ for which $\cos \theta = -\sqrt{3}/2$. Do not use a calculator.

In Exercises 9 and 10, find the exact values of the remaining five trigonometric functions of θ satisfying the given conditions.

9. $\cos \theta = \frac{3}{5}, \quad \tan \theta < 0$

10. $\sec \theta = -\frac{29}{20}, \quad \sin \theta > 0$

In Exercises 11–13, sketch the graph of the function. (Include two full periods.)

11. $g(x) = -2 \sin\!\left(x - \dfrac{\pi}{4}\right)$

12. $f(t) = \cos\!\left(t + \dfrac{\pi}{2}\right) - 1$

13. $f(x) = \frac{1}{2} \tan 2x$

In Exercises 14 and 15, use a graphing utility to graph the function. If the function is periodic, find its period. If not, describe the behavior of the function as x increases without bound.

14. $y = \sin 2\pi x + 2 \cos \pi x$

15. $y = 6e^{-0.12x} \cos(0.25x)$

16. Find a, b, and c for the function $f(x) = a \sin(bx + c)$ such that the graph of f matches the figure.

17. Find the exact value of $\cot\!\left(\arcsin \frac{3}{8}\right)$.

18. Sketch the graph of the function $f(x) = 2 \arcsin\!\left(\frac{1}{2}x\right)$.

19. An airplane is 90 miles south and 110 miles east of an airport. What bearing should the pilot take to fly directly to the airport?

20. A ball on a spring starts at its lowest point of 6 inches below equilibrium, bounces to its maximum height of 6 inches above equilibrium, and returns to its lowest point in a total of 2 seconds. Write an equation for the simple harmonic motion of the ball.

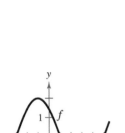

$(-2, 6)$

θ

y

x

Figure for 5

y

f

1

$-\pi$ π 2π

-1

-2

x

Figure for 16

Proofs in Mathematics ■ ■ ■ ■ ■ ■ ■ ■ ■ ■ ■ ■ ■ ■ ■ ■ ■

The Pythagorean Theorem

The Pythagorean Theorem is one of the most famous theorems in mathematics. More than 350 different proofs now exist. James A. Garfield, the twentieth president of the United States, developed a proof of the Pythagorean Theorem in 1876. His proof, shown below, involves the fact that two congruent right triangles and an isosceles right triangle can form a trapezoid.

The Pythagorean Theorem

In a right triangle, the sum of the squares of the lengths of the legs is equal to the square of the length of the hypotenuse, where a and b are the lengths of the legs and c is the length of the hypotenuse.

$$a^2 + b^2 = c^2$$

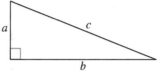

Proof

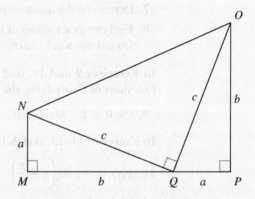

$$\text{Area of trapezoid } MNOP = \text{Area of } \triangle MNQ + \text{Area of } \triangle PQO + \text{Area of } \triangle NOQ$$

$$\frac{1}{2}(a + b)(a + b) = \frac{1}{2}ab + \frac{1}{2}ab + \frac{1}{2}c^2$$

$$\frac{1}{2}(a + b)(a + b) = ab + \frac{1}{2}c^2$$

$$(a + b)(a + b) = 2ab + c^2$$

$$a^2 + 2ab + b^2 = 2ab + c^2$$

$$a^2 + b^2 = c^2$$

344

P.S. Problem Solving

1. Angle of Rotation The restaurant at the top of the Space Needle in Seattle, Washington, is circular and has a radius of 47.25 feet. The dining part of the restaurant revolves, making about one complete revolution every 48 minutes. A dinner party, seated at the edge of the revolving restaurant at 6:45 P.M., finishes at 8:57 P.M.

(a) Find the angle through which the dinner party rotated.

(b) Find the distance the party traveled during dinner.

2. Bicycle Gears A bicycle's gear ratio is the number of times the freewheel turns for every one turn of the chainwheel (see figure). The table shows the numbers of teeth in the freewheel and chainwheel for the first five gears of an 18-speed touring bicycle. The chainwheel completes one rotation for each gear. Find the angle through which the freewheel turns for each gear. Give your answers in both degrees and radians.

DATA	Gear Number	Number of Teeth in Freewheel	Number of Teeth in Chainwheel
	1	32	24
	2	26	24
	3	22	24
	4	32	40
	5	19	24

Spreadsheet at LarsonPrecalculus.com

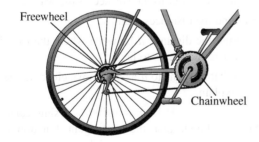

Freewheel

Chainwheel

3. Height of a Ferris Wheel Car A model for the height h (in feet) of a Ferris wheel car is

$h = 50 + 50 \sin 8\pi t$

where t is the time (in minutes). (The Ferris wheel has a radius of 50 feet.) This model yields a height of 50 feet when $t = 0$. Alter the model so that the height of the car is 1 foot when $t = 0$.

4. Periodic Function The function f is periodic, with period c. So, $f(t + c) = f(t)$. Determine whether each statement is true or false. Explain.

(a) $f(t - 2c) = f(t)$ (b) $f\left(t + \frac{1}{2}c\right) = f\left(\frac{1}{2}t\right)$

(c) $f\left(\frac{1}{2}[t + c]\right) = f\left(\frac{1}{2}t\right)$ (d) $f\left(\frac{1}{2}[t + 4c]\right) = f\left(\frac{1}{2}t\right)$

5. Surveying A surveyor in a helicopter is determining the width of an island, as shown in the figure.

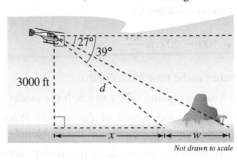

Not drawn to scale

(a) What is the shortest distance d the helicopter must travel to land on the island?

(b) What is the horizontal distance x the helicopter must travel before it is directly over the nearer end of the island?

(c) Find the width w of the island. Explain how you found your answer.

6. Similar Triangles and Trigonometric Functions Use the figure below.

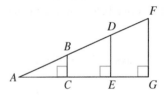

(a) Explain why $\triangle ABC$, $\triangle ADE$, and $\triangle AFG$ are similar triangles.

(b) What does similarity imply about the ratios

$$\frac{BC}{AB}, \quad \frac{DE}{AD}, \quad \text{and} \quad \frac{FG}{AF}?$$

(c) Does the value of $\sin A$ depend on which triangle from part (a) is used to calculate it? Does the value of $\sin A$ change when you use a different right triangle similar to the three given triangles?

(d) Do your conclusions from part (c) apply to the other five trigonometric functions? Explain.

7. Using Technology Use a graphing utility to graph h, and use the graph to determine whether h is even, odd, or neither.

(a) $h(x) = \cos^2 x$ (b) $h(x) = \sin^2 x$

8. Squares of Even and Odd Functions Given that f is an even function and g is an odd function, use the results of Exercise 7 to make a conjecture about each function h.

(a) $h(x) = [f(x)]^2$ (b) $h(x) = [g(x)]^2$

9. Blood Pressure The pressure P (in millimeters of mercury) against the walls of the blood vessels of a patient is modeled by

$$P = 100 - 20 \cos \frac{8\pi t}{3}$$

where t is the time (in seconds).

(a) Use a graphing utility to graph the model.

(b) What is the period of the model? What does it represent in the context of the problem?

(c) What is the amplitude of the model? What does it represent in the context of the problem?

(d) If one cycle of this model is equivalent to one heartbeat, what is the pulse of the patient?

(e) A physician wants the patient's pulse rate to be 64 beats per minute or less. What should the period be? What should the coefficient of t be?

10. Biorhythms A popular theory that attempts to explain the ups and downs of everyday life states that each person has three cycles, called biorhythms, which begin at birth. These three cycles can be modeled by the sine functions below, where t is the number of days since birth.

Physical (23 days): $P = \sin \dfrac{2\pi t}{23}, \quad t \geq 0$

Emotional (28 days): $E = \sin \dfrac{2\pi t}{28}, \quad t \geq 0$

Intellectual (33 days): $I = \sin \dfrac{2\pi t}{33}, \quad t \geq 0$

Consider a person who was born on July 20, 1995.

(a) Use a graphing utility to graph the three models in the same viewing window for $7300 \leq t \leq 7380$.

(b) Describe the person's biorhythms during the month of September 2015.

(c) Calculate the person's three energy levels on September 22, 2015.

11. Graphical Reasoning

(a) Use a graphing utility to graph the functions
$$f(x) = 2 \cos 2x + 3 \sin 3x$$
and
$$g(x) = 2 \cos 2x + 3 \sin 4x.$$

(b) Use the graphs from part (a) to find the period of each function.

(c) Is the function $h(x) = A \cos \alpha x + B \sin \beta x$, where α and β are positive integers, periodic? Explain.

12. Analyzing Trigonometric Functions Two trigonometric functions f and g have periods of 2, and their graphs intersect at $x = 5.35$.

(a) Give one positive value of x less than 5.35 and one value of x greater than 5.35 at which the functions have the same value.

(b) Determine one negative value of x at which the graphs intersect.

(c) Is it true that $f(13.35) = g(-4.65)$? Explain.

13. Refraction When you stand in shallow water and look at an object below the surface of the water, the object will look farther away from you than it really is. This is because when light rays pass between air and water, the water refracts, or bends, the light rays. The index of refraction for water is 1.333. This is the ratio of the sine of θ_1 and the sine of θ_2 (see figure).

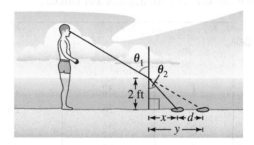

(a) While standing in water that is 2 feet deep, you look at a rock at angle $\theta_1 = 60°$ (measured from a line perpendicular to the surface of the water). Find θ_2.

(b) Find the distances x and y.

(c) Find the distance d between where the rock is and where it appears to be.

(d) What happens to d as you move closer to the rock? Explain.

14. Polynomial Approximation Using calculus, it can be shown that the arctangent function can be approximated by the polynomial

$$\arctan x \approx x - \frac{x^3}{3} + \frac{x^5}{5} - \frac{x^7}{7}$$

where x is in radians.

(a) Use a graphing utility to graph the arctangent function and its polynomial approximation in the same viewing window. How do the graphs compare?

(b) Study the pattern in the polynomial approximation of the arctangent function and predict the next term. Then repeat part (a). How does the accuracy of the approximation change when an additional term is added?

5 Analytic Trigonometry

Standing Waves *(Exercise 80, page 379)*

Projectile Motion
(Example 10, page 387)

Ferris Wheel *(Exercise 94, page 373)*

Shadow Length
(Exercise 62, page 361)

Friction *(Exercise 65, page 354)*

5.1 Using Fundamental Identities

- ■ Recognize and write the fundamental trigonometric identities.
- ■ Use the fundamental trigonometric identities to evaluate trigonometric functions, simplify trigonometric expressions, and rewrite trigonometric expressions.

Introduction

In Chapter 4, you studied the basic definitions, properties, graphs, and applications of the individual trigonometric functions. In this chapter, you will learn how to use the fundamental identities to perform the four tasks listed below.

1. Evaluate trigonometric functions.
2. Simplify trigonometric expressions.
3. Develop additional trigonometric identities.
4. Solve trigonometric equations.

Fundamental trigonometric identities are useful in simplifying trigonometric expressions. For example, in Exercise 65 on page 354, you will use trigonometric identities to simplify an expression for the coefficient of friction.

Fundamental Trigonometric Identities

Reciprocal Identities

$$\sin u = \frac{1}{\csc u} \qquad \cos u = \frac{1}{\sec u} \qquad \tan u = \frac{1}{\cot u}$$

$$\csc u = \frac{1}{\sin u} \qquad \sec u = \frac{1}{\cos u} \qquad \cot u = \frac{1}{\tan u}$$

Quotient Identities

$$\tan u = \frac{\sin u}{\cos u} \qquad \cot u = \frac{\cos u}{\sin u}$$

Pythagorean Identities

$$\sin^2 u + \cos^2 u = 1 \qquad 1 + \tan^2 u = \sec^2 u \qquad 1 + \cot^2 u = \csc^2 u$$

Cofunction Identities

$$\sin\left(\frac{\pi}{2} - u\right) = \cos u \qquad \cos\left(\frac{\pi}{2} - u\right) = \sin u$$

$$\tan\left(\frac{\pi}{2} - u\right) = \cot u \qquad \cot\left(\frac{\pi}{2} - u\right) = \tan u$$

$$\sec\left(\frac{\pi}{2} - u\right) = \csc u \qquad \csc\left(\frac{\pi}{2} - u\right) = \sec u$$

Even/Odd Identities

$$\sin(-u) = -\sin u \qquad \cos(-u) = \cos u \qquad \tan(-u) = -\tan u$$

$$\csc(-u) = -\csc u \qquad \sec(-u) = \sec u \qquad \cot(-u) = -\cot u$$

• • REMARK You should learn the fundamental trigonometric identities well, because you will use them frequently in trigonometry and they will also appear in calculus. Note that u can be an angle, a real number, or a variable.

Pythagorean identities are sometimes used in radical form such as

$$\sin u = \pm\sqrt{1 - \cos^2 u}$$

or

$$\tan u = \pm\sqrt{\sec^2 u - 1}$$

where the sign depends on the choice of u.

Using the Fundamental Identities

One common application of trigonometric identities is to use given information about trigonometric functions to evaluate other trigonometric functions.

EXAMPLE 1 **Using Identities to Evaluate a Function**

Use the conditions $\sec u = -\frac{3}{2}$ and $\tan u > 0$ to find the values of all six trigonometric functions.

Solution Using a reciprocal identity, you have

$$\cos u = \frac{1}{\sec u} = \frac{1}{-3/2} = -\frac{2}{3}.$$

Using a Pythagorean identity, you have

$$\sin^2 u = 1 - \cos^2 u \qquad \text{Pythagorean identity}$$
$$= 1 - \left(-\frac{2}{3}\right)^2 \qquad \text{Substitute } -\frac{2}{3} \text{ for } \cos u.$$
$$= \frac{5}{9}. \qquad \text{Simplify.}$$

Because $\sec u < 0$ and $\tan u > 0$, it follows that u lies in Quadrant III. Moreover, $\sin u$ is negative when u is in Quadrant III, so choose the negative root and obtain $\sin u = -\sqrt{5}/3$. Knowing the values of the sine and cosine enables you to find the values of the remaining trigonometric functions.

$$\sin u = -\frac{\sqrt{5}}{3} \qquad\qquad \csc u = \frac{1}{\sin u} = -\frac{3}{\sqrt{5}} = -\frac{3\sqrt{5}}{5}$$

$$\cos u = -\frac{2}{3} \qquad\qquad \sec u = -\frac{3}{2}$$

$$\tan u = \frac{\sin u}{\cos u} = \frac{-\sqrt{5}/3}{-2/3} = \frac{\sqrt{5}}{2} \qquad \cot u = \frac{1}{\tan u} = \frac{2}{\sqrt{5}} = \frac{2\sqrt{5}}{5}$$

 **✓ Checkpoint** Audio-video solution in English & Spanish at LarsonPrecalculus.com

Use the conditions $\tan x = \frac{1}{3}$ and $\cos x < 0$ to find the values of all six trigonometric functions.

EXAMPLE 2 **Simplifying a Trigonometric Expression**

Simplify the expression.

$$\sin x \cos^2 x - \sin x$$

Solution First factor out the common monomial factor $\sin x$ and then use a Pythagorean identity.

$$\sin x \cos^2 x - \sin x = \sin x(\cos^2 x - 1) \qquad \text{Factor out common monomial factor.}$$
$$= -\sin x(1 - \cos^2 x) \qquad \text{Factor out } -1.$$
$$= -\sin x(\sin^2 x) \qquad \text{Pythagorean identity}$$
$$= -\sin^3 x \qquad \text{Multiply.}$$

 **✓ Checkpoint** Audio-video solution in English & Spanish at LarsonPrecalculus.com

Simplify the expression.

$$\cos^2 x \csc x - \csc x$$

▷ **TECHNOLOGY** Use a graphing utility to check the result of Example 2. To do this, enter

$$Y1 = -(\sin(X))^3$$

and

$$Y2 = \sin(X)(\cos(X))^2$$
$$- \sin(X).$$

Select the *line* style for Y1 and the *path* style for Y2, then graph both equations in the same viewing window. The two graphs *appear* to coincide, so it is reasonable to assume that their expressions are equivalent. Note that the actual equivalence of the expressions can only be verified algebraically, as in Example 2. This graphical approach is only to check your work.

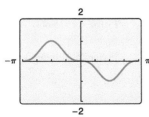

When factoring trigonometric expressions, it is helpful to find a polynomial form that fits the expression, as shown in Example 3.

EXAMPLE 3   **Factoring Trigonometric Expressions**

Factor each expression.

a. $\sec^2\theta - 1$ **b.** $4\tan^2\theta + \tan\theta - 3$

Solution

a. This expression has the polynomial form $u^2 - v^2$, which is the difference of two squares. It factors as

$$\sec^2\theta - 1 = (\sec\theta + 1)(\sec\theta - 1).$$

b. This expression has the polynomial form $ax^2 + bx + c$, and it factors as

$$4\tan^2\theta + \tan\theta - 3 = (4\tan\theta - 3)(\tan\theta + 1).$$

✓ **Checkpoint**))) *Audio-video solution in English & Spanish at LarsonPrecalculus.com*

Factor each expression.

a. $1 - \cos^2\theta$ **b.** $2\csc^2\theta - 7\csc\theta + 6$

In some cases, when factoring or simplifying a trigonometric expression, it is helpful to first rewrite the expression in terms of just *one* trigonometric function or in terms of *sine and cosine only*. These strategies are demonstrated in Examples 4 and 5.

EXAMPLE 4   **Factoring a Trigonometric Expression**

Factor $\csc^2 x - \cot x - 3$.

Solution   Use the identity $\csc^2 x = 1 + \cot^2 x$ to rewrite the expression.

$$\csc^2 x - \cot x - 3 = (1 + \cot^2 x) - \cot x - 3 \qquad \text{Pythagorean identity}$$

$$= \cot^2 x - \cot x - 2 \qquad \text{Combine like terms.}$$

$$= (\cot x - 2)(\cot x + 1) \qquad \text{Factor.}$$

✓ **Checkpoint**))) *Audio-video solution in English & Spanish at LarsonPrecalculus.com*

Factor $\sec^2 x + 3\tan x + 1$.

EXAMPLE 5   **Simplifying a Trigonometric Expression**

See LarsonPrecalculus.com for an interactive version of this type of example.

$$\sin t + \cot t \cos t = \sin t + \left(\frac{\cos t}{\sin t}\right)\cos t \qquad \text{Quotient identity}$$

$$= \frac{\sin^2 t + \cos^2 t}{\sin t} \qquad \text{Add fractions.}$$

$$= \frac{1}{\sin t} \qquad \text{Pythagorean identity}$$

$$= \csc t \qquad \text{Reciprocal identity}$$

✓ **Checkpoint**))) *Audio-video solution in English & Spanish at LarsonPrecalculus.com*

Simplify $\csc x - \cos x \cot x$.

▷ **ALGEBRA HELP**   In Example 3, you factor the difference of two squares and you factor a trinomial. To review the techniques for factoring polynomials, see Appendix A.3.

REMARK   Remember that when adding rational expressions, you must first find the least common denominator (LCD). In Example 5, the LCD is $\sin t$.

EXAMPLE 6   **Adding Trigonometric Expressions**

Perform the addition and simplify: $\dfrac{\sin\theta}{1+\cos\theta}+\dfrac{\cos\theta}{\sin\theta}$.

Solution

$$\dfrac{\sin\theta}{1+\cos\theta}+\dfrac{\cos\theta}{\sin\theta}=\dfrac{(\sin\theta)(\sin\theta)+(\cos\theta)(1+\cos\theta)}{(1+\cos\theta)(\sin\theta)}$$

$$=\dfrac{\sin^2\theta+\cos^2\theta+\cos\theta}{(1+\cos\theta)(\sin\theta)}\qquad\text{Multiply.}$$

$$=\dfrac{1+\cos\theta}{(1+\cos\theta)(\sin\theta)}\qquad\text{Pythagorean identity}$$

$$=\dfrac{1}{\sin\theta}\qquad\text{Divide out common factor.}$$

$$=\csc\theta\qquad\text{Reciprocal identity}$$

✓ *Checkpoint* ◀))) *Audio-video solution in English & Spanish at LarsonPrecalculus.com*

Perform the addition and simplify: $\dfrac{1}{1+\sin\theta}+\dfrac{1}{1-\sin\theta}$.

The next two examples involve techniques for rewriting expressions in forms that are used in calculus.

EXAMPLE 7   **Rewriting a Trigonometric Expression**

Rewrite $\dfrac{1}{1+\sin x}$ so that it is *not* in fractional form.

Solution   From the Pythagorean identity

$$\cos^2 x=1-\sin^2 x=(1-\sin x)(1+\sin x)$$

multiplying both the numerator and the denominator by $(1-\sin x)$ will produce a monomial denominator.

$$\dfrac{1}{1+\sin x}=\dfrac{1}{1+\sin x}\cdot\dfrac{1-\sin x}{1-\sin x}\qquad\begin{array}{l}\text{Multiply numerator and}\\\text{denominator by }(1-\sin x).\end{array}$$

$$=\dfrac{1-\sin x}{1-\sin^2 x}\qquad\text{Multiply.}$$

$$=\dfrac{1-\sin x}{\cos^2 x}\qquad\text{Pythagorean identity}$$

$$=\dfrac{1}{\cos^2 x}-\dfrac{\sin x}{\cos^2 x}\qquad\text{Write as separate fractions.}$$

$$=\dfrac{1}{\cos^2 x}-\dfrac{\sin x}{\cos x}\cdot\dfrac{1}{\cos x}\qquad\text{Product of fractions}$$

$$=\sec^2 x-\tan x\sec x\qquad\text{Reciprocal and quotient identities}$$

✓ *Checkpoint* ◀))) *Audio-video solution in English & Spanish at LarsonPrecalculus.com*

Rewrite $\dfrac{\cos^2\theta}{1-\sin\theta}$ so that it is *not* in fractional form.

Trigonometric Substitution

Use the substitution $x = 2 \tan \theta$, $0 < \theta < \pi/2$, to write $\sqrt{4 + x^2}$ as a trigonometric function of θ.

Solution Begin by letting $x = 2 \tan \theta$. Then, you obtain

$$\sqrt{4 + x^2} = \sqrt{4 + (2 \tan \theta)^2} \qquad \text{Substitute } 2 \tan \theta \text{ for } x.$$

$$= \sqrt{4 + 4 \tan^2 \theta} \qquad \text{Property of exponents}$$

$$= \sqrt{4(1 + \tan^2 \theta)} \qquad \text{Factor.}$$

$$= \sqrt{4 \sec^2 \theta} \qquad \text{Pythagorean identity}$$

$$= 2 \sec \theta. \qquad \sec \theta > 0 \text{ for } 0 < \theta < \frac{\pi}{2}$$

✓ *Checkpoint* ◀))) Audio-video solution in English & Spanish at LarsonPrecalculus.com

Use the substitution $x = 3 \sin \theta$, $0 < \theta < \pi/2$, to write $\sqrt{9 - x^2}$ as a trigonometric function of θ.

Figure 5.1 shows the right triangle illustration of the trigonometric substitution $x = 2 \tan \theta$ in Example 8. You can use this triangle to check the solution to Example 8. For $0 < \theta < \pi/2$, you have

$$\text{opp} = x, \quad \text{adj} = 2, \quad \text{and} \quad \text{hyp} = \sqrt{4 + x^2}.$$

Using these expressions,

$$\sec \theta = \frac{\text{hyp}}{\text{adj}} = \frac{\sqrt{4 + x^2}}{2}.$$

So, $2 \sec \theta = \sqrt{4 + x^2}$, and the solution checks.

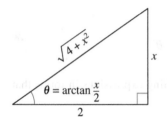

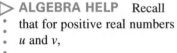

Figure 5.1

Rewriting a Logarithmic Expression

Rewrite $\ln|\csc \theta| + \ln|\tan \theta|$ as a single logarithm and simplify the result.

Solution

$$\ln|\csc \theta| + \ln|\tan \theta| = \ln|\csc \theta \tan \theta| \qquad \text{Product Property of Logarithms}$$

$$= \ln\left|\frac{1}{\sin \theta} \cdot \frac{\sin \theta}{\cos \theta}\right| \qquad \text{Reciprocal and quotient identities}$$

$$= \ln\left|\frac{1}{\cos \theta}\right| \qquad \text{Simplify.}$$

$$= \ln|\sec \theta| \qquad \text{Reciprocal identity}$$

▷ **ALGEBRA HELP** Recall that for positive real numbers u and v,

$$\ln u + \ln v = \ln(uv).$$

To review the properties of logarithms, see Section 3.3.

✓ *Checkpoint* ◀))) Audio-video solution in English & Spanish at LarsonPrecalculus.com

Rewrite $\ln|\sec x| + \ln|\sin x|$ as a single logarithm and simplify the result.

Summarize (Section 5.1)

1. State the fundamental trigonometric identities (*page 348*).

2. Explain how to use the fundamental trigonometric identities to evaluate trigonometric functions, simplify trigonometric expressions, and rewrite trigonometric expressions (*pages 349–352*). For examples of these concepts, see Examples 1–9.

5.1 Exercises

Vocabulary: Fill in the blank to complete the trigonometric identity.

1. $\dfrac{\sin u}{\cos u} = $ _____

2. $\dfrac{1}{\sin u} = $ _____

3. $\dfrac{1}{\tan u} = $ _____

4. $\sec\left(\dfrac{\pi}{2} - u\right) = $ _____

5. $\sin^2 u + \cos^2 u = $ _____

6. $\sin(-u) = $ _____

Skills and Applications

 Using Identities to Evaluate a Function
In Exercises 7–12, use the given conditions to find the values of all six trigonometric functions.

7. $\sec x = -\dfrac{5}{2}$, $\tan x < 0$
8. $\csc x = -\dfrac{7}{6}$, $\tan x > 0$
9. $\sin \theta = -\dfrac{3}{4}$, $\cos \theta > 0$
10. $\cos \theta = \dfrac{2}{3}$, $\sin \theta < 0$
11. $\tan x = \dfrac{2}{3}$, $\cos x > 0$
12. $\cot x = \dfrac{7}{4}$, $\sin x < 0$

Matching Trigonometric Expressions In Exercises 13–18, match the trigonometric expression with its simplified form.

(a) $\csc x$ (b) -1 (c) 1

(d) $\sin x \tan x$ (e) $\sec^2 x$ (f) $\sec x$

13. $\sec x \cos x$
14. $\cot^2 x - \csc^2 x$
15. $\cos x(1 + \tan^2 x)$
16. $\cot x \sec x$
17. $\dfrac{\sec^2 x - 1}{\sin^2 x}$
18. $\dfrac{\cos^2[(\pi/2) - x]}{\cos x}$

 Simplifying a Trigonometric Expression
In Exercises 19–22, use the fundamental identities to simplify the expression. (There is more than one correct form of each answer).

19. $\dfrac{\tan \theta \cot \theta}{\sec \theta}$
20. $\cos\left(\dfrac{\pi}{2} - x\right) \sec x$
21. $\tan^2 x - \tan^2 x \sin^2 x$
22. $\sin^2 x \sec^2 x - \sin^2 x$

 Factoring a Trigonometric Expression
In Exercises 23–32, factor the expression. Use the fundamental identities to simplify, if necessary. (There is more than one correct form of each answer.)

23. $\dfrac{\sec^2 x - 1}{\sec x - 1}$
24. $\dfrac{\cos x - 2}{\cos^2 x - 4}$
25. $1 - 2\cos^2 x + \cos^4 x$
26. $\sec^4 x - \tan^4 x$
27. $\cot^3 x + \cot^2 x + \cot x + 1$
28. $\sec^3 x - \sec^2 x - \sec x + 1$
29. $3 \sin^2 x - 5 \sin x - 2$
30. $6 \cos^2 x + 5 \cos x - 6$
31. $\cot^2 x + \csc x - 1$
32. $\sin^2 x + 3 \cos x + 3$

 Simplifying a Trigonometric Expression
In Exercises 33–40, use the fundamental identities to simplify the expression. (There is more than one correct form of each answer.)

33. $\tan \theta \csc \theta$
34. $\tan(-x) \cos x$
35. $\sin \phi(\csc \phi - \sin \phi)$
36. $\cos x(\sec x - \cos x)$
37. $\sin \beta \tan \beta + \cos \beta$
38. $\cot u \sin u + \tan u \cos u$
39. $\dfrac{1 - \sin^2 x}{\csc^2 x - 1}$
40. $\dfrac{\cos^2 y}{1 - \sin y}$

Multiplying Trigonometric Expressions In Exercises 41 and 42, perform the multiplication and use the fundamental identities to simplify. (There is more than one correct form of each answer.)

41. $(\sin x + \cos x)^2$
42. $(2 \csc x + 2)(2 \csc x - 2)$

 Adding or Subtracting Trigonometric Expressions In Exercises 43–48, perform the addition or subtraction and use the fundamental identities to simplify. (There is more than one correct form of each answer.)

43. $\dfrac{1}{1 + \cos x} + \dfrac{1}{1 - \cos x}$
44. $\dfrac{1}{\sec x + 1} - \dfrac{1}{\sec x - 1}$
45. $\dfrac{\cos x}{1 + \sin x} - \dfrac{\cos x}{1 - \sin x}$
46. $\dfrac{\sin x}{1 + \cos x} + \dfrac{\sin x}{1 - \cos x}$
47. $\tan x - \dfrac{\sec^2 x}{\tan x}$
48. $\dfrac{\cos x}{1 + \sin x} + \dfrac{1 + \sin x}{\cos x}$

♩ Rewriting a Trigonometric Expression In Exercises 49 and 50, rewrite the expression so that it is *not* in fractional form. (There is more than one correct form of each answer.)

49. $\dfrac{\sin^2 y}{1 - \cos y}$
50. $\dfrac{5}{\tan x + \sec x}$

Trigonometric Functions and Expressions In Exercises 51 and 52, use a graphing utility to determine which of the six trigonometric functions is equal to the expression. Verify your answer algebraically.

51. $\dfrac{\tan x + 1}{\sec x + \csc x}$

52. $\dfrac{1}{\sin x}\left(\dfrac{1}{\cos x} - \cos x\right)$

Trigonometric Substitution In Exercises 53–56, use the trigonometric substitution to write the algebraic expression as a trigonometric function of θ, where $0 < \theta < \pi/2$.

53. $\sqrt{9 - x^2}$, $x = 3 \cos \theta$

54. $\sqrt{49 - x^2}$, $x = 7 \sin \theta$

55. $\sqrt{x^2 - 4}$, $x = 2 \sec \theta$

56. $\sqrt{9x^2 + 25}$, $3x = 5 \tan \theta$

Trigonometric Substitution In Exercises 57 and 58, use the trigonometric substitution to write the algebraic equation as a trigonometric equation of θ, where $-\pi/2 < \theta < \pi/2$. Then find $\sin \theta$ and $\cos \theta$.

57. $\sqrt{2} = \sqrt{4 - x^2}$, $x = 2 \sin \theta$

58. $5\sqrt{3} = \sqrt{100 - x^2}$, $x = 10 \cos \theta$

Solving a Trigonometric Equation In Exercises 59 and 60, use a graphing utility to solve the equation for θ, where $0 \le \theta < 2\pi$.

59. $\sin \theta = \sqrt{1 - \cos^2 \theta}$ **60.** $\sec \theta = \sqrt{1 + \tan^2 \theta}$

Rewriting a Logarithmic Expression In Exercises 61–64, rewrite the expression as a single logarithm and simplify the result.

61. $\ln|\sin x| + \ln|\cot x|$ **62.** $\ln|\cos x| - \ln|\sin x|$

63. $\ln|\tan t| - \ln(1 - \cos^2 t)$

64. $\ln(\cos^2 t) + \ln(1 + \tan^2 t)$

• • 65. Friction • • • • • • • • • • •

The forces acting on an object weighing W units on an inclined plane positioned at an angle of θ with the horizontal (see figure) are modeled by

$\mu W \cos \theta = W \sin \theta$,

where μ is the coefficient of friction. Solve the equation for μ and simplify the result.

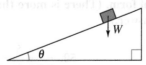

66. Rate of Change The rate of change of the function $f(x) = \sec x + \cos x$ is given by the expression $\sec x \tan x - \sin x$. Show that this expression can also be written as $\sin x \tan^2 x$.

Exploration

True or False? In Exercises 67 and 68, determine whether the statement is true or false. Justify your answer.

67. The quotient identities and reciprocal identities can be used to write any trigonometric function in terms of sine and cosine.

68. A cofunction identity can transform a tangent function into a cosecant function.

Analyzing Trigonometric Functions In Exercises 69 and 70, fill in the blanks. (*Note:* The notation $x \to c^+$ indicates that x approaches c from the right and $x \to c^-$ indicates that x approaches c from the left.)

69. As $x \to \left(\dfrac{\pi}{2}\right)^-$, $\tan x \to$ ▨ and $\cot x \to$ ▨ .

70. As $x \to \pi^+$, $\sin x \to$ ▨ and $\csc x \to$ ▨ .

71. Error Analysis Describe the error.

$$\dfrac{\sin \theta}{\cos(-\theta)} = \dfrac{\sin \theta}{-\cos \theta}$$
$$= -\tan \theta$$

72. Trigonometric Substitution Use the trigonometric substitution $u = a \tan \theta$, where $-\pi/2 < \theta < \pi/2$ and $a > 0$, to simplify the expression $\sqrt{a^2 + u^2}$.

73. Writing Trigonometric Functions in Terms of Sine Write each of the other trigonometric functions of θ in terms of $\sin \theta$.

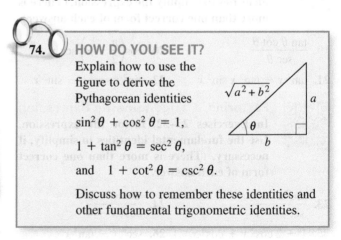

74. **HOW DO YOU SEE IT?** Explain how to use the figure to derive the Pythagorean identities

$\sin^2 \theta + \cos^2 \theta = 1$,

$1 + \tan^2 \theta = \sec^2 \theta$,

and $1 + \cot^2 \theta = \csc^2 \theta$.

Discuss how to remember these identities and other fundamental trigonometric identities.

75. Rewriting a Trigonometric Expression Rewrite the expression below in terms of $\sin \theta$ and $\cos \theta$.

$$\dfrac{\sec \theta(1 + \tan \theta)}{\sec \theta + \csc \theta}$$

5.2 Verifying Trigonometric Identities

Trigonometric identities enable you to rewrite trigonometric equations that model real-life situations. For example, in Exercise 62 on page 361, trigonometric identities can help you simplify an equation that models the length of a shadow cast by a gnomon (a device used to tell time).

■ Verify trigonometric identities.

Verifying Trigonometric Identities

In this section, you will study techniques for verifying trigonometric identities. In the next section, you will study techniques for solving trigonometric equations. The key to both verifying identities *and* solving equations is your ability to use the fundamental identities and the rules of algebra to rewrite trigonometric expressions.

Remember that a *conditional equation* is an equation that is true for only some of the values in the domain of the variable. For example, the conditional equation

$$\sin x = 0 \qquad \text{Conditional equation}$$

is true only for

$$x = n\pi$$

where n is an integer. When you are finding the values of the variable for which the equation is true, you are *solving* the equation.

On the other hand, an equation that is true for all real values in the domain of the variable is an *identity*. For example, the familiar equation

$$\sin^2 x = 1 - \cos^2 x \qquad \text{Identity}$$

is true for all real numbers x. So, it is an identity.

Although there are similarities, verifying that a trigonometric equation is an identity is quite different from solving an equation. There is no well-defined set of rules to follow in verifying trigonometric identities, the process is best learned through practice.

Guidelines for Verifying Trigonometric Identities

1. Work with one side of the equation at a time. It is often better to work with the more complicated side first.

2. Look for opportunities to factor an expression, add fractions, square a binomial, or create a monomial denominator.

3. Look for opportunities to use the fundamental identities. Note which functions are in the final expression you want. Sines and cosines pair up well, as do secants and tangents, and cosecants and cotangents.

4. When the preceding guidelines do not help, try converting all terms to sines and cosines.

5. Always try *something*. Even making an attempt that leads to a dead end can provide insight.

Verifying trigonometric identities is a useful process when you need to convert a trigonometric expression into a form that is more useful algebraically. When you verify an identity, you cannot *assume* that the two sides of the equation are equal because you are trying to verify that they *are* equal. As a result, when verifying identities, you cannot use operations such as adding the same quantity to each side of the equation or cross multiplication.

| EXAMPLE 1 | **Verifying a Trigonometric Identity** |

Verify the identity $\dfrac{\sec^2 \theta - 1}{\sec^2 \theta} = \sin^2 \theta$.

Solution Start with the left side because it is more complicated.

$$\frac{\sec^2 \theta - 1}{\sec^2 \theta} = \frac{\tan^2 \theta}{\sec^2 \theta} \qquad \text{Pythagorean identity}$$

$$= \tan^2 \theta(\cos^2 \theta) \qquad \text{Reciprocal identity}$$

$$= \frac{\sin^2 \theta}{(\cos^2 \theta)} (\cos^2 \theta) \qquad \text{Quotient identity}$$

$$= \sin^2 \theta \qquad \text{Simplify.}$$

· · REMARK Remember that an identity is only true for all real values in the domain of the variable. For instance, in Example 1 the identity is not true when $\theta = \pi/2$ because $\sec^2 \theta$ is undefined when $\theta = \pi/2$.

Notice that you verify the identity by starting with the left side of the equation (the more complicated side) and using the fundamental trigonometric identities to simplify it until you obtain the right side.

✓ **Checkpoint** ◀))) Audio-video solution in English & Spanish at LarsonPrecalculus.com

Verify the identity $\dfrac{\sin^2 \theta + \cos^2 \theta}{\cos^2 \theta \sec^2 \theta} = 1$.

There can be more than one way to verify an identity. Here is another way to verify the identity in Example 1.

$$\frac{\sec^2 \theta - 1}{\sec^2 \theta} = \frac{\sec^2 \theta}{\sec^2 \theta} - \frac{1}{\sec^2 \theta} \qquad \text{Write as separate fractions.}$$

$$= 1 - \cos^2 \theta \qquad \text{Reciprocal identity}$$

$$= \sin^2 \theta \qquad \text{Pythagorean identity}$$

| EXAMPLE 2 | **Verifying a Trigonometric Identity** |

Verify the identity $2 \sec^2 \alpha = \dfrac{1}{1 - \sin \alpha} + \dfrac{1}{1 + \sin \alpha}$.

Algebraic Solution

Start with the right side because it is more complicated.

$$\frac{1}{1 - \sin \alpha} + \frac{1}{1 + \sin \alpha} = \frac{1 + \sin \alpha + 1 - \sin \alpha}{(1 - \sin \alpha)(1 + \sin \alpha)} \qquad \text{Add fractions.}$$

$$= \frac{2}{1 - \sin^2 \alpha} \qquad \text{Simplify.}$$

$$= \frac{2}{\cos^2 \alpha} \qquad \text{Pythagorean identity}$$

$$= 2 \sec^2 \alpha \qquad \text{Reciprocal identity}$$

Numerical Solution

Use a graphing utility to create a table that shows the values of $y_1 = 2/\cos^2 x$ and $y_2 = [1/(1 - \sin x)] + [1/(1 + \sin x)]$ for different values of x.

X	Y1	Y2
-.5	2.5969	2.5969
-.25	2.1304	2.1304
0	2	2
.25	2.1304	2.1304
.5	2.5969	2.5969
.75	3.7357	3.7357
1	6.851	6.851

X=-.5

The values in the table for y_1 and y_2 appear to be identical, so the equation appears to be an identity.

✓ **Checkpoint** Audio-video solution in English & Spanish at LarsonPrecalculus.com

Verify the identity $2 \csc^2 \beta = \dfrac{1}{1 - \cos \beta} + \dfrac{1}{1 + \cos \beta}$.

EXAMPLE 3 **Verifying a Trigonometric Identity**

Verify the identity

$$(\tan^2 x + 1)(\cos^2 x - 1) = -\tan^2 x.$$

Algebraic Solution

Apply Pythagorean identities before multiplying.

$$(\tan^2 x + 1)(\cos^2 x - 1) = (\sec^2 x)(-\sin^2 x) \qquad \text{Pythagorean identities}$$

$$= -\frac{\sin^2 x}{\cos^2 x} \qquad \text{Reciprocal identity}$$

$$= -\left(\frac{\sin x}{\cos x}\right)^2 \qquad \text{Property of exponents}$$

$$= -\tan^2 x \qquad \text{Quotient identity}$$

Graphical Solution

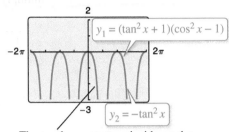

The graphs appear to coincide, so the given equation appears to be an identity.

✓ **Checkpoint** ◄))) *Audio-video solution in English & Spanish at LarsonPrecalculus.com*

Verify the identity $(\sec^2 x - 1)(\sin^2 x - 1) = -\sin^2 x.$

EXAMPLE 4 **Converting to Sines and Cosines**

Verify each identity.

a. $\tan x \csc x = \sec x$

b. $\tan x + \cot x = \sec x \csc x$

Solution

a. Convert the left side into sines and cosines.

$$\tan x \csc x = \frac{\sin x}{\cos x} \cdot \frac{1}{\sin x} \qquad \text{Quotient and reciprocal identities}$$

$$= \frac{1}{\cos x} \qquad \text{Simplify.}$$

$$= \sec x \qquad \text{Reciprocal identity}$$

b. Convert the left side into sines and cosines.

$$\tan x + \cot x = \frac{\sin x}{\cos x} + \frac{\cos x}{\sin x} \qquad \text{Quotient identities}$$

$$= \frac{\sin^2 x + \cos^2 x}{\cos x \sin x} \qquad \text{Add fractions.}$$

$$= \frac{1}{\cos x \sin x} \qquad \text{Pythagorean identity}$$

$$= \frac{1}{\cos x} \cdot \frac{1}{\sin x} \qquad \text{Product of fractions}$$

$$= \sec x \csc x \qquad \text{Reciprocal identities}$$

✓ **Checkpoint** ◄))) *Audio-video solution in English & Spanish at LarsonPrecalculus.com*

Verify each identity.

a. $\cot x \sec x = \csc x$

b. $\csc x - \sin x = \cos x \cot x$

▷ **ALGEBRA HELP**   To review the techniques for rationalizing a denominator, see Appendix A.2.

Recall from algebra that *rationalizing the denominator* using conjugates is, on occasion, a powerful simplification technique. A related form of this technique works for simplifying trigonometric expressions as well. For example, to simplify

$$\frac{1}{1 - \cos x}$$

multiply the numerator and the denominator by $1 + \cos x$.

$$\frac{1}{1 - \cos x} = \frac{1}{1 - \cos x}\left(\frac{1 + \cos x}{1 + \cos x}\right)$$

$$= \frac{1 + \cos x}{1 - \cos^2 x}$$

$$= \frac{1 + \cos x}{\sin^2 x}$$

$$= \csc^2 x(1 + \cos x)$$

The expression $\csc^2 x(1 + \cos x)$ is considered a simplified form of

$$\frac{1}{1 - \cos x}$$

because $\csc^2 x(1 + \cos x)$ does not contain fractions.

EXAMPLE 5   **Verifying a Trigonometric Identity**

See LarsonPrecalculus.com for an interactive version of this type of example.

Verify the identity $\sec x + \tan x = \dfrac{\cos x}{1 - \sin x}$.

Algebraic Solution

Begin with the *right* side and create a monomial denominator by multiplying the numerator and the denominator by $1 + \sin x$.

$$\frac{\cos x}{1 - \sin x} = \frac{\cos x}{1 - \sin x}\left(\frac{1 + \sin x}{1 + \sin x}\right) \qquad \text{Multiply numerator and denominator by } 1 + \sin x.$$

$$= \frac{\cos x + \cos x \sin x}{1 - \sin^2 x} \qquad \text{Multiply.}$$

$$= \frac{\cos x + \cos x \sin x}{\cos^2 x} \qquad \text{Pythagorean identity}$$

$$= \frac{\cos x}{\cos^2 x} + \frac{\cos x \sin x}{\cos^2 x} \qquad \text{Write as separate fractions.}$$

$$= \frac{1}{\cos x} + \frac{\sin x}{\cos x} \qquad \text{Simplify.}$$

$$= \sec x + \tan x \qquad \text{Identities}$$

Graphical Solution

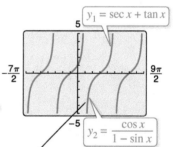

The graphs appear to coincide, so the given equation appears to be an identity.

 ✓ **Checkpoint**   *Audio-video solution in English & Spanish at LarsonPrecalculus.com*

Verify the identity $\csc x + \cot x = \dfrac{\sin x}{1 - \cos x}$.

In Examples 1 through 5, you have been verifying trigonometric identities by working with one side of the equation and converting to the form given on the other side. On occasion, it is practical to work with each side *separately*, to obtain one common form that is equivalent to both sides. This is illustrated in Example 6.

EXAMPLE 6   **Working with Each Side Separately**

Verify the identity $\dfrac{\cot^2 \theta}{1 + \csc \theta} = \dfrac{1 - \sin \theta}{\sin \theta}$.

Algebraic Solution	**Numerical Solution**
Working with the left side, you have	Use a graphing utility to create a table that shows the values of

Algebraic Solution

Working with the left side, you have

$$\dfrac{\cot^2 \theta}{1 + \csc \theta} = \dfrac{\csc^2 \theta - 1}{1 + \csc \theta} \qquad \text{Pythagorean identity}$$

$$= \dfrac{(\csc \theta - 1)(\csc \theta + 1)}{1 + \csc \theta} \qquad \text{Factor.}$$

$$= \csc \theta - 1. \qquad \text{Simplify.}$$

Now, simplifying the right side, you have

$$\dfrac{1 - \sin \theta}{\sin \theta} = \dfrac{1}{\sin \theta} - \dfrac{\sin \theta}{\sin \theta} = \csc \theta - 1.$$

This verifies the identity because both sides are equal to $\csc \theta - 1$.

Numerical Solution

Use a graphing utility to create a table that shows the values of

$$y_1 = \dfrac{\cot^2 x}{1 + \csc x} \quad \text{and} \quad y_2 = \dfrac{1 - \sin x}{\sin x}$$

for different values of x.

X	Y1	Y2
-.5	-3.086	-3.086
-.25	-5.042	-5.042
0	ERROR	ERROR
.25	3.042	3.042
.5	1.0858	1.0858
.75	.46705	.46705
1	.1884	.1884
X=1		

The values for y_1 and y_2 appear to be identical, so the equation appears to be an identity.

✓ *Checkpoint* ◄))) *Audio-video solution in English & Spanish at LarsonPrecalculus.com*

Verify the identity $\dfrac{\tan^2 \theta}{1 + \sec \theta} = \dfrac{1 - \cos \theta}{\cos \theta}$.

Example 7 shows powers of trigonometric functions rewritten as more complicated sums of products of trigonometric functions. This is a common procedure used in calculus.

EXAMPLE 7   **Two Examples from Calculus**

Verify each identity.

a. $\tan^4 x = \tan^2 x \sec^2 x - \tan^2 x$ **b.** $\csc^4 x \cot x = \csc^2 x(\cot x + \cot^3 x)$

Solution

a. $\tan^4 x = (\tan^2 x)(\tan^2 x)$ Write as separate factors.

       $= \tan^2 x(\sec^2 x - 1)$ Pythagorean identity

       $= \tan^2 x \sec^2 x - \tan^2 x$ Multiply.

b. $\csc^4 x \cot x = \csc^2 x \csc^2 x \cot x$ Write as separate factors.

       $= \csc^2 x(1 + \cot^2 x) \cot x$ Pythagorean identity

       $= \csc^2 x(\cot x + \cot^3 x)$ Multiply.

✓ *Checkpoint* ◄))) *Audio-video solution in English & Spanish at LarsonPrecalculus.com*

Verify each identity.

a. $\tan^3 x = \tan x \sec^2 x - \tan x$ **b.** $\sin^3 x \cos^4 x = (\cos^4 x - \cos^6 x) \sin x$

Summarize   *(Section 5.2)*

1. State the guidelines for verifying trigonometric identities *(page 355)*. For examples of verifying trigonometric identities, see Examples 1–7.

5.2 Exercises

See **CalcChat.com** for tutorial help and worked-out solutions to odd-numbered exercises.

Vocabulary

In Exercises 1 and 2, fill in the blanks.

1. An equation that is true for all real values in the domain of the variable is an _____.

2. An equation that is true for only some values in the domain of the variable is a _____ _____.

In Exercises 3–8, fill in the blank to complete the fundamental trigonometric identity.

3. $\dfrac{1}{\cot u} = $ _____

4. $\dfrac{\cos u}{\sin u} = $ _____

5. $\cos\left(\dfrac{\pi}{2} - u\right) = $ _____

6. $1 + $ _____ $= \csc^2 u$

7. $\csc(-u) = $ _____

8. $\sec(-u) = $ _____

Skills and Applications

 Verifying a Trigonometric Identity   In Exercises 9–18, verify the identity.

9. $\tan t \cot t = 1$

10. $\dfrac{\tan x \cot x}{\cos x} = \sec x$

11. $(1 + \sin \alpha)(1 - \sin \alpha) = \cos^2 \alpha$

12. $\cos^2 \beta - \sin^2 \beta = 2\cos^2 \beta - 1$

13. $\cos^2 \beta - \sin^2 \beta = 1 - 2\sin^2 \beta$

14. $\sin^2 \alpha - \sin^4 \alpha = \cos^2 \alpha - \cos^4 \alpha$

15. $\tan\left(\dfrac{\pi}{2} - \theta\right)\tan \theta = 1$

16. $\dfrac{\cos[(\pi/2) - x]}{\sin[(\pi/2) - x]} = \tan x$

17. $\sin t \csc\left(\dfrac{\pi}{2} - t\right) = \tan t$

18. $\sec^2 y - \cot^2\left(\dfrac{\pi}{2} - y\right) = 1$

 Verifying a Trigonometric Identity In Exercises 19–24, verify the identity algebraically. Use the *table* feature of a graphing utility to check your result numerically.

19. $\dfrac{1}{\tan x} + \dfrac{1}{\cot x} = \tan x + \cot x$

20. $\dfrac{1}{\sin x} - \dfrac{1}{\csc x} = \csc x - \sin x$

21. $\dfrac{1 + \sin \theta}{\cos \theta} + \dfrac{\cos \theta}{1 + \sin \theta} = 2\sec \theta$

22. $\dfrac{\cos \theta \cot \theta}{1 - \sin \theta} - 1 = \csc \theta$

23. $\dfrac{1}{\cos x + 1} + \dfrac{1}{\cos x - 1} = -2\csc x \cot x$

24. $\cos x - \dfrac{\cos x}{1 - \tan x} = \dfrac{\sin x \cos x}{\sin x - \cos x}$

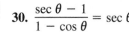 **Verifying a Trigonometric Identity   In Exercises 25–30, verify the identity algebraically. Use a graphing utility to check your result graphically.**

25. $\sec y \cos y = 1$

26. $\cot^2 y(\sec^2 y - 1) = 1$

27. $\dfrac{\tan^2 \theta}{\sec \theta} = \sin \theta \tan \theta$

28. $\dfrac{\cot^3 t}{\csc t} = \cos t(\csc^2 t - 1)$

29. $\dfrac{1}{\tan \beta} + \tan \beta = \dfrac{\sec^2 \beta}{\tan \beta}$

30. $\dfrac{\sec \theta - 1}{1 - \cos \theta} = \sec \theta$

Converting to Sines and Cosines   In Exercises 31–36, verify the identity by converting the left side into sines and cosines.

31. $\dfrac{\cot^2 t}{\csc t} = \dfrac{1 - \sin^2 t}{\sin t}$

32. $\cos x + \sin x \tan x = \sec x$

33. $\sec x - \cos x = \sin x \tan x$

34. $\cot x - \tan x = \sec x(\csc x - 2\sin x)$

35. $\dfrac{\cot x}{\sec x} = \csc x - \sin x$

36. $\dfrac{\csc(-x)}{\sec(-x)} = -\cot x$

Verifying a Trigonometric Identity   In Exercises 37–42, verify the identity.

37. $\sin^{1/2} x \cos x - \sin^{5/2} x \cos x = \cos^3 x \sqrt{\sin x}$

38. $\sec^6 x(\sec x \tan x) - \sec^4 x(\sec x \tan x) = \sec^5 x \tan^3 x$

39. $(1 + \sin y)[1 + \sin(-y)] = \cos^2 y$

40. $\dfrac{\tan x + \tan y}{1 - \tan x \tan y} = \dfrac{\cot x + \cot y}{\cot x \cot y - 1}$

41. $\sqrt{\dfrac{1 + \sin \theta}{1 - \sin \theta}} = \dfrac{1 + \sin \theta}{|\cos \theta|}$

42. $\dfrac{\cos x - \cos y}{\sin x + \sin y} + \dfrac{\sin x - \sin y}{\cos x + \cos y} = 0$

Error Analysis   **In Exercises 43 and 44, describe the error(s).**

43. $\dfrac{1}{\tan x} + \cot(-x) = \cot x + \cot x = 2\cot x$

44. $\dfrac{1 + \sec(-\theta)}{\sin(-\theta) + \tan(-\theta)} = \dfrac{1 - \sec\theta}{\sin\theta - \tan\theta}$

$$= \dfrac{1 - \sec\theta}{(\sin\theta)[1 - (1/\cos\theta)]}$$

$$= \dfrac{1 - \sec\theta}{\sin\theta(1 - \sec\theta)}$$

$$= \dfrac{1}{\sin\theta}$$

$$= \csc\theta$$

 Determining Trigonometric Identities   **In Exercises 45–50, (a) use a graphing utility to graph each side of the equation to determine whether the equation is an identity, (b) use the *table* feature of the graphing utility to determine whether the equation is an identity, and (c) confirm the results of parts (a) and (b) algebraically.**

45. $(1 + \cot^2 x)(\cos^2 x) = \cot^2 x$

46. $\csc x(\csc x - \sin x) + \dfrac{\sin x - \cos x}{\sin x} + \cot x = \csc^2 x$

47. $2 + \cos^2 x - 3\cos^4 x = \sin^2 x(3 + 2\cos^2 x)$

48. $\tan^4 x + \tan^2 x - 3 = \sec^2 x(4\tan^2 x - 3)$

49. $\dfrac{1 + \cos x}{\sin x} = \dfrac{\sin x}{1 - \cos x}$ **50.** $\dfrac{\cot\alpha}{\csc\alpha + 1} = \dfrac{\csc\alpha + 1}{\cot\alpha}$

 Verifying a Trigonometric Identity   **In Exercises 51–54, verify the identity.**

51. $\tan^5 x = \tan^3 x \sec^2 x - \tan^3 x$

52. $\sec^4 x \tan^2 x = (\tan^2 x + \tan^4 x)\sec^2 x$

53. $\cos^3 x \sin^2 x = (\sin^2 x - \sin^4 x)\cos x$

54. $\sin^4 x + \cos^4 x = 1 - 2\cos^2 x + 2\cos^4 x$

Using Cofunction Identities   **In Exercises 55 and 56, use the cofunction identities to evaluate the expression without using a calculator.**

55. $\sin^2 25° + \sin^2 65°$

56. $\tan^2 63° + \cot^2 16° - \sec^2 74° - \csc^2 27°$

 Verifying a Trigonometric Identity   **In Exercises 57–60, verify the identity.**

57. $\tan(\sin^{-1} x) = \dfrac{x}{\sqrt{1 - x^2}}$ **58.** $\cos(\sin^{-1} x) = \sqrt{1 - x^2}$

59. $\tan\left(\sin^{-1}\dfrac{x - 1}{4}\right) = \dfrac{x - 1}{\sqrt{16 - (x - 1)^2}}$

60. $\tan\left(\cos^{-1}\dfrac{x + 1}{2}\right) = \dfrac{\sqrt{4 - (x + 1)^2}}{x + 1}$

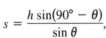 **61. Rate of Change**   The rate of change of the function $f(x) = \sin x + \csc x$ is given by the expression $\cos x - \csc x \cot x$. Show that the expression for the rate of change can also be written as $-\cos x \cot^2 x$.

62. Shadow Length

The length s of a shadow cast by a vertical gnomon (a device used to tell time) of height h when the angle of the sun above the horizon is θ can be modeled by the equation

$$s = \dfrac{h\sin(90° - \theta)}{\sin\theta},$$

$0° < \theta \le 90°$.

(a) Verify that the expression for s is equal to $h\cot\theta$.

(b) Use a graphing utility to create a table of the lengths s for different values of θ. Let $h = 5$ feet.

(c) Use your table from part (b) to determine the angle of the sun that results in the minimum length of the shadow.

(d) Based on your results from part (c), what time of day do you think it is when the angle of the sun above the horizon is 90°?

Exploration

True or False?   **In Exercises 63–65, determine whether the statement is true or false. Justify your answer.**

63. $\tan x^2 = \tan^2 x$ **64.** $\cos\left(\theta - \dfrac{\pi}{2}\right) = \sin\theta$

65. The equation $\sin^2\theta + \cos^2\theta = 1 + \tan^2\theta$ is an identity because $\sin^2(0) + \cos^2(0) = 1$ and $1 + \tan^2(0) = 1$.

66. 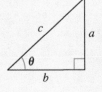 **HOW DO YOU SEE IT?**   Explain how to use the figure to derive the identity

$$\dfrac{\sec^2\theta - 1}{\sec^2\theta} = \sin^2\theta$$

given in Example 1.

Think About It   **In Exercises 67–70, explain why the equation is not an identity and find one value of the variable for which the equation is not true.**

67. $\sin\theta = \sqrt{1 - \cos^2\theta}$ **68.** $\tan\theta = \sqrt{\sec^2\theta - 1}$

69. $1 - \cos\theta = \sin\theta$ **70.** $1 + \tan\theta = \sec\theta$

5.3   Solving Trigonometric Equations

- ■ Use standard algebraic techniques to solve trigonometric equations.
- ■ Solve trigonometric equations of quadratic type.
- ■ Solve trigonometric equations involving multiple angles.
- ■ Use inverse trigonometric functions to solve trigonometric equations.

Trigonometric equations have many applications in circular motion. For example, in Exercise 94 on page 373, you will solve a trigonometric equation to determine when a person riding a Ferris wheel will be at certain heights above the ground.

Introduction

To solve a trigonometric equation, use standard algebraic techniques (when possible) such as collecting like terms, extracting square roots, and factoring. Your preliminary goal in solving a trigonometric equation is to *isolate* the trigonometric function on one side of the equation. For example, to solve the equation $2 \sin x = 1$, divide each side by 2 to obtain

$$\sin x = \frac{1}{2}.$$

To solve for x, note in the graph of $y = \sin x$ below that the equation $\sin x = \frac{1}{2}$ has solutions $x = \pi/6$ and $x = 5\pi/6$ in the interval $[0, 2\pi)$. Moreover, because $\sin x$ has a period of 2π, there are infinitely many other solutions, which can be written as

$$x = \frac{\pi}{6} + 2n\pi \quad \text{and} \quad x = \frac{5\pi}{6} + 2n\pi \qquad \text{General solution}$$

where n is an integer. Notice the solutions for $n = \pm 1$ in the graph of $y = \sin x$.

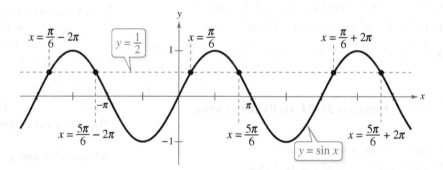

The figure below illustrates another way to show that the equation $\sin x = \frac{1}{2}$ has infinitely many solutions. Any angles that are coterminal with $\pi/6$ or $5\pi/6$ are also solutions of the equation.

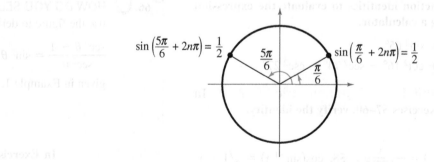

When solving trigonometric equations, write your answer(s) using exact values (when possible) rather than decimal approximations.

EXAMPLE 1 **Collecting Like Terms**

Solve

$$\sin x + \sqrt{2} = -\sin x.$$

Solution Begin by isolating $\sin x$ on one side of the equation.

$\sin x + \sqrt{2} = -\sin x$	Write original equation.
$\sin x + \sin x + \sqrt{2} = 0$	Add $\sin x$ to each side.
$\sin x + \sin x = -\sqrt{2}$	Subtract $\sqrt{2}$ from each side.
$2\sin x = -\sqrt{2}$	Combine like terms.
$\sin x = -\dfrac{\sqrt{2}}{2}$	Divide each side by 2.

The period of $\sin x$ is 2π, so first find all solutions in the interval $[0, 2\pi)$. These solutions are $x = 5\pi/4$ and $x = 7\pi/4$. Finally, add multiples of 2π to each of these solutions to obtain the general form

$$x = \frac{5\pi}{4} + 2n\pi \quad \text{and} \quad x = \frac{7\pi}{4} + 2n\pi \qquad \text{General solution}$$

where n is an integer.

✓ **Checkpoint**))) Audio-video solution in English & Spanish at LarsonPrecalculus.com

Solve $\sin x - \sqrt{2} = -\sin x$.

EXAMPLE 2 **Extracting Square Roots**

Solve

$$3\tan^2 x - 1 = 0.$$

Solution Begin by isolating $\tan x$ on one side of the equation.

$3\tan^2 x - 1 = 0$	Write original equation.
$3\tan^2 x = 1$	Add 1 to each side.
$\tan^2 x = \dfrac{1}{3}$	Divide each side by 3.
$\tan x = \pm\dfrac{1}{\sqrt{3}}$	Extract square roots.
$\tan x = \pm\dfrac{\sqrt{3}}{3}$	Rationalize the denominator.

> •• REMARK When you extract square roots, make sure you account for both the positive and negative solutions.

The period of $\tan x$ is π, so first find all solutions in the interval $[0, \pi)$. These solutions are $x = \pi/6$ and $x = 5\pi/6$. Finally, add multiples of π to each of these solutions to obtain the general form

$$x = \frac{\pi}{6} + n\pi \quad \text{and} \quad x = \frac{5\pi}{6} + n\pi \qquad \text{General solution}$$

where n is an integer.

✓ **Checkpoint**))) Audio-video solution in English & Spanish at LarsonPrecalculus.com

Solve $4\sin^2 x - 3 = 0$.

The equations in Examples 1 and 2 involved only one trigonometric function. When two or more functions occur in the same equation, collect all terms on one side and try to separate the functions by factoring or by using appropriate identities. This may produce factors that yield no solutions, as illustrated in Example 3.

EXAMPLE 3   Factoring

Solve $\cot x \cos^2 x = 2 \cot x$.

Solution   Begin by collecting all terms on one side of the equation and factoring.

$$\cot x \cos^2 x = 2 \cot x \qquad \text{Write original equation.}$$

$$\cot x \cos^2 x - 2 \cot x = 0 \qquad \text{Subtract } 2 \cot x \text{ from each side.}$$

$$\cot x(\cos^2 x - 2) = 0 \qquad \text{Factor.}$$

Set each factor equal to zero and isolate the trigonometric function, if necessary.

$$\cot x = 0 \quad \text{or} \quad \cos^2 x - 2 = 0$$

$$\cos^2 x = 2$$

$$\cos x = \pm\sqrt{2}$$

In the interval $(0, \pi)$, the equation $\cot x = 0$ has the solution

$$x = \frac{\pi}{2}.$$

No solution exists for $\cos x = \pm\sqrt{2}$ because $\pm\sqrt{2}$ are outside the range of the cosine function. The period of $\cot x$ is π, so add multiples of π to $x = \pi/2$ to get the general form

$$x = \frac{\pi}{2} + n\pi \qquad \text{General solution}$$

where n is an integer. Confirm this graphically by sketching the graph of $y = \cot x \cos^2 x - 2 \cot x$.

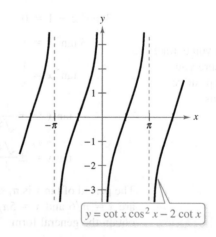

$$y = \cot x \cos^2 x - 2 \cot x$$

Notice that the x-intercepts occur at

$$-\frac{3\pi}{2}, \quad -\frac{\pi}{2}, \quad \frac{\pi}{2}, \quad \frac{3\pi}{2}$$

and so on. These x-intercepts correspond to the solutions of $\cot x \cos^2 x = 2 \cot x$.

✓ **Checkpoint** ◄))) *Audio-video solution in English & Spanish at LarsonPrecalculus.com*

Solve $\sin^2 x = 2 \sin x$.

▷ **ALGEBRA HELP**   To
review the techniques for
solving quadratic equations,
see Appendix A.5.

Equations of Quadratic Type

Below are two examples of trigonometric equations of quadratic type

$$ax^2 + bx + c = 0.$$

To solve equations of this type, use factoring (when possible) or use the Quadratic Formula.

Quadratic in sin x	**Quadratic in sec x**
$2 \sin^2 x - \sin x - 1 = 0$	$\sec^2 x - 3 \sec x - 2 = 0$
$2(\sin x)^2 - (\sin x) - 1 = 0$	$(\sec x)^2 - 3(\sec x) - 2 = 0$

EXAMPLE 4   **Solving an Equation of Quadratic Type**

Find all solutions of $2 \sin^2 x - \sin x - 1 = 0$ in the interval $[0, 2\pi)$.

Algebraic Solution

Treat the equation as quadratic in $\sin x$ and factor.

$$2 \sin^2 x - \sin x - 1 = 0 \qquad \text{Write original equation.}$$
$$(2 \sin x + 1)(\sin x - 1) = 0 \qquad \text{Factor.}$$

Setting each factor equal to zero, you obtain the following solutions in the interval $[0, 2\pi)$.

$$2 \sin x + 1 = 0 \qquad \text{or} \quad \sin x - 1 = 0$$

$$\sin x = -\frac{1}{2} \qquad\qquad \sin x = 1$$

$$x = \frac{7\pi}{6}, \frac{11\pi}{6} \qquad\qquad x = \frac{\pi}{2}$$

Graphical Solution

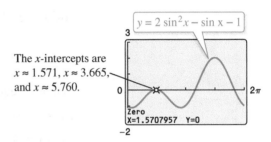

The x-intercepts are $x \approx 1.571$, $x \approx 3.665$, and $x \approx 5.760$.

Use the x-intercepts to conclude that the approximate solutions of $2 \sin^2 x - \sin x - 1 = 0$ in the interval $[0, 2\pi)$ are

$$x \approx 1.571 \approx \frac{\pi}{2}, \quad x \approx 3.665 \approx \frac{7\pi}{6}, \quad \text{and} \quad x \approx 5.760 \approx \frac{11\pi}{6}.$$

✓ *Checkpoint*  Audio-video solution in English & Spanish at LarsonPrecalculus.com

Find all solutions of $2 \sin^2 x - 3 \sin x + 1 = 0$ in the interval $[0, 2\pi)$.

EXAMPLE 5   **Rewriting with a Single Trigonometric Function**

Solve $2 \sin^2 x + 3 \cos x - 3 = 0$.

Solution   This equation contains both sine and cosine functions. Rewrite the equation so that it has only cosine functions by using the identity $\sin^2 x = 1 - \cos^2 x$.

$$2 \sin^2 x + 3 \cos x - 3 = 0 \qquad \text{Write original equation.}$$
$$2(1 - \cos^2 x) + 3 \cos x - 3 = 0 \qquad \text{Pythagorean identity}$$
$$2 \cos^2 x - 3 \cos x + 1 = 0 \qquad \text{Multiply each side by } -1.$$
$$(2 \cos x - 1)(\cos x - 1) = 0 \qquad \text{Factor.}$$

Setting each factor equal to zero, you obtain the solutions $x = 0$, $x = \pi/3$, and $x = 5\pi/3$ in the interval $[0, 2\pi)$. Because $\cos x$ has a period of 2π, the general solution is

$$x = 2n\pi, \quad x = \frac{\pi}{3} + 2n\pi, \quad \text{and} \quad x = \frac{5\pi}{3} + 2n\pi \qquad \text{General solution}$$

where n is an integer.

✓ *Checkpoint*  Audio-video solution in English & Spanish at LarsonPrecalculus.com

Solve $3 \sec^2 x - 2 \tan^2 x - 4 = 0$.

Sometimes you square each side of an equation to obtain an equation of quadratic type, as demonstrated in the next example. This procedure can introduce extraneous solutions, so check any solutions in the original equation to determine whether they are valid or extraneous.

• • REMARK You square each side of the equation in Example 6 because the squares of the sine and cosine functions are related by a Pythagorean identity. The same is true for the squares of the secant and tangent functions and for the squares of the cosecant and cotangent functions.
• • • • • • • • • • • • • • • • • ▷

EXAMPLE 6 Squaring and Converting to Quadratic Type

See LarsonPrecalculus.com for an interactive version of this type of example.

Find all solutions of $\cos x + 1 = \sin x$ in the interval $[0, 2\pi)$.

Solution It is not clear how to rewrite this equation in terms of a single trigonometric function. Notice what happens when you square each side of the equation.

$$\cos x + 1 = \sin x \qquad \text{Write original equation.}$$

$$\cos^2 x + 2\cos x + 1 = \sin^2 x \qquad \text{Square each side.}$$

$$\cos^2 x + 2\cos x + 1 = 1 - \cos^2 x \qquad \text{Pythagorean identity}$$

$$\cos^2 x + \cos^2 x + 2\cos x + 1 - 1 = 0 \qquad \text{Rewrite equation.}$$

$$2\cos^2 x + 2\cos x = 0 \qquad \text{Combine like terms.}$$

$$2\cos x(\cos x + 1) = 0 \qquad \text{Factor.}$$

Set each factor equal to zero and solve for x.

$$2\cos x = 0 \qquad \text{or} \quad \cos x + 1 = 0$$

$$\cos x = 0 \qquad\qquad\qquad \cos x = -1$$

$$x = \frac{\pi}{2}, \frac{3\pi}{2} \qquad\qquad\qquad x = \pi$$

Because you squared the original equation, check for extraneous solutions.

Check $x = \dfrac{\pi}{2}$

$$\cos\frac{\pi}{2} + 1 \stackrel{?}{=} \sin\frac{\pi}{2} \qquad \text{Substitute } \tfrac{\pi}{2} \text{ for } x.$$

$$0 + 1 = 1 \qquad \text{Solution checks.} \checkmark$$

Check $x = \dfrac{3\pi}{2}$

$$\cos\frac{3\pi}{2} + 1 \stackrel{?}{=} \sin\frac{3\pi}{2} \qquad \text{Substitute } \tfrac{3\pi}{2} \text{ for } x.$$

$$0 + 1 \neq -1 \qquad \text{Solution does not check.}$$

Check $x = \pi$

$$\cos\pi + 1 \stackrel{?}{=} \sin\pi \qquad \text{Substitute } \pi \text{ for } x.$$

$$-1 + 1 = 0 \qquad \text{Solution checks.} \checkmark$$

Of the three possible solutions, $x = 3\pi/2$ is extraneous. So, in the interval $[0, 2\pi)$, the only two solutions are

$$x = \frac{\pi}{2} \quad \text{and} \quad x = \pi.$$

✓ **Checkpoint** ◀))) *Audio-video solution in English & Spanish at LarsonPrecalculus.com*

Find all solutions of $\sin x + 1 = \cos x$ in the interval $[0, 2\pi)$.

Functions Involving Multiple Angles

The next two examples involve trigonometric functions of multiple angles of the forms cos ku and tan ku. To solve equations involving these forms, first solve the equation for ku, and then divide your result by k.

EXAMPLE 7   Solving a Multiple-Angle Equation

Solve $2 \cos 3t - 1 = 0$.

Solution

$2 \cos 3t - 1 = 0$	Write original equation.
$2 \cos 3t = 1$	Add 1 to each side.
$\cos 3t = \dfrac{1}{2}$	Divide each side by 2.

In the interval $[0, 2\pi)$, you know that $3t = \pi/3$ and $3t = 5\pi/3$ are the only solutions, so, in general, you have

$$3t = \frac{\pi}{3} + 2n\pi \quad \text{and} \quad 3t = \frac{5\pi}{3} + 2n\pi.$$

Dividing these results by 3, you obtain the general solution

$$t = \frac{\pi}{9} + \frac{2n\pi}{3} \quad \text{and} \quad t = \frac{5\pi}{9} + \frac{2n\pi}{3} \qquad \text{General solution}$$

where n is an integer.

✓ **Checkpoint** 🔊))) *Audio-video solution in English & Spanish at LarsonPrecalculus.com*

Solve $2 \sin 2t - \sqrt{3} = 0$.

EXAMPLE 8   Solving a Multiple-Angle Equation

$3 \tan \dfrac{x}{2} + 3 = 0$	Original equation
$3 \tan \dfrac{x}{2} = -3$	Subtract 3 from each side.
$\tan \dfrac{x}{2} = -1$	Divide each side by 3.

In the interval $[0, \pi)$, you know that $x/2 = 3\pi/4$ is the only solution, so, in general, you have

$$\frac{x}{2} = \frac{3\pi}{4} + n\pi.$$

Multiplying this result by 2, you obtain the general solution

$$x = \frac{3\pi}{2} + 2n\pi \qquad \text{General solution}$$

where n is an integer.

✓ **Checkpoint** 🔊))) *Audio-video solution in English & Spanish at LarsonPrecalculus.com*

Solve $2 \tan \dfrac{x}{2} - 2 = 0$.

Using Inverse Functions

EXAMPLE 9 Using Inverse Functions

$$\sec^2 x - 2\tan x = 4 \qquad \text{Original equation}$$

$$1 + \tan^2 x - 2\tan x - 4 = 0 \qquad \text{Pythagorean identity}$$

$$\tan^2 x - 2\tan x - 3 = 0 \qquad \text{Combine like terms.}$$

$$(\tan x - 3)(\tan x + 1) = 0 \qquad \text{Factor.}$$

Setting each factor equal to zero, you obtain two solutions in the interval $(-\pi/2, \pi/2)$. [Recall that the range of the inverse tangent function is $(-\pi/2, \pi/2)$.]

$$x = \arctan 3 \quad \text{and} \quad x = \arctan(-1) = -\pi/4$$

Finally, $\tan x$ has a period of π, so add multiples of π to obtain

$$x = \arctan 3 + n\pi \quad \text{and} \quad x = (-\pi/4) + n\pi \qquad \text{General solution}$$

where n is an integer. You can use a calculator to approximate the value of arctan 3.

✓ **Checkpoint** Audio-video solution in English & Spanish at LarsonPrecalculus.com

Solve $4\tan^2 x + 5\tan x - 6 = 0$.

EXAMPLE 10 Using the Quadratic Formula

Find all solutions of $\sin^2 x - 3\sin x - 2 = 0$ in the interval $[0, 2\pi)$.

Solution

The expression $\sin^2 x - 3\sin x - 2$ cannot be factored, so use the Quadratic Formula.

$$\sin^2 x - 3\sin x - 2 = 0 \qquad \text{Write original equation.}$$

$$\sin x = \frac{-(-3) \pm \sqrt{(-3)^2 - 4(1)(-2)}}{2(1)} \qquad \text{Quadratic Formula}$$

$$\sin x = \frac{3 \pm \sqrt{17}}{2} \qquad \text{Simplify.}$$

So, $\sin x = \dfrac{3 + \sqrt{17}}{2} \approx 3.5616$ or $\sin x = \dfrac{3 - \sqrt{17}}{2} \approx -0.5616$. The range of the sine function is $[-1, 1]$, so $\sin x = \dfrac{3 + \sqrt{17}}{2}$ has no solution for x. Use a calculator to approximate a solution of $\sin x = \dfrac{3 - \sqrt{17}}{2}$.

$$x = \arcsin\left(\frac{3 - \sqrt{17}}{2}\right) \approx -0.5963$$

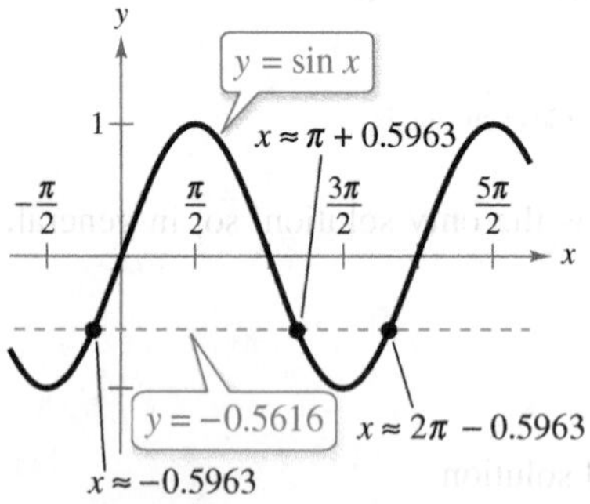

Figure 5.2

Note that this solution is not in the interval $[0, 2\pi)$. To find the solutions in $[0, 2\pi)$, sketch the graphs of $y = \sin x$ and $y = -0.5616$, as shown in Figure 5.2. From the graph, it appears that $\sin x \approx -0.5616$ on the interval $[0, 2\pi)$ when

$$x \approx \pi + 0.5963 \approx 3.7379 \quad \text{and} \quad x \approx 2\pi - 0.5963 \approx 5.6869.$$

So, the solutions of $\sin^2 x - 3\sin x - 2 = 0$ in $[0, 2\pi)$ are $x \approx 3.7379$ and $x \approx 5.6869$.

✓ **Checkpoint** Audio-video solution in English & Spanish at LarsonPrecalculus.com

Find all solutions of $\sin^2 x + 2\sin x - 1 = 0$ in the interval $[0, 2\pi)$.

EXAMPLE 11   **Surface Area of a Honeycomb Cell**

The surface area S (in square inches) of a honeycomb cell is given by

$$S = 6hs + 1.5s^2\left(\frac{\sqrt{3} - \cos\theta}{\sin\theta}\right), \quad 0° < \theta \le 90°$$

where $h = 2.4$ inches, $s = 0.75$ inch, and θ is the angle shown in the figure at the right. What value of θ gives the minimum surface area?

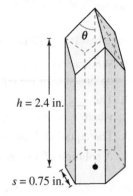

$h = 2.4$ in.

$s = 0.75$ in.

Solution

Letting $h = 2.4$ and $s = 0.75$, you obtain

$$S = 10.8 + 0.84375\left(\frac{\sqrt{3} - \cos\theta}{\sin\theta}\right).$$

Graph this function using a graphing utility set in *degree* mode. Use the *minimum* feature to approximate the minimum point on the graph, as shown in the figure below.

$$y = 10.8 + 0.84375\left(\frac{\sqrt{3} - \cos x}{\sin x}\right)$$

14

0   150

11

Minimum
X=54.735623__Y=11.993243_

So, the minimum surface area occurs when

$$\theta \approx 54.7356°.$$

REMARK By using calculus, it can be shown that the *exact* minimum surface area occurs when

$$\theta = \arccos\left(\frac{1}{\sqrt{3}}\right).$$

✓ *Checkpoint* ◀))) *Audio-video solution in English & Spanish at LarsonPrecalculus.com*

Use the equation for the surface area of a honeycomb cell given in Example 11 with $h = 3.2$ inches and $s = 0.75$ inch. What value of θ gives the minimum surface area?

Summarize (Section 5.3)

1. Explain how to use standard algebraic techniques to solve trigonometric equations *(page 362)*. For examples of using standard algebraic techniques to solve trigonometric equations, see Examples 1–3.

2. Explain how to solve a trigonometric equation of quadratic type *(page 365)*. For examples of solving trigonometric equations of quadratic type, see Examples 4–6.

3. Explain how to solve a trigonometric equation involving multiple angles *(page 367)*. For examples of solving trigonometric equations involving multiple angles, see Examples 7 and 8.

4. Explain how to use inverse trigonometric functions to solve trigonometric equations *(page 368)*. For examples of using inverse trigonometric functions to solve trigonometric equations, see Examples 9–11.

5.3 Exercises

See **CalcChat.com** for tutorial help and worked-out solutions to odd-numbered exercises.

Vocabulary: Fill in the blanks.

1. When solving a trigonometric equation, the preliminary goal is to _____ the trigonometric function on one side of the equation.

2. The _____ solution of the equation $2 \sin \theta + 1 = 0$ is $\theta = \frac{7\pi}{6} + 2n\pi$ and $\theta = \frac{11\pi}{6} + 2n\pi$, where n is an integer.

3. The equation $2 \tan^2 x - 3 \tan x + 1 = 0$ is a trigonometric equation of _____ type.

4. A solution of an equation that does not satisfy the original equation is an _____ solution.

Skills and Applications

Verifying Solutions In Exercises 5–10, verify that each x-value is a solution of the equation.

5. $\tan x - \sqrt{3} = 0$

 (a) $x = \frac{\pi}{3}$

 (b) $x = \frac{4\pi}{3}$

6. $\sec x - 2 = 0$

 (a) $x = \frac{\pi}{3}$

 (b) $x = \frac{5\pi}{3}$

7. $3 \tan^2 2x - 1 = 0$

 (a) $x = \frac{\pi}{12}$

 (b) $x = \frac{5\pi}{12}$

8. $2 \cos^2 4x - 1 = 0$

 (a) $x = \frac{\pi}{16}$

 (b) $x = \frac{3\pi}{16}$

9. $2 \sin^2 x - \sin x - 1 = 0$

 (a) $x = \frac{\pi}{2}$

 (b) $x = \frac{7\pi}{6}$

10. $\csc^4 x - 4 \csc^2 x = 0$

 (a) $x = \frac{\pi}{6}$

 (b) $x = \frac{5\pi}{6}$

 Solving a Trigonometric Equation In Exercises 11–28, solve the equation.

11. $\sqrt{3} \csc x - 2 = 0$

12. $\tan x + \sqrt{3} = 0$

13. $\cos x + 1 = -\cos x$

14. $3 \sin x + 1 = \sin x$

15. $3 \sec^2 x - 4 = 0$

16. $3 \cot^2 x - 1 = 0$

17. $4 \cos^2 x - 1 = 0$

18. $2 - 4 \sin^2 x = 0$

19. $\sin x(\sin x + 1) = 0$

20. $(2 \sin^2 x - 1)(\tan^2 x - 3) = 0$

21. $\cos^3 x - \cos x = 0$

22. $\sec^2 x - 1 = 0$

23. $3 \tan^3 x = \tan x$

24. $\sec x \csc x = 2 \csc x$

25. $2 \cos^2 x + \cos x - 1 = 0$

26. $2 \sin^2 x + 3 \sin x + 1 = 0$

27. $\sec^2 x - \sec x = 2$

28. $\csc^2 x + \csc x = 2$

Solving a Trigonometric Equation In Exercises 29–38, find all solutions of the equation in the interval $[0, 2\pi)$.

29. $\sin x - 2 = \cos x - 2$

30. $\cos x + \sin x \tan x = 2$

31. $2 \sin^2 x = 2 + \cos x$

32. $\tan^2 x = \sec x - 1$

33. $\sin^2 x = 3 \cos^2 x$

34. $2 \sec^2 x + \tan^2 x - 3 = 0$

35. $2 \sin x + \csc x = 0$

36. $3 \sec x - 4 \cos x = 0$

37. $\csc x + \cot x = 1$

38. $\sec x + \tan x = 1$

Solving a Multiple-Angle Equation In Exercises 39–46, solve the multiple-angle equation.

39. $2 \cos 2x - 1 = 0$

40. $2 \sin 2x + \sqrt{3} = 0$

41. $\tan 3x - 1 = 0$

42. $\sec 4x - 2 = 0$

43. $2 \cos \frac{x}{2} - \sqrt{2} = 0$

44. $2 \sin \frac{x}{2} + \sqrt{3} = 0$

45. $3 \tan \frac{x}{2} - \sqrt{3} = 0$

46. $\tan \frac{x}{2} + \sqrt{3} = 0$

Finding x-Intercepts In Exercises 47 and 48, find the x-intercepts of the graph.

47. $y = \sin \frac{\pi x}{2} + 1$

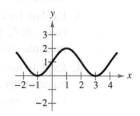

48. $y = \sin \pi x + \cos \pi x$

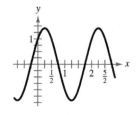

Approximating Solutions In Exercises 49–58, use a graphing utility to approximate (to three decimal places) the solutions of the equation in the interval $[0, 2\pi)$.

49. $5 \sin x + 2 = 0$

50. $2 \tan x + 7 = 0$

51. $\sin x - 3 \cos x = 0$

52. $\sin x + 4 \cos x = 0$

53. $\cos x = x$

54. $\tan x = \csc x$

55. $\sec^2 x - 3 = 0$

56. $\csc^2 x - 5 = 0$

57. $2 \tan^2 x = 15$

58. $6 \sin^2 x = 5$

 Using Inverse Functions In Exercises 59–70, solve the equation.

59. $\tan^2 x + \tan x - 12 = 0$

60. $\tan^2 x - \tan x - 2 = 0$

61. $\sec^2 x - 6 \tan x = -4$

62. $\sec^2 x + \tan x = 3$

63. $2 \sin^2 x + 5 \cos x = 4$

64. $2 \cos^2 x + 7 \sin x = 5$

65. $\cot^2 x - 9 = 0$

66. $\cot^2 x - 6 \cot x + 5 = 0$

67. $\sec^2 x - 4 \sec x = 0$

68. $\sec^2 x + 2 \sec x - 8 = 0$

69. $\csc^2 x + 3 \csc x - 4 = 0$

70. $\csc^2 x - 5 \csc x = 0$

Using the Quadratic Formula In Exercises 71–74, use the Quadratic Formula to find all solutions of the equation in the interval $[0, 2\pi)$. Round your result to four decimal places.

71. $12 \sin^2 x - 13 \sin x + 3 = 0$

72. $3 \tan^2 x + 4 \tan x - 4 = 0$

73. $\tan^2 x + 3 \tan x + 1 = 0$

74. $4 \cos^2 x - 4 \cos x - 1 = 0$

Approximating Solutions In Exercises 75–78, use a graphing utility to approximate (to three decimal places) the solutions of the equation in the given interval.

75. $3 \tan^2 x + 5 \tan x - 4 = 0$, $\left[-\dfrac{\pi}{2}, \dfrac{\pi}{2}\right]$

76. $\cos^2 x - 2 \cos x - 1 = 0$, $[0, \pi]$

77. $4 \cos^2 x - 2 \sin x + 1 = 0$, $\left[-\dfrac{\pi}{2}, \dfrac{\pi}{2}\right]$

78. $2 \sec^2 x + \tan x - 6 = 0$, $\left[-\dfrac{\pi}{2}, \dfrac{\pi}{2}\right]$

Approximating Maximum and Minimum Points In Exercises 79–84, (a) use a graphing utility to graph the function and approximate the maximum and minimum points on the graph in the interval $[0, 2\pi)$, and (b) solve the trigonometric equation and verify that its solutions are the x-coordinates of the maximum and minimum points of f. (Calculus is required to find the trigonometric equation.)

Function	Trigonometric Equation
79. $f(x) = \sin^2 x + \cos x$	$2 \sin x \cos x - \sin x = 0$
80. $f(x) = \cos^2 x - \sin x$	$-2 \sin x \cos x - \cos x = 0$
81. $f(x) = \sin x + \cos x$	$\cos x - \sin x = 0$
82. $f(x) = 2 \sin x + \cos 2x$	$2 \cos x - 4 \sin x \cos x = 0$
83. $f(x) = \sin x \cos x$	$-\sin^2 x + \cos^2 x = 0$
84. $f(x) = \sec x + \tan x - x$	$\sec x \tan x + \sec^2 x = 1$

Number of Points of Intersection In Exercises 85 and 86, use the graph to approximate the number of points of intersection of the graphs of y_1 and y_2.

85. $y_1 = 2 \sin x$
$y_2 = 3x + 1$

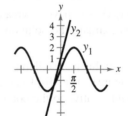

86. $y_1 = 2 \sin x$
$y_2 = \frac{1}{2}x + 1$

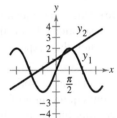

87. Graphical Reasoning Consider the function

$$f(x) = \frac{\sin x}{x}$$

and its graph, shown in the figure below.

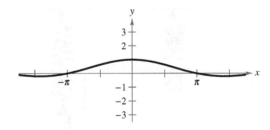

(a) What is the domain of the function?

(b) Identify any symmetry and any asymptotes of the graph.

(c) Describe the behavior of the function as $x \to 0$.

(d) How many solutions does the equation

$$\frac{\sin x}{x} = 0$$

have in the interval $[-8, 8]$? Find the solutions.

88. Graphical Reasoning Consider the function

$$f(x) = \cos \frac{1}{x}$$

and its graph, shown in the figure below.

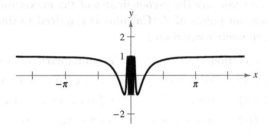

(a) What is the domain of the function?

(b) Identify any symmetry and any asymptotes of the graph.

(c) Describe the behavior of the function as $x \to 0$.

(d) How many solutions does the equation

$$\cos \frac{1}{x} = 0$$

have in the interval $[-1, 1]$? Find the solutions.

(e) Does the equation $\cos(1/x) = 0$ have a greatest solution? If so, then approximate the solution. If not, then explain why.

89. Harmonic Motion A weight is oscillating on the end of a spring (see figure). The displacement from equilibrium of the weight relative to the point of equilibrium is given by

$$y = \tfrac{1}{12}(\cos 8t - 3 \sin 8t)$$

where y is the displacement (in meters) and t is the time (in seconds). Find the times when the weight is at the point of equilibrium $(y = 0)$ for $0 \le t \le 1$.

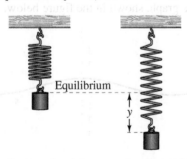

Equilibrium

90. Damped Harmonic Motion The displacement from equilibrium of a weight oscillating on the end of a spring is given by

$$y = 1.56e^{-0.22t} \cos 4.9t$$

where y is the displacement (in feet) and t is the time (in seconds). Use a graphing utility to graph the displacement function for $0 \le t \le 10$. Find the time beyond which the distance between the weight and equilibrium does not exceed 1 foot.

91. Equipment Sales The monthly sales S (in hundreds of units) of skiing equipment at a sports store are approximated by

$$S = 58.3 + 32.5 \cos \frac{\pi t}{6}$$

where t is the time (in months), with $t = 1$ corresponding to January. Determine the months in which sales exceed 7500 units.

92. Projectile Motion A baseball is hit at an angle of θ with the horizontal and with an initial velocity of $v_0 = 100$ feet per second. An outfielder catches the ball 300 feet from home plate (see figure). Find θ when the range r of a projectile is given by

$$r = \frac{1}{32}v_0^2 \sin 2\theta.$$

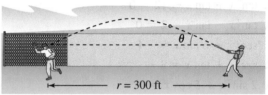

$r = 300$ ft

Not drawn to scale

93. Meteorology The table shows the normal daily high temperatures C in Chicago (in degrees Fahrenheit) for month t, with $t = 1$ corresponding to January. (*Source: NOAA*)

DATA	Month, t	Chicago, C
	1	31.0
	2	35.3
	3	46.6
	4	59.0
	5	70.0
	6	79.7
	7	84.1
	8	81.9
	9	74.8
	10	62.3
	11	48.2
	12	34.8

Spreadsheet at LarsonPrecalculus.com

(a) Use a graphing utility to create a scatter plot of the data.

(b) Find a cosine model for the temperatures.

(c) Graph the model and the scatter plot in the same viewing window. How well does the model fit the data?

(d) What is the overall normal daily high temperature?

(e) Use the graphing utility to determine the months during which the normal daily high temperature is above 72°F and below 72°F.

94. Ferris Wheel

The height h (in feet) above ground of a seat on a Ferris wheel at time t (in minutes) can be modeled by

$$h(t) = 53 + 50 \sin\left(\frac{\pi}{16}t - \frac{\pi}{2}\right).$$

The wheel makes one revolution every 32 seconds. The ride begins when $t = 0$.

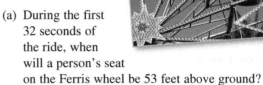

(a) During the first 32 seconds of the ride, when will a person's seat on the Ferris wheel be 53 feet above ground?

(b) When will a person's seat be at the top of the Ferris wheel for the first time during the ride? For a ride that lasts 160 seconds, how many times will a person's seat be at the top of the ride, and at what times?

95. Geometry The area of a rectangle inscribed in one arc of the graph of $y = \cos x$ (see figure) is given by

$$A = 2x \cos x, \quad 0 < x < \pi/2.$$

(a) Use a graphing utility to graph the area function, and approximate the area of the largest inscribed rectangle.

(b) Determine the values of x for which $A \geq 1$.

96. Quadratic Approximation Consider the function

$$f(x) = 3 \sin(0.6x - 2).$$

(a) Approximate the zero of the function in the interval $[0, 6]$.

(b) A quadratic approximation agreeing with f at $x = 5$ is

$$g(x) = -0.45x^2 + 5.52x - 13.70.$$

Use a graphing utility to graph f and g in the same viewing window. Describe the result.

(c) Use the Quadratic Formula to find the zeros of g. Compare the zero of g in the interval $[0, 6]$ with the result of part (a).

Fixed Point In Exercises 97 and 98, find the least positive fixed point of the function f. [A *fixed point* of a function f is a real number c such that $f(c) = c$.]

97. $f(x) = \tan(\pi x/4)$

98. $f(x) = \cos x$

Exploration

True or False? In Exercises 99 and 100, determine whether the statement is true or false. Justify your answer.

99. The equation $2 \sin 4t - 1 = 0$ has four times the number of solutions in the interval $[0, 2\pi)$ as the equation $2 \sin t - 1 = 0$.

100. The trigonometric equation $\sin x = 3.4$ can be solved using an inverse trigonometric function.

101. Think About It Explain what happens when you divide each side of the equation $\cot x \cos^2 x = 2 \cot x$ by $\cot x$. Is this a correct method to use when solving equations?

102. **HOW DO YOU SEE IT?** Explain how to use the figure to solve the equation $2 \cos x - 1 = 0$.

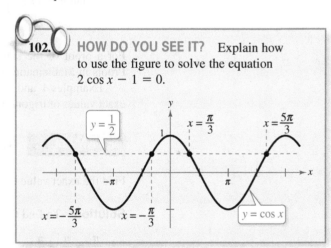

103. Graphical Reasoning Use a graphing utility to confirm the solutions found in Example 6 in two different ways.

(a) Graph both sides of the equation and find the x-coordinates of the points at which the graphs intersect.

Left side: $y = \cos x + 1$

Right side: $y = \sin x$

(b) Graph the equation $y = \cos x + 1 - \sin x$ and find the x-intercepts of the graph.

(c) Do both methods produce the same x-values? Which method do you prefer? Explain.

Project: Meteorology To work an extended application analyzing the normal daily high temperatures in Phoenix, Arizona, and in Seattle, Washington, visit this text's website at *LarsonPrecalculus.com*. *(Source: NOAA)*

5.4 Sum and Difference Formulas

Sum and difference formulas are used to model standing waves, such as those produced in a guitar string. For example, in Exercise 80 on page 379, you will use a sum formula to write the equation of a standing wave.

■ Use sum and difference formulas to evaluate trigonometric functions, verify identities, and solve trigonometric equations.

Using Sum and Difference Formulas

In this section and the next, you will study the uses of several trigonometric identities and formulas.

Sum and Difference Formulas

$$\sin(u + v) = \sin u \cos v + \cos u \sin v$$

$$\sin(u - v) = \sin u \cos v - \cos u \sin v$$

$$\cos(u + v) = \cos u \cos v - \sin u \sin v$$

$$\cos(u - v) = \cos u \cos v + \sin u \sin v$$

$$\tan(u + v) = \frac{\tan u + \tan v}{1 - \tan u \tan v} \qquad \tan(u - v) = \frac{\tan u - \tan v}{1 + \tan u \tan v}$$

For a proof of the sum and difference formulas for $\cos(u \pm v)$ and $\tan(u \pm v)$, see Proofs in Mathematics on page 395.

Examples 1 and 2 show how **sum and difference formulas** enable you to find exact values of trigonometric functions involving sums or differences of special angles.

EXAMPLE 1   **Evaluating a Trigonometric Function**

Find the exact value of $\sin \dfrac{\pi}{12}$.

Solution   To find the *exact* value of $\sin(\pi/12)$, use the fact that

$$\frac{\pi}{12} = \frac{\pi}{3} - \frac{\pi}{4}$$

with the formula for $\sin(u - v)$.

$$\sin \frac{\pi}{12} = \sin\left(\frac{\pi}{3} - \frac{\pi}{4}\right)$$

$$= \sin \frac{\pi}{3} \cos \frac{\pi}{4} - \cos \frac{\pi}{3} \sin \frac{\pi}{4}$$

$$= \frac{\sqrt{3}}{2}\left(\frac{\sqrt{2}}{2}\right) - \frac{1}{2}\left(\frac{\sqrt{2}}{2}\right)$$

$$= \frac{\sqrt{6} - \sqrt{2}}{4}$$

Check this result on a calculator by comparing its value to $\sin(\pi/12) \approx 0.2588$.

✓ **Checkpoint** ◀))) Audio-video solution in English & Spanish at LarsonPrecalculus.com

Find the exact value of $\cos \dfrac{\pi}{12}$.

•• REMARK Another way to solve Example 2 is to use the fact that $75° = 120° - 45°$ with the formula for $\cos(u - v)$.

EXAMPLE 2 **Evaluating a Trigonometric Function**

Find the exact value of $\cos 75°$.

Solution Use the fact that $75° = 30° + 45°$ with the formula for $\cos(u + v)$.

$$\cos 75° = \cos(30° + 45°)$$

$$= \cos 30° \cos 45° - \sin 30° \sin 45°$$

$$= \frac{\sqrt{3}}{2}\left(\frac{\sqrt{2}}{2}\right) - \frac{1}{2}\left(\frac{\sqrt{2}}{2}\right)$$

$$= \frac{\sqrt{6} - \sqrt{2}}{4}$$

✓ **Checkpoint** ◀))) Audio-video solution in English & Spanish at LarsonPrecalculus.com

Find the exact value of $\sin 75°$.

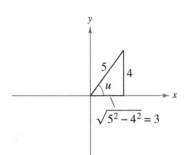

Figure 5.3

EXAMPLE 3 **Evaluating a Trigonometric Expression**

Find the exact value of $\sin(u + v)$ given $\sin u = 4/5$, where $0 < u < \pi/2$, and $\cos v = -12/13$, where $\pi/2 < v < \pi$.

Solution Because $\sin u = 4/5$ and u is in Quadrant I, $\cos u = 3/5$, as shown in Figure 5.3. Because $\cos v = -12/13$ and v is in Quadrant II, $\sin v = 5/13$, as shown in Figure 5.4. Use these values in the formula for $\sin(u + v)$.

$$\sin(u + v) = \sin u \cos v + \cos u \sin v$$

$$= \frac{4}{5}\left(-\frac{12}{13}\right) + \frac{3}{5}\left(\frac{5}{13}\right)$$

$$= -\frac{33}{65}$$

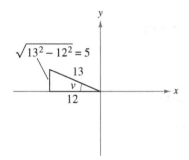

Figure 5.4

✓ **Checkpoint** ◀))) Audio-video solution in English & Spanish at LarsonPrecalculus.com

Find the exact value of $\cos(u + v)$ given $\sin u = 12/13$, where $0 < u < \pi/2$, and $\cos v = -3/5$, where $\pi/2 < v < \pi$.

EXAMPLE 4 **An Application of a Sum Formula**

Write $\cos(\arctan 1 + \arccos x)$ as an algebraic expression.

Solution This expression fits the formula for $\cos(u + v)$. Figure 5.5 shows angles $u = \arctan 1$ and $v = \arccos x$.

$$\cos(u + v) = \cos(\arctan 1)\cos(\arccos x) - \sin(\arctan 1)\sin(\arccos x)$$

$$= \frac{1}{\sqrt{2}} \cdot x - \frac{1}{\sqrt{2}} \cdot \sqrt{1 - x^2}$$

$$= \frac{x - \sqrt{1 - x^2}}{\sqrt{2}}$$

$$= \frac{\sqrt{2}x - \sqrt{2 - 2x^2}}{2}$$

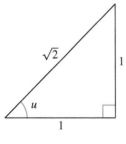

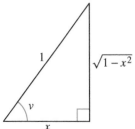

Figure 5.5

✓ **Checkpoint** ◀))) Audio-video solution in English & Spanish at LarsonPrecalculus.com

Write $\sin(\arctan 1 + \arccos x)$ as an algebraic expression.

Hipparchus, considered the most important of the Greek astronomers, was born about 190 B.C. in Nicaea. He is credited with the invention of trigonometry, and his work contributed to the derivation of the sum and difference formulas for $\sin(A \pm B)$ and $\cos(A \pm B)$.

EXAMPLE 5   **Verifying a Cofunction Identity**

See LarsonPrecalculus.com for an interactive version of this type of example.

Verify the cofunction identity $\cos\left(\dfrac{\pi}{2} - x\right) = \sin x$.

Solution   Use the formula for $\cos(u - v)$.

$$\cos\left(\frac{\pi}{2} - x\right) = \cos \frac{\pi}{2} \cos x + \sin \frac{\pi}{2} \sin x$$

$$= (0)(\cos x) + (1)(\sin x)$$

$$= \sin x$$

✓ **Checkpoint** ◀))) *Audio-video solution in English & Spanish at LarsonPrecalculus.com*

Verify the cofunction identity $\sin\left(x - \dfrac{\pi}{2}\right) = -\cos x$.

Sum and difference formulas can be used to derive **reduction formulas** for rewriting expressions such as

$$\sin\left(\theta + \frac{n\pi}{2}\right) \quad \text{and} \quad \cos\left(\theta + \frac{n\pi}{2}\right), \quad \text{where } n \text{ is an integer}$$

as trigonometric functions of only θ.

EXAMPLE 6   **Deriving Reduction Formulas**

Write each expression as a trigonometric function of only θ.

a. $\cos\left(\theta - \dfrac{3\pi}{2}\right)$

b. $\tan(\theta + 3\pi)$

Solution

a. Use the formula for $\cos(u - v)$.

$$\cos\left(\theta - \frac{3\pi}{2}\right) = \cos \theta \cos \frac{3\pi}{2} + \sin \theta \sin \frac{3\pi}{2}$$

$$= (\cos \theta)(0) + (\sin \theta)(-1)$$

$$= -\sin \theta$$

b. Use the formula for $\tan(u + v)$.

$$\tan(\theta + 3\pi) = \frac{\tan \theta + \tan 3\pi}{1 - \tan \theta \tan 3\pi}$$

$$= \frac{\tan \theta + 0}{1 - (\tan \theta)(0)}$$

$$= \tan \theta$$

✓ **Checkpoint** ◀))) *Audio-video solution in English & Spanish at LarsonPrecalculus.com*

Write each expression as a trigonometric function of only θ.

a. $\sin\left(\dfrac{3\pi}{2} - \theta\right)$ **b.** $\tan\left(\theta - \dfrac{\pi}{4}\right)$

EXAMPLE 7   **Solving a Trigonometric Equation**

Find all solutions of $\sin[x + (\pi/4)] + \sin[x - (\pi/4)] = -1$ in the interval $[0, 2\pi)$.

Algebraic Solution

Use sum and difference formulas to rewrite the equation.

$$\sin x \cos \frac{\pi}{4} + \cos x \sin \frac{\pi}{4} + \sin x \cos \frac{\pi}{4} - \cos x \sin \frac{\pi}{4} = -1$$

$$2 \sin x \cos \frac{\pi}{4} = -1$$

$$2(\sin x)\left(\frac{\sqrt{2}}{2}\right) = -1$$

$$\sin x = -\frac{1}{\sqrt{2}}$$

$$\sin x = -\frac{\sqrt{2}}{2}$$

So, the solutions in the interval $[0, 2\pi)$ are $x = \dfrac{5\pi}{4}$ and $x = \dfrac{7\pi}{4}$.

Graphical Solution

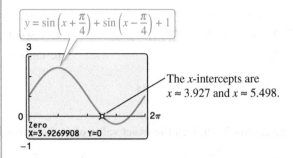

$y = \sin\left(x + \frac{\pi}{4}\right) + \sin\left(x - \frac{\pi}{4}\right) + 1$

The x-intercepts are $x \approx 3.927$ and $x \approx 5.498$.

Zero
X=3.9269908   Y=0

Use the x-intercepts of

$$y = \sin[x + (\pi/4)] + \sin[x - (\pi/4)] + 1$$

to conclude that the approximate solutions in the interval $[0, 2\pi)$ are

$$x \approx 3.927 \approx \frac{5\pi}{4} \quad \text{and} \quad x \approx 5.498 \approx \frac{7\pi}{4}.$$

✓ **Checkpoint** ◀))) Audio-video solution in English & Spanish at LarsonPrecalculus.com

Find all solutions of $\sin[x + (\pi/2)] + \sin[x - (3\pi/2)] = 1$ in the interval $[0, 2\pi)$. ▪

The next example is an application from calculus.

EXAMPLE 8   **An Application from Calculus**   ∫

Verify that $\dfrac{\sin(x + h) - \sin x}{h} = (\cos x)\left(\dfrac{\sin h}{h}\right) - (\sin x)\left(\dfrac{1 - \cos h}{h}\right)$, where $h \neq 0$.

Solution   Use the formula for $\sin(u + v)$.

$$\frac{\sin(x + h) - \sin x}{h} = \frac{\sin x \cos h + \cos x \sin h - \sin x}{h}$$

$$= \frac{\cos x \sin h - \sin x(1 - \cos h)}{h}$$

$$= (\cos x)\left(\frac{\sin h}{h}\right) - (\sin x)\left(\frac{1 - \cos h}{h}\right)$$

✓ **Checkpoint** ◀))) Audio-video solution in English & Spanish at LarsonPrecalculus.com

Verify that $\dfrac{\cos(x + h) - \cos x}{h} = (\cos x)\left(\dfrac{\cos h - 1}{h}\right) - (\sin x)\left(\dfrac{\sin h}{h}\right)$, where $h \neq 0$.

▪

Summarize   (Section 5.4)

1. State the sum and difference formulas for sine, cosine, and tangent (*page 374*). For examples of using the sum and difference formulas to evaluate trigonometric functions, verify identities, and solve trigonometric equations, see Examples 1–8.

5.4 Exercises

Vocabulary: Fill in the blank.

1. $\sin(u - v) =$ _____
2. $\cos(u + v) =$ _____
3. $\tan(u + v) =$ _____
4. $\sin(u + v) =$ _____
5. $\cos(u - v) =$ _____
6. $\tan(u - v) =$ _____

Skills and Applications

Evaluating Trigonometric Expressions In Exercises 7–10, find the exact value of each expression.

7. (a) $\cos\left(\dfrac{\pi}{4} + \dfrac{\pi}{3}\right)$ (b) $\cos\dfrac{\pi}{4} + \cos\dfrac{\pi}{3}$

8. (a) $\sin\left(\dfrac{7\pi}{6} - \dfrac{\pi}{3}\right)$ (b) $\sin\dfrac{7\pi}{6} - \sin\dfrac{\pi}{3}$

9. (a) $\sin(135° - 30°)$ (b) $\sin 135° - \cos 30°$

10. (a) $\cos(120° + 45°)$ (b) $\cos 120° + \cos 45°$

 Evaluating Trigonometric Functions In Exercises 11–26, find the exact values of the sine, cosine, and tangent of the angle.

11. $\dfrac{11\pi}{12} = \dfrac{3\pi}{4} + \dfrac{\pi}{6}$
12. $\dfrac{7\pi}{12} = \dfrac{\pi}{3} + \dfrac{\pi}{4}$

13. $\dfrac{17\pi}{12} = \dfrac{9\pi}{4} - \dfrac{5\pi}{6}$
14. $-\dfrac{\pi}{12} = \dfrac{\pi}{6} - \dfrac{\pi}{4}$

15. $105° = 60° + 45°$
16. $165° = 135° + 30°$

17. $-195° = 30° - 225°$
18. $255° = 300° - 45°$

19. $\dfrac{13\pi}{12}$
20. $\dfrac{19\pi}{12}$

21. $-\dfrac{5\pi}{12}$
22. $-\dfrac{7\pi}{12}$

23. $285°$
24. $15°$

25. $-165°$
26. $-105°$

Rewriting a Trigonometric Expression In Exercises 27–34, write the expression as the sine, cosine, or tangent of an angle.

27. $\sin 3 \cos 1.2 - \cos 3 \sin 1.2$

28. $\cos\dfrac{\pi}{7} \cos\dfrac{\pi}{5} - \sin\dfrac{\pi}{7} \sin\dfrac{\pi}{5}$

29. $\sin 60° \cos 15° + \cos 60° \sin 15°$

30. $\cos 130° \cos 40° - \sin 130° \sin 40°$

31. $\dfrac{\tan(\pi/15) + \tan(2\pi/5)}{1 - \tan(\pi/15) \tan(2\pi/5)}$

32. $\dfrac{\tan 1.1 - \tan 4.6}{1 + \tan 1.1 \tan 4.6}$

33. $\cos 3x \cos 2y + \sin 3x \sin 2y$

34. $\sin x \cos 2x + \cos x \sin 2x$

 Evaluating a Trigonometric Expression In Exercises 35–40, find the exact value of the expression.

35. $\sin\dfrac{\pi}{12} \cos\dfrac{\pi}{4} + \cos\dfrac{\pi}{12} \sin\dfrac{\pi}{4}$

36. $\cos\dfrac{\pi}{16} \cos\dfrac{3\pi}{16} - \sin\dfrac{\pi}{16} \sin\dfrac{3\pi}{16}$

37. $\cos 130° \cos 10° + \sin 130° \sin 10°$

38. $\sin 100° \cos 40° - \cos 100° \sin 40°$

39. $\dfrac{\tan(9\pi/8) - \tan(\pi/8)}{1 + \tan(9\pi/8) \tan(\pi/8)}$

40. $\dfrac{\tan 25° + \tan 110°}{1 - \tan 25° \tan 110°}$

Evaluating a Trigonometric Expression In Exercises 41–46, find the exact value of the trigonometric expression given that $\sin u = -\dfrac{3}{5}$, where $3\pi/2 < u < 2\pi$, and $\cos v = \dfrac{15}{17}$, where $0 < v < \pi/2$.

41. $\sin(u + v)$
42. $\cos(u - v)$

43. $\tan(u + v)$
44. $\csc(u - v)$

45. $\sec(v - u)$
46. $\cot(u + v)$

Evaluating a Trigonometric Expression In Exercises 47–52, find the exact value of the trigonometric expression given that $\sin u = -\dfrac{7}{25}$ and $\cos v = -\dfrac{4}{5}$. (Both u and v are in Quadrant III.)

47. $\cos(u + v)$
48. $\sin(u + v)$

49. $\tan(u - v)$
50. $\cot(v - u)$

51. $\csc(u - v)$
52. $\sec(v - u)$

 An Application of a Sum or Difference Formula In Exercises 53–56, write the trigonometric expression as an algebraic expression.

53. $\sin(\arcsin x + \arccos x)$

54. $\sin(\arctan 2x - \arccos x)$

55. $\cos(\arccos x + \arcsin x)$

56. $\cos(\arccos x - \arctan x)$

Verifying a Trigonometric Identity In Exercises 57–64, verify the identity.

57. $\sin\left(\dfrac{\pi}{2} - x\right) = \cos x$ **58.** $\sin\left(\dfrac{\pi}{2} + x\right) = \cos x$

59. $\sin\left(\dfrac{\pi}{6} + x\right) = \dfrac{1}{2}(\cos x + \sqrt{3} \sin x)$

60. $\cos\left(\dfrac{5\pi}{4} - x\right) = -\dfrac{\sqrt{2}}{2}(\cos x + \sin x)$

61. $\tan(\theta + \pi) = \tan\theta$ **62.** $\tan\left(\dfrac{\pi}{4} - \theta\right) = \dfrac{1 - \tan\theta}{1 + \tan\theta}$

63. $\cos(\pi - \theta) + \sin\left(\dfrac{\pi}{2} + \theta\right) = 0$

64. $\cos(x + y)\cos(x - y) = \cos^2 x - \sin^2 y$

Deriving a Reduction Formula In Exercises 65–68, write the expression as a trigonometric function of only θ, and use a graphing utility to confirm your answer graphically.

65. $\cos\left(\dfrac{3\pi}{2} - \theta\right)$ **66.** $\sin(\pi + \theta)$

67. $\csc\left(\dfrac{3\pi}{2} + \theta\right)$ **68.** $\cot(\theta - \pi)$

Solving a Trigonometric Equation In Exercises 69–74, find all solutions of the equation in the interval $[0, 2\pi)$.

69. $\sin(x + \pi) - \sin x + 1 = 0$

70. $\cos(x + \pi) - \cos x - 1 = 0$

71. $\cos\left(x + \dfrac{\pi}{4}\right) - \cos\left(x - \dfrac{\pi}{4}\right) = 1$

72. $\sin\left(x + \dfrac{\pi}{6}\right) - \sin\left(x - \dfrac{7\pi}{6}\right) = \dfrac{\sqrt{3}}{2}$

73. $\tan(x + \pi) + 2\sin(x + \pi) = 0$

74. $\sin\left(x + \dfrac{\pi}{2}\right) - \cos^2 x = 0$

Approximating Solutions In Exercises 75–78, use a graphing utility to approximate the solutions of the equation in the interval $[0, 2\pi)$.

75. $\cos\left(x + \dfrac{\pi}{4}\right) + \cos\left(x - \dfrac{\pi}{4}\right) = 1$

76. $\tan(x + \pi) - \cos\left(x + \dfrac{\pi}{2}\right) = 0$

77. $\sin\left(x + \dfrac{\pi}{2}\right) + \cos^2 x = 0$

78. $\cos\left(x - \dfrac{\pi}{2}\right) - \sin^2 x = 0$

79. Harmonic Motion A weight is attached to a spring suspended vertically from a ceiling. When a driving force is applied to the system, the weight moves vertically from its equilibrium position, and this motion is modeled by

$$y = \dfrac{1}{3}\sin 2t + \dfrac{1}{4}\cos 2t$$

where y is the displacement (in feet) from equilibrium of the weight and t is the time (in seconds).

(a) Use the identity

$$a \sin B\theta + b \cos B\theta = \sqrt{a^2 + b^2}\,\sin(B\theta + C)$$

where $C = \arctan(b/a)$, $a > 0$, to write the model in the form

$$y = \sqrt{a^2 + b^2}\,\sin(Bt + C).$$

(b) Find the amplitude of the oscillations of the weight.

(c) Find the frequency of the oscillations of the weight.

80. Standing Waves

The equation of a standing wave is obtained by adding the displacements of two waves traveling in opposite directions (see figure). Assume that each of the waves has amplitude A, period T, and wavelength λ. The models for two such waves are

$$y_1 = A\cos 2\pi\left(\dfrac{t}{T} - \dfrac{x}{\lambda}\right) \quad \text{and} \quad y_2 = A\cos 2\pi\left(\dfrac{t}{T} + \dfrac{x}{\lambda}\right).$$

Show that

$$y_1 + y_2 = 2A\cos\dfrac{2\pi t}{T}\cos\dfrac{2\pi x}{\lambda}.$$

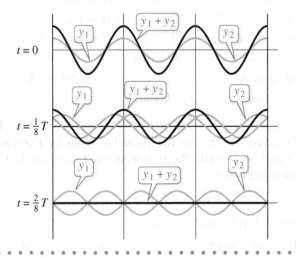

Exploration

True or False? In Exercises 81–84, determine whether the statement is true or false. Justify your answer.

81. $\sin(u \pm v) = \sin u \cos v \pm \cos u \sin v$

82. $\cos(u \pm v) = \cos u \cos v \pm \sin u \sin v$

83. When α and β are supplementary,
$$\sin \alpha \cos \beta = \cos \alpha \sin \beta.$$

84. When A, B, and C form $\triangle ABC$, $\cos(A + B) = -\cos C$.

85. Error Analysis Describe the error.
$$\tan\left(x - \frac{\pi}{4}\right) = \frac{\tan x - \tan(\pi/4)}{1 - \tan x \tan(\pi/4)}$$
$$= \frac{\tan x - 1}{1 - \tan x}$$
$$= -1$$

86. **HOW DO YOU SEE IT?** Explain how to use the figure to justify each statement.

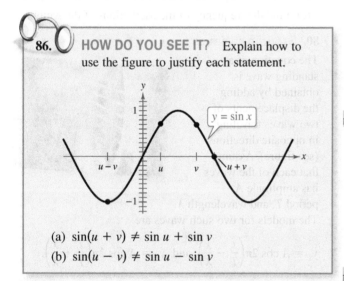

(a) $\sin(u + v) \neq \sin u + \sin v$

(b) $\sin(u - v) \neq \sin u - \sin v$

Verifying an Identity In Exercises 87–90, verify the identity.

87. $\cos(n\pi + \theta) = (-1)^n \cos \theta$, n is an integer

88. $\sin(n\pi + \theta) = (-1)^n \sin \theta$, n is an integer

89. $a \sin B\theta + b \cos B\theta = \sqrt{a^2 + b^2} \sin(B\theta + C)$,
where $C = \arctan(b/a)$ and $a > 0$

90. $a \sin B\theta + b \cos B\theta = \sqrt{a^2 + b^2} \cos(B\theta - C)$,
where $C = \arctan(a/b)$ and $b > 0$

Rewriting a Trigonometric Expression In Exercises 91–94, use the formulas given in Exercises 89 and 90 to write the trigonometric expression in the following forms.

(a) $\sqrt{a^2 + b^2} \sin(B\theta + C)$

(b) $\sqrt{a^2 + b^2} \cos(B\theta - C)$

91. $\sin \theta + \cos \theta$ **92.** $3 \sin 2\theta + 4 \cos 2\theta$

93. $12 \sin 3\theta + 5 \cos 3\theta$ **94.** $\sin 2\theta + \cos 2\theta$

Rewriting a Trigonometric Expression In Exercises 95 and 96, use the formulas given in Exercises 89 and 90 to write the trigonometric expression in the form $a \sin B\theta + b \cos B\theta$.

95. $2 \sin[\theta + (\pi/4)]$ **96.** $5 \cos[\theta - (\pi/4)]$

Angle Between Two Lines In Exercises 97 and 98, use the figure, which shows two lines whose equations are $y_1 = m_1 x + b_1$ and $y_2 = m_2 x + b_2$. Assume that both lines have positive slopes. Derive a formula for the angle between the two lines. Then use your formula to find the angle between the given pair of lines.

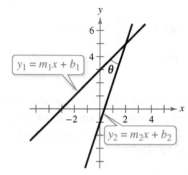

97. $y = x$ and $y = \sqrt{3}x$ **98.** $y = x$ and $y = x/\sqrt{3}$

Graphical Reasoning In Exercises 99 and 100, use a graphing utility to graph y_1 and y_2 in the same viewing window. Use the graphs to determine whether $y_1 = y_2$. Explain your reasoning.

99. $y_1 = \cos(x + 2)$, $y_2 = \cos x + \cos 2$

100. $y_1 = \sin(x + 4)$, $y_2 = \sin x + \sin 4$

101. Proof Write a proof of the formula for $\sin(u + v)$. Write a proof of the formula for $\sin(u - v)$.

102. An Application from Calculus Let $x = \pi/3$ in the identity in Example 8 and define the functions f and g as follows.

$$f(h) = \frac{\sin[(\pi/3) + h] - \sin(\pi/3)}{h}$$

$$g(h) = \cos\frac{\pi}{3}\left(\frac{\sin h}{h}\right) - \sin\frac{\pi}{3}\left(\frac{1 - \cos h}{h}\right)$$

(a) What are the domains of the functions f and g?

(b) Use a graphing utility to complete the table.

h	0.5	0.2	0.1	0.05	0.02	0.01
$f(h)$						
$g(h)$						

(c) Use the graphing utility to graph the functions f and g.

(d) Use the table and the graphs to make a conjecture about the values of the functions f and g as $h \to 0^+$.

5.5 Multiple-Angle and Product-to-Sum Formulas

A variety of trigonometric formulas enable you to rewrite trigonometric equations in more convenient forms. For example, in Exercise 71 on page 389, you will use a half-angle formula to rewrite an equation relating the Mach number of a supersonic airplane to the apex angle of the cone formed by the sound waves behind the airplane.

- ◾ Use multiple-angle formulas to rewrite and evaluate trigonometric functions.
- ◾ Use power-reducing formulas to rewrite trigonometric expressions.
- ◾ Use half-angle formulas to rewrite and evaluate trigonometric functions.
- ◾ Use product-to-sum and sum-to-product formulas to rewrite and evaluate trigonometric expressions.
- ◾ Use trigonometric formulas to rewrite real-life models.

Multiple-Angle Formulas

In this section, you will study four other categories of trigonometric identities.

1. The first category involves *functions of multiple angles* such as sin ku and cos ku.

2. The second category involves *squares of trigonometric functions* such as $\sin^2 u$.

3. The third category involves *functions of half-angles* such as $\sin(u/2)$.

4. The fourth category involves *products of trigonometric functions* such as sin u cos v.

You should learn the **double-angle formulas** because they are used often in trigonometry and calculus. For proofs of these formulas, see Proofs in Mathematics on page 395.

Double-Angle Formulas

$$\sin 2u = 2 \sin u \cos u \qquad \cos 2u = \cos^2 u - \sin^2 u$$

$$\tan 2u = \frac{2 \tan u}{1 - \tan^2 u} \qquad \qquad = 2 \cos^2 u - 1$$

$$= 1 - 2 \sin^2 u$$

EXAMPLE 1   Solving a Multiple-Angle Equation

Solve $2 \cos x + \sin 2x = 0$.

Solution   Begin by rewriting the equation so that it involves trigonometric functions of only x. Then factor and solve.

$$2 \cos x + \sin 2x = 0 \qquad \text{Write original equation.}$$

$$2 \cos x + 2 \sin x \cos x = 0 \qquad \text{Double-angle formula}$$

$$2 \cos x(1 + \sin x) = 0 \qquad \text{Factor.}$$

$$2 \cos x = 0 \quad \text{and} \quad 1 + \sin x = 0 \qquad \text{Set factors equal to zero.}$$

$$x = \frac{\pi}{2}, \frac{3\pi}{2} \qquad \qquad x = \frac{3\pi}{2} \qquad \text{Solutions in } [0, 2\pi)$$

So, the general solution is

$$x = \frac{\pi}{2} + 2n\pi \quad \text{and} \quad x = \frac{3\pi}{2} + 2n\pi$$

where n is an integer. Verify these solutions graphically.

✓ **Checkpoint** ◀))) *Audio-video solution in English & Spanish at LarsonPrecalculus.com*

Solve $\cos 2x + \cos x = 0$.

EXAMPLE 2   **Evaluating Functions Involving Double Angles**

Use the conditions below to find $\sin 2\theta$, $\cos 2\theta$, and $\tan 2\theta$.

$$\cos \theta = \frac{5}{13}, \quad \frac{3\pi}{2} < \theta < 2\pi$$

Solution   From Figure 5.6,

$$\sin \theta = \frac{y}{r} = -\frac{12}{13} \quad \text{and} \quad \tan \theta = \frac{y}{x} = -\frac{12}{5}.$$

Use these values with each of the double-angle formulas.

$$\sin 2\theta = 2 \sin \theta \cos \theta = 2\left(-\frac{12}{13}\right)\left(\frac{15}{13}\right) = -\frac{120}{169}$$

$$\cos 2\theta = 2 \cos^2 \theta - 1 = 2\left(\frac{25}{169}\right) - 1 = -\frac{119}{169}$$

$$\tan 2\theta = \frac{2 \tan \theta}{1 - \tan^2 \theta} = \frac{2\left(-\dfrac{12}{5}\right)}{1 - \left(-\dfrac{12}{5}\right)^2} = \frac{120}{119}$$

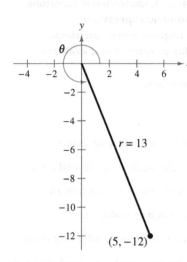

Figure 5.6

✓ *Checkpoint* ◀))) *Audio-video solution in English & Spanish at LarsonPrecalculus.com*

Use the conditions below to find $\sin 2\theta$, $\cos 2\theta$, and $\tan 2\theta$.

$$\sin \theta = \frac{3}{5}, \quad 0 < \theta < \frac{\pi}{2}$$

The double-angle formulas are not restricted to the angles 2θ and θ. Other *double* combinations, such as 4θ and 2θ or 6θ and 3θ, are also valid. Here are two examples.

$$\sin 4\theta = 2 \sin 2\theta \cos 2\theta \quad \text{and} \quad \cos 6\theta = \cos^2 3\theta - \sin^2 3\theta$$

By using double-angle formulas together with the sum formulas given in the preceding section, you can derive other multiple-angle formulas.

EXAMPLE 3   **Deriving a Triple-Angle Formula**

Rewrite $\sin 3x$ in terms of $\sin x$.

Solution

$\sin 3x = \sin(2x + x)$	Rewrite the angle as a sum.
$= \sin 2x \cos x + \cos 2x \sin x$	Sum formula
$= 2 \sin x \cos x \cos x + (1 - 2 \sin^2 x) \sin x$	Double-angle formulas
$= 2 \sin x \cos^2 x + \sin x - 2 \sin^3 x$	Distributive Property
$= 2 \sin x(1 - \sin^2 x) + \sin x - 2 \sin^3 x$	Pythagorean identity
$= 2 \sin x - 2 \sin^3 x + \sin x - 2 \sin^3 x$	Distributive Property
$= 3 \sin x - 4 \sin^3 x$	Simplify.

✓ *Checkpoint* ◀))) *Audio-video solution in English & Spanish at LarsonPrecalculus.com*

Rewrite $\cos 3x$ in terms of $\cos x$.

Power-Reducing Formulas

The double-angle formulas can be used to obtain the **power-reducing formulas.**

Power-Reducing Formulas

$$\sin^2 u = \frac{1 - \cos 2u}{2}$$

$$\cos^2 u = \frac{1 + \cos 2u}{2}$$

$$\tan^2 u = \frac{1 - \cos 2u}{1 + \cos 2u}$$

For a proof of the power-reducing formulas, see Proofs in Mathematics on page 396. Example 4 shows a typical power reduction used in calculus.

EXAMPLE 4   **Reducing a Power**

Rewrite $\sin^4 x$ in terms of first powers of the cosines of multiple angles.

Solution   Note the repeated use of power-reducing formulas.

$$\sin^4 x = (\sin^2 x)^2 \qquad \text{Property of exponents}$$

$$= \left(\frac{1 - \cos 2x}{2}\right)^2 \qquad \text{Power-reducing formula}$$

$$= \frac{1}{4}(1 - 2\cos 2x + \cos^2 2x) \qquad \text{Expand.}$$

$$= \frac{1}{4}\left(1 - 2\cos 2x + \frac{1 + \cos 4x}{2}\right) \qquad \text{Power-reducing formula}$$

$$= \frac{1}{4} - \frac{1}{2}\cos 2x + \frac{1}{8} + \frac{1}{8}\cos 4x \qquad \text{Distributive Property}$$

$$= \frac{3}{8} - \frac{1}{2}\cos 2x + \frac{1}{8}\cos 4x \qquad \text{Simplify.}$$

$$= \frac{1}{8}(3 - 4\cos 2x + \cos 4x) \qquad \text{Factor out common factor.}$$

Use a graphing utility to check this result, as shown below. Notice that the graphs coincide.

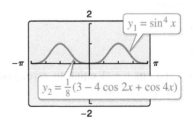

✓ Checkpoint))) *Audio-video solution in English & Spanish at LarsonPrecalculus.com*

Rewrite $\tan^4 x$ in terms of first powers of the cosines of multiple angles.

Half-Angle Formulas

You can derive some useful alternative forms of the power-reducing formulas by replacing u with $u/2$. The results are called **half-angle formulas.**

•• REMARK To find the exact value of a trigonometric function with an angle measure in D° M′ S″ form using a half-angle formula, first convert the angle measure to decimal degree form. Then multiply the resulting angle measure by 2.
▷

Half-Angle Formulas

$$\sin \frac{u}{2} = \pm \sqrt{\frac{1 - \cos u}{2}} \qquad \cos \frac{u}{2} = \pm \sqrt{\frac{1 + \cos u}{2}}$$

$$\tan \frac{u}{2} = \frac{1 - \cos u}{\sin u} = \frac{\sin u}{1 + \cos u}$$

The signs of $\sin \dfrac{u}{2}$ and $\cos \dfrac{u}{2}$ depend on the quadrant in which $\dfrac{u}{2}$ lies.

•• REMARK Use your calculator to verify the result obtained in Example 5. That is, evaluate $\sin 105°$ and $\left(\sqrt{2 + \sqrt{3}}\right)/2$. Note that both values are approximately 0.9659258.
▷

EXAMPLE 5 **Using a Half-Angle Formula**

Find the exact value of $\sin 105°$.

Solution Begin by noting that $105°$ is half of $210°$. Then, use the half-angle formula for $\sin(u/2)$ and the fact that $105°$ lies in Quadrant II.

$$\sin 105° = \sqrt{\frac{1 - \cos 210°}{2}} = \sqrt{\frac{1 + \left(\sqrt{3}/2\right)}{2}} = \frac{\sqrt{2 + \sqrt{3}}}{2}$$

The positive square root is chosen because $\sin \theta$ is positive in Quadrant II.

✓ *Checkpoint*  *Audio-video solution in English & Spanish at LarsonPrecalculus.com*

Find the exact value of $\cos 105°$.

EXAMPLE 6 **Solving a Trigonometric Equation**

Find all solutions of $1 + \cos^2 x = 2 \cos^2 \dfrac{x}{2}$ in the interval $[0, 2\pi)$.

Algebraic Solution

$$1 + \cos^2 x = 2 \cos^2 \frac{x}{2} \qquad \text{Write original equation.}$$

$$1 + \cos^2 x = 2\left(\pm \sqrt{\frac{1 + \cos x}{2}}\right)^2 \qquad \text{Half-angle formula}$$

$$1 + \cos^2 x = 1 + \cos x \qquad \text{Simplify.}$$

$$\cos^2 x - \cos x = 0 \qquad \text{Simplify.}$$

$$\cos x(\cos x - 1) = 0 \qquad \text{Factor.}$$

By setting the factors $\cos x$ and $\cos x - 1$ equal to zero, you find that the solutions in the interval $[0, 2\pi)$ are

$$x = \frac{\pi}{2}, \quad x = \frac{3\pi}{2}, \quad \text{and} \quad x = 0.$$

Graphical Solution

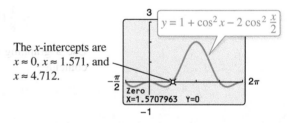

The x-intercepts are $x \approx 0$, $x \approx 1.571$, and $x \approx 4.712$.

Use the x-intercepts of $y = 1 + \cos^2 x - 2 \cos^2(x/2)$ to conclude that the approximate solutions of $1 + \cos^2 x = 2 \cos^2(x/2)$ in the interval $[0, 2\pi)$ are

$$x = 0, \quad x \approx 1.571 \approx \frac{\pi}{2}, \quad \text{and} \quad x \approx 4.712 \approx \frac{3\pi}{2}.$$

✓ *Checkpoint*  *Audio-video solution in English & Spanish at LarsonPrecalculus.com*

Find all solutions of $\cos^2 x = \sin^2(x/2)$ in the interval $[0, 2\pi)$.

Product-to-Sum and Sum-to-Product Formulas

Each of the **product-to-sum formulas** can be proved using the sum and difference formulas discussed in the preceding section.

Product-to-Sum Formulas

$$\sin u \sin v = \frac{1}{2}[\cos(u - v) - \cos(u + v)]$$

$$\cos u \cos v = \frac{1}{2}[\cos(u - v) + \cos(u + v)]$$

$$\sin u \cos v = \frac{1}{2}[\sin(u + v) + \sin(u - v)]$$

$$\cos u \sin v = \frac{1}{2}[\sin(u + v) - \sin(u - v)]$$

Product-to-sum formulas are used in calculus to solve problems involving the products of sines and cosines of two different angles.

EXAMPLE 7   Writing Products as Sums

Rewrite the product $\cos 5x \sin 4x$ as a sum or difference.

Solution   Using the appropriate product-to-sum formula, you obtain

$$\cos 5x \sin 4x = \frac{1}{2}[\sin(5x + 4x) - \sin(5x - 4x)]$$

$$= \frac{1}{2}\sin 9x - \frac{1}{2}\sin x.$$

✓ **Checkpoint** *Audio-video solution in English & Spanish at LarsonPrecalculus.com*

Rewrite the product $\sin 5x \cos 3x$ as a sum or difference.

Occasionally, it is useful to reverse the procedure and write a sum of trigonometric functions as a product. This can be accomplished with the **sum-to-product formulas.**

Sum-to-Product Formulas

$$\sin u + \sin v = 2 \sin\left(\frac{u + v}{2}\right)\cos\left(\frac{u - v}{2}\right)$$

$$\sin u - \sin v = 2 \cos\left(\frac{u + v}{2}\right)\sin\left(\frac{u - v}{2}\right)$$

$$\cos u + \cos v = 2 \cos\left(\frac{u + v}{2}\right)\cos\left(\frac{u - v}{2}\right)$$

$$\cos u - \cos v = -2 \sin\left(\frac{u + v}{2}\right)\sin\left(\frac{u - v}{2}\right)$$

For a proof of the sum-to-product formulas, see Proofs in Mathematics on page 396.

EXAMPLE 8   Using a Sum-to-Product Formula

Find the exact value of $\cos 195° + \cos 105°$.

Solution   Use the appropriate sum-to-product formula.

$$\cos 195° + \cos 105° = 2 \cos\left(\frac{195° + 105°}{2}\right) \cos\left(\frac{195° - 105°}{2}\right)$$

$$= 2 \cos 150° \cos 45°$$

$$= 2\left(-\frac{\sqrt{3}}{2}\right)\left(\frac{\sqrt{2}}{2}\right)$$

$$= -\frac{\sqrt{6}}{2}$$

✓ *Checkpoint* Audio-video solution in English & Spanish at LarsonPrecalculus.com

Find the exact value of $\sin 195° + \sin 105°$.

EXAMPLE 9   Solving a Trigonometric Equation

See LarsonPrecalculus.com for an interactive version of this type of example.

Solve $\sin 5x + \sin 3x = 0$.

Solution

$$\sin 5x + \sin 3x = 0 \qquad \text{Write original equation.}$$

$$2 \sin\left(\frac{5x + 3x}{2}\right) \cos\left(\frac{5x - 3x}{2}\right) = 0 \qquad \text{Sum-to-product formula}$$

$$2 \sin 4x \cos x = 0 \qquad \text{Simplify.}$$

Set the factor $2 \sin 4x$ equal to zero. The solutions in the interval $[0, 2\pi)$ are

$$x = 0, \frac{\pi}{4}, \frac{\pi}{2}, \frac{3\pi}{4}, \pi, \frac{5\pi}{4}, \frac{3\pi}{2}, \frac{7\pi}{4}.$$

The equation $\cos x = 0$ yields no additional solutions, so the solutions are of the form $x = n\pi/4$, where n is an integer. To confirm this graphically, sketch the graph of $y = \sin 5x + \sin 3x$, as shown below.

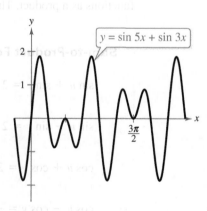

Notice from the graph that the x-intercepts occur at multiples of $\pi/4$.

✓ *Checkpoint* Audio-video solution in English & Spanish at LarsonPrecalculus.com

Solve $\sin 4x - \sin 2x = 0$.

Application

EXAMPLE 10 **Projectile Motion**

Kicking a football with an initial velocity of 80 feet per second at an angle of 45° with the horizontal results in a distance traveled of 200 feet.

Ignoring air resistance, the range of a projectile fired at an angle θ with the horizontal and with an initial velocity of v_0 feet per second is given by

$$r = \frac{1}{16} v_0^2 \sin \theta \cos \theta$$

where r is the horizontal distance (in feet) that the projectile travels. A football player can kick a football from ground level with an initial velocity of 80 feet per second.

a. Rewrite the projectile motion model in terms of the first power of the sine of a multiple angle.

b. At what angle must the player kick the football so that the football travels 200 feet?

Solution

a. Use a double-angle formula to rewrite the projectile motion model as

$$r = \frac{1}{32} v_0^2 (2 \sin \theta \cos \theta) \qquad \text{Write original model.}$$

$$= \frac{1}{32} v_0^2 \sin 2\theta. \qquad \text{Double-angle formula}$$

b. $$r = \frac{1}{32} v_0^2 \sin 2\theta \qquad \text{Write projectile motion model.}$$

$$200 = \frac{1}{32} (80)^2 \sin 2\theta \qquad \text{Substitute 200 for } r \text{ and 80 for } v_0.$$

$$200 = 200 \sin 2\theta \qquad \text{Simplify.}$$

$$1 = \sin 2\theta \qquad \text{Divide each side by 200.}$$

You know that $2\theta = \pi/2$. Dividing this result by 2 produces $\theta = \pi/4$, or 45°. So, the player must kick the football at an angle of 45° so that the football travels 200 feet.

✓ *Checkpoint* 🔊))) *Audio-video solution in English & Spanish at LarsonPrecalculus.com*

In Example 10, for what angle is the horizontal distance the football travels a maximum?

Summarize (Section 5.5)

1. State the double-angle formulas *(page 381)*. For examples of using multiple-angle formulas to rewrite and evaluate trigonometric functions, see Examples 1–3.

2. State the power-reducing formulas *(page 383)*. For an example of using power-reducing formulas to rewrite a trigonometric expression, see Example 4.

3. State the half-angle formulas *(page 384)*. For examples of using half-angle formulas to rewrite and evaluate trigonometric functions, see Examples 5 and 6.

4. State the product-to-sum and sum-to-product formulas *(page 385)*. For an example of using a product-to-sum formula to rewrite a trigonometric expression, see Example 7. For examples of using sum-to-product formulas to rewrite and evaluate trigonometric functions, see Examples 8 and 9.

5. Describe an example of how to use a trigonometric formula to rewrite a real-life model *(page 387, Example 10)*.

5.5   Exercises

See CalcChat.com for tutorial help and worked-out solutions to odd-numbered exercises.

Vocabulary: Fill in the blank to complete the trigonometric formula.

1. $\sin 2u = $ _____

2. $\cos 2u = $ _____

3. $\sin u \cos v = $ _____

4. $\dfrac{1 - \cos 2u}{1 + \cos 2u} = $ _____

5. $\sin \dfrac{u}{2} = $ _____

6. $\cos u - \cos v = $ _____

Skills and Applications

 Solving a Multiple-Angle Equation In Exercises 7–14, solve the equation.

7. $\sin 2x - \sin x = 0$

8. $\sin 2x \sin x = \cos x$

9. $\cos 2x - \cos x = 0$

10. $\cos 2x + \sin x = 0$

11. $\sin 4x = -2 \sin 2x$

12. $(\sin 2x + \cos 2x)^2 = 1$

13. $\tan 2x - \cot x = 0$

14. $\tan 2x - 2 \cos x = 0$

Using a Double-Angle Formula In Exercises 15–20, use a double-angle formula to rewrite the expression.

15. $6 \sin x \cos x$

16. $\sin x \cos x$

17. $6 \cos^2 x - 3$

18. $\cos^2 x - \frac{1}{2}$

19. $4 - 8 \sin^2 x$

20. $10 \sin^2 x - 5$

 Evaluating Functions Involving Double Angles In Exercises 21–24, use the given conditions to find the exact values of $\sin 2u$, $\cos 2u$, and $\tan 2u$ using the double-angle formulas.

21. $\sin u = -3/5$, $\quad 3\pi/2 < u < 2\pi$

22. $\cos u = -4/5$, $\quad \pi/2 < u < \pi$

23. $\tan u = 3/5$, $\quad 0 < u < \pi/2$

24. $\sec u = -2$, $\quad \pi < u < 3\pi/2$

25. Deriving a Multiple-Angle Formula Rewrite $\cos 4x$ in terms of $\cos x$.

26. Deriving a Multiple-Angle Formula Rewrite $\tan 3x$ in terms of $\tan x$.

 Reducing Powers In Exercises 27–34, use the power-reducing formulas to rewrite the expression in terms of first powers of the cosines of multiple angles.

27. $\cos^4 x$

28. $\sin^8 x$

29. $\sin^4 2x$

30. $\cos^4 2x$

31. $\tan^4 2x$

32. $\tan^2 2x \cos^4 2x$

33. $\sin^2 2x \cos^2 2x$

34. $\sin^4 x \cos^2 x$

 Using Half-Angle Formulas In Exercises 35–40, use the half-angle formulas to determine the exact values of the sine, cosine, and tangent of the angle.

35. $75°$

36. $165°$

37. $112° \, 30'$

38. $67° \, 30'$

39. $\pi/8$

40. $7\pi/12$

 Using Half-Angle Formulas In Exercises 41–44, use the given conditions to (a) determine the quadrant in which $u/2$ lies, and (b) find the exact values of $\sin(u/2)$, $\cos(u/2)$, and $\tan(u/2)$ using the half-angle formulas.

41. $\cos u = 7/25$, $\quad 0 < u < \pi/2$

42. $\sin u = 5/13$, $\quad \pi/2 < u < \pi$

43. $\tan u = -5/12$, $\quad 3\pi/2 < u < 2\pi$

44. $\cot u = 3$, $\quad \pi < u < 3\pi/2$

 Solving a Trigonometric Equation In Exercises 45–48, find all solutions of the equation in the interval $[0, 2\pi)$. Use a graphing utility to graph the equation and verify the solutions.

45. $\sin \dfrac{x}{2} + \cos x = 0$

46. $\sin \dfrac{x}{2} + \cos x - 1 = 0$

47. $\cos \dfrac{x}{2} - \sin x = 0$

48. $\tan \dfrac{x}{2} - \sin x = 0$

 Using Product-to-Sum Formulas In Exercises 49–52, use the product-to-sum formulas to rewrite the product as a sum or difference.

49. $\sin 5\theta \sin 3\theta$

50. $7 \cos(-5\beta) \sin 3\beta$

51. $\cos 2\theta \cos 4\theta$

52. $\sin(x + y) \cos(x - y)$

Using Sum-to-Product Formulas In Exercises 53–56, use the sum-to-product formulas to rewrite the sum or difference as a product.

53. $\sin 5\theta - \sin 3\theta$

54. $\sin 3\theta + \sin \theta$

55. $\cos 6x + \cos 2x$

56. $\cos x + \cos 4x$

Using Sum-to-Product Formulas **In Exercises 57–60, use the sum-to-product formulas to find the exact value of the expression.**

57. $\sin 75° + \sin 15°$

58. $\cos 120° + \cos 60°$

59. $\cos \dfrac{3\pi}{4} - \cos \dfrac{\pi}{4}$

60. $\sin \dfrac{5\pi}{4} - \sin \dfrac{3\pi}{4}$

 Solving a Trigonometric Equation **In Exercises 61–64, find all solutions of the equation in the interval $[0, 2\pi)$. Use a graphing utility to graph the equation and verify the solutions.**

61. $\sin 6x + \sin 2x = 0$

62. $\cos 2x - \cos 6x = 0$

63. $\dfrac{\cos 2x}{\sin 3x - \sin x} - 1 = 0$

64. $\sin^2 3x - \sin^2 x = 0$

 Verifying a Trigonometric Identity **In Exercises 65–70, verify the identity.**

65. $\csc 2\theta = \dfrac{\csc \theta}{2 \cos \theta}$

66. $\cos^4 x - \sin^4 x = \cos 2x$

67. $(\sin x + \cos x)^2 = 1 + \sin 2x$

68. $\tan \dfrac{u}{2} = \csc u - \cot u$

69. $\dfrac{\sin x \pm \sin y}{\cos x + \cos y} = \tan \dfrac{x \pm y}{2}$

70. $\cos\left(\dfrac{\pi}{3} + x\right) + \cos\left(\dfrac{\pi}{3} - x\right) = \cos x$

• • **71. Mach Number** • • • • • • • • • • • • • • • •

The Mach number M of a supersonic airplane is the ratio of its speed to the speed of sound. When an airplane travels faster than the speed of sound, the sound waves form a cone behind the airplane. The Mach number is related to the apex angle θ of the cone by $\sin(\theta/2) = 1/M$.

(a) Use a half-angle formula to rewrite the equation in terms of $\cos \theta$.

(b) Find the angle θ that corresponds to a Mach number of 2.

(c) Find the angle θ that corresponds to a Mach number of 4.5.

(d) The speed of sound is about 760 miles per hour. Determine the speed of an object with the Mach numbers from parts (b) and (c).

72. Projectile Motion   The range of a projectile fired at an angle θ with the horizontal and with an initial velocity of v_0 feet per second is

$$r = \dfrac{1}{32} v_0^2 \sin 2\theta$$

where r is the horizontal distance (in feet) the projectile travels. An athlete throws a javelin at 75 feet per second. At what angle must the athlete throw the javelin so that the javelin travels 130 feet?

73. Railroad Track   When two railroad tracks merge, the overlapping portions of the tracks are in the shapes of circular arcs (see figure). The radius r (in feet) of each arc and the angle θ are related by

$$\dfrac{x}{2} = 2r \sin^2 \dfrac{\theta}{2}.$$

Write a formula for x in terms of $\cos \theta$.

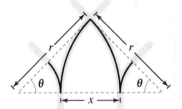

74.   **HOW DO YOU SEE IT?**   Explain how to use the figure to verify the double-angle formulas (a) $\sin 2u = 2 \sin u \cos u$ and (b) $\cos 2u = \cos^2 u - \sin^2 u$.

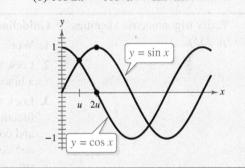

Exploration

True or False?   **In Exercises 75 and 76, determine whether the statement is true or false. Justify your answer.**

75. The sine function is an odd function, so

$$\sin(-2x) = -2 \sin x \cos x.$$

76. $\sin \dfrac{u}{2} = -\sqrt{\dfrac{1 - \cos u}{2}}$ when u is in the second quadrant.

77. Complementary Angles   Verify each identity for complementary angles ϕ and θ.

(a) $\sin(\phi - \theta) = \cos 2\theta$

(b) $\cos(\phi - \theta) = \sin 2\theta$

Chapter Summary

What Did You Learn?	Explanation/Examples	Review Exercises
Section 5.1 Recognize and write the fundamental trigonometric identities *(p. 348)*.	**Reciprocal Identities** $\sin u = 1/\csc u \qquad \cos u = 1/\sec u \qquad \tan u = 1/\cot u$ $\csc u = 1/\sin u \qquad \sec u = 1/\cos u \qquad \cot u = 1/\tan u$ **Quotient Identities:** $\tan u = \dfrac{\sin u}{\cos u}, \quad \cot u = \dfrac{\cos u}{\sin u}$ **Pythagorean Identities:** $\sin^2 u + \cos^2 u = 1,$ $1 + \tan^2 u = \sec^2 u, \quad 1 + \cot^2 u = \csc^2 u$ **Cofunction Identities** $\sin[(\pi/2) - u] = \cos u \qquad \cos[(\pi/2) - u] = \sin u$ $\tan[(\pi/2) - u] = \cot u \qquad \cot[(\pi/2) - u] = \tan u$ $\sec[(\pi/2) - u] = \csc u \qquad \csc[(\pi/2) - u] = \sec u$ **Even/Odd Identities** $\sin(-u) = -\sin u \qquad \cos(-u) = \cos u \qquad \tan(-u) = -\tan u$ $\csc(-u) = -\csc u \qquad \sec(-u) = \sec u \qquad \cot(-u) = -\cot u$	1–4
Use the fundamental trigonometric identities to evaluate trigonometric functions, simplify trigonometric expressions, and rewrite trigonometric expressions *(p. 349)*.	In some cases, when factoring or simplifying a trigonometric expression, it is helpful to rewrite the expression in terms of just *one* trigonometric function or in terms of *sine and cosine only*.	5–18
Section 5.2 Verify trigonometric identities *(p. 355)*.	**Guidelines for Verifying Trigonometric Identities** **1.** Work with one side of the equation at a time. **2.** Look to factor an expression, add fractions, square a binomial, or create a monomial denominator. **3.** Look to use the fundamental identities. Note which functions are in the final expression you want. Sines and cosines pair up well, as do secants and tangents, and cosecants and cotangents. **4.** When the preceding guidelines do not help, try converting all terms to sines and cosines. **5.** Always try *something*.	19–26
Section 5.3 Use standard algebraic techniques to solve trigonometric equations *(p. 362)*.	Use standard algebraic techniques (when possible) such as collecting like terms, extracting square roots, and factoring to solve trigonometric equations.	27–32
Solve trigonometric equations of quadratic type *(p. 365)*.	To solve trigonometric equations of quadratic type $ax^2 + bx + c = 0$, use factoring (when possible) or use the Quadratic Formula.	33–36
Solve trigonometric equations involving multiple angles *(p. 367)*.	To solve equations that contain forms such as $\sin ku$ or $\cos ku$, first solve the equation for ku, and then divide your result by k.	37–42
Use inverse trigonometric functions to solve trigonometric equations *(p. 368)*.	After factoring an equation, you may get an equation such as $(\tan x - 3)(\tan x + 1) = 0$. In such cases, use inverse trigonometric functions to solve. (See Example 9.)	43–46

What Did You Learn?	**Explanation/Examples**	**Review Exercises**
Section 5.4 Use sum and difference formulas to evaluate trigonometric functions, verify identities, and solve trigonometric equations (*p. 374*).	**Sum and Difference Formulas** $$\sin(u+v)=\sin u\cos v+\cos u\sin v$$ $$\sin(u-v)=\sin u\cos v-\cos u\sin v$$ $$\cos(u+v)=\cos u\cos v-\sin u\sin v$$ $$\cos(u-v)=\cos u\cos v+\sin u\sin v$$ $$\tan(u+v)=\frac{\tan u+\tan v}{1-\tan u\tan v}$$ $$\tan(u-v)=\frac{\tan u-\tan v}{1+\tan u\tan v}$$	47–62
Use multiple-angle formulas to rewrite and evaluate trigonometric functions (*p. 381*).	**Double-Angle Formulas** $$\sin 2u=2\sin u\cos u \qquad \cos 2u=\cos^2 u-\sin^2 u$$ $$\tan 2u=\frac{2\tan u}{1-\tan^2 u} \qquad \begin{aligned}&=2\cos^2 u-1\\&=1-2\sin^2 u\end{aligned}$$	63–66
Use power-reducing formulas to rewrite trigonometric expressions (*p. 383*).	**Power-Reducing Formulas** $$\sin^2 u=\frac{1-\cos 2u}{2},\quad \cos^2 u=\frac{1+\cos 2u}{2}$$ $$\tan^2 u=\frac{1-\cos 2u}{1+\cos 2u}$$	67, 68
Use half-angle formulas to rewrite and evaluate trigonometric functions (*p. 384*).	**Half-Angle Formulas** $$\sin\frac{u}{2}=\pm\sqrt{\frac{1-\cos u}{2}},\quad \cos\frac{u}{2}=\pm\sqrt{\frac{1+\cos u}{2}}$$ $$\tan\frac{u}{2}=\frac{1-\cos u}{\sin u}=\frac{\sin u}{1+\cos u}$$ The signs of $\sin\frac{u}{2}$ and $\cos\frac{u}{2}$ depend on the quadrant in which $u/2$ lies.	69–74
Section 5.5 Use product-to-sum and sum-to-product formulas to rewrite and evaluate trigonometric expressions (*p. 385*).	**Product-to-Sum Formulas** $$\sin u\sin v=(1/2)[\cos(u-v)-\cos(u+v)]$$ $$\cos u\cos v=(1/2)[\cos(u-v)+\cos(u+v)]$$ $$\sin u\cos v=(1/2)[\sin(u+v)+\sin(u-v)]$$ $$\cos u\sin v=(1/2)[\sin(u+v)-\sin(u-v)]$$ **Sum-to-Product Formulas** $$\sin u+\sin v=2\sin\left(\frac{u+v}{2}\right)\cos\left(\frac{u-v}{2}\right)$$ $$\sin u-\sin v=2\cos\left(\frac{u+v}{2}\right)\sin\left(\frac{u-v}{2}\right)$$ $$\cos u+\cos v=2\cos\left(\frac{u+v}{2}\right)\cos\left(\frac{u-v}{2}\right)$$ $$\cos u-\cos v=-2\sin\left(\frac{u+v}{2}\right)\sin\left(\frac{u-v}{2}\right)$$	75–78
Use trigonometric formulas to rewrite real-life models (*p. 387*).	A trigonometric formula can be used to rewrite the projectile motion model $r=(1/16)\,v_0^2\sin\theta\cos\theta$. (See Example 10.)	79, 80

Review Exercises See CalcChat.com for tutorial help and worked-out solutions to odd-numbered exercises.

5.1 Recognizing a Fundamental Identity In Exercises 1–4, name the trigonometric function that is equivalent to the expression.

1. $\dfrac{\cos x}{\sin x}$

2. $\dfrac{1}{\cos x}$

3. $\sin\left(\dfrac{\pi}{2} - x\right)$

4. $\sqrt{\cot^2 x + 1}$

Using Identities to Evaluate a Function In Exercises 5 and 6, use the given conditions and fundamental trigonometric identities to find the values of all six trigonometric functions.

5. $\cos\theta = -\frac{2}{5}$, $\tan\theta > 0$ 6. $\cot x = -\frac{2}{3}$, $\cos x < 0$

Simplifying a Trigonometric Expression In Exercises 7–16, use the fundamental trigonometric identities to simplify the expression. (There is more than one correct form of each answer.)

7. $\dfrac{1}{\cot^2 x + 1}$

8. $\dfrac{\tan\theta}{1 - \cos^2\theta}$

9. $\tan^2 x(\csc^2 x - 1)$

10. $\cot^2 x(\sin^2 x)$

11. $\dfrac{\cot\left(\dfrac{\pi}{2} - u\right)}{\cos u}$

12. $\dfrac{\sec^2(-\theta)}{\csc^2\theta}$

13. $\cos^2 x + \cos^2 x \cot^2 x$

14. $(\tan x + 1)^2 \cos x$

15. $\dfrac{1}{\csc\theta + 1} - \dfrac{1}{\csc\theta - 1}$

16. $\dfrac{\tan^2 x}{1 + \sec x}$

Trigonometric Substitution In Exercises 17 and 18, use the trigonometric substitution to write the algebraic expression as a trigonometric function of θ, where $0 < \theta < \pi/2$.

17. $\sqrt{25 - x^2}$, $x = 5\sin\theta$ 18. $\sqrt{x^2 - 16}$, $x = 4\sec\theta$

5.2 Verifying a Trigonometric Identity In Exercises 19–26, verify the identity.

19. $\cos x(\tan^2 x + 1) = \sec x$

20. $\sec^2 x \cot x - \cot x = \tan x$

21. $\sin\left(\dfrac{\pi}{2} - \theta\right)\tan\theta = \sin\theta$

22. $\cot\left(\dfrac{\pi}{2} - x\right)\csc x = \sec x$

23. $\dfrac{1}{\tan\theta\csc\theta} = \cos\theta$ 24. $\dfrac{1}{\tan x \csc x \sin x} = \cot x$

25. $\sin^5 x \cos^2 x = (\cos^2 x - 2\cos^4 x + \cos^6 x)\sin x$

26. $\cos^3 x \sin^2 x = (\sin^2 x - \sin^4 x)\cos x$

5.3 Solving a Trigonometric Equation In Exercises 27–32, solve the equation.

27. $\sin x = \sqrt{3} - \sin x$ 28. $4\cos\theta = 1 + 2\cos\theta$

29. $3\sqrt{3}\tan u = 3$ 30. $\frac{1}{2}\sec x - 1 = 0$

31. $3\csc^2 x = 4$ 32. $4\tan^2 u - 1 = \tan^2 u$

Solving a Trigonometric Equation In Exercises 33–42, find all solutions of the equation in the interval $[0, 2\pi)$.

33. $\sin^3 x = \sin x$ 34. $2\cos^2 x + 3\cos x = 0$

35. $\cos^2 x + \sin x = 1$ 36. $\sin^2 x + 2\cos x = 2$

37. $2\sin 2x - \sqrt{2} = 0$ 38. $2\cos\dfrac{x}{2} + 1 = 0$

39. $3\tan^2\left(\dfrac{x}{3}\right) - 1 = 0$ 40. $\sqrt{3}\tan 3x = 0$

41. $\cos 4x(\cos x - 1) = 0$ 42. $3\csc^2 5x = -4$

Using Inverse Functions In Exercises 43–46, solve the equation.

43. $\tan^2 x - 2\tan x = 0$

44. $2\tan^2 x - 3\tan x = -1$

45. $\tan^2\theta + \tan\theta - 6 = 0$

46. $\sec^2 x + 6\tan x + 4 = 0$

5.4 Evaluating Trigonometric Functions In Exercises 47–50, find the exact values of the sine, cosine, and tangent of the angle.

47. $75° = 120° - 45°$ 48. $375° = 135° + 240°$

49. $\dfrac{25\pi}{12} = \dfrac{11\pi}{6} + \dfrac{\pi}{4}$ 50. $\dfrac{19\pi}{12} = \dfrac{11\pi}{6} - \dfrac{\pi}{4}$

Rewriting a Trigonometric Expression In Exercises 51 and 52, write the expression as the sine, cosine, or tangent of an angle.

51. $\sin 60°\cos 45° - \cos 60°\sin 45°$

52. $\dfrac{\tan 68° - \tan 115°}{1 + \tan 68°\tan 115°}$

Evaluating a Trigonometric Expression In Exercises 53–56, find the exact value of the trigonometric expression given that $\tan u = \frac{3}{4}$ and $\cos v = -\frac{4}{5}$. (u is in Quadrant I and v is in Quadrant III.)

53. $\sin(u + v)$

54. $\tan(u + v)$

55. $\cos(u - v)$

56. $\sin(u - v)$

Verifying a Trigonometric Identity In Exercises 57–60, verify the identity.

57. $\cos\left(x + \dfrac{\pi}{2}\right) = -\sin x$ **58.** $\tan\left(x - \dfrac{\pi}{2}\right) = -\cot x$

59. $\tan(\pi - x) = -\tan x$ **60.** $\sin(x - \pi) = -\sin x$

Solving a Trigonometric Equation In Exercises 61 and 62, find all solutions of the equation in the interval $[0, 2\pi)$.

61. $\sin\left(x + \dfrac{\pi}{4}\right) - \sin\left(x - \dfrac{\pi}{4}\right) = 1$

62. $\cos\left(x + \dfrac{\pi}{6}\right) - \cos\left(x - \dfrac{\pi}{6}\right) = 1$

5.5 Evaluating Functions Involving Double Angles In Exercises 63 and 64, use the given conditions to find the exact values of $\sin 2u$, $\cos 2u$, and $\tan 2u$ using the double-angle formulas.

63. $\sin u = \frac{4}{5}$, $0 < u < \pi/2$

64. $\cos u = -2/\sqrt{5}$, $\pi/2 < u < \pi$

Verifying a Trigonometric Identity In Exercises 65 and 66, use the double-angle formulas to verify the identity algebraically and use a graphing utility to confirm your result graphically.

65. $\sin 4x = 8\cos^3 x \sin x - 4\cos x \sin x$

66. $\tan^2 x = \dfrac{1 - \cos 2x}{1 + \cos 2x}$

Reducing Powers In Exercises 67 and 68, use the power-reducing formulas to rewrite the expression in terms of first powers of the cosines of multiple angles.

67. $\tan^2 3x$ **68.** $\sin^2 x \cos^2 x$

Using Half-Angle Formulas In Exercises 69 and 70, use the half-angle formulas to determine the exact values of the sine, cosine, and tangent of the angle.

69. $-75°$ **70.** $5\pi/12$

Using Half-Angle Formulas In Exercises 71–74, use the given conditions to (a) determine the quadrant in which $u/2$ lies, and (b) find the exact values of $\sin(u/2)$, $\cos(u/2)$, and $\tan(u/2)$ using the half-angle formulas.

71. $\tan u = \dfrac{4}{3}$, $\pi < u < \dfrac{3\pi}{2}$

72. $\sin u = \dfrac{3}{5}$, $0 < u < \dfrac{\pi}{2}$

73. $\cos u = -\dfrac{2}{7}$, $\dfrac{\pi}{2} < u < \pi$

74. $\tan u = -\dfrac{\sqrt{21}}{2}$, $\dfrac{3\pi}{2} < u < 2\pi$

Using Product-to-Sum Formulas In Exercises 75 and 76, use the product-to-sum formulas to rewrite the product as a sum or difference.

75. $\cos 4\theta \sin 6\theta$

76. $2\sin 7\theta \cos 3\theta$

Using Sum-to-Product Formulas In Exercises 77 and 78, use the sum-to-product formulas to rewrite the sum or difference as a product.

77. $\cos 6\theta + \cos 5\theta$

78. $\sin 3x - \sin x$

79. Projectile Motion A baseball leaves the hand of a player at first base at an angle of θ with the horizontal and at an initial velocity of $v_0 = 80$ feet per second. A player at second base 100 feet away catches the ball. Find θ when the range r of a projectile is

$$r = \dfrac{1}{32}v_0^2 \sin 2\theta.$$

80. Geometry A trough for feeding cattle is 4 meters long and its cross sections are isosceles triangles with the two equal sides being $\frac{1}{2}$ meter (see figure). The angle between the two sides is θ.

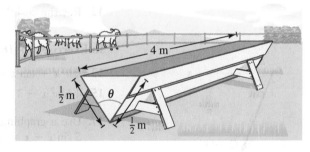

(a) Write the volume of the trough as a function of $\theta/2$.

(b) Write the volume of the trough as a function of θ and determine the value of θ such that the volume is maximized.

Exploration

True or False? In Exercises 81–84, determine whether the statement is true or false. Justify your answer.

81. If $\dfrac{\pi}{2} < \theta < \pi$, then $\cos\dfrac{\theta}{2} < 0$.

82. $\cot x \sin^2 x = \cos x \sin x$

83. $4\sin(-x)\cos(-x) = -2\sin 2x$

84. $4\sin 45° \cos 15° = 1 + \sqrt{3}$

85. Think About It Is it possible for a trigonometric equation that is not an identity to have an infinite number of solutions? Explain.

Chapter Test

See **CalcChat.com** for tutorial help and worked-out solutions to odd-numbered exercises.

Take this test as you would take a test in class. When you are finished, check your work against the answers given in the back of the book.

1. Use the conditions $\csc \theta = \frac{5}{2}$ and $\tan \theta < 0$ to find the values of all six trigonometric functions.

2. Use the fundamental identities to simplify $\csc^2 \beta (1 - \cos^2 \beta)$.

3. Factor and simplify $\dfrac{\sec^4 x - \tan^4 x}{\sec^2 x + \tan^2 x}$.

4. Add and simplify $\dfrac{\cos \theta}{\sin \theta} + \dfrac{\sin \theta}{\cos \theta}$.

In Exercises 5–10, verify the identity.

5. $\sin \theta \sec \theta = \tan \theta$

6. $\sec^2 x \tan^2 x + \sec^2 x = \sec^4 x$

7. $\dfrac{\csc \alpha + \sec \alpha}{\sin \alpha + \cos \alpha} = \cot \alpha + \tan \alpha$

8. $\tan\left(x + \dfrac{\pi}{2}\right) = -\cot x$

9. $1 + \cos 10y = 2 \cos^2 5y$

10. $\sin \dfrac{\alpha}{3} \cos \dfrac{\alpha}{3} = \dfrac{1}{2} \sin \dfrac{2\alpha}{3}$

11. Rewrite $4 \sin 3\theta \cos 2\theta$ as a sum or difference.

12. Rewrite $\cos\left(\theta + \dfrac{\pi}{2}\right) - \cos\left(\theta - \dfrac{\pi}{2}\right)$ as a product.

In Exercises 13–16, find all solutions of the equation in the interval $[0, 2\pi)$.

13. $\tan^2 x + \tan x = 0$

14. $\sin 2\alpha - \cos \alpha = 0$

15. $4 \cos^2 x - 3 = 0$

16. $\csc^2 x - \csc x - 2 = 0$

17. Use a graphing utility to approximate (to three decimal places) the solutions of $5 \sin x - x = 0$ in the interval $[0, 2\pi)$.

18. Find the exact value of $\cos 105°$ using the fact that $105° = 135° - 30°$.

19. Use the figure to find the exact values of $\sin 2u$, $\cos 2u$, and $\tan 2u$.

20. Cheyenne, Wyoming, has a latitude of 41°N. At this latitude, the number of hours of daylight D can be modeled by

$$D = 2.914 \sin(0.017t - 1.321) + 12.134$$

where t represents the day, with $t = 1$ corresponding to January 1. Use a graphing utility to determine the days on which there are more than 10 hours of daylight. *(Source: U.S. Naval Observatory)*

21. The heights h_1 and h_2 (in feet) above ground of two people in different seats on a Ferris wheel can be modeled by

$$h_1 = 28 \cos 10t + 38$$

and

$$h_2 = 28 \cos\left[10\left(t - \dfrac{\pi}{6}\right)\right] + 38, \quad 0 \le t \le 2$$

where t represents the time (in minutes). When are the two people at the same height?

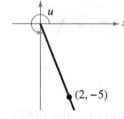

Figure for 19

Proofs in Mathematics ■ ■ ■ ■ ■ ■ ■ ■ ■ ■ ■ ■ ■ ■ ■

Sum and Difference Formulas *(p. 374)*

$$\sin(u + v) = \sin u \cos v + \cos u \sin v \qquad \tan(u + v) = \frac{\tan u + \tan v}{1 - \tan u \tan v}$$

$$\sin(u - v) = \sin u \cos v - \cos u \sin v$$

$$\cos(u + v) = \cos u \cos v - \sin u \sin v \qquad \tan(u - v) = \frac{\tan u - \tan v}{1 + \tan u \tan v}$$

$$\cos(u - v) = \cos u \cos v + \sin u \sin v$$

Proof

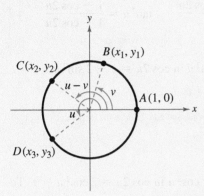

In the proofs of the formulas for $\cos(u \pm v)$, assume that $0 < v < u < 2\pi$. The top figure at the left uses u and v to locate the points $B(x_1, y_1)$, $C(x_2, y_2)$, and $D(x_3, y_3)$ on the unit circle. So, $x_i^2 + y_i^2 = 1$ for $i = 1, 2$, and 3. In the bottom figure, arc lengths AC and BD are equal, so segment lengths AC and BD are also equal. This leads to the following.

$$\sqrt{(x_2 - 1)^2 + (y_2 - 0)^2} = \sqrt{(x_3 - x_1)^2 + (y_3 - y_1)^2}$$

$$x_2^2 - 2x_2 + 1 + y_2^2 = x_3^2 - 2x_1x_3 + x_1^2 + y_3^2 - 2y_1y_3 + y_1^2$$

$$(x_2^2 + y_2^2) + 1 - 2x_2 = (x_3^2 + y_3^2) + (x_1^2 + y_1^2) - 2x_1x_3 - 2y_1y_3$$

$$1 + 1 - 2x_2 = 1 + 1 - 2x_1x_3 - 2y_1y_3$$

$$x_2 = x_3x_1 + y_3y_1$$

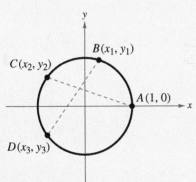

Substitute the values $x_2 = \cos(u - v)$, $x_3 = \cos u$, $x_1 = \cos v$, $y_3 = \sin u$, and $y_1 = \sin v$ to obtain $\cos(u - v) = \cos u \cos v + \sin u \sin v$. To establish the formula for $\cos(u + v)$, consider $u + v = u - (-v)$ and use the formula just derived to obtain

$$\cos(u + v) = \cos[u - (-v)]$$

$$= \cos u \cos(-v) + \sin u \sin(-v)$$

$$= \cos u \cos v - \sin u \sin v.$$

You can use the sum and difference formulas for sine and cosine to prove the formulas for $\tan(u \pm v)$.

$$\tan(u \pm v) = \frac{\sin(u \pm v)}{\cos(u \pm v)} = \frac{\sin u \cos v \pm \cos u \sin v}{\cos u \cos v \mp \sin u \sin v}$$

$$= \frac{\dfrac{\sin u \cos v \pm \cos u \sin v}{\cos u \cos v}}{\dfrac{\cos u \cos v \mp \sin u \sin v}{\cos u \cos v}} = \frac{\dfrac{\sin u \cos v}{\cos u \cos v} \pm \dfrac{\cos u \sin v}{\cos u \cos v}}{\dfrac{\cos u \cos v}{\cos u \cos v} \mp \dfrac{\sin u \sin v}{\cos u \cos v}}$$

$$= \frac{\dfrac{\sin u}{\cos u} \pm \dfrac{\sin v}{\cos v}}{1 \mp \dfrac{\sin u}{\cos u} \cdot \dfrac{\sin v}{\cos v}} = \frac{\tan u \pm \tan v}{1 \mp \tan u \tan v} \qquad \blacksquare$$

Double-Angle Formulas *(p. 381)*

$$\sin 2u = 2 \sin u \cos u \qquad\qquad \cos 2u = \cos^2 u - \sin^2 u$$

$$\tan 2u = \frac{2 \tan u}{1 - \tan^2 u} \qquad\qquad\quad = 2 \cos^2 u - 1$$

$$= 1 - 2 \sin^2 u$$

TRIGONOMETRY AND ASTRONOMY

Early astronomers used trigonometry to calculate measurements in the universe. For instance, they used trigonometry to calculate the circumference of Earth and the distance from Earth to the moon. Another major accomplishment in astronomy using trigonometry was computing distances to stars.

Proof Prove each Double-Angle Formula by letting $v = u$ in the corresponding sum formula.

$$\sin 2u = \sin(u + u) = \sin u \cos u + \cos u \sin u = 2 \sin u \cos u$$

$$\cos 2u = \cos(u + u) = \cos u \cos u - \sin u \sin u = \cos^2 u - \sin^2 u$$

$$\tan 2u = \tan(u + u) = \frac{\tan u + \tan u}{1 - \tan u \tan u} = \frac{2 \tan u}{1 - \tan^2 u}$$

Power-Reducing Formulas   *(p. 383)*

$$\sin^2 u = \frac{1 - \cos 2u}{2} \qquad \cos^2 u = \frac{1 + \cos 2u}{2} \qquad \tan^2 u = \frac{1 - \cos 2u}{1 + \cos 2u}$$

Proof Prove the first formula by solving for $\sin^2 u$ in $\cos 2u = 1 - 2 \sin^2 u$.

$$\cos 2u = 1 - 2 \sin^2 u \qquad \text{Write double-angle formula.}$$

$$2 \sin^2 u = 1 - \cos 2u \qquad \text{Subtract } \cos 2u \text{ from, and add } 2 \sin^2 u \text{ to, each side.}$$

$$\sin^2 u = \frac{1 - \cos 2u}{2} \qquad \text{Divide each side by 2.}$$

Similarly, to prove the second formula, solve for $\cos^2 u$ in $\cos 2u = 2 \cos^2 u - 1$. To prove the third formula, use a quotient identity.

$$\tan^2 u = \frac{\sin^2 u}{\cos^2 u} = \frac{\dfrac{1 - \cos 2u}{2}}{\dfrac{1 + \cos 2u}{2}} = \frac{1 - \cos 2u}{1 + \cos 2u}$$

Sum-to-Product Formulas   *(p. 385)*

$$\sin u + \sin v = 2 \sin\left(\frac{u + v}{2}\right) \cos\left(\frac{u - v}{2}\right)$$

$$\sin u - \sin v = 2 \cos\left(\frac{u + v}{2}\right) \sin\left(\frac{u - v}{2}\right)$$

$$\cos u + \cos v = 2 \cos\left(\frac{u + v}{2}\right) \cos\left(\frac{u - v}{2}\right)$$

$$\cos u - \cos v = -2 \sin\left(\frac{u + v}{2}\right) \sin\left(\frac{u - v}{2}\right)$$

Proof To prove the first formula, let $x = u + v$ and $y = u - v$. Then substitute $u = (x + y)/2$ and $v = (x - y)/2$ in the product-to-sum formula.

$$\sin u \cos v = \frac{1}{2}[\sin(u + v) + \sin(u - v)]$$

$$\sin\left(\frac{x + y}{2}\right) \cos\left(\frac{x - y}{2}\right) = \frac{1}{2}(\sin x + \sin y)$$

$$2 \sin\left(\frac{x + y}{2}\right) \cos\left(\frac{x - y}{2}\right) = \sin x + \sin y$$

The other sum-to-product formulas can be proved in a similar manner.

P.S. Problem Solving ■ ■ ■ ■ ■ ■ ■ ■ ■ ■ ■ ■ ■ ■ ■

1. **Writing Trigonometric Functions in Terms of Cosine** Write each of the other trigonometric functions of θ in terms of $\cos \theta$.

2. **Verifying a Trigonometric Identity** Verify that for all integers n,
$$\cos\left[\frac{(2n + 1)\pi}{2}\right] = 0.$$

3. **Verifying a Trigonometric Identity** Verify that for all integers n,
$$\sin\left[\frac{(12n + 1)\pi}{6}\right] = \frac{1}{2}.$$

4. **Sound Wave** A sound wave is modeled by
$$p(t) = \frac{1}{4\pi}[p_1(t) + 30p_2(t) + p_3(t) + p_5(t) + 30p_6(t)]$$

where $p_n(t) = \frac{1}{n}\sin(524n\pi t)$, and t represents the time (in seconds).

   (a) Find the sine components $p_n(t)$ and use a graphing utility to graph the components. Then verify the graph of p shown below.

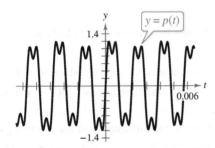

   (b) Find the period of each sine component of p. Is p periodic? If so, then what is its period?

   (c) Use the graphing utility to find the t-intercepts of the graph of p over one cycle.

   (d) Use the graphing utility to approximate the absolute maximum and absolute minimum values of p over one cycle.

5. **Geometry** Three squares of side length s are placed side by side (see figure). Make a conjecture about the relationship between the sum $u + v$ and w. Prove your conjecture by using the identity for the tangent of the sum of two angles.

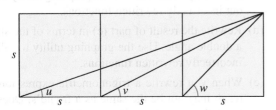

6. **Projectile Motion** The path traveled by an object (neglecting air resistance) that is projected at an initial height of h_0 feet, an initial velocity of v_0 feet per second, and an initial angle θ is given by
$$y = -\frac{16}{v_0^2 \cos^2 \theta}x^2 + (\tan \theta)x + h_0$$

where the horizontal distance x and the vertical distance y are measured in feet. Find a formula for the maximum height of an object projected from ground level at velocity v_0 and angle θ. To do this, find half of the horizontal distance
$$\frac{1}{32}v_0^2 \sin 2\theta$$

and then substitute it for x in the model for the path of a projectile (where $h_0 = 0$).

7. **Geometry** The length of each of the two equal sides of an isosceles triangle is 10 meters (see figure). The angle between the two sides is θ.

   (a) Write the area of the triangle as a function of $\theta/2$.

   (b) Write the area of the triangle as a function of θ. Determine the value of θ such that the area is a maximum.

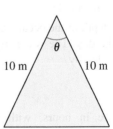   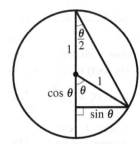

Figure for 7   Figure for 8

8. **Geometry** Use the figure to derive the formulas for
$$\sin\frac{\theta}{2}, \cos\frac{\theta}{2}, \text{ and } \tan\frac{\theta}{2}$$

where θ is an acute angle.

9. **Force** The force F (in pounds) on a person's back when he or she bends over at an angle θ from an upright position is modeled by
$$F = \frac{0.6W \sin(\theta + 90°)}{\sin 12°}$$

where W represents the person's weight (in pounds).

   (a) Simplify the model.

   (b) Use a graphing utility to graph the model, where $W = 185$ and $0° < \theta < 90°$.

   (c) At what angle is the force maximized? At what angle is the force minimized?

10. Hours of Daylight The number of hours of daylight that occur at any location on Earth depends on the time of year and the latitude of the location. The equations below model the numbers of hours of daylight in Seward, Alaska (60° latitude), and New Orleans, Louisiana (30° latitude).

$$D = 12.2 - 6.4 \cos\left[\frac{\pi(t + 0.2)}{182.6}\right] \quad \text{Seward}$$

$$D = 12.2 - 1.9 \cos\left[\frac{\pi(t + 0.2)}{182.6}\right] \quad \text{New Orleans}$$

In these models, D represents the number of hours of daylight and t represents the day, with $t = 0$ corresponding to January 1.

(a) Use a graphing utility to graph both models in the same viewing window. Use a viewing window of $0 \le t \le 365$.

(b) Find the days of the year on which both cities receive the same amount of daylight.

(c) Which city has the greater variation in the number of hours of daylight? Which constant in each model would you use to determine the difference between the greatest and least numbers of hours of daylight?

(d) Determine the period of each model.

11. Ocean Tide The tide, or depth of the ocean near the shore, changes throughout the day. The water depth d (in feet) of a bay can be modeled by

$$d = 35 - 28 \cos\frac{\pi}{6.2}t$$

where t represents the time in hours, with $t = 0$ corresponding to 12:00 A.M.

(a) Algebraically find the times at which the high and low tides occur.

(b) If possible, algebraically find the time(s) at which the water depth is 3.5 feet.

(c) Use a graphing utility to verify your results from parts (a) and (b).

12. Piston Heights The heights h (in inches) of pistons 1 and 2 in an automobile engine can be modeled by

$$h_1 = 3.75 \sin 733t + 7.5$$

and

$$h_2 = 3.75 \sin 733\left(t + \frac{4\pi}{3}\right) + 7.5$$

respectively, where t is measured in seconds.

(a) Use a graphing utility to graph the heights of these pistons in the same viewing window for $0 \le t \le 1$.

(b) How often are the pistons at the same height?

13. Index of Refraction The index of refraction n of a transparent material is the ratio of the speed of light in a vacuum to the speed of light in the material. Some common materials and their indices of refraction are air (1.00), water (1.33), and glass (1.50). Triangular prisms are often used to measure the index of refraction based on the formula

$$n = \frac{\sin\left(\dfrac{\theta}{2} + \dfrac{\alpha}{2}\right)}{\sin\dfrac{\theta}{2}}.$$

For the prism shown in the figure, $\alpha = 60°$.

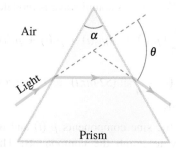

(a) Write the index of refraction as a function of $\cot(\theta/2)$.

(b) Find θ for a prism made of glass.

14. Sum Formulas

(a) Write a sum formula for $\sin(u + v + w)$.

(b) Write a sum formula for $\tan(u + v + w)$.

15. Solving Trigonometric Inequalities Find the solution of each inequality in the interval $[0, 2\pi)$.

(a) $\sin x \ge 0.5$ (b) $\cos x \le -0.5$

(c) $\tan x < \sin x$ (d) $\cos x \ge \sin x$

16. Sum of Fourth Powers Consider the function $f(x) = \sin^4 x + \cos^4 x$.

(a) Use the power-reducing formulas to write the function in terms of cosine to the first power.

(b) Determine another way of rewriting the original function. Use a graphing utility to rule out incorrectly rewritten functions.

(c) Add a trigonometric term to the original function so that it becomes a perfect square trinomial. Rewrite the function as a perfect square trinomial minus the term that you added. Use the graphing utility to rule out incorrectly rewritten functions.

(d) Rewrite the result of part (c) in terms of the sine of a double angle. Use the graphing utility to rule out incorrectly rewritten functions.

(e) When you rewrite a trigonometric expression, the result may not be the same as a friend's. Does this mean that one of you is wrong? Explain.

6 Additional Topics in Trigonometry

Work *(page 434)*

Ohm's Law
(Exercise 95, page 453)

Air Navigation *(Example 11, page 424)*

Mechanical Engineering
(Exercise 56, page 415)

Surveying *(page 401)*

6.1 Law of Sines

The Law of Sines is a useful tool for solving real-life problems involving oblique triangles. For example, in Exercise 46 on page 407, you will use the Law of Sines to determine the distance from a boat to a shoreline.

- Use the Law of Sines to solve oblique triangles (AAS or ASA).
- Use the Law of Sines to solve oblique triangles (SSA).
- Find the areas of oblique triangles.
- Use the Law of Sines to model and solve real-life problems.

Introduction

In Chapter 4, you studied techniques for solving right triangles. In this section and the next, you will solve **oblique triangles**—triangles that have no right angles. As standard notation, the angles of a triangle are labeled A, B, and C, and their opposite sides are labeled a, b, and c, as shown in the figure.

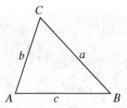

To solve an oblique triangle, you need to know the measure of at least one side and any two other measures of the triangle—the other two sides, two angles, or one angle and one other side. So, there are four cases.

1. Two angles and any side (AAS or ASA)
2. Two sides and an angle opposite one of them (SSA)
3. Three sides (SSS)
4. Two sides and their included angle (SAS)

The first two cases can be solved using the **Law of Sines,** whereas the last two cases require the Law of Cosines (see Section 6.2).

Law of Sines

If ABC is a triangle with sides a, b, and c, then

$$\frac{a}{\sin A} = \frac{b}{\sin B} = \frac{c}{\sin C}.$$

A is acute.

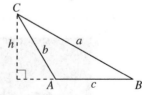

A is obtuse.

The Law of Sines can also be written in the reciprocal form

$$\frac{\sin A}{a} = \frac{\sin B}{b} = \frac{\sin C}{c}.$$

For a proof of the Law of Sines, see Proofs in Mathematics on page 462.

$b = 28$ ft
C
$102°$
a
$29°$
A
c
B

Figure 6.1

EXAMPLE 1 **Given Two Angles and One Side—AAS**

For the triangle in Figure 6.1, $C = 102°$, $B = 29°$, and $b = 28$ feet. Find the remaining angle and sides.

Solution The third angle of the triangle is

$$A = 180° - B - C = 180° - 29° - 102° = 49°.$$

By the Law of Sines, you have

$$\frac{a}{\sin A} = \frac{b}{\sin B} = \frac{c}{\sin C}.$$

Using $b = 28$ produces

$$a = \frac{b}{\sin B}(\sin A) = \frac{28}{\sin 29°}(\sin 49°) \approx 43.59 \text{ feet}$$

and

$$c = \frac{b}{\sin B}(\sin C) = \frac{28}{\sin 29°}(\sin 102°) \approx 56.49 \text{ feet}.$$

✓ **Checkpoint**  Audio-video solution in English & Spanish at LarsonPrecalculus.com

For the triangle shown, $A = 30°$, $B = 45°$, and $a = 32$ centimeters. Find the remaining angle and sides.

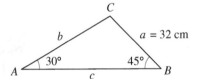

C
b
$a = 32$ cm
$30°$
A
$45°$
c
B

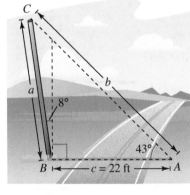

In the 1850s, surveyors used the Law of Sines to calculate the height of Mount Everest. Their calculation was within 30 feet of the currently accepted value.

EXAMPLE 2 **Given Two Angles and One Side—ASA**

A pole tilts toward the sun at an 8° angle from the vertical, and it casts a 22-foot shadow. (See Figure 6.2.) The angle of elevation from the tip of the shadow to the top of the pole is 43°. How tall is the pole?

Solution In Figure 6.2, $A = 43°$ and

$$B = 90° + 8° = 98°.$$

So, the third angle is

$$C = 180° - A - B = 180° - 43° - 98° = 39°.$$

By the Law of Sines, you have

$$\frac{a}{\sin A} = \frac{c}{\sin C}.$$

The shadow length c is $c = 22$ feet, so the height of the pole is

$$a = \frac{c}{\sin C}(\sin A) = \frac{22}{\sin 39°}(\sin 43°) \approx 23.84 \text{ feet}.$$

✓ **Checkpoint**  Audio-video solution in English & Spanish at LarsonPrecalculus.com

Find the height of the tree shown in the figure.

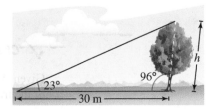

C
a
$8°$
b
$43°$
B
$c = 22$ ft
A

Figure 6.2

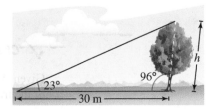

$23°$
$96°$
h
30 m

The Ambiguous Case (SSA)

In Examples 1 and 2, you saw that two angles and one side determine a unique triangle. However, if two sides and one opposite angle are given, then three possible situations can occur: (1) no such triangle exists, (2) one such triangle exists, or (3) two distinct triangles exist that satisfy the conditions.

The Ambiguous Case (SSA)

Consider a triangle in which a, b, and A are given.   ($h = b \sin A$)

	A is acute.	A is acute.	A is acute.	A is acute.	A is obtuse.	A is obtuse.
Sketch						
Necessary condition	$a < h$	$a = h$	$a \geq b$	$h < a < b$	$a \leq b$	$a > b$
Triangles possible	None	One	One	Two	None	One

EXAMPLE 3   **Single-Solution Case—SSA**

See LarsonPrecalculus.com for an interactive version of this type of example.

For the triangle in Figure 6.3, $a = 22$ inches, $b = 12$ inches, and $A = 42°$. Find the remaining side and angles.

Solution   By the Law of Sines, you have

$$\frac{\sin B}{b} = \frac{\sin A}{a}$$ Reciprocal form

$$\sin B = b\left(\frac{\sin A}{a}\right)$$ Multiply each side by b.

$$\sin B = 12\left(\frac{\sin 42°}{22}\right)$$ Substitute for A, a, and b.

$$B \approx 21.41°.$$ Solve for acute angle B.

Next, subtract to determine that $C \approx 180° - 42° - 21.41° = 116.59°$. Then find the remaining side.

$$\frac{c}{\sin C} = \frac{a}{\sin A}$$ Law of Sines

$$c = \frac{a}{\sin A}(\sin C)$$ Multiply each side by $\sin C$.

$$c \approx \frac{22}{\sin 42°}(\sin 116.59°)$$ Substitute for a, A, and C.

$$c \approx 29.40 \text{ inches}$$ Simplify.

C
$b = 12$ in. $a = 22$ in.
$42°$
A c B
One solution: $a \geq b$
Figure 6.3

✓ **Checkpoint**   Audio-video solution in English & Spanish at LarsonPrecalculus.com

Given $A = 31°$, $a = 12$ inches, and $b = 5$ inches, find the remaining side and angles of the triangle.

No solution: $a < h$
Figure 6.4

EXAMPLE 4 **No-Solution Case—SSA**

Show that there is no triangle for which $a = 15$ feet, $b = 25$ feet, and $A = 85°$.

Solution Begin by making the sketch shown in Figure 6.4. From this figure, it appears that no triangle is possible. Verify this using the Law of Sines.

$$\frac{\sin B}{b} = \frac{\sin A}{a} \qquad \text{Reciprocal form}$$

$$\sin B = b\left(\frac{\sin A}{a}\right) \qquad \text{Multiply each side by } b.$$

$$\sin B = 25\left(\frac{\sin 85°}{15}\right) \approx 1.6603 > 1$$

This contradicts the fact that $|\sin B| \leq 1$. So, no triangle can be formed with sides $a = 15$ feet and $b = 25$ feet and angle $A = 85°$.

✓ **Checkpoint** ◀))) *Audio-video solution in English & Spanish at LarsonPrecalculus.com*

Show that there is no triangle for which $a = 4$ feet, $b = 14$ feet, and $A = 60°$.

EXAMPLE 5 **Two-Solution Case—SSA**

Find two triangles for which $a = 12$ meters, $b = 31$ meters, and $A = 20.50°$.

Solution Because $h = b \sin A = 31(\sin 20.50°) \approx 10.86$ meters and $h < a < b$, there are two possible triangles. By the Law of Sines, you have

$$\frac{\sin B}{b} = \frac{\sin A}{a} \qquad \text{Reciprocal form}$$

$$\sin B = b\left(\frac{\sin A}{a}\right) = 31\left(\frac{\sin 20.50°}{12}\right) \approx 0.9047.$$

There are two angles between $0°$ and $180°$ whose sine is approximately 0.9047, $B_1 \approx 64.78°$ and $B_2 \approx 180° - 64.78° = 115.22°$. For $B_1 \approx 64.78°$, you obtain

$$C \approx 180° - 20.50° - 64.78° = 94.72°$$

$$c = \frac{a}{\sin A}(\sin C) \approx \frac{12}{\sin 20.50°}(\sin 94.72°) \approx 34.15 \text{ meters.}$$

For $B_2 \approx 115.22°$, you obtain

$$C \approx 180° - 20.50° - 115.22° = 44.28°$$

$$c = \frac{a}{\sin A}(\sin C) \approx \frac{12}{\sin 20.50°}(\sin 44.28°) \approx 23.92 \text{ meters.}$$

The resulting triangles are shown below.

Two solutions: $h < a < b$

✓ **Checkpoint** ◀))) *Audio-video solution in English & Spanish at LarsonPrecalculus.com*

Find two triangles for which $a = 4.5$ feet, $b = 5$ feet, and $A = 58°$.

Area of an Oblique Triangle

•• **REMARK** To obtain the
height of the obtuse triangle,
use the reference angle
$180° − A$ and the difference
formula for sine:

$$h = b \sin(180° − A)$$

$$= b(\sin 180° \cos A$$

$$- \cos 180° \sin A)$$

$$= b[0 \cdot \cos A − (−1) \cdot \sin A]$$

$$= b \sin A.$$

The procedure used to prove the Law of Sines leads to a formula for the area of an oblique triangle. Consider the two triangles below.

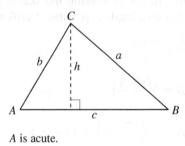

A is acute.

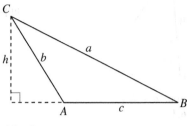

A is obtuse.

Note that each triangle has a height of $h = b \sin A$. Consequently, the area of each triangle is

$$\text{Area} = \frac{1}{2}(\text{base})(\text{height})$$

$$= \frac{1}{2}(c)(b \sin A)$$

$$= \frac{1}{2}bc \sin A.$$

By similar arguments, you can develop the other two forms shown below.

Area of an Oblique Triangle

The area of any triangle is one-half the product of the lengths of two sides times the sine of their included angle. That is,

$$\text{Area} = \frac{1}{2}bc \sin A = \frac{1}{2}ab \sin C = \frac{1}{2}ac \sin B.$$

Note that when angle A is 90°, the formula gives the area of a right triangle:

$$\text{Area} = \frac{1}{2}bc(\sin 90°) = \frac{1}{2}bc = \frac{1}{2}(\text{base})(\text{height}). \qquad \text{sin } 90° = 1$$

You obtain similar results for angles C and B equal to 90°.

EXAMPLE 6 **Finding the Area of a Triangular Lot**

Find the area of a triangular lot with two sides of lengths 90 meters and 52 meters and an included angle of 102°, as shown in Figure 6.5.

Solution The area is

$$\text{Area} = \frac{1}{2}ab \sin C = \frac{1}{2}(90)(52)(\sin 102°) \approx 2289 \text{ square meters.}$$

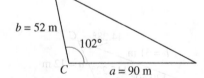

Figure 6.5

✓ **Checkpoint** ◀))) *Audio-video solution in English & Spanish at LarsonPrecalculus.com*

Find the area of a triangular lot with two sides of lengths 24 yards and 18 yards and an included angle of 80°.

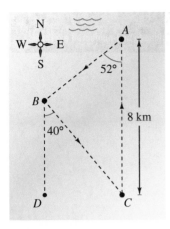

Figure 6.6

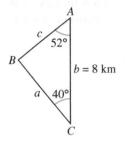

Figure 6.7

Application

| EXAMPLE 7 | An Application of the Law of Sines

The course for a boat race starts at point A and proceeds in the direction S 52° W to point B, then in the direction S 40° E to point C, and finally back to point A, as shown in Figure 6.6. Point C lies 8 kilometers directly south of point A. Approximate the total distance of the race course.

Solution The lines BD and AC are parallel, so $\angle BCA \cong \angle CBD$. Consequently, triangle ABC has the measures shown in Figure 6.7. The measure of angle B is $180° - 52° - 40° = 88°$. Using the Law of Sines,

$$\frac{a}{\sin 52°} = \frac{8}{\sin 88°} = \frac{c}{\sin 40°}.$$

Solving for a and c, you have

$$a = \frac{8}{\sin 88°}(\sin 52°) \approx 6.31 \quad \text{and} \quad c = \frac{8}{\sin 88°}(\sin 40°) \approx 5.15.$$

So, the total distance of the course is approximately

$$8 + 6.31 + 5.15 = 19.46 \text{ kilometers.}$$

✓ **Checkpoint** ◀))) *Audio-video solution in English & Spanish at LarsonPrecalculus.com*

On a small lake, you swim from point A to point B at a bearing of N 28° E, then to point C at a bearing of N 58° W, and finally back to point A, as shown in the figure below. Point C lies 800 meters directly north of point A. Approximate the total distance that you swim.

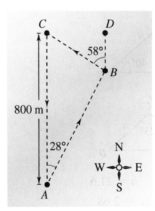

Summarize (Section 6.1)

1. State the Law of Sines *(page 400)*. For examples of using the Law of Sines to solve oblique triangles (AAS or ASA), see Examples 1 and 2.

2. List the necessary conditions and the corresponding numbers of possible triangles for the ambiguous case (SSA) *(page 402)*. For examples of using the Law of Sines to solve oblique triangles (SSA), see Examples 3–5.

3. State the formulas for the area of an oblique triangle *(page 404)*. For an example of finding the area of an oblique triangle, see Example 6.

4. Describe a real-life application of the Law of Sines *(page 405, Example 7)*.

6.1 Exercises

See **CalcChat.com** for tutorial help and worked-out solutions to odd-numbered exercises.

Vocabulary: Fill in the blanks.

1. An _____ triangle is a triangle that has no right angle.

2. For triangle ABC, the Law of Sines is $\dfrac{a}{\sin A} = $ _____ $ = \dfrac{c}{\sin C}$.

3. Two _____ and one _____ determine a unique triangle.

4. The area of an oblique triangle ABC is $\frac{1}{2}bc \sin A = \frac{1}{2}ab \sin C = $ _____.

Skills and Applications

 Using the Law of Sines **In Exercises 5–22, use the Law of Sines to solve the triangle. Round your answers to two decimal places.**

5.
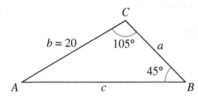

6.
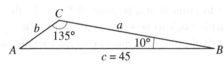

7.
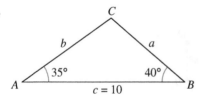

8.
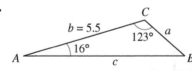

9. $A = 102.4°$, $C = 16.7°$, $a = 21.6$
10. $A = 24.3°$, $C = 54.6°$, $c = 2.68$
11. $A = 83° \, 20'$, $C = 54.6°$, $c = 18.1$
12. $A = 5° \, 40'$, $B = 8° \, 15'$, $b = 4.8$
13. $A = 35°$, $B = 65°$, $c = 10$
14. $A = 120°$, $B = 45°$, $c = 16$
15. $A = 55°$, $B = 42°$, $c = \frac{3}{4}$
16. $B = 28°$, $C = 104°$, $a = 3\frac{5}{8}$
17. $A = 36°$, $a = 8$, $b = 5$
18. $A = 60°$, $a = 9$, $c = 7$
19. $A = 145°$, $a = 14$, $b = 4$
20. $A = 100°$, $a = 125$, $c = 10$
21. $B = 15° \, 30'$, $a = 4.5$, $b = 6.8$
22. $B = 2° \, 45'$, $b = 6.2$, $c = 5.8$

 Using the Law of Sines **In Exercises 23–32, use the Law of Sines to solve (if possible) the triangle. If two solutions exist, find both. Round your answers to two decimal places.**

23. $A = 110°$, $a = 125$, $b = 100$
24. $A = 110°$, $a = 125$, $b = 200$
25. $A = 76°$, $a = 18$, $b = 20$
26. $A = 76°$, $a = 34$, $b = 21$
27. $A = 58°$, $a = 11.4$, $b = 12.8$
28. $A = 58°$, $a = 4.5$, $b = 12.8$
29. $A = 120°$, $a = b = 25$
30. $A = 120°$, $a = 25$, $b = 24$
31. $A = 45°$, $a = b = 1$
32. $A = 25° \, 4'$, $a = 9.5$, $b = 22$

 Using the Law of Sines **In Exercises 33–36, find values for b such that the triangle has (a) one solution, (b) two solutions (if possible), and (c) no solution.**

33. $A = 36°$, $a = 5$
34. $A = 60°$, $a = 10$
35. $A = 105°$, $a = 80$
36. $A = 132°$, $a = 215$

 Finding the Area of a Triangle **In Exercises 37–44, find the area of the triangle. Round your answers to one decimal place.**

37. $A = 125°$, $b = 9$, $c = 6$
38. $C = 150°$, $a = 17$, $b = 10$
39. $B = 39°$, $a = 25$, $c = 12$
40. $A = 72°$, $b = 31$, $c = 44$
41. $C = 103° \, 15'$, $a = 16$, $b = 28$
42. $B = 54° \, 30'$, $a = 62$, $c = 35$
43. $A = 67°$, $B = 43°$, $a = 8$
44. $B = 118°$, $C = 29°$, $a = 52$

45. Height A tree grows at an angle of 4° from the vertical due to prevailing winds. At a point 40 meters from the base of the tree, the angle of elevation to the top of the tree is 30° (see figure).

(a) Write an equation that you can use to find the height h of the tree.

(b) Find the height of the tree.

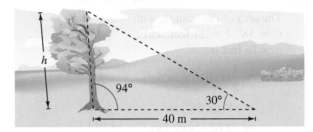

46. Distance

A boat is traveling due east parallel to the shoreline at a speed of 10 miles per hour. At a given time, the bearing to a lighthouse is S 70° E, and 15 minutes later the bearing is S 63° E (see figure). The lighthouse is located at the shoreline. What is the distance from the boat to the shoreline?

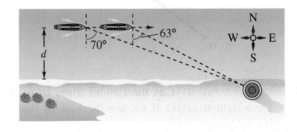

47. Environmental Science The bearing from the Pine Knob fire tower to the Colt Station fire tower is N 65° E, and the two towers are 30 kilometers apart. A fire spotted by rangers in each tower has a bearing of N 80° E from Pine Knob and S 70° E from Colt Station (see figure). Find the distance of the fire from each tower.

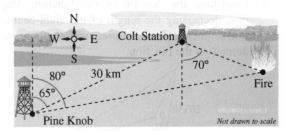

48. Bridge Design A bridge is built across a small lake from a gazebo to a dock (see figure). The bearing from the gazebo to the dock is S 41° W. From a tree 100 meters from the gazebo, the bearings to the gazebo and the dock are S 74° E and S 28° E, respectively. Find the distance from the gazebo to the dock.

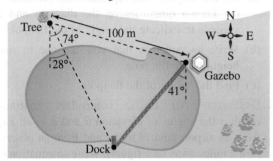

49. Angle of Elevation A 10-meter utility pole casts a 17-meter shadow directly down a slope when the angle of elevation of the sun is 42° (see figure). Find θ, the angle of elevation of the ground.

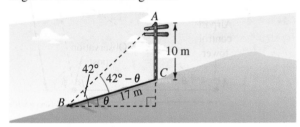

50. Flight Path A plane flies 500 kilometers with a bearing of 316° from Naples to Elgin (see figure). The plane then flies 720 kilometers from Elgin to Canton (Canton is due west of Naples). Find the bearing of the flight from Elgin to Canton.

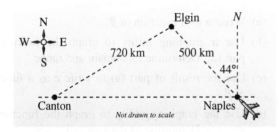

Not drawn to scale

51. Altitude The angles of elevation to an airplane from two points A and B on level ground are 55° and 72°, respectively. The points A and B are 2.2 miles apart, and the airplane is east of both points in the same vertical plane.

(a) Draw a diagram that represents the problem. Show the known quantities on the diagram.

(b) Find the distance between the plane and point B.

(c) Find the altitude of the plane.

(d) Find the distance the plane must travel before it is directly above point A.

52. Height   A flagpole at a right angle to the horizontal is located on a slope that makes an angle of 12° with the horizontal. The flagpole's shadow is 16 meters long and points directly up the slope. The angle of elevation from the tip of the shadow to the sun is 20°.

(a) Draw a diagram that represents the problem. Show the known quantities on the diagram and use a variable to indicate the height of the flagpole.

(b) Write an equation that you can use to find the height of the flagpole.

(c) Find the height of the flagpole.

53. Distance   Air traffic controllers continuously monitor the angles of elevation θ and ϕ to an airplane from an airport control tower and from an observation post 2 miles away (see figure). Write an equation giving the distance d between the plane and the observation post in terms of θ and ϕ.

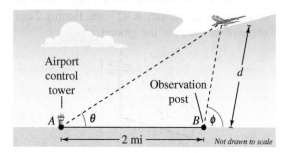

Not drawn to scale

54. Numerical Analysis   In the figure, α and β are positive angles.

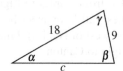

(a) Write α as a function of β.

(b) Use a graphing utility to graph the function in part (a). Determine its domain and range.

(c) Use the result of part (a) to write c as a function of β.

(d) Use the graphing utility to graph the function in part (c). Determine its domain and range.

(e) Complete the table. What can you infer?

β	0.4	0.8	1.2	1.6	2.0	2.4	2.8
α							
c							

Exploration

True or False?   **In Exercises 55–58, determine whether the statement is true or false. Justify your answer.**

55. If a triangle contains an obtuse angle, then it must be oblique.

56. Two angles and one side of a triangle do not necessarily determine a unique triangle.

57. When you know the three angles of an oblique triangle, you can solve the triangle.

58. The ratio of any two sides of a triangle is equal to the ratio of the sines of the opposite angles of the two sides.

59. Error Analysis   Describe the error.

The area of the triangle with $C = 58°$, $b = 11$ feet, and $c = 16$ feet is

$$\text{Area} = \frac{1}{2}(11)(16)(\sin 58°)$$

$$= 88(\sin 58°)$$

$$\approx 74.63 \text{ square feet.}$$

60.   **HOW DO YOU SEE IT?**   In the figure, a triangle is to be formed by drawing a line segment of length a from $(4, 3)$ to the positive x-axis. For what value(s) of a can you form (a) one triangle, (b) two triangles, and (c) no triangles? Explain.

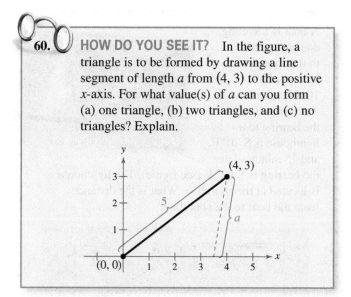

61. Think About It   Can the Law of Sines be used to solve a right triangle? If so, use the Law of Sines to solve the triangle with

$$B = 50°, \quad C = 90°, \quad \text{and} \quad a = 10.$$

Is there another way to solve the triangle? Explain.

62. Using Technology

(a) Write the area A of the shaded region in the figure as a function of θ.

(b) Use a graphing utility to graph the function.

(c) Determine the domain of the function. Explain how decreasing the length of the eight-centimeter line segment affects the area of the region and the domain of the function.

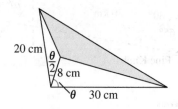

6.2 Law of Cosines

The Law of Cosines is a useful tool for solving real-life problems involving oblique triangles. For example, in Exercise 56 on page 415, you will use the Law of Cosines to determine the total distance a piston moves in an engine.

- Use the Law of Cosines to solve oblique triangles (SSS or SAS).
- Use the Law of Cosines to model and solve real-life problems.
- Use Heron's Area Formula to find areas of triangles.

Introduction

Two cases remain in the list of conditions needed to solve an oblique triangle—SSS and SAS. When you are given three sides (SSS), or two sides and their included angle (SAS), you cannot solve the triangle using the Law of Sines alone. In such cases, use the **Law of Cosines**.

Law of Cosines

Standard Form	**Alternative Form**
$a^2 = b^2 + c^2 - 2bc \cos A$	$\cos A = \dfrac{b^2 + c^2 - a^2}{2bc}$
$b^2 = a^2 + c^2 - 2ac \cos B$	$\cos B = \dfrac{a^2 + c^2 - b^2}{2ac}$
$c^2 = a^2 + b^2 - 2ab \cos C$	$\cos C = \dfrac{a^2 + b^2 - c^2}{2ab}$

For a proof of the Law of Cosines, see Proofs in Mathematics on page 462.

EXAMPLE 1   **Given Three Sides—SSS**

Find the three angles of the triangle shown below.

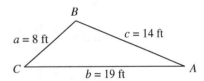

Solution   It is a good idea to find the angle opposite the longest side first—side b in this case. Using the alternative form of the Law of Cosines,

$$\cos B = \frac{a^2 + c^2 - b^2}{2ac} = \frac{8^2 + 14^2 - 19^2}{2(8)(14)} \approx -0.4509.$$

Because $\cos B$ is negative, B is an *obtuse* angle given by $B \approx 116.80°$. At this point, use the Law of Sines to determine A.

$$\sin A = a\left(\frac{\sin B}{b}\right) \approx 8\left(\frac{\sin 116.80°}{19}\right) \approx 0.3758$$

The angle B is obtuse and a triangle can have at most one obtuse angle, so you know that A must be acute. So, $A \approx 22.07°$ and $C \approx 180° - 22.07° - 116.80° = 41.13°$.

✓ **Checkpoint**))) Audio-video solution in English & Spanish at LarsonPrecalculus.com

Find the three angles of the triangle whose sides have lengths $a = 6$ centimeters, $b = 8$ centimeters, and $c = 12$ centimeters.

Smart-foto/Shutterstock.com

Do you see why it was wise to find the largest angle *first* in Example 1? Knowing the cosine of an angle, you can determine whether the angle is acute or obtuse. That is,

$$\cos \theta > 0 \quad \text{for} \quad 0° < \theta < 90° \qquad \text{Acute}$$

$$\cos \theta < 0 \quad \text{for} \quad 90° < \theta < 180°. \qquad \text{Obtuse}$$

So, in Example 1, after you find that angle B is obtuse, you know that angles A and C must both be acute. Furthermore, if the largest angle is acute, then the remaining two angles must also be acute.

> **EXAMPLE 2** **Given Two Sides and Their Included Angle—SAS**

See LarsonPrecalculus.com for an interactive version of this type of example.

Find the remaining angles and side of the triangle shown below.

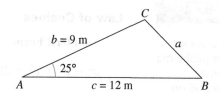

• • **REMARK** When solving
an oblique triangle given three
sides, use the alternative form
of the Law of Cosines to solve
for an angle. When solving
an oblique triangle given two
sides and their included angle,
use the standard form of the
Law of Cosines to solve for the
remaining side.

▷ **Solution** Use the standard form of the Law of Cosines to find side a.

$$a^2 = b^2 + c^2 - 2bc \cos A$$

$$a^2 = 9^2 + 12^2 - 2(9)(12) \cos 25°$$

$$a^2 \approx 29.2375$$

$$a \approx 5.4072 \text{ meters}$$

Next, use the ratio $(\sin A)/a$, the given value of b, and the reciprocal form of the Law of Sines to find B.

$$\frac{\sin B}{b} = \frac{\sin A}{a} \qquad \text{Reciprocal form}$$

$$\sin B = b\left(\frac{\sin A}{a}\right) \qquad \text{Multiply each side by } b.$$

$$\sin B \approx 9\left(\frac{\sin 25°}{5.4072}\right) \qquad \text{Substitute for } A, a, \text{ and } b.$$

$$\sin B \approx 0.7034 \qquad \text{Use a calculator.}$$

There are two angles between $0°$ and $180°$ whose sine is approximately 0.7034, $B_1 \approx 44.70°$ and $B_2 \approx 180° - 44.70° = 135.30°$.

For $B_1 \approx 44.70°$,

$$C_1 \approx 180° - 25° - 44.70° = 110.30°.$$

For $B_2 \approx 135.30°$,

$$C_2 \approx 180° - 25° - 135.30° = 19.70°.$$

Side c is the longest side of the triangle, which means that angle C is the largest angle of the triangle. So, $C \approx 110.30°$ and $B \approx 44.70°$.

✓ *Checkpoint* ◀))) Audio-video solution in English & Spanish at LarsonPrecalculus.com

Given $A = 80°$, $b = 16$ meters, and $c = 12$ meters, find the remaining angles and side of the triangle.

Applications

EXAMPLE 3 An Application of the Law of Cosines

The pitcher's mound on a women's softball field is 43 feet from home plate and the distance between the bases is 60 feet, as shown in Figure 6.8. (The pitcher's mound is *not* halfway between home plate and second base.) How far is the pitcher's mound from first base?

Solution In triangle *HPF*, $H = 45°$ (line segment *HP* bisects the right angle at *H*), $f = 43$, and $p = 60$. Using the standard form of the Law of Cosines for this SAS case,

$$h^2 = f^2 + p^2 - 2fp \cos H$$
$$= 43^2 + 60^2 - 2(43)(60) \cos 45°$$
$$\approx 1800.3290.$$

So, the approximate distance from the pitcher's mound to first base is

$$h \approx \sqrt{1800.3290} \approx 42.43 \text{ feet.}$$

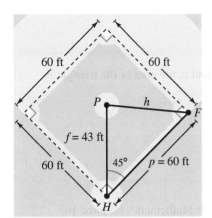

Figure 6.8

✓ *Checkpoint*))) *Audio-video solution in English & Spanish at LarsonPrecalculus.com*

In a softball game, a batter hits a ball to dead center field, a distance of 240 feet from home plate. The center fielder then throws the ball to third base and gets a runner out. The distance between the bases is 60 feet. How far is the center fielder from third base?

EXAMPLE 4 An Application of the Law of Cosines

A ship travels 60 miles due north and then adjusts its course, as shown in Figure 6.9. After traveling 80 miles in this new direction, the ship is 139 miles from its point of departure. Describe the bearing from point *B* to point *C*.

Solution You have $a = 80$, $b = 139$, and $c = 60$. So, using the alternative form of the Law of Cosines,

$$\cos B = \frac{a^2 + c^2 - b^2}{2ac}$$
$$= \frac{80^2 + 60^2 - 139^2}{2(80)(60)}$$
$$\approx -0.9709.$$

So, $B \approx 166.14°$, and the bearing measured from due north from point *B* to point *C* is approximately $180° - 166.14° = 13.86°$, or N 13.86° W.

✓ *Checkpoint* ))) *Audio-video solution in English & Spanish at LarsonPrecalculus.com*

A ship travels 40 miles due east and then changes direction, as shown at the right. After traveling 30 miles in this new direction, the ship is 56 miles from its point of departure. Describe the bearing from point *B* to point *C*.

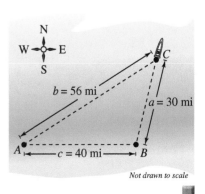

Not drawn to scale

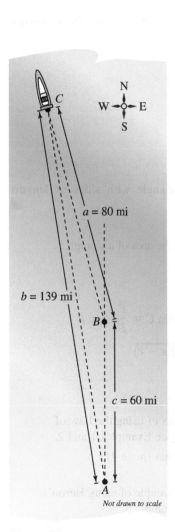

Not drawn to scale

Figure 6.9

Heron's Area Formula

The Law of Cosines can be used to establish a formula for the area of a triangle. This formula is called **Heron's Area Formula** after the Greek mathematician Heron (ca. 10–75 A.D.).

Heron's Area Formula

Given any triangle with sides of lengths a, b, and c, the area of the triangle is

$$\text{Area} = \sqrt{s(s - a)(s - b)(s - c)}$$

where

$$s = \frac{a + b + c}{2}.$$

For a proof of Heron's Area Formula, see Proofs in Mathematics on page 463.

EXAMPLE 5   **Using Heron's Area Formula**

Use Heron's Area Formula to find the area of a triangle with sides of lengths $a = 43$ meters, $b = 53$ meters, and $c = 72$ meters.

Solution   First, determine that $s = (a + b + c)/2 = 168/2 = 84$. Then Heron's Area Formula yields

$$\begin{aligned}
\text{Area} &= \sqrt{s(s - a)(s - b)(s - c)} \\
&= \sqrt{84(84 - 43)(84 - 53)(84 - 72)} \\
&= \sqrt{84(41)(31)(12)} \\
&\approx 1131.89 \text{ square meters.}
\end{aligned}$$

 ✓ **Checkpoint** 🔊))) *Audio-video solution in English & Spanish at LarsonPrecalculus.com*

Use Heron's Area Formula to find the area of a triangle with sides of lengths $a = 5$ inches, $b = 9$ inches, and $c = 8$ inches. ■

You have now studied three different formulas for the area of a triangle.

Standard Formula: $\text{Area} = \dfrac{1}{2}bh$

Oblique Triangle: $\text{Area} = \dfrac{1}{2}bc \sin A = \dfrac{1}{2}ab \sin C = \dfrac{1}{2}ac \sin B$

Heron's Area Formula: $\text{Area} = \sqrt{s(s - a)(s - b)(s - c)}$

Summarize   (Section 6.2)

1. State the Law of Cosines *(page 409)*. For examples of using the Law of Cosines to solve oblique triangles (SSS or SAS), see Examples 1 and 2.

2. Describe real-life applications of the Law of Cosines *(page 411, Examples 3 and 4)*.

3. State Heron's Area Formula *(page 412)*. For an example of using Heron's Area Formula to find the area of a triangle, see Example 5.

6.2 Exercises

See **CalcChat.com** for tutorial help and worked-out solutions to odd-numbered exercises.

Vocabulary: Fill in the blanks.

1. The standard form of the Law of Cosines for $\cos B = \dfrac{a^2 + c^2 - b^2}{2ac}$ is _____.

2. When solving an oblique triangle given three sides, use the _____ form of the Law of Cosines to solve for an angle.

3. When solving an oblique triangle given two sides and their included angle, use the _____ form of the Law of Cosines to solve for the remaining side.

4. The Law of Cosines can be used to establish a formula for the area of a triangle called _____ _____ Formula.

Skills and Applications

 Using the Law of Cosines **In Exercises 5–24, use the Law of Cosines to solve the triangle. Round your answers to two decimal places.**

5.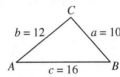
6.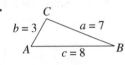
7.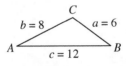
8.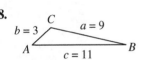
9.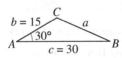
10.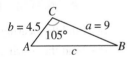
11.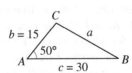
12.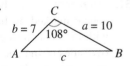

13. $a = 11$, $b = 15$, $c = 21$
14. $a = 55$, $b = 25$, $c = 72$
15. $a = 2.5$, $b = 1.8$, $c = 0.9$
16. $a = 75.4$, $b = 52.5$, $c = 52.5$
17. $A = 120°$, $b = 6$, $c = 7$
18. $A = 48°$, $b = 3$, $c = 14$
19. $B = 10° 35'$, $a = 40$, $c = 30$
20. $B = 75° 20'$, $a = 9$, $c = 6$
21. $B = 125° 40'$, $a = 37$, $c = 37$
22. $C = 15° 15'$, $a = 7.45$, $b = 2.15$
23. $C = 43°$, $a = \frac{4}{9}$, $b = \frac{7}{9}$
24. $C = 101°$, $a = \frac{3}{8}$, $b = \frac{3}{4}$

 Finding Measures in a Parallelogram **In Exercises 25–30, find the missing values by solving the parallelogram shown in the figure. (The lengths of the diagonals are given by c and d.)**

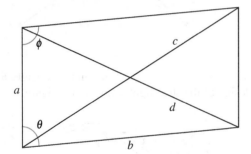

	a	b	c	d	θ	ϕ
25.	5	8			45°	
26.	25	35				120°
27.	10	14	20			
28.	40	60		80		
29.	15		25	20		
30.		25	50	35		

 Solving a Triangle **In Exercises 31–36, determine whether the Law of Cosines is needed to solve the triangle. Then solve (if possible) the triangle. If two solutions exist, find both. Round your answers to two decimal places.**

31. $a = 8$, $c = 5$, $B = 40°$
32. $a = 10$, $b = 12$, $C = 70°$
33. $A = 24°$, $a = 4$, $b = 18$
34. $a = 11$, $b = 13$, $c = 7$
35. $A = 42°$, $B = 35°$, $c = 1.2$
36. $B = 12°$, $a = 160$, $b = 63$

Using Heron's Area Formula In Exercises 37–44, use Heron's Area Formula to find the area of the triangle.

37. $a = 6$, $b = 12$, $c = 17$

38. $a = 33$, $b = 36$, $c = 21$

39. $a = 2.5$, $b = 10.2$, $c = 8$

40. $a = 12.32$, $b = 8.46$, $c = 15.9$

41. $a = 1$, $b = \frac{1}{2}$, $c = \frac{5}{4}$

42. $a = \frac{3}{5}$, $b = \frac{4}{3}$, $c = \frac{7}{8}$

43. $A = 80°$, $b = 75$, $c = 41$

44. $C = 109°$, $a = 16$, $b = 3.5$

45. **Surveying**   To approximate the length of a marsh, a surveyor walks 250 meters from point A to point B, then turns 75° and walks 220 meters to point C (see figure). Approximate the length AC of the marsh.

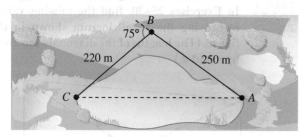

46. **Streetlight Design**   Determine the angle θ in the design of the streetlight shown in the figure.

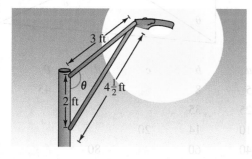

47. **Baseball**   A baseball player in center field is approximately 330 feet from a television camera that is behind home plate. A batter hits a fly ball that goes to the wall 420 feet from the camera (see figure). The camera turns 8° to follow the play. Approximately how far does the center fielder have to run to make the catch?

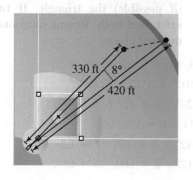

48. **Baseball**   On a baseball diamond with 90-foot sides, the pitcher's mound is 60.5 feet from home plate. How far is the pitcher's mound from third base?

49. **Length**   A 100-foot vertical tower is built on the side of a hill that makes a 6° angle with the horizontal (see figure). Find the length of each of the two guy wires that are anchored 75 feet uphill and downhill from the base of the tower.

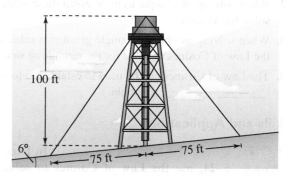

50. **Navigation**   On a map, Minneapolis is 165 millimeters due west of Albany, Phoenix is 216 millimeters from Minneapolis, and Phoenix is 368 millimeters from Albany (see figure).

(a) Find the bearing of Minneapolis from Phoenix.

(b) Find the bearing of Albany from Phoenix.

51. **Navigation**   A boat race runs along a triangular course marked by buoys A, B, and C. The race starts with the boats headed west for 3700 meters. The other two sides of the course lie to the north of the first side, and their lengths are 1700 meters and 3000 meters. Draw a diagram that gives a visual representation of the problem. Then find the bearings for the last two legs of the race.

52. **Air Navigation**   A plane flies 810 miles from Franklin to Centerville with a bearing of 75°. Then it flies 648 miles from Centerville to Rosemount with a bearing of 32°. Draw a diagram that gives a visual representation of the problem. Then find the straight-line distance and bearing from Franklin to Rosemount.

53. **Surveying**   A triangular parcel of land has 115 meters of frontage, and the other boundaries have lengths of 76 meters and 92 meters. What angles does the frontage make with the two other boundaries?

54. Surveying A triangular parcel of ground has sides of lengths 725 feet, 650 feet, and 575 feet. Find the measure of the largest angle.

55. Distance Two ships leave a port at 9 A.M. One travels at a bearing of N 53° W at 12 miles per hour, and the other travels at a bearing of S 67° W at s miles per hour.

(a) Use the Law of Cosines to write an equation that relates s and the distance d between the two ships at noon.

(b) Find the speed s that the second ship must travel so that the ships are 43 miles apart at noon.

• • **56. Mechanical Engineering** • • • • • • • • • •

An engine has a seven-inch connecting rod fastened to a crank (see figure).

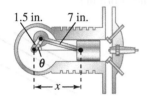

(a) Use the Law of Cosines to write an equation giving the relationship between x and θ.

(b) Write x as a function of θ. (Select the sign that yields positive values of x.)

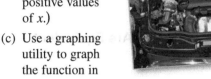

(c) Use a graphing utility to graph the function in part (b).

(d) Use the graph in part (c) to determine the total distance the piston moves in one cycle.

57. Geometry A triangular parcel of land has sides of lengths 200 feet, 500 feet, and 600 feet. Find the area of the parcel.

58. Geometry A parking lot has the shape of a parallelogram (see figure). The lengths of two adjacent sides are 70 meters and 100 meters. The angle between the two sides is 70°. What is the area of the parking lot?

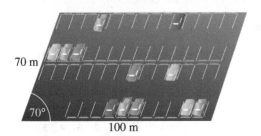

70 m

70°

100 m

59. Geometry You want to buy a triangular lot measuring 510 yards by 840 yards by 1120 yards. The price of the land is $2000 per acre. How much does the land cost? (*Hint:* 1 acre = 4840 square yards)

60. Geometry You want to buy a triangular lot measuring 1350 feet by 1860 feet by 2490 feet. The price of the land is $2200 per acre. How much does the land cost? (*Hint:* 1 acre = 43,560 square feet)

Exploration

True or False? In Exercises 61 and 62, determine whether the statement is true or false. Justify your answer.

61. In Heron's Area Formula, s is the average of the lengths of the three sides of the triangle.

62. In addition to SSS and SAS, the Law of Cosines can be used to solve triangles with AAS conditions.

63. Think About It What familiar formula do you obtain when you use the standard form of the Law of Cosines, $c^2 = a^2 + b^2 - 2ab \cos C$, and you let $C = 90°$? What is the relationship between the Law of Cosines and this formula?

64. Writing Describe how the Law of Cosines can be used to solve the ambiguous case of the oblique triangle ABC, where $a = 12$ feet, $b = 30$ feet, and $A = 20°$. Is the result the same as when the Law of Sines is used to solve the triangle? Describe the advantages and the disadvantages of each method.

65. Writing In Exercise 64, the Law of Cosines was used to solve a triangle in the two-solution case of SSA. Can the Law of Cosines be used to solve the no-solution and single-solution cases of SSA? Explain.

66. HOW DO YOU SEE IT? To solve the triangle, would you begin by using the Law of Sines or the Law of Cosines? Explain.

(a)

C

$b = 16$ $a = 12$

A $c = 18$ B

(b)

A

$35°$ c

b

$55°$

C $a = 18$ B

67. Proof Use the Law of Cosines to prove each identity.

(a) $\dfrac{1}{2}bc(1 + \cos A) = \dfrac{a + b + c}{2} \cdot \dfrac{-a + b + c}{2}$

(b) $\dfrac{1}{2}bc(1 - \cos A) = \dfrac{a - b + c}{2} \cdot \dfrac{a + b - c}{2}$

6.3 Vectors in the Plane

Vectors are useful tools for modeling and solving real-life problems involving magnitude and direction. For instance, in Exercise 94 on page 428, you will use vectors to determine the speed and true direction of a commercial jet.

- ❑ **Represent vectors as directed line segments.**
- ❑ **Write component forms of vectors.**
- ❑ **Perform basic vector operations and represent vector operations graphically.**
- ❑ **Write vectors as linear combinations of unit vectors.**
- ❑ **Find direction angles of vectors.**
- ❑ **Use vectors to model and solve real-life problems.**

Introduction

Quantities such as force and velocity involve both *magnitude* and *direction* and cannot be completely characterized by a single real number. To represent such a quantity, you can use a **directed line segment,** as shown in Figure 6.10. The directed line segment $\overrightarrow{PQ}$ has **initial point** P and **terminal point** Q. Its **magnitude** (or **length**) is denoted by $\|\overrightarrow{PQ}\|$ and can be found using the Distance Formula.

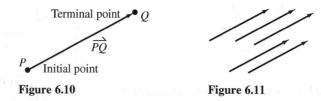

Figure 6.10   **Figure 6.11**

Two directed line segments that have the same magnitude and direction are *equivalent*. For example, the directed line segments in Figure 6.11 are all equivalent. The set of all directed line segments that are equivalent to the directed line segment $\overrightarrow{PQ}$ is a **vector v in the plane,** written $\mathbf{v} = \overrightarrow{PQ}$. Vectors are denoted by lowercase, boldface letters such as $\mathbf{u}$, $\mathbf{v}$, and $\mathbf{w}$.

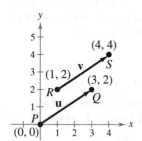

Figure 6.12

EXAMPLE 1   Showing That Two Vectors Are Equivalent

Show that $\mathbf{u}$ and $\mathbf{v}$ in Figure 6.12 are equivalent.

Solution   From the Distance Formula, $\overrightarrow{PQ}$ and $\overrightarrow{RS}$ have the *same magnitude*.

$$\|\overrightarrow{PQ}\| = \sqrt{(3 - 0)^2 + (2 - 0)^2} = \sqrt{13}$$
$$\|\overrightarrow{RS}\| = \sqrt{(4 - 1)^2 + (4 - 2)^2} = \sqrt{13}$$

Moreover, both line segments have the *same direction* because they are both directed toward the upper right on lines with a slope of

$$\frac{4 - 2}{4 - 1} = \frac{2 - 0}{3 - 0} = \frac{2}{3}.$$

Because $\overrightarrow{PQ}$ and $\overrightarrow{RS}$ have the same magnitude and direction, $\mathbf{u}$ and $\mathbf{v}$ are equivalent.

✓ *Checkpoint* 🔊)) Audio-video solution in English & Spanish at LarsonPrecalculus.com

Show that $\mathbf{u}$ and $\mathbf{v}$ in the figure at the right are equivalent.

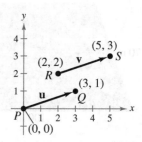

Component Form of a Vector

The directed line segment whose initial point is the origin is often the most convenient representative of a set of equivalent directed line segments. This representative of the vector **v** is in **standard position.**

A vector whose initial point is the origin $(0, 0)$ can be uniquely represented by the coordinates of its terminal point (v_1, v_2). This is the **component form of a vector v,** written as $\mathbf{v} = \langle v_1, v_2 \rangle$. The coordinates v_1 and v_2 are the *components* of **v**. If both the initial point and the terminal point lie at the origin, then **v** is the **zero vector** and is denoted by $\mathbf{0} = \langle 0, 0 \rangle$.

▷ **TECHNOLOGY** Consult the user's guide for your graphing utility for specific instructions on how to use your graphing utility to graph vectors.

> ### Component Form of a Vector
>
> The component form of the vector with initial point $P(p_1, p_2)$ and terminal point $Q(q_1, q_2)$ is given by
>
> $$\overrightarrow{PQ} = \langle q_1 - p_1, q_2 - p_2 \rangle = \langle v_1, v_2 \rangle = \mathbf{v}.$$
>
> The **magnitude** (or **length**) of **v** is given by
>
> $$\|\mathbf{v}\| = \sqrt{(q_1 - p_1)^2 + (q_2 - p_2)^2} = \sqrt{v_1^2 + v_2^2}.$$
>
> If $\|\mathbf{v}\| = 1$, then **v** is a **unit vector.** Moreover, $\|\mathbf{v}\| = 0$ if and only if **v** is the zero vector **0**.

Two vectors $\mathbf{u} = \langle u_1, u_2 \rangle$ and $\mathbf{v} = \langle v_1, v_2 \rangle$ are *equal* if and only if $u_1 = v_1$ and $u_2 = v_2$. For instance, in Example 1, the vector **u** from $P(0, 0)$ to $Q(3, 2)$ is $\mathbf{u} = \overrightarrow{PQ} = \langle 3 - 0, 2 - 0 \rangle = \langle 3, 2 \rangle$, and the vector **v** from $R(1, 2)$ to $S(4, 4)$ is $\mathbf{v} = \overrightarrow{RS} = \langle 4 - 1, 4 - 2 \rangle = \langle 3, 2 \rangle$. So, the vectors **u** and **v** in Example 1 are equal.

EXAMPLE 2 **Finding the Component Form of a Vector**

Find the component form and magnitude of the vector **v** that has initial point $(4, -7)$ and terminal point $(-1, 5)$.

Algebraic Solution

Let

$$P(4, -7) = (p_1, p_2)$$

and

$$Q(-1, 5) = (q_1, q_2).$$

Then, the components of $\mathbf{v} = \langle v_1, v_2 \rangle$ are

$$v_1 = q_1 - p_1 = -1 - 4 = -5$$

$$v_2 = q_2 - p_2 = 5 - (-7) = 12.$$

So, $\mathbf{v} = \langle -5, 12 \rangle$ and the magnitude of **v** is

$$\|\mathbf{v}\| = \sqrt{(-5)^2 + 12^2}$$

$$= \sqrt{169}$$

$$= 13.$$

Graphical Solution

Use centimeter graph paper to plot the points $P(4, -7)$ and $Q(-1, 5)$. Carefully sketch the vector **v**. Use the sketch to find the components of $\mathbf{v} = \langle v_1, v_2 \rangle$. Then use a centimeter ruler to find the magnitude of **v**. The figure at the right shows that the components of **v** are $v_1 = -5$ and $v_2 = 12$, so $\mathbf{v} = \langle -5, 12 \rangle$. The figure also shows that the magnitude of **v** is $\|\mathbf{v}\| = 13$.

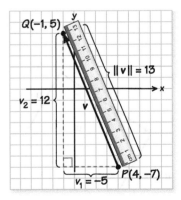

✓ *Checkpoint* ◀)) *Audio-video solution in English & Spanish at LarsonPrecalculus.com*

Find the component form and magnitude of the vector **v** that has initial point $(-2, 3)$ and terminal point $(-7, 9)$.

Vector Operations

The two basic vector operations are **scalar multiplication** and **vector addition.** In operations with vectors, numbers are usually referred to as **scalars.** In this text, scalars will always be real numbers. Geometrically, the product of a vector **v** and a scalar k is the vector that is $|k|$ times as long as **v**. When k is positive, $k\mathbf{v}$ has the same direction as **v**, and when k is negative, $k\mathbf{v}$ has the direction opposite that of **v**, as shown in Figure 6.13.

To add two vectors **u** and **v** geometrically, first position them (without changing their lengths or directions) so that the initial point of the second vector **v** coincides with the terminal point of the first vector **u**. The sum **u** + **v** is the vector formed by joining the initial point of the first vector **u** with the terminal point of the second vector **v**, as shown in the next two figures. This technique is called the **parallelogram law** for vector addition because the vector **u** + **v**, often called the **resultant** of vector addition, is the diagonal of a parallelogram with adjacent sides **u** and **v**.

Figure 6.13

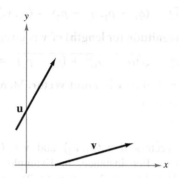
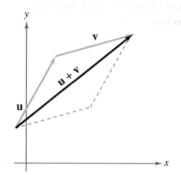

Definitions of Vector Addition and Scalar Multiplication

Let $\mathbf{u} = \langle u_1, u_2 \rangle$ and $\mathbf{v} = \langle v_1, v_2 \rangle$ be vectors and let k be a scalar (a real number). Then the **sum** of **u** and **v** is the vector

$$\mathbf{u} + \mathbf{v} = \langle u_1 + v_1, u_2 + v_2 \rangle \qquad \text{Sum}$$

and the **scalar multiple** of k times **u** is the vector

$$k\mathbf{u} = k\langle u_1, u_2 \rangle = \langle ku_1, ku_2 \rangle. \qquad \text{Scalar multiple}$$

The **negative** of $\mathbf{v} = \langle v_1, v_2 \rangle$ is

$$-\mathbf{v} = (-1)\mathbf{v}$$
$$= \langle -v_1, -v_2 \rangle \qquad \text{Negative}$$

and the **difference** of **u** and **v** is

$$\mathbf{u} - \mathbf{v} = \mathbf{u} + (-\mathbf{v}) \qquad \text{Add } (-\mathbf{v}). \text{ See Figure 6.14.}$$
$$= \langle u_1 - v_1, u_2 - v_2 \rangle. \qquad \text{Difference}$$

To represent **u** − **v** geometrically, use directed line segments with the *same* initial point. The difference **u** − **v** is the vector from the terminal point of **v** to the terminal point of **u**, which is equal to

$$\mathbf{u} + (-\mathbf{v})$$

$\mathbf{u} - \mathbf{v} = \mathbf{u} + (-\mathbf{v})$
Figure 6.14

as shown in Figure 6.14.

Example 3 illustrates the component definitions of vector addition and scalar multiplication. In this example, note the geometrical interpretations of each of the vector operations.

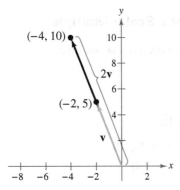

Figure 6.15

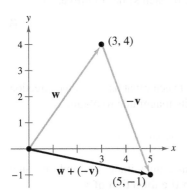

Figure 6.16

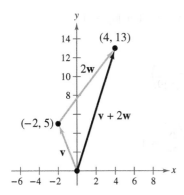

Figure 6.17

•• **REMARK** Property 9 can be stated as: The magnitude of a scalar multiple $c\mathbf{v}$ is the absolute value of c times the magnitude of $\mathbf{v}$.

EXAMPLE 3 **Vector Operations**

See LarsonPrecalculus.com for an interactive version of this type of example.

Let $\mathbf{v} = \langle -2, 5 \rangle$ and $\mathbf{w} = \langle 3, 4 \rangle$. Find each vector.

a. $2\mathbf{v}$ **b.** $\mathbf{w} - \mathbf{v}$ **c.** $\mathbf{v} + 2\mathbf{w}$

Solution

a. Multiplying $\mathbf{v} = \langle -2, 5 \rangle$ by the scalar 2, you have

$$2\mathbf{v} = 2\langle -2, 5 \rangle$$
$$= \langle 2(-2), 2(5) \rangle$$
$$= \langle -4, 10 \rangle.$$

Figure 6.15 shows a sketch of $2\mathbf{v}$.

b. The difference of $\mathbf{w}$ and $\mathbf{v}$ is

$$\mathbf{w} - \mathbf{v} = \langle 3, 4 \rangle - \langle -2, 5 \rangle$$
$$= \langle 3 - (-2), 4 - 5 \rangle$$
$$= \langle 5, -1 \rangle.$$

Figure 6.16 shows a sketch of $\mathbf{w} - \mathbf{v}$. Note that the figure shows the vector difference $\mathbf{w} - \mathbf{v}$ as the sum $\mathbf{w} + (-\mathbf{v})$.

c. The sum of $\mathbf{v}$ and $2\mathbf{w}$ is

$$\mathbf{v} + 2\mathbf{w} = \langle -2, 5 \rangle + 2\langle 3, 4 \rangle$$
$$= \langle -2, 5 \rangle + \langle 2(3), 2(4) \rangle$$
$$= \langle -2, 5 \rangle + \langle 6, 8 \rangle$$
$$= \langle -2 + 6, 5 + 8 \rangle$$
$$= \langle 4, 13 \rangle.$$

Figure 6.17 shows a sketch of $\mathbf{v} + 2\mathbf{w}$.

✓ **Checkpoint** 🔊))) *Audio-video solution in English & Spanish at LarsonPrecalculus.com*

Let $\mathbf{u} = \langle 1, 4 \rangle$ and $\mathbf{v} = \langle 3, 2 \rangle$. Find each vector.

a. $\mathbf{u} + \mathbf{v}$ **b.** $\mathbf{u} - \mathbf{v}$ **c.** $2\mathbf{u} - 3\mathbf{v}$

Vector addition and scalar multiplication share many of the properties of ordinary arithmetic.

Properties of Vector Addition and Scalar Multiplication

Let $\mathbf{u}$, $\mathbf{v}$, and $\mathbf{w}$ be vectors and let c and d be scalars. Then the properties listed below are true.

1. $\mathbf{u} + \mathbf{v} = \mathbf{v} + \mathbf{u}$

2. $(\mathbf{u} + \mathbf{v}) + \mathbf{w} = \mathbf{u} + (\mathbf{v} + \mathbf{w})$

3. $\mathbf{u} + \mathbf{0} = \mathbf{u}$

4. $\mathbf{u} + (-\mathbf{u}) = \mathbf{0}$

5. $c(d\mathbf{u}) = (cd)\mathbf{u}$

6. $(c + d)\mathbf{u} = c\mathbf{u} + d\mathbf{u}$

7. $c(\mathbf{u} + \mathbf{v}) = c\mathbf{u} + c\mathbf{v}$

8. $1(\mathbf{u}) = \mathbf{u}$, $0(\mathbf{u}) = \mathbf{0}$

9. $\|c\mathbf{v}\| = |c|\|\mathbf{v}\|$

William Rowan Hamilton (1805–1865), an Irish mathematician, did some of the earliest work with vectors. Hamilton spent many years developing a system of vector-like quantities called quaternions. Although Hamilton was convinced of the benefits of quaternions, the operations he defined did not produce good models for physical phenomena. It was not until the latter half of the nineteenth century that the Scottish physicist James Maxwell (1831–1879) restructured Hamilton's quaternions in a form that is useful for representing physical quantities such as force, velocity, and acceleration.

EXAMPLE 4   **Finding the Magnitude of a Scalar Multiple**

Let $\mathbf{u} = \langle 1, 3 \rangle$ and $\mathbf{v} = \langle -2, 5 \rangle$. Find the magnitude of each scalar multiple.

a. $\|2\mathbf{u}\|$ **b.** $\|-5\mathbf{u}\|$ **c.** $\|3\mathbf{v}\|$

Solution

a. $\|2\mathbf{u}\| = |2|\|\mathbf{u}\| = |2|\|\langle 1, 3 \rangle\| = |2|\sqrt{1^2 + 3^2} = 2\sqrt{10}$

b. $\|-5\mathbf{u}\| = |-5|\|\mathbf{u}\| = |-5|\|\langle 1, 3 \rangle\| = |-5|\sqrt{1^2 + 3^2} = 5\sqrt{10}$

c. $\|3\mathbf{v}\| = |3|\|\mathbf{v}\| = |3|\|\langle -2, 5 \rangle\| = |3|\sqrt{(-2)^2 + 5^2} = 3\sqrt{29}$

✓ *Checkpoint*))) Audio-video solution in English & Spanish at LarsonPrecalculus.com

Let $\mathbf{u} = \langle 4, -1 \rangle$ and $\mathbf{v} = \langle 3, 2 \rangle$. Find the magnitude of each scalar multiple.

a. $\|3\mathbf{u}\|$ **b.** $\|-2\mathbf{v}\|$ **c.** $\|5\mathbf{v}\|$

Unit Vectors

In many applications of vectors, it is useful to find a unit vector that has the same direction as a given nonzero vector $\mathbf{v}$. To do this, divide $\mathbf{v}$ by its magnitude to obtain

$$\mathbf{u} = \text{unit vector} = \frac{\mathbf{v}}{\|\mathbf{v}\|} = \left(\frac{1}{\|\mathbf{v}\|}\right)\mathbf{v}. \qquad \text{Unit vector in direction of } \mathbf{v}$$

Note that $\mathbf{u}$ is a scalar multiple of $\mathbf{v}$. The vector $\mathbf{u}$ has a magnitude of 1 and the same direction as $\mathbf{v}$. The vector $\mathbf{u}$ is called a **unit vector in the direction of $\mathbf{v}$.**

EXAMPLE 5   **Finding a Unit Vector**

Find a unit vector $\mathbf{u}$ in the direction of $\mathbf{v} = \langle -2, 5 \rangle$. Verify that $\|\mathbf{u}\| = 1$.

Solution   The unit vector $\mathbf{u}$ in the direction of $\mathbf{v}$ is

$$\frac{\mathbf{v}}{\|\mathbf{v}\|} = \frac{\langle -2, 5 \rangle}{\sqrt{(-2)^2 + 5^2}} = \frac{1}{\sqrt{29}}\langle -2, 5 \rangle = \left\langle \frac{-2}{\sqrt{29}}, \frac{5}{\sqrt{29}} \right\rangle.$$

This vector has a magnitude of 1 because

$$\sqrt{\left(\frac{-2}{\sqrt{29}}\right)^2 + \left(\frac{5}{\sqrt{29}}\right)^2} = \sqrt{\frac{4}{29} + \frac{25}{29}} = \sqrt{\frac{29}{29}} = 1.$$

✓ *Checkpoint*))) Audio-video solution in English & Spanish at LarsonPrecalculus.com

Find a unit vector $\mathbf{u}$ in the direction of $\mathbf{v} = \langle 6, -1 \rangle$. Verify that $\|\mathbf{u}\| = 1$.

The unit vectors $\langle 1, 0 \rangle$ and $\langle 0, 1 \rangle$ are the **standard unit vectors** and are denoted by

$$\mathbf{i} = \langle 1, 0 \rangle \quad \text{and} \quad \mathbf{j} = \langle 0, 1 \rangle$$

as shown in Figure 6.18. (Note that the lowercase letter $\mathbf{i}$ is in boldface and not italicized to distinguish it from the imaginary unit $i = \sqrt{-1}$.) These vectors can be used to represent any vector $\mathbf{v} = \langle v_1, v_2 \rangle$, because

$$\mathbf{v} = \langle v_1, v_2 \rangle = v_1\langle 1, 0 \rangle + v_2\langle 0, 1 \rangle = v_1\mathbf{i} + v_2\mathbf{j}.$$

The scalars v_1 and v_2 are the **horizontal** and **vertical components of $\mathbf{v}$,** respectively. The vector sum $v_1\mathbf{i} + v_2\mathbf{j}$ is a **linear combination** of the vectors $\mathbf{i}$ and $\mathbf{j}$. Any vector in the plane can be written as a linear combination of the standard unit vectors $\mathbf{i}$ and $\mathbf{j}$.

Figure 6.18

$\mathbf{j} = \langle 0, 1 \rangle$

$\mathbf{i} = \langle 1, 0 \rangle$

EXAMPLE 6   **Writing a Linear Combination of Unit Vectors**

Let **u** be the vector with initial point $(2, -5)$ and terminal point $(-1, 3)$. Write **u** as a linear combination of the standard unit vectors **i** and **j**.

Solution   Begin by writing the component form of the vector **u**. Then write the component form in terms of **i** and **j**.

$$\mathbf{u} = \langle -1 - 2, 3 - (-5) \rangle = \langle -3, 8 \rangle = -3\mathbf{i} + 8\mathbf{j}$$

This result is shown graphically below.

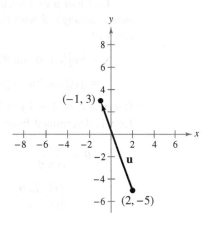

✓ **Checkpoint** 🔊))) *Audio-video solution in English & Spanish at LarsonPrecalculus.com*

Let **u** be the vector with initial point $(-2, 6)$ and terminal point $(-8, 3)$. Write **u** as a linear combination of the standard unit vectors **i** and **j**.

EXAMPLE 7   **Vector Operations**

Let $\mathbf{u} = -3\mathbf{i} + 8\mathbf{j}$ and $\mathbf{v} = 2\mathbf{i} - \mathbf{j}$. Find $2\mathbf{u} - 3\mathbf{v}$.

Solution   It is not necessary to convert **u** and **v** to component form to solve this problem. Just perform the operations with the vectors in unit vector form.

$$\begin{aligned} 2\mathbf{u} - 3\mathbf{v} &= 2(-3\mathbf{i} + 8\mathbf{j}) - 3(2\mathbf{i} - \mathbf{j}) \\ &= -6\mathbf{i} + 16\mathbf{j} - 6\mathbf{i} + 3\mathbf{j} \\ &= -12\mathbf{i} + 19\mathbf{j} \end{aligned}$$

✓ **Checkpoint** 🔊))) *Audio-video solution in English & Spanish at LarsonPrecalculus.com*

Let $\mathbf{u} = \mathbf{i} - 2\mathbf{j}$ and $\mathbf{v} = -3\mathbf{i} + 2\mathbf{j}$. Find $5\mathbf{u} - 2\mathbf{v}$.

In Example 7, you could perform the operations in component form by writing

$$\mathbf{u} = -3\mathbf{i} + 8\mathbf{j} = \langle -3, 8 \rangle \quad \text{and} \quad \mathbf{v} = 2\mathbf{i} - \mathbf{j} = \langle 2, -1 \rangle.$$

The difference of $2\mathbf{u}$ and $3\mathbf{v}$ is

$$\begin{aligned} 2\mathbf{u} - 3\mathbf{v} &= 2\langle -3, 8 \rangle - 3\langle 2, -1 \rangle \\ &= \langle -6, 16 \rangle - \langle 6, -3 \rangle \\ &= \langle -6 - 6, 16 - (-3) \rangle \\ &= \langle -12, 19 \rangle. \end{aligned}$$

Compare this result with the solution to Example 7.

Direction Angles

If **u** is a unit vector such that θ is the angle (measured counterclockwise) from the positive x-axis to **u**, then the terminal point of **u** lies on the unit circle and you have

$$\mathbf{u} = \langle x, y \rangle$$
$$= \langle \cos \theta, \sin \theta \rangle$$
$$= (\cos \theta)\mathbf{i} + (\sin \theta)\mathbf{j}$$

as shown in Figure 6.19. The angle θ is the **direction angle** of the vector **u**.

Consider a unit vector **u** with direction angle θ. If $\mathbf{v} = a\mathbf{i} + b\mathbf{j}$ is any vector that makes an angle θ with the positive x-axis, then it has the same direction as **u** and you can write

$$\mathbf{v} = \|\mathbf{v}\|\langle \cos \theta, \sin \theta \rangle$$
$$= \|\mathbf{v}\|(\cos \theta)\mathbf{i} + \|\mathbf{v}\|(\sin \theta)\mathbf{j}.$$

Because $\mathbf{v} = a\mathbf{i} + b\mathbf{j} = \|\mathbf{v}\|(\cos \theta)\mathbf{i} + \|\mathbf{v}\|(\sin \theta)\mathbf{j}$, it follows that the direction angle θ for **v** is determined from

$$\tan \theta = \frac{\sin \theta}{\cos \theta} \qquad \text{Quotient identity}$$

$$= \frac{\|\mathbf{v}\| \sin \theta}{\|\mathbf{v}\| \cos \theta} \qquad \text{Multiply numerator and denominator by } \|\mathbf{v}\|.$$

$$= \frac{b}{a}. \qquad \text{Simplify.}$$

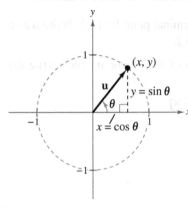

$\|\mathbf{u}\| = 1$
Figure 6.19

| EXAMPLE 8 | **Finding Direction Angles of Vectors** |

Find the direction angle of each vector.

a. $\mathbf{u} = 3\mathbf{i} + 3\mathbf{j}$ **b.** $\mathbf{v} = 3\mathbf{i} - 4\mathbf{j}$

Solution

a. The direction angle is determined from

$$\tan \theta = \frac{b}{a} = \frac{3}{3} = 1.$$

So, $\theta = 45°$, as shown in Figure 6.20.

b. The direction angle is determined from

$$\tan \theta = \frac{b}{a} = \frac{-4}{3}.$$

Moreover, $\mathbf{v} = 3\mathbf{i} - 4\mathbf{j}$ lies in Quadrant IV, so θ lies in Quadrant IV, and its reference angle is

$$\theta' = \left| \arctan\left(-\frac{4}{3}\right) \right| \approx |-0.9273 \text{ radian}| \approx |-53.13°| = 53.13°.$$

It follows that $\theta \approx 360° - 53.13° = 306.87°$, as shown in Figure 6.21.

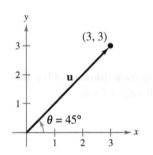

Figure 6.20

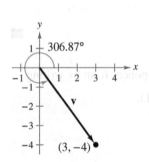

Figure 6.21

✓ *Checkpoint* ◄))) Audio-video solution in English & Spanish at LarsonPrecalculus.com

Find the direction angle of each vector.

a. $\mathbf{v} = -6\mathbf{i} + 6\mathbf{j}$ **b.** $\mathbf{v} = -7\mathbf{i} - 4\mathbf{j}$

Applications

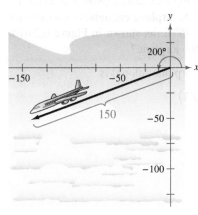

Figure 6.22

EXAMPLE 9 **Finding the Component Form of a Vector**

Find the component form of the vector that represents the velocity of an airplane descending at a speed of 150 miles per hour at an angle 20° below the horizontal, as shown in Figure 6.22.

Solution The velocity vector **v** has a magnitude of 150 and a direction angle of $\theta = 200°$.

$$\mathbf{v} = \|\mathbf{v}\|(\cos \theta)\mathbf{i} + \|\mathbf{v}\|(\sin \theta)\mathbf{j}$$
$$= 150(\cos 200°)\mathbf{i} + 150(\sin 200°)\mathbf{j}$$
$$\approx 150(-0.9397)\mathbf{i} + 150(-0.3420)\mathbf{j}$$
$$\approx -140.96\mathbf{i} - 52.30\mathbf{j}$$
$$= \langle -140.96, -52.30 \rangle$$

Check that **v** has a magnitude of 150.

$$\|\mathbf{v}\| \approx \sqrt{(-140.96)^2 + (-51.30)^2} \approx \sqrt{22{,}501.41} \approx 150 \qquad \text{Solution checks.}$$

✓ **Checkpoint**  Audio-video solution in English & Spanish at LarsonPrecalculus.com

Find the component form of the vector that represents the velocity of an airplane descending at a speed of 100 miles per hour at an angle 15° below the horizontal ($\theta = 195°$).

EXAMPLE 10 **Using Vectors to Determine Weight**

A force of 600 pounds is required to pull a boat and trailer up a ramp inclined at 15° from the horizontal. Find the combined weight of the boat and trailer.

Solution Use Figure 6.23 to make the observations below.

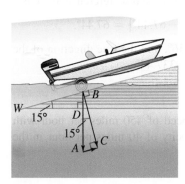

Figure 6.23

$\|\overrightarrow{BA}\|$ = force of gravity = combined weight of boat and trailer

$\|\overrightarrow{BC}\|$ = force against ramp

$\|\overrightarrow{AC}\|$ = force required to move boat up ramp = 600 pounds

Note that $\overrightarrow{AC}$ is parallel to the ramp. So, by construction, triangles *BWD* and *ABC* are similar and angle *ABC* is 15°. In triangle *ABC*, you have

$$\sin 15° = \frac{\|\overrightarrow{AC}\|}{\|\overrightarrow{BA}\|}$$
$$\sin 15° = \frac{600}{\|\overrightarrow{BA}\|}$$
$$\|\overrightarrow{BA}\| = \frac{600}{\sin 15°}$$
$$\|\overrightarrow{BA}\| \approx 2318.$$

So, the combined weight is approximately 2318 pounds.

✓ **Checkpoint**  Audio-video solution in English & Spanish at LarsonPrecalculus.com

A force of 500 pounds is required to pull a boat and trailer up a ramp inclined at 12° from the horizontal. Find the combined weight of the boat and trailer. ◼

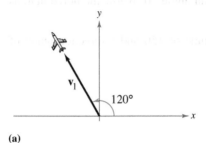

(a)

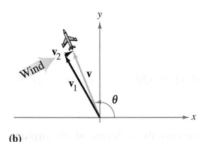

(b)
Figure 6.24

Pilots can take advantage of
fast-moving air currents called jet
streams to decrease travel time.

EXAMPLE 11   Using Vectors to Find Speed and Direction

An airplane travels at a speed of 500 miles per hour with a bearing of 330° at a fixed altitude with a negligible wind velocity, as shown in Figure 6.24(a). (Note that a bearing of 330° corresponds to a direction angle of 120°.) The airplane encounters a wind with a velocity of 70 miles per hour in the direction N 45° E, as shown in Figure 6.24(b). What are the resultant speed and true direction of the airplane?

Solution   Using Figure 6.24, the velocity of the airplane (alone) is

$$\mathbf{v}_1 = 500\langle\cos 120°, \sin 120°\rangle = \langle-250, 250\sqrt{3}\rangle$$

and the velocity of the wind is

$$\mathbf{v}_2 = 70\langle\cos 45°, \sin 45°\rangle = \langle 35\sqrt{2}, 35\sqrt{2}\rangle.$$

So, the velocity of the airplane (in the wind) is

$$\mathbf{v} = \mathbf{v}_1 + \mathbf{v}_2$$
$$= \langle-250 + 35\sqrt{2}, 250\sqrt{3} + 35\sqrt{2}\rangle$$
$$\approx \langle-200.5, 482.5\rangle$$

and the resultant speed of the airplane is

$$\|\mathbf{v}\| \approx \sqrt{(-200.5)^2 + (482.5)^2} \approx 522.5 \text{ miles per hour.}$$

To find the direction angle θ of the flight path, you have

$$\tan \theta \approx \frac{482.5}{-200.5} \approx -2.4065.$$

The flight path lies in Quadrant II, so θ lies in Quadrant II, and its reference angle is

$$\theta' \approx |\arctan(-2.4065)| \approx |-1.1770 \text{ radians}| \approx |-67.44°| = 67.44°.$$

So, the direction angle is $\theta \approx 180° - 67.44° = 112.56°$, and the true direction of the airplane is approximately $270° + (180° - 112.56°) = 337.44°$.

✓ **Checkpoint** ◀))) *Audio-video solution in English & Spanish at LarsonPrecalculus.com*

Repeat Example 11 for an airplane traveling at a speed of 450 miles per hour with a bearing of 300° that encounters a wind with a velocity of 40 miles per hour in the direction N 30° E.

Summarize   (Section 6.3)

1. Explain how to represent a vector as a directed line segment *(page 416)*. For an example involving vectors represented as directed line segments, see Example 1.

2. Explain how to find the component form of a vector *(page 417)*. For an example of finding the component form of a vector, see Example 2.

3. Explain how to perform basic vector operations *(page 418)*. For an example of performing basic vector operations, see Example 3.

4. Explain how to write a vector as a linear combination of unit vectors *(page 420)*. For examples involving unit vectors, see Examples 5–7.

5. Explain how to find the direction angle of a vector *(page 422)*. For an example of finding direction angles of vectors, see Example 8.

6. Describe real-life applications of vectors *(pages 423 and 424, Examples 9–11)*.

6.3 Exercises

Vocabulary: Fill in the blanks.

1. You can use a _____ _____ _____ to represent a quantity that involves both magnitude and direction.

2. The directed line segment $\overrightarrow{PQ}$ has _____ point P and _____ point Q.

3. The set of all directed line segments that are equivalent to a given directed line segment $\overrightarrow{PQ}$ is a _____ **v** in the plane.

4. Two vectors are equivalent when they have the same _____ and the same _____.

5. The directed line segment whose initial point is the origin is in _____ _____.

6. A vector that has a magnitude of 1 is a _____ _____.

7. The two basic vector operations are scalar _____ and vector _____.

8. The vector sum $v_1\mathbf{i} + v_2\mathbf{j}$ is a _____ _____ of the vectors **i** and **j**, and the scalars v_1 and v_2 are the _____ and _____ components of **v**, respectively.

Skills and Applications

 Determining Whether Two Vectors Are Equivalent In Exercises 9–14, determine whether **u** and **v** are equivalent. Explain.

9.
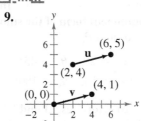

10.
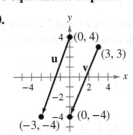

	Vector	Initial Point	Terminal Point
11.	**u**	(2, 2)	(−1, 4)
	v	(−3, −1)	(−5, 2)
12.	**u**	(2, 0)	(7, 4)
	v	(−8, 1)	(2, 9)
13.	**u**	(2, −1)	(5, −10)
	v	(6, 1)	(9, −8)
14.	**u**	(8, 1)	(13, −1)
	v	(−2, 4)	(−7, 6)

 Finding the Component Form of a Vector In Exercises 15–24, find the component form and magnitude of the vector **v**.

15.
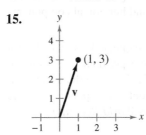

16.
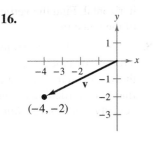

17.
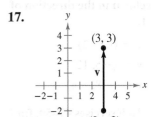

18.
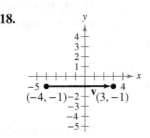

	Initial Point	Terminal Point
19.	(−3, −5)	(−11, 1)
20.	(−2, 7)	(5, −17)
21.	(1, 3)	(−8, −9)
22.	(17, −5)	(9, 3)
23.	(−1, 5)	(15, −21)
24.	(−3, 11)	(9, 40)

Sketching the Graph of a Vector In Exercises 25–30, use the figure to sketch a graph of the specified vector. To print an enlarged copy of the graph, go to *MathGraphs.com*.

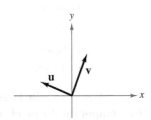

25. $-\mathbf{v}$

26. $5\mathbf{v}$

27. $\mathbf{u} + \mathbf{v}$

28. $\mathbf{u} + 2\mathbf{v}$

29. $\mathbf{u} - \mathbf{v}$

30. $\mathbf{v} - \dfrac{1}{2}\mathbf{u}$

Vector Operations In Exercises 31–36, find (a) **u** + **v**, (b) **u** − **v**, and (c) 2**u** − 3**v**. Then sketch each resultant vector.

31. $\mathbf{u} = \langle 2, 1 \rangle$, $\mathbf{v} = \langle 1, 3 \rangle$
32. $\mathbf{u} = \langle 2, 3 \rangle$, $\mathbf{v} = \langle 4, 0 \rangle$
33. $\mathbf{u} = \langle -5, 3 \rangle$, $\mathbf{v} = \langle 0, 0 \rangle$
34. $\mathbf{u} = \langle 0, 0 \rangle$, $\mathbf{v} = \langle 2, 1 \rangle$
35. $\mathbf{u} = \langle 0, -7 \rangle$, $\mathbf{v} = \langle 1, -2 \rangle$
36. $\mathbf{u} = \langle -3, 1 \rangle$, $\mathbf{v} = \langle 2, -5 \rangle$

 Finding the Magnitude of a Scalar Multiple In Exercises 37–40, find the magnitude of the scalar multiple, where $\mathbf{u} = \langle 2, 0 \rangle$ and $\mathbf{v} = \langle -3, 6 \rangle$.

37. $\|5\mathbf{u}\|$
38. $\|4\mathbf{v}\|$
39. $\|-3\mathbf{v}\|$
40. $\left\|-\frac{3}{4}\mathbf{u}\right\|$

 Finding a Unit Vector In Exercises 41–46, find a unit vector **u** in the direction of **v**. Verify that $\|\mathbf{u}\| = 1$.

41. $\mathbf{v} = \langle 3, 0 \rangle$
42. $\mathbf{v} = \langle 0, -2 \rangle$
43. $\mathbf{v} = \langle -2, 2 \rangle$
44. $\mathbf{v} = \langle -5, 12 \rangle$
45. $\mathbf{v} = \langle 1, -6 \rangle$
46. $\mathbf{v} = \langle -8, -4 \rangle$

 Finding a Vector In Exercises 47–50, find the vector **v** with the given magnitude and the same direction as **u**.

47. $\|\mathbf{v}\| = 10$, $\mathbf{u} = \langle -3, 4 \rangle$
48. $\|\mathbf{v}\| = 3$, $\mathbf{u} = \langle -12, -5 \rangle$
49. $\|\mathbf{v}\| = 9$, $\mathbf{u} = \langle 2, 5 \rangle$
50. $\|\mathbf{v}\| = 8$, $\mathbf{u} = \langle 3, 3 \rangle$

Writing a Linear Combination of Unit Vectors In Exercises 51–54, the initial and terminal points of a vector are given. Write the vector as a linear combination of the standard unit vectors **i** and **j**.

Initial Point	Terminal Point
51. $(-2, 1)$	$(3, -2)$
52. $(0, -2)$	$(3, 6)$
53. $(0, 1)$	$(-6, 4)$
54. $(2, 3)$	$(-1, -5)$

Vector Operations In Exercises 55–60, find the component form of **v** and sketch the specified vector operations geometrically, where $\mathbf{u} = 2\mathbf{i} - \mathbf{j}$ and $\mathbf{w} = \mathbf{i} + 2\mathbf{j}$.

55. $\mathbf{v} = \frac{3}{2}\mathbf{u}$
56. $\mathbf{v} = \frac{3}{4}\mathbf{w}$
57. $\mathbf{v} = \mathbf{u} + 2\mathbf{w}$
58. $\mathbf{v} = -\mathbf{u} + \mathbf{w}$
59. $\mathbf{v} = \mathbf{u} - 2\mathbf{w}$
60. $\mathbf{v} = \frac{1}{2}(3\mathbf{u} + \mathbf{w})$

 Finding the Direction Angle of a Vector In Exercises 61–64, find the magnitude and direction angle of the vector **v**.

61. $\mathbf{v} = 6\mathbf{i} - 6\mathbf{j}$
62. $\mathbf{v} = -5\mathbf{i} + 4\mathbf{j}$
63. $\mathbf{v} = 3(\cos 60°\mathbf{i} + \sin 60°\mathbf{j})$
64. $\mathbf{v} = 8(\cos 135°\mathbf{i} + \sin 135°\mathbf{j})$

Finding the Component Form of a Vector In Exercises 65–70, find the component form of **v** given its magnitude and the angle it makes with the positive x-axis. Then sketch **v**.

Magnitude	Angle
65. $\|\mathbf{v}\| = 3$	$\theta = 0°$
66. $\|\mathbf{v}\| = 4\sqrt{3}$	$\theta = 90°$
67. $\|\mathbf{v}\| = \frac{7}{2}$	$\theta = 150°$
68. $\|\mathbf{v}\| = 2\sqrt{3}$	$\theta = 45°$
69. $\|\mathbf{v}\| = 3$	**v** in the direction $3\mathbf{i} + 4\mathbf{j}$
70. $\|\mathbf{v}\| = 2$	**v** in the direction $\mathbf{i} + 3\mathbf{j}$

Finding the Component Form of a Vector In Exercises 71 and 72, find the component form of the sum of **u** and **v** with direction angles θ_u and θ_v.

71. $\|\mathbf{u}\| = 4$, $\theta_u = 60°$
$\|\mathbf{v}\| = 4$, $\theta_v = 90°$
72. $\|\mathbf{u}\| = 20$, $\theta_u = 45°$
$\|\mathbf{v}\| = 50$, $\theta_v = 180°$

Using the Law of Cosines In Exercises 73 and 74, use the Law of Cosines to find the angle α between the vectors. (Assume $0° \leq \alpha \leq 180°$.)

73. $\mathbf{v} = \mathbf{i} + \mathbf{j}$, $\mathbf{w} = 2\mathbf{i} - 2\mathbf{j}$
74. $\mathbf{v} = \mathbf{i} + 2\mathbf{j}$, $\mathbf{w} = 2\mathbf{i} - \mathbf{j}$

Resultant Force In Exercises 75 and 76, find the angle between the forces given the magnitude of their resultant. (*Hint:* Write force 1 as a vector in the direction of the positive x-axis and force 2 as a vector at an angle θ with the positive x-axis.)

	Force 1	Force 2	Resultant Force
75.	45 pounds	60 pounds	90 pounds
76.	3000 pounds	1000 pounds	3750 pounds

77. Velocity A gun with a muzzle velocity of 1200 feet per second is fired at an angle of 6° above the horizontal. Find the vertical and horizontal components of the velocity.

78. Velocity Pitcher Aroldis Chapman threw a pitch with a recorded velocity of 105 miles per hour. Assuming he threw the pitch at an angle of 3.5° below the horizontal, find the vertical and horizontal components of the velocity. (*Source: Guinness World Records*)

79. Resultant Force Forces with magnitudes of 125 newtons and 300 newtons act on a hook (see figure). The angle between the two forces is 45°. Find the direction and magnitude of the resultant of these forces. (*Hint:* Write the vector representing each force in component form, then add the vectors.)

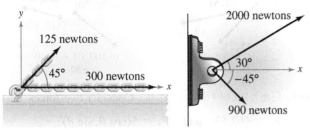

Figure for 79   Figure for 80

80. Resultant Force Forces with magnitudes of 2000 newtons and 900 newtons act on a machine part at angles of 30° and −45°, respectively, with the positive x-axis (see figure). Find the direction and magnitude of the resultant of these forces.

81. Resultant Force Three forces with magnitudes of 75 pounds, 100 pounds, and 125 pounds act on an object at angles of 30°, 45°, and 120°, respectively, with the positive x-axis. Find the direction and magnitude of the resultant of these forces.

82. Resultant Force Three forces with magnitudes of 70 pounds, 40 pounds, and 60 pounds act on an object at angles of −30°, 45°, and 135°, respectively, with the positive x-axis. Find the direction and magnitude of the resultant of these forces.

83. Cable Tension The cranes shown in the figure are lifting an object that weighs 20,240 pounds. Find the tension (in pounds) in the cable of each crane.

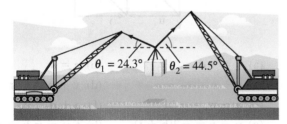

84. Cable Tension Repeat Exercise 83 for $\theta_1 = 35.6°$ and $\theta_2 = 40.4°$.

85. Rope Tension A tetherball weighing 1 pound is pulled outward from the pole by a horizontal force **u** until the rope makes a 45° angle with the pole (see figure). Determine the resulting tension (in pounds) in the rope and the magnitude of **u**.

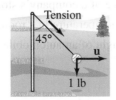

86. Physics Use the figure to determine the tension (in pounds) in each cable supporting the load.

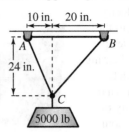

87. Tow Line Tension Two tugboats are towing a loaded barge and the magnitude of the resultant is 6000 pounds directed along the axis of the barge (see figure). Find the tension (in pounds) in the tow lines when they each make an 18° angle with the axis of the barge.

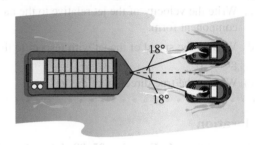

88. Rope Tension To carry a 100-pound cylindrical weight, two people lift on the ends of short ropes that are tied to an eyelet on the top center of the cylinder. Each rope makes a 20° angle with the vertical. Draw a diagram that gives a visual representation of the problem. Then find the tension (in pounds) in the ropes.

Inclined Ramp **In Exercises 89–92, a force of F pounds is required to pull an object weighing W pounds up a ramp inclined at θ degrees from the horizontal.**

89. Find F when $W = 100$ pounds and $\theta = 12°$.

90. Find W when $F = 600$ pounds and $\theta = 14°$.

91. Find θ when $F = 5000$ pounds and $W = 15,000$ pounds.

92. Find F when $W = 5000$ pounds and $\theta = 26°$.

93. Air Navigation An airplane travels in the direction of 148° with an airspeed of 875 kilometers per hour. Due to the wind, its groundspeed and direction are 800 kilometers per hour and 140°, respectively (see figure). Find the direction and speed of the wind.

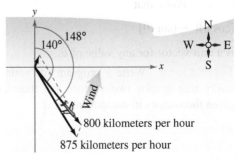

94. Air Navigation

A commercial jet travels from Miami to Seattle. The jet's velocity with respect to the air is 580 miles per hour, and its bearing is 332°. The jet encounters a wind with a velocity of 60 miles per hour from the southwest.

(a) Draw a diagram that gives a visual representation of the problem.

(b) Write the velocity of the wind as a vector in component form.

(c) Write the velocity of the jet relative to the air in component form.

(d) What is the speed of the jet with respect to the ground?

(e) What is the true direction of the jet?

Exploration

True or False?   In Exercises 95–98, determine whether the statement is true or false. Justify your answer.

95. If **u** and **v** have the same magnitude and direction, then **u** and **v** are equivalent.

96. If **u** is a unit vector in the direction of **v**, then $\mathbf{v} = \|\mathbf{v}\|\mathbf{u}$.

97. If $\mathbf{v} = a\mathbf{i} + b\mathbf{j} = \mathbf{0}$, then $a = -b$.

98. If $\mathbf{u} = a\mathbf{i} + b\mathbf{j}$ is a unit vector, then $a^2 + b^2 = 1$.

99. Error Analysis   Describe the error in finding the component form of the vector **u** that has initial point $(-3, 4)$ and terminal point $(6, -1)$.

The components are $u_1 = -3 - 6 = -9$ and $u_2 = 4 - (-1) = 5$. So, $\mathbf{u} = \langle -9, 5 \rangle$.

100. Error Analysis   Describe the error in finding the direction angle θ of the vector $\mathbf{v} = -5\mathbf{i} + 8\mathbf{j}$.

Because $\tan \theta = \dfrac{b}{a} = \dfrac{8}{-5}$, the reference angle is $\theta' = \left| \arctan\left(-\dfrac{8}{5}\right) \right| \approx |-57.99°| = 57.99°$ and $\theta \approx 360° - 57.99° = 302.01°$.

101. Proof   Prove that

$$(\cos \theta)\mathbf{i} + (\sin \theta)\mathbf{j}$$

is a unit vector for any value of θ.

102. Technology   Write a program for your graphing utility that graphs two vectors and their difference given the vectors in component form.

Finding the Difference of Two Vectors   In Exercises 103 and 104, use the program in Exercise 102 to find the difference of the vectors shown in the figure.

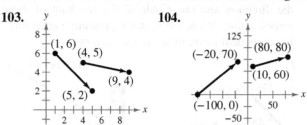

105. Graphical Reasoning   Consider two forces

$$\mathbf{F}_1 = \langle 10, 0 \rangle \quad \text{and} \quad \mathbf{F}_2 = 5\langle \cos \theta, \sin \theta \rangle.$$

(a) Find $\|\mathbf{F}_1 + \mathbf{F}_2\|$ as a function of θ.

(b) Use a graphing utility to graph the function in part (a) for $0 \le \theta < 2\pi$.

(c) Use the graph in part (b) to determine the range of the function. What is its maximum, and for what value of θ does it occur? What is its minimum, and for what value of θ does it occur?

(d) Explain why the magnitude of the resultant is never 0.

106. **HOW DO YOU SEE IT?**   Use the figure to determine whether each statement is true or false. Justify your answer.

(a) $\mathbf{a} = -\mathbf{d}$ (b) $\mathbf{c} = \mathbf{s}$

(c) $\mathbf{a} + \mathbf{u} = \mathbf{c}$ (d) $\mathbf{v} + \mathbf{w} = -\mathbf{s}$

(e) $\mathbf{a} + \mathbf{w} = -2\mathbf{d}$ (f) $\mathbf{a} + \mathbf{d} = \mathbf{0}$

(g) $\mathbf{u} - \mathbf{v} = -2(\mathbf{b} + \mathbf{t})$ (h) $\mathbf{t} - \mathbf{w} = \mathbf{b} - \mathbf{a}$

107. Writing   Give geometric descriptions of (a) vector addition and (b) scalar multiplication.

108. Writing   Identify the quantity as a scalar or as a vector. Explain.

(a) The muzzle velocity of a bullet

(b) The price of a company's stock

(c) The air temperature in a room

(d) The weight of an automobile

6.4 Vectors and Dot Products

- Find the dot product of two vectors and use the properties of the dot product.
- Find the angle between two vectors and determine whether two vectors are orthogonal.
- Write a vector as the sum of two vector components.
- Use vectors to determine the work done by a force.

The Dot Product of Two Vectors

So far, you have studied two vector operations—vector addition and multiplication by a scalar—each of which yields another vector. In this section, you will study a third vector operation, the **dot product.** This operation yields a scalar, rather than a vector.

The dot product of two vectors has many real-life applications. For example, in Exercise 74 on page 436, you will use the dot product to find the force necessary to keep a sport utility vehicle from rolling down a hill.

Definition of the Dot Product

The **dot product** of $\mathbf{u} = \langle u_1, u_2 \rangle$ and $\mathbf{v} = \langle v_1, v_2 \rangle$ is $\mathbf{u} \cdot \mathbf{v} = u_1 v_1 + u_2 v_2$.

Properties of the Dot Product

Let $\mathbf{u}$, $\mathbf{v}$, and $\mathbf{w}$ be vectors in the plane or in space and let c be a scalar.

1. $\mathbf{u} \cdot \mathbf{v} = \mathbf{v} \cdot \mathbf{u}$

2. $\mathbf{0} \cdot \mathbf{v} = 0$

3. $\mathbf{u} \cdot (\mathbf{v} + \mathbf{w}) = \mathbf{u} \cdot \mathbf{v} + \mathbf{u} \cdot \mathbf{w}$

4. $\mathbf{v} \cdot \mathbf{v} = \|\mathbf{v}\|^2$

5. $c(\mathbf{u} \cdot \mathbf{v}) = c\mathbf{u} \cdot \mathbf{v} = \mathbf{u} \cdot c\mathbf{v}$

For proofs of the properties of the dot product, see Proofs in Mathematics on page 464.

> **REMARK** In Example 1, be sure you see that the dot product of two vectors is a scalar (a real number), not a vector. Moreover, notice that the dot product can be positive, zero, or negative.

EXAMPLE 1 **Finding Dot Products**

a. $\langle 4, 5 \rangle \cdot \langle 2, 3 \rangle = 4(2) + 5(3)$

$= 8 + 15$

$= 23$

b. $\langle 2, -1 \rangle \cdot \langle 1, 2 \rangle = 2(1) + (-1)(2)$

$= 2 - 2$

$= 0$

c. $\langle 0, 3 \rangle \cdot \langle 4, -2 \rangle = 0(4) + 3(-2)$

$= 0 - 6$

$= -6$

✓ *Checkpoint* ◀))) *Audio-video solution in English & Spanish at LarsonPrecalculus.com*

Find each dot product.

a. $\langle 3, 4 \rangle \cdot \langle 2, -3 \rangle$ **b.** $\langle -3, -5 \rangle \cdot \langle 1, -8 \rangle$ **c.** $\langle -6, 5 \rangle \cdot \langle 5, 6 \rangle$

Using Properties of the Dot Product

Let $\mathbf{u} = \langle -1, 3 \rangle$, $\mathbf{v} = \langle 2, -4 \rangle$, and $\mathbf{w} = \langle 1, -2 \rangle$. Find each quantity.

a. $(\mathbf{u} \cdot \mathbf{v})\mathbf{w}$ **b.** $\mathbf{u} \cdot 2\mathbf{v}$ **c.** $\|\mathbf{u}\|$

Solution Begin by finding the dot product of $\mathbf{u}$ and $\mathbf{v}$ and the dot product of $\mathbf{u}$ and $\mathbf{u}$.

$$\mathbf{u} \cdot \mathbf{v} = \langle -1, 3 \rangle \cdot \langle 2, -4 \rangle = -1(2) + 3(-4) = -14$$

$$\mathbf{u} \cdot \mathbf{u} = \langle -1, 3 \rangle \cdot \langle -1, 3 \rangle = -1(-1) + 3(3) = 10$$

a. $(\mathbf{u} \cdot \mathbf{v})\mathbf{w} = -14\langle 1, -2 \rangle = \langle -14, 28 \rangle$

b. $\mathbf{u} \cdot 2\mathbf{v} = 2(\mathbf{u} \cdot \mathbf{v}) = 2(-14) = -28$

c. Because $\|\mathbf{u}\|^2 = \mathbf{u} \cdot \mathbf{u} = 10$, it follows that $\|\mathbf{u}\| = \sqrt{\mathbf{u} \cdot \mathbf{u}} = \sqrt{10}$.

Notice that the product in part (a) is a vector, whereas the product in part (b) is a scalar. Can you see why?

✓ **Checkpoint**)))) *Audio-video solution in English & Spanish at LarsonPrecalculus.com*

Let $\mathbf{u} = \langle 3, 4 \rangle$ and $\mathbf{v} = \langle -2, 6 \rangle$. Find each quantity.

a. $(\mathbf{u} \cdot \mathbf{v})\mathbf{v}$ **b.** $\mathbf{u} \cdot (\mathbf{u} + \mathbf{v})$ **c.** $\|\mathbf{v}\|$

The Angle Between Two Vectors

The **angle between two nonzero vectors** is the angle θ, $0 \le \theta \le \pi$, between their respective standard position vectors, as shown in Figure 6.25. This angle can be found using the dot product.

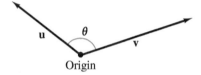

Figure 6.25

Angle Between Two Vectors

If θ is the angle between two nonzero vectors $\mathbf{u}$ and $\mathbf{v}$, then

$$\cos \theta = \frac{\mathbf{u} \cdot \mathbf{v}}{\|\mathbf{u}\| \, \|\mathbf{v}\|}.$$

For a proof of the angle between two vectors, see Proofs in Mathematics on page 464.

Finding the Angle Between Two Vectors

See LarsonPrecalculus.com for an interactive version of this type of example.

Find the angle θ between $\mathbf{u} = \langle 4, 3 \rangle$ and $\mathbf{v} = \langle 3, 5 \rangle$ (see Figure 6.26).

Solution

$$\cos \theta = \frac{\mathbf{u} \cdot \mathbf{v}}{\|\mathbf{u}\| \, \|\mathbf{v}\|} = \frac{\langle 4, 3 \rangle \cdot \langle 3, 5 \rangle}{\|\langle 4, 3 \rangle\| \, \|\langle 3, 5 \rangle\|} = \frac{4(3) + 3(5)}{\sqrt{4^2 + 3^2}\sqrt{3^2 + 5^2}} = \frac{27}{5\sqrt{34}}$$

This implies that the angle between the two vectors is

$$\theta = \cos^{-1} \frac{27}{5\sqrt{34}} \approx 0.3869 \text{ radian} \approx 22.17°. \qquad \text{Use a calculator.}$$

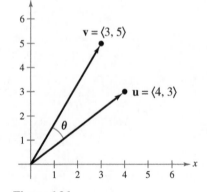

Figure 6.26

✓ **Checkpoint**)))) *Audio-video solution in English & Spanish at LarsonPrecalculus.com*

Find the angle θ between $\mathbf{u} = \langle 2, 1 \rangle$ and $\mathbf{v} = \langle 1, 3 \rangle$.

Rewriting the expression for the angle between two vectors in the form

$$\mathbf{u} \cdot \mathbf{v} = \|\mathbf{u}\| \, \|\mathbf{v}\| \cos \theta \qquad \text{Alternative form of dot product}$$

produces an alternative way to calculate the dot product. This form shows that $\mathbf{u} \cdot \mathbf{v}$ and $\cos \theta$ always have the same sign, because $\|\mathbf{u}\|$ and $\|\mathbf{v}\|$ are always positive. The figures below show the five possible orientations of two vectors.

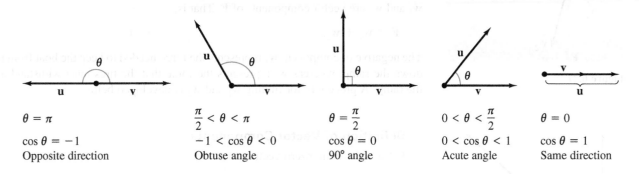

$\theta = \pi$	$\dfrac{\pi}{2} < \theta < \pi$	$\theta = \dfrac{\pi}{2}$	$0 < \theta < \dfrac{\pi}{2}$	$\theta = 0$
$\cos \theta = -1$	$-1 < \cos \theta < 0$	$\cos \theta = 0$	$0 < \cos \theta < 1$	$\cos \theta = 1$
Opposite direction	Obtuse angle	90° angle	Acute angle	Same direction

---

Definition of Orthogonal Vectors

The vectors $\mathbf{u}$ and $\mathbf{v}$ are **orthogonal** if and only if $\mathbf{u} \cdot \mathbf{v} = 0$.

The terms *orthogonal* and *perpendicular* have essentially the same meaning—meeting at right angles. Even though the angle between the zero vector and another vector is not defined, it is convenient to extend the definition of orthogonality to include the zero vector. In other words, the zero vector is orthogonal to every vector $\mathbf{u}$, because $\mathbf{0} \cdot \mathbf{u} = 0$.

▷ **TECHNOLOGY**
A graphing utility program that graphs two vectors and finds the angle between them is available at *CengageBrain.com*. Use this program, called "Finding the Angle Between Two Vectors," to verify the solutions to Examples 3 and 4.

EXAMPLE 4   **Determining Orthogonal Vectors**

Determine whether the vectors $\mathbf{u} = \langle 2, -3 \rangle$ and $\mathbf{v} = \langle 6, 4 \rangle$ are orthogonal.

Solution   Find the dot product of the two vectors.

$$\mathbf{u} \cdot \mathbf{v} = \langle 2, -3 \rangle \cdot \langle 6, 4 \rangle = 2(6) + (-3)(4) = 0$$

The dot product is 0, so the two vectors are orthogonal (see figure below).

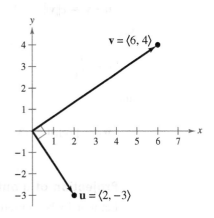

✓ *Checkpoint* ◀))) Audio-video solution in English & Spanish at LarsonPrecalculus.com

Determine whether the vectors $\mathbf{u} = \langle 6, 10 \rangle$ and $\mathbf{v} = \left\langle -\frac{1}{3}, \frac{1}{5} \right\rangle$ are orthogonal.

Finding Vector Components

You have seen applications in which you add two vectors to produce a resultant vector. Many applications in physics and engineering pose the reverse problem—decomposing a given vector into the sum of two **vector components.**

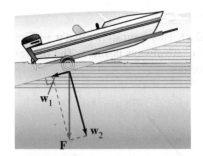

Figure 6.27

Consider a boat on an inclined ramp, as shown in Figure 6.27. The force **F** due to gravity pulls the boat *down* the ramp and *against* the ramp. These two orthogonal forces $\mathbf{w}_1$ and $\mathbf{w}_2$ are vector components of **F**. That is,

$$\mathbf{F} = \mathbf{w}_1 + \mathbf{w}_2.$$ Vector components of **F**

The negative of component $\mathbf{w}_1$ represents the force needed to keep the boat from rolling down the ramp, whereas $\mathbf{w}_2$ represents the force that the tires must withstand against the ramp. A procedure for finding $\mathbf{w}_1$ and $\mathbf{w}_2$ is developed below.

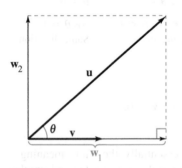

θ is acute.

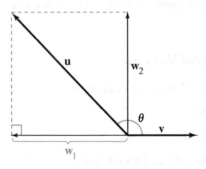

θ is obtuse.
Figure 6.28

Definition of Vector Components

Let **u** and **v** be nonzero vectors such that

$$\mathbf{u} = \mathbf{w}_1 + \mathbf{w}_2$$

where $\mathbf{w}_1$ and $\mathbf{w}_2$ are orthogonal and $\mathbf{w}_1$ is parallel to (or a scalar multiple of) **v**, as shown in Figure 6.28. The vectors $\mathbf{w}_1$ and $\mathbf{w}_2$ are **vector components** of **u**. The vector $\mathbf{w}_1$ is the **projection** of **u** onto **v** and is denoted by

$$\mathbf{w}_1 = \text{proj}_{\mathbf{v}}\mathbf{u}.$$

The vector $\mathbf{w}_2$ is given by

$$\mathbf{w}_2 = \mathbf{u} - \mathbf{w}_1.$$

To find the component $\mathbf{w}_2$, first find the projection of **u** onto **v**. To find the projection, use the dot product.

$$\mathbf{u} = \mathbf{w}_1 + \mathbf{w}_2$$

$$\mathbf{u} = c\mathbf{v} + \mathbf{w}_2 \qquad \mathbf{w}_1 \text{ is a scalar multiple of } \mathbf{v}.$$

$$\mathbf{u} \cdot \mathbf{v} = (c\mathbf{v} + \mathbf{w}_2) \cdot \mathbf{v} \qquad \text{Dot product of each side with } \mathbf{v}$$

$$\mathbf{u} \cdot \mathbf{v} = c\mathbf{v} \cdot \mathbf{v} + \mathbf{w}_2 \cdot \mathbf{v} \qquad \text{Property 3 of the dot product}$$

$$\mathbf{u} \cdot \mathbf{v} = c\|\mathbf{v}\|^2 + 0 \qquad \mathbf{w}_2 \text{ and } \mathbf{v} \text{ are orthogonal.}$$

So,

$$c = \frac{\mathbf{u} \cdot \mathbf{v}}{\|\mathbf{v}\|^2}$$

and

$$\mathbf{w}_1 = \text{proj}_{\mathbf{v}}\mathbf{u} = c\mathbf{v} = \left(\frac{\mathbf{u} \cdot \mathbf{v}}{\|\mathbf{v}\|^2}\right)\mathbf{v}.$$

Projection of u onto v

Let **u** and **v** be nonzero vectors. The projection of **u** onto **v** is given by

$$\text{proj}_{\mathbf{v}}\mathbf{u} = \left(\frac{\mathbf{u} \cdot \mathbf{v}}{\|\mathbf{v}\|^2}\right)\mathbf{v}.$$

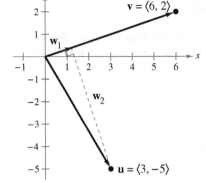

Figure 6.29

| EXAMPLE 5 | **Decomposing a Vector into Components** |

Find the projection of $\mathbf{u} = \langle 3, -5 \rangle$ onto $\mathbf{v} = \langle 6, 2 \rangle$. Then write $\mathbf{u}$ as the sum of two orthogonal vectors, one of which is $\text{proj}_{\mathbf{v}}\mathbf{u}$.

Solution The projection of $\mathbf{u}$ onto $\mathbf{v}$ is

$$\mathbf{w}_1 = \text{proj}_{\mathbf{v}}\mathbf{u} = \left(\frac{\mathbf{u} \cdot \mathbf{v}}{\|\mathbf{v}\|^2}\right)\mathbf{v} = \left(\frac{8}{40}\right)\langle 6, 2 \rangle = \left\langle \frac{6}{5}, \frac{2}{5} \right\rangle$$

as shown in Figure 6.29. The component $\mathbf{w}_2$ is

$$\mathbf{w}_2 = \mathbf{u} - \mathbf{w}_1 = \langle 3, -5 \rangle - \left\langle \frac{6}{5}, \frac{2}{5} \right\rangle = \left\langle \frac{9}{5}, -\frac{27}{5} \right\rangle.$$

So,

$$\mathbf{u} = \mathbf{w}_1 + \mathbf{w}_2 = \left\langle \frac{6}{5}, \frac{2}{5} \right\rangle + \left\langle \frac{9}{5}, -\frac{27}{5} \right\rangle = \langle 3, -5 \rangle.$$

✓ *Checkpoint* Audio-video solution in English & Spanish at LarsonPrecalculus.com

Find the projection of $\mathbf{u} = \langle 3, 4 \rangle$ onto $\mathbf{v} = \langle 8, 2 \rangle$. Then write $\mathbf{u}$ as the sum of two orthogonal vectors, one of which is $\text{proj}_{\mathbf{v}}\mathbf{u}$.

| EXAMPLE 6 | **Finding a Force** |

A 200-pound cart is on a ramp inclined at 30°, as shown in Figure 6.30. What force is required to keep the cart from rolling down the ramp?

Solution The force due to gravity is vertical and downward, so use the vector

$$\mathbf{F} = -200\mathbf{j} \qquad \text{Force due to gravity}$$

to represent the gravitational force. To find the force required to keep the cart from rolling down the ramp, project $\mathbf{F}$ onto a unit vector $\mathbf{v}$ in the direction of the ramp, where

$$\mathbf{v} = (\cos 30°)\mathbf{i} + (\sin 30°)\mathbf{j}$$

$$= \frac{\sqrt{3}}{2}\mathbf{i} + \frac{1}{2}\mathbf{j}. \qquad \text{Unit vector along ramp}$$

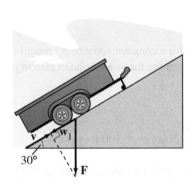

Figure 6.30

So, the projection of $\mathbf{F}$ onto $\mathbf{v}$ is

$$\mathbf{w}_1 = \text{proj}_{\mathbf{v}}\mathbf{F}$$

$$= \left(\frac{\mathbf{F} \cdot \mathbf{v}}{\|\mathbf{v}\|^2}\right)\mathbf{v}$$

$$= (\mathbf{F} \cdot \mathbf{v})\mathbf{v} \qquad \|\mathbf{v}\|^2 = 1$$

$$= (-200)\left(\frac{1}{2}\right)\mathbf{v}$$

$$= -100\left(\frac{\sqrt{3}}{2}\mathbf{i} + \frac{1}{2}\mathbf{j}\right).$$

The magnitude of this force is 100. So, a force of 100 pounds is required to keep the cart from rolling down the ramp.

✓ *Checkpoint*  Audio-video solution in English & Spanish at LarsonPrecalculus.com

Rework Example 6 for a 150-pound cart that is on a ramp inclined at 15°.

Force acts along the line of motion.
Figure 6.31

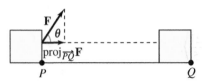

Force acts at angle θ with the line of motion.
Figure 6.32

Figure 6.33

Work

The work W done by a constant force $\mathbf{F}$ acting along the line of motion of an object is given by

$$W = (\text{magnitude of force})(\text{distance}) = \|\mathbf{F}\|\|\overrightarrow{PQ}\|$$

as shown in Figure 6.31. When the constant force $\mathbf{F}$ is *not* directed along the line of motion, as shown in Figure 6.32, the work W done by the force is given by

$$W = \|\text{proj}_{\overrightarrow{PQ}}\mathbf{F}\|\ \|\overrightarrow{PQ}\| \qquad \text{Projection form for work}$$
$$= (\cos \theta)\|\mathbf{F}\|\ \|\overrightarrow{PQ}\| \qquad \|\text{proj}_{PQ}\mathbf{F}\| = (\cos \theta)\|\mathbf{F}\|$$
$$= \mathbf{F} \cdot \overrightarrow{PQ}. \qquad \text{Alternate form of dot product}$$

The definition below summarizes the concept of work.

Definition of Work

The **work** W done by a constant force $\mathbf{F}$ as its point of application moves along the vector $\overrightarrow{PQ}$ is given by either formula below.

1. $W = \|\text{proj}_{\overrightarrow{PQ}}\mathbf{F}\|\ \|\overrightarrow{PQ}\|$ Projection form

2. $W = \mathbf{F} \cdot \overrightarrow{PQ}$ Dot product form

---

EXAMPLE 7   **Determining Work**

To close a sliding barn door, a person pulls on a rope with a constant force of 50 pounds at a constant angle of 60°, as shown in Figure 6.33. Determine the work done in moving the barn door 12 feet to its closed position.

Solution   Use a projection to find the work.

$$W = \|\text{proj}_{\overrightarrow{PQ}}\mathbf{F}\|\ \|\overrightarrow{PQ}\| = (\cos 60°)\|\mathbf{F}\|\ \|\overrightarrow{PQ}\| = \frac{1}{2}(50)(12) = 300 \text{ foot-pounds}$$

So, the work done is 300 foot-pounds. Verify this result by finding the vectors $\mathbf{F}$ and $\overrightarrow{PQ}$ and calculating their dot product.

✓ **Checkpoint**  Audio-video solution in English & Spanish at LarsonPrecalculus.com

A person pulls a wagon by exerting a constant force of 35 pounds on a handle that makes a 30° angle with the horizontal. Determine the work done in pulling the wagon 40 feet.

Work is done only when an object is moved. It does not matter how much force is applied—if an object does not move, then no work is done.

Summarize (Section 6.4)

1. State the definition of the dot product and list the properties of the dot product (*page 429*). For examples of finding dot products and using the properties of the dot product, see Examples 1 and 2.

2. Explain how to find the angle between two vectors and how to determine whether two vectors are orthogonal (*page 430*). For examples involving the angle between two vectors, see Examples 3 and 4.

3. Explain how to write a vector as the sum of two vector components (*page 432*). For examples involving vector components, see Examples 5 and 6.

4. State the definition of work (*page 434*). For an example of determining work, see Example 7.

6.4 Exercises

See CalcChat.com for tutorial help and worked-out solutions to odd-numbered exercises.

Vocabulary: Fill in the blanks.

1. The _____ _____ of two vectors yields a scalar, rather than a vector.
2. The dot product of $\mathbf{u} = \langle u_1, u_2 \rangle$ and $\mathbf{v} = \langle v_1, v_2 \rangle$ is $\mathbf{u} \cdot \mathbf{v} =$ _____.
3. If θ is the angle between two nonzero vectors $\mathbf{u}$ and $\mathbf{v}$, then $\cos \theta =$ _____.
4. The vectors $\mathbf{u}$ and $\mathbf{v}$ are _____ if and only if $\mathbf{u} \cdot \mathbf{v} = 0$.
5. The projection of $\mathbf{u}$ onto $\mathbf{v}$ is given by $\text{proj}_\mathbf{v}\mathbf{u} =$ _____.
6. The work W done by a constant force $\mathbf{F}$ as its point of application moves along the vector $\overrightarrow{PQ}$ is given by $W =$ _____ or $W =$ _____.

Skills and Applications

Finding a Dot Product In Exercises 7–12, find $\mathbf{u} \cdot \mathbf{v}$.

7. $\mathbf{u} = \langle 7, 1 \rangle$
   $\mathbf{v} = \langle -3, 2 \rangle$
8. $\mathbf{u} = \langle 6, 10 \rangle$
   $\mathbf{v} = \langle -2, 3 \rangle$
9. $\mathbf{u} = \langle -6, 2 \rangle$
   $\mathbf{v} = \langle 1, 3 \rangle$
10. $\mathbf{u} = \langle -2, 5 \rangle$
    $\mathbf{v} = \langle -1, -8 \rangle$
11. $\mathbf{u} = 4\mathbf{i} - 2\mathbf{j}$
    $\mathbf{v} = \mathbf{i} - \mathbf{j}$
12. $\mathbf{u} = \mathbf{i} - 2\mathbf{j}$
    $\mathbf{v} = -2\mathbf{i} - \mathbf{j}$

Using Properties of the Dot Product In Exercises 13–22, use the vectors $\mathbf{u} = \langle 3, 3 \rangle$, $\mathbf{v} = \langle -4, 2 \rangle$, and $\mathbf{w} = \langle 3, -1 \rangle$ to find the quantity. State whether the result is a vector or a scalar.

13. $\mathbf{u} \cdot \mathbf{u}$
14. $3\mathbf{u} \cdot \mathbf{v}$
15. $(\mathbf{u} \cdot \mathbf{v})\mathbf{v}$
16. $(\mathbf{u} \cdot 2\mathbf{v})\mathbf{w}$
17. $(\mathbf{v} \cdot \mathbf{0})\mathbf{w}$
18. $(\mathbf{u} + \mathbf{v}) \cdot \mathbf{0}$
19. $\|\mathbf{w}\| - 1$
20. $2 - \|\mathbf{u}\|$
21. $(\mathbf{u} \cdot \mathbf{v}) - (\mathbf{u} \cdot \mathbf{w})$
22. $(\mathbf{v} \cdot \mathbf{u}) - (\mathbf{w} \cdot \mathbf{v})$

Finding the Magnitude of a Vector In Exercises 23–28, use the dot product to find the magnitude of $\mathbf{u}$.

23. $\mathbf{u} = \langle -8, 15 \rangle$
24. $\mathbf{u} = \langle 4, -6 \rangle$
25. $\mathbf{u} = 20\mathbf{i} + 25\mathbf{j}$
26. $\mathbf{u} = 12\mathbf{i} - 16\mathbf{j}$
27. $\mathbf{u} = 6\mathbf{j}$
28. $\mathbf{u} = -21\mathbf{i}$

Finding the Angle Between Two Vectors In Exercises 29–38, find the angle θ (in radians) between the vectors.

29. $\mathbf{u} = \langle 1, 0 \rangle$
    $\mathbf{v} = \langle 0, -2 \rangle$
30. $\mathbf{u} = \langle 3, 2 \rangle$
    $\mathbf{v} = \langle 4, 0 \rangle$
31. $\mathbf{u} = 3\mathbf{i} + 4\mathbf{j}$
    $\mathbf{v} = -2\mathbf{j}$
32. $\mathbf{u} = 2\mathbf{i} - 3\mathbf{j}$
    $\mathbf{v} = \mathbf{i} - 2\mathbf{j}$

33. $\mathbf{u} = 2\mathbf{i} - \mathbf{j}$
    $\mathbf{v} = 6\mathbf{i} - 3\mathbf{j}$
34. $\mathbf{u} = 5\mathbf{i} + 5\mathbf{j}$
    $\mathbf{v} = -6\mathbf{i} + 6\mathbf{j}$
35. $\mathbf{u} = -6\mathbf{i} - 3\mathbf{j}$
    $\mathbf{v} = -8\mathbf{i} + 4\mathbf{j}$
36. $\mathbf{u} = 2\mathbf{i} - 3\mathbf{j}$
    $\mathbf{v} = 4\mathbf{i} + 3\mathbf{j}$
37. $\mathbf{u} = \cos\left(\dfrac{\pi}{3}\right)\mathbf{i} + \sin\left(\dfrac{\pi}{3}\right)\mathbf{j}$
    $\mathbf{v} = \cos\left(\dfrac{3\pi}{4}\right)\mathbf{i} + \sin\left(\dfrac{3\pi}{4}\right)\mathbf{j}$
38. $\mathbf{u} = \cos\left(\dfrac{\pi}{4}\right)\mathbf{i} + \sin\left(\dfrac{\pi}{4}\right)\mathbf{j}$
    $\mathbf{v} = \cos\left(\dfrac{5\pi}{4}\right)\mathbf{i} + \sin\left(\dfrac{5\pi}{4}\right)\mathbf{j}$

Finding the Angle Between Two Vectors In Exercises 39–42, find the angle θ (in degrees) between the vectors.

39. $\mathbf{u} = 3\mathbf{i} + 4\mathbf{j}$
    $\mathbf{v} = -7\mathbf{i} + 5\mathbf{j}$
40. $\mathbf{u} = 6\mathbf{i} - 3\mathbf{j}$
    $\mathbf{v} = -4\mathbf{i} - 4\mathbf{j}$
41. $\mathbf{u} = -5\mathbf{i} - 5\mathbf{j}$
    $\mathbf{v} = -8\mathbf{i} + 8\mathbf{j}$
42. $\mathbf{u} = 2\mathbf{i} - 3\mathbf{j}$
    $\mathbf{v} = 8\mathbf{i} + 3\mathbf{j}$

Finding the Angles in a Triangle In Exercises 43–46, use vectors to find the interior angles of the triangle with the given vertices.

43. $(1, 2), (3, 4), (2, 5)$
44. $(-3, -4), (1, 7), (8, 2)$
45. $(-3, 0), (2, 2), (0, 6)$
46. $(-3, 5), (-1, 9), (7, 9)$

Using the Angle Between Two Vectors In Exercises 47–50, find $\mathbf{u} \cdot \mathbf{v}$, where θ is the angle between $\mathbf{u}$ and $\mathbf{v}$.

47. $\|\mathbf{u}\| = 4, \quad \|\mathbf{v}\| = 10, \quad \theta = 2\pi/3$
48. $\|\mathbf{u}\| = 4, \quad \|\mathbf{v}\| = 12, \quad \theta = \pi/3$
49. $\|\mathbf{u}\| = 100, \quad \|\mathbf{v}\| = 250, \quad \theta = \pi/6$
50. $\|\mathbf{u}\| = 9, \quad \|\mathbf{v}\| = 36, \quad \theta = 3\pi/4$

Determining Orthogonal Vectors **In Exercises 51–56, determine whether u and v are orthogonal.**

51. $u = \langle 3, 15 \rangle$
$v = \langle -1, 5 \rangle$

52. $u = \langle 30, 12 \rangle$
$v = \langle \frac{1}{2}, -\frac{5}{4} \rangle$

53. $u = 2i - 2j$
$v = -i - j$

54. $u = \frac{1}{4}(3i - j)$
$v = 5i + 6j$

55. $u = i$
$v = -2i + 2j$

56. $u = \langle \cos\theta, \sin\theta \rangle$
$v = \langle \sin\theta, -\cos\theta \rangle$

Decomposing a Vector into Components **In Exercises 57–60, find the projection of u onto v. Then write u as the sum of two orthogonal vectors, one of which is proj$_v$u.**

57. $u = \langle 2, 2 \rangle$
$v = \langle 6, 1 \rangle$

58. $u = \langle 0, 3 \rangle$
$v = \langle 2, 15 \rangle$

59. $u = \langle 4, 2 \rangle$
$v = \langle 1, -2 \rangle$

60. $u = \langle -3, -2 \rangle$
$v = \langle -4, -1 \rangle$

Finding the Projection of u onto v **In Exercises 61–64, use the graph to find the projection of u onto v. (The terminal points of the vectors in standard position are given.) Use the formula for the projection of u onto v to verify your result.**

61.
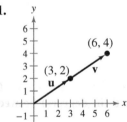

62.
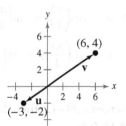

63.
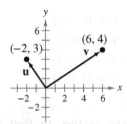

64.
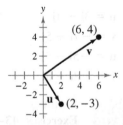

Finding Orthogonal Vectors **In Exercises 65–68, find two vectors in opposite directions that are orthogonal to the vector u. (There are many correct answers.)**

65. $u = \langle 3, 5 \rangle$

66. $u = \langle -8, 3 \rangle$

67. $u = \frac{1}{2}i - \frac{2}{3}j$

68. $u = -\frac{5}{2}i - 3j$

Work **In Exercises 69 and 70, determine the work done in moving a particle from P to Q when the magnitude and direction of the force are given by v.**

69. $P(0, 0)$, $Q(4, 7)$, $v = \langle 1, 4 \rangle$

70. $P(1, 3)$, $Q(-3, 5)$, $v = -2i + 3j$

71. Business The vector $u = \langle 1225, 2445 \rangle$ gives the numbers of hours worked by employees of a temporary work agency at two pay levels. The vector $v = \langle 12.20, 8.50 \rangle$ gives the hourly wage (in dollars) paid at each level, respectively.

(a) Find the dot product $u \cdot v$ and interpret the result in the context of the problem.

(b) Identify the vector operation used to increase wages by 2%.

72. Revenue The vector $u = \langle 3140, 2750 \rangle$ gives the numbers of hamburgers and hot dogs, respectively, sold at a fast-food stand in one month. The vector $v = \langle 2.25, 1.75 \rangle$ gives the prices (in dollars) of the food items, respectively.

(a) Find the dot product $u \cdot v$ and interpret the result in the context of the problem.

(b) Identify the vector operation used to increase the prices by 2.5%.

73. Physics A truck with a gross weight of 30,000 pounds is parked on a slope of $d°$ (see figure). Assume that the only force to overcome is the force of gravity.

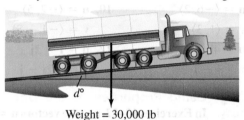

Weight = 30,000 lb

(a) Find the force required to keep the truck from rolling down the hill in terms of d.

(b) Use a graphing utility to complete the table.

d	0°	1°	2°	3°	4°	5°
Force						

d	6°	7°	8°	9°	10°
Force					

(c) Find the force perpendicular to the hill when $d = 5°$.

74. Braking Load

A sport utility vehicle with a gross weight of 5400 pounds is parked on a slope of 10°. Assume that the only force to overcome is the force of gravity. Find the force required to keep the vehicle from rolling down the hill. Find the force perpendicular to the hill.

75. Work Determine the work done by a person lifting a 245-newton bag of sugar 3 meters.

76. Work Determine the work done by a crane lifting a 2400-pound car 5 feet.

77. Work A constant force of 45 pounds, exerted at an angle of 30° with the horizontal, is required to slide a table across a floor. Determine the work done in sliding the table 20 feet.

78. Work A constant force of 50 pounds, exerted at an angle of 25° with the horizontal, is required to slide a desk across a floor. Determine the work done in sliding the desk 15 feet.

79. Work A tractor pulls a log 800 meters, and the tension in the cable connecting the tractor and the log is approximately 15,691 newtons. The direction of the constant force is 35° above the horizontal. Determine the work done in pulling the log.

80. Work One of the events in a strength competition is to pull a cement block 100 feet. One competitor pulls the block by exerting a constant force of 250 pounds on a rope attached to the block at an angle of 30° with the horizontal (see figure). Determine the work done in pulling the block.

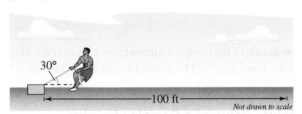

81. Work A child pulls a toy wagon by exerting a constant force of 25 pounds on a handle that makes a 20° angle with the horizontal (see figure). Determine the work done in pulling the wagon 50 feet.

82. Work A ski patroller pulls a rescue toboggan across a flat snow surface by exerting a constant force of 35 pounds on a handle that makes a 22° angle with the horizontal (see figure). Determine the work done in pulling the toboggan 200 feet.

Exploration

True or False? **In Exercises 83 and 84, determine whether the statement is true or false. Justify your answer.**

83. The work W done by a constant force **F** acting along the line of motion of an object is represented by a vector.

84. A sliding door moves along the line of vector $\overrightarrow{PQ}$. If a force is applied to the door along a vector that is orthogonal to $\overrightarrow{PQ}$, then no work is done.

Error Analysis **In Exercises 85 and 86, describe the error in finding the quantity when u = $\langle 2, -1 \rangle$ and v = $\langle -3, 5 \rangle$.**

85. $\mathbf{v} \cdot \mathbf{0} = \langle 0, 0 \rangle$ ✗

86. $\mathbf{u} \cdot 2\mathbf{v} = \langle 2, -1 \rangle \cdot \langle -6, 10 \rangle$
$$= 2(-6) - (-1)(10)$$
$$= -12 + 10$$
$$= -2$$

Finding an Unknown Vector Component **In Exercises 87 and 88, find the value of k such that vectors u and v are orthogonal.**

87. $\mathbf{u} = 8\mathbf{i} + 4\mathbf{j}$
$\mathbf{v} = 2\mathbf{i} - k\mathbf{j}$

88. $\mathbf{u} = -3k\mathbf{i} + 5\mathbf{j}$
$\mathbf{v} = 2\mathbf{i} - 4\mathbf{j}$

89. Think About It Let **u** be a unit vector. What is the value of $\mathbf{u} \cdot \mathbf{u}$? Explain.

90. 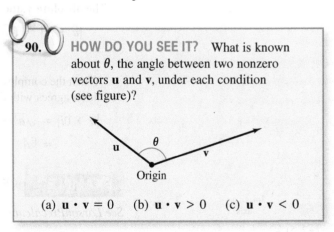 **HOW DO YOU SEE IT?** What is known about θ, the angle between two nonzero vectors **u** and **v**, under each condition (see figure)?

(a) $\mathbf{u} \cdot \mathbf{v} = 0$ (b) $\mathbf{u} \cdot \mathbf{v} > 0$ (c) $\mathbf{u} \cdot \mathbf{v} < 0$

91. Think About It What can be said about the vectors **u** and **v** under each condition?

(a) The projection of **u** onto **v** equals **u**.

(b) The projection of **u** onto **v** equals **0**.

92. Proof Use vectors to prove that the diagonals of a rhombus are perpendicular.

93. Proof Prove that

$$\|\mathbf{u} - \mathbf{v}\|^2 = \|\mathbf{u}\|^2 + \|\mathbf{v}\|^2 - 2\mathbf{u} \cdot \mathbf{v}.$$

6.5 The Complex Plane

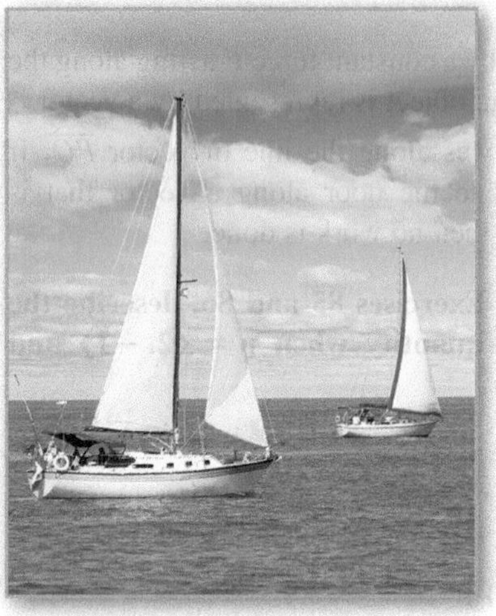

The complex plane has many practical applications. For example, in Exercise 49 on page 444, you will use the complex plane to write complex numbers that represent the positions of two ships.

- **Plot complex numbers in the complex plane and find absolute values of complex numbers.**
- **Perform operations with complex numbers in the complex plane.**
- **Use the Distance and Midpoint Formulas in the complex plane.**

The Complex Plane

Just as a real number can be represented by a point on the real number line, a complex number $z = a + bi$ can be represented by the point (a, b) in a coordinate plane (the **complex plane**). In the complex plane, the horizontal axis is the **real axis** and the vertical axis is the **imaginary axis,** as shown in the figure below.

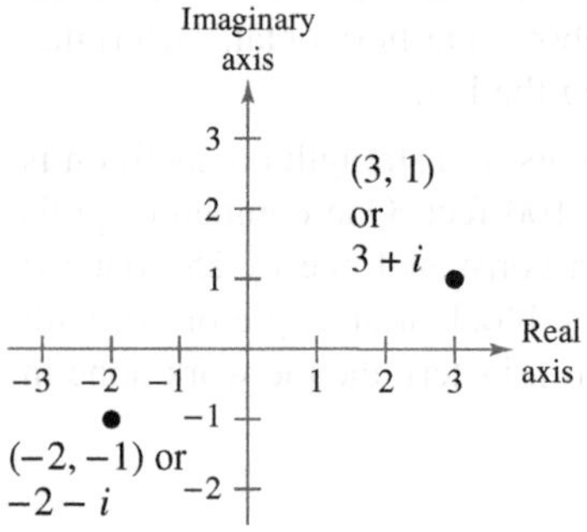

The **absolute value,** or **modulus,** of the complex number $z = a + bi$ is the distance between the origin $(0, 0)$ and the point (a, b). (The plural of modulus is *moduli.*)

Definition of the Absolute Value of a Complex Number

The **absolute value** of the complex number $z = a + bi$ is

$$|a + bi| = \sqrt{a^2 + b^2}.$$

When the complex number $z = a + bi$ is a real number (that is, when $b = 0$), this definition agrees with that given for the absolute value of a real number

$$|a + 0i| = \sqrt{a^2 + 0^2}$$
$$= |a|.$$

EXAMPLE 1 Finding the Absolute Value of a Complex Number

See LarsonPrecalculus.com for an interactive version of this type of example.

Plot $z = -2 + 5i$ in the complex plane and find its absolute value.

Solution The number is plotted in Figure 6.34. It has an absolute value of

$$|z| = \sqrt{(-2)^2 + 5^2}$$
$$= \sqrt{29}.$$

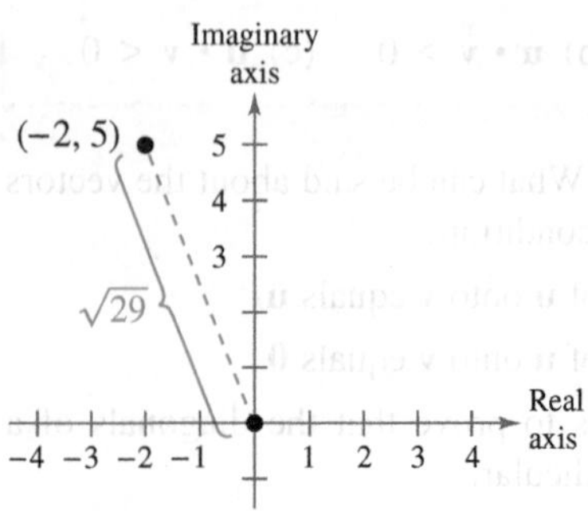

Figure 6.34

✓ ***Checkpoint*** *Audio-video solution in English & Spanish at LarsonPrecalculus.com*

Plot $z = 3 - 4i$ in the complex plane and find its absolute value.

Michael C. Gray/Shutterstock.com

Operations with Complex Numbers in the Complex Plane

In Section 6.3, you learned how to add and subtract vectors geometrically in the coordinate plane. In a similar way, you can add and subtract complex numbers geometrically in the complex plane.

The complex number $z = a + bi$ can be represented by the vector $\mathbf{u} = \langle a, b \rangle$. For example, the complex number $z = 1 + 2i$ can be represented by the vector $\mathbf{u} = \langle 1, 2 \rangle$. To add two complex numbers geometrically, first represent them as vectors $\mathbf{u}$ and $\mathbf{v}$. Then add the vectors, as shown in the next two figures. The sum of the vectors represents the sum of the complex numbers.

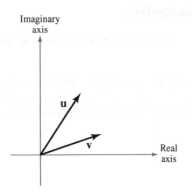
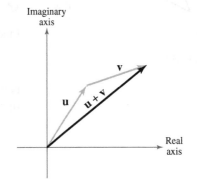

EXAMPLE 2 Adding in the Complex Plane

Find $(1 + 3i) + (2 + i)$ in the complex plane.

Solution

Let the vectors $\mathbf{u} = \langle 1, 3 \rangle$ and $\mathbf{v} = \langle 2, 1 \rangle$ represent the complex numbers $1 + 3i$ and $2 + i$, respectively. Graph the vectors $\mathbf{u}$, $\mathbf{v}$, and $\mathbf{u} + \mathbf{v}$, as shown at the right. From the graph, $\mathbf{u} + \mathbf{v} = \langle 3, 4 \rangle$, which implies that

$$(1 + 3i) + (2 + i) = 3 + 4i.$$

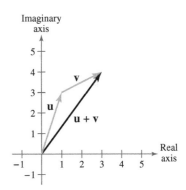

✓ **Checkpoint** ◀))) *Audio-video solution in English & Spanish at LarsonPrecalculus.com*

Find $(3 + i) + (1 + 2i)$ in the complex plane.

To subtract two complex numbers geometrically, first represent them as vectors $\mathbf{u}$ and $\mathbf{v}$. Then subtract the vectors, as shown in the figure below. The difference of the vectors represents the difference of the complex numbers.

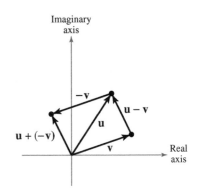

Imaginary
axis

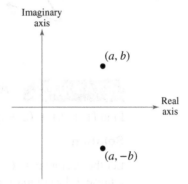

Figure 6.35

EXAMPLE 3 **Subtracting in the Complex Plane**

Find $(4 + 2i) - (3 - i)$ in the complex plane.

Solution

Let the vectors $\mathbf{u} = \langle 4, 2 \rangle$ and $\mathbf{v} = \langle 3, -1 \rangle$ represent the complex numbers $4 + 2i$ and $3 - i$, respectively. Graph the vectors $\mathbf{u}$, $-\mathbf{v}$, and $\mathbf{u} + (-\mathbf{v})$, as shown in Figure 6.35. From the graph, $\mathbf{u} - \mathbf{v} = \mathbf{u} + (-\mathbf{v}) = \langle 1, 3 \rangle$, which implies that

$$(4 + 2i) - (3 - i) = 1 + 3i.$$

✓ *Checkpoint* ◄))) *Audio-video solution in English & Spanish at LarsonPrecalculus.com*

Find $(2 - 4i) - (1 + i)$ in the complex plane.

Recall that the complex numbers $a + bi$ and $a - bi$ are *complex conjugates*. The points (a, b) and $(a, -b)$ are reflections of each other in the real axis, as shown in the figure below. This information enables you to find a complex conjugate geometrically.

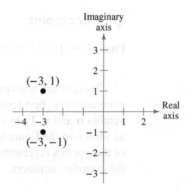

EXAMPLE 4 **Complex Conjugates in the Complex Plane**

Plot $z = -3 + i$ and its complex conjugate in the complex plane. Write the conjugate as a complex number.

Solution

The figure below shows the point $(-3, 1)$ and its reflection in the real axis, $(-3, -1)$. So, the complex conjugate of $-3 + i$ is $-3 - i$.

✓ *Checkpoint* ◄))) *Audio-video solution in English & Spanish at LarsonPrecalculus.com*

Plot $z = 2 - 3i$ and its complex conjugate in the complex plane. Write the conjugate as a complex number.

Distance and Midpoint Formulas in the Complex Plane

For two points in the complex plane, the distance between the points is the modulus (or absolute value) of the difference of the two corresponding complex numbers. Let (a, b) and (s, t) be points in the complex plane. One way to write the difference of the corresponding complex numbers is $(s + ti) - (a + bi) = (s - a) + (t - b)i$. The modulus of the difference is

$$|(s - a) + (t - b)i| = \sqrt{(s - a)^2 + (t - b)^2}.$$

So, $d = \sqrt{(s - a)^2 + (t - b)^2}$ is the distance between the points in the complex plane.

Imaginary axis

Real axis

Figure 6.36

> ### Distance Formula in the Complex Plane
> The distance d between the points (a, b) and (s, t) in the complex plane is
> $$d = \sqrt{(s - a)^2 + (t - b)^2}.$$

Figure 6.36 shows the points represented as vectors. The magnitude of the vector $\mathbf{u} - \mathbf{v}$ is the distance between (a, b) and (s, t).

$$\mathbf{u} - \mathbf{v} = \langle s - a, t - b \rangle$$
$$\|\mathbf{u} - \mathbf{v}\| = \sqrt{(s - a)^2 + (t - b)^2}$$

EXAMPLE 5   **Finding Distance in the Complex Plane**

Find the distance between $2 + 3i$ and $5 - 2i$ in the complex plane.

Solution

Let $a + bi = 2 + 3i$ and $s + ti = 5 - 2i$. The distance is

$$d = \sqrt{(s - a)^2 + (t - b)^2}$$
$$= \sqrt{(5 - 2)^2 + (-2 - 3)^2}$$
$$= \sqrt{3^2 + (-5)^2}$$
$$= \sqrt{34}$$
$$\approx 5.83 \text{ units}$$

as shown in the figure below.

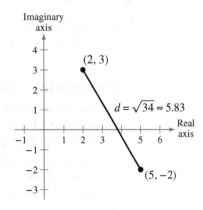

✓ *Checkpoint* ◀))) *Audio-video solution in English & Spanish at LarsonPrecalculus.com*

Find the distance between $5 - 4i$ and $6 + 5i$ in the complex plane.

To find the midpoint of the line segment joining two points in the complex plane, find the average values of the respective coordinates of the two endpoints.

> ### Midpoint Formula in the Complex Plane
>
> The midpoint of the line segment joining the points (a, b) and (s, t) in the complex plane is
>
> $$\text{Midpoint} = \left(\frac{a + s}{2}, \frac{b + t}{2} \right).$$

EXAMPLE 6 **Finding a Midpoint in the Complex Plane**

Find the midpoint of the line segment joining the points corresponding to $4 - 3i$ and $2 + 2i$ in the complex plane.

Solution

Let the points $(4, -3)$ and $(2, 2)$ represent the complex numbers $4 - 3i$ and $2 + 2i$, respectively. Apply the Midpoint Formula.

$$\text{Midpoint} = \left(\frac{a + s}{2}, \frac{b + t}{2} \right) = \left(\frac{4 + 2}{2}, \frac{-3 + 2}{2} \right) = \left(3, -\frac{1}{2} \right)$$

The midpoint is $\left(3, -\frac{1}{2} \right)$, as shown in the figure below.

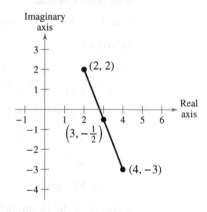

✓ *Checkpoint* ◄))) *Audio-video solution in English & Spanish at LarsonPrecalculus.com*

Find the midpoint of the line segment joining the points corresponding to $2 + i$ and $5 - 5i$ in the complex plane.

> ### Summarize (Section 6.5)
>
> 1. State the definition of the absolute value, or modulus, of a complex number *(page 438)*. For an example of finding the absolute value of a complex number, see Example 1.
> 2. Explain how to add, subtract, and find complex conjugates of complex numbers in the complex plane *(page 439)*. For examples of performing operations with complex numbers in the complex plane, see Examples 2–4.
> 3. Explain how to use the Distance and Midpoint Formulas in the complex plane *(page 441)*. For examples of using the Distance and Midpoint Formulas in the complex plane, see Examples 5 and 6.

6.5 Exercises

See **CalcChat.com** for tutorial help and worked-out solutions to odd-numbered exercises.

Vocabulary: Fill in the blanks.

1. In the complex plane, the horizontal axis is the _____ axis.
2. In the complex plane, the vertical axis is the _____ axis.
3. The _____ _____ of the complex number $a + bi$ is the distance between the origin and (a, b).
4. To subtract two complex numbers geometrically, first represent them as _____.
5. The points that represent a complex number and its complex conjugate are _____ of each other in the real axis.
6. The distance between two points in the complex plane is the _____ of the difference of the two corresponding complex numbers.

Skills and Applications

Matching In Exercises 7–14, match the complex number with its representation in the complex plane. [The representations are labeled (a)–(h).]

(a)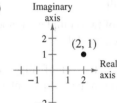

(b)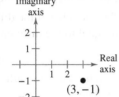

(c)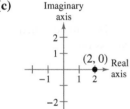

(d)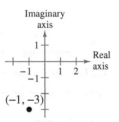

(e)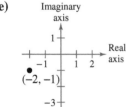

(f)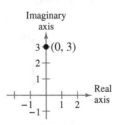

(g)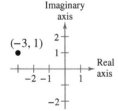

(h)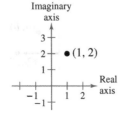

7. 2
8. $3i$
9. $1 + 2i$
10. $2 + i$
11. $3 - i$
12. $-3 + i$
13. $-2 - i$
14. $-1 - 3i$

 Finding the Absolute Value of a Complex Number In Exercises 15–20, plot the complex number and find its absolute value.

15. $-7i$
16. -7
17. $-6 + 8i$
18. $5 - 12i$
19. $4 - 6i$
20. $-8 + 3i$

 Adding in the Complex Plane In Exercises 21–28, find the sum of the complex numbers in the complex plane.

21. $(3 + i) + (2 + 5i)$
22. $(5 + 2i) + (3 + 4i)$
23. $(8 - 2i) + (2 + 6i)$
24. $(3 - i) + (-1 + 2i)$
25.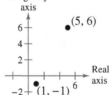
26.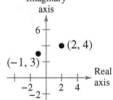
27.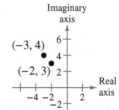
28.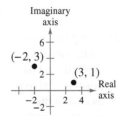

 Subtracting in the Complex Plane In Exercises 29–36, find the difference of the complex numbers in the complex plane.

29. $(4 + 2i) - (6 + 4i)$
30. $(-3 + i) - (3 + i)$
31. $(5 - i) - (-5 + 2i)$
32. $(2 - 3i) - (3 + 2i)$
33. $2 - (2 + 6i)$
34. $-3 - (2 + 2i)$
35. $-2i - (3 - 5i)$
36. $3i - (-3 + 7i)$

Complex Conjugates in the Complex Plane **In Exercises 37–40, plot the complex number and its complex conjugate. Write the conjugate as a complex number.**

37. $2 + 3i$

38. $5 - 4i$

39. $-1 - 2i$

40. $-7 + 3i$

Finding Distance in the Complex Plane **In Exercises 41–44, find the distance between the complex numbers in the complex plane.**

41. $1 + 2i, -1 + 4i$

42. $-5 + i, -2 + 5i$

43. $6i, 3 - 4i$

44. $-7 - 3i, 3 + 5i$

Finding a Midpoint in the Complex Plane **In Exercises 45–48, find the midpoint of the line segment joining the points corresponding to the complex numbers in the complex plane.**

45. $2 + i, 6 + 5i$

46. $-3 + 4i, 1 - 2i$

47. $7i, 9 - 10i$

48. $-1 - \frac{3}{4}i, \frac{1}{2} + \frac{1}{4}i$

49. Sailing

Ship A is 3 miles east and 4 miles north of port. Ship B is 5 miles west and 2 miles north of port (see figure).

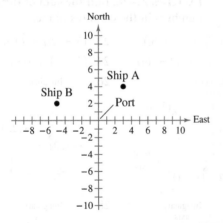

(a) Using the positive imaginary axis as north and the positive real axis as east, write complex numbers that represent the positions of Ship A and Ship B relative to port.

(b) How can you use the complex numbers in part (a) to find the distance between Ship A and Ship B?

50. Force Two forces are acting on a point. The first force has a horizontal component of 5 newtons and a vertical component of 3 newtons. The second force has a horizontal component of 4 newtons and a vertical component of 2 newtons.

(a) Plot the vectors that represent the two forces in the complex plane.

(b) Find the horizontal and vertical components of the resultant force acting on the point using the complex plane.

Exploration

True or False? **In Exercises 51–54, determine whether the statement is true or false. Justify your answer.**

51. The modulus of a complex number can be real or imaginary.

52. The distance between two points in the complex plane is always real.

53. The modulus of the sum of two complex numbers is equal to the sum of their moduli.

54. The modulus of the difference of two complex numbers is equal to the difference of their moduli.

55. Think About It What does the set of all points with the same modulus represent in the complex plane? Explain.

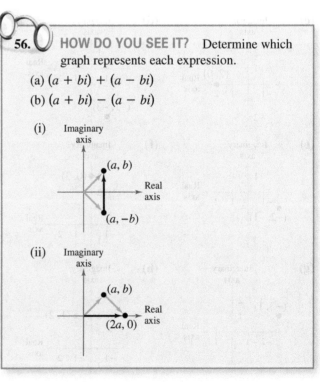

56. **HOW DO YOU SEE IT?** Determine which graph represents each expression.

(a) $(a + bi) + (a - bi)$

(b) $(a + bi) - (a - bi)$

(i)   Imaginary axis

(a, b)

Real axis

$(a, -b)$

(ii)   Imaginary axis

(a, b)

Real axis

$(2a, 0)$

57. Think About It The points corresponding to a complex number and its complex conjugate are plotted in the complex plane. What type of triangle do these points form with the origin?

6.6 Trigonometric Form of a Complex Number

Trigonometric forms of complex numbers have applications in circuit analysis. For example, in Exercise 95 on page 453, you will use trigonometric forms of complex numbers to find the voltage of an alternating current circuit.

- Write trigonometric forms of complex numbers.
- Multiply and divide complex numbers written in trigonometric form.
- Use DeMoivre's Theorem to find powers of complex numbers.
- Find nth roots of complex numbers.

Trigonometric Form of a Complex Number

In Section 2.4, you learned how to add, subtract, multiply, and divide complex numbers. To work effectively with *powers* and *roots* of complex numbers, it is helpful to write complex numbers in trigonometric form. Consider the nonzero complex number $a + bi$, plotted at the right. By letting θ be the angle from the positive real axis (measured counterclockwise) to the line segment connecting the origin and the point (a, b), you can write $a = r \cos \theta$ and $b = r \sin \theta$, where $r = \sqrt{a^2 + b^2}$. Consequently, you have $a + bi = (r \cos \theta) + (r \sin \theta)i$, from which you can obtain the **trigonometric form of a complex number.**

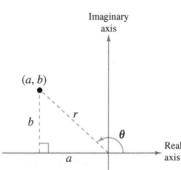

Trigonometric Form of a Complex Number

The **trigonometric form** of the complex number $z = a + bi$ is

$$z = r(\cos \theta + i \sin \theta)$$

where $a = r \cos \theta$, $b = r \sin \theta$, $r = \sqrt{a^2 + b^2}$, and $\tan \theta = b/a$. The number r is the **modulus** of z, and θ is an **argument** of z.

· · REMARK   For $0 \le \theta < 2\pi$,
use the guidelines below.
When z lies in Quadrant I,
$\theta = \arctan(b/a)$. When z lies
in Quadrant II or Quadrant III,
$\theta = \pi + \arctan(b/a)$.
When z lies in Quadrant IV,
$\theta = 2\pi + \arctan(b/a)$.

The trigonometric form of a complex number is also called the *polar form*. There are infinitely many choices for θ, so the trigonometric form of a complex number is not unique. Normally, θ is restricted to the interval $0 \le \theta < 2\pi$, although on occasion it is convenient to use $\theta < 0$.

EXAMPLE 1   Trigonometric Form of a Complex Number

Write the complex number $z = -2 - 2\sqrt{3}i$ in trigonometric form.

Solution   The modulus of z is $r = \sqrt{(-2)^2 + (-2\sqrt{3})^2} = \sqrt{16} = 4$, and the argument θ is determined from

$$\tan \theta = \frac{b}{a} = \frac{-2\sqrt{3}}{-2} = \sqrt{3}.$$

Because $z = -2 - 2\sqrt{3}i$ lies in Quadrant III, as shown in Figure 6.37, you have $\theta = \pi + \arctan\sqrt{3} = \pi + (\pi/3) = 4\pi/3$. So, the trigonometric form of z is

$$z = r(\cos \theta + i \sin \theta) = 4\left(\cos \frac{4\pi}{3} + i \sin \frac{4\pi}{3}\right).$$

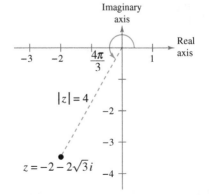

Figure 6.37

✓ **Checkpoint** ◀))) *Audio-video solution in English & Spanish at LarsonPrecalculus.com*

Write the complex number $z = 6 - 6i$ in trigonometric form.

| EXAMPLE 2 | **Writing a Complex Number in Standard Form** |

Write $z = \sqrt{8}[\cos(-\pi/3) + i \sin(-\pi/3)]$ in standard form $a + bi$.

Solution   Because $\cos(-\pi/3) = 1/2$ and $\sin(-\pi/3) = -\sqrt{3}/2$, you can write

$$z = \sqrt{8}\left[\cos\left(-\frac{\pi}{3}\right) + i \sin\left(-\frac{\pi}{3}\right)\right] = 2\sqrt{2}\left(\frac{1}{2} - \frac{\sqrt{3}}{2}i\right) = \sqrt{2} - \sqrt{6}i.$$

✓ **Checkpoint** ◀))) *Audio-video solution in English & Spanish at LarsonPrecalculus.com*

Write $z = 8[\cos(2\pi/3) + i \sin(2\pi/3)]$ in standard form $a + bi$. ∎

Multiplication and Division of Complex Numbers

The trigonometric form adapts nicely to multiplication and division of complex numbers. Consider two complex numbers $z_1 = r_1(\cos \theta_1 + i \sin \theta_1)$ and $z_2 = r_2(\cos \theta_2 + i \sin \theta_2)$. The product of z_1 and z_2 is

$$z_1 z_2 = r_1 r_2 (\cos \theta_1 + i \sin \theta_1)(\cos \theta_2 + i \sin \theta_2)$$
$$= r_1 r_2 [(\cos \theta_1 \cos \theta_2 - \sin \theta_1 \sin \theta_2) + i(\sin \theta_1 \cos \theta_2 + \cos \theta_1 \sin \theta_2)].$$

Using the sum and difference formulas for cosine and sine, this equation is equivalent to

$$z_1 z_2 = r_1 r_2 [\cos(\theta_1 + \theta_2) + i \sin(\theta_1 + \theta_2)].$$

This establishes the first part of the rule below. The second part is left for you to verify (see Exercise 99).

▷

Product and Quotient of Two Complex Numbers

Let $z_1 = r_1(\cos \theta_1 + i \sin \theta_1)$ and $z_2 = r_2(\cos \theta_2 + i \sin \theta_2)$ be complex numbers.

$$z_1 z_2 = r_1 r_2 [\cos(\theta_1 + \theta_2) + i \sin(\theta_1 + \theta_2)] \qquad \text{Product}$$

$$\frac{z_1}{z_2} = \frac{r_1}{r_2}[\cos(\theta_1 - \theta_2) + i \sin(\theta_1 - \theta_2)], \quad z_2 \neq 0 \qquad \text{Quotient}$$

| EXAMPLE 3 | **Multiplying Complex Numbers** |

Find the product $z_1 z_2$ of $z_1 = 3\left(\cos \dfrac{\pi}{4} + i \sin \dfrac{\pi}{4}\right)$ and $z_2 = 2\left(\cos \dfrac{3\pi}{4} + i \sin \dfrac{3\pi}{4}\right)$.

Solution

$$z_1 z_2 = 3\left(\cos \frac{\pi}{4} + i \sin \frac{\pi}{4}\right) \cdot 2\left(\cos \frac{3\pi}{4} + i \sin \frac{3\pi}{4}\right)$$

$$= 6\left[\cos\left(\frac{\pi}{4} + \frac{3\pi}{4}\right) + i \sin\left(\frac{\pi}{4} + \frac{3\pi}{4}\right)\right] \qquad \text{Multiply moduli and add arguments.}$$

$$= 6(\cos \pi + i \sin \pi)$$

$$= 6[-1 + i(0)]$$

$$= -6$$

✓ **Checkpoint** ◀))) *Audio-video solution in English & Spanish at LarsonPrecalculus.com*

Find the product $z_1 z_2$ of $z_1 = 2\left(\cos \dfrac{5\pi}{6} + i \sin \dfrac{5\pi}{6}\right)$ and $z_2 = 5\left(\cos \dfrac{7\pi}{6} + i \sin \dfrac{7\pi}{6}\right)$. ∎

EXAMPLE 4 **Multiplying Complex Numbers**

$$2\left(\cos\frac{2\pi}{3} + i\sin\frac{2\pi}{3}\right) \cdot 8\left(\cos\frac{11\pi}{6} + i\sin\frac{11\pi}{6}\right)$$

$$= 16\left[\cos\left(\frac{2\pi}{3} + \frac{11\pi}{6}\right) + i\sin\left(\frac{2\pi}{3} + \frac{11\pi}{6}\right)\right]$$ Multiply moduli and add arguments.

$$= 16\left(\cos\frac{5\pi}{2} + i\sin\frac{5\pi}{2}\right)$$

$$= 16\left(\cos\frac{\pi}{2} + i\sin\frac{\pi}{2}\right)$$ $\frac{5\pi}{2}$ and $\frac{\pi}{2}$ are coterminal.

$$= 16i$$

✓ **Checkpoint**))) Audio-video solution in English & Spanish at LarsonPrecalculus.com

Find the product z_1z_2 of $z_1 = 3\left(\cos\frac{\pi}{3} + i\sin\frac{\pi}{3}\right)$ and $z_2 = 4\left(\cos\frac{\pi}{6} + i\sin\frac{\pi}{6}\right)$.

EXAMPLE 5 **Dividing Complex Numbers**

$$\frac{24(\cos 300° + i\sin 300°)}{8(\cos 75° + i\sin 75°)} = 3[\cos(300° - 75°) + i\sin(300° - 75°)]$$

$$= 3(\cos 225° + i\sin 225°)$$

$$= -\frac{3\sqrt{2}}{2} - \frac{3\sqrt{2}}{2}i$$

✓ **Checkpoint**))) Audio-video solution in English & Spanish at LarsonPrecalculus.com

Find the quotient z_1/z_2 of $z_1 = \cos 40° + i\sin 40°$ and $z_2 = \cos 10° + i\sin 10°$. ■

In Section 6.5, you added, subtracted, and found complex conjugates of complex numbers geometrically in the complex plane. In a similar way, you can multiply complex numbers geometrically in the complex plane.

EXAMPLE 6 **Multiplying in the Complex Plane**

Find the product z_1z_2 of $z_1 = 2\left(\cos\frac{\pi}{6} + i\sin\frac{\pi}{6}\right)$ and $z_2 = 2\left(\cos\frac{\pi}{3} + i\sin\frac{\pi}{3}\right)$ in the complex plane.

Solution

Let $\mathbf{u} = 2\langle\cos(\pi/6), \sin(\pi/6)\rangle = \langle\sqrt{3}, 1\rangle$ and $\mathbf{v} = 2\langle\cos(\pi/3), \sin(\pi/3)\rangle = \langle1, \sqrt{3}\rangle$. Then $\|\mathbf{u}\| = \sqrt{(\sqrt{3})^2 + 1^2} = \sqrt{4} = 2$ and $\|\mathbf{v}\| = \sqrt{1^2 + (\sqrt{3})^2} = \sqrt{4} = 2$. So, the magnitude of the product vector is $2(2) = 4$. The sum of the direction angles is $(\pi/6) + (\pi/3) = \pi/2$. So, the product vector lies on the imaginary axis and is represented in vector form as $\langle0, 4\rangle$, as shown in Figure 6.38. This implies that $z_1z_2 = 4i$.

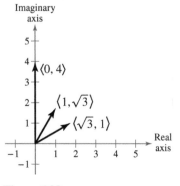

Figure 6.38

✓ **Checkpoint**))) Audio-video solution in English & Spanish at LarsonPrecalculus.com

Find the product z_1z_2 of $z_1 = 2\left(\cos\frac{\pi}{4} + i\sin\frac{\pi}{4}\right)$ and $z_2 = 4\left(\cos\frac{3\pi}{4} + i\sin\frac{3\pi}{4}\right)$ in the complex plane. ■

Powers of Complex Numbers

The trigonometric form of a complex number is used to raise a complex number to a power. To accomplish this, consider repeated use of the multiplication rule.

$$z = r(\cos \theta + i \sin \theta)$$

$$z^2 = r(\cos \theta + i \sin \theta)r(\cos \theta + i \sin \theta) = r^2(\cos 2\theta + i \sin 2\theta)$$

$$z^3 = r^2(\cos 2\theta + i \sin 2\theta)r(\cos \theta + i \sin \theta) = r^3(\cos 3\theta + i \sin 3\theta)$$

$$z^4 = r^4(\cos 4\theta + i \sin 4\theta)$$

$$z^5 = r^5(\cos 5\theta + i \sin 5\theta)$$

$$\vdots$$

This pattern leads to **DeMoivre's Theorem,** which is named after the French mathematician Abraham DeMoivre (1667–1754).

Abraham DeMoivre (1667–1754) is remembered for his work in probability theory and DeMoivre's Theorem. His book *The Doctrine of Chances* (published in 1718) includes the theory of recurring series and the theory of partial fractions.

DeMoivre's Theorem

If $z = r(\cos \theta + i \sin \theta)$ is a complex number and n is a positive integer, then

$$z^n = [r(\cos \theta + i \sin \theta)]^n$$

$$= r^n(\cos n\theta + i \sin n\theta).$$

EXAMPLE 7   **Finding a Power of a Complex Number**

Use DeMoivre's Theorem to find $\left(-1 + \sqrt{3}i\right)^{12}$.

Solution   The modulus of $z = -1 + \sqrt{3}i$ is

$$r = \sqrt{(-1)^2 + \left(\sqrt{3}\right)^2} = 2$$

and the argument θ is determined from $\tan \theta = \sqrt{3}/(-1)$. Because $z = -1 + \sqrt{3}i$ lies in Quadrant II,

$$\theta = \pi + \arctan \frac{\sqrt{3}}{-1} = \pi + \left(-\frac{\pi}{3}\right) = \frac{2\pi}{3}.$$

So, the trigonometric form of z is

$$z = -1 + \sqrt{3}i = 2\left(\cos \frac{2\pi}{3} + i \sin \frac{2\pi}{3}\right).$$

Then, by DeMoivre's Theorem, you have

$$\left(-1 + \sqrt{3}i\right)^{12} = \left[2\left(\cos \frac{2\pi}{3} + i \sin \frac{2\pi}{3}\right)\right]^{12}$$

$$= 2^{12}\left[\cos \frac{12(2\pi)}{3} + i \sin \frac{12(2\pi)}{3}\right]$$

$$= 4096(\cos 8\pi + i \sin 8\pi)$$

$$= 4096(1 + 0)$$

$$= 4096.$$

✓ **Checkpoint**))) *Audio-video solution in English & Spanish at LarsonPrecalculus.com*

Use DeMoivre's Theorem to find $(-1 - i)^4$.   ▪

Roots of Complex Numbers

Recall that a consequence of the Fundamental Theorem of Algebra is that a polynomial equation of degree n has n solutions in the complex number system. For example, the equation $x^6 = 1$ has six solutions. To find these solutions, use factoring and the Quadratic Formula.

$$x^6 - 1 = 0$$

$$(x^3 - 1)(x^3 + 1) = 0$$

$$(x - 1)(x^2 + x + 1)(x + 1)(x^2 - x + 1) = 0$$

Consequently, the solutions are

$$x = \pm 1, \quad x = \frac{-1 \pm \sqrt{3}i}{2}, \quad \text{and} \quad x = \frac{1 \pm \sqrt{3}i}{2}.$$

Each of these numbers is a sixth root of 1. In general, an **nth root of a complex number** is defined as follows.

Definition of an nth Root of a Complex Number

The complex number $u = a + bi$ is an **nth root** of the complex number z when

$$z = u^n$$

$$= (a + bi)^n.$$

To find a formula for an nth root of a complex number, let u be an nth root of z, where

$$u = s(\cos \beta + i \sin \beta)$$

and

$$z = r(\cos \theta + i \sin \theta).$$

By DeMoivre's Theorem and the fact that $u^n = z$, you have

$$s^n(\cos n\beta + i \sin n\beta) = r(\cos \theta + i \sin \theta).$$

Taking the absolute value of each side of this equation, it follows that $s^n = r$. Substituting back into the previous equation and dividing by r gives

$$\cos n\beta + i \sin n\beta = \cos \theta + i \sin \theta.$$

So, it follows that

$$\cos n\beta = \cos \theta$$

and

$$\sin n\beta = \sin \theta.$$

Both sine and cosine have a period of 2π, so these last two equations have solutions if and only if the angles differ by a multiple of 2π. Consequently, there must exist an integer k such that

$$n\beta = \theta + 2\pi k$$

$$\beta = \frac{\theta + 2\pi k}{n}.$$

Substituting this value of β and $s = \sqrt[n]{r}$ into the trigonometric form of u gives the result stated on the next page.

Finding *n*th Roots of a Complex Number

For a positive integer n, the complex number $z = r(\cos \theta + i \sin \theta)$ has exactly n distinct nth roots given by

$$z_k = \sqrt[n]{r}\left(\cos\frac{\theta + 2\pi k}{n} + i \sin\frac{\theta + 2\pi k}{n}\right).$$

where $k = 0, 1, 2, \ldots, n - 1$.

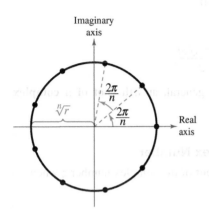

Figure 6.39

When $k > n - 1$, the roots begin to repeat. For example, when $k = n$, the angle

$$\frac{\theta + 2\pi n}{n} = \frac{\theta}{n} + 2\pi$$

is coterminal with θ/n, which is also obtained when $k = 0$.

The formula for the nth roots of a complex number z has a geometrical interpretation, as shown in Figure 6.39. Note that the nth roots of z all have the same magnitude $\sqrt[n]{r}$, so they all lie on a circle of radius $\sqrt[n]{r}$ with center at the origin. Furthermore, successive nth roots have arguments that differ by $2\pi/n$, so the n roots are equally spaced around the circle.

You have already found the sixth roots of 1 by factoring and using the Quadratic Formula. Example 8 shows how to solve the same problem with the formula for nth roots.

EXAMPLE 8 **Finding the *n*th Roots of a Real Number**

Find all sixth roots of 1.

Solution First, write 1 in the trigonometric form $z = 1(\cos 0 + i \sin 0)$. Then, by the nth root formula with $n = 6$, $r = 1$, and $\theta = 0$, the roots have the form

$$z_k = \sqrt[6]{1}\left(\cos\frac{0 + 2\pi k}{6} + i \sin\frac{0 + 2\pi k}{6}\right) = \cos\frac{\pi k}{3} + i \sin\frac{\pi k}{3}.$$

So, for $k = 0, 1, 2, 3, 4$, and 5, the roots are as listed below. (See Figure 6.40.)

$$z_0 = \cos 0 + i \sin 0 = 1$$

$$z_1 = \cos\frac{\pi}{3} + i \sin\frac{\pi}{3} = \frac{1}{2} + \frac{\sqrt{3}}{2}i \qquad \text{Increment by } \frac{2\pi}{n} = \frac{2\pi}{6} = \frac{\pi}{3}$$

$$z_2 = \cos\frac{2\pi}{3} + i \sin\frac{2\pi}{3} = -\frac{1}{2} + \frac{\sqrt{3}}{2}i$$

$$z_3 = \cos\pi + i \sin\pi = -1$$

$$z_4 = \cos\frac{4\pi}{3} + i \sin\frac{4\pi}{3} = -\frac{1}{2} - \frac{\sqrt{3}}{2}i$$

$$z_5 = \cos\frac{5\pi}{3} + i \sin\frac{5\pi}{3} = \frac{1}{2} - \frac{\sqrt{3}}{2}i$$

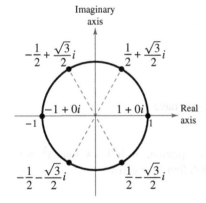

Figure 6.40

✓ *Checkpoint* ◉))) *Audio-video solution in English & Spanish at LarsonPrecalculus.com*

Find all fourth roots of 1.

In Figure 6.40, notice that the roots obtained in Example 8 all have a magnitude of 1 and are equally spaced around the unit circle. Also notice that the complex roots occur in conjugate pairs, as discussed in Section 2.5. The n distinct nth roots of 1 are called the ***n*th roots of unity.**

EXAMPLE 9 **Finding the *n*th Roots of a Complex Number**

See LarsonPrecalculus.com for an interactive version of this type of example.

Find the three cube roots of $z = -2 + 2i$.

Solution The modulus of z is

$$r = \sqrt{(-2)^2 + 2^2} = \sqrt{8}$$

and the argument θ is determined from

$$\tan \theta = \frac{b}{a} = \frac{2}{-2} = -1.$$

Because z lies in Quadrant II, the trigonometric form of z is

$$z = -2 + 2i = \sqrt{8}(\cos 135° + i \sin 135°). \qquad \theta = \pi + \arctan(-1) = 3\pi/4 = 135°$$

By the *n*th root formula, the roots have the form

$$z_k = \sqrt[6]{8}\left(\cos \frac{135° + 360°k}{3} + i \sin \frac{135° + 360°k}{3}\right).$$

So, for $k = 0$, 1, and 2, the roots are as listed below. (See Figure 6.41.)

$$z_0 = \sqrt[6]{8}\left(\cos \frac{135° + 360°(0)}{3} + i \sin \frac{135° + 360°(0)}{3}\right)$$

$$= \sqrt{2}(\cos 45° + i \sin 45°)$$

$$= 1 + i$$

$$z_1 = \sqrt[6]{8}\left(\cos \frac{135° + 360°(1)}{3} + i \sin \frac{135° + 360°(1)}{3}\right)$$

$$= \sqrt{2}(\cos 165° + i \sin 165°)$$

$$\approx -1.3660 + 0.3660i$$

$$z_2 = \sqrt[6]{8}\left(\cos \frac{135° + 360°(2)}{3} + i \sin \frac{135° + 360°(2)}{3}\right)$$

$$= \sqrt{2}(\cos 285° + i \sin 285°)$$

$$\approx 0.3660 - 1.3660i.$$

REMARK In Example 9, $r = \sqrt{8}$, so it follows that

$$\sqrt[n]{r} = \sqrt[3]{\sqrt{8}}$$

$$= \sqrt[3 \cdot 2]{8}$$

$$= \sqrt[6]{8}.$$

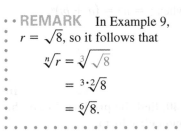

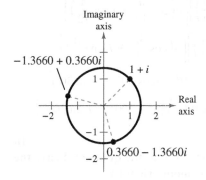

Figure 6.41

✓ *Checkpoint* 🔊)) *Audio-video solution in English & Spanish at LarsonPrecalculus.com*

Find the three cube roots of $z = -6 + 6i$.

Summarize (Section 6.6)

1. State the trigonometric form of a complex number *(page 445)*. For examples of writing complex numbers in trigonometric form and standard form, see Examples 1 and 2.

2. Explain how to multiply and divide complex numbers written in trigonometric form *(page 446)*. For examples of multiplying and dividing complex numbers written in trigonometric form, see Examples 3–6.

3. Explain how to use DeMoivre's Theorem to find a power of a complex number *(page 448)*. For an example of using DeMoivre's Theorem, see Example 7.

4. Explain how to find the *n*th roots of a complex number *(page 449)*. For examples of finding *n*th roots of complex numbers, see Examples 8 and 9.

6.6 Exercises

See **CalcChat.com** for tutorial help and worked-out solutions to odd-numbered exercises.

Vocabulary: Fill in the blanks.

1. The _____ _____ of the complex number $z = a + bi$ is $z = r(\cos\theta + i\sin\theta)$, where r is the _____ of z and θ is an _____ of z.

2. _____ Theorem states that if $z = r(\cos\theta + i\sin\theta)$ is a complex number and n is a positive integer, then $z^n = r^n(\cos n\theta + i\sin n\theta)$.

3. The complex number $u = a + bi$ is an _____ _____ of the complex number z when $z = u^n = (a + bi)^n$.

4. Successive nth roots of a complex number have arguments that differ by _____.

Skills and Applications

 Trigonometric Form of a Complex Number In Exercises 5–24, plot the complex number. Then write the trigonometric form of the complex number.

5. $1 + i$ 6. $5 - 5i$

7. $1 - \sqrt{3}i$ 8. $4 - 4\sqrt{3}i$

9. $-2(1 + \sqrt{3}i)$ 10. $\frac{5}{2}(\sqrt{3} - i)$

11. $-5i$ 12. $12i$

13. 2 14. 4

15. $-7 + 4i$ 16. $3 - i$

17. $2\sqrt{2} - i$ 18. $-3 - i$

19. $5 + 2i$ 20. $8 + 3i$

21. $3 + \sqrt{3}i$ 22. $3\sqrt{2} - 7i$

23. $-8 - 5\sqrt{3}i$ 24. $-9 - 2\sqrt{10}i$

 Writing a Complex Number in Standard Form In Exercises 25–32, write the standard form of the complex number. Then plot the complex number.

25. $2(\cos 60° + i\sin 60°)$ 26. $5(\cos 135° + i\sin 135°)$

27. $\sqrt{48}[\cos(-30°) + i\sin(-30°)]$

28. $\sqrt{8}(\cos 225° + i\sin 225°)$

29. $\frac{9}{4}\left(\cos\frac{3\pi}{4} + i\sin\frac{3\pi}{4}\right)$

30. $6\left(\cos\frac{5\pi}{12} + i\sin\frac{5\pi}{12}\right)$

31. $5[\cos(198° \, 45') + i\sin(198° \, 45')]$

32. $9.75[\cos(280° \, 30') + i\sin(280° \, 30')]$

Writing a Complex Number in Standard Form In Exercises 33–36, use a graphing utility to write the complex number in standard form.

33. $5\left(\cos\frac{\pi}{9} + i\sin\frac{\pi}{9}\right)$ 34. $10\left(\cos\frac{2\pi}{5} + i\sin\frac{2\pi}{5}\right)$

35. $2(\cos 155° + i\sin 155°)$ 36. $9(\cos 58° + i\sin 58°)$

 Multiplying Complex Numbers In Exercises 37–40, find the product. Leave the result in trigonometric form.

37. $\left[2\left(\cos\frac{\pi}{4} + i\sin\frac{\pi}{4}\right)\right]\left[6\left(\cos\frac{\pi}{12} + i\sin\frac{\pi}{12}\right)\right]$

38. $\left[\frac{3}{4}\left(\cos\frac{\pi}{3} + i\sin\frac{\pi}{3}\right)\right]\left[4\left(\cos\frac{3\pi}{4} + i\sin\frac{3\pi}{4}\right)\right]$

39. $\left[\frac{5}{3}(\cos 120° + i\sin 120°)\right]\left[\frac{2}{3}(\cos 30° + i\sin 30°)\right]$

40. $\left[\frac{1}{2}(\cos 100° + i\sin 100°)\right]\left[\frac{4}{5}(\cos 300° + i\sin 300°)\right]$

 Dividing Complex Numbers In Exercises 41–44, find the quotient. Leave the result in trigonometric form.

41. $\dfrac{3(\cos 50° + i\sin 50°)}{9(\cos 20° + i\sin 20°)}$ 42. $\dfrac{\cos 120° + i\sin 120°}{2(\cos 40° + i\sin 40°)}$

43. $\dfrac{\cos\pi + i\sin\pi}{\cos(\pi/3) + i\sin(\pi/3)}$ 44. $\dfrac{5(\cos 4.3 + i\sin 4.3)}{4(\cos 2.1 + i\sin 2.1)}$

Multiplying or Dividing Complex Numbers In Exercises 45–50, (a) write the trigonometric forms of the complex numbers, (b) perform the operation using the trigonometric forms, and (c) perform the operation using the standard forms, and check your result with that of part (b).

45. $(2 + 2i)(1 - i)$ 46. $(\sqrt{3} + i)(1 + i)$

47. $-2i(1 + i)$ 48. $3i(1 - \sqrt{2}i)$

49. $\dfrac{3 + 4i}{1 - \sqrt{3}i}$ 50. $\dfrac{1 + \sqrt{3}i}{6 - 3i}$

Multiplying in the Complex Plane In Exercises 51 and 52, find the product in the complex plane.

51. $\left[2\left(\cos\frac{2\pi}{3} + i\sin\frac{2\pi}{3}\right)\right]\left[\frac{1}{2}\left(\cos\frac{\pi}{3} + i\sin\frac{\pi}{3}\right)\right]$

52. $\left[2\left(\cos\frac{\pi}{4} + i\sin\frac{\pi}{4}\right)\right]\left[3\left(\cos\frac{\pi}{4} + i\sin\frac{\pi}{4}\right)\right]$

Finding a Power of a Complex Number
In Exercises 53–68, use DeMoivre's Theorem to find the power of the complex number. Write the result in standard form.

53. $[5(\cos 20° + i \sin 20°)]^3$ **54.** $[3(\cos 60° + i \sin 60°)]^4$

55. $\left(\cos \dfrac{\pi}{4} + i \sin \dfrac{\pi}{4}\right)^{12}$ **56.** $\left[2\left(\cos \dfrac{\pi}{2} + i \sin \dfrac{\pi}{2}\right)\right]^8$

57. $[5(\cos 3.2 + i \sin 3.2)]^4$ **58.** $(\cos 0 + i \sin 0)^{20}$

59. $[3(\cos 15° + i \sin 15°)]^4$ **60.** $\left[2\left(\cos \dfrac{\pi}{8} + i \sin \dfrac{\pi}{8}\right)\right]^6$

61. $(1 + i)^5$ **62.** $(2 + 2i)^6$

63. $(-1 + i)^6$ **64.** $(3 - 2i)^8$

65. $2(\sqrt{3} + i)^{10}$ **66.** $4(1 - \sqrt{3}i)^3$

67. $(3 - 2i)^5$ **68.** $(\sqrt{5} - 4i)^3$

Graphing Powers of a Complex Number In Exercises 69 and 70, represent the powers z, z^2, z^3, and z^4 graphically. Describe the pattern.

69. $z = \dfrac{\sqrt{2}}{2}(1 + i)$ **70.** $z = \dfrac{1}{2}(1 + \sqrt{3}i)$

Finding the nth Roots of a Complex Number In Exercises 71–86, (a) use the formula on page 450 to find the roots of the complex number, (b) write each of the roots in standard form, and (c) represent each of the roots graphically.

71. Square roots of $5(\cos 120° + i \sin 120°)$

72. Square roots of $16(\cos 60° + i \sin 60°)$

73. Cube roots of $8\left(\cos \dfrac{2\pi}{3} + i \sin \dfrac{2\pi}{3}\right)$

74. Fifth roots of $32\left(\cos \dfrac{5\pi}{6} + i \sin \dfrac{5\pi}{6}\right)$

75. Cube roots of $-\dfrac{125}{2}(1 + \sqrt{3}i)$

76. Cube roots of $-4\sqrt{2}(-1 + i)$

77. Square roots of $-25i$

78. Fourth roots of $625i$

79. Fourth roots of 16 **80.** Fourth roots of i

81. Fifth roots of 1 **82.** Cube roots of 1000

83. Cube roots of -125 **84.** Fourth roots of -4

85. Fifth roots of $4(1 - i)$ **86.** Sixth roots of $64i$

Solving an Equation In Exercises 87–94, use the formula on page 450 to find all solutions of the equation and represent the solutions graphically.

87. $x^4 + i = 0$ **88.** $x^3 + 1 = 0$

89. $x^5 + 243 = 0$ **90.** $x^3 - 27 = 0$

91. $x^4 + 16i = 0$ **92.** $x^6 + 64i = 0$

93. $x^3 - (1 - i) = 0$ **94.** $x^4 + (1 + i) = 0$

95. Ohm's Law

Ohm's law for alternating current circuits is $E = IZ$, where E is the voltage in volts, I is the current in amperes, and Z is the impedance in ohms. Each variable is a complex number.

(a) Write E in trigonometric form when $I = 6(\cos 41° + i \sin 41°)$ amperes and $Z = 4[\cos(-11°) + i \sin(-11°)]$ ohms.

(b) Write the voltage from part (a) in standard form.

(c) A voltmeter measures the magnitude of the voltage in a circuit. What would be the reading on a voltmeter for the circuit described in part (a)?

96. **HOW DO YOU SEE IT?**
The figure shows one of the fourth roots of a complex number z.

(a) How many roots are not shown?

(b) Describe the other roots.

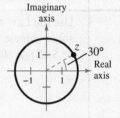

Exploration

True or False? In Exercises 97 and 98, determine whether the statement is true or false. Justify your answer.

97. Geometrically, the nth roots of any complex number z are all equally spaced around the unit circle.

98. The product of two complex numbers is zero only when the modulus of one (or both) of the complex numbers is zero.

99. Quotient of Two Complex Numbers Given two complex numbers $z_1 = r_1(\cos \theta_1 + i \sin \theta_1)$ and $z_2 = r_2(\cos \theta_2 + i \sin \theta_2)$, $z_2 \neq 0$, show that

$$\dfrac{z_1}{z_2} = \dfrac{r_1}{r_2}[\cos(\theta_1 - \theta_2) + i \sin(\theta_1 - \theta_2)].$$

100. Negative of a Complex Number Show that the negative of $z = r(\cos \theta + i \sin \theta)$ is $-z = r[\cos(\theta + \pi) + i \sin(\theta + \pi)]$.

101. Complex Conjugates Show that

$$\bar{z} = r[\cos(-\theta) + i \sin(-\theta)]$$

is the complex conjugate of $z = r(\cos \theta + i \sin \theta)$. Then find (a) $z\bar{z}$ and (b) $z/\bar{z}$, $\bar{z} \neq 0$.

Chapter Summary

	What Did You Learn?	**Explanation/Examples**	**Review Exercises**
Section 6.1	Use the Law of Sines to solve oblique triangles (AAS or ASA) *(p. 400)*.	**Law of Sines** If ABC is a triangle with sides a, b, and c, then $$\frac{a}{\sin A} = \frac{b}{\sin B} = \frac{c}{\sin C}.$$	1–12
	Use the Law of Sines to solve oblique triangles (SSA) *(p. 402)*.	If two sides and one opposite angle are given, then three possible situations can occur: (1) no such triangle exists, (2) one such triangle exists, or (3) two distinct triangles exist that satisfy the conditions.	1–12
	Find the areas of oblique triangles *(p. 404)*.	$\text{Area} = \frac{1}{2}bc \sin A = \frac{1}{2}ab \sin C = \frac{1}{2}ac \sin B$	13–16
	Use the Law of Sines to model and solve real-life problems *(p. 405)*.	The Law of Sines can be used to approximate the total distance of a boat race course. (See Example 7.)	17, 18
Section 6.2	Use the Law of Cosines to solve oblique triangles (SSS or SAS) *(p. 409)*.	**Law of Cosines** **Standard Form** $a^2 = b^2 + c^2 - 2bc \cos A$ $b^2 = a^2 + c^2 - 2ac \cos B$ $c^2 = a^2 + b^2 - 2ab \cos C$ **Alternative Form** $\cos A = \dfrac{b^2 + c^2 - a^2}{2bc}$ $\cos B = \dfrac{a^2 + c^2 - b^2}{2ac}$ $\cos C = \dfrac{a^2 + b^2 - c^2}{2ab}$	19–30
	Use the Law of Cosines to model and solve real-life problems *(p. 411)*.	The Law of Cosines can be used to find the distance between the pitcher's mound and first base on a women's softball field. (See Example 3.)	31, 32
	Use Heron's Area Formula to find areas of triangles *(p. 412)*.	**Heron's Area Formula:** Given any triangle with sides of lengths a, b, and c, the area of the triangle is $\text{Area} = \sqrt{s(s-a)(s-b)(s-c)}$, where $s = (a+b+c)/2$.	33–36
Section 6.3	Represent vectors as directed line segments *(p. 416)*.		37, 38
	Write component forms of vectors *(p. 417)*.	The component form of the vector with initial point $P(p_1, p_2)$ and terminal point $Q(q_1, q_2)$ is given by $\overrightarrow{PQ} = \langle q_1 - p_1, q_2 - p_2 \rangle = \langle v_1, v_2 \rangle = \mathbf{v}.$	39, 40
	Perform basic vector operations and represent vector operations graphically *(p. 418)*.	Let $\mathbf{u} = \langle u_1, u_2 \rangle$ and $\mathbf{v} = \langle v_1, v_2 \rangle$ be vectors and let k be a scalar (a real number). $\mathbf{u} + \mathbf{v} = \langle u_1 + v_1, u_2 + v_2 \rangle \quad k\mathbf{u} = \langle ku_1, ku_2 \rangle$ $-\mathbf{v} = \langle -v_1, -v_2 \rangle \quad \mathbf{u} - \mathbf{v} = \langle u_1 - v_1, u_2 - v_2 \rangle$	41–48, 53–58
	Write vectors as linear combinations of unit vectors *(p. 420)*.	The vector sum $\mathbf{v} = \langle v_1, v_2 \rangle = v_1 \langle 1, 0 \rangle + v_2 \langle 0, 1 \rangle = v_1 \mathbf{i} + v_2 \mathbf{j}$ is a linear combination of the vectors $\mathbf{i}$ and $\mathbf{j}$.	49–52

	What Did You Learn?	Explanation/Examples	Review Exercises
Section 6.3	Find direction angles of vectors (p. 422).	If $\mathbf{u} = a\mathbf{i} + b\mathbf{j}$, then the direction angle is determined from $\tan\theta = b/a$.	59–66
	Use vectors to model and solve real-life problems (p. 423).	Vectors can be used to find the resultant speed and true direction of an airplane. (See Example 11.)	67, 68
Section 6.4	Find the dot product of two vectors and use the properties of the dot product (p. 429).	The dot product of $\mathbf{u} = \langle u_1, u_2 \rangle$ and $\mathbf{v} = \langle v_1, v_2 \rangle$ is $\mathbf{u} \cdot \mathbf{v} = u_1 v_1 + u_2 v_2$.	69–80
	Find the angle between two vectors and determine whether two vectors are orthogonal (p. 430).	If θ is the angle between two nonzero vectors $\mathbf{u}$ and $\mathbf{v}$, then $\cos\theta = \dfrac{\mathbf{u} \cdot \mathbf{v}}{\|\mathbf{u}\|\|\mathbf{v}\|}$. The vectors $\mathbf{u}$ and $\mathbf{v}$ are orthogonal if and only if $\mathbf{u} \cdot \mathbf{v} = 0$.	81–88
	Write a vector as the sum of two vector components (p. 432).	Many applications in physics and engineering require the decomposition of a given vector into the sum of two vector components. (See Example 6.)	89–92
	Use vectors to determine the work done by a force (p. 434).	The work W done by a constant force $\mathbf{F}$ as its point of application moves along the vector $\overrightarrow{PQ}$ is given by **1.** $W = \|\text{proj}_{\overrightarrow{PQ}}\,\mathbf{F}\|\|\overrightarrow{PQ}\|$ or **2.** $W = \mathbf{F} \cdot \overrightarrow{PQ}$.	93–96
Section 6.5	Plot complex numbers in the complex plane and find absolute values of complex numbers (p. 438).	A complex number $z = a + bi$ can be represented by the point (a, b) in the complex plane. The horizontal axis is the real axis and the vertical axis is the imaginary axis. The absolute value, or modulus, of $z = a + bi$ is $|a + bi| = \sqrt{a^2 + b^2}$.	97–100
	Perform operations with complex numbers in the complex plane (p. 439).	Complex numbers can be added and subtracted geometrically in the complex plane. The points representing the complex conjugates $a + bi$ and $a - bi$ are reflections of each other in the real axis.	101–106
	Use the Distance and Midpoint Formulas in the complex plane (p. 441).	Let (a, b) and (s, t) be points in the complex plane. **Distance Formula** $d = \sqrt{(s - a)^2 + (t - b)^2}$ **Midpoint Formula** $\text{Midpoint} = \left(\dfrac{a + s}{2}, \dfrac{b + t}{2} \right)$	107–110
Section 6.6	Write trigonometric forms of complex numbers (p. 445).	The trigonometric form of the complex number $z = a + bi$ is $z = r(\cos\theta + i\sin\theta)$, where $a = r\cos\theta$, $b = r\sin\theta$, $r = \sqrt{a^2 + b^2}$, and $\tan\theta = b/a$.	111–116
	Multiply and divide complex numbers written in trigonometric form (p. 446).	Let $z_1 = r_1(\cos\theta_1 + i\sin\theta_1)$ and $z_2 = r_2(\cos\theta_2 + i\sin\theta_2)$ be complex numbers. $z_1 z_2 = r_1 r_2 [\cos(\theta_1 + \theta_2) + i\sin(\theta_1 + \theta_2)]$ $z_1/z_2 = (r_1/r_2)[\cos(\theta_1 - \theta_2) + i\sin(\theta_1 - \theta_2)], \quad z_2 \neq 0$	117–120
	Use DeMoivre's Theorem to find powers of complex numbers (p. 448).	**DeMoivre's Theorem:** If $z = r(\cos\theta + i\sin\theta)$ is a complex number and n is a positive integer, then $z^n = [r(\cos\theta + i\sin\theta)]^n = r^n(\cos n\theta + i\sin n\theta)$.	121–124
	Find nth roots of complex numbers (p. 449).	The complex number $u = a + bi$ is an nth root of the complex number z when $z = u^n = (a + bi)^n$.	125–132

Review Exercises   See CalcChat.com for tutorial help and worked-out solutions to odd-numbered exercises.

6.1 **Using the Law of Sines** In Exercises 1–12, use the Law of Sines to solve (if possible) the triangle. If two solutions exist, find both. Round your answers to two decimal places.

1. $A = 38°$, $B = 70°$, $a = 8$
2. $A = 22°$, $B = 121°$, $a = 19$
3. $B = 72°$, $C = 82°$, $b = 54$
4. $B = 10°$, $C = 20°$, $c = 33$
5. $A = 16°$, $B = 98°$, $c = 8.4$
6. $A = 95°$, $B = 45°$, $c = 104.8$
7. $A = 24°$, $C = 48°$, $b = 27.5$
8. $B = 64°$, $C = 36°$, $a = 367$
9. $B = 150°$, $b = 30$, $c = 10$
10. $B = 150°$, $a = 10$, $b = 3$
11. $A = 75°$, $a = 51.2$, $b = 33.7$
12. $B = 25°$, $a = 6.2$, $b = 4$

Finding the Area of a Triangle In Exercises 13–16, find the area of the triangle. Round your answers to one decimal place.

13. $A = 33°$, $b = 7$, $c = 10$
14. $B = 80°$, $a = 4$, $c = 8$
15. $C = 119°$, $a = 18$, $b = 6$
16. $A = 11°$, $b = 22$, $c = 21$

17. **Height** From a certain distance, the angle of elevation to the top of a building is 17°. At a point 50 meters closer to the building, the angle of elevation is 31°. Find the height of the building.

18. **River Width** A surveyor finds that a tree on the opposite bank of a river flowing due east has a bearing of N 22° 30′ E from a certain point and a bearing of N 15° W from a point 400 feet downstream. Find the width of the river.

6.2 **Using the Law of Cosines** In Exercises 19–26, use the Law of Cosines to solve the triangle. Round your answers to two decimal places.

19. $a = 6$, $b = 9$, $c = 14$
20. $a = 75$, $b = 50$, $c = 110$
21. $a = 2.5$, $b = 5.0$, $c = 4.5$
22. $a = 16.4$, $b = 8.8$, $c = 12.2$
23. $B = 108°$, $a = 11$, $c = 11$
24. $B = 150°$, $a = 10$, $c = 20$
25. $C = 43°$, $a = 22.5$, $b = 31.4$
26. $A = 62°$, $b = 11.34$, $c = 19.52$

Solving a Triangle In Exercises 27–30, determine whether the Law of Cosines is needed to solve the triangle. Then solve (if possible) the triangle. If two solutions exist, find both. Round your answers to two decimal places.

27. $C = 64°$, $b = 9$, $c = 13$
28. $B = 52°$, $a = 4$, $c = 5$
29. $a = 13$, $b = 15$, $c = 24$
30. $A = 44°$, $B = 31°$, $c = 2.8$

31. **Geometry** The lengths of the diagonals of a parallelogram are 10 feet and 16 feet. Find the lengths of the sides of the parallelogram when the diagonals intersect at an angle of 28°.

32. **Air Navigation** Two planes leave an airport at approximately the same time. One flies 425 miles per hour at a bearing of 355°, and the other flies 530 miles per hour at a bearing of 67°. Draw a diagram that gives a visual representation of the problem and determine the distance between the planes after they fly for 2 hours.

Using Heron's Area Formula In Exercises 33–36, use Heron's Area Formula to find the area of the triangle.

33. $a = 3$, $b = 6$, $c = 8$
34. $a = 15$, $b = 8$, $c = 10$
35. $a = 12.3$, $b = 15.8$, $c = 3.7$
36. $a = \dfrac{4}{5}$, $b = \dfrac{3}{4}$, $c = \dfrac{5}{8}$

6.3 **Determining Whether Two Vectors Are Equivalent** In Exercises 37 and 38, determine whether u and v are equivalent. Explain.

37. 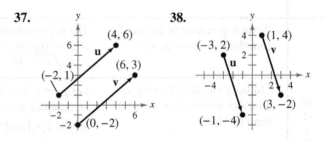 38.

Finding the Component Form of a Vector In Exercises 39 and 40, find the component form and magnitude of the vector v.

39. Initial point: $(0, 10)$

    Terminal point: $(7, 3)$

40. Initial point: $(1, 5)$

    Terminal point: $(15, 9)$

Vector Operations In Exercises 41–48, find (a) $\mathbf{u} + \mathbf{v}$, (b) $\mathbf{u} - \mathbf{v}$, (c) $4\mathbf{u}$, and (d) $3\mathbf{v} + 5\mathbf{u}$. Then sketch each resultant vector.

41. $\mathbf{u} = \langle -1, -3 \rangle$, $\mathbf{v} = \langle -3, 6 \rangle$

42. $\mathbf{u} = \langle 4, 5 \rangle$, $\mathbf{v} = \langle 0, -1 \rangle$

43. $\mathbf{u} = \langle -5, 2 \rangle$, $\mathbf{v} = \langle 4, 4 \rangle$

44. $\mathbf{u} = \langle 1, -8 \rangle$, $\mathbf{v} = \langle 3, -2 \rangle$

45. $\mathbf{u} = 2\mathbf{i} - \mathbf{j}$, $\mathbf{v} = 5\mathbf{i} + 3\mathbf{j}$

46. $\mathbf{u} = -7\mathbf{i} - 3\mathbf{j}$, $\mathbf{v} = 4\mathbf{i} - \mathbf{j}$

47. $\mathbf{u} = 4\mathbf{i}$, $\mathbf{v} = -\mathbf{i} + 6\mathbf{j}$

48. $\mathbf{u} = -6\mathbf{j}$, $\mathbf{v} = \mathbf{i} + \mathbf{j}$

Writing a Linear Combination of Unit Vectors In Exercises 49–52, the initial and terminal points of a vector are given. Write the vector as a linear combination of the standard unit vectors i and j.

Initial Point	Terminal Point
49. $(2, 3)$	$(1, 8)$
50. $(4, -2)$	$(-2, -10)$
51. $(3, 4)$	$(9, 8)$
52. $(-2, 7)$	$(5, -9)$

Vector Operations In Exercises 53–58, find the component form of w and sketch the specified vector operations geometrically, where $\mathbf{u} = 6\mathbf{i} - 5\mathbf{j}$ and $\mathbf{v} = 10\mathbf{i} + 3\mathbf{j}$.

53. $\mathbf{w} = 3\mathbf{v}$

54. $\mathbf{w} = \frac{1}{2}\mathbf{v}$

55. $\mathbf{w} = 2\mathbf{u} + \mathbf{v}$

56. $\mathbf{w} = 4\mathbf{u} - 5\mathbf{v}$

57. $\mathbf{w} = 5\mathbf{u} - 4\mathbf{v}$

58. $\mathbf{w} = -3\mathbf{u} + 2\mathbf{v}$

Finding the Direction Angle of a Vector In Exercises 59–64, find the magnitude and direction angle of the vector v.

59. $\mathbf{v} = 5\mathbf{i} + 4\mathbf{j}$

60. $\mathbf{v} = -4\mathbf{i} + 7\mathbf{j}$

61. $\mathbf{v} = -3\mathbf{i} - 3\mathbf{j}$

62. $\mathbf{v} = 8\mathbf{i} - \mathbf{j}$

63. $\mathbf{v} = 7(\cos 60°\mathbf{i} + \sin 60°\mathbf{j})$

64. $\mathbf{v} = 3(\cos 150°\mathbf{i} + \sin 150°\mathbf{j})$

Finding the Component Form of a Vector In Exercises 65 and 66, find the component form of v given its magnitude and the angle it makes with the positive x-axis. Then sketch v.

Magnitude	Angle
65. $\|\mathbf{v}\| = 8$	$\theta = 120°$
66. $\|\mathbf{v}\| = \frac{1}{2}$	$\theta = 225°$

67. Resultant Force Forces with magnitudes of 85 pounds and 50 pounds act on a single point at angles of 45° and 60°, respectively, with the positive x-axis. Find the direction and magnitude of the resultant of these forces.

68. Rope Tension Two ropes support a 180-pound weight, as shown in the figure. Find the tension in each rope.

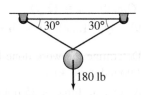

180 lb

6.4 **Finding a Dot Product** In Exercises 69–72, find $\mathbf{u} \cdot \mathbf{v}$.

69. $\mathbf{u} = \langle 6, 7 \rangle$
     $\mathbf{v} = \langle -3, 9 \rangle$

70. $\mathbf{u} = \langle -7, 12 \rangle$
     $\mathbf{v} = \langle -4, -14 \rangle$

71. $\mathbf{u} = 3\mathbf{i} + 7\mathbf{j}$
     $\mathbf{v} = 11\mathbf{i} - 5\mathbf{j}$

72. $\mathbf{u} = -7\mathbf{i} + 2\mathbf{j}$
     $\mathbf{v} = 16\mathbf{i} - 12\mathbf{j}$

Using Properties of the Dot Product In Exercises 73–80, use the vectors $\mathbf{u} = \langle -4, 2 \rangle$ and $\mathbf{v} = \langle 5, 1 \rangle$ to find the quantity. State whether the result is a vector or a scalar.

73. $2\mathbf{u} \cdot \mathbf{u}$

74. $3\mathbf{u} \cdot \mathbf{v}$

75. $4 - \|\mathbf{u}\|$

76. $\|\mathbf{v}\|^2$

77. $\mathbf{u}(\mathbf{u} \cdot \mathbf{v})$

78. $(\mathbf{u} \cdot \mathbf{v})\mathbf{v}$

79. $(\mathbf{u} \cdot \mathbf{u}) - (\mathbf{u} \cdot \mathbf{v})$

80. $(\mathbf{v} \cdot \mathbf{v}) - (\mathbf{v} \cdot \mathbf{u})$

Finding the Angle Between Two Vectors In Exercises 81–84, find the angle θ (in degrees) between the vectors.

81. $\mathbf{u} = \langle 2\sqrt{2}, -4 \rangle$, $\mathbf{v} = \langle -\sqrt{2}, 1 \rangle$

82. $\mathbf{u} = \langle 3, \sqrt{3} \rangle$, $\mathbf{v} = \langle 4, 3\sqrt{3} \rangle$

83. $\mathbf{u} = \cos \dfrac{7\pi}{4}\mathbf{i} + \sin \dfrac{7\pi}{4}\mathbf{j}$, $\mathbf{v} = \cos \dfrac{5\pi}{6}\mathbf{i} + \sin \dfrac{5\pi}{6}\mathbf{j}$

84. $\mathbf{u} = \cos 45°\mathbf{i} + \sin 45°\mathbf{j}$, $\mathbf{v} = \cos 300°\mathbf{i} + \sin 300°\mathbf{j}$

Determining Orthogonal Vectors In Exercises 85–88, determine whether u and v are orthogonal.

85. $\mathbf{u} = \langle -3, 8 \rangle$
     $\mathbf{v} = \langle 8, 3 \rangle$

86. $\mathbf{u} = \langle \frac{1}{4}, -\frac{1}{2} \rangle$
     $\mathbf{v} = \langle -2, 4 \rangle$

87. $\mathbf{u} = -\mathbf{i}$
     $\mathbf{v} = \mathbf{i} + 2\mathbf{j}$

88. $\mathbf{u} = -2\mathbf{i} + \mathbf{j}$
     $\mathbf{v} = 3\mathbf{i} + 6\mathbf{j}$

Decomposing a Vector into Components In Exercises 89–92, find the projection of u onto v. Then write u as the sum of two orthogonal vectors, one of which is $\text{proj}_{\mathbf{v}}\,\mathbf{u}$.

89. $\mathbf{u} = \langle -4, 3 \rangle$, $\mathbf{v} = \langle -8, -2 \rangle$

90. $\mathbf{u} = \langle 5, 6 \rangle$, $\mathbf{v} = \langle 10, 0 \rangle$

91. $\mathbf{u} = \langle 2, 7 \rangle$, $\mathbf{v} = \langle 1, -1 \rangle$

92. $\mathbf{u} = \langle -3, 5 \rangle$, $\mathbf{v} = \langle -5, 2 \rangle$

Work In Exercises 93 and 94, determine the work done in moving a particle from P to Q when the magnitude and direction of the force are given by v.

93. $P(5, 3)$, $Q(8, 9)$, $\mathbf{v} = \langle 2, 7 \rangle$

94. $P(-2, -9)$, $Q(-12, 8)$, $\mathbf{v} = 3\mathbf{i} - 6\mathbf{j}$

95. Work Determine the work done by a crane lifting an 18,000-pound truck 4 feet.

96. Work A constant force of 25 pounds, exerted at an angle of 20° with the horizontal, is required to slide a crate across a floor. Determine the work done in sliding the crate 12 feet.

6.5 Finding the Absolute Value of a Complex Number In Exercises 97–100, plot the complex number and find its absolute value.

97. $7i$ 　　　　**98.** $-6i$

99. $5 + 3i$ 　　**100.** $-10 - 4i$

Adding in the Complex Plane In Exercises 101 and 102, find the sum of the complex numbers in the complex plane.

101. $(2 + 3i) + (1 - 2i)$ 　**102.** $(-4 + 2i) + (2 + i)$

Subtracting in the Complex Plane In Exercises 103 and 104, find the difference of the complex numbers in the complex plane.

103. $(1 + 2i) - (3 + i)$ 　**104.** $(-2 + i) - (1 + 4i)$

Complex Conjugates in the Complex Plane In Exercises 105 and 106, plot the complex number and its complex conjugate. Write the conjugate as a complex number.

105. $3 + i$ 　　　**106.** $2 - 5i$

Finding Distance in the Complex Plane In Exercises 107 and 108, find the distance between the complex numbers in the complex plane.

107. $3 + 2i, 2 - i$ 　**108.** $1 + 5i, -1 + 3i$

Finding a Midpoint in the Complex Plane In Exercises 109 and 110, find the midpoint of the line segment joining the points corresponding to the complex numbers in the complex plane.

109. $1 + i, 4 + 3i$ 　**110.** $2 - i, 1 + 4i$

6.6 Trigonometric Form of a Complex Number In Exercises 111–116, plot the complex number. Then write the trigonometric form of the complex number.

111. $4i$ 　　　**112.** -7

113. $7 - 7i$ 　　**114.** $5 + 12i$

115. $-5 - 12i$ 　**116.** $-3\sqrt{3} + 3i$

Multiplying Complex Numbers In Exercises 117 and 118, find the product. Leave the result in trigonometric form.

117. $\left[2\left(\cos\frac{\pi}{4} + i \sin\frac{\pi}{4}\right)\right]\left[2\left(\cos\frac{\pi}{3} + i \sin\frac{\pi}{3}\right)\right]$

118. $\left[4\left(\cos\frac{\pi}{3} + i \sin\frac{\pi}{3}\right)\right]\left[3\left(\cos\frac{5\pi}{6} + i \sin\frac{5\pi}{6}\right)\right]$

Dividing Complex Numbers In Exercises 119 and 120, find the quotient. Leave the result in trigonometric form.

119. $\dfrac{2(\cos 60° + i \sin 60°)}{3(\cos 15° + i \sin 15°)}$ 　**120.** $\dfrac{\cos 150° + i \sin 150°}{2(\cos 50° + i \sin 50°)}$

Finding a Power of a Complex Number In Exercises 121–124, use DeMoivre's Theorem to find the power of the complex number. Write the result in standard form.

121. $\left[5\left(\cos\frac{\pi}{12} + i \sin\frac{\pi}{12}\right)\right]^4$

122. $\left[2\left(\cos\frac{4\pi}{15} + i \sin\frac{4\pi}{15}\right)\right]^5$

123. $(2 + 3i)^6$

124. $(1 - i)^8$

Finding the nth Roots of a Complex Number In Exercises 125–128, (a) use the formula on page 450 to find the roots of the complex number, (b) write each of the roots in standard form, and (c) represent each of the roots graphically.

125. Sixth roots of $-729i$ 　**126.** Fourth roots of $256i$

127. Cube roots of 8 　　**128.** Fifth roots of -1024

Solving an Equation In Exercises 129–132, use the formula on page 450 to find all solutions of the equation and represent the solutions graphically.

129. $x^4 + 81 = 0$

130. $x^5 - 32 = 0$

131. $x^3 + 8i = 0$

132. $x^4 - 64i = 0$

Exploration

True or False? In Exercises 133 and 134, determine whether the statement is true or false. Justify your answer.

133. The Law of Sines is true when one of the angles in the triangle is a right angle.

134. When the Law of Sines is used, the solution is always unique.

135. Writing What characterizes a vector in the plane?

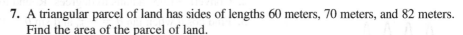

Chapter Test

See CalcChat.com for tutorial help and worked-out solutions to odd-numbered exercises.

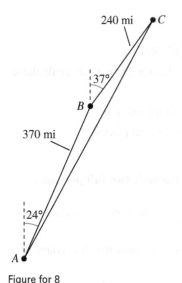

240 mi   •C

37°

B •

370 mi

24°

A •

Figure for 8

Take this test as you would take a test in class. When you are finished, check your work against the answers given in the back of the book.

In Exercises 1–6, determine whether the Law of Cosines is needed to solve the triangle. Then solve (if possible) the triangle. If two solutions exist, find both. Round your answers to two decimal places.

1. $A = 24°$, $B = 68°$, $a = 12.2$ 2. $B = 110°$, $C = 28°$, $a = 15.6$
3. $A = 24°$, $a = 11.2$, $b = 13.4$ 4. $a = 6.0$, $b = 7.3$, $c = 12.4$
5. $B = 100°$, $a = 23$, $b = 15$ 6. $C = 121°$, $a = 34$, $b = 55$

7. A triangular parcel of land has sides of lengths 60 meters, 70 meters, and 82 meters. Find the area of the parcel of land.

8. An airplane flies 370 miles from point A to point B with a bearing of 24°. Then it flies 240 miles from point B to point C with a bearing of 37° (see figure). Find the straight-line distance and bearing from point A to point C.

In Exercises 9 and 10, find the component form of the vector v.

9. Initial point of **v**: $(-3, 7)$; terminal point of **v**: $(11, -16)$
10. Magnitude of **v**: $\|\mathbf{v}\| = 12$; direction of **v**: $\mathbf{u} = \langle 3, -5 \rangle$

In Exercises 11–14, $\mathbf{u} = \langle 2, 7 \rangle$ and $\mathbf{v} = \langle -6, 5 \rangle$. Find the resultant vector and sketch its graph.

11. $\mathbf{u} + \mathbf{v}$ 12. $\mathbf{u} - \mathbf{v}$
13. $5\mathbf{u} - 3\mathbf{v}$ 14. $4\mathbf{u} + 2\mathbf{v}$

15. Find the distance between $4 + 3i$ and $1 - i$ in the complex plane.

16. Forces with magnitudes of 250 pounds and 130 pounds act on an object at angles of 45° and $-60°$, respectively, with the positive x-axis. Find the direction and magnitude of the resultant of these forces.

17. Find the angle θ (in degrees) between the vectors $\mathbf{u} = \langle -1, 5 \rangle$ and $\mathbf{v} = \langle 3, -2 \rangle$.

18. Determine whether the vectors $\mathbf{u} = \langle 6, -10 \rangle$ and $\mathbf{v} = \langle 5, 3 \rangle$ are orthogonal.

19. Find the projection of $\mathbf{u} = \langle 6, 7 \rangle$ onto $\mathbf{v} = \langle -5, -1 \rangle$. Then write $\mathbf{u}$ as the sum of two orthogonal vectors, one of which is proj$_\mathbf{v}\mathbf{u}$.

20. A 500-pound motorcycle is stopped at a red light on a hill inclined at 12°. Find the force required to keep the motorcycle from rolling down the hill.

21. Write the complex number $z = 4 - 4i$ in trigonometric form.

22. Write the complex number $z = 6(\cos 120° + i \sin 120°)$ in standard form.

In Exercises 23 and 24, use DeMoivre's Theorem to find the power of the complex number. Write the result in standard form.

23. $\left[3\left(\cos \dfrac{7\pi}{6} + i \sin \dfrac{7\pi}{6} \right) \right]^8$ 24. $(3 - 3i)^6$

25. Find the fourth roots of 256.

26. Find all solutions of the equation $x^3 - 27i = 0$ and represent the solutions graphically.

Cumulative Test for Chapters 4–6

See CalcChat.com for tutorial help and worked-out solutions to odd-numbered exercises.

Take this test as you would take a test in class. When you are finished, check your work against the answers given in the back of the book.

1. Consider the angle $\theta = -120°$.
   (a) Sketch the angle in standard position.
   (b) Determine a coterminal angle in the interval $[0°, 360°)$.
   (c) Rewrite the angle in radian measure as a multiple of π. Do not use a calculator.
   (d) Find the reference angle θ'.
   (e) Find the exact values of the six trigonometric functions of θ.

2. Convert -1.45 radians to degrees. Round to three decimal places.

3. Find $\cos \theta$ when $\tan \theta = -\frac{21}{20}$ and $\sin \theta < 0$.

In Exercises 4–6, sketch the graph of the function. (Include two full periods.)

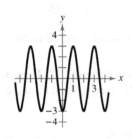

Figure for 7

4. $f(x) = 3 - 2 \sin \pi x$

5. $g(x) = \frac{1}{2} \tan\left(x - \frac{\pi}{2}\right)$

6. $h(x) = -\sec(x + \pi)$

7. Find a, b, and c for the function $h(x) = a \cos(bx + c)$ such that the graph of h matches the figure.

8. Sketch the graph of the function $f(x) = \frac{1}{2}x \sin x$ on the interval $[-3\pi, 3\pi]$.

In Exercises 9 and 10, find the exact value of the expression.

9. $\tan(\arctan 4.9)$

10. $\tan\left(\arcsin \frac{3}{5}\right)$

11. Write an algebraic expression that is equivalent to $\sin(\arccos 2x)$.

12. Use the fundamental identities to simplify: $\cos\left(\frac{\pi}{2} - x\right)\csc x$.

13. Subtract and simplify: $\dfrac{\sin \theta - 1}{\cos \theta} - \dfrac{\cos \theta}{\sin \theta - 1}$.

In Exercises 14–16, verify the identity.

14. $\cot^2 \alpha(\sec^2 \alpha - 1) = 1$

15. $\sin(x + y) \sin(x - y) = \sin^2 x - \sin^2 y$

16. $\sin^2 x \cos^2 x = \frac{1}{8}(1 - \cos 4x)$

In Exercises 17 and 18, find all solutions of the equation in the interval $[0, 2\pi)$.

17. $2 \cos^2 \beta - \cos \beta = 0$

18. $3 \tan \theta - \cot \theta = 0$

19. Use the Quadratic Formula to find all solutions of the equation in the interval $[0, 2\pi)$: $\sin^2 x + 2 \sin x + 1 = 0$.

20. Given that $\sin u = \frac{12}{13}$, $\cos v = \frac{3}{5}$, and angles u and v are both in Quadrant I, find $\tan(u - v)$.

21. Given that $\tan u = \dfrac{1}{2}$ and $0 < u < \dfrac{\pi}{2}$, find the exact value of $\tan(2u)$.

22. Given that $\tan u = \dfrac{4}{3}$ and $0 < u < \dfrac{\pi}{2}$, find the exact value of $\sin \dfrac{u}{2}$.

23. Rewrite $5 \sin \dfrac{3\pi}{4} \cdot \cos \dfrac{7\pi}{4}$ as a sum or difference.

24. Rewrite $\cos 9x - \cos 7x$ as a product.

In Exercises 25–30, determine whether the Law of Cosines is needed to solve the triangle at the left, then solve the triangle. Round your answers to two decimal places.

25. $A = 30°$, $a = 9$, $b = 8$

26. $A = 30°$, $b = 8$, $c = 10$

27. $A = 30°$, $C = 90°$, $b = 10$

28. $a = 4.7$, $b = 8.1$, $c = 10.3$

29. $A = 45°$, $B = 26°$, $c = 20$

30. $C = 80°$, $a = 1.2$, $b = 10$

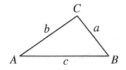

Figure for 25–30

31. Find the area of a triangle with two sides of lengths 7 inches and 12 inches and an included angle of 99°.

32. Use Heron's Area Formula to find the area of a triangle with sides of lengths 30 meters, 41 meters, and 45 meters.

33. Write the vector with initial point $(-1, 2)$ and terminal point $(6, 10)$ as a linear combination of the standard unit vectors $\mathbf{i}$ and $\mathbf{j}$.

34. Find a unit vector $\mathbf{u}$ in the direction of $\mathbf{v} = \mathbf{i} + \mathbf{j}$.

35. Find $\mathbf{u} \cdot \mathbf{v}$ for $\mathbf{u} = 3\mathbf{i} + 4\mathbf{j}$ and $\mathbf{v} = \mathbf{i} - 2\mathbf{j}$.

36. Find the projection of $\mathbf{u} = \langle 8, -2 \rangle$ onto $\mathbf{v} = \langle 1, 5 \rangle$. Then write $\mathbf{u}$ as the sum of two orthogonal vectors, one of which is $\text{proj}_{\mathbf{v}}\mathbf{u}$.

37. Plot $3 - 2i$ and its complex conjugate. Write the conjugate as a complex number.

38. Write the complex number $-2 + 2i$ in trigonometric form.

39. Find the product of $[4(\cos 30° + i \sin 30°)]$ and $[6(\cos 120° + i \sin 120°)]$. Leave the result in trigonometric form.

40. Find the three cube roots of 1.

41. Find all solutions of the equation $x^4 + 625 = 0$ and represent the solutions graphically.

42. A ceiling fan with 21-inch blades makes 63 revolutions per minute. Find the angular speed of the fan in radians per minute. Find the linear speed (in inches per minute) of the tips of the blades.

43. Find the area of the sector of a circle with a radius of 12 yards and a central angle of 105°.

44. From a point 200 feet from a flagpole, the angles of elevation to the bottom and top of the flag are 16° 45′ and 18°, respectively. Approximate the height of the flag to the nearest foot.

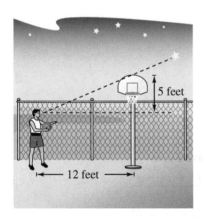

Figure for 45

45. To determine the angle of elevation of a star in the sky, you align the star and the top of the backboard of a basketball hoop that is 5 feet higher than your eyes in your line of vision (see figure). Your horizontal distance from the backboard is 12 feet. What is the angle of elevation of the star?

46. Find a model for a particle in simple harmonic motion with a displacement (at $t = 0$) of 4 inches, an amplitude of 4 inches, and a period of 8 seconds.

47. An airplane has a speed of 500 kilometers per hour at a bearing of 30°. The wind velocity is 50 kilometers per hour in the direction N 60° E. Find the resultant speed and true direction of the airplane.

48. A constant force of 85 pounds, exerted at an angle of 60° with the horizontal, is required to slide an object across a floor. Determine the work done in sliding the object 10 feet.

Proofs in Mathematics ■ ■ ■ ■ ■ ■ ■ ■ ■ ■ ■ ■ ■ ■ ■

Law of Sines (p. 400)

If ABC is a triangle with sides a, b, and c, then

$$\frac{a}{\sin A} = \frac{b}{\sin B} = \frac{c}{\sin C}.$$

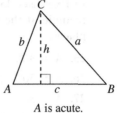

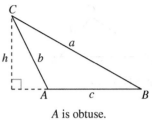

A is acute. A is obtuse.

Proof

For either triangle shown above, you have

$$\sin A = \frac{h}{b} \implies h = b \sin A \quad \text{and} \quad \sin B = \frac{h}{a} \implies h = a \sin B$$

where h is an altitude. Equating these two values of h, you have

$$a \sin B = b \sin A \quad \text{or} \quad \frac{a}{\sin A} = \frac{b}{\sin B}.$$

Note that $\sin A \neq 0$ and $\sin B \neq 0$ because no angle of a triangle can have a measure of $0°$ or $180°$. In a similar manner, construct an altitude h from vertex B to side AC (extended in the obtuse triangle), as shown at the left. Then you have

$$\sin A = \frac{h}{c} \implies h = c \sin A \quad \text{and} \quad \sin C = \frac{h}{a} \implies h = a \sin C.$$

Equating these two values of h, you have

$$a \sin C = c \sin A \quad \text{or} \quad \frac{a}{\sin A} = \frac{c}{\sin C}.$$

By the Transitive Property of Equality,

$$\frac{a}{\sin A} = \frac{b}{\sin B} = \frac{c}{\sin C}.$$

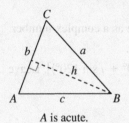

A is acute.

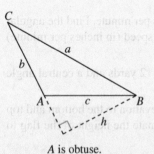

A is obtuse.

Law of Cosines (p. 409)

Standard Form	Alternative Form
$a^2 = b^2 + c^2 - 2bc \cos A$	$\cos A = \dfrac{b^2 + c^2 - a^2}{2bc}$
$b^2 = a^2 + c^2 - 2ac \cos B$	$\cos B = \dfrac{a^2 + c^2 - b^2}{2ac}$
$c^2 = a^2 + b^2 - 2ab \cos C$	$\cos C = \dfrac{a^2 + b^2 - c^2}{2ab}$

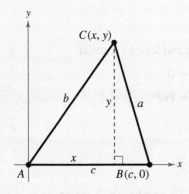

$C(x, y)$

b

y

a

x

A c $B(c, 0)$

Proof

To prove the first formula, consider the triangle at the left, which has three acute angles. Note that vertex B has coordinates $(c, 0)$. Furthermore, C has coordinates (x, y), where $x = b \cos A$ and $y = b \sin A$. Because a is the distance from C to B, it follows that

$$a = \sqrt{(x - c)^2 + (y - 0)^2}$$ Distance Formula

$$a^2 = (x - c)^2 + (y - 0)^2$$ Square each side.

$$a^2 = (b \cos A - c)^2 + (b \sin A)^2$$ Substitute for x and y.

$$a^2 = b^2 \cos^2 A - 2bc \cos A + c^2 + b^2 \sin^2 A$$ Expand.

$$a^2 = b^2(\sin^2 A + \cos^2 A) + c^2 - 2bc \cos A$$ Factor out b^2.

$$a^2 = b^2 + c^2 - 2bc \cos A.$$ $\sin^2 A + \cos^2 A = 1$

Similar arguments are used to establish the second and third formulas. ■

> **Heron's Area Formula** *(p. 412)*
>
> Given any triangle with sides of lengths a, b, and c, the area of the triangle is
>
> $$\text{Area} = \sqrt{s(s - a)(s - b)(s - c)}, \quad \text{where } s = \frac{a + b + c}{2}.$$

· · REMARK

$$\frac{1}{2}bc(1 + \cos A)$$

$$= \frac{1}{2}bc\left(1 + \frac{b^2 + c^2 - a^2}{2bc}\right)$$

$$= \frac{1}{2}bc\left(\frac{2bc + b^2 + c^2 - a^2}{2bc}\right)$$

$$= \frac{1}{4}(2bc + b^2 + c^2 - a^2)$$

$$= \frac{1}{4}(b^2 + 2bc + c^2 - a^2)$$

$$= \frac{1}{4}[(b + c)^2 - a^2]$$

$$= \frac{1}{4}(b + c + a)(b + c - a)$$

$$= \frac{a + b + c}{2} \cdot \frac{-a + b + c}{2}$$

▷

Proof

From Section 6.1, you know that

$$\text{Area} = \frac{1}{2}bc \sin A$$ Formula for the area of an oblique triangle

$$= \sqrt{\frac{1}{4}b^2c^2 \sin^2 A}$$ Square each side and then take the square root of each side.

$$= \sqrt{\frac{1}{4}b^2c^2(1 - \cos^2 A)}$$ Pythagorean identity

$$= \sqrt{\left[\frac{1}{2}bc(1 + \cos A)\right]\left[\frac{1}{2}bc(1 - \cos A)\right]}.$$ Factor.

Using the alternate form of the Law of Cosines,

$$\frac{1}{2}bc(1 + \cos A) = \frac{a + b + c}{2} \cdot \frac{-a + b + c}{2}$$

and

$$\frac{1}{2}bc(1 - \cos A) = \frac{a - b + c}{2} \cdot \frac{a + b - c}{2}.$$

Letting $s = (a + b + c)/2$, rewrite these two equations as

$$\frac{1}{2}bc(1 + \cos A) = s(s - a) \quad \text{and} \quad \frac{1}{2}bc(1 - \cos A) = (s - b)(s - c).$$

Substitute into the last formula for area to conclude that

$$\text{Area} = \sqrt{s(s - a)(s - b)(s - c)}.$$ ■

Properties of the Dot Product   *(p. 429)*

Let **u**, **v**, and **w** be vectors in the plane or in space and let c be a scalar.

1. $\mathbf{u} \cdot \mathbf{v} = \mathbf{v} \cdot \mathbf{u}$ 2. $\mathbf{0} \cdot \mathbf{v} = 0$

3. $\mathbf{u} \cdot (\mathbf{v} + \mathbf{w}) = \mathbf{u} \cdot \mathbf{v} + \mathbf{u} \cdot \mathbf{w}$ 4. $\mathbf{v} \cdot \mathbf{v} = \|\mathbf{v}\|^2$

5. $c(\mathbf{u} \cdot \mathbf{v}) = c\mathbf{u} \cdot \mathbf{v} = \mathbf{u} \cdot c\mathbf{v}$

Proof

Let $\mathbf{u} = \langle u_1, u_2 \rangle$, $\mathbf{v} = \langle v_1, v_2 \rangle$, $\mathbf{w} = \langle w_1, w_2 \rangle$, $\mathbf{0} = \langle 0, 0 \rangle$, and let c be a scalar.

1. $\mathbf{u} \cdot \mathbf{v} = u_1 v_1 + u_2 v_2 = v_1 u_1 + v_2 u_2 = \mathbf{v} \cdot \mathbf{u}$

2. $\mathbf{0} \cdot \mathbf{v} = 0 \cdot v_1 + 0 \cdot v_2 = 0$

3. $\mathbf{u} \cdot (\mathbf{v} + \mathbf{w}) = \mathbf{u} \cdot \langle v_1 + w_1, v_2 + w_2 \rangle$

$\qquad = u_1(v_1 + w_1) + u_2(v_2 + w_2)$

$\qquad = u_1 v_1 + u_1 w_1 + u_2 v_2 + u_2 w_2$

$\qquad = (u_1 v_1 + u_2 v_2) + (u_1 w_1 + u_2 w_2)$

$\qquad = \mathbf{u} \cdot \mathbf{v} + \mathbf{u} \cdot \mathbf{w}$

4. $\mathbf{v} \cdot \mathbf{v} = v_1^2 + v_2^2 = \left(\sqrt{v_1^2 + v_2^2}\right)^2 = \|\mathbf{v}\|^2$

5. $c(\mathbf{u} \cdot \mathbf{v}) = c(\langle u_1, u_2 \rangle \cdot \langle v_1, v_2 \rangle)$

$\qquad = c(u_1 v_1 + u_2 v_2)$

$\qquad = (cu_1)v_1 + (cu_2)v_2$

$\qquad = \langle cu_1, cu_2 \rangle \cdot \langle v_1, v_2 \rangle$

$\qquad = c\mathbf{u} \cdot \mathbf{v}$

Angle Between Two Vectors   *(p. 430)*

If θ is the angle between two nonzero vectors **u** and **v**, then $\cos \theta = \dfrac{\mathbf{u} \cdot \mathbf{v}}{\|\mathbf{u}\|\|\mathbf{v}\|}$.

Proof

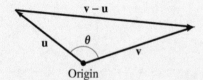

Consider the triangle determined by vectors **u**, **v**, and $\mathbf{v} - \mathbf{u}$, as shown at the left. By the Law of Cosines,

$$\|\mathbf{v} - \mathbf{u}\|^2 = \|\mathbf{u}\|^2 + \|\mathbf{v}\|^2 - 2\|\mathbf{u}\|\|\mathbf{v}\| \cos \theta$$

$$(\mathbf{v} - \mathbf{u}) \cdot (\mathbf{v} - \mathbf{u}) = \|\mathbf{u}\|^2 + \|\mathbf{v}\|^2 - 2\|\mathbf{u}\|\|\mathbf{v}\| \cos \theta$$

$$(\mathbf{v} - \mathbf{u}) \cdot \mathbf{v} - (\mathbf{v} - \mathbf{u}) \cdot \mathbf{u} = \|\mathbf{u}\|^2 + \|\mathbf{v}\|^2 - 2\|\mathbf{u}\|\|\mathbf{v}\| \cos \theta$$

$$\mathbf{v} \cdot \mathbf{v} - \mathbf{u} \cdot \mathbf{v} - \mathbf{v} \cdot \mathbf{u} + \mathbf{u} \cdot \mathbf{u} = \|\mathbf{u}\|^2 + \|\mathbf{v}\|^2 - 2\|\mathbf{u}\|\|\mathbf{v}\| \cos \theta$$

$$\|\mathbf{v}\|^2 - 2\mathbf{u} \cdot \mathbf{v} + \|\mathbf{u}\|^2 = \|\mathbf{u}\|^2 + \|\mathbf{v}\|^2 - 2\|\mathbf{u}\|\|\mathbf{v}\| \cos \theta$$

$$\cos \theta = \frac{\mathbf{u} \cdot \mathbf{v}}{\|\mathbf{u}\|\|\mathbf{v}\|}.$$

P.S. Problem Solving ▪ ▪ ▪ ▪ ▪ ▪ ▪ ▪ ▪ ▪ ▪ ▪ ▪ ▪

1. Distance In the figure, a beam of light is directed at the blue mirror, reflected to the red mirror, and then reflected back to the blue mirror. Find PT, the distance that the light travels from the red mirror back to the blue mirror.

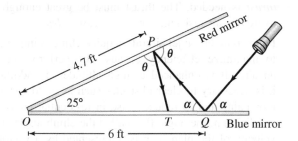

2. Correcting a Course A triathlete sets a course to swim S 25° E from a point on shore to a buoy $\frac{3}{4}$ mile away. After swimming 300 yards through a strong current, the triathlete is off course at a bearing of S 35° E. Find the bearing and distance the triathlete needs to swim to correct her course.

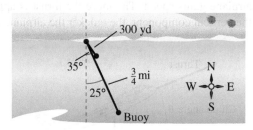

3. Locating Lost Hikers A group of hikers is lost in a national park. Two ranger stations receive an emergency SOS signal from the hikers. Station B is 75 miles due east of station A. The bearing from station A to the signal is S 60° E and the bearing from station B to the signal is S 75° W.

(a) Draw a diagram that gives a visual representation of the problem.

(b) Find the distance from each station to the SOS signal.

(c) A rescue party is in the park 20 miles from station A at a bearing of S 80° E. Find the distance and the bearing the rescue party must travel to reach the lost hikers.

4. Seeding a Courtyard You are seeding a triangular courtyard. One side of the courtyard is 52 feet long and another side is 46 feet long. The angle opposite the 52-foot side is 65°.

(a) Draw a diagram that gives a visual representation of the problem.

(b) How long is the third side of the courtyard?

(c) One bag of grass seed covers an area of 50 square feet. How many bags of grass seed do you need to cover the courtyard?

5. Finding Magnitudes For each pair of vectors, find the value of each expression.

(i) $\|\mathbf{u}\|$ (ii) $\|\mathbf{v}\|$ (iii) $\|\mathbf{u} + \mathbf{v}\|$

(iv) $\left\|\dfrac{\mathbf{u}}{\|\mathbf{u}\|}\right\|$ (v) $\left\|\dfrac{\mathbf{v}}{\|\mathbf{v}\|}\right\|$ (vi) $\left\|\dfrac{\mathbf{u} + \mathbf{v}}{\|\mathbf{u} + \mathbf{v}\|}\right\|$

(a) $\mathbf{u} = \langle 1, -1 \rangle$
  $\mathbf{v} = \langle -1, 2 \rangle$

(b) $\mathbf{u} = \langle 0, 1 \rangle$
  $\mathbf{v} = \langle 3, -3 \rangle$

(c) $\mathbf{u} = \langle 1, \frac{1}{2} \rangle$
  $\mathbf{v} = \langle 2, 3 \rangle$

(d) $\mathbf{u} = \langle 2, -4 \rangle$
  $\mathbf{v} = \langle 5, 5 \rangle$

6. Writing a Vector in Terms of Other Vectors Write the vector $\mathbf{w}$ in terms of $\mathbf{u}$ and $\mathbf{v}$, given that the terminal point of $\mathbf{w}$ bisects the line segment (see figure).

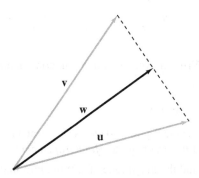

7. Proof Prove that if $\mathbf{u}$ is orthogonal to $\mathbf{v}$ and $\mathbf{w}$, then $\mathbf{u}$ is orthogonal to

$$c\mathbf{v} + d\mathbf{w}$$

for any scalars c and d.

8. Comparing Work Two forces of the same magnitude $\mathbf{F}_1$ and $\mathbf{F}_2$ act at angles θ_1 and θ_2, respectively. Use a diagram to compare the work done by $\mathbf{F}_1$ with the work done by $\mathbf{F}_2$ in moving along the vector $\overrightarrow{PQ}$ when

(a) $\theta_1 = -\theta_2$

(b) $\theta_1 = 60°$ and $\theta_2 = 30°$.

9. Think About It For each graph of the roots of a complex number, write each of the roots in trigonometric form.

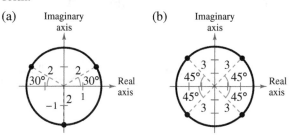

10. Skydiving A skydiver falls at a constant downward velocity of 120 miles per hour. In the figure, vector **u** represents the skydiver's velocity. A steady breeze pushes the skydiver to the east at 40 miles per hour. Vector **v** represents the wind velocity.

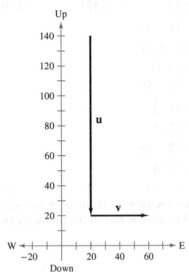

(a) Write the vectors **u** and **v** in component form.

(b) Let

$$\mathbf{s} = \mathbf{u} + \mathbf{v}.$$

Use the figure to sketch **s**. To print an enlarged copy of the graph, go to *MathGraphs.com*.

(c) Find the magnitude of **s**. What information does the magnitude give you about the skydiver's fall?

(d) Without wind, the skydiver would fall in a path perpendicular to the ground. At what angle to the ground is the path of the skydiver when affected by the 40-mile-per-hour wind from due west?

(e) The next day, the skydiver falls at a constant downward velocity of 120 miles per hour and a steady breeze pushes the skydiver to the west at 30 miles per hour. Draw a new figure that gives a visual representation of the problem and find the skydiver's new velocity.

11. Think About It The vectors **u** and **v** have the same magnitudes in the two figures. In which figure is the magnitude of the sum greater? Explain.

(a)   (b)

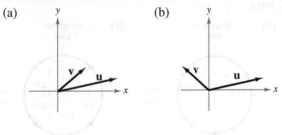

12. Speed and Velocity of an Airplane Four basic forces are in action during flight: weight, lift, thrust, and drag. To fly through the air, an object must overcome its own *weight*. To do this, it must create an upward force called *lift*. To generate lift, a forward motion called *thrust* is needed. The thrust must be great enough to overcome air resistance, which is called *drag*.

For a commercial jet aircraft, a quick climb is important to maximize efficiency because the performance of an aircraft is enhanced at high altitudes. In addition, it is necessary to clear obstacles such as buildings and mountains and to reduce noise in residential areas. In the diagram, the angle θ is called the climb angle. The velocity of the plane can be represented by a vector **v** with a vertical component $\|\mathbf{v}\| \sin \theta$ (called climb speed) and a horizontal component $\|\mathbf{v}\| \cos \theta$, where $\|\mathbf{v}\|$ is the speed of the plane.

When taking off, a pilot must decide how much of the thrust to apply to each component. The more the thrust is applied to the horizontal component, the faster the airplane gains speed. The more the thrust is applied to the vertical component, the quicker the airplane climbs.

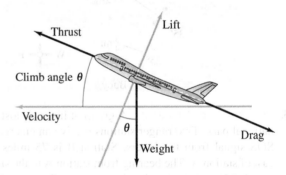

(a) Complete the table for an airplane that has a speed of $\|\mathbf{v}\| = 100$ miles per hour.

θ	0.5°	1.0°	1.5°	2.0°	2.5°	3.0°
$\|\mathbf{v}\| \sin \theta$						
$\|\mathbf{v}\| \cos \theta$						

(b) Does an airplane's speed equal the sum of the vertical and horizontal components of its velocity? If not, how could you find the speed of an airplane whose velocity components were known?

(c) Use the result of part (b) to find the speed of an airplane with the given velocity components.

    (i) $\|\mathbf{v}\| \sin \theta = 5.235$ miles per hour

        $\|\mathbf{v}\| \cos \theta = 149.909$ miles per hour

    (ii) $\|\mathbf{v}\| \sin \theta = 10.463$ miles per hour

        $\|\mathbf{v}\| \cos \theta = 149.634$ miles per hour

Appendix A   Review of Fundamental Concepts of Algebra

A.1   Real Numbers and Their Properties

Real numbers can represent many real-life quantities. For example, in Exercises 49–52 on page A12, you will use real numbers to represent the federal surplus or deficit.

■ **Represent and classify real numbers.**
■ **Order real numbers and use inequalities.**
■ **Find the absolute values of real numbers and find the distance between two real numbers.**
■ **Evaluate algebraic expressions.**
■ **Use the basic rules and properties of algebra.**

Real Numbers

Real numbers can describe quantities in everyday life such as age, miles per gallon, and population. Symbols such as

$$-5, \ 9, \ 0, \ \tfrac{4}{3}, \ 0.666\ldots, \ 28.21, \ \sqrt{2}, \ \pi, \ \text{and} \ \sqrt[3]{-32}$$

represent real numbers. Here are some important **subsets** (each member of a subset B is also a member of a set A) of the real numbers. The three dots, or *ellipsis points,* tell you that the pattern continues indefinitely.

$$\{1, 2, 3, 4, \ldots\} \qquad \text{Set of natural numbers}$$

$$\{0, 1, 2, 3, 4, \ldots\} \qquad \text{Set of whole numbers}$$

$$\{\ldots, -3, -2, -1, 0, 1, 2, 3, \ldots\} \qquad \text{Set of integers}$$

A real number is **rational** when it can be written as the ratio p/q of two integers, where $q \neq 0$. For example, the numbers

$$\tfrac{1}{3} = 0.3333\ldots = 0.\overline{3}, \quad \tfrac{1}{8} = 0.125, \quad \text{and} \quad \tfrac{125}{111} = 1.126126\ldots = 1.\overline{126}$$

are rational. The decimal representation of a rational number either repeats $\left(\text{as in } \tfrac{173}{55} = 3.1\overline{45}\right)$ or terminates $\left(\text{as in } \tfrac{1}{2} = 0.5\right)$. A real number that cannot be written as the ratio of two integers is **irrational.** The decimal representation of an irrational number neither terminates nor repeats. For example, the numbers

$$\sqrt{2} = 1.4142135\ldots \approx 1.41 \quad \text{and} \quad \pi = 3.1415926\ldots \approx 3.14$$

are irrational. (The symbol $\approx$ means "is approximately equal to.") Figure A.1 shows subsets of the real numbers and their relationships to each other.

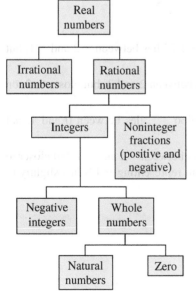

Subsets of the real numbers
Figure A.1

EXAMPLE 1   Classifying Real Numbers

Determine which numbers in the set $\left\{-13, -\sqrt{5}, -1, -\tfrac{1}{3}, 0, \tfrac{5}{8}, \sqrt{2}, \pi, 7\right\}$ are (a) natural numbers, (b) whole numbers, (c) integers, (d) rational numbers, and (e) irrational numbers.

Solution

a. Natural numbers: $\{7\}$ **b.** Whole numbers: $\{0, 7\}$

c. Integers: $\{-13, -1, 0, 7\}$ **d.** Rational numbers: $\left\{-13, -1, -\tfrac{1}{3}, 0, \tfrac{5}{8}, 7\right\}$

e. Irrational numbers: $\left\{-\sqrt{5}, \sqrt{2}, \pi\right\}$

✓ *Checkpoint*  *Audio-video solution in English & Spanish at LarsonPrecalculus.com*

Repeat Example 1 for the set $\left\{-\pi, -\tfrac{1}{4}, \tfrac{6}{3}, \tfrac{1}{2}\sqrt{2}, -7.5, -1, 8, -22\right\}$.

Real numbers are represented graphically on the **real number line.** When you draw a point on the real number line that corresponds to a real number, you are **plotting** the real number. The point representing 0 on the real number line is the **origin.** Numbers to the right of 0 are positive, and numbers to the left of 0 are negative, as shown in the figure below. The term **nonnegative** describes a number that is either positive or zero.

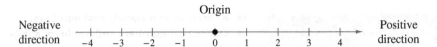

As the next two number lines illustrate, there is a *one-to-one correspondence* between real numbers and points on the real number line.

Every real number corresponds to exactly one point on the real number line.

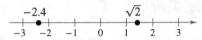

Every point on the real number line corresponds to exactly one real number.

EXAMPLE 2   **Plotting Points on the Real Number Line**

Plot the real numbers on the real number line.

a. $-\dfrac{7}{4}$

b. 2.3

c. $\dfrac{2}{3}$

d. -1.8

Solution   The figure below shows all four points.

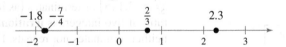

a. The point representing the real number $-\frac{7}{4} = -1.75$ lies between -2 and -1, but closer to -2, on the real number line.

b. The point representing the real number 2.3 lies between 2 and 3, but closer to 2, on the real number line.

c. The point representing the real number $\frac{2}{3} = 0.666\ldots$ lies between 0 and 1, but closer to 1, on the real number line.

d. The point representing the real number -1.8 lies between -2 and -1, but closer to -2, on the real number line. Note that the point representing -1.8 lies slightly to the left of the point representing $-\frac{7}{4}$.

✓ **Checkpoint** 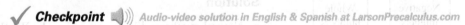 Audio-video solution in English & Spanish at LarsonPrecalculus.com

Plot the real numbers on the real number line.

a. $\dfrac{5}{2}$ **b.** -1.6

c. $-\dfrac{3}{4}$ **d.** 0.7

Ordering Real Numbers

One important property of real numbers is that they are *ordered*.

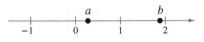

$a < b$ if and only if a lies to the left of b.

Figure A.2

> ### Definition of Order on the Real Number Line
>
> If a and b are real numbers, then a is *less than b* when $b - a$ is positive. The **inequality** $a < b$ denotes the **order** of a and b. This relationship can also be described by saying that b is *greater than a* and writing $b > a$. The inequality $a \le b$ means that a is *less than or equal to b*, and the inequality $b \ge a$ means that b is *greater than or equal to a*. The symbols $<$, $>$, $\le$, and $\ge$ are *inequality symbols*.

Geometrically, this definition implies that $a < b$ if and only if a lies to the *left* of b on the real number line, as shown in Figure A.2.

EXAMPLE 3   **Ordering Real Numbers**

Place the appropriate inequality symbol ($<$ or $>$) between the pair of real numbers.

a. $-3, 0$ **b.** $-2, -4$ **c.** $\frac{1}{4}, \frac{1}{3}$

Solution

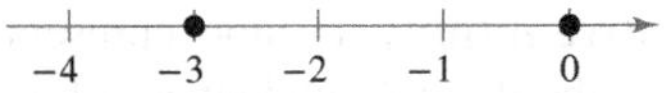

Figure A.3

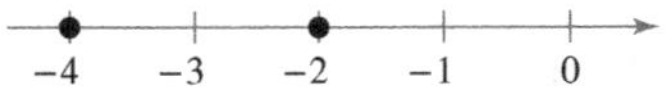

a. On the real number line, -3 lies to the left of 0, as shown in Figure A.3. So, you can say that -3 is *less than* 0, and write $-3 < 0$.

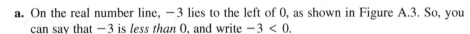

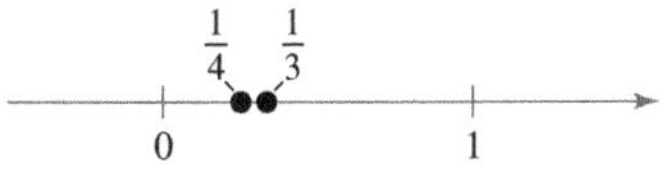

Figure A.4

b. On the real number line, -2 lies to the right of -4, as shown in Figure A.4. So, you can say that -2 is *greater than* -4, and write $-2 > -4$.

Figure A.5

c. On the real number line, $\frac{1}{4}$ lies to the left of $\frac{1}{3}$, as shown in Figure A.5. So, you can say that $\frac{1}{4}$ is *less than* $\frac{1}{3}$, and write $\frac{1}{4} < \frac{1}{3}$.

✓ *Checkpoint* ◀))) Audio-video solution in English & Spanish at LarsonPrecalculus.com

Place the appropriate inequality symbol ($<$ or $>$) between the pair of real numbers.

a. $1, -5$ **b.** $\frac{3}{2}, 7$ **c.** $-\frac{2}{3}, -\frac{3}{4}$

EXAMPLE 4   **Interpreting Inequalities**

See LarsonPrecalculus.com for an interactive version of this type of example.

Describe the subset of real numbers that the inequality represents.

a. $x \le 2$ **b.** $-2 \le x < 3$

Solution

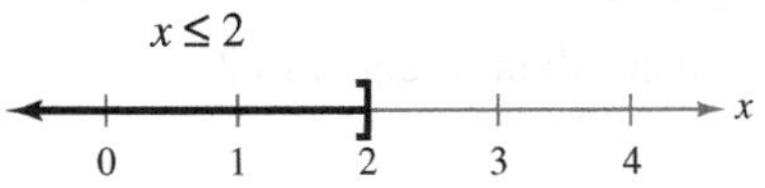

Figure A.6

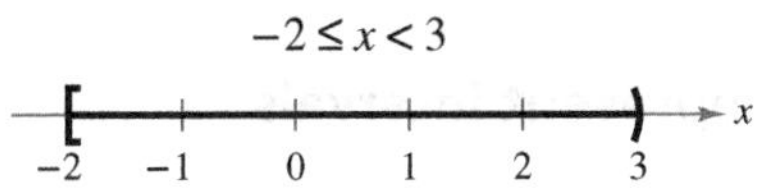

a. The inequality $x \le 2$ denotes all real numbers less than or equal to 2, as shown in Figure A.6.

Figure A.7

b. The inequality $-2 \le x < 3$ means that $x \ge -2$ *and* $x < 3$. This "double inequality" denotes all real numbers between -2 and 3, including -2 but not including 3, as shown in Figure A.7.

✓ *Checkpoint* ◀))) Audio-video solution in English & Spanish at LarsonPrecalculus.com

Describe the subset of real numbers that the inequality represents.

a. $x > -3$ **b.** $0 < x \le 4$

Inequalities can describe subsets of real numbers called **intervals.** In the bounded intervals below, the real numbers a and b are the **endpoints** of each interval. The endpoints of a closed interval are included in the interval, whereas the endpoints of an open interval are not included in the interval.

REMARK The reason that the four types of intervals at the right are called *bounded* is that each has a finite length. An interval that does not have a finite length is *unbounded* (see below).

Bounded Intervals on the Real Number Line

Notation	Interval Type	Inequality	Graph
$[a, b]$	Closed	$a \le x \le b$	
(a, b)	Open	$a < x < b$	
$[a, b)$		$a \le x < b$	
$(a, b]$		$a < x \le b$	

The symbols ∞, **positive infinity,** and $-\infty$, **negative infinity,** do not represent real numbers. They are convenient symbols used to describe the unboundedness of an interval such as $(1, \infty)$ or $(-\infty, 3]$.

REMARK Whenever you write an interval containing ∞ or $-\infty$, always use a parenthesis and never a bracket next to these symbols. This is because ∞ and $-\infty$ are never included in the interval.

Unbounded Intervals on the Real Number Line

Notation	Interval Type	Inequality	Graph
$[a, \infty)$		$x \ge a$	
(a, ∞)	Open	$x > a$	
$(-\infty, b]$		$x \le b$	
$(-\infty, b)$	Open	$x < b$	
$(-\infty, \infty)$	Entire real line	$-\infty < x < \infty$	

EXAMPLE 5 **Interpreting Intervals**

a. The interval $(-1, 0)$ consists of all real numbers greater than -1 and less than 0.

b. The interval $[2, \infty)$ consists of all real numbers greater than or equal to 2.

✓ **Checkpoint** 🔊))) Audio-video solution in English & Spanish at LarsonPrecalculus.com

Give a verbal description of the interval $[-2, 5)$.

EXAMPLE 6 **Using Inequalities to Represent Intervals**

a. The inequality $c \le 2$ can represent the statement "c is at most 2."

b. The inequality $-3 < x \le 5$ can represent "all x in the interval $(-3, 5]$."

✓ **Checkpoint** 🔊))) Audio-video solution in English & Spanish at LarsonPrecalculus.com

Use inequality notation to represent the statement "x is less than 4 and at least -2." ■

Absolute Value and Distance

The **absolute value** of a real number is its *magnitude,* or the distance between the origin and the point representing the real number on the real number line.

Definition of Absolute Value

If a is a real number, then the **absolute value** of a is

$$|a| = \begin{cases} a, & a \geq 0 \\ -a, & a < 0 \end{cases}.$$

Notice in this definition that the absolute value of a real number is never negative. For example, if $a = -5$, then $|-5| = -(-5) = 5$. The absolute value of a real number is either positive or zero. Moreover, 0 is the only real number whose absolute value is 0. So, $|0| = 0$.

Properties of Absolute Values

1. $|a| \geq 0$ **2.** $|-a| = |a|$

3. $|ab| = |a||b|$ **4.** $\left|\dfrac{a}{b}\right| = \dfrac{|a|}{|b|}, \quad b \neq 0$

EXAMPLE 7   **Finding Absolute Values**

a. $|-15| = 15$ **b.** $\left|\dfrac{2}{3}\right| = \dfrac{2}{3}$

c. $|-4.3| = 4.3$ **d.** $-|-6| = -(6) = -6$

✓ **Checkpoint** 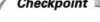 Audio-video solution in English & Spanish at LarsonPrecalculus.com

Evaluate each expression.

a. $|1|$ **b.** $-\left|\dfrac{3}{4}\right|$ **c.** $\dfrac{2}{|-3|}$ **d.** $-|0.7|$

EXAMPLE 8   **Evaluating an Absolute Value Expression**

Evaluate $\dfrac{|x|}{x}$ for (a) $x > 0$ and (b) $x < 0$.

Solution

a. If $x > 0$, then x is positive and $|x| = x$. So, $\dfrac{|x|}{x} = \dfrac{x}{x} = 1$.

b. If $x < 0$, then x is negative and $|x| = -x$. So, $\dfrac{|x|}{x} = \dfrac{-x}{x} = -1$.

✓ **Checkpoint** 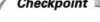 Audio-video solution in English & Spanish at LarsonPrecalculus.com

Evaluate $\dfrac{|x + 3|}{x + 3}$ for (a) $x > -3$ and (b) $x < -3$.

The **Law of Trichotomy** states that for any two real numbers a and b, *precisely* one of three relationships is possible:

$$a = b, \quad a < b, \quad \text{or} \quad a > b. \qquad \text{Law of Trichotomy}$$

EXAMPLE 9 Comparing Real Numbers

Place the appropriate symbol ($<$, $>$, or $=$) between the pair of real numbers.

a. $|-4| \quad\quad |3|$ **b.** $|-10| \quad\quad |10|$ **c.** $-|-7| \quad\quad |-7|$

Solution

a. $|-4| > |3|$ because $|-4| = 4$ and $|3| = 3$, and 4 is greater than 3.

b. $|-10| = |10|$ because $|-10| = 10$ and $|10| = 10$.

c. $-|-7| < |-7|$ because $-|-7| = -7$ and $|-7| = 7$, and -7 is less than 7.

✓ **Checkpoint**))) *Audio-video solution in English & Spanish at LarsonPrecalculus.com*

Place the appropriate symbol ($<$, $>$, or $=$) between the pair of real numbers.

a. $|-3| \quad\quad |4|$

b. $-|-4| \quad\quad -|4|$

c. $|-3| \quad\quad -|-3|$

Absolute value can be used to find the distance between two points on the real number line. For example, the distance between -3 and 4 is

$$|-3 - 4| = |-7|$$
$$= 7$$

as shown in Figure A.8.

The distance between -3 and 4 is 7.

Figure A.8

Distance Between Two Points on the Real Number Line

Let a and b be real numbers. The **distance between a and b** is

$$d(a, b) = |b - a| = |a - b|.$$

EXAMPLE 10 Finding a Distance

Find the distance between -25 and 13.

Solution

The distance between -25 and 13 is

$$|-25 - 13| = |-38| = 38. \qquad \text{Distance between } -25 \text{ and } 13$$

The distance can also be found as follows.

$$|13 - (-25)| = |38| = 38 \qquad \text{Distance between } -25 \text{ and } 13$$

✓ **Checkpoint**))) *Audio-video solution in English & Spanish at LarsonPrecalculus.com*

a. Find the distance between 35 and -23.

b. Find the distance between -35 and -23.

c. Find the distance between 35 and 23.

Algebraic Expressions

One characteristic of algebra is the use of letters to represent numbers. The letters are **variables,** and combinations of letters and numbers are **algebraic expressions.** Here are a few examples of algebraic expressions.

$$5x, \qquad 2x - 3, \qquad \frac{4}{x^2 + 2}, \qquad 7x + y$$

Definition of an Algebraic Expression

An **algebraic expression** is a collection of letters (**variables**) and real numbers (**constants**) combined using the operations of addition, subtraction, multiplication, division, and exponentiation.

The **terms** of an algebraic expression are those parts that are separated by *addition.* For example, $x^2 - 5x + 8 = x^2 + (-5x) + 8$ has three terms: x^2 and $-5x$ are the **variable terms** and 8 is the **constant term.** For terms such as x^2, $-5x$, and 8, the numerical factor is the **coefficient.** Here, the coefficients are 1, -5, and 8.

EXAMPLE 11   **Identifying Terms and Coefficients**

Algebraic Expression	Terms	Coefficients
a. $5x - \dfrac{1}{7}$	$5x, -\dfrac{1}{7}$	$5, -\dfrac{1}{7}$
b. $2x^2 - 6x + 9$	$2x^2, -6x, 9$	$2, -6, 9$
c. $\dfrac{3}{x} + \dfrac{1}{2}x^4 - y$	$\dfrac{3}{x}, \dfrac{1}{2}x^4, -y$	$3, \dfrac{1}{2}, -1$

✓ *Checkpoint* 🔊))) Audio-video solution in English & Spanish at LarsonPrecalculus.com

Identify the terms and coefficients of $-2x + 4$.

The **Substitution Principle** states, "If $a = b$, then b can replace a in any expression involving a." Use the Substitution Principle to **evaluate** an algebraic expression by substituting numerical values for each of the variables in the expression. The next example illustrates this.

EXAMPLE 12   **Evaluating Algebraic Expressions**

Expression	Value of Variable	Substitute.	Value of Expression
a. $-3x + 5$	$x = 3$	$-3(3) + 5$	$-9 + 5 = -4$
b. $3x^2 + 2x - 1$	$x = -1$	$3(-1)^2 + 2(-1) - 1$	$3 - 2 - 1 = 0$
c. $\dfrac{2x}{x + 1}$	$x = -3$	$\dfrac{2(-3)}{-3 + 1}$	$\dfrac{-6}{-2} = 3$

Note that you must substitute the value for *each* occurrence of the variable.

✓ *Checkpoint* 🔊))) Audio-video solution in English & Spanish at LarsonPrecalculus.com

Evaluate $4x - 5$ when $x = 0$.

Basic Rules of Algebra

There are four arithmetic operations with real numbers: *addition, multiplication, subtraction,* and *division,* denoted by the symbols $+$, $\times$ or $\cdot$, $-$, and $\div$ or $/$, respectively. Of these, addition and multiplication are the two primary operations. Subtraction and division are the inverse operations of addition and multiplication, respectively.

Definitions of Subtraction and Division

Subtraction: Add the opposite.   **Division:** Multiply by the reciprocal.

$$a - b = a + (-b)$$ $$\text{If } b \neq 0, \text{ then } a/b = a\left(\frac{1}{b}\right) = \frac{a}{b}.$$

In these definitions, $-b$ is the **additive inverse** (or opposite) of b, and $1/b$ is the **multiplicative inverse** (or reciprocal) of b. In the fractional form a/b, a is the **numerator** of the fraction and b is the **denominator.**

The properties of real numbers below are true for variables and algebraic expressions as well as for real numbers, so they are often called the **Basic Rules of Algebra.** Formulate a verbal description of each of these properties. For example, the first property states that *the order in which two real numbers are added does not affect their sum.*

Basic Rules of Algebra

Let a, b, and c be real numbers, variables, or algebraic expressions.

Property		Example
Commutative Property of Addition:	$a + b = b + a$	$4x + x^2 = x^2 + 4x$
Commutative Property of Multiplication:	$ab = ba$	$(4 - x)x^2 = x^2(4 - x)$
Associative Property of Addition:	$(a + b) + c = a + (b + c)$	$(x + 5) + x^2 = x + (5 + x^2)$
Associative Property of Multiplication:	$(ab)c = a(bc)$	$(2x \cdot 3y)(8) = (2x)(3y \cdot 8)$
Distributive Properties:	$a(b + c) = ab + ac$	$3x(5 + 2x) = 3x \cdot 5 + 3x \cdot 2x$
	$(a + b)c = ac + bc$	$(y + 8)y = y \cdot y + 8 \cdot y$
Additive Identity Property:	$a + 0 = a$	$5y^2 + 0 = 5y^2$
Multiplicative Identity Property:	$a \cdot 1 = a$	$(4x^2)(1) = 4x^2$
Additive Inverse Property:	$a + (-a) = 0$	$5x^3 + (-5x^3) = 0$
Multiplicative Inverse Property:	$a \cdot \dfrac{1}{a} = 1, \quad a \neq 0$	$(x^2 + 4)\left(\dfrac{1}{x^2 + 4}\right) = 1$

Subtraction is defined as "adding the opposite," so the Distributive Properties are also true for subtraction. For example, the "subtraction form" of $a(b + c) = ab + ac$ is $a(b - c) = ab - ac$. Note that the operations of subtraction and division are neither commutative nor associative. The examples

$$7 - 3 \neq 3 - 7 \quad \text{and} \quad 20 \div 4 \neq 4 \div 20$$

show that subtraction and division are not commutative. Similarly

$$5 - (3 - 2) \neq (5 - 3) - 2 \quad \text{and} \quad 16 \div (4 \div 2) \neq (16 \div 4) \div 2$$

demonstrate that subtraction and division are not associative.

EXAMPLE 13   **Identifying Rules of Algebra**

Identify the rule of algebra illustrated by the statement.

a. $(5x^3)2 = 2(5x^3)$ **b.** $(4x + 3) - (4x + 3) = 0$

c. $7x \cdot \dfrac{1}{7x} = 1, \quad x \neq 0$ **d.** $(2 + 5x^2) + x^2 = 2 + (5x^2 + x^2)$

Solution

a. This statement illustrates the Commutative Property of Multiplication. In other words, you obtain the same result whether you multiply $5x^3$ by 2, or 2 by $5x^3$.

b. This statement illustrates the Additive Inverse Property. In terms of subtraction, this property states that when any expression is subtracted from itself, the result is 0.

c. This statement illustrates the Multiplicative Inverse Property. Note that x must be a nonzero number. The reciprocal of x is undefined when x is 0.

d. This statement illustrates the Associative Property of Addition. In other words, to form the sum $2 + 5x^2 + x^2$, it does not matter whether 2 and $5x^2$, or $5x^2$ and x^2 are added first.

✓ *Checkpoint* *Audio-video solution in English & Spanish at LarsonPrecalculus.com*

Identify the rule of algebra illustrated by the statement.

a. $x + 9 = 9 + x$ **b.** $5(x^3 \cdot 2) = (5x^3)2$ **c.** $(2 + 5x^2)y^2 = 2 \cdot y^2 + 5x^2 \cdot y^2$

• • REMARK   Notice the difference between the *opposite of a number* and a *negative number*. If a is already negative, then its opposite, $-a$, is positive. For example, if $a = -5$, then

$$-a = -(-5) = 5.$$

Properties of Negation and Equality

Let a, b, and c be real numbers, variables, or algebraic expressions.

Property	Example
1. $(-1)a = -a$	$(-1)7 = -7$
2. $-(-a) = a$	$-(-6) = 6$
3. $(-a)b = -(ab) = a(-b)$	$(-5)3 = -(5 \cdot 3) = 5(-3)$
4. $(-a)(-b) = ab$	$(-2)(-x) = 2x$
5. $-(a + b) = (-a) + (-b)$	$-(x + 8) = (-x) + (-8)$
	$\qquad\quad = -x - 8$
6. If $a = b$, then $a \pm c = b \pm c$.	$\frac{1}{2} + 3 = 0.5 + 3$
7. If $a = b$, then $ac = bc$.	$4^2 \cdot 2 = 16 \cdot 2$
8. If $a \pm c = b \pm c$, then $a = b$.	$1.4 - 1 = \frac{7}{5} - 1 \implies 1.4 = \frac{7}{5}$
9. If $ac = bc$ and $c \neq 0$, then $a = b$.	$3x = 3 \cdot 4 \implies x = 4$

• • REMARK   The "or" in the Zero-Factor Property includes the possibility that either or both factors may be zero. This is an *inclusive or*, and it is generally the way the word "or" is used in mathematics.

Properties of Zero

Let a and b be real numbers, variables, or algebraic expressions.

1. $a + 0 = a$ and $a - 0 = a$ **2.** $a \cdot 0 = 0$

3. $\dfrac{0}{a} = 0, \quad a \neq 0$ **4.** $\dfrac{a}{0}$ is undefined.

5. Zero-Factor Property: If $ab = 0$, then $a = 0$ or $b = 0$.

Properties and Operations of Fractions

Let a, b, c, and d be real numbers, variables, or algebraic expressions such that $b \neq 0$ and $d \neq 0$.

1. **Equivalent Fractions:** $\dfrac{a}{b} = \dfrac{c}{d}$ if and only if $ad = bc$.

2. **Rules of Signs:** $-\dfrac{a}{b} = \dfrac{-a}{b} = \dfrac{a}{-b}$ and $\dfrac{-a}{-b} = \dfrac{a}{b}$

3. **Generate Equivalent Fractions:** $\dfrac{a}{b} = \dfrac{ac}{bc}$, $c \neq 0$

4. **Add or Subtract with Like Denominators:** $\dfrac{a}{b} \pm \dfrac{c}{b} = \dfrac{a \pm c}{b}$

5. **Add or Subtract with Unlike Denominators:** $\dfrac{a}{b} \pm \dfrac{c}{d} = \dfrac{ad \pm bc}{bd}$

6. **Multiply Fractions:** $\dfrac{a}{b} \cdot \dfrac{c}{d} = \dfrac{ac}{bd}$

7. **Divide Fractions:** $\dfrac{a}{b} \div \dfrac{c}{d} = \dfrac{a}{b} \cdot \dfrac{d}{c} = \dfrac{ad}{bc}$, $c \neq 0$

> **REMARK** In Property 1, the phrase "if and only if" implies two statements. One statement is: If $a/b = c/d$, then $ad = bc$. The other statement is: If $ad = bc$, where $b \neq 0$ and $d \neq 0$, then $a/b = c/d$.

EXAMPLE 14 **Properties and Operations of Fractions**

a. $\dfrac{x}{5} = \dfrac{3 \cdot x}{3 \cdot 5} = \dfrac{3x}{15}$

b. $\dfrac{7}{x} \div \dfrac{3}{2} = \dfrac{7}{x} \cdot \dfrac{2}{3} = \dfrac{14}{3x}$

✓ **Checkpoint** ◀))) Audio-video solution in English & Spanish at LarsonPrecalculus.com

a. Multiply fractions: $\dfrac{3}{5} \cdot \dfrac{x}{6}$.

b. Add fractions: $\dfrac{x}{10} + \dfrac{2x}{5}$.

> **REMARK** The number 1 is neither prime nor composite.

If a, b, and c are integers such that $ab = c$, then a and b are **factors** or **divisors** of c. A **prime number** is an integer that has exactly two positive factors—itself and 1—such as 2, 3, 5, 7, and 11. The numbers 4, 6, 8, 9, and 10 are **composite** because each can be written as the product of two or more prime numbers. The **Fundamental Theorem of Arithmetic** states that every positive integer greater than 1 is a prime number or can be written as the product of prime numbers in precisely one way (disregarding order). For example, the *prime factorization* of 24 is $24 = 2 \cdot 2 \cdot 2 \cdot 3$.

Summarize (Appendix A.1)

1. Explain how to represent and classify real numbers *(pages A1 and A2)*. For examples of representing and classifying real numbers, see Examples 1 and 2.

2. Explain how to order real numbers and use inequalities *(pages A3 and A4)*. For examples of ordering real numbers and using inequalities, see Examples 3–6.

3. State the definition of the absolute value of a real number *(page A5)*. For examples of using absolute value, see Examples 7–10.

4. Explain how to evaluate an algebraic expression *(page A7)*. For examples involving algebraic expressions, see Examples 11 and 12.

5. State the basic rules and properties of algebra *(pages A8–A10)*. For examples involving the basic rules and properties of algebra, see Examples 13 and 14.

A.1   Exercises

See CalcChat.com for tutorial help and worked-out solutions to odd-numbered exercises.

Vocabulary: Fill in the blanks.

1. The decimal representation of an _____ number neither terminates nor repeats.
2. The point representing 0 on the real number line is the _____.
3. The distance between the origin and a point representing a real number on the real number line is the _____ _____ of the real number.
4. A number that can be written as the product of two or more prime numbers is a _____ number.
5. The _____ of an algebraic expression are those parts that are separated by addition.
6. The _____ _____ states that if $ab = 0$, then $a = 0$ or $b = 0$.

Skills and Applications

Classifying Real Numbers In Exercises 7–10, determine which numbers in the set are (a) natural numbers, (b) whole numbers, (c) integers, (d) rational numbers, and (e) irrational numbers.

7. $\left\{-9, -\frac{7}{2}, 5, \frac{2}{3}, \sqrt{2}, 0, 1, -4, 2, -11\right\}$
8. $\left\{\sqrt{5}, -7, -\frac{7}{3}, 0, 3.14, \frac{5}{4}, -3, 12, 5\right\}$
9. $\{2.01, 0.\overline{6}, -13, 0.010110111 \ldots, 1, -6\}$
10. $\left\{25, -17, -\frac{12}{5}, \sqrt{9}, 3.12, \frac{1}{2}\pi, 7, -11.1, 13\right\}$

Plotting Points on the Real Number Line In Exercises 11 and 12, plot the real numbers on the real number line.

11. (a) 3   (b) $\frac{7}{2}$ (c) $-\frac{5}{2}$ (d) -5.2
12. (a) 8.5   (b) $\frac{4}{3}$ (c) -4.75 (d) $-\frac{8}{3}$

Plotting and Ordering Real Numbers In Exercises 13–16, plot the two real numbers on the real number line. Then place the appropriate inequality symbol (< or >) between them.

13. $-4, -8$
14. $1, \frac{16}{3}$
15. $\frac{5}{6}, \frac{2}{3}$
16. $-\frac{8}{7}, -\frac{3}{7}$

Interpreting an Inequality or an Interval In Exercises 17–24, (a) give a verbal description of the subset of real numbers represented by the inequality or the interval, (b) sketch the subset on the real number line, and (c) state whether the subset is bounded or unbounded.

17. $x \le 5$
18. $x < 0$
19. $-2 < x < 2$
20. $0 < x \le 6$
21. $[4, \infty)$
22. $(-\infty, 2)$
23. $[-5, 2)$
24. $(-1, 2]$

Using Inequality and Interval Notation In Exercises 25–28, use inequality notation and interval notation to describe the set.

25. y is nonnegative.
26. y is no more than 25.
27. t is at least 10 and at most 22.
28. k is less than 5 but no less than -3.

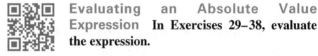

Evaluating an Absolute Value Expression In Exercises 29–38, evaluate the expression.

29. $|-10|$
30. $|0|$
31. $|3 - 8|$
32. $|6 - 2|$
33. $|-1| - |-2|$
34. $-3 - |-3|$
35. $5|-5|$
36. $-4|-4|$
37. $\dfrac{|x + 2|}{x + 2}, \quad x < -2$
38. $\dfrac{|x - 1|}{x - 1}, \quad x > 1$

Comparing Real Numbers In Exercises 39–42, place the appropriate symbol (<, >, or =) between the pair of real numbers.

39. $|-4|$ ▨ $|4|$
40. -5 ▨ $-|5|$
41. $-|-6|$ ▨ $|-6|$
42. $-|-2|$ ▨ $-|2|$

Finding a Distance In Exercises 43–46, find the distance between a and b.

43. $a = 126, b = 75$
44. $a = -20, b = 30$
45. $a = -\frac{5}{2}, b = 0$
46. $a = -\frac{1}{4}, b = -\frac{11}{4}$

Using Absolute Value Notation In Exercises 47 and 48, use absolute value notation to represent the situation.

47. The distance between x and 5 is no more than 3.
48. The distance between x and -10 is at least 6.

• • Federal Deficit • • • • • • • • • • • • • • • • •

In Exercises 49–52, use the bar graph, which shows the receipts of the federal government (in billions of dollars) for selected years from 2008 through 2014. In each exercise, you are given the expenditures of the federal government. Find the magnitude of the surplus or deficit for the year. (*Source: U.S. Office of Management and Budget*)

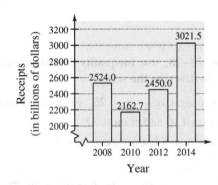

Year	Receipts, R	Expenditures, E	$\lvert R - E \rvert$
49. 2008		$2982.5 billion	
50. 2010		$3457.1 billion	
51. 2012		$3537.0 billion	
52. 2014		$3506.1 billion	

 Identifying Terms and Coefficients In Exercises 53–58, identify the terms. Then identify the coefficients of the variable terms of the expression.

53. $7x + 4$ **54.** $2x - 3$

55. $6x^3 - 5x$ **56.** $4x^3 + 0.5x - 5$

57. $3\sqrt{3}x^2 + 1$ **58.** $2\sqrt{2}x^2 - 3$

Evaluating an Algebraic Expression In Exercises 59–64, evaluate the expression for each value of x. (If not possible, state the reason.)

59. $4x - 6$ (a) $x = -1$ (b) $x = 0$

60. $9 - 7x$ (a) $x = -3$ (b) $x = 3$

61. $x^2 - 3x + 2$ (a) $x = 0$ (b) $x = -1$

62. $-x^2 + 5x - 4$ (a) $x = -1$ (b) $x = 1$

63. $\dfrac{x + 1}{x - 1}$ (a) $x = 1$ (b) $x = -1$

64. $\dfrac{x - 2}{x + 2}$ (a) $x = 2$ (b) $x = -2$

Identifying Rules of Algebra In Exercises 65–68, identify the rule(s) of algebra illustrated by the statement.

65. $\dfrac{1}{h + 6}(h + 6) = 1, \quad h \neq -6$

66. $(x + 3) - (x + 3) = 0$

67. $x(3y) = (x \cdot 3)y = (3x)y$

68. $\frac{1}{7}(7 \cdot 12) = \left(\frac{1}{7} \cdot 7\right)12 = 1 \cdot 12 = 12$

Operations with Fractions In Exercises 69–72, perform the operation. (Write fractional answers in simplest form.)

69. $\dfrac{2x}{3} - \dfrac{x}{4}$ **70.** $\dfrac{3x}{4} + \dfrac{x}{5}$

71. $\dfrac{3x}{10} \cdot \dfrac{5}{6}$ **72.** $\dfrac{2x}{3} \div \dfrac{6}{7}$

Exploration

True or False? In Exercises 73–75, determine whether the statement is true or false. Justify your answer.

73. Every nonnegative number is positive.

74. If $a > 0$ and $b < 0$, then $ab > 0$.

75. If $a < 0$ and $b < 0$, then $ab > 0$.

76. **HOW DO YOU SEE IT?** Match each description with its graph. Which types of real numbers shown in Figure A.1 on page A1 may be included in a range of prices? a range of lengths? Explain.

(i)

| 1.87 | 1.88 | 1.89 | 1.90 | 1.91 | 1.92 | 1.93 |

(ii)

| 1.87 | 1.88 | 1.89 | 1.90 | 1.91 | 1.92 | 1.93 |

(a) The price of an item is within $0.03 of $1.90.

(b) The distance between the prongs of an electric plug may not differ from 1.9 centimeters by more than 0.03 centimeter.

77. Conjecture

(a) Use a calculator to complete the table.

n	0.0001	0.01	1	100	10,000
$\dfrac{5}{n}$					

(b) Use the result from part (a) to make a conjecture about the value of $5/n$ as n (i) approaches 0, and (ii) increases without bound.

A.2 Exponents and Radicals

Real numbers and algebraic expressions are often written with exponents and radicals. For example, in Exercise 69 on page A24, you will use an expression involving rational exponents to find the times required for a funnel to empty for different water heights.

- Use properties of exponents.
- Use scientific notation to represent real numbers.
- Use properties of radicals.
- Simplify and combine radical expressions.
- Rationalize denominators and numerators.
- Use properties of rational exponents.

Integer Exponents and Their Properties

Repeated *multiplication* can be written in **exponential form.**

Repeated Multiplication	Exponential Form
$a \cdot a \cdot a \cdot a \cdot a$	a^5
$(-4)(-4)(-4)$	$(-4)^3$
$(2x)(2x)(2x)(2x)$	$(2x)^4$

Exponential Notation

If a is a real number and n is a positive integer, then

$$a^n = \underbrace{a \cdot a \cdot a \cdots a}_{n \text{ factors}}$$

where n is the **exponent** and a is the **base.** You read a^n as "a to the nth **power.**"

An exponent can also be negative or zero. Properties 3 and 4 below show how to use negative and zero exponents.

Properties of Exponents

Let a and b be real numbers, variables, or algebraic expressions, and let m and n be integers. (All denominators and bases are nonzero.)

Property	Example
1. $a^m a^n = a^{m+n}$	$3^2 \cdot 3^4 = 3^{2+4} = 3^6 = 729$
2. $\dfrac{a^m}{a^n} = a^{m-n}$	$\dfrac{x^7}{x^4} = x^{7-4} = x^3$
3. $a^{-n} = \dfrac{1}{a^n} = \left(\dfrac{1}{a}\right)^n$	$y^{-4} = \dfrac{1}{y^4} = \left(\dfrac{1}{y}\right)^4$
4. $a^0 = 1$	$(x^2 + 1)^0 = 1$
5. $(ab)^m = a^m b^m$	$(5x)^3 = 5^3 x^3 = 125x^3$
6. $(a^m)^n = a^{mn}$	$(y^3)^{-4} = y^{3(-4)} = y^{-12} = \dfrac{1}{y^{12}}$
7. $\left(\dfrac{a}{b}\right)^m = \dfrac{a^m}{b^m}$	$\left(\dfrac{2}{x}\right)^3 = \dfrac{2^3}{x^3} = \dfrac{8}{x^3}$
8. $|a^2| = |a|^2 = a^2$	$|(-2)^2| = |-2|^2 = 2^2 = 4 = (-2)^2$

The properties of exponents listed on the preceding page apply to *all* integers m and n, not just to positive integers, as shown in Examples 1–4.

It is important to recognize the difference between expressions such as $(-2)^4$ and -2^4. In $(-2)^4$, the parentheses tell you that the exponent applies to the negative sign as well as to the 2, but in $-2^4 = -(2^4)$, the exponent applies only to the 2. So, $(-2)^4 = 16$ and $-2^4 = -16$.

EXAMPLE 1 Evaluating Exponential Expressions

a. $(-5)^2 = (-5)(-5) = 25$ Negative sign is part of the base.

b. $-5^2 = -(5)(5) = -25$ Negative sign is *not* part of the base.

c. $2 \cdot 2^4 = 2^{1+4} = 2^5 = 32$ Property 1

d. $\dfrac{4^4}{4^6} = 4^{4-6} = 4^{-2} = \dfrac{1}{4^2} = \dfrac{1}{16}$ Properties 2 and 3

✓ **Checkpoint** *Audio-video solution in English & Spanish at LarsonPrecalculus.com*

Evaluate each expression.

a. -3^4 **b.** $(-3)^4$

c. $3^2 \cdot 3$ **d.** $\dfrac{3^5}{3^8}$

▷ **TECHNOLOGY** When using a calculator to evaluate exponential expressions, it is important to know when to use parentheses because the calculator follows the order of operations. For example, here is how you would evaluate $(-2)^4$ on a graphing utility.

$(\,)\ (\,(-)\,)\ 2\ (\,)\ \boxed{\wedge}\ 4\ \boxed{\text{ENTER}}$

The display will be 16. If you omit the parentheses, the display will be -16.

EXAMPLE 2 Evaluating Algebraic Expressions

Evaluate each algebraic expression when $x = 3$.

a. $5x^{-2}$ **b.** $\dfrac{1}{3}(-x)^3$

Solution

a. When $x = 3$, the expression $5x^{-2}$ has a value of

$$5x^{-2} = 5(3)^{-2} = \frac{5}{3^2} = \frac{5}{9}.$$

b. When $x = 3$, the expression $\dfrac{1}{3}(-x)^3$ has a value of

$$\frac{1}{3}(-x)^3 = \frac{1}{3}(-3)^3 = \frac{1}{3}(-27) = -9.$$

✓ **Checkpoint** *Audio-video solution in English & Spanish at LarsonPrecalculus.com*

Evaluate each algebraic expression when $x = 4$.

a. $-x^{-2}$ **b.** $\dfrac{1}{4}(-x)^4$

EXAMPLE 3   **Using Properties of Exponents**

Use the properties of exponents to simplify each expression.

a. $(-3ab^4)(4ab^{-3})$ **b.** $(2xy^2)^3$ **c.** $3a(-4a^2)^0$ **d.** $\left(\dfrac{5x^3}{y}\right)^2$

Solution

a. $(-3ab^4)(4ab^{-3}) = (-3)(4)(a)(a)(b^4)(b^{-3}) = -12a^2b$

b. $(2xy^2)^3 = 2^3(x)^3(y^2)^3 = 8x^3y^6$

c. $3a(-4a^2)^0 = 3a(1) = 3a$

d. $\left(\dfrac{5x^3}{y}\right)^2 = \dfrac{5^2(x^3)^2}{y^2} = \dfrac{25x^6}{y^2}$

✓ **Checkpoint**  *Audio-video solution in English & Spanish at LarsonPrecalculus.com*

Use the properties of exponents to simplify each expression.

a. $(2x^{-2}y^3)(-x^4y)$ **b.** $(4a^2b^3)^0$ **c.** $(-5z)^3(z^2)$ **d.** $\left(\dfrac{3x^4}{x^2y^2}\right)^2$

EXAMPLE 4   **Rewriting with Positive Exponents**

a. $x^{-1} = \dfrac{1}{x}$ Property 3

b. $\dfrac{1}{3x^{-2}} = \dfrac{1(x^2)}{3}$ Property 3 (The exponent -2 does not apply to 3.)

$= \dfrac{x^2}{3}$ Simplify.

c. $\dfrac{12a^3b^{-4}}{4a^{-2}b} = \dfrac{12a^3 \cdot a^2}{4b \cdot b^4}$ Property 3

$= \dfrac{3a^5}{b^5}$ Property 1

d. $\left(\dfrac{3x^2}{y}\right)^{-2} = \dfrac{3^{-2}(x^2)^{-2}}{y^{-2}}$ Properties 5 and 7

$= \dfrac{3^{-2}x^{-4}}{y^{-2}}$ Property 6

$= \dfrac{y^2}{3^2x^4}$ Property 3

$= \dfrac{y^2}{9x^4}$ Simplify.

✓ **Checkpoint** *Audio-video solution in English & Spanish at LarsonPrecalculus.com*

Rewrite each expression with positive exponents. Simplify, if possible.

a. $2a^{-2}$ **b.** $\dfrac{3a^{-3}b^4}{15ab^{-1}}$

c. $\left(\dfrac{x}{10}\right)^{-1}$ **d.** $(-2x^2)^3(4x^3)^{-1}$

•• REMARK   Rarely in algebra is there only one way to solve a problem. Do not be concerned when the steps you use to solve a problem are not exactly the same as the steps presented in this text. It is important to use steps that you understand and, of course, steps that are justified by the rules of algebra. For example, the fractional form of Property 3 is

$$\left(\dfrac{a}{b}\right)^{-m} = \left(\dfrac{b}{a}\right)^m.$$

So, you might prefer the steps below for Example 4(d).

$$\left(\dfrac{3x^2}{y}\right)^{-2} = \left(\dfrac{y}{3x^2}\right)^2 = \dfrac{y^2}{9x^4}$$

Scientific Notation

Exponents provide an efficient way of writing and computing with very large (or very small) numbers. For example, there are about 366 billion billion gallons of water on Earth—that is, 366 followed by 18 zeros.

$$366{,}000{,}000{,}000{,}000{,}000{,}000$$

It is convenient to write such numbers in **scientific notation.** This notation has the form $\pm c \times 10^n$, where $1 \le c < 10$ and n is an integer. So, the number of gallons of water on Earth, written in scientific notation, is

$$3.66 \times 100{,}000{,}000{,}000{,}000{,}000{,}000 = 3.66 \times 10^{20}.$$

The *positive* exponent 20 tells you that the number is *large* (10 or more) and that the decimal point has been moved 20 places. A *negative* exponent tells you that the number is *small* (less than 1). For example, the mass (in grams) of one electron is approximately

$$9.1 \times 10^{-28} = 0.00000000000000000000000000091.$$

28 decimal places

EXAMPLE 5   **Scientific Notation**

a. $0.0000782 = 7.82 \times 10^{-5}$

b. $836{,}100{,}000 = 8.361 \times 10^8$

✓ **Checkpoint**))) Audio-video solution in English & Spanish at LarsonPrecalculus.com

Write 45,850 in scientific notation.

EXAMPLE 6   **Decimal Notation**

a. $-9.36 \times 10^{-6} = -0.00000936$

b. $1.345 \times 10^2 = 134.5$

✓ **Checkpoint**))) Audio-video solution in English & Spanish at LarsonPrecalculus.com

Write -2.718×10^{-3} in decimal notation.

▷ **TECHNOLOGY** Most calculators automatically switch to scientific notation when showing large (or small) numbers that exceed the display range.
     To enter numbers in scientific notation, your calculator should have an exponential entry key labeled

⎡EE⎤ or ⎡EXP⎤.

Consult the user's guide for instructions on keystrokes and how your calculator displays numbers in scientific notation.

EXAMPLE 7   **Using Scientific Notation**

Evaluate $\dfrac{(2{,}400{,}000{,}000)(0.0000045)}{(0.00003)(1500)}$.

Solution   Begin by rewriting each number in scientific notation. Then simplify.

$$\frac{(2{,}400{,}000{,}000)(0.0000045)}{(0.00003)(1500)} = \frac{(2.4 \times 10^9)(4.5 \times 10^{-6})}{(3.0 \times 10^{-5})(1.5 \times 10^3)}$$

$$= \frac{(2.4)(4.5)(10^3)}{(4.5)(10^{-2})}$$

$$= (2.4)(10^5)$$

$$= 240{,}000$$

✓ **Checkpoint**))) Audio-video solution in English & Spanish at LarsonPrecalculus.com

Evaluate $(24{,}000{,}000{,}000)(0.00000012)(300{,}000)$.

Radicals and Their Properties

A **square root** of a number is one of its two equal factors. For example, 5 is a square root of 25 because 5 is one of the two equal factors of 25. In a similar way, a **cube root** of a number is one of its three equal factors, as in $125 = 5^3$.

Definition of *n*th Root of a Number

Let a and b be real numbers and let $n \geq 2$ be a positive integer. If

$$a = b^n$$

then b is an **nth root of a.** If $n = 2$, then the root is a **square root.** If $n = 3$, then the root is a **cube root.**

Some numbers have more than one *n*th root. For example, both 5 and -5 are square roots of 25. The *principal square* root of 25, written as $\sqrt{25}$, is the positive root, 5.

Principal *n*th Root of a Number

Let a be a real number that has at least one *n*th root. The **principal nth root of** a is the *n*th root that has the same sign as a. It is denoted by a **radical symbol**

$$\sqrt[n]{a}. \qquad \text{Principal } n\text{th root}$$

The positive integer $n \geq 2$ is the **index** of the radical, and the number a is the **radicand.** When $n = 2$, omit the index and write $\sqrt{a}$ rather than $\sqrt[2]{a}$. (The plural of index is *indices.*)

A common misunderstanding is that the square root sign implies both negative and positive roots. This is not correct. The square root sign implies only a positive root. When a negative root is needed, you must use the negative sign with the square root sign.

$$\text{Incorrect: } \sqrt{4} = \pm 2 \quad \text{✗} \qquad \text{Correct: } -\sqrt{4} = -2 \quad \text{and} \quad \sqrt{4} = 2$$

EXAMPLE 8   **Evaluating Radical Expressions**

a. $\sqrt{36} = 6$ because $6^2 = 36$.

b. $-\sqrt{36} = -6$ because $-\left(\sqrt{36}\right) = -\left(\sqrt{6^2}\right) = -(6) = -6$.

c. $\sqrt[3]{\dfrac{125}{64}} = \dfrac{5}{4}$ because $\left(\dfrac{5}{4}\right)^3 = \dfrac{5^3}{4^3} = \dfrac{125}{64}$.

d. $\sqrt[5]{-32} = -2$ because $(-2)^5 = -32$.

e. $\sqrt[4]{-81}$ is not a real number because no real number raised to the fourth power produces -81.

✓ **Checkpoint** ◀))) *Audio-video solution in English & Spanish at LarsonPrecalculus.com*

Evaluate each expression, if possible.

a. $-\sqrt{144}$ b. $\sqrt{-144}$

c. $\sqrt{\dfrac{25}{64}}$ d. $-\sqrt[3]{\dfrac{8}{27}}$

Here are some generalizations about the nth roots of real numbers.

Generalizations About nth Roots of Real Numbers			
Real Number a	Integer $n > 0$	Root(s) of a	Example
$a > 0$	n is even.	$\sqrt[n]{a}, \ -\sqrt[n]{a}$	$\sqrt[4]{81} = 3, \ -\sqrt[4]{81} = -3$
$a > 0$ or $a < 0$	n is odd.	$\sqrt[n]{a}$	$\sqrt[3]{-8} = -2$
$a < 0$	n is even.	No real roots	$\sqrt{-4}$ is not a real number.
$a = 0$	n is even or odd.	$\sqrt[n]{0} = 0$	$\sqrt[5]{0} = 0$

Integers such as 1, 4, 9, 16, 25, and 36 are **perfect squares** because they have integer square roots. Similarly, integers such as 1, 8, 27, 64, and 125 are **perfect cubes** because they have integer cube roots.

Properties of Radicals

Let a and b be real numbers, variables, or algebraic expressions such that the roots below are real numbers, and let m and n be positive integers.

Property	Example
1. $\sqrt[n]{a^m} = \left(\sqrt[n]{a}\right)^m$	$\sqrt[3]{8^2} = \left(\sqrt[3]{8}\right)^2 = (2)^2 = 4$
2. $\sqrt[n]{a} \cdot \sqrt[n]{b} = \sqrt[n]{ab}$	$\sqrt{5} \cdot \sqrt{7} = \sqrt{5 \cdot 7} = \sqrt{35}$
3. $\dfrac{\sqrt[n]{a}}{\sqrt[n]{b}} = \sqrt[n]{\dfrac{a}{b}}, \quad b \neq 0$	$\dfrac{\sqrt[4]{27}}{\sqrt[4]{9}} = \sqrt[4]{\dfrac{27}{9}} = \sqrt[4]{3}$
4. $\sqrt[m]{\sqrt[n]{a}} = \sqrt[mn]{a}$	$\sqrt[3]{\sqrt{10}} = \sqrt[6]{10}$
5. $\left(\sqrt[n]{a}\right)^n = a$	$\left(\sqrt{3}\right)^2 = 3$
6. For n even, $\sqrt[n]{a^n} = |a|$.	$\sqrt{(-12)^2} = |-12| = 12$
For n odd, $\sqrt[n]{a^n} = a$.	$\sqrt[3]{(-12)^3} = -12$

A common use of Property 6 is $\sqrt{a^2} = |a|$.

EXAMPLE 9 **Using Properties of Radicals**

Use the properties of radicals to simplify each expression.

a. $\sqrt{8} \cdot \sqrt{2}$ **b.** $\left(\sqrt[3]{5}\right)^3$

c. $\sqrt[3]{x^3}$ **d.** $\sqrt[6]{y^6}$

Solution

a. $\sqrt{8} \cdot \sqrt{2} = \sqrt{8 \cdot 2} = \sqrt{16} = 4$ **b.** $\left(\sqrt[3]{5}\right)^3 = 5$

c. $\sqrt[3]{x^3} = x$ **d.** $\sqrt[6]{y^6} = |y|$

✓ **Checkpoint**)))) Audio-video solution in English & Spanish at LarsonPrecalculus.com

Use the properties of radicals to simplify each expression.

a. $\dfrac{\sqrt{125}}{\sqrt{5}}$ **b.** $\sqrt[3]{125^2}$

c. $\sqrt[3]{x^2} \cdot \sqrt[3]{x}$ **d.** $\sqrt{\sqrt{x}}$

Simplifying Radical Expressions

An expression involving radicals is in **simplest form** when the three conditions below are satisfied.

1. All possible factors are removed from the radical.

2. All fractions have radical-free denominators (a process called *rationalizing the denominator* accomplishes this).

3. The index of the radical is reduced.

To simplify a radical, factor the radicand into factors whose exponents are multiples of the index. Write the roots of these factors outside the radical. The "leftover" factors make up the new radicand.

· · REMARK   When you simplify a radical, it is important that both the original and the simplified expressions are defined for the same values of the variable. For instance, in Example 10(c), $\sqrt{75x^3}$ and $5x\sqrt{3x}$ are both defined only for nonnegative values of x. Similarly, in Example 10(e), $\sqrt[4]{(5x)^4}$ and $5|x|$ are both defined for all real values of x.

EXAMPLE 10   **Simplifying Radical Expressions**

Perfect cube   Leftover factor

a. $\sqrt[3]{24} = \sqrt[3]{8 \cdot 3} = \sqrt[3]{2^3 \cdot 3} = 2\sqrt[3]{3}$

Perfect   Leftover
4th power   factor

b. $\sqrt[4]{48} = \sqrt[4]{16 \cdot 3} = \sqrt[4]{2^4 \cdot 3} = 2\sqrt[4]{3}$

c. $\sqrt{75x^3} = \sqrt{25x^2 \cdot 3x} = \sqrt{(5x)^2 \cdot 3x} = 5x\sqrt{3x}$

d. $\sqrt[3]{24a^4} = \sqrt[3]{8a^3 \cdot 3a} = \sqrt[3]{(2a)^3 \cdot 3a} = 2a\sqrt[3]{3a}$

e. $\sqrt[4]{(5x)^4} = |5x| = 5|x|$

✓ *Checkpoint* ◀))) *Audio-video solution in English & Spanish at LarsonPrecalculus.com*

Simplify each radical expression.

a. $\sqrt{32}$ **b.** $\sqrt[3]{250}$ **c.** $\sqrt{24a^5}$ **d.** $\sqrt[3]{-135x^3}$

Radical expressions can be combined (added or subtracted) when they are **like radicals**—that is, when they have the same index and radicand. For example, $\sqrt{2}$, $3\sqrt{2}$, and $\frac{1}{2}\sqrt{2}$ are like radicals, but $\sqrt{3}$ and $\sqrt{2}$ are unlike radicals. To determine whether two radicals can be combined, first simplify each radical.

EXAMPLE 11   **Combining Radical Expressions**

a. $2\sqrt{48} - 3\sqrt{27} = 2\sqrt{16 \cdot 3} - 3\sqrt{9 \cdot 3}$ Find square factors.

$\qquad\qquad\qquad = 8\sqrt{3} - 9\sqrt{3}$ Find square roots and multiply by coefficients.

$\qquad\qquad\qquad = (8 - 9)\sqrt{3}$ Combine like radicals.

$\qquad\qquad\qquad = -\sqrt{3}$ Simplify.

b. $\sqrt[3]{16x} - \sqrt[3]{54x^4} = \sqrt[3]{8 \cdot 2x} - \sqrt[3]{27x^3 \cdot 2x}$ Find cube factors.

$\qquad\qquad\qquad = 2\sqrt[3]{2x} - 3x\sqrt[3]{2x}$ Find cube roots.

$\qquad\qquad\qquad = (2 - 3x)\sqrt[3]{2x}$ Combine like radicals.

✓ *Checkpoint* ◀))) *Audio-video solution in English & Spanish at LarsonPrecalculus.com*

Simplify each radical expression.

a. $3\sqrt{8} + \sqrt{18}$ **b.** $\sqrt[3]{81x^5} - \sqrt[3]{24x^2}$

Rationalizing Denominators and Numerators

To rationalize a denominator or numerator of the form $a - b\sqrt{m}$ or $a + b\sqrt{m}$, multiply both numerator and denominator by a **conjugate:** $a + b\sqrt{m}$ and $a - b\sqrt{m}$ are conjugates of each other. If $a = 0$, then the rationalizing factor for $\sqrt{m}$ is itself, $\sqrt{m}$. For cube roots, choose a rationalizing factor that produces a perfect cube radicand.

EXAMPLE 12   **Rationalizing Single-Term Denominators**

a. $\dfrac{5}{2\sqrt{3}} = \dfrac{5}{2\sqrt{3}} \cdot \dfrac{\sqrt{3}}{\sqrt{3}}$ $\sqrt{3}$ is rationalizing factor.

$= \dfrac{5\sqrt{3}}{2(3)}$ Multiply.

$= \dfrac{5\sqrt{3}}{6}$ Simplify.

b. $\dfrac{2}{\sqrt[3]{5}} = \dfrac{2}{\sqrt[3]{5}} \cdot \dfrac{\sqrt[3]{5^2}}{\sqrt[3]{5^2}}$ $\sqrt[3]{5^2}$ is rationalizing factor.

$= \dfrac{2\sqrt[3]{5^2}}{\sqrt[3]{5^3}}$ Multiply.

$= \dfrac{2\sqrt[3]{25}}{5}$ Simplify.

✓ **Checkpoint** 🔊))) *Audio-video solution in English & Spanish at LarsonPrecalculus.com*

Rationalize the denominator of each expression.

a. $\dfrac{5}{3\sqrt{2}}$ **b.** $\dfrac{1}{\sqrt[3]{25}}$

EXAMPLE 13   **Rationalizing a Denominator with Two Terms**

$\dfrac{2}{3 + \sqrt{7}} = \dfrac{2}{3 + \sqrt{7}} \cdot \dfrac{3 - \sqrt{7}}{3 - \sqrt{7}}$ Multiply numerator and denominator by conjugate of denominator.

$= \dfrac{2(3 - \sqrt{7})}{3(3 - \sqrt{7}) + \sqrt{7}(3 - \sqrt{7})}$ Distributive Property

$= \dfrac{2(3 - \sqrt{7})}{3(3) - 3(\sqrt{7}) + \sqrt{7}(3) - \sqrt{7}(\sqrt{7})}$ Distributive Property

$= \dfrac{2(3 - \sqrt{7})}{(3)^2 - (\sqrt{7})^2}$ Simplify.

$= \dfrac{2(3 - \sqrt{7})}{2}$ Simplify.

$= 3 - \sqrt{7}$ Divide out common factor.

✓ **Checkpoint** 🔊))) *Audio-video solution in English & Spanish at LarsonPrecalculus.com*

Rationalize the denominator: $\dfrac{8}{\sqrt{6} - \sqrt{2}}$.

Sometimes it is necessary to rationalize the numerator of an expression. For instance, in Appendix A.4 you will use the technique shown in Example 14 on the next page to rationalize the numerator of an expression from calculus.

EXAMPLE 14 Rationalizing a Numerator

$$\frac{\sqrt{5} - \sqrt{7}}{2} = \frac{\sqrt{5} - \sqrt{7}}{2} \cdot \frac{\sqrt{5} + \sqrt{7}}{\sqrt{5} + \sqrt{7}}$$ Multiply numerator and denominator by conjugate of numerator.

$$= \frac{\left(\sqrt{5}\right)^2 - \left(\sqrt{7}\right)^2}{2\left(\sqrt{5} + \sqrt{7}\right)}$$ Simplify.

$$= \frac{5 - 7}{2\left(\sqrt{5} + \sqrt{7}\right)}$$ Property 5 of radicals

$$= \frac{-2}{2\left(\sqrt{5} + \sqrt{7}\right)}$$ Simplify.

$$= \frac{-1}{\sqrt{5} + \sqrt{7}}$$ Divide out common factor.

> **• • REMARK** Do not confuse the expression $\sqrt{5} + \sqrt{7}$ with the expression $\sqrt{5 + 7}$. In general, $\sqrt{x + y}$ does not equal $\sqrt{x} + \sqrt{y}$. Similarly, $\sqrt{x^2 + y^2}$ does not equal $x + y$.

✓ **Checkpoint** ◀))) Audio-video solution in English & Spanish at LarsonPrecalculus.com

Rationalize the numerator: $\dfrac{2 - \sqrt{2}}{3}$.

Rational Exponents and Their Properties

> **Definition of Rational Exponents**
>
> If a is a real number and n is a positive integer such that the principal nth root of a exists, then $a^{1/n}$ is defined as
>
> $$a^{1/n} = \sqrt[n]{a}.$$
>
> Moreover, if m is a positive integer, then
>
> $$a^{m/n} = \left(a^{1/n}\right)^m.$$
>
> $1/n$ and m/n are called **rational exponents** of a.

> **• • REMARK** If m and n have no common factors, then it is also true that $a^{m/n} = (a^m)^{1/n}$.

The numerator of a rational exponent denotes the *power* to which the base is raised, and the denominator denotes the *index* or the *root* to be taken.

$$b^{m/n} = \left(\sqrt[n]{b}\right)^m = \sqrt[n]{b^m}$$

(with labels: Power over the m; Index over the n)

When you are working with rational exponents, the properties of integer exponents still apply. For example, $2^{1/2}2^{1/3} = 2^{(1/2)+(1/3)} = 2^{5/6}$.

EXAMPLE 15 Changing From Radical to Exponential Form

a. $\sqrt{3} = 3^{1/2}$

b. $\sqrt{(3xy)^5} = \sqrt[2]{(3xy)^5} = (3xy)^{5/2}$

c. $2x\sqrt[4]{x^3} = (2x)(x^{3/4}) = 2x^{1+(3/4)} = 2x^{7/4}$

✓ **Checkpoint** ◀))) Audio-video solution in English & Spanish at LarsonPrecalculus.com

Write (a) $\sqrt[3]{27}$, (b) $\sqrt{x^3y^5z}$, and (c) $3x\sqrt[3]{x^2}$ in exponential form.

▷ TECHNOLOGY There are four
methods of evaluating radicals on
most graphing utilities. For square
roots, you can use the *square root
key* ⎣√⎦. For cube roots, you can
use the *cube root key* ⎣∛⎦. For
other roots, first convert the radical
to exponential form and then use
the *exponential key* ⎣^⎦, or use the
*x*th *root key* ⎣ˣ√⎦ (or menu choice).
Consult the user's guide for
your graphing utility for specific
keystrokes.

EXAMPLE 16   Changing From Exponential to Radical Form

See LarsonPrecalculus.com for an interactive version of this type of example.

a. $(x^2 + y^2)^{3/2} = \left(\sqrt{x^2 + y^2}\right)^3 = \sqrt{(x^2 + y^2)^3}$

b. $2y^{3/4}z^{1/4} = 2(y^3z)^{1/4} = 2\sqrt[4]{y^3z}$

c. $a^{-3/2} = \dfrac{1}{a^{3/2}} = \dfrac{1}{\sqrt{a^3}}$

d. $x^{0.2} = x^{1/5} = \sqrt[5]{x}$

✓ **Checkpoint** ◀))) *Audio-video solution in English & Spanish at LarsonPrecalculus.com*

Write each expression in radical form.

a. $(x^2 - 7)^{-1/2}$ **b.** $-3b^{1/3}c^{2/3}$

c. $a^{0.75}$ **d.** $(x^2)^{2/5}$

Rational exponents are useful for evaluating roots of numbers on a calculator, for reducing the index of a radical, and for simplifying expressions in calculus.

EXAMPLE 17   Simplifying with Rational Exponents

a. $(-32)^{-4/5} = \left(\sqrt[5]{-32}\right)^{-4} = (-2)^{-4} = \dfrac{1}{(-2)^4} = \dfrac{1}{16}$

b. $(-5x^{5/3})(3x^{-3/4}) = -15x^{(5/3)-(3/4)} = -15x^{11/12}, \quad x \neq 0$

c. $\sqrt[9]{a^3} = a^{3/9} = a^{1/3} = \sqrt[3]{a}$ Reduce index.

d. $\sqrt[3]{\sqrt{125}} = \sqrt[6]{125} = \sqrt[6]{(5)^3} = 5^{3/6} = 5^{1/2} = \sqrt{5}$

e. $(2x - 1)^{4/3}(2x - 1)^{-1/3} = (2x - 1)^{(4/3)-(1/3)} = 2x - 1, \quad x \neq \frac{1}{2}$

· · · · · · · · · · · · · · ▷

·· REMARK The expression in
Example 17(b) is not defined
when $x = 0$ because $0^{-3/4}$ is
not a real number. Similarly, the
expression in Example 17(e) is
not defined when $x = \frac{1}{2}$ because

$$\left(2 \cdot \tfrac{1}{2} - 1\right)^{-1/3} = (0)^{-1/3}$$

is not a real number.

✓ **Checkpoint** ◀))) *Audio-video solution in English & Spanish at LarsonPrecalculus.com*

Simplify each expression.

a. $(-125)^{-2/3}$ **b.** $(4x^2y^{3/2})(-3x^{-1/3}y^{-3/5})$

c. $\sqrt[3]{\sqrt[4]{27}}$ **d.** $(3x + 2)^{5/2}(3x + 2)^{-1/2}$

Summarize   (Appendix A.2)

1. Make a list of the properties of exponents (*page A13*). For examples that use these properties, see Examples 1–4.

2. State the definition of scientific notation (*page A16*). For examples involving scientific notation, see Examples 5–7.

3. Make a list of the properties of radicals (*page A18*). For examples involving radicals, see Examples 8 and 9.

4. Explain how to simplify a radical expression (*page A19*). For examples of simplifying radical expressions, see Examples 10 and 11.

5. Explain how to rationalize a denominator or a numerator (*page A20*). For examples of rationalizing denominators and numerators, see Examples 12–14.

6. State the definition of a rational exponent (*page A21*). For examples involving rational exponents, see Examples 15–17.

A.2   Exercises

See **CalcChat.com** for tutorial help and worked-out solutions to odd-numbered exercises.

Vocabulary: Fill in the blanks.

1. In the exponential form a^n, n is the _____ and a is the _____.
2. A convenient way of writing very large or very small numbers is _____ _____.
3. One of the two equal factors of a number is a _____ _____ of the number.
4. In the radical form $\sqrt[n]{a}$, the positive integer n is the _____ of the radical and the number a is the _____.
5. Radical expressions can be combined (added or subtracted) when they are _____ _____.
6. The expressions $a + b\sqrt{m}$ and $a - b\sqrt{m}$ are _____ of each other.
7. The process used to create a radical-free denominator is known as _____ the denominator.
8. In the expression $b^{m/n}$, m denotes the _____ to which the base is raised and n denotes the _____ or root to be taken.

Skills and Applications

Evaluating Exponential Expressions In Exercises 9–14, evaluate each expression.

9. (a) $5 \cdot 5^3$ (b) $\dfrac{5^2}{5^4}$
10. (a) $(3^3)^0$ (b) -3^2
11. (a) $(2^3 \cdot 3^2)^2$ (b) $\left(-\dfrac{3}{5}\right)^3\left(\dfrac{5}{3}\right)^2$
12. (a) $\dfrac{3}{3^{-4}}$ (b) $48(-4)^{-3}$
13. (a) $\dfrac{4 \cdot 3^{-2}}{2^{-2} \cdot 3^{-1}}$ (b) $(-2)^0$
14. (a) $3^{-1} + 2^{-2}$ (b) $(3^{-2})^2$

Evaluating an Algebraic Expression In Exercises 15–20, evaluate the expression for the given value of x.

15. $-3x^3$, $x = 2$
16. $7x^{-2}$, $x = 4$
17. $6x^0$, $x = 10$
18. $2x^3$, $x = -3$
19. $-3x^4$, $x = -2$
20. $12(-x)^3$, $x = -\frac{1}{3}$

Using Properties of Exponents In Exercises 21–26, simplify each expression.

21. (a) $(5z)^3$ (b) $5x^4(x^2)$
22. (a) $(-2x)^2$ (b) $(4x^3)^0$
23. (a) $6y^2(2y^0)^2$ (b) $(-z)^3(3z^4)$
24. (a) $\dfrac{7x^2}{x^3}$ (b) $\dfrac{12(x + y)^3}{9(x + y)}$
25. (a) $\left(\dfrac{4}{y}\right)^3\left(\dfrac{3}{y}\right)^4$ (b) $\left(\dfrac{b^{-2}}{a^{-2}}\right)\left(\dfrac{b}{a}\right)^2$
26. (a) $[(x^2y^{-2})^{-1}]^{-1}$ (b) $(5x^2z^6)^3(5x^2z^6)^{-3}$

Rewriting with Positive Exponents In Exercises 27–30, rewrite each expression with positive exponents. Simplify, if possible.

27. (a) $(x + 5)^0$ (b) $(2x^2)^{-2}$
28. (a) $(4y^{-2})(8y^4)$ (b) $(z + 2)^{-3}(z + 2)^{-1}$
29. (a) $\left(\dfrac{x^{-3}y^4}{5}\right)^{-3}$ (b) $\left(\dfrac{a^{-2}}{b^{-2}}\right)\left(\dfrac{b}{a}\right)^3$
30. (a) $\dfrac{3^n \cdot 3^{2n}}{3^{3n} \cdot 3^2}$ (b) $\dfrac{x^2 \cdot x^n}{x^3 \cdot x^n}$

Scientific Notation In Exercises 31 and 32, write the number in scientific notation.

31. 10,250.4   32. -0.000125

Decimal Notation In Exercises 33–36, write the number in decimal notation.

33. 3.14×10^{-4} 34. -2.058×10^6
35. Light year: 9.46×10^{12} kilometers
36. Diameter of a human hair: 9.0×10^{-6} meter

Using Scientific Notation In Exercises 37 and 38, evaluate each expression without using a calculator.

37. (a) $(2.0 \times 10^9)(3.4 \times 10^{-4})$
    (b) $(1.2 \times 10^7)(5.0 \times 10^{-3})$
38. (a) $\dfrac{6.0 \times 10^8}{3.0 \times 10^{-3}}$ (b) $\dfrac{2.5 \times 10^{-3}}{5.0 \times 10^2}$

Evaluating Radical Expressions In Exercises 39 and 40, evaluate each expression without using a calculator.

39. (a) $\sqrt{9}$ (b) $\sqrt[3]{\dfrac{27}{8}}$ 40. (a) $\sqrt[3]{27}$ (b) $(\sqrt{36})^3$

Using Properties of Radicals In Exercises 41 and 42, use the properties of radicals to simplify each expression.

41. (a) $\left(\sqrt[5]{2}\right)^5$ (b) $\sqrt[5]{32x^5}$
42. (a) $\sqrt{12} \cdot \sqrt{3}$ (b) $\sqrt[4]{(3x^2)^4}$

 Simplifying Radical Expressions In Exercises 43–50, simplify each radical expression.

43. (a) $\sqrt{20}$ (b) $\sqrt[3]{128}$
44. (a) $\sqrt[3]{\frac{16}{27}}$ (b) $\sqrt{\frac{75}{4}}$
45. (a) $\sqrt{72x^3}$ (b) $\sqrt{54xy^4}$
46. (a) $\sqrt{\frac{18^2}{z^3}}$ (b) $\sqrt{\frac{32a^4}{b^2}}$
47. (a) $\sqrt[3]{16x^5}$ (b) $\sqrt{75x^2y^{-4}}$
48. (a) $\sqrt[4]{3x^4y^2}$ (b) $\sqrt[5]{160x^8z^4}$
49. (a) $2\sqrt{20x^2} + 5\sqrt{125x^2}$
    (b) $8\sqrt{147x} - 3\sqrt{48x}$
50. (a) $3\sqrt[3]{54x^3} + \sqrt[3]{16x^3}$
    (b) $\sqrt[3]{64x} - \sqrt[3]{27x^4}$

 Rationalizing a Denominator In Exercises 51–54, rationalize the denominator of the expression. Then simplify your answer.

51. $\dfrac{1}{\sqrt{3}}$ 52. $\dfrac{8}{\sqrt[3]{2}}$

53. $\dfrac{5}{\sqrt{14} - 2}$ 54. $\dfrac{3}{\sqrt{5} + \sqrt{6}}$

Rationalizing a Numerator In Exercises 55 and 56, rationalize the numerator of the expression. Then simplify your answer.

55. $\dfrac{\sqrt{5} + \sqrt{3}}{3}$ 56. $\dfrac{\sqrt{7} - 3}{4}$

 Writing Exponential and Radical Forms In Exercises 57–60, fill in the missing form of the expression.

Radical Form	Rational Exponent Form
57. $\sqrt[3]{64}$	
58. $x^2\sqrt{x}$	
59.	$3x^{-2/3}$
60.	$a^{0.4}$

 Simplifying Expressions In Exercises 61–68, simplify each expression.

61. (a) $32^{-3/5}$ (b) $\left(\frac{16}{81}\right)^{-3/4}$
62. (a) $100^{-3/2}$ (b) $\left(\frac{9}{4}\right)^{-1/2}$

63. (a) $\sqrt[4]{3^2}$ (b) $\sqrt[6]{(x + 1)^4}$
64. (a) $\sqrt[6]{x^3}$ (b) $\sqrt[4]{(3x^2)^4}$
65. (a) $\sqrt{\sqrt{32}}$ (b) $\sqrt{\sqrt[4]{2x}}$
66. (a) $\sqrt{\sqrt{243(x + 1)}}$
    (b) $\sqrt{\sqrt[3]{10a^7b}}$
67. (a) $(x - 1)^{1/3}(x - 1)^{2/3}$
    (b) $(x - 1)^{1/3}(x - 1)^{-4/3}$
68. (a) $(4x + 3)^{5/2}(4x + 3)^{-5/3}$
    (b) $(4x + 3)^{-5/2}(4x + 3)^{2/3}$

• • **69. Mathematical Modeling** • • • • • • • • •
A funnel is filled with water to a height of h centimeters. The formula

$$t = 0.03[12^{5/2} - (12 - h)^{5/2}], \quad 0 \le h \le 12$$

represents the amount of time t (in seconds) that it will take for the funnel to empty. Use the *table* feature of a graphing utility to find the times required for the funnel to empty for water heights of $h = 0, h = 1, h = 2, \ldots, h = 12$ centimeters.

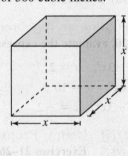 **HOW DO YOU SEE IT?** Package A is a cube with a volume of 500 cubic inches. Package B is a cube with a volume of 250 cubic inches. Is the length x of a side of package A greater than, less than, or equal to twice the length of a side of package B? Explain.

Exploration

True or False? In Exercises 71–74, determine whether the statement is true or false. Justify your answer.

71. $\dfrac{x^{k+1}}{x} = x^k$ 72. $(a^n)^k = a^{nk}$

73. $(a + b)^2 = a^2 + b^2$

74. $\dfrac{a}{\sqrt{b}} = \dfrac{a^2}{\left(\sqrt{b}\right)^2} = \dfrac{a^2}{b}$

A.3   Polynomials and Factoring

Polynomial factoring has many real-life applications. For example, in Exercise 84 on page A34, you will use polynomial factoring to write an alternative form of an expression that models the rate of change of an autocatalytic chemical reaction.

- Write polynomials in standard form.
- Add, subtract, and multiply polynomials.
- Use special products to multiply polynomials.
- Factor out common factors from polynomials.
- Factor special polynomial forms.
- Factor trinomials as the product of two binomials.
- Factor polynomials by grouping.

Polynomials

One of the most common types of algebraic expressions is the **polynomial.** Some examples are $2x + 5$, $3x^4 - 7x^2 + 2x + 4$, and $5x^2y^2 - xy + 3$. The first two are *polynomials in x* and the third is a *polynomial in x and y*. The terms of a polynomial in x have the form ax^k, where a is the **coefficient** and k is the **degree** of the term. For example, the polynomial $2x^3 - 5x^2 + 1 = 2x^3 + (-5)x^2 + (0)x + 1$ has coefficients 2, -5, 0, and 1.

> **Definition of a Polynomial in x**
>
> Let $a_0, a_1, a_2, \ldots, a_n$ be real numbers and let n be a nonnegative integer. A polynomial in x is an expression of the form
>
> $$a_n x^n + a_{n-1} x^{n-1} + \cdots + a_1 x + a_0$$
>
> where $a_n \neq 0$. The polynomial is of **degree** n, a_n is the **leading coefficient,** and a_0 is the **constant term.**

▷ **•• REMARK**   Expressions are not polynomials when a variable is underneath a radical or when a polynomial expression (with degree greater than 0) is in the denominator of a term. For example, the expressions $x^3 - \sqrt{3x} = x^3 - (3x)^{1/2}$ and $x^2 + (5/x) = x^2 + 5x^{-1}$ are not polynomials.

Polynomials with one, two, and three terms are **monomials, binomials,** and **trinomials,** respectively. A polynomial written with descending powers of x is in **standard form.**

EXAMPLE 1   **Writing Polynomials in Standard Form**

Polynomial	Standard Form	Degree	Leading Coefficient
a. $4x^2 - 5x^7 - 2 + 3x$	$-5x^7 + 4x^2 + 3x - 2$	7	-5
b. $4 - 9x^2$	$-9x^2 + 4$	2	-9
c. 8	8 or $8x^0$	0	8

✓ *Checkpoint* ◀))) *Audio-video solution in English & Spanish at LarsonPrecalculus.com*

Write the polynomial $6 - 7x^3 + 2x$ in standard form. Then identify the degree and leading coefficient of the polynomial.

A polynomial that has all zero coefficients is called the **zero polynomial,** denoted by 0. No degree is assigned to the zero polynomial. For polynomials in more than one variable, the degree of a *term* is the sum of the exponents of the variables in the term. The degree of the *polynomial* is the highest degree of its terms. For example, the degree of the polynomial $-2x^3y^6 + 4xy - x^7y^4$ is 11 because the sum of the exponents in the last term is the greatest. The leading coefficient of the polynomial is the coefficient of the highest-degree term.

Operations with Polynomials

You can add and subtract polynomials in much the same way you add and subtract real numbers. Add or subtract the *like terms* (terms having the same variables to the same powers) by adding or subtracting their coefficients. For example, $-3xy^2$ and $5xy^2$ are like terms and their sum is

$$-3xy^2 + 5xy^2 = (-3 + 5)xy^2 = 2xy^2.$$

EXAMPLE 2   Adding or Subtracting Polynomials

a. $(5x^3 - 7x^2 - 3) + (x^3 + 2x^2 - x + 8)$

$\qquad = (5x^3 + x^3) + (-7x^2 + 2x^2) + (-x) + (-3 + 8)$ ⟶ Group like terms.

$\qquad = 6x^3 - 5x^2 - x + 5$ ⟶ Combine like terms.

b. $(7x^4 - x^2 - 4x + 2) - (3x^4 - 4x^2 + 3x)$

$\qquad = 7x^4 - x^2 - 4x + 2 - 3x^4 + 4x^2 - 3x$ ⟶ Distributive Property

$\qquad = (7x^4 - 3x^4) + (-x^2 + 4x^2) + (-4x - 3x) + 2$ ⟶ Group like terms.

$\qquad = 4x^4 + 3x^2 - 7x + 2$ ⟶ Combine like terms.

✓ *Checkpoint*))) *Audio-video solution in English & Spanish at LarsonPrecalculus.com*

Find the difference $(2x^3 - x + 3) - (x^2 - 2x - 3)$ and write the resulting polynomial in standard form.

▷

•• **REMARK** When a negative sign precedes an expression inside parentheses, remember to distribute the negative sign to each term inside the parentheses. In other words, multiply each term by -1.

$$-(3x^4 - 4x^2 + 3x)$$
$$= -3x^4 + 4x^2 - 3x$$

To find the *product* of two polynomials, use the right and left Distributive Properties. For example, you can find the product of $3x - 2$ and $5x + 7$ by first treating $5x + 7$ as a single quantity.

$$(3x - 2)(5x + 7) = 3x(5x + 7) - 2(5x + 7)$$

$$= (3x)(5x) + (3x)(7) - (2)(5x) - (2)(7)$$

$$= 15x^2 + 21x - 10x - 14$$

| Product of First terms | Product of Outer terms | Product of Inner terms | Product of Last terms |

$$= 15x^2 + 11x - 14$$

Note that when using the **FOIL Method** above (which can be used only to multiply two binomials), some of the terms in the product may be like terms that can be combined into one term.

EXAMPLE 3   Finding a Product by the FOIL Method

Use the FOIL Method to find the product of $2x - 4$ and $x + 5$.

Solution

$$\overset{F}{\quad}\ \overset{O}{\quad}\ \overset{I}{\quad}\ \overset{L}{\quad}$$
$$(2x - 4)(x + 5) = 2x^2 + 10x - 4x - 20 = 2x^2 + 6x - 20$$

✓ *Checkpoint*))) *Audio-video solution in English & Spanish at LarsonPrecalculus.com*

Use the FOIL Method to find the product of $3x - 1$ and $x - 5$.

Special Products

Some binomial products have special forms that occur frequently in algebra. You do not need to memorize these formulas because you can use the Distributive Property to multiply. However, becoming familiar with these formulas will enable you to manipulate the algebra more quickly.

Special Products

Let u and v be real numbers, variables, or algebraic expressions.

Special Product	Example
Sum and Difference of Same Terms	
$(u + v)(u - v) = u^2 - v^2$	$(x + 4)(x - 4) = x^2 - 4^2$
	$= x^2 - 16$
Square of a Binomial	
$(u + v)^2 = u^2 + 2uv + v^2$	$(x + 3)^2 = x^2 + 2(x)(3) + 3^2$
	$= x^2 + 6x + 9$
$(u - v)^2 = u^2 - 2uv + v^2$	$(3x - 2)^2 = (3x)^2 - 2(3x)(2) + 2^2$
	$= 9x^2 - 12x + 4$
Cube of a Binomial	
$(u + v)^3 = u^3 + 3u^2v + 3uv^2 + v^3$	$(x + 2)^3 = x^3 + 3x^2(2) + 3x(2^2) + 2^3$
	$= x^3 + 6x^2 + 12x + 8$
$(u - v)^3 = u^3 - 3u^2v + 3uv^2 - v^3$	$(x - 1)^3 = x^3 - 3x^2(1) + 3x(1^2) - 1^3$
	$= x^3 - 3x^2 + 3x - 1$

EXAMPLE 4   **Sum and Difference of Same Terms**

Find each product.

a. $(5x + 9)(5x - 9)$ **b.** $(x + y - 2)(x + y + 2)$

Solution

a. The product of a sum and a difference of the *same* two terms has no middle term and takes the form $(u + v)(u - v) = u^2 - v^2$.

$$(5x + 9)(5x - 9) = (5x)^2 - 9^2 = 25x^2 - 81$$

b. One way to find this product is to group $x + y$ and form a special product.

$$
\begin{aligned}
(x + y - 2)(x + y + 2) &= [(x + y) - 2][(x + y) + 2] \\
&= (x + y)^2 - 2^2 \quad \text{Sum and difference of same terms} \\
&= x^2 + 2xy + y^2 - 4 \quad \text{Square of a binomial.}
\end{aligned}
$$

✓ **Checkpoint**))) Audio-video solution in English & Spanish at LarsonPrecalculus.com

Find each product.

a. $(3x - 2)(3x + 2)$ **b.** $(x - 2 + 3y)(x - 2 - 3y)$

Polynomials with Common Factors

The process of writing a polynomial as a product is called **factoring.** It is an important tool for solving equations and for simplifying rational expressions.

Unless noted otherwise, when you are asked to factor a polynomial, assume that you are looking for factors that have integer coefficients. If a polynomial does not factor using integer coefficients, then it is **prime** or **irreducible over the integers.** For example, the polynomial

$$x^2 - 3$$

is irreducible over the integers. Over the *real numbers,* this polynomial factors as

$$x^2 - 3 = (x + \sqrt{3})(x - \sqrt{3}).$$

A polynomial is **completely factored** when each of its factors is prime. For example,

$$x^3 - x^2 + 4x - 4 = (x - 1)(x^2 + 4) \qquad \text{Completely factored}$$

is completely factored, but

$$x^3 - x^2 - 4x + 4 = (x - 1)(x^2 - 4) \qquad \text{Not completely factored}$$

is not completely factored. Its complete factorization is

$$x^3 - x^2 - 4x + 4 = (x - 1)(x + 2)(x - 2).$$

The simplest type of factoring involves a polynomial that can be written as the product of a monomial and another polynomial. The technique used here is the Distributive Property, $a(b + c) = ab + ac$, in the *reverse* direction.

$$ab + ac = a(b + c) \qquad \qquad a \text{ is a common factor.}$$

Factoring out any common factors is the first step in completely factoring a polynomial.

EXAMPLE 5 **Factoring Out Common Factors**

Factor each expression.

a. $6x^3 - 4x$

b. $-4x^2 + 12x - 16$

c. $(x - 2)(2x) + (x - 2)(3)$

Solution

a. $6x^3 - 4x = 2x(3x^2) - 2x(2)$ $\qquad \qquad 2x$ is a common factor.

$\qquad \qquad \quad = 2x(3x^2 - 2)$

b. $-4x^2 + 12x - 16 = -4(x^2) + (-4)(-3x) + (-4)4$ $\qquad -4$ is a common factor.

$\qquad \qquad \qquad \qquad \quad = -4(x^2 - 3x + 4)$

c. $(x - 2)(2x) + (x - 2)(3) = (x - 2)(2x + 3)$ $\qquad (x - 2)$ is a common factor.

✓ **Checkpoint**))) *Audio-video solution in English & Spanish at LarsonPrecalculus.com*

Factor each expression.

a. $5x^3 - 15x^2$

b. $-3 + 6x - 12x^3$

c. $(x + 1)(x^2) - (x + 1)(2)$

Factoring Special Polynomial Forms

Some polynomials have special forms that arise from the special product forms on page A27. You should learn to recognize these forms.

Factoring Special Polynomial Forms

Factored Form	**Example**

Difference of Two Squares

$$u^2 - v^2 = (u + v)(u - v)$$

$$9x^2 - 4 = (3x)^2 - 2^2 = (3x + 2)(3x - 2)$$

Perfect Square Trinomial

$$u^2 + 2uv + v^2 = (u + v)^2$$

$$x^2 + 6x + 9 = x^2 + 2(x)(3) + 3^2 = (x + 3)^2$$

$$u^2 - 2uv + v^2 = (u - v)^2$$

$$x^2 - 6x + 9 = x^2 - 2(x)(3) + 3^2 = (x - 3)^2$$

Sum or Difference of Two Cubes

$$u^3 + v^3 = (u + v)(u^2 - uv + v^2)$$

$$x^3 + 8 = x^3 + 2^3 = (x + 2)(x^2 - 2x + 4)$$

$$u^3 - v^3 = (u - v)(u^2 + uv + v^2)$$

$$27x^3 - 1 = (3x)^3 - 1^3 = (3x - 1)(9x^2 + 3x + 1)$$

The factored form of a difference of two squares is always a set of **conjugate pairs.**

$$u^2 - v^2 = (u + v)(u - v)$$

Conjugate pairs

Difference   Opposite signs

To recognize perfect square terms, look for coefficients that are squares of integers and variables raised to *even powers.*

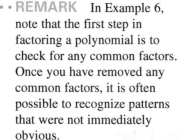

EXAMPLE 6   **Factoring Out a Common Factor First**

$$3 - 12x^2 = 3(1 - 4x^2)$$ 3 is a common factor.

$$= 3[1^2 - (2x)^2]$$ Rewrite $1 - 4x^2$ as the difference of two squares.

$$= 3(1 + 2x)(1 - 2x)$$ Factor.

✓ **Checkpoint** 🔊)) *Audio-video solution in English & Spanish at LarsonPrecalculus.com*

Factor $100 - 4y^2$.

REMARK In Example 6, note that the first step in factoring a polynomial is to check for any common factors. Once you have removed any common factors, it is often possible to recognize patterns that were not immediately obvious.

EXAMPLE 7   **Factoring the Difference of Two Squares**

a. $(x + 2)^2 - y^2 = [(x + 2) + y][(x + 2) - y] = (x + 2 + y)(x + 2 - y)$

b. $16x^4 - 81 = (4x^2)^2 - 9^2$ Rewrite as the difference of two squares.

$$= (4x^2 + 9)(4x^2 - 9)$$ Factor.

$$= (4x^2 + 9)[(2x)^2 - 3^2]$$ Rewrite $4x^2 - 9$ as the difference of two squares.

$$= (4x^2 + 9)(2x + 3)(2x - 3)$$ Factor.

✓ **Checkpoint** 🔊)) *Audio-video solution in English & Spanish at LarsonPrecalculus.com*

Factor $(x - 1)^2 - 9y^4$.

A **perfect square trinomial** is the square of a binomial, and it has the form

$$u^2 + 2uv + v^2 = (u + v)^2 \quad \text{or} \quad u^2 - 2uv + v^2 = (u - v)^2.$$

Like signs   Like signs

Note that the first and last terms are squares and the middle term is twice the product of u and v.

EXAMPLE 8   Factoring Perfect Square Trinomials

Factor each trinomial.

a. $x^2 - 10x + 25$ **b.** $16x^2 + 24x + 9$

Solution

a. $x^2 - 10x + 25 = x^2 - 2(x)(5) + 5^2 = (x - 5)^2$

b. $16x^2 + 24x + 9 = (4x)^2 + 2(4x)(3) + 3^2 = (4x + 3)^2$

✓ **Checkpoint** ◀))) Audio-video solution in English & Spanish at LarsonPrecalculus.com

Factor $9x^2 - 30x + 25$.

The next two formulas show the sum and difference of two cubes. Pay special attention to the signs of the terms.

Like signs   Like signs

$$u^3 + v^3 = (u + v)(u^2 - uv + v^2) \quad u^3 - v^3 = (u - v)(u^2 + uv + v^2)$$

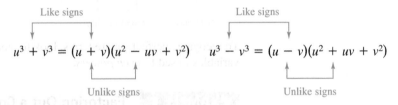

Unlike signs   Unlike signs

EXAMPLE 9   Factoring the Difference of Two Cubes

$x^3 - 27 = x^3 - 3^3$ Rewrite 27 as 3^3.

$\qquad\quad = (x - 3)(x^2 + 3x + 9)$ Factor.

✓ **Checkpoint** ◀))) Audio-video solution in English & Spanish at LarsonPrecalculus.com

Factor $64x^3 - 1$.

EXAMPLE 10   Factoring the Sum of Two Cubes

a. $y^3 + 8 = y^3 + 2^3$ Rewrite 8 as 2^3.

$\qquad\quad = (y + 2)(y^2 - 2y + 4)$ Factor.

b. $3x^3 + 192 = 3(x^3 + 64)$ 3 is a common factor.

$\qquad\qquad\quad = 3(x^3 + 4^3)$ Rewrite 64 as 4^3.

$\qquad\qquad\quad = 3(x + 4)(x^2 - 4x + 16)$ Factor.

✓ **Checkpoint** ◀))) Audio-video solution in English & Spanish at LarsonPrecalculus.com

Factor each expression.

a. $x^3 + 216$ **b.** $5y^3 + 135$

Trinomials with Binomial Factors

To factor a trinomial of the form $ax^2 + bx + c$, use the pattern below.

Factors of a

$$ax^2 + bx + c = (\boxed{}x + \boxed{})(\boxed{}x + \boxed{})$$

Factors of c

The goal is to find a combination of factors of a and c such that the sum of the outer and inner products is the middle term bx. For example, for the trinomial $6x^2 + 17x + 5$, you can write all possible factorizations and determine which one has outer and inner products whose sum is $17x$.

$$(6x + 5)(x + 1),\ (6x + 1)(x + 5),\ (2x + 1)(3x + 5),\ (2x + 5)(3x + 1)$$

The correct factorization is $(2x + 5)(3x + 1)$ because the sum of the outer (O) and inner (I) products is $17x$.

$$\begin{array}{ccccc} \text{F} & \text{O} & \text{I} & \text{L} & \text{O} + \text{I} \\ \downarrow & \downarrow & \downarrow & \downarrow & \downarrow \end{array}$$

$$(2x + 5)(3x + 1) = 6x^2 + 2x + 15x + 5 = 6x^2 + 17x + 5$$

REMARK   Factoring a trinomial can involve trial and error. However, it is relatively easy to check your answer by multiplying the factors. The product should be the original trinomial. For instance, in Example 11, verify that $(x - 3)(x - 4) = x^2 - 7x + 12$.

▷

EXAMPLE 11   Factoring a Trinomial: Leading Coefficient Is 1

Factor $x^2 - 7x + 12$.

Solution   For this trinomial, $a = 1$, $b = -7$, and $c = 12$. Because b is negative and c is positive, both factors of 12 must be negative. So, the possible factorizations of $x^2 - 7x + 12$ are

$$(x - 1)(x - 12),\quad (x - 2)(x - 6),\quad \text{and}\quad (x - 3)(x - 4).$$

Testing the middle term, you will find the correct factorization to be

$$x^2 - 7x + 12 = (x - 3)(x - 4).\qquad \text{O} + \text{I} = -4x - 3x = -7x$$

✓ **Checkpoint** ◀))) Audio-video solution in English & Spanish at LarsonPrecalculus.com

Factor $x^2 + x - 6$.

EXAMPLE 12   Factoring a Trinomial: Leading Coefficient Is Not 1

See LarsonPrecalculus.com for an interactive version of this type of example.

Factor $2x^2 + x - 15$.

Solution   For this trinomial, $a = 2$, $b = 1$, and $c = -15$. Because c is negative, its factors must have unlike signs. The eight possible factorizations are below.

$$(2x - 1)(x + 15)\quad (2x + 1)(x - 15)\quad (2x - 3)(x + 5)\quad (2x + 3)(x - 5)$$

$$(2x - 5)(x + 3)\quad (2x + 5)(x - 3)\quad (2x - 15)(x + 1)\quad (2x + 15)(x - 1)$$

Testing the middle term, you will find the correct factorization to be

$$2x^2 + x - 15 = (2x - 5)(x + 3).\qquad \text{O} + \text{I} = 6x - 5x = x$$

✓ **Checkpoint** ◀))) Audio-video solution in English & Spanish at LarsonPrecalculus.com

Factor $2x^2 - 5x + 3$.

Factoring by Grouping

Sometimes, polynomials with more than three terms can be **factored by grouping.**

EXAMPLE 13   **Factoring by Grouping**

$$x^3 - 2x^2 - 3x + 6 = (x^3 - 2x^2) - (3x - 6)$$ Group terms.

$$= x^2(x - 2) - 3(x - 2)$$ Factor each group.

$$= (x - 2)(x^2 - 3)$$ $(x - 2)$ is a common factor.

✓ **Checkpoint** 🔊))) *Audio-video solution in English & Spanish at LarsonPrecalculus.com*

Factor $x^3 + x^2 - 5x - 5$.

> ▷
>
> •• REMARK   Sometimes, more than one grouping will work. For instance, another way to factor the polynomial in Example 13 is
>
> $$x^3 - 2x^2 - 3x + 6$$
> $$= (x^3 - 3x) - (2x^2 - 6)$$
> $$= x(x^2 - 3) - 2(x^2 - 3)$$
> $$= (x^2 - 3)(x - 2).$$
>
> Notice that this is the same result as in Example 13.

Factoring by grouping can eliminate some of the trial and error involved in factoring a trinomial. To factor a trinomial of the form $ax^2 + bx + c$ by grouping, choose factors of the product ac that sum to b and use these factors to rewrite the middle term. Example 14 illustrates this technique.

EXAMPLE 14   **Factoring a Trinomial by Grouping**

In the trinomial $2x^2 + 5x - 3$, $a = 2$ and $c = -3$, so the product ac is -6. Now, -6 factors as $(6)(-1)$ and $6 + (-1) = 5 = b$. So, rewrite the middle term as $5x = 6x - x$ and factor by grouping.

$$2x^2 + 5x - 3 = 2x^2 + 6x - x - 3$$ Rewrite middle term.

$$= (2x^2 + 6x) - (x + 3)$$ Group terms.

$$= 2x(x + 3) - (x + 3)$$ Factor $2x^2 + 6x$.

$$= (x + 3)(2x - 1)$$ $(x + 3)$ is a common factor.

✓ **Checkpoint** 🔊))) *Audio-video solution in English & Spanish at LarsonPrecalculus.com*

Use factoring by grouping to factor $2x^2 + 5x - 12$.

Summarize   (Appendix A.3)

1. State the definition of a polynomial in x and explain what is meant by the standard form of a polynomial *(page A25)*. For an example of writing polynomials in standard form, see Example 1.

2. Explain how to add and subtract polynomials *(page A26)*. For an example of adding and subtracting polynomials, see Example 2.

3. Explain the FOIL Method *(page A26)*. For an example of finding a product using the FOIL Method, see Example 3.

4. Explain how to find binomial products that have special forms *(page A27)*. For an example of binomial products that have special forms, see Example 4.

5. Explain what it means to completely factor a polynomial *(page A28)*. For an example of factoring out common factors, see Example 5.

6. Make a list of the special polynomial forms of factoring *(page A29)*. For examples of factoring these special forms, see Examples 6–10.

7. Explain how to factor a trinomial of the form $ax^2 + bx + c$ *(page A31)*. For examples of factoring trinomials of this form, see Examples 11 and 12.

8. Explain how to factor a polynomial by grouping *(page A32)*. For examples of factoring by grouping, see Examples 13 and 14.

A.3   Exercises

See **CalcChat.com** for tutorial help and worked-out solutions to odd-numbered exercises.

Vocabulary: Fill in the blanks.

1. For the polynomial $a_nx^n + a_{n-1}x^{n-1} + \cdots + a_1x + a_0$, $a_n \neq 0$, the degree is _____, the leading coefficient is _____, and the constant term is _____.

2. A polynomial with one term is a _____, while a polynomial with two terms is a _____ and a polynomial with three terms is a _____.

3. To add or subtract polynomials, add or subtract the _____ _____ by adding or subtracting their coefficients.

4. The letters in "FOIL" stand for F _____, O _____, I _____, and L _____.

5. The process of writing a polynomial as a product is called _____.

6. A polynomial is _____ _____ when each of its factors is prime.

7. A _____ _____ _____ is the square of a binomial, and it has the form $u^2 + 2uv + v^2$ or $u^2 - 2uv + v^2$.

8. Sometimes, polynomials with more than three terms can be factored by _____.

Skills and Applications

 Writing Polynomials in Standard Form In Exercises 9–14, (a) write the polynomial in standard form, (b) identify the degree and leading coefficient of the polynomial, and (c) state whether the polynomial is a monomial, a binomial, or a trinomial.

9. $7x$

10. 3

11. $14x - \frac{1}{2}x^5$

12. $3 + 2x$

13. $1 + 6x^4 - 4x^5$

14. $-y + 25y^2 + 1$

 Adding or Subtracting Polynomials In Exercises 15–18, add or subtract and write the result in standard form.

15. $(6x + 5) - (8x + 15)$

16. $(2x^2 + 1) - (x^2 - 2x + 1)$

17. $(15x^2 - 6) + (-8.3x^3 - 14.7x^2 - 17)$

18. $(15.6w^4 - 14w - 17.4) + (16.9w^4 - 9.2w + 13)$

 Multiplying Polynomials In Exercises 19–36, multiply the polynomials.

19. $3x(x^2 - 2x + 1)$

20. $y^2(4y^2 + 2y - 3)$

21. $-5z(3z - 1)$

22. $-3x(5x + 2)$

23. $(3x - 5)(2x + 1)$

24. $(7x - 2)(4x - 3)$

25. $(x^2 - x + 2)(x^2 + x + 1)$

26. $(2x^2 - x + 4)(x^2 + 3x + 2)$

27. $(x + 10)(x - 10)$

28. $(4a + 5b)(4a - 5b)$

29. $(2x + 3)^2$

30. $(8x + 3)^2$

31. $(x + 3)^3$

32. $(3x + 2y)^3$

33. $[(x - 3) + y]^2$

34. $[(x + 1) - y]^2$

35. $[(m - 3) + n][(m - 3) - n]$

36. $[(x - 3y) + z][(x - 3y) - z]$

 Factoring Out a Common Factor In Exercises 37–40, factor out the common factor.

37. $2x^3 - 6x$

38. $3z^3 - 6z^2 + 9z$

39. $3x(x - 5) + 8(x - 5)$

40. $(x + 3)^2 - 4(x + 3)$

 Factoring the Difference of Two Squares In Exercises 41–44, completely factor the difference of two squares.

41. $25y^2 - 4$

42. $81 - 36z^2$

43. $(x - 1)^2 - 4$

44. $25 - (z + 5)^2$

 Factoring a Perfect Square Trinomial In Exercises 45–50, factor the perfect square trinomial.

45. $x^2 - 4x + 4$

46. $4t^2 + 4t + 1$

47. $25z^2 - 30z + 9$

48. $36y^2 + 84y + 49$

49. $4y^2 - 12y + 9$

50. $9u^2 + 24uv + 16v^2$

Factoring the Sum or Difference of Two Cubes In Exercises 51–54, factor the sum or difference of two cubes.

51. $x^3 + 125$

52. $x^3 - 8$

53. $8t^3 - 1$

54. $27t^3 + 8$

Factoring a Trinomial **In Exercises 55–62, factor the trinomial.**

55. $x^2 + x - 2$ **56.** $s^2 - 5s + 6$

57. $3x^2 + 10x - 8$ **58.** $2x^2 - 3x - 27$

59. $5x^2 + 31x + 6$ **60.** $8x^2 + 51x + 18$

61. $-5y^2 - 8y + 4$ **62.** $-6z^2 + 17z + 3$

 Factoring by Grouping **In Exercises 63–68, factor by grouping.**

63. $x^3 - x^2 + 2x - 2$ **64.** $x^3 + 5x^2 - 5x - 25$

65. $2x^3 - x^2 - 6x + 3$ **66.** $3x^3 + x^2 - 15x - 5$

67. $3x^5 + 6x^3 - 2x^2 - 4$ **68.** $8x^5 - 6x^2 + 12x^3 - 9$

 Factoring a Trinomial by Grouping **In Exercises 69–72, factor the trinomial by grouping.**

69. $2x^2 + 9x + 9$ **70.** $6x^2 + x - 2$

71. $6x^2 - x - 15$ **72.** $12x^2 - 13x + 1$

 Factoring Completely **In Exercises 73–82, completely factor the expression.**

73. $6x^2 - 54$ **74.** $12x^2 - 48$

75. $x^3 - x^2$ **76.** $x^3 - 16x$

77. $2x^2 + 4x - 2x^3$ **78.** $9x^2 + 12x - 3x^3$

79. $5 - x + 5x^2 - x^3$ **80.** $3u - 2u^2 + 6 - u^3$

81. $2(x - 2)(x + 1)^2 - 3(x - 2)^2(x + 1)$

82. $2(x + 1)(x - 3)^2 - 3(x + 1)^2(x - 3)$

83. Geometry The cylindrical shell shown in the figure has a volume of

$$V = \pi R^2 h - \pi r^2 h.$$

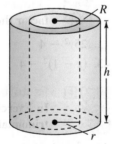

(a) Factor the expression for the volume.

(b) From the result of part (a), show that the volume is 2π(average radius)(thickness of the shell)h.

84. Chemistry

The rate of change of an autocatalytic chemical reaction is $kQx - kx^2$, where Q is the amount of the original substance, x is the amount of substance formed, and k is a constant of proportionality. Factor the expression.

Exploration

True or False? In Exercises 85–87, determine whether the statement is true or false. Justify your answer.

85. The product of two binomials is always a second-degree polynomial.

86. The sum of two binomials is always a binomial.

87. The difference of two perfect squares can be factored as the product of conjugate pairs.

88. Error Analysis Describe the error.

$$9x^2 - 9x - 54 = (3x + 6)(3x - 9)$$
$$= 3(x + 2)(x - 3)$$

89. Degree of a Product Find the degree of the product of two polynomials of degrees m and n.

90. Degree of a Sum Find the degree of the sum of two polynomials of degrees m and n, where $m < n$.

91. Think About It When the polynomial

$$-x^3 + 3x^2 + 2x - 1$$

is subtracted from an unknown polynomial, the difference is $5x^2 + 8$. Find the unknown polynomial.

92. Logical Reasoning Verify that $(x + y)^2$ is not equal to $x^2 + y^2$ by letting $x = 3$ and $y = 4$ and evaluating both expressions. Are there any values of x and y for which $(x + y)^2 = x^2 + y^2$? Explain.

93. Think About It Give an example of a polynomial that is prime.

94. 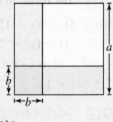 **HOW DO YOU SEE IT?** The figure shows a large square with an area of a^2 that contains a smaller square with an area of b^2.

(a) Describe the regions that represent $a^2 - b^2$. How can you rearrange these regions to show that $a^2 - b^2 = (a - b)(a + b)$?

(b) How can you use the figure to show that $(a - b)^2 = a^2 - 2ab + b^2$?

(c) Draw another figure to show that $(a + b)^2 = a^2 + 2ab + b^2$. Explain how the figure shows this.

Factoring with Variables in the Exponents In Exercises 95 and 96, factor the expression as completely as possible.

95. $x^{2n} - y^{2n}$ **96.** $x^{3n} + y^{3n}$

A.4 Rational Expressions

■ Find domains of algebraic expressions.
■ Simplify rational expressions.
■ Add, subtract, multiply, and divide rational expressions.
■ Simplify complex fractions and rewrite difference quotients.

Domain of an Algebraic Expression

The set of real numbers for which an algebraic expression is defined is the **domain** of the expression. Two algebraic expressions are **equivalent** when they have the same domain and yield the same values for all numbers in their domain. For example,

$$(x + 1) + (x + 2) \quad \text{and} \quad 2x + 3$$

are equivalent because

$$(x + 1) + (x + 2) = x + 1 + x + 2$$
$$= x + x + 1 + 2$$
$$= 2x + 3.$$

Rational expressions have many real-life applications. For example, in Exercise 71 on page A43, you will work with a rational expression that models the temperature of food in a refrigerator.

EXAMPLE 1 Finding Domains of Algebraic Expressions

a. The domain of the polynomial

$$2x^3 + 3x + 4$$

is the set of all real numbers. In fact, the domain of any polynomial is the set of all real numbers, unless the domain is specifically restricted.

b. The domain of the radical expression

$$\sqrt{x - 2}$$

is the set of real numbers greater than or equal to 2, because the square root of a negative number is not a real number.

c. The domain of the expression

$$\frac{x + 2}{x - 3}$$

is the set of all real numbers except $x = 3$, which would result in division by zero, which is undefined.

✓ *Checkpoint* Audio-video solution in English & Spanish at LarsonPrecalculus.com

Find the domain of each expression.

a. $4x^3 + 3, \quad x \geq 0$ **b.** $\sqrt{x + 7}$ **c.** $\dfrac{1 - x}{x}$

The quotient of two algebraic expressions is a *fractional expression*. Moreover, the quotient of two *polynomials* such as

$$\frac{1}{x}, \quad \frac{2x - 1}{x + 1}, \quad \text{or} \quad \frac{x^2 - 1}{x^2 + 1}$$

is a **rational expression.**

Simplifying Rational Expressions

Recall that a fraction is in simplest form when its numerator and denominator have no factors in common other than ± 1. To write a fraction in simplest form, divide out common factors.

$$\frac{a \cdot \cancel{c}}{b \cdot \cancel{c}} = \frac{a}{b}, \quad c \neq 0$$

The key to success in simplifying rational expressions lies in your ability to *factor* polynomials. When simplifying rational expressions, factor each polynomial completely to determine whether the numerator and denominator have factors in common.

EXAMPLE 2 Simplifying a Rational Expression

$$\frac{x^2 + 4x - 12}{3x - 6} = \frac{(x + 6)(\cancel{x - 2})}{3(\cancel{x - 2})} \qquad \text{Factor completely.}$$

$$= \frac{x + 6}{3}, \quad x \neq 2 \qquad \text{Divide out common factor.}$$

Note that the original expression is undefined when $x = 2$ (because division by zero is undefined). To make the simplified expression *equivalent* to the original expression, you must restrict the domain of the simplified expression by excluding the value of $x = 2$.

✓ **Checkpoint**  Audio-video solution in English & Spanish at LarsonPrecalculus.com

Write $\dfrac{4x + 12}{x^2 - 3x - 18}$ in simplest form. ■

Sometimes it may be necessary to change the sign of a factor by factoring out (-1) to simplify a rational expression, as shown in Example 3.

EXAMPLE 3 Simplifying a Rational Expression

$$\frac{12 + x - x^2}{2x^2 - 9x + 4} = \frac{(4 - x)(3 + x)}{(2x - 1)(x - 4)} \qquad \text{Factor completely.}$$

$$= \frac{-(\cancel{x - 4})(3 + x)}{(2x - 1)(\cancel{x - 4})} \qquad (4 - x) = -(x - 4)$$

$$= -\frac{3 + x}{2x - 1}, \quad x \neq 4 \qquad \text{Divide out common factor.}$$

✓ **Checkpoint** ◄))) Audio-video solution in English & Spanish at LarsonPrecalculus.com

Write $\dfrac{3x^2 - x - 2}{5 - 4x - x^2}$ in simplest form. ■

In this text, the domain is usually not listed with a rational expression. It is *implied* that the real numbers that make the denominator zero are excluded from the domain. Also, when performing operations with rational expressions, this text follows the convention of listing *by the simplified expression* all values of x that must be specifically excluded from the domain to make the domains of the simplified and original expressions agree. Example 3, for instance, lists the restriction $x \neq 4$ with the simplified expression to make the two domains agree. Note that the value $x = \frac{1}{2}$ is excluded from *both* domains, so it is not necessary to list this value.

• • REMARK In Example 2, do not make the mistake of trying to simplify further by dividing out terms.

$$\frac{x + 6}{3} = \frac{x + \cancel{6}}{\cancel{3}}$$

$$= x + 2$$

To simplify fractions, divide out common *factors,* not terms.

Operations with Rational Expressions

To multiply or divide rational expressions, use the properties of fractions discussed in Appendix A.1. Recall that to divide fractions, you invert the divisor and multiply.

EXAMPLE 4   **Multiplying Rational Expressions**

$$\frac{2x^2 + x - 6}{x^2 + 4x - 5} \cdot \frac{x^3 - 3x^2 + 2x}{4x^2 - 6x} = \frac{(2x - 3)(x + 2)}{(x + 5)(x - 1)} \cdot \frac{x(x - 2)(x - 1)}{2x(2x - 3)}$$

$$= \frac{(x + 2)(x - 2)}{2(x + 5)}, \quad x \neq 0, x \neq 1, x \neq \tfrac{3}{2}$$

✓ *Checkpoint* ◀))) *Audio-video solution in English & Spanish at LarsonPrecalculus.com*

Multiply and simplify: $\dfrac{15x^2 + 5x}{x^3 - 3x^2 - 18x} \cdot \dfrac{x^2 - 2x - 15}{3x^2 - 8x - 3}$.

> •• **REMARK** Note that Example 4 lists the restrictions $x \neq 0$, $x \neq 1$, and $x \neq \tfrac{3}{2}$ with the simplified expression to make the two domains agree. Also note that the value $x = -5$ is excluded from both domains, so it is not necessary to list this value.

EXAMPLE 5   **Dividing Rational Expressions**

$$\frac{x^3 - 8}{x^2 - 4} \div \frac{x^2 + 2x + 4}{x^3 + 8} = \frac{x^3 - 8}{x^2 - 4} \cdot \frac{x^3 + 8}{x^2 + 2x + 4} \qquad \text{Invert and multiply.}$$

$$= \frac{(x - 2)(x^2 + 2x + 4)}{(x + 2)(x - 2)} \cdot \frac{(x + 2)(x^2 - 2x + 4)}{(x^2 + 2x + 4)}$$

$$= x^2 - 2x + 4, \quad x \neq \pm 2 \qquad \text{Divide out common factors.}$$

✓ *Checkpoint* ◀))) *Audio-video solution in English & Spanish at LarsonPrecalculus.com*

Divide and simplify: $\dfrac{x^3 - 1}{x^2 - 1} \div \dfrac{x^2 + x + 1}{x^2 + 2x + 1}$. ■

To add or subtract rational expressions, use the LCD (least common denominator) method or the *basic definition*

$$\frac{a}{b} \pm \frac{c}{d} = \frac{ad \pm bc}{bd}, \quad b \neq 0, d \neq 0. \qquad \text{Basic definition}$$

This definition provides an efficient way of adding or subtracting two fractions that have no common factors in their denominators.

EXAMPLE 6   **Subtracting Rational Expressions**

$$\frac{x}{x - 3} - \frac{2}{3x + 4} = \frac{x(3x + 4) - 2(x - 3)}{(x - 3)(3x + 4)} \qquad \text{Basic definition}$$

$$= \frac{3x^2 + 4x - 2x + 6}{(x - 3)(3x + 4)} \qquad \text{Distributive Property}$$

$$= \frac{3x^2 + 2x + 6}{(x - 3)(3x + 4)} \qquad \text{Combine like terms.}$$

> •• **REMARK** When subtracting rational expressions, remember to distribute the negative sign to *all* the terms in the quantity that is being subtracted.

✓ *Checkpoint* ◀))) *Audio-video solution in English & Spanish at LarsonPrecalculus.com*

Subtract and simplify: $\dfrac{x}{2x - 1} - \dfrac{1}{x + 2}$. ■

For three or more fractions, or for fractions with a repeated factor in the denominators, the LCD method works well. Recall that the least common denominator of several fractions consists of the product of all prime factors in the denominators, with each factor given the highest power of its occurrence in any denominator. Here is a numerical example.

$$\frac{1}{6} + \frac{3}{4} - \frac{2}{3} = \frac{1 \cdot 2}{6 \cdot 2} + \frac{3 \cdot 3}{4 \cdot 3} - \frac{2 \cdot 4}{3 \cdot 4} \qquad \text{The LCD is 12.}$$

$$= \frac{2}{12} + \frac{9}{12} - \frac{8}{12}$$

$$= \frac{3}{12}$$

$$= \frac{1}{4}$$

Sometimes, the numerator of the answer has a factor in common with the denominator. In such cases, simplify the answer, as shown in the example above.

EXAMPLE 7   Combining Rational Expressions: The LCD Method

See LarsonPrecalculus.com for an interactive version of this type of example.

Perform the operations and simplify.

$$\frac{3}{x - 1} - \frac{2}{x} + \frac{x + 3}{x^2 - 1}$$

Solution   Use the factored denominators

$$(x - 1), \quad x, \quad \text{and} \quad (x + 1)(x - 1)$$

to determine that the LCD is $x(x + 1)(x - 1)$.

$$\frac{3}{x - 1} - \frac{2}{x} + \frac{x + 3}{(x + 1)(x - 1)}$$

$$= \frac{3(x)(x + 1)}{x(x + 1)(x - 1)} - \frac{2(x + 1)(x - 1)}{x(x + 1)(x - 1)} + \frac{(x + 3)(x)}{x(x + 1)(x - 1)}$$

$$= \frac{3(x)(x + 1) - 2(x + 1)(x - 1) + (x + 3)(x)}{x(x + 1)(x - 1)}$$

$$= \frac{3x^2 + 3x - 2x^2 + 2 + x^2 + 3x}{x(x + 1)(x - 1)} \qquad \text{Multiply.}$$

$$= \frac{(3x^2 - 2x^2 + x^2) + (3x + 3x) + 2}{x(x + 1)(x - 1)} \qquad \text{Group like terms.}$$

$$= \frac{2x^2 + 6x + 2}{x(x + 1)(x - 1)} \qquad \text{Combine like terms.}$$

$$= \frac{2(x^2 + 3x + 1)}{x(x + 1)(x - 1)} \qquad \text{Factor.}$$

✓ **Checkpoint** ◀))) *Audio-video solution in English & Spanish at LarsonPrecalculus.com*

Perform the operations and simplify.

$$\frac{4}{x} - \frac{x + 5}{x^2 - 4} + \frac{4}{x + 2}$$

Complex Fractions and the Difference Quotient

Complex fractions are fractional expressions with separate fractions in the numerator, denominator, or both. Here are two examples.

$$\frac{\left(\dfrac{1}{x}\right)}{x^2 + 1} \quad \text{and} \quad \frac{\left(\dfrac{1}{x}\right)}{\left(\dfrac{1}{x^2 + 1}\right)}$$

One way to simplify a complex fraction is to combine the fractions in the numerator into a single fraction and then combine the fractions in the denominator into a single fraction. Then invert the denominator and multiply. Example 8 shows this method.

EXAMPLE 8   **Simplifying a Complex Fraction**

$$\frac{\left(\dfrac{2}{x} - 3\right)}{\left(1 - \dfrac{1}{x-1}\right)} = \frac{\left[\dfrac{2 - 3(x)}{x}\right]}{\left[\dfrac{1(x-1) - 1}{x-1}\right]} \qquad \text{Combine fractions.}$$

$$= \frac{\left(\dfrac{2 - 3x}{x}\right)}{\left(\dfrac{x - 2}{x - 1}\right)} \qquad \text{Simplify.}$$

$$= \frac{2 - 3x}{x} \cdot \frac{x - 1}{x - 2} \qquad \text{Invert and multiply.}$$

$$= \frac{(2 - 3x)(x - 1)}{x(x - 2)}, \quad x \neq 1$$

✓ **Checkpoint** 🔊 *Audio-video solution in English & Spanish at LarsonPrecalculus.com*

Simplify the complex fraction $\dfrac{\left(\dfrac{1}{x + 2} + 1\right)}{\left(\dfrac{x}{3} - 1\right)}$.

Another way to simplify a complex fraction is to multiply its numerator and denominator by the LCD of all fractions in its numerator and denominator. This method, applied to the fraction in Example 8, is shown below. Notice that both methods yield the same result.

$$\frac{\left(\dfrac{2}{x} - 3\right)}{\left(1 - \dfrac{1}{x-1}\right)} \cdot \frac{x(x - 1)}{x(x - 1)} = \frac{\left(\dfrac{2}{x}\right)(x)(x - 1) - (3)(x)(x - 1)}{(1)(x)(x - 1) - \left(\dfrac{1}{x-1}\right)(x)(x - 1)} \qquad \begin{array}{l}\text{LCD is}\\ x(x - 1).\end{array}$$

$$= \frac{2(x - 1) - 3x(x - 1)}{x(x - 1) - x} \qquad \text{Simplify.}$$

$$= \frac{(2 - 3x)(x - 1)}{x(x - 2)}, \quad x \neq 1 \qquad \text{Factor.}$$

The next three examples illustrate some methods for simplifying rational expressions involving negative exponents and radicals. These types of expressions occur frequently in calculus.

To simplify an expression with negative exponents, one method is to begin by factoring out the common factor with the *lesser* exponent. Remember that when factoring, you *subtract* exponents. For example, in $3x^{-5/2} + 2x^{-3/2}$, the lesser exponent is $-\frac{5}{2}$ and the common factor is $x^{-5/2}$.

$$3x^{-5/2} + 2x^{-3/2} = x^{-5/2}[3 + 2x^{-3/2-(-5/2)}]$$

$$= x^{-5/2}(3 + 2x^1)$$

$$= \frac{3 + 2x}{x^{5/2}}$$

EXAMPLE 9 **Simplifying an Expression**

Simplify $x(1 - 2x)^{-3/2} + (1 - 2x)^{-1/2}$.

Solution Begin by factoring out the common factor with the lesser exponent.

$$x(1 - 2x)^{-3/2} + (1 - 2x)^{-1/2} = (1 - 2x)^{-3/2}[x + (1 - 2x)^{(-1/2)-(-3/2)}]$$

$$= (1 - 2x)^{-3/2}[x + (1 - 2x)^1]$$

$$= \frac{1 - x}{(1 - 2x)^{3/2}}$$

✓ **Checkpoint** ◀))) Audio-video solution in English & Spanish at LarsonPrecalculus.com

Simplify $(x - 1)^{-1/3} - x(x - 1)^{-4/3}$.

The next example shows a complex fraction with a negative exponent and a second method for simplifying an expression with negative exponents.

EXAMPLE 10 **Simplifying an Expression**

$$\frac{\left[\dfrac{1}{(4 - x^2)^{-1/2}} + \dfrac{x^2}{(4 - x^2)^{1/2}}\right]}{4 - x^2} = \frac{(4 - x^2)^{1/2} + x^2(4 - x^2)^{-1/2}}{4 - x^2}$$

$$= \frac{(4 - x^2)^{1/2} + x^2(4 - x^2)^{-1/2}}{4 - x^2} \cdot \frac{(4 - x^2)^{1/2}}{(4 - x^2)^{1/2}}$$

$$= \frac{(4 - x^2)^1 + x^2(4 - x^2)^0}{(4 - x^2)^{3/2}}$$

$$= \frac{4 - x^2 + x^2}{(4 - x^2)^{3/2}}$$

$$= \frac{4}{(4 - x^2)^{3/2}}$$

✓ **Checkpoint** ◀))) Audio-video solution in English & Spanish at LarsonPrecalculus.com

Simplify

$$\frac{\left[\dfrac{x^2}{(x^2 - 2)^{1/2}} + \dfrac{1}{(x^2 - 2)^{-1/2}}\right]}{x^2 - 2}.$$

Difference quotients, such as

$$\frac{\sqrt{x + h} - \sqrt{x}}{h}$$

occur frequently in calculus. Often, they need to be rewritten in an equivalent form, as shown in Example 11.

EXAMPLE 11   **Rewriting a Difference Quotient**

Rewrite the difference quotient

$$\frac{\sqrt{x + h} - \sqrt{x}}{h}$$

by rationalizing its numerator.

Solution

$$\frac{\sqrt{x + h} - \sqrt{x}}{h} = \frac{\sqrt{x + h} - \sqrt{x}}{h} \cdot \frac{\sqrt{x + h} + \sqrt{x}}{\sqrt{x + h} + \sqrt{x}}$$

$$= \frac{\left(\sqrt{x + h}\right)^2 - \left(\sqrt{x}\right)^2}{h\left(\sqrt{x + h} + \sqrt{x}\right)}$$

$$= \frac{x + h - x}{h\left(\sqrt{x + h} + \sqrt{x}\right)}$$

$$= \frac{h}{h\left(\sqrt{x + h} + \sqrt{x}\right)}$$

$$= \frac{1}{\sqrt{x + h} + \sqrt{x}}, \quad h \neq 0$$

✓ *Checkpoint* *Audio-video solution in English & Spanish at LarsonPrecalculus.com*

Rewrite the difference quotient

$$\frac{\sqrt{9 + h} - 3}{h}$$

by rationalizing its numerator.

Summarize   (Appendix A.4)

1. State the definition of the domain of an algebraic expression *(page A35)*. For an example of finding domains of algebraic expressions, see Example 1.

2. State the definition of a rational expression and explain how to simplify a rational expression *(pages A35 and A36)*. For examples of simplifying rational expressions, see Examples 2 and 3.

3. Explain how to multiply, divide, add, and subtract rational expressions *(page A37)*. For examples of operations with rational expressions, see Examples 4–7.

4. State the definition of a complex fraction *(page A39)*. For an example of simplifying a complex fraction, see Example 8.

5. Explain how to rewrite a difference quotient *(page A41)*. For an example of rewriting a difference quotient, see Example 11.

A.4 Exercises

See **CalcChat.com** for tutorial help and worked-out solutions to odd-numbered exercises.

Vocabulary: **Fill in the blanks.**

1. The set of real numbers for which an algebraic expression is defined is the _____ of the expression.

2. The quotient of two algebraic expressions is a fractional expression, and the quotient of two polynomials is a _____ _____.

3. Fractional expressions with separate fractions in the numerator, denominator, or both are _____ fractions.

4. Two algebraic expressions that have the same domain and yield the same values for all numbers in their domains are _____.

Skills and Applications

 Finding the Domain of an Algebraic Expression **In Exercises 5–16, find the domain of the expression.**

5. $3x^2 - 4x + 7$

6. $6x^2 - 9, \quad x > 0$

7. $\dfrac{1}{3 - x}$

8. $\dfrac{1}{x + 5}$

9. $\dfrac{x + 6}{3x + 2}$

10. $\dfrac{x - 4}{1 - 2x}$

11. $\dfrac{x^2 - 5x + 6}{x^2 + 6x + 8}$

12. $\dfrac{x^2 - 1}{x^2 + 3x - 10}$

13. $\sqrt{x - 7}$

14. $\sqrt{2x - 5}$

15. $\dfrac{1}{\sqrt{x - 3}}$

16. $\dfrac{1}{\sqrt{x + 2}}$

 Simplifying a Rational Expression **In Exercises 17–30, write the rational expression in simplest form.**

17. $\dfrac{15x^2}{10x}$

18. $\dfrac{18y^2}{60y^5}$

19. $\dfrac{x - 5}{10 - 2x}$

20. $\dfrac{12 - 4x}{x - 3}$

21. $\dfrac{y^2 - 16}{y + 4}$

22. $\dfrac{x^2 - 25}{5 - x}$

23. $\dfrac{6y + 9y^2}{12y + 8}$

24. $\dfrac{4y - 8y^2}{10y - 5}$

25. $\dfrac{x^2 + 4x - 5}{x^2 + 8x + 15}$

26. $\dfrac{x^2 + 8x - 20}{x^2 + 11x + 10}$

27. $\dfrac{x^2 - x - 2}{10 - 3x - x^2}$

28. $\dfrac{4 + 3x - x^2}{2x^2 - 7x - 4}$

29. $\dfrac{x^2 - 16}{x^3 + x^2 - 16x - 16}$

30. $\dfrac{x^2 - 1}{x^3 + x^2 + 9x + 9}$

31. **Error Analysis** Describe the error.

$$\dfrac{5x^3}{2x^3 + 4} = \dfrac{5x^3}{2x^3 + 4} = \dfrac{5}{2 + 4} = \dfrac{5}{6} \quad \times$$

32. **Evaluating a Rational Expression** Complete the table. What can you conclude?

x	0	1	2	3	4	5	6
$\dfrac{x - 3}{x^2 - x - 6}$							
$\dfrac{1}{x + 2}$							

 Multiplying or Dividing Rational Expressions **In Exercises 33–38, perform the multiplication or division and simplify.**

33. $\dfrac{5}{x - 1} \cdot \dfrac{x - 1}{25(x - 2)}$

34. $\dfrac{r}{r - 1} \div \dfrac{r^2}{r^2 - 1}$

35. $\dfrac{x^2 - 4}{12} \div \dfrac{2 - x}{2x + 4}$

36. $\dfrac{t^2 - t - 6}{t^2 + 6t + 9} \cdot \dfrac{t + 3}{t^2 - 4}$

37. $\dfrac{x^2 + xy - 2y^2}{x^3 + x^2y} \cdot \dfrac{x}{x^2 + 3xy + 2y^2}$

38. $\dfrac{x^2 - 14x + 49}{x^2 - 49} \div \dfrac{3x - 21}{x + 7}$

 Adding or Subtracting Rational Expressions **In Exercises 39–46, perform the addition or subtraction and simplify.**

39. $\dfrac{x - 1}{x + 2} - \dfrac{x - 4}{x + 2}$

40. $\dfrac{2x - 1}{x + 3} + \dfrac{1 - x}{x + 3}$

41. $\dfrac{1}{3x + 2} + \dfrac{x}{x + 1}$

42. $\dfrac{x}{x + 4} - \dfrac{6}{x - 1}$

43. $\dfrac{3}{2x + 4} - \dfrac{x}{x + 2}$

44. $\dfrac{2}{x^2 - 9} + \dfrac{4}{x + 3}$

45. $-\dfrac{1}{x} + \dfrac{2}{x^2 + 1} + \dfrac{1}{x^3 + x}$

46. $\dfrac{2}{x + 1} + \dfrac{2}{x - 1} + \dfrac{1}{x^2 - 1}$

Error Analysis In Exercises 47 and 48, describe the error.

47.
$$\frac{x+4}{x+2} - \frac{3x-8}{x+2} = \frac{x+4-3x-8}{x+2}$$
$$= \frac{-2x-4}{x+2}$$
$$= \frac{-2(x+2)}{x+2}$$
$$= -2, \quad x \neq -2$$

48.
$$\frac{6-x}{x(x+2)} + \frac{x+2}{x^2} + \frac{8}{x^2(x+2)}$$
$$= \frac{6-x+(x+2)^2+8}{x^2(x+2)}$$
$$= \frac{6-x+x^2+4x+4+8}{x^2(x+2)}$$
$$= \frac{x^2+3x+18}{x^2(x+2)}$$

 Simplifying a Complex Fraction In Exercises 49–54, simplify the complex fraction.

49.
$$\frac{\left(\dfrac{x}{2}-1\right)}{x-2}$$

50.
$$\frac{x+5}{\left(\dfrac{x}{5}-5\right)}$$

51.
$$\frac{\left[\dfrac{x^2}{(x+1)^2}\right]}{\left[\dfrac{x}{(x+1)^3}\right]}$$

52.
$$\frac{\left(\dfrac{x^2-1}{x}\right)}{\left[\dfrac{(x-1)^2}{x}\right]}$$

53.
$$\frac{\left(\sqrt{x}-\dfrac{1}{2\sqrt{x}}\right)}{\sqrt{x}}$$

54.
$$\frac{\left(\dfrac{t^2}{\sqrt{t^2+1}}-\sqrt{t^2+1}\right)}{t^2}$$

Factoring an Expression In Exercises 55–58, factor the expression by factoring out the common factor with the lesser exponent.

55. $x^2(x^2+3)^{-4} + (x^2+3)^3$

56. $2x(x-5)^{-3} - 4x^2(x-5)^{-4}$

57. $2x^2(x-1)^{1/2} - 5(x-1)^{-1/2}$

58. $4x^3(x+1)^{-3/2} - x(x+1)^{-1/2}$

 Simplifying an Expression In Exercises 59 and 60, simplify the expression.

59.
$$\frac{3x^{1/3} - x^{-2/3}}{3x^{-2/3}}$$

60.
$$\frac{-x^3(1-x^2)^{-1/2} - 2x(1-x^2)^{1/2}}{x^4}$$

Simplifying a Difference Quotient In Exercises 61–64, simplify the difference quotient.

61.
$$\frac{\left(\dfrac{1}{x+h}-\dfrac{1}{x}\right)}{h}$$

62.
$$\frac{\left[\dfrac{1}{(x+h)^2}-\dfrac{1}{x^2}\right]}{h}$$

63.
$$\frac{\left(\dfrac{1}{x+h-4}-\dfrac{1}{x-4}\right)}{h}$$

64.
$$\frac{\left(\dfrac{x+h}{x+h+1}-\dfrac{x}{x+1}\right)}{h}$$

 Rewriting a Difference Quotient In Exercises 65–70, rewrite the difference quotient by rationalizing the numerator.

65.
$$\frac{\sqrt{x+2}-\sqrt{x}}{2}$$

66.
$$\frac{\sqrt{z-3}-\sqrt{z}}{-3}$$

67.
$$\frac{\sqrt{t+3}-\sqrt{3}}{t}$$

68.
$$\frac{\sqrt{x+5}-\sqrt{5}}{x}$$

69.
$$\frac{\sqrt{x+h+1}-\sqrt{x+1}}{h}$$

70.
$$\frac{\sqrt{x+h-2}-\sqrt{x-2}}{h}$$

71. Refrigeration

After placing food (at room temperature) in a refrigerator, the time required for the food to cool depends on the amount of food, the air circulation in the refrigerator, the original temperature of the food, and the temperature of the refrigerator. The model that gives the temperature of food that has an original temperature of 75°F and is placed in a 40°F refrigerator is

$$T = 10\left(\frac{4t^2 + 16t + 75}{t^2 + 4t + 10}\right)$$

where T is the temperature (in degrees Fahrenheit) and t is the time (in hours).

(a) Complete the table.

t	0	2	4	6	8	10	12
T							

t	14	16	18	20	22
T					

(b) What value of T does the mathematical model appear to be approaching?

72. Rate   A copier copies at a rate of 50 pages per minute.

(a) Find the time required to copy one page.

(b) Find the time required to copy x pages.

(c) Find the time required to copy 120 pages.

Probability   In Exercises 73 and 74, consider an experiment in which a marble is tossed into a box whose base is shown in the figure. The probability that the marble will come to rest in the shaded portion of the base is equal to the ratio of the shaded area to the total area of the figure. Find the probability.

73.

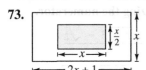

74.

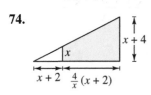

75. Interactive Money Management   The table shows the numbers of U.S. households (in millions) using online banking and mobile banking from 2011 through 2014.   (*Source: Fiserv, Inc.*)

Year	Online Banking	Mobile Banking
2011	79	18
2012	81	24
2013	83	30
2014	86	35

Mathematical models for the data are

$$\text{Number using online banking} = \frac{-2.9709t + 70.517}{-0.0474t + 1}$$

$$\text{Number using mobile banking} = \frac{0.661t^2 - 47}{0.007t^2 + 1}$$

where t represents the year, with $t = 11$ corresponding to 2011.

(a) Using the models, create a table showing the numbers of households using online banking and the numbers of households using mobile banking for the given years.

(b) Compare the values from the models with the actual data.

(c) Determine a model for the ratio of the number of households using mobile banking to the number of households using online banking.

(d) Use the model from part (c) to find the ratios for the given years. Interpret your results.

76. Finance   The formula that approximates the annual interest rate r of a monthly installment loan is

$$r = \frac{24(NM - P)}{N} \div \left(P + \frac{NM}{12} \right)$$

where N is the total number of payments, M is the monthly payment, and P is the amount financed.

(a) Approximate the annual interest rate for a five-year car loan of $28,000 that has monthly payments of $525.

(b) Simplify the expression for the annual interest rate r, and then rework part (a).

77. Electrical Engineering   The formula for the total resistance R_T (in ohms) of two resistors connected in parallel is

$$R_T = \frac{1}{\left(\dfrac{1}{R_1} + \dfrac{1}{R_2} \right)}$$

where R_1 and R_2 are the resistance values of the first and second resistors, respectively. Simplify the expression for the total resistance R_T.

78. **HOW DO YOU SEE IT?**   The mathematical model

$$P = 100 \left(\frac{t^2 - t + 1}{t^2 + 1} \right), \quad t \geq 0$$

gives the percent P of the normal level of oxygen in a pond, where t is the time (in weeks) after organic waste is dumped into the pond. The bar graph shows the situation. What conclusions can you draw from the bar graph?

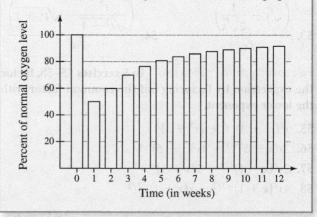

Exploration

True or False?   In Exercises 79 and 80, determine whether the statement is true or false. Justify your answer.

79. $\dfrac{x^{2n} - 1^{2n}}{x^n - 1^n} = x^n + 1^n$

80. $\dfrac{x^2 - 3x + 2}{x - 1} = x - 2$, for all values of x

A.5 Solving Equations

- ▢ **Identify different types of equations.**
- ▢ **Solve linear equations in one variable and rational equations that lead to linear equations.**
- ▢ **Solve quadratic equations by factoring, extracting square roots, completing the square, and using the Quadratic Formula.**
- ▢ **Solve polynomial equations of degree three or greater.**
- ▢ **Solve radical equations.**
- ▢ **Solve absolute value equations.**
- ▢ **Use common formulas to solve real-life problems.**

Linear equations have many real-life applications, such as in forensics. For example, in Exercises 95 and 96 on page A57, you will use linear equations to determine height from femur length.

Equations and Solutions of Equations

An **equation** in x is a statement that two algebraic expressions are equal. For example, $3x - 5 = 7$, $x^2 - x - 6 = 0$, and $\sqrt{2x} = 4$ are equations. To **solve** an equation in x means to find all values of x for which the equation is true. Such values are **solutions.** For example, $x = 4$ is a solution of the equation $3x - 5 = 7$ because $3(4) - 5 = 7$ is a true statement.

The solutions of an equation depend on the kinds of numbers being considered. For example, in the set of rational numbers, $x^2 = 10$ has no solution because there is no rational number whose square is 10. However, in the set of real numbers, the equation has the two solutions $x = \sqrt{10}$ and $x = -\sqrt{10}$.

An equation that is true for *every* real number in the domain of the variable is an **identity.** For example,

$$x^2 - 9 = (x + 3)(x - 3) \qquad \text{Identity}$$

is an identity because it is a true statement for any real value of x. The equation

$$\frac{x}{3x^2} = \frac{1}{3x} \qquad \text{Identity}$$

is an identity because it is true for any nonzero real value of x.

An equation that is true for just *some* (but not all) of the real numbers in the domain of the variable is a **conditional equation.** For example, the equation

$$x^2 - 9 = 0 \qquad \text{Conditional equation}$$

is conditional because $x = 3$ and $x = -3$ are the only values in the domain that satisfy the equation.

A **contradiction** is an equation that is *false* for *every* real number in the domain of the variable. For example, the equation

$$2x - 4 = 2x + 1 \qquad \text{Contradiction}$$

is a contradiction because there are no real values of x for which the equation is true.

Linear and Rational Equations

> ### Definition of a Linear Equation in One Variable
>
> A **linear equation in one variable** x is an equation that can be written in the standard form
>
> $$ax + b = 0$$
>
> where a and b are real numbers with $a \neq 0$.

A linear equation in one variable has exactly one solution. To see this, consider the steps below. (Remember that $a \neq 0$.)

$$ax + b = 0 \qquad \text{Write original equation.}$$

$$ax = -b \qquad \text{Subtract } b \text{ from each side.}$$

$$x = -\frac{b}{a} \qquad \text{Divide each side by } a.$$

The above suggests that to solve a conditional equation in x, you isolate x on one side of the equation using a sequence of **equivalent equations,** each having the same solution as the original equation. The operations that yield equivalent equations come from the properties of equality reviewed in Appendix A.1.

Generating Equivalent Equations

An equation can be transformed into an *equivalent equation* by one or more of the steps listed below.

	Given Equation	Equivalent Equation
1. Remove symbols of grouping, combine like terms, or simplify fractions on one or both sides of the equation.	$2x - x = 4$	$x = 4$
2. Add (or subtract) the same quantity to (from) *each* side of the equation.	$x + 1 = 6$	$x = 5$
3. Multiply (or divide) *each* side of the equation by the same *nonzero* quantity.	$2x = 6$	$x = 3$
4. Interchange the two sides of the equation.	$2 = x$	$x = 2$

In Example 1, you will use these steps to solve linear equations in one variable x.

EXAMPLE 1   Solving Linear Equations

a. $3x - 6 = 0 \qquad$ Original equation

$\quad\quad 3x = 6 \qquad$ Add 6 to each side.

$\quad\quad\quad x = 2 \qquad$ Divide each side by 3.

b. $5x + 4 = 3x - 8 \qquad$ Original equation

$\quad\quad 2x + 4 = -8 \qquad$ Subtract $3x$ from each side.

$\quad\quad\quad 2x = -12 \qquad$ Subtract 4 from each side.

$\quad\quad\quad\quad x = -6 \qquad$ Divide each side by 2.

✓ *Checkpoint*))) *Audio-video solution in English & Spanish at LarsonPrecalculus.com*

Solve each equation.

a. $7 - 2x = 15$

b. $7x - 9 = 5x + 7$

> •• **REMARK** After solving an equation, you should check each solution in the original equation. For instance, here is a check of the solution in Example 1(a).
>
> $3x - 6 = 0 \qquad$ Write original equation.
>
> $3(2) - 6 \overset{?}{=} 0 \qquad$ Substitute 2 for x.
>
> $\quad\quad 0 = 0 \qquad$ Solution checks. ✓
>
> Check the solution in Example 1(b) on your own.

• • • • • • • • • • • • • • • • ▷

REMARK An equation with a single fraction on each side can be cleared of denominators by *cross multiplying*. To do this, multiply the left numerator by the right denominator and the right numerator by the left denominator.

$$\frac{a}{b} = \frac{c}{d} \qquad \text{Original equation}$$

$$ad = cb \qquad \text{Cross multiply.}$$

A **rational equation** involves one or more rational expressions. To solve a rational equation, multiply every term by the least common denominator (LCD) of all the terms. This clears the original equation of fractions and produces a simpler equation.

EXAMPLE 2 **Solving a Rational Equation**

Solve $\dfrac{x}{3} + \dfrac{3x}{4} = 2$.

Solution

$$\frac{x}{3} + \frac{3x}{4} = 2 \qquad \text{Write original equation.}$$

$$(12)\frac{x}{3} + (12)\frac{3x}{4} = (12)2 \qquad \text{Multiply each term by the LCD.}$$

$$4x + 9x = 24 \qquad \text{Simplify.}$$

$$13x = 24 \qquad \text{Combine like terms.}$$

$$x = \frac{24}{13} \qquad \text{Divide each side by 13.}$$

The solution is $x = \frac{24}{13}$. Check this in the original equation.

✓ *Checkpoint*  Audio-video solution in English & Spanish at LarsonPrecalculus.com

Solve $\dfrac{4x}{9} - \dfrac{1}{3} = x + \dfrac{5}{3}$. ◼

When multiplying or dividing an equation by a *variable expression*, it is possible to introduce an **extraneous solution,** which is a solution that does not satisfy the original equation.

EXAMPLE 3 **An Equation with an Extraneous Solution**

See LarsonPrecalculus.com for an interactive version of this type of example.

Solve $\dfrac{1}{x - 2} = \dfrac{3}{x + 2} - \dfrac{6x}{x^2 - 4}$.

Solution The LCD is $x^2 - 4 = (x + 2)(x - 2)$. Multiply each term by the LCD.

$$\frac{1}{x - 2}(x + 2)(x - 2) = \frac{3}{x + 2}(x + 2)(x - 2) - \frac{6x}{x^2 - 4}(x + 2)(x - 2)$$

$$x + 2 = 3(x - 2) - 6x, \quad x \neq \pm 2$$

$$x + 2 = 3x - 6 - 6x$$

$$x + 2 = -3x - 6$$

$$4x = -8$$

$$x = -2 \qquad \text{Extraneous solution}$$

In the original equation, $x = -2$ yields a denominator of zero. So, $x = -2$ is an extraneous solution, and the original equation has *no solution*.

✓ *Checkpoint*  Audio-video solution in English & Spanish at LarsonPrecalculus.com

Solve $\dfrac{3x}{x - 4} = 5 + \dfrac{12}{x - 4}$. ◼

Quadratic Equations

A **quadratic equation** in x is an equation that can be written in the general form

$$ax^2 + bx + c = 0$$

where a, b, and c are real numbers with $a \neq 0$. A quadratic equation in x is also called a **second-degree polynomial equation** in x.

You should be familiar with the four methods for solving quadratic equations listed below.

Solving a Quadratic Equation

Factoring

If $ab = 0$, then $a = 0$ or $b = 0$. Zero-Factor Property

Example: $x^2 - x - 6 = 0$

$$(x - 3)(x + 2) = 0$$

$$x - 3 = 0 \implies x = 3$$

$$x + 2 = 0 \implies x = -2$$

Extracting Square Roots

If $u^2 = c$, where $c > 0$, then $u = \pm\sqrt{c}$. Square Root Principle

Example: $(x + 3)^2 = 16$

$$x + 3 = \pm 4$$

$$x = -3 \pm 4$$

$$x = 1 \quad \text{or} \quad x = -7$$

Completing the Square

If $x^2 + bx = c$, then

$$x^2 + bx + \left(\frac{b}{2}\right)^2 = c + \left(\frac{b}{2}\right)^2$$ Add $\left(\frac{b}{2}\right)^2$ to each side.

$$\left(x + \frac{b}{2}\right)^2 = c + \frac{b^2}{4}.$$

Example: $x^2 + 6x = 5$

$$x^2 + 6x + 3^2 = 5 + 3^2$$ Add $\left(\frac{6}{2}\right)^2$ to each side.

$$(x + 3)^2 = 14$$

$$x + 3 = \pm\sqrt{14}$$

$$x = -3 \pm \sqrt{14}$$

Quadratic Formula

If $ax^2 + bx + c = 0$, then $x = \dfrac{-b \pm \sqrt{b^2 - 4ac}}{2a}$.

Example: $2x^2 + 3x - 1 = 0$

$$x = \frac{-3 \pm \sqrt{3^2 - 4(2)(-1)}}{2(2)}$$

$$= \frac{-3 \pm \sqrt{17}}{4}$$

> **• • REMARK** It is possible to solve every quadratic equation by completing the square or using the Quadratic Formula.

EXAMPLE 4   **Solving Quadratic Equations by Factoring**

a. $2x^2 + 9x + 7 = 3$ Original equation

$2x^2 + 9x + 4 = 0$ Write in general form.

$(2x + 1)(x + 4) = 0$ Factor.

$2x + 1 = 0 \implies x = -\frac{1}{2}$ Set 1st factor equal to 0 and solve.

$x + 4 = 0 \implies x = -4$ Set 2nd factor equal to 0 and solve.

The solutions are $x = -\frac{1}{2}$ and $x = -4$. Check these in the original equation.

b. $6x^2 - 3x = 0$ Original equation

$3x(2x - 1) = 0$ Factor.

$3x = 0 \implies x = 0$ Set 1st factor equal to 0 and solve.

$2x - 1 = 0 \implies x = \frac{1}{2}$ Set 2nd factor equal to 0 and solve.

The solutions are $x = 0$ and $x = \frac{1}{2}$. Check these in the original equation.

✓ *Checkpoint* Audio-video solution in English & Spanish at LarsonPrecalculus.com

Solve $2x^2 - 3x + 1 = 6$ by factoring.

Note that the method of solution in Example 4 is based on the Zero-Factor Property from Appendix A.1. This property applies only to equations written in general form (in which the right side of the equation is zero). So, collect all terms on one side *before* factoring. For example, in the equation $(x - 5)(x + 2) = 8$, it is *incorrect* to set each factor equal to 8. Solve this equation correctly on your own. Then check the solutions in the original equation.

EXAMPLE 5   **Extracting Square Roots**

Solve each equation by extracting square roots.

a. $4x^2 = 12$

b. $(x - 3)^2 = 7$

Solution

a. $4x^2 = 12$ Write original equation.

$x^2 = 3$ Divide each side by 4.

$x = \pm\sqrt{3}$ Extract square roots.

The solutions are $x = \sqrt{3}$ and $x = -\sqrt{3}$. Check these in the original equation.

b. $(x - 3)^2 = 7$ Write original equation.

$x - 3 = \pm\sqrt{7}$ Extract square roots.

$x = 3 \pm \sqrt{7}$ Add 3 to each side.

The solutions are $x = 3 \pm \sqrt{7}$. Check these in the original equation.

✓ *Checkpoint* Audio-video solution in English & Spanish at LarsonPrecalculus.com

Solve each equation by extracting square roots.

a. $3x^2 = 36$

b. $(x - 1)^2 = 10$

When solving quadratic equations by completing the square, you must add $(b/2)^2$ to *each side* in order to maintain equality. When the leading coefficient is *not* 1, divide each side of the equation by the leading coefficient *before* completing the square, as shown in Example 7.

EXAMPLE 6   Completing the Square: Leading Coefficient Is 1

Solve $x^2 + 2x - 6 = 0$ by completing the square.

Solution

$x^2 + 2x - 6 = 0$	Write original equation.
$x^2 + 2x = 6$	Add 6 to each side.
$x^2 + 2x + 1^2 = 6 + 1^2$	Add 1^2 to each side.

(Half of 2)^2

$(x + 1)^2 = 7$	Simplify.
$x + 1 = \pm\sqrt{7}$	Extract square roots.
$x = -1 \pm \sqrt{7}$	Subtract 1 from each side.

The solutions are

$$x = -1 \pm \sqrt{7}.$$

Check these in the original equation.

✓ **Checkpoint** ◀))) *Audio-video solution in English & Spanish at LarsonPrecalculus.com*

Solve $x^2 - 4x - 1 = 0$ by completing the square.

EXAMPLE 7   Completing the Square: Leading Coefficient Is Not 1

Solve $3x^2 - 4x - 5 = 0$ by completing the square.

Solution

$3x^2 - 4x - 5 = 0$	Write original equation.
$3x^2 - 4x = 5$	Add 5 to each side.
$x^2 - \dfrac{4}{3}x = \dfrac{5}{3}$	Divide each side by 3.
$x^2 - \dfrac{4}{3}x + \left(-\dfrac{2}{3}\right)^2 = \dfrac{5}{3} + \left(-\dfrac{2}{3}\right)^2$	Add $\left(-\dfrac{2}{3}\right)^2$ to each side.

$\left(\text{Half of } -\dfrac{4}{3}\right)^2$

$\left(x - \dfrac{2}{3}\right)^2 = \dfrac{19}{9}$	Simplify.
$x - \dfrac{2}{3} = \pm\dfrac{\sqrt{19}}{3}$	Extract square roots.
$x = \dfrac{2}{3} \pm \dfrac{\sqrt{19}}{3}$	Add $\dfrac{2}{3}$ to each side.

✓ **Checkpoint** ◀))) *Audio-video solution in English & Spanish at LarsonPrecalculus.com*

Solve $3x^2 - 10x - 2 = 0$ by completing the square. ∎

| EXAMPLE 8 | **The Quadratic Formula: Two Distinct Solutions** |

Use the Quadratic Formula to solve $x^2 + 3x = 9$.

Solution

$$x^2 + 3x = 9 \qquad \text{Write original equation.}$$

$$x^2 + 3x - 9 = 0 \qquad \text{Write in general form.}$$

$$x = \frac{-b \pm \sqrt{b^2 - 4ac}}{2a} \qquad \text{Quadratic Formula}$$

$$x = \frac{-3 \pm \sqrt{(3)^2 - 4(1)(-9)}}{2(1)} \qquad \text{Substitute } a = 1, b = 3, \text{ and } c = -9.$$

$$x = \frac{-3 \pm \sqrt{45}}{2} \qquad \text{Simplify.}$$

$$x = \frac{-3 \pm 3\sqrt{5}}{2} \qquad \text{Simplify.}$$

The two solutions are

$$x = \frac{-3 + 3\sqrt{5}}{2} \quad \text{and} \quad x = \frac{-3 - 3\sqrt{5}}{2}.$$

Check these in the original equation.

✓ **Checkpoint** ◀))) *Audio-video solution in English & Spanish at LarsonPrecalculus.com*

Use the Quadratic Formula to solve $3x^2 + 2x = 10$.

> **REMARK** When you use the Quadratic Formula, remember that *before* applying the formula, you must first write the quadratic equation in general form.

| EXAMPLE 9 | **The Quadratic Formula: One Solution** |

Use the Quadratic Formula to solve $8x^2 - 24x + 18 = 0$.

Solution

$$8x^2 - 24x + 18 = 0 \qquad \text{Write original equation.}$$

$$4x^2 - 12x + 9 = 0 \qquad \text{Divide out common factor of 2.}$$

$$x = \frac{-b \pm \sqrt{b^2 - 4ac}}{2a} \qquad \text{Quadratic Formula}$$

$$x = \frac{-(-12) \pm \sqrt{(-12)^2 - 4(4)(9)}}{2(4)} \qquad \text{Substitute } a = 4, b = -12, \text{ and } c = 9.$$

$$x = \frac{12 \pm \sqrt{0}}{8} \qquad \text{Simplify.}$$

$$x = \frac{3}{2} \qquad \text{Simplify.}$$

This quadratic equation has only one solution: $x = \frac{3}{2}$. Check this in the original equation.

✓ **Checkpoint** ◀))) *Audio-video solution in English & Spanish at LarsonPrecalculus.com*

Use the Quadratic Formula to solve $18x^2 - 48x + 32 = 0$. ∎

Note that you could have solved Example 9 without first dividing out a common factor of 2. Substituting $a = 8$, $b = -24$, and $c = 18$ into the Quadratic Formula produces the same result.

Polynomial Equations of Higher Degree

Sometimes, the methods used to solve quadratic equations can be extended to solve polynomial equations of higher degrees.

• • • • • • • • • • • • • • • ▷

• • REMARK   A common mistake when solving an equation such as that in Example 10 is to divide each side of the equation by the variable factor x^2. This loses the solution $x = 0$. When solving a polynomial equation, always write the equation in general form, then factor the polynomial and set each factor equal to zero. Do not divide each side of an equation by a variable factor in an attempt to simplify the equation.

EXAMPLE 10   Solving a Polynomial Equation by Factoring

Solve $3x^4 = 48x^2$ and check your solution(s).

Solution   First write the polynomial equation in general form. Then factor the polynomial, set each factor equal to zero, and solve.

$$3x^4 = 48x^2 \qquad \text{Write original equation.}$$
$$3x^4 - 48x^2 = 0 \qquad \text{Write in general form.}$$
$$3x^2(x^2 - 16) = 0 \qquad \text{Factor out common factor.}$$
$$3x^2(x + 4)(x - 4) = 0 \qquad \text{Factor completely.}$$
$$3x^2 = 0 \ \Longrightarrow \ x = 0 \qquad \text{Set 1st factor equal to 0 and solve.}$$
$$x + 4 = 0 \ \Longrightarrow \ x = -4 \qquad \text{Set 2nd factor equal to 0 and solve.}$$
$$x - 4 = 0 \ \Longrightarrow \ x = 4 \qquad \text{Set 3rd factor equal to 0 and solve.}$$

Check these solutions by substituting in the original equation.

Check

$$3(0)^4 \overset{?}{=} 48(0)^2 \ \Longrightarrow \ 0 = 0 \qquad \text{0 checks.} \ \checkmark$$
$$3(-4)^4 \overset{?}{=} 48(-4)^2 \ \Longrightarrow \ 768 = 768 \qquad \text{-4 checks.} \ \checkmark$$
$$3(4)^4 \overset{?}{=} 48(4)^2 \ \Longrightarrow \ 768 = 768 \qquad \text{4 checks.} \ \checkmark$$

So, the solutions are

$$x = 0, \quad x = -4, \quad \text{and} \quad x = 4.$$

✓ **Checkpoint** ◀))) *Audio-video solution in English & Spanish at LarsonPrecalculus.com*

Solve $9x^4 - 12x^2 = 0$ and check your solution(s).

EXAMPLE 11   Solving a Polynomial Equation by Factoring

Solve $x^3 - 3x^2 - 3x + 9 = 0$.

Solution

$$x^3 - 3x^2 - 3x + 9 = 0 \qquad \text{Write original equation.}$$
$$x^2(x - 3) - 3(x - 3) = 0 \qquad \text{Group terms and factor.}$$
$$(x - 3)(x^2 - 3) = 0 \qquad \text{$(x - 3)$ is a common factor.}$$
$$x - 3 = 0 \ \Longrightarrow \ x = 3 \qquad \text{Set 1st factor equal to 0 and solve.}$$
$$x^2 - 3 = 0 \ \Longrightarrow \ x = \pm\sqrt{3} \qquad \text{Set 2nd factor equal to 0 and solve.}$$

The solutions are $x = 3$, $x = \sqrt{3}$, and $x = -\sqrt{3}$. Check these in the original equation.

✓ **Checkpoint** ◀))) *Audio-video solution in English & Spanish at LarsonPrecalculus.com*

Solve each equation.

a. $x^3 - 5x^2 - 2x + 10 = 0$

b. $6x^3 - 27x^2 - 54x = 0$

Radical Equations

REMARK When squaring each side of an equation or raising each side of an equation to a rational power, it is possible to introduce extraneous solutions. So when using such operations, checking your solutions is crucial.

A **radical equation** is an equation that involves one or more radical expressions. Examples 12 and 13 demonstrate how to solve radical equations.

EXAMPLE 12 **Solving Radical Equations**

a. $\sqrt{2x + 7} - x = 2$ Original equation

$\sqrt{2x + 7} = x + 2$ Isolate radical.

$2x + 7 = x^2 + 4x + 4$ Square each side.

$0 = x^2 + 2x - 3$ Write in general form.

$0 = (x + 3)(x - 1)$ Factor.

$x + 3 = 0 \implies x = -3$ Set 1st factor equal to 0 and solve.

$x - 1 = 0 \implies x = 1$ Set 2nd factor equal to 0 and solve.

Checking these values shows that the only solution is $x = 1$.

b. $\sqrt{2x - 5} - \sqrt{x - 3} = 1$ Original equation

$\sqrt{2x - 5} = \sqrt{x - 3} + 1$ Isolate $\sqrt{2x - 5}$.

$2x - 5 = x - 3 + 2\sqrt{x - 3} + 1$ Square each side.

$x - 3 = 2\sqrt{x - 3}$ Isolate $2\sqrt{x - 3}$.

$x^2 - 6x + 9 = 4(x - 3)$ Square each side.

$x^2 - 10x + 21 = 0$ Write in general form.

$(x - 3)(x - 7) = 0$ Factor.

$x - 3 = 0 \implies x = 3$ Set 1st factor equal to 0 and solve.

$x - 7 = 0 \implies x = 7$ Set 2nd factor equal to 0 and solve.

REMARK When an equation contains two radical expressions, it may not be possible to isolate both of them in the first step. In such cases, you may have to isolate radical expressions at *two* different stages in the solution, as shown in Example 12(b).

The solutions are $x = 3$ and $x = 7$. Check these in the original equation.

✓ *Checkpoint*  Audio-video solution in English & Spanish at LarsonPrecalculus.com

Solve $-\sqrt{40 - 9x} + 2 = x$.

EXAMPLE 13 **Solving an Equation Involving a Rational Exponent**

Solve $(x - 4)^{2/3} = 25$.

Solution

$(x - 4)^{2/3} = 25$ Write original equation.

$\sqrt[3]{(x - 4)^2} = 25$ Rewrite in radical form.

$(x - 4)^2 = 15{,}625$ Cube each side.

$x - 4 = \pm 125$ Extract square roots.

$x = 129, \ x = -121$ Add 4 to each side.

The solutions are $x = 129$ and $x = -121$. Check these in the original equation.

✓ *Checkpoint* Audio-video solution in English & Spanish at LarsonPrecalculus.com

Solve $(x - 5)^{2/3} = 16$.

Absolute Value Equations

An **absolute value equation** is an equation that involves one or more absolute value expressions. To solve an absolute value equation, remember that the expression inside the absolute value bars can be positive or negative. This results in *two* separate equations, each of which must be solved. For example, the equation

$$|x - 2| = 3$$

results in the two equations

$$x - 2 = 3$$

and

$$-(x - 2) = 3$$

which implies that the original equation has two solutions: $x = 5$ and $x = -1$.

EXAMPLE 14   **Solving an Absolute Value Equation**

Solve $|x^2 - 3x| = -4x + 6$ and check your solution(s).

Solution   Solve the two equations below.

First Equation

$x^2 - 3x = -4x + 6$	Use positive expression.
$x^2 + x - 6 = 0$	Write in general form.
$(x + 3)(x - 2) = 0$	Factor.
$x + 3 = 0 \implies x = -3$	Set 1st factor equal to 0 and solve.
$x - 2 = 0 \implies x = 2$	Set 2nd factor equal to 0 and solve.

Second Equation

$-(x^2 - 3x) = -4x + 6$	Use negative expression.
$x^2 - 7x + 6 = 0$	Write in general form.
$(x - 1)(x - 6) = 0$	Factor.
$x - 1 = 0 \implies x = 1$	Set 1st factor equal to 0 and solve.
$x - 6 = 0 \implies x = 6$	Set 2nd factor equal to 0 and solve.

Check

$|(-3)^2 - 3(-3)| \overset{?}{=} -4(-3) + 6$	Substitute -3 for x.
$18 = 18$	-3 checks. ✓
$|(2)^2 - 3(2)| \overset{?}{=} -4(2) + 6$	Substitute 2 for x.
$2 \neq -2$	2 does not check.
$|(1)^2 - 3(1)| \overset{?}{=} -4(1) + 6$	Substitute 1 for x.
$2 = 2$	1 checks. ✓
$|(6)^2 - 3(6)| \overset{?}{=} -4(6) + 6$	Substitute 6 for x.
$18 \neq -18$	6 does not check.

The solutions are $x = -3$ and $x = 1$.

✓ *Checkpoint* ◀))) Audio-video solution in English & Spanish at LarsonPrecalculus.com

Solve $|x^2 + 4x| = 7x + 18$ and check your solution(s). ∎

Common Formulas

You will use the geometric formulas listed below at various times throughout this course. For your convenience, some of these formulas along with several others are also given on the inside cover of this text.

Common Formulas for Area A, Perimeter P, Circumference C, and Volume V

Rectangle	Circle	Rectangular Solid	Circular Cylinder	Sphere
$A = lw$	$A = \pi r^2$	$V = lwh$	$V = \pi r^2 h$	$V = \frac{4}{3}\pi r^3$
$P = 2l + 2w$	$C = 2\pi r$			

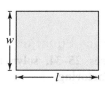 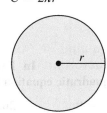 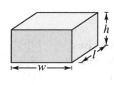 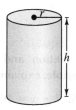 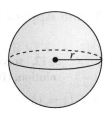

Figure A.9

REMARK To check that the answer in Example 15 is reasonable, substitute $h = 3.98$ into the formula for the volume of a cylinder and simplify.

$$V = \pi r^2 h$$
$$\approx \pi(4)^2(3.98)$$
$$\approx 200$$

EXAMPLE 15 **Using a Geometric Formula**

The cylindrical can shown in Figure A.9 has a volume of 200 cubic centimeters (cm^3). Find the height of the can.

Solution The formula for the *volume of a cylinder* is $V = \pi r^2 h$. To find the height of the can, solve for h. Then, using $V = 200$ and $r = 4$, find the height.

$$V = \pi r^2 h \quad \Longrightarrow \quad h = \frac{V}{\pi r^2} = \frac{200}{\pi(4)^2} = \frac{200}{16\pi} \approx 3.98$$

So, the height of the can is about 3.98 centimeters.

 **Checkpoint** Audio-video solution in English & Spanish at LarsonPrecalculus.com

A cylindrical container has a volume of 84 cubic inches and a radius of 3 inches. Find the height of the container.

Summarize   (Appendix A.5)

1. State the definitions of an identity, a conditional equation, and a contradiction *(page A45)*.

2. State the definition of a linear equation in one variable *(page A45)*. For examples of solving linear equations and rational equations that lead to linear equations, see Examples 1–3.

3. List the four methods for solving quadratic equations discussed in this section *(page A48)*. For examples of solving quadratic equations, see Examples 4–9.

4. Explain how to solve a polynomial equation of degree three or greater by factoring *(page A52)*. For examples of solving polynomial equations by factoring, see Examples 10 and 11.

5. Explain how to solve a radical equation *(page A53)*. For an example of solving radical equations, see Example 12.

6. Explain how to solve an absolute value equation *(page A54)*. For an example of solving an absolute value equation, see Example 14.

7. State the common geometric formulas listed in this section *(page A55)*. For an example that uses a volume formula, see Example 15.

A.5 Exercises

See **CalcChat.com** for tutorial help and worked-out solutions to odd-numbered exercises.

Vocabulary: Fill in the blanks.

1. An _____ is a statement that equates two algebraic expressions.
2. A linear equation in one variable x is an equation that can be written in the standard form _____.
3. An _____ solution is a solution that does not satisfy the original equation.
4. Four methods for solving quadratic equations are _____, extracting _____ _____, _____ the _____, and the _____ _____.

Skills and Applications

 Solving a Linear Equation **In Exercises 5–12, solve the equation and check your solution. (If not possible, explain why.)**

5. $x + 11 = 15$
6. $7 - x = 19$
7. $7 - 2x = 25$
8. $7x + 2 = 23$
9. $3x - 5 = 2x + 7$
10. $4y + 2 - 5y = 7 - 6y$
11. $x - 3(2x + 3) = 8 - 5x$
12. $9x - 10 = 5x + 2(2x - 5)$

 Solving a Rational Equation **In Exercises 13–24, solve the equation and check your solution. (If not possible, explain why.)**

13. $\dfrac{3x}{8} - \dfrac{4x}{3} = 4$
14. $\dfrac{5x}{4} + \dfrac{1}{2} = x - \dfrac{1}{2}$

15. $\dfrac{5x - 4}{5x + 4} = \dfrac{2}{3}$

16. $\dfrac{10x + 3}{5x + 6} = \dfrac{1}{2}$

17. $10 - \dfrac{13}{x} = 4 + \dfrac{5}{x}$

18. $\dfrac{1}{x} + \dfrac{2}{x - 5} = 0$

19. $\dfrac{x}{x + 4} + \dfrac{4}{x + 4} + 2 = 0$

20. $\dfrac{7}{2x + 1} - \dfrac{8x}{2x - 1} = -4$

21. $\dfrac{2}{(x - 4)(x - 2)} = \dfrac{1}{x - 4} + \dfrac{2}{x - 2}$

22. $\dfrac{12}{(x - 1)(x + 3)} = \dfrac{3}{x - 1} + \dfrac{2}{x + 3}$

23. $\dfrac{1}{x - 3} + \dfrac{1}{x + 3} = \dfrac{10}{x^2 - 9}$

24. $\dfrac{1}{x - 2} + \dfrac{3}{x + 3} = \dfrac{4}{x^2 + x - 6}$

 Solving a Quadratic Equation by Factoring **In Exercises 25–34, solve the quadratic equation by factoring.**

25. $6x^2 + 3x = 0$
26. $8x^2 - 2x = 0$
27. $x^2 + 10x + 25 = 0$
28. $x^2 - 2x - 8 = 0$
29. $3 + 5x - 2x^2 = 0$
30. $4x^2 + 12x + 9 = 0$
31. $16x^2 - 9 = 0$
32. $-x^2 + 8x = 12$
33. $\frac{3}{4}x^2 + 8x + 20 = 0$
34. $\frac{1}{8}x^2 - x - 16 = 0$

 Extracting Square Roots **In Exercises 35–42, solve the equation by extracting square roots. When a solution is irrational, list both the exact solution and its approximation rounded to two decimal places.**

35. $x^2 = 49$
36. $x^2 = 43$
37. $3x^2 = 81$
38. $9x^2 = 36$
39. $(x - 4)^2 = 49$
40. $(x + 9)^2 = 24$
41. $(2x - 1)^2 = 18$
42. $(x - 7)^2 = (x + 3)^2$

 Completing the Square **In Exercises 43–50, solve the quadratic equation by completing the square.**

43. $x^2 + 4x - 32 = 0$
44. $x^2 - 2x - 3 = 0$
45. $x^2 + 4x + 2 = 0$
46. $x^2 + 8x + 14 = 0$
47. $6x^2 - 12x = -3$
48. $4x^2 - 4x = 1$
49. $2x^2 + 5x - 8 = 0$
50. $3x^2 - 4x - 7 = 0$

 Using the Quadratic Formula **In Exercises 51–64, use the Quadratic Formula to solve the equation.**

51. $2x^2 + x - 1 = 0$
52. $2x^2 - x - 1 = 0$
53. $9x^2 + 30x + 25 = 0$
54. $28x - 49x^2 = 4$
55. $2x^2 - 7x + 1 = 0$
56. $3x + x^2 - 1 = 0$
57. $12x - 9x^2 = -3$
58. $9x^2 - 37 = 6x$
59. $2 + 2x - x^2 = 0$
60. $x^2 + 10 + 8x = 0$
61. $8t = 5 + 2t^2$
62. $25h^2 + 80h = -61$
63. $(y - 5)^2 = 2y$
64. $(z + 6)^2 = -2z$

Choosing a Method In Exercises 65–72, solve the equation using any convenient method.

65. $x^2 - 2x - 1 = 0$

66. $14x^2 + 42x = 0$

67. $(x + 2)^2 = 64$

68. $x^2 - 14x + 49 = 0$

69. $x^2 - x - \frac{11}{4} = 0$

70. $x^2 + 3x - \frac{3}{4} = 0$

71. $3x + 4 = 2x^2 - 7$

72. $(x + 1)^2 = x^2$

 Solving a Polynomial Equation In Exercises 73–76, solve the equation. Check your solutions.

73. $6x^4 - 54x^2 = 0$

74. $5x^3 + 30x^2 + 45x = 0$

75. $x^3 + 2x^2 - 8x = 16$

76. $x^3 - 3x^2 - x = -3$

 Solving a Radical Equation In Exercises 77–84, solve the equation. Check your solutions.

77. $\sqrt{5x} - 10 = 0$

78. $\sqrt{x + 8} - 5 = 0$

79. $4 + \sqrt[3]{2x - 9} = 0$

80. $\sqrt[3]{12 - x} - 3 = 0$

81. $\sqrt{x + 8} = 2 + x$

82. $2x = \sqrt{-5x + 24} - 3$

83. $\sqrt{x - 3} + 1 = \sqrt{x}$

84. $2\sqrt{x + 1} - \sqrt{2x + 3} = 1$

 Solving an Equation Involving a Rational Exponent In Exercises 85–88, solve the equation. Check your solutions.

85. $(x - 5)^{3/2} = 8$

86. $(x^2 - x - 22)^{3/2} = 27$

87. $3x(x - 1)^{1/2} + 2(x - 1)^{3/2} = 0$

88. $4x^2(x - 1)^{1/3} + 6x(x - 1)^{4/3} = 0$

 Solving an Absolute Value Function In Exercises 89–92, solve the equation. Check your solutions.

89. $|2x - 5| = 11$

90. $|3x + 2| = 7$

91. $|x + 1| = x^2 - 5$

92. $|x^2 + 6x| = 3x + 18$

93. Volume of a Billiard Ball A billiard ball has a volume of 5.96 cubic inches. Find the radius of the billiard ball.

94. Length of a Tank The diameter of a cylindrical propane gas tank is 4 feet. The total volume of the tank is 603.2 cubic feet. Find the length of the tank.

• • **Forensics** • • • • • • • • • • • • • • • •

In Exercises 95 and 96, use the following information. The relationship between the length of an adult's femur (thigh bone) and the height of the adult can be approximated by the linear equations

$y = 0.514x - 14.75$ Female

$y = 0.532x - 17.03$ Male

where y is the length of the femur in inches and x is the height of the adult in inches (see figure).

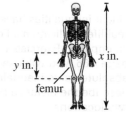

95. A crime scene investigator discovers a femur belonging to an adult human female. The bone is 18 inches long. Estimate the height of the female.

96. Officials search a forest for a missing man who is 6 feet 2 inches tall. They find an adult male femur that is 23 inches long. Is it possible that the femur belongs to the missing man?

Exploration

True or False? In Exercises 97–99, determine whether the statement is true or false. Justify your answer.

97. An equation can never have more than one extraneous solution.

98. The equation $2(x - 3) + 1 = 2x - 5$ has no solution.

99. The equation

$$\sqrt{x + 10} - \sqrt{x - 10} = 0$$

has no solution.

100. 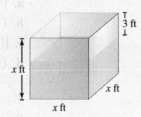 **HOW DO YOU SEE IT?** The figure shows a glass cube partially filled with water.

(a) What does the expression $x^2(x - 3)$ represent?

(b) Given $x^2(x - 3) = 320$, explain how to find the capacity of the cube.

A.6 Linear Inequalities in One Variable

Linear inequalities have many real-life applications. For example, in Exercise 104 on page A66, you will use an absolute value inequality to describe the distance between two locations.

- ■ Represent solutions of linear inequalities in one variable.
- ■ Use properties of inequalities to write equivalent inequalities.
- ■ Solve linear inequalities in one variable.
- ■ Solve absolute value inequalities.
- ■ Use linear inequalities to model and solve real-life problems.

Introduction

Simple inequalities were discussed in Appendix A.1. There, the inequality symbols $<$, $\le$, $>$, and $\ge$ were used to compare two numbers and to denote subsets of real numbers. For example, the simple inequality

$$x \ge 3$$

denotes all real numbers x that are greater than or equal to 3.

Now, you will expand your work with inequalities to include more involved statements such as

$$5x - 7 < 3x + 9 \quad \text{and} \quad -3 \le 6x - 1 < 3.$$

As with an equation, you **solve an inequality** in the variable x by finding all values of x for which the inequality is true. Such values are **solutions** that **satisfy** the inequality. The set of all real numbers that are solutions of an inequality is the **solution set** of the inequality. For example, the solution set of

$$x + 1 < 4$$

is all real numbers that are less than 3.

The set of all points on the real number line that represents the solution set is the **graph of the inequality.** Graphs of many types of inequalities consist of intervals on the real number line. See Appendix A.1 to review the nine basic types of intervals on the real number line. Note that each type of interval can be classified as *bounded* or *unbounded*.

EXAMPLE 1 Intervals and Inequalities

Write an inequality that represents each interval. Then state whether the interval is bounded or unbounded.

a. $(-3, 5]$ **b.** $(-3, \infty)$

c. $[0, 2]$ **d.** $(-\infty, \infty)$

Solution

a. $(-3, 5]$ corresponds to $-3 < x \le 5$.   Bounded

b. $(-3, \infty)$ corresponds to $x > -3$.   Unbounded

c. $[0, 2]$ corresponds to $0 \le x \le 2$.   Bounded

d. $(-\infty, \infty)$ corresponds to $-\infty < x < \infty$.   Unbounded

✓ **Checkpoint** ◀))) *Audio-video solution in English & Spanish at LarsonPrecalculus.com*

Write an inequality that represents each interval. Then state whether the interval is bounded or unbounded.

a. $[-1, 3]$ **b.** $(-1, 6)$ **c.** $(-\infty, 4)$ **d.** $[0, \infty)$ ■

Properties of Inequalities

The procedures for solving linear inequalities in one variable are similar to those for solving linear equations. To isolate the variable, use the **properties of inequalities.** These properties are similar to the properties of equality, but there are two important exceptions. When you multiply or divide each side of an inequality by a negative number, you must reverse the direction of the inequality symbol. Here is an example.

$$-2 < 5 \qquad \text{Original inequality}$$
$$(-3)(-2) > (-3)(5) \qquad \text{Multiply each side by } -3 \text{ and reverse inequality symbol.}$$
$$6 > -15 \qquad \text{Simplify.}$$

Notice that when you do not reverse the inequality symbol in the example above, you obtain the false statement

$$6 < -15. \qquad \text{False statement}$$

Two inequalities that have the same solution set are **equivalent.** For example, the inequalities

$$x + 2 < 5$$

and

$$x < 3$$

are equivalent. To obtain the second inequality from the first, subtract 2 from each side of the inequality. The list below describes operations used to write equivalent inequalities.

Properties of Inequalities

Let a, b, c, and d be real numbers.

1. Transitive Property

$$a < b \text{ and } b < c \implies a < c$$

2. Addition of Inequalities

$$a < b \text{ and } c < d \implies a + c < b + d$$

3. Addition of a Constant

$$a < b \implies a + c < b + c$$

4. Multiplication by a Constant

$$\text{For } c > 0, a < b \implies ac < bc$$
$$\text{For } c < 0, a < b \implies ac > bc \qquad \text{Reverse the inequality symbol.}$$

Each of the properties above is true when you replace the symbol $<$ with $\leq$ and you replace the symbol $>$ with $\geq$. For example, another form of the multiplication property is shown below.

$$\text{For } c > 0, a \leq b \implies ac \leq bc$$
$$\text{For } c < 0, a \leq b \implies ac \geq bc$$

On your own, verify each of the properties of inequalities by using several examples with real numbers.

Solving Linear Inequalities in One Variable

The simplest type of inequality to solve is a **linear inequality** in one variable. For example, $2x + 3 > 4$ is a linear inequality in x.

EXAMPLE 2   Solving a Linear Inequality

Solve $5x - 7 > 3x + 9$. Then graph the solution set.

Solution

$$5x - 7 > 3x + 9 \qquad \text{Write original inequality.}$$
$$2x - 7 > 9 \qquad \text{Subtract } 3x \text{ from each side.}$$
$$2x > 16 \qquad \text{Add 7 to each side.}$$
$$x > 8 \qquad \text{Divide each side by 2.}$$

The solution set is all real numbers that are greater than 8, denoted by $(8, \infty)$. The graph of this solution set is shown below. Note that a parenthesis at 8 on the real number line indicates that 8 *is not* part of the solution set.

Solution interval: $(8, \infty)$

✓ *Checkpoint* 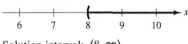 *Audio-video solution in English & Spanish at LarsonPrecalculus.com*

Solve $7x - 3 \le 2x + 7$. Then graph the solution set.

REMARK Checking the solution set of an inequality is not as simple as checking the solution(s) of an equation. However, to get an indication of the validity of a solution set, substitute a few convenient values of x. For instance, in Example 2, substitute $x = 5$ and $x = 10$ into the original inequality.

EXAMPLE 3   Solving a Linear Inequality

See LarsonPrecalculus.com for an interactive version of this type of example.

Solve $1 - \frac{3}{2}x \ge x - 4$.

Algebraic Solution

$$1 - \tfrac{3}{2}x \ge x - 4 \qquad \text{Write original inequality.}$$
$$2 - 3x \ge 2x - 8 \qquad \text{Multiply each side by 2.}$$
$$2 - 5x \ge -8 \qquad \text{Subtract } 2x \text{ from each side.}$$
$$-5x \ge -10 \qquad \text{Subtract 2 from each side.}$$
$$x \le 2 \qquad \begin{array}{l}\text{Divide each side by } -5 \text{ and}\\ \text{reverse the inequality symbol.}\end{array}$$

The solution set is all real numbers that are less than or equal to 2, denoted by $(-\infty, 2]$. The graph of this solution set is shown below. Note that a bracket at 2 on the real number line indicates that 2 *is* part of the solution set.

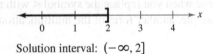

Solution interval: $(-\infty, 2]$

Graphical Solution

Use a graphing utility to graph $y_1 = 1 - \frac{3}{2}x$ and $y_2 = x - 4$ in the same viewing window.

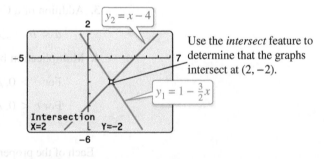

Use the *intersect* feature to determine that the graphs intersect at $(2, -2)$.

The graph of y_1 lies above the graph of y_2 to the left of their point of intersection, which implies that $y_1 \ge y_2$ for all $x \le 2$.

✓ *Checkpoint* *Audio-video solution in English & Spanish at LarsonPrecalculus.com*

Solve $2 - \frac{5}{3}x > x - 6$ (a) algebraically and (b) graphically.

Sometimes it is possible to write two inequalities as a **double inequality.** For example, you can write the two inequalities

$$-4 \le 5x - 2$$

and

$$5x - 2 < 7$$

as

$$-4 \le 5x - 2 < 7. \qquad \text{Double inequality}$$

This form allows you to solve the two inequalities together, as demonstrated in Example 4.

EXAMPLE 4   Solving a Double Inequality

Solve $-3 \le 6x - 1 < 3$. Then graph the solution set.

Solution   One way to solve this double inequality is to isolate x as the middle term.

$$-3 \le 6x - 1 < 3 \qquad \text{Write original inequality.}$$

$$-3 + 1 \le 6x - 1 + 1 < 3 + 1 \qquad \text{Add 1 to each part.}$$

$$-2 \le 6x < 4 \qquad \text{Simplify.}$$

$$\frac{-2}{6} \le \frac{6x}{6} < \frac{4}{6} \qquad \text{Divide each part by 6.}$$

$$-\frac{1}{3} \le x < \frac{2}{3} \qquad \text{Simplify.}$$

The solution set is all real numbers that are greater than or equal to $-\frac{1}{3}$ and less than $\frac{2}{3}$, denoted by $\left[-\frac{1}{3}, \frac{2}{3}\right)$. The graph of this solution set is shown below.

Solution interval: $\left[-\frac{1}{3}, \frac{2}{3}\right)$

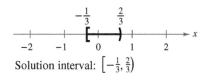 **Checkpoint** ◀))) *Audio-video solution in English & Spanish at LarsonPrecalculus.com*

Solve $1 < 2x + 7 < 11$. Then graph the solution set.

Another way to solve the double inequality in Example 4 is to solve it in two parts.

$$-3 \le 6x - 1 \quad \text{and} \quad 6x - 1 < 3$$

$$-2 \le 6x \qquad\qquad\quad 6x < 4$$

$$-\frac{1}{3} \le x \qquad\qquad\quad x < \frac{2}{3}$$

The solution set consists of all real numbers that satisfy *both* inequalities. In other words, the solution set is the set of all values of x for which

$$-\frac{1}{3} \le x < \frac{2}{3}.$$

When combining two inequalities to form a double inequality, be sure that the inequalities satisfy the Transitive Property. For example, it is *incorrect* to combine the inequalities $3 < x$ and $x \le -1$ as $3 < x \le -1$. This "inequality" is wrong because 3 is not less than -1.

Absolute Value Inequalities

▷ **TECHNOLOGY** A graphing utility can help you identify the solution set of an inequality. For instance, to find the solution set of $|x - 5| < 2$ (see Example 5a), rewrite the inequality as $|x - 5| - 2 < 0$, enter

$$Y1 = abs(X - 5) - 2$$

and press the *graph* key. The graph should resemble the one shown below.

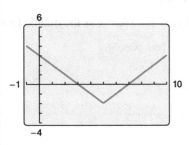

Notice that the graph is below the *x*-axis on the interval $(3, 7)$.

Solving an Absolute Value Inequality

Let *u* be an algebraic expression and let *a* be a real number such that $a > 0$.

1. $|u| < a$ if and only if $-a < u < a$.
2. $|u| \le a$ if and only if $-a \le u \le a$.
3. $|u| > a$ if and only if $u < -a$ or $u > a$.
4. $|u| \ge a$ if and only if $u \le -a$ or $u \ge a$.

EXAMPLE 5 **Solving Absolute Value Inequalities**

Solve each inequality. Then graph the solution set.

a. $|x - 5| < 2$ **b.** $|x + 3| \ge 7$

Solution

a. $|x - 5| < 2$ Write original inequality.

     $-2 < x - 5 < 2$ Write related double inequality.

  $-2 + 5 < x - 5 + 5 < 2 + 5$ Add 5 to each part.

        $3 < x < 7$ Simplify.

The solution set is all real numbers that are greater than 3 and less than 7, denoted by $(3, 7)$. The graph of this solution set is shown below. Note that the graph of the inequality can be described as all real numbers less than two units from 5.

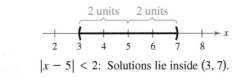

$|x - 5| < 2$: Solutions lie inside $(3, 7)$.

b. $|x + 3| \ge 7$ Write original inequality.

     $x + 3 \le -7$ or $x + 3 \ge 7$ Write related inequalities.

$x + 3 - 3 \le -7 - 3$ $x + 3 - 3 \ge 7 - 3$ Subtract 3 from each side.

      $x \le -10$ $x \ge 4$ Simplify.

The solution set is all real numbers that are less than or equal to -10 *or* greater than or equal to 4, denoted by $(-\infty, -10] \cup [4, \infty)$. The symbol $\cup$ is the *union* symbol, which denotes the combining of two sets. The graph of this solution set is shown below. Note that the graph of the inequality can be described as all real numbers at least seven units from -3.

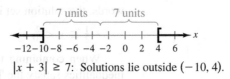

$|x + 3| \ge 7$: Solutions lie outside $(-10, 4)$.

✓ **Checkpoint** ◀))) Audio-video solution in English & Spanish at LarsonPrecalculus.com

Solve $|x - 20| \le 4$. Then graph the solution set.

Applications

EXAMPLE 6   **Comparative Shopping**

A car sharing company offers two plans, as shown in Figure A.10. How many hours must you use a car in one month for plan B to cost more than plan A?

Solution   Let h represent the number of hours you use the car. Write and solve an inequality.

$$10.25h + 8 > 8.75h + 50$$
$$1.5h > 42$$
$$h > 28$$

Plan B costs more when you use the car for more than 28 hours in one month.

 **Checkpoint**  *Audio-video solution in English & Spanish at LarsonPrecalculus.com*

Rework Example 6 when plan A costs $25 per month plus $10 per hour.

Car Sharing Company

Plan A:
$50.00 per month
plus $8.75 per hour

Plan B:
$8.00 per month
plus $10.25 per hour

Figure A.10

EXAMPLE 7   **Accuracy of a Measurement**

You buy a bag of chocolates that cost $9.89 per pound. The scale used to weigh the bag is accurate to within $\frac{1}{32}$ pound. According to the scale, the bag weighs $\frac{1}{2}$ pound and costs $4.95. How much might you have been undercharged or overcharged?

Solution   Let x represent the actual weight of the bag. The difference of the actual weight and the weight shown on the scale is at most $\frac{1}{32}$ pound. That is, $\left|x - \frac{1}{2}\right| \le \frac{1}{32}$. Solve this inequality.

$$-\frac{1}{32} \le x - \frac{1}{2} \le \frac{1}{32}$$
$$\frac{15}{32} \le x \le \frac{17}{32}$$

The least the bag can weigh is $\frac{15}{32}$ pound, which would have cost $4.64. The most the bag can weigh is $\frac{17}{32}$ pound, which would have cost $5.25. So, you might have been overcharged by as much as $0.31 or undercharged by as much as $0.30.

 Checkpoint *Audio-video solution in English & Spanish at LarsonPrecalculus.com*

Rework Example 7 when the scale is accurate to within $\frac{1}{64}$ pound.

Summarize   (Appendix A.6)

1. Explain how to use inequalities to represent intervals *(page A58)*. For an example of writing inequalities that represent intervals, see Example 1.

2. State the properties of inequalities *(page A59)*.

3. Explain how to solve a linear inequality in one variable *(page A60)*. For examples of solving linear inequalities in one variable, see Examples 2–4.

4. Explain how to solve an absolute value inequality *(page A62)*. For an example of solving absolute value inequalities, see Example 5.

5. Describe real-life applications of linear inequalities in one variable *(page A63, Examples 6 and 7)*.

A.6 Exercises

See **CalcChat.com** for tutorial help and worked-out solutions to odd-numbered exercises.

Vocabulary: Fill in the blanks.

1. The set of all real numbers that are solutions of an inequality is the _____ _____ of the inequality.
2. The set of all points on the real number line that represents the solution set of an inequality is the _____ of the inequality.
3. It is sometimes possible to write two inequalities as a _____ inequality.
4. The symbol $\cup$ is the _____ symbol, which denotes the combining of two sets.

Skills and Applications

 Intervals and Inequalities In Exercises 5–12, write an inequality that represents the interval. Then state whether the interval is bounded or unbounded.

5. $[-2, 6)$ 6. $(-7, 4)$
7. $[-1, 5]$ 8. $(2, 10]$
9. $(11, \infty)$ 10. $[-5, \infty)$
11. $(-\infty, -2)$ 12. $(-\infty, 7]$

 Solving a Linear Inequality In Exercises 13–30, solve the inequality. Then graph the solution set.

13. $4x < 12$ 14. $10x < -40$
15. $-2x > -3$ 16. $-6x > 15$
17. $x - 5 \geq 7$ 18. $x + 7 \leq 12$
19. $2x + 7 < 3 + 4x$ 20. $3x + 1 \geq 2 + x$
21. $3x - 4 \geq 4 - 5x$ 22. $6x - 4 \leq 2 + 8x$
23. $4 - 2x < 3(3 - x)$ 24. $4(x + 1) < 2x + 3$
25. $\frac{3}{4}x - 6 \leq x - 7$ 26. $3 + \frac{2}{7}x > x - 2$
27. $\frac{1}{2}(8x + 1) \geq 3x + \frac{5}{2}$ 28. $9x - 1 < \frac{3}{4}(16x - 2)$
29. $3.6x + 11 \geq -3.4$ 30. $15.6 - 1.3x < -5.2$

 Solving a Double Inequality In Exercises 31–42, solve the inequality. Then graph the solution set.

31. $1 < 2x + 3 < 9$ 32. $-9 \leq -2x - 7 < 5$
33. $0 < 3(x + 7) \leq 20$ 34. $-1 \leq -(x - 4) < 7$
35. $-4 < \dfrac{2x - 3}{3} < 4$ 36. $0 \leq \dfrac{x + 3}{2} < 5$
37. $-1 < \dfrac{-x - 2}{3} \leq 1$ 38. $-1 \leq \dfrac{-3x + 5}{7} \leq 2$
39. $\dfrac{3}{4} > x + 1 > \dfrac{1}{4}$
40. $-1 < 2 - \dfrac{x}{3} < 1$
41. $3.2 \leq 0.4x - 1 \leq 4.4$
42. $1.6 < 0.3x + 1 < 2.8$

Solving an Absolute Value Inequality In Exercises 43–58, solve the inequality. Then graph the solution set. (Some inequalities have no solution.)

43. $|x| < 5$ 44. $|x| \geq 8$
45. $\left|\dfrac{x}{2}\right| > 1$ 46. $\left|\dfrac{x}{3}\right| < 2$
47. $|x - 5| < -1$ 48. $|x - 7| < -5$
49. $|x - 20| \leq 6$ 50. $|x - 8| \geq 0$
51. $|7 - 2x| \geq 9$ 52. $|1 - 2x| < 5$
53. $\left|\dfrac{x - 3}{2}\right| \geq 4$ 54. $\left|1 - \dfrac{2x}{3}\right| < 1$
55. $|9 - 2x| - 2 < -1$ 56. $|x + 14| + 3 > 17$
57. $2|x + 10| \geq 9$ 58. $3|4 - 5x| \leq 9$

Using Technology In Exercises 59–68, use a graphing utility to graph the inequality and identify the solution set.

59. $7x > 21$ 60. $-4x \leq 9$
61. $8 - 3x \geq 2$ 62. $20 < 6x - 1$
63. $4(x - 3) \leq 8 - x$ 64. $3(x + 1) < x + 7$
65. $|x - 8| \leq 14$ 66. $|2x + 9| > 13$
67. $2|x + 7| \geq 13$ 68. $\frac{1}{2}|x + 1| \leq 3$

Using Technology In Exercises 69–74, use a graphing utility to graph the equation. Use the graph to approximate the values of x that satisfy each inequality.

Equation	Inequalities	
69. $y = 3x - 1$	(a) $y \geq 2$	(b) $y \leq 0$
70. $y = \frac{2}{3}x + 1$	(a) $y \leq 5$	(b) $y \geq 0$
71. $y = -\frac{1}{2}x + 2$	(a) $0 \leq y \leq 3$	(b) $y \geq 0$
72. $y = -3x + 8$	(a) $-1 \leq y \leq 3$	(b) $y \leq 0$
73. $y = |x - 3|$	(a) $y \leq 2$	(b) $y \geq 4$
74. $y = \left|\frac{1}{2}x + 1\right|$	(a) $y \leq 4$	(b) $y \geq 1$

75. **Think About It** The graph of $|x - 5| < 3$ can be described as all real numbers less than three units from 5. Give a similar description of $|x - 10| < 8$.

76. Think About It The graph of $|x - 2| > 5$ can be described as all real numbers more than five units from 2. Give a similar description of $|x - 8| > 4$.

Using Absolute Value In Exercises 77–84, use absolute value notation to define the interval (or pair of intervals) on the real number line.

77.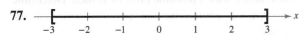

78.

79.

80.

81. All real numbers at least three units from 7

82. All real numbers more than five units from 8

83. All real numbers less than four units from -3

84. All real numbers no more than seven units from -6

Writing an Inequality In Exercises 85–88, write an inequality to describe the situation.

85. During a trading day, the price P of a stock is no less than $7.25 and no more than $7.75.

86. During a month, a person's weight w is greater than 180 pounds but less than 185.5 pounds.

87. The expected return r on an investment is no more than 8%.

88. The expected net income I of a company is no less than $239 million.

Physiology One formula that relates a person's maximum heart rate r (in beats per minute) to the person's age A (in years) is

$r = 220 - A.$

In Exercises 89 and 90, determine the interval in which the person's heart rate is from 50% to 85% of the maximum heart rate. *(Source: American Heart Association)*

89. a 20-year-old

90. a 40-year-old

91. Job Offers You are considering two job offers. The first job pays $13.50 per hour. The second job pays $9.00 per hour plus $0.75 per unit produced per hour. How many units must you produce per hour for the second job to pay more per hour than the first job?

92. Job Offers You are considering two job offers. The first job pays $13.75 per hour. The second job pays $10.00 per hour plus $1.25 per unit produced per hour. How many units must you produce per hour for the second job to pay more than the first job?

93. Investment What annual interest rates yield a balance of more than $2000 on a 10-year investment of $1000? $[A = P(1 + rt)]$

94. Investment What annual interest rates yield a balance of more than $750 on a 5-year investment of $500? $[A = P(1 + rt)]$

95. Cost, Revenue, and Profit The revenue from selling x units of a product is $R = 115.95x$. The cost of producing x units is $C = 95x + 750$. To obtain a profit, the revenue must be greater than the cost. For what values of x does this product return a profit?

96. Cost, Revenue, and Profit The revenue from selling x units of a product is $R = 24.55x$. The cost of producing x units is $C = 15.4x + 150,000$. To obtain a profit, the revenue must be greater than the cost. For what values of x does this product return a profit?

97. Daily Sales A doughnut shop sells a dozen doughnuts for $7.95. Beyond the fixed costs (rent, utilities, and insurance) of $165 per day, it costs $1.45 for enough materials and labor to produce a dozen doughnuts. The daily profit from doughnut sales varies between $400 and $1200. Between what levels (in dozens of doughnuts) do the daily sales vary?

98. Weight Loss Program A person enrolls in a diet and exercise program that guarantees a loss of at least $1\frac{1}{2}$ pounds per week. The person's weight at the beginning of the program is 164 pounds. Find the maximum number of weeks before the person attains a goal weight of 128 pounds.

99. GPA An equation that relates the college grade-point averages y and high school grade-point averages x of the students at a college is $y = 0.692x + 0.988$.

(a) Use a graphing utility to graph the model.

(b) Use the graph to estimate the values of x that predict a college grade-point average of at least 3.0.

(c) Verify your estimate from part (b) algebraically.

(d) List other factors that may influence college GPA.

100. Weightlifting The 6RM load for a weightlifting exercise is the maximum weight at which a person can perform six repetitions. An equation that relates an athlete's 6RM bench press load x (in kilograms) and the athlete's 6RM barbell curl load y (in kilograms) is $y = 0.33x + 6.20$. *(Source: Journal of Sports Science & Medicine)*

(a) Use a graphing utility to graph the model.

(b) Use the graph to estimate the values of x that predict a 6RM barbell curl load of no more than 80 kilograms.

(c) Verify your estimate from part (b) algebraically.

(d) List other factors that may influence an athlete's 6RM barbell curl load.

101. Chemists' Wages The mean hourly wage W (in dollars) of chemists in the United States from 2000 through 2014 can be modeled by

$$W = 0.903t + 26.08, \quad 0 \le t \le 14$$

where t represents the year, with $t = 0$ corresponding to 2000. *(Source: U.S. Bureau of Labor Statistics)*

(a) According to the model, when was the mean hourly wage at least \$30, but no more than \$32?

(b) Use the model to predict when the mean hourly wage will exceed \$45.

102. Milk Production Milk production M (in billions of pounds) in the United States from 2000 through 2014 can be modeled by

$$M = 3.00t + 163.3, \quad 0 \le t \le 14$$

where t represents the year, with $t = 0$ corresponding to 2000. *(Source: U.S. Department of Agriculture)*

(a) According to the model, when was the annual milk production greater than 180 billion pounds, but no more than 190 billion pounds?

(b) Use the model to predict when milk production will exceed 230 billion pounds.

103. Time Study The times required to perform a task in a manufacturing process by approximately two-thirds of the workers in a study satisfy the inequality

$$|t - 15.6| \le 1.9$$

where t is time in minutes. Determine the interval in which these times lie.

104. Geography

A geographic information system reports that the distance between two locations is 206 meters. The system is accurate to within 3 meters.

(a) Write an absolute value inequality for the possible distances between the locations.

(b) Graph the solution set.

105. Accuracy of Measurement You buy 6 T-bone steaks that cost \$8.99 per pound. The weight that is listed on the package is 5.72 pounds. The scale that weighed the package is accurate to within $\frac{1}{32}$ pound. How much might you be undercharged or overcharged?

106. Accuracy of Measurement You stop at a self-service gas station to buy 15 gallons of 87-octane gasoline at \$2.22 per gallon. The gas pump is accurate to within $\frac{1}{10}$ gallon. How much might you be undercharged or overcharged?

107. Geometry The side length of a square is 10.4 inches with a possible error of $\frac{1}{16}$ inch. Determine the interval containing the possible areas of the square.

108. Geometry The side length of a square is 24.2 centimeters with a possible error of 0.25 centimeter. Determine the interval containing the possible areas of the square.

Exploration

True or False? **In Exercises 109–112, determine whether the statement is true or false. Justify your answer.**

109. If a, b, and c are real numbers, and $a < b$, then $a + c < b + c$.

110. If $a, b,$ and c are real numbers, and $a \le b$, then $ac \le bc$.

111. If $-10 \le x \le 8$, then $-10 \ge -x$ and $-x \ge -8$.

112. If $-2 < x < -1$, then $1 < -x < 2$.

113. Think About It Give an example of an inequality whose solution set is $(-\infty, \infty)$.

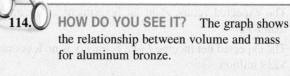

114. **HOW DO YOU SEE IT?** The graph shows the relationship between volume and mass for aluminum bronze.

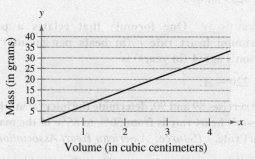

(a) Estimate the mass when the volume is 2 cubic centimeters.

(b) Approximate the interval for the mass when the volume is greater than or equal to 0 cubic centimeters and less than 4 cubic centimeters.

115. Think About It Find sets of values of a, b, and c such that $0 \le x \le 10$ is a solution of the inequality $|ax - b| \le c$.

A.7 Errors and the Algebra of Calculus

■ Avoid common algebraic errors.
■ Recognize and use algebraic techniques that are common in calculus.

Algebraic Errors to Avoid

This section contains five lists of common algebraic errors: errors involving parentheses, errors involving fractions, errors involving exponents, errors involving radicals, and errors involving dividing out common factors. Many of these errors occur because they seem to be the *easiest* things to do. For example, students often believe that the operations of subtraction and division are commutative and associative. The examples below illustrate the fact that subtraction and division are neither commutative nor associative.

Not commutative	Not associative
$4 - 3 \neq 3 - 4$	$8 - (6 - 2) \neq (8 - 6) - 2$
$15 \div 5 \neq 5 \div 15$	$20 \div (4 \div 2) \neq (20 \div 4) \div 2$

Errors Involving Parentheses

Potential Error	Correct Form	Comment
$a - (x - b) = a - x - b$ ✗	$a - (x - b) = a - x + b$	Distribute negative sign to each term in parentheses.
$(a + b)^2 = a^2 + b^2$ ✗	$(a + b)^2 = a^2 + 2ab + b^2$	Remember the middle term when squaring binomials.
$\left(\frac{1}{2}a\right)\left(\frac{1}{2}b\right) = \frac{1}{2}(ab)$ ✗	$\left(\frac{1}{2}a\right)\left(\frac{1}{2}b\right) = \frac{1}{4}(ab) = \frac{ab}{4}$	$\frac{1}{2}$ occurs twice as a factor.
$(3x + 6)^2 = 3(x + 2)^2$ ✗	$(3x + 6)^2 = [3(x + 2)]^2$ $= 3^2(x + 2)^2$	When factoring, raise all factors to the power.

Errors Involving Fractions

Potential Error	Correct Form	Comment
$\frac{2}{x + 4} = \frac{2}{x} + \frac{2}{4}$ ✗	Leave as $\frac{2}{x + 4}$.	The fraction is already in simplest form.
$\frac{\left(\frac{x}{a}\right)}{b} = \frac{bx}{a}$ ✗	$\frac{\left(\frac{x}{a}\right)}{b} = \left(\frac{x}{a}\right)\left(\frac{1}{b}\right) = \frac{x}{ab}$	Multiply by the reciprocal when dividing fractions.
$\frac{1}{a} + \frac{1}{b} = \frac{1}{a + b}$ ✗	$\frac{1}{a} + \frac{1}{b} = \frac{b + a}{ab}$	Use the property for adding fractions with unlike denominators.
$\frac{1}{3x} = \frac{1}{3}x$ ✗	$\frac{1}{3x} = \frac{1}{3} \cdot \frac{1}{x}$	Use the property for multiplying fractions.
$(1/3)x = \frac{1}{3x}$ ✗	$(1/3)x = \frac{1}{3} \cdot x = \frac{x}{3}$	Be careful when expressing fractions in the form $1/a$.
$(1/x) + 2 = \frac{1}{x + 2}$ ✗	$(1/x) + 2 = \frac{1}{x} + 2 = \frac{1 + 2x}{x}$	Be careful when expressing fractions in the form $1/a$. Be sure to find a common denominator before adding fractions.

Errors Involving Exponents

Potential Error	Correct Form	Comment
$(x^2)^3 = x^5$ ✗	$(x^2)^3 = x^{2 \cdot 3} = x^6$	Multiply exponents when raising a power to a power.
$x^2 \cdot x^3 = x^6$ ✗	$x^2 \cdot x^3 = x^{2+3} = x^5$	Add exponents when multiplying powers with like bases.
$(2x)^3 = 2x^3$ ✗	$(2x)^3 = 2^3 x^3 = 8x^3$	Raise each factor to the power.
$\dfrac{1}{x^2 - x^3} = x^{-2} - x^{-3}$ ✗	Leave as $\dfrac{1}{x^2 - x^3}$.	Do not move term-by-term from denominator to numerator.

Errors Involving Radicals

Potential Error	Correct Form	Comment
$\sqrt{5x} = 5\sqrt{x}$ ✗	$\sqrt{5x} = \sqrt{5}\sqrt{x}$	Radicals apply to every factor inside the radical.
$\sqrt{x^2 + a^2} = x + a$ ✗	Leave as $\sqrt{x^2 + a^2}$.	Do not apply radicals term-by-term when adding or subtracting terms.
$\sqrt{-x + a} = -\sqrt{x - a}$ ✗	Leave as $\sqrt{-x + a}$.	Do not factor negative signs out of square roots.

Errors Involving Dividing Out

Potential Error	Correct Form	Comment
$\dfrac{a + bx}{a} = 1 + bx$ ✗	$\dfrac{a + bx}{a} = \dfrac{a}{a} + \dfrac{bx}{a} = 1 + \dfrac{b}{a}x$	Divide out common factors, not common terms.
$\dfrac{a + ax}{a} = a + x$ ✗	$\dfrac{a + ax}{a} = \dfrac{a(1 + x)}{a} = 1 + x$	Factor before dividing out common factors.
$1 + \dfrac{x}{2x} = 1 + \dfrac{1}{x}$ ✗	$1 + \dfrac{x}{2x} = 1 + \dfrac{1}{2} = \dfrac{3}{2}$	Divide out common factors.

A good way to avoid errors is to *work slowly, write neatly,* and *think about each step.* Each time you write a step, ask yourself why the step is algebraically legitimate. For example, the step below is legitimate because *dividing the numerator and denominator by the same nonzero number produces an equivalent fraction.*

$$\frac{2x}{6} = \frac{\cancel{2} \cdot x}{\cancel{2} \cdot 3} = \frac{x}{3}$$

EXAMPLE 1 **Describing and Correcting an Error**

Describe and correct the error. $\quad \dfrac{1}{2x} + \dfrac{1}{3x} = \dfrac{1}{5x}$ ✗

Solution Use the property for adding fractions with unlike denominators.

$$\frac{1}{2x} + \frac{1}{3x} = \frac{3x + 2x}{6x^2} = \frac{5x}{6x^2} = \frac{5}{6x}$$

✓ *Checkpoint* ◄))) *Audio-video solution in English & Spanish at LarsonPrecalculus.com*

Describe and correct the error. $\quad \sqrt{x^2 + 4} = x + 2$ ✗

Some Algebra of Calculus

In calculus it is often necessary to rewrite a simplified algebraic expression. See the following lists, which are from a standard calculus text.

Unusual Factoring

Expression	Useful Calculus Form	Comment
$\dfrac{5x^4}{8}$	$\dfrac{5}{8}x^4$	Write with fractional coefficient.
$\dfrac{x^2 + 3x}{-6}$	$-\dfrac{1}{6}(x^2 + 3x)$	Write with fractional coefficient.
$2x^2 - x - 3$	$2\left(x^2 - \dfrac{x}{2} - \dfrac{3}{2}\right)$	Factor out the leading coefficient.
$\dfrac{x}{2}(x + 1)^{-1/2} + (x + 1)^{1/2}$	$\dfrac{(x + 1)^{-1/2}}{2}[x + 2(x + 1)]$	Factor out the fractional coefficient and the variable expression with the lesser exponent.

Writing with Negative Exponents

Expression	Useful Calculus Form	Comment
$\dfrac{9}{5x^3}$	$\dfrac{9}{5}x^{-3}$	Move the factor to the numerator and change the sign of the exponent.
$\dfrac{7}{\sqrt{2x - 3}}$	$7(2x - 3)^{-1/2}$	Move the factor to the numerator and change the sign of the exponent.

Writing a Fraction as a Sum

Expression	Useful Calculus Form	Comment
$\dfrac{x + 2x^2 + 1}{\sqrt{x}}$	$x^{1/2} + 2x^{3/2} + x^{-1/2}$	Divide each term of the numerator by $x^{1/2}$.
$\dfrac{1 + x}{x^2 + 1}$	$\dfrac{1}{x^2 + 1} + \dfrac{x}{x^2 + 1}$	Rewrite the fraction as a sum of fractions.
$\dfrac{2x}{x^2 + 2x + 1}$	$\dfrac{2x + 2 - 2}{x^2 + 2x + 1}$	Add and subtract the same term.
	$= \dfrac{2x + 2}{x^2 + 2x + 1} - \dfrac{2}{(x + 1)^2}$	Rewrite the fraction as a difference of fractions.
$\dfrac{x^2 - 2}{x + 1}$	$x - 1 - \dfrac{1}{x + 1}$	Use polynomial long division. (See Section 2.3.)
$\dfrac{x + 7}{x^2 - x - 6}$	$\dfrac{2}{x - 3} - \dfrac{1}{x + 2}$	Use the method of partial fractions. (See Section 7.4.)

Inserting Factors and Terms

Expression	Useful Calculus Form	Comment
$(2x - 1)^3$	$\dfrac{1}{2}(2x - 1)^3(2)$	Multiply and divide by 2.
$7x^2(4x^3 - 5)^{1/2}$	$\dfrac{7}{12}(4x^3 - 5)^{1/2}(12x^2)$	Multiply and divide by 12.
$\dfrac{4x^2}{9} - 4y^2 = 1$	$\dfrac{x^2}{9/4} - \dfrac{y^2}{1/4} = 1$	Write with fractional denominators.
$\dfrac{x}{x + 1}$	$\dfrac{x + 1 - 1}{x + 1} = 1 - \dfrac{1}{x + 1}$	Add and subtract the same term.

The next five examples demonstrate many of the steps in the preceding lists.

EXAMPLE 2   **Factors Involving Negative Exponents**

Factor $x(x + 1)^{-1/2} + (x + 1)^{1/2}$.

Solution   When multiplying powers with like bases, you add exponents. When factoring, you are undoing multiplication, and so you *subtract* exponents.

$$x(x + 1)^{-1/2} + (x + 1)^{1/2} = (x + 1)^{-1/2}[x(x + 1)^0 + (x + 1)^1]$$
$$= (x + 1)^{-1/2}[x + (x + 1)]$$
$$= (x + 1)^{-1/2}(2x + 1)$$

✓ **Checkpoint**  Audio-video solution in English & Spanish at LarsonPrecalculus.com

Factor $x(x - 2)^{-1/2} + 6(x - 2)^{1/2}$.

Another way to simplify the expression in Example 2 is to multiply the expression by a fractional form of 1 and then use the Distributive Property.

$$[x(x + 1)^{-1/2} + (x + 1)^{1/2}] \cdot \frac{(x + 1)^{1/2}}{(x + 1)^{1/2}} = \frac{x(x + 1)^0 + (x + 1)^1}{(x + 1)^{1/2}}$$
$$= \frac{2x + 1}{\sqrt{x + 1}}$$

EXAMPLE 3   **Inserting Factors in an Expression**

Insert the required factor: $\dfrac{x + 2}{(x^2 + 4x - 3)^2} = ()\dfrac{1}{(x^2 + 4x - 3)^2}(2x + 4)$.

Solution   The expression on the right side of the equation is twice the expression on the left side. To make both sides equal, insert a factor of $\frac{1}{2}$.

$$\frac{x + 2}{(x^2 + 4x - 3)^2} = \left(\frac{1}{2}\right)\frac{1}{(x^2 + 4x - 3)^2}(2x + 4)$$

✓ **Checkpoint**   Audio-video solution in English & Spanish at LarsonPrecalculus.com

Insert the required factor: $\dfrac{6x - 3}{(x^2 - x + 4)^2} = ()\dfrac{1}{(x^2 - x + 4)^2}(2x - 1)$.

EXAMPLE 4   **Rewriting Fractions**

Show that the two expressions are equivalent.

$$\frac{16x^2}{25} - 9y^2 = \frac{x^2}{25/16} - \frac{y^2}{1/9}$$

Solution   To write the expression on the left side of the equation in the form given on the right side, multiply the numerator and denominator of the first term by $1/16$ and multiply the numerator and denominator of the second term by $1/9$.

$$\frac{16x^2}{25} - 9y^2 = \frac{16x^2}{25}\left(\frac{1/16}{1/16}\right) - 9y^2\left(\frac{1/9}{1/9}\right) = \frac{x^2}{25/16} - \frac{y^2}{1/9}$$

✓ *Checkpoint* *Audio-video solution in English & Spanish at LarsonPrecalculus.com*

Show that the two expressions are equivalent.

$$\frac{9x^2}{16} + 25y^2 = \frac{x^2}{16/9} + \frac{y^2}{1/25}$$

EXAMPLE 5   **Rewriting with Negative Exponents**

Rewrite each expression using negative exponents.

a. $\dfrac{-4x}{(1 - 2x^2)^2}$ **b.** $\dfrac{2}{5x^3} - \dfrac{1}{\sqrt{x}} + \dfrac{3}{5(4x)^2}$

Solution

a. $\dfrac{-4x}{(1 - 2x^2)^2} = -4x(1 - 2x^2)^{-2}$

b. $\dfrac{2}{5x^3} - \dfrac{1}{\sqrt{x}} + \dfrac{3}{5(4x)^2} = \dfrac{2}{5x^3} - \dfrac{1}{x^{1/2}} + \dfrac{3}{5(4x)^2} = \dfrac{2}{5}x^{-3} - x^{-1/2} + \dfrac{3}{5}(4x)^{-2}$

✓ *Checkpoint* *Audio-video solution in English & Spanish at LarsonPrecalculus.com*

Rewrite $\dfrac{-6x}{(1 - 3x^2)^2} + \dfrac{1}{\sqrt[3]{x}}$ using negative exponents.

EXAMPLE 6   **Rewriting Fractions as Sums of Terms**

Rewrite each fraction as the sum of three terms.

a. $\dfrac{x^2 - 4x + 8}{2x}$ **b.** $\dfrac{x + 2x^2 + 1}{\sqrt{x}}$

Solution

a. $\dfrac{x^2 - 4x + 8}{2x} = \dfrac{x^2}{2x} - \dfrac{4x}{2x} + \dfrac{8}{2x} = \dfrac{x}{2} - 2 + \dfrac{4}{x}$

b. $\dfrac{x + 2x^2 + 1}{\sqrt{x}} = \dfrac{x}{x^{1/2}} + \dfrac{2x^2}{x^{1/2}} + \dfrac{1}{x^{1/2}} = x^{1/2} + 2x^{3/2} + x^{-1/2}$

✓ *Checkpoint*  *Audio-video solution in English & Spanish at LarsonPrecalculus.com*

Rewrite each fraction as the sum of three terms.

a. $\dfrac{x^4 - 2x^3 + 5}{x^3}$ **b.** $\dfrac{x^2 - x + 5}{\sqrt{x}}$

A.7   Exercises

See **CalcChat.com** for tutorial help and worked-out solutions to odd-numbered exercises.

Vocabulary: Fill in the blanks.

1. To rewrite the expression $\dfrac{3}{x^5}$ using negative exponents, move x^5 to the _____ and change the sign of the exponent.

2. When dividing fractions, multiply by the _____.

Skills and Applications

Describing and Correcting an Error In Exercises 3–12, describe and correct the error.

3. $(x + 3)^2 = x^2 + 9$ $\times$

4. $5z + 3(x - 2) = 5z + 3x - 2$ $\times$

5. $\sqrt{x + 9} = \sqrt{x} + 3$ $\times$ 6. $\sqrt{25 - x^2} = 5 - x$ $\times$

7. $\dfrac{2x^2 + 1}{5x} = \dfrac{2x + 1}{5}$ $\times$ 8. $\dfrac{6x + y}{6x - y} = \dfrac{x + y}{x - y}$ $\times$

9. $(4x)^2 = 4x^2$ $\times$

10. $\dfrac{1}{a^{-1} + b^{-1}} = \left(\dfrac{1}{a + b}\right)^{-1}$ $\times$

11. $\dfrac{3}{x} + \dfrac{4}{y} = \dfrac{7}{x + y}$ $\times$ 12. $5 + (1/y) = \dfrac{1}{5 + y}$ $\times$

Factors Involving Negative Exponents In Exercises 13–16, factor the expression.

13. $2x(x + 2)^{-1/2} + (x + 2)^{1/2}$

14. $x^2(x^2 + 1)^{-5} - (x^2 + 1)^{-4}$

15. $4x^3(2x - 1)^{3/2} - 2x(2x - 1)^{-1/2}$

16. $x(x + 1)^{-4/3} + (x + 1)^{2/3}$

Unusual Factoring In Exercises 17–24, complete the factored form of the expression.

17. $\dfrac{5x + 3}{4} = \dfrac{1}{4}(\boxed{})$ 18. $\dfrac{7x^2}{10} = \dfrac{7}{10}(\boxed{})$

19. $\frac{2}{3}x^2 + \frac{1}{3}x + 5 = \frac{1}{3}(\boxed{})$ 20. $\frac{3}{4}x + \frac{1}{2} = \frac{1}{4}(\boxed{})$

21. $x^{1/3} - 5x^{4/3} = x^{1/3}(\boxed{})$

22. $3(2x + 1)x^{1/2} + 4x^{3/2} = x^{1/2}(\boxed{})$

23. $\dfrac{1}{10}(2x + 1)^{5/2} - \dfrac{1}{6}(2x + 1)^{3/2} = \dfrac{(2x + 1)^{3/2}}{15}(\boxed{})$

24. $\dfrac{3}{7}(t + 1)^{7/3} - \dfrac{3}{4}(t + 1)^{4/3} = \dfrac{3(t + 1)^{4/3}}{28}(\boxed{})$

Inserting Factors in an Expression In Exercises 25–28, insert the required factor in the parentheses.

25. $x^2(x^3 - 1)^4 = (\boxed{})(x^3 - 1)^4(3x^2)$

26. $x(1 - 2x^2)^3 = (\boxed{})(1 - 2x^2)^3(-4x)$

27. $\dfrac{4x + 6}{(x^2 + 3x + 7)^3} = (\boxed{})\dfrac{1}{(x^2 + 3x + 7)^3}(2x + 3)$

28. $\dfrac{x + 1}{(x^2 + 2x - 3)^2} = (\boxed{})\dfrac{1}{(x^2 + 2x - 3)^2}(2x + 2)$

Rewriting Fractions In Exercises 29–34, show that the two expressions are equivalent.

29. $4x^2 + \dfrac{6y^2}{10} = \dfrac{x^2}{1/4} + \dfrac{3y^2}{5}$

30. $\dfrac{4x^2}{14} - 2y^2 = \dfrac{2x^2}{7} - \dfrac{y^2}{1/2}$

31. $\dfrac{25x^2}{36} + \dfrac{4y^2}{9} = \dfrac{x^2}{36/25} + \dfrac{y^2}{9/4}$

32. $\dfrac{5x^2}{9} - \dfrac{16y^2}{49} = \dfrac{x^2}{9/5} - \dfrac{y^2}{49/16}$

33. $\dfrac{x^2}{3/10} - \dfrac{y^2}{4/5} = \dfrac{10x^2}{3} - \dfrac{5y^2}{4}$

34. $\dfrac{x^2}{5/8} + \dfrac{y^2}{6/11} = \dfrac{8x^2}{5} + \dfrac{11y^2}{6}$

Rewriting with Negative Exponents In Exercises 35–40, rewrite the expression using negative exponents.

35. $\dfrac{7}{(x + 3)^5}$ 36. $\dfrac{2 - x}{(x + 1)^{3/2}}$

37. $\dfrac{2x^5}{(3x + 5)^4}$ 38. $\dfrac{x + 1}{x(6 - x)^{1/2}}$

39. $\dfrac{4}{3x} + \dfrac{4}{x^4} - \dfrac{7x}{\sqrt[3]{2x}}$ 40. $\dfrac{x}{x - 2} + \dfrac{1}{x^2} + \dfrac{8}{3(9x)^3}$

Rewriting a Fraction as a Sum of Terms In Exercises 41–46, rewrite the fraction as a sum of two or more terms.

41. $\dfrac{x^2 + 6x + 12}{3x}$ 42. $\dfrac{x^3 - 5x^2 + 4}{x^2}$

43. $\dfrac{4x^3 - 7x^2 + 1}{x^{1/3}}$ 44. $\dfrac{2x^5 - 3x^3 + 5x - 1}{x^{3/2}}$

45. $\dfrac{3 - 5x^2 - x^4}{\sqrt{x}}$ 46. $\dfrac{x^3 - 5x^4}{3x^2}$

Simplifying an Expression **In Exercises 47–58, simplify the expression.**

47. $\dfrac{-2(x^2 - 3)^{-3}(2x)(x + 1)^3 - 3(x + 1)^2(x^2 - 3)^{-2}}{[(x + 1)^3]^2}$

48. $\dfrac{x^5(-3)(x^2 + 1)^{-4}(2x) - (x^2 + 1)^{-3}(5)x^4}{(x^5)^2}$

49. $\dfrac{(6x + 1)^3(27x^2 + 2) - (9x^3 + 2x)(3)(6x + 1)^2(6)}{[(6x + 1)^3]^2}$

50. $\dfrac{(4x^2 + 9)^{1/2}(2) - (2x + 3)\left(\frac{1}{2}\right)(4x^2 + 9)^{-1/2}(8x)}{[(4x^2 + 9)^{1/2}]^2}$

51. $\dfrac{(x + 2)^{3/4}(x + 3)^{-2/3} - (x + 3)^{1/3}(x + 2)^{-1/4}}{[(x + 2)^{3/4}]^2}$

52. $(2x - 1)^{1/2} - (x + 2)(2x - 1)^{-1/2}$

53. $\dfrac{2(3x - 1)^{1/3} - (2x + 1)\left(\frac{1}{3}\right)(3x - 1)^{-2/3}(3)}{(3x - 1)^{2/3}}$

54. $\dfrac{(x + 1)\left(\frac{1}{2}\right)(2x - 3x^2)^{-1/2}(2 - 6x) - (2x - 3x^2)^{1/2}}{(x + 1)^2}$

55. $\dfrac{1}{(x^2 + 4)^{1/2}} \cdot \dfrac{1}{2}(x^2 + 4)^{-1/2}(2x)$

56. $\dfrac{1}{x^2 - 6}(2x) + \dfrac{1}{2x + 5}(2)$

57. $(x^2 + 5)^{1/2}\left(\dfrac{3}{2}\right)(3x - 2)^{1/2}(3)$

    $+ (3x - 2)^{3/2}\left(\dfrac{1}{2}\right)(x^2 + 5)^{-1/2}(2x)$

58. $(3x + 2)^{-1/2}(3)(x - 6)^{1/2}(1)$

    $+ (x - 6)^3\left(-\dfrac{1}{2}\right)(3x + 2)^{-3/2}(3)$

59. **Verifying an Equation**

    (a) Verify that $y_1 = y_2$ analytically.

    $$y_1 = x^2\left(\frac{1}{3}\right)(x^2 + 1)^{-2/3}(2x) + (x^2 + 1)^{1/3}(2x)$$

    $$y_2 = \frac{2x(4x^2 + 3)}{3(x^2 + 1)^{2/3}}$$

    (b) Complete the table and demonstrate the equality in part (a) numerically.

x	-2	-1	$-\frac{1}{2}$	0	1	2	$\frac{5}{2}$
y_1							
y_2							

    (c) Use a graphing utility to verify the equality in part (a) graphically.

60. **Athletics** An athlete has set up a course in which she is dropped off by a boat 2 miles from the nearest point on shore. Once she reaches the shore, she must run to a point 4 miles down the coast and 2 miles inland (see figure). She can swim 2 miles per hour and run 6 miles per hour. The time t (in hours) required for her to complete the course can be approximated by the model

    $$t = \frac{\sqrt{x^2 + 4}}{2} + \frac{\sqrt{(4 - x)^2 + 4}}{6}$$

    where x is the distance (in miles) down the coast from her starting point to the point at which she leaves the water to start her run.

    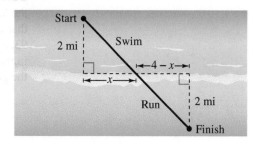

    (a) Use a table to approximate the distance down the coast that will yield the minimum amount of time required for the athlete to complete the course.

    (b) The expression below was obtained using calculus. It can be used to find the minimum amount of time required for the triathlete to reach the finish line. Simplify the expression.

    $$\frac{1}{2}x(x^2 + 4)^{-1/2} + \frac{1}{6}(x - 4)(x^2 - 8x + 20)^{-1/2}$$

Exploration

61. **Writing** Write a paragraph explaining to a classmate why

    $$\frac{1}{(x - 2)^{1/2} + x^4} \neq (x - 2)^{-1/2} + x^{-4}.$$

62. **Think About It** You are taking a course in calculus, and for one of the homework problems you obtain the following answer.

    $$\frac{2}{3}x(2x - 3)^{3/2} - \frac{2}{15}(2x - 3)^{5/2}$$

    The answer in the back of the book is

    $$\frac{2}{5}(2x - 3)^{3/2}(x + 1).$$

    Show how the second answer can be obtained from the first. Then use the same technique to simplify the expression

    $$\frac{2}{3}x(4 + x)^{3/2} - \frac{2}{15}(4 + x)^{5/2}.$$

Line Plots

A **line plot** uses a portion of a real number line to order numbers. Line plots ar·
especially useful for ordering small sets of numbers (about 50 or less) by hand.

EXAMPLE 2   **Constructing a Line Plot**

Use a line plot to organize the test scores listed below. Which score occurs with th·
greatest frequency?   *(Spreadsheet at LarsonPrecalculus.com)*

DATA

93, 70, 76, 67, 86, 93, 82, 78, 83, 86, 64, 78, 76, 66, 83,
83, 96, 74, 69, 76, 64, 74, 79, 76, 88, 76, 81, 82, 74, 70

Solution   Begin by determining the least and greatest data values. For this data se·
the least value is 64 and the greatest is 96. Next, draw a portion of a real number lin·
that includes the interval $[64, 96]$. To create the line plot, start with the first number
93, and enter a ● above 93 on the number line. Continue recording ●'s for the number
in the list until you obtain the line plot below. The line plot shows that 76 occurs wit·
the greatest frequency.

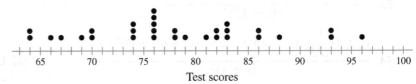

Test scores

✓ **Checkpoint** 🔊))) *Audio-video solution in English & Spanish at LarsonPrecalculus.com*

Use a line plot to organize the test scores listed below. Which score occurs with th·
greatest frequency?   *(Spreadsheet at LarsonPrecalculus.com)*

DATA

68, 73, 67, 95, 71, 82, 85, 74, 82, 61, 87, 92, 78, 74, 64,
71, 74, 82, 71, 83, 92, 82, 78, 72, 82, 64, 85, 67, 71, 62

EXAMPLE 3   **Analyzing a Line Plot**

The line plot shows the daily high temperatures (in degrees Fahrenheit) in a city durin·
the month of June.

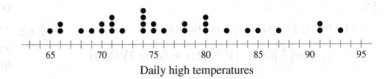

Daily high temperatures

a. What is the range of daily high temperatures?

b. On how many days was the high temperature in the 80s?

Solution

a. The line plot shows that the maximum daily high temperature was 93°F and th·
minimum daily high temperature was 65°F. So, the range is $93 - 65 = 28°F$.

b. There are 7 ●'s in the interval $[80, 90)$. So, the high temperature was in the 80s o·
7 days.

✓ **Checkpoint** 🔊))) *Audio-video solution in English & Spanish at LarsonPrecalculus.com*

In Example 3, on how many days was the high temperature less than 80°F?

2. Test Scores The histograms represent the test scores of two classes of a college course in mathematics. Which histogram shows the lesser standard deviation? Explain.

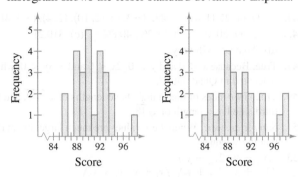

Finding Mean and Standard Deviation In **Exercises 23 and 24, each line plot represents a data set. Find the mean and standard deviation of each data set.**

23. (a)

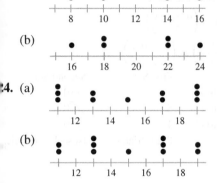

(b)

24. (a)

(b)

Finding Mean, Variance, and Standard Deviation In Exercises 25–30, find the mean ($\bar{x}$), variance (v), and standard deviation (σ) of the data set.

25. 4, 10, 8, 2   **26.** 2, 12, 4, 7, 5

27. 0, 1, 1, 2, 2, 2, 3, 3, 4

28. 2, 2, 2, 2, 2, 2

29. 42, 50, 61, 47, 56, 68

30. 1.2, 0.6, 2.8, 1.7, 0.9

Using the Alternative Formula for Standard Deviation In Exercises 31–36, use the alternative formula for standard deviation to find the standard deviation of the data set.

31. 3, 5, 7, 7, 18, 8

32. 24, 20, 38, 15, 52, 33, 46

33. 246, 336, 473, 167, 219

34. 5.1, 6.7, 4.5, 10.2, 9.1

35. 8.6, 6.4, 2.9, 5.0, 6.7

36. 9.2, 10.6, 7.2, 4.3, 7.0

Quartiles and Box-and-Whisker Plots
In Exercises 37–40, find the lower and upper quartiles and sketch a box-and-whisker plot for each data set.

37. 11, 10, 11, 14, 17, 16, 14, 11, 8, 14, 20

38. 46, 48, 48, 50, 52, 47, 51, 47, 49, 53

39. 19, 12, 14, 9, 14, 15, 17, 13, 19, 11, 10, 19

40. 20.1, 43.4, 34.9, 23.9, 33.5, 24.1, 22.5, 42.4, 25.7, 17.4, 23.8, 33.3, 17.3, 36.4, 21.8

41. Price of Gold The data represent the average prices of gold (in dollars per troy ounce) for the years 1996 through 2015. Use a graphing utility to find the mean, variance, and standard deviation of the data. What percent of the data lies within one standard deviation of the mean? *(Source: U.S. Bureau of Mines and U.S. Geological Survey)* *(Spreadsheet at LarsonPrecalculus.com)*

 DATA

389,	332,	295,	280,	280,
272,	311,	365,	411,	446,
606,	699,	874,	975,	1227,
1572,	1673,	1415,	1269,	1170

42. Product Lifetime

A manufacturer redesigns a machine part to increase its average lifetime. The two data sets list the lifetimes (in months) of 20 randomly selected parts of each design.
(Spreadsheet at LarsonPrecalculus.com)

Original Design

15.1	78.3	56.3	68.9	30.6
27.2	12.5	42.7	72.7	20.2
53.0	13.5	11.0	18.4	85.2
10.8	38.3	85.1	10.0	12.6

New Design

55.8	71.5	25.6	19.0	23.1
37.2	60.0	35.3	18.9	80.5
46.7	31.1	67.9	23.5	99.5
54.0	23.2	45.5	24.8	87.8

(a) Construct a box-and-whisker plot for each data set.

(b) Explain the differences between the box-and-whisker plots you constructed in part (a).

Answers to Odd-Numbered Exercises and Tests

Chapter 1

Section 1.1 *(page 8)*

1. Cartesian   **3.** Distance Formula

5.

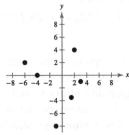

7. $(-3, 4)$ **9.** Quadrant IV   **11.** Quadrant II

13. Quadrant II or IV

15.

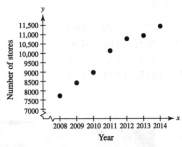

17. 13   **19.** $\sqrt{61}$ **21.** $\dfrac{\sqrt{277}}{6}$

23. (a) 5, 12, 13   (b) $5^2 + 12^2 = 13^2$

25. $\left(\sqrt{5}\right)^2 + \left(\sqrt{45}\right)^2 = \left(\sqrt{50}\right)^2$

27. Distances between the points: $\sqrt{29}$, $\sqrt{58}$, $\sqrt{29}$

29. (a)

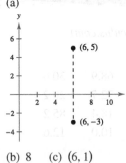

(b) 8   (c) $(6, 1)$

31. (a)

(b) 10   (c) $(5, 4)$

33. (a)

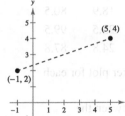

(b) $2\sqrt{10}$
(c) $(2, 3)$

35. (a)

(b) $\sqrt{556.52}$
(c) $(-5.6, 8.6)$

37. $30\sqrt{41} \approx 192$ km   **39.** \$40,560.5 million

41. $(0, 1), (4, 2), (1, 4)$ **43.** $(-3, 6), (2, 10), (2, 4), (-3, 4)$

45. (a) 2000–2010   (b) 53.7%; 40.8%   (c) \$10.21

(d) Answers will vary.

47. True. Because $x < 0$ and $y > 0$, $2x < 0$ and $-3y < 0$, which is located in Quadrant III.

49. True. Two sides of the triangle have lengths of $\sqrt{149}$, and the third side has a length of $\sqrt{18}$.

51. *Sample answer:* When the x-values are much larger or smaller that the y-values.

53. $(2x_m - x_1, 2y_m - y_1)$

55. $\left(\dfrac{3x_1 + x_2}{4}, \dfrac{3y_1 + y_2}{4}\right), \left(\dfrac{x_1 + x_2}{2}, \dfrac{y_1 + y_2}{2}\right),$
$\left(\dfrac{x_1 + 3x_2}{4}, \dfrac{y_1 + 3y_2}{4}\right)$

57. Use the Midpoint Formula to prove that the diagonals of the parallelogram bisect each other.

$$\left(\frac{b + a}{2}, \frac{c + 0}{2}\right) = \left(\frac{a + b}{2}, \frac{c}{2}\right)$$
$$\left(\frac{a + b + 0}{2}, \frac{c + 0}{2}\right) = \left(\frac{a + b}{2}, \frac{c}{2}\right)$$

59. (a)

	First Set	Second Set
Distance A to B	3	$\sqrt{10}$
Distance B to C	5	$\sqrt{10}$
Distance A to C	4	$\sqrt{40}$
	Right triangle	Isosceles triangle

(b)

The first set of points is not collinear. The second set of points is collinear.

(c) A set of three points is collinear when the sum of two distances among the points is exactly equal to the third distance.

Section 1.2 *(page 19)*

1. solution or solution point   **3.** intercepts

5. circle; (h, k); r **7.** (a) Yes   (b) Yes

9. (a) Yes   (b) No   **11.** (a) No   (b) Yes

13. (a) No   (b) Yes

5.

x	-1	0	1	2	$\frac{5}{2}$
y	7	5	3	1	0
(x, y)	$(-1, 7)$	$(0, 5)$	$(1, 3)$	$(2, 1)$	$\left(\frac{5}{2}, 0\right)$

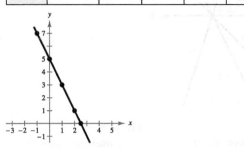

7.

x	-1	0	1	2	3
y	4	0	-2	-2	0
(x, y)	$(-1, 4)$	$(0, 0)$	$(1, -2)$	$(2, -2)$	$(3, 0)$

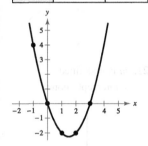

19. x-intercept: $(3, 0)$
y-intercept: $(0, 9)$

23. x-intercept: $\left(\frac{6}{5}, 0\right)$
y-intercept: $(0, -6)$

27. x-intercept: $\left(\frac{7}{3}, 0\right)$
y-intercept: $(0, 7)$

31. x-intercept: $(6, 0)$
y-intercepts: $\left(0, \pm\sqrt{6}\right)$

33. y-axis symmetry

37. Origin symmetry

41.

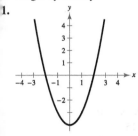

21. x-intercept: $(-2, 0)$
y-intercept: $(0, 2)$

25. x-intercept: $(-4, 0)$
y-intercept: $(0, 2)$

29. x-intercepts: $(0, 0)$, $(2, 0)$
y-intercept: $(0, 0)$

35. Origin symmetry

39. x-axis symmetry

43.

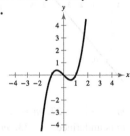

45. x-intercept: $\left(\frac{1}{3}, 0\right)$
y-intercept: $(0, 1)$
No symmetry

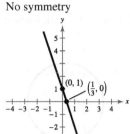

47. x-intercepts: $(0, 0)$, $(2, 0)$
y-intercept: $(0, 0)$
No symmetry

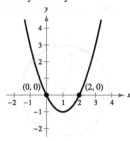

49. x-intercept: $\left(\sqrt[3]{-3}, 0\right)$
y-intercept: $(0, 3)$
No symmetry

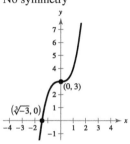

51. x-intercept: $(3, 0)$
y-intercept: None
No symmetry

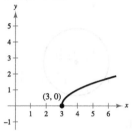

53. x-intercept: $(6, 0)$
y-intercept: $(0, 6)$
No symmetry

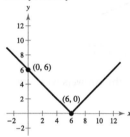

55. x-intercept: $(-1, 0)$
y-intercepts: $(0, \pm 1)$
x-axis symmetry

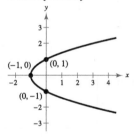

57.

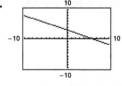

Intercepts: $(6, 0)$, $(0, 3)$

59.

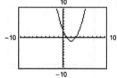

Intercepts:
$(3, 0)$, $(1, 0)$, $(0, 3)$

61.

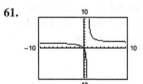

Intercept: $(0, 0)$

63.

Intercepts: $(-1, 0)$, $(0, 1)$

65.

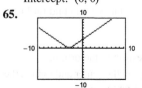

Intercepts: $(-3, 0)$, $(0, 3)$

67. $x^2 + y^2 = 9$ **69.** $(x + 4)^2 + (y - 5)^2 = 4$

71. $(x - 3)^2 + (y - 8)^2 = 169$

73. $(x + 3)^2 + (y + 3)^2 = 61$

75. Center: $(0, 0)$; Radius: 5   **77.** Center: $(1, -3)$; Radius: 3

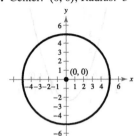 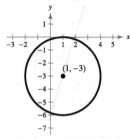

79. Center: $\left(\frac{1}{2}, \frac{1}{2}\right)$; Radius: $\frac{3}{2}$

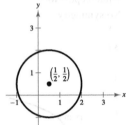

81.

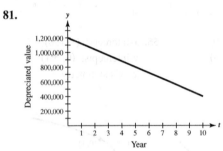

83. (a)   (b) Answers will vary.

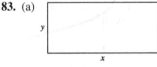

(c)   (d) $x = 86\frac{2}{3}, y = 86\frac{2}{3}$

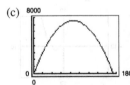

(e) A regulation NFL playing field is 120 yards long and $53\frac{1}{3}$ yards wide. The actual area is 6400 square yards.

85. (a)   The model fits the data well;

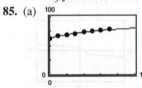

Each data value is close to the graph of the model.

(b) 74.7 yr   (c) 1964

(d) $(0, 63.6)$; In 1940, the life expectancy of a child (at birth) was 63.6 years.

(e) Answers will vary.

87. False. $y = x$ is symmetric with respect to the origin.

89. True. *Sample answer:* Depending on the center and radius, the graph could intersect one, both, or neither axis.

91. (a) $a = 1, b = 0$ (b) $a = 0, b = 1$

Section 1.3   *(page 31)*

1. linear   **3.** point-slope   **5.** perpendicular

7. Linear extrapolation   **9.** (a) L_2 (b) L_3 (c) L_1

11.   **13.** $\frac{3}{2}$

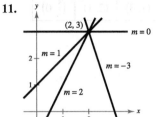

15. $m = 5$ **17.** $m = -\frac{3}{4}$

    y-intercept: $(0, 3)$ y-intercept: $(0, -1)$

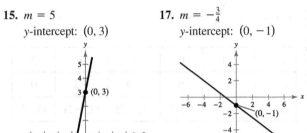

19. $m = 0$ **21.** m is undefined.

    y-intercept: $(0, 5)$ y-intercept: none

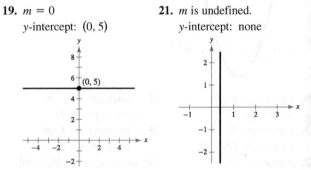

23. $m = \frac{7}{6}$

    y-intercept: $(0, -5)$

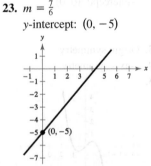

25. $m = -\frac{3}{2}$ **27.** $m = 2$ **29.** $m = 0$

31. m is undefined.   **33.** $m = 0.15$

35. $(-1, 7), (0, 7), (4, 7)$ **37.** $(-4, 6), (-3, 8), (-2, 10)$

39. $(-2, 7), \left(0, \frac{19}{3}\right), (1, 6)$ **41.** $(-4, -5), (-4, 0), (-4, 2)$

43. $y = 3x - 2$

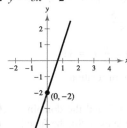

45. $y = -2x$

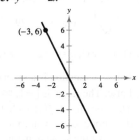

47. $y = -\frac{1}{3}x + \frac{4}{3}$

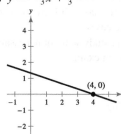

49. $y = -\frac{1}{2}x - 2$

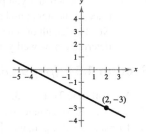

51. $y = \frac{5}{2}$

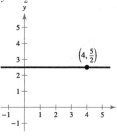

53. $y = 5x + 27.3$

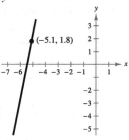

55. $y = -\frac{3}{5}x + 2$

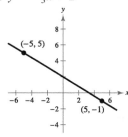

57. $x = -7$

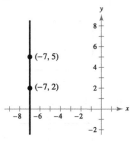

59. $y = -\frac{1}{2}x + \frac{3}{2}$

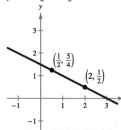

61. $y = 0.4x + 0.2$

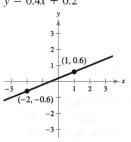

63. $y = -1$

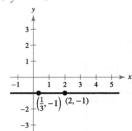

65. Parallel   **67.** Neither   **69.** Perpendicular
71. Parallel   **73.** (a) $y = 2x - 3$ (b) $y = -\frac{1}{2}x + 2$
75. (a) $y = -\frac{3}{4}x + \frac{3}{8}$ (b) $y = \frac{4}{3}x + \frac{127}{72}$
77. (a) $y = 4$ (b) $x = -2$
79. (a) $y = x + 4.3$ (b) $y = -x + 9.3$
81. $5x + 3y - 15 = 0$ **83.** $12x + 3y + 2 = 0$
85. $x + y - 3 = 0$
87. (a) Sales increasing 135 units/yr
      (b) No change in sales
      (c) Sales decreasing 40 units/yr
89. 12 ft
91. $V(t) = -150t + 5400, \quad 16 \le t \le 21$
93. C-intercept: fixed initial cost; Slope: cost of producing an additional laptop bag
95. $V(t) = -175t + 875, \quad 0 \le t \le 5$
97. $F = 1.8C + 32$ or $C = \frac{5}{9}F - \frac{160}{9}$
99. (a) $C = 21t + 42{,}000$ (b) $R = 45t$
      (c) $P = 24t - 42{,}000$ (d) 1750 h
101. False. The slope with the greatest magnitude corresponds to the steepest line.
103. Find the slopes of the lines containing each two points and use the relationship $m_1 = -\dfrac{1}{m_2}$.
105. The scale on the y-axis is unknown, so the slopes of the lines cannot be determined.
107. No. The slopes of two perpendicular lines have opposite signs (assume that neither line is vertical or horizontal).
109. The line $y = 4x$ rises most quickly, and the line $y = -4x$ falls most quickly. The greater the magnitude of the slope (the absolute value of the slope), the faster the line rises or falls.
111. $3x - 2y - 1 = 0$ **113.** $80x + 12y + 139 = 0$

Section 1.4 *(page 44)*

1. domain; range; function   **3.** implied domain
5. Function   **7.** Not a function
9. (a) Function
      (b) Not a function, because the element 1 in A corresponds to two elements, -2 and 1, in B.
      (c) Function
      (d) Not a function, because not every element in A is matched with an element in B.
11. Not a function   **13.** Function   **15.** Function
17. Function   **19.** (a) -2 (b) -14 (c) $3x + 1$
21. (a) 15   (b) $4t^2 - 19t + 27$ (c) $4t^2 - 3t - 10$
23. (a) 1   (b) 2.5   (c) $3 - 2|x|$
25. (a) $-\dfrac{1}{9}$ (b) Undefined   (c) $\dfrac{1}{y^2 + 6y}$

27. (a) 1 (b) −1 (c) $\dfrac{|x-1|}{x-1}$

29. (a) −1 (b) 2 (c) 6

31.

x	−2	−1	0	1	2
$f(x)$	1	4	5	4	1

33.

x	−2	−1	0	1	2
$f(x)$	5	$\frac{9}{2}$	4	1	0

35. 5   **37.** $\frac{4}{3}$ **39.** ±9   **41.** 0, ±1   **43.** −1, 2

45. 0, ±2   **47.** All real numbers x

49. All real numbers y such that $y \geq -6$

51. All real numbers x except $x = 0, -2$

53. All real numbers s such that $s \geq 1$ except $s = 4$

55. All real numbers x such that $x > 0$

57. (a) The maximum volume is 1024 cubic centimeters.

(b)

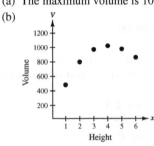

Yes, V is a function of x.

(c) $V = x(24 - 2x)^2$, $0 < x < 12$

59. $A = \dfrac{P^2}{16}$ **61.** No, the ball will be at a height of 18.5 feet.

63. $A = \dfrac{x^2}{2(x-2)}$, $x > 2$

65. 2008: 67.36%
2009: 70.13%
2010: 72.90%
2011: 75.67%
2012: 79.30%
2013: 81.25%
2014: 83.20%

67. (a) $C = 12.30x + 98,000$
(b) $R = 17.98x$
(c) $P = 5.68x - 98,000$

69. (a)

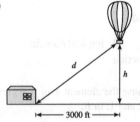

(b) $h = \sqrt{d^2 - 3000^2}$, $d \geq 3000$

71. (a) $R = \dfrac{240n - n^2}{20}$, $n \geq 80$

(b)

n	90	100	110	120	130	140	150
$R(n)$	$675	$700	$715	$720	$715	$700	$675

The revenue is maximum when 120 people take the trip.

73. $2 + h$, $h \neq 0$ **75.** $3x^2 + 3xh + h^2 + 3$, $h \neq 0$

77. $-\dfrac{x+3}{9x^2}$, $x \neq 3$ **79.** $\dfrac{\sqrt{5x-5}}{x-5}$

81. $g(x) = cx^2$; $c = -2$ **83.** $r(x) = \dfrac{c}{x}$; $c = 32$

85. False. A function is a special type of relation.

87. False. The range is $[-1, \infty)$.

89. The domain of $f(x)$ includes $x = 1$ and the domain of $g(x)$ does not because you cannot divide by 0. So, the functions do not have the same domain.

91. No; x is the independent variable, f is the name of the function.

93. (a) Yes. The amount you pay in sales tax will increase as the price of the item purchased increases.

(b) No. The length of time that you study will not necessarily determine how well you do on an exam.

Section 1.5   (page 56)

1. Vertical Line Test   **3.** decreasing

5. average rate of change; secant

7. Domain: $(-2, 2]$; range: $[-1, 8]$
(a) −1 (b) 0 (c) −1 (d) 8

9. Domain: $(-\infty, \infty)$; range: $(-2, \infty)$
(a) 0 (b) 1 (c) 2 (d) 3

11. Function   **13.** Not a function   **15.** −6   **17.** $-\frac{5}{2}, 6$

19. −3   **21.** 0, $\pm\sqrt{6}$ **23.** ±3, 4   **25.** $\frac{1}{2}$

27. (a)

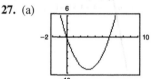

0, 6

(b) 0, 6

29. (a)

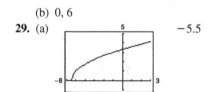

−5.5

(b) $-\frac{11}{2}$

31. (a)

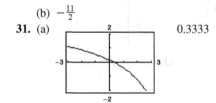

0.3333

(b) $\frac{1}{3}$

33. Decreasing on $(-\infty, \infty)$

35. Increasing on $(1, \infty)$; Decreasing on $(-\infty, -1)$

37. Increasing on $(1, \infty)$; Decreasing on $(-\infty, -1)$
Constant on $(-1, 1)$

39. Increasing on $(-\infty, -1)$, $(0, \infty)$; Decreasing on $(-1, 0)$

41.

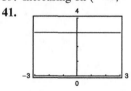

Constant on $(-\infty, \infty)$

43.

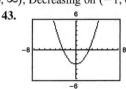

Decreasing on $(-\infty, 0)$
Increasing on $(0, \infty)$

45.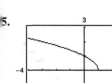

47.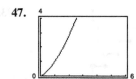

Decreasing on $(-\infty, 1)$ Increasing on $(0, \infty)$

49. Relative minimum: $(-1.5, -2.25)$
51. Relative maximum: $(0, 15)$
  Relative minimum: $(4, -17)$
53. Relative minimum: $(0.33, -0.38)$

55. 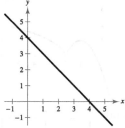   **57.**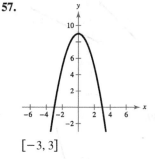

$(-\infty, 4]$ $[-3, 3]$

59.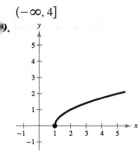

$[1, \infty)$

61. -2 **63.** -1

65. (a)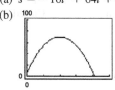

(b) About -6.14; The amount the U.S. federal government
  spent on research and development for defense decreased
  by about \$6.14 billion each year from 2010 to 2014.

67. (a) $s = -16t^2 + 64t + 6$

(b) 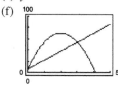   (c) 16 ft/sec

(d) The slope of the secant line is positive.
(e) $y = 16t + 6$
(f)

69. (a) $s = -16t^2 + 120t$

(b) 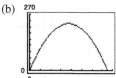   (c) -8 ft/sec

(d) The slope of the secant line is negative.
(e) $y = -8t + 240$
(f)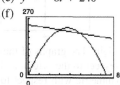

71. Even; y-axis symmetry   **73.** Neither; no symmetry
75. Neither; no symmetry

77. 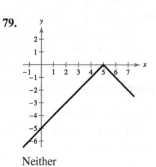   **79.**

Even   Neither

81.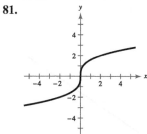

Odd

83. $h = 3 - 4x + x^2$ **85.** $L = 2 - \sqrt[3]{2y}$
87. The negative symbol should be divided out of each term,
  which yields $f(-x) = -(2x^3 + 5)$. So, the function is neither
  even nor odd.
89. (a) Ten thousands   (b) Ten millions   (c) Tens
  (d) Ones
91. False. The function $f(x) = \sqrt{x^2 + 1}$ has a domain of all real
  numbers.
93. True. A graph that is symmetric with respect to the y-axis
  cannot be increasing on its entire domain.
95. (a) $\left(\frac{5}{3}, -7\right)$ (b) $\left(\frac{5}{3}, 7\right)$
97. (a) 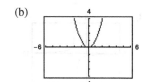   (b)

(c)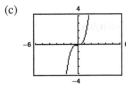

(d)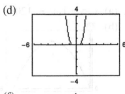

(e)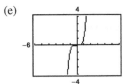

(f)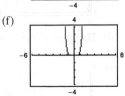

All the graphs pass through the origin. The graphs of the odd powers of x are symmetric with respect to the origin, and the graphs of the even powers are symmetric with respect to the y-axis. As the powers increase, the graphs become flatter in the interval $-1 < x < 1$.

99. (a) Even. The graph is a reflection in the x-axis.
   (b) Even. The graph is a reflection in the y-axis.
   (c) Even. The graph is a downward shift of f.
   (d) Neither. The graph is a right shift of f.

Section 1.6   (page 65)

1. Greatest integer function   **3.** Reciprocal function
5. Square root function   **7.** Absolute value function
9. Linear function
11. (a) $f(x) = -2x + 6$ **13.** (a) $f(x) = \frac{2}{3}x - 2$
   (b)   (b)

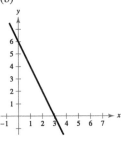   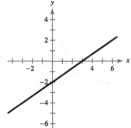

15.   **17.**

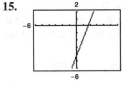   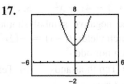

19.   **21.**

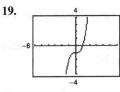   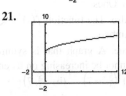

23.   **25.**

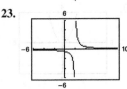   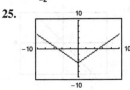

27. (a) 2   (b) 2   (c) -4 (d) 3
29. (a) 1   (b) -4 (c) 3   (d) 2

31. 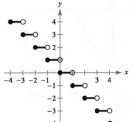   **33.**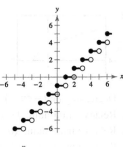

35. 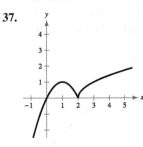   **37.**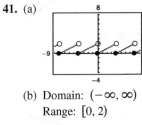

39. 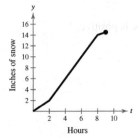   **41.** (a)

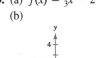

   (b) Domain: $(-\infty, \infty)$
      Range: $[0, 2)$

43. (a) $W(30) = 420$; $W(40) = 560$;
      $W(45) = 665$; $W(50) = 770$
   (b) $W(h) = \begin{cases} 14h, & 0 < h \le 36 \\ 21(h - 36) + 504, & h > 36 \end{cases}$
   (c) $W(h) = \begin{cases} 16h, & 0 < h \le 40 \\ 24(h - 40) + 640, & h > 40 \end{cases}$

45.

Interval	Input Pipe	Drain Pipe 1	Drain Pipe
$[0, 5]$	Open	Closed	Closed
$[5, 10]$	Open	Open	Closed
$[10, 20]$	Closed	Closed	Closed
$[20, 30]$	Closed	Closed	Open
$[30, 40]$	Open	Open	Open
$[40, 45]$	Open	Closed	Open
$[45, 50]$	Open	Open	Open
$[50, 60]$	Open	Open	Closed

47. $f(t) = \begin{cases} t, & 0 \le t \le 2 \\ 2t - 2, & 2 < t \le 8 \\ \frac{1}{2}t + 10, & 8 < t \le 9 \end{cases}$

Total accumulation = 14.5 in.

49. False. A piecewise-defined function is a function that is defined by two or more equations over a specified domain. That domain may or may not include x- and y-intercepts.

Section 1.7 *(page 72)*

1. rigid   3. vertical stretch; vertical shrink

5. (a) 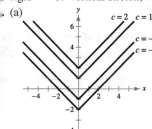   (b)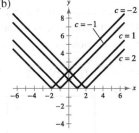

7. (a) 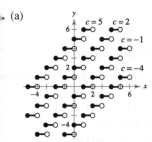   (b)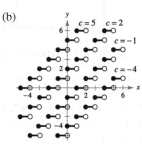

9. (a) 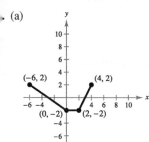   (b)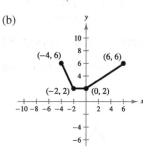

(c) 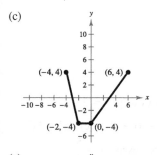   (d)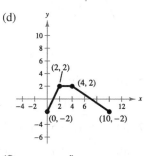

(e) 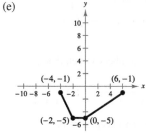   (f)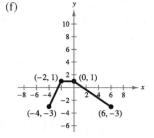

(g)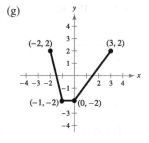

11. (a) $y = x^2 - 1$ (b) $y = -(x + 1)^2 + 1$

13. (a) $y = -|x + 3|$ (b) $y = |x - 2| - 4$

15. Right shift of $y = x^3$; $y = (x - 2)^3$

17. Reflection in the x-axis of $y = x^2$; $y = -x^2$

19. Reflection in the x-axis and upward shift of $y = \sqrt{x}$;
    $y = 1 - \sqrt{x}$

21. (a) $f(x) = x^2$
    (b) Upward shift of six units
    (c) 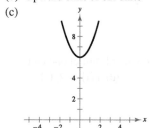   (d) $g(x) = f(x) + 6$

23. (a) $f(x) = x^3$
    (b) Reflection in the x-axis and a right shift of two units
    (c) 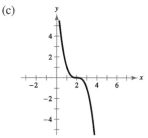   (d) $g(x) = -f(x - 2)$

25. (a) $f(x) = x^2$
    (b) Reflection in the x-axis, a left shift of one unit, and a
    downward shift of three units
    (c) 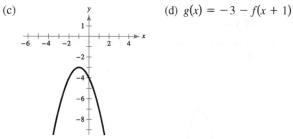   (d) $g(x) = -3 - f(x + 1)$

27. (a) $f(x) = |x|$
    (b) Right shift of one unit and an upward shift of two units
    (c) 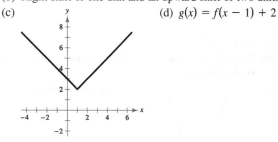   (d) $g(x) = f(x - 1) + 2$

29. (a) $f(x) = \sqrt{x}$ (b) Vertical stretch
(c)   (d) $g(x) = 2f(x)$

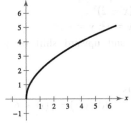

31. (a) $f(x) = [\![x]\!]$
(b) Vertical stretch and a downward shift of one unit
(c)   (d) $g(x) = 2f(x) - 1$

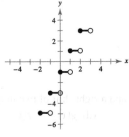

33. (a) $f(x) = |x|$
(b) Horizontal shrink
(c)   (d) $g(x) = f(2x)$

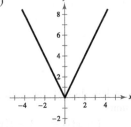

35. (a) $f(x) = x^2$
(b) Reflection in the x-axis, a vertical stretch, and an upward shift of one unit
(c)   (d) $g(x) = -2f(x) + 1$

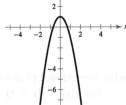

37. (a) $f(x) = |x|$
(b) Vertical stretch, a right shift of one unit, and an upward shift of two units
(c)   (d) $g(x) = 3f(x - 1) + 2$

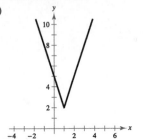

39. $g(x) = (x - 3)^2 - 7$ **41.** $g(x) = (x - 13)^3$

43. $g(x) = -|x| - 12$ **45.** $g(x) = -\sqrt{-x + 6}$
47. (a) $y = -3x^2$ (b) $y = 4x^2 + 3$
49. (a) $y = -\frac{1}{2}|x|$ (b) $y = 3|x| - 3$
51. Vertical stretch of $y = x^3$; $y = 2x^3$
53. Reflection in the x-axis and vertical shrink of $y = x^2$; $y = -\frac{1}{2}x^2$
55. Reflection in the y-axis and vertical shrink of $y = \sqrt{x}$; $y = \frac{1}{2}\sqrt{-x}$
57. $y = -(x - 2)^3 + 2$ **59.** $y = -\sqrt{x} - 3$
61. (a)

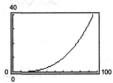

(b) $H\left(\dfrac{x}{1.6}\right) = 0.00001132x^3$; Horizontal stretch
63. False. The graph of $y = f(-x)$ is a reflection of the graph of $f(x)$ in the y-axis.
65. True. $|-x| = |x|$ **67.** $(-2, 0), (-1, 1), (0, 2)$
69. The equation should be $g(x) = (x - 1)^3$.
71. (a) $g(t) = \frac{3}{4}f(t)$ (b) $g(t) = f(t) + 10,000$
(c) $g(t) = f(t - 2)$

Section 1.8   *(page 81)*

1. addition; subtraction; multiplication; division
3.

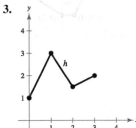

5. (a) $2x$ (b) 4 (c) $x^2 - 4$
(d) $\dfrac{x + 2}{x - 2}$; all real numbers x except $x = 2$
7. (a) $x^2 + 4x - 5$ (b) $x^2 - 4x + 5$ (c) $4x^3 - 5x^2$
(d) $\dfrac{x^2}{4x - 5}$; all real numbers x except $x = \dfrac{5}{4}$
9. (a) $x^2 + 6 + \sqrt{1 - x}$ (b) $x^2 + 6 - \sqrt{1 - x}$
(c) $(x^2 + 6)\sqrt{1 - x}$
(d) $\dfrac{(x^2 + 6)\sqrt{1 - x}}{1 - x}$; all real numbers x such that $x < 1$
11. (a) $\dfrac{x^4 + x^3 + x}{x + 1}$ (b) $\dfrac{-x^4 - x^3 + x}{x + 1}$ (c) $\dfrac{x^4}{x + 1}$
(d) $\dfrac{1}{x^2(x + 1)}$; all real numbers x except $x = 0, -1$
13. 7   **15.** 5   **17.** $-9t^2 + 3t + 5$ **19.** 306
21. $\frac{8}{23}$ **23.** -9
25.   **27.**

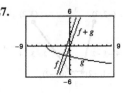

$f(x), g(x)$ $f(x), f(x)$

9. (a) $x + 5$ (b) $x + 5$ (c) $x - 6$

1. (a) $(x - 1)^2$ (b) $x^2 - 1$ (c) $x - 2$

3. (a) x (b) x (c) $x^9 + 3x^6 + 3x^3 + 2$

5. (a) $\sqrt{x^2 + 4}$ (b) $x + 4$
Domains of f and $g \circ f$: all real numbers x such that $x \geq -4$
Domains of g and $f \circ g$: all real numbers x

7. (a) x^2 (b) x^2
Domains of $f, g, f \circ g$, and $g \circ f$: all real numbers x

9. (a) $|x + 6|$ (b) $|x| + 6$
Domains of $f, g, f \circ g$, and $g \circ f$: all real numbers x

1. (a) $\dfrac{1}{x + 3}$ (b) $\dfrac{1}{x} + 3$
Domains of f and $g \circ f$: all real numbers x except $x = 0$
Domain of g: all real numbers x
Domain of $f \circ g$: all real numbers x except $x = -3$

3. (a)   (b)

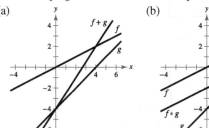

5. (a) 3   (b) 0   **47.** (a) 0   (b) 4

9. $f(x) = x^2,\ g(x) = 2x + 1$ **51.** $f(x) = \sqrt[3]{x},\ g(x) = x^2 - 4$

3. $f(x) = \dfrac{1}{x},\ g(x) = x + 2$ **55.** $f(x) = \dfrac{x + 3}{4 + x},\ g(x) = -x^2$

7. (a) $T = \frac{3}{4}x + \frac{1}{15}x^2$
(b)

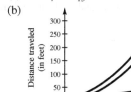

(c) The braking function $B(x)$; As x increases, $B(x)$ increases at a faster rate than $R(x)$.

9. (a) $c(t) = \dfrac{b(t) - d(t)}{p(t)} \times 100$

(b) $c(16)$ is the percent change in the population due to births and deaths in the year 2016.

1. (a) $r(x) = \dfrac{x}{2}$ (b) $A(r) = \pi r^2$

(c) $(A \circ r)(x) = \pi\left(\dfrac{x}{2}\right)^2$;

(A ∘ r)(x) represents the area of the circular base of the tank on the square foundation with side length x.

3. $g(f(x))$ represents 3 percent of an amount over \$500,000.

5. False. $(f \circ g)(x) = 6x + 1$ and $(g \circ f)(x) = 6x + 6$

7. (a) $O(M(Y)) = 2\left(6 + \frac{1}{2}Y\right) = 12 + Y$
(b) Middle child is 8 years old; youngest child is 4 years old.

9. Proof

1. (a) *Sample answer:* $f(x) = x + 1,\ g(x) = x + 3$
(b) *Sample answer:* $f(x) = x^2,\ g(x) = x^3$

73. (a) Proof
(b) $\frac{1}{2}[f(x) + f(-x)] + \frac{1}{2}[f(x) - f(-x)]$
$\qquad = \frac{1}{2}[f(x) + f(-x) + f(x) - f(-x)]$
$\qquad = \frac{1}{2}[2f(x)]$
$\qquad = f(x)$

(c) $f(x) = (x^2 + 1) + (-2x)$
$k(x) = \dfrac{-1}{(x + 1)(x - 1)} + \dfrac{x}{(x + 1)(x - 1)}$

Section 1.9 *(page 90)*

1. inverse   **3.** range; domain   **5.** one-to-one

7. $f^{-1}(x) = \frac{1}{6}x$ **9.** $f^{-1}(x) = \dfrac{x - 1}{3}$

11. $f^{-1}(x) = \sqrt{x + 4}$ **13.** $f^{-1}(x) = \sqrt[3]{x - 1}$

15. $f(g(x)) = f(4x + 9) = \dfrac{(4x + 9) - 9}{4} = x$

$g(f(x)) = g\left(\dfrac{x - 9}{4}\right) = 4\left(\dfrac{x - 9}{4}\right) + 9 = x$

17. $f(g(x)) = f(\sqrt[3]{4x}) = \dfrac{(\sqrt[3]{4x})^3}{4} = x$

$g(f(x)) = g\left(\dfrac{x^3}{4}\right) = \sqrt[3]{4\left(\dfrac{x^3}{4}\right)} = x$

19.

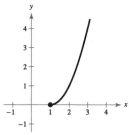

21. (a) $f(g(x)) = f(x + 5) = (x + 5) - 5 = x$
$g(f(x)) = g(x - 5) = (x - 5) + 5 = x$
(b)

23. (a) $f(g(x)) = f\left(\dfrac{x - 1}{7}\right) = 7\left(\dfrac{x - 1}{7}\right) + 1 = x$

$g(f(x)) = g(7x + 1) = \dfrac{(7x + 1) - 1}{7} = x$

(b)

25. (a) $f(g(x)) = f(\sqrt[3]{x}) = (\sqrt[3]{x})^3 = x$
$g(f(x)) = g(x^3) = \sqrt[3]{(x^3)} = x$

(b)

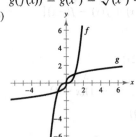

27. (a) $f(g(x)) = f(x^2 - 5) = \sqrt{(x^2 - 5) + 5} = x,\ x \geq 0$
$g(f(x)) = g(\sqrt{x + 5}) = (\sqrt{x + 5})^2 - 5 = x$

(b)

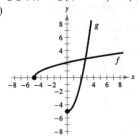

29. (a) $f(g(x)) = f\left(\dfrac{1}{x}\right) = \dfrac{1}{(1/x)} = x$

$g(f(x)) = g\left(\dfrac{1}{x}\right) = \dfrac{1}{(1/x)} = x$

(b)

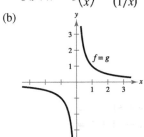

31. (a) $f(g(x)) = f\left(-\dfrac{5x + 1}{x - 1}\right) = \dfrac{-\left(\dfrac{5x + 1}{x - 1}\right) - 1}{-\left(\dfrac{5x + 1}{x - 1}\right) + 5}$

$= \dfrac{-5x - 1 - x + 1}{-5x - 1 + 5x - 5} = x$

$g(f(x)) = g\left(\dfrac{x - 1}{x + 5}\right) = \dfrac{-5\left(\dfrac{x - 1}{x + 5}\right) - 1}{\dfrac{x - 1}{x + 5} - 1}$

$= \dfrac{-5x + 5 - x - 5}{x - 1 - x - 5} = x$

(b)

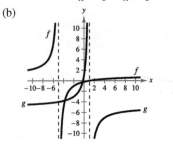

33. No

35.

x	3	5	7	9	11	13
$f^{-1}(x)$	-1	0	1	2	3	4

37. Yes   **39.** No

41.

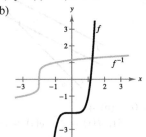

The function does not have an inverse function.

43.

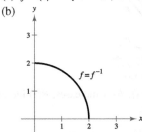

The function does not have an inverse function.

45. (a) $f^{-1}(x) = \sqrt[5]{x + 2}$

(b)

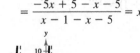

(c) The graph of f^{-1} is the reflection of the graph of f in the line $y = x$.

(d) The domains and ranges of f and f^{-1} are all real numbers x.

47. (a) $f^{-1}(x) = \sqrt{4 - x^2},\ 0 \leq x \leq 2$

(b)

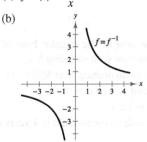

(c) The graph of f^{-1} is the same as the graph of f.

(d) The domains and ranges of f and f^{-1} are all real numbers x such that $0 \leq x \leq 2$.

49. (a) $f^{-1}(x) = \dfrac{4}{x}$

(b)

(c) The graph of f^{-1} is the same as the graph of f.

(d) The domains and ranges of f and f^{-1} are all real numbers x except $x = 0$.

1. (a) $f^{-1}(x) = \dfrac{2x + 1}{x - 1}$

(b)

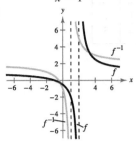

(c) The graph of f^{-1} is the reflection of the graph of f in the line $y = x$.

(d) The domain of f and the range of f^{-1} are all real numbers x except $x = 2$. The domain of f^{-1} and the range of f are all real numbers x except $x = 1$.

3. (a) $f^{-1}(x) = x^3 + 1$

(b)

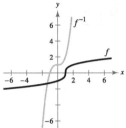

(c) The graph of f^{-1} is the reflection of the graph of f in the line $y = x$.

(d) The domains and ranges of f and f^{-1} are all real numbers x.

5. No inverse function   **57.** $g^{-1}(x) = 6x - 1$

9. No inverse function   **61.** $f^{-1}(x) = \sqrt{x} - 3$

3. No inverse function   **65.** No inverse function

7. $f^{-1}(x) = \dfrac{x^2 - 3}{2}, \ x \geq 0$ **69.** $f^{-1}(x) = \dfrac{5x - 4}{6 - 4x}$

1. $f^{-1}(x) = x - 2$

The domain of f and the range of f^{-1} are all real numbers x such that $x \geq -2$. The domain of f^{-1} and the range of f are all real numbers x such that $x \geq 0$.

3. $f^{-1}(x) = \sqrt{x} - 6$

The domain of f and the range of f^{-1} are all real numbers x such that $x \geq -6$. The domain of f^{-1} and the range of f are all real numbers x such that $x \geq 0$.

5. $f^{-1}(x) = \dfrac{\sqrt{-2(x - 5)}}{2}$

The domain of f and the range of f^{-1} are all real numbers x such that $x \geq 0$. The domain of f^{-1} and the range of f are all real numbers x such that $x \leq 5$.

7. $f^{-1}(x) = x + 3$

The domain of f and the range of f^{-1} are all real numbers x such that $x \geq 4$. The domain of f^{-1} and the range of f are all real numbers x such that $x \geq 1$.

9. 32   **81.** 472   **83.** $2\sqrt[3]{x + 3}$

5. $\dfrac{x + 1}{2}$ **87.** $\dfrac{x + 1}{2}$

9. (a) $y = \dfrac{x - 10}{0.75}$

$x = $ hourly wage; $y = $ number of units produced

(b) 19 units

91. False. $f(x) = x^2$ has no inverse function.

93.

x	1	3	4	6
y	1	2	6	7

x	1	2	6	7
$f^{-1}(x)$	1	3	4	6

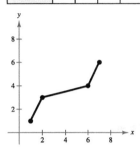

95. Proof   **97.** $k = \dfrac{1}{4}$

99.

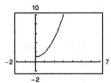

There is an inverse function $f^{-1}(x) = \sqrt{x - 1}$ because the domain of f is equal to the range of f^{-1} and the range of f is equal to the domain of f^{-1}.

101. This situation could be represented by a one-to-one function if the runner does not stop to rest. The inverse function would represent the time in hours for a given number of miles completed.

Section 1.10 *(page 100)*

1. variation; regression   **3.** least squares regression

5. directly proportional   **7.** combined

9.

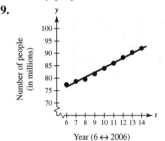

The model is a good fit for the data.

11.

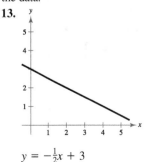

$y = \frac{1}{4}x + 3$

13.

$y = -\frac{1}{2}x + 3$

15.

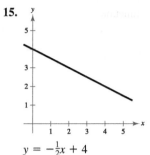

$y = -\frac{1}{2}x + 4$

17. (a) and (b)

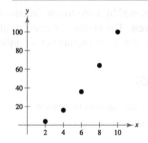

Year (84 ↔ 1984)

Sample answer: $y = -0.1t + 64$

(c) $y = -0.097t + 63.72$ (d) The models are similar.

19. $y = 7x$ **21.** $y = \frac{1}{5}x$ **23.** $y = 2\pi x$

25.

x	2	4	6	8	10
$y = x^2$	4	16	36	64	100

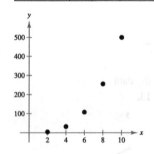

27.

x	2	4	6	8	10
$y = \frac{1}{2}x^3$	4	32	108	256	500

29.

x	2	4	6	8	10
$y = \dfrac{2}{x}$	1	$\frac{1}{2}$	$\frac{1}{3}$	$\frac{1}{4}$	$\frac{1}{5}$

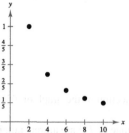

31.

x	2	4	6	8	10
$y = \dfrac{10}{x^2}$	$\frac{5}{2}$	$\frac{5}{8}$	$\frac{5}{18}$	$\frac{5}{32}$	$\frac{1}{10}$

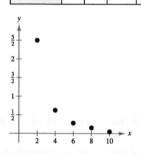

33. Inversely; Answers will vary.   **35.** $y = \dfrac{5}{x}$

37. $y = -\dfrac{7}{10}x$ **39.** $A = kr^2$ **41.** $y = \dfrac{k}{x^2}$

43. $F = \dfrac{kg}{r^2}$ **45.** $R = k(T - T_e)$ **47.** $P = kVI$

49. y is directly proportional to the square of x.

51. A is jointly proportional to b and h.

53. $y = 18x$ **55.** $y = \dfrac{75}{x}$ **57.** $z = 2xy$ **59.** $P = \dfrac{18x}{y^2}$

61. $I = 0.035P$ **63.** Model: $y = \frac{33}{13}x$; 25.4 cm, 50.8 cm

65. $293\frac{1}{3}$N   **67.** About 39.47 lb   **69.** About 0.61 mi/h

71. (a)

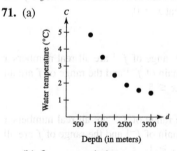

Depth (in meters)

(b) Inverse variation

(c) About 4919.9

(d) 1640 m

73. (a) 200 Hz   (b) 50 Hz   (c) 100 Hz

75. True. If $y = k_1 x$ and $x = k_2 z$, then $y = k_1(k_2 z) = (k_1 k_2)z$.

77. π is a constant, not a variable.   **79.** Direct; $y = 2t$

eview Exercises *(page 106)*

1.

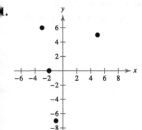

3. Quadrant IV

5.

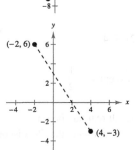

Distance: $3\sqrt{13}$
Midpoint: $\left(1, \frac{3}{2}\right)$

7.

x	-2	-1	0	1	2
y	-11	-8	-5	-2	1

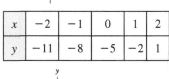

9.

x	-1	0	1	2	3	4
y	4	0	-2	-2	0	4

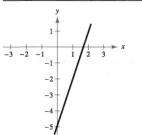

11. x-intercept: $\left(-\frac{7}{2}, 0\right)$
y-intercept: $(0, 7)$

15. x-intercept: $\left(\frac{1}{4}, 0\right)$
y-intercept: $(0, 1)$
No symmetry

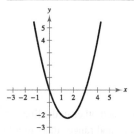

13. x-intercepts: $(1, 0), (5, 0)$
y-intercept: $(0, 5)$

17. x-intercepts: $\left(\pm\sqrt{6}, 0\right)$
y-intercept: $(0, 6)$
y-axis symmetry

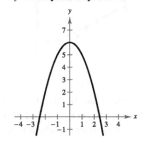

19. x-intercept: $\left(\sqrt[3]{-5}, 0\right)$
y-intercept: $(0, 5)$
No symmetry

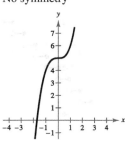

21. x-intercept: $(-5, 0)$
y-intercept: $\left(0, \sqrt{5}\right)$
No symmetry

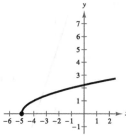

23. Center: $(0, 0)$
Radius: 3

25. Center: $(-2, 0)$
Radius: 4

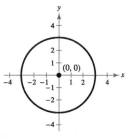

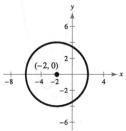

27. $(x - 2)^2 + (y + 3)^2 = 13$

29. $m = -\frac{1}{2}$
y-intercept: $(0, 1)$

31. $m = 0$
y-intercept: $(0, 1)$

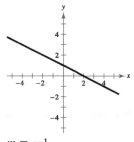

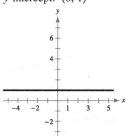

33. $m = -1$

35. $y = \frac{1}{3}x - 7$

37. $y = \frac{1}{2}x + 7$

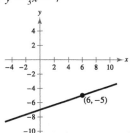

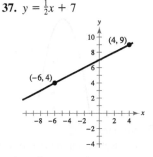

39. (a) $y = \frac{5}{4}x - \frac{23}{4}$ (b) $y = -\frac{4}{5}x + \frac{2}{5}$ **41.** $S = 0.80L$

43. Not a function   **45.** Function

47. (a) 5 (b) 17 (c) $t^4 + 1$ (d) $t^2 + 2t + 2$

49. All real numbers x such that $-5 \le x \le 5$

51. 16 ft/sec   **53.** $4x + 2h + 3, \ h \ne 0$

55. Function   **57.** $\frac{7}{3}, 3$

59.

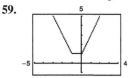

Increasing on $(0, \infty)$
Decreasing on $(-\infty, -1)$
Constant on $(-1, 0)$

61. Relative maximum: $(1, 2)$ **63.** 4

65. Even; y-axis symmetry

67. (a) $f(x) = -3x$ **69.**

(b)

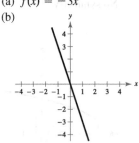

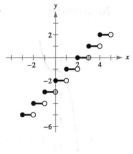

71. (a) $f(x) = x^2$ (b) Downward shift of nine units

(c)   (d) $h(x) = f(x) - 9$

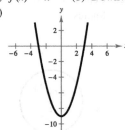

73. (a) $f(x) = \sqrt{x}$

(b) Reflection in the x-axis and an upward shift of four units

(c)   (d) $h(x) = -f(x) + 4$

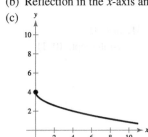

75. (a) $f(x) = x^2$

(b) Reflection in the x-axis, a left shift of two units, and an upward shift of three units

(c)   (d) $h(x) = -f(x + 2) + 3$

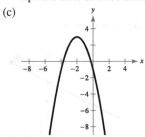

77. (a) $f(x) = [\![x]\!]$

(b) Reflection in the x-axis and an upward shift of six units

(c)   (d) $h(x) = -f(x) + 6$

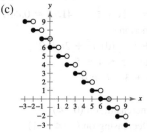

79. (a) $f(x) = [\![x]\!]$

(b) Right shift of nine units and a vertical stretch

(c)   (d) $h(x) = 5f(x - 9)$

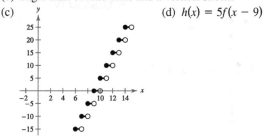

81. (a) $x^2 + 2x + 2$ (b) $x^2 - 2x + 4$

(c) $2x^3 - x^2 + 6x - 3$

(d) $\dfrac{x^2 + 3}{2x - 1}$; all real numbers x except $x = \dfrac{1}{2}$

83. (a) $x - \frac{8}{3}$ (b) $x - 8$

Domains of f, g, $f \circ g$, and $g \circ f$: all real numbers x

85. $f(g(x)) = 0.95x - 100$; $(f \circ g)(x)$ represents the 5% discount before the $100 rebate.

87. $f^{-1}(x) = \dfrac{x - 8}{3}$

$$f(f^{-1}(x)) = 3\left(\dfrac{x - 8}{3}\right) + 8 = x$$

$$f^{-1}(f(x)) = \dfrac{3x + 8 - 8}{3} = x$$

89.

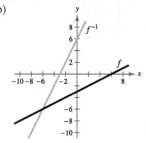

The function does not have an inverse function.

91. (a) $f^{-1}(x) = 2x + 6$

(b)

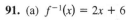

(c) The graphs are reflections of each other in the line $y = x$

(d) Both f and f^{-1} have domains and ranges that are all real number x.

93. $x > 4$; $f^{-1}(x) = \sqrt{\dfrac{x}{2} + 4}$, $x \neq 0$

95. (a) and (b)   **97.** $44.80

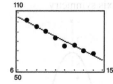

$B = -5.02t + 135.6$

The model fits the data well.

99. True. If $f(x) = x^3$ and $g(x) = \sqrt[3]{x}$, then the domain of g is all real numbers x, which is equal to the range of f, and vice versa

Chapter Test *(page 109)*

1.

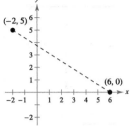

2. $V(h) = 16\pi h$

Midpoint: $\left(2, \frac{5}{2}\right)$; Distance: $\sqrt{89}$

3. x-intercept: $\left(\frac{3}{5}, 0\right)$
y-intercept: $(0, 3)$
No symmetry

4. x-intercepts: $(\pm 4, 0)$
y-intercept: $(0, 4)$
y-axis symmetry

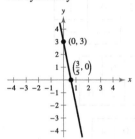

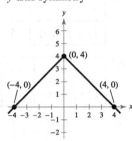

5. x-intercepts: $(\pm 1, 0)$
y-intercept: $(0, -1)$
y-axis symmetry

6. $(x - 1)^2 + (y - 3)^2 = 16$

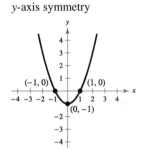

7. $y = -4x - 3$

8. $y = \frac{5}{3}x - \frac{1}{3}$

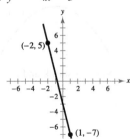

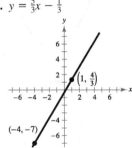

9. (a) $y = -\frac{5}{2}x + 4$ (b) $y = \frac{2}{5}x + 4$

10. (a) $-\dfrac{1}{8}$ (b) $-\dfrac{1}{28}$ (c) $\dfrac{\sqrt{x}}{x^2 - 18x}$ **11.** $x \le 3$

12. (a) -5

(b)

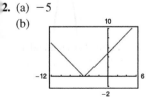

(c) Increasing on $(-5, \infty)$
Decreasing on $(-\infty, -5)$

(d) Neither

13. (a) $0, 3$

(b)

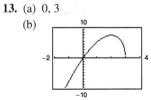

(c) Increasing on $(-\infty, 2)$
Decreasing on $(2, 3)$

(d) Neither

14. (a) $0, \pm 0.4314$

(b)

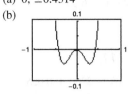

(c) Increasing on $(-0.31, 0)$,
$(0.31, \infty)$
Decreasing on $(-\infty, -0.31)$,
$(0, 0.31)$

(d) Even

15.

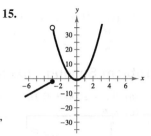

16. (a) $f(x) = [\![x]\!]$ (b) Vertical stretch

(c)

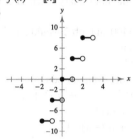

17. (a) $y = \sqrt{x}$

(b) Left shift of five units and an upward shift of eight units

(c)

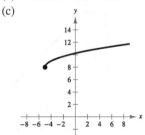

18. (a) $y = x^3$

(b) Reflection in the x-axis, a vertical stretch, a right shift of five units, and an upward shift of three units

19. (a) $2x^2 - 4x - 2$ (b) $4x^2 + 4x - 12$

(c) $-3x^4 - 12x^3 + 22x^2 + 28x - 35$

(d) $\dfrac{3x^2 - 7}{-x^2 - 4x + 5}$

(e) $3x^4 + 24x^3 + 18x^2 - 120x + 68$

(f) $-9x^4 + 30x^2 - 16$

20. (a) $\dfrac{1 + 2x^{3/2}}{x}$ (b) $\dfrac{1 - 2x^{3/2}}{x}$

(c) $\dfrac{2\sqrt{x}}{x}$ (d) $\dfrac{1}{2x^{3/2}}$

(e) $\dfrac{\sqrt{x}}{2x}$ (f) $\dfrac{2\sqrt{x}}{x}$

21. $f^{-1}(x) = \sqrt[3]{x - 8}$ **22.** No inverse

23. $f^{-1}(x) = \left(\frac{1}{3}x\right)^{2/3}, \quad x \geq 0$ **24.** $v = 6\sqrt{s}$

25. $A = \dfrac{25}{6}xy$ **26.** $b = \dfrac{48}{a}$

Problem Solving *(page 111)*

1. (a) $W_1 = 2000 + 0.07S$ (b) $W_2 = 2300 + 0.05S$

(c)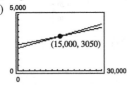

    Both jobs pay the same monthly salary when sales equal
    $15,000.

(d) No. Job 1 would pay $3400 and job 2 would pay $3300.

3. (a) The function will be even.

(b) The function will be odd.

(c) The function will be neither even nor odd.

5. $f(x) = a_{2n}x^{2n} + a_{2n-2}x^{2n-2} + \cdots + a_2x^2 + a_0$

   $f(-x) = a_{2n}(-x)^{2n} + a_{2n-2}(-x)^{2n-2}$

$$+ \cdots + a_2(-x)^2 + a_0$$

$$= f(x)$$

7. (a) $81\frac{2}{3}$ h

(b) $25\frac{5}{7}$ mi/h

(c) $y = \dfrac{-180}{7}x + 3400$

    Domain: $0 \leq x \leq \dfrac{1190}{9}$

    Range: $0 \leq y \leq 3400$

(d)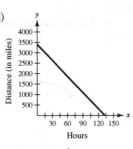

9. (a) $(f \circ g)(x) = 4x + 24$ (b) $(f \circ g)^{-1}(x) = \frac{1}{4}x - 6$

(c) $f^{-1}(x) = \frac{1}{4}x; \; g^{-1}(x) = x - 6$

(d) $(g^{-1} \circ f^{-1})(x) = \frac{1}{4}x - 6$; They are the same.

(e) $(f \circ g)(x) = 8x^3 + 1; \; (f \circ g)^{-1}(x) = \frac{1}{2}\sqrt[3]{x - 1};$

    $f^{-1}(x) = \sqrt[3]{x - 1}; \; g^{-1}(x) = \frac{1}{2}x;$

    $(g^{-1} \circ f^{-1})(x) = \frac{1}{2}\sqrt[3]{x - 1}$

(f) Answers will vary.

(g) $(f \circ g)^{-1}(x) = (g^{-1} \circ f^{-1})(x)$

11. (a)   (b)

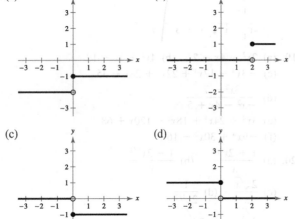

(c)   (d)

(e)   (f)

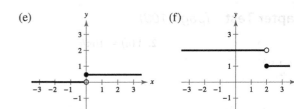

13. Proof

15. (a)

x	-4	-2	0	4
$f(f^{-1}(x))$	-4	-2	0	4

(b)

x	-3	-2	0	1
$(f + f^{-1})(x)$	5	1	-3	-5

(c)

x	-3	-2	0	1
$(f \cdot f^{-1})(x)$	4	0	2	6

(d)

x	-4	-3	0	4
$\lvert f^{-1}(x) \rvert$	2	1	1	3

Chapter 2

Section 2.1 *(page 120)*

1. polynomial   **3.** quadratic; parabola

5. b   **6.** a   **7.** c   **8.** d

9. (a)   (b)

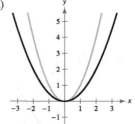 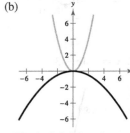

    Vertical shrink   Vertical shrink and a
                              reflection in the x-axis

(c)   (d)

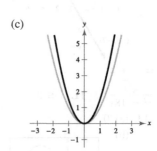 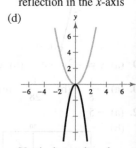

    Vertical stretch   Vertical stretch and a
                              reflection in the x-axis

1. (a)

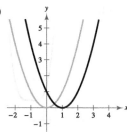

Right shift of one unit

(b)

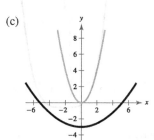

Horizontal shrink and an
upward shift of one unit

(c)

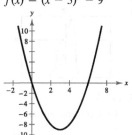

Horizontal stretch and a
downward shift of three
units

(d)

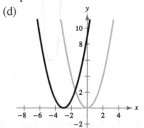

Left shift of three units

3. $f(x) = (x - 3)^2 - 9$

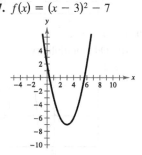

Vertex: $(3, -9)$
Axis of symmetry: $x = 3$
x-intercepts: $(0, 0), (6, 0)$

7. $f(x) = (x - 3)^2 - 7$

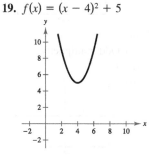

Vertex: $(3, -7)$
Axis of symmetry: $x = 3$
x-intercepts: $\left(3 \pm \sqrt{7}, 0\right)$

15. $h(x) = (x - 4)^2$

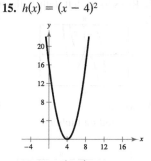

Vertex: $(4, 0)$
Axis of symmetry: $x = 4$
x-intercept: $(4, 0)$

19. $f(x) = (x - 4)^2 + 5$

Vertex: $(4, 5)$
Axis of symmetry: $x = 4$
No x-intercept

21. $f(x) = \left(x - \frac{1}{2}\right)^2 + 1$

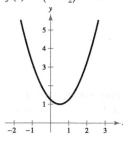

Vertex: $\left(\frac{1}{2}, 1\right)$
Axis of symmetry: $x = \frac{1}{2}$
No x-intercept

25. $h(x) = 4\left(x - \frac{1}{2}\right)^2 + 20$

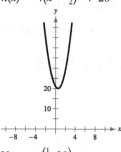

Vertex: $\left(\frac{1}{2}, 20\right)$
Axis of symmetry: $x = \frac{1}{2}$
No x-intercept

23. $f(x) = -(x - 1)^2 + 6$

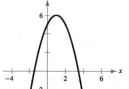

Vertex: $(1, 6)$
Axis of symmetry: $x = 1$
x-intercepts: $\left(1 \pm \sqrt{6}, 0\right)$

27.

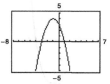

Vertex: $(-1, 4)$
Axis of symmetry: $x = -1$
x-intercepts: $(1, 0), (-3, 0)$

29.

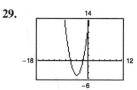

Vertex: $(-4, -5)$
Axis of symmetry: $x = -4$
x-intercepts: $\left(-4 \pm \sqrt{5}, 0\right)$

31.

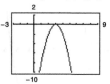

Vertex: $(3, 0)$
Axis of symmetry: $x = 3$
x-intercept: $(3, 0)$

33.

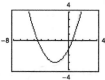

Vertex: $(-2, -3)$
Axis of symmetry: $x = -2$
x-intercepts: $\left(-2 \pm \sqrt{6}, 0\right)$

35. $f(x) = (x + 2)^2 - 1$ **37.** $f(x) = (x + 2)^2 + 5$

39. $f(x) = 4(x - 1)^2 - 2$ **41.** $f(x) = \frac{3}{4}(x - 5)^2 + 12$

43. $f(x) = -\frac{24}{49}\left(x + \frac{1}{4}\right)^2 + \frac{3}{2}$ **45.** $f(x) = -\frac{16}{3}\left(x + \frac{5}{2}\right)^2$

47. $(-1, 0), (3, 0)$ **49.** $(-3, 0), \left(\frac{1}{2}, 0\right)$

51.

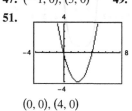

$(0, 0), (4, 0)$

53.

$(3, 0), (6, 0)$

55.

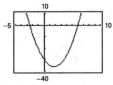

$\left(-\frac{5}{2}, 0\right), (6, 0)$

57. $f(x) = x^2 - 9$
$\quad g(x) = -x^2 + 9$

59. $f(x) = x^2 - 3x - 4$
$\quad g(x) = -x^2 + 3x + 4$

61. $f(x) = 2x^2 + 7x + 3$
$\quad g(x) = -2x^2 - 7x - 3$

63. 55, 55   **65.** 12, 6   **67.** 16 ft   **69.** 20 fixtures

71. (a) $14,000,000; $14,375,000; $13,500,000
$\quad$ (b) $24; $14,400,000; Answers will vary.

73. (a) $A = \dfrac{8x(50 - x)}{3}$ (b) $x = 25$ ft, $y = 33\frac{1}{3}$ ft

75. True. The equation has no real solutions, so the graph has no x-intercepts.

77. $b = \pm 20$ **79.** $f(x) = a\left(x + \dfrac{b}{2a}\right)^2 + \dfrac{4ac - b^2}{4a}$

81. Proof

Section 2.2   *(page 132)*

1. continuous   **3.** $n; n - 1$ **5.** touches; crosses

7. standard   **9.** c   **10.** f   **11.** a   **12.** e

13. d   **14.** b

15. (a)

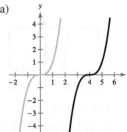

(b)

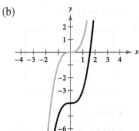

(c)

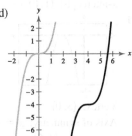

(d)

17. (a)

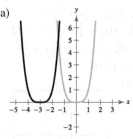

(b)

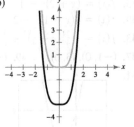

(c)

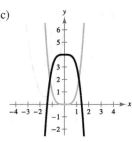

(d)

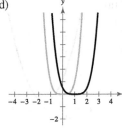

(e)

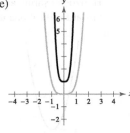

(f)

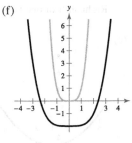

19. Falls to the left, rises to the right

21. Falls to the left and to the right

23. Rises to the left, falls to the right

25. Rises to the left and to the right

27. Rises to the left, falls to the right

29.

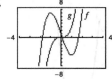

31.

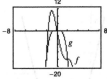

33. (a) ± 6
$\quad$ (b) Odd multiplicity
$\quad$ (c) 1
$\quad$ (d)

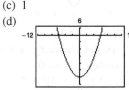

35. (a) 3
$\quad$ (b) Even multiplicity
$\quad$ (c) 1
$\quad$ (d)

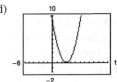

37. (a) $-2, 1$
$\quad$ (b) Odd multiplicity
$\quad$ (c) 1
$\quad$ (d)

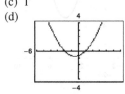

39. (a) $0, 1 \pm \sqrt{2}$
$\quad$ (b) Odd multiplicity
$\quad$ (c) 2
$\quad$ (d)

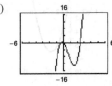

41. (a) $0, 2 \pm \sqrt{3}$
$\quad$ (b) Odd multiplicity
$\quad$ (c) 2
$\quad$ (d)

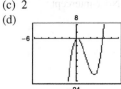

43. (a) $0, \pm\sqrt{3}$
$\quad$ (b) 0, odd multiplicity;
$\quad\quad$ $\pm\sqrt{3}$, even multiplicity
$\quad$ (c) 4
$\quad$ (d)

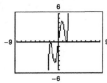

45. (a) No real zero
(b) No multiplicity
(c) 1
(d)

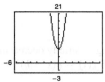

47. (a) $\pm 2, -3$
(b) Odd multiplicity
(c) 2
(d)

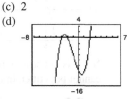

49. (a)

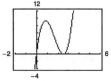

(b) and (c) $0, \frac{5}{2}$
(d) The answers are the same.

51. (a)

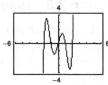

(b) and (c) $0, \pm 1, \pm 2$
(d) The answers are the same.

53. $f(x) = x^2 - 7x$ **55.** $f(x) = x^3 + 6x^2 + 8x$

57. $f(x) = x^4 - 4x^3 - 9x^2 + 36x$ **59.** $f(x) = x^2 - 2x - 1$

61. $f(x) = x^3 - 6x^2 + 7x + 2$ **63.** $f(x) = x^2 + 6x + 9$

65. $f(x) = x^3 + 4x^2 - 5x$

67. $f(x) = x^4 + x^3 - 15x^2 + 23x - 10$ **69.** $f(x) = x^5 - 3x^3$

71. (a) Rises to the left and to the right
(b) No zeros   (c) Answers will vary.
(d)

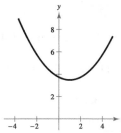

73. (a) Falls to the left, rises to the right
(b) $0, 5, -5$ (c) Answers will vary.
(d)

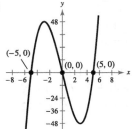

75. (a) Rises to the left and to the right
(b) $-2, 2$ (c) Answers will vary.
(d)

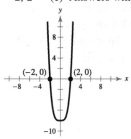

77. (a) Falls to the left, rises to the right
(b) 0, 2, 3   (c) Answers will vary.
(d)

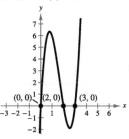

79. (a) Rises to the left, falls to the right
(b) $-5, 0$ (c) Answers will vary.
(d)

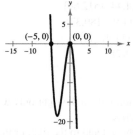

81. (a) Falls to the left, rises to the right
(b) $-2, 0$ (c) Answers will vary.
(d)

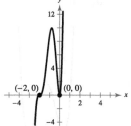

83. (a) Falls to the left and to the right
(b) ± 2 (c) Answers will vary.
(d)

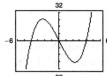

85.

Zeros: $0, \pm 4$,
odd multiplicity

87.

Zeros: -1,
even multiplicity;
$3, \frac{9}{2}$, odd multiplicity

89. (a) $[-1, 0], [1, 2], [2, 3]$ (b) $-0.879, 1.347, 2.532$

91. (a) $[-2, -1], [0, 1]$ (b) $-1.585, 0.779$

93. (a) $V(x) = x(36 - 2x)^2$ (b) Domain: $0 < x < 18$

(c)

6 in. × 24 in. × 24 in.

(d)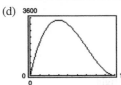

$x = 6$; The results are the same.

95. (a) Relative maximum: $(4.44, 1512.60)$
Relative minimum: $(11.97, 189.37)$
(b) Increasing: $(3, 4.44), (11.97, 16)$
Decreasing: $(4.44, 11.97)$
(c) Answers will vary.

97. $x \approx 200$

99. True. A polynomial function falls to the right only when the leading coefficient is negative.

101. False. The graph falls to the left and to the right or the graph rises to the left and to the right.

103. Answers will vary. *Sample answers:*

$a_4 < 0$ $a_4 > 0$

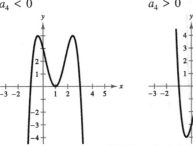 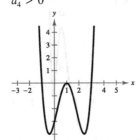

105.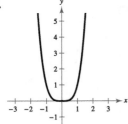

(a) Upward shift of two units; Even
(b) Left shift of two units; Neither
(c) Reflection in the y-axis; Even
(d) Reflection in the x-axis; Even
(e) Horizontal stretch; Even   (f) Vertical shrink; Even
(g) $g(x) = x^3, x \geq 0$; Neither   (h) $g(x) = x^{16}$; Even

107.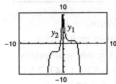

(a) y_1 is decreasing, y_2 is increasing.
(b) Yes; a; If $a > 0$, then the graph is increasing, and if $a < 0$, then the graph is decreasing.

(c)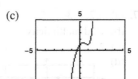

No; f is not strictly increasing or strictly decreasing, so f cannot be written in the form $f(x) = a(x - h)^5 + k$.

Section 2.3   *(page 142)*

1. $f(x)$: dividend; $d(x)$: divisor;
$q(x)$: quotient; $r(x)$: remainder

3. improper   **5.** Factor   **7.** Answers will vary.

9. (a) and (b)

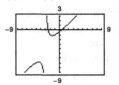

(c) Answers will vary.

11. $2x + 4$, $x \neq -3$ **13.** $x^2 - 3x + 1$, $x \neq -\frac{5}{4}$

15. $x^3 + 3x^2 - 1$, $x \neq -2$ **17.** $6 - \dfrac{1}{x + 1}$

19. $x - \dfrac{x + 9}{x^2 + 1}$ **21.** $2x - 8 + \dfrac{x - 1}{x^2 + 1}$

23. $x + 3 + \dfrac{6x^2 - 8x + 3}{(x - 1)^3}$ **25.** $2x^2 - 2x + 6$, $x \neq 4$

27. $6x^2 + 25x + 74 + \dfrac{248}{x - 3}$ **29.** $4x^2 - 9$, $x \neq -2$

31. $-x^2 + 10x - 25$, $x \neq -10$ **33.** $x^2 + x + 4 + \dfrac{21}{x - 4}$

35. $10x^3 + 10x^2 + 60x + 360 + \dfrac{1360}{x - 6}$

37. $x^2 - 8x + 64$, $x \neq -8$

39. $-3x^3 - 6x^2 - 12x - 24 - \dfrac{48}{x - 2}$

41. $-x^3 - 6x^2 - 36x - 36 - \dfrac{216}{x - 6}$

43. $4x^2 + 14x - 30$, $x \neq -\frac{1}{2}$
45. $f(x) = (x - 3)(x^2 + 2x - 4) - 5$, $f(3) = -5$
47. $f(x) = \left(x + \frac{2}{3}\right)(15x^3 - 6x + 4) + \frac{34}{3}$, $f\left(-\frac{2}{3}\right) = \frac{34}{3}$
49. $f(x) = \left(x - 1 + \sqrt{3}\right)\left[-4x^2 + \left(2 + 4\sqrt{3}\right)x + \left(2 + 2\sqrt{3}\right)\right]$,
$f\left(1 - \sqrt{3}\right) = 0$

51. (a) -2 (b) 1 (c) 36 (d) 5
53. (a) -35 (b) $-\frac{5}{8}$ (c) -10 (d) -211
55. $(x + 3)(x + 2)(x + 1)$; Solutions: $-3, -2, -1$
57. $(2x - 1)(x - 5)(x - 2)$; Solutions: $\frac{1}{2}, 5, 2$
59. $\left(x + \sqrt{3}\right)\left(x - \sqrt{3}\right)(x + 2)$; Solutions: $-\sqrt{3}, \sqrt{3}, -2$
61. $(x - 1)\left(x - 1 - \sqrt{3}\right)\left(x - 1 + \sqrt{3}\right)$;
Solutions: $1, 1 + \sqrt{3}, 1 - \sqrt{3}$

3. (a) Answers will vary.   (b) $2x - 1$
(c) $f(x) = (2x - 1)(x + 2)(x - 1)$ (d) $\frac{1}{2}, -2, 1$
(e)

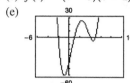

5. (a) Answers will vary.   (b) $(x - 4)(x - 1)$
(c) $f(x) = (x - 5)(x + 2)(x - 4)(x - 1)$ (d) $5, -2, 4, 1$
(e)

7. (a) Answers will vary.   (b) $x + 7$
(c) $f(x) = (x + 7)(2x + 1)(3x - 2)$ (d) $-7, -\frac{1}{2}, \frac{2}{3}$
(e)

9. (a) Answers will vary.   (b) $x - \sqrt{5}$
(c) $f(x) = (x - \sqrt{5})(x + \sqrt{5})(2x - 1)$ (d) $\pm\sqrt{5}, \frac{1}{2}$
(e)

11. (a) $2, \pm 2.236$ (b) 2
(c) $f(x) = (x - 2)(x - \sqrt{5})(x + \sqrt{5})$
73. (a) $-2, 0.268, 3.732$ (b) -2
(c) $h(t) = (t + 2)[t - (2 + \sqrt{3})][t - (2 - \sqrt{3})]$
75. (a) $0, 3, 4, \pm 1.414$ (b) 0
(c) $h(x) = x(x - 4)(x - 3)(x + \sqrt{2})(x - \sqrt{2})$
77. $x^2 - 7x - 8, \ x \neq -8$ **79.** $x^2 + 3x, \ x \neq -2, -1$
81. (a) 3,200,000

(b) \$250,366
(c) Answers will vary.
83. False. $-\frac{4}{7}$ is a zero of f.
85. True. The degree of the numerator is greater than the degree of the denominator.
87. $x^{2n} + 6x^n + 9, \ x^n \neq -3$ **89.** $k = -1$, not 1.
91. $c = -210$ **93.** $k = 7$

Section 2.4   *(page 150)*

1. real   **3.** pure imaginary   **5.** principal square
7. $a = 9, b = 8$ **9.** $a = 8, b = 4$ **11.** $2 + 5i$
13. $1 - 2\sqrt{3}i$ **15.** $2\sqrt{10}i$ **17.** 23 **19.** $-1 - 6i$
21. $0.2i$ **23.** $7 + 4i$ **25.** 1 **27.** $3 - 3\sqrt{2}i$
29. $-14 + 20i$ **31.** $5 + i$ **33.** $108 + 12i$ **35.** 11

37. $-13 + 84i$ **39.** $9 - 2i, 85$ **41.** $-1 + \sqrt{5}i, 6$
43. $-2\sqrt{5}i, 20$ **45.** $\sqrt{6}, 6$ **47.** $\frac{8}{41} + \frac{10}{41}i$
49. $\frac{12}{13} + \frac{5}{13}i$ **51.** $-4 - 9i$ **53.** $-\frac{120}{1681} - \frac{27}{1681}i$
55. $-\frac{1}{2} - \frac{5}{2}i$ **57.** $\frac{62}{949} + \frac{297}{949}i$ **59.** $-2\sqrt{3}$ **61.** -15
63. $7\sqrt{2}i$ **65.** $(21 + 5\sqrt{2}) + (7\sqrt{5} - 3\sqrt{10})i$
67. $1 \pm i$ **69.** $-2 \pm \frac{1}{2}i$ **71.** $-2 \pm \frac{\sqrt{5}}{2}i$
73. $2 \pm \sqrt{2}i$ **75.** $\frac{5}{7} \pm \frac{5\sqrt{13}}{7}i$ **77.** $-1 + 6i$
79. $-14i$ **81.** $-432\sqrt{2}i$ **83.** i **85.** 81
87. (a) $z_1 = 9 + 16i, z_2 = 20 - 10i$

(b) $z = \dfrac{11{,}240}{877} + \dfrac{4630}{877}i$

89. False. *Sample answer:* $(1 + i) + (3 + i) = 4 + 2i$
91. True. $x^4 - x^2 + 14 = 56$
$$(-i\sqrt{6})^4 - (-i\sqrt{6})^2 + 14 \overset{?}{=} 56$$
$$36 + 6 + 14 \overset{?}{=} 56$$
$$56 = 56$$
93. $i, -1, -i, 1, i, -1, -i, 1$; The pattern repeats the first four results. Divide the exponent by 4.
When the remainder is 1, the result is i.
When the remainder is 2, the result is -1.
When the remainder is 3, the result is $-i$.
When the remainder is 0, the result is 1.
95. $\sqrt{-6}\sqrt{-6} = \sqrt{6}i\sqrt{6}i = 6i^2 = -6$ **97.** Proof

Section 2.5   *(page 162)*

1. Fundamental Theorem of Algebra   **3.** Rational Zero
5. linear; quadratic; quadratic   **7.** Descartes's Rule of Signs
9. 3   **11.** 5   **13.** 2   **15.** $\pm 1, \pm 2$
17. $\pm 1, \pm 3, \pm 5, \pm 9, \pm 15, \pm 45, \pm\frac{1}{2}, \pm\frac{3}{2}, \pm\frac{5}{2}, \pm\frac{9}{2}, \pm\frac{15}{2}, \pm\frac{45}{2}$
19. $-2, -1, 3$ **21.** No rational zeros   **23.** $-6, -1$
25. $-1, \frac{1}{2}$ **27.** $-2, 3, \pm\frac{2}{3}$ **29.** $1, \frac{3}{5} \pm \frac{\sqrt{19}}{5}$
31. $-3, 1, -2 \pm \sqrt{6}$
33. (a) $\pm 1, \pm 2, \pm 4$
(b)   (c) $-2, -1, 2$

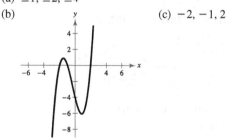

35. (a) $\pm 1, \pm 3, \pm\frac{1}{2}, \pm\frac{3}{2}, \pm\frac{1}{4}, \pm\frac{3}{4}$
(b)   (c) $-\frac{1}{4}, 1, 3$

37. (a) $\pm 1, \pm 2, \pm 4, \pm 8, \pm \frac{1}{2}$

(b)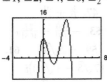

(c) $-\frac{1}{2}, 1, 2, 4$

39. (a) $\pm 1, \pm 3, \pm \frac{1}{2}, \pm \frac{3}{2}, \pm \frac{1}{4}, \pm \frac{3}{4}, \pm \frac{1}{8}, \pm \frac{3}{8}, \pm \frac{1}{16}, \pm \frac{3}{16}, \pm \frac{1}{32}, \pm \frac{3}{32}$

(b)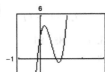

(c) $1, \frac{3}{4}, -\frac{1}{8}$

41. $f(x) = x^3 - x^2 + 25x - 25$

43. $f(x) = x^4 - 6x^3 + 14x^2 - 16x + 8$

45. $f(x) = 3x^4 - 17x^3 + 25x^2 + 23x - 22$

47. $f(x) = 2x^4 + 2x^3 - 2x^2 + 2x - 4$

49. $f(x) = x^3 + x^2 - 2x + 12$

51. (a) $(x^2 + 4)(x^2 - 2)$ (b) $(x^2 + 4)(x + \sqrt{2})(x - \sqrt{2})$
(c) $(x + 2i)(x - 2i)(x + \sqrt{2})(x - \sqrt{2})$

53. (a) $(x^2 - 6)(x^2 - 2x + 3)$
(b) $(x + \sqrt{6})(x - \sqrt{6})(x^2 - 2x + 3)$
(c) $(x + \sqrt{6})(x - \sqrt{6})(x - 1 - \sqrt{2}i)(x - 1 + \sqrt{2}i)$

55. $\pm 2i, 1$ **57.** $2, 3 \pm 2i$ **59.** $1, 3, 1 \pm \sqrt{2}i$

61. $(x + 6i)(x - 6i); \pm 6i$

63. $(x - 1 - 4i)(x - 1 + 4i); 1 \pm 4i$

65. $(x - 2)(x + 2)(x - 2i)(x + 2i); \pm 2, \pm 2i$

67. $(z - 1 + i)(z - 1 - i); 1 \pm i$

69. $(x + 1)(x - 2 + i)(x - 2 - i); -1, 2 \pm i$

71. $(x - 2)^2(x + 2i)(x - 2i); 2, \pm 2i$

73. $-10, -7 \pm 5i$ **75.** $-\frac{3}{4}, 1 \pm \frac{1}{2}i$ **77.** $-2, -\frac{1}{2}, \pm i$

79. One positive real zero, no negative real zeros

81. No positive real zeros, one negative real zero

83. Two or no positive real zeros, two or no negative real zeros

85. Two or no positive real zeros, one negative real zero

87–89. Answers will vary.   **91.** $\frac{3}{4}, \pm \frac{1}{2}$ **93.** $-\frac{3}{4}$

95. $\pm 2, \pm \frac{3}{2}$ **97.** $\pm 1, \frac{1}{4}$

99. d   **100.** a   **101.** b   **102.** c

103. (a)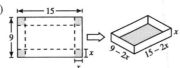

(b) $V(x) = x(9 - 2x)(15 - 2x)$
Domain: $0 < x < \frac{9}{2}$

(c)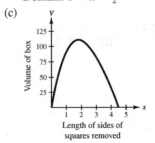

1.82 cm × 5.36 cm × 11.36 cm

(d) $\frac{1}{2}, \frac{7}{2}, 8$; 8 is not in the domain of V.

105. (a) $V(x) = x^3 + 9x^2 + 26x + 24 = 120$
(b) 4 ft × 5 ft × 6 ft

107. False. The most complex zeros it can have is two, and the Linear Factorization Theorem guarantees that there are three linear factors, so one zero must be real.

109. r_1, r_2, r_3 **111.** $5 + r_1, 5 + r_2, 5 + r_3$

113. The zeros cannot be determined.

115. Answers will vary. There are infinitely many possible functions for f. Sample equation and graph:
$f(x) = -2x^3 + 3x^2 + 11x - 6$

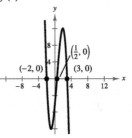

117. $f(x) = x^3 - 3x^2 + 4x - 2$

119. The function should be
$f(x) = (x + 2)(x - 3.5)(x + i)(x - i)$.

121. $f(x) = x^4 + 5x^2 + 4$

123. (a) $x^2 + b$ (b) $x^2 - 2ax + a^2 + b^2$

Section 2.6   *(page 175)*

1. rational functions   **3.** horizontal asymptote

5. Domain: all real numbers x except $x = 1$
$f(x) \to -\infty$ as $x \to 1^-$, $f(x) \to \infty$ as $x \to 1^+$

7. Domain: all real numbers x except $x = \pm 1$
$f(x) \to \infty$ as $x \to -1^-$ and as $x \to 1^+$,
$f(x) \to -\infty$ as $x \to -1^+$ and as $x \to 1^-$

9. Vertical asymptote: $x = 0$
Horizontal asymptote: $y = 0$

11. Vertical asymptote: $x = 5$
Horizontal asymptote: $y = -1$

13. Vertical asymptote: $x = 1$

15. Vertical asymptote: $x = \frac{1}{2}$
Horizontal asymptote: $y = \frac{1}{2}$

17. (a) Domain: all real numbers x except $x = -1$
(b) y-intercept: $(0, 1)$
(c) Vertical asymptote: $x = -1$
Horizontal asymptote: $y = 0$
(d)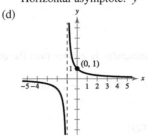

9. (a) Domain: all real numbers x except $x = -4$

(b) y-intercept: $\left(0, -\frac{1}{4}\right)$

(c) Vertical asymptote: $x = -4$

Horizontal asymptote: $y = 0$

(d)

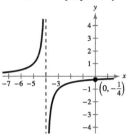

11. (a) Domain: all real numbers x except $x = -2$

(b) x-intercept: $\left(-\frac{3}{2}, 0\right)$

y-intercept: $\left(0, \frac{3}{2}\right)$

(c) Vertical asymptote: $x = -2$

Horizontal asymptote: $y = 2$

(d)

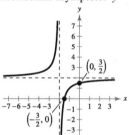

13. (a) Domain: all real numbers x (b) Intercept: $(0, 0)$

(c) Horizontal asymptote: $y = 1$

(d)

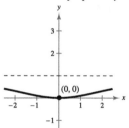

15. (a) Domain: all real numbers s (b) Intercept: $(0, 0)$

(c) Horizontal asymptote: $y = 0$

(d)

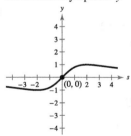

27. (a) Domain: all real numbers x except $x = 4, -1$

(b) Intercept: $(0, 0)$

(c) Vertical asymptotes: $x = -1, x = 4$

Horizontal asymptote: $y = 0$

(d)

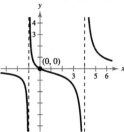

29. (a) Domain: all real numbers x except $x = \pm 4$

(b) y-intercept: $\left(0, \frac{1}{4}\right)$

(c) Vertical asymptote: $x = -4$

Horizontal asymptote: $y = 0$

(d)

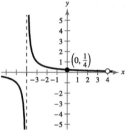

31. (a) Domain: all real numbers t except $t = 1$

(b) t-intercept: $(-1, 0)$

y-intercept: $(0, 1)$

(c) Vertical asymptote: None

Horizontal asymptote: None

(d)

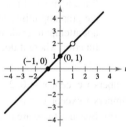

33. (a) Domain: all real numbers x except $x = -1, 5$

(b) x-intercept: $(-5, 0)$

y-intercept: $(0, 5)$

(c) Vertical asymptote: $x = -1$

Horizontal asymptote: $y = 1$

(d)

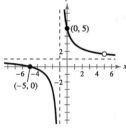

35. (a) Domain: all real numbers x except $x = 2, -3$

(b) Intercept: $(0, 0)$

(c) Vertical asymptote: $x = 2$

Horizontal asymptote: $y = 1$

(d)

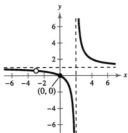

37. (a) Domain: all real numbers x except $x = \pm 1, 2$

(b) x-intercepts: $(3, 0), \left(-\frac{1}{2}, 0\right)$

y-intercept: $\left(0, -\frac{3}{2}\right)$

(c) Vertical asymptotes: $x = 2, x = \pm 1$

Horizontal asymptote: $y = 0$

(d)

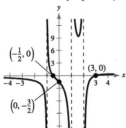

39. d   **40.** a   **41.** c   **42.** b

43. (a) Domain of f: all real numbers x except $x = -1$

Domain of g: all real numbers x

(b)

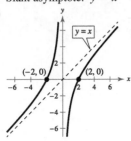

(c) Because there are only a finite number of pixels, the graphing utility may not attempt to evaluate the function where it does not exist.

45. (a) Domain of f: all real numbers x except $x = 0, 2$

Domain of g: all real numbers x except $x = 0$

(b)

(c) Because there are only a finite number of pixels, the graphing utility may not attempt to evaluate the function where it does not exist.

47. (a) Domain: all real numbers x except $x = 0$

(b) x-intercepts: $(\pm 2, 0)$

(c) Vertical asymptote: $x = 0$

Slant asymptote: $y = x$

(d)

49. (a) Domain: all real numbers x except $x = 0$

(b) No intercepts

(c) Vertical asymptote: $x = 0$

Slant asymptote: $y = 2x$

(d)

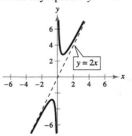

51. (a) Domain: all real numbers x except $x = 0$

(b) No intercepts

(c) Vertical asymptote: $x = 0$

Slant asymptote: $y = x$

(d)

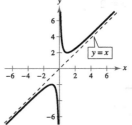

53. (a) Domain: all real numbers t except $t = -5$

(b) y-intercept: $\left(0, -\frac{1}{5}\right)$

(c) Vertical asymptote: $t = -5$

Slant asymptote: $y = -t + 5$

(d)

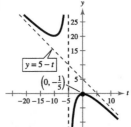

55. (a) Domain: all real numbers x except $x = \pm 2$

(b) Intercept: $(0, 0)$

(c) Vertical asymptotes: $x = \pm 2$

Slant asymptote: $y = x$

(d)

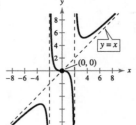

. (a) Domain: all real numbers x except $x = 1$
(b) y-intercept: $(0, -1)$
(c) Vertical asymptote: $x = 1$
Slant asymptote: $y = x$
(d)

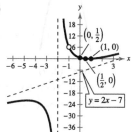

. (a) Domain: all real numbers x except $x = -1, -2$
(b) y-intercept: $\left(0, \frac{1}{2}\right)$
x-intercepts: $\left(\frac{1}{2}, 0\right), (1, 0)$
(c) Vertical asymptote: $x = -2$
Slant asymptote: $y = 2x - 7$
(d)

.

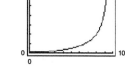

63.

Domain: all real numbers x
except $x = -2$
Vertical asymptote:
$x = -2$
Slant asymptote: $y = x$
$y = x$

Domain: all real numbers x
except $x = 0$
Vertical asymptote: $x = 0$
Slant asymptote:
$y = -x + 3$
$y = -x + 3$

. (a) $(-1, 0)$ (b) -1 **67.** (a) $(1, 0), (-1, 0)$ (b) ± 1

. (a)

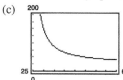

(b) 4411.76; $25,000$; $225,000$
(c) No. The function is undefined at $p = 100$.

. 12.8 in. × 8.5 in.

. (a) Answers will vary.
(b) Vertical asymptote: $x = 25$
Horizontal asymptote: $y = 25$
(c)

(d)

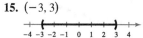

x	30	35	40	45	50	55	60
y	150	87.5	66.7	56.3	50	45.8	42.9

(e) *Sample answer:* No. You might expect the average speed for the round trip to be the average of the average speeds for the two parts of the trip.
(f) No. At 20 miles per hour you would use more time in one direction than is required for the round trip at an average speed of 50 miles per hour.

75. False. Polynomials do not have vertical asymptotes.

77. False. If the degree of the numerator is greater than the degree of the denominator, then no horizontal asymptote exists. However, a slant asymptote exists only if the degree of the numerator is one greater than the degree of the denominator.

79. Yes; No; Every rational function is the ratio of two polynomial functions of the form $f(x) = \dfrac{N(x)}{D(x)}$.

81. $f(x) = \dfrac{x^3}{(x + 2)(x - 1)}$

Section 2.7 *(page 185)*

1. positive; negative **3.** zeros; undefined values

5. (a) No (b) Yes (c) Yes (d) No

7. (a) Yes (b) No (c) No (d) Yes

9. $-3, 6$ **11.** $4, 5$

13. $(-2, 0)$ **15.** $(-3, 3)$

17. $[-7, 3]$ **19.** $(-\infty, -4] \cup [-2, \infty)$

21. $(-3, 2)$ **23.** $(-3, 1)$

25. $\left(-\infty, -\frac{4}{3}\right) \cup (5, \infty)$ **27.** $(-1, 1) \cup (3, \infty)$

29. $(-\infty, -3) \cup (3, 7)$ **31.** $(-\infty, 0) \cup \left(0, \frac{3}{2}\right)$

33. $[-2, 0] \cup [2, \infty)$ **35.** $[-2, \infty)$

37. The solution set consists of the single real number $\frac{1}{2}$.

39. The solution set is empty.

41. $(-\infty, 0) \cup \left(\frac{1}{4}, \infty\right)$ **43.** $(-7, 1)$

45. $(-5, 3) \cup (11, \infty)$ **47.** $\left(-\frac{3}{4}, 3\right) \cup [6, \infty)$

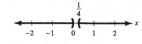

49. $(-3, -2] \cup [0, 3)$

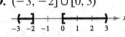

51. $(-\infty, -1) \cup (1, \infty)$

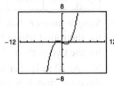

53.

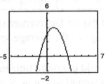

(a) $x \leq -1, x \geq 3$
(b) $0 \leq x \leq 2$

55.

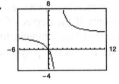

(a) $-2 \leq x \leq 0$,
$2 \leq x \leq \infty$
(b) $x \leq 4$

57.

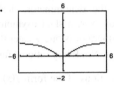

(a) $0 \leq x < 2$
(b) $2 < x \leq 4$

59.

(a) $|x| \geq 2$
(b) $-\infty < x < \infty$

61. $(-3.89, 3.89)$ **63.** $(-0.13, 25.13)$ **65.** $(2.26, 2.39)$
67. (a) $t = 10$ sec (b) 4 sec $< t < 6$ sec
69. $40,000 \leq x \leq 50,000$; $\$50.00 \leq p \leq \$55.00$
71. $[-2, 2]$ **73.** $(-\infty, 4] \cup [5, \infty)$ **75.** $(-5, 0] \cup (7, \infty)$
77. (a) and (c)

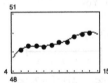

(b) $N = -0.001231t^4 + 0.04723t^3 - 0.6452t^2$
$\qquad + 3.783t + 41.21$

(d) 2017

(e) *Sample answer:* No. For $t > 15$, the model rapidly
decreases.

79. 13.8 m $\leq L \leq 36.2$ m **81.** $R_1 \geq 2$ ohms
83. False. There are four test intervals.
85.

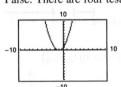

For part (b), the y-values that are
less than or equal to 0 occur only
at $x = -1$.

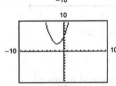

For part (c), there are no y-values
that are less than 0.

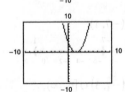

For part (d), the y-values that are
greater than 0 occur for all values
of x except 2.

87. (a) $(-\infty, -6] \cup [6, \infty)$
(b) When $a > 0$ and $c > 0$, $b \leq -2\sqrt{ac}$ or $b \geq 2\sqrt{ac}$.
89. (a) $\left(-\infty, -2\sqrt{30}\right] \cup \left[2\sqrt{30}, \infty\right)$
(b) When $a > 0$ and $c > 0$, $b \leq -2\sqrt{ac}$ or $b \geq 2\sqrt{ac}$.

Review Exercises *(page 190)*

1. (a)

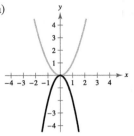

Vertical stretch and a
reflection in the x-axis

(b)

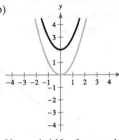

Upward shift of two units

3. $g(x) = (x - 1)^2 - 1$

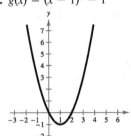

Vertex: $(1, -1)$
Axis of symmetry: $x = 1$
x-intercepts: $(0, 0), (2, 0)$

5. $h(x) = -(x - 2)^2 + 7$

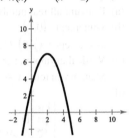

Vertex: $(2, 7)$
Axis of symmetry: $x = 2$
x-intercepts: $\left(2 \pm \sqrt{7}, 0\right)$

7. $h(x) = 4\left(x + \frac{1}{2}\right)^2 + 12$

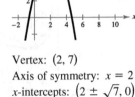

Vertex: $\left(-\frac{1}{2}, 12\right)$
Axis of symmetry: $x = -\frac{1}{2}$
No x-intercept

9. (a) $y = 500 - x$
$\qquad A(x) = 500x - x^2$
(b) $x = 250, y = 250$

11.

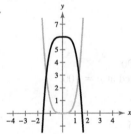

13. Falls to the left and to the right
15. Falls to the left, rises to the right

27. (a) Falls to the left, rises to the right
(b) $-2, 0$ (c) Answers will vary.
(d)
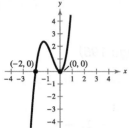

29. (a) Rises to the left, falls to the right
(b) -1 (c) Answers will vary.
(d)
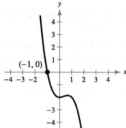

21. (a) $[-1, 0]$ (b) -0.900 **23.** $6x + 3 + \dfrac{17}{5x - 3}$

25. $2x^2 - 9x - 6, \ x \neq 8$

27. (a) Answers will vary.   (b) $(2x + 5), (x - 3)$
(c) $f(x) = (2x + 5)(x - 3)(x + 6)$ (d) $-\frac{5}{2}, 3, -6$
(e)
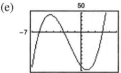

29. $4 + 3i$ **31.** $-3 - 3i$ **33.** $15 + 6i$ **35.** $\frac{4}{5} + \frac{8}{5}i$
37. $\frac{21}{13} - \frac{1}{13}i$ **39.** $1 \pm 3i$ **41.** 2 **43.** $-1, \frac{7}{4}, 6$
45. $(x + 2)(x - 3)(x - 6); -2, 3, 6$
47. One or three positive real zeros, two or no negative real zeros
49. Domain: all real numbers x except $x = -10$
Vertical asymptote: $x = -10$
Horizontal asymptote: $y = 3$
51. (a) Domain: all real numbers x except $x = 0$
(b) No intercepts
(c) Vertical asymptote: $x = 0$
Horizontal asymptote: $y = 0$
(d)
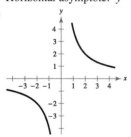

53. (a) Domain: all real numbers x except $x = \pm 4$
(b) Intercept: $(0, 0)$
(c) Vertical asymptotes: $x = \pm 4$
Horizontal asymptote: $y = 0$
(d)
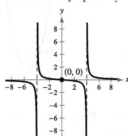

55. (a) Domain: all real numbers x except $x = 0, \frac{1}{3}$
(b) x-intercept: $\left(\frac{3}{2}, 0\right)$
(c) Vertical asymptote: $x = 0$
Horizontal asymptote: $y = 2$
(d)
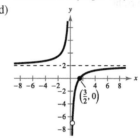

57. (a) Domain: all real numbers x
(b) Intercept: $(0, 0)$ (c) Slant asymptote: $y = 2x$
(d)
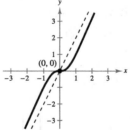

59. (a)
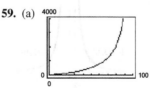

(b) \$176 million; \$528 million; \$1584 million
(or \$1.584 billion)
(c) No; The function is undefined when $p = 100$.

61. $\left(-\frac{2}{3}, \frac{1}{4}\right)$ **63.** $[-5, -1) \cup (1, \infty)$
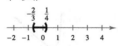

65. 9 days

67. False. The domain of $f(x) = \dfrac{1}{x^2 + 1}$ is the set of all real numbers.

Chapter Test   *(page 192)*

1. (a) 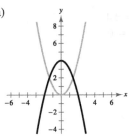   (b)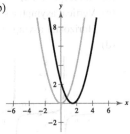

Reflection in the x-axis and an upward shift of four units

Right shift of $\frac{3}{2}$ units

2. $y = (x - 3)^2 - 6$

3. (a) 50 ft
 (b) 5. Yes, changing the constant term results in a vertical shift of the graph, so the maximum height changes.

4. Rises to the left, falls to the right

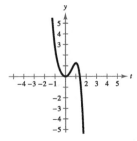

5. $3x + \dfrac{x - 1}{x^2 + 1}$ **6.** $2x^3 - 4x^2 + 5x - 6 + \dfrac{11}{x + 2}$

7. $(2x - 5)(x + \sqrt{3})(x - \sqrt{3})$; Zeros: $\frac{5}{2}, \pm \sqrt{3}$

8. (a) -14 (b) $19 + 17i$ **9.** $\frac{8}{5} - \frac{16}{5}i$

10. $f(x) = x^4 - 2x^3 + 9x^2 - 18x$

11. $f(x) = x^4 - 6x^3 + 16x^2 - 18x + 7$

12. $-5, -\frac{2}{3}, 1$ **13.** $-2, 4, -1 \pm \sqrt{2}i$

14. x-intercepts: $\left(\pm \sqrt{3}, 0\right)$
 Vertical asymptote: $x = 0$
 Horizontal asymptote: $y = -1$

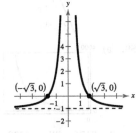

15. x-intercept: $\left(-\frac{3}{2}, 0\right)$
 y-intercept: $\left(0, \frac{3}{4}\right)$
 Vertical asymptote: $x = -4$
 Horizontal asymptote: $y = 2$

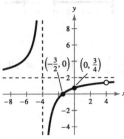

16. y-intercept: $(0, -2)$
 Vertical asymptote: $x = 1$
 Slant asymptote: $y = x + 1$

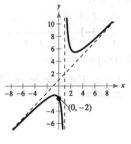

17. $x < -4$ or $x > \frac{3}{2}$ **18.** $x \le -12$ or $-6 < x < 0$

Problem Solving   *(page 195)*

1. Answers will vary.

3. 2 in. × 2 in. × 5 in.

5. (a) and (b) $y = -x^2 + 5x - 4$

7. (a) $f(x) = (x - 2)x^2 + 5 = x^3 - 2x^2 + 5$
 (b) $f(x) = -(x + 3)x^2 + 1 = -x^3 - 3x^2 + 1$

9. $(a + bi)(a - bi) = a^2 + abi - abi - b^2i^2$
 $= a^2 + b^2$

11. (a) As $|a|$ increases, the graph stretches vertically. For $a < 0$ the graph is reflected in the x-axis.
 (b) As $|b|$ increases, the vertical asymptote is translated. For $b > 0$, the graph is translated to the right. For $b < 0$, the graph is reflected in the x-axis and is translated to the left.

13. No. Complex zeros always occur in conjugate pairs.

Chapter 3

Section 3.1   *(page 206)*

1. algebraic   **3.** One-to-One   **5.** $A = P\left(1 + \dfrac{r}{n}\right)^{nt}$

7. 0.863   **9.** 1.552   **11.** 1767.767

13. d   **14.** c   **15.** a   **16.** b

17.

x	-2	-1	0	1	2
$f(x)$	0.020	0.143	1	7	49

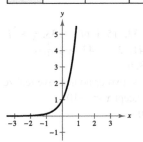

19.

x	-2	-1	0	1	2
$f(x)$	0.063	0.25	1	4	16

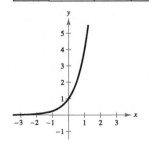

1.

x	−2	−1	0	1	2
f(x)	0.016	0.063	0.25	1	4

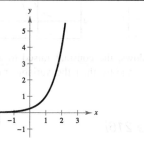

3.

x	−3	−2	−1	0	1
f(x)	3.25	3.5	4	5	7

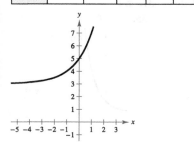

25. 2   **27.** −5   **29.** Shift the graph of f one unit up.
31. Reflect the graph of f in the y-axis and shift three units to the right.
33. 6.686   **35.** 7166.647
37.

x	−8	−7	−6	−5	−4
f(x)	0.055	0.149	0.406	1.104	3

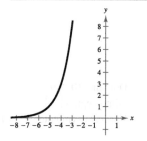

39.

x	−2	−1	0	1	2
f(x)	4.037	4.100	4.271	4.736	6

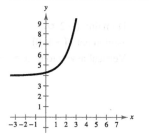

41. 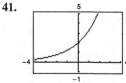   **43.**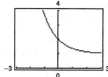

45. $\frac{1}{3}$ **47.** 3, −1

49.

n	1	2	4	12
A	$1828.49	$1830.29	$1831.19	$1831.80

n	365	Continuous
A	$1832.09	$1832.10

51.

n	1	2	4	12
A	$5477.81	$5520.10	$5541.79	$5556.46

n	365	Continuous
A	$5563.61	$5563.85

53.

t	10	20	30
A	$17,901.90	$26,706.49	$39,841.40

t	40	50
A	$59,436.39	$88,668.67

55.

t	10	20	30
A	$22,986.49	$44,031.56	$84,344.25

t	40	50
A	$161,564.86	$309,484.08

57. $104,710.29   **59.** $44.23

61. (a)

(b)

t	25	26	27	28
P (in millions)	350.281	352.107	353.943	355.788

t	29	30	31	32
P (in millions)	357.643	359.508	361.382	363.266

t	33	34	35	36
P (in millions)	365.160	367.064	368.977	370.901

t	37	38	39	40
P (in millions)	372.835	374.779	376.732	378.697

t	41	42	43	44
P (in millions)	380.671	382.656	384.651	386.656

t	45	46	47	48
P (in millions)	388.672	390.698	392.735	394.783

t	49	50	51	52
P (in millions)	396.841	398.910	400.989	403.080

t	53	54	55
P (in millions)	405.182	407.294	409.417

(c) 2064

63. (a) 16 g   (b) 1.85 g
(c)
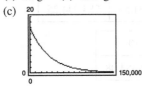

65. (a) $V(t) = 49,810\left(\frac{7}{8}\right)^t$ (b) \$29,197.71
67. True. As $x \to -\infty$, $f(x) \to -2$ but never reaches -2.
69. $f(x) = h(x)$ **71.** $f(x) = g(x) = h(x)$
73.
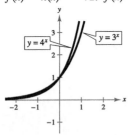

(a) $x < 0$ (b) $x > 0$
75.
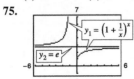

As the x-value increases, y_1 approaches the value of e.

77. (a)
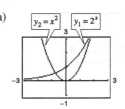
(b)
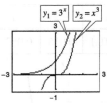

In both viewing windows, the constant raised to a variable power increases more rapidly than the variable raised to constant power.
79. c, d

Section 3.2 *(page 216)*

1. logarithmic   **3.** natural; e **5.** $x = y$ **7.** $4^2 = 16$
9. $12^1 = 12$ **11.** $\log_5 125 = 3$ **13.** $\log_4 \frac{1}{64} = -3$
15. 6   **17.** 0   **19.** -2 **21.** -0.058 **23.** 1.097
25. 1   **27.** 0   **29.** 5   **31.** ± 2
33.
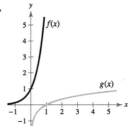
35.
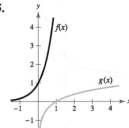

37. a; Upward shift of two units
38. d; Right shift of one unit
39. b; Reflection in the y-axis and a right shift of one unit
40. c; Reflection in the x-axis
41.
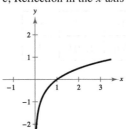
Domain: $(0, \infty)$
x-intercept: $(1, 0)$
Vertical asymptote: $x = 0$

43.
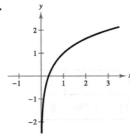
Domain: $(0, \infty)$
x-intercept: $\left(\frac{1}{3}, 0\right)$
Vertical asymptote: $x = 0$

45.
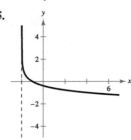
Domain: $(-2, \infty)$
x-intercept: $(-1, 0)$
Vertical asymptote: $x = -$

47.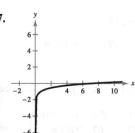
Domain: $(0, \infty)$
x-intercept: $(7, 0)$
Vertical asymptote: $x = 0$

49. $e^{-0.693\ldots} = \frac{1}{2}$ **51.** $e^{5.521\ldots} = 250$
53. $\ln 7.3890\ldots = 2$ **55.** $\ln \frac{1}{2} = -4x$ **57.** 2.913
59. 6.438   **61.** 4   **63.** 0   **65.** 1

67.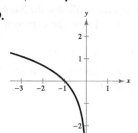
Domain: $(4, \infty)$
x-intercept: $(5, 0)$
Vertical asymptote: $x = 4$

69.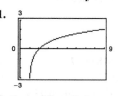
Domain: $(-\infty, 0)$
x-intercept: $(-1, 0)$
Vertical asymptote: $x = 0$

71. 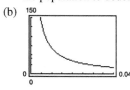   **73.**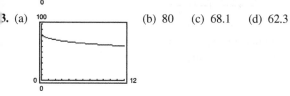

75. 8   **77.** $-2, 3$
79. (a) 30 yr; 10 yr
(b) \$323,179; \$199,109; \$173,179; \$49,109
(c) $x = 750$; The monthly payment must be greater than \$750.
81. (a)

r	0.005	0.010	0.015	0.020	0.025	0.030
t	138.6	69.3	46.2	34.7	27.7	23.1

As the rate of increase r increases, the time t in years for the population to double decreases.
(b)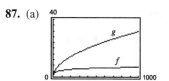

83. (a) 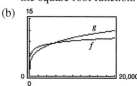   (b) 80   (c) 68.1   (d) 62.3

85. False. Reflecting $g(x)$ in the line $y = x$ will determine the graph of $f(x)$.

87. (a)

$g(x)$; The natural log function grows at a slower rate than the square root function.
(b)

$g(x)$; The natural log function grows at a slower rate than the fourth root function.
89. $y = \log_2 x$, so y is a logarithmic function of x.
91. (a)

x	1	5	10	10^2
$f(x)$	0	0.322	0.230	0.046

x	10^4	10^6
$f(x)$	0.00092	0.0000138

(b) 0
(c)

Section 3.3 *(page 223)*

1. change-of-base   **3.** $\dfrac{1}{\log_b a}$ **5.** (a) $\dfrac{\log 16}{\log 5}$ (b) $\dfrac{\ln 16}{\ln 5}$
7. (a) $\dfrac{\log \frac{3}{10}}{\log x}$ (b) $\dfrac{\ln \frac{3}{10}}{\ln x}$ **9.** 2.579   **11.** -0.606
13. $\log_3 5 + \log_3 7$ **15.** $\log_3 7 - 2 \log_3 5$
17. $1 + \log_3 7 - \log_3 5$ **19.** 2   **21.** $-\frac{1}{3}$
23. -2 is not in the domain of $\log_2 x$.   **25.** $\frac{3}{4}$ **27.** 7
29. 2   **31.** $\frac{3}{2}$ **33.** 1.1833   **35.** -1.6542 **37.** 1.9563
39. -2.7124 **41.** $\ln 7 + \ln x$ **43.** $4 \log_8 x$
45. $1 - \log_5 x$ **47.** $\frac{1}{2} \ln z$ **49.** $\ln x + \ln y + 2 \ln z$
51. $\ln z + 2 \ln(z - 1)$
53. $\frac{1}{2} \log_2 (a + 2) + \frac{1}{2} \log_2 (a - 2) - \log_2 7$
55. $2 \log_5 x - 2 \log_5 y - 3 \log_5 z$ **57.** $\frac{1}{3} \ln y + \frac{1}{3} \ln z - \frac{2}{3} \ln x$
59. $\frac{3}{4} \ln x + \frac{1}{4} \ln(x^2 + 3)$ **61.** $\ln 3x$ **63.** $\log_7 (z - 2)^{2/3}$
65. $\log_3 \dfrac{5}{x^3}$ **67.** $\log x(x + 1)^2$ **69.** $\log \dfrac{xz^3}{y^2}$
71. $\ln \dfrac{x}{(x + 1)(x - 1)}$ **73.** $\ln \sqrt{\dfrac{x(x + 3)^2}{x^2 - 1}}$
75. $\log_8 \dfrac{\sqrt[3]{y(y + 4)^2}}{y - 1}$
77. $\log_2 \frac{32}{4} = \log_2 32 - \log_2 4$; Property 2
79. $\beta = 10(\log I + 12)$; 60 dB   **81.** 70 dB
83. $\ln y = \frac{1}{4} \ln x$ **85.** $\ln y = -\frac{1}{4} \ln x + \ln \frac{5}{2}$
87. $\ln y = -0.14 \ln x + 5.7$

89. (a) and (b)   (c)

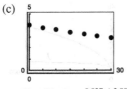

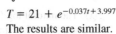

$T = 21 + e^{-0.037t + 3.997}$
The results are similar.

(d)

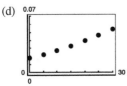

$$T = 21 + \dfrac{1}{0.001t + 0.016}$$

91. False; $\ln 1 = 0$ **93.** False; $\ln(x - 2) \neq \ln x - \ln 2$
95. False; $u = v^2$

97. $f(x) = \dfrac{\log x}{\log 2} = \dfrac{\ln x}{\ln 2}$ **99.** $f(x) = \dfrac{\log x}{\log \frac{1}{4}} = \dfrac{\ln x}{\ln \frac{1}{4}}$

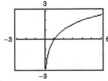   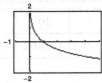

101. The Power Property cannot be used because $\ln e$ is raised to the second power, not just e.

103.

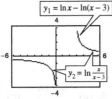

No; $\dfrac{x}{x - 3} > 0$ when $x < 0$.

105. $\ln 1 = 0$ $\ln 9 \approx 2.1972$
$\ln 2 \approx 0.6931$ $\ln 10 \approx 2.3025$
$\ln 3 \approx 1.0986$ $\ln 12 \approx 2.4848$
$\ln 4 \approx 1.3862$ $\ln 15 \approx 2.7080$
$\ln 5 \approx 1.6094$ $\ln 16 \approx 2.7724$
$\ln 6 \approx 1.7917$ $\ln 18 \approx 2.8903$
$\ln 8 \approx 2.0793$ $\ln 20 \approx 2.9956$

Section 3.4   *(page 233)*

1. (a) $x = y$ (b) $x = y$ (c) x (d) x
3. (a) Yes   (b) No   (c) Yes
5. (a) Yes   (b) No   (c) No   **7.** 2
9. 2   **11.** $\ln 2 \approx 0.693$ **13.** $e^{-1} \approx 0.368$
15. 64   **17.** (3, 8)   **19.** 2, -1

21. $\dfrac{\ln 5}{\ln 3} \approx 1.465$ **23.** $\ln 39 \approx 3.664$ **25.** $\dfrac{\ln 80}{2 \ln 3} \approx 1.994$

27. $2 - \dfrac{\ln 400}{\ln 3} \approx -3.454$ **29.** $\dfrac{1}{3} \log \dfrac{3}{2} \approx 0.059$

31. $\dfrac{\ln 12}{3} \approx 0.828$ **33.** 0   **35.** $\dfrac{\ln \frac{8}{3}}{3 \ln 2} + \dfrac{1}{3} \approx 0.805$

37. $-\dfrac{\ln 2}{\ln 3 - \ln 2} \approx -1.710$ **39.** 0, $\dfrac{\ln 4}{\ln 5} \approx 0.861$

41. $\ln 5 \approx 1.609$ **43.** $\ln \frac{4}{5} \approx -0.223$

45. $\dfrac{\ln 4}{365 \ln\left(1 + \dfrac{0.065}{365}\right)} \approx 21.330$ **47.** $e^{-3} \approx 0.050$

49. $\dfrac{e^{2.1}}{6} \approx 1.361$ **51.** $e^{-2} \approx 0.135$ **53.** $2(3^{11/6}) \approx 14.98$

55. No solution   **57.** No solution   **59.** No solution
61. 2   **63.** 3.328   **65.** -0.478 **67.** 20.086
69. 1.482   **71.** (a) 27.73 yr   (b) 43.94 yr   **73.** $-1, 0$
75. 1   **77.** $e^{-1} \approx 0.368$ **79.** $e^{-1/2} \approx 0.607$
81. (a) $y = 100$ and $y = 0$; The range falls between 0% and 100%
(b) Males: 69.51 in.   Females: 64.49 in.
83. 5 years   **85.** 2011   **87.** About 3.039 min
89. $\log_b uv = \log_b u + \log_b v$
True by Property 1 in Section 3.3.
91. $\log_b(u - v) = \log_b u - \log_b v$
False.
$1.95 \approx \log(100 - 10) \neq \log 100 - \log 10 = 1$
93. Yes. See Exercise 57.
95. For $rt < \ln 2$ years, double the amount you invest. For $rt > \ln 2$ years, double your interest rate or double the number of years, because either of these will double the exponent in the exponential function.
97. (a) 7%   (b) 7.25%   (c) 7.19%   (d) 7.45%
The investment plan with the greatest effective yield and the highest balance after 5 years is plan (d).

Section 3.5   *(page 243)*

1. $y = ae^{bx}$; $y = ae^{-bx}$ **3.** normally distributed

5. (a) $P = \dfrac{A}{e^{rt}}$ (b) $t = \dfrac{\ln\left(\dfrac{A}{P}\right)}{r}$

7. 19.8 yr; $1419.07   **9.** 8.9438%; $1834.37
11. $6376.28; 15.4 yr   **13.** $303,580.52
15. (a) 7.27 yr   (b) 6.96 yr   (c) 6.93 yr   (d) 6.93 yr
17. (a)

r	2%	4%	6%	8%	10%	12%
t	54.93	27.47	18.31	13.73	10.99	9.16

(b)

r	2%	4%	6%	8%	10%	12%
t	55.48	28.01	18.85	14.27	11.53	9.69

19.

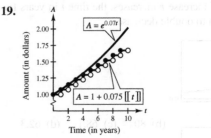

Continuous compounding

21. 6.48 g   **23.** 2.26 g   **25.** $y = e^{0.7675x}$
27. $y = 5e^{-0.4024x}$

9. (a)

Year	Population
1980	104,752
1990	143,251
2000	195,899
2010	267,896

(b) 2019

(c) *Sample answer:* No; As t increases, the population increases rapidly.

31. $k = 0.2988$; About 5,309,734 hits **33.** About 800 bacteria

35. (a) $V = -150t + 575$ (b) $V = 575e^{-0.3688t}$

(c)

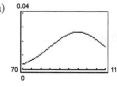

The exponential model depreciates faster.

(d) Linear model: \$425; \$125

Exponential model: \$397.65; \$190.18

(e) Answers will vary.

37. About 12,180 yr old

39. (a)

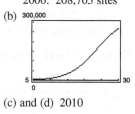

(b) 100

41. (a) 1998: 63,922 sites

2003: 149,805 sites

2006: 208,705 sites

(b)

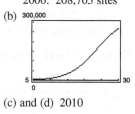

(c) and (d) 2010

43. (a) 203 animals (b) 13 mo

(c)

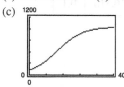

Horizontal asymptotes: $p = 0$, $p = 1000$. The population size will approach 1000 as time increases.

45. (a) $10^{7.6} \approx 39,810,717$

(b) $10^{5.6} \approx 398,107$ (c) $10^{6.6} \approx 3,981,072$

47. (a) 20 dB (b) 70 dB (c) 40 dB (d) 90 dB

49. 95% **51.** 4.64 **53.** 1.58×10^{-6} moles/L

55. $10^{5.1}$ **57.** 3:00 A.M.

59. (a)

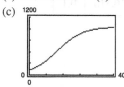

(b) $t \approx 21$ yr; Yes

61. False. The domain can be the set of real numbers for a logistic growth function.

63. False. The graph of $f(x)$ is the graph of $g(x)$ shifted five units up.

65. Answers will vary.

Review Exercises *(page 250)*

1. 0.164 **3.** 1.587 **5.** 1456.529

7.

x	-1	0	1	2	3
$f(x)$	8	5	4.25	4.063	4.016

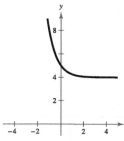

9.

x	-1	0	1	2	3
$f(x)$	4.008	4.04	4.2	5	9

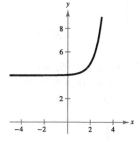

11.

x	-2	-1	0	1	2
$f(x)$	3.25	3.5	4	5	7

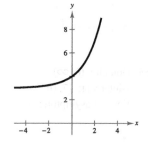

13. 1 **15.** 4 **17.** Shift the graph of f one unit up.

19. Reflect f in the x-axis and shift one unit up.

21. 29.964 **23.** 1.822

25.

x	-2	-1	0	1	2
$h(x)$	2.72	1.65	1	0.61	0.37

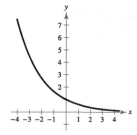

27.

x	-3	-2	-1	0	1
$f(x)$	0.37	1	2.72	7.39	20.09

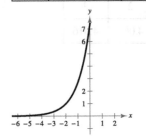

29. (a) 0.283   (b) 0.487   (c) 0.811

31.

n	1	2	4	12
A	\$6719.58	\$6734.28	\$6741.74	\$6746.77

n	365	Continuous
A	\$6749.21	\$6749.29

33. $\log_3 27 = 3$ **35.** $\ln 2.2255 \ldots = 0.8$ **37.** 3
39. -2 **41.** 7   **43.** -5
45. Domain: $(0, \infty)$ **47.** Domain: $(-5, \infty)$
x-intercept: $(1, 0)$ x-intercept: $(9995, 0)$
Vertical asymptote: $x = 0$ Vertical asymptote: $x = -5$

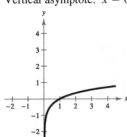

49. 3.118   **51.** 0.25
53. Domain: $(0, \infty)$ **55.** Domain: $(6, \infty)$
x-intercept: $(e^{-6}, 0)$ x-intercept: $(7, 0)$
Vertical asymptote: $x = 0$ Vertical asymptote: $x = 6$

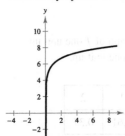

57. About 14.32 parsecs   **59.** (a) and (b) 2.585
61. (a) and (b) -2.322 **63.** $\log_2 5 - \log_2 3$
65. $2 \log_2 3 - \log_2 5$ **67.** $\log 7 + 2 \log x$
69. $2 - \frac{1}{2} \log_3 x$ **71.** $2 \ln x + 2 \ln y + \ln z$
73. $\ln 7x$ **75.** $\log \dfrac{x}{\sqrt{y}}$ **77.** $\log_3 \dfrac{\sqrt{x}}{(y + 8)^2}$

79. (a) $0 \le h < 18{,}000$
(b)
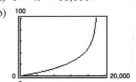
Vertical asymptote: $h = 18{,}000$
(c) The plane is climbing at a slower rate, so the tim
required increases.
(d) 5.46 min
81. 3   **83.** $\ln 3 \approx 1.099$ **85.** $e^4 \approx 54.598$ **87.** 1, 3
89. $\dfrac{\ln 32}{\ln 2} = 5$ **91.** $\frac{1}{3} e^{8.2} \approx 1213.650$
93. $\dfrac{3}{2} + \dfrac{\sqrt{9 + 4e}}{2} \approx 3.729$ **95.** No solution   **97.** 0.900
99. 2.447   **101.** 1.482   **103.** 73.2 yr
105. e   **106.** b   **107.** f   **108.** d   **109.** a   **110.** c
111. $y = 2e^{0.1014x}$
113.
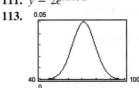
71
115. (a) 10^{-6} W/m²   (b) $10\sqrt{10}$ W/m²
(c) 1.259×10^{-12} W/m²
117. True by the inverse properties.

Chapter Test   (page 253)

1. 0.410   **2.** 0.032   **3.** 0.497   **4.** 22.198
5.

x	-1	$-\frac{1}{2}$	0	$\frac{1}{2}$	1
$f(x)$	10	3.162	1	0.316	0.1

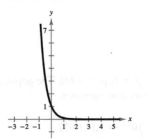

6.

x	-1	0	1	2	3
$f(x)$	-0.005	-0.028	-0.167	-1	-6

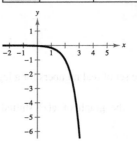

7.

x	-1	$-\frac{1}{2}$	0	$\frac{1}{2}$	1
$f(x)$	0.865	0.632	0	-1.718	-6.389

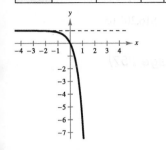

8. (a) -0.89 (b) 9.2

9. Domain: $(0, \infty)$
x-intercept: $(10^{-4}, 0)$
Vertical asymptote: $x = 0$

10. Domain: $(4, \infty)$
x-intercept: $(5, 0)$
Vertical asymptote: $x = 4$

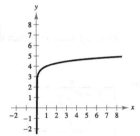
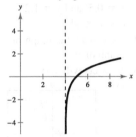

11. Domain: $(-6, \infty)$
x-intercept: $(e^{-1} - 6, 0)$
Vertical asymptote: $x = -6$

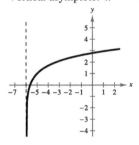

12. 2.209   **13.** -0.167 **14.** -11.047
15. $\log_2 3 + 4 \log_2 a$ **16.** $\frac{1}{2} \ln x - \ln 7$
17. $1 + 2 \log x - 3 \log y$ **18.** $\log_3 13y$ **19.** $\ln \dfrac{x^4}{y^4}$
20. $\ln \dfrac{x^3 y^2}{x + 3}$ **21.** -2 **22.** $\dfrac{\ln 44}{-5} \approx -0.757$
23. $\dfrac{\ln 197}{4} \approx 1.321$ **24.** $e^{1/2} \approx 1.649$
25. $e^{-11/4} \approx 0.0639$ **26.** 20
27. $y = 2745 e^{0.1570t}$ **28.** 55%

29. (a)

x	$\frac{1}{4}$	1	2	4	5	6
H	58.720	75.332	86.828	103.43	110.59	117.38

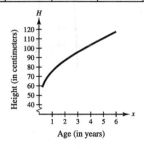

(b) 103 cm; 103.43 cm

Cumulative Test for Chapters 1–3   *(page 254)*

1.

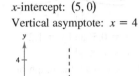

Midpoint: $\left(\frac{1}{2}, 2\right)$
Distance: $\sqrt{61}$

2.

3.

4.

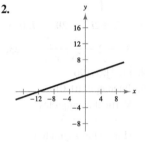
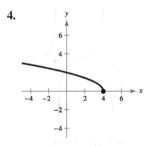

5. $y = 2x + 2$
6. For some values of x there correspond two values of y.
7. (a) $\dfrac{3}{2}$ (b) Division by 0 is undefined.   (c) $\dfrac{s + 2}{s}$
8. (a) Vertical shrink   (b) Upward shift of two units
   (c) Left shift of two units
9. (a) $4x - 3$ (b) $-2x - 5$ (c) $3x^2 - 11x - 4$
   (d) $\dfrac{x - 4}{3x + 1}$; Domain: all real numbers x except $x = -\dfrac{1}{3}$
10. (a) $\sqrt{x - 1} + x^2 + 1$ (b) $\sqrt{x - 1} - x^2 - 1$
   (c) $x^2 \sqrt{x - 1} + \sqrt{x - 1}$
   (d) $\dfrac{\sqrt{x - 1}}{x^2 + 1}$; Domain: all real numbers x such that $x \geq 1$
11. (a) $2x + 12$ (b) $\sqrt{2x^2 + 6}$
   Domain of $f \circ g$: all real numbers x such that $x \geq -6$
   Domain of $g \circ f$: all real numbers x
12. (a) $|x| - 2$ (b) $|x - 2|$
   Domains of $f \circ g$ and $g \circ f$: all real numbers x
13. $h(x)^{-1} = \frac{1}{3}(x + 4)$ **14.** 2438.64 kW
15. $y = -\frac{3}{4}(x + 8)^2 + 5$

16.
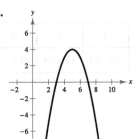

17.
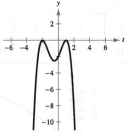

18.
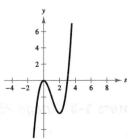

19. $-2, \pm 2i$ **20.** $-7, 0, 3$ **21.** $4, -\frac{1}{2}, 1 \pm 3i$

22. $3x - 2 - \dfrac{3x - 2}{2x^2 + 1}$ **23.** $3x^3 + 6x^2 + 14x + 23 + \dfrac{49}{x - 2}$

24. $[1, 2]$; 1.196

25. Intercept: $(0, 0)$
Vertical asymptotes:
$x = 1, x = -3$
Horizontal asymptote:
$y = 0$

26. y-intercept: $(0, 2)$
x-intercept: $(2, 0)$
Vertical asymptote: $x = 1$
Horizontal asymptote:
$y = 1$

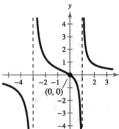
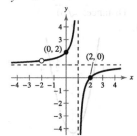

27. y-intercept: $(0, 6)$
x-intercepts: $(2, 0), (3, 0)$
Vertical asymptote: $x = -1$
Slant asymptote: $y = x - 6$

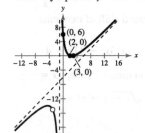

28. $x \le -3$ or $0 \le x \le 3$ **29.** All real numbers x such that
$x < -5$ or $x > -1$

30. Reflect f in the x-axis and y-axis, and shift three units to the right.

31. Reflect f in the x-axis and shift four units up.

32. 1.991 **33.** -0.067 **34.** 1.717 **35.** 0.390

36. $\ln(x + 5) + \ln(x - 5) - 4 \ln x$ **37.** $\ln \dfrac{x^2}{\sqrt{x + 5}}, \ x > 0$

38. $\dfrac{\ln 12}{2} \approx 1.242$ **39.** $\ln 6 \approx 1.792$ or $\ln 7 \approx 1.946$

40. $e^6 - 2 \approx 401.429$ **41.** \$16,302.05

42. 6.3 h   **43.** 2023

Problem Solving   *(page 257)*

1.
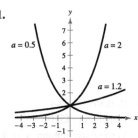
$y = 0.5^x$ and $y = 1.2^x$
$0 < a \le e^{1/e}$

3. As $x \to \infty$, the graph of e^x increases at a greater rate than th
graph of x^n.

5. Answers will vary.

7. (a)
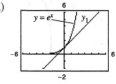
(b)
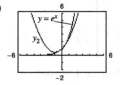

(c)
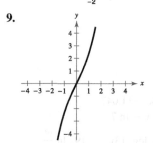

9.
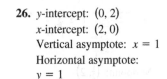
$f^{-1}(x) = \ln\!\left(\dfrac{x + \sqrt{x^2 + 4}}{2}\right)$

11. c

13. $t = \dfrac{\ln c_1 - \ln c_2}{\left(\dfrac{1}{k_2} - \dfrac{1}{k_1}\right)\ln \dfrac{1}{2}}$

15. (a) $y_1 = 252{,}606(1.0310)^t$
(b) $y_2 = 400.88t^2 - 1464.6t + 291{,}782$
(c)
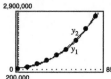

(d) The exponential model is a better fit. No, because th
model is rapidly approaching infinity.

7. $1, e^2$

9. $y_4 = (x - 1) - \frac{1}{2}(x - 1)^2 + \frac{1}{3}(x - 1)^3 - \frac{1}{4}(x - 1)^4$

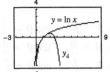

The pattern implies that
$$\ln x = (x - 1) - \frac{1}{2}(x - 1)^2 + \frac{1}{3}(x - 1)^3 - \cdots.$$

21.

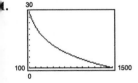

17.7 ft³/min

23. (a)

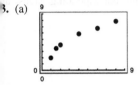

25. (a)

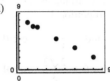

(b)–(e) Answers will vary.   (b)–(e) Answers will vary.

Chapter 4

Section 4.1   (page 267)

1. coterminal   **3.** complementary; supplementary
5. linear; angular   **7.** 1 rad   **9.** −3 rad
11. (a) Quadrant I   (b) Quadrant II
13. (a)   (b)

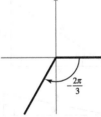

15. *Sample answers:* (a) $\frac{13\pi}{6}, -\frac{11\pi}{6}$ (b) $\frac{7\pi}{6}, -\frac{17\pi}{6}$

17. (a) Complement: $\frac{5\pi}{12}$; Supplement: $\frac{11\pi}{12}$

(b) Complement: none; Supplement: $\frac{\pi}{12}$

19. (a) Complement: $\frac{\pi}{2} - 1 \approx 0.57$;

Supplement: $\pi - 1 \approx 2.14$
(b) Complement: none; Supplement: $\pi - 2 \approx 1.14$

21. 210°   **23.** −60°

25. (a) Quadrant II   (b) Quadrant IV

27. (a)   (b)

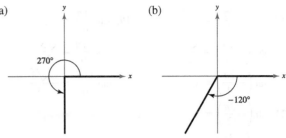

29. (a) 480°, −240°   (b) 150°, −570°
31. (a) Complement: 72°; Supplement: 162°
(b) Complement: 5°; Supplement: 95°
33. (a) Complement: 66°; Supplement: 156°
(b) Complement: none; Supplement: 54°
35. (a) $\frac{2\pi}{3}$ (b) $-\frac{\pi}{9}$ **37.** (a) 270°   (b) −210°
39. 0.785   **41.** −0.009   **43.** 81.818°   **45.** −756.000°
47. (a) 54.75°   (b) −128.5°
49. (a) 240° 36′   (b) −145° 48′   **51.** 10π in. ≈ 31.42 in.
53. $\frac{15}{8}$ rad   **55.** 4 rad   **57.** About 18.85 in.²
59. 20° should be multiplied by $\frac{\pi \text{ rad}}{180 \text{ deg}}$.
61. About 592 mi   **63.** About 23.87°
65. (a) 8π rad/min ≈ 25.13 rad/min
(b) 200π ft/min ≈ 628.3 ft/min
67. (a) About 35.70 mi/h   (b) About 739.50 revolutions/min
69.

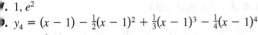

$A = 93.75\pi$ m² ≈ 294.52 m²

71. False. $\frac{180°}{\pi}$ is in degree measure.
73. True. Let α and β represent coterminal angles, and let n represent an integer.
$$\alpha = \beta + n(360°)$$
$$\alpha - \beta = n(360°)$$
75. When θ is constant, the length of the arc is proportional to the radius ($s = r\theta$).
77. The speed increases. The linear velocity is proportional to the radius.
79. Proof

Section 4.2   (page 275)

1. unit circle   **3.** period
5. $\sin t = \frac{5}{13}$ $\csc t = \frac{13}{5}$ **7.** $\sin t = -\frac{3}{5}$ $\csc t = -\frac{5}{3}$
$\cos t = \frac{12}{13}$ $\sec t = \frac{13}{12}$ $\cos t = -\frac{4}{5}$ $\sec t = -\frac{5}{4}$
$\tan t = \frac{5}{12}$ $\cot t = \frac{12}{5}$ $\tan t = \frac{3}{4}$ $\cot t = \frac{4}{3}$

9. $(0, 1)$ **11.** $\left(-\frac{\sqrt{3}}{2}, \frac{1}{2}\right)$

13. $\sin \frac{\pi}{4} = \frac{\sqrt{2}}{2}$ **15.** $\sin\left(-\frac{\pi}{6}\right) = -\frac{1}{2}$

$\cos \frac{\pi}{4} = \frac{\sqrt{2}}{2}$ $\cos\left(-\frac{\pi}{6}\right) = \frac{\sqrt{3}}{2}$

$\tan \frac{\pi}{4} = 1$ $\tan\left(-\frac{\pi}{6}\right) = -\frac{\sqrt{3}}{3}$

17. $\sin\left(-\dfrac{7\pi}{4}\right) = \dfrac{\sqrt{2}}{2}$

$\cos\left(-\dfrac{7\pi}{4}\right) = \dfrac{\sqrt{2}}{2}$

$\tan\left(-\dfrac{7\pi}{4}\right) = 1$

19. $\sin\dfrac{11\pi}{6} = -\dfrac{1}{2}$

$\cos\dfrac{11\pi}{6} = \dfrac{\sqrt{3}}{2}$

$\tan\dfrac{11\pi}{6} = -\dfrac{\sqrt{3}}{3}$

21. $\sin\left(-\dfrac{3\pi}{2}\right) = 1$

$\cos\left(-\dfrac{3\pi}{2}\right) = 0$

$\tan\left(-\dfrac{3\pi}{2}\right)$ is undefined.

23. $\sin\dfrac{2\pi}{3} = \dfrac{\sqrt{3}}{2}$ $\csc\dfrac{2\pi}{3} = \dfrac{2\sqrt{3}}{3}$

$\cos\dfrac{2\pi}{3} = -\dfrac{1}{2}$ $\sec\dfrac{2\pi}{3} = -2$

$\tan\dfrac{2\pi}{3} = -\sqrt{3}$ $\cot\dfrac{2\pi}{3} = -\dfrac{\sqrt{3}}{3}$

25. $\sin\dfrac{4\pi}{3} = -\dfrac{\sqrt{3}}{2}$ $\csc\dfrac{4\pi}{3} = -\dfrac{2\sqrt{3}}{3}$

$\cos\dfrac{4\pi}{3} = -\dfrac{1}{2}$ $\sec\dfrac{4\pi}{3} = -2$

$\tan\dfrac{4\pi}{3} = \sqrt{3}$ $\cot\dfrac{4\pi}{3} = \dfrac{\sqrt{3}}{3}$

27. $\sin\left(-\dfrac{5\pi}{3}\right) = \dfrac{\sqrt{3}}{2}$ $\csc\left(-\dfrac{5\pi}{3}\right) = \dfrac{2\sqrt{3}}{3}$

$\cos\left(-\dfrac{5\pi}{3}\right) = \dfrac{1}{2}$ $\sec\left(-\dfrac{5\pi}{3}\right) = 2$

$\tan\left(-\dfrac{5\pi}{3}\right) = \sqrt{3}$ $\cot\left(-\dfrac{5\pi}{3}\right) = \dfrac{\sqrt{3}}{3}$

29. $\sin\left(-\dfrac{\pi}{2}\right) = -1$ $\csc\left(-\dfrac{\pi}{2}\right) = -1$

$\cos\left(-\dfrac{\pi}{2}\right) = 0$ $\sec\left(-\dfrac{\pi}{2}\right)$ is undefined.

$\tan\left(-\dfrac{\pi}{2}\right)$ is undefined. $\cot\left(-\dfrac{\pi}{2}\right) = 0$

31. 0   **33.** $\frac{1}{2}$ **35.** $-\frac{1}{2}$ **37.** (a) $-\frac{1}{2}$ (b) -2

39. (a) $-\frac{1}{5}$ (b) -5 **41.** (a) $\frac{4}{5}$ (b) $-\frac{4}{5}$

43. 0.5646   **45.** 0.4142   **47.** -1.0009

49. (a) 0.50 ft   (b) About 0.04 ft   (c) About -0.49 ft

51. False. $\sin(-t) = -\sin(t)$ means that the function is odd, not that the sine of a negative angle is a negative number.

53. True. The tangent function has a period of π.

55. (a) y-axis symmetry   (b) $\sin t_1 = \sin(\pi - t_1)$
(c) $\cos(\pi - t_1) = -\cos t_1$

57. The calculator was in degree mode instead of radian mode.

59. (a)

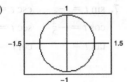

Circle of radius 1 centered at $(0, 0)$

(b) The t-values represent the central angle in radians. The x- and y-values represent the location in the coordinate plane.

(c) $-1 \le x \le 1, -1 \le y \le 1$

61. It is an odd function.

Section 4.3   *(page 284)*

1. (a) v   (b) iv   (c) vi   (d) iii   (e) i   (f) ii

3. complementary

5. $\sin\theta = \frac{3}{5}$ $\csc\theta = \frac{5}{3}$ **7.** $\sin\theta = \frac{9}{41}$ $\csc\theta = \frac{41}{9}$

$\cos\theta = \frac{4}{5}$ $\sec\theta = \frac{5}{4}$ $\cos\theta = \frac{40}{41}$ $\sec\theta = \frac{41}{40}$

$\tan\theta = \frac{3}{4}$ $\cot\theta = \frac{4}{3}$ $\tan\theta = \frac{9}{40}$ $\cot\theta = \frac{40}{9}$

9. $\sin\theta = \dfrac{\sqrt{2}}{2}$ $\csc\theta = \sqrt{2}$

$\cos\theta = \dfrac{\sqrt{2}}{2}$ $\sec\theta = \sqrt{2}$

$\tan\theta = 1$ $\cot\theta = 1$

11. $\sin\theta = \frac{8}{17}$ $\csc\theta = \frac{17}{8}$

$\cos\theta = \frac{15}{17}$ $\sec\theta = \frac{17}{15}$

$\tan\theta = \frac{8}{15}$ $\cot\theta = \frac{15}{8}$

The triangles are similar, and corresponding sides are proportional.

13. $\sin\theta = \dfrac{1}{3}$ $\csc\theta = 3$

$\cos\theta = \dfrac{2\sqrt{2}}{3}$ $\sec\theta = \dfrac{3\sqrt{2}}{4}$

$\tan\theta = \dfrac{\sqrt{2}}{4}$ $\cot\theta = 2\sqrt{2}$

The triangles are similar, and corresponding sides are proportional.

15. $\sin\theta = \frac{8}{17}$ $\csc\theta = \frac{17}{8}$

$\sec\theta = \frac{17}{15}$

$\tan\theta = \frac{8}{15}$ $\cot\theta = \frac{15}{8}$

17. $\sin\theta = \dfrac{\sqrt{11}}{6}$ $\csc\theta = \dfrac{6\sqrt{11}}{11}$

$\cos\theta = \dfrac{5}{6}$

$\tan\theta = \dfrac{\sqrt{11}}{5}$ $\cot\theta = \dfrac{5\sqrt{11}}{11}$

19. $\csc\theta = 5$

$\cos\theta = \dfrac{2\sqrt{6}}{5}$ $\sec\theta = \dfrac{5\sqrt{6}}{12}$

$\tan\theta = \dfrac{\sqrt{6}}{12}$ $\cot\theta = 2\sqrt{6}$

21. $\sin\theta = \dfrac{\sqrt{10}}{10}$ $\csc\theta = \sqrt{10}$

$\cos\theta = \dfrac{3\sqrt{10}}{10}$ $\sec\theta = \dfrac{\sqrt{10}}{3}$

$\tan\theta = \dfrac{1}{3}$

23. $\dfrac{\pi}{6}; \dfrac{\sqrt{3}}{3}$ **25.** $45°; \dfrac{\sqrt{2}}{2}$ **27.** $45°; \sqrt{2}$

29. (a) 0.3420   (b) 0.3420   **31.** (a) 0.2455   (b) 4.0737

33. (a) 0.9964   (b) 1.0036   **35.** (a) 3.2205   (b) 0.3105

37. (a) $\dfrac{1}{2}$ (b) $\dfrac{\sqrt{3}}{2}$ (c) $\sqrt{3}$ (d) $\dfrac{\sqrt{3}}{3}$

39. (a) $\dfrac{2\sqrt{2}}{3}$ (b) $2\sqrt{2}$ (c) 3   (d) 3

41. (a) $\dfrac{1}{3}$ (b) $\sqrt{10}$ (c) $\dfrac{1}{3}$ (d) $\dfrac{\sqrt{10}}{10}$

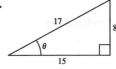

3–51. Answers will vary.   **53.** (a) $30° = \frac{\pi}{6}$ (b) $30° = \frac{\pi}{6}$

55. (a) $60° = \frac{\pi}{3}$ (b) $45° = \frac{\pi}{4}$

57. (a) $60° = \frac{\pi}{3}$ (b) $45° = \frac{\pi}{4}$ **59.** $x = 9, y = 9\sqrt{3}$

61. $x = \frac{32\sqrt{3}}{3}, r = \frac{64\sqrt{3}}{3}$

63. About 443.2 m; about 323.3 m   **65.** $30° = \frac{\pi}{6}$

67. (a) About 219.9 ft   (b) About 160.9 ft

69. $(x_1, y_1) = (28\sqrt{3}, 28)$
$(x_2, y_2) = (28, 28\sqrt{3})$

71. $\sin 20° \approx 0.34$, $\cos 20° \approx 0.94$, $\tan 20° \approx 0.36$,
$\csc 20° \approx 2.92$, $\sec 20° \approx 1.06$, $\cot 20° \approx 2.75$

73. (a) About 519.33 ft
(b) About 1174.17 ft
(c) About 173.11 ft/min

75. True. $\csc x = \frac{1}{\sin x}$ **77.** False. $\frac{\sqrt{2}}{2} + \frac{\sqrt{2}}{2} \neq 1$

79. False. $1.7321 \neq 0.0349$

81. Yes, $\tan \theta$ is equal to opp/adj. You can find the value of the hypotenuse by the Pythagorean Theorem. Then you can find $\sec \theta$, which is equal to hyp/adj.

83.

θ	0.1	0.2	0.3	0.4	0.5
$\sin \theta$	0.0998	0.1987	0.2955	0.3894	0.4794

(a) θ is greater.

(b) As $\theta \to 0$, $\sin \theta \to 0$ and $\frac{\theta}{\sin \theta} \to 1$.

Section 4.4   (page 294)

1. $\frac{y}{r}$ **3.** $\frac{y}{x}$ **5.** $\cos \theta$ **7.** zero; defined

9. (a) $\sin \theta = \frac{3}{5}$ $\csc \theta = \frac{5}{3}$
$\cos \theta = \frac{4}{5}$ $\sec \theta = \frac{5}{4}$
$\tan \theta = \frac{3}{4}$ $\cot \theta = \frac{4}{3}$
(b) $\sin \theta = \frac{15}{17}$ $\csc \theta = \frac{17}{15}$
$\cos \theta = -\frac{8}{17}$ $\sec \theta = -\frac{17}{8}$
$\tan \theta = -\frac{15}{8}$ $\cot \theta = -\frac{8}{15}$

11. (a) $\sin \theta = -\frac{1}{2}$ $\csc \theta = -2$
$\cos \theta = -\frac{\sqrt{3}}{2}$ $\sec \theta = -\frac{2\sqrt{3}}{3}$
$\tan \theta = \frac{\sqrt{3}}{3}$ $\cot \theta = \sqrt{3}$
(b) $\sin \theta = -\frac{\sqrt{17}}{17}$ $\csc \theta = -\sqrt{17}$
$\cos \theta = \frac{4\sqrt{17}}{17}$ $\sec \theta = \frac{\sqrt{17}}{4}$
$\tan \theta = -\frac{1}{4}$ $\cot \theta = -4$

13. $\sin \theta = \frac{12}{13}$ $\csc \theta = \frac{13}{12}$
$\cos \theta = \frac{5}{13}$ $\sec \theta = \frac{13}{5}$
$\tan \theta = \frac{12}{5}$ $\cot \theta = \frac{5}{12}$

15. $\sin \theta = -\frac{2\sqrt{29}}{29}$ $\csc \theta = -\frac{\sqrt{29}}{2}$
$\cos \theta = -\frac{5\sqrt{29}}{29}$ $\sec \theta = -\frac{\sqrt{29}}{5}$
$\tan \theta = \frac{2}{5}$ $\cot \theta = \frac{5}{2}$

17. $\sin \theta = \frac{4}{5}$ $\csc \theta = \frac{5}{4}$
$\cos \theta = -\frac{3}{5}$ $\sec \theta = -\frac{5}{3}$
$\tan \theta = -\frac{4}{3}$ $\cot \theta = -\frac{3}{4}$

19. Quadrant I   **21.** Quadrant II

23. $\sin \theta = \frac{15}{17}$ $\csc \theta = \frac{17}{15}$
$\cos \theta = \frac{8}{17}$ $\sec \theta = \frac{17}{8}$
$\cot \theta = \frac{8}{15}$

25. $\csc \theta = \frac{5}{3}$
$\cos \theta = -\frac{4}{5}$ $\sec \theta = -\frac{5}{4}$
$\tan \theta = -\frac{3}{4}$ $\cot \theta = -\frac{4}{3}$

27. $\sin \theta = -\frac{\sqrt{10}}{10}$ $\csc \theta = -\sqrt{10}$
$\cos \theta = \frac{3\sqrt{10}}{10}$ $\sec \theta = \frac{\sqrt{10}}{3}$
$\tan \theta = -\frac{1}{3}$

29. $\sin \theta = 1$
    $\sec \theta$ is undefined.
$\tan \theta$ is undefined. $\cot \theta = 0$

31. $\sin \theta = 0$ $\csc \theta$ is undefined.
$\cos \theta = -1$ $\sec \theta = -1$
$\tan \theta = 0$ $\cot \theta$ is undefined.

33. $\sin \theta = \frac{\sqrt{2}}{2}$ $\csc \theta = \sqrt{2}$
$\cos \theta = -\frac{\sqrt{2}}{2}$ $\sec \theta = -\sqrt{2}$
$\tan \theta = -1$ $\cot \theta = -1$

35. $\sin \theta = \frac{2\sqrt{5}}{5}$ $\csc \theta = \frac{\sqrt{5}}{2}$
$\cos \theta = \frac{\sqrt{5}}{5}$ $\sec \theta = \sqrt{5}$
$\tan \theta = 2$ $\cot \theta = \frac{1}{2}$

37. 0   **39.** Undefined   **41.** 1
43. Undefined   **45.** 0
47. $\theta' = 20°$ **49.** $\theta' = 55°$

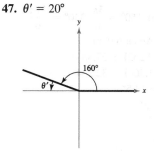

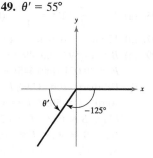

51. $\theta' = \dfrac{\pi}{3}$

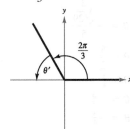

53. $\theta' = 2\pi - 4.8$

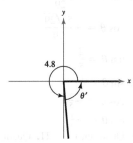

55. $\sin 225° = -\dfrac{\sqrt{2}}{2}$

$\cos 225° = -\dfrac{\sqrt{2}}{2}$

$\tan 225° = 1$

57. $\sin 750° = \dfrac{1}{2}$

$\cos 750° = \dfrac{\sqrt{3}}{2}$

$\tan 750° = \dfrac{\sqrt{3}}{3}$

59. $\sin(-120°) = -\dfrac{\sqrt{3}}{2}$

$\cos(-120°) = -\dfrac{1}{2}$

$\tan(-120°) = \sqrt{3}$

61. $\sin \dfrac{2\pi}{3} = \dfrac{\sqrt{3}}{2}$

$\cos \dfrac{2\pi}{3} = -\dfrac{1}{2}$

$\tan \dfrac{2\pi}{3} = -\sqrt{3}$

63. $\sin\left(-\dfrac{\pi}{6}\right) = -\dfrac{1}{2}$

$\cos\left(-\dfrac{\pi}{6}\right) = \dfrac{\sqrt{3}}{2}$

$\tan\left(-\dfrac{\pi}{6}\right) = -\dfrac{\sqrt{3}}{3}$

65. $\sin \dfrac{11\pi}{4} = \dfrac{\sqrt{2}}{2}$

$\cos \dfrac{11\pi}{4} = -\dfrac{\sqrt{2}}{2}$

$\tan \dfrac{11\pi}{4} = -1$

67. $\sin\left(-\dfrac{17\pi}{6}\right) = -\dfrac{1}{2}$

$\cos\left(-\dfrac{17\pi}{6}\right) = -\dfrac{\sqrt{3}}{2}$

$\tan\left(-\dfrac{17\pi}{6}\right) = \dfrac{\sqrt{3}}{3}$

69. $\dfrac{4}{5}$ **71.** $-\dfrac{\sqrt{13}}{12}$ **73.** $\dfrac{8\sqrt{39}}{39}$ **75.** 0.1736

77. -0.3420 **79.** -28.6363 **81.** 1.4142 **83.** 0.3640

85. -2.6131 **87.** -0.6052 **89.** 1.8382

91. (a) $30° = \dfrac{\pi}{6}, 150° = \dfrac{5\pi}{6}$ (b) $210° = \dfrac{7\pi}{6}, 330° = \dfrac{11\pi}{6}$

93. (a) $60° = \dfrac{\pi}{3}, 300° = \dfrac{5\pi}{3}$ (b) $60° = \dfrac{\pi}{3}, 300° = \dfrac{5\pi}{3}$

95. (a) $45° = \dfrac{\pi}{4}, 225° = \dfrac{5\pi}{4}$ (b) $150° = \dfrac{5\pi}{6}, 330° = \dfrac{11\pi}{6}$

97. (a) 12 mi (b) 6 mi (c) About 6.9 mi

99. (a) $B = 24.593 \sin(0.495t - 2.262) + 57.387$

$F = 39.071 \sin(0.448t - 1.366) + 32.204$

(b) February: $B \approx 33.9°, F \approx 14.5°$

April: $B \approx 50.5°, F \approx 48.3°$

May: $B \approx 62.6°, F \approx 62.2°$

July: $B \approx 80.3°, F \approx 70.5°$

September: $B \approx 77.4°, F \approx 50.1°$

October: $B \approx 68.2°, F \approx 33.3°$

December: $B \approx 44.8°, F \approx 2.4°$

(c)

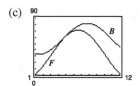

Answers will vary.

101. About 0.79 amp

103. False. In each of the four quadrants, the signs of the secant function and the cosine function are the same because these functions are reciprocals of each other.

105. Answers will vary.

Section 4.5 *(page 304)*

1. cycle **3.** phase shift **5.** Period: $\dfrac{2\pi}{5}$; Amplitude: 2

7. Period: 4; Amplitude: $\dfrac{3}{4}$ **9.** Period: $\dfrac{8\pi}{5}$; Amplitude: $\dfrac{1}{2}$

11. Period: 24; Amplitude: $\dfrac{5}{3}$

13. The period of g is one-fifth the period of f.

15. g is a reflection of f in the x-axis.

17. g is a shift of f π units to the right.

19. g is a shift of f three units up.

21. The graph of g has twice the amplitude of the graph of f.

23. The graph of g is a horizontal shift of the graph of f π units to the right.

25.

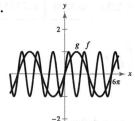

27.

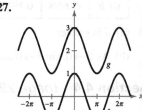

29.

31.

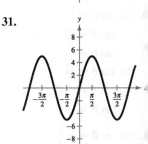

33.

35.

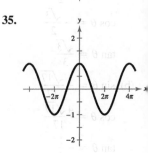

39.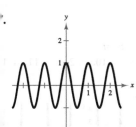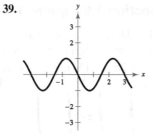

43.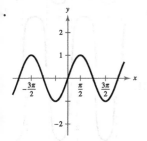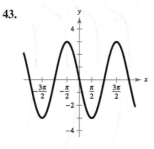

47.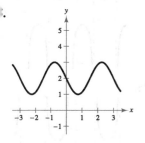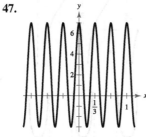

51.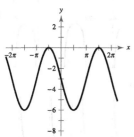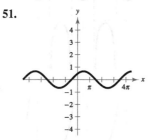

. (a) Horizontal shrink and a phase shift $\pi/4$ unit right

(b) 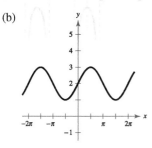   (c) $g(x) = f(4x - \pi)$

. (a) Shift two units up and a phase shift $\pi/2$ units right

(b)   (c) $g(x) = f\left(x - \dfrac{\pi}{2}\right) + 2$

57. (a) Horizontal shrink, a vertical stretch, a shift three units down, and a phase shift $\pi/4$ unit right

(b)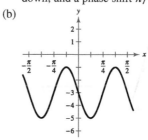

(c) $g(x) = 2f(4x - \pi) - 3$

59. 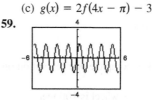   **61.**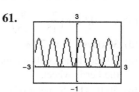

63.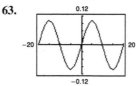

65. $a = 2, d = 1$ **67.** $a = -4, d = 4$

69. $a = -3, b = 2, c = 0$ **71.** $a = 2, b = 1, c = -\dfrac{\pi}{4}$

73.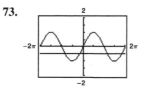

$x = -\dfrac{\pi}{6}, -\dfrac{5\pi}{6}, \dfrac{7\pi}{6}, \dfrac{11\pi}{6}$

75. $y = 1 + 2\sin(2x - \pi)$ **77.** $y = \cos(2x + 2\pi) - \dfrac{3}{2}$

79. (a) 4 sec
(b) 15 cycles/min
(c)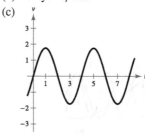

81. (a) $\dfrac{6}{5}$ sec   (b) 50 heartbeats/min

83. (a) and (c)

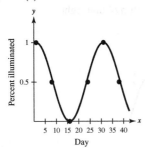

The model fits the data well.

(b) $y = 0.5 \cos\left(\dfrac{\pi x}{15} - \dfrac{\pi}{15}\right) + 0.5$

(d) 30 days   (e) 25%

85. (a) 20 sec; It takes 20 seconds to complete one revolution on the Ferris wheel.

(b) 50 ft; The diameter of the Ferris wheel is 100 feet.

(c)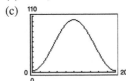

87. False. The graph of g is shifted one period to the left.

89. True. $-\sin\left(x + \dfrac{\pi}{2}\right) = -\cos x$

91.

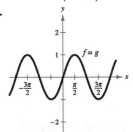

Conjecture:

$\sin x = \cos\left(x - \dfrac{\pi}{2}\right)$

93.

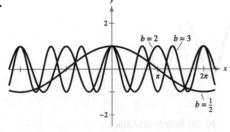

The value of b affects the period of the graph.

$b = \frac{1}{2} \rightarrow \frac{1}{2}$ cycle

$b = 2 \rightarrow 2$ cycles

$b = 3 \rightarrow 3$ cycles

95. (a) 0.4794, 0.4794   (b) 0.8417, 0.8415   (c) 0.5, 0.5

(d) 0.8776, 0.8776   (e) 0.5417, 0.5403

(f) 0.7074, 0.7071

The error increases as x moves farther away from 0.

Section 4.6   (page 315)

1. odd; origin   **3.** reciprocal   **5.** π

7. $(-\infty, -1] \cup [1, \infty)$ **9.** e, π **10.** c, 2π **11.** a, 1

12. d, 2π **13.** f, 4   **14.** b, 4

15. 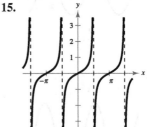   **17.**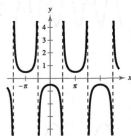

19. 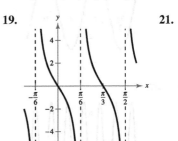   **21.**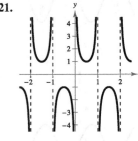

23. 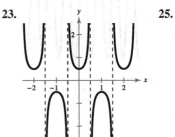   **25.**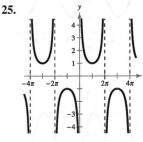

27. 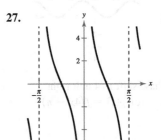   **29.**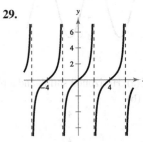

31. 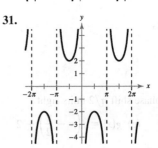   **33.**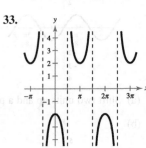

35. 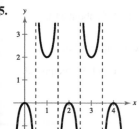   **37.**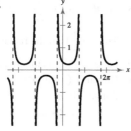

39. 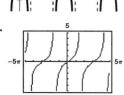   **41.**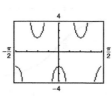

43. 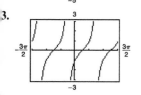   **45.**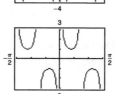

47.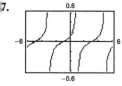

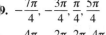

49. $-\dfrac{7\pi}{4}, -\dfrac{3\pi}{4}, \dfrac{\pi}{4}, \dfrac{5\pi}{4}$ **51.** $-\dfrac{7\pi}{6}, -\dfrac{\pi}{6}, \dfrac{5\pi}{6}, \dfrac{11\pi}{6}$

53. $-\dfrac{4\pi}{3}, -\dfrac{2\pi}{3}, \dfrac{2\pi}{3}, \dfrac{4\pi}{3}$ **55.** $-\dfrac{7\pi}{4}, -\dfrac{5\pi}{4}, \dfrac{\pi}{4}, \dfrac{3\pi}{4}$

57. Even   **59.** Odd   **61.** Odd   **63.** Even

65. d, $f \to 0$ as $x \to 0$.   **66.** a, $f \to 0$ as $x \to 0$.

67. b, $g \to 0$ as $x \to 0$.   **68.** c, $g \to 0$ as $x \to 0$.

69. 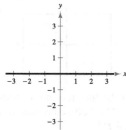   **71.**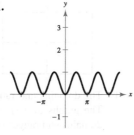

The functions are equal.   The functions are equal.

73. 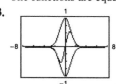   **75.**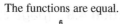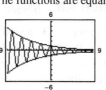

As $x \to \infty$, $g(x) \to 0$.   As $x \to \infty$, $f(x) \to 0$.

77. 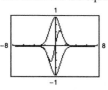   **79.**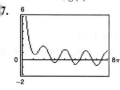

As $x \to 0$, $y \to \infty$.   As $x \to 0$, $g(x) \to 1$.

81.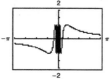

As $x \to 0$, $f(x)$ oscillates between 1 and -1.

83. (a) Period of $H(t)$: 12 mo
Period of $L(t)$: 12 mo
(b) Summer; winter   (c) About 0.5 mo

85. $d = 27 \sec x$

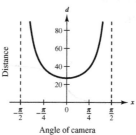

87. True. For a given value of x, the y-coordinate of $\csc x$ is the reciprocal of the y-coordinate of $\sin x$.

89. (a)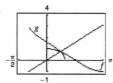

0.7391

(b) 1, 0.5403, 0.8576, 0.6543, 0.7935, 0.7014, 0.7640, 0.7221, 0.7504, 0.7314, . . . ; 0.7391

91. (a) $f(x) \to \infty$ (b) $f(x) \to -\infty$
(c) $f(x) \to \infty$ (d) $f(x) \to -\infty$

93. (a) $f(x) \to -\infty$ (b) $f(x) \to \infty$
(c) $f(x) \to -\infty$ (d) $f(x) \to \infty$

Section 4.7 *(page 324)*

1. $y = \sin^{-1} x$; $-1 \le x \le 1$

3. $y = \tan^{-1} x$; $-\infty < x < \infty$; $-\dfrac{\pi}{2} < y < \dfrac{\pi}{2}$ **5.** $\dfrac{\pi}{6}$ **7.** $\dfrac{\pi}{3}$

9. $\dfrac{\pi}{6}$ **11.** Not possible   **13.** $-\dfrac{\pi}{3}$ **15.** $\dfrac{2\pi}{3}$ **17.** $-\dfrac{\pi}{3}$

19.

21. 1.19   **23.** -0.85 **25.** -1.25 **27.** Not possible
29. 1.99   **31.** 0.74   **33.** 1.07   **35.** -1.50

37. $-\dfrac{\pi}{3}, -\dfrac{\sqrt{3}}{3}, 1$ **39.** $\theta = \arctan \dfrac{x}{4}$ **41.** $\theta = \arcsin \dfrac{x+2}{5}$

43. $\theta = \arccos \dfrac{x+3}{2x}$ **45.** 0.3   **47.** Not possible

49. $\dfrac{\pi}{4}$ **51.** $\dfrac{3}{5}$ **53.** $\dfrac{\sqrt{5}}{5}$ **55.** $\dfrac{13}{12}$

57. $-\dfrac{5}{3}$ **59.** $-\dfrac{\sqrt{5}}{2}$ **61.** 2   **63.** $\sqrt{1 - 4x^2}$

65. $\dfrac{1}{x}$ **67.** $\sqrt{1-x^2}$ **69.** $\dfrac{\sqrt{9-x^2}}{x}$ **71.** $\dfrac{\sqrt{x^2+a^2}}{x}$

73.

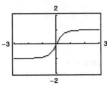

Asymptotes: $y = \pm 1$

75. $\dfrac{9}{\sqrt{x^2+81}}$ **77.** $\dfrac{|x-1|}{\sqrt{x^2-2x+10}}$

79.

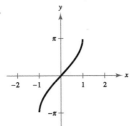

81.

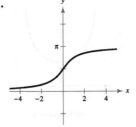

Vertical stretch   Shift $\dfrac{\pi}{2}$ units up

83.

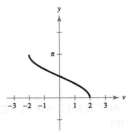

85.

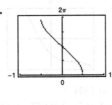

Horizontal stretch

87.

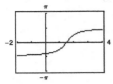

89.

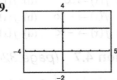

91. $3\sqrt{2}\,\sin\!\left(2t+\dfrac{\pi}{4}\right)$

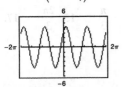

The graph implies that the identity is true.

93. $\dfrac{\pi}{2}$ **95.** $\dfrac{\pi}{2}$ **97.** π

99. (a) $\theta = \arcsin\dfrac{5}{s}$ (b) About 0.13, about 0.25

101. (a) About 32.9°   (b) About 6.5 m

103. (a)

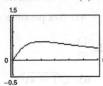

(b) 2 ft   (c) $\beta = 0$; As x increases, β approaches 0.

105. (a) $\theta = \arctan\dfrac{x}{20}$ (b) About 14.0°, about 31.0°

107. True. $-\dfrac{\pi}{4}$ is in the range of the arctangent function.

109. False. $\sin^{-1} x$ is the inverse of $\sin x$, not the reciprocal.

111. Domain: $(-\infty, \infty)$
Range: $(0, \pi)$

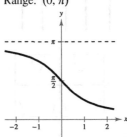

113. Domain: $(-\infty, -1] \cup [1, \infty)$
Range: $[-\pi/2, 0) \cup (0, \pi/2]$

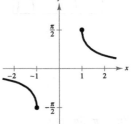

115. $\dfrac{\pi}{4}$ **117.** $\dfrac{3\pi}{4}$ **119.** $-\dfrac{\pi}{2}$ **121.** 1.17

123. -0.12 **125.** 0.19

127. (a) $\dfrac{\pi}{4}$ (b) $\dfrac{\pi}{2}$ (c) About 1.25   (d) About 2.03

129. (a) $f \circ f^{-1}$ $f^{-1} \circ f$

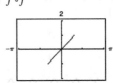   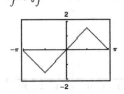

(b) The domains and ranges of the functions are restricted.
The graphs of $f \circ f^{-1}$ and $f^{-1} \circ f$ differ because of the
domains and ranges of f and f^{-1}.

Section 4.8   *(page 334)*

1. bearing   **3.** period

5. $a \approx 10.39$ **7.** $b \approx 14.21$
$b = 6$ $c \approx 14.88$
$B = 30°$ $A = 17.2°$

9. $c = 5$ **11.** $a \approx 52.88$
$A \approx 36.87°$ $A \approx 73.46°$
$B \approx 53.13°$ $B \approx 16.54°$

13. 3.00   **15.** 2.50   **17.** About 214.45 ft

19. About 19.7 ft   **21.** About 20.5 ft   **23.** About 11.8 km

25. About 56.3°   **27.** About 75.97°

29. About 3.23 mi or about 17,054 ft

31. (a) $l = 250$ ft, $A \approx 36.87°$, $B \approx 53.13°$
(b) About 4.87 sec

33. About 508 mi north, about 650 mi east

5. (a) About 104.95 nmi south, about 58.18 nmi west
(b) S 36.7° W; about 130.9 nmi

7. N 56.31° W   **39.** (a) N 58° E   (b) About 68.82 m

41. About 35.3°   **43.** About 29.4 in.   **45.** $d = 4 \sin \pi t$

47. $d = 3 \cos \dfrac{4\pi t}{3}$ **49.** $\omega = 524\pi$

51. (a) 9   (b) $\dfrac{3}{5}$ (c) 9   (d) $\dfrac{5}{12}$

53. (a) $\dfrac{1}{4}$ (b) 3   (c) 0   (d) $\dfrac{1}{6}$

55. (a)   (b) $\dfrac{\pi}{8}$ (c) $\dfrac{\pi}{32}$

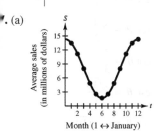

57. (a)

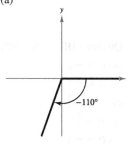

(b) $S = 8 + 6.3 \cos \dfrac{\pi}{6} t$ or $S = 8 + 6.3 \sin\left(\dfrac{\pi}{6} t + \dfrac{\pi}{2}\right)$

The model is a good fit.
(c) 12. Yes, sales of outerwear are seasonal.
(d) Maximum displacement from average sales of $8 million

59. False. The scenario does not create a right triangle because the tower is not vertical.

Review Exercises   *(page 340)*

1. (a)   **3.** (a)

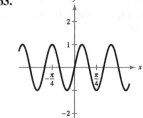

(b) Quadrant IV   (b) Quadrant III
(c) $\dfrac{23\pi}{4}, -\dfrac{\pi}{4}$ (c) 250°, −470°

5. 7.854   **7.** −0.279   **9.** 54°   **11.** −200.535°

13. 198° 24'   **15.** About 48.17 in.   **17.** About 523.60 in.²

19. $\left(-\dfrac{1}{2}, \dfrac{\sqrt{3}}{2}\right)$ **21.** $\left(-\dfrac{\sqrt{3}}{2}, -\dfrac{1}{2}\right)$

23. $\sin \dfrac{3\pi}{4} = \dfrac{\sqrt{2}}{2}$ $\csc \dfrac{3\pi}{4} = \sqrt{2}$

$\cos \dfrac{3\pi}{4} = -\dfrac{\sqrt{2}}{2}$ $\sec \dfrac{3\pi}{4} = -\sqrt{2}$

$\tan \dfrac{3\pi}{4} = -1$ $\cot \dfrac{3\pi}{4} = -1$

25. $\dfrac{\sqrt{2}}{2}$ **27.** $-\dfrac{\sqrt{3}}{2}$ **29.** 3.2361   **31.** −75.3130

33. $\sin \theta = \dfrac{4\sqrt{41}}{41}$ $\csc \theta = \dfrac{\sqrt{41}}{4}$

$\cos \theta = \dfrac{5\sqrt{41}}{41}$ $\sec \theta = \dfrac{\sqrt{41}}{5}$

$\tan \theta = \dfrac{4}{5}$ $\cot \theta = \dfrac{5}{4}$

35. 0.6494   **37.** 3.6722

39. (a) 3   (b) $\dfrac{2\sqrt{2}}{3}$ (c) $\dfrac{3\sqrt{2}}{4}$ (d) $\dfrac{\sqrt{2}}{4}$

41. About 73.3 m

43. $\sin \theta = \dfrac{4}{5}$ $\csc \theta = \dfrac{5}{4}$

$\cos \theta = \dfrac{3}{5}$ $\sec \theta = \dfrac{5}{3}$

$\tan \theta = \dfrac{4}{3}$ $\cot \theta = \dfrac{3}{4}$

45. $\sin \theta = \dfrac{4}{5}$ $\csc \theta = \dfrac{5}{4}$

$\cos \theta = \dfrac{3}{5}$ $\sec \theta = \dfrac{5}{3}$

$\tan \theta = \dfrac{4}{3}$ $\cot \theta = \dfrac{3}{4}$

47. $\sin \theta = -\dfrac{\sqrt{11}}{6}$ $\csc \theta = -\dfrac{6\sqrt{11}}{11}$

$\cos \theta = \dfrac{5}{6}$

$\tan \theta = -\dfrac{\sqrt{11}}{5}$ $\cot \theta = -\dfrac{5\sqrt{11}}{11}$

49. $\sin \theta = \dfrac{\sqrt{21}}{5}$ $\csc \theta = \dfrac{5\sqrt{21}}{21}$

$\sec \theta = -\dfrac{5}{2}$

$\tan \theta = -\dfrac{\sqrt{21}}{2}$ $\cot \theta = -\dfrac{2\sqrt{21}}{21}$

51. $\theta' = 84°$ **53.** $\theta' = \dfrac{\pi}{5}$

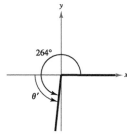   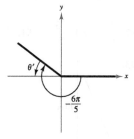

55. $\sin(-150°) = -\dfrac{1}{2}$; $\cos(-150°) = -\dfrac{\sqrt{3}}{2}$; $\tan(-150°) = \dfrac{\sqrt{3}}{3}$

57. $\sin \dfrac{\pi}{3} = \dfrac{\sqrt{3}}{2}$; $\cos \dfrac{\pi}{3} = \dfrac{1}{2}$; $\tan \dfrac{\pi}{3} = \sqrt{3}$

59. 0.9613   **61.** −0.4452

63.   **65.**

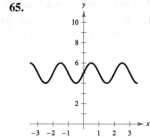

67.

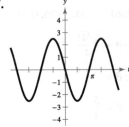

69. (a) $y = 2 \sin 528\pi x$ (b) 264 cycles/sec

71.   **73.**

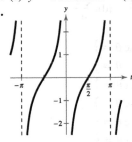   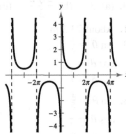

75.

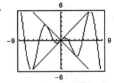

As $x \to +\infty$, $f(x)$ oscillates.

77. $-\dfrac{\pi}{2}$ **79.** $\dfrac{\pi}{6}$ **81.** -0.92 **83.** 0.06

85.

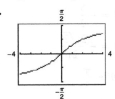

87. $\dfrac{4}{5}$ **89.** $\dfrac{13}{5}$ **91.** $\dfrac{\sqrt{4 - x^2}}{x}$

93.

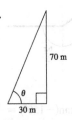

$\theta \approx 66.8°$

95. About 1221 mi, about 85.6°
97. False. For each θ there corresponds exactly one value of y.
99. The function is undefined because $\sec \theta = 1/\cos \theta$.
101. The ranges of the other four trigonometric functions are $(-\infty, \infty)$ or $(-\infty, -1] \cup [1, \infty)$.

Chapter Test   *(page 343)*

1. (a)

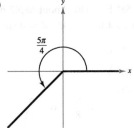

(b) $\dfrac{13\pi}{4}, \ -\dfrac{3\pi}{4}$

(c) $225°$

2. 3500 rad/min   **3.** About 709.04 ft²

4. $\sin \theta = \dfrac{3\sqrt{13}}{13}$ $\csc \theta = \dfrac{\sqrt{13}}{3}$

$\cos \theta = \dfrac{2\sqrt{13}}{13}$ $\sec \theta = \dfrac{\sqrt{13}}{2}$

$\cot \theta = \dfrac{2}{3}$

5. $\sin \theta = \dfrac{3\sqrt{10}}{10}$ $\csc \theta = \dfrac{\sqrt{10}}{3}$

$\cos \theta = -\dfrac{\sqrt{10}}{10}$ $\sec \theta = -\sqrt{10}$

$\tan \theta = -3$ $\cot \theta = -\dfrac{1}{3}$

6. $\theta' = 25°$

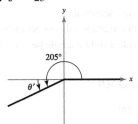

7. Quadrant III   **8.** 150°, 210°

9. $\sin \theta = -\frac{4}{5}$ **10.** $\sin \theta = \frac{21}{29}$

$\tan \theta = -\frac{4}{3}$ $\cos \theta = -\frac{20}{29}$

$\csc \theta = -\frac{5}{4}$ $\tan \theta = -\frac{21}{20}$

$\sec \theta = \frac{5}{3}$ $\csc \theta = \frac{29}{21}$

$\cot \theta = -\frac{3}{4}$ $\cot \theta = -\frac{20}{21}$

11.   **12.**

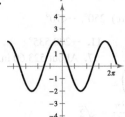   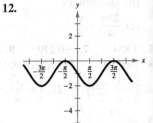

3.

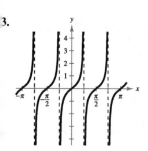

4.

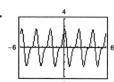

Period: 2

15.

Not periodic;
As $x \to \infty$, $y \to 0$.

6. $a = -2$, $b = \dfrac{1}{2}$, $c = -\dfrac{\pi}{4}$ **17.** $\dfrac{\sqrt{55}}{3}$

8.

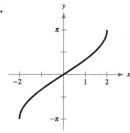

19. About 309.3°   **20.** $d = -6 \cos \pi t$

Problem Solving   *(page 345)*

1. (a) $\dfrac{11\pi}{2}$ rad or 990°   (b) About 816.42 ft

3. $h = 51 - 50 \sin\left(8\pi t + \dfrac{\pi}{2}\right)$

5. (a) About 4767 ft   (b) About 3705 ft

(c) $w \approx 2183$ ft, $\tan 63° = \dfrac{w + 3705}{3000}$

7. (a)

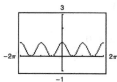

Even

(b)

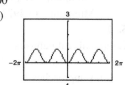

Even

9. (a)

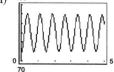

(b) Period = $\dfrac{3}{4}$ sec; Answers will vary.

(c) 20 mm; Answers will vary.   (d) 80 beats/min

(e) Period = $\dfrac{15}{16}$ sec; $\dfrac{32\pi}{15}$

11. (a)

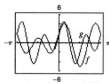

(b) Period of f: 2π
Period of g: π

(c) Yes, because the sine and cosine functions are periodic.

13. (a) About 40.5°   (b) $x \approx 1.71$ ft; $y \approx 3.46$ ft

(c) About 1.75 ft

(d) As you move closer to the rock, d must get smaller and smaller. The angles θ_1 and θ_2 will decrease along with the distance y, so d will decrease.

Chapter 5

Section 5.1   *(page 353)*

1. $\tan u$ **3.** $\cot u$ **5.** 1

7. $\sin x = \dfrac{\sqrt{21}}{5}$ $\csc x = \dfrac{5\sqrt{21}}{21}$

$\cos x = -\dfrac{2}{5}$ $\sec x = -\dfrac{5}{2}$

$\tan x = -\dfrac{\sqrt{21}}{2}$ $\cot x = -\dfrac{2\sqrt{21}}{21}$

9. $\sin \theta = -\dfrac{3}{4}$ $\csc \theta = -\dfrac{4}{3}$

$\cos \theta = \dfrac{\sqrt{7}}{4}$ $\sec \theta = \dfrac{4\sqrt{7}}{7}$

$\tan \theta = -\dfrac{3\sqrt{7}}{7}$ $\cot \theta = -\dfrac{\sqrt{7}}{3}$

11. $\sin x = \dfrac{2\sqrt{13}}{13}$ $\csc x = \dfrac{\sqrt{13}}{2}$

$\cos x = \dfrac{3\sqrt{13}}{13}$ $\sec x = \dfrac{\sqrt{13}}{3}$

$\tan x = \dfrac{2}{3}$ $\cot x = \dfrac{3}{2}$

13. c   **14.** b   **15.** f   **16.** a   **17.** e   **18.** d

19. $\cos \theta$ **21.** $\sin^2 x$ **23.** $\sec x + 1$ **25.** $\sin^4 x$

27. $\csc^2 x(\cot x + 1)$ **29.** $(3 \sin x + 1)(\sin x - 2)$

31. $(\csc x - 1)(\csc x + 2)$ **33.** $\sec \theta$ **35.** $\cos^2 \phi$

37. $\sec \beta$ **39.** $\sin^2 x$ **41.** $1 + 2 \sin x \cos x$

43. $2 \csc^2 x$ **45.** $-2 \tan x$ **47.** $-\cot x$ **49.** $1 + \cos y$

51. $\sin x$ **53.** $3 \sin \theta$ **55.** $2 \tan \theta$

57. $2 \cos \theta = \sqrt{2}$; $\sin \theta = \pm\dfrac{\sqrt{2}}{2}$; $\cos \theta = \dfrac{\sqrt{2}}{2}$

59. $0 \le \theta \le \pi$ **61.** $\ln|\cos x|$

63. $\ln|\csc t \sec t|$ **65.** $\mu = \tan \theta$

67. True. $\csc x = \dfrac{1}{\sin x}$, $\sec x = \dfrac{1}{\cos x}$, $\tan x = \dfrac{\sin x}{\cos x}$, $\cot x = \dfrac{\cos x}{\sin x}$

69. $\infty, 0$ **71.** $\cos(-\theta) = \cos \theta$

73. $\cos \theta = \pm\sqrt{1 - \sin^2 \theta}$ **75.** $\dfrac{\sin \theta}{\cos \theta}$

$\tan \theta = \pm\dfrac{\sin \theta}{\sqrt{1 - \sin^2 \theta}}$

$\cot \theta = \pm\dfrac{\sqrt{1 - \sin^2 \theta}}{\sin \theta}$

$\sec \theta = \pm\dfrac{1}{\sqrt{1 - \sin^2 \theta}}$

$\csc \theta = \dfrac{1}{\sin \theta}$

Section 5.2 *(page 360)*

1. identity   **3.** tan u **5.** sin u **7.** $-\csc u$

9–41. Answers will vary.

43. $\cot(-x) = -\cot x$

45. (a)

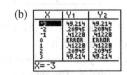

(b)

X	Y₁	Y₂

X = -3   Identity

(c) Answers will vary.

47. (a)

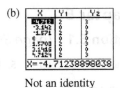

(b)

X	Y₁	Y₂

X = -4.71238898038   Not an identity

(c) Answers will vary.

49. (a)

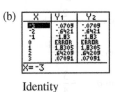

(b)

X	Y₁	Y₂

X = -3   Identity

(c) Answers will vary.

51–53. Answers will vary.   **55.** 1

57–61. Answers will vary.

63. False. $\tan x^2 = \tan(x \cdot x) \neq \tan^2 x = (\tan x)(\tan x)$

65. False. An identity is an equation that is true for all real values of θ.

67. The equation is not an identity because $\sin \theta = \pm\sqrt{1 - \cos^2 \theta}$.

Sample answer: $\dfrac{7\pi}{4}$

69. The equation is not an identity because $1 - \cos^2 \theta = \sin^2 \theta$.

Sample answer: $-\dfrac{\pi}{2}$

Section 5.3 *(page 370)*

1. isolate   **3.** quadratic   **5–9.** Answers will vary.

11. $\dfrac{\pi}{3} + 2n\pi, \dfrac{2\pi}{3} + 2n\pi$ **13.** $\dfrac{2\pi}{3} + 2n\pi, \dfrac{4\pi}{3} + 2n\pi$

15. $\dfrac{\pi}{6} + n\pi, \dfrac{5\pi}{6} + n\pi$ **17.** $\dfrac{\pi}{3} + n\pi, \dfrac{2\pi}{3} + n\pi$

19. $n\pi, \dfrac{3\pi}{2} + 2n\pi$ **21.** $\dfrac{n\pi}{2}$ **23.** $n\pi, \dfrac{\pi}{6} + n\pi, \dfrac{5\pi}{6} + n\pi$

25. $\pi + 2n\pi, \dfrac{\pi}{3} + 2n\pi, \dfrac{5\pi}{3} + 2n\pi$

27. $\dfrac{\pi}{3} + 2n\pi, \pi + 2n\pi, \dfrac{5\pi}{3} + 2n\pi$ **29.** $\dfrac{\pi}{4}, \dfrac{5\pi}{4}$

31. $\dfrac{\pi}{2}, \dfrac{3\pi}{2}, \dfrac{2\pi}{3}, \dfrac{4\pi}{3}$ **33.** $\dfrac{\pi}{3}, \dfrac{2\pi}{3}, \dfrac{4\pi}{3}, \dfrac{5\pi}{3}$ **35.** No solution

37. $\dfrac{\pi}{2}$ **39.** $\dfrac{\pi}{6} + n\pi, \dfrac{5\pi}{6} + n\pi$ **41.** $\dfrac{\pi}{12} + \dfrac{n\pi}{3}$

43. $\dfrac{\pi}{2} + 4n\pi, \dfrac{7\pi}{2} + 4n\pi$ **45.** $\dfrac{\pi}{3} + 2n\pi$ **47.** $3 + 4n$

49. 3.553, 5.872   **51.** 1.249, 4.391   **53.** 0.739

55. 0.955, 2.186, 4.097, 5.328   **57.** 1.221, 1.921, 4.362, 5.062

59. $\arctan(-4) + n\pi, \arctan 3 + n\pi$

61. $\dfrac{\pi}{4} + n\pi, \arctan 5 + n\pi$ **63.** $\dfrac{\pi}{3} + 2n\pi, \dfrac{5\pi}{3} + 2n\pi$

65. $\arctan\frac{1}{3} + n\pi, \arctan\left(-\frac{1}{3}\right) + n\pi$

67. $\arccos\frac{1}{4} + 2n\pi, -\arccos\frac{1}{4} + 2n\pi$

69. $\dfrac{\pi}{2} + 2n\pi, \arcsin\left(-\dfrac{1}{4}\right) + 2n\pi, \arcsin\dfrac{1}{4} + 2n\pi$

71. 0.3398, 0.8481, 2.2935, 2.8018

73. 1.9357, 2.7767, 5.0773, 5.9183

75. $-1.154, 0.534$ **77.** 1.110

79. (a)

(b) $\dfrac{\pi}{3} \approx 1.0472$

$\dfrac{5\pi}{3} \approx 5.2360$

0

$\pi \approx 3.1416$

Maximum: $(1.0472, 1.25)$
Maximum: $(5.2360, 1.25)$
Minimum: $(0, 1)$
Minimum: $(3.1416, -1)$

81. (a)

(b) $\dfrac{\pi}{4} \approx 0.7854$

$\dfrac{5\pi}{4} \approx 3.9270$

Maximum: $(0.7854, 1.4142)$
Minimum: $(3.9270, -1.4142)$

83. (a)

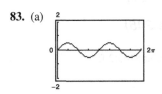

(b) $\dfrac{\pi}{4} \approx 0.7854$

$\dfrac{5\pi}{4} \approx 3.9270$

$\dfrac{3\pi}{4} \approx 2.3562$

$\dfrac{7\pi}{4} \approx 5.4978$

Maximum: $(0.7854. 0.5)$
Maximum: $(3.9270, 0.5)$
Minimum: $(2.3562, -0.5)$
Minimum: $(5.4978, -0.5)$

85. 1

87. (a) All real numbers x except $x = 0$

(b) y-axis symmetry; Horizontal asymptote: $y = 0$

(c) y approaches 1.

(d) Four solutions: $\pm\pi, \pm 2\pi$

89. 0.04 sec, 0.43 sec, 0.83 sec

91. January, November, December

93. (a) and (c)

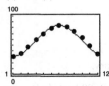

The model fits the data well.

(b) $C = 26.55 \cos\left(\dfrac{\pi t}{6} - \dfrac{7\pi}{6}\right) + 57.55$ (d) $57.55°F$

(e) Above $72°F$: June through September

Below $72°F$: October through May

5. (a)

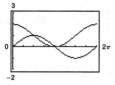

$A \approx 1.12$

(b) $0.6 < x < 1.1$

7. 1

9. True. The first equation has a smaller period than the second equation, so it will have more solutions in the interval $[0, 2\pi)$.

1. The equation would become $\cos^2 x = 2$; this is not the correct method to use when solving equations.

3. (a)

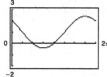

Graphs intersect when $x = \dfrac{\pi}{2}$ and $x = \pi$.

(b)

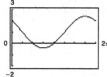

x-intercepts: $\left(\dfrac{\pi}{2}, 0\right), (\pi, 0)$

(c) Yes; Answers will vary.

ection 5.4   *(page 378)*

1. $\sin u \cos v - \cos u \sin v$ **3.** $\dfrac{\tan u + \tan v}{1 - \tan u \tan v}$

5. $\cos u \cos v + \sin u \sin v$

7. (a) $\dfrac{\sqrt{2} - \sqrt{6}}{4}$ (b) $\dfrac{\sqrt{2} + 1}{2}$

9. (a) $\dfrac{\sqrt{6} + \sqrt{2}}{4}$ (b) $\dfrac{\sqrt{2} - \sqrt{3}}{2}$

1. $\sin \dfrac{11\pi}{12} = \dfrac{\sqrt{2}}{4}\left(\sqrt{3} - 1\right)$

$\cos \dfrac{11\pi}{12} = -\dfrac{\sqrt{2}}{4}\left(\sqrt{3} + 1\right)$

$\tan \dfrac{11\pi}{12} = -2 + \sqrt{3}$

3. $\sin \dfrac{17\pi}{12} = -\dfrac{\sqrt{2}}{4}\left(\sqrt{3} + 1\right)$ **15.** $\sin 105° = \dfrac{\sqrt{2}}{4}\left(\sqrt{3} + 1\right)$

$\cos \dfrac{17\pi}{12} = \dfrac{\sqrt{2}}{4}\left(1 - \sqrt{3}\right)$ $\cos 105° = \dfrac{\sqrt{2}}{4}\left(1 - \sqrt{3}\right)$

$\tan \dfrac{17\pi}{12} = 2 + \sqrt{3}$ $\tan 105° = -2 - \sqrt{3}$

7. $\sin(-195°) = \dfrac{\sqrt{2}}{4}\left(\sqrt{3} - 1\right)$

$\cos(-195°) = -\dfrac{\sqrt{2}}{4}\left(\sqrt{3} + 1\right)$

$\tan(-195°) = -2 + \sqrt{3}$

19. $\sin \dfrac{13\pi}{12} = \dfrac{\sqrt{2}}{4}\left(1 - \sqrt{3}\right)$

$\cos \dfrac{13\pi}{12} = -\dfrac{\sqrt{2}}{4}\left(1 + \sqrt{3}\right)$

$\tan \dfrac{13\pi}{12} = 2 - \sqrt{3}$

21. $\sin\left(-\dfrac{5\pi}{12}\right) = -\dfrac{\sqrt{2}}{4}\left(1 + \sqrt{3}\right)$

$\cos\left(-\dfrac{5\pi}{12}\right) = \dfrac{\sqrt{2}}{4}\left(\sqrt{3} - 1\right)$

$\tan\left(-\dfrac{5\pi}{12}\right) = -2 - \sqrt{3}$

23. $\sin 285° = -\dfrac{\sqrt{2}}{4}\left(\sqrt{3} + 1\right)$

$\cos 285° = \dfrac{\sqrt{2}}{4}\left(\sqrt{3} - 1\right)$

$\tan 285° = -\left(2 + \sqrt{3}\right)$

25. $\sin(-165°) = -\dfrac{\sqrt{2}}{4}\left(\sqrt{3} - 1\right)$

$\cos(-165°) = -\dfrac{\sqrt{2}}{4}\left(1 + \sqrt{3}\right)$

$\tan(-165°) = 2 - \sqrt{3}$

27. $\sin 1.8$ **29.** $\sin 75°$ **31.** $\tan \dfrac{7\pi}{15}$ **33.** $\cos(3x - 2y)$

35. $\dfrac{\sqrt{3}}{2}$ **37.** $-\dfrac{1}{2}$ **39.** 0   **41.** $-\dfrac{13}{85}$ **43.** $-\dfrac{13}{84}$

45. $\dfrac{85}{36}$ **47.** $\dfrac{3}{5}$ **49.** $-\dfrac{44}{117}$ **51.** $-\dfrac{125}{44}$ **53.** 1   **55.** 0

57–63. Answers will vary.   **65.** $-\sin \theta$ **67.** $-\sec \theta$

69. $\dfrac{\pi}{6}, \dfrac{5\pi}{6}$ **71.** $\dfrac{5\pi}{4}, \dfrac{7\pi}{4}$ **73.** $0, \dfrac{\pi}{3}, \pi, \dfrac{5\pi}{3}$ **75.** $\dfrac{\pi}{4}, \dfrac{7\pi}{4}$

77. $\dfrac{\pi}{2}, \pi, \dfrac{3\pi}{2}$

79. (a) $y = \dfrac{5}{12}\sin(2t + 0.6435)$

(b) $\dfrac{5}{12}$ ft   (c) $\dfrac{1}{\pi}$ cycle/sec

81. True. $\sin(u \pm v) = \sin u \cos v \pm \cos u \sin v$

83. False. $\sin \alpha \cos \beta = -\cos \alpha \sin \beta$

85. The denominator should be $1 + \tan x \tan \dfrac{\pi}{4} = 1 + \tan x$.

87–89. Answers will vary.

91. (a) $\sqrt{2}\sin\left(\theta + \dfrac{\pi}{4}\right)$ (b) $\sqrt{2}\cos\left(\theta - \dfrac{\pi}{4}\right)$

93. (a) $13\sin(3\theta + 0.3948)$ (b) $13\cos(3\theta - 1.1760)$

95. $\sqrt{2}\sin \theta + \sqrt{2}\cos \theta$ **97.** 15°

99.   **101.** Proof

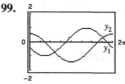

No, $y_1 \neq y_2$ because their graphs are different.

Section 5.5 (page 388)

1. $2 \sin u \cos u$ **3.** $\frac{1}{2}[\sin(u+v) + \sin(u-v)]$

5. $\pm\sqrt{\dfrac{1-\cos u}{2}}$ **7.** $n\pi, \dfrac{\pi}{3} + 2n\pi, \dfrac{5\pi}{3} + 2n\pi$ **9.** $\dfrac{2n\pi}{3}$

11. $\dfrac{n\pi}{2}$ **13.** $\dfrac{\pi}{6} + n\pi, \dfrac{\pi}{2} + n\pi, \dfrac{5\pi}{6} + n\pi$ **15.** $3 \sin 2x$

17. $3 \cos 2x$ **19.** $4 \cos 2x$

21. $\sin 2u = -\frac{24}{25}, \cos 2u = \frac{7}{25}, \tan 2u = -\frac{24}{7}$

23. $\sin 2u = \frac{15}{17}, \cos 2u = \frac{8}{17}, \tan 2u = \frac{15}{8}$

25. $8 \cos^4 x - 8 \cos^2 x + 1$ **27.** $\frac{1}{8}(3 + 4\cos 2x + \cos 4x)$

29. $\frac{1}{8}(3 - 4\cos 4x + \cos 8x)$ **31.** $\dfrac{(3 - 4\cos 4x + \cos 8x)}{(3 + 4\cos 4x + \cos 8x)}$

33. $\frac{1}{8}(1 - \cos 8x)$

35. $\sin 75° = \frac{1}{2}\sqrt{2 + \sqrt{3}}$

$\cos 75° = \frac{1}{2}\sqrt{2 - \sqrt{3}}$

$\tan 75° = 2 + \sqrt{3}$

37. $\sin 112° \, 30' = \frac{1}{2}\sqrt{2 + \sqrt{2}}$

$\cos 112° \, 30' = -\frac{1}{2}\sqrt{2 - \sqrt{2}}$

$\tan 112° \, 30' = -1 - \sqrt{2}$

39. $\sin \dfrac{\pi}{8} = \frac{1}{2}\sqrt{2 - \sqrt{2}}$

$\cos \dfrac{\pi}{8} = \frac{1}{2}\sqrt{2 + \sqrt{2}}$

$\tan \dfrac{\pi}{8} = \sqrt{2} - 1$

41. (a) Quadrant I

(b) $\sin \dfrac{u}{2} = \dfrac{3}{5}, \cos \dfrac{u}{2} = \dfrac{4}{5}, \tan \dfrac{u}{2} = \dfrac{3}{4}$

43. (a) Quadrant II

(b) $\sin \dfrac{u}{2} = \dfrac{\sqrt{26}}{26}, \cos \dfrac{u}{2} = -\dfrac{5\sqrt{26}}{26}, \tan \dfrac{u}{2} = -\dfrac{1}{5}$

45. π **47.** $\dfrac{\pi}{3}, \pi, \dfrac{5\pi}{3}$ **49.** $\frac{1}{2}(\cos 2\theta - \cos 8\theta)$

51. $\frac{1}{2}(\cos(-2\theta) + \cos 6\theta)$ **53.** $2 \cos 4\theta \sin \theta$

55. $2 \cos 4x \cos 2x$ **57.** $\dfrac{\sqrt{6}}{2}$ **59.** $-\sqrt{2}$

61. $0, \dfrac{\pi}{4}, \dfrac{\pi}{2}, \dfrac{3\pi}{4}, \pi, \dfrac{5\pi}{4}, \dfrac{3\pi}{2}, \dfrac{7\pi}{4}$ **63.** $\dfrac{\pi}{6}, \dfrac{5\pi}{6}$

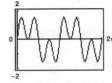
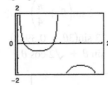

65–69. Answers will vary.

71. (a) $\cos \theta = \dfrac{M^2 - 2}{M^2}$ (b) $\dfrac{\pi}{3}$ (c) 0.4482

(d) 1520 mi/h; 3420 mi/h

73. $x = 2r(1 - \cos \theta)$

75. True. $\sin(-2x) = 2 \sin(-x) \cos(-x) = -2 \sin x \cos x$.

77. Answers will vary.

Review Exercises (page 392)

1. $\cot x$ **3.** $\cos x$

5. $\sin \theta = -\dfrac{\sqrt{21}}{5}$ $\csc \theta = -\dfrac{5\sqrt{21}}{21}$

$\cos \theta = -\dfrac{2}{5}$ $\sec \theta = -\dfrac{5}{2}$

$\tan \theta = \dfrac{\sqrt{21}}{2}$ $\cot \theta = \dfrac{2\sqrt{21}}{21}$

7. $\sin^2 x$ **9.** 1 **11.** $\tan u \sec u$ **13.** $\cot^2 x$

15. $-2 \tan^2 \theta$ **17.** $5 \cos \theta$ **19–25.** Answers will vary.

27. $\dfrac{\pi}{3} + 2n\pi, \dfrac{2\pi}{3} + 2n\pi$ **29.** $\dfrac{\pi}{6} + n\pi$

31. $\dfrac{\pi}{3} + n\pi, \dfrac{2\pi}{3} + n\pi$ **33.** $0, \dfrac{\pi}{2}, \pi, \dfrac{3\pi}{2}$ **35.** $0, \dfrac{\pi}{2}, \pi$

37. $\dfrac{\pi}{8}, \dfrac{3\pi}{8}, \dfrac{9\pi}{8}, \dfrac{11\pi}{8}$ **39.** $\dfrac{\pi}{2}$

41. $0, \dfrac{\pi}{8}, \dfrac{3\pi}{8}, \dfrac{5\pi}{8}, \dfrac{7\pi}{8}, \dfrac{9\pi}{8}, \dfrac{11\pi}{8}, \dfrac{13\pi}{8}, \dfrac{15\pi}{8}$

43. $n\pi, \arctan 2 + n\pi$ **45.** $\arctan(-3) + n\pi, \arctan 2 + n\pi$

47. $\sin 75° = \dfrac{\sqrt{2}}{4}(1 + \sqrt{3})$ **49.** $\sin \dfrac{25\pi}{12} = \dfrac{\sqrt{2}}{4}(\sqrt{3} - 1)$

$\cos 75° = \dfrac{\sqrt{2}}{4}(\sqrt{3} - 1)$ $\cos \dfrac{25\pi}{12} = \dfrac{\sqrt{2}}{4}(\sqrt{3} + 1)$

$\tan 75° = 2 + \sqrt{3}$ $\tan \dfrac{25\pi}{12} = 2 - \sqrt{3}$

51. $\sin 15°$ **53.** $-\frac{24}{25}$ **55.** -1

57–59. Answers will vary.   **61.** $\dfrac{\pi}{4}, \dfrac{7\pi}{4}$

63. $\sin 2u = \frac{24}{25}$

$\cos 2u = -\frac{7}{25}$

$\tan 2u = -\frac{24}{7}$

65. Answers will vary.   **67.** $\dfrac{1 - \cos 6x}{1 + \cos 6x}$

69. $\sin(-75°) = -\frac{1}{2}\sqrt{2 + \sqrt{3}}$

$\cos(-75°) = \frac{1}{2}\sqrt{2 - \sqrt{3}}$

$\tan(-75°) = -2 - \sqrt{3}$

71. (a) Quadrant II

(b) $\sin \dfrac{u}{2} = \dfrac{2\sqrt{5}}{5}, \cos \dfrac{u}{2} = -\dfrac{\sqrt{5}}{5}, \tan \dfrac{u}{2} = -2$

73. (a) Quadrant I

(b) $\sin \dfrac{u}{2} = \dfrac{3\sqrt{14}}{14}, \cos \dfrac{u}{2} = \dfrac{\sqrt{70}}{14}, \tan \dfrac{u}{2} = \dfrac{3\sqrt{5}}{5}$

75. $\frac{1}{2}[\sin 10\theta - \sin(-2\theta)]$

77. $2 \cos \dfrac{11\theta}{2} \cos \dfrac{\theta}{2}$ **79.** $\theta = 15°$ or $\dfrac{\pi}{12}$

81. False. If $(\pi/2) < \theta < \pi$, then $\theta/2$ lies in Quadrant I.

83. True. $4 \sin(-x) \cos(-x) = 4(-\sin x) \cos x$

$= -4 \sin x \cos x$

$= -2(2 \sin x \cos x)$

$= -2 \sin 2x$

85. Yes. *Sample answer:* $\sin x = \frac{1}{2}$ has an infinite number of solutions.

Chapter Test   *(page 394)*

1. $\sin \theta = \dfrac{2}{5}$ $\csc \theta = \dfrac{5}{2}$

  $\cos \theta = -\dfrac{\sqrt{21}}{5}$ $\sec \theta = -\dfrac{5\sqrt{21}}{21}$

  $\tan \theta = -\dfrac{2\sqrt{21}}{21}$ $\cot \theta = -\dfrac{\sqrt{21}}{2}$

2. 1   **3.** 1   **4.** $\csc \theta \sec \theta$ **5–10.** Answers will vary.

11. $2(\sin 5\theta + \sin \theta)$ **12.** $-2 \sin \theta$

13. $0, \dfrac{3\pi}{4}, \pi, \dfrac{7\pi}{4}$ **14.** $\dfrac{\pi}{6}, \dfrac{\pi}{2}, \dfrac{5\pi}{6}, \dfrac{3\pi}{2}$ **15.** $\dfrac{\pi}{6}, \dfrac{5\pi}{6}, \dfrac{7\pi}{6}, \dfrac{11\pi}{6}$

16. $\dfrac{\pi}{6}, \dfrac{5\pi}{6}, \dfrac{3\pi}{2}$ **17.** $0, 2.596$ **18.** $\dfrac{\sqrt{2} - \sqrt{6}}{4}$

19. $\sin 2u = -\dfrac{20}{29}, \cos 2u = -\dfrac{21}{29}, \tan 2u = \dfrac{20}{21}$

20. Day 30 to day 310

21. 0.26 min, 0.58 min, 0.89 min, 1.20 min, 1.52 min, 1.83 min

Problem Solving   *(page 397)*

1. $\sin \theta = \pm\sqrt{1 - \cos^2 \theta}$

  $\tan \theta = \pm\dfrac{\sqrt{1 - \cos^2 \theta}}{\cos \theta}$

  $\csc \theta = \pm\dfrac{1}{\sqrt{1 - \cos^2 \theta}}$

  $\sec \theta = \dfrac{1}{\cos \theta}$

  $\cot \theta = \pm\dfrac{\cos \theta}{\sqrt{1 - \cos^2 \theta}}$

3. Answers will vary.

5. $u + v = w$; Proof

7. (a) $A = 100 \sin \dfrac{\theta}{2} \cos \dfrac{\theta}{2}$ (b) $A = 50 \sin \theta; \theta = \dfrac{\pi}{2}$

9. (a) $F = \dfrac{0.6W \cos \theta}{\sin 12°}$

  (b) 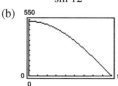   (c) Maximum: $\theta = 0°$
                                  Minimum: $\theta = 90°$

11. (a) High tides: 6:12 A.M., 6:36 P.M.
     Low tides: 12:00 A.M., 12:24 P.M.
  (b) The water depth never falls below 7 feet.
  (c)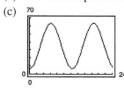

13. (a) $n = \dfrac{1}{2}\left(\cot \dfrac{\theta}{2} + \sqrt{3}\right)$ (b) $\theta \approx 76.5°$

15. (a) $\dfrac{\pi}{6} \le x \le \dfrac{5\pi}{6}$ (b) $\dfrac{2\pi}{3} \le x \le \dfrac{4\pi}{3}$

  (c) $\dfrac{\pi}{2} < x < \pi, \dfrac{3\pi}{2} < x < 2\pi$

  (d) $0 \le x \le \dfrac{\pi}{4}, \dfrac{5\pi}{4} \le x < 2\pi$

Chapter 6

Section 6.1   *(page 406)*

1. oblique   **3.** angles; side

5. $A = 30°, a \approx 14.14, c \approx 27.32$

7. $C = 105°, a \approx 5.94, b \approx 6.65$

9. $B = 60.9°, b \approx 19.32, c \approx 6.36$

11. $B = 42° 4', a \approx 22.05, b \approx 14.88$

13. $C = 80°, a \approx 5.82, b \approx 9.20$

15. $C = 83°, a \approx 0.62, b \approx 0.51$

17. $B \approx 21.55°, C \approx 122.45°, c \approx 11.49$

19. $B \approx 9.43°, C \approx 25.57°, c \approx 10.53$

21. $A \approx 10° 11', C \approx 154° 19', c \approx 11.03$

23. $B \approx 48.74°, C \approx 21.26°, c \approx 48.23$

25. No solution

27. Two solutions:
  $B \approx 72.21°, C \approx 49.79°, c \approx 10.27$
  $B \approx 107.79°, C \approx 14.21°, c \approx 3.30$

29. No solution   **31.** $B = 45°, C = 90°, c \approx 1.41$

33. (a) $b \le 5, b = \dfrac{5}{\sin 36°}$ (b) $5 < b < \dfrac{5}{\sin 36°}$

  (c) $b > \dfrac{5}{\sin 36°}$

35. (a) $b < 80$ (b) Not possible   (c) $b \ge 80$

37. 22.1   **39.** 94.4   **41.** 218.0   **43.** 22.3

45. (a) $\dfrac{h}{\sin 30°} = \dfrac{40}{\sin 56°}$ (b) About 24.1 m

47. From Pine Knob: about 42.4 km
   From Colt Station: about 15.5 km

49. About 16.1°

51. (a) 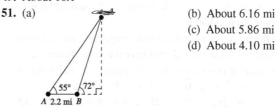   (b) About 6.16 mi
                                  (c) About 5.86 mi
                                  (d) About 4.10 mi

53. $d = \dfrac{2 \sin \theta}{\sin(\phi - \theta)}$

55. True. If an angle of a triangle is obtuse, then the other two angles must be acute.

57. False. When just three angles are known, the triangle cannot be solved.

59. The formula is $A = \dfrac{1}{2}ab \sin C$, so solve the triangle to find the value of a.

61. Yes. $A = 40°, b \approx 11.9, c \approx 15.6$; Yes; *Sample answer:* Use the definitions of cosine and tangent.

Section 6.2   *(page 413)*

1. $b^2 = a^2 + c^2 - 2ac \cos B$ **3.** standard

5. $A \approx 38.62°, B \approx 48.51°, C \approx 92.87°$

7. $A \approx 26.38°, B \approx 36.34°, C \approx 117.28°$

9. $B \approx 23.79°, C \approx 126.21°, a \approx 18.59$

11. $B \approx 29.44°, C \approx 100.56°, a \approx 23.38$

13. $A \approx 30.11°, B \approx 43.16°, C \approx 106.73°$

15. $A \approx 132.77°, B \approx 31.91°, C \approx 15.32°$

17. $B \approx 27.46°, C \approx 32.54°, a \approx 11.27$

19. $A \approx 141° 45', C \approx 27° 40', b \approx 11.87$

21. $A = 27° 10', C = 27° 10', b \approx 65.84$

23. $A \approx 33.80°, B \approx 103.20°, c \approx 0.54$ **25.** $12.07; 5.69; 135°$

27. $13.86; 68.2°; 111.8°$ **29.** $16.96; 77.2°; 102.8°$

31. Yes; $A \approx 102.44°, C \approx 37.56°, b \approx 5.26$

33. No; No solution   **35.** No; $C = 103°, a \approx 0.82, b \approx 0.71$

37. 23.53 **39.** 5.35 **41.** 0.24 **43.** 1514.14

45. About 373.3 m   **47.** About 103.9 ft

49. About 131.1 ft, about 118.6 ft

51.

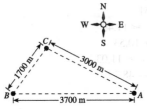

N 37.1° E, S 63.1° E

53. $41.2°, 52.9°$

55. (a) $d = \sqrt{9s^2 - 108s + 1296}$ (b) About 15.87 mi/h

57. About 46,837.5 ft²   **59.** \$83,336.37

61. False. For s to be the average of the lengths of the three sides of the triangle, s would be equal to $(a + b + c)/3$.

63. $c^2 = a^2 + b^2$; The Pythagorean Theorem is a special case of the Law of Cosines.

65. The Law of Cosines can be used to solve the single-solution case of SSA. There is no method that can solve the no-solution case of SSA.

67. Proof

Section 6.3   *(page 425)*

1. directed line segment   **3.** vector   **5.** standard position

7. multiplication; addition

9. Equivalent; **u** and **v** have the same magnitude and direction.

11. Not equivalent; **u** and **v** do not have the same direction.

13. Equivalent; **u** and **v** have the same magnitude and direction.

15. $\mathbf{v} = \langle 1, 3 \rangle; \|\mathbf{v}\| = \sqrt{10}$ **17.** $\mathbf{v} = \langle 0, 5 \rangle; \|\mathbf{v}\| = 5$

19. $\mathbf{v} = \langle -8, 6 \rangle; \|\mathbf{v}\| = 10$ **21.** $\mathbf{v} = \langle -9, -12 \rangle; \|\mathbf{v}\| = 15$

23. $\mathbf{v} = \langle 16, -26 \rangle; \|\mathbf{v}\| = 2\sqrt{233}$

25.

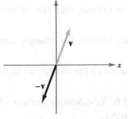

27.

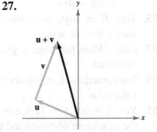

29.

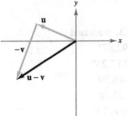

31. (a) $\langle 3, 4 \rangle$ (b) $\langle 1, -2 \rangle$

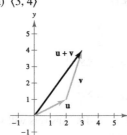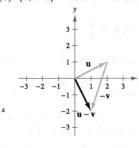

(c) $\langle 1, -7 \rangle$

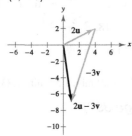

33. (a) $\langle -5, 3 \rangle$ (b) $\langle -5, 3 \rangle$

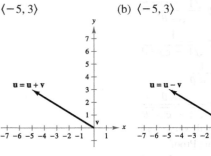

(c) $\langle -10, 6 \rangle$

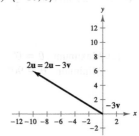

5. (a) $\langle 1, -9 \rangle$ (b) $\langle -1, -5 \rangle$

(c) $\langle -3, -8 \rangle$

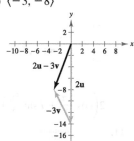

7. 10   **39.** $9\sqrt{5}$ **41.** $\langle 1, 0 \rangle$ **43.** $\left\langle -\dfrac{\sqrt{2}}{2}, \dfrac{\sqrt{2}}{2} \right\rangle$

5. $\left\langle \dfrac{\sqrt{37}}{37}, -\dfrac{6\sqrt{37}}{37} \right\rangle$ **47.** $\mathbf{v} = \langle -6, 8 \rangle$

9. $\mathbf{v} = \left\langle \dfrac{18\sqrt{29}}{29}, \dfrac{45\sqrt{29}}{29} \right\rangle$ **51.** $5\mathbf{i} - 3\mathbf{j}$ **53.** $-6\mathbf{i} + 3\mathbf{j}$

5. $\mathbf{v} = \langle 3, -\tfrac{3}{2} \rangle$ **57.** $\mathbf{v} = \langle 4, 3 \rangle$

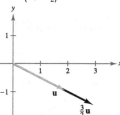

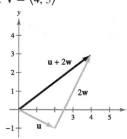

?. $\mathbf{v} = \langle 0, -5 \rangle$

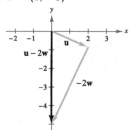

. $\|\mathbf{v}\| = 6\sqrt{2}; \theta = 315°$ **63.** $\|\mathbf{v}\| = 3; \theta = 60°$

. $\mathbf{v} = \langle 3, 0 \rangle$ **67.** $\mathbf{v} = \left\langle -\dfrac{7\sqrt{3}}{4}, \dfrac{7}{4} \right\rangle$

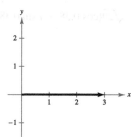

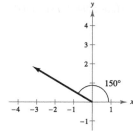

69. $\mathbf{v} = \langle \tfrac{9}{5}, \tfrac{12}{5} \rangle$

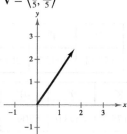

71. $\langle 2, 4 + 2\sqrt{3} \rangle$ **73.** 90°   **75.** About 62.7°
77. Vertical $\approx$ 125.4 ft/sec, horizontal $\approx$ 1193.4 ft/sec
79. About 12.8°; about 398.32 N
81. About 71.3°; about 228.5 lb
83. $T_L \approx$ 15,484 lb
    $T_R \approx$ 19,786 lb
85. $\sqrt{2}$ lb; 1 lb   **87.** About 3154.4 lb   **89.** About 20.8 lb
91. About 19.5°   **93.** N 21.4° E; about 138.7 km/h
95. True. See Example 1.   **97.** True. $a = b = 0$
99. $u_1 = 6 - (-3) = 9$ and $u_2 = -1 - 4 = -5$, so $\mathbf{u} = \langle 9, -5 \rangle$.
101. Proof   **103.** $\langle 1, 3 \rangle$ or $\langle -1, -3 \rangle$
105. (a) $5\sqrt{5 + 4\cos\theta}$

(b)
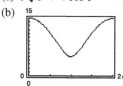

(c) Range: $[5, 15]$
    Maximum is 15 when $\theta = 0$.
    Minimum is 5 when $\theta = \pi$.
(d) The magnitudes of $\mathbf{F}_1$ and $\mathbf{F}_2$ are not the same.
107. (a) and (b) Answers will vary.

Section 6.4   (page 435)

1. dot product   **3.** $\dfrac{\mathbf{u} \cdot \mathbf{v}}{\|\mathbf{u}\|\|\mathbf{v}\|}$ **5.** $\left(\dfrac{\mathbf{u} \cdot \mathbf{v}}{\|\mathbf{v}\|^2} \right)\mathbf{v}$ **7.** -19
9. 0   **11.** 6   **13.** 18; scalar   **15.** $\langle 24, -12 \rangle$; vector
17. 0; vector   **19.** $\sqrt{10} - 1$; scalar   **21.** -12; scalar
23. 17   **25.** $5\sqrt{41}$ **27.** 6   **29.** $\dfrac{\pi}{2}$ **31.** About 2.50
33. 0   **35.** About 0.93   **37.** $\dfrac{5\pi}{12}$ **39.** About 91.33°
41. 90°   **43.** 26.57°, 63.43°, 90°
45. 41.63°, 53.13°, 85.24°   **47.** -20 **49.** $12{,}500\sqrt{3}$
51. Not orthogonal   **53.** Orthogonal   **55.** Not orthogonal
57. $\tfrac{1}{37}\langle 84, 14 \rangle, \tfrac{1}{37}\langle -10, 60 \rangle$ **59.** $\langle 0, 0 \rangle, \langle 4, 2 \rangle$ **61.** $\langle 3, 2 \rangle$
63. $\langle 0, 0 \rangle$ **65.** $\langle -5, 3 \rangle, \langle 5, -3 \rangle$
67. $\tfrac{2}{3}\mathbf{i} + \tfrac{1}{2}\mathbf{j}, -\tfrac{2}{3}\mathbf{i} - \tfrac{1}{2}\mathbf{j}$ **69.** 32
71. (a) \$35,727.50
        This value gives the total amount paid to the employees.
    (b) Multiply $\mathbf{v}$ by 1.02.

73. (a) Force $= 30,000 \sin d$

(b)

d	0°	1°	2°	3°	4°	5°
Force	0	523.6	1047.0	1570.1	2092.7	2614.7

d	6°	7°	8°	9°	10°
Force	3135.9	3656.1	4175.2	4693.0	5209.4

(c) About 29,885.8 lb

75. 735 N-m   **77.** About 779.4 ft-lb

79. About 10,282,651.78 N-m   **81.** About 1174.62 ft-lb

83. False. Work is represented by a scalar.

85. The dot product is the scalar 0.   **87.** 4

89. 1; $\mathbf{u} \cdot \mathbf{u} = \|\mathbf{u}\|^2$

91. (a) $\mathbf{u}$ and $\mathbf{v}$ are parallel.   (b) $\mathbf{u}$ and $\mathbf{v}$ are orthogonal.

93. Proof

Section 6.5   *(page 443)*

1. real   **3.** absolute value   **5.** reflections   **7.** c   **8.** f

9. h   **10.** a   **11.** b   **12.** g   **13.** e   **14.** d

15.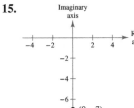

7

17.

10

19.

$2\sqrt{13}$

21. $5 + 6i$ **23.** $10 + 4i$ **25.** $6 + 5i$ **27.** $-5 + 7i$

29. $-2 - 2i$ **31.** $10 - 3i$ **33.** $-6i$ **35.** $-3 + 3i$

37.

$2 - 3i$

39.

$-1 + 2i$

41. $2\sqrt{2} \approx 2.83$ **43.** $\sqrt{109} \approx 10.44$

45. $(4, 3)$ **47.** $\left(\frac{9}{2}, -\frac{3}{2}\right)$

49. (a) Ship A: $3 + 4i$, Ship B: $-5 + 2i$

(b) *Sample answer:* Find the modulus of the difference of the complex numbers.

51. False. The modulus is always real.

53. False. $|1 + i| + |1 - i| = 2\sqrt{2}$ and $|(1 + i) + (1 - i)| =$

55. A circle; The modulus represents the distance from the origin

57. Isosceles triangle; The moduli are equal.

Section 6.6   *(page 452)*

1. trigonometric form; modulus; argument   **3.** nth root

5. 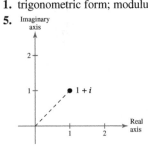   **7.**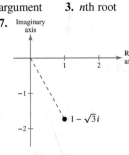

$\sqrt{2}\left(\cos \dfrac{\pi}{4} + i \sin \dfrac{\pi}{4}\right)$ $2\left(\cos \dfrac{5\pi}{3} + i \sin \dfrac{5\pi}{3}\right)$

9. 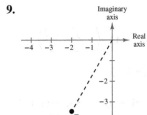   **11.**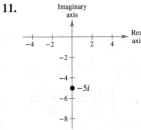

$4\left(\cos \dfrac{4\pi}{3} + i \sin \dfrac{4\pi}{3}\right)$ $5\left(\cos \dfrac{3\pi}{2} + i \sin \dfrac{3\pi}{2}\right)$

13. 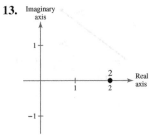   **15.**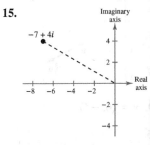

$2(\cos 0 + i \sin 0)$ $\sqrt{65}(\cos 2.62 + i \sin 2.62)$

17. 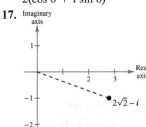   **19.**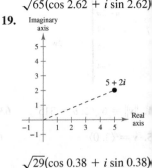

$3(\cos 5.94 + i \sin 5.94)$ $\sqrt{29}(\cos 0.38 + i \sin 0.38)$

1.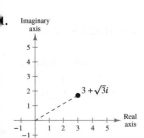

$2\sqrt{3}\left(\cos\dfrac{\pi}{6} + i\sin\dfrac{\pi}{6}\right)$

23.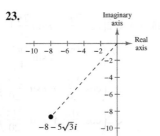

$\sqrt{139}(\cos 3.97 + i\sin 3.97)$

25. $1 + \sqrt{3}i$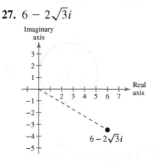

27. $6 - 2\sqrt{3}i$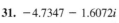

29. $-\dfrac{9\sqrt{2}}{8} + \dfrac{9\sqrt{2}}{8}i$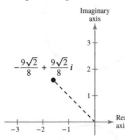

31. $-4.7347 - 1.6072i$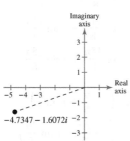

33. $4.6985 + 1.7101i$ **35.** $-1.8126 + 0.8452i$

37. $12\left(\cos\dfrac{\pi}{3} + i\sin\dfrac{\pi}{3}\right)$ **39.** $\dfrac{10}{9}(\cos 150° + i\sin 150°)$

41. $\dfrac{1}{3}(\cos 30° + i\sin 30°)$ **43.** $\cos\dfrac{2\pi}{3} + i\sin\dfrac{2\pi}{3}$

45. (a) $\left[2\sqrt{2}\left(\cos\dfrac{\pi}{4} + i\sin\dfrac{\pi}{4}\right)\right]\left[\sqrt{2}\left(\cos\dfrac{7\pi}{4} + i\sin\dfrac{7\pi}{4}\right)\right]$
(b) $4(\cos 0 + i\sin 0) = 4$ (c) 4

47. (a) $\left[2\left(\cos\dfrac{3\pi}{2} + i\sin\dfrac{3\pi}{2}\right)\right]\left[\sqrt{2}\left(\cos\dfrac{\pi}{4} + i\sin\dfrac{\pi}{4}\right)\right]$
(b) $2\sqrt{2}\left(\cos\dfrac{7\pi}{4} + i\sin\dfrac{7\pi}{4}\right) = 2 - 2i$
(c) $-2i - 2i^2 = -2i + 2 = 2 - 2i$

49. (a) $[5(\cos 0.93 + i\sin 0.93)] \div \left[2\left(\cos\dfrac{5\pi}{3} + i\sin\dfrac{5\pi}{3}\right)\right]$
(b) $\dfrac{5}{2}(\cos 1.97 + i\sin 1.97) \approx -0.982 + 2.299i$
(c) About $-0.982 + 2.299i$

51. -1 **53.** $\dfrac{125}{2} + \dfrac{125\sqrt{3}}{2}i$ **55.** -1

57. $608.0 + 144.7i$ **59.** $\dfrac{81}{2} + \dfrac{81\sqrt{3}}{2}i$ **61.** $-4 - 4i$

63. $8i$ **65.** $1024 - 1024\sqrt{3}i$ **67.** $-597 - 122i$

69.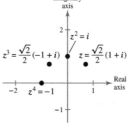

The absolute value of each is 1, and the consecutive powers of z are each 45° apart.

71. (a) $\sqrt{5}(\cos 60° + i\sin 60°)$
$\sqrt{5}(\cos 240° + i\sin 240°)$
(b) $\dfrac{\sqrt{5}}{2} + \dfrac{\sqrt{15}}{2}i, \ -\dfrac{\sqrt{5}}{2} - \dfrac{\sqrt{15}}{2}i$
(c)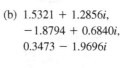

73. (a) $2\left(\cos\dfrac{2\pi}{9} + i\sin\dfrac{2\pi}{9}\right)$
$2\left(\cos\dfrac{8\pi}{9} + i\sin\dfrac{8\pi}{9}\right)$
$2\left(\cos\dfrac{14\pi}{9} + i\sin\dfrac{14\pi}{9}\right)$
(b) $1.5321 + 1.2856i,$
$-1.8794 + 0.6840i,$
$0.3473 - 1.9696i$
(c)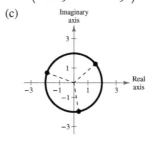

75. (a) $5\left(\cos\dfrac{4\pi}{9} + i\sin\dfrac{4\pi}{9}\right)$
$5\left(\cos\dfrac{10\pi}{9} + i\sin\dfrac{10\pi}{9}\right)$
$5\left(\cos\dfrac{16\pi}{9} + i\sin\dfrac{16\pi}{9}\right)$
(b) $0.8682 + 4.9240i,$
$-4.6985 - 1.7101i,$
$3.8302 - 3.2139i$
(c)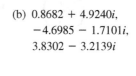

77. (a) $5\left(\cos\dfrac{3\pi}{4} + i\sin\dfrac{3\pi}{4}\right)$
$5\left(\cos\dfrac{7\pi}{4} + i\sin\dfrac{7\pi}{4}\right)$
(b) $-\dfrac{5\sqrt{2}}{2} + \dfrac{5\sqrt{2}}{2}i,$
$\dfrac{5\sqrt{2}}{2} - \dfrac{5\sqrt{2}}{2}i$

(c)

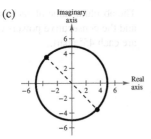

79. (a) $2(\cos 0 + i \sin 0)$
$2\left(\cos \dfrac{\pi}{2} + i \sin \dfrac{\pi}{2}\right)$
$2(\cos \pi + i \sin \pi)$
$2\left(\cos \dfrac{3\pi}{2} + i \sin \dfrac{3\pi}{2}\right)$

(b) $2, 2i, -2, -2i$

(c)

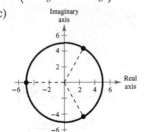

81. (a) $\cos 0 + i \sin 0$
$\cos \dfrac{2\pi}{5} + i \sin \dfrac{2\pi}{5}$
$\cos \dfrac{4\pi}{5} + i \sin \dfrac{4\pi}{5}$
$\cos \dfrac{6\pi}{5} + i \sin \dfrac{6\pi}{5}$
$\cos \dfrac{8\pi}{5} + i \sin \dfrac{8\pi}{5}$

(b) $1, 0.3090 + 0.9511i,$
$-0.8090 + 0.5878i,$
$-0.8090 - 0.5878i,$
$0.3090 - 0.9511i$

(c)

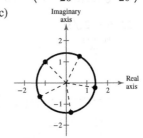

83. (a) $5\left(\cos \dfrac{\pi}{3} + i \sin \dfrac{\pi}{3}\right)$
$5(\cos \pi + i \sin \pi)$
$5\left(\cos \dfrac{5\pi}{3} + i \sin \dfrac{5\pi}{3}\right)$

(b) $\dfrac{5}{2} + \dfrac{5\sqrt{3}}{2}i, -5,$
$\dfrac{5}{2} - \dfrac{5\sqrt{3}}{2}i$

(c)

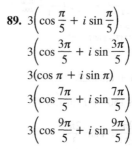

85. (a) $\sqrt{2}\left(\cos \dfrac{7\pi}{20} + i \sin \dfrac{7\pi}{20}\right)$
$\sqrt{2}\left(\cos \dfrac{3\pi}{4} + i \sin \dfrac{3\pi}{4}\right)$
$\sqrt{2}\left(\cos \dfrac{23\pi}{20} + i \sin \dfrac{23\pi}{20}\right)$
$\sqrt{2}\left(\cos \dfrac{31\pi}{20} + i \sin \dfrac{31\pi}{20}\right)$
$\sqrt{2}\left(\cos \dfrac{39\pi}{20} + i \sin \dfrac{39\pi}{20}\right)$

(b) $0.6420 + 1.2601i,$
$-1 + i,$
$-1.2601 - 0.6420i,$
$0.2212 - 1.3968i,$
$1.3968 - 0.2212i$

(c)

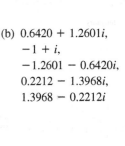

87. $\cos \dfrac{3\pi}{8} + i \sin \dfrac{3\pi}{8}$
$\cos \dfrac{7\pi}{8} + i \sin \dfrac{7\pi}{8}$
$\cos \dfrac{11\pi}{8} + i \sin \dfrac{11\pi}{8}$
$\cos \dfrac{15\pi}{8} + i \sin \dfrac{15\pi}{8}$

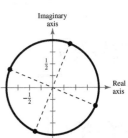

89. $3\left(\cos \dfrac{\pi}{5} + i \sin \dfrac{\pi}{5}\right)$
$3\left(\cos \dfrac{3\pi}{5} + i \sin \dfrac{3\pi}{5}\right)$
$3(\cos \pi + i \sin \pi)$
$3\left(\cos \dfrac{7\pi}{5} + i \sin \dfrac{7\pi}{5}\right)$
$3\left(\cos \dfrac{9\pi}{5} + i \sin \dfrac{9\pi}{5}\right)$

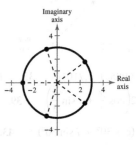

91. $2\left(\cos \dfrac{3\pi}{8} + i \sin \dfrac{3\pi}{8}\right)$
$2\left(\cos \dfrac{7\pi}{8} + i \sin \dfrac{7\pi}{8}\right)$
$2\left(\cos \dfrac{11\pi}{8} + i \sin \dfrac{11\pi}{8}\right)$
$2\left(\cos \dfrac{15\pi}{8} + i \sin \dfrac{15\pi}{8}\right)$

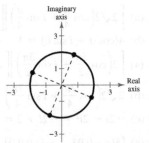

93. $\sqrt[6]{2}\left(\cos \dfrac{7\pi}{12} + i \sin \dfrac{7\pi}{12}\right)$
$\sqrt[6]{2}\left(\cos \dfrac{5\pi}{4} + i \sin \dfrac{5\pi}{4}\right)$
$\sqrt[6]{2}\left(\cos \dfrac{23\pi}{12} + i \sin \dfrac{23\pi}{12}\right)$

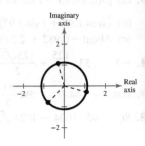

5. (a) $E = 24(\cos 30° + i \sin 30°)$ volts
   (b) $E = 12\sqrt{3} + 12i$ volts   (c) $|E| = 24$ volts
7. False. They are equally spaced around the circle centered at the origin with radius $\sqrt[n]{r}$.
9. Answers will vary.
1. Answers will vary; (a) r^2 (b) $\cos 2\theta + i \sin 2\theta$

eview Exercises *(page 456)*

1. $C = 72°, b \approx 12.21, c \approx 12.36$
3. $A = 26°, a \approx 24.89, c \approx 56.23$
5. $C = 66°, a \approx 2.53, b \approx 9.11$
7. $B = 108°, a \approx 11.76, c \approx 21.49$
9. $A \approx 20.41°, C \approx 9.59°, a \approx 20.92$
1. $B \approx 39.48°, C \approx 65.52°, c \approx 48.24$ **13.** 19.1
5. 47.2   **17.** About 31.1 m
9. $A \approx 16.99°, B \approx 26.00°, C \approx 137.01°$
1. $A \approx 29.92°, B \approx 86.18°, C \approx 63.90°$
3. $A = 36°, C = 36°, b \approx 17.80$
5. $A \approx 45.76°, B \approx 91.24°, c \approx 21.42$
7. No; $A \approx 77.52°, B \approx 38.48°, a \approx 14.12$
9. Yes; $A \approx 28.62°, B \approx 33.56°, C \approx 117.82°$
1. About 4.3 ft, about 12.6 ft   **33.** 7.64   **35.** 8.36
7. Equivalent; **u** and **v** have the same magnitude and direction.
9. $\langle 7, -7 \rangle; 7\sqrt{2}$
1. (a) $\langle -4, 3 \rangle$ (b) $\langle 2, -9 \rangle$

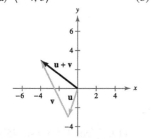

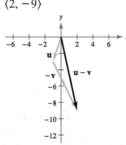

(c) $\langle -4, -12 \rangle$ (d) $\langle -14, 3 \rangle$

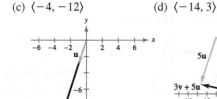

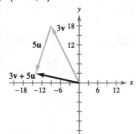

3. (a) $\langle -1, 6 \rangle$ (b) $\langle -9, -2 \rangle$

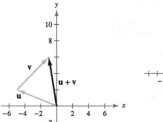

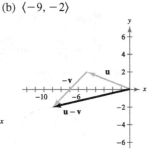

(c) $\langle -20, 8 \rangle$ (d) $\langle -13, 22 \rangle$

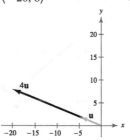

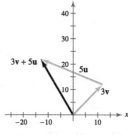

45. (a) $7\mathbf{i} + 2\mathbf{j}$ (b) $-3\mathbf{i} - 4\mathbf{j}$

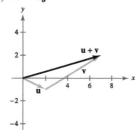

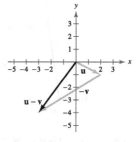

(c) $8\mathbf{i} - 4\mathbf{j}$ (d) $25\mathbf{i} + 4\mathbf{j}$

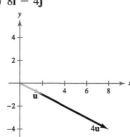

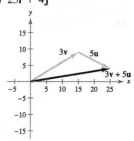

47. (a) $3\mathbf{i} + 6\mathbf{j}$ (b) $5\mathbf{i} - 6\mathbf{j}$

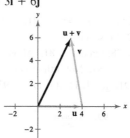

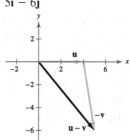

(c) $16\mathbf{i}$ (d) $17\mathbf{i} + 18\mathbf{j}$

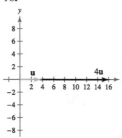

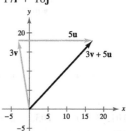

49. $-\mathbf{i} + 5\mathbf{j}$ **51.** $6\mathbf{i} + 4\mathbf{j}$

53. $\langle 30, 9 \rangle$

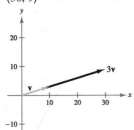

55. $\langle 22, -7 \rangle$

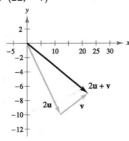

57. $\langle -10, -37 \rangle$

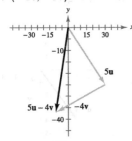

59. $\|\mathbf{v}\| = \sqrt{41}$; $\theta \approx 38.7°$ **61.** $\|\mathbf{v}\| = 3\sqrt{2}$; $\theta = 225°$
63. $\|\mathbf{v}\| = 7$; $\theta = 60°$
65. $\mathbf{v} = \langle -4, 4\sqrt{3} \rangle$

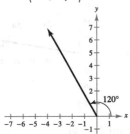

67. About 50.5°; about 133.92 lb   **69.** 45   **71.** -2
73. 40; scalar   **75.** $4 - 2\sqrt{5}$; scalar
77. $\langle 72, -36 \rangle$; vector   **79.** 38; scalar   **81.** About 160.5°
83. 165°   **85.** Orthogonal   **87.** Not orthogonal
89. $-\frac{13}{17}\langle 4, 1 \rangle$, $\frac{16}{17}\langle -1, 4 \rangle$ **91.** $\frac{5}{2}\langle -1, 1 \rangle$, $\frac{9}{2}\langle 1, 1 \rangle$ **93.** 48
95. 72,000 ft-lb

97.

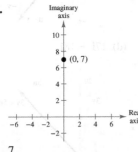

7

99.

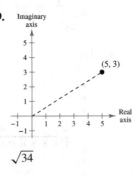

$\sqrt{34}$

101. $3 + i$ **103.** $-2 + i$
105.

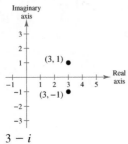

$3 - i$

107. $\sqrt{10}$ **109.** $\left(\frac{5}{2}, 2\right)$
111.

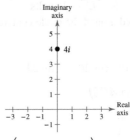

$4\left(\cos\frac{\pi}{2} + i\sin\frac{\pi}{2}\right)$

113.

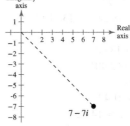

$7\sqrt{2}\left(\cos\frac{7\pi}{4} + i\sin\frac{7\pi}{4}\right)$

115.

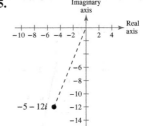

$13(\cos 4.32 + i\sin 4.32)$

117. $4\left(\cos\frac{7\pi}{12} + i\sin\frac{7\pi}{12}\right)$ **119.** $\frac{2}{3}(\cos 45° + i\sin 45°)$

121. $\frac{625}{2} + \frac{625\sqrt{3}}{2}i$ **123.** $2035 - 828i$

125. (a) $3\left(\cos\frac{\pi}{4} + i\sin\frac{\pi}{4}\right)$

$3\left(\cos\frac{7\pi}{12} + i\sin\frac{7\pi}{12}\right)$

$3\left(\cos\frac{11\pi}{12} + i\sin\frac{11\pi}{12}\right)$

$3\left(\cos\frac{5\pi}{4} + i\sin\frac{5\pi}{4}\right)$

$3\left(\cos\frac{19\pi}{12} + i\sin\frac{19\pi}{12}\right)$

$3\left(\cos\frac{23\pi}{12} + i\sin\frac{23\pi}{12}\right)$

(b) $\frac{3\sqrt{2}}{2} + \frac{3\sqrt{2}}{2}i$,

$-0.7765 + 2.8978i$,

$-2.8978 + 0.7765i$,

$-\frac{3\sqrt{2}}{2} - \frac{3\sqrt{2}}{2}i$,

$0.7765 - 2.8978i$,

$2.8978 - 0.7765i$

(c)

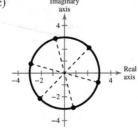

27. (a) $2(\cos 0 + i \sin 0)$
$2\left(\cos \dfrac{2\pi}{3} + i \sin \dfrac{2\pi}{3}\right)$
$2\left(\cos \dfrac{4\pi}{3} + i \sin \dfrac{4\pi}{3}\right)$

(b) $2, -1 + \sqrt{3}\,i,$
$-1 - \sqrt{3}\,i$

(c)

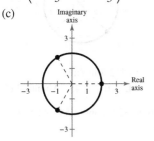

29. $3\left(\cos \dfrac{\pi}{4} + i \sin \dfrac{\pi}{4}\right) = \dfrac{3\sqrt{2}}{2} + \dfrac{3\sqrt{2}}{2}i;$
$3\left(\cos \dfrac{3\pi}{4} + i \sin \dfrac{3\pi}{4}\right) = -\dfrac{3\sqrt{2}}{2} + \dfrac{3\sqrt{2}}{2}i;$
$3\left(\cos \dfrac{5\pi}{4} + i \sin \dfrac{5\pi}{4}\right) = -\dfrac{3\sqrt{2}}{2} - \dfrac{3\sqrt{2}}{2}i;$
$3\left(\cos \dfrac{7\pi}{4} + i \sin \dfrac{7\pi}{4}\right) = \dfrac{3\sqrt{2}}{2} - \dfrac{3\sqrt{2}}{2}i$

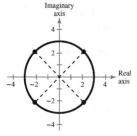

31. $2\left(\cos \dfrac{\pi}{2} + i \sin \dfrac{\pi}{2}\right) = 2i$
$2\left(\cos \dfrac{7\pi}{6} + i \sin \dfrac{7\pi}{6}\right) = -\sqrt{3} - i$
$2\left(\cos \dfrac{11\pi}{6} + i \sin \dfrac{11\pi}{6}\right) = \sqrt{3} - i$

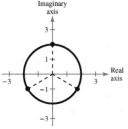

33. True. $\sin 90°$ is defined in the Law of Sines.
35. Direction and magnitude

Chapter Test   *(page 459)*

1. No; $C = 88°, b \approx 27.81, c \approx 29.98$
2. No; $A = 42°, b \approx 21.91, c \approx 10.95$
3. No; Two solutions:
  $B \approx 29.12°, C \approx 126.88°, c \approx 22.03$
  $B \approx 150.88°, C \approx 5.12°, c \approx 2.46$
4. Yes; $A \approx 19.12°, B \approx 23.49°, C \approx 137.39°$
5. No; No solution

6. Yes; $A \approx 21.90°, B \approx 37.10°, c \approx 78.15$
7. 2052.5 m²   **8.** 606.3 mi; 29.1°   **9.** $\langle 14, -23 \rangle$
10. $\left\langle \dfrac{18\sqrt{34}}{17}, -\dfrac{30\sqrt{34}}{17} \right\rangle$
11. $\langle -4, 12 \rangle$ **12.** $\langle 8, 2 \rangle$

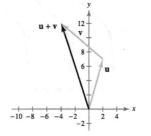
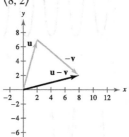

13. $\langle 28, 20 \rangle$ **14.** $\langle -4, 38 \rangle$

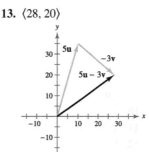
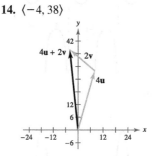

15. 5   **16.** About 14.9°; about 250.15 lb   **17.** 135°
18. Orthogonal   **19.** $\dfrac{37}{26}\langle 5, 1 \rangle; \dfrac{29}{26}\langle -1, 5 \rangle$ **20.** About 104 lb
21. $4\sqrt{2}\left(\cos \dfrac{7\pi}{4} + i \sin \dfrac{7\pi}{4}\right)$ **22.** $-3 + 3\sqrt{3}\,i$
23. $-\dfrac{6561}{2} - \dfrac{6561\sqrt{3}}{2}i$ **24.** $5832i$ **25.** $4, -4, 4i, -4i$

26. $3\left(\cos \dfrac{\pi}{6} + i \sin \dfrac{\pi}{6}\right)$
$3\left(\cos \dfrac{5\pi}{6} + i \sin \dfrac{5\pi}{6}\right)$
$3\left(\cos \dfrac{3\pi}{2} + i \sin \dfrac{3\pi}{2}\right)$

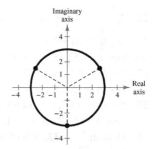

Cumulative Test for Chapters 4–6   *(page 460)*

1. (a)

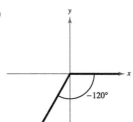

(b) 240°
(c) $-\dfrac{2\pi}{3}$
(d) 60°

(e) $\sin(-120°) = -\dfrac{\sqrt{3}}{2}$ $\csc(-120°) = -\dfrac{2\sqrt{3}}{3}$

$\cos(-120°) = -\dfrac{1}{2}$ $\sec(-120°) = -2$

$\tan(-120°) = \sqrt{3}$ $\cot(-120°) = \dfrac{\sqrt{3}}{3}$

2. $-83.079°$ **3.** $\frac{20}{29}$

4.

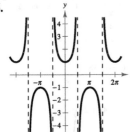

5.

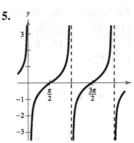

6.

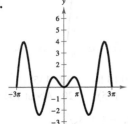

7. $a = -3, b = \pi, c = 0$

8.

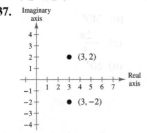

9. 4.9

10. $\frac{3}{4}$

11. $\sqrt{1 - 4x^2}$

12. 1

13. $2 \tan \theta$

14–16. Answers will vary.

17. $\frac{\pi}{3}, \frac{\pi}{2}, \frac{3\pi}{2}, \frac{5\pi}{3}$

18. $\frac{\pi}{6}, \frac{5\pi}{6}, \frac{7\pi}{6}, \frac{11\pi}{6}$ **19.** $\frac{3\pi}{2}$ **20.** $\frac{16}{63}$ **21.** $\frac{4}{3}$

22. $\frac{\sqrt{5}}{5}$ **23.** $\frac{5}{2}\left(\sin \frac{5\pi}{2} - \sin \pi\right)$ **24.** $-2 \sin 8x \sin x$

25. No; $B \approx 26.39°, C \approx 123.61°, c \approx 14.99$

26. Yes; $B \approx 52.48°, C \approx 97.52°, a \approx 5.04$

27. No; $B = 60°, a \approx 5.77, c \approx 11.55$

28. Yes; $A \approx 26.28°, B \approx 49.74°, C \approx 103.98°$

29. No; $C = 109°, a \approx 14.96, b \approx 9.27$

30. Yes; $A \approx 6.88°, B \approx 93.12°, c \approx 9.86$

31. 41.48 in.^{2} **32.** 599.09 m^2 **33.** $7\mathbf{i} + 8\mathbf{j}$

34. $\left\langle \frac{\sqrt{2}}{2}, \frac{\sqrt{2}}{2} \right\rangle$ **35.** -5 **36.** $-\frac{1}{13}\langle 1, 5 \rangle; \frac{21}{13}\langle 5, -1 \rangle$

37.

$3 + 2i$

38. $2\sqrt{2}\left(\cos \frac{3\pi}{4} + i \sin \frac{3\pi}{4}\right)$ **39.** $24(\cos 150° + i \sin 150°)$

40. $\cos 0 + i \sin 0 = 1$

$\cos \frac{2\pi}{3} + i \sin \frac{2\pi}{3} = -\frac{1}{2} + \frac{\sqrt{3}}{2}i$

$\cos \frac{4\pi}{3} + i \sin \frac{4\pi}{3} = -\frac{1}{2} - \frac{\sqrt{3}}{2}i$

41. $\frac{5\sqrt{2}}{2} + \frac{5\sqrt{2}}{2}i$

$-\frac{5\sqrt{2}}{2} + \frac{5\sqrt{2}}{2}i$

$-\frac{5\sqrt{2}}{2} - \frac{5\sqrt{2}}{2}i$

$\frac{5\sqrt{2}}{2} - \frac{5\sqrt{2}}{2}i$

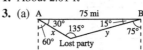

42. About 395.8 rad/min; about 8312.7 in./min

43. 42π yd$^2 \approx 131.95$ yd^2 **44.** 5 ft   **45.** About 22.6°

46. $d = 4 \cos \frac{\pi}{4}t$ **47.** About 32.6°; about 543.9 km/h

48. 425 ft-lb

Problem Solving   *(page 465)*

1. About 2.01 ft

3. (a)

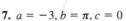

(b) Station A: about 27.45 mi
Station B: about 53.03 mi

(c) About 11.03 mi; S 21.7° E

5. (a) (i) $\sqrt{2}$ (ii) $\sqrt{5}$ (iii) 1
(iv) 1   (v) 1   (vi) 1

(b) (i) 1   (ii) $3\sqrt{2}$ (iii) $\sqrt{13}$
(iv) 1   (v) 1   (vi) 1

(c) (i) $\frac{\sqrt{5}}{2}$ (ii) $\sqrt{13}$ (iii) $\frac{\sqrt{85}}{2}$
(iv) 1   (v) 1   (vi) 1

(d) (i) $2\sqrt{5}$ (ii) $5\sqrt{2}$ (iii) $5\sqrt{2}$
(iv) 1   (v) 1   (vi) 1

7. Proof

9. (a) $2(\cos 30° + i \sin 30°)$ (b) $3(\cos 45° + i \sin 45°)$
$2(\cos 150° + i \sin 150°)$ $3(\cos 135° + i \sin 135°)$
$2(\cos 270° + i \sin 270°)$ $3(\cos 225° + i \sin 225°)$
                           $3(\cos 315° + i \sin 315°)$

11. a; The angle between the vectors is acute.

Chapter 7

Section 7.1   *(page 475)*

1. solution   **3.** points; intersection

5. (a) No   (b) No   (c) No   (d) Yes   **7.** $(2, 2)$

9. $(2, 6), (-1, 3)$ **11.** $(0, 0), (2, -4)$

13. $(0, 1), (1, -1), (3, 1)$ **15.** $(6, 4)$ **17.** $\left(\frac{1}{2}, 3\right)$

19. $(1, 1)$ **21.** $\left(\frac{20}{3}, \frac{40}{3}\right)$ **23.** No solution

25. \$5500 at 2%; \$6500 at 6%

27. \$6000 at 2.8%; \$6000 at 3.8%

29. $(-2, 4), (0, 0)$ **31.** No solution   **33.** $(6, 2)$

35. $\left(-\frac{3}{2}, \frac{1}{2}\right)$ **37.** $(2, 2), (4, 0)$ **39.** No solution

41. $(4, 3), (-4, 3)$ **43.** $(0, 1)$ **45.** $(5.31, -0.54)$

47. $(1, 2)$ **49.** No solution   **51.** $(0.287, 1.751)$

53. $\left(\frac{1}{2}, 2\right), \left(-4, -\frac{1}{4}\right)$ **55.** 293 units

57. (a) 344 units   (b) 2495 units

. (a) 8 weeks

(b)

	1	2	3	4
$360 - 24x$	336	312	288	264
$24 + 18x$	42	60	78	96

	5	6	7	8
$360 - 24x$	240	216	192	168
$24 + 18x$	114	132	150	168

. $y = 2x - 2$, not $-2x + 2$.　**63.** 12 m × 16 m

. 10 km × 12 km

. False. You can solve for either variable in either equation and then back-substitute.

. *Sample answer:* After substituting, the resulting equation may be a contradiction or have imaginary solutions.

. (a)–(c) Answers will vary.

Section 7.2 (page 486)

. elimination　　**3.** consistent; inconsistent

. $(1, 5)$　　　　　　　**7.** $(1, -1)$

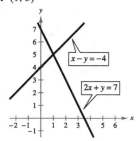

. No solution　　　　**11.** $\left(a, \frac{3}{2}a - \frac{5}{2}\right)$

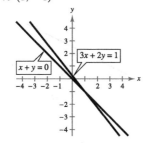

. $(4, 1)$　**15.** $\left(\frac{3}{2}, -\frac{1}{2}\right)$　**17.** $\left(-3, \frac{5}{3}\right)$　**19.** $(4, -1)$

. $\left(-\frac{6}{35}, \frac{43}{35}\right)$　**23.** $(101, 96)$　**25.** No solution

. Infinitely many solutions: $\left(a, -\frac{1}{2} + \frac{5}{6}a\right)$　**29.** $(5, -2)$

. a; infinitely many solutions; consistent

. c; one solution; consistent

. d; no solutions; inconsistent

. b; one solution; consistent　　**35.** $(4, 1)$　　**37.** $(10, 5)$

. $(19, -55)$　　**41.** 660 mi/h; 60 mi/h

. Cheeseburger: 550 calories; fries: 320 calories

. $(240, 404)$　　**47.** $(2,000,000, 100)$

49. (a) $\begin{cases} x + y = 30 \\ 0.25x + 0.5y = 12 \end{cases}$

(b)

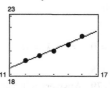

Decreases

(c) 25% solution: 12 L; 50% solution: 18 L

51. $18,000

53. (a) Pharmacy A: $P = 0.52t + 12.9$

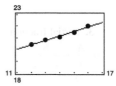

Pharmacy B: $P = 0.39t + 15.7$

(b) Yes. 2021

55. $y = 0.97x + 2.1$

57. (a) $y = 14x + 19$　　(b) 41.4 bushels/acre

59. False. Two lines that coincide have infinitely many points of intersection.

61. $k = -4$

63. No. Two lines will intersect only once or will coincide, and if they coincide the system will have infinitely many solutions.

65. Answers will vary.

67. $(39,600, 398)$. It is necessary to change the scale on the axes to see the point of intersection.

69. $u = 1, v = -\tan x$

Section 7.3 (page 498)

1. row-echelon　　**3.** Gaussian　　**5.** nonsquare

7. (a) No　(b) No　(c) No　(d) Yes

9. (a) No　(b) No　(c) Yes　(d) No

11. $(-13, -10, 8)$　　**13.** $(3, 10, 2)$　　**15.** $\left(\frac{11}{4}, 7, 11\right)$

17. $\begin{cases} x - 2y + 3z = 5 \\ y - 2z = 9 \\ 2x - 3z = 0 \end{cases}$

First step in putting the system in row-echelon form.

19. $(-2, 2)$　　**21.** $(4, 3)$　　**23.** $(4, 1, 2)$　　**25.** $\left(1, \frac{1}{2}, -3\right)$

27. No solution　　**29.** $\left(\frac{1}{8}, -\frac{5}{8}, -\frac{1}{2}\right)$　　**31.** $(0, 0, 0)$

33. No solution　　**35.** $(-a + 3, a + 1, a)$

37. $(-3a + 10, 5a - 7, a)$　　**39.** $(1, 1, 1, 1)$

41. $(2a, 21a - 1, 8a)$　　**43.** $\left(-\frac{3}{2}a + \frac{1}{2}, -\frac{2}{3}a + 1, a\right)$

45. $s = -16t^2 + 144$　　**47.** $y = \frac{1}{2}x^2 - 2x$

49. $y = x^2 - 6x + 8$　　**51.** $y = 4x^2 - 2x + 1$

53. $x^2 + y^2 - 10x = 0$　　**55.** $x^2 + y^2 + 6x - 8y = 0$

57. The leading coefficient of the third equation is not 1, so the system is not in row-echelon form.

59. $300,000 at 8%
$400,000 at 9%
$75,000 at 10%

61. $x = 60°, y = 67°, z = 53°$ **63.** 75 ft, 63 ft, 42 ft

65. $I_1 = 1, I_2 = 2, I_3 = 1$ **67.** $y = x^2 - x$

69. (a) $y = 0.0514x^2 + 0.8771x + 1.8857$

(b)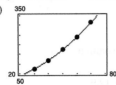

The model fits the data well.

(c) About 356 ft

71. $x = \pm\dfrac{\sqrt{2}}{2}, y = \dfrac{1}{2}, \lambda = 1$ or $x = 0, y = 0, \lambda = 0$

73. False. See Example 6 on page 495.

75. No. Answers will vary.   **77–79.** Answers will vary.

Section 7.4   *(page 508)*

1. partial fraction decomposition   **3.** partial fraction

5. b   **6.** c   **7.** d   **8.** a   **9.** $\dfrac{A}{x} + \dfrac{B}{x-2}$

11. $\dfrac{A}{x+2} + \dfrac{B}{(x+2)^2} + \dfrac{C}{(x+2)^3} + \dfrac{D}{(x+2)^4}$

13. $\dfrac{A}{x} + \dfrac{Bx+C}{x^2+10}$ **15.** $\dfrac{A}{x} + \dfrac{B}{x^2} + \dfrac{Cx+D}{x^2+3} + \dfrac{Ex+F}{(x^2+3)^2}$

17. $\dfrac{1}{x} - \dfrac{1}{x+1}$ **19.** $\dfrac{1}{x-1} - \dfrac{1}{x+2}$

21. $\dfrac{1}{2}\left(\dfrac{1}{x-1} - \dfrac{1}{x+1}\right)$ **23.** $-\dfrac{3}{x} - \dfrac{1}{x+2} + \dfrac{5}{x-2}$

25. $\dfrac{3}{x-3} + \dfrac{9}{(x-3)^2}$ **27.** $\dfrac{3}{x} - \dfrac{1}{x^2} + \dfrac{1}{x+1}$

29. $\dfrac{3}{x} - \dfrac{2x-2}{x^2+1}$ **31.** $-\dfrac{1}{x-1} + \dfrac{x+2}{x^2-2}$

33. $\dfrac{1}{8}\left(\dfrac{1}{2x+1} + \dfrac{1}{2x-1} - \dfrac{4x}{4x^2+1}\right)$

35. $\dfrac{1}{x+1} + \dfrac{2}{x^2-2x+3}$ **37.** $\dfrac{2}{x^2+4} + \dfrac{x}{(x^2+4)^2}$

39. $\dfrac{5}{(x^2+3)^2} - \dfrac{17}{(x^2+3)^3}$ **41.** $\dfrac{2}{x} - \dfrac{3}{x^2} - \dfrac{2x-3}{x^2+2} - \dfrac{4x-6}{(x^2+2)^2}$

43. $1 - \dfrac{2x+1}{x^2+x+1}$ **45.** $2x - 7 + \dfrac{17}{x+2} + \dfrac{1}{x+1}$

47. $x + 3 + \dfrac{6}{x-1} + \dfrac{4}{(x-1)^2} + \dfrac{1}{(x-1)^3}$

49. $x + \dfrac{2}{x} + \dfrac{1}{x+1} + \dfrac{3}{(x+1)^2}$ **51.** $\dfrac{3}{2x-1} - \dfrac{2}{x+1}$

53. $\dfrac{2}{x} + \dfrac{4}{x+1} - \dfrac{3}{x-1}$ **55.** $\dfrac{1}{x^2+2} + \dfrac{x}{(x^2+2)^2}$

57. $2x + \dfrac{1}{2}\left(\dfrac{3}{x-4} - \dfrac{1}{x+2}\right)$ **59.** $\dfrac{60}{100-p} - \dfrac{60}{100+p}$

61. False. The partial fraction decomposition is

$\dfrac{A}{x+10} + \dfrac{B}{x-10} + \dfrac{C}{(x-10)^2}.$

63. False. The degrees could be equal.

65. The expression is improper, so first divide the denominator

into the numerator to obtain $1 + \dfrac{x+1}{x^2-x}$.

Section 7.5   *(page 517)*

1. solution   **3.** solution

5. 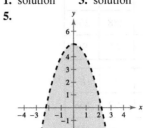   **7.**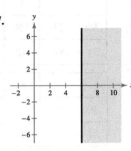

9. 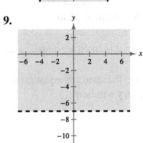   **11.**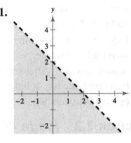

13. 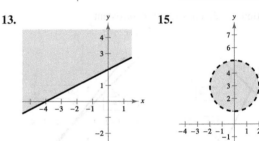   **15.**

17.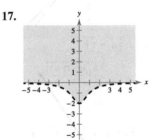

19. 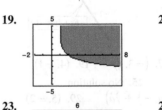   **21.**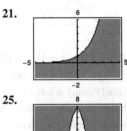

23. 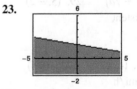   **25.**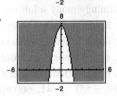

27. $y < 5x + 5$ **29.** $y \geq x^2 - 4$

31.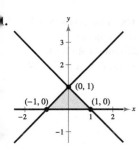

33.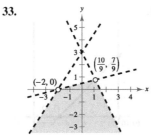

35.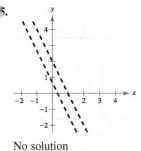

37.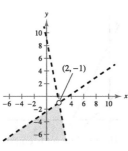

No solution

39.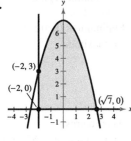

41.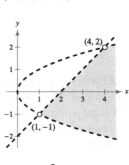

43.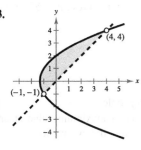

45.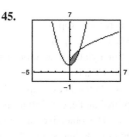

47.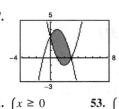

49.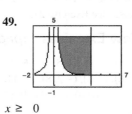

51. $\begin{cases} x \geq 0 \\ y \geq 0 \\ y \leq 6 - x \end{cases}$

53. $\begin{cases} x \geq 0 \\ y \geq 0 \\ x^2 + y^2 < 64 \end{cases}$

55. $\begin{cases} x \geq 4 \\ x \leq 9 \\ y \geq 3 \\ y \leq 9 \end{cases}$

57. $\begin{cases} y \geq 0 \\ y \leq 5x \\ y \leq -x + 6 \end{cases}$

59. (a)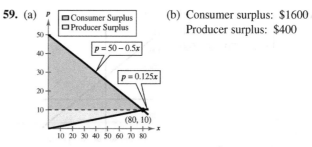

(b) Consumer surplus: $1600
Producer surplus: $400

61. (a)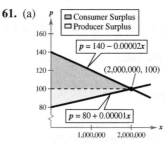

(b) Consumer surplus:
$40,000,000
Producer surplus:
$20,000,000

63. $\begin{cases} x + y \leq 20,000 \\ y \geq 2x \\ x \geq 5,000 \\ y \geq 5,000 \end{cases}$

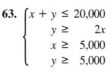

65. $\begin{cases} x + \frac{3}{2}y \leq 12 \\ \frac{4}{3}x + \frac{3}{2}y \leq 15 \\ x \geq 0 \\ y \geq 0 \end{cases}$

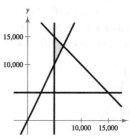

67. (a) $\begin{cases} 180x + 100y \geq 1000 \\ 6x + y \geq 18 \\ 220x + 40y \geq 400 \\ x \geq 0 \\ y \geq 0 \end{cases}$

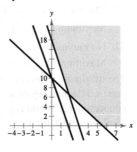

(b) Answers will vary.

69. (a) $\begin{cases} x \ge 50 \\ y \ge 40 \\ 55x + 70y \le 7500 \end{cases}$

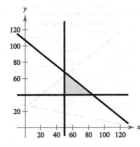

(b) Answers will vary.

71. True. The figure is a rectangle with a length of 9 units and a width of 11 units.

73. Test a point on each side of the line.

75. (a) iv   (b) ii   (c) iii   (d) i

Section 7.6   *(page 526)*

1. optimization   **3.** objective   **5.** inside; on

7. Minimum at $(0, 0)$: 0   **9.** Minimum at $(1, 0)$: 2
Maximum at $(5, 0)$: 20   Maximum at $(3, 4)$: 26

11. Minimum at $(0, 20)$: 140
Maximum at $(60, 20)$: 740

13.

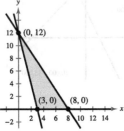

Minimum at $(3, 0)$: 9
Maximum at any point on
the line segment connecting
$(0, 12)$ and $(8, 0)$: 24

15.

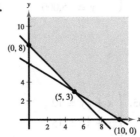

Minimum at $(5, 3)$: 35
No maximum

17.

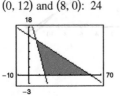

Minimum at $(7.2, 13.2)$: 34.8
Maximum at $(60, 0)$: 180

19.

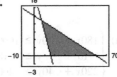

Minimum at $(7.2, 13.2)$: 7.2
Maximum at $(60, 0)$: 60

21. Minimum at $(0, 0)$: 0
Maximum at $(0, 5)$: 25

23. Minimum at $(0, 0)$: 0
Maximum at $\left(\frac{22}{3}, \frac{19}{6}\right)$: $\frac{271}{6}$

25. Minimum at $(4, 3)$: 10
No maximum

27. No minimum
Maximum at $(12, 5)$: 7
The maximum, 5, occurs at any
point on the line segment
connecting $(2, 0)$ and $\left(\frac{20}{19}, \frac{45}{19}\right)$.
Minimum at $(0, 0)$: 0

29.

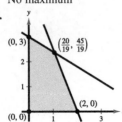

31.

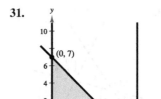

The constraint $x \le 10$
is extraneous.
Minimum at $(7, 0)$: -7
Maximum at $(0, 7)$: 14

33.

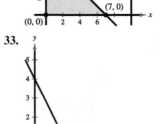

The feasible set is empty.

35.

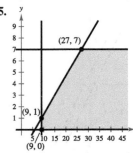

The solution region is
unbounded.
Minimum at $(9, 0)$: 9
No maximum

37. 230 units of the $225 model
45 units of the $250 model
Optimal profit: $8295

39. 2 bottles of brand X   **41.** 13 audits
5 bottles of brand Y   0 tax returns
Optimal revenue: $20,800

43. 60 acres for crop A
90 acres for crop B
Optimal yield: 63,000 bushels

45. $0 on TV ads
$1,000,000 on newspaper ads
Optimal audience: 250 million people

47. True. The objective function has a maximum value at an
point on the line segment connecting the two vertices.

49. False. See Exercise 27.

Review Exercises   *(page 531)*

1. $(1, 1)$ **3.** $\left(\frac{3}{2}, 5\right)$ **5.** $(0.25, 0.625)$ **7.** $(5, 4)$

9. $(0, 0), (2, 8), (-2, 8)$ **11.** $(4, -2)$

13. $(1.41, -0.66), (-1.41, 10.66)$ **15.** $(0, -2)$

17. No solution

19.

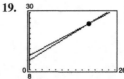

The BMI for males exceeds the BMI for females after age 1

21. 16 ft × 18 ft   **23.** $\left(\frac{5}{2}, 3\right)$ **25.** $(0, 0)$ **27.** $\left(\frac{8}{5}a + \frac{14}{5}, \right)$

29. d, one solution, consistent

30. c, infinitely many solutions, consistent

31. b, no solution, inconsistent

32. a, one solution, consistent   **33.** $(100,000, 23)$

35. $(2, -4, -5)$ **37.** $(-6, 7, 10)$ **39.** $\left(\frac{24}{5}, \frac{22}{5}, -\frac{8}{5}\right)$

41. $\left(-\frac{3}{4}, 0, -\frac{5}{4}\right)$ **43.** $(a - 4, a - 3, a)$

. $y = 2x^2 + x - 5$ **47.** $x^2 + y^2 - 4x + 4y - 1 = 0$

. 10 gal of spray X   **51.** $16,000 at 7%
  5 gal of spray Y $13,000 at 9%
  12 gal of spray Z $11,000 at 11%

. $s = -16t^2 + 150$ **55.** $\dfrac{A}{x} + \dfrac{B}{x + 20}$

. $\dfrac{A}{x} + \dfrac{B}{x^2} + \dfrac{C}{x - 5}$ **59.** $\dfrac{3}{x + 2} - \dfrac{4}{x + 4}$

. $1 - \dfrac{25}{8(x + 5)} + \dfrac{9}{8(x - 3)}$ **63.** $\dfrac{1}{2}\left(\dfrac{3}{x - 1} - \dfrac{x - 3}{x^2 + 1}\right)$

. $\dfrac{3}{x^2 + 1} + \dfrac{4x - 3}{(x^2 + 1)^2}$

.   **69.**
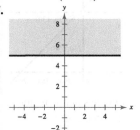
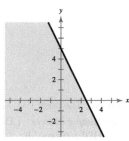

.   **73.**
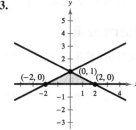
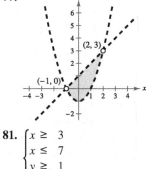

.   **77.**
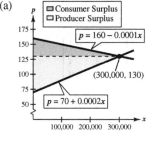

.   **81.** $\begin{cases} x \ge 3 \\ x \le 7 \\ y \ge 1 \\ y \le 10 \end{cases}$

. (a)

  (b) Consumer surplus:
      $4,500,000
      Producer surplus:
      $9,000,000

85. $\begin{cases} 20x + 30y \le 24,000 \\ 12x + 8y \le 12,400 \\ x \ge 0 \\ y \ge 0 \end{cases}$

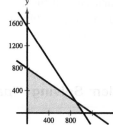

87.   **89.**
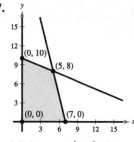
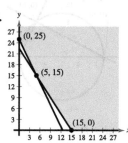

Minimum at $(0, 0)$: 0   Minimum at $(15, 0)$: 26.25
Maximum at $(5, 8)$: 47   No maximum

91. 72 haircuts
       0 permanents
       Optimal revenue: $1800

93. True. The nonparallel sides of the trapezoid are equal in length.

95. $\begin{cases} 4x + y = -22 \\ \frac{1}{2}x + y = 6 \end{cases}$ **97.** $\begin{cases} 3x + y = 7 \\ -6x + 3y = 1 \end{cases}$

99. $\begin{cases} x + y + z = 6 \\ x + y - z = 0 \\ x - y - z = 2 \end{cases}$ **101.** $\begin{cases} 2x + 2y - 3z = 7 \\ x - 2y + z = 4 \\ -x + 4y - z = -1 \end{cases}$

103. An inconsistent system of linear equations has no solution.

Chapter Test *(page 535)*

1. $(-4, -5)$ **2.** $(0, -1), (1, 0), (2, 1)$ **3.** $(8, 4), (2, -2)$
4. $(4, 2)$ **5.** $(-3, 0), (2, 5)$ **6.** $(1, 4), (0.034, 0.619)$
7. $(-2, -5)$ **8.** $(10, -3)$ **9.** $(2, -3, 1)$

10. No solution   **11.** $-\dfrac{1}{x + 1} + \dfrac{3}{x - 2}$ **12.** $\dfrac{2}{x^2} + \dfrac{3}{2 - x}$

13. $x - \dfrac{5}{x} + \dfrac{3}{x + 1} + \dfrac{3}{x - 1}$ **14.** $-\dfrac{2}{x} + \dfrac{3x}{x^2 + 2}$

15.   **16.**
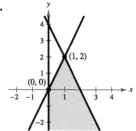
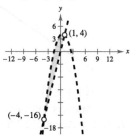

17.
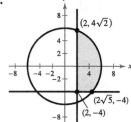

18. Minimum at $(0, 0)$: 0
Maximum at $(12, 0)$: 240

20. $y = -\frac{1}{2}x^2 + x + 6$

19. $24,000 in 4% fund
$26,000 in 5.5% fund

21. 900 units of model I
4400 units of model II
Optimal profit: $203,000

Problem Solving *(page 537)*

1.

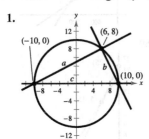

$a = 8\sqrt{5}, b = 4\sqrt{5}, c = 20$
$\left(8\sqrt{5}\right)^2 + \left(4\sqrt{5}\right)^2 = 20^2$
Therefore, the triangle is a right triangle.

3. $ad \neq bc$

5. (a)

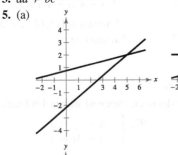

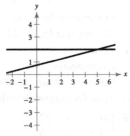

$(5, 2)$
Answers will vary.

(b)

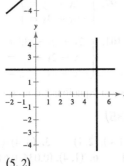

$\left(\frac{3}{2}a + \frac{7}{2}, a\right)$
Answers will vary.

7. 10.1 ft; About 252.7 ft   **9.** $12.00

11. (a) $(3, -4)$ (b) $\left(\dfrac{2}{-a + 5}, \dfrac{1}{4a - 1}, \dfrac{1}{a}\right)$

13. (a) $\left(\dfrac{-5a + 16}{6}, \dfrac{5a - 16}{6}, a\right)$

(b) $\left(\dfrac{-11a + 36}{14}, \dfrac{13a - 40}{14}, a\right)$

(c) $(-a + 3, a - 3, a)$ (d) Infinitely many

15. $\begin{cases} a + \quad t \le \quad 32 \\ 0.15a \quad\quad \ge \quad 1.9 \\ 193a + 772t \ge 11,000 \end{cases}$

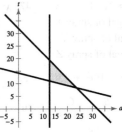

17. (a) $\begin{cases} 0 < y < 130 \\ x \ge 60 \\ x + y \le 200 \end{cases}$ (b)

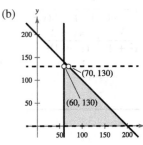

(c) No. The point $(90, 120)$ is not in the solution region.
(d) *Sample answer:* LDL/VLDL: 135 mg/dL;
   HDL: 65 mg/dL
(e) *Sample answer:* $(75, 90)$; $\frac{165}{75} = 2.2 < 4$

Chapter 8

Section 8.1 *(page 549)*

1. square   **3.** augmented   **5.** row-equivalent
7. 1×2 **9.** 3×1 **11.** 2×2 **13.** 3×3

15. $\begin{bmatrix} 2 & -1 & \vdots & 7 \\ 1 & 1 & \vdots & 2 \end{bmatrix}$ **17.** $\begin{bmatrix} 1 & -1 & 2 & \vdots & 2 \\ 4 & -3 & 1 & \vdots & -1 \\ 2 & 1 & 0 & \vdots & 0 \end{bmatrix}$

19. $\begin{bmatrix} 3 & -5 & 2 & \vdots & 12 \\ 12 & 0 & -7 & \vdots & 10 \end{bmatrix}$ **21.** $\begin{cases} x + y = 3 \\ 5x - 3y = -1 \end{cases}$

23. $\begin{cases} 2x \quad\quad + 5z = -12 \\ \quad y - 2z = 7 \\ 6x + 3y \quad\quad = 2 \end{cases}$

25. $\begin{cases} 9x + 12y + 3z \quad\quad = 0 \\ -2x + 18y + 5z + 2w = 10 \\ x + 7y - 8z \quad\quad = -4 \\ 3x \quad\quad + 2z \quad\quad = -10 \end{cases}$

27. Add 5 times Row 2 to Row 1.
29. Interchange Row 1 and Row 2.
Add 4 times new Row 1 to Row 3.

31. $\begin{bmatrix} 1 & 2 & \frac{8}{3} \\ 4 & -3 & 6 \end{bmatrix}$ **33.** $\begin{bmatrix} 1 & 1 & 1 \\ 0 & -7 & -1 \end{bmatrix}$

35. $\begin{bmatrix} 1 & 0 & 14 & -11 \\ 0 & 1 & -2 & 2 \\ 0 & 0 & 1 & -7 \end{bmatrix}$

37. $\begin{bmatrix} 1 & 1 & 4 & -1 \\ 0 & 5 & -2 & 6 \\ 0 & 3 & 20 & 4 \end{bmatrix}; \begin{bmatrix} 1 & 1 & 4 & -1 \\ 0 & 1 & -\frac{2}{5} & \frac{6}{5} \\ 0 & 3 & 20 & 4 \end{bmatrix}$

1. (a) (i) $\begin{bmatrix} 3 & 0 & \vdots & -6 \\ 6 & -4 & \vdots & -28 \end{bmatrix}$

(ii) $\begin{bmatrix} 3 & 0 & \vdots & -6 \\ 0 & -4 & \vdots & -16 \end{bmatrix}$

(iii) $\begin{bmatrix} 3 & 0 & \vdots & -6 \\ 0 & 1 & \vdots & 4 \end{bmatrix}$

(iv) $\begin{bmatrix} 1 & 0 & \vdots & -2 \\ 0 & 1 & \vdots & 4 \end{bmatrix}$

(b) $\begin{cases} -3x + 4y = 22 \\ 6x - 4y = -28 \end{cases}$

Solution: $(-2, 4)$

(c) Answers will vary.

41. Reduced row-echelon form   **43.** Not in row-echelon form

45. $\begin{bmatrix} 1 & 1 & 0 & 5 \\ 0 & 1 & 2 & 0 \\ 0 & 0 & 1 & -1 \end{bmatrix}$ **47.** $\begin{bmatrix} 1 & -1 & -1 & 1 \\ 0 & 1 & 6 & 3 \\ 0 & 0 & 0 & 0 \end{bmatrix}$

49. $\begin{bmatrix} 1 & 0 & 0 \\ 0 & 1 & 0 \\ 0 & 0 & 1 \end{bmatrix}$ **51.** $\begin{bmatrix} 1 & 2 & 0 & 0 \\ 0 & 0 & 1 & 0 \\ 0 & 0 & 0 & 1 \\ 0 & 0 & 0 & 0 \end{bmatrix}$

53. $\begin{bmatrix} 1 & 0 & 3 & 16 \\ 0 & 1 & 2 & 12 \end{bmatrix}$ **55.** $\begin{cases} x - 2y = 4 \\ y = -1 \end{cases}$
$(2, -1)$

57. $\begin{cases} x - y + 2z = 4 \\ \phantom{x - {}}y - z = 2 \\ z = -2 \end{cases}$
$(8, 0, -2)$

59. $(-3, 5)$ **61.** $(-5, 6)$ **63.** $(-4, -3, 6)$

65. No solution   **67.** $(3, -2, 5, 0)$ **69.** $(3, -4)$

71. $(-1, -4)$ **73.** $(5a + 4, -3a + 2, a)$ **75.** $(4, -3, 2)$

77. $(7, -3, 4)$ **79.** $(0, 2 - 4a, a)$ **81.** $(1, 0, 4, -2)$

83. $(-2a, a, a, 0)$ **85.** The dimension is 4×1.

87. $f(x) = -x^2 + x + 1$ **89.** $f(x) = -9x^2 - 5x + 11$

91. $f(x) = x^2 + 2x + 5$

93. $y = 7.5t + 28$; About 141 cases; Yes, because the data values increase in a linear pattern.

95. \$1,200,000 at 8%
\$200,000 at 9%
\$600,000 at 12%

97. False. It is a 2×4 matrix.   **99.** They are the same.

Section 8.2   (page 564)

1. equal   **3.** zero; O **5.** $x = -4, y = 23$

7. $x = -1, y = 3$

9. (a) $\begin{bmatrix} 3 & -2 \\ 1 & 7 \end{bmatrix}$ (b) $\begin{bmatrix} -1 & 0 \\ 3 & -9 \end{bmatrix}$ (c) $\begin{bmatrix} 3 & -3 \\ 6 & -3 \end{bmatrix}$

(d) $\begin{bmatrix} -1 & -1 \\ 8 & -19 \end{bmatrix}$

11. (a), (b), and (d) Not possible   (c) $\begin{bmatrix} 18 & 0 & 9 \\ -3 & -12 & 0 \end{bmatrix}$

13. (a) $\begin{bmatrix} 9 & 5 \\ 1 & -2 \\ -3 & 15 \end{bmatrix}$ (b) $\begin{bmatrix} 7 & -7 \\ 3 & 8 \\ -5 & -5 \end{bmatrix}$ (c) $\begin{bmatrix} 24 & -3 \\ 6 & 9 \\ -12 & 15 \end{bmatrix}$

(d) $\begin{bmatrix} 22 & -15 \\ 8 & 19 \\ -14 & -5 \end{bmatrix}$

15. (a) $\begin{bmatrix} 5 & 5 & -2 & 4 & 4 \\ -5 & 10 & 0 & -4 & -7 \end{bmatrix}$

(b) $\begin{bmatrix} 3 & 5 & 0 & 2 & 4 \\ 7 & -6 & -4 & 2 & 7 \end{bmatrix}$

(c) $\begin{bmatrix} 12 & 15 & -3 & 9 & 12 \\ 3 & 6 & -6 & -3 & 0 \end{bmatrix}$

(d) $\begin{bmatrix} 10 & 15 & -1 & 7 & 12 \\ 15 & -10 & -10 & 3 & 14 \end{bmatrix}$

17. $\begin{bmatrix} -8 & -7 \\ 15 & -1 \end{bmatrix}$ **19.** $\begin{bmatrix} -24 & -4 & 12 \\ -12 & 32 & 12 \end{bmatrix}$ **21.** $\begin{bmatrix} 10 & 8 \\ -59 & 9 \end{bmatrix}$

23. $\begin{bmatrix} -17.12 & 2.2 \\ 11.56 & 10.24 \end{bmatrix}$ **25.** $\begin{bmatrix} -10.81 & -5.36 & 0.4 \\ -14.04 & 10.69 & -14.76 \end{bmatrix}$

27. $\begin{bmatrix} -4 & 6 & -2 \\ 4 & 0 & 10 \end{bmatrix}$ **29.** $\begin{bmatrix} -2 & 0 & 5 \\ -\frac{5}{2} & 0 & \frac{7}{2} \end{bmatrix}$

31. $\begin{bmatrix} 3 & -\frac{1}{2} & -\frac{13}{2} \\ 3 & 0 & -\frac{11}{2} \end{bmatrix}$ **33.** $\begin{bmatrix} 2 & -5 & 5 \\ -5 & 0 & -6 \end{bmatrix}$

35. $\begin{bmatrix} -2 & 51 \\ -8 & 33 \\ 0 & 27 \end{bmatrix}$ **37.** Not possible   **39.** $\begin{bmatrix} 1 & 0 & 0 \\ 0 & 1 & 0 \\ 0 & 0 & \frac{7}{2} \end{bmatrix}$
3×2 3×3

41. $\begin{bmatrix} 70 & -17 & 73 \\ 32 & 11 & 6 \\ 16 & -38 & 70 \end{bmatrix}$ **43.** $\begin{bmatrix} 151 & 25 & 48 \\ 516 & 279 & 387 \\ 47 & -20 & 87 \end{bmatrix}$

45. (a) $\begin{bmatrix} 0 & 15 \\ 6 & 12 \end{bmatrix}$ (b) $\begin{bmatrix} -2 & 2 \\ 31 & 14 \end{bmatrix}$ (c) $\begin{bmatrix} 9 & 6 \\ 12 & 12 \end{bmatrix}$

47. (a) $\begin{bmatrix} 5 & -9 & 0 \\ 3 & 0 & -8 \\ -1 & 4 & 11 \end{bmatrix}$ (b) $\begin{bmatrix} 5 & -9 & 0 \\ 3 & 0 & -8 \\ -1 & 4 & 11 \end{bmatrix}$

(c) $\begin{bmatrix} -2 & -45 & 72 \\ 23 & -59 & -88 \\ -4 & 53 & 89 \end{bmatrix}$

49. (a) $\begin{bmatrix} 19 \\ 48 \end{bmatrix}$ (b) Not possible   (c) $\begin{bmatrix} 14 & -8 \\ 16 & 142 \end{bmatrix}$

51. (a) $\begin{bmatrix} 7 & 7 & 14 \\ 8 & 8 & 16 \\ -1 & -1 & -2 \end{bmatrix}$ (b) $[13]$ (c) Not possible

53. $\begin{bmatrix} 5 & 8 \\ -4 & -16 \end{bmatrix}$ **55.** $\begin{bmatrix} -4 & 10 \\ 3 & 14 \end{bmatrix}$

57. (a) $\langle 4, 7 \rangle$ (b) $\langle -2, 3 \rangle$ (c) $\langle 8, 1 \rangle$

59. (a) $\langle 3, 6 \rangle$ (b) $\langle -7, -2 \rangle$ (c) $\langle 17, 10 \rangle$

61. $\langle 4, -2 \rangle$; Reflection in the x-axis

63. $\langle 2, 4 \rangle$; Reflection in the line $y = x$

65. $\langle 8, 2 \rangle$; Horizontal stretch

67. (a) $\begin{bmatrix} 2 & 3 \\ 1 & 4 \end{bmatrix}\begin{bmatrix} x_1 \\ x_2 \end{bmatrix} = \begin{bmatrix} 5 \\ 10 \end{bmatrix}$ (b) $\begin{bmatrix} -2 \\ 3 \end{bmatrix}$

69. (a) $\begin{bmatrix} 1 & -2 & 3 \\ -1 & 3 & -1 \\ 2 & -5 & 5 \end{bmatrix}\begin{bmatrix} x_1 \\ x_2 \\ x_3 \end{bmatrix} = \begin{bmatrix} 9 \\ -6 \\ 17 \end{bmatrix}$ (b) $\begin{bmatrix} 1 \\ -1 \\ 2 \end{bmatrix}$

71. (a) $\begin{bmatrix} 1 & -5 & 2 \\ -3 & 1 & -1 \\ 0 & -2 & 5 \end{bmatrix}\begin{bmatrix} x_1 \\ x_2 \\ x_3 \end{bmatrix} = \begin{bmatrix} -20 \\ 8 \\ -16 \end{bmatrix}$ (b) $\begin{bmatrix} -1 \\ 3 \\ -2 \end{bmatrix}$

73. $\begin{bmatrix} 110 & 99 & 77 & 33 \\ 44 & 22 & 66 & 66 \end{bmatrix}$

75. [$1037.50 $1400 $1012.50]

The entries represent the profits from both crops at each of the three outlets.

77. $\begin{bmatrix} \$23.20 & \$20.50 \\ \$38.20 & \$33.80 \\ \$76.90 & \$68.50 \end{bmatrix}$

The entries represent the labor costs at each plant for each size of boat.

79. $\begin{bmatrix} 0.40 & 0.15 & 0.15 \\ 0.28 & 0.53 & 0.17 \\ 0.32 & 0.32 & 0.68 \end{bmatrix}$

P^2 gives the proportions of the voting population that changed parties or remained loyal to their parties from the first election to the third.

81. True. The sum of two matrices of different dimensions is undefined.

83. $\begin{bmatrix} 1 & 0 \\ 2 & 1 \end{bmatrix} \neq \begin{bmatrix} 0 & 0 \\ 3 & 2 \end{bmatrix}$ **85.** $\begin{bmatrix} 3 & -2 \\ 4 & 3 \end{bmatrix} \neq \begin{bmatrix} 2 & -2 \\ 5 & 4 \end{bmatrix}$

87. $AC = BC = \begin{bmatrix} 2 & 3 \\ 2 & 3 \end{bmatrix}$ **89.** Answers will vary.

91. AB is a diagonal matrix whose entries are the products of the corresponding entries of A and B.

Section 8.3   *(page 574)*

1. inverse   **3.** determinant   **5–11.** $AB = I$ and $BA = I$

13. $\begin{bmatrix} 3 & -1 \\ -5 & 2 \end{bmatrix}$ **15.** $\begin{bmatrix} -3 & 2 \\ -2 & 1 \end{bmatrix}$ **17.** $\begin{bmatrix} 1 & -\frac{1}{2} \\ -2 & \frac{3}{2} \end{bmatrix}$

19. $\begin{bmatrix} 1 & 1 & -1 \\ -3 & 2 & -1 \\ 3 & -3 & 2 \end{bmatrix}$ **21.** Not possible

23. $\begin{bmatrix} -\frac{1}{8} & 0 & 0 & 0 \\ 0 & 1 & 0 & 0 \\ 0 & 0 & \frac{1}{4} & 0 \\ 0 & 0 & 0 & -\frac{1}{5} \end{bmatrix}$ **25.** $\begin{bmatrix} -175 & 37 & -13 \\ 95 & -20 & 7 \\ 14 & -3 & 1 \end{bmatrix}$

27. $\begin{bmatrix} -12 & -5 & -9 \\ -4 & -2 & -4 \\ -8 & -4 & -6 \end{bmatrix}$ **29.** $\begin{bmatrix} 0 & -1.\overline{81} & 0.\overline{90} \\ -10 & 5 & 5 \\ 10 & -2.\overline{72} & -3.\overline{63} \end{bmatrix}$

31. $\begin{bmatrix} 1 & 0 & 1 & 0 \\ 0 & 1 & 0 & 1 \\ 2 & 0 & 1 & 0 \\ 0 & 1 & 0 & 2 \end{bmatrix}$ **33.** $\begin{bmatrix} \frac{5}{13} & -\frac{3}{13} \\ \frac{1}{13} & \frac{2}{13} \end{bmatrix}$

35. Not possible   **37.** $\begin{bmatrix} -4 & 2 \\ 10 & -\frac{10}{3} \end{bmatrix}$ **39.** $(5, 0)$

41. $(-8, -6)$ **43.** $(3, 8, -11)$ **45.** $(2, 1, 0, 0)$
47. $(-1, 1)$ **49.** No solution   **51.** $(-4, -8)$
53. $(-1, 3, 2)$ **55.** $\left(\frac{13}{16}, \frac{11}{16}, 0 \right)$
57. $3684.21 in AAA-rated bonds
      $2105.26 in A-rated bonds
      $4210.53 in B-rated bonds
59. $I_1 = 0.5$ amp   **61.** $I_1 = 4$ amps
      $I_2 = 3$ amps $I_2 = 1$ amp
      $I_3 = 3.5$ amps $I_3 = 5$ amps

63. 100 bags of potting soil for seedlings
      100 bags of potting soil for general potting
      100 bags of potting soil for hardwood plants

65. (a) $\begin{cases} 2.5r + 4l + 2i = 300 \\ -r + 2l + 2i = 0 \\ r + l + i = 120 \end{cases}$

$\begin{bmatrix} 2.5 & 4 & 2 \\ -1 & 2 & 2 \\ 1 & 1 & 1 \end{bmatrix} \begin{bmatrix} r \\ l \\ i \end{bmatrix} = \begin{bmatrix} 300 \\ 0 \\ 120 \end{bmatrix}$

(b) 80 roses, 10 lilies, 30 irises

67. True. If B is the inverse of A, then $AB = I = BA$.
69. Answers will vary.   **71.** $k \neq -\frac{3}{2}; k = -\frac{3}{2}$
73. (a) Answers will vary.

(b) $A^{-1} = \begin{bmatrix} 1/a_{11} & 0 & 0 & \cdots & 0 \\ 0 & 1/a_{22} & 0 & \cdots & 0 \\ 0 & 0 & 1/a_{33} & \cdots & 0 \\ \vdots & \vdots & \vdots & & \vdots \\ 0 & 0 & 0 & \cdots & 1/a_{nn} \end{bmatrix}$

75. Answers will vary.

Section 8.4   *(page 582)*

1. determinant   **3.** cofactor   **5.** 4   **7.** 16   **9.** -3
11. 0   **13.** 6   **15.** 0   **17.** -23 **19.** -24
21. $\frac{11}{6}$ **23.** 11   **25.** -1924 **27.** 0.08
29. (a) $M_{11} = -6, M_{12} = 3, M_{21} = 5, M_{22} = 4$
      (b) $C_{11} = -6, C_{12} = -3, C_{21} = -5, C_{22} = 4$
31. (a) $M_{11} = 3, M_{12} = -4, M_{13} = 1, M_{21} = 2, M_{22} = 2,$
      $M_{23} = -4, M_{31} = -4, M_{32} = 10, M_{33} = 8$
      (b) $C_{11} = 3, C_{12} = 4, C_{13} = 1, C_{21} = -2, C_{22} = 2,$
      $C_{23} = 4, C_{31} = -4, C_{32} = -10, C_{33} = 8$
33. (a) $M_{11} = 10, M_{12} = -43, M_{13} = 2, M_{21} = -30, M_{22} = 1$
      $M_{23} = -6, M_{31} = 54, M_{32} = -53, M_{33} = -34$
      (b) $C_{11} = 10, C_{12} = 43, C_{13} = 2, C_{21} = 30, C_{22} = 17,$
      $C_{23} = 6, C_{31} = 54, C_{32} = 53, C_{33} = -34$
35. (a) and (b) -36 **37.** (a) and (b) 96
39. (a) and (b) -75 **41.** (a) and (b) 0
43. (a) and (b) 225   **45.** -9 **47.** 0   **49.** 0
51. -58 **53.** 72   **55.** 0   **57.** 412
59. -126 **61.** -336

63. (a) -3 (b) -2 (c) $\begin{bmatrix} -2 & 0 \\ 0 & -3 \end{bmatrix}$ (d) 6

65. (a) -8 (b) 0   (c) $\begin{bmatrix} -4 & 4 \\ 1 & -1 \end{bmatrix}$ (d) 0

67. (a) 2   (b) -6 (c) $\begin{bmatrix} 1 & 4 & 3 \\ -1 & 0 & 3 \\ 0 & 2 & 0 \end{bmatrix}$ (d) -12

69. $A = \begin{bmatrix} 3 & 3 \\ 1 & 2 \end{bmatrix}$ **71.** $A = \begin{bmatrix} 4 & 2 & -1 \\ 2 & 1 & 0 \\ 1 & 1 & 3 \end{bmatrix}$

73. $A = \begin{bmatrix} 2 & 3 \\ 8 & 12 \end{bmatrix}$ **75–79.** Answers will vary.   **81.** ± 2

83. $-2, 1$ **85.** $-1, -4$ **87.** $8uv - 1$ **89.** e^{5x}
91. $1 - \ln x$
93. True. If an entire row is zero, then each cofactor in the expansion is multiplied by zero.

55. Answers will vary.

57. The signs of the cofactors should be −, +, −.

59. (a) Columns 2 and 3 of A were interchanged.
$$|A| = -115 = -|B|$$
(b) Rows 1 and 3 of A were interchanged.
$$|A| = -40 = -|B|$$

61. (a) Multiply Row 1 by 5.
(b) Multiply Column 2 by 4 and Column 3 by 3.

63. (a) 28 (b) −10 (c) −12
The determinant of a diagonal matrix is the product of the entries on the main diagonal.

Section 8.5 *(page 595)*

1. Cramer's Rule

3. $A = \pm\dfrac{1}{2}\begin{vmatrix} x_1 & y_1 & 1 \\ x_2 & y_2 & 1 \\ x_3 & y_3 & 1 \end{vmatrix}$

5. uncoded; coded **7.** $(1, -1)$ **9.** Not possible

11. $(-1, 3, 2)$ **13.** $(-2, 1, -1)$ **15.** 7 **17.** 14

19. $y = \frac{16}{5}$ or $y = 0$ **21.** $250\ \text{mi}^2$ **23.** Collinear

25. Not collinear **27.** Collinear **29.** $y = -3$

31. $3x - 5y = 0$ **33.** $x + 3y - 5 = 0$

35. $2x + 3y - 8 = 0$

37. $(0, 0), (0, 3), (6, 0), (6, 3)$

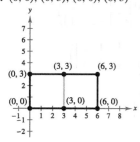

39. $(-4, 3), (-5, 3), (-4, 4), (-5, 4)$

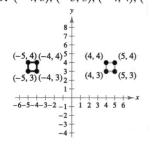

41. 2 square units **43.** 10 square units

45. (a) Uncoded: $[3\quad 15], [13\quad 5], [0\quad 8], [15\quad 13], [5\quad 0],$
$[19\quad 15], [15\quad 14]$
(b) Encoded: 48 81 28 51 24 40 54 95 5
10 64 113 57 100

47. (a) Uncoded: $[3\quad 1\quad 12], [12\quad 0\quad 13], [5\quad 0\quad 20],$
$[15\quad 13\quad 15], [18\quad 18\quad 15], [23\quad 0\quad 0]$
(b) Encoded: −68 21 35 −66 14 39 −115
35 60 −62 15 32 −54 12 27 23 −23 0

49. 1 −25 −65 17 15 −9 −12 −62 −119
27 51 48 43 67 48 57 111 117

51. −5 −41 −87 91 207 257 11 −5 −41 40
80 84 76 177 227

53. HAPPY NEW YEAR **55.** CLASS IS CANCELED

57. SEND PLANES **59.** MEET ME TONIGHT RON

61. $I_1 = -0.5$ amp
$I_2 = 1$ amp
$I_3 = 0.5$ amp

63. False. The denominator is the determinant of the coefficient matrix.

65. The system has either no solutions or infinitely many solutions.

67. 12

Review Exercises *(page 600)*

1. 1×2 **3.** 2×5 **5.** $\begin{bmatrix} 3 & -10 & \vdots & 15 \\ 5 & 4 & \vdots & 22 \end{bmatrix}$

7. $\begin{cases} x & + 2z = -8 \\ 2x - 2y + 3z = 12 \\ 4x + 7y + z = 3 \end{cases}$ **9.** $\begin{bmatrix} 1 & 2 & 3 \\ 0 & 1 & 1 \\ 0 & 0 & 1 \end{bmatrix}$

11. $\begin{cases} x + 2y + 3z = 9 \\ y - 2z = 2 \\ z = -1 \end{cases}$ **13.** $\begin{cases} x + 3y + 4z = 1 \\ y + 2z = 3 \\ z = 4 \end{cases}$

$(12, 0, -1)$ $(0, -5, 4)$

15. $(10, -12)$ **17.** $\left(-\frac{1}{5}, \frac{7}{10}\right)$ **19.** No solution

21. $(1, -2, 2)$ **23.** $\left(-2a + \frac{3}{2}, 2a + 1, a\right)$ **25.** $(5, 2, -6)$

27. $(1, 2, 2)$ **29.** $(2, -3, 3)$ **31.** $(2, 6, -10, -3)$

33. $x = 12, y = 11$ **35.** $x = 1, y = 11$

37. (a) $\begin{bmatrix} -1 & 8 \\ 15 & 13 \end{bmatrix}$ (b) $\begin{bmatrix} 5 & -12 \\ -9 & -3 \end{bmatrix}$
(c) $\begin{bmatrix} 8 & -8 \\ 12 & 20 \end{bmatrix}$ (d) $\begin{bmatrix} -2 & 16 \\ 30 & 26 \end{bmatrix}$

39. (a) $\begin{bmatrix} 5 & 7 \\ -3 & 14 \\ 31 & 42 \end{bmatrix}$ (b) $\begin{bmatrix} 5 & 1 \\ -11 & -10 \\ -9 & -38 \end{bmatrix}$
(c) $\begin{bmatrix} 20 & 16 \\ -28 & 8 \\ 44 & 8 \end{bmatrix}$ (d) $\begin{bmatrix} 10 & 14 \\ -6 & 28 \\ 62 & 84 \end{bmatrix}$

41. $\begin{bmatrix} 22 & -17 \\ 14 & 11 \end{bmatrix}$ **43.** $\begin{bmatrix} -16 & -6 \\ -12 & 4 \\ -14 & -8 \end{bmatrix}$ **45.** $\begin{bmatrix} -11 & -6 \\ 8 & -13 \\ -18 & -8 \end{bmatrix}$

47. $\begin{bmatrix} 3 & \frac{2}{3} \\ -\frac{4}{3} & \frac{11}{3} \\ \frac{10}{3} & 0 \end{bmatrix}$ **49.** $\begin{bmatrix} -30 & 4 \\ 51 & 70 \end{bmatrix}; 2 \times 2$

51. $\begin{bmatrix} 100 & 220 \\ 12 & -4 \\ 84 & 212 \end{bmatrix}; 3 \times 2$ **53.** $\begin{bmatrix} 14 & -22 & 22 \\ 19 & -41 & 80 \\ 42 & -66 & 66 \end{bmatrix}$

55. Not possible

57. (a) $\begin{bmatrix} -1 & -1 \\ 18 & -4 \end{bmatrix}$ (b) $\begin{bmatrix} 1 & 14 \\ -2 & -6 \end{bmatrix}$ (c) $\begin{bmatrix} 13 & 6 \\ 8 & 13 \end{bmatrix}$

59. $\langle 2, -5 \rangle$; Reflection in the x-axis

61. $\langle 1, 5 \rangle$; Horizontal shrink **63.** $\begin{bmatrix} 76 & 114 & 133 \\ 38 & 95 & 76 \end{bmatrix}$

65–67. $AB = I$ and $BA = I$ **69.** $\begin{bmatrix} 4 & -5 \\ 5 & -6 \end{bmatrix}$

71. $\begin{bmatrix} \frac{1}{2} & -1 & -\frac{1}{2} \\ \frac{1}{2} & -\frac{2}{3} & -\frac{5}{6} \\ 0 & \frac{2}{3} & \frac{1}{3} \end{bmatrix}$ **73.** $\begin{bmatrix} 13 & 6 & -4 \\ -12 & -5 & 3 \\ 5 & 2 & -1 \end{bmatrix}$

75. $\begin{bmatrix} 1 & -1 \\ 4 & -\frac{7}{2} \end{bmatrix}$ **77.** Not possible   **79.** $(36, 11)$

81. $(-6, -1)$ **83.** $(2, 3)$ **85.** $(-8, 18)$

87. $(2, -1, -2)$ **89.** $(-3, 1)$ **91.** $\left(\frac{1}{6}, -\frac{7}{4}\right)$

93. 26   **95.** 116

97. (a) $M_{11} = 4, M_{12} = 7, M_{21} = -1, M_{22} = 2$

(b) $C_{11} = 4, C_{12} = -7, C_{21} = 1, C_{22} = 1$

99. (a) $M_{11} = 30, M_{12} = -12, M_{13} = -21, M_{21} = 20,$
$M_{22} = 19, M_{23} = 22, M_{31} = 5, M_{32} = -2, M_{33} = 19$

(b) $C_{11} = 30, C_{12} = 12, C_{13} = -21, C_{21} = -20,$
$C_{22} = 19, C_{23} = -22, C_{31} = 5, C_{32} = 2, C_{33} = 19$

101. -6 **103.** 15   **105.** 130   **107.** $(4, 7)$

109. $(-1, 4, 5)$ **111.** 16   **113.** Collinear

115. $x - 2y + 4 = 0$ **117.** $2x + 6y - 13 = 0$

119. 8 square units   **121.** SEE YOU FRIDAY

123. False. The matrix must be square.

125. If A is a square matrix, then the cofactor C_{ij} of the entry a_{ij} is $(-1)^{i+j}M_{ij}$, where M_{ij} is the determinant obtained by deleting the ith row and jth column of A. The determinant of A is the sum of the entries of any row or column of A multiplied by their respective cofactors.

Chapter Test   *(page 604)*

1. $\begin{bmatrix} 1 & 0 & 0 \\ 0 & 1 & 0 \\ 0 & 0 & 1 \end{bmatrix}$ **2.** $\begin{bmatrix} 1 & 0 & -1 & 2 \\ 0 & 1 & 0 & -1 \\ 0 & 0 & 0 & 0 \\ 0 & 0 & 0 & 0 \end{bmatrix}$

3. $\begin{bmatrix} 4 & 3 & -2 & \vdots & 14 \\ -1 & -1 & 2 & \vdots & -5 \\ 3 & 1 & -4 & \vdots & 8 \end{bmatrix}, \left(1, 3, -\frac{1}{2}\right)$

4. (a) $\begin{bmatrix} 1 & 5 \\ 0 & -4 \end{bmatrix}$

(b) $\begin{bmatrix} 6 & -3 & 12 \\ 0 & 18 & -9 \end{bmatrix}$

(c) $\begin{bmatrix} 8 & 15 \\ -5 & -13 \end{bmatrix}$

(d) $\begin{bmatrix} 10 & -5 & 20 \\ -10 & -1 & -17 \end{bmatrix}$

(e) Not possible

5. $\langle -3, -2 \rangle$; Reflection in the line $y = -x$

6. $\begin{bmatrix} \frac{2}{7} & \frac{3}{7} \\ \frac{5}{7} & \frac{4}{7} \end{bmatrix}$ **7.** $\begin{bmatrix} -\frac{5}{2} & 4 & -3 \\ 5 & -7 & 6 \\ 4 & -6 & 5 \end{bmatrix}$ **8.** $(12, 18)$

9. -112 **10.** 0   **11.** 43   **12.** $(-3, 5)$

13. $(-2, 4, 6)$ **14.** 7

15. Uncoded: $\begin{bmatrix} 11 & 14 & 15 \end{bmatrix}, \begin{bmatrix} 3 & 11 & 0 \end{bmatrix}, \begin{bmatrix} 15 & 14 & 0 \end{bmatrix}, \begin{bmatrix} 23 & 15 & 15 \end{bmatrix},$
$\begin{bmatrix} 4 & 0 & 0 \end{bmatrix}$

Encoded: $115 \ -41 \ -59 \ 14 \ -3 \ -11 \ 29 \ -15$
$-14 \ 128 \ -53 \ -60 \ 4 \ -4 \ 0$

16. 75 L of 60% solution, 25 L of 20% solution

Problem Solving   *(page 607)*

1. (a) $AT = \begin{bmatrix} -1 & -4 & -2 \\ 1 & 2 & 3 \end{bmatrix}, AAT = \begin{bmatrix} -1 & -2 & -3 \\ -1 & -4 & -2 \end{bmatrix}$

A represents a counterclockwise rotation.

(b) AAT is rotated clockwise 90° to obtain AT. AT is the rotated clockwise 90° to obtain T.

3. (a) Yes   (b) No   (c) No   (d) No   (e) No   (f) N

5. (a) $A^2 - 2A + 5I = \begin{bmatrix} 1 & 2 \\ -2 & 1 \end{bmatrix}^2 - 2\begin{bmatrix} 1 & 2 \\ -2 & 1 \end{bmatrix} + 5\begin{bmatrix} 1 & 0 \\ 0 & 1 \end{bmatrix}$

$= \begin{bmatrix} -3 & 4 \\ -4 & -3 \end{bmatrix} - \begin{bmatrix} 2 & 4 \\ -4 & 2 \end{bmatrix} + \begin{bmatrix} 5 & 0 \\ 0 & 5 \end{bmatrix}$

$= \begin{bmatrix} 0 & 0 \\ 0 & 0 \end{bmatrix}$

(b) $A^{-1} = \frac{1}{5}(2I - A)$

$\begin{bmatrix} 1 & 2 \\ -2 & 1 \end{bmatrix}^{-1} = \frac{1}{5}\left(2\begin{bmatrix} 1 & 0 \\ 0 & 1 \end{bmatrix} - \begin{bmatrix} 1 & 2 \\ -2 & 1 \end{bmatrix}\right)$

$\begin{bmatrix} \frac{1}{5} & -\frac{2}{5} \\ \frac{2}{5} & \frac{1}{5} \end{bmatrix} = \frac{1}{5}\left(\begin{bmatrix} 2 & 0 \\ 0 & 2 \end{bmatrix} - \begin{bmatrix} 1 & 2 \\ -2 & 1 \end{bmatrix}\right)$

$\begin{bmatrix} \frac{1}{5} & -\frac{2}{5} \\ \frac{2}{5} & \frac{1}{5} \end{bmatrix} = \frac{1}{5}\begin{bmatrix} 1 & -2 \\ 2 & 1 \end{bmatrix}$

$\begin{bmatrix} \frac{1}{5} & -\frac{2}{5} \\ \frac{2}{5} & \frac{1}{5} \end{bmatrix} = \begin{bmatrix} \frac{1}{5} & -\frac{2}{5} \\ \frac{2}{5} & \frac{1}{5} \end{bmatrix}$

(c) Answers will vary.

7. $A^T = \begin{bmatrix} -1 & 2 \\ 1 & 0 \\ -2 & 1 \end{bmatrix}, B^T = \begin{bmatrix} -3 & 1 & 1 \\ 0 & 2 & -1 \end{bmatrix}$

$(AB)^T = \begin{bmatrix} 2 & -5 \\ 4 & -1 \end{bmatrix} = B^T A^T$

9. $x = 6$ **11.** Answers will vary.

13. $\begin{vmatrix} x & 0 & 0 & d \\ -1 & x & 0 & c \\ 0 & -1 & x & b \\ 0 & 0 & -1 & a \end{vmatrix}$

15. Sulfur: 32 atomic mass units
Nitrogen: 14 atomic mass units
Fluorine: 19 atomic mass units

17. REMEMBER SEPTEMBER THE ELEVENTH

19. $A^{-1} = \begin{bmatrix} 0.0625 & -0.4375 & 0.625 \\ 0.1875 & 0.6875 & -1.125 \\ -0.125 & -0.125 & 0.75 \end{bmatrix}$

$|A^{-1}| = \frac{1}{16}, |A| = 16$

$|A^{-1}| = \frac{1}{|A|}$

12. (a) The upper graph represents the first investment (compound interest), and the lower graph represents the second investment (simple interest).

(b) $y_1 = 500(1.07)^t$

$y_2 = 35t + 500$

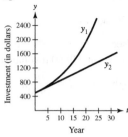

(c) You should choose compound interest because the earnings would be higher.

13. $t = \dfrac{\ln c_1 - \ln c_2}{\left(\dfrac{1}{k_2} - \dfrac{1}{k_1}\right)\ln\dfrac{1}{2}}$ **14.** $B = 500\left(\dfrac{2}{5}\right)^{t/2}$

15. (a) $y_1 = 252{,}606(1.0310)^t$

(b) $y_2 = 400.88t^2 - 1464.6t + 291{,}782$

(c)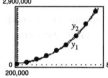

(d) The exponential model is a better fit. No, because the model is rapidly approaching infinity.

16. Answers will vary.   **17.** $1, e^2$

18. (a) 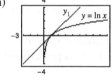   (b)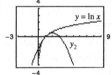

(c)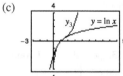

19. $y_4 = (x - 1) - \frac{1}{2}(x - 1)^2 + \frac{1}{3}(x - 1)^3 - \frac{1}{4}(x - 1)^4$

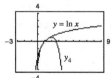

The pattern implies that

$\ln x = (x - 1) - \frac{1}{2}(x - 1)^2 + \frac{1}{3}(x - 1)^3 - \cdots$.

20. (a) Slope $= \ln b$; y-intercept: $(0, \ln a)$

(b) Slope $= b$; y-intercept: $(0, \ln a)$

21.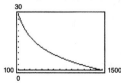

$17.7 \text{ ft}^3/\text{min}$

22. (a) $15 \text{ ft}^3/\text{min}$ (b) 382.0 ft^3 (c) 382.0 ft^2

23. (a) 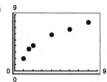   **24.** (a)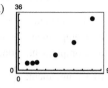

(b)–(e) Answers will vary.   (b)–(e) Answers will vary.

25. (a) 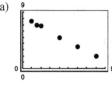   **26.** (a)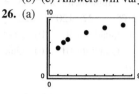

(b)–(e) Answers will vary.   (b)–(e) Answers will vary.

Chapter 4

Section 4.1   *(page 267)*

1. coterminal   **2.** radian

3. complementary; supplementary   **4.** degree

5. linear; angular   **6.** $A = \frac{1}{2}r^2\theta$ **7.** 1 rad

8. 5.5 rad   **9.** -3 rad   **10.** 6.5 rad

11. (a) Quadrant I   (b) Quadrant II

12. (a) Quadrant IV   (b) Quadrant III

13. (a)   (b)

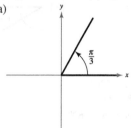 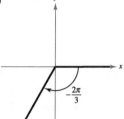

14. (a)   (b)

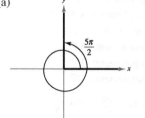 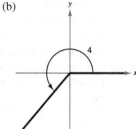

15. *Sample answers:* (a) $\dfrac{13\pi}{6}, -\dfrac{11\pi}{6}$ (b) $\dfrac{7\pi}{6}, -\dfrac{17\pi}{6}$

16. *Sample answers:* (a) $\dfrac{8\pi}{3}, -\dfrac{4\pi}{3}$ (b) $\dfrac{7\pi}{4}, -\dfrac{\pi}{4}$

17. (a) Complement: $\dfrac{5\pi}{12}$; Supplement: $\dfrac{11\pi}{12}$

(b) Complement: none; Supplement: $\dfrac{\pi}{12}$

18. (a) Complement: $\dfrac{\pi}{6}$; Supplement: $\dfrac{2\pi}{3}$

(b) Complement: $\dfrac{\pi}{4}$; Supplement: $\dfrac{3\pi}{4}$

19. (a) Complement: $\frac{\pi}{2} - 1 \approx 0.57$;

Supplement: $\pi - 1 \approx 2.14$

(b) Complement: none; Supplement: $\pi - 2 \approx 1.14$

20. (a) Complement: none; Supplement: $\pi - 3 \approx 0.14$

(b) Complement: $\frac{\pi}{2} - 1.5 \approx 0.07$;

Supplement: $\pi - 1.5 \approx 1.64$

21. $210°$　**22.** $120°$　**23.** $-60°$　**24.** $-330°$

25. (a) Quadrant II　(b) Quadrant IV

26. (a) Quadrant III　(b) Quadrant I

27. (a)　　　　　　　　(b)

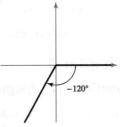

28. (a)　　　　　　　　(b)

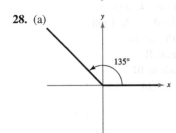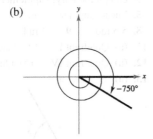

29. (a) $480°, -240°$　(b) $150°, -570°$

30. (a) $405°, -315°$　(b) $300°, -60°$

31. (a) Complement: $72°$; Supplement: $162°$

(b) Complement: $5°$; Supplement: $95°$

32. (a) Complement: $44°$; Supplement: $134°$

(b) Complement: none; Supplement: $87°$

33. (a) Complement: $66°$; Supplement: $156°$

(b) Complement: none; Supplement: $54°$

34. (a) Complement: none; Supplement: $50°$

(b) Complement: none; Supplement: $10°$

35. (a) $\frac{2\pi}{3}$　(b) $-\frac{\pi}{9}$　**36.** (a) $-\frac{\pi}{3}$　(b) $\frac{4\pi}{5}$

37. (a) $270°$　(b) $-210°$　**38.** (a) $-105°$　(b) $225°$

39. 0.785　**40.** -0.842　**41.** -0.009　**42.** 6.021

43. $81.818°$　**44.** $337.500°$　**45.** $-756.000°$

46. $-32.659°$　**47.** (a) $54.75°$　(b) $-128.5°$

48. (a) $135.177°$　(b) $-408.272°$

49. (a) $240° 36'$　(b) $-145° 48'$

50. (a) $345° 7' 12''$　(b) $-3° 34' 48''$

51. 10π in. ≈ 31.42 in.　**52.** 2.5π m ≈ 7.85 m　**53.** $\frac{15}{8}$ rad

54. $\frac{4}{7}$ rad　**55.** 4 rad　**56.** $-\frac{4}{5}$ rad

57. About 18.85 in.^{2}　**58.** About 12.27 ft^2

59. $20°$ should be multiplied by $\frac{\pi \text{ rad}}{180 \text{ deg}}$.

60. $72°$ should be converted to radians first.

61. About 592 mi　**62.** About 686 mi　**63.** About $23.87°$

64. (a) $10,400\pi \approx 32,672.6$ rad/min

(b) $\frac{9425\pi}{3}$ ft/min ≈ 9869.8 ft/min

65. (a) 8π rad/min ≈ 25.13 rad/min

(b) 200π ft/min ≈ 628.3 ft/min

66. (a) $\frac{1000\pi}{3}$ rad/sec ≈ 1047.20 rad/sec

(b) 20π m/sec ≈ 62.83 m/sec

67. (a) About 35.70 mi/h　(b) About 739.50 revolutions/min

68. (a) $\frac{14\pi}{3}$ ft/sec; About 10 mi/h　(b) $d = \frac{7\pi}{7920}n$

(c) $d = \frac{7\pi}{7920}t$

69.

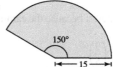

$A = 93.75\pi$ m$^2 \approx 294.52$ m^2

70. $A = 175\pi$ in.$^2 \approx 549.8$ in.^{2}

71. False. $\frac{180°}{\pi}$ is in degree measure.

72. False. A measurement of 4π radians corresponds to two complete revolutions from the initial to the terminal side of an angle.

73. True. Let α and β represent coterminal angles, and let n represent an integer.

$\alpha = \beta + n(360°)$

$\alpha - \beta = n(360°)$

74. False. The terminal side of the angle lies on the x-axis.

75. When θ is constant, the length of the arc is proportional to the radius ($s = r\theta$).

76. B and C have the same initial and terminal sides as A.

77. The speed increases. The linear velocity is proportional to the radius.

78. Radian. 1 rad $\approx 57.3°$　**79.** Proof

Section 4.2　(page 275)

1. unit circle　**2.** periodic　**3.** period　**4.** odd; even

5. $\sin t = \frac{5}{13}$　$\csc t = \frac{13}{5}$　**6.** $\sin t = \frac{15}{17}$　$\csc t = \frac{17}{15}$

$\cos t = \frac{12}{13}$　$\sec t = \frac{13}{12}$　$\cos t = -\frac{8}{17}$　$\sec t = -\frac{17}{8}$

$\tan t = \frac{5}{12}$　$\cot t = \frac{12}{5}$　$\tan t = -\frac{15}{8}$　$\cot t = -\frac{8}{15}$

7. $\sin t = -\frac{3}{5}$　$\csc t = -\frac{5}{3}$　**8.** $\sin t = -\frac{5}{13}$　$\csc t = -\frac{13}{5}$

$\cos t = -\frac{4}{5}$　$\sec t = -\frac{5}{4}$　$\cos t = \frac{12}{13}$　$\sec t = \frac{13}{12}$

$\tan t = \frac{3}{4}$　$\cot t = \frac{4}{3}$　$\tan t = -\frac{5}{12}$　$\cot t = -\frac{12}{5}$

9. $(0, 1)$　**10.** $\left(\frac{\sqrt{2}}{2}, \frac{\sqrt{2}}{2}\right)$

11. $\left(-\frac{\sqrt{3}}{2}, \frac{1}{2}\right)$　**12.** $\left(-\frac{1}{2}, -\frac{\sqrt{3}}{2}\right)$

13. $\sin \frac{\pi}{4} = \frac{\sqrt{2}}{2}$　　**14.** $\sin \frac{\pi}{3} = \frac{\sqrt{3}}{2}$

$\cos \frac{\pi}{4} = \frac{\sqrt{2}}{2}$　　$\cos \frac{\pi}{3} = \frac{1}{2}$

$\tan \frac{\pi}{4} = 1$　　$\tan \frac{\pi}{3} = \sqrt{3}$

15. $\sin\left(-\dfrac{\pi}{6}\right) = -\dfrac{1}{2}$

$\cos\left(-\dfrac{\pi}{6}\right) = \dfrac{\sqrt{3}}{2}$

$\tan\left(-\dfrac{\pi}{6}\right) = -\dfrac{\sqrt{3}}{3}$

16. $\sin\left(-\dfrac{\pi}{4}\right) = -\dfrac{\sqrt{2}}{2}$

$\cos\left(-\dfrac{\pi}{4}\right) = \dfrac{\sqrt{2}}{2}$

$\tan\left(-\dfrac{\pi}{4}\right) = -1$

17. $\sin\left(-\dfrac{7\pi}{4}\right) = \dfrac{\sqrt{2}}{2}$

$\cos\left(-\dfrac{7\pi}{4}\right) = \dfrac{\sqrt{2}}{2}$

$\tan\left(-\dfrac{7\pi}{4}\right) = 1$

18. $\sin\left(-\dfrac{4\pi}{3}\right) = \dfrac{\sqrt{3}}{2}$

$\cos\left(-\dfrac{4\pi}{3}\right) = -\dfrac{1}{2}$

$\tan\left(-\dfrac{4\pi}{3}\right) = -\sqrt{3}$

19. $\sin\dfrac{11\pi}{6} = -\dfrac{1}{2}$

$\cos\dfrac{11\pi}{6} = \dfrac{\sqrt{3}}{2}$

$\tan\dfrac{11\pi}{6} = -\dfrac{\sqrt{3}}{3}$

20. $\sin\dfrac{5\pi}{3} = -\dfrac{\sqrt{3}}{2}$

$\cos\dfrac{5\pi}{3} = \dfrac{1}{2}$

$\tan\dfrac{5\pi}{3} = -\sqrt{3}$

21. $\sin\left(-\dfrac{3\pi}{2}\right) = 1$

$\cos\left(-\dfrac{3\pi}{2}\right) = 0$

$\tan\left(-\dfrac{3\pi}{2}\right)$ is undefined.

22. $\sin(-2\pi) = 0$

$\cos(-2\pi) = 1$

$\tan(-2\pi) = 0$

23. $\sin\dfrac{2\pi}{3} = \dfrac{\sqrt{3}}{2}$

$\cos\dfrac{2\pi}{3} = -\dfrac{1}{2}$

$\tan\dfrac{2\pi}{3} = -\sqrt{3}$

$\csc\dfrac{2\pi}{3} = \dfrac{2\sqrt{3}}{3}$

$\sec\dfrac{2\pi}{3} = -2$

$\cot\dfrac{2\pi}{3} = -\dfrac{\sqrt{3}}{3}$

24. $\sin\dfrac{5\pi}{6} = \dfrac{1}{2}$

$\cos\dfrac{5\pi}{6} = -\dfrac{\sqrt{3}}{2}$

$\tan\dfrac{5\pi}{6} = -\dfrac{\sqrt{3}}{3}$

$\csc\dfrac{5\pi}{6} = 2$

$\sec\dfrac{5\pi}{6} = -\dfrac{2\sqrt{3}}{3}$

$\cot\dfrac{5\pi}{6} = -\sqrt{3}$

25. $\sin\dfrac{4\pi}{3} = -\dfrac{\sqrt{3}}{2}$

$\cos\dfrac{4\pi}{3} = -\dfrac{1}{2}$

$\tan\dfrac{4\pi}{3} = \sqrt{3}$

$\csc\dfrac{4\pi}{3} = -\dfrac{2\sqrt{3}}{3}$

$\sec\dfrac{4\pi}{3} = -2$

$\cot\dfrac{4\pi}{3} = \dfrac{\sqrt{3}}{3}$

26. $\sin\dfrac{7\pi}{4} = -\dfrac{\sqrt{2}}{2}$

$\cos\dfrac{7\pi}{4} = \dfrac{\sqrt{2}}{2}$

$\tan\dfrac{7\pi}{4} = -1$

$\csc\dfrac{7\pi}{4} = -\sqrt{2}$

$\sec\dfrac{7\pi}{4} = \sqrt{2}$

$\cot\dfrac{7\pi}{4} = -1$

27. $\sin\left(-\dfrac{5\pi}{3}\right) = \dfrac{\sqrt{3}}{2}$

$\cos\left(-\dfrac{5\pi}{3}\right) = \dfrac{1}{2}$

$\tan\left(-\dfrac{5\pi}{3}\right) = \sqrt{3}$

$\csc\left(-\dfrac{5\pi}{3}\right) = \dfrac{2\sqrt{3}}{3}$

$\sec\left(-\dfrac{5\pi}{3}\right) = 2$

$\cot\left(-\dfrac{5\pi}{3}\right) = \dfrac{\sqrt{3}}{3}$

28. $\sin\left(-\dfrac{3\pi}{2}\right) = 1$ $\csc\left(-\dfrac{3\pi}{2}\right) = 1$

$\cos\left(-\dfrac{3\pi}{2}\right) = 0$ $\sec\left(-\dfrac{3\pi}{2}\right)$ is undefined.

$\tan\left(-\dfrac{3\pi}{2}\right)$ is undefined. $\cot\left(-\dfrac{3\pi}{2}\right) = 0$

29. $\sin\left(-\dfrac{\pi}{2}\right) = -1$ $\csc\left(-\dfrac{\pi}{2}\right) = -1$

$\cos\left(-\dfrac{\pi}{2}\right) = 0$ $\sec\left(-\dfrac{\pi}{2}\right)$ is undefined.

$\tan\left(-\dfrac{\pi}{2}\right)$ is undefined. $\cot\left(-\dfrac{\pi}{2}\right) = 0$

30. $\sin(-\pi) = 0$ $\csc(-\pi)$ is undefined.

$\cos(-\pi) = -1$ $\sec(-\pi) = -1$

$\tan(-\pi) = 0$ $\cot(-\pi)$ is undefined.

31. 0 **32.** -1 **33.** $\dfrac{1}{2}$

34. $\dfrac{\sqrt{2}}{2}$ **35.** $-\dfrac{1}{2}$ **36.** $-\dfrac{\sqrt{3}}{2}$

37. (a) $-\dfrac{1}{2}$ (b) -2 **38.** (a) $-\dfrac{3}{8}$ (b) $-\dfrac{8}{3}$

39. (a) $-\dfrac{1}{5}$ (b) -5 **40.** (a) $-\dfrac{3}{4}$ (b) $-\dfrac{4}{3}$

41. (a) $\dfrac{4}{5}$ (b) $-\dfrac{4}{5}$ **42.** (a) $-\dfrac{4}{5}$ (b) $-\dfrac{4}{5}$

43. 0.5646 **44.** -0.9422 **45.** 0.4142 **46.** -1.2540

47. -1.0009 **48.** -0.5090

49. (a) 0.50 ft   (b) About 0.04 ft   (c) About -0.49 ft

50. (a)

t	0	$\dfrac{1}{4}$	$\dfrac{1}{2}$	$\dfrac{3}{4}$	1
y	0.5	0.0275	-0.3002	-0.0498	0.1766

(b) $t \approx 0.26$ (c) The displacement approaches 0.

51. False. $\sin(-t) = -\sin(t)$ means that the function is odd, not that the sine of a negative angle is a negative number.

52. False. The real number 0 corresponds to the point $(1, 0)$.

53. True. The tangent function has a period of π.   **54.** True

55. (a) y-axis symmetry   (b) $\sin t_1 = \sin(\pi - t_1)$
(c) $\cos(\pi - t_1) = -\cos t_1$

56. Answers will vary.

57. The calculator was in degree mode instead of radian mode.

58. Answers will vary.

59. (a)

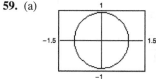

Circle of radius 1 centered at $(0, 0)$

(b) The t-values represent the central angle in radians. The x- and y-values represent the location in the coordinate plane.

(c) $-1 \le x \le 1, -1 \le y \le 1$

60. (a) Yes; x and y are both nonzero.

(b) Positive: $\tan t$, $\cot t$
Negative: $\sin t$, $\cos t$, $\csc t$, $\sec t$
Answers will vary.

61. It is an odd function.

62. It is an even function.

Section 4.3 *(page 284)*

1. (a) v   (b) iv   (c) vi   (d) iii   (e) i   (f) ii

2. opposite; adjacent; hypotenuse   **3.** complementary

4. elevation; depression

5. $\sin \theta = \frac{3}{5}$ $\csc \theta = \frac{5}{3}$ **6.** $\sin \theta = \frac{5}{13}$ $\csc \theta = \frac{13}{5}$

$\cos \theta = \frac{4}{5}$ $\sec \theta = \frac{5}{4}$ $\cos \theta = \frac{12}{13}$ $\sec \theta = \frac{13}{12}$

$\tan \theta = \frac{3}{4}$ $\cot \theta = \frac{4}{3}$ $\tan \theta = \frac{5}{12}$ $\cot \theta = \frac{12}{5}$

7. $\sin \theta = \frac{9}{41}$ $\csc \theta = \frac{41}{9}$ **8.** $\sin \theta = \frac{\sqrt{2}}{2}$ $\csc \theta = \sqrt{2}$

$\cos \theta = \frac{40}{41}$ $\sec \theta = \frac{41}{40}$ $\cos \theta = \frac{\sqrt{2}}{2}$ $\sec \theta = \sqrt{2}$

$\tan \theta = \frac{9}{40}$ $\cot \theta = \frac{40}{9}$ $\tan \theta = 1$ $\cot \theta = 1$

9. $\sin \theta = \frac{\sqrt{2}}{2}$ $\csc \theta = \sqrt{2}$ **10.** $\sin \theta = \frac{7}{25}$ $\csc \theta = \frac{25}{7}$

$\cos \theta = \frac{\sqrt{2}}{2}$ $\sec \theta = \sqrt{2}$ $\cos \theta = \frac{24}{25}$ $\sec \theta = \frac{25}{24}$

$\tan \theta = 1$ $\cot \theta = 1$ $\tan \theta = \frac{7}{24}$ $\cot \theta = \frac{24}{7}$

11. $\sin \theta = \frac{8}{17}$ $\csc \theta = \frac{17}{8}$

$\cos \theta = \frac{15}{17}$ $\sec \theta = \frac{17}{15}$

$\tan \theta = \frac{8}{15}$ $\cot \theta = \frac{15}{8}$

The triangles are similar, and corresponding sides are proportional.

12. $\sin \theta = \frac{3}{5}$ $\csc \theta = \frac{5}{3}$

$\cos \theta = \frac{4}{5}$ $\sec \theta = \frac{5}{4}$

$\tan \theta = \frac{3}{4}$ $\cot \theta = \frac{4}{3}$

The triangles are similar, and corresponding sides are proportional.

13. $\sin \theta = \frac{1}{3}$ $\csc \theta = 3$

$\cos \theta = \frac{2\sqrt{2}}{3}$ $\sec \theta = \frac{3\sqrt{2}}{4}$

$\tan \theta = \frac{\sqrt{2}}{4}$ $\cot \theta = 2\sqrt{2}$

The triangles are similar, and corresponding sides are proportional.

14. $\sin \theta = \frac{\sqrt{5}}{5}$ $\csc \theta = \sqrt{5}$

$\cos \theta = \frac{2\sqrt{5}}{5}$ $\sec \theta = \frac{\sqrt{5}}{2}$

$\tan \theta = \frac{1}{2}$ $\cot \theta = 2$

The triangles are similar, and corresponding sides are proportional.

15.

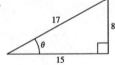

$\sin \theta = \frac{8}{17}$ $\csc \theta = \frac{17}{8}$

$\sec \theta = \frac{17}{15}$

$\tan \theta = \frac{8}{15}$ $\cot \theta = \frac{15}{8}$

16.

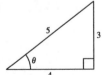

$\csc \theta = \frac{5}{3}$

$\cos \theta = \frac{4}{5}$ $\sec \theta = \frac{5}{4}$

$\tan \theta = \frac{3}{4}$ $\cot \theta = \frac{4}{3}$

17.

$\sin \theta = \frac{\sqrt{11}}{6}$ $\csc \theta = \frac{6\sqrt{11}}{11}$

$\cos \theta = \frac{5}{6}$

$\tan \theta = \frac{\sqrt{11}}{5}$ $\cot \theta = \frac{5\sqrt{11}}{11}$

18.

$\sin \theta = \frac{4\sqrt{41}}{41}$ $\csc \theta = \frac{\sqrt{41}}{4}$

$\cos \theta = \frac{5\sqrt{41}}{41}$ $\sec \theta = \frac{\sqrt{41}}{5}$

$\cot \theta = \frac{5}{4}$

19.

$\csc \theta = 5$

$\cos \theta = \frac{2\sqrt{6}}{5}$ $\sec \theta = \frac{5\sqrt{6}}{12}$

$\tan \theta = \frac{\sqrt{6}}{12}$ $\cot \theta = 2\sqrt{6}$

20.

$\sin \theta = \frac{4\sqrt{15}}{17}$ $\csc \theta = \frac{17\sqrt{15}}{60}$

$\cos \theta = \frac{7}{17}$

$\tan \theta = \frac{4\sqrt{15}}{7}$ $\cot \theta = \frac{7\sqrt{15}}{60}$

21.

$\sin \theta = \frac{\sqrt{10}}{10}$ $\csc \theta = \sqrt{10}$

$\cos \theta = \frac{3\sqrt{10}}{10}$ $\sec \theta = \frac{\sqrt{10}}{3}$

$\tan \theta = \frac{1}{3}$

22.

$\sin \theta = \frac{1}{9}$

$\cos \theta = \frac{4\sqrt{5}}{9}$ $\sec \theta = \frac{9\sqrt{5}}{20}$

$\tan \theta = \frac{\sqrt{5}}{20}$ $\cot \theta = 4\sqrt{5}$

23. $\frac{\pi}{6}, \frac{\sqrt{3}}{3}$ **24.** $\frac{\pi}{4}, \frac{\sqrt{2}}{2}$ **25.** $45°; \frac{\sqrt{2}}{2}$

26. $60°; \sqrt{3}$ **27.** $45°; \sqrt{2}$ **28.** $30°; 2$

29. (a) 0.3420   (b) 0.3420   **30.** (a) 0.4348   (b) 0.4348

31. (a) 0.2455   (b) 4.0737   **32.** (a) 0.1843   (b) 5.5186

33. (a) 0.9964   (b) 1.0036   **34.** (a) 1.3499   (b) 1.3432

35. (a) 3.2205   (b) 0.3105   **36.** (a) 1.7946   (b) 0.5572

37. (a) $\frac{1}{2}$ (b) $\frac{\sqrt{3}}{2}$ (c) $\sqrt{3}$ (d) $\frac{\sqrt{3}}{3}$

38. (a) 2   (b) $\frac{\sqrt{3}}{3}$ (c) $\frac{\sqrt{3}}{2}$ (d) $\sqrt{3}$

39. (a) $\frac{2\sqrt{2}}{3}$ (b) $2\sqrt{2}$ (c) 3   (d) 3

40. (a) $\frac{1}{5}$ (b) $\frac{\sqrt{6}}{12}$ (c) $2\sqrt{6}$ (d) $\frac{2\sqrt{6}}{5}$

41. (a) $\frac{1}{3}$ (b) $\sqrt{10}$ (c) $\frac{1}{3}$ (d) $\frac{\sqrt{10}}{10}$

42. (a) $\frac{4\sqrt{7}}{7}$ (b) $\frac{3}{4}$ (c) $\frac{\sqrt{7}}{3}$ (d) $\frac{\sqrt{7}}{4}$

43–52. Answers will vary.

53. (a) $30° = \dfrac{\pi}{6}$ (b) $30° = \dfrac{\pi}{6}$

54. (a) $45° = \dfrac{\pi}{4}$ (b) $45° = \dfrac{\pi}{4}$

55. (a) $60° = \dfrac{\pi}{3}$ (b) $45° = \dfrac{\pi}{4}$

56. (a) $60° = \dfrac{\pi}{3}$ (b) $45° = \dfrac{\pi}{4}$

57. (a) $60° = \dfrac{\pi}{3}$ (b) $45° = \dfrac{\pi}{4}$

58. (a) $60° = \dfrac{\pi}{3}$ (b) $45° = \dfrac{\pi}{4}$ **59.** $x = 9, y = 9\sqrt{3}$

60. $x = 30\sqrt{3}, r = 60$ **61.** $x = \dfrac{32\sqrt{3}}{3}, r = \dfrac{64\sqrt{3}}{3}$

62. $x = 20, r = 20\sqrt{2}$ **63.** About 443.2 m; about 323.3 m

64. (a) 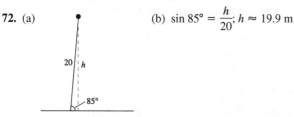   (b) $\tan \theta = \dfrac{6}{3} = \dfrac{h}{135}$

(c) 270 ft

Not drawn to scale

65. $30° = \dfrac{\pi}{6}$ **66.** About 137.6 ft

67. (a) About 219.9 ft   (b) About 160.9 ft

68. About 1.3 mi

69. $(x_1, y_1) = \left(28\sqrt{3}, 28\right)$
$(x_2, y_2) = \left(28, 28\sqrt{3}\right)$

70. About 6.57 cm

71. $\sin 20° \approx 0.34$, $\cos 20° \approx 0.94$, $\tan 20° \approx 0.36$,
$\csc 20° \approx 2.92$, $\sec 20° \approx 1.06$, $\cot 20° \approx 2.75$

72. (a)   (b) $\sin 85° = \dfrac{h}{20}$; $h \approx 19.9$ m

(c) The side of the triangle labeled h will become shorter.

(d)

Angle, θ	80°	70°	60°	50°
Height	19.7	18.8	17.3	15.3

Angle, θ	40°	30°	20°	10°
Height	12.9	10.0	6.8	3.5

(e) As $\theta \to 0°$, $h \to 0$.

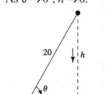

73. (a) About 519.33 ft
(b) About 1174.17 ft
(c) About 173.11 ft/min

74. $\cos 60° = \dfrac{adj}{hyp} = \dfrac{1}{2}$ **75.** True. $\csc x = \dfrac{1}{\sin x}$

76. False. $\dfrac{2\sqrt{3}}{3} \neq 2$ **77.** False. $\dfrac{\sqrt{2}}{2} + \dfrac{\sqrt{2}}{2} \neq 1$

78. True. $\dfrac{1}{2} - \dfrac{1}{2} = 0$ **79.** False. $1.7321 \neq 0.0349$

80. False. $\tan 25° \neq (\tan 5°)(\tan 5°)$

81. Yes, $\tan \theta$ is equal to opp/adj. You can find the value of the hypotenuse by the Pythagorean Theorem. Then you can find $\sec \theta$, which is equal to hyp/adj.

82. (a) y (b) y
(c) The side opposite θ is adjacent to $90° - \theta$.

83.

θ	0.1	0.2	0.3	0.4	0.5
$\sin \theta$	0.0998	0.1987	0.2955	0.3894	0.4794

(a) θ is greater.

(b) As $\theta \to 0$, $\sin \theta \to 0$ and $\dfrac{\theta}{\sin \theta} \to 1$.

84.

θ	0°	18°	36°	54°	72°	90°
$\sin \theta$	0	0.3090	0.5878	0.8090	0.9511	1
$\cos \theta$	1	0.9511	0.8090	0.5878	0.3090	0

(a) Increasing function   (b) Decreasing function
(c) As the angle increases, the length of the side opposite the angle increases relative to the length of the hypotenuse, and the length of the side adjacent to the angle decreases relative to the length of the hypotenuse. So, as the sine function increases, the cosine function decreases.

Section 4.4   *(page 294)*

1. $\dfrac{y}{r}$ **2.** $\csc \theta$ **3.** $\dfrac{y}{x}$ **4.** $\dfrac{r}{x}$ **5.** $\cos \theta$ **6.** $\cot \theta$

7. zero; defined   **8.** reference

9. (a) $\sin \theta = \dfrac{3}{5}$ $\csc \theta = \dfrac{5}{3}$
     $\cos \theta = \dfrac{4}{5}$ $\sec \theta = \dfrac{5}{4}$
     $\tan \theta = \dfrac{3}{4}$ $\cot \theta = \dfrac{4}{3}$

(b) $\sin \theta = \dfrac{15}{17}$ $\csc \theta = \dfrac{17}{15}$
     $\cos \theta = -\dfrac{8}{17}$ $\sec \theta = -\dfrac{17}{8}$
     $\tan \theta = -\dfrac{15}{8}$ $\cot \theta = -\dfrac{8}{15}$

10. (a) $\sin \theta = -\dfrac{5}{13}$ $\csc \theta = -\dfrac{13}{5}$
     $\cos \theta = -\dfrac{12}{13}$ $\sec \theta = -\dfrac{13}{12}$
     $\tan \theta = \dfrac{5}{12}$ $\cot \theta = \dfrac{12}{5}$

(b) $\sin \theta = -\dfrac{\sqrt{2}}{2}$ $\csc \theta = -\sqrt{2}$
     $\cos \theta = \dfrac{\sqrt{2}}{2}$ $\sec \theta = \sqrt{2}$
     $\tan \theta = -1$ $\cot \theta = -1$

CHAPTER 4

11. (a) $\sin \theta = -\dfrac{1}{2}$ $\qquad$ $\csc \theta = -2$

$\cos \theta = -\dfrac{\sqrt{3}}{2}$ $\qquad$ $\sec \theta = -\dfrac{2\sqrt{3}}{3}$

$\tan \theta = \dfrac{\sqrt{3}}{3}$ $\qquad$ $\cot \theta = \sqrt{3}$

(b) $\sin \theta = -\dfrac{\sqrt{17}}{17}$ $\qquad$ $\csc \theta = -\sqrt{17}$

$\cos \theta = \dfrac{4\sqrt{17}}{17}$ $\qquad$ $\sec \theta = \dfrac{\sqrt{17}}{4}$

$\tan \theta = -\dfrac{1}{4}$ $\qquad$ $\cot \theta = -4$

12. (a) $\sin \theta = \dfrac{\sqrt{10}}{10}$ $\qquad$ $\csc \theta = \sqrt{10}$

$\cos \theta = \dfrac{3\sqrt{10}}{10}$ $\qquad$ $\sec \theta = \dfrac{\sqrt{10}}{3}$

$\tan \theta = \dfrac{1}{3}$ $\qquad$ $\cot \theta = 3$

(b) $\sin \theta = \dfrac{\sqrt{21}}{5}$ $\qquad$ $\csc \theta = \dfrac{5\sqrt{21}}{21}$

$\cos \theta = -\dfrac{2}{5}$ $\qquad$ $\sec \theta = -\dfrac{5}{2}$

$\tan \theta = -\dfrac{\sqrt{21}}{2}$ $\qquad$ $\cot \theta = -\dfrac{2\sqrt{21}}{21}$

13. $\sin \theta = \dfrac{12}{13}$ $\qquad$ $\csc \theta = \dfrac{13}{12}$

$\cos \theta = \dfrac{5}{13}$ $\qquad$ $\sec \theta = \dfrac{13}{5}$

$\tan \theta = \dfrac{12}{5}$ $\qquad$ $\cot \theta = \dfrac{5}{12}$

14. $\sin \theta = \dfrac{15}{17}$ $\qquad$ $\csc \theta = \dfrac{17}{15}$

$\cos \theta = \dfrac{8}{17}$ $\qquad$ $\sec \theta = \dfrac{17}{8}$

$\tan \theta = \dfrac{15}{8}$ $\qquad$ $\cot \theta = \dfrac{8}{15}$

15. $\sin \theta = -\dfrac{2\sqrt{29}}{29}$ $\qquad$ $\csc \theta = -\dfrac{\sqrt{29}}{2}$

$\cos \theta = -\dfrac{5\sqrt{29}}{29}$ $\qquad$ $\sec \theta = -\dfrac{\sqrt{29}}{5}$

$\tan \theta = \dfrac{2}{5}$ $\qquad$ $\cot \theta = \dfrac{5}{2}$

16. $\sin \theta = \dfrac{5\sqrt{29}}{29}$ $\qquad$ $\csc \theta = \dfrac{\sqrt{29}}{5}$

$\cos \theta = -\dfrac{2\sqrt{29}}{29}$ $\qquad$ $\sec \theta = -\dfrac{\sqrt{29}}{2}$

$\tan \theta = -\dfrac{5}{2}$ $\qquad$ $\cot \theta = -\dfrac{2}{5}$

17. $\sin \theta = \dfrac{4}{5}$ $\qquad$ $\csc \theta = \dfrac{5}{4}$

$\cos \theta = -\dfrac{3}{5}$ $\qquad$ $\sec \theta = -\dfrac{5}{3}$

$\tan \theta = -\dfrac{4}{3}$ $\qquad$ $\cot \theta = -\dfrac{3}{4}$

18. $\sin \theta = -\dfrac{4\sqrt{15}}{17}$ $\qquad$ $\csc \theta = -\dfrac{17\sqrt{15}}{60}$

$\cos \theta = \dfrac{7}{17}$ $\qquad$ $\sec \theta = \dfrac{17}{7}$

$\tan \theta = -\dfrac{4\sqrt{15}}{7}$ $\qquad$ $\cot \theta = -\dfrac{7\sqrt{15}}{60}$

19. Quadrant I $\qquad$ **20.** Quadrant III

21. Quadrant II $\qquad$ **22.** Quadrant IV

23. $\sin \theta = \dfrac{15}{17}$ $\qquad$ $\csc \theta = \dfrac{17}{15}$

$\cos \theta = \dfrac{8}{17}$ $\qquad$ $\sec \theta = \dfrac{17}{8}$

$\cot \theta = \dfrac{8}{15}$

24. $\sin \theta = -\dfrac{15}{17}$ $\qquad$ $\csc \theta = -\dfrac{17}{15}$

$\sec \theta = \dfrac{17}{8}$

$\tan \theta = -\dfrac{15}{8}$ $\qquad$ $\cot \theta = -\dfrac{8}{15}$

25. $\qquad$ $\csc \theta = \dfrac{5}{3}$

$\cos \theta = -\dfrac{4}{5}$ $\qquad$ $\sec \theta = -\dfrac{5}{4}$

$\tan \theta = -\dfrac{3}{4}$ $\qquad$ $\cot \theta = -\dfrac{4}{3}$

26. $\sin \theta = -\dfrac{3}{5}$ $\qquad$ $\csc \theta = -\dfrac{5}{3}$

$\sec \theta = -\dfrac{5}{4}$

$\tan \theta = \dfrac{3}{4}$ $\qquad$ $\cot \theta = \dfrac{4}{3}$

27. $\sin \theta = -\dfrac{\sqrt{10}}{10}$ $\qquad$ $\csc \theta = -\sqrt{10}$

$\cos \theta = \dfrac{3\sqrt{10}}{10}$ $\qquad$ $\sec \theta = \dfrac{\sqrt{10}}{3}$

$\tan \theta = -\dfrac{1}{3}$

28. $\sin \theta = \dfrac{1}{4}$

$\cos \theta = -\dfrac{\sqrt{15}}{4}$ $\qquad$ $\sec \theta = -\dfrac{4\sqrt{15}}{15}$

$\tan \theta = -\dfrac{\sqrt{15}}{15}$ $\qquad$ $\cot \theta = -\sqrt{15}$

29. $\sin \theta = 1$ $\qquad$ $\sec \theta$ is undefined.

$\tan \theta$ is undefined. $\qquad$ $\cot \theta = 0$

30. $\qquad$ $\csc \theta$ is undefined.

$\cos \theta = -1$

$\tan \theta = 0$ $\qquad$ $\cot \theta$ is undefined.

31. $\sin \theta = 0$ $\qquad$ $\csc \theta$ is undefined.

$\cos \theta = -1$ $\qquad$ $\sec \theta = -1$

$\tan \theta = 0$ $\qquad$ $\cot \theta$ is undefined.

32. $\sin \theta = -1$ $\qquad$ $\csc \theta = -1$

$\cos \theta = 0$ $\qquad$ $\sec \theta$ is undefined.

$\tan \theta$ is undefined. $\qquad$ $\cot \theta = 0$

33. $\sin \theta = \dfrac{\sqrt{2}}{2}$ $\qquad$ $\csc \theta = \sqrt{2}$

$\cos \theta = -\dfrac{\sqrt{2}}{2}$ $\qquad$ $\sec \theta = -\sqrt{2}$

$\tan \theta = -1$ $\qquad$ $\cot \theta = -1$

34. $\sin \theta = -\dfrac{\sqrt{10}}{10}$ $\qquad$ $\csc \theta = -\sqrt{10}$

$\cos \theta = -\dfrac{3\sqrt{10}}{10}$ $\qquad$ $\sec \theta = -\dfrac{\sqrt{10}}{3}$

$\tan \theta = \dfrac{1}{3}$ $\qquad$ $\cot \theta = 3$

35. $\sin \theta = \dfrac{2\sqrt{5}}{5}$ $\qquad$ $\csc \theta = \dfrac{\sqrt{5}}{2}$

$\cos \theta = \dfrac{\sqrt{5}}{5}$ $\qquad$ $\sec \theta = \sqrt{5}$

$\tan \theta = 2$ $\qquad$ $\cot \theta = \dfrac{1}{2}$

36. $\sin \theta = -\dfrac{4}{5}$ $\qquad$ $\csc \theta = -\dfrac{5}{4}$

$\cos \theta = \dfrac{3}{5}$ $\qquad$ $\sec \theta = \dfrac{5}{3}$

$\tan \theta = -\dfrac{4}{3}$ $\qquad$ $\cot \theta = -\dfrac{3}{4}$

37. 0 $\qquad$ **38.** -1 $\qquad$ **39.** Undefined $\qquad$ **40.** -1 $\qquad$ **41.** 1

42. Undefined $\qquad$ **43.** Undefined $\qquad$ **44.** 0

45. 0   **46.** Undefined

47. $\theta' = 20°$

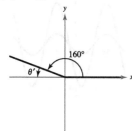

48. $\theta' = 51°$

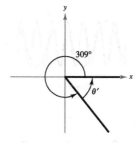

49. $\theta' = 55°$

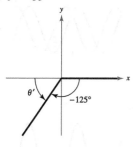

50. $\theta' = 35°$

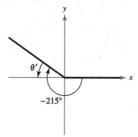

51. $\theta' = \dfrac{\pi}{3}$

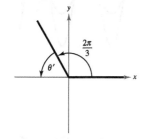

52. $\theta' = \dfrac{\pi}{6}$

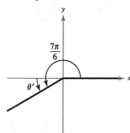

53. $\theta' = 2\pi - 4.8$

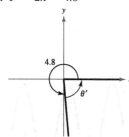

54. $\theta' = 12.9 - 4\pi$

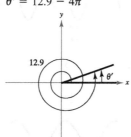

55. $\sin 225° = -\dfrac{\sqrt{2}}{2}$

$\cos 225° = -\dfrac{\sqrt{2}}{2}$

$\tan 225° = 1$

56. $\sin 300° = -\dfrac{\sqrt{3}}{2}$

$\cos 300° = \dfrac{1}{2}$

$\tan 300° = -\sqrt{3}$

57. $\sin 750° = \dfrac{1}{2}$

$\cos 750° = \dfrac{\sqrt{3}}{2}$

$\tan 750° = \dfrac{\sqrt{3}}{3}$

58. $\sin 675° = -\dfrac{\sqrt{2}}{2}$

$\cos 675° = \dfrac{\sqrt{2}}{2}$

$\tan 675° = -1$

59. $\sin(-120°) = -\dfrac{\sqrt{3}}{2}$

$\cos(-120°) = -\dfrac{1}{2}$

$\tan(-120°) = \sqrt{3}$

60. $\sin(-570°) = \dfrac{1}{2}$

$\cos(-570°) = -\dfrac{\sqrt{3}}{2}$

$\tan(-570°) = -\dfrac{\sqrt{3}}{3}$

61. $\sin \dfrac{2\pi}{3} = \dfrac{\sqrt{3}}{2}$

$\cos \dfrac{2\pi}{3} = -\dfrac{1}{2}$

$\tan \dfrac{2\pi}{3} = -\sqrt{3}$

62. $\sin \dfrac{3\pi}{4} = \dfrac{\sqrt{2}}{2}$

$\cos \dfrac{3\pi}{4} = -\dfrac{\sqrt{2}}{2}$

$\tan \dfrac{3\pi}{4} = -1$

63. $\sin\left(-\dfrac{\pi}{6}\right) = -\dfrac{1}{2}$

$\cos\left(-\dfrac{\pi}{6}\right) = \dfrac{\sqrt{3}}{2}$

$\tan\left(-\dfrac{\pi}{6}\right) = -\dfrac{\sqrt{3}}{3}$

64. $\sin\left(-\dfrac{2\pi}{3}\right) = -\dfrac{\sqrt{3}}{2}$

$\cos\left(-\dfrac{2\pi}{3}\right) = -\dfrac{1}{2}$

$\tan\left(-\dfrac{2\pi}{3}\right) = \sqrt{3}$

65. $\sin \dfrac{11\pi}{4} = \dfrac{\sqrt{2}}{2}$

$\cos \dfrac{11\pi}{4} = -\dfrac{\sqrt{2}}{2}$

$\tan \dfrac{11\pi}{4} = -1$

66. $\sin \dfrac{13\pi}{6} = \dfrac{1}{2}$

$\cos \dfrac{13\pi}{6} = \dfrac{\sqrt{3}}{2}$

$\tan \dfrac{13\pi}{6} = \dfrac{\sqrt{3}}{3}$

67. $\sin\left(-\dfrac{17\pi}{6}\right) = -\dfrac{1}{2}$

$\cos\left(-\dfrac{17\pi}{6}\right) = -\dfrac{\sqrt{3}}{2}$

$\tan\left(-\dfrac{17\pi}{6}\right) = \dfrac{\sqrt{3}}{3}$

68. $\sin\left(-\dfrac{23\pi}{4}\right) = \dfrac{\sqrt{2}}{2}$

$\cos\left(-\dfrac{23\pi}{4}\right) = \dfrac{\sqrt{2}}{2}$

$\tan\left(-\dfrac{23\pi}{4}\right) = 1$

69. $\dfrac{4}{5}$ **70.** $\sqrt{10}$ **71.** $-\dfrac{\sqrt{13}}{12}$ **72.** $-\sqrt{3}$

73. $\dfrac{8\sqrt{39}}{39}$ **74.** $\dfrac{4\sqrt{65}}{65}$ **75.** 0.1736

76. -1.4826 **77.** -0.3420 **78.** 0.5

79. -28.6363 **80.** 3.2361   **81.** 1.4142

82. -2.7475 **83.** 0.3640   **84.** 0.6235

85. -2.6131 **86.** -1.4142 **87.** -0.6052

88. 0.2190   **89.** 1.8382   **90.** -8.9164

91. (a) $30° = \dfrac{\pi}{6},\ 150° = \dfrac{5\pi}{6}$ (b) $210° = \dfrac{7\pi}{6},\ 330° = \dfrac{11\pi}{6}$

92. (a) $45° = \dfrac{\pi}{4},\ 315° = \dfrac{7\pi}{4}$ (b) $135° = \dfrac{3\pi}{4},\ 225° = \dfrac{5\pi}{4}$

93. (a) $60° = \dfrac{\pi}{3},\ 300° = \dfrac{5\pi}{3}$ (b) $60° = \dfrac{\pi}{3},\ 300° = \dfrac{5\pi}{3}$

94. (a) $60° = \dfrac{\pi}{3},\ 120° = \dfrac{2\pi}{3}$ (b) $60° = \dfrac{\pi}{3},\ 120° = \dfrac{2\pi}{3}$

95. (a) $45° = \dfrac{\pi}{4},\ 225° = \dfrac{5\pi}{4}$ (b) $150° = \dfrac{5\pi}{6},\ 330° = \dfrac{11\pi}{6}$

96. (a) $90° = \dfrac{\pi}{2},\ 270° = \dfrac{3\pi}{2}$ (b) $135° = \dfrac{3\pi}{4},\ 225° = \dfrac{5\pi}{4}$

97. (a) 12 mi   (b) 6 mi   (c) About 6.9 mi

98. (a) 2 cm   (b) About 0.14 cm   (c) About -1.98 cm

99. (a) $B = 24.593 \sin(0.495t - 2.262) + 57.387$
$F = 39.071 \sin(0.448t - 1.366) + 32.204$

(b) February: $B \approx 33.9°$, $F \approx 14.5°$
April: $B \approx 50.5°$, $F \approx 48.3°$
May: $B \approx 62.6°$, $F \approx 62.2°$
July: $B \approx 80.3°$, $F \approx 70.5°$
September: $B \approx 77.4°$, $F \approx 50.1°$
October: $B \approx 68.2°$, $F \approx 33.3°$
December: $B \approx 44.8°$, $F \approx 2.4°$

(c)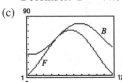
Answers will vary.

100. (a) 26,134 units   (b) 31,438 units   (c) 21,452 units
(d) 26,765 units

101. About 0.79 amp

102. As θ increases from 0° to 90°, x decreases from 12 cm to 0 cm and y increases from 0 cm to 12 cm. Therefore, $\sin \theta = y/12$ increases from 0 to 1 and $\cos \theta = x/12$ decreases from 1 to 0. Thus, $\tan \theta = y/x$ increases without bound. When $\theta = 90°$, the tangent is undefined.

103. False. In each of the four quadrants, the signs of the secant function and the cosine function are the same because these functions are reciprocals of each other.

104. False. For θ in Quadrant II, $\theta' = 180° - \theta$. For θ in Quadrant III, $\theta' = \theta - 180°$. For θ in Quadrant IV, $\theta' = 360° - \theta$.

105. Answers will vary.

106. (a) $\sin t = y$ (b) $r = 1$ because it is a unit circle.
$\cos t = x$
(c) $\sin \theta = y$ (d) $\sin t = \sin \theta$ and $\cos t = \cos \theta$
$\cos \theta = x$

Section 4.5   (page 304)

1. cycle   **2.** amplitude   **3.** phase shift

4. vertical translation   **5.** Period: $\dfrac{2\pi}{5}$; Amplitude: 2

6. Period: π; Amplitude: 3   **7.** Period: 4; Amplitude: $\frac{3}{4}$

8. Period: 6; Amplitude: 5   **9.** Period: $\dfrac{8\pi}{5}$; Amplitude: $\frac{1}{2}$

10. Period: 12π; Amplitude: $\frac{1}{4}$

11. Period: 24; Amplitude: $\frac{5}{3}$ **12.** Period: $\frac{1}{5}$; Amplitude: $\frac{2}{5}$

13. The period of g is one-fifth the period of f.

14. The amplitude of g is twice the amplitude of f.

15. g is a reflection of f in the x-axis.

16. g is a reflection of f in the y-axis.

17. g is a shift of f π units to the right.

18. g is a shift of f π units to the left.

19. g is a shift of f three units up.

20. g is a shift of f two units down.

21. The graph of g has twice the amplitude of the graph of f.

22. The period of g is one-third the period of f.

23. The graph of g is a horizontal shift of the graph of f π units to the right.

24. g is a shift of f two units up.

25.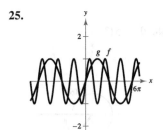

26.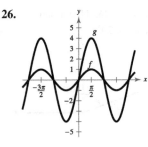

27.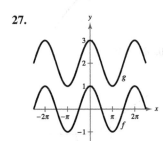

28.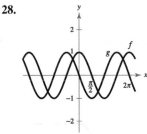

29.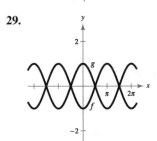

30.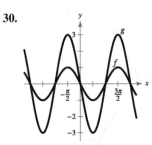

31.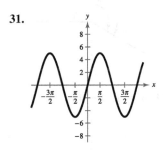

32.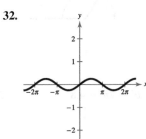

33.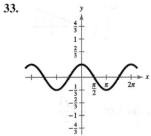

34.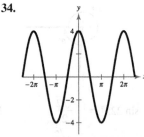

35.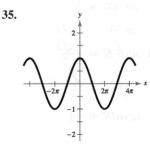

36.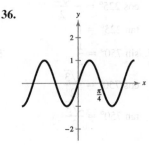

37.

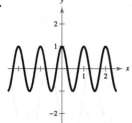

38.

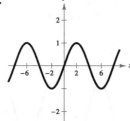

49.

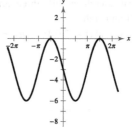

50.

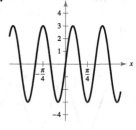

39.

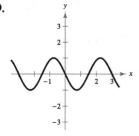

40.

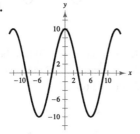

51.

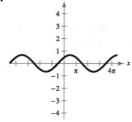

52.

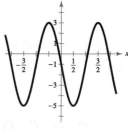

41.

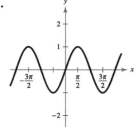

42.

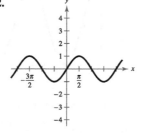

53. (a) Horizontal shrink and a phase shift $\pi/4$ unit right

(b)   (c) $g(x) = f(4x - \pi)$

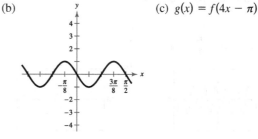

43.

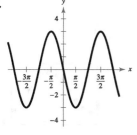

44.

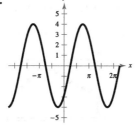

54. (a) Horizontal shrink and a phase shift $\pi/2$ units left

(b)   (c) $g(x) = f(2x + \pi)$

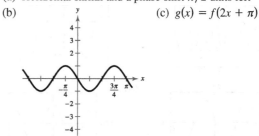

45.

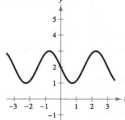

46.

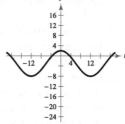

55. (a) Shift two units up and a phase shift $\pi/2$ units right

(b)   (c) $g(x) = f\left(x - \dfrac{\pi}{2}\right) + 2$

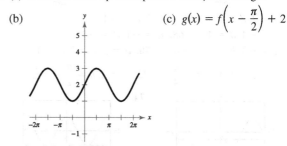

47.

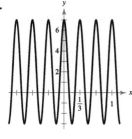

48.

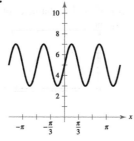

56. (a) Shift one unit up and a phase shift π units left

(b)   (c) $g(x) = 1 + f(x + \pi)$

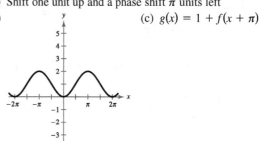

CHAPTER 4

57. (a) Horizontal shrink, a vertical stretch, a shift three units down, and a phase shift $\pi/4$ unit right

(b)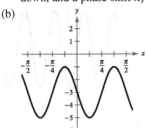

(c) $g(x) = 2f(4x - \pi) - 3$

58. (a) Reflection in the x-axis, a horizontal shrink, a shift four units up, and a phase shift $\pi/4$ unit left

(b)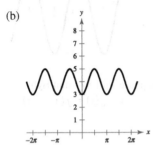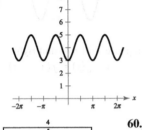

(c) $g(x) = 4 - f\left(2x + \dfrac{\pi}{2}\right)$

59. 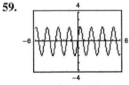   **60.**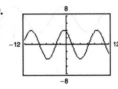

61. 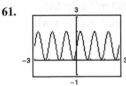   **62.**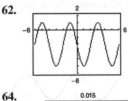

63. 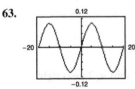   **64.**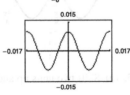

65. $a = 2, d = 1$ **66.** $a = 2, d = -1$

67. $a = -4, d = 4$ **68.** $a = -1, d = -3$

69. $a = -3, b = 2, c = 0$ **70.** $a = 2, b = \frac{1}{2}, c = 0$

71. $a = 2, b = 1, c = -\dfrac{\pi}{4}$ **72.** $a = 2, b = \dfrac{\pi}{2}, c = \dfrac{\pi}{2}$

73. 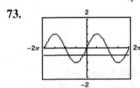   **74.**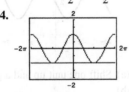

$x = -\dfrac{\pi}{6}, -\dfrac{5\pi}{6}, \dfrac{7\pi}{6}, \dfrac{11\pi}{6}$ $x = -\pi, \pi$

75. $y = 1 + 2\sin(2x - \pi)$ **76.** $y = 3\sin\left(\dfrac{1}{2}x + \dfrac{\pi}{8}\right) - 1$

77. $y = \cos(2x + 2\pi) - \dfrac{3}{2}$ **78.** $y = 2 + 3\cos\left(\dfrac{1}{2}x - \dfrac{\pi}{4}\right)$

79. (a) 4 sec
(b) 15 cycles/min
(c)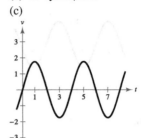

80. (a) 6 sec
(b) 10 cycles/min
(c)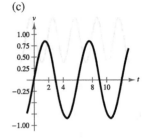

Sample answer: $t = 0, t = 6$

81. (a) $\frac{6}{5}$ sec   (b) 50 heartbeats/min

82. (a) $\frac{1}{440}$ sec   (b) 440 cycles/sec

83. (a) and (c)

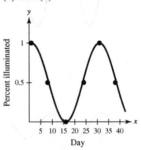

The model fits the data well.

(b) $y = 0.5\cos\left(\dfrac{\pi x}{15} - \dfrac{\pi}{15}\right) + 0.5$

(d) 30 days   (e) 25%

84. (a) $I(t) = 46.2 + 32.4\cos\left(\dfrac{\pi t}{6} - 3.67\right)$

(b) 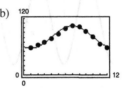   (c)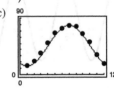

The model fits the data well.   The model fits the data well.

(d) Las Vegas: 80.6°; International Falls: 46.2°
The constant term gives the annual average maximum temperature.

(e) 12; yes; One full period is one year.

(f) International Falls; 23.5, 32.4; The greater the amplitude, the greater the variability in temperature.

85. (a) 20 sec; It takes 20 seconds to complete one revolution on the Ferris wheel.

(b) 50 ft; The diameter of the Ferris wheel is 100 feet.

(c)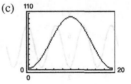

86. (a) 365; Yes, because there are 365 days in a year.

(b) 30.3 gal; the constant term

(c)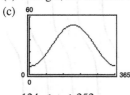

124 < t < 252

87. False. The graph of g is shifted one period to the left.

88. False. The function $y = \frac{1}{2} \cos 2x$ has an amplitude that is one-half that of $y = \cos x$. For $y = a \cos bx$, the amplitude is $|a|$.

89. True. $-\sin\left(x + \frac{\pi}{2}\right) = -\cos x$

90. (a) For $c \neq 0$, the value of c shifts the graph of $f(x) = \sin x$ c units to the left when $c < 0$ and c units to the right when $c > 0$.

(b) The green graph

91.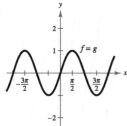

Conjecture:

$$\sin x = \cos\left(x - \frac{\pi}{2}\right)$$

92.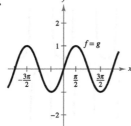

Conjecture:

$$\sin x = -\cos\left(x + \frac{\pi}{2}\right)$$

93.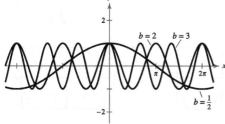

The value of b affects the period of the graph.

$b = \frac{1}{2} \rightarrow \frac{1}{2}$ cycle

$b = 2 \rightarrow 2$ cycles

$b = 3 \rightarrow 3$ cycles

94. (a) 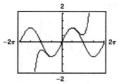   The graphs appear to coincide from $-\frac{\pi}{2}$ to $\frac{\pi}{2}$.

(b) 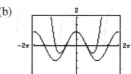   The graphs appear to coincide from $-\frac{\pi}{2}$ to $\frac{\pi}{2}$.

(c) $-\frac{x^7}{7!}, -\frac{x^6}{6!}$

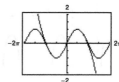   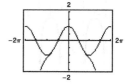

The interval of accuracy increased.

95. (a) 0.4794, 0.4794   (b) 0.8417, 0.8415   (c) 0.5, 0.5

(d) 0.8776, 0.8776   (e) 0.5417, 0.5403

(f) 0.7074, 0.7071

The error increases as x moves farther away from 0.

Section 4.6   *(page 315)*

1. odd; origin   **2.** vertical   **3.** reciprocal   **4.** damping

5. π **6.** $x \neq n\pi$ **7.** $(-\infty, -1] \cup [1, \infty)$ **8.** 2π

9. e, π **10.** c, 2π **11.** a, 1   **12.** d, 2π

13. f, 4   **14.** b, 4

15. 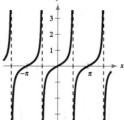   **16.**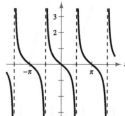

17. 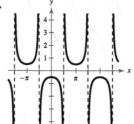   **18.**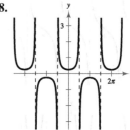

19. 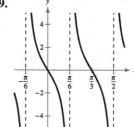   **20.**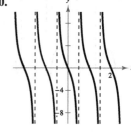

21.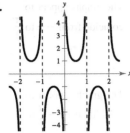

22.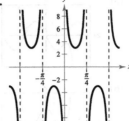

23.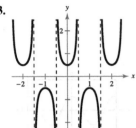

24.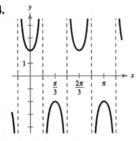

25.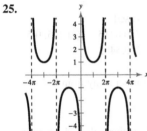

26.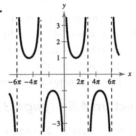

27.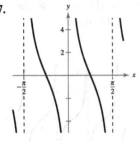

28.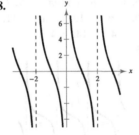

29.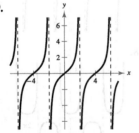

30.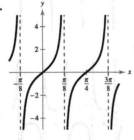

31.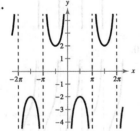

32.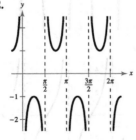

33.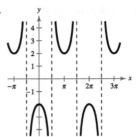

34.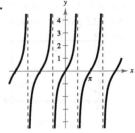

35.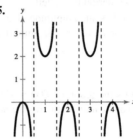

36.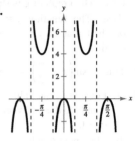

37.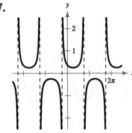

38.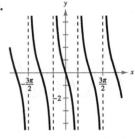

39.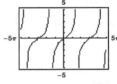

40.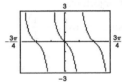

41.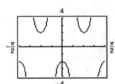

42.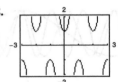

43.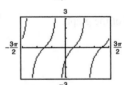

44.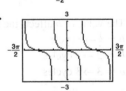

45.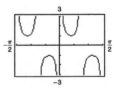

46.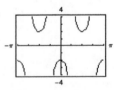

47.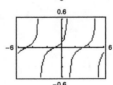

48.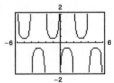

49. $-\dfrac{7\pi}{4}, -\dfrac{3\pi}{4}, \dfrac{\pi}{4}, \dfrac{5\pi}{4}$

50. $-\dfrac{5\pi}{3}, -\dfrac{2\pi}{3}, \dfrac{\pi}{3}, \dfrac{4\pi}{3}$

51. $-\dfrac{7\pi}{6}, -\dfrac{\pi}{6}, \dfrac{5\pi}{6}, \dfrac{11\pi}{6}$ **52.** $-\dfrac{7\pi}{4}, -\dfrac{3\pi}{4}, \dfrac{\pi}{4}, \dfrac{5\pi}{4}$

53. $-\dfrac{4\pi}{3}, -\dfrac{2\pi}{3}, \dfrac{2\pi}{3}, \dfrac{4\pi}{3}$ **54.** $-\dfrac{5\pi}{3}, -\dfrac{\pi}{3}, \dfrac{\pi}{3}, \dfrac{5\pi}{3}$

55. $-\dfrac{7\pi}{4}, -\dfrac{5\pi}{4}, \dfrac{\pi}{4}, \dfrac{3\pi}{4}$ **56.** $-\dfrac{5\pi}{6}, -\dfrac{\pi}{6}, \dfrac{7\pi}{6}, \dfrac{11\pi}{6}$

57. Even   **58.** Odd   **59.** Odd   **60.** Odd

61. Odd   **62.** Even   **63.** Even   **64.** Odd

65. d, $f \to 0$ as $x \to 0$.   **66.** a, $f \to 0$ as $x \to 0$.

67. b, $g \to 0$ as $x \to 0$.   **68.** c, $g \to 0$ as $x \to 0$.

69.   **70.**

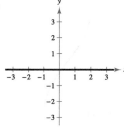   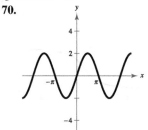

The functions are equal.   The functions are equal.

71.   **72.**

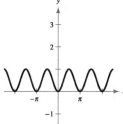   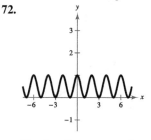

The functions are equal.   The functions are equal.

73.   **74.**

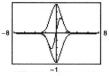   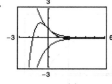

As $x \to \infty$, $g(x) \to 0$.   As $x \to \infty$, $f(x) \to 0$.

75.   **76.**

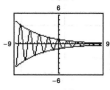   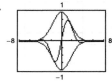

As $x \to \infty$, $f(x) \to 0$.   As $x \to \infty$, $h(x) \to 0$.

77.   **78.**

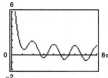   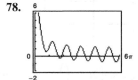

As $x \to 0$, $y \to \infty$.   As $x \to 0$, $y \to \infty$.

79.   **80.**

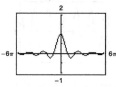   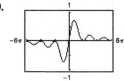

As $x \to 0$, $g(x) \to 1$.   As $x \to 0$, $f(x) \to 0$.

81.   **82.**

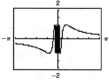   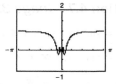

As $x \to 0$, $f(x)$ oscillates   As $x \to 0$, $h(x) \to 0$.
between 1 and -1.

83. (a) Period of $H(t)$: 12 mo
   Period of $L(t)$: 12 mo

   (b) Summer; winter   (c) About 0.5 mo

84. (a)   (b) 132,000 units

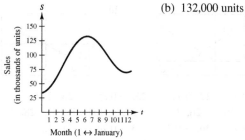

Month (1 ↔ January)

85. $d = 27 \sec x$ **86.** $d = 7 \cot x$

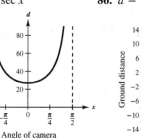   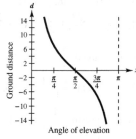

Angle of camera   Angle of elevation

87. True. For a given value of x, the y-coordinate of $\csc x$ is the reciprocal of the y-coordinate of $\sin x$.

88. True. $y = \sec x$ is equal to $y = 1/\cos x$, and if the reciprocal of $y = \sin x$ is translated $\pi/2$ units to the left, then

$$\frac{1}{\sin\left(x + \dfrac{\pi}{2}\right)} = \frac{1}{\cos x} = \sec x.$$

89. (a)

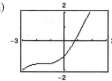

0.7391

   (b) 1, 0.5403, 0.8576, 0.6543, 0.7935, 0.7014, 0.7640, 0.7221, 0.7504, 0.7314, . . . ; 0.7391

90. (a) (i) $f(x) = \tan 2x$; The period is $\pi/2$ and the graph is increasing on $(-\pi/4, \pi/4)$.

   (b) (ii) $f(x) = \csc 4x$; The period is $\pi/2$ and the graph has vertical asymptotes at multiples of $\pi/4$.

91. (a) $f(x) \to \infty$ (b) $f(x) \to -\infty$
   (c) $f(x) \to \infty$ (d) $f(x) \to -\infty$

92. (a) $f(x) \to \infty$ (b) $f(x) \to -\infty$
   (c) $f(x) \to -\infty$ (d) $f(x) \to \infty$

93. (a) $f(x) \to -\infty$ (b) $f(x) \to \infty$
   (c) $f(x) \to -\infty$ (d) $f(x) \to \infty$

94. (a) $f(x) \to -\infty$ (b) $f(x) \to \infty$
   (c) $f(x) \to \infty$ (d) $f(x) \to -\infty$

CHAPTER 4

Section 4.7 (page 324)

1. $y = \sin^{-1} x; -1 \le x \le 1$ 2. $y = \arccos x; 0 \le y \le \pi$

3. $y = \tan^{-1} x; -\infty < x < \infty; -\dfrac{\pi}{2} < y < \dfrac{\pi}{2}$ 4. inverse

5. $\dfrac{\pi}{6}$ 6. 0 7. $\dfrac{\pi}{3}$ 8. $\dfrac{\pi}{2}$ 9. $\dfrac{\pi}{6}$ 10. $\dfrac{\pi}{4}$

11. Not possible   12. $\dfrac{\pi}{3}$ 13. $-\dfrac{\pi}{3}$ 14. Not possible

15. $\dfrac{2\pi}{3}$ 16. $\dfrac{\pi}{4}$ 17. $-\dfrac{\pi}{3}$ 18. $-\dfrac{\pi}{6}$

19.

20.

21. 1.19 22. 0.71 23. -0.85 24. 2.35
25. -1.25 26. 1.53 27. Not possible   28. 1.31
29. 1.99 30. -0.13 31. 0.74 32. 1.23
33. 1.07 34. Not possible
35. -1.50 36. -1.52

37. $-\dfrac{\pi}{3}, -\dfrac{\sqrt{3}}{3}, 1$ 38. $\pi, \dfrac{2\pi}{3}, \dfrac{\sqrt{3}}{2}$ 39. $\theta = \arctan \dfrac{x}{4}$

40. $\theta = \arccos \dfrac{4}{x}$ 41. $\theta = \arcsin \dfrac{x+2}{5}$

42. $\theta = \arctan \dfrac{x+1}{10}$ 43. $\theta = \arccos \dfrac{x+3}{2x}$

44. $\theta = \arcsin \dfrac{1}{x+1}, x \ne 1$ 45. 0.3 46. 45

47. Not possible   48. -0.2 49. $\dfrac{\pi}{4}$ 50. $\dfrac{\pi}{2}$ 51. $\dfrac{3}{5}$

52. $\dfrac{3}{5}$ 53. $\dfrac{\sqrt{5}}{5}$ 54. Not possible   55. $\dfrac{13}{12}$

56. $-\dfrac{13}{5}$ 57. $-\dfrac{5}{3}$ 58. $-\dfrac{4}{3}$ 59. $-\dfrac{\sqrt{5}}{2}$ 60. $\dfrac{8}{5}$

61. 2 62. -1 63. $\sqrt{1-4x^2}$ 64. $\dfrac{x}{\sqrt{x^2+1}}$

65. $\dfrac{1}{x}$ 66. $\sqrt{9x^2+1}$ 67. $\sqrt{1-x^2}$ 68. $\dfrac{1}{\sqrt{2x-x^2}}$

69. $\dfrac{\sqrt{9-x^2}}{x}$ 70. x 71. $\dfrac{\sqrt{x^2+a^2}}{x}$

72. $\dfrac{\sqrt{r^2-(x-h)^2}}{r}$

73.

Asymptotes: $y = \pm 1$

74.

Asymptote: $x = 0$

75. $\dfrac{9}{\sqrt{x^2+81}}$ 76. $\dfrac{x}{6}$ 77. $\dfrac{|x-1|}{\sqrt{x^2-2x+10}}$

78. $\dfrac{\sqrt{4x-x^2}}{x-2}$

79.

Vertical stretch

80.

Horizontal shrink

81.

Shift $\dfrac{\pi}{2}$ units up

82.

Shift two units left

83.

Horizontal stretch

84.

Horizontal stretch

85.

86.

87.

88.

89.

90.

91. $3\sqrt{2} \sin\left(2t + \dfrac{\pi}{4}\right)$

The graph implies that the identity is true.

92. $5 \sin\left(\pi t + \arctan \dfrac{4}{3}\right)$

The graph implies that the identity is true.

93. $\dfrac{\pi}{2}$ **94.** 0 **95.** $\dfrac{\pi}{2}$ **96.** $-\dfrac{\pi}{2}$ **97.** π **98.** $-\dfrac{\pi}{2}$

99. (a) $\theta = \arcsin \dfrac{5}{s}$ (b) About 0.13, about 0.25

100. (a) $\theta = \arctan \dfrac{s}{750}$ (b) About 21.8°, about 58.0°

101. (a) About 32.9°   (b) About 6.5 m

102. (a)

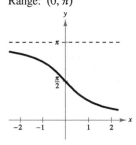

20 ft
47 ft

(b) About 23.1°   (c) About 12.8 ft

103. (a)

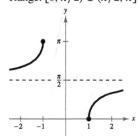

(b) 2 ft   (c) $\beta = 0$; As x increases, β approaches 0.

104. (a) $\theta = \arcsin \dfrac{6}{x}$ (b) 30°; about 59.0°

105. (a) $\theta = \arctan \dfrac{x}{20}$ (b) About 14.0°, about 31.0°

106. False. $\dfrac{5\pi}{6}$ is not in the range of the arcsine function.

107. True. $-\dfrac{\pi}{4}$ is in the range of the arctangent function.

108. False. $\arctan(-1) = -\dfrac{\pi}{4} \neq \dfrac{\arcsin(-1)}{\arccos(-1)} = \dfrac{-\pi/2}{\pi} = -\dfrac{1}{2}$

109. False. $\sin^{-1} x$ is the inverse of $\sin x$, not the reciprocal.

110. (a) $-1 \le x < \dfrac{\sqrt{2}}{2}$ (b) $x = \dfrac{\sqrt{2}}{2}$ (c) $\dfrac{\sqrt{2}}{2} < x \le 1$

111. Domain: $(-\infty, \infty)$
Range: $(0, \pi)$

112. Domain: $(-\infty, -1] \cup [1, \infty)$
Range: $[0, \pi/2) \cup (\pi/2, \pi]$

113. Domain: $(-\infty, -1] \cup [1, \infty)$
Range: $[-\pi/2, 0) \cup (0, \pi/2]$

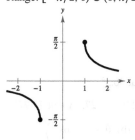

114. Answers will vary.   **115.** $\dfrac{\pi}{4}$ **116.** 0 **117.** $\dfrac{3\pi}{4}$

118. $\dfrac{5\pi}{6}$ **119.** $-\dfrac{\pi}{2}$ **120.** $\dfrac{\pi}{3}$

121. 1.17   **122.** 2.29   **123.** -0.12 **124.** -0.08

125. 0.19   **126.** 2.73

127. (a) $\dfrac{\pi}{4}$ (b) $\dfrac{\pi}{2}$ (c) About 1.25   (d) About 2.03

128.

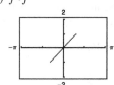

As x increases to infinity, g approaches 3π, but f has no maximum.
$a \approx 87.54$

129. (a) $f \circ f^{-1}$ $f^{-1} \circ f$

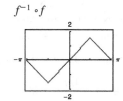

(b) The domains and ranges of the functions are restricted. The graphs of $f \circ f^{-1}$ and $f^{-1} \circ f$ differ because of the domains and ranges of f and f^{-1}.

130. (a)–(e) Proofs

Section 4.8   (page 334)

1. bearing   **2.** harmonic motion
3. period   **4.** frequency
5. $a \approx 10.39$ **6.** $a \approx 8.58$ **7.** $b \approx 14.21$
$b = 6$ $c \approx 9.46$ $c \approx 14.88$
$B = 30°$ $A = 65°$ $A = 17.2°$
8. $b \approx 274.27$ **9.** $c = 5$ **10.** $b \approx 24.49$
$c \approx 277.24$ $A \approx 36.87°$ $A \approx 45.58°$
$B = 81.6°$ $B \approx 53.13°$ $B \approx 44.42°$
11. $a \approx 52.88$ **12.** $a \approx 9.36$
$A \approx 73.46°$ $A \approx 81.97°$
$B \approx 16.54°$ $B \approx 8.03°$
13. 3.00   **14.** 2.83   **15.** 2.50   **16.** About 2.80
17. About 214.45 ft   **18.** About 32.97 ft
19. About 19.7 ft   **20.** About 81.2 ft   **21.** About 20.5 ft
22. About 1933.3 ft   **23.** About 11.8 km

CHAPTER 4

24. (a)

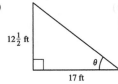

(b) $\tan \theta = \dfrac{12\frac{1}{2}}{17}$ (c) About 36.3°

25. About 56.3°   **26.** About 2.0°   **27.** About 75.97°

28. (a) $\sqrt{h^2 + 34h + 10{,}289}$ (b) $\theta = \arccos \dfrac{100}{l}$

(c) About 53.02 ft

29. About 3.23 mi or about 17,054 ft

30. (a) About 25.98 ft   (b) $\arctan \dfrac{25.98}{d}$

(c) 45.0 ft $\le d \le$ 55.7 ft

31. (a) $l = 250$ ft, $A \approx 36.87°$, $B \approx 53.13°$

(b) About 4.87 sec

32. (a) About 4820.7 ft   (b) About 124.5 sec

33. About 508 mi north, about 650 mi east

34. (a) About 429.26 mi north, about 2434.44 mi west

(b) 280°

35. (a) About 104.95 nmi south, about 58.18 nmi west

(b) S 36.7° W; about 130.9 nmi

36. (a) About 21.4 h

(b) About 5.86 nmi east; about 239.93 nmi south

(c) About 178.6°

37. N 56.31° W   **38.** 208°

39. (a) N 58° E   (b) About 68.82 m   **40.** $d \approx 5.46$ km

41. About 35.3°   **42.** About 54.7°   **43.** About 29.4 in.

44. 25 in.   **45.** $d = 4 \sin \pi t$ **46.** $d = 3 \sin \dfrac{\pi t}{3}$

47. $d = 3 \cos \dfrac{4\pi t}{3}$ **48.** $d = 2 \cos \dfrac{\pi t}{5}$

49. $\omega = 524\pi$ **50.** $d = \dfrac{7}{4} \cos \dfrac{\pi t}{5}$

51. (a) 9   (b) $\frac{3}{5}$ (c) 9   (d) $\frac{5}{12}$

52. (a) $\frac{1}{2}$ (b) 10   (c) $\frac{1}{2}$ (d) $\frac{1}{40}$

53. (a) $\frac{1}{4}$ (b) 3   (c) 0   (d) $\frac{1}{6}$

54. (a) $\frac{1}{64}$ (b) 396   (c) 0   (d) $\frac{1}{792}$

55. (a)

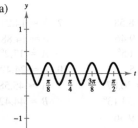

(b) $\dfrac{\pi}{8}$ (c) $\dfrac{\pi}{32}$

56. (a)

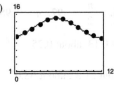

(b) 12; Yes, there are 12 months in a year.

(c) 2.77; The maximum change in the number of hours of daylight

57. (a)

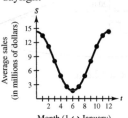

Month (1 ↔ January)

(b) $S = 8 + 6.3 \cos \dfrac{\pi}{6}t$ or $S = 8 + 6.3 \sin\left(\dfrac{\pi}{6}t + \dfrac{\pi}{2}\right)$

The model is a good fit.

(c) 12. Yes, sales of outerwear are seasonal.

(d) Maximum displacement from average sales of $8 million

58. (a) 4 cm   (b) 3 sec   (c) $d = a \cos \omega t$

59. False. The scenario does not create a right triangle because the tower is not vertical.

60. False. The bearing N 24° E means 24 degrees east of north.

Review Exercises   *(page 340)*

1. (a)

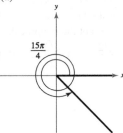

(b) Quadrant IV

(c) $\dfrac{23\pi}{4}, -\dfrac{\pi}{4}$

2. (a)

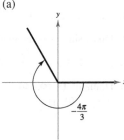

(b) Quadrant II

(c) $\dfrac{2\pi}{3}, -\dfrac{10\pi}{3}$

3. (a)

(b) Quadrant III

(c) 250°, −470°

4. (a)

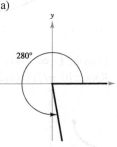

(b) Quadrant IV

(c) 640°, −80°

5. 7.854   **6.** 3.316   **7.** −0.279   **8.** −1.955

9. 54°   **10.** −330°   **11.** −200.535°

12. 326.586°   **13.** 198° 24′   **14.** −5° 57′ 36″

15. About 48.17 in.

16. (a) $66\frac{2}{3}\pi$ rad/min   (b) 400π in./min

17. About 523.60 in.^{2} **18.** About 58.90 mm^2

19. $\left(-\frac{1}{2}, \frac{\sqrt{3}}{2}\right)$ **20.** $\left(\frac{\sqrt{2}}{2}, -\frac{\sqrt{2}}{2}\right)$

21. $\left(-\frac{\sqrt{3}}{2}, -\frac{1}{2}\right)$ **22.** $\left(-\frac{1}{2}, \frac{\sqrt{3}}{2}\right)$

23. $\sin\frac{3\pi}{4} = \frac{\sqrt{2}}{2}$ $\csc\frac{3\pi}{4} = \sqrt{2}$

$\cos\frac{3\pi}{4} = -\frac{\sqrt{2}}{2}$ $\sec\frac{3\pi}{4} = -\sqrt{2}$

$\tan\frac{3\pi}{4} = -1$ $\cot\frac{3\pi}{4} = -1$

24. $\sin\left(-\frac{2\pi}{3}\right) = -\frac{\sqrt{3}}{2}$ $\csc\left(-\frac{2\pi}{3}\right) = -\frac{2\sqrt{3}}{3}$

$\cos\left(-\frac{2\pi}{3}\right) = -\frac{1}{2}$ $\sec\left(-\frac{2\pi}{3}\right) = -2$

$\tan\left(-\frac{2\pi}{3}\right) = \sqrt{3}$ $\cot\left(-\frac{2\pi}{3}\right) = \frac{\sqrt{3}}{3}$

25. $\frac{\sqrt{2}}{2}$ **26.** 1   **27.** $-\frac{\sqrt{3}}{2}$ **28.** $-\frac{\sqrt{3}}{2}$ **29.** 3.2361

30. -0.3420 **31.** -75.3130 **32.** -1.1368

33. $\sin\theta = \frac{4\sqrt{41}}{41}$ $\csc\theta = \frac{\sqrt{41}}{4}$

$\cos\theta = \frac{5\sqrt{41}}{41}$ $\sec\theta = \frac{\sqrt{41}}{5}$

$\tan\theta = \frac{4}{5}$ $\cot\theta = \frac{5}{4}$

34. $\sin\theta = \frac{\sqrt{3}}{2}$ $\csc\theta = \frac{2\sqrt{3}}{3}$

$\cos\theta = \frac{1}{2}$ $\sec\theta = 2$

$\tan\theta = \sqrt{3}$ $\cot\theta = \frac{\sqrt{3}}{3}$

35. 0.6494   **36.** 5.3860   **37.** 3.6722   **38.** 0.2045

39. (a) 3   (b) $\frac{2\sqrt{2}}{3}$ (c) $\frac{3\sqrt{2}}{4}$ (d) $\frac{\sqrt{2}}{4}$

40. (a) $\frac{1}{5}$ (b) $2\sqrt{6}$ (c) $\frac{\sqrt{6}}{12}$ (d) 5

41. About 73.3 m   **42.** About 19.5 ft

43. $\sin\theta = \frac{4}{5}$ $\csc\theta = \frac{5}{4}$

$\cos\theta = \frac{3}{5}$ $\sec\theta = \frac{5}{3}$

$\tan\theta = \frac{4}{3}$ $\cot\theta = \frac{3}{4}$

44. $\sin\theta = -\frac{4}{5}$ $\csc\theta = -\frac{5}{4}$

$\cos\theta = \frac{3}{5}$ $\sec\theta = \frac{5}{3}$

$\tan\theta = -\frac{4}{3}$ $\cot\theta = -\frac{5}{4}$

45. $\sin\theta = \frac{4}{5}$ $\csc\theta = \frac{5}{4}$

$\cos\theta = \frac{3}{5}$ $\sec\theta = \frac{5}{3}$

$\tan\theta = \frac{4}{3}$ $\cot\theta = \frac{3}{4}$

46. $\sin\theta = -\frac{\sqrt{26}}{26}$ $\csc\theta = -\sqrt{26}$

$\cos\theta = -\frac{5\sqrt{26}}{26}$ $\sec\theta = -\frac{\sqrt{26}}{5}$

$\tan\theta = \frac{1}{5}$ $\cot\theta = 5$

47. $\sin\theta = -\frac{\sqrt{11}}{6}$ $\csc\theta = -\frac{6\sqrt{11}}{11}$

$\cos\theta = \frac{5}{6}$

$\tan\theta = -\frac{\sqrt{11}}{5}$ $\cot\theta = -\frac{5\sqrt{11}}{11}$

48. $\sin\theta = \frac{2}{3}$

$\cos\theta = -\frac{\sqrt{5}}{3}$ $\sec\theta = -\frac{3\sqrt{5}}{5}$

$\tan\theta = -\frac{2\sqrt{5}}{5}$ $\cot\theta = -\frac{\sqrt{5}}{2}$

49. $\sin\theta = \frac{\sqrt{21}}{5}$ $\csc\theta = \frac{5\sqrt{21}}{21}$

$\sec\theta = -\frac{5}{2}$

$\tan\theta = -\frac{\sqrt{21}}{2}$ $\cot\theta = -\frac{2\sqrt{21}}{21}$

50. $\csc\theta = -2$

$\cos\theta = \frac{\sqrt{3}}{2}$ $\sec\theta = \frac{2\sqrt{3}}{3}$

$\tan\theta = -\frac{\sqrt{3}}{3}$ $\cot\theta = -\sqrt{3}$

51. $\theta' = 84°$ **52.** $\theta' = 85°$

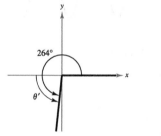

53. $\theta' = \frac{\pi}{5}$ **54.** $\theta' = \frac{\pi}{3}$

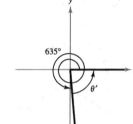

55. $\sin(-150°) = -\frac{1}{2}$; $\cos(-150°) = -\frac{\sqrt{3}}{2}$; $\tan(-150°) = \frac{\sqrt{3}}{3}$

56. $\sin 495° = \frac{\sqrt{2}}{2}$; $\cos 495° = -\frac{\sqrt{2}}{2}$; $\tan 495° = -1$

57. $\sin\frac{\pi}{3} = \frac{\sqrt{3}}{2}$; $\cos\frac{\pi}{3} = \frac{1}{2}$; $\tan\frac{\pi}{3} = \sqrt{3}$

58. $\sin\left(-\frac{5\pi}{4}\right) = \frac{\sqrt{2}}{2}$; $\cos\left(-\frac{5\pi}{4}\right) = -\frac{\sqrt{2}}{2}$; $\tan\left(-\frac{5\pi}{4}\right) = -1$

59. 0.9613   **60.** 0.7536   **61.** -0.4452 **62.** 0.2225

63.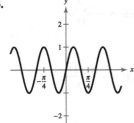

64.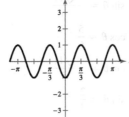

65.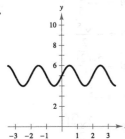

66.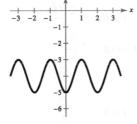

67.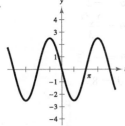

68.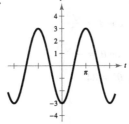

69. (a) $y = 2 \sin 528\pi x$ (b) 264 cycles/sec

70. (a)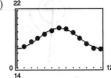

(b) 12; Yes. One period is one year.

(c) 1.41; 1.41 represents the maximum change in time from the average time ($d = 18.10$) of sunset.

71.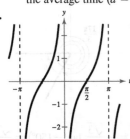

72.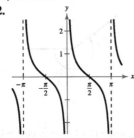

73.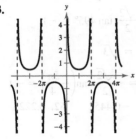

74.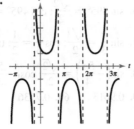

75.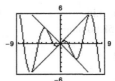

76.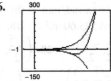

As $x \to +\infty$, $f(x)$ oscillates.   As $x \to \infty$, $f(x)$ oscillates.

77. $-\dfrac{\pi}{2}$ **78.** 0   **79.** $\dfrac{\pi}{6}$ **80.** $\dfrac{3\pi}{4}$ **81.** -0.92

82. 1.19   **83.** 0.06   **84.** -0.25

85.

86.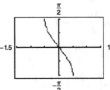

87. $\frac{4}{5}$ **88.** $\frac{4}{3}$ **89.** $\frac{13}{5}$ **90.** $-\frac{5}{12}$

91. $\dfrac{\sqrt{4 - x^2}}{x}$ **92.** $\dfrac{1}{\sqrt{1 - (x - 1)^2}}$

93.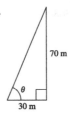

$\theta \approx 66.8°$

94. About 9.6 ft   **95.** About 1221 mi, about 85.6°

96. $d = 0.75 \cos \dfrac{2\pi t}{3}$

97. False. For each θ there corresponds exactly one value of y.

98. False. $3\pi/4$ is not in the range of the arctangent function.

99. The function is undefined because $\sec \theta = 1/\cos \theta$.

100. (a)

θ	0.1	0.4	0.7
$\tan\left(\theta - \dfrac{\pi}{2}\right)$	-9.9666	-2.3652	-1.1872
$-\cot \theta$	-9.9666	-2.3652	-1.1872

θ	1.0	1.3
$\tan\left(\theta - \dfrac{\pi}{2}\right)$	-0.6421	-0.2776
$-\cot \theta$	-0.6421	-0.2776

(b) $\tan\left(\theta - \dfrac{\pi}{2}\right) = -\cot \theta$

101. The ranges of the other four trigonometric functions are $(-\infty, \infty)$ or $(-\infty, -1] \cup [1, \infty)$.

102. (a) The displacement is increased.

(b) The friction damps the oscillations more quickly.

(c) The frequency of the oscillations increases.

Chapter Test *(page 343)*

1. (a)

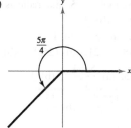

(b) $\dfrac{13\pi}{4}, -\dfrac{3\pi}{4}$

(c) $225°$

2. 3500 rad/min   **3.** About 709.04 ft²

4. $\sin \theta = \dfrac{3\sqrt{13}}{13}$ $\csc \theta = \dfrac{\sqrt{13}}{3}$

$\cos \theta = \dfrac{2\sqrt{13}}{13}$ $\sec \theta = \dfrac{\sqrt{13}}{2}$

$\cot \theta = \dfrac{2}{3}$

5. $\sin \theta = \dfrac{3\sqrt{10}}{10}$ $\csc \theta = \dfrac{\sqrt{10}}{3}$

$\cos \theta = -\dfrac{\sqrt{10}}{10}$ $\sec \theta = -\sqrt{10}$

$\tan \theta = -3$ $\cot \theta = -\dfrac{1}{3}$

6. $\theta' = 25°$

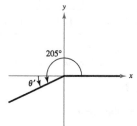

7. Quadrant III   **8.** $150°, 210°$

9. $\sin \theta = -\dfrac{4}{5}$ **10.** $\sin \theta = \dfrac{21}{29}$

$\tan \theta = -\dfrac{4}{3}$ $\cos \theta = -\dfrac{20}{29}$

$\csc \theta = -\dfrac{5}{4}$ $\tan \theta = -\dfrac{21}{20}$

$\sec \theta = \dfrac{5}{3}$ $\csc \theta = \dfrac{29}{21}$

$\cot \theta = -\dfrac{3}{4}$ $\cot \theta = -\dfrac{20}{21}$

11.

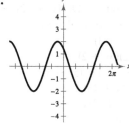

12.

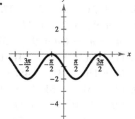

13.

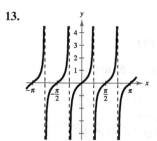

14.

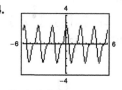

Period: 2

15.

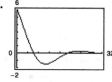

Not periodic;
As $x \to \infty$, $y \to 0$.

16. $a = -2, b = \dfrac{1}{2}, c = -\dfrac{\pi}{4}$ **17.** $\dfrac{\sqrt{55}}{3}$

18.

19. About $309.3°$

20. $d = -6 \cos \pi t$

Problem Solving *(page 345)*

1. (a) $\dfrac{11\pi}{2}$ rad or $990°$ (b) About 816.42 ft

2. Gear 1: $270°, \dfrac{3\pi}{2}$ rad

Gear 2: About $332.3°, \dfrac{24\pi}{13}$ rad

Gear 3: About $392.7°, \dfrac{24\pi}{11}$ rad

Gear 4: $450°, \dfrac{5\pi}{2}$ rad

Gear 5: About $454.7°, \dfrac{48\pi}{19}$ rad

3. $h = 51 - 50 \sin\left(8\pi t + \dfrac{\pi}{2}\right)$

4. (a) True. $f(t - 2c) = f(t - 2c + 2c) = f(t)$

(b) False. $f\left(t + \dfrac{1}{2}c\right)$ has a period of c and $f\left(\dfrac{1}{2}t\right)$ has a period of $2c$.

(c) False. $f\left(\dfrac{1}{2}[t + c]\right) = f\left(\dfrac{1}{2}t + \dfrac{1}{2}c\right)$ and the period of $f\left(\dfrac{1}{2}t\right)$ is $2c$.

(d) True. $f\left(\dfrac{1}{2}[t + 4c]\right) = f\left(\dfrac{1}{2}t + 2c\right)$ and the period of $f\left(\dfrac{1}{2}t\right)$ is $2c$.

5. (a) About 4767 ft   (b) About 3705 ft

(c) $w \approx 2183$ ft, $\tan 63° = \dfrac{w + 3705}{3000}$

6. (a) Answers will vary.   (b) The ratios are equal.

(c) No; No

(d) Yes, because all six trigonometric functions are ratios of right triangles.

CHAPTER 4

Appendix A

Appendix A.1   *(page A11)*

1. irrational   **3.** absolute value   **5.** terms

7. (a) 5, 1, 2   (b) 0, 5, 1, 2   (c) $-9, 5, 0, 1, -4, 2, -11$
(d) $-9, -\frac{7}{2}, 5, \frac{2}{3}, 0, 1, -4, 2, -11$ (e) $\sqrt{2}$

9. (a) 1   (b) 1   (c) $-13, 1, -6$
(d) $2.01, 0.\overline{6}, -13, 1, -6$ (e) $0.010110111\ldots$

11. (a) ![number line] x (b) ![number line] $\frac{7}{2}$ x

(c) $-\frac{5}{2}$![number line] x (d) -5.2 ![number line] x

13. ![number line] x **15.** $\frac{2}{3}\ \frac{5}{6}$![number line] x
$-4 > -8$

$\frac{5}{6} > \frac{2}{3}$

17. (a) $x \le 5$ denotes the set of all real numbers less than or equal to 5.
(b) ![number line] x (c) Unbounded

19. (a) $-2 < x < 2$ denotes the set of all real numbers greater than -2 and less than 2.
(b) ![number line] x (c) Bounded

21. (a) $[4, \infty)$ denotes the set of all real numbers greater than or equal to 4.
(b) ![number line] x (c) Unbounded

23. (a) $[-5, 2)$ denotes the set of all real numbers greater than or equal to -5 and less than 2.
(b) ![number line] x (c) Bounded

Inequality	Interval
25. $y \ge 0$	$[0, \infty)$
27. $10 \le t \le 22$	$[10, 22]$

29. 10   **31.** 5   **33.** -1 **35.** 25   **37.** -1
39. $|-4| = |4|$ **41.** $-|-6| < |-6|$ **43.** 51   **45.** $\frac{5}{2}$
47. $|x - 5| \le 3$ **49.** \$2524.0 billion; \$458.5 billion
51. \$2450.0 billion; \$1087.0 billion
53. $7x$ and 4 are the terms; 7 is the coefficient.
55. $6x^3$ and $-5x$ are the terms; 6 and -5 are the coefficients.
57. $3\sqrt{3}x^2$ and 1 are the terms; $3\sqrt{3}$ is the coefficient.
59. (a) -10 (b) -6 **61.** (a) 2   (b) 6
63. (a) Division by 0 is undefined.   (b) 0
65. Multiplicative Inverse Property
67. Associative and Commutative Properties of Multiplication
69. $\frac{5x}{12}$ **71.** $\frac{x}{4}$
73. False. Zero is nonnegative, but not positive.
75. True. The product of two negative numbers is positive.
77. (a)

n	0.0001	0.01	1	100	10,000
$\dfrac{5}{n}$	50,000	500	5	0.05	0.0005

(b) (i) The value of $5/n$ approaches infinity as n approaches 0.
(ii) The value of $5/n$ approaches 0 as n increases without bound.

Appendix A.2   *(page A23)*

1. exponent; base   **3.** square root   **5.** like radicals
7. rationalizing   **9.** (a) 625   (b) $\frac{1}{25}$
11. (a) 5184   (b) $-\frac{3}{5}$ **13.** (a) $\frac{16}{3}$ (b) 1
15. -24 **17.** 6   **19.** -48 **21.** (a) $125z^3$ (b) $5x^6$
23. (a) $24y^2$ (b) $-3z^7$ **25.** (a) $\dfrac{5184}{y^7}$ (b) 1
27. (a) 1   (b) $\dfrac{1}{4x^4}$ **29.** (a) $\dfrac{125x^9}{y^{12}}$ (b) $\dfrac{b^5}{a^5}$
31. 1.02504×10^4 **33.** 0.000314
35. 9,460,000,000,000 km   **37.** (a) 6.8×10^5 (b) 6.0×10^4
39. (a) 3   (b) $\frac{3}{2}$ **41.** (a) 2   (b) $2x$
43. (a) $2\sqrt{5}$ (b) $4\sqrt[3]{2}$ **45.** (a) $6x\sqrt{2x}$ (b) $3y^2\sqrt{6x}$
47. (a) $2x\sqrt[3]{2x^2}$ (b) $\dfrac{5|x|\sqrt{3}}{y^2}$
49. (a) $29|x|\sqrt{5}$ (b) $44\sqrt{3x}$ **51.** $\dfrac{\sqrt{3}}{3}$ **53.** $\dfrac{\sqrt{14} + 2}{2}$
55. $\dfrac{2}{3(\sqrt{5} - \sqrt{3})}$ **57.** $64^{1/3}$ **59.** $\dfrac{3}{\sqrt[3]{x^2}}$
61. (a) $\frac{1}{8}$ (b) $\frac{27}{8}$ **63.** (a) $\sqrt{3}$ (b) $\sqrt[3]{(x + 1)^2}$
65. (a) $2\sqrt[4]{2}$ (b) $\sqrt[8]{2x}$ **67.** (a) $x - 1$ (b) $\dfrac{1}{x - 1}$
69.

h	0	1	2	3	4	5	6
t	0	2.93	5.48	7.67	9.53	11.08	12.32

h	7	8	9	10	11	12
t	13.29	14.00	14.50	14.80	14.93	14.96

71. False. When $x = 0$, the expressions are not equal.
73. False. For instance, $(3 + 5)^2 = 8^2 = 64 \ne 34 = 3^2 + 5^2$.

Appendix A.3   *(page A33)*

1. $n; a_n; a_0$ **3.** like terms   **5.** factoring
7. perfect square trinomial
9. (a) $7x$ (b) Degree: 1; Leading coefficient: 7
(c) Monomial
11. (a) $-\frac{1}{2}x^5 + 14x$ (b) Degree: 5; Leading coefficient: $-\frac{1}{2}$
(c) Binomial
13. (a) $-4x^5 + 6x^4 + 1$
(b) Degree: 5; Leading coefficient: -4 (c) Trinomial
15. $-2x - 10$ **17.** $-8.3x^3 + 0.3x^2 - 23$
19. $3x^3 - 6x^2 + 3x$ **21.** $-15z^2 + 5z$ **23.** $6x^2 - 7x - 5$
25. $x^4 + 2x^2 + x + 2$ **27.** $x^2 - 100$ **29.** $4x^2 + 12x + 9$
31. $x^3 + 9x^2 + 27x + 27$ **33.** $x^2 + 2xy + y^2 - 6x - 6y + 9$
35. $m^2 - n^2 - 6m + 9$ **37.** $2x(x^2 - 3)$
39. $(x - 5)(3x + 8)$ **41.** $(5y - 2)(5y + 2)$
43. $(x + 1)(x - 3)$ **45.** $(x - 2)^2$ **47.** $(5z - 3)^2$
49. $(2y - 3)^2$ **51.** $(x + 5)(x^2 - 5x + 25)$
53. $(2t - 1)(4t^2 + 2t + 1)$ **55.** $(x + 2)(x - 1)$
57. $(3x - 2)(x + 4)$ **59.** $(5x + 1)(x + 6)$
61. $-(5y - 2)(y + 2)$ **63.** $(x - 1)(x^2 + 2)$
65. $(2x - 1)(x^2 - 3)$ **67.** $(3x^3 - 2)(x^2 + 2)$
69. $(x + 3)(2x + 3)$ **71.** $(2x + 3)(3x - 5)$
73. $6(x + 3)(x - 3)$ **75.** $x^2(x - 1)$ **77.** $-2x(x + 1)(x - 2)$
79. $(5 - x)(1 + x^2)$ **81.** $-(x - 2)(x + 1)(x - 8)$

83. (a) $\pi h(R + r)(R - r)$ (b) $V = 2\pi\left[\left(\dfrac{R + r}{2}\right)(R - r)\right]h$

85. False. $(4x^2 + 1)(3x + 1) = 12x^3 + 4x^2 + 3x + 1$

87. True. $a^2 - b^2 = (a + b)(a - b)$ **89.** $m + n$

91. $-x^3 + 8x^2 + 2x + 7$

93. Answers will vary. *Sample answer:* $x^2 - 3$

95. $(x^n + y^n)(x^n - y^n)$

Appendix A.4   (page A42)

1. domain   **3.** complex   **5.** All real numbers x

7. All real numbers x such that $x \neq 3$

9. All real numbers x such that $x \neq -\frac{2}{3}$

11. All real numbers x such that $x \neq -4, -2$

13. All real numbers x such that $x \geq 7$

15. All real numbers x such that $x > 3$ **17.** $\dfrac{3x}{2}, \ x \neq 0$

19. $-\dfrac{1}{2}, \ x \neq 5$ **21.** $y - 4, \ y \neq -4$ **23.** $\dfrac{3y}{4}, \ y \neq -\dfrac{2}{3}$

25. $\dfrac{x - 1}{x + 3}, \ x \neq -5$ **27.** $-\dfrac{x + 1}{x + 5}, \ x \neq 2$ **29.** $\dfrac{1}{x + 1}, \ x \neq \pm 4$

31. When simplifying fractions, only common factors can be divided out, not terms.

33. $\dfrac{1}{5(x - 2)}, \ x \neq 1$ **35.** $-\dfrac{(x + 2)^2}{6}, \ x \neq \pm 2$

37. $\dfrac{x - y}{x(x + y)^2}, \ x \neq -2y$ **39.** $\dfrac{3}{x + 2}$ **41.** $\dfrac{3x^2 + 3x + 1}{(x + 1)(3x + 2)}$

43. $\dfrac{3 - 2x}{2(x + 2)}$ **45.** $\dfrac{2 - x}{x^2 + 1}, \ x \neq 0$

47. The minus sign should be distributed to each term in the numerator of the second fraction to yield $x + 4 - 3x + 8$.

49. $\dfrac{1}{2}, \ x \neq 2$ **51.** $x(x + 1), \ x \neq -1, 0$ **53.** $\dfrac{2x - 1}{2x}, \ x > 0$

55. $\dfrac{x^2 + (x^2 + 3)^7}{(x^2 + 3)^4}$ **57.** $\dfrac{2x^3 - 2x^2 - 5}{(x - 1)^{1/2}}$ **59.** $\dfrac{3x - 1}{3}, \ x \neq 0$

61. $\dfrac{-1}{x(x + h)}, \ h \neq 0$ **63.** $\dfrac{-1}{(x - 4)(x + h - 4)}, \ h \neq 0$

65. $\dfrac{1}{\sqrt{x + 2} + \sqrt{x}}$ **67.** $\dfrac{1}{\sqrt{t + 3} + \sqrt{3}}, \ t \neq 0$

69. $\dfrac{1}{\sqrt{x + h + 1} + \sqrt{x + 1}}, \ h \neq 0$

71. (a)

t	0	2	4	6	8	10	12
T	75	55.9	48.3	45	43.3	42.3	41.7

t	14	16	18	20	22
T	41.3	41.1	40.9	40.7	40.6

(b) The model is approaching a T-value of 40.

73. $\dfrac{x}{2(2x + 1)}, \ x \neq 0$

75. (a)

Year, t	Online Banking	Mobile Banking
11	79.1	17.9
12	80.9	24.0
13	83.1	29.6
14	86.0	34.8

(b) The values from the models are close to the actual data.

(c) Ratio $= \dfrac{0.0313t^3 - 0.661t^2 - 2.23t + 47}{0.0208t^3 - 0.494t^2 + 2.97t - 70.5}$

(d)

Year	Ratio
2011	0.2267
2012	0.2977
2013	0.3578
2014	0.4061

Answers will vary.

77. $\dfrac{R_1 R_2}{R_2 + R_1}$

79. False. In order for the simplified expression to be equivalent to the original expression, the domain of the simplified expression needs to be restricted. If n is even, $x \neq -1, 1$. If n is odd, $x \neq 1$.

Appendix A.5   (page A56)

1. equation   **3.** extraneous   **5.** 4   **7.** -9 **9.** 12

11. No solution   **13.** $-\dfrac{96}{23}$ **15.** 4   **17.** 3

19. No solution. The variable is divided out.

21. No solution. The solution is extraneous.

23. 5   **25.** $0, -\dfrac{1}{2}$ **27.** -5 **29.** $-\dfrac{1}{2}, 3$ **31.** $\pm\dfrac{3}{4}$

33. $-\dfrac{20}{3}, -4$ **35.** ± 7 **37.** $\pm 3\sqrt{3} \approx 5.20$ **39.** $-3, 11$

41. $\dfrac{1 \pm 3\sqrt{2}}{2} \approx 2.62, -1.62$ **43.** $4, -8$ **45.** $-2 \pm \sqrt{2}$

47. $-1 \pm \dfrac{\sqrt{2}}{2}$ **49.** $\dfrac{-5 \pm \sqrt{89}}{4}$ **51.** $\dfrac{1}{2}, -1$ **53.** $-\dfrac{5}{3}$

55. $\dfrac{7}{4} \pm \dfrac{\sqrt{41}}{4}$ **57.** $\dfrac{2}{3} \pm \dfrac{\sqrt{7}}{3}$ **59.** $1 \pm \sqrt{3}$ **61.** $2 \pm \dfrac{\sqrt{6}}{2}$

63. $6 \pm \sqrt{11}$ **65.** $1 \pm \sqrt{2}$ **67.** $-10, 6$ **69.** $\dfrac{1}{2} \pm \sqrt{3}$

71. $\dfrac{3}{4} \pm \dfrac{\sqrt{97}}{4}$ **73.** $0, \pm 3$ **75.** $-2, \pm 2\sqrt{2}$ **77.** 20

79. $-\dfrac{55}{2}$ **81.** 1   **83.** 4   **85.** 9   **87.** 1   **89.** $8, -3$

91. $-\dfrac{1}{2} - \dfrac{\sqrt{17}}{2}, 3$ **93.** $\sqrt[3]{\dfrac{4.47}{\pi}} \approx 1.12$ in.   **95.** 63.7 in.

97. False. See Example 14 on page A54.

99. True. There is no value that satisfies this equation.

Appendix A.6   (page A64)

1. solution set   **3.** double   **5.** $-2 \leq x < 6$; Bounded

7. $-1 \leq x \leq 5$; Bounded   **9.** $x > 11$; Unbounded

11. $x < -2$; Unbounded

13. $x < 3$ **15.** $x < \dfrac{3}{2}$

17. $x \geq 12$ **19.** $x > 2$

21. $x \geq 1$ **23.** $x < 5$

25. $x \geq 4$ **27.** $x \geq 2$

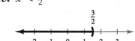

29. $x \geq -4$

31. $-1 < x < 3$

33. $-7 < x \leq -\frac{1}{3}$

35. $-\frac{9}{2} < x < \frac{15}{2}$

37. $-5 \leq x < 1$

39. $-\frac{3}{4} < x < -\frac{1}{4}$

41. $10.5 \leq x \leq 13.5$

43. $-5 < x < 5$

45. $x < -2, x > 2$

47. No solution

49. $14 \leq x \leq 26$

51. $x \leq -1, x \geq 8$

53. $x \leq -5, x \geq 11$

55. $4 < x < 5$

57. $x \leq -\frac{29}{2}, x \geq -\frac{11}{2}$

59.

$x > 3$

61.

$x \leq 2$

63.

$x \leq 4$

65.

$-6 \leq x \leq 22$

67.

$x \leq -\frac{27}{2}, x \geq -\frac{1}{2}$

69.

(a) $x \geq 1$　(b) $x \leq \frac{1}{3}$

71.

(a) $-2 \leq x \leq 4$
(b) $x \leq 4$

73.

(a) $1 \leq x \leq 5$
(b) $x \leq -1, x \geq 7$

75. All real numbers less than eight units from 10
77. $|x| \leq 3$　**79.** $|x - 7| \geq 3$　**81.** $|x - 7| \geq 3$
83. $|x + 3| < 4$　**85.** $7.25 \leq P \leq 7.75$　**87.** $r < 0.08$
89. $100 \leq r \leq 170$　**91.** More than 6 units per hour
93. Greater than 10%　**95.** $x \geq 36$　**97.** $87 \leq x \leq 210$
99. (a)

(b) $x \geq 2.9$
(c) $x \geq 2.908$
(d) Answers will vary.

101. (a) $4.34 \leq t \leq 6.56$ (Between 2004 and 2006)
(b) $t > 20.95$ (2020)
103. $13.7 < t < 17.5$　**105.** \$0.28
107. $106.864 \text{ in.}^2 \leq \text{area} \leq 109.464 \text{ in.}^2$
109. True by the Addition of a Constant Property of Inequalities.
111. False. If $-10 \leq x \leq 8$, then $10 \geq -x$ and $-x \geq -8$.
113. *Sample answer:* $x < x + 1$
115. *Sample answer:* $a = 1, b = 5, c = 5$

Appendix A.7　*(page A72)*

1. numerator
3. The middle term needs to be included.
$(x + 3)^2 = x^2 + 6x + 9$
5. $\sqrt{x + 9}$ cannot be simplified.
7. Divide out common factors, not common terms.
$\dfrac{2x^2 + 1}{5x}$ cannot be simplified.
9. The exponent also applies to the coefficient. $(4x)^2 = 16x^2$
11. To add fractions, first find a common denominator.
$\dfrac{3}{x} + \dfrac{4}{y} = \dfrac{3y + 4x}{xy}$
13. $(x + 2)^{-1/2}(3x + 2)$
15. $2x(2x - 1)^{-1/2}[2x^2(2x - 1)^2 - 1]$　**17.** $5x + 3$
19. $2x^2 + x + 15$　**21.** $1 - 5x$　**23.** $3x - 1$　**25.** $\frac{1}{3}$
27. 2　**29–33.** Answers will vary.　**35.** $7(x + 3)^{-5}$
37. $2x^5(3x + 5)^{-4}$　**39.** $\frac{4}{3}x^{-1} + 4x^{-4} - 7x(2x)^{-1/3}$
41. $\dfrac{x}{3} + 2 + \dfrac{4}{x}$　**43.** $4x^{8/3} - 7x^{5/3} + \dfrac{1}{x^{1/3}}$
45. $\dfrac{3}{x^{1/2}} - 5x^{3/2} - x^{7/2}$　**47.** $\dfrac{-7x^2 - 4x + 9}{(x^2 - 3)^3(x + 1)^4}$
49. $\dfrac{27x^2 - 24x + 2}{(6x + 1)^4}$　**51.** $\dfrac{-1}{(x + 3)^{2/3}(x + 2)^{7/4}}$
53. $\dfrac{4x - 3}{(3x - 1)^{4/3}}$　**55.** $\dfrac{x}{x^2 + 4}$
57. $\dfrac{(3x - 2)^{1/2}(15x^2 - 4x + 45)}{2(x^2 + 5)^{1/2}}$
59. (a) Answers will vary.
(b)

x	-2	-1	$-\frac{1}{2}$	0	1	2	$\frac{5}{2}$
y_1	-8.7	-2.9	-1.1	0	2.9	8.7	12.5
y_2	-8.7	-2.9	-1.1	0	2.9	8.7	12.5

(c) Answers will vary.
61. You cannot move term-by-term from the denominator to the numerator.

Technology

Chapter 1   *(page 3)*

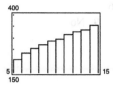

(page 27)

The lines appear perpendicular on the square setting.

(page 40)

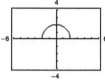   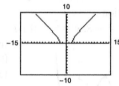

Domain: $[-2, 2]$ Domain: $(-\infty, -2] \cup [2, \infty)$
                        Yes, for -2 and 2.

(page 63)

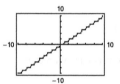   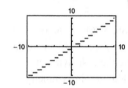

The graph in *dot* mode illustrates that the range is the set of all integers.

Chapter 3   *(page 237)*

$S = 0.00036(2.130)^t$

The exponential regression model has the same coefficient as the model in Example 1. However, the model given in Example 1 contains the natural exponential function.

Chapter 4   *(page 298)*

No graph is visible. $-\pi \le x \le \pi$ and $-0.5 \le y \le 0.5$ displays a good view of the graph.

Chapter 7   *(page 470)*

The point of intersection (7000, 5000) agrees with the solution.

(page 472)

$(2, 0)$; $(0, -1)$, $(2, 1)$; None

Chapter 8   *(page 556)*

$$\begin{bmatrix} 1 & 1 \\ 1 & -5 \end{bmatrix}$$

Chapter 10   *(page 740)*

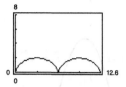

Checkpoints

Chapter 1

Section 1.1

1.
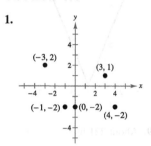

2.
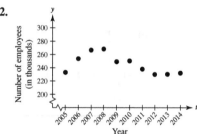

3. $\sqrt{37} \approx 6.08$
4. $d_1 = \sqrt{45}$, $d_2 = \sqrt{20}$, $d_3 = \sqrt{65}$
   $\left(\sqrt{45}\right)^2 + \left(\sqrt{20}\right)^2 = \left(\sqrt{65}\right)^2$
5. $(1, -1)$ 6. $\sqrt{709} \approx 27$ yd   7. \$4.8 billion
8. $(1, 2), (1, -2), (-1, 0), (-1, -4)$

Section 1.2

1. (a) No   (b) Yes
2. (a)   (b)

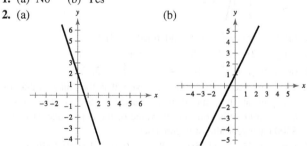

3. (a) 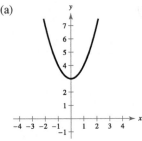 (b)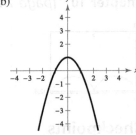

4. x-intercepts: $(0, 0), (-5, 0),$
y-intercept: $(0, 0)$

5. x-axis symmetry

6. 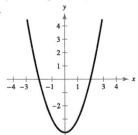 **7.**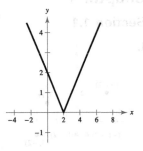

8. $(x + 3)^2 + (y + 5)^2 = 25$ **9.** About 221 lb

Section 1.3

1. (a) 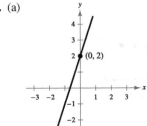 (b)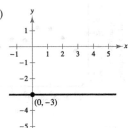

(c)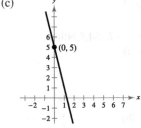

2. (a) 2   (b) $-\frac{3}{2}$ (c) Undefined   (d) 0
3. (a) $y = 2x - 13$ (b) $y = -\frac{2}{3}x + \frac{5}{3}$ (c) $y = 1$
4. (a) $y = \frac{5}{3}x + \frac{23}{3}$ (b) $y = -\frac{3}{5}x - \frac{7}{5}$ **5.** Yes
6. The y-intercept, $(0, 1500)$, tells you that the initial value of
a copier at the time it is purchased is $1500. The slope,
$m = -300$, tells you that the value of the copier decreases by
$300 each year after it is purchased.
7. $y = -4125x + 24{,}750$ **8.** $y = 0.7t + 4.4;$ $9.3 billion

Section 1.4

1. (a) Not a function   (b) Function
2. (a) Not a function   (b) Function
3. (a) -2 (b) -38 (c) $-3x^2 + 6x + 7$
4. $f(-2) = 5, f(2) = 1, f(3) = 2$ **5.** ± 4 **6.** $-4, 3$

7. (a) $\{-2, -1, 0, 1, 2\}$
(b) All real numbers x except $x = 3$
(c) All real numbers r such that $r > 0$
(d) All real numbers x such that $x \geq 16$
8. (a) $S(r) = 10\pi r^2$ (b) $S(h) = \frac{5}{8}\pi h^2$ **9.** No
10. 2009: 772   2013: 1277
      2010: 841   2014: 1437
      2011: 910   2015: 1597
      2012: 1117
11. $2x + h + 2, h \neq 0$

Section 1.5

1. (a) All real numbers x except $x = -3$
(b) $f(0) = 3; f(3) = -6$ (c) $(-\infty, 3]$
2. Function
3. (a) $x = -8, x = \frac{3}{2}$ (b) $t = 25$ (c) $x = \pm\sqrt{2}$
4. 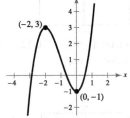   Increasing on $(-\infty, -2)$ and
$(0, \infty)$
Decreasing on $(-2, 0)$

5. $(-0.88, 6.06)$ **6.** (a) -3 (b) 0
7. (a) 20 ft/sec   (b) $\frac{140}{3}$ ft/sec
8. (a) Neither; No symmetry   (b) Even; y-axis symmetry
(c) Odd; Origin symmetry

Section 1.6

1. $f(x) = -\frac{5}{2}x + 1$ **2.** $f\left(-\frac{3}{2}\right) = 0, f(1) = 3, f\left(-\frac{5}{2}\right) = -1$
3.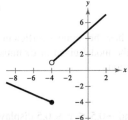

Section 1.7

1. (a) 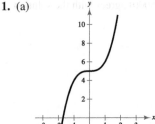 (b)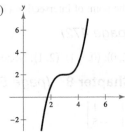

2. $j(x) = -(x + 3)^4$
3. (a) The graph of g is a reflection in the x-axis of the graph
of f.
(b) The graph of h is a reflection in the y-axis of the graph
of f.

4. (a) The graph of g is a vertical stretch of the graph of f.
   (b) The graph of h is a vertical shrink of the graph of f.
5. (a) The graph of g is a horizontal shrink of the graph of f.
   (b) The graph of h is a horizontal stretch of the graph of f.

Section 1.8

1. $x^2 - x + 1$; 3   **2.** $x^2 + x - 1$; 11   **3.** $x^2 - x^3$; -18

4. $\left(\dfrac{f}{g}\right)(x) = \dfrac{\sqrt{x-3}}{\sqrt{16-x^2}}$; Domain: $[3, 4)$

   $\left(\dfrac{g}{f}\right)(x) = \dfrac{\sqrt{16-x^2}}{\sqrt{x-3}}$; Domain: $(3, 4]$

5. (a) $8x^2 + 7$ (b) $16x^2 + 80x + 101$ (c) 9
6. All real numbers x **7.** $f(x) = \frac{1}{5}\sqrt[3]{x},\ g(x) = 8 - x$
8. (a) $(N \circ T)(t) = 32t^2 + 36t + 204$ (b) About 4.5 h

Section 1.9

1. $f^{-1}(x) = 5x, f(f^{-1}(x)) = \frac{1}{5}(5x) = x, f^{-1}(f(x)) = 5\left(\frac{1}{5}x\right) = x$
2. $g(x) = 7x + 4$
3. 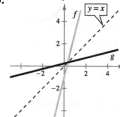   **4.**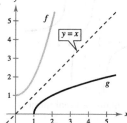

5. (a) Yes   (b) No   **6.** $f^{-1}(x) = \dfrac{5 - 2x}{x + 3}$

7. $f^{-1}(x) = x^3 - 10$

Section 1.10

1. 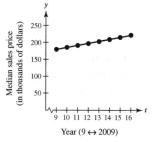   The model is a good fit for the data.

2. 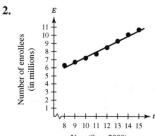 $E = 0.65t + 0.8$
   The model is a good fit for the data.

3. $I = 0.075P$ **4.** 576 ft   **5.** 508 units
6. About 1314 ft   **7.** 14,000 joules

Chapter 2

Section 2.1

1. (a) 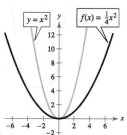   The graph of $f(x) = \frac{1}{4}x^2$ is broader than the graph of $y = x^2$.

   (b) 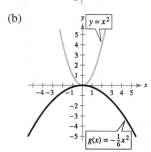   The graph of $g(x) = -\frac{1}{6}x^2$ is a reflection in the x-axis and is broader than the graph of $y = x^2$.

   (c) 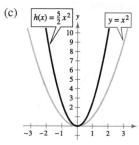   The graph of $h(x) = \frac{5}{2}x^2$ is narrower than the graph of $y = x^2$.

   (d) 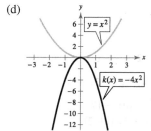   The graph of $k(x) = -4x^2$ is a reflection in the x-axis and is narrower than the graph of $y = x^2$.

2.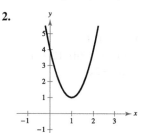
   Vertex: $(1, 1)$
   Axis: $x = 1$

3.
   Vertex: $(2, -1)$
   x-intercepts: $(1, 0), (3, 0)$

4. $y = (x + 4)^2 + 11$ **5.** About 39.7 ft

Section 2.2

1. (a) 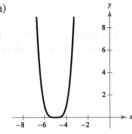 (b)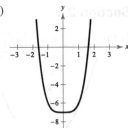

(c) 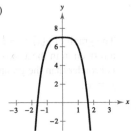 (d)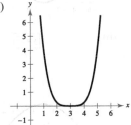

2. (a) Falls to the left, rises to the right
   (b) Rises to the left, falls to the right
3. Real zeros: $x = 0, x = 6$; Turning points: 2
4. 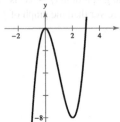 5.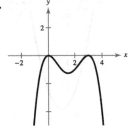

6. $x \approx 3.196$

Section 2.3

1. $(x + 4)(3x + 7)(3x - 7)$ 2. $x^2 + x + 3, x \neq 3$
3. $2x^3 - x^2 + 3 - \dfrac{6}{3x + 1}$ 4. $5x^2 - 2x + 3$
5. (a) 1   (b) 396   (c) $-\frac{13}{2}$ (d) -17
6. $f(-3) = 0, f(x) = (x + 3)(x - 5)(x + 2)$

Section 2.4

1. (a) $12 - i$ (b) $-2 + 7i$ (c) i (d) 0
2. (a) $-15 + 10i$ (b) $18 - 6i$ (c) 41   (d) $12 + 16i$
3. (a) 45   (b) 29   4. $\frac{3}{5} + \frac{4}{5}i$ 5. $-2\sqrt{7}$
6. $-\dfrac{7}{8} \pm \dfrac{\sqrt{23}}{8}i$

Section 2.5

1. 4   2. No rational zeros   3. 5   4. $-3, \frac{1}{2}, 2$
5. $-1, \dfrac{-3 - 3\sqrt{17}}{4} \approx -3.8423, \dfrac{-3 + 3\sqrt{17}}{4} \approx 2.3423$
6. $f(x) = x^4 + 45x^2 - 196$
7. $f(x) = -x^4 - x^3 - 2x^2 - 4x + 8$ 8. $\frac{2}{3}, \pm 4i$
9. $f(x) = (x - 1)(x + 1)(x - 3i)(x + 3i); \pm 1, \pm 3i$
10. No positive real zeros, three or no negative real zeros
11. $\frac{1}{2}$ 12. 7 in. × 7 in. × 9 in.

Section 2.6

1. Domain: all real numbers x such that $x \neq 1$
   $f(x)$ decreases without bound as x approaches 1 from the left and increases without bound as x approaches 1 from the right.
2. Vertical asymptote: $x = -1$
   Horizontal asymptote: $y = 3$

3. 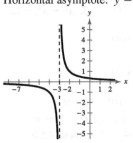 4.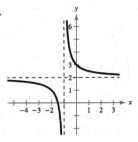

Domain: all real numbers x except $x = -3$

Domain: all real numbers x except $x = -1$

5. 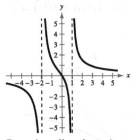 6.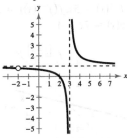

Domain: all real numbers x except $x = -2, 1$

Domain: all real numbers x except $x = -2, 3$

7.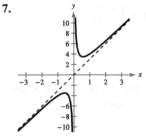

Domain: all real numbers x except $x = 0$
8. (a) \$63.75 million; \$208.64 million; \$1020 million
   (b) No. The function is undefined at $p = 100$.
9. 12.9 in. by 6.5 in.

Section 2.7

1. $(-4, 5)$

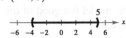

2. (a) $\left(-\frac{5}{2}, 1\right)$
   (b) 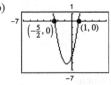   The graph is below the x-axis when x is greater than $-\frac{5}{2}$ and less than 1. So, the solution set is $\left(-\frac{5}{2}, 1\right)$.

3. $\left(-2, \frac{1}{3}\right) \cup (2, \infty)$

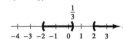

4. (a) The solution set is empty.

(b) The solution set consists of the single real number $\{-2\}$.

(c) The solution set consists of all real numbers except $x = 3$.

(d) The solution set is all real numbers.

5. (a) $\left(-\infty, \frac{11}{4}\right] \cup (3, \infty)$ (b) $(-\infty, -17) \cup (6, \infty)$

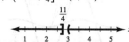

6. $180{,}000 \le x \le 300{,}000$ **7.** $(-\infty, 2] \cup [5, \infty)$

Chapter 3

Section 3.1

1. 0.0528248

2. 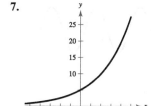 $g(x) = 9^x$ $f(x) = 3^x$

3. $g(x) = 9^{-x}$ $f(x) = 3^{-x}$

4. (a) 2   (b) 3

5. (a) Shift the graph of f two units to the right.

(b) Shift the graph of f three units up.

(c) Reflect the graph of f in the y-axis and shift three units down.

6. (a) 1.3498588   (b) 0.3011942   (c) 492.7490411

7.

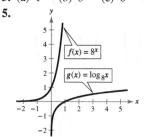

8. (a) \$7927.75   (b) \$7935.08   (c) \$7938.78

9. About 9.970 lb; about 0.275 lb

Section 3.2

1. (a) 0   (b) -3 (c) 3

2. (a) 2.4393327   (b) Error or complex number

(c) -0.3010300

3. (a) 1   (b) 3   (c) 0   **4.** ± 3

5. 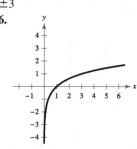 $f(x) = 8^x$ $g(x) = \log_8 x$

6.

Vertical asymptote: $x = 0$

7. (a) 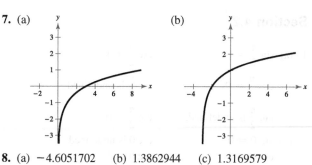   (b)

8. (a) -4.6051702 (b) 1.3862944   (c) 1.3169579

(d) Error or complex number

9. (a) $\frac{1}{3}$ (b) 0   (c) $\frac{3}{4}$ (d) 7   **10.** $(-3, \infty)$

11. (a) 70.84   (b) 61.18   (c) 59.61

Section 3.3

1. 3.5850   **2.** 3.5850

3. (a) $\log 3 + 2 \log 5$ (b) $2 \log 3 - 3 \log 5$ **4.** 4

5. $\log_3 4 + 2 \log_3 x - \frac{1}{2} \log_3 y$ **6.** $\log \dfrac{(x+3)^2}{(x-2)^4}$

7. $\ln y = \frac{2}{3} \ln x$

Section 3.4

1. (a) 9   (b) 216   (c) $\ln 5$ (d) $-\frac{1}{2}$

2. (a) $4, -2$ (b) 1.723   **3.** 3.401   **4.** -4.778

5. 1.099, 1.386   **6.** (a) $e^{2/3}$ (b) 7   (c) 12

7. 0.513   **8.** $\frac{32}{3}$ **9.** 10

10. About 13.2 years; It takes longer for your money to double at a lower interest rate.

11. 2010

Section 3.5

1. 2018   **2.** 400 bacteria   **3.** About 38,000 yr

4.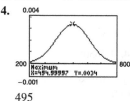

495

5. About 7 days   **6.** (a) 1,000,000   (b) About 80,000,000

Chapter 4

Section 4.1

1. (a) $\dfrac{\pi}{4}, -\dfrac{7\pi}{4}$ (b) $\dfrac{5\pi}{3}, -\dfrac{7\pi}{3}$

2. (a) Complement: $\dfrac{\pi}{3}$; Supplement: $\dfrac{5\pi}{6}$

(b) Complement: none; Supplement: $\dfrac{\pi}{6}$

3. (a) $\dfrac{\pi}{3}$ (b) $\dfrac{16\pi}{9}$ **4.** (a) 30°   (b) 300°

5. 24π in. ≈ 75.40 in.   **6.** About 0.84 cm/sec

7. (a) 4800π rad/min   (b) About 60.319 in./min

8. About 1117 ft^2

Section 4.2

1. (a) $\sin \frac{\pi}{2} = 1$ $\qquad$ $\csc \frac{\pi}{2} = 1$

$\cos \frac{\pi}{2} = 0$ $\qquad$ $\sec \frac{\pi}{2}$ is undefined.

$\tan \frac{\pi}{2}$ is undefined. $\qquad$ $\cot \frac{\pi}{2} = 0$

(b) $\sin 0 = 0$ $\qquad$ $\csc 0$ is undefined.

$\cos 0 = 1$ $\qquad$ $\sec 0 = 1$

$\tan 0 = 0$ $\qquad$ $\cot 0$ is undefined.

(c) $\sin\left(-\frac{5\pi}{6}\right) = -\frac{1}{2}$ $\qquad$ $\csc\left(-\frac{5\pi}{6}\right) = -2$

$\cos\left(-\frac{5\pi}{6}\right) = -\frac{\sqrt{3}}{2}$ $\qquad$ $\sec\left(-\frac{5\pi}{6}\right) = -\frac{2\sqrt{3}}{3}$

$\tan\left(-\frac{5\pi}{6}\right) = \frac{\sqrt{3}}{3}$ $\qquad$ $\cot\left(-\frac{5\pi}{6}\right) = \sqrt{3}$

(d) $\sin\left(-\frac{3\pi}{4}\right) = -\frac{\sqrt{2}}{2}$ $\qquad$ $\csc\left(-\frac{3\pi}{4}\right) = -\sqrt{2}$

$\cos\left(-\frac{3\pi}{4}\right) = -\frac{\sqrt{2}}{2}$ $\qquad$ $\sec\left(-\frac{3\pi}{4}\right) = -\sqrt{2}$

$\tan\left(-\frac{3\pi}{4}\right) = 1$ $\qquad$ $\cot\left(-\frac{3\pi}{4}\right) = 1$

2. (a) 0 $\quad$ (b) $-\frac{\sqrt{3}}{1}$ $\quad$ (c) 0.3

3. (a) 0.78183148 $\quad$ (b) 1.0997502

Section 4.3

1. $\sin \theta = \frac{1}{2}$ $\qquad$ $\csc \theta = 2$

$\cos \theta = \frac{\sqrt{3}}{2}$ $\qquad$ $\sec \theta = \frac{2\sqrt{3}}{3}$

$\tan \theta = \frac{\sqrt{3}}{3}$ $\qquad$ $\cot \theta = \sqrt{3}$

2. $\cot 45° = 1$, $\sec 45° = \sqrt{2}$, $\csc 45° = \sqrt{2}$

3. $\tan 60° = \sqrt{3}$, $\tan 30° = \frac{\sqrt{3}}{3}$ $\qquad$ **4.** 1.7650691

5. (a) 0.28 $\quad$ (b) 0.2917 $\qquad$ **6.** (a) $\frac{1}{2}$ $\quad$ (b) $\sqrt{5}$

7. Answers will vary. $\qquad$ **8.** About 40 ft

9. 60° $\qquad$ **10.** About 17.6 ft; about 17.2 ft

Section 4.4

1. $\sin \theta = \frac{3\sqrt{13}}{13}$, $\cos \theta = -\frac{2\sqrt{13}}{13}$, $\tan \theta = -\frac{3}{2}$

2. $\cos \theta = -\frac{3}{5}$, $\tan \theta = -\frac{4}{3}$

3. $-1; 0$ $\qquad$ **4.** (a) 33° $\quad$ (b) $\frac{4\pi}{9}$ $\quad$ (c) $\frac{\pi}{5}$

5. (a) $-\frac{\sqrt{2}}{2}$ $\quad$ (b) $-\frac{1}{2}$ $\quad$ (c) $-\frac{\sqrt{3}}{3}$ $\qquad$ **6.** (a) $-\frac{3}{5}$ $\quad$ (b) $\frac{4}{3}$

7. (a) -1.8040478 $\quad$ (b) -1.0428352 $\quad$ (c) 0.8090170

Section 4.5

1.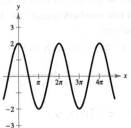

2.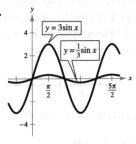

3.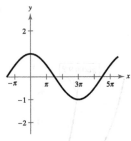

4. Period: 2π

Amplitude: 2

$\left[\frac{\pi}{2}, \frac{5\pi}{2}\right]$ corresponds to one cycle.

Key points: $\left(\frac{\pi}{2}, 2\right)$, $(\pi, 0)$, $\left(\frac{3\pi}{2}, -2\right)$, $(2\pi, 0)$, $\left(\frac{5\pi}{2}, 2\right)$

5.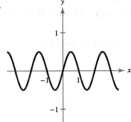

6.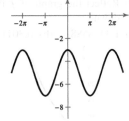

7. $y = 5.6 \sin(0.524t - 0.524) + 5.7$

Section 4.6

1.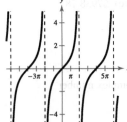

2.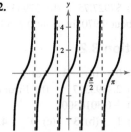

3.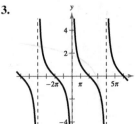

4.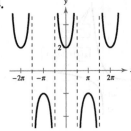

5.

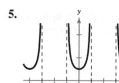

6.

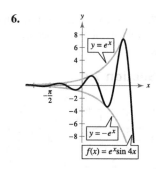

$f(x) = e^x \sin 4x$

Section 4.7

1. (a) $\dfrac{\pi}{2}$

(b) Not possible

2.

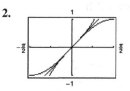

3. π

4. (a) 1.3670516 (b) Not possible   (c) 1.9273001

5. (a) -14 (b) $-\dfrac{\pi}{4}$ (c) 0.54 **6.** $\dfrac{4}{5}$ **7.** $\sqrt{x^2 + 1}$

Section 4.8

1. $a \approx 5.46, c \approx 15.96, B = 70°$ **2.** About 15.8 ft
3. About 15.1 ft   **4.** $3.58°$
5. Bearing: N $53°$ W, Distance: about 3.2 nmi

6. $d = 6 \sin \dfrac{2\pi}{3} t$

7. (a) 4 (b) 3 cycles per unit of time   (c) 4 (d) $\dfrac{1}{12}$

Chapter 5

Section 5.1

1. $\sin x = -\dfrac{\sqrt{10}}{10}, \cos x = -\dfrac{3\sqrt{10}}{10}, \tan x = \dfrac{1}{3}, \csc x = -\sqrt{10}$

$\sec x = -\dfrac{\sqrt{10}}{3}, \cot x = 3$

2. $-\sin x$
3. (a) $(1 + \cos \theta)(1 - \cos \theta)$ (b) $(2 \csc \theta - 3)(\csc \theta - 2)$
4. $(\tan x + 1)(\tan x + 2)$ **5.** $\sin x$ **6.** $2 \sec^2 \theta$
7. $1 + \sin \theta$ **8.** $3 \cos \theta$ **9.** $\ln|\tan x|$

Section 5.2

1–7. Answers will vary.

Section 5.3

1. $\dfrac{\pi}{4} + 2n\pi, \dfrac{3\pi}{4} + 2n\pi$

2. $\dfrac{\pi}{3} + 2n\pi, \dfrac{2\pi}{3} + 2n\pi, \dfrac{4\pi}{3} + 2n\pi, \dfrac{5\pi}{3} + 2n\pi$ **3.** $n\pi$

4. $\dfrac{\pi}{6}, \dfrac{\pi}{2}, \dfrac{5\pi}{6}$ **5.** $\dfrac{\pi}{4} + n\pi, \dfrac{3\pi}{4} + n\pi$ **6.** $0, \dfrac{3\pi}{2}$

7. $\dfrac{\pi}{6} + n\pi, \dfrac{\pi}{3} + n\pi$ **8.** $\dfrac{\pi}{2} + 2n\pi$

9. $\arctan \dfrac{3}{4} + n\pi, \arctan(-2) + n\pi$ **10.** $0.4271, 2.7145$
11. $\theta \approx 54.7356°$

Section 5.4

1. $\dfrac{\sqrt{2} + \sqrt{6}}{4}$ **2.** $\dfrac{\sqrt{2} + \sqrt{6}}{4}$ **3.** $-\dfrac{63}{65}$

4. $\dfrac{x + \sqrt{1 - x^2}}{\sqrt{2}}$ **5.** Answers will vary.

6. (a) $-\cos \theta$ (b) $\dfrac{\tan \theta - 1}{\tan \theta + 1}$ **7.** $\dfrac{\pi}{3}, \dfrac{5\pi}{3}$

8. Answers will vary.

Section 5.5

1. $\dfrac{\pi}{3} + 2n\pi, \pi + 2n\pi, \dfrac{5\pi}{3} + 2n\pi$

2. $\sin 2\theta = \dfrac{24}{25}, \cos 2\theta = \dfrac{7}{25}, \tan 2\theta = \dfrac{24}{7}$

3. $4 \cos^3 x - 3 \cos x$ **4.** $\dfrac{\cos 4x - 4 \cos 2x + 3}{\cos 4x + 4 \cos 2x + 3}$

5. $\dfrac{-\sqrt{2 - \sqrt{3}}}{2}$ **6.** $\dfrac{\pi}{3}, \pi, \dfrac{5\pi}{3}$ **7.** $\dfrac{1}{2} \sin 8x + \dfrac{1}{2} \sin 2x$

8. $\dfrac{\sqrt{2}}{2}$ **9.** $\dfrac{\pi}{6} + \dfrac{2n\pi}{3}, \dfrac{\pi}{2} + \dfrac{2n\pi}{3}, n\pi$ **10.** $45°$

Chapter 6

Section 6.1

1. $C = 105°, b \approx 45.25$ cm, $c \approx 61.82$ cm   **2.** 13.40 m
3. $B \approx 12.39°, C \approx 136.61°, c \approx 16.01$ in.
4. $\sin B \approx 3.0311 > 1$
5. Two solutions:
  $B \approx 70.4°, C \approx 51.6°, c \approx 4.16$ ft
  $B \approx 109.6°, C \approx 12.4°, c \approx 1.14$ ft
6. About 213 yd^2 **7.** About 1856.59 m

Section 6.2

1. $A \approx 26.38°, B \approx 36.34°, C \approx 117.28°$
2. $B \approx 59.66°, C \approx 40.34°, a \approx 18.26$ m   **3.** About 202 ft
4. N $15.37°$ E   **5.** About 19.90 in.^{2}

Section 6.3

1. $\|\overrightarrow{PQ}\| = \|\overrightarrow{RS}\| = \sqrt{10}$, slope$_{\overrightarrow{PQ}}$ = slope$_{\overrightarrow{RS}}$ = $\dfrac{1}{3}$
$\overrightarrow{PQ}$ and $\overrightarrow{RS}$ have the same magnitude and direction, so they are equivalent.
2. $\mathbf{v} = \langle -5, 6 \rangle, \|\mathbf{v}\| = \sqrt{61}$
3. (a) $\langle 4, 6 \rangle$ (b) $\langle -2, 2 \rangle$ (c) $\langle -7, 2 \rangle$
4. (a) $3\sqrt{17}$ (b) $2\sqrt{13}$ (c) $5\sqrt{13}$

5. $\left\langle \dfrac{6}{\sqrt{37}}, -\dfrac{1}{\sqrt{37}} \right\rangle$ **6.** $-6\mathbf{i} - 3\mathbf{j}$ **7.** $11\mathbf{i} - 14\mathbf{j}$

8. (a) $135°$ (b) About $209.74°$ **9.** $\langle -96.59, -25.88 \rangle$
10. About 2405 lb   **11.** $\|\mathbf{v}\| \approx 451.8$ mi/h, $\theta \approx 305.1°$

Section 6.4

1. (a) -6 (b) 37 (c) 0
2. (a) $\langle -36, 108 \rangle$ (b) 43 (c) $2\sqrt{10}$ **3.** $45°$
4. Yes   **5.** $\dfrac{1}{17}\langle 64, 16 \rangle; \dfrac{1}{17}\langle -13, 52 \rangle$ **6.** About 38.8 lb
7. About 1212 ft-lb

Section 6.5

1.

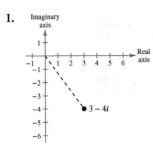

5

2. $4 + 3i$　　**3.** $1 - 5i$

4.

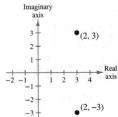

$2 + 3i$

5. $\sqrt{82} \approx 9.06$ units　　**6.** $\left(\frac{7}{2}, -2\right)$

Section 6.6

1. $6\sqrt{2}\left(\cos\dfrac{7\pi}{4} + i\sin\dfrac{7\pi}{4}\right)$　　**2.** $-4 + 4\sqrt{3}i$　　**3.** 10

4. $12i$　　**5.** $\dfrac{\sqrt{3}}{2} + \dfrac{1}{2}i$　　**6.** -8　　**7.** -4　　**8.** $1, i, -1, -i$

9. $\sqrt[3]{3} + \sqrt[3]{3}i, -1.9707 + 0.5279i, 0.5279 - 1.9701i$

Chapter 7

Section 7.1

1. $(3, 3)$　　**2.** \$6250 at 6.5%, \$18,750 at 8.5%

3. $(-3, -1), (2, 9)$　　**4.** No solution　　**5.** $(1, 3)$

6. About 5172 pairs　　**7.** 5 weeks

Section 7.2

1. $\left(\frac{3}{4}, \frac{5}{2}\right)$　　**2.** $(4, 3)$　　**3.** $(3, -1)$　　**4.** $(9, 12)$

5.

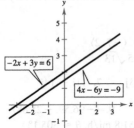

No solution; inconsistent

6. No solution　　**7.** Infinitely many solutions: $(a, 4a + 3)$

8. About 471.18 mi/h; about 16.63 mi/h　　**9.** $(1,500,000, 537)$

Section 7.3

1. $(4, -3, 3)$　　**2.** $(1, 1)$　　**3.** $(1, 2, 3)$　　**4.** No solution

5. Infinitely many solutions: $(-23a + 22, 15a - 13, a)$

6. Infinitely many solutions: $\left(\frac{1}{4}a, \frac{17}{4}a - 3, a\right)$

7. $s = -16t^2 + 20t + 100$; The object was thrown upward at a velocity of 20 feet per second from a height of 100 feet.

8. $y = \frac{1}{3}x^2 - 2x$

Section 7.4

1. $-\dfrac{3}{2x + 1} + \dfrac{2}{x - 1}$　　**2.** $x - \dfrac{3}{x} + \dfrac{4}{x^2} + \dfrac{3}{x + 1}$

3. $-\dfrac{5}{x} + \dfrac{7x}{x^2 + 1}$　　**4.** $\dfrac{x + 3}{x^2 + 4} - \dfrac{6x + 5}{(x^2 + 4)^2}$

5. $\dfrac{1}{x} - \dfrac{2}{x^2} + \dfrac{2 - x}{x^2 + 2} + \dfrac{4 - 2x}{(x^2 + 2)^2}$

Section 7.5

1.
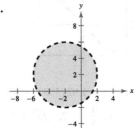

2.
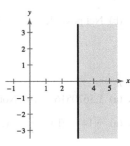

3.
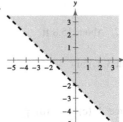

4.
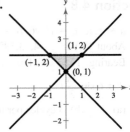

5.
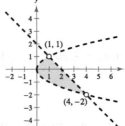

6.
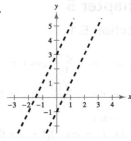

No solution

7.
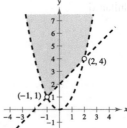

8. Consumer surplus: \$22,500,000

Producer surplus: \$33,750,000

9. $\begin{cases} 8x + 2y \geq 16 & \text{Nutrient A} \\ x + y \geq 5 & \text{Nutrient B} \\ 2x + 7y \geq 20 & \text{Nutrient C} \\ x \geq 0 \\ y \geq 0 \end{cases}$

Section 7.6

1. Maximum at $(0, 6)$: 30   2. Minimum at $(0, 0)$: 0
3. Maximum at $(60, 20)$: 880   4. Minimum at $(10, 0)$: 30
5. $2925; 1050 boxes of chocolate-covered creams, 150 boxes of chocolate-covered nuts
6. 3 bottles of brand X, 2 bottles of brand Y

Chapter 8

Section 8.1

1. 2×3 2. $\begin{bmatrix} 1 & 1 & 1 & \vdots & 2 \\ 2 & -1 & 3 & \vdots & -1 \\ -1 & 2 & -1 & \vdots & 4 \end{bmatrix}$; 3×4

3. Add -3 times Row 1 to Row 2.
4. Answers will vary. Solution: $(-1, 0, 1)$
5. Reduced row-echelon form   6. $(4, -2, 1)$
7. No solution   8. $(7, 4, -3)$ 9. $(3a + 8, 2a - 5, a)$

Section 8.2

1. $a_{11} = 6, a_{12} = 3, a_{21} = -2, a_{22} = 4$

2. (a) $\begin{bmatrix} 6 & -2 \\ 2 & 3 \end{bmatrix}$ (b) $\begin{bmatrix} 0 & 0 \\ 0 & 0 \\ 0 & 0 \end{bmatrix}$ (c) Not possible   (d) $\begin{bmatrix} 0 \\ 0 \\ 2 \end{bmatrix}$

3. (a) $\begin{bmatrix} 4 & -5 \\ 1 & 1 \\ -4 & 1 \end{bmatrix}$ (b) $\begin{bmatrix} 12 & -3 \\ 0 & 12 \\ -9 & 24 \end{bmatrix}$ (c) $\begin{bmatrix} 12 & -11 \\ 2 & 6 \\ -11 & 10 \end{bmatrix}$

4. $\begin{bmatrix} 1 & 2 \\ 10 & -4 \end{bmatrix}$ 5. $\begin{bmatrix} -6 & 6 \\ -10 & 6 \end{bmatrix}$ 6. $\begin{bmatrix} 5 & 0 \\ -1 & 4 \end{bmatrix}$

7. $\begin{bmatrix} -1 & 30 \\ 2 & -4 \\ 1 & 12 \end{bmatrix}$ 8. $\begin{bmatrix} -3 & -22 \\ 3 & 10 \\ -5 & 10 \end{bmatrix}$

9. (a) $\begin{bmatrix} 3 & -1 \\ -9 & 3 \end{bmatrix}$ (b) $[6]$ (c) Not possible

10. $\begin{bmatrix} 7 & 0 \\ 0 & 7 \end{bmatrix}$ 11. (a) $\langle -5, 1 \rangle$ (b) $\langle 17, 23 \rangle$

12. $\langle -3, 1 \rangle$; Reflection in the y-axis

13. (a) $\begin{bmatrix} -2 & -3 \\ 6 & 1 \end{bmatrix} \begin{bmatrix} x_1 \\ x_2 \end{bmatrix} = \begin{bmatrix} -4 \\ -36 \end{bmatrix}$ (b) $\begin{bmatrix} -7 \\ 6 \end{bmatrix}$

14. Total cost for women's team: $2310
    Total cost for men's team: $2719

Section 8.3

1. $AB = I$ and $BA = I$ 2. $\begin{bmatrix} 3 & 2 \\ 1 & 1 \end{bmatrix}$

3. $\begin{bmatrix} -4 & -2 & 5 \\ -2 & -1 & 2 \\ -1 & 0 & 1 \end{bmatrix}$ 4. $\begin{bmatrix} \frac{4}{23} & \frac{1}{23} \\ -\frac{3}{23} & \frac{5}{23} \end{bmatrix}$ 5. $(2, -1, -2)$

Section 8.4

1. (a) -7 (b) 10 (c) 0
2. $M_{11} = -9, M_{12} = -10, M_{13} = 2, M_{21} = 5, M_{22} = -2,$
   $M_{23} = -3, M_{31} = 13, M_{32} = 5, M_{33} = -1$
   $C_{11} = -9, C_{12} = 10, C_{13} = 2, C_{21} = -5, C_{22} = -2,$
   $C_{23} = 3, C_{31} = 13, C_{32} = -5, C_{33} = -1$

3. -31 4. 704

Section 8.5

1. $(3, -2)$ 2. $(2, -3, 1)$ 3. 9 square units
4. Collinear   5. $x - y + 2 = 0$
6. $(0, 0), (2, 0), (0, 4), (2, 4)$ 7. 10 square units
8. $[15 \quad 23 \quad 12] [19 \quad 0 \quad 1] [18 \quad 5 \quad 0] [14 \quad 15 \quad 3]$
   $[20 \quad 21 \quad 18] [14 \quad 1 \quad 12]$
9. $110 \quad -39 \quad -59 \quad 25 \quad -21 \quad -3 \quad 23 \quad -18 \quad -5 \quad 47$
   $-20 \quad -24 \quad 149 \quad -56 \quad -75 \quad 87 \quad -38 \quad -37$
10. OWLS ARE NOCTURNAL

Chapter 9

Section 9.1

1. $3, 5, 7, 9$ 2. $1, \frac{3}{2}, \frac{1}{3}, \frac{3}{4}$
3. (a) $a_n = 4n - 3$ (b) $a_n = (-1)^{n+1}(2n)$
4. $6, 7, 8, 9, 10$ 5. $1, 3, 4, 7, 11$ 6. $2, 4, 5, \frac{14}{3}, \frac{41}{12}$
7. $4(n + 1)$ 8. 44 9. (a) 0.5555 (b) $\frac{5}{9}$
10. (a) $1000, $1002.50, $1005.01   (b) $1127.33

Section 9.2

1. $2, 5, 8, 11; d = 3$ 2. $a_n = 5n - 6$
3. $-3, 1, 5, 9, 13, 17, 21, 25, 29, 33, 37$ 4. 79 5. 217
6. (a) 630 (b) $N(1 + 2N)$ 7. $43,560$ 8. 1470
9. $2,500,000

Section 9.3

1. $-12, 24, -48, 96; r = -2$
2. $2, 8, 32, 128, 512$

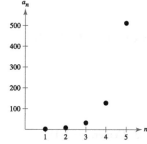

3. 104.02 4. $a_n = 4(5)^{n-1}$; $195,312,500$ 5. $\frac{2187}{32}$
6. 2.667 7. (a) 10 (b) 6.25 8. $3500.85

Section 9.4

1. (a) $\dfrac{6}{(k + 1)(k + 4)}$ (b) $k + 3 \le 3k^2$
   (c) $2^{4k+2} + 1 > 5k + 5$
2–5. Proofs   6. $S_k = k(2k + 1)$; Proof
7. (a) 210 (b) 785 8. $a_n = n^2 - n - 2$

Section 9.5

1. (a) 462 (b) 36 (c) 1 (d) 1
2. (a) 21 (b) 21 (c) 14 (d) 14
3. $1, 9, 36, 84, 126, 126, 84, 36, 9, 1$
4. $x^4 + 8x^3 + 24x^2 + 32x + 16$

5. (a) $y^4 - 8y^3 + 24y^2 - 32y + 16$
(b) $32x^5 - 80x^4y + 80x^3y^2 - 40x^2y^3 + 10xy^4 - y^5$
6. $125 + 75y^2 + 15y^4 + y^6$
7. (a) $1120a^4b^4$ (b) $-3,421,440$

Section 9.6

1. 3 ways   **2.** 2 ways   **3.** 27,000 combinations
4. 2,600,000 numbers   **5.** 24 permutations   **6.** 20 ways
7. 1260 ways   **8.** 21 ways
9. 22,100 three-card poker hands   **10.** 1,051,050 teams

Section 9.7

1. $\{HH1, HH2, HH3, HH4, HH5, HH6, HT1, HT2, HT3, HT4,$
$HT5, HT6, TH1, TH2, TH3, TH4, TH5, TH6, TT1, TT2, TT3$
$TT4, TT5, TT6\}$
2. (a) $\frac{1}{8}$ (b) $\frac{1}{4}$ **3.** $\frac{1}{9}$ **4.** Answers will vary.
5. $\frac{320}{2311} \approx 0.138$ **6.** $\frac{1}{962,598}$ **7.** $\frac{4}{13} \approx 0.308$
8. $\frac{66}{529} \approx 0.125$ **9.** $\frac{121}{900} \approx 0.134$ **10.** About 0.116
11. 0.452

Chapter 10
Section 10.1

1. (a) $0.6747 \approx 38.7°$ (b) $\frac{3\pi}{4} \approx 135°$ **2.** $1.4841 \approx 85.0°$
3. $\frac{12}{\sqrt{10}} \approx 3.79$ units   **4.** $\frac{3}{\sqrt{34}} \approx 0.51$ unit
5. (a) $\frac{8}{\sqrt{5}} \approx 3.58$ units   (b) 12 square units

Section 10.2

1. $x^2 = \frac{3}{2}y$ **2.** $(y + 3)^2 = 8(x - 2)$ **3.** $(2, -3)$
4. $y = 6x - 3$

Section 10.3

1. $\frac{(x - 2)^2}{7} + \frac{(y - 3)^2}{16} = 1$
2.
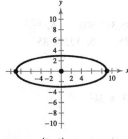
Center: $(0, 0)$
Vertices: $(-9, 0), (9, 0)$

3. Center: $(-2, 1)$
Vertices: $(-2, -2), (-2, 4)$
Foci: $\left(-2, 1 \pm \sqrt{5}\right)$
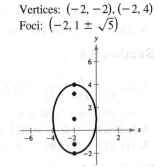

4. Center: $(-1, 3)$
Vertices: $(-4, 3), (2, 3)$
Foci: $(-3, 3), (1, 3)$
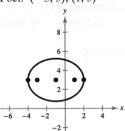

5. Aphelion:
4.080 astronomical units
Perihelion:
0.340 astronomical unit
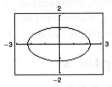

Section 10.4

1. $\frac{(y + 1)^2}{9} - \frac{(x - 2)^2}{7} = 1$
2.
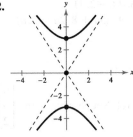
3.
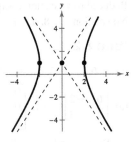

4. $\frac{(x - 6)^2}{9} - \frac{(y - 2)^2}{4} = 1$
5. The explosion occurred somewhere on the right branch of the hyperbola
$$\frac{x^2}{4,840,000} - \frac{y^2}{2,129,600} = 1.$$
6. (a) Circle   (b) Hyperbola   (c) Ellipse   (d) Parabola

Section 10.5

1. $\frac{(y')^2}{12} - \frac{(x')^2}{12} = 1$ **2.** $\frac{(x')^2}{1} - \frac{(y')^2}{9} = 1$
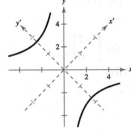
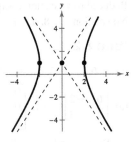

3. $(x')^2 = -2(y' - 3)$ **4.** Parabola;
$$y = \frac{2x \pm \sqrt{-6x - 10}}{4}$$
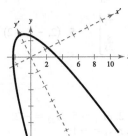
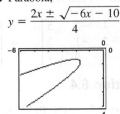

Section 10.6

1.

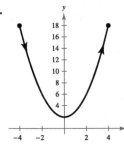

2.

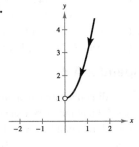

The curve starts at $(-4, 18)$
and ends at $(4, 18)$.

3. (a)

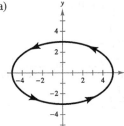

(b)

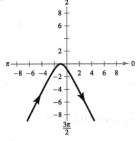

4. (a) $x = t, y = t^2 + 2$ (b) $x = 2 - t, y = t^2 - 4t + 6$
5. $x = a(\theta + \sin \theta), y = a(1 + \cos \theta)$

Section 10.7

1. (a)

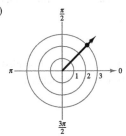

(b)

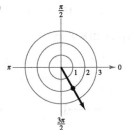

(c)

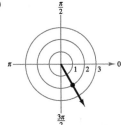

2.

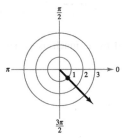

$\left(-1, -\dfrac{5\pi}{4}\right), \left(1, \dfrac{7\pi}{4}\right), \left(1, -\dfrac{\pi}{4}\right)$

3. $(-2, 0)$ **4.** $\left(2, \dfrac{\pi}{2}\right)$

5. (a) The graph consists of all points seven units from the pole;
   $x^2 + y^2 = 49$

   (b) The graph consists of all points on the line that makes an
   angle of $\pi/4$ with the polar axis and passes through the
   pole; $y = x$

   (c) The graph is a circle with center $(0, 3)$ and radius 3;
   $x^2 + (y - 3)^2 = 9$

Section 10.8

1.

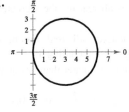

2.

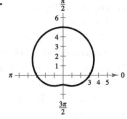

3.

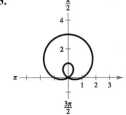

4.

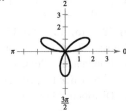

5.

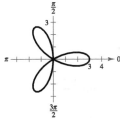

6.

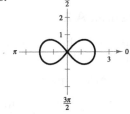

Section 10.9

1. Hyperbola
2. Hyperbola

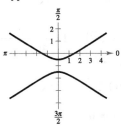

3. $r = \dfrac{2}{1 - \cos \theta}$

4. $r = \dfrac{0.625}{1 + 0.847 \sin \theta}$; about 0.340 astronomical unit

CHECKPOINTS

Appendix A

Appendix A.1

1. (a) Natural numbers: $\left\{\frac{6}{3}, 8\right\}$　(b) Whole numbers: $\left\{\frac{6}{3}, 8\right\}$
(c) Integers: $\left\{\frac{6}{3}, -1, 8, -22\right\}$
(d) Rational numbers: $\left\{-\frac{1}{4}, \frac{6}{3}, -7.5, -1, 8, -22\right\}$
(e) Irrational numbers: $\left\{-\pi, \frac{1}{2}\sqrt{2}\right\}$

2.
```
    -1.6  -3/4   0.7   5/2
    +--●--●--+--●--+--●--+---→ x
    -2  -1   0   1   2   3   4
```

3. (a) $1 > -5$　(b) $\frac{3}{2} < 7$　(c) $-\frac{2}{3} > -\frac{3}{4}$
4. (a) The inequality $x > -3$ denotes all real numbers greater than -3.
(b) The inequality $0 < x \le 4$ denotes all real numbers between 0 and 4, not including 0, but including 4.
5. The interval consists of all real numbers greater than or equal to -2 and less than 5.
6. $-2 \le x < 4$　7. (a) 1　(b) $-\frac{3}{4}$　(c) $\frac{2}{3}$　(d) -0.7
8. (a) 1　(b) -1
9. (a) $|-3| < |4|$　(b) $-|-4| = -|4|$
(c) $|-3| > -|-3|$
10. (a) 58　(b) 12　(c) 12
11. Terms: $-2x, 4$; Coefficients: $-2, 4$　12. -5
13. (a) Commutative Property of Addition
(b) Associative Property of Multiplication
(c) Distributive Property
14. (a) $\dfrac{x}{10}$　(b) $\dfrac{x}{2}$

Appendix A.2

1. (a) -81　(b) 81　(c) 27　(d) $\frac{1}{27}$
2. (a) $-\frac{1}{16}$　(b) 64
3. (a) $-2x^2y^4$　(b) 1　(c) $-125z^5$　(d) $\dfrac{9x^4}{y^4}$
4. (a) $\dfrac{2}{a^2}$　(b) $\dfrac{b^5}{5a^4}$　(c) $\dfrac{10}{x}$　(d) $-2x^3$
5. 4.585×10^4　6. -0.002718　7. $864{,}000{,}000$
8. (a) -12　(b) Not a real number　(c) $\frac{5}{8}$　(d) $-\frac{2}{3}$
9. (a) 5　(b) 25　(c) x　(d) $\sqrt[4]{x}$
10. (a) $4\sqrt{2}$　(b) $5\sqrt[3]{2}$　(c) $2a^2\sqrt{6a}$　(d) $-3x\sqrt[3]{5}$
11. (a) $9\sqrt{2}$　(b) $(3x - 2)\sqrt[3]{3x^2}$
12. (a) $\dfrac{5\sqrt{2}}{6}$　(b) $\dfrac{\sqrt[3]{5}}{5}$　13. $2\left(\sqrt{6} + \sqrt{2}\right)$　14. $\dfrac{2}{3\left(2 + \sqrt{2}\right)}$
15. (a) $27^{1/3}$　(b) $x^{3/2}y^{5/2}z^{1/2}$　(c) $3x^{5/3}$
16. (a) $\dfrac{1}{\sqrt{x^2 - 7}}$　(b) $-3\sqrt[3]{bc^2}$　(c) $\sqrt[4]{a^3}$　(d) $\sqrt[5]{x^4}$
17. (a) $\frac{1}{25}$　(b) $-12x^{5/3}y^{9/10}, x \ne 0, y \ne 0$
(c) $\sqrt[4]{3}$　(d) $(3x + 2)^2, x \ne -\frac{2}{3}$

Appendix A.3

1. Standard form: $-7x^3 + 2x + 6$
Degree: 3; Leading coefficient: -7
2. $2x^3 - x^2 + x + 6$　3. $3x^2 - 16x + 5$
4. (a) $9x^2 - 4$　(b) $x^2 - 4x + 4 - 9y^2$
5. (a) $5x^2(x - 3)$　(b) $-3(1 - 2x + 4x^3)$
(c) $(x + 1)(x^2 - 2)$

6. $4(5 + y)(5 - y)$　7. $(x - 1 + 3y^2)(x - 1 - 3y^2)$
8. $(3x - 5)^2$　9. $(4x - 1)(16x^2 + 4x + 1)$
10. (a) $(x + 6)(x^2 - 6x + 36)$　(b) $5(y + 3)(y^2 - 3y + 9)$
11. $(x + 3)(x - 2)$　12. $(2x - 3)(x - 1)$
13. $(x^2 - 5)(x + 1)$　14. $(2x - 3)(x + 4)$

Appendix A.4

1. (a) All nonnegative real numbers x
(b) All real numbers x such that $x \ge -7$
(c) All real numbers x such that $x \ne 0$
2. $\dfrac{4}{x - 6}, x \ne -3$　3. $-\dfrac{3x + 2}{5 + x}, x \ne 1$
4. $\dfrac{5(x - 5)}{(x - 6)(x - 3)}, x \ne -3, x \ne -\dfrac{1}{3}, x \ne 0$
5. $x + 1, x \ne \pm1$　6. $\dfrac{x^2 + 1}{(2x - 1)(x + 2)}$
7. $\dfrac{7x^2 - 13x - 16}{x(x + 2)(x - 2)}$　8. $\dfrac{3(x + 3)}{(x - 3)(x + 2)}$　9. $-\dfrac{1}{(x - 1)^{4/3}}$
10. $\dfrac{2(x + 1)(x - 1)}{(x^2 - 2)^{3/2}}$　11. $\dfrac{1}{\sqrt{9 + h} + 3}, h \ne 0$

Appendix A.5

1. (a) -4　(b) 8　2. $-\dfrac{18}{5}$　3. No solution
4. $-1, \dfrac{5}{2}$　5. (a) $\pm2\sqrt{3}$　(b) $1 \pm \sqrt{10}$　6. $2 \pm \sqrt{5}$
7. $\dfrac{5}{3} \pm \dfrac{\sqrt{31}}{3}$　8. $-\dfrac{1}{3} \pm \dfrac{\sqrt{31}}{3}$　9. $\dfrac{4}{3}$　10. $0, \pm\dfrac{2\sqrt{3}}{3}$
11. (a) $5, \pm\sqrt{2}$　(b) $0, -\dfrac{3}{2}, 6$　12. -9　13. $-59, 69$
14. $-2, 6$　15. About 2.97 in.

Appendix A.6

1. (a) $1 \le x \le 3$; Bounded　(b) $-1 < x < 6$; Bounded
(c) $x < 4$; Unbounded　(d) $x \ge 0$; Unbounded
2. $x \le 2$
```
    ←----------●--+--+--→ x
       0   1   2   3   4
```
3. (a) $x < 3$
(b)
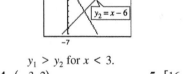

$y_1 > y_2$ for $x < 3$.
4. $(-3, 2)$
```
    +--(--+--+--+--+--+--)--+--→ x
    -4 -3 -2 -1  0  1  2  3
```
5. $[16, 24]$
```
    +--+--[--+--+--+--+--]--+--→ x
    12 14 16 18 20 22 24 26 28
```
6. More than 68 hours
7. You might have been overcharged by as much as \$0.16 or undercharged by as much as \$0.15.

Appendix A.7

1. Do not apply radicals term-by-term. Leave as $\sqrt{x^2 + 4}$.
2. $(x - 2)^{-1/2}(7x - 12)$　3. 3
4. Answers will vary.　5. $-6x(1 - 3x^2)^{-2} + x^{-1/3}$
6. (a) $x - 2 + \dfrac{5}{x^3}$　(b) $x^{3/2} - x^{1/2} + 5x^{-1/2}$

Definition of the Six Trigonometric Functions

Right triangle definitions, where $0 < \theta < \pi/2$

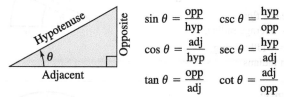

$$\sin \theta = \frac{\text{opp}}{\text{hyp}} \qquad \csc \theta = \frac{\text{hyp}}{\text{opp}}$$

$$\cos \theta = \frac{\text{adj}}{\text{hyp}} \qquad \sec \theta = \frac{\text{hyp}}{\text{adj}}$$

$$\tan \theta = \frac{\text{opp}}{\text{adj}} \qquad \cot \theta = \frac{\text{adj}}{\text{opp}}$$

Circular function definitions, where θ is any angle

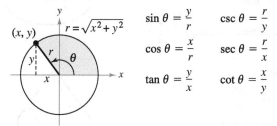

$$\sin \theta = \frac{y}{r} \qquad \csc \theta = \frac{r}{y}$$

$$\cos \theta = \frac{x}{r} \qquad \sec \theta = \frac{r}{x}$$

$$\tan \theta = \frac{y}{x} \qquad \cot \theta = \frac{x}{y}$$

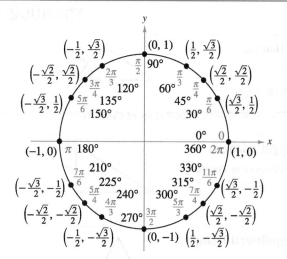

Reciprocal Identities

$$\sin u = \frac{1}{\csc u} \qquad \cos u = \frac{1}{\sec u} \qquad \tan u = \frac{1}{\cot u}$$

$$\csc u = \frac{1}{\sin u} \qquad \sec u = \frac{1}{\cos u} \qquad \cot u = \frac{1}{\tan u}$$

Quotient Identities

$$\tan u = \frac{\sin u}{\cos u} \qquad \cot u = \frac{\cos u}{\sin u}$$

Pythagorean Identities

$$\sin^2 u + \cos^2 u = 1$$

$$1 + \tan^2 u = \sec^2 u \qquad 1 + \cot^2 u = \csc^2 u$$

Cofunction Identities

$$\sin\left(\frac{\pi}{2} - u\right) = \cos u \qquad \cot\left(\frac{\pi}{2} - u\right) = \tan u$$

$$\cos\left(\frac{\pi}{2} - u\right) = \sin u \qquad \sec\left(\frac{\pi}{2} - u\right) = \csc u$$

$$\tan\left(\frac{\pi}{2} - u\right) = \cot u \qquad \csc\left(\frac{\pi}{2} - u\right) = \sec u$$

Even/Odd Identities

$$\sin(-u) = -\sin u \qquad \cot(-u) = -\cot u$$

$$\cos(-u) = \cos u \qquad \sec(-u) = \sec u$$

$$\tan(-u) = -\tan u \qquad \csc(-u) = -\csc u$$

Sum and Difference Formulas

$$\sin(u \pm v) = \sin u \cos v \pm \cos u \sin v$$

$$\cos(u \pm v) = \cos u \cos v \mp \sin u \sin v$$

$$\tan(u \pm v) = \frac{\tan u \pm \tan v}{1 \mp \tan u \tan v}$$

Double-Angle Formulas

$$\sin 2u = 2 \sin u \cos u$$

$$\cos 2u = \cos^2 u - \sin^2 u = 2 \cos^2 u - 1 = 1 - 2 \sin^2 u$$

$$\tan 2u = \frac{2 \tan u}{1 - \tan^2 u}$$

Power-Reducing Formulas

$$\sin^2 u = \frac{1 - \cos 2u}{2}$$

$$\cos^2 u = \frac{1 + \cos 2u}{2}$$

$$\tan^2 u = \frac{1 - \cos 2u}{1 + \cos 2u}$$

Sum-to-Product Formulas

$$\sin u + \sin v = 2 \sin\left(\frac{u + v}{2}\right)\cos\left(\frac{u - v}{2}\right)$$

$$\sin u - \sin v = 2 \cos\left(\frac{u + v}{2}\right)\sin\left(\frac{u - v}{2}\right)$$

$$\cos u + \cos v = 2 \cos\left(\frac{u + v}{2}\right)\cos\left(\frac{u - v}{2}\right)$$

$$\cos u - \cos v = -2 \sin\left(\frac{u + v}{2}\right)\sin\left(\frac{u - v}{2}\right)$$

Product-to-Sum Formulas

$$\sin u \sin v = \frac{1}{2}[\cos(u - v) - \cos(u + v)]$$

$$\cos u \cos v = \frac{1}{2}[\cos(u - v) + \cos(u + v)]$$

$$\sin u \cos v = \frac{1}{2}[\sin(u + v) + \sin(u - v)]$$

$$\cos u \sin v = \frac{1}{2}[\sin(u + v) - \sin(u - v)]$$

FORMULAS FROM GEOMETRY

Triangle:

$h = a \sin \theta$

$\text{Area} = \frac{1}{2}bh$

$c^2 = a^2 + b^2 - 2ab \cos \theta$ (Law of Cosines)

Right Triangle:

Pythagorean Theorem
$c^2 = a^2 + b^2$

Equilateral Triangle:

$h = \frac{\sqrt{3}s}{2}$

$\text{Area} = \frac{\sqrt{3}s^2}{4}$

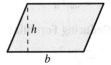

Parallelogram:

$\text{Area} = bh$

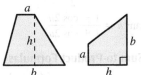

Trapezoid:

$\text{Area} = \frac{h}{2}(a + b)$

Circle:

$\text{Area} = \pi r^2$

$\text{Circumference} = 2\pi r$

Sector of Circle:

$\text{Area} = \frac{\theta r^2}{2}$

$s = r\theta$

θ in radians

Circular Ring:

$\text{Area} = \pi(R^2 - r^2)$

$\quad\quad = 2\pi pw$

$p = $ average radius,

$w = $ width of ring

Sector of Circular Ring:

$\text{Area} = \theta pw$

$p = $ average radius,

$w = $ width of ring,

θ in radians

Ellipse:

$\text{Area} = \pi ab$

$\text{Circumference} \approx 2\pi \sqrt{\dfrac{a^2 + b^2}{2}}$

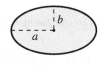

Cone:

$\text{Volume} = \dfrac{Ah}{3}$

$A = $ area of base

Right Circular Cone:

$\text{Volume} = \dfrac{\pi r^2 h}{3}$

$\text{Lateral Surface Area} = \pi r \sqrt{r^2 + h^2}$

Frustum of Right Circular Cone:

$\text{Volume} = \dfrac{\pi(r^2 + rR + R^2)h}{3}$

$\text{Lateral Surface Area} = \pi s(R + r)$

Right Circular Cylinder:

$\text{Volume} = \pi r^2 h$

$\text{Lateral Surface Area} = 2\pi rh$

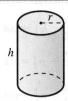

Sphere:

$\text{Volume} = \dfrac{4}{3}\pi r^3$

$\text{Surface Area} = 4\pi r^2$

Wedge:

$A = B \sec \theta$

$A = $ area of upper face,

$B = $ area of base

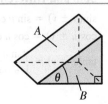

ALGEBRA

Factors and Zeros of Polynomials: Given the polynomial $p(x) = a_n x^n + a_{n-1}x^{n-1} + \cdots + a_1 x + a_0$. If $p(b) = 0$, then b is a *zero* of p and a *solution* of the equation $p(x) = 0$. Furthermore, $(x - b)$ is a *factor* of the polynomial.

Fundamental Theorem of Algebra: If $f(x)$ is a polynomial function of degree n, where $n > 0$, then f has at least one zero in the complex number system.

Quadratic Formula: If $p(x) = ax^2 + bx + c$, $a \neq 0$ and $b^2 - 4ac \geq 0$, then the real zeros of p are $x = \dfrac{-b \pm \sqrt{b^2 - 4ac}}{2a}$.

Special Factors:

$x^2 - a^2 = (x - a)(x + a)$

$x^3 - a^3 = (x - a)(x^2 + ax + a^2)$

$x^3 + a^3 = (x + a)(x^2 - ax + a^2)$

$x^4 - a^4 = (x - a)(x + a)(x^2 + a^2)$

$x^4 + a^4 = (x^2 + \sqrt{2}ax + a^2)(x^2 - \sqrt{2}ax + a^2)$

$x^n - a^n = (x - a)(x^{n-1} + ax^{n-2} + \cdots + a^{n-1})$

$x^n + a^n = (x + a)(x^{n-1} - ax^{n-2} + \cdots + a^{n-1})$, for n odd

$x^{2n} - a^{2n} = (x^n - a^n)(x^n + a^n)$

Examples

$x^2 - 9 = (x - 3)(x + 3)$

$x^3 - 8 = (x - 2)(x^2 + 2x + 4)$

$x^3 + 4 = \left(x + \sqrt[3]{4}\right)\left(x^2 - \sqrt[3]{4}x + \sqrt[3]{16}\right)$

$x^4 - 4 = \left(x - \sqrt{2}\right)\left(x + \sqrt{2}\right)(x^2 + 2)$

$x^4 + 4 = (x^2 + 2x + 2)(x^2 - 2x + 2)$

$x^5 - 1 = (x - 1)(x^4 + x^3 + x^2 + x + 1)$

$x^7 + 1 = (x + 1)(x^6 - x^5 + x^4 - x^3 + x^2 - x + 1)$

$x^6 - 1 = (x^3 - 1)(x^3 + 1)$

Binomial Theorem:

$(x + a)^2 = x^2 + 2ax + a^2$

$(x - a)^2 = x^2 - 2ax + a^2$

$(x + a)^3 = x^3 + 3ax^2 + 3a^2x + a^3$

$(x - a)^3 = x^3 - 3ax^2 + 3a^2x - a^3$

$(x + a)^4 = x^4 + 4ax^3 + 6a^2x^2 + 4a^3x + a^4$

$(x - a)^4 = x^4 - 4ax^3 + 6a^2x^2 - 4a^3x + a^4$

$(x + a)^n = x^n + nax^{n-1} + \dfrac{n(n-1)}{2!}a^2x^{n-2} + \cdots + na^{n-1}x + a^n$

$(x - a)^n = x^n - nax^{n-1} + \dfrac{n(n-1)}{2!}a^2x^{n-2} - \cdots \pm na^{n-1}x \mp a^n$

Examples

$(x + 3)^2 = x^2 + 6x + 9$

$(x^2 - 5)^2 = x^4 - 10x^2 + 25$

$(x + 2)^3 = x^3 + 6x^2 + 12x + 8$

$(x - 1)^3 = x^3 - 3x^2 + 3x - 1$

$\left(x + \sqrt{2}\right)^4 = x^4 + 4\sqrt{2}x^3 + 12x^2 + 8\sqrt{2}x + 4$

$(x - 4)^4 = x^4 - 16x^3 + 96x^2 - 256x + 256$

$(x + 1)^5 = x^5 + 5x^4 + 10x^3 + 10x^2 + 5x + 1$

$(x - 1)^6 = x^6 - 6x^5 + 15x^4 - 20x^3 + 15x^2 - 6x + 1$

Rational Zero Test: If $p(x) = a_n x^n + a_{n-1}x^{n-1} + \cdots + a_1 x + a_0$ has integer coefficients, then every *rational* zero of p is of the form $x = r/s$, where r is a factor of a_0 and s is a factor of a_n.

Exponents and Radicals:

$a^0 = 1, \ a \neq 0$

$a^{-x} = \dfrac{1}{a^x}$

$a^x a^y = a^{x+y}$

$\dfrac{a^x}{a^y} = a^{x-y}$

$(a^x)^y = a^{xy}$

$(ab)^x = a^x b^x$

$\left(\dfrac{a}{b}\right)^x = \dfrac{a^x}{b^x}$

$\sqrt{a} = a^{1/2}$

$\sqrt[n]{a} = a^{1/n}$

$\sqrt[n]{a^m} = a^{m/n} = \left(\sqrt[n]{a}\right)^m$

$\sqrt[n]{ab} = \sqrt[n]{a}\,\sqrt[n]{b}$

$\sqrt[n]{\dfrac{a}{b}} = \dfrac{\sqrt[n]{a}}{\sqrt[n]{b}}$

Conversion Table:

1 centimeter $\approx$ 0.394 inch

1 meter $\approx$ 39.370 inches

$\qquad \approx$ 3.281 feet

1 kilometer $\approx$ 0.621 mile

1 liter $\approx$ 0.264 gallon

1 newton $\approx$ 0.225 pound

1 joule $\approx$ 0.738 foot-pound

1 gram $\approx$ 0.035 ounce

1 kilogram $\approx$ 2.205 pounds

1 inch $=$ 2.54 centimeters

1 foot $=$ 30.48 centimeters

$\qquad \approx$ 0.305 meter

1 mile $\approx$ 1.609 kilometers

1 gallon $\approx$ 3.785 liters

1 pound $\approx$ 4.448 newtons

1 foot-pound $\approx$ 1.356 joules

1 ounce $\approx$ 28.350 grams

1 pound $\approx$ 0.454 kilogram